FACIES MODELS 4

Edited by

Noel P. James
Department of Geological Sciences
and Geological Engineering
Queen's University
Kingston, ON, K7L 3N6, Canada

and

Robert W. Dalrymple
Department of Geological Sciences
and Geological Engineering
Queen's University
Kingston, ON, K7L 3N6, Canada

GEOtext 6
Geological Association of Canada
2010

CONTENTS

Page

PART ONE - PRINCIPLES, TOOLS AND CONCEPTS

PART TWO - TERRIGENOUS CLASTIC FACIES MODELS

PART THREE - CHEMICAL AND BIOCHEMICAL FACIES MODELS

1. INTRODUCTION TO THE VOLUME

Noel P. James, Department of Geological Sciences and Geological Engineering, Queen's University, Kingston, ON, K7L 3N6, Canada

Robert W. Dalrymple, Department of Geological Sciences and Geological Engineering, Queen's University, Kingston, ON, K7L 3N6, Canada

Facies models are encapsulated summaries of sedimentation in a depositional environment. Since the initial formulation of this concept (Walker, 1979a), facies models have become the cornerstone of modern sedimentology, and will continue to be a vital starting point for whatever new areas of research develop in the future.

The originator of the concept, Roger Walker, envisioned facies models as living entities that would evolve by a feedback process, through the continual incorporation of new information and insights from both modern environments and the deposits of their ancient counterparts (Walker, 1979b; *see* Chapter 2). This process is abundantly clear in this new edition, which is a legacy of Roger Walker's fundamental contribution to sedimentary geology. In the 19 years since the last edition of this book was published, the attention of the sedimentological community has shifted from the creation of facies models as an end in itself to a focus on controls, that is, how external variables influence the operation of sedimentary environments. As a result, we now know much more about how changes in such variables as tectonic setting, relative sea level, climate, ocean-water chemistry, and the composition of the biosphere influence the nature of the sedimentary record of past depositional systems. These aspects are featured much more in this edition than ever before, continuing the trend, started in previous editions, which featured discussion of the influence of sea-level change.

Many of the chapters contained herein bear little resemblance to those in the last version because our knowledge of sedimentary systems has increased dramatically since 1992. The important points stressed before have become common knowledge over the intervening time (*e.g.*, tidal flats were considered at length in the siliciclastic tidal chapter in the previous edition, but are not given the same extended treatment here), whereas other aspects, some of which were not even discussed before, have risen to prominence. In the siliciclastic section, for example, the dispersal of fine-grained suspended sediment from rivers, which is one of the most important aspects of sediment delivery from the land to the sea, was barely mentioned in the 1992 edition. Now it is the subject of considerable active research in modern continental-margin settings, and the new insights are beginning to find application in the study of ancient successions. Increasing interest in lakes and bioelemental sediments (iron formations, phosphates, and siliceous sediments) has led to the inclusion of entirely new chapters on these deposits. Insights gained from recent research on the modern and past global ocean, including the importance of the role of nutrient variation, the role of climate change, and the evolving biosphere, are pervasive in the chapters on biochemical sediments.

As our knowledge of depositional environments has increased, many of the 'classic' ancient examples that were featured in previous editions have been questioned or reinterpreted, and are no longer featured here. Such evolution of thought is a common and expected aspect of the discipline, and students should not be surprised or disoriented by such changes of mind.

There is a close genetic linkage between facies models and sequence stratigraphy, not the least of which is the fact that the identification of the facies stacking patterns (*i.e.*, progradational or retrogradational) that are used to identify systems tracts requires a thorough knowledge of facies models in order to interpret the environments correctly. Similarly, the placement of such key sequence-stratigraphic surfaces as sequence boundaries and maximum flooding surfaces can only be done with confidence if the changing environments in the succession are interpreted correctly. Therefore, each chapter explores, in greater or lesser detail, the way in which the environment in question responds to changes in accommodation.

To get the maximum benefit from this volume, the level of required background preparation is about that of a 2nd- to 3rd-year undergraduate course in the principles of sedimentation and stratigraphy (Fig. 1). The book could be used for advanced undergraduate courses, although it is perhaps most suitable at the graduate level. Practicing professionals will also find much that is useful because presentations in the various chapters incorporate the very latest, cutting-edge research. The text is, however, written in the approachable fashion that is the hallmark of previous editions. Important terms with which students might not be familiar are defined when used first and are highlighted in italics. Although the book contains the most up-to-date information, the coverage of each environment is purposely condensed, so as to capture the 'essence' of that depositional system. All authors have, as a consequence, tried to keep the number of references to a minimum, and have not necessarily cited the origin of every idea presented. The references at the end of each chapter begin with the basic sources of information and these references are annotated. The remaining references are for those who wish to pursue a specific topic in greater depth.

We would like to thank all the

Figure 1. A group of upper level undergraduate and graduate students from Queen's University learning about carbonate facies models on a sunny autumn afternoon, while sitting on Ordovician limestone along the banks of the Oureau River, Quebec.

authors for their scholarship and for framing their chapters in the context of modern sedimentology. There is much original research in these contributions. A great deal of this information has come from research supported by the Natural Sciences and Engineering Research Council of Canada, and for this the authors and editors are very grateful. In many ways this volume stands as a testament to the effectiveness and wisdom of the NSERC system.

This book has been made possible because the Geological Association of Canada (GAC) has thrown its full support behind all aspects of the volume's preparation. This support has been augmented by the Canadian Society of Petroleum Geologists which has enthusiastically backed the volume throughout. The current edition would not have been possible without the continuous support of, and oversight by, Sandra Barr, Books Editor for GAC. The book itself reflects the efforts of many people, in addition to the authors. The drafting and editorial skills of Isabelle Malcolm at Queen's University are reflected throughout. The GAC staff in St. John's is as much responsible for the book as the authors and editors. Bev Strickland formulated the layout, while Karen Dawe skillfully controlled all publishing matters.

The study of depositional environments is an international endeavor, and yet we have chosen to make the book unabashedly Canadian. All of the authors or one of their co-authors is either in Canada today or have studied and worked in Canada in the past. The approach is, however, not parochial but international, with many writers choosing to involve colleagues from around the world. The result is a global approach but with a Canadian flavor. All of the authors are recognized experts on the environment about which they write, bringing a wealth of personal experience to their presentation. As editors, we are excited and pleased by the high quality and insight provided by the following chapters. We hope you find the book useful. Enjoy!

REFERENCES

Walker, R.G., *ed.*, 1979a, Facies Models: Geoscience Canada Reprint Series, 1, (1st edition), Geological Association of Canada, 211 p.

Walker, R.G., 1979b, Facies and facies models 1) General Introduction *in* Walker, R.G., *ed.*, Facies Models: Geoscience Canada Reprint Series, 1, (1st edition), Geological Association of Canada, p. 1-7.

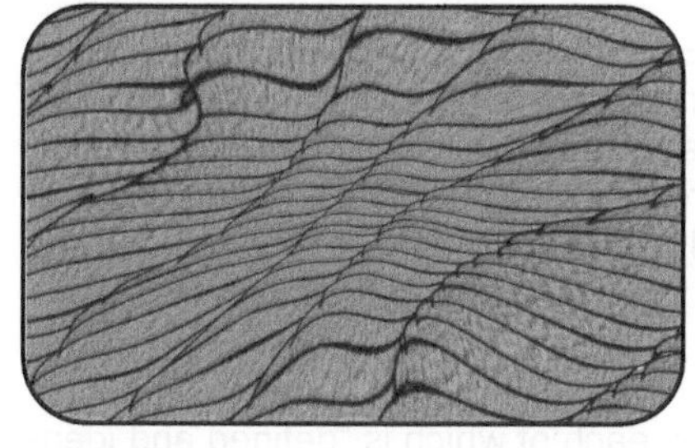

2. INTERPRETING SEDIMENTARY SUCCESSIONS: FACIES, FACIES ANALYSIS AND FACIES MODELS

Robert W. Dalrymple, Department of Geological Sciences and Geological Engineering, Queen's University, Kingston, ON, K7L 3N6, Canada

PREFACE

The goal of this chapter is the description of the facies-analysis approach to the interpretation of sedimentary successions. The use of facies and facies models is now so widespread that there is a tendency to forget that the formalization of this approach occurred quite recently. The popularity of this method is due to its effectiveness as a means of teasing information from the stratigraphic record, but it is due also, in no small measure, to the elegance of its presentation by Roger Walker in previous editions of this chapter. Indeed, Roger Walker's description of the creation and use of facies models is undoubtedly the definitive statement on the subject. As a result, the present version, while containing updated material and an expanded discussion of the methods used, relies very heavily on the version that appeared in the 1992 edition of this chapter (Walker, 1992). To make the presentation as smooth as possible, quotes have not been used extensively, even though significant parts of the text have been taken, with only minor revision, from the 1992 edition. This reflects the fact that it is difficult to improve on what has already been written.

INTRODUCTION

The surface of the Earth is a complex place. Many different biological, chemical and physical processes operate here, usually in intricate, nonlinear combinations, to generate sedimentary deposits. Unraveling the record contained in them is difficult. Over the last thirty years or so, there has been an exponential increase in our ability to extract information from sedimentary successions. The increased sophistication of our environmental interpretations is largely a result of the widespread use of the techniques of *facies analysis* and *facies models* (italicized words in this chapter are defined in Table 1) that are the subject of this volume.

Facies analysis and environmental interpretation rely on the observation that every biological, chemical and physical process produces a specific record of its action – a particular sedimentary structure, texture or type of fossil. Thus, each observable feature of a deposit can be used to infer something about the process(es) responsible for that feature if we understand the *process-response relationship(s)* in question. From the assemblage of processes that are deduced in this way, we can then infer the *depositional environment* in which the deposit was formed (Fig. 1). Without a thorough understanding of the linkage between environments, processes and sedimentary products, rigorous interpretations of ancient deposits would be impossible.

Whereas the *depositional environment* determines the processes that operate and hence the nature of the resulting deposits, the environments are themselves a function of the larger context in which they sit. Such contextual factors include the tectonic setting, sea-level change, climate and the geological age of the deposits (*see* Chapters 4 and 13 for more detailed discussions). All of these factors influence the nature of the environment and are recorded by the deposits, albeit in an indirect manner. Therefore, if a researcher is interested in making deductions about such things, it is necessary to start with a correct interpretation of the depositional processes and depositional environments; jumping too quickly to a sequence-stratigraphic or paleoclimatic interpretation is dangerous.

Over the years, sedimentary geologists have developed a rigorous methodology for the interpretation of depositional processes and environments. This approach (Fig. 1) consists of the differentiation and interpretation of successively larger bodies of sediment or sedimentary rock including *facies*, *facies associations* and *facies successions*. This approach relies heavily on the *facies models* presented in this book. Such environmental interpretations can then be used to move to more advanced and larger scale objectives, such as the sequence-stratigraphic subdivision of the succession, or the determination of the role of tectonism or climate on sedimentation. The purpose of this chapter is to outline this practical approach, and to present the general concepts that will be used in the following chapters.

FACIES ANALYSIS

Facies

The most widely used process for interpreting the genesis of a sedimentary succession begins with its subdivision into its fundamental building blocks (Fig. 1). These building blocks, which are internally homogeneous in some basic way, are termed *facies* (from the Latin word for the aspect, or 'appearance of' something; Table 1). As Walker (1992) has noted, the modern geological usage of the term *facies* was introduced by Gressly, in 1838, who used it to imply the sum total of all lithological and paleontological aspects of a stratigraphic unit. A translation of Gressly's extended definition is given by Middleton (1978). The term has been defined in many ways, but it is now generally used to apply to a body of rock, rather than to an abstract set of characteristics. It can be used in a purely descriptive manner (mudstone facies), and also in an interpretative sense (fluvial facies), but only after careful study of a succession. Succinct discussions of the problems and controversies associated with the term facies have been given by Middleton (1978), Anderton (1985) and

Table 1. Glossary of terms used in this chapter and throughout the book

Accommodation – the space available for potential sediment accumulation. In marine environments it is generally taken as being equivalent to the water depth, whereas in fluvial environments, it is the distance between the fluvial equilibrium profile and the instantaneous depositional surface. A similar equilibrium-profile approach has been applied to marine and turbidite systems, but is not widely used. Changes in accommodation are produced by eustatic sea-level change, tectonic movement of the sediment surface and changes in the depositional energy of the system that lead to sediment deposition or erosion.

Allostratigraphy – Subdivision of the stratigraphic record into mappable rock bodies, each of which is "defined and identified on the basis of its bounding discontinuities" (NACSN, 2005, p. 1578).

Architectural Element – a morphological subdivision of a particular depositional system characterized by a distinctive bedding geometry and assemblage of facies. They are directly related to the main landscape elements (*e.g.*, channels, bars, levees, *etc.*) of the original depositional system.

Base Level – the lowest point to which a fluvial system can erode. It is equal to sea (or lake) level and is the downstream tie-point for the fluvial equilibrium profile.

Bounding Discontinuity – a laterally traceable discontinuity of any kind; can be an unconformity, ravinement surface, onlap or downlap surface, condensed section or hardground.

Condensed Section – a deposit formed in deep water by very slow sedimentation such that the timelines are close together. It is characterized by hemipelagic or pelagic deposits, shell accumulations and/or biochemical deposits.

Correlative Conformity – the conformable surface within the basin that is temporally equivalent with the tip of the subaerial unconformity.

Depositional Environment – geographic and/or geomorphic area characterized by a distinct assemblage of depositional processes.

Depositional System – "three dimensional assemblage of lithofacies, genetically linked by active or inferred processes and environments" (Posamentier *et al.* 1988, p. 110). It embraces depositional environments and the processes acting therein over a specific interval of time. The deposits of contemporaneous depositional systems form systems tracts.

Eustasy – a world-wide change of absolute sea level, as measured relative to a fixed point such as the centre of the earth. Eustatic changes result from variations in the volume of water in the ocean basins (glacial control), or a change in the volume of the basins themselves (related to rates of mid-ocean ridge building and seafloor spreading).

Facies – a body of rock characterized by a particular combination of lithology and physical and biological structures that bestow an aspect ("facies") that is different from the bodies of rock above, below and laterally adjacent. The characteristics used to define facies are generally those that have genetic significance.

Facies Analysis- The widely used approach to the interpretation of sedimentary rocks that is based on the interpretation of the attributes of *facies*, *facies associations* and *facies successions* in terms of the processes responsible for their genesis, followed by the deduction of the most likely depositional environments in which the inferred processes may have operated. An iterative approach that uses preliminary environmental interpretations, in combination with facies models, to make predictions that are tested by additional observations, is a powerful way to arrive at sophisticated interpretations.

Facies Association- "groups of facies genetically related to one another and which have some environmental significance" (Collinson, 1969, p. 207).

Facies Model – a general summary of a particular depositional system, based on many individual examples from recent sediments and ancient rocks. This term should not be used for the environmental synthesis of an individual deposit.

Facies Succession – a vertical succession of facies characterized by a progressive change in one or more parameters such as the abundance of sand, grain size, sedimentary structures, bed thickness or faunal composition.

Flooding Surface – a surface produced by transgression that is placed at an abrupt upward transition from shallow-water deposits into those that accumulated in deeper water (typically offshore shale in marine settings). It is commonly marked by erosion caused by landward migration of the shoreline and is, thus, a ravinement surface.

Table 1. Continued

Maximum Flooding Surface (MFS) – a surface separating a transgressive systems tract (below) from a highstand systems tract (above). It is commonly characterized by a condensed horizon reflecting very slow deposition (*i.e.*, sediment starvation), or may be marked by biochemical sedimentation in an otherwise siliciclastic succession. Markers in the overlying systems tract typically downlap onto the MFS. It corresponds to the time when the shoreline is at its most landward location (*i.e.*, at the time of maximum transgression). In proximal areas, it corresponds approximately to the time of deepest water, but this need not be the case in more distal marine areas where subsidence continues during the time of sediment starvation.

Parasequence – "a relatively conformable succession of genetically related beds or bedsets bounded by marine flooding surfaces and their correlative surfaces" (Posamentier *et al.*, 1988, p. 110). Parasequences display an upward shallowing succession of facies.

Process-Response Relationship: the idea that a process (physical, chemical or biological) operating within a depositional environment with a certain intensity, will generate a limited range of physical responses that can be recorded in the resulting deposit. The direct and predictable linkages between processes and products allow a sedimentologist to deduce the processes that operated in an ancient environment from the observed attributes of the deposit. This concept is at the core of facies analysis.

Ravinement Surface – an erosion surface produced in the coastal zone, either by wave erosion within the shoreface or by tidal scour in channels within estuaries, during marine transgression of a formerly subaerial environment.

Sequence – "a succession of strata deposited during a full cycle of change in accommodation or sediment supply" (Catuneanu *et al.*, 2009, p. 19). Such cycles can range in duration for a few thousand years to many millions of years. Usually, they are taken to be bounded by unconformities and their correlative conformities.

Sequence Boundary – the surface separating two sequences. Different workers place this surface at different locations within a depositional cycle (Catuneanu *et al.*, 2009). It is most commonly placed at the subaerial erosion surface and at the correlative conformity that occurs in the area seaward of the lowstand shoreline. When defined in this way, the sequence boundary falls close to the lowest point in the relative sea-level cycle.

Sequence Stratigraphy – a variant of allostratigraphy that examines the stratigraphic stacking patterns that result from changes in accommodation.

Systems Tract – a linkage of contemporaneous depositional systems, forming the subdivision of a sequence. Systems tracts consist of conformable strata that were deposited during a particular segment of an accommodation (or base-level) cycle. Up to four systems tracts can be identified within a sequence: highstand systems tract (HST), falling-stage systems tract (FSST), lowstand systems tract (LST) and transgressive systems tract (TST).

Unconformity – a surface at which there is a 'significant' gap in time, caused by a combination of non-deposition and erosion. Most such surfaces are formed in subaerial settings during a relative lowstand of sea level. However, they can form in other ways, such as by non-deposition (with or without erosion) in distal marine environments. Subaerial unconformity surfaces and their correlative conformities are used as sequence boundaries. In this context, 'significant' necessitates that the temporal gap is a substantial fraction of a cycle of base-level change.

Reading (1986, 2003).

The most widely accepted definition of facies is given by Middleton (1978), who wrote that:

> "*the more common (modern) usage is exemplified by De Raaf et al. (1965) who subdivided a group of three formations into a cyclical repetition of a number of facies distinguished by "lithological, structural and organic aspects detectable in the field". The facies may be given informal designations ("facies A," etc.) or brief descriptive designations (e.g., "laminated siltstone facies"); and it is understood that they are units that will ultimately be given an environmental interpretation; but the facies definition is quite objective and based on the total field aspect of the rocks themselves… The key to the interpretation of facies is to combine observations made on their spatial relations and internal characteristics (lithology and sedimentary structures) with comparative information from other well-studied stratigraphic units, and particularly from studies of modern sedimentary environments*".

The subdivision of a succession into facies must be based on a careful description of the deposits (Fig. 2): this is what Middleton (1978) meant when he said that the establishment of facies is 'objective'. Facies are normally distinguished on the basis of attributes that are genetically significant because the focus of facies analysis is generally to ascertain some aspect of the origin of a succession. Thus, diagenetic and weathering characteristics (such as rock color and the type of cement) are generally of lesser value than primary attributes such as grain size

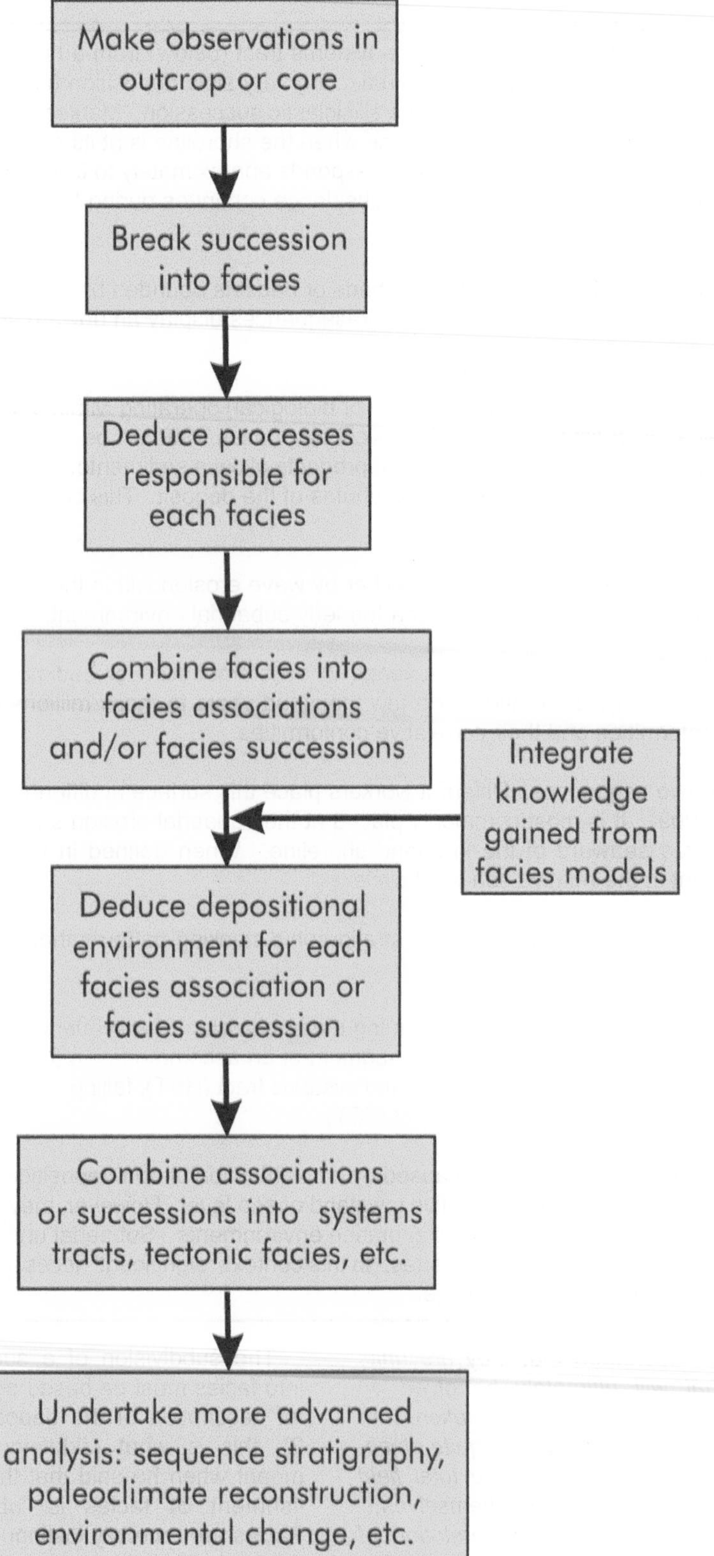

Figure 1. Relationship between facies, facies associations, facies successions, depositional environments and systems tracts as used in this book. This flow chart also illustrates the logical progression that a worker should take in the interpretation of a deposit. Only after the depositional processes that created the individual facies are known should one deduce the depositional environment. More advanced and larger scale analyses, such as the sequence-stratigraphic subdivision of the succession, are then built on the succession of environments that are present. Failure to follow this interpretive process may lead to erroneous interpretations.

and sorting, physical structures, fossil content, composition of syndepositional authigenic minerals, *etc.* The choice of characteristics to use in the designation of facies is generally evident only after the deposit has been examined in detail: the subdivision of the succession into facies is one of the last things that is done before the interpretation phase begins, because it is only then that the full range of deposit characteristics is known and the number of facies needed to describe the succession adequately can be assessed.

In the case of carbonate, fluvial and deep-marine deposits, some workers have suggested that there is a standard set of deposit types formed within each environment. They have created standardized facies schemes that are presumed to apply to all examples of such deposits (carbonates – Folk, 1959; Dunham, 1962 (*see* Chapters 13 to 18); fluvial – Miall 1977, 1985 (*see* Chapter 6); deep-marine – Mutti and Ricci Lucchi, 1972; Mutti, 1992 (*see* Chapter 12)). The use of such standardized schemes can speed up logging, and helps to point out key attributes of the succession worthy of recording. They should never be used, however, without critical examination to ensure that unexpected attributes are not overlooked. For example, the facies scheme for fluvial deposits does not include the tidal features that will be present in river deposits formed near a coast. Undue reliance on the standardized facies scheme might lead an unobservant worker to miss these tidal features.

Most studies have defined facies using qualitatively assessed combinations of distinctive sedimentary and organic structures (*e.g.*, De Raaf *et al.*, 1965; Cant and Walker, 1976), as observed in outcrops and cores. Facies can also be defined seismically, using such attributes as reflection amplitude, continuity and geometry (Roksandić, 1978; Tebo and Hart, 2005). If borehole data are available, facies and facies successions can be created using geophysical well-log attributes (*e.g.*, Saggaf and Nebrija, 2000). Statistical methods have been used to define facies (*e.g.*, Klovan, 1964), but such methods have generally fallen out of favor because it is

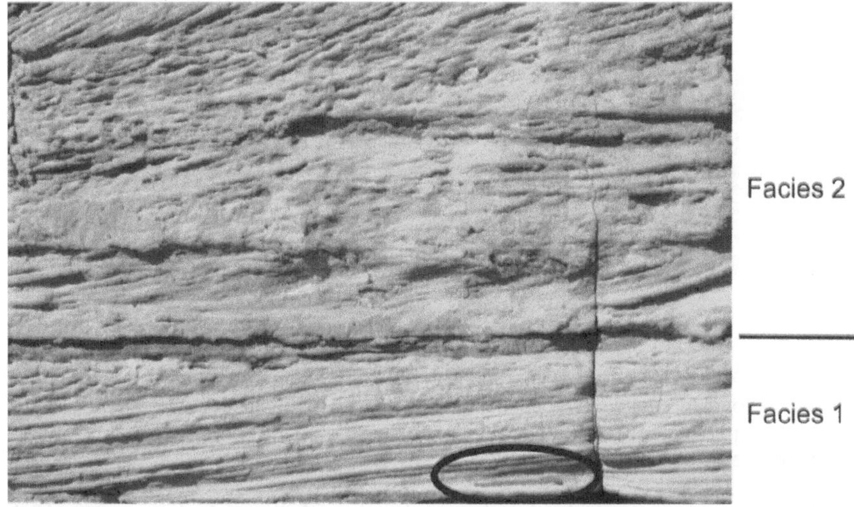

Figure 2. Outcrop of the Shannon Sandstone, Wyoming, containing hummocky cross stratification in fine-grained sandstone (base) and crossbedding in medium-grained sandstone (above). This difference in sedimentary structures and grain size forms the basis for the creation of two facies. Pen at base (circled) for scale.

Figure 3. Shannon Sandstone (top) gradationally overlying uniform mudstones of the Cody Shale. In a regional study with limited data or time, the separation of this succession into only two facies (A and B) may be the only distinction that is possible. Figure 2, however, comes from the upper left-hand portion of this outcrop and illustrates the finer scale detail that exists within Facies B. If time and the nature of the study warrant documentation of the detail shown in Figure 2, Facies B here could be designated a facies association consisting of all sandstones facies.

commonly difficult to quantify the genetically important information contained in physical and biological structures. The introduction of knowledge-based decision algorithms (*e.g.*, fuzzy logic and neural networks) into sedimentary geology has, however, brought about a resurgence of interest in statistical approaches, especially for use with subsurface data (*e.g.*, Saggaf and Nebrija, 2000; Tebo and Hart, 2005; Qi *et al.*, 2007).

Facies can be defined on many different physical scales, ranging from individual beds or groupings of similar beds (Fig. 2) to large-scale bodies that are comparable to lithostratigraphic units such as members or even formations (Fig. 3). The degree of subdivision is governed, in part, by the objectives of the study. If the objective is routine description on a regional scale, a broad subdivision may be sufficient (*e.g.*, all fluvial deposits as distinct from all marine deposits). If, however, the objective is more detailed, such as the teasing out all of the processes and environmental parameters responsible for a complex succession, or the characterization of a hydrocarbon reservoir when planning an expensive development program, then the facies subdivision must be done at a much finer scale. Attention to details is crucial if the best possible environmental interpretation is to be obtained (Fig. 4).

The scale of subdivision is also dependant on the time available, the degree of preservation of primary features, and the abundance of physical and biological structures. A thick succession of homogeneous deposits (*e.g.*, the lower part of Figure 3) will be difficult to subdivide, but a similar thickness of interbedded, coastal and shallow-marine sandstones and shales (with abundant and varied physical and biogenic structures) might be divisible into a large number of distinct facies. It is always better to create an unduly fine-scale subdivision in the field or when logging core – facies can always be recombined in the office, but a crude initial subdivision cannot be refined later. Anderton (1985) and Miall (2000) discuss strategies for erecting a set of facies for an individual deposit.

Facies Associations and Facies Successions

Many, if not most, individual, small-scale facies can have ambiguous environmental interpretations. For example, a medium-grained sandstone containing decimeter-scale crossbeds could have formed in any one of several depositional environments including a meandering or braided river, a tidal inlet, a coarse-grained shoreface, or a shelf dominated by tidal currents, because the process capable of forming subaqueous dunes is not unique to one setting. Diamict facies (*see* Chapter 5) are particularly difficult to interpret unambiguously, and the same applies to many carbonate lithofacies (*e.g.*, a bed of fossiliferous wacke-

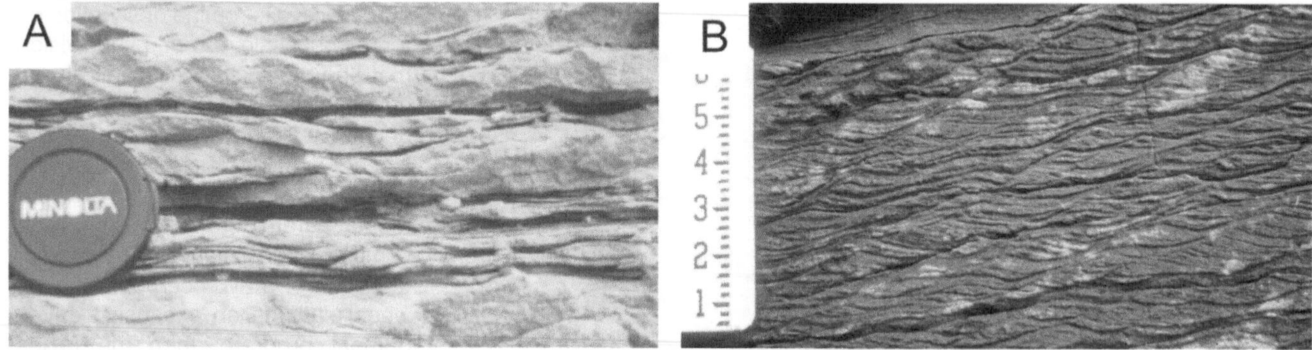

Figure 4. These two examples of current ripples might be lumped into one facies because they contain the same sedimentary structure. However, those in **A** contain mud drapes and bi-directional foreset dip directions, implying deposition in a tidal environment. Those in **B** contain no mud drapes, are uni-directional and are climbing steeply; they were deposited on the upper part of a fluvial point bar. Details such as these can be crucial in arriving at a correct environmental interpretation.

stone). Indeed, many facies defined descriptively in the field or core might, at first, suggest no particular environmental interpretation at all. The key to interpretation is to analyze all of the facies communally, in context, thereby obtaining important information that the facies, considered individually, might not provide. Two related approaches can be used.

The first is to combine closely related facies into *facies associations* (Figs. 2 and 3) that are defined as "groups of facies genetically related to one another and which have some environmental significance" (Collinson, 1969). The assemblage of facies comprising a facies association should occur in intimate physical association with each other, although not every occurrence of the association need contain all elements. Facies associations will more likely correspond to a unique depositional environment than any of the facies by themselves because the associations are larger bodies. Relatively detailed environmental reconstructions have become routine and many excellent studies exist for any given depositional environment. Therefore, there is less need now than there was10 to 20 years ago to describe facies in intricate detail. Thus, whereas papers that described dozens of facies in (excruciating) detail were commonplace in the 1960s and 70s, facies are now typically documented at a coarser level, with greater amalgamation than before: today's facies were yesterday's facies associations. In fact, many studies jump immediately to facies associations with little or no description of the individual facies, the latter being documented in tabular form to save precious space in a journal (*e.g.*, MacNaughton *et al.*, 1997). This is acceptable as long as the necessary detailed observations have been made to justify the process and environmental interpretations that are presented.

The second approach is to use any sequential ordering of facies (or facies associations) to define *facies successions*. The concept of a succession implies that certain facies properties change progressively in a specific direction (vertically or laterally), although an unordered 'random' succession is a possibility. The properties that might change regularly include the proportion of sand (sandier-upward succession; Figs. 3 and 5), the grain size of the sand (hence a coarsening-upward succession), the amount of bioturbation, or the type and/or abundance of fossils present. The term succession can also be used in an interpretive way, such as for an upward-shoaling shoreface or tidal-flat succession (*see* Chapters 8, 9 and 16). In older work, it was common to see the term 'facies sequence', but this wording should be avoided because the word *sequence* has now been given a very specific definition (Table 1). Consequently, the term *facies succession* should be used to avoid ambiguity and the potential for misinterpretation. The only exception that has been retained here is the 'Bouma sequence' because this term is deeply entrenched in the literature.

The relationship between *depositional systems* in space, and the resulting stratigraphic successions developed through time, was first emphasized by Johannes Walther (1894, *in* Middleton, 1973) in his Law of the Correlation of Facies, also known as Walther's Law. Walther stated that,

> "*it is a basic statement of far-reaching significance that only those facies and facies areas can be superimposed primarily which can be observed beside each other at the present time*".

Application of the law suggests that, in a vertical succession of facies, a *gradational* transition from one facies to another (Fig. 5) implies that the two facies represent environments that were once adjacent laterally. If the contact between two facies or facies associations is sharp and/or erosional (Fig. 6), there is no way of knowing whether they represent environments that were once laterally adjacent. Indeed, sharp breaks between facies (marked, for example, by channel scours, or by intensely bioturbated horizons or shell concentrations that imply a lengthy period of non-deposition) may signify fundamental changes in the depositional environment and the beginning of a new cycle of sedimentation. These sharp breaks, or *bounding discontinuities*, are now used to separate stratigraphic *sequences*, *systems tracts* and *allostratigraphic* units (Fig. 5).

Although Walther's Law has proven to be highly useful, especially in terrigenous-clastic deposits, the lateral migration of environments that forms the basis of the law is not the only way in which a vertical succession of facies may be formed. Walther's Law assumes, implicitly,

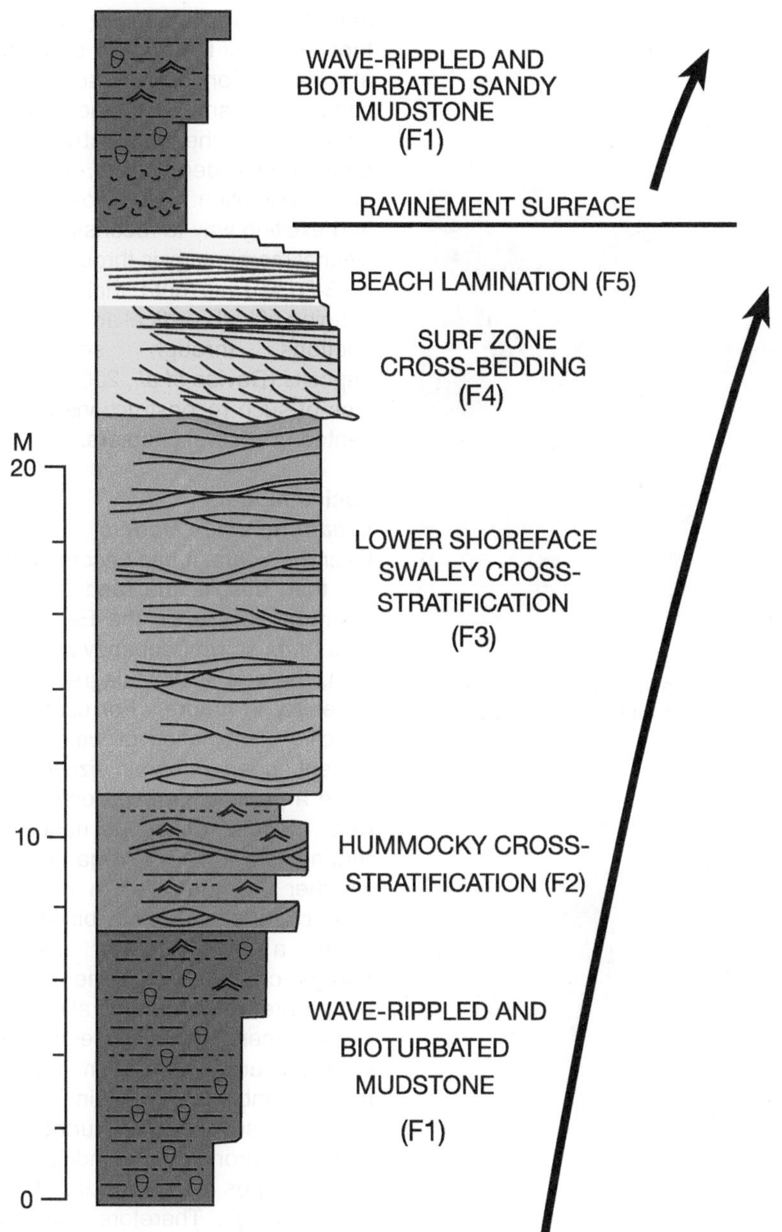

Figure 5. Gradational upward-coarsening succession of facies (F1–F5) formed by progradation of a beach-shoreface-offshore environment (*see* Chapter 8). The *ravinement surface*, which marks an abrupt shift back to offshore shale (F1), is a discontinuity surface produced by transgressive coastal erosion. Such discontinuities mark the boundaries between facies successions. Here, this surface is the boundary between two *parasequences*. (Modified after Plint, 1988, Figure 3).

that the larger environment does not change substantially through time and that all facies coexist throughout the accumulation of the succession. This assumption need not be true in all cases (Kendall, 1988). For example, a small sea-level fall on a rimmed carbonate platform can lead to restriction of water circulation in the back-reef area and an increase of salinity (*see* Chapters 14 and 20). This will cause a change from open-marine limestone having a diverse fauna to a restricted dolo-lime mudstone with little or no fauna. In this case, the vertical juxtaposition of these two facies reflects a secular evolution of water chemistry rather than the superposition of coexisting subenvironments.

Architectural Elements and Paleogeomorphological Reconstruction

The concept of *architectural elements* (Allen, 1983; Miall, 1985; *see* Chapter 6) is separate from, but related to, the concepts of facies, facies associations and facies successions. Architectural elements are related to the geometry of the deposits, and are distinguished first and foremost on the basis of bedding geometry; thus, successions in which the bedding is horizontal are distinct from successions in which the bedding possesses an original dip. The later group of architectural elements is subdivided further, on the basis of the direction that the master bedding planes dip relative to the local current (downstream, upstream or transverse). The nature of the deposits within the element allows further subdivision, especially for those elements that are horizontally bedded. The idea that a hierarchy of surfaces exists within a deposit is closely related to that of architectural elements: surfaces with limited lateral extent are truncated by or lap onto surfaces that are more extensive and, thus, of higher order. Architectural-element analysis is performed by careful examination of outcrops using photomosaics, LIDAR and mapping to determine the geometry and extent of every surface (Fig. 7). This allows the worker to reconstruct the morphology of the depositional environment at a range of scales. Although architectural elements were first defined, and are most commonly used, in the study of fluvial deposits, the concepts can be applied to the deposits of any depositional environment. They are most useful, however, with the deposits of environments where there is channelized flow (*e.g.*, fluvial, tidal and submarine-fan settings) because this is where the bedding geometry is most complex.

The relationship of architectural elements to facies, facies associations and facies successions is not simple because it depends on the scale at which the various sediment bodies are recognized. Bedding geometry is an important descriptive attribute of a deposit and can be used to define or distinguish between facies elements at a wide range of scales. Thus, a particular architectural element could be a component of a facies, a facies in its own right, or even a facies succession (*e.g.*, the upward-fining succession that typi-

Figure 6. Shannon Sandstone resting erosively on the Cody Shale. Compare with the gradational contact shown in Figure 3. The two successions have very different interpretations because of the presence of the discontinuity. (For more on the origin of the enigmatic Shannon Sandstone, *see* Suter and Clifton, 1998).

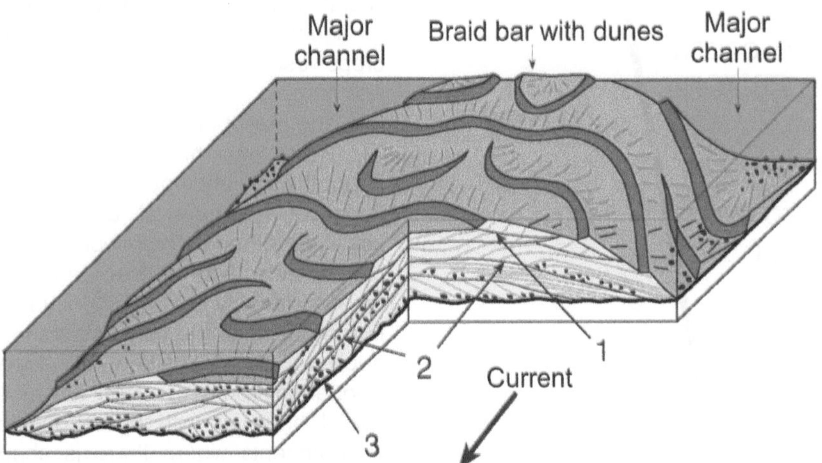

Figure 7. Reconstruction of a fluvial braid bar, based on the architecture of the deposits. The numbers give the surface hierarchies following Miall (1985): 1 – boundaries between individual crossbed or ripple sets; 2 – master bedding planes, the dips of which define lateral-accretion and forward-accretion architectural elements (in flow-transverse and flow-parallel sections, respectively); and 3 – horizontal erosion surface produced by migration of a channel thalweg. (After Allen, 1983, Figure 19a).

fies the lateral-accretion architectural elements generated by point bars).

The principal use of architectural-element analysis and surface hierarchies is to reconstruct the geomorphology of the depositional environment, within the limits of what is preserved. In fact, architectural elements define the landscape elements that were originally present (*e.g.*, channels, bars and levees in fluvial deposits). The reconstruction of paleo-geomorphology has been a topic of increasing interest over the last 15 to 20 years. Not only is this possible at a relatively small scale by means of architectural-element analysis, but it is increasingly possible at a regional or basin scale using sequence-stratigraphic techniques. Both techniques allow geomorphic reconstructions because they are based on the identification of surfaces (*i.e.*, bedding planes) that approximate time lines and follow closely the topography of the original depositional surface. Conventional seismic sections and seismic sequence analysis also permit the reconstruction of depositional geometries, but at a lower resolution. The most detailed and exciting way to reconstruct depositional morphology is through the use of 3-D seismic data that allow the imaging of both vertical and horizontal sections through sedimentary deposits (Davies *et al.*, 2007). Examples of such reconstructions are presented in several chapters.

Facies Models

Creating Facies Models

Over the years, it has become apparent that, despite the randomness of many processes on the Earth's surface (*e.g.*, storm frequency and intensity), there is a high degree of predictability in nature. For example, a specific combination of water depth, current speed and grain size will generate a predictable bed configuration (*e.g.*, ripples, dunes or upper flow regime plane bed) regardless of whether the current is in a river or tidal channel. In the biochemical realm, a specific water chemistry always creates the same suite of evaporate minerals or allows the same general faunal assemblage to inhabit an area (the exact members of the assemblage depending on the age of the succession). Furthermore, certain environmental conditions and deposit types typify specific depositional settings. Therefore, river channels commonly contain dunes and form crossbedded sandstone, shorefaces are typified by wave-generated structures and especially hummocky cross stratification, and carbonate tidal flats are characterized by frequent exposure and environmental stress that lead to the proliferation of microbial mats. As a result, there is a tendency for a given environment to be associated with a particular assemblage of facies.

Individual environments also tend to behave in the same manner regardless of contextual differences. Thus, meander bends in rivers migrate sideways, producing upward fining facies successions, regardless of whether the sediment is gravel or fine sand, and when carbonate tidal flats prograde, they generate upward-shallowing successions with a gradual upward decrease in faunal diversity and an increase in the effects of exposure, no matter what the tectonic setting or age. There will, of course, be differences between any two river bends or tidal flats, but it will be variation on a common theme. Such common themes are the basis of what is one of most important tools at the disposal of sedimentary geologists, namely *facies models*.

As stated succinctly by Walker (1992, p. 6), "*a facies model can be defined as a general summary of a given depositional system*". The process by which they are created, and the uses to which they may be put, are illustrated in Figure 8, using turbidites as an example (*see* also Chapter 3, Figure 7). It is assumed that, if enough modern and ancient turbidites can be studied using a diversity of techniques, *and* we learn enough about how turbidity currents behave using analogue and numerical experiments, we should be able to make some *general* statements about the nature of the deposits created by them (*see* Chapter 12). Generally, the creation of a facies model involves the subjective assessment of all of the individual case studies to determine which attributes are common to all examples and which ones are of a purely local nature as a result of the unique context and history of each one. In essence, the wealth of information obtained from the many examples is then *distilled*, boiling away the local details while concentrating the important features that they all have in common. Those general characteristics, including the spatial distribution of processes and facies, together with the longer term behavior of the system, comprise the facies model for that environment. But what constitutes local detail, and what is general? Which aspects do we dismiss, and which do we extract and consider important?

As Walker (1992) notes,

"*answering these questions involves experience, judgment, knowledge and argument among sedimentologists*".

Consequently, facies models are dynamic constructs that must be continuously updated to reflect our current understanding of an environment. This is shown in Figure 8 by the feedback loop between newly described examples and the model itself. For example, the recent appreciation that the mud and sand issuing from the mouth of a river tends to move alongshore has given rise to revised models for deltas and shelves that include a degree of asymmetry that was not present before (*see* Chapters 8 and 10).

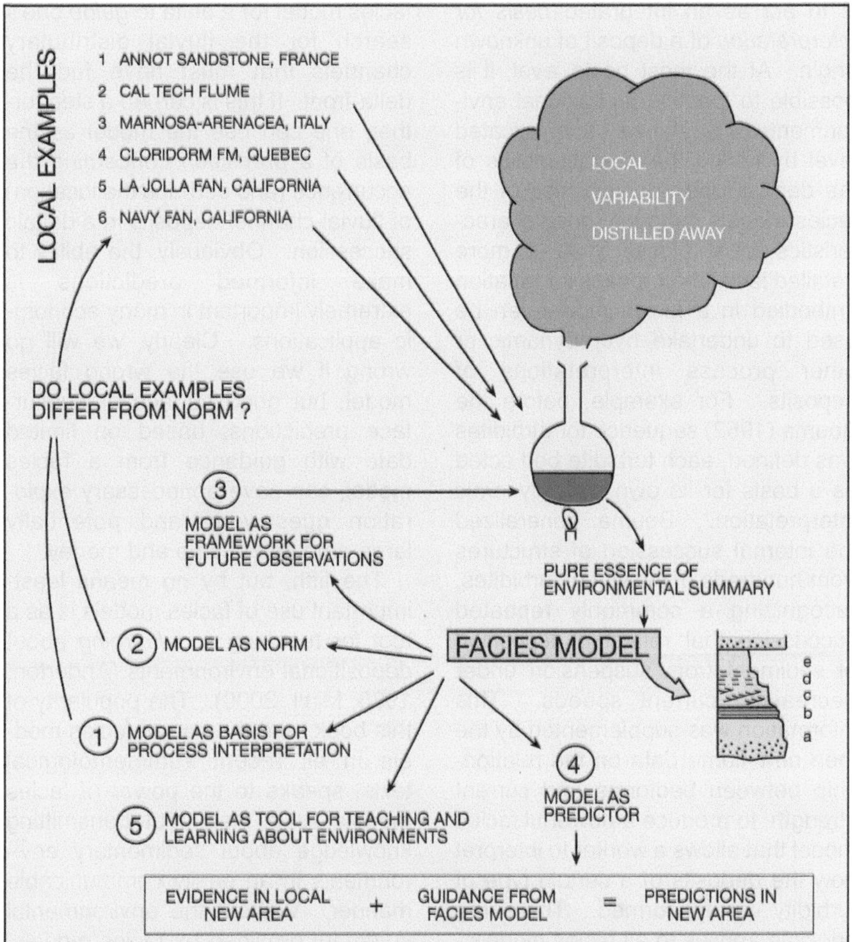

Figure 8. Distillation process leading to the generation of a facies model, with the uses to which a facies model may be put. (Modified after Walker, 1984, Figure. 9).

It follows from the preceding discussion that a facies model (in the sense used here) cannot be created from the interpretation of a single example. It is common, however, to see in the literature the environmental reconstruction of a given deposit referred to as a 'facies model'. This is inconsistent with the sense of generality that is an explicit element of facies models. The reconstruction or summary of a given deposit, regardless of how elegant it is, must retain the unique aspects of that deposit. This prevents it from fulfilling the five functions of facies models discussed below.

The Five uses of Facies Models

In previous editions of this chapter (*e.g.*, Walker, 1992), facies models were said to have four inter-related functions (Fig. 8):

1. It can act as an integrated basis for *interpretation* for the system that it represents;
2. It can act as a *norm*, for purposes of comparison;
3. It can act as a *framework* and *guide* for further observations; and
4. It can act as a *predictor* in new geological situations.

To these, a fifth is added:

5. It can be used as an aid for *teaching* and *learning* about the environment in question.

Much of the following discussion is taken from Walker (1992).

The first function of facies models

is to act as an integrated *basis for interpretation* of a deposit of unknown origin. At the most basic level, it is possible to identify depositional environments to a relatively sophisticated level by noting the characteristics of the deposit and seeing which of the facies models matches those characteristics most closely. At a more detailed level, the process information embodied in a facies model can be used to undertake hydrodynamic or other process interpretations of deposits. For example, before the Bouma (1962) sequence for turbidites was defined, each turbidite bed acted as a basis for its own hydrodynamic interpretation. Bouma generalized the internal succession of structures from hundreds of individual turbidites, recognizing a commonly repeated succession that reflected deposition of sediment from suspension under decreasing current speeds. This information was supplemented by the then new flume data on the relationship between bedforms and current strength, to produce a powerful facies model that allows a worker to interpret how the deposits of a certain type of turbidity current formed. The same principle applies to all facies models.

Second, the model may act as a *norm*, against which an unknown deposit can be compared (Fig. 8). Without a norm, we are unable to say whether a particular deposit contains any unusual features. If the deposit conforms closely to the facies model, its interpretation is simplified. If the new example differs, we can specify exactly how it differs, and then ask questions about the new example that could not have been asked without the norm. For example, compared with the norm, why are the beds in a coarse-grained deep-water deposit thicker and why do they contain successions of structures that do not match the classical Bouma sequence? These questions can open up new avenues of productive thought that may lead to an enhanced understanding of the deposit under study; without the norm, such questions cannot be asked and important characteristics might be overlooked.

The third and fourth functions of the model are closely linked. If previous work indicates that the deposit is deltaic, then it is logical to use the facies model for a delta to *guide* one's search for the fluvial distributary channels that must have fed the delta-front. If this is carried a step further, one can use the model as the basis of a *prediction* concerning the occurrence (and perhaps the location) of fluvial-channel deposits in a deltaic succession. Obviously, the ability to make informed predictions is extremely important in many economic applications. Clearly, we will go wrong if we use the wrong facies model, but good surface or subsurface predictions, based on limited data with guidance from a facies model, can save unnecessary exploration guesswork, and potentially large amounts of time and money.

The fifth, but by no means least-important use of facies models is as a tool for *teaching* and *learning* about depositional environments (Anderton, 1985; Miall, 2000). The popularity of this book, and the use of facies models in all recent sedimentological texts, speaks to the power of facies models as a means of transmitting knowledge about sedimentary environments in an easily communicable manner. Without the environmental summary provided by facies models, students at any level would be overwhelmed by the complex and unstructured information that is available for each depositional setting. Therefore, facies models are an essential element of a good sedimentary education.

SEQUENCE STRATIGRAPHY AND ITS RELATIONSHIP TO FACIES MODELS

Facies models encapsulate the innate, autogenic behavior of a depositional environment or, in the case of the Bouma sequence (*see* Chapter 12), the hydrodynamic processes responsible for the deposition of an event bed by a turbidity current. *Sequence stratigraphy*, on the other hand, is a tool for examining the response of large *depositional systems* to allogenic forcing. Despite the fundamental difference between these two approaches to interpreting sedimentary deposits, there are important relationships between them. As a result, modern-day facies modeling cannot be undertaken without reference to at least some sequence-stratigraphic concepts such as changing *accommodation*. It is necessary, therefore, to review briefly the basics of sequence stratigraphy, to explore how they relate to facies modeling. This is not a comprehensive examination of sequence stratigraphy, and interested readers are encouraged to consult such recent references as Postamentier and Allen (1999), Schlager (2005), Catuneanu (2006) and Catuneanu *et al.* (2009) for more detailed discussions.

Sequence stratigraphy is used to examine and interpret the stratigraphic organization produced by the interplay between changes in accommodation and sedimentation/erosion. *Accommodation*, which is the space available for possible sediment accumulation, is one of the most important and pervasive controls on sedimentation: positive accommodation allows sediment accumulation to take place, whereas negative accommodation causes sedimentation to cease and creates the potential for erosion and the production of an *unconformity*. In marine environments, the change in accommodation is given by the vector summation of subsidence/uplift of the seafloor and eustatic (*i.e.*, global) sea-level fall/rise. Thus, subsidence and a eustatic rise create accommodation, whereas uplift and eustatic fall reduce accommodation. Under most circumstances, it is not possible to decouple the influences of *eustasy* and tectonic movements. Therefore, the combined result is referred to as a change of relative sea level. In terrestrial environments far removed from the influence of the sea, the factors influencing accommodation are more complex. In fluvial settings, vertical movement of the fluvial equilibrium profile, which is defined as the fluvial gradient that allows complete bypassing of the sediment supplied from upstream (*i.e.*, there is neither erosion or deposition averaged over many river floods), either creates accommodation (rise of the equilibrium profile) or removes it (fall of the equilibrium profile). Changes in the position of the fluvial equilibrium profile may be caused by changes of *base level* (*i.e.*, lake or sea level at the downstream end of a river), by tectonic changes in the elevation of the land surface, or by changes in the sediment-transport capacity of the river as a result of climate change. The controls on accommodation in aeolian and glacial environments have not been explored systematically but are related to the factors that determine the balance between sedimentation and erosion (*e.g.*, wind strength, elevation of the water table, the erosiveness of the glacier, *etc.*).

It is now recognized that accommodation changes cyclically at frequencies ranging from Milankovitch time scales (tens of thousands to a few hundred thousand years) to multi-million-year-long tectonic cycles (Schwarzacher, 2000). These repeated changes in accommodation (Fig. 9) lead to a systematic change in the nature of sediment accumulation. The resulting sedimentary response reflects the extent to which sediment fills, or does not fill, the accommodation that is available. Thus, if the rate of sediment supply by physical, chemical or biological processes exceeds the rate at which accommodation is produced, then the local accommodation will be decreased (*i.e.*, the water depth will become less in subaqueous environments) and there will be aggradation and/or progradation. In coastal and shallow-marine, terrigenous-clastic environments where much of the river-supplied sediment is deposited relatively close to the coast (*see* Chapters 8 and 10), this leads to progradation (*i.e.*, regression–a seaward movement of the coast and the associated marginal-marine facies). Two types of regression are recognized (Fig. 9), normal regression which occurs with a rising sea level (*i.e.,* regression with a rising trajectory for the shoreline or platform margin) as a result of sediment supply outpacing the rate of creation of accommodation, and forced regression, which occurs as sea level falls (*i.e.*, the shoreline trajectory is downward). Conversely, if the rate of sediment supply is less than the rate of generation of accommodation at the coast, there is transgression (*i.e., a* landward movement of environments). In carbonate environments, the spatial distribution of sedimentation rate is different from that in terrigenous-clastic settings because of

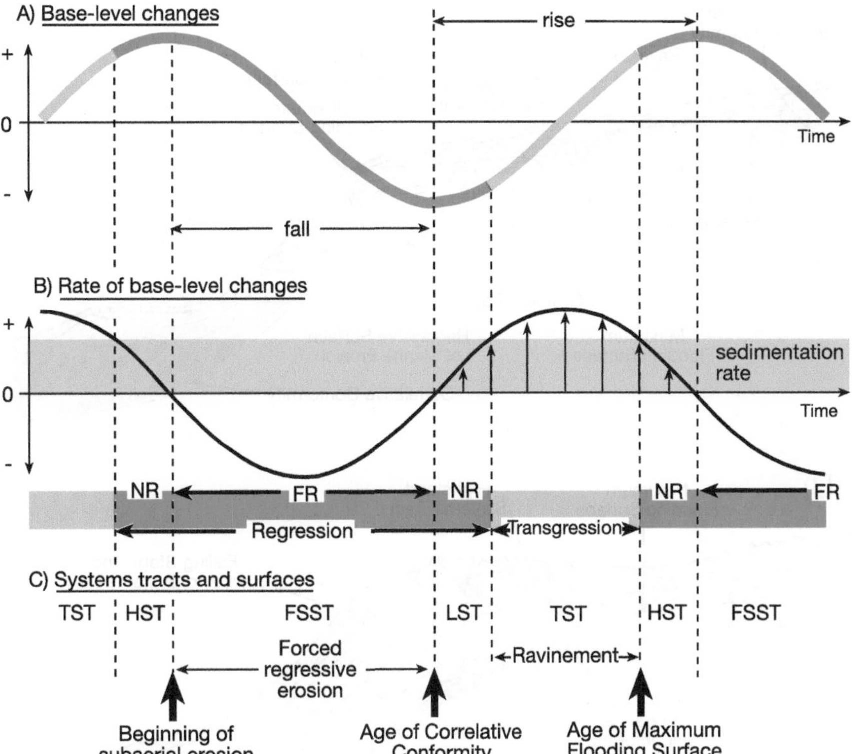

Figure 9. Schematic representation of a cycle of accommodation change resulting from changes of base level (= sea level). This cycle is shown as being symmetric, although this need not be the case. For example, eustatic sea-level cycles during the late Quaternary have been markedly asymmetric, with a long falling stage and a very abrupt rise. (Based on Catuneanu, 2006, Figure. 3.19).

the ability of sediment to be generated over wide areas (*see* Chapter 13). In warm-water (tropical) settings where photozoan assemblages predominate, sediment production is concentrated in shallow water, but may be spread more or less evenly over broad platforms, rather than being concentrated near the coast. This leads to vertical aggradation of the carbonate platform as long as accommodation continues to be created, with progradation of the platform margin if sediment is exported from shallow water into the adjacent deeper basins, a situation that occurs most vigorously at times of limited accommodation across the top of the platform. In cool- to cold-water settings, the sedimentation rate is not so strongly depth controlled and more uniform aggradation/progradation of seaward-sloping ramps occurs whenever there is available accommodation and sufficient sediment production.

The repeated rises and falls of relative sea level (or, more generally, of base level), and the associated transgressions and regressions, are widely believed to produce a predictable organization of the stratigraphic record (Fig. 10). Sequence stratigraphy is the approach that most workers now adopt to describe this organization. Several types of sediment bodies are recognized, including, from smallest to largest, *parasequences*, *systems tracts* and *sequences*. The sequence is the fundamental unit of sequence stratigraphy and is generally regarded as the deposits associated with a single accommodation cycle (Fig. 9). There are various schemes for the placement of *sequence boundaries* (*see* discussion in Catuneanu *et* al., 2009), but in this book, it is generally placed at the subaerial unconformity that forms when relative sea level is low, and is extended seaward along the *correlative conformity* in areas where sedimentation was continuous. Systems tracts are the component building blocks of sequences. They are defined by their position on the relative base-level curve (Fig. 9) and are recognized by the stacking

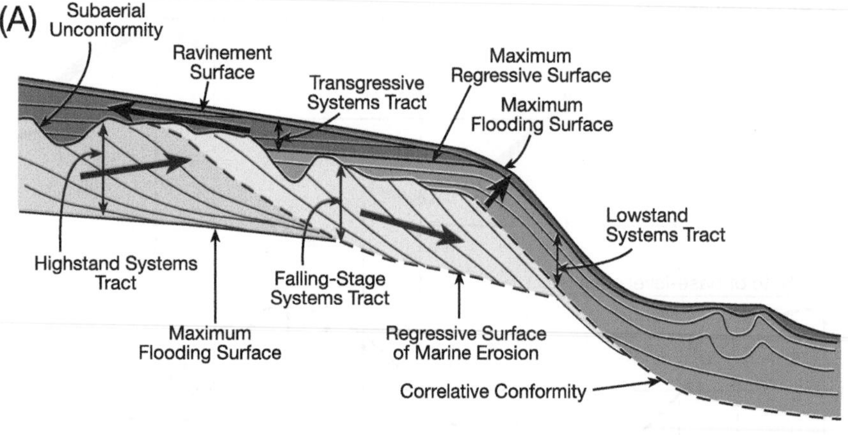

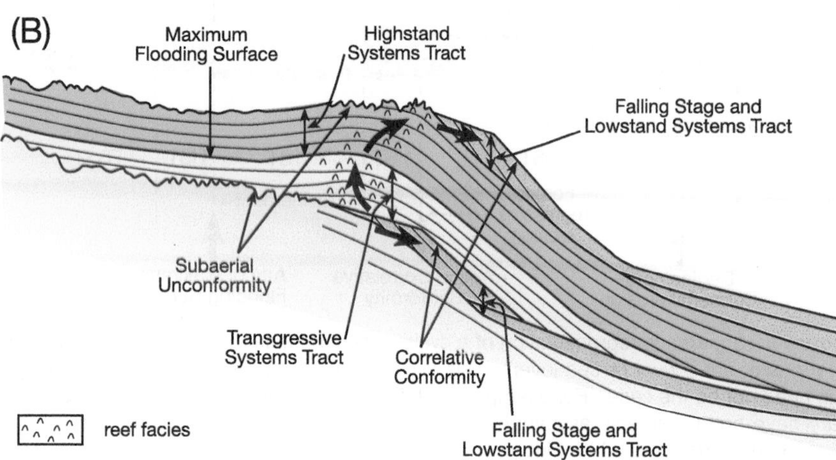

Figure 10. Schematic dip sections for **(A)** a siliciclastic system and **(B)** a rimmed carbonate platform, showing the stratal architecture, systems tracts and major surfaces produced over an accommodation cycle. (**A**, based on many sources and especially Willis and Gabel, 2001, Figure 8a; and **B**) based on Handford and Loucks, 1993, Figure 6).

patterns of the deposits. In this book, up to four systems tracts are recognized within an idealized, complete sequence (Fig. 10):

1. Highstand systems tract (HST)—progradation with a rising base level (an upward shoreline trajectory; *i.e.*, normal regression) when base level is near or at its highest elevation;
2. Falling-stage systems tract (FSST)—progradation with a falling shoreline trajectory (*i.e.*, forced regression);
3. Lowstand systems tract (LST)—normal progradation with a rising shoreline trajectory when base level is near or at its lowest elevation; and the
4. Transgressive systems tract (TST)—transgression caused when base level rises more rapidly than sediment accumulates.

The sequence boundary is typically defined as lying on top of the FSST and beneath the LST. The TST is separated from the HST by the *maximum flooding surface*, which forms at the time when the shoreline is at its most landward position at the end of a transgression. In distal, deeper water areas, it is typically marked by a *condensed section* (*i.e.*, a thin succession of fine-grained deposits that accumulated very slowly), although it is also possible for it to be marked by biochemical deposition in an otherwise siliciclastic succession.

A *parasequence* is an upward-shallowing, progradational succession of strata that is bounded by *flooding surfaces*, across which an abrupt deepening caused by a transgression is evident (Fig. 5). Two possible origins exist for parasequences. In deltaic successions, progradation of an individual delta lobe generates a parasequence, with the bounding flooding surfaces representing transgression of the abandoned lobe (*see* Chapter 10). Parasequences may also be created by the interplay of two or more orders of eustatic sea-level change and tectonics. In this case, parasequences can form only during times when the rate of rise produced by a longer duration cycle, supplemented by tectonic subsidence, more than compensates for the rate of fall of a shorter duration cycle, leading to a stepped continuous rise of relative sea level: parasequences are deposited during the episodes of slow rise, and the flooding surfaces form during periods of rapid rise. Unfortunately, many sequences can masquerade as a parasequence because the evidence for the base-level fall has been removed or is not otherwise evident in the study area. This may be the case with many carbonate upward-shallowing successions. Therefore, if the term 'parasequence' is used at all, it should only be used in an observational sense, with no genetic connotation, until such time as the detailed accommodation history has been deciphered.

Although facies modeling and sequence stratigraphy are distinct, they are related in that the changes in accommodation can have a direct affect on the nature of sedimentary environments. For example, changes in accommodation without commensurate changes in the sedimentation rate will lead to changes in water depth that are expressed in the nature of the facies. Accommodation changes cause the equilibrium profile of channels (whether fluvial, tidal or deep-sea) to be raised or lowered, which, in turn, influences the degree of amalgamation of the channel deposits. In siliciclastic environments, the amount of accommodation in areas up-system from the site in question exerts a profound influence on the grain size of the sediment available for deposition: limited accommodation allows coarser sediment to reach a particular area than does higher accommodation along the transport path (*see* discussion in Chapter 4). In shallow-marine carbonate settings, the local production of sediment isolates the environment from changes in accommodation elsewhere, but shallowing nevertheless occurs if production exceeds the rate of creation of accommodation. In the case of the deep sea, the existence of unfilled accommodation on the adjacent shelf or carbonate platform means that little or no sediment will be exported to deep water and this depositional system is largely inactive.

As a result of the important influence of accommodation, most facies models now take this into account. Thus, all chapters in this book include at least some discussion of how the environment responds to changes in accommodation, and provide insight into how key sequence-stratigraphic surfaces, such as sequence boundaries and maximum flooding surfaces, can be recognized. In the extreme case, separate facies models exist for different accommodation scenarios. Thus, there are distinct facies models for prograding coasts (*see* Chapters 8,10 and 16, in particular) and transgressing coasts (*see* Chapter 11). The sequence-stratigraphic subdivision of a succession must be based on the application of facies-analysis techniques because the identification of systems tracts and key sequence-stratigraphic surfaces requires evaluation of the facies stacking patterns and shoreline/platform-margin trajectories, discussed above. Thus, a sequence-stratigraphic analysis of a succession cannot be done without the application of the information provided in this book (Fig. 1).

METHODOLOGY OF STRATIGRAPHIC ANALYSIS

The stratigraphic record is exceedingly complex and care must be taken when approaching the analysis of any succession, regardless of how simple it may appear at first glance (*see*, for example, Figure 35 in Chapter 8). This section provides some guidelines for undertaking this challenging task. It applies equally to interpretations at the local or basin scale, and to the application of both facies and sequence-stratigraphic models, although the former is emphasized in the examples used. Additional discussions can be found in Anderton (1985) and Miall (2000).

Process-response Relationships

Facies analysis and facies modeling are based on the presumption that there is a known relationship between each depositional process operating in an environment and the sedimentary record of that process. This understanding is at the heart of the approach outlined in Figure 1. As indicated there, the approach least likely to result in an erroneous interpretation is first to interpret the processes that were responsible for the observed attributes of the succession and then, from the assemblage of inferred processes, to deduce the most likely environment in which the deposits formed. Only when the environments comprising a succession have been reconstructed, is it reasonable to move to more synthetic interpretations, such as the sequence-stratigraphic subdivision of the succession. As one gains experience, it is possible to move more rapidly through this interpretation process, although it is usually necessary to return to basics when beginning to work on deposits with which the worker may have limited experience.

To undertake facies analysis in this way, it is necessary to have a thorough understanding of the relevant processes (biological, chemical or physical). Unfortunately, such basic knowledge is too often passed over quickly in introductory courses, to get to the exciting new, large-scale concepts. It is also necessary to remember that there might be no one-to-one relationship between a process and the resulting feature in the deposit. Thus, while subaqueous dunes produce crossbedding, not every crossbed was formed by a subaqueous dune. This occurs because several processes generate crossbedding. Yet another example: flute marks typify the bases of turbidite beds, but flutes can also be formed by wave action or tidal currents acting on a layer of mud. Such 'non-unique solutions' are common and make interpretation of sedimentary successions difficult. In most cases, it is the total assemblage of attributes that allows one to choose between alternatives and to eliminate the ambiguity that exists in such cases.

Use of Multiple Working Hypotheses

Because there may be more than one interpretation for a sedimentary feature or a succession, particularly at an early stage in the analysis, it is recommended that conscious use be made of the idea of multiple working hypotheses (Chamberlin, 1890; Railsback, 2004; Fig. 11). Thus, once a certain body of observations has been assembled, a worker should try to think of all of the possible ways to interpret them rather than jumping to a single conclusion about their origin. So, a crossbedded sandstone might represent the deposits of several possible environments, the possibilities including a river channel, a tidal channel, a tidal inlet, a coarse-grained upper shoreface, a current-reworked transgressive-shelf lag, and so on. From there, one can begin to think about ways to eliminate some of the possibilities. For example, the presence of bimodal crossbed orientations in close proximity (and not separated by a pronounced erosion surface that would suggest that the oppositely oriented crossbeds are not related to each other genetically) would reduce the probability that the deposit was formed in a river while giving support for a tidal origin. Additional observations should then be sought to restrict the interpretation further.

In the end, this approach should lead to the most robust interpretation possible with the available data. Not only will all possible interpretations have been considered, thereby avoiding potential surprises later, but you will be able to argue logically why some were excluded from further consideration while the one that is ultimately chosen is supported by the observations.

This approach represents the application of the formal scientific method whereby you: 1) create a 'hypothesis' (an initial interpretation), 2) test that hypothesis by the collection of additional observations, and 3) ultimately arrive at the equivalent of a 'theory' (the final interpretation) that explains the observations best (Fig. 11). Thus, one can hypothesize that a crossbedded sandstone is a river deposit. From this preliminary interpretation it follows that the cross

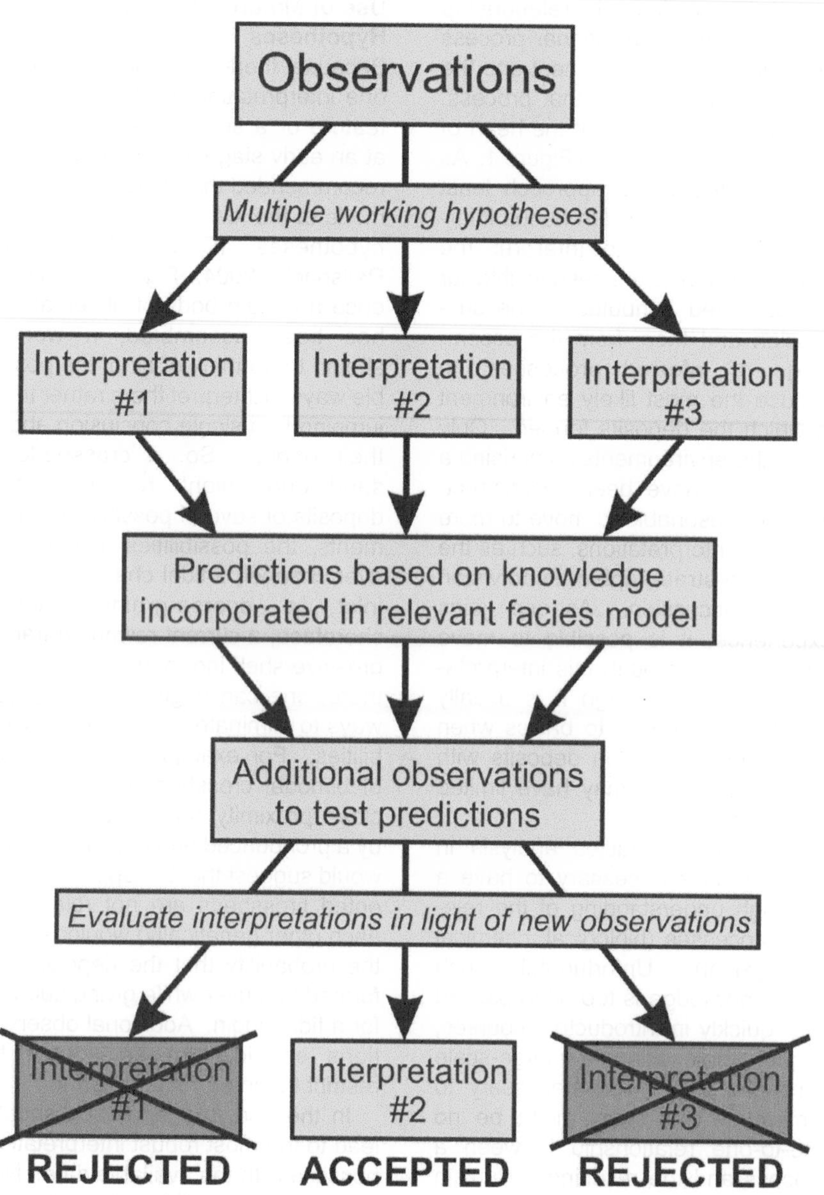

Figure 11. Flow chart illustrating the recommended method for the use of facies models in environmental interpretation. The use of multiple working hypotheses and their testing using predictions created using information provided by facies models allows a worker to accept, or reject, hypotheses on a more solid basis than would be the case without the use of this method.

bedding should show a unidirectional paleocurrent pattern. This can then be tested by looking for paleocurrent indicators. The finding of abundant evidence of a bimodal paleocurrent pattern, accompanied by closely spaced reactivation surfaces (*see* Chapter 9), would logically falsify the fluvial interpretation. The great value of facies models is their ability to suggest the types of tests that might be undertaken. Similar tests can be devised to evaluate the other possible hypotheses for the interpretation of the crossbedded sandstone. The one of the several preliminary interpretations (the 'hypotheses') that survives all possible tests would then represent, in formal scientific terminology, the 'theory' derived from the observations. As such, it would not be proven conclusively, but would be the best possible interpretation until additional data and further tests were available.

CONCLUDING REMARKS

As the title of this book indicates, the following chapters are devoted to presenting the facies models for each of the environments covered. As explained here, these 'distilled' environmental summaries are a crucial tool in the reconstruction of ancient sedimentary successions, and for making predictions about the occurrence of specific types of deposits, such as those that might form hydrocarbon reservoirs or ground-water aquifers. As powerful as these models are, they must be used carefully to avoid problems.

A worker should never expect his or her deposit to match a facies model exactly. Every depositional setting is unique in some way, and should be expected to show features that are not shown in the distillation for the environment in question. The differences should not be treated as noise; they potentially represent an important signal that might tell the worker something useful about the setting in which the deposit formed. Undue reliance on a model can also lead to missing crucial information about a deposit: we see what we expect to see unless we are careful. It is for this reason that the method shown in Figure 11 is advocated, because the incorporation of observations and insights gained from facies models, coupled with rigorous testing of multiple interpretations, minimizes the potential to be mislead by a preconceived idea or by an elegant model that superficially resembles the deposit in question. Interpreting ancient sedimentary successions is a challenging but exciting endeavor. The information contained in the subsequent chapters, when used with care, should allow one to make highly sophisticated environmental reconstructions and will serve as the basis for larger scale analyses of the stratigraphic record.

ACKNOWLEDGEMENTS

First and foremost, I thank Roger Walker for his outstanding articulation of the concepts presented in this chapter. Without his clear writing and teaching on the subject, much of the approach advocated here would not exist. I also thank the many generations of unwitting students who were the guinea pigs on whom the ideas presented here were honed. This chapter benefited greatly from the comments of Octavian Catuneanu, Noel James and Andrew Miall.

REFERENCES

Basic sources of information

Anderton, R., 1985, Clastic facies models and facies analysis, *in* Brenchley, P.J. and Williams, B.J.P., *eds.*, Sedimentology: Recent Developments and Applied Aspects: Oxford, Blackwell Scientific Publications, p. 31-47.
A review of the practical methodology for undertaking a facies analysis, with a thoughtful discussion of the creation and use of facies models.

Catuneanu, O., 2006, Principles of Sequence Stratigraphy: Amsterdam, Elsevier, 375 p.
The most recent textbook dealing with this important subject. Concentrates primarily on siliciclastic environments, but does consider carbonates too.

Miall, A.D., 2000, Principles of Sedimentary Basin Analysis, 3rd edition: New York, Springer-Verlag Inc., 616 p.
Chapter 4 of this book contains a thorough review of the subject of facies and facies analysis, with practical tips on how to approach the analysis of an individual succession.

Middleton, G.V., 1973, Johannes Walther's Law of the Correlation of Facies: Geological Society of America, Bulletin, v. 84, p. 979-988.
An excellent discussion of the use, misuse and application of Walther's Law.

Middleton, G.V., 1978, Facies, *in* Fairbridge, R.W. and Bourgeois, J., *eds.*, Encyclopedia of Sedimentology: Stroudsbury, Pennsylvania, Dowden, Hutchinson and Ross, p. 323-325.
One of the best and most concise statements of the facies concept, discussing various ways in which the term has been used.

Reading, H.G., *ed.*, 1986. Sedimentary Environments and Facies, 2nd Edition: Oxford, Blackwell Scientific Publications, 615 p.
The second edition of this book remains an indispensable reference on depositional environments and facies models.

Reading, H.G., 2003, Facies models, *in* Middleton, G.V., *ed.*, Encyclopedia of Sediments and Sedimentary Rocks: Boston, Kluwer Academic Publishers, p. 268-272.
The most recent review of the subject.

Schlager, W., 2005, Carbonate sedimentology and sequence stratigraphy: SEPM, Concepts in Sedimentology and Paleontology 8, 200 p.
An excellent introduction to all aspects of carbonate sedimentary rocks.

Other references

Allen, J.R.L., 1983, Studies in fluviatile sedimentation: bar complexes and sandstone sheets (low sinuosity braided streams) in the Brownstones (L. Devonian), Welsh Borders: Sedimentary Geology, v. 33, p. 237-293.

Bouma, A.H., 1962, Sedimentology of Some Flysch Deposits: Amsterdam, Elsevier, 168 p.

Cant, D.J. and Walker, R.G., 1976, Development of a braided fluvial facies model for the Devonian Battery Point Sandstone, Quebec: Canadian Journal of Earth Sciences, v. 13, p. 102-119.

Catuneanu, O., Abreu, V., Bhattacharya, J.P., Blum, M.D., Dalrymple, R.W., Eriksson, P.G., Fielding, C.R., Fisher, W.L., Galloway, W.E., Gibling, M.R., Giles, K.A., Holbrook, J.M., Jordan, R., Kendall, C.G.St.C., Macurda, B., Martinsen, O.J., Miall, A.D., Neal, J.E., Nummedal, D., Pomar, L., Posamentier, H.W., Pratt, B.R., Sarg, J.F., Shanley, K.W., Steel, R.J., Strasser, A., Tucker, M.E. and Winker C., 2009, Towards the standardization of sequence stratigraphy: Earth-Science Reviews, v. 92, p. 1-33.

Chamberlin, T.C., 1890, The method of multiple working hypotheses: Science (old series) v. 15, p. 92-96. (Reprinted in Science (new series) 1965, v. 148, p. 754-759).

Collinson, J.D., 1969, The sedimentology of the Grindslow Shales and the Kinderscout Grit: a deltaic complex in the Namurian of northern England: Journal of Sedimentary Petrology, v. 39, p. 194-221.

Davies, R. J., Posamentier H. W., Wood L. J. and Cartwright J. A., *eds.*, 2007, Seismic Geomorphology: Applications to Hydrocarbon Exploration and Production: The Geological Society of London Special Publication 277, 274 p.

De Raaf, J.F.M., Reading, H.G. and Walker, R.G., 1965, Cyclic sedimentation in the Lower Westphalian of North Devon, England: Sedimentology, v. 4, p. 1-52.

Dunham, R.J., 1962, Classification of carbonate rocks according to depositional texture: American Association of Petroleum Geologists Memoir 1, pp. 108-121.

Folk, R.L., 1959, Practical petrographic classification of limestones: Bulletin of the American Association of Petroleum Geologists, v. 43, p. 1-38.

Handford, C.R. and Loucks, R.G., 1993, Carbonate depositional sequences and systems tracts—Responses of carbonate platforms to relative sea-level changes, *in* Loucks, R.G., and Sarg, J.F., *eds.*, Carbonate Sequence Stratigraphy—Recent Developments and Applications: American Association of Petroleum Geologists, Memoir 57, p. 3-41.

Kendall, A.C., 1988, Aspects of evaporate basin stratigraphy, *in* Schrieber, B.C., *ed.*, Evaporites and Hydrocarbons: New York, Columbia University Press, p. 11-65.

Klovan, J.E., 1964, Facies analysis of the Redwater Reef complex Alberta, Canada: Bulletin of Canadian Petroleum Geology, v. 12, p. 1-100.

MacNaughton, R.B., Dalrymple, R.W. and Narbonne, G.M., 1997, Early Cambrian braid-delta deposits, MacKenzie Mountains, north-western Canada: Sedimentology, v. 44, p. 587-609.

Miall, A.D., 1977, A review of the braided river depositional environment: Earth-Science Reviews, v. 13, p. 1-62.

Miall, A.D., 1985, Architectural element analysis: a new method of facies analysis applied to fluvial deposits: Earth-Science Reviews, v. 22, p. 261-308.

Miall, A.D., 1999, In defense of facies classifications and models: Journal of Sedimentary Research, v. 69, p. 2-5.

Moore, R.C., 1949, Meaning of facies, *in* Longwell, C.R., *ed.*, Sedimentary Facies in Geological History: Geological Society of America, Memoir 39, p. 1-34.

Mutti, E., 1992, Turbidite Sandstones: AGIP, Instituto di Geologia, Universitá di Parma. Milan, 275 p.

Mutti, E. and Ricci Lucchi, F., 1972, Le torbiditi dell'Appennino settentrionale: introduzione all'analisi di facies: Memorie della Societa Geologica Italiana, v. 11, p. 161-199. English translation by T.H. Nilsen, 1978, International Geology Review, v. 20, p. 125-166.

North American Commission on Stratigraphic Nomenclature (NACSN), 2005, North American Stratigraphic Code: American Association of Petroleum Geologists Bulletin, v. 89, p. 1547-1591.

Plint, A.G., 1988, Sharp-based shoreface sequences and "offshore bars" in the Cardium Formation: Their relationship to relative changes in sea level, *in* Wilgus, C.K. *et al., eds.,* Sea Level Changes: An Integrated Approach: SEPM Special Publication 42, p. 357-370.

Postamentier, H.W. and Allen, G.P., 1999, Siliciclastic Sequence Stratigraphy: SEPM Concepts in Sedimentology and Paleontology 7, 210 p.

Posamentier, H.W., Jervey, M.T. and Vail, P.R., 1988. Eustatic controls on clastic deposition I — conceptual framework, *in* Wilgus, C.K., Hastings, B.S., Kendall, C.G.St.C., Posamentier, H.W., Ross, C.A. and Van Wagoner, J.C., *eds.*, Sea Level Changes — An Integrated Approach: SEPM, Special Publication 42, p. 110–124.

Qi, L, Carr, T.R. and Goldstein, R.H., 2007, Geostatistical three-dimensional modeling of oolites shoals, St. Louis Limestone, southwest Kansas: American Association of Petroleum Geologists Bulletin, v. 91, p. 69-96.

Railsback, L.B., 2004, T. C. Chamberlin's

"Method of Multiple Working Hypotheses": An encapsulation for modern students: Houston Geological Society Bulletin, v. 47, p. 68-69. (Also available at http://www.gly.uga.edu/railsback/railsback_chamberlin.html)

Roksandić, M.M., 1978, Seismic facies analysis concepts: Geophysical Prospecting, v. 26, p. 383-398.

Saggaf, M.M. and Nebrija, E.L., 2000, Estimation of lithologies and depositional facies from wire-long logs: American Association of Petroleum Geologists Bulletin, v., 84, p. 1633-1646.

Schwarzacher, A., 2000, Repetition and cycles in stratigraphy: Earth-Science Reviews, v. 50, p. 51-75.

Suter, J. and Clifton, E., 1998, The Shannon Sandstone and isolated linear sand bodies: Interpretations and Realizations, *in* Isolated Shallow Marine Sandbodies: Sequence Stratigraphic Analysis and Sedimentologic Interpretation, Bergman, K.M. and Snedden, J.W., *eds.*: SEPM Special Publication 64, p. 321-356.

Tebo, J.M. and Hart, B.S., 2005, Use of volume-based 3-D seismic attribute analysis to characterize physical-property distribution: A case study to delineate sedimentological heterogeneity at the Appleton Field, southwestern Alabama, U.S.A.: Journal of Sedimentary Research, v. 75, p. 723-735.

Walker, R.G., 1984, General introduction: facies, facies sequences and facies models, *in* Walker, R.G., *ed.*, Facies Models, 2nd edition: St. Johns, Geological Association of Canada, p. 1-9.

Walker, R.G., 1992, Facies, facies models and modern stratigraphic concepts, *in* Walker, R.G. and James, N.P., *eds.*, Facies Models—Response to Sea-level Change: St. Johns, Geological Association of Canada, p. 1-14.

Willis, B.J. and Gabel, S., 2001, Sharp-based, tide-dominated deltas of the Sego Sandstone: Sedimentology, v. 48, p. 479-506.

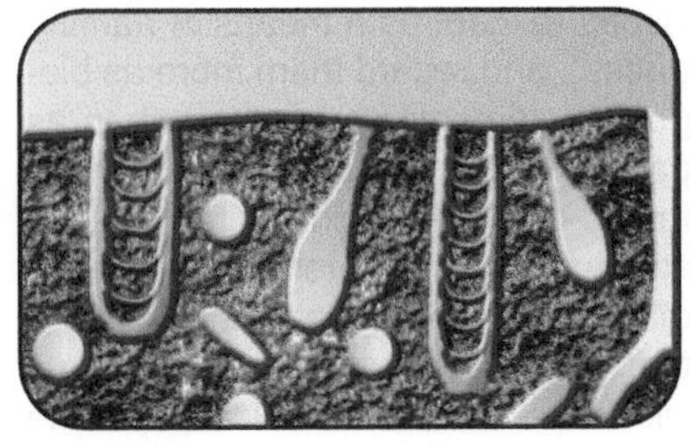

3. ICHNOLOGY AND FACIES MODELS

James A. MacEachern, ARISE, Department of Earth Sciences, Simon Fraser University, Burnaby, BC, V5A 1S6, Canada

S. George Pemberton, IRG, Department of Earth and Atmospheric Sciences, University of Alberta, Edmonton, AB, T6G 2E3, Canada

Murray K. Gingras, IRG, Department of Earth and Atmospheric Sciences, University of Alberta, Edmonton, AB, T6G 2E3, Canada

Kerrie L. Bann, Ichnofacies Analysis Inc., 9 Sienna Hills Court, Calgary, AB, T3H 2W3, Canada

This chapter builds upon the previous ichnology chapter (Pemberton *et al.*, 1992) in *Facies Models: Response to Sea Level Change*, and introduces the reader to the principles of ichnology and their applications to facies analysis and depositional environment interpretation. The chapter combines aspects of modern animal–sediment relationships, the preserved biogenic structures in the ancient record, and sedimentology to demonstrate the utility of undertaking integrated ichnological–sedimentological facies analysis. General techniques are outlined to facilitate development of practical skills in ichnological data collection. The ichnological analysis showcased here is conceptually underpinned by the ichnofacies concept.

WHAT IS ICHNOLOGY?

Ichnology is the study of traces of organisms, which are a record of their activities within the environment they occupy. These traces include a myriad of structure types such as tracks, trails, burrows, feeding structures, and other disruptions of sedimentary media owing to the activities of animals. The study of ichnology typically falls into two sub-disciplines: the study of modern trace-making organisms and present-day animal–sediment relationships (*i.e.*, neoichnology), and the study of the traces of organisms (trace fossils or ichnofossils) as preserved in the ancient record (*i.e.*, paleoichnology or palichnology). The relationship of the two sub-disciplines parallels that of process sedimentology and sedimentary facies analysis in the lithofacies realm (*see* Chapter 2). Neoichnology, or 'process ichnology' (Gingras *et al.*, 2007), addresses animal–sediment relationships from the perspective of *known* environments of occupation and *identified* behaviors, with the aim of assessing or determining the sedimentological significance of the resulting biogenic structures. Paleoichnology evaluates the *preserved record*, and concentrates on identifying ichnofossils, tiering relationships and recurring groupings of trace fossils, integrating them with the sedimentary facies to infer organism behaviors and interpret the original depositional environments. The two approaches provide positive feedback – paleoichnology deriving most of its usable inferences as to organism behavior (the study of ethology) from neoichnological analysis, and neoichnology deriving its insights about the preservation potential of structures and their resulting morphology based on the preserved ancient record. Integration of the two sub-disciplines provides insights that expand the role of ichnology in sedimentary facies analysis.

Trace fossils are primary features of the sedimentary succession, comprise the predominant expression of macroscopic animal life (particularly in clastic successions), and are strongly facies controlled. Organisms are very sensitive to the environment; as such, their traces can provide insight into a wide range of environmental conditions, some of which are poorly expressed, if at all, by physical sedimentary characteristics (*e.g.*, oxygenation, salinity, and deposition rate). Ichnological datasets proffer insights into depositional conditions operating at various temporal scales, including daily, seasonally, and (through assessment of tiering relationships and stacked trace-fossil suites) long-term changes in the environment and their cumulative effects on successive generations of fauna. The integration of ichnological data with sedimentological analyses ultimately leads to more refined facies models.

There are some important caveats to the above statements. First, one must remember that trace fossils are not the organisms themselves – they are indications of the activities of the trace makers and must not be treated as if they were animals.

Second, trace-fossil assemblages record a complex set of environmental conditions, and not just a single factor (*e.g.*, water depth). In the past, some workers erroneously focused on the role of paleobathymetry on distributions of trace-fossil suites. Trace-fossil distributions are determined by environmental factors such as substrate consistency, sediment caliber, food-resource types and their distributions, energy conditions, salinity, oxygenation, water turbidity, and deposition rate, but *not* water depth *per se*. Although the depositional parameters that control trace-fossil distributions *tend* to change with water depth, the relationship is passive (Frey *et al.*, 1990). Facies analysts must endeavor not to assign bathymetric interpretations to trace-fossil suites, without careful assessment of the sedimentology and facies architecture of the unit.

Third, trace fossils, although temporally long ranging and environmentally controlled, do show some progressive temporal variations, recording the evolution and extinction of different trace-making organisms, as well as changing organism responses to environmental stresses. Ichnological suites associated with continental deposits have changed markedly during the Phanerozoic. So, also, have suites associated with the progressive invasion of brackish-water settings (Beynon *et al.,* 1988; Gingras *et al.,* 1999; Buatois *et al.,* 2005). In general, organisms evolved in the marine realm and have gradually invaded reduced salinity and freshwater environments over time. Additionally, the development and diversification of various terrestrial plants have helped to drive changes in the availability of food resources in continental settings. These changes, coupled with adaptive strategies and the evolutionary development of particular trace makers (*e.g.,* extinction of trilobites in the Paleozoic; evolution of decapod shrimp in the Mesozoic, *etc.*) have helped to produce predictable changes in the character of trace-fossil suites through time. For example, certain ichnogenera are far more prevalent in the Paleozoic (*e.g., Phycodes, Cruziana,* and *Rusophycus*), whereas others are more characteristic of the Mesozoic and Cenozoic eras (*e.g., Ophiomorpha, Scolicia,* and *Taphrhelminthopsis*). In the continental realm, nesting structures such as *Coprinisphaera, Celliforma* and *Termitichnus* are more typical of Upper Cretaceous and Cenozoic strata.

Terminology and Classification

The study of ichnology requires some understanding of the terms used to describe and characterize biogenic structures. To date, the most ready single source of such information is Bromley (1996), in which are outlined the principles and varying approaches taken by specialists in the study of ichnology. Also in that volume, terms such as 'spreite', 'meniscae', 'shaft', 'tunnel', 'lining', 'tiering', *etc.,* are discussed at length. These terms are not discussed here and the reader is referred to that source. Numerous volumes since then have included introductory chapters or papers that discuss the conceptual framework of ichnology, interspersed with case studies that outline the application of ichnology (*e.g.,* Pemberton, 1992; McIlroy, 2004a; MacEachern *et al.,* 2007a; Miller, 2007). Seilacher (2007) has recently produced a volume that outlines a rigorous paleontologic treatment of distinctive and representative ichnogenera, accompanied by handsome drawings and insightful discussions of trace-fossil morphologic elements. This chapter concentrates on the practical applications of ichnology to facies analysis and facies models, and so taxonomic aspects of the science are not emphasized.

Depending upon the aspect of ichnology in which one is interested, various trace-fossil classification schemes are available. Historically, classifications of biogenic structures have been descriptive, preservational (toponomic), taxonomic, or ethological (based on interpreted behavior). In the neoichnological realm, a philosophy of naming structures for the inferred trace-maker is apparent (*e.g.,* Gingras *et al.,* 1999, 2004; Hasiotis, 2002, 2007). This approach can be confusing, particularly given that trace fossils are named using the same Linnaean system of genus and species employed for naming organisms and their body fossils. Neoichnological analyses show us that different trace makers are capable of generating the same trace when performing the same behavior, and the same trace maker can produce a wide variety of biogenic structures, depending upon what behaviors are being performed. A philosophy of phylogenetic classification potentially can be misleading, therefore, and students of the discipline must harden themselves against the tendency to treat ichnofossils *as* the trace makers themselves. Remember that *'Planolites'* did not *do* anything; it is merely a structure left behind by an organism performing some behavior.

From the perspective of paleoichnology and sedimentary facies analysis, the most common schemes for trace-fossil classification are a combination of taxonomic (based on overall morphology) and ethologic (inferred organism behavior; Fig. 1). This is because most 'applied ichnologists' are not paleontologists but are formally trained as sedimentary geologists who have been brought to ichnology for the purposes of solving sedimentological and stratigraphic problems. Most sedimentologist–ichnologists are less concerned with taxonomic issues (except as a means of naming traces), and regard them more as biogenic sedimentary structures than as paleontologic entities. The strong facies-controlled character of the structures and long temporal duration of most ichnogenera easily lead to that perception, and allow trace fossils to be wielded in much the same way as primary physical structures for the interpretation of depositional environments. Nevertheless, trace fossils *are* paleontologic entities and not physical structures, and pitfalls can be encountered when they are not recognized as such (*e.g.,* progressive evolutionary adaptations to brackish-water settings during the Phanerozoic). The way forward lies in the continued collaboration of such specialists with one another.

'Doing' Ichnology – Applied Ichnological Analysis

The importance of ichnology to the fields of sedimentology and genetic stratigraphy stem largely from the following facts about trace fossils:

1. They have long temporal ranges, which, although a disadvantage in high-resolution biostratigraphy, greatly facilitate paleoecological comparisons of rocks that differ in age;
2. Many ichnogenera possess relatively narrow facies ranges that reflect similar responses by different trace-making organisms to particular paleoecological factors;
3. They are characterized by largely *in situ* preservation, which guarantees that most suites are genetically linked to the depositional environment of the enclosing facies or to those of an overlying discontinuity, if they are palimpsest (post-depositional);
4. They commonly occur in otherwise unfossiliferous rocks, especially in siliciclastic sediments; and
5. They are generated by predominantly non-preservable soft-bodied biota, which, nevertheless, commonly represent the bulk of the environment's biomass (Ekdale *et al.,* l984).

Such characteristics are exceedingly useful from the perspective of facies analysis, including reconstruction of paleoenvironments, assessment of paleoecological factors, and documenting local and regional temporally significant facies changes across stratigraphic discontinuities (*e.g., see* Chapter 2).

From the perspective of sedimentary facies analysis (*i.e.,* reconstruction of depositional environments), the application of ichnology requires more than the mere identification of a few readily identifiable trace fossils. Certainly, the accurate identification of trace-fossil genera and, in some cases, ichnospecies, is paramount, but this constitutes only one aspect of a thorough ichnological analysis. The identification of trace fossils to ichnogenus and ichnospecies levels necessarily requires considerable experience that is usually acquired during graduate studies or one's ensuing career. It is a steep learning curve, but highly rewarding. This chapter cannot hope to give the reader the necessary criteria upon which to identify the many recurring trace fossils. There are useful manuals, however, wherein the reader can learn how to identify trace fossils. Although not revised since 1975, the Treatise of Invertebrate Paleontology (Häntzschel, 1975, Part W, Supplement 1) remains the anchor point for the taxonomic treatment of trace fossils. It is, however, not the sole, or even the best repository of trace-fossil taxonomy. Seilacher (2007) has taken the first credible step in grouping ichnogenera into broader associations, based upon generally similar constructional morphology. There are also a number of guidebooks, short-course volumes and the like, which serve the purpose of showcasing the diagnoses for identifying various ichnogenera (*e.g.,* Frey, 1975; Basan, 1978; Ekdale *et al.,* 1984; Curran, 1985; Pemberton, 1992; Donovan, 1994). More recently, Hasiotis (2002) has provided a much-needed review of continental trace fossils.

Atlases referring to trace fossils associated with marine and marine-influenced strata are more common, and vary from those concerned with outcrop expressions to those concerned with core expressions. The identification of trace fossils in core is particularly challenging. Chamberlain (1978) probably stands as the earliest attempt to establish a rigorous

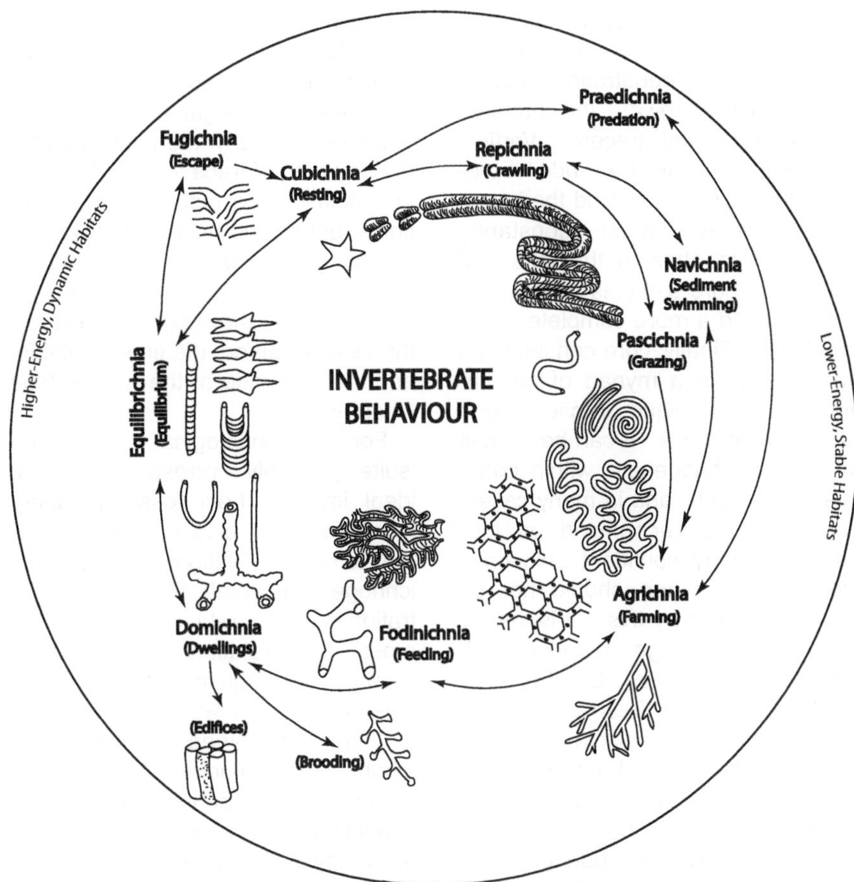

Figure 1. The ethology of trace fossils can be represented as a continuum of behaviors that may combine or grade with each other. In this scheme, the ethological patterns are grouped so that the families of feeding behaviors are evident. The resulting categories (*i.e.,* fodinichnia, domichnia, fugichnia, cubichnia, repichnia, pascichnia, and agrichnia) represent the highest level of trace-fossil classification, in some cases referred to as 'ichnofamilies'. Most of these behaviors are generally associated with particular depositional processes and those associations are the general basis for ichnofacies classification (modified from Müller, 1962 and Gingras *et al.,* 2007).

approach. Pemberton and Frey (1984) and Frey and Pemberton (1985) remain two of the most influential manuscripts illustrating the utility of trace-fossil analysis using core. Students of ichnology should avail themselves of the large number of published manuscripts and special volumes (*e.g.,* Pemberton, 1992; McIlroy, 2004a; Miller, 2007; Bromley *et al.,* 2007; MacEachern *et al.,* 2007a), wherein trace fossils are annotated on photographic plates.

The study of ichnology in subsurface core is quite different from that of outcrop. Outcrop areas give more extensive exposure, which allows large structures to be identified, displays the spatial variability in associated traces, and yields bedding plane views of the trace fossils. As such, certain traces are better expressed in outcrop than in core, including most surface trails and resting structures, graphoglyptids (deep-sea patterned grazing, farming and feeding structures), robust forms such as *Piscichnus*, and *Conichnus*, and complex forms such as *Rhizocorallium*. Additionally, because more of the morphology is preserved in outcrop exposure, identification of ichnospecies is facilitated. Core, by contrast, limits one's ability to discern such details. Ichnospecies, unless distinctive in their small-scale morphology (*e.g.,* pellet shape in *Ophiomorpha*), or constitute ichnotaxa that are small and fully captured by core-scale intervals (*e.g., Phycosiphon* and *Schaubcylindrichnus freyi*), are not readily discernible in core datasets. Cored intervals also tend to lack bedding-plane expressions, inhibiting the differentiation of various grazing structures and identification of surface locomotion and resting structures. Core, owing to its general vertical penetration, also is strongly biased toward the intersection of horizontal traces. Vertical structures tend to be under-represented in most cases, and the ichnological worker must constantly remember to observe the tops and bottoms of core samples in the hope of generating a more complete trace-fossil suite. Finally, core can intersect trace fossils in a myriad of orientations, giving oblique transects and partial expressions. It can be difficult to identify ichnogenera within cored intervals reliably, and in many cases, structures are recognized as biogenic, but defy confident taxonomic identification – one should not be overly frustrated by this – it is merely part of the challenge of working with subsurface datasets. On the other hand, core yields fresh exposures, and in many cases, displays delicate aspects of burrow infill and ornamentation, and preserves details illustrating how trace was constructed. Additionally, core provides superior ichnological expressions of mudstones and shales, which tend to weather recessively in outcrop. In outcrop, many details of a mudstone's ichnological suite come from studying casts of traces on the soles of sandstones (*e.g.,* low-energy turbidites and some tempestites). In thick muddy successions, outcrops tend to yield incomplete suites, unless blocks can be cut and polished in the lab. Additionally, sandy successions may suffer from extensive surficial weathering, which obscures ichnological detail. In any case, core is what subsurface facies analysts have to work with, and so they must work around the limitations and exploit the benefits to the fullest degree.

Ichnological Suites – Ethology and Diversity

The identification of ichnogenera leads to the recognition of recurring associations of trace fossils within the facies. Depending upon the skill of the worker, trace fossils may be combined into broad assemblages (wherein all ichnogenera in a unit are combined, even if they are not directly associated in time or depositional conditions), or ichnological suites (groupings of genetically associated trace fossils). In modern settings, such genetically associated biogenic structures are regarded as 'ichnocoenoses'. Clearly, the more detailed the grouping of trace fossils, the higher resolution the resulting interpretation – but this is no different than in lithofacies analysis. The level of analysis undertaken ultimately depends upon the scope of the study, the detail required, the time available, and the expertise of the worker (*see* Chapter 2).

For most ichnological analysts, the 'suite' probably corresponds to the ideal level of trace-fossil grouping. Some workers also concentrate on the temporal relationships between ichnogenera, depth of burrow penetration into the strata, and the ensuing crosscutting relationships – termed 'tiering' – although in many cases, the different tiers represent the activities of contemporaneous organisms inhabiting the same depositional environment but responding to different aspects of the ecosystem. At the facies level, they all impart important insights into the depositional environment. For highly detailed ichnological analyses, the relative proportions of the different ichnotaxa may be assessed. This gives a superior appraisal of the trace-fossil suite by allowing dominant elements to be differentiated from subordinate or rare elements, although it can be a time-consuming effort. In complex, commonly heterolithic facies, recognizing variations in traces associated with lithologically distinct beds (*e.g.,* sandy *versus* muddy interbeds) is also important, particularly in the characterization of event-style deposition (*e.g.,* tempestites and turbidites; *cf.* Chapters 8 and 12, respectively).

The establishment of the suite permits assessment of two distinct but related aspects of the facies. The first is the characterization of the ethological (behavioral) activity of the trace makers (Figs. 1 and 2). Ekdale *et al.* (1984) recognized seven basic, but overlapping, behavioral groups: resting traces, locomotion traces, dwelling structures, feeding burrows, grazing structures, farming systems, and escape traces. To these have been added, with varying degrees of practical success, predation traces, re-equilibration (or multiple adjustment) structures, and sediment-swimming structures or navichnia (*see* Gingras *et al.,* 2007). The behaviors recorded by trace fossils give insight into organism responses to a number of environmental parameters, such as substrate consistency, sediment caliber, food resource type, depositional energy, salinity (particularly reduced salinity or fluctuating salinity), water turbidity, soil moisture, position of the water table, and climate, depending upon whether the ichnological suites record marine or continental conditions, and whether the traces formed in subaqueous, intertidal or subaerial settings. The *range* of behaviors recorded in a single deposit gives some insight into how dynamic the depositional setting was! Environments typified by largely unchanging depositional conditions tend toward limited ethological variability. Settings prone to temporal changes in depositional conditions lend themselves to more variable ethological responses by trace makers. The more marked the degree of dynamic changes, the more discrete the trace-fossil ethological groupings become. In extreme cases, variations lead to the generation of discrete ichnological suites. One such extreme is associated with turbidite emplacement on the abyssal seafloor. In such cases, there may be virtually no overlap between suites associated with the sediment gravity event bed and those in the intervening mudstone units. Storm beds emplaced onto the shelf likewise may show suites discrete from the fair-weather mudstones. In shallower water conditions, the distinction between event suites of tempestites and the fair-weather suites tend to be less distinct (*e.g.,* tempestites in proximal lower shoreface and middle shoreface environments). It must be emphasized, however, that these variations record in-place changes in the dynamic processes intrinsic to the depositional setting itself (autogenic changes), and do not record longer term changes in the depositional environment.

Narrow ethological ranges typify relatively stable environments. High-energy, shallow-water environments, for example, tend to be characterized by persistent wave agitation at the

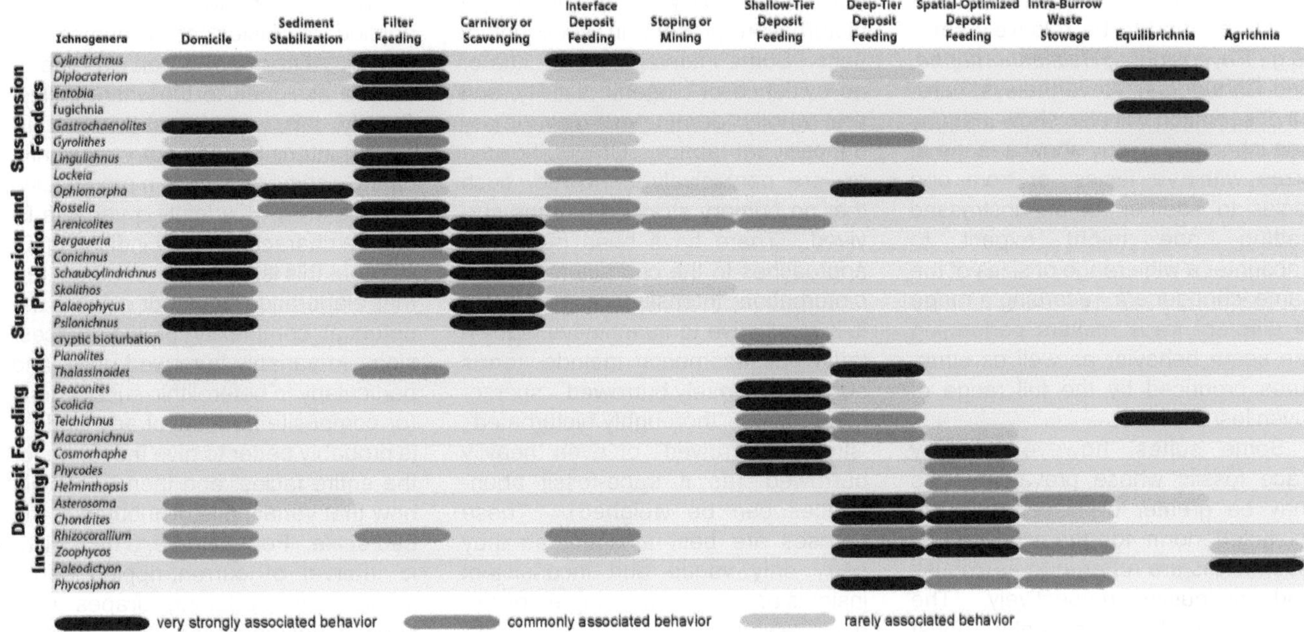

Figure 2. Table of commonly observed trace fossils and their interpreted behaviors (modified from Gingras *et al.*, 2007). Behaviors range from sessile living and filter feeding (on the left) to generally motile deposit feeding (to the right). Suspension-feeding structures dominantly comprise elements characteristic of the *Skolithos* Ichnofacies, whereas deposit-feeding traces represent ichnogenera typical of the *Cruziana* through *Zoophycos* ichnofacies. Agrichnia dominantly represent the *Nereites* Ichnofacies. Many trace fossils can be interpreted to represent more than a single behavior and are commonly regarded as facies-crossing forms. Additionally, shallow-tier deposit feeding is a useful strategy for collecting food in dynamic sedimentary environments; consequently, such structures also occur in more than one ichnofacies.

sediment–water interface. Deep-water settings may be characterized by slow gradual hemipelagic settling of silt and clay. Persistent conditions favor uniform sedimentary parameters that allow for ecological optimization, marked specialization, and lead to a reduction of overall diversity.

Proper identification of ichnogenera also permits an assessment of *ichnological diversity*. To do this, it is not sufficient to identify a few, easily recognized ichnogenera within the facies. Highly diverse suites (showing a number of ichnogenera attributable to a number of ethological groupings) tend to be typical of depositional settings with limited physico-chemical stress. Such settings in the marine realm tend to be characterized by normal-marine salinities, variable substrates, abundant food, and well-oxygenated water columns. In continental settings, optimal conditions tend to be associated with reduced sedimentation rate, persistence of soil moisture, variable but abundant food resources, and variable climatic factors (*e.g.*, ranges in temperature and annual precipitation, periods of snow or ice cover, *etc.*). As physico-chemical stresses are imposed on a setting, organism diversity declines, and specialized structures decrease in favor of more generalized, opportunistic behaviors (*cf.* MacEachern *et al.*, 2007b).

Recognition of environmental stress can be instrumental in the accurate identification of the sedimentary environment. This commonly requires the careful documentation of the ichnological suites, changes in the suite within a succession, and comparisons to unstressed successions in associated units. Under extreme conditions of environmental stress, such departures in the ichnological character are readily apparent, and may lead to monospecific suites (*e.g.*, brackish-water settings: Beynon *et al.*, 1988; Gingras *et al.*, 1999, or dysaerobic settings: Savrda and Bottjer, 1989; Martin, 2004). It is also helpful to recognize the presence of ichnogenera that tend to be restricted to settings less prone to physico-chemical stress (*e.g.*, *Scolicia, Spirophyton, Cosmorhaphe, Asterosoma, Conichnus,* and *Bergaueria*), and suites dominated by what are regarded as 'facies-crossing' elements (*e.g.*, *Planolites, Teichichnus, Cylindrichnus, Palaeophycus, Skolithos, Thalassinoides,* and *Ophiomorpha*). Traces very commonly associated with fully marine conditions tend to be those with strong phylogenetic linkages to euryhaline trace makers (*e.g.*, echinoderms, anemones), and those associated with elaborate feeding/farming/chemosymbiosis strategies (particularly bedding-plane confined geometrically patterned trace fossils – graphoglyptids – typical of the deep sea). Facies-crossing structures tend to be morphologically simple, and comprise recurring elements in numerous suites, typically present within units of varying lithology. Such structures commonly record generalized feeding styles of organisms able to exploit a variety of resource types (trophic generalists). This is particularly the case for salinity-stressed regimes.

Trace-Fossil Size

Trace fossils vary significantly in size. Some ichnogenera are predisposed to large sizes (*e.g.*, *Piscichnus,* a fish-generated feeding structure; *Psilonichnus,* commonly a crab-dwelling structure; *Conichnus,* an anemone-dwelling structure). Other structures tend to be relatively small (*e.g., Phycosiphon, Helminthorhaphe,* and *Chondrites*). Nevertheless, careful observation will also show that any one ichnogenus may show a range of sizes within a facies, or from one facies to another. In an unstressed setting, one might expect to encounter a wide range of sizes of the same ichnogenera, recording a range of different trace makers performing the same behavior as well as structures produced by the full range of juveniles to adults.

Some suites, however, display trace fossils whose prevalent sizes may be greater or lesser than the expected norm for the ichnogenus. Such forms are referred to as 'robust' and 'diminutive', respectively. The implications for persistently robust expressions have yet to be adequately explained, although preliminary work suggests that it may be, in part, a response to cold-water conditions. Diminution, however, has been readily encountered in modern studies, and is closely tied to physico-chemical stress. Stresses that are particularly well expressed through trace diminution are salinity reduction (brackish water), lowered oxygenation, and waters sufficiently warm that oxygenation levels are suppressed. In such stressful settings, some organisms may be predisposed to small sizes to facilitate reduction of volume to surface area ratios, permitting exchange of salt, oxygen or heat across their body membranes. Other organisms grow more slowly in stressed settings. Finally, environmentally stressed conditions typically lead to high organism mortality rates, leading to a predominance of traces generated by juvenile organisms. Whatever the cause of the response, the outcome tends to be the same: persistently diminutive traces typically record depositional settings characterized by physico-chemical stress. It is the responsibility of the facies analyst to discern the particular stresses that operated, through a combination of sedimentological, ichnological and micropaleontological analyses.

Assessing Bioturbation Intensity and Burrowing Uniformity

Bioturbation Intensity There is a marked variability in the degree of bioturbation present in sedimentary units. Some intervals display little or no evidence of biogenic activity, and are wholly dominated by primary depositional fabrics. Others, however, are pervasively burrowed, such that no primary stratification has survived. There have been numerous approaches to the characterization of bioturbation intensity, ranging from fully qualitative to quantitative. Qualitative assessments include terms such as 'intensely burrowed', 'sparsely burrowed', 'thoroughly bioturbated', 'strongly burrowed', or even 'heavily burrowed' (as if trace-fossil abundances can be weighed). These phrases are best avoided, as they carry only vague and inconsistent insights as to the actual degree of biogenic reworking.

Early attempts at quantifying bioturbation arose from the efforts of neoichnological workers. They were able to show from box cores, vibracores and x-radiographs that certain grades of bioturbation (assessed by the percent of the area in a vertical profile occupied by burrows) could be discerned by eye (*see* Reineck and Singh, 1980). Despite the inherent challenges in reliably assigning percentages, a scale consisting of seven grades (Grades 0-6) has stood the test of time, and forms the underpinning of currently employed bioturbation-intensity schemes.

Taylor and Goldring (1993) adopted this approach, renaming it 'Bioturbation Index', and assigned each of the original grades a BI number (BI 0-6); BI 0 recording unburrowed media and BI 6 reflecting 100% bioturbation. The BI system purports to be a fully quantitative assessment of bioturbation, based on the percentage of original fabric reworked by biogenic activity (as determined by constituent diagrams). The inaccuracies inherent in arriving at percentages of burrowing, and the inability to consistently reproduce them have led workers to adopt a semi-quantitative flash-card approach to BI for the purposes of visual comparison (Fig. 3; *cf,.* Bann *et al.,* 2008). For an alternative framework for quantifying bioturbation intensity, see the earlier ichnofabric-index scheme proposed by Droser and Bottjer (1986).

The Bioturbation Index can be applied to facies at a number of scales. At its broadest, an average BI can be assigned to the entire facies. Clearly, this works best where burrowing intensity does not vary significantly, or where there is a progressive change in BI. Certainly a variable BI can be characterized by indicating 'BI 0-3', but this is vague, and without further elaboration does not capture the potential complexity of the succession. In variably burrowed units, and particularly in heterolithic successions (or composite bedsets of any kind), it is probably better to give the range for the entire facies, and then to explain how that variation is distributed at the bed scale. For example, a heterolithic interval of current-rippled sandstone with mudstone drapes may show sandstone beds that have a BI of 2 or 3, and mudstone drapes with a BI of 0 or 1 (or vice versa in some shallow-marine successions). As the proportion of sand and mud changes in the succession, the general character of burrowing might shift, and the combination of these two scales (facies level and bed level) of BI assessment would capture the observed complexity. At a more detailed scale of resolution, one might assign BI values to each bed, rather than to the facies themselves. While such an approach lends itself to higher resolution, for most purposes (especially thinly bedded units), facies characterization will require that this detail ultimately be synthesized into the recurring expression of the facies for the purposes of interpretation. In our experience, BI variation at the facies scale, coupled with average variability in burrowing expressed by the component beds of heterolithic successions, provides the detail required for most facies evaluations. For very high-resolution studies, however, particularly for facies characterized by complex tier relationships, BI assessment of individual tiers may be necessary (*cf.* Bromley, 1996).

Generally, elevated BI values tend to be associated with units recording some combination of slow deposition and/or high infaunal biomass (Fig. 4); the interplay of these two can normally be discerned through integration with the sedimentology and other paleontological evaluations (*e.g.,*

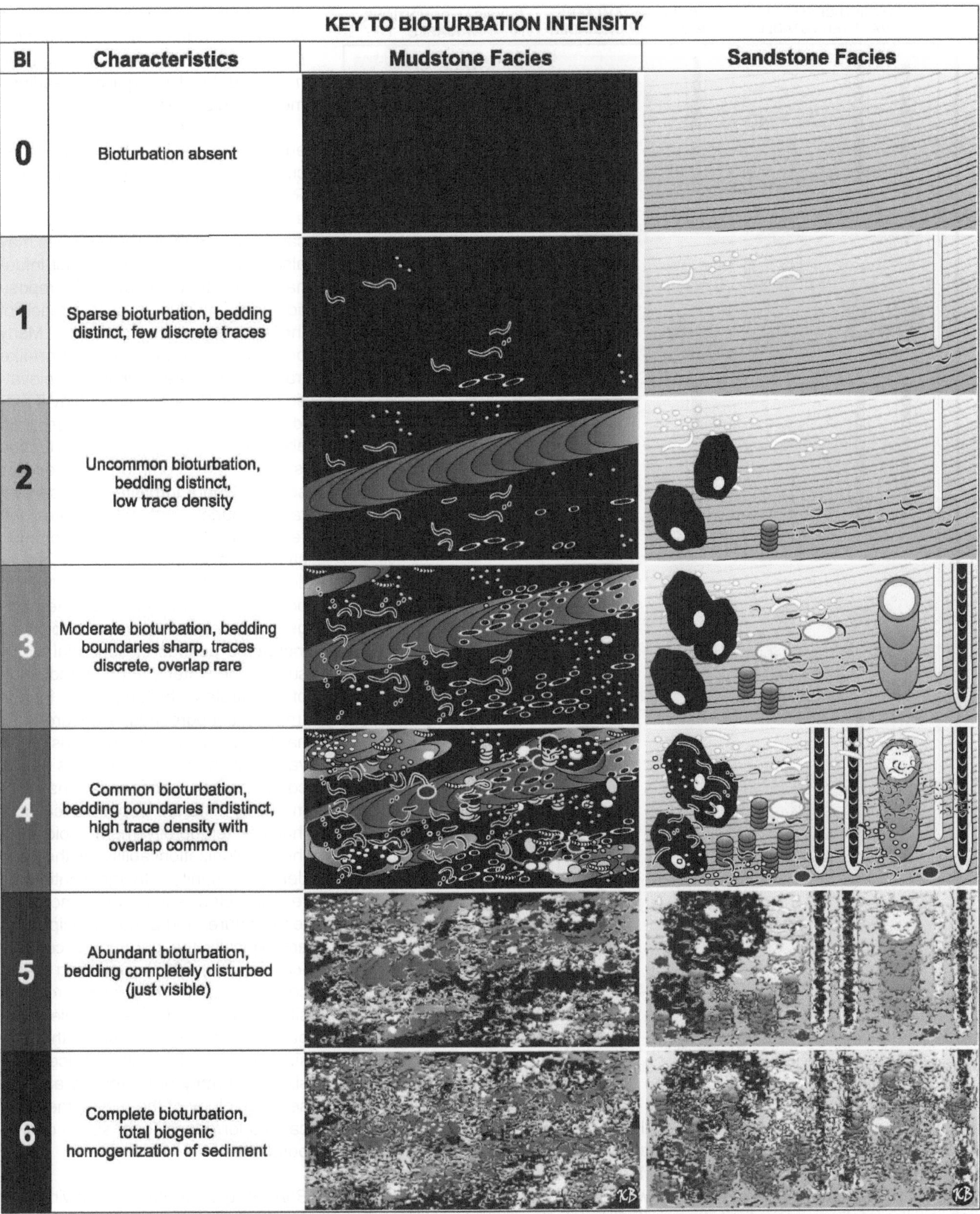

Figure 3. The Bioturbation Index (BI) for sand- and mud-dominated facies (modified from Bann *et al.,* 2008).

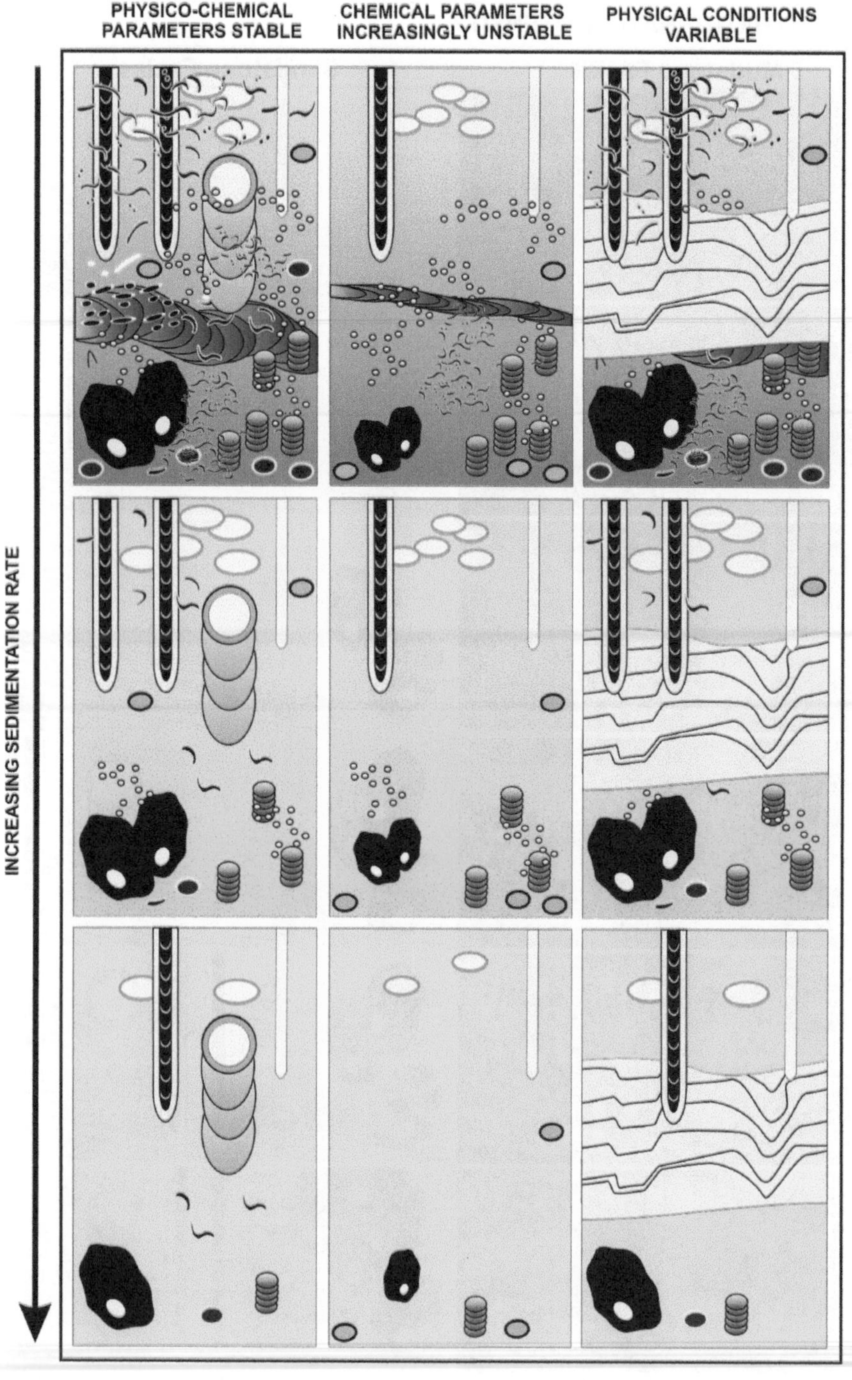

Figure 4. Burrow distributions as a function of changing sedimentation rates and environmental stability.

microfossil studies). Bioturbation intensities tend to decline in response to increasing sedimentation rates, and smaller populations of infauna. High sedimentation rates limit the range of deployed ethologies (dwelling and near-surface deposit-feeding/detritus feeding are favored over deep-tier deposit-feeding and grazing), makes (re)colonization of the substrate challenging, and/or limits the ability of organisms to reposition their burrows following sediment accumulation. Elevated deposition rates are commonly also associated with abundant escape structures (fugichnia), dwelling structures showing spreiten lying beneath a dwelling structure (indicating that the organism has shifted its burrow upward in response to sediment accumulation) and re-equilibrated structures (leading to vertically stacked burrows). The latter is particularly well expressed in ichnogenera such as *Rosselia* (Fig. 5), *Teichichnus, Diplocraterion, Conichnus, Lingulichnus,* and bivalve equilibrium-adjustment structures (*see* Fig. 2).

Low BI values, then, are associated with high sedimentation rates (discussed above) and/or low biomass. Reduced infaunal biomass is typically a function of elevated physico-chemical stress. High deposition rates may also limit biomass, and restrict infauna to those that are capable of repositioning themselves to within reach of the sediment–water interface. Many polychaetes and other worm-like trace makers are inhibited by elevated deposition rates, particularly if rapidly emplaced beds are more than about a decimeter thick. This situation is particularly common in deep-sea settings where turbidite emplacement rapidly buries and exterminates the resident benthic community. Common stresses also include salinity reduction and/or fluctuations, reduction in bottom-water and/or substrate oxygenation, elevated water turbidities, increased periods of subaerial exposure at the water's edge, introduction of substrates challenging to inhabit (muddy soupgrounds, woodgrounds, hardgrounds, and firmgrounds), and variations in energy conditions and/or concomitant accumulations of markedly different substrate calibers that are temporally more rapid than the recolonization ability of the resident community. In the continental realm, factors such as soil moisture, temperature, and annual precipitation are typically the paramount controls (*cf.* Hasiotis, 2007). Selecting which combination of stresses were responsible for a given facies is challenging, but by combining BI values with diversity, ethological variation, trace-fossil size, uniformity of burrowing and the sedimentology of the facies, the likely causes for depositional stress can be postulated.

Burrowing Uniformity Burrowing uniformity between beds can generally be taken as a reflection of environmental stability (Fig. 4). This stability can be entirely independent of bioturbation intensity, however, and should be evaluated separately. For example, a stable setting could display a BI of any value, depending upon what aspect of the setting is stable - a setting typified by persistently stressed conditions may lead to a virtually unburrowed facies, whereas an unstressed marine succession with persistently slow sedimentation can result in pervasive bioturbation. These clearly form extremes in a continuum, although the latter conditions are certainly common and widespread in many calm-water offshore, shelf and deep-marine settings. Lack of burrowing uniformity results from changes in bioturbation intensity, owing to temporal variations in depositional conditions! These changes characteristically coincide with event-style deposition, the most common being storm-bed deposition in the shallow-marine realm, turbidite emplacement in the deep sea, or river-flood deposits in the terrestrial realm. In each scenario, prevailing ambient or fair-weather sedimentation is interrupted by the abrupt introduction of an anomalous sedimentary body related to a short-lived episodic event.

The introduction of event beds may lead not only to changes in bioturbation intensity, but also to the development of discrete trace-fossil suites, depending upon the sediment caliber and food-resource contrasts between the ambient depositional regime and the event bed, as well as the amount of time available for recolonization (Pemberton and MacEachern, 1997). These suites may crosscut one another, leading to complex or mixed suites. For example, sandy event beds may display dwelling structures such as *Ophiomorpha* or *Diplocraterion,* whereas the fair-weather beds may contain mainly deposit-feeding structures such as *Chondrites, Zoophycos,* and *Asterosoma.* In the past, such recurring overprints were referred to as the 'mixed *Skolithos–Cruziana* Ichnofacies' but this is undesirable (see discussion on the Ichnofacies Paradigm).

Event beds tend to display overall reductions in bioturbation intensities relative to the background deposits. In settings prone to high-frequency events, this reduction in bioturbation is largely a function of a reduced colonization window (*i.e.,* diminished time for larval recruitment). Some

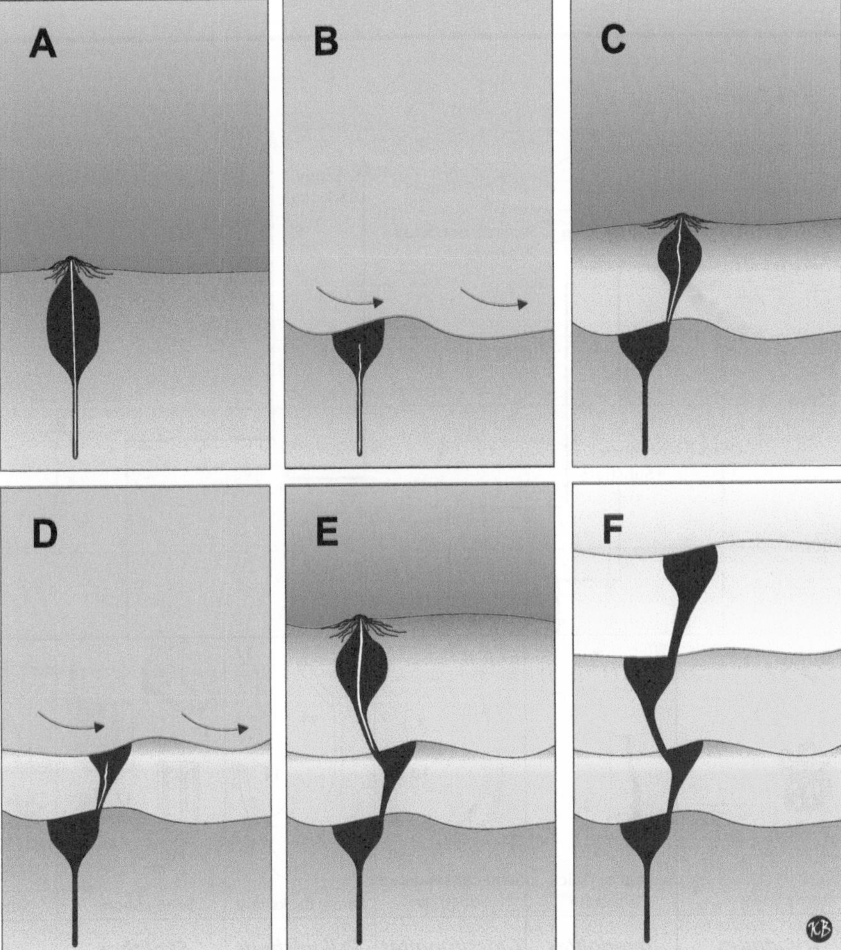

Figure 5. Readjustment behaviors expressed by *Rosselia* in response to deposition and erosion. Trace makers shift the position of the burrow to maintain or re-establish their connection to the sediment–water interface. **A.** Initial colonization; **B.** erosional truncation preceding a depositional event; **C.** the trace maker repositions its burrow following deposition; **D–F.** iterations of erosional truncation and burrow re-establishment (modified from Gingras *et al.,* 2007).

reduction in burrowing, however, may also reflect the degree of unsuitability of the introduced sediment for the needs of the normal inhabitants of the setting. Turbidites deposited in the deep sea, for example, are particularly prone to this, and many beds may entirely escape biogenic disruption due to the absence of larval recruitment into an unfamiliar substrate and infaunal avoidance of the anomalous sediment. For these reasons, recolonization of the area may be delayed until the event bed is buried by ambient deposition. Rapid bed emplacement may be corroborated by the presence of escape structures generated by organisms entrained in the flow or buried by the event bed itself. Thinner, episodically emplaced beds may be marked by readjustment or re-equilibration structures.

Seasonal variation in sedimentary conditions can also lead to predictable variation in the ichnological record. Inclined heterolithic stratification in tidal point-bar deposits, for example, commonly displays repeated colonization events. The resulting suites reflect the contrasting seasonal conditions and can comprise unburrowed and burrowed units, or can present two suites reflecting the two (dominant) ambient seasonal conditions (*e.g.,* Dalrymple *et al.,* 1991; Pearson and Gingras, 2006).

THE ICHNOFACIES PARADIGM AND FACIES MODELS

The ichnofacies paradigm (Fig. 6) is an elegant, unifying framework manifest by recurring, strongly facies-controlled (*i.e.,* environmentally relat-

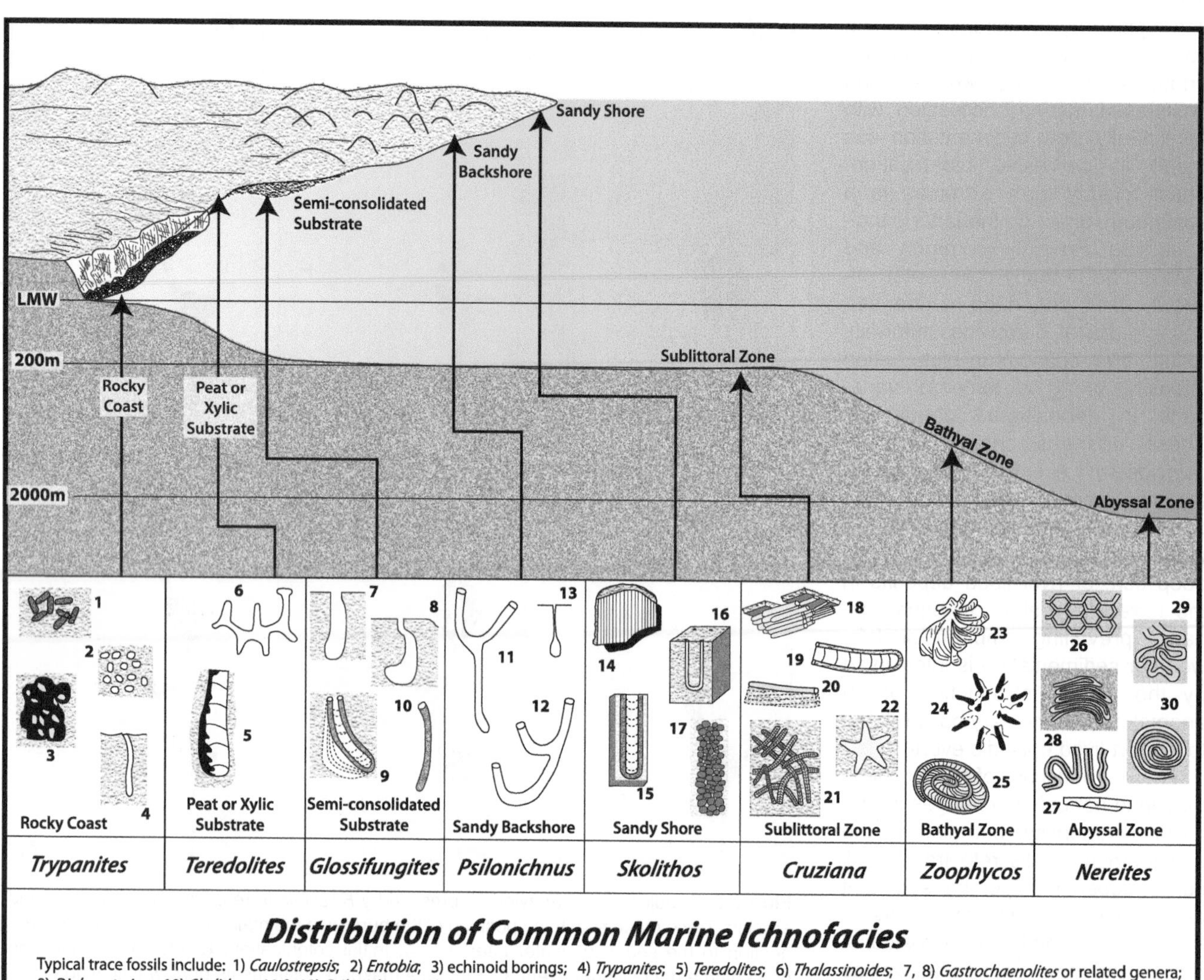

Figure 6. An idealized proximal–distal trend for recurring Seilacherian ichnofacies, reflecting progressive changes in depositional conditions (modified after Pemberton *et al.*, 1992). This schematic trend is not intended to express the complete distribution range for each ichnofacies (see text for explanation).

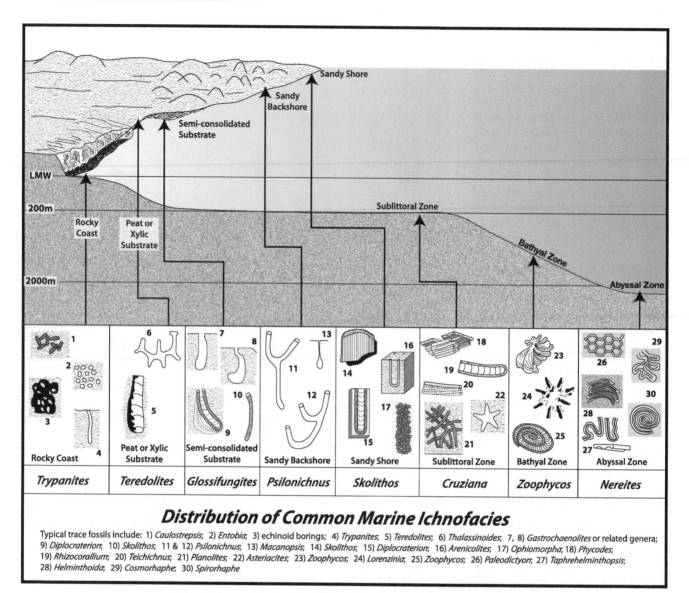

ed) groupings of trace fossils that reflect specific combinations of organism behavior (ethology; Seilacher, 1967). The namesake ichnogenera represent commonly present significant elements of the suites upon which the original ichnofacies were described. These ichnogenera selected also represent the dominant ethological grouping distinctive of the ichnofacies. In practice, the namesake ichnogenera need not be present, so long as the suite itself conforms to the ichnofacies' ethological grouping.

Ichnofacies are part of the total aspect of the rock, and consist of the primary biogenic structures created by organisms that inhabited the depositional environment. Insights into the depositional environment are derived from the fact that organisms respond in predictable ways to variations in environmental conditions. Although in the marine realm many of these conditions change progressively with increasing water depth, ichnogenera display, at most, a passive relationship to bathymetry. This passive relationship *does* tend to lead to predictable proximal–distal trends in some successions, but there are numerous exceptions; most are associated with dynamic depositional regimes (*e.g.,* submarine fans, fan deltas, sandy tide-swept shelves). In the past, the bathymetric aspect of ichnofacies was overemphasized, and care must be taken when assigning trace-fossil suites to relative water depths. Given these caveats, ichnofacies, like lithofacies, are subject to Walther's Law (*see* Chapter 2). The utility of ichnofacies to paleoenvironmental reconstruction, therefore, also lies in their lateral continuity and predictable vertical succession, leading to mappable units. Accurate interpretations of depositional environments favor reliable predictions of laterally adjacent settings and their associated ichnofacies. Like all facies analyses, interpretations of ichnofaunas are improved substantially when they are evaluated in the context of the host rocks and their sedimentologic (*i.e.,* lithofacies) and stratigraphic implications.

Archetypal ichnofacies are especially effective for characterizing deep-marine through to shallow-marine settings, although more recent studies have investigated and expanded their utility in continental deposits as well. Intergradations between the archetypal ichnofacies are also common and demonstrate a continuum of changing depositional conditions. As a result, very high-resolution analyses can be achieved. Departures from the archetypal ichnofacies are also common, but their recognition and interpretation are only possible by comparison with these established temporally and globally recurring groupings (MacEachern *et al.,* 2007b). By their very nature, such anomalous ichnological suites yield important insights into the specific characteristics of the depositional setting, highlighting animal–sediment interactions in response to imposed environmental stresses. In this way, brackish-water environments, anoxic to dysaerobic settings, and areas of fluviodeltaic deposition can be readily recognized.

In spite of being one of the most important concepts in ichnology, confusion continues to surround the nature and utility of the ichnofacies paradigm. For numerous workers, the paradigm is viewed as a restrictive framework, limiting the expression of observed variability in modern and ancient settings. For others, ichnofacies are regarded as nothing more than repeated suites or assemblages of trace fossils observed at the local scale; ichnofacies are neither.

Ichnofacies are *facies models* that address animal–sediment responses in the depositional environment. They have been built through the distillation of recurring characteristics of trace-fossil suites from numerous temporally and geographically ranging case studies (Fig. 7). The distillation of observations from these studies allows the ichnofacies to retain the recurring characteristics of the assigned suites (*e.g.,* diversity, ethological grouping and ethological

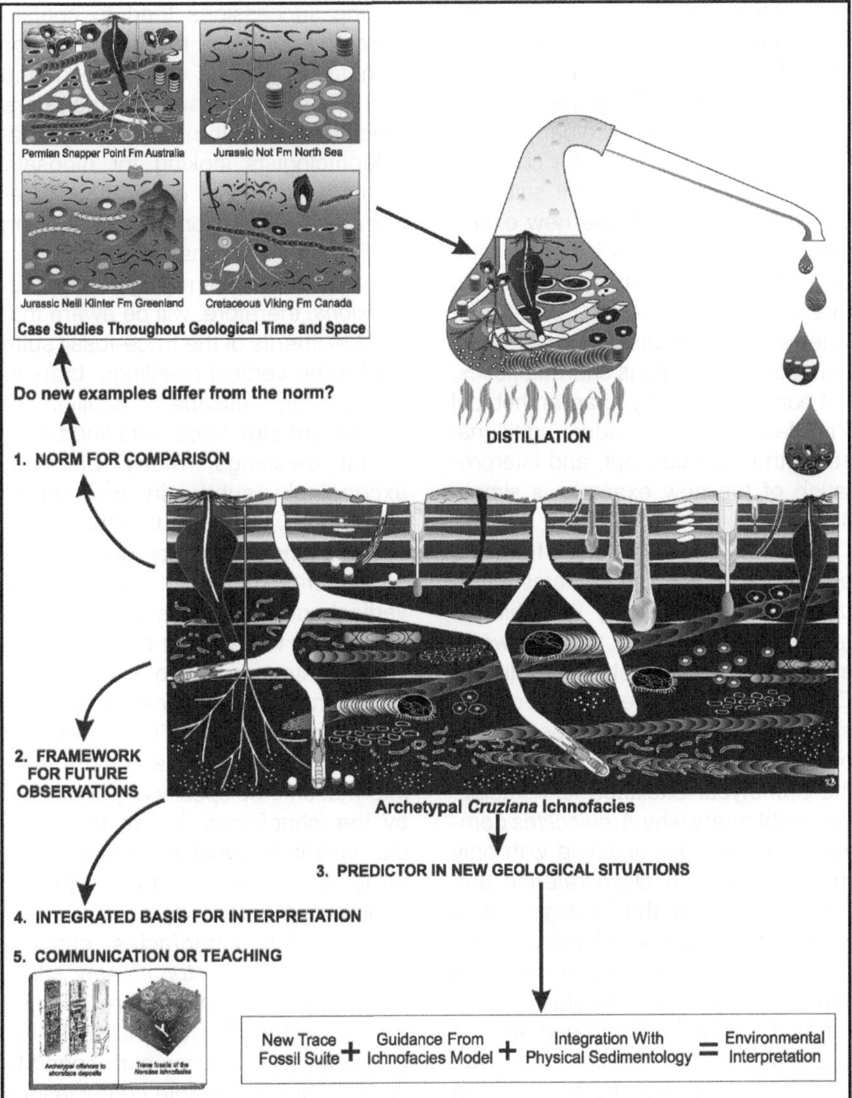

Figure 7. The ichnofacies paradigm as a facies model (concept after Walker, 1992; see text for details).

range, response to substrate), independent of their case-specific characteristics. This distillation also serves to minimize (to some extent) the changes in suites that have occurred as a result of biologic evolution – although some of the ichnogenera are different, their ethologic affinities are similar and therefore represent similar animal–sediment responses to the environment. The unique characteristics of the suites are retained, however, for evaluation of the specific case study. Just as importantly, ichnofacies are validated and, indeed, have been refined via neoichnological studies from modern environments, and through our *'theoretical'* understanding of how organisms live and behave in response to physico-chemical conditions. This 'process ichnology' (Gingras *et al.,* 2007) underpins the ichnofacies concept in a manner comparable to that served by process sedimentology for the new generation of lithologic facies models (*see* Chapter 2).

Ichnofacies models serve the same five functions (Fig. 7) as do lithofacies models (*e.g.,* Walker, 1992; *see* also Chapter 2).

First, an ichnofacies acts as a norm for the purposes of comparison. An archetypal ichnofacies comprises the features shared by all suites attributable to the ichnofacies. Any individual suite need not contain all the ichnogenera characteristic of the ichnofacies, or even the namesake element. Rather, the suite must contain traces that are consistent with the ethological grouping that defines the ichnofacies – this is what all suites, regardless of age and locality, have in common. Archetypal ichnofacies, by acting as norms for comparison, allow new trace-fossil suites to be compared – does the new example meet the criteria for inclusion within a particular ichnofacies? Without this norm, one cannot determine whether a new example contains any unusual or anomalous characteristics. If it conforms exactly to an archetypal ichnofacies, then it adds additional strength to the concept, and interpretation of the new example is simplified.

If a new example differs from the established norm and contains attributes that do not easily fit within an archetypal ichnofacies, then the worker can specify precisely *how* it differs. One can now ask questions about the suite that could not have been posed before. For example, compared to the archetypal *Skolithos* Ichnofacies, we might query why *Arenicolites* dominates a particular example with only minor associated *Diplocraterion* and *Lockeia*. Rather than assign a new ichnofacies to account for this example, we can explain *why* it contrasts with the archetypal ichnofacies, and perhaps assign an interpretation of opportunistic colonization of newly available sandy substrates, such as might be encountered in tempestites deposited in the offshore realm. These settings yield characteristics consistent with the *Skolithos* Ichnofacies (shifting particulate substrates, with generally high-energy conditions), but with limits imposed upon it (predominantly facies-crossing elements typical of rapid opportunistic colonization of newly available substrates). The interpretation of a high abundance but low diversity, and mainly facies-crossing expression of the *Skolithos* Ichnofacies leads to precise paleoenvironmental interpretations, without the unnecessary creation of a new ichnofacies to explain it. Without the archetypal ichnofacies, however, such questions cannot be asked, and avenues of research are left unexplored.

Second, the ichnofacies concept acts as a *framework* or *guide for future observations*. Sand-prone marine successions tend to contain suites attributable to the *Skolithos* Ichnofacies. When one is working in such units, mental search criteria are erected, much the same way that paleontologists looking for dinosaur bones have search criteria that differ from those of paleontologists looking for fish fossils – bone shapes, scales, etc. The worker studying *sandy successions*, therefore, will be aware that most elements of the trace-fossil suite tend to be vertical dwellings, branching dwellings, escape or equilibrium-adjustment structures, and lined horizontal dwellings, many of them exceedingly subtle in well-sorted sandstones. Realizing that the bioturbation intensities in such units tend to be underestimated in exposures of vertical orientation, it is useful for the researcher to seek out particular outcrop orientations and rock types, or to look at the bedding planes of cores as often as possible. This does not exclude the careful search for new information not specifically indicated by the ichnofacies, but it does help the worker to avoid overlooking features that might require specific approaches or techniques.

Third, the ichnofacies concept *serves as a predictor in new situations*. The paradigm erected by Seilacher (Fig. 6) demonstrates that the various ichnofacies correspond to different but predictable combinations of environmental parameters. The archetypal *Skolithos* Ichnofacies reflects high-energy, shifting particulate substrates, generally in marine water. The archetypal *Cruziana* Ichnofacies records lower energy, generally more cohesive substrates, with mainly sediment accumulation via suspension selling. We may then, with some justification, consider that if we have a locality containing suites attributable to the *Skolithos* Ichnofacies, and with other facies indicators consistent with an open marine, upper shoreface environment, by moving in a more distal direction, we can expect to encounter mud-prone, open-marine facies carrying suites attributable to the *Cruziana* Ichnofacies. In contrast, should we move in what we believe to be the seaward direction, but encounter sand-prone environments having trace suites corresponding to the *Psilonichnus* Ichnofacies, we might then predict that we have erred in our paleogeographic reconstruction. In actuality, however, it is far more complex than this simple example suggests. In a wave-dominated estuary (*see* Chapter 11), for example, one might encounter a low-diversity trace-fossil suite of the *Skolithos* Ichnofacies, which gradually passes seaward into a low-diversity expression of the *Cruziana* Ichnofacies reflecting a stressed infaunal community. Given what we know of wave-dominated estuaries, we might logically interpret the transition to indicate a shift from the bayhead delta to the central basin. We might then predict that by continuing in the same direction, we will encounter sand-prone facies with higher diversity trace-fossil suites of the *Skolithos* Ichnofacies, corresponding to the estuary mouth complex (*see* Chapter 11). The close relationship between ichnofacies, lithofacies, facies models, and Walther's Law permits such predictable distributions. Prediction is very important when dealing with limited datasets (subsurface studies or less extensive outcrop exposures), but can also assist with the selection of field localities in studies with extensive exposure that could not possibly be fully covered in the time frame of the study.

A fourth use of ichnofacies — as with physically based facies models — is that it may *act as an integrated basis for interpretation* of the system that it represents (Walker, 1992). Such a use is achieved once the model is mature, and is based on the combined features of many case studies. The turbidite model, for example, has reached this level: it has a hydrodynamic basis for interpretation — deposition from the deceleration of turbulent sediment-gravity flows (*see* Chapter 12). The ichnofacies concept is only starting to approach this use, owing to the continued clarification of the environmental parameters that constrain the faunal behaviors and the types and diversities of the resulting ichnogenera. This is particularly true for the *Skolithos, Cruziana* and *Zoophycos* ichnofacies, because of the great abundance of shallow-marine successions that have been studied. The *Skolithos* Ichnofacies, where manifest by suites containing elements such as *Conichnus, Bergaueria, Diplocraterion* and *Skolithos*, record persistent, high-energy depositional conditions, shifting particulate substrates, clean (low-turbidity) and well-oxygenated water columns, normal marine salinities, and abundant food material held in suspension. No matter what specific paleoenvironment is ultimately determined for the unit, it *must conform to these physico-chemical parameters.* The essence of this fourth usage is that it permits the reconstruction of the processes that formed the deposits in question.

Finally, ichnofacies serve a fifth role, as expounded in Chapter 2: it *facilitates teaching and communication.* The ichnofacies concept provides a basis for understanding the ways in which organisms respond to their environment and forms a theoretical basis for evaluating observed trace-fossil suites. Using ichnofacies, these relationships can be effectively presented to students. Indeed, the presence of a volume such as this and the inclusion of this chapter attest to the teaching role of the ichnofacies paradigm.

Opponents to the ichnofacies approach criticize its utility by implying that trace-fossil suites and ichnofabrics are 'lumped' into broad entities that then are assigned to a limited number of environments (*e.g.,* Taylor *et al.,* 2003; McIlroy, 2004b). Their concern is that the details imparted by the observed ichnological suites and ichnofabrics become downplayed in the interpretation of the facies. A similar criticism has been leveled against sedimentological facies models (*see* Chapter 2). However, to turn the phrase of Walker (1992), it is the generality embodied by the ichnofacies, as opposed to the summary of one particular case study, which enables the ichnofacies paradigm to serve its most valuable functions.

Competent workers do not look at a sandstone interval and say: "this has suites attributable to the *Skolithos* Ichnofacies and therefore the unit is a shoreface deposit!" The *Skolithos* Ichnofacies is *not* diagnostic of middle and upper shoreface environments. Rather, it is a theoretical construct characterized by dwelling structures made by inferred suspension feeders, surface-detritus feeders, and carnivores favoring high-energy, shifting, sandy substrates with clean, fully marine water columns. Suites attributable to the *Skolithos* Ichnofacies *may*, indeed, correspond to middle or upper shoreface settings, but they may alternatively correspond to proximal delta fronts with minimal water turbidity, to estuary-mouth shoals with strong tidal exchange, tidal channels or inlets, sandy open bays, or any other depositional environment that honors the particular physico-chemical parameters upon which the ichnofacies is founded. The ichnofacies concept's strength lies in the environmental validity of each of its ethological groupings. Assigning a trace-fossil suite to a particular ichnofacies does not result in a pigeonhole identification of a depositional environment.

Instead, the paradigm, as it stands, permits insightful assessments and interpretations of the observed trace suite, so that the most reasonable depositional environment selected for the facies adheres to the depositional characteristics exemplified by the ichnofacies. Whether one chooses to employ trace-fossil suites integrated with lithofacies analysis and genetic stratigraphic frameworks, or ichnofabric analysis, the ultimate paleoenvironmental interpretation of that ichnological data requires understanding of animal–sediment interactions as constrained by the environment. The ichnofacies paradigm comprises the unifying framework within which ichnological observations can be accurately interpreted in a depositional context.

The Archetypal Ichnofacies

Eleven temporally and geographically recurring archetypal ichnofacies have been defined, each named for a characteristic ichnotaxon. These fall into five softground marine types (*Psilonichnus, Skolithos, Cruziana, Zoophycos*, and *Nereites*), three substrate-controlled types (*Trypanites, Glossifungites,* and *Teredolites*), and three softground continental types (*Scoyenia, Mermia,* and *Coprinisphaera*). Recent summaries of the ichnofacies concept can be found in MacEachern *et al.* (2007b,d).

Continental Ichnofacies

The *Coprinisphaera, Scoyenia* and *Mermia* ichnofacies encompass ichnological suites found in continental settings (*see* Chapters 6, 7 and 21). The systematic ichnological analysis of the continental realm is relatively recent, with an ever-increasing number of case studies being added to the developing models. As a result, refinements to these ichnofacies are to be expected. Workers have demonstrated that the continental regime is far more diverse and complex, ichnologically, than previously considered. At present, there is the sense that there are insufficient examples described in the literature to permit the degree of synthesis and distillation that has been achieved in the marine realm. Hasiotis (2002, 2007), for example, has conducted studies from an actualistic perspective, employing ichnocoenoses that demonstrate the overall inadequacy of current continental ichnofacies models for global application.

Subaerially exposed areas are characterized by animal responses to factors such as soil-moisture content, stability of the groundwater table, high and low extremes of temperature, variability of climate, and soil/sediment composition. Likewise, variability in lacustrine and intermittently subaqueous settings is a result of factors such as organic content of the sediment, dissolved oxygen concentrations, sediment caliber, climatic seasonality, temperature, and salinity.

Scoyenia *Ichnofacies*

Buatois and Mángano (1995) refined the *Scoyenia* Ichnofacies (Figs. 8 and 9A, B), and ascribed it to low-energy continental settings characterized by periodically subaerial conditions. Most settings are inundated intermittently with freshwater, particularly during river floods. Incipient paleosols may be locally developed, associated with intervals of floodplain mudstones (*see* Chapter 6). Upper parts of point bars and channel banks are also typical sites for suites attributable to the *Scoyenia* Ichnofa-

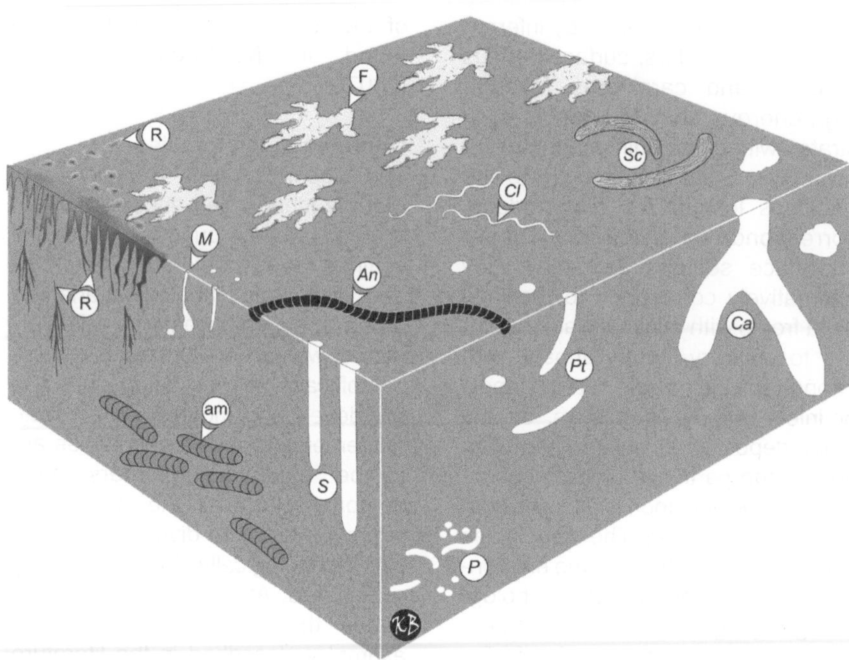

Figure 8. Schematic block diagram of the *Scoyenia* Ichnofacies. Trace-fossil abbreviations are as follows: *An: Ancorichnus; Ca: Camborygma; Cl: Cochlichnus*; F: footprints; M: *Macanopsis; Na: Naktodemasis; P: Planolites; Pt: Palaeophycus tubularis*; Rt: rhizoliths; *S: Skolithos;* and *Sc: Scoyenia* (figure modified after MacEachern *et al.* (2007d).

cies. Event-style fluvial deposits, particularly as crevasse splays, are common.

The *Scoyenia* Ichnofacies encompasses trace-fossil suites that consist of:

1. Horizontal, meniscate (backfilled) structures made by mobile deposit-feeding organisms;
2. Horizontal mobile deposit-feeding structures; locomotion traces, including tracks and trails;
3. Fish-fin markings;
4. Vertical dwelling structures;
5. Horizontal dwelling structures;
6. A mixture of invertebrate (predominantly arthropod) and vertebrate structures (dwelling and footprints); and
7. Plant-root traces.

Suites tend to display low to moderate diversities and localized high abundances. Ornamented structures and scratch marks are more typical of burrow excavation during subaerial exposure and substrate desiccation. The *Scoyenia* Ichnofacies may comprise two distinct expressions: one characterized by meniscate, backfilled structures without ornamentation developed in a soft substrate, and the second characterized by striated traces developed in a firm substrate, which commonly crosscut the former (Buatois and Mángano, 2004) as a result of progressive desiccation.

Mermia *Ichnofacies*

The *Mermia* Ichnofacies (Figs. 9C, D and 10) encompasses freshwater trace-fossil suites associated with low-energy, permanently subaqueous conditions, characterized by a high degree of environmental stability (Buatois and Mángano, 1995). These conditions are met by freshwater lakes, particularly shallow lake margins where oxygenation of the water column is readily maintained (*see* Chapter 21). Suites attributable to the *Mermia* Ichnofacies may also occur in some glacial lakes (*see* Chapter 5) and freshwater delta complexes, although case studies of these settings are uncommon. Recent studies identifying the *Mermia* Ichnofacies show it to be dominated by horizontal to sub-horizontal grazing and feeding traces produced by mobile detritus and deposit feeders, as well as by a subordinate number of locomotion traces. Fish, amphibian, reptile, and possibly even mammal feeding structures can be present. Suites display high to moderate diversities and abundances, particularly in lacustrine settings that are hydrologically open. Nevertheless, freshwater biomes are dominated by seasonal changes in water chemistry, hydraulic energy and temperature, and a propensity (in lakes) to establish anaerobic conditions. These issues are coupled with dynamic lake levels over longer time frames, such that most freshwater bodies are hostile to animal colonization at times. In semi-arid and arid climates, constant evaporation of water produces high salinities that are not inhabited by infaunal invertebrates. Correspondingly, long-term inland seas possess no sediment dwellers, as freshwater invertebrates have not radiated into land-locked saline basins.

Coprinisphaera *Ichnofacies*

The *Coprinisphaera* Ichnofacies (Figs. 9E, F and 11) was erected by Genise *et al.* (2000) to accommodate suites associated with more or less permanently subaerially exposed continental settings (*i.e.*, paleosols; *see* Chapter 6) that vary from dry and cold to warm and humid. Ethologically, the principal grouping is nesting/breeding (calichnia), however, such behavior has not been extended beyond the Upper Cretaceous. Accordingly, employing calichnia as the ethological basis for the ichnofacies limits its temporal extent. Suites also include dwellings employed as refugia, aestivation (high temperature-induced torpor), and ambush predation (including large vertebrate-generated domiciles), as well as mobile deposit-feeding structures (particularly earthworms and other oligochaetes), and rhizoliths produced by plants. These ethological groupings persist well into the Paleozoic, and would constitute a superior basis for the erection of a paleosol-related ichnofacies. The predominant trace makers forming calichnia include bees, ants, wasps, beetles (particularly dung beetles), termites, and other insects, which show a marked evolutionary diversification linked to the increase in importance of flowering and seed-bearing plants (*i.e.*, angiosperms). These trace-fossil suites are prone to complex tiering patterns, particularly in mature soils, reflecting the variable depths of emplacement of the nests. Suites show moderate to relatively high diversity, and generally high

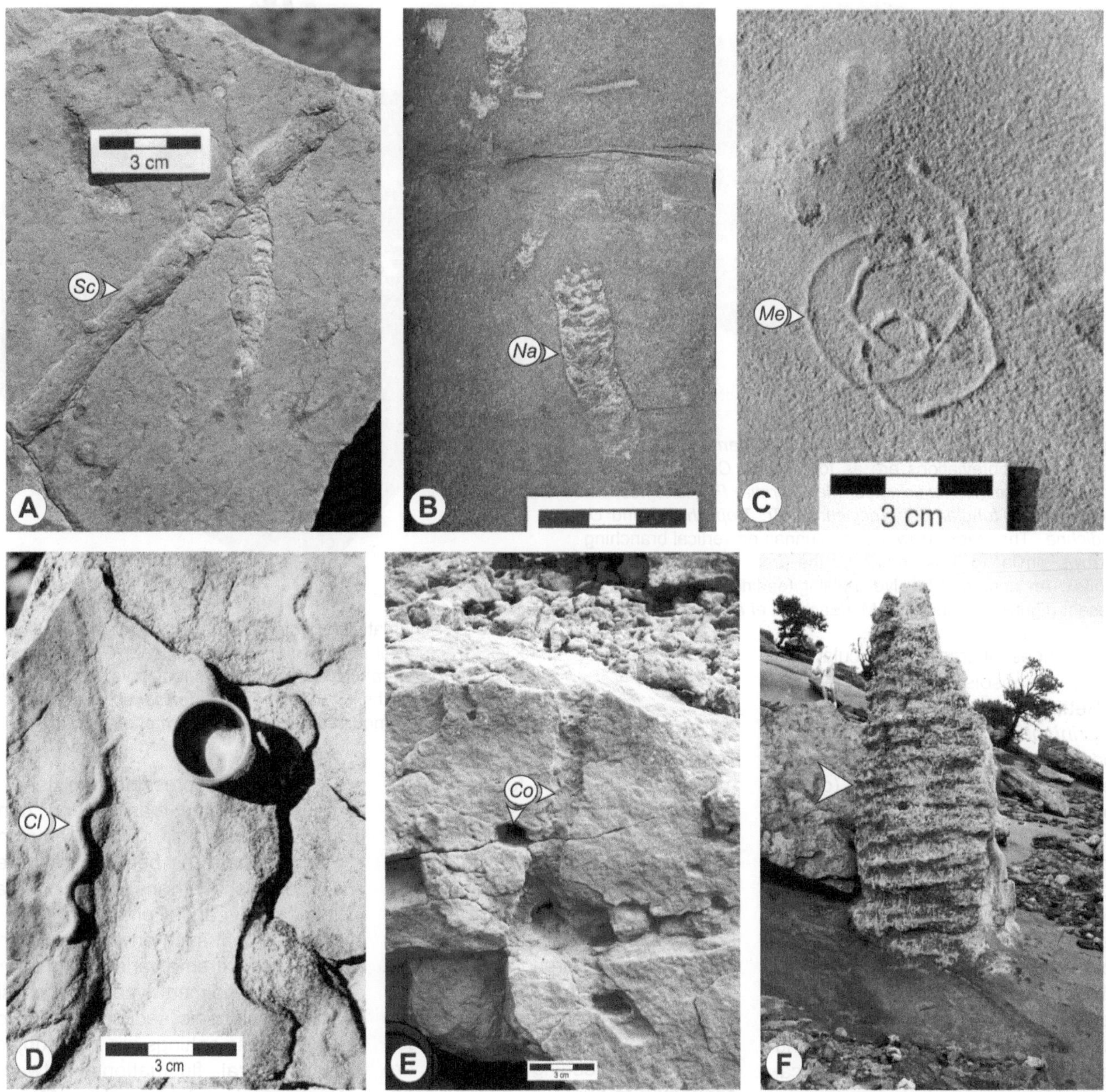

Figure 9. Example photos of trace fossils typical of continental suites. **A.** The trace fossil *Scoyenia (Sc)* with meniscae and scratch marks, from the Lower Triassic of the Karoo Basin, South Africa. This is the namesake ichnogenus of the *Scoyenia* Ichnofacies. Photo is courtesy of Fiona J. MacEachern. **B.** The trace fossil *Naktodemasis (Na)* from the Lower Cretaceous Mannville Group, Alberta, interpreted as a beetle larve burrow generated in unconsolidated sediment. This is a common element of the *Scoyenia* Ichnofacies. **C.** The trace fossil *Mermia (Me)*, namesake of the *Mermia* Ichnofacies, from the Permian of the Karoo Basin, South Africa. This was misidentified in MacEachern *et al.* (2007d) as *Gordia*. **D.** The trace fossil *Cochlichnus (Cl)*, a common constituent of the *Mermia* Ichnofacies, from Triassic Bowen Basin, Queensland, Australia. **E.** The trace fossil *Coprinisphaera (Co)*, interpreted as a dung-beetle nest from a paleosol unit, Eocene– Oligocene beds, Ebro Basin, Spain. This ichnogenus is the namesake of the *Coprinisphaera* Ichnofacies. **F.** The trace fossil *Termitichnus* (arrow) from a paleosol unit of the Upper Jurassic Morrison Formation, Fort Wingate, New Mexico. This ichnogenus is a common element of the *Coprinisphaera* Ichnofacies. Photos **E** and **F** are courtesy of Stephen Hasiotis, but his kind permission to use these images should not be taken as his agreement with their ichnofacies designations.

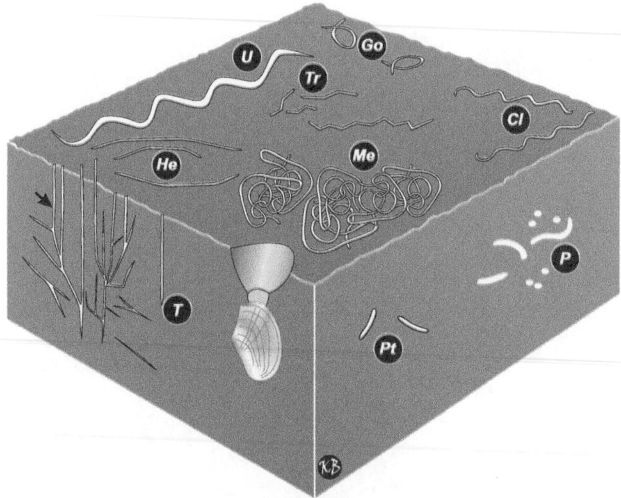

Figure 10. Schematic block diagram of the *Mermia* Ichnofacies. Trace-fossil abbreviations are as follows: *Cl: Cochlichnus; Go: Gordia*; He: *Helminthoidichnites; Me: Mermia; P: Planolites; Pt: Palaeophycus tubularis; T: Trichichnus; Tr: Treptichnus*, and *U: Undichna*. The black arrow points to unnamed vertical branching burrows similar to those made by the present-day oligochaete *Tubifex*. An unnamed bivalve dwelling/feeding structure is also present (figure modified after MacEachern *et al.*, 2007d).

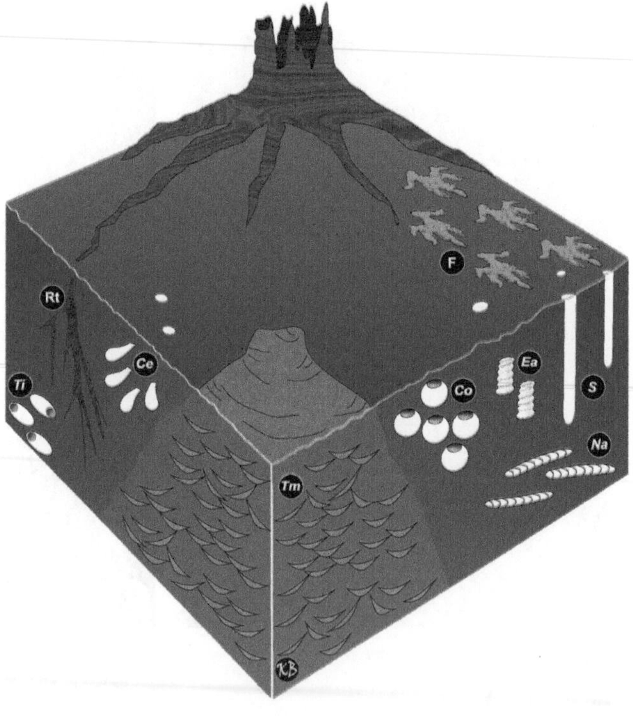

Figure 11. Schematic block diagram of the *Coprinisphaera* Ichnofacies. Trace-fossil abbreviations are as follows: *Ce: Celliforma; Co: Coprinisphaera; Ea: Eatonichnus*; F: footprints; *Na: Naktodemasis;* Rt: rhizoliths; *S: Skolithos; Ti: Teisseirei*; and *Tm: Termitichnus* (figure modified after MacEachern *et al.*, 2007b).

abundances of traces, particularly in mature paleosols.

Settings characteristic of the *Coprinisphaera* Ichnofacies correspond to paleosols developed in paleoecosystems of herbaceous (grass) communities. This may effectively limit the ichnofacies to units ranging from Late Cretaceous to the Holocene as there were no herbaceous communities earlier in Earth history. Climatically, settings range from arid and cold steppes (dominated by nests of hymenopterous insects; *i.e.*, insects with 4 membranous wings, such as bees, wasps and ants) to humid and hot subtropical savannas (dominated by termite nests). Paleosol settings occupy alluvial plains, desiccated floodplains, crevasse splays, levees, abandoned point bars, and vegetated eolian environments (Genise *et al.*, 2000). These settings are strongly controlled by local variations in temperature, radiation, humidity, and wind speed near the ground associated with vegetation, topography, and overall climatic conditions (Hasiotis, 2007).

Softground Marine Ichnofacies

Five marine-softground ichnofacies have global and temporal recurrence: the *Psilonichnus, Skolithos, Cruziana, Zoophycos*, and *Nereites* ichnofacies. These characteristic ichnofacies primarily reflect different feeding behaviors and responses to substrate consistency, with the *Psilonichnus* and *Skolithos* ichnofacies containing scavenging and filter-feeding/suspension-feeding traces, the *Cruziana* and *Zoophycos* ichnofacies representing deposit-feeding through to sediment grazing, and the *Nereites* Ichnofacies representing grazing and farming trace fossils in low food-resource settings. These feeding strategies are connected to hydraulic energy; associated substrates range from semi-consolidated to unconsolidated, and coarse grained to fine grained, respectively (*i.e., Psilonichnus* through to *Nereites* ichnofacies). Due to a broad range of modern and outcrop-based studies, these marine-softground ichnofacies are well established, and, where fully developed, indicate normal-marine conditions. Recent research has demonstrated that marine softground ichnofacies form a continuum along the depositional profile, adding precision to paleoenvironmental interpretations.

Psilonichnus *Ichnofacies*

The *Psilonichnus* Ichnofacies (Figs. 12 and 13A) represents a mixture of marine, marginal-marine, and nonmarine conditions. Typical environments include the beach backshore, dunes, washover fans, and supratidal flats. Frey and Pemberton (1987) indicated that such environments are subject to extreme variations in energy levels, sediment types, and physical and biogenic sedimentary structures. These settings are also subject to temporal fluctuations between fresh and saline water. Marine processes generally dominate during spring tides and storm surges, whereas eolian processes and subaerial exposure of supratidal flats predominate during neap tides and non-storm periods.

The topographically elevated position of this setting and its extended periods of subaerial exposure preclude colonization by most benthic marine animals. The only persistent, notable exceptions are amphibious crabs of the Family Ocypodidae, which include both scavengers and surficial deposit feeders; these animals typically excavate J-, Y-, or U-shaped dwelling burrows referable to the trace fossil *Psilonichnus*. Other

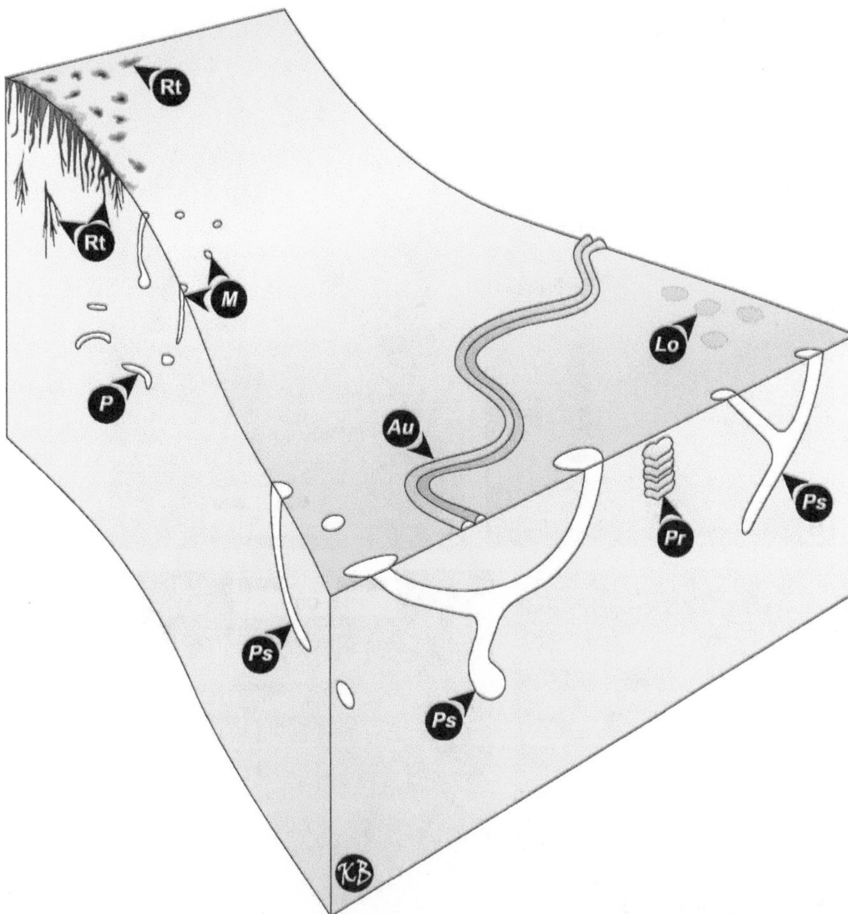

Figure 12. Schematic block diagram of the *Psilonichnus* Ichnofacies. Trace-fossil abbreviations are as follows: *Au: Aulichnites; Lo: Lockeia; M: Macanopsis; P: Planolites; Pr: Protovirgularia; Ps: Psilonichnus*; and Rt: rhizoliths (figure after MacEachern *et al.*, 2007d).

biogenic structures are generated by essentially terrestrial organisms and include the vertical shafts of insects and spiders, the horizontal tunnels of other insects and tetrapods, and the ephemeral tracks, trails, and fecal pellets of insects, reptiles, birds, and mammals. Plant roots constitute another biogenic structure common to the *Psilonichnus* Ichnofacies. The types of plants able to exploit these substrates range from intertidal halophytes on the distal margins of some washover fans, to maritime or terrestrial grasses, weeds, vines, shrubs, bushes, and trees on dunes.

The burrows of amphibious crabs may appear in the uppermost foreshore or the upper part of estuarine point bars, and so the *Psilonichnus* Ichnofacies overlaps the *Skolithos* Ichnofacies in occurrence; however, the boundary between these two ichnofacies is normally distinct. In contrast, because of the potentially large number of terrestrial traces in the *Psilonichnus* ichnofacies, it may be broadly intergradational with the *Scoyenia* Ichnofacies.

Skolithos *Ichnofacies*

The *Skolithos* Ichnofacies (Figs. 13B–D and 14) is indicative of high levels of wave or current energy, and typically is developed in clean, well-sorted, loose or shifting substrates. Abrupt changes in the rates of deposition, erosion, and physical reworking are common. Owing to frequent water agitation, little food is deposited on the substrate and is, instead, held in suspension within the water column. Correspondingly, most trace makers found here are suspension feeders and substrates serve mainly as an anchoring and sheltering medium. Thus, in both subtidal and intertidal examples, the organisms typically construct deeply penetrating, more or less permanent domiciles. Such conditions commonly occur on the foreshore and shoreface of beaches, bars, and spits: associated stratification features typically consist of parallel to subparallel, gently seaward-dipping laminae, large- and small-scale trough crossbeds, and current ripple laminae. Sandy subaqueous settings, such as sandy bays, can also possess oscillation ripples. Suites attributable to the *Skolithos* Ichnofacies also occur within crossbedded sands of some tidal-fluvial estuarine channels, tidal inlets and tidal channels, provided water turbidity is not excessive (*see* Chapter 9). In more seaward parts of the shoreline, suites attributable to the ichnofacies occur in swaley cross-stratified (SCS) and hummocky cross-stratified (HCS) tempestites of the middle shoreface and proximal lower shoreface (*see* Chapter 8). The *Skolithos* Ichnofacies may appear in slightly to substantially deeper water deposits wherever energy levels, food supplies, and substrate characteristics are suitable (Crimes and Fedonkin, 1994). Potential examples include submarine canyons, deep-sea fans, and bathyal slopes swept by strong contour currents (*see* Chapter 12). Such suites, however, will be spatially associated with low-energy suites typical of deep-water ambient conditions (*e.g., Zoophycos* and *Nereites* ichnofacies). Therefore, as emphasized previously, paleobathymetric interpretations cannot be based solely on checklists of trace-fossil names: evaluation of associated physical sedimentary structures, stratigraphic position, and other evidence is essential, even in normal beach-to-offshore successions.

The *Skolithos* Ichnofacies is characterized by;

- Predominantly cylindrical vertical or U-shaped burrows;
- Spreiten lying below or above some domiciles, indicating aggradation or degradation at the sediment-water interface, respectively;
- Few horizontal structures;
- Few deposit-feeding structures produced by mobile organisms (*Macaronichnus* being an exception);
- Low-diversity suites, although individual forms may be abundant;

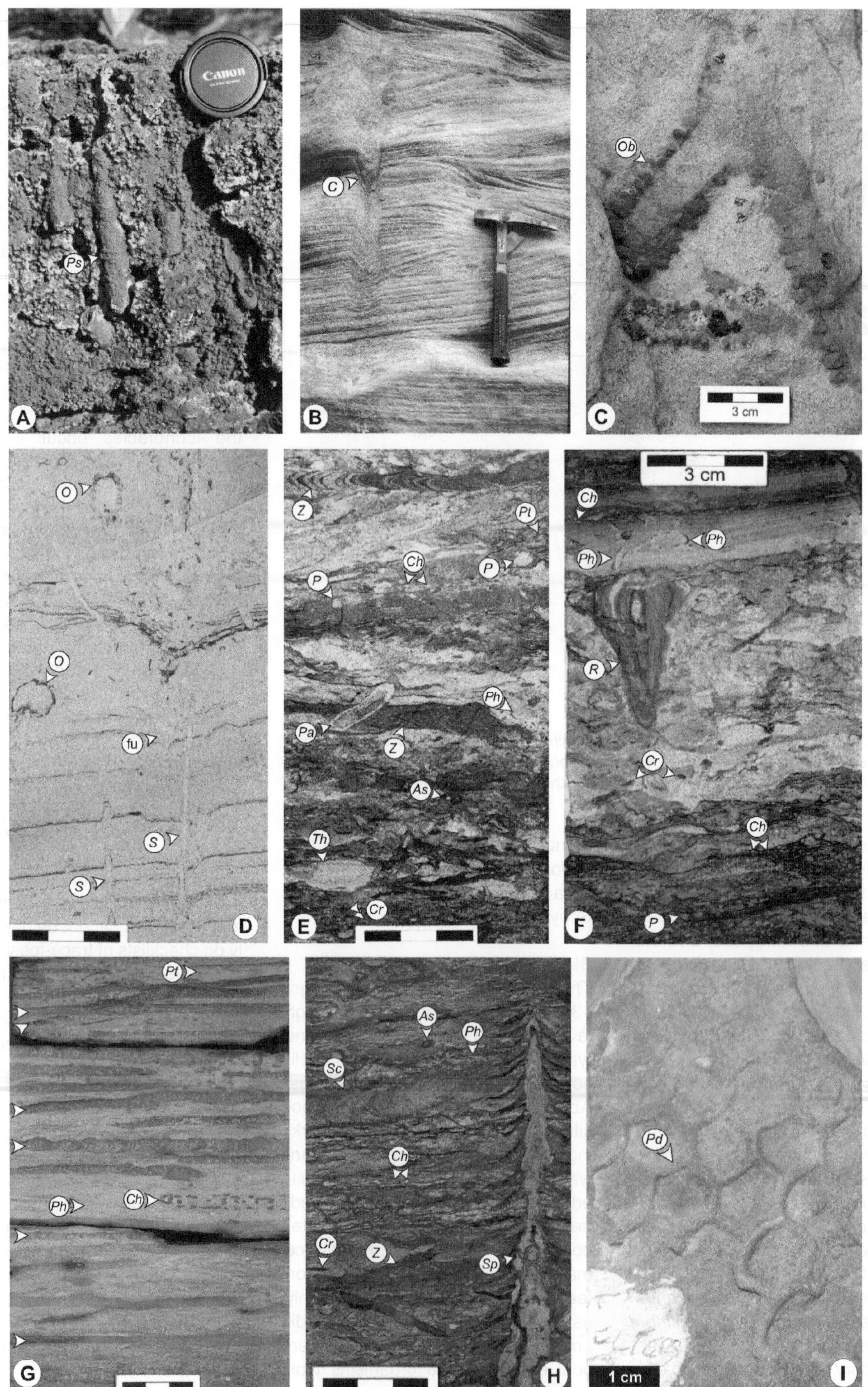

Figure 13. Caption on opposite page.

Figure 13. *(opposite page)* Photos of marine softground trace-fossil suites and their ichnofacies designations. **A.** The ichnogenus *Psilonichnus (Ps)* from a backshore setting in the Upper Cretaceous Woodbine Formation, north-central Texas recording the *Psilonichnus* Ichnofacies. **B–D.** Suites common to the *Skolithos* Ichnofacies. **B.** *Conichnus (C)* in an upper shoreface, crossbedded sandstone of the Upper Cretaceous Blackhawk Formation (Spring Canyon Member), Utah. **C.** Robust *Ophiomorpha borneensis (Ob)* of the lower shoreface, Upper Cretaceous Sego Sandstone, Utah. **D.** *Skolithos (S), Ophiomorpha (O)*, and fugichnia (fu) in cross-stratified sandstones (BI 3) of the Upper Cretaceous Ferron Sandstone, Utah. **E** and **F.** Suites attributed to the *Cruziana* Ichnofacies. **E.** Bioturbated (BI 5) sandy mudstone of the proximal upper offshore, with a diverse suite that includes *Zoophycos (Z), Planolites (P), Chondrites (Ch), Phycosiphon (Ph), Thalassinoides (Th), Asterosoma (As), Cosmorhaphe (Cr), Palaeophycus tubularis (Pt)*, and *Palaeophycus heberti (Pa)*. The unit is from the Upper Cretaceous Cardium Formation, Alberta. **F.** Diverse suite showing BI 4, with *Rosselia (R), Cosmorhaphe (Cr), Planolites (P), Chondrites (Ch)*, and *Phycosiphon (Ph)*. Unit is from the Lower Jurassic Plover Fm, Australia. **G** and **H.** Suites attributable to the *Zoophycos* Ichnofacies. Both units are from the Cretaceous Nise Formation, Norwegian Shelf. **G.** Silty and sandy mudstone of the slope with BI 5, dominated by *Chondrites (Ch), Phycosiphon (Ph)*, rare *Palaeophycus tubularis (Pt)* and other unnamed grazing structures, crosscut by deeper tier *Zoophycos* (white arrows). Note the well-developed spreiten in the *Zoophycos*. **H.** Bioturbated (BI 5) muddy sandstone unit of the upper slope, with *Spirophyton (Sp), Zoophycos (Z), Scolicia (Sc), Asterosoma (As), Cosmorhaphe (Cr), Phycosiphon (Ph)*, and *Chondrites (Ch)*. **I.** Bedding plane of a sandstone from the Miocene Waitemata Group, Auckland, New Zealand, showing the graphoglyptid *Paleodictyon (Pd)*, a unique constituent of the *Nereites* Ichnofacies.

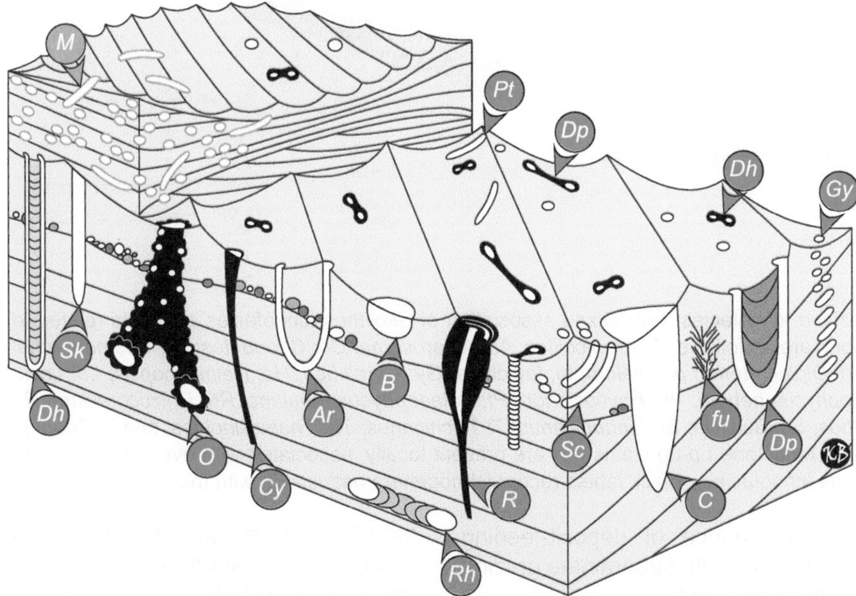

Figure 14. Schematic block diagram of the *Skolithos* Ichnofacies. Trace-fossil abbreviations are as follows: *Ar: Arenicolites; B: Bergaueria; C: Conichnus; Cy: Cylindrichnus; Dh: Diplocraterion habichi; Dp: Diplocraterion parallelum*; fu: fugichnia; *Gy: Gyrolithes; M: Macaronichnus; O: Ophiomorpha; Pt: Palaeophycus; R: Rosselia; Rh: Rhizocorallium; Sc: Schaubcylindrichnus*, and *Sk: Skolithos*. Red labels reflect common constituents. Green labels show facies-crossing forms that are less common, and which occur in the *Cruziana* Ichnofacies as well. Blue label marks assemblage that replaces *Skolithos* Ichnofacies in some settings. Figure after MacEachern *et al.* (2007b).

- Predominance of dwelling burrows constructed by suspension feeders or passive carnivores; and
- Vertebrate traces in some low-energy intertidal settings.

Cruziana *Ichnofacies*

The *Cruziana* Ichnofacies (Figs. 13E, F and 15) is characteristic of subtidal, poorly sorted or heterolithic, cohesive to semi-cohesive substrates. Depositional conditions range from long-term moderate-energy levels in shallow water below fair-weather wave base but above storm wave base, to generally low-energy levels in deeper, quieter waters (*see* Chapter 8). Such settings are subject to short-lived periods of high-energy event-style deposition, particularly associated with storm deposits. Sediment textures and bedding styles exhibit considerable diversity, including thin-bedded, well-sorted silt and sand, discrete mud and shell layers, interbedded muddy silt, clean silt and sand, and extremely poorly sorted beds derived from any of the above through intense bioturbation. Complex tier relationships may permit differentiation between thinly interstratified heterolithic bedding that has been subsequently biogenically homogenized and accumulations of admixed silty and sandy mudstone. Physical sedimentary structures, where not modified or destroyed through bioturbation, include parallel and subparallel lamination, HCS, oscillation-ripple lamination, or trough crossbedded sand.

With fluctuating energy levels, food supplies consist of both suspended and deposited components; either fraction may predominate locally, the two may be intermixed, or there may be marked temporal variability in the availability of the two food resources. Characteristic organisms therefore include suspension and deposit feeders, as well as mobile carnivores and scavengers. In the case of temporal variability in the supply of suspended and deposited food, discrete suites may be established, showing stacked tiers. In storm-dominated settings, this yields suites dominated by opportunistic vertical to subvertical dwelling-structures associated with event beds, passing into horizontal dwellings, feeding, foraging and grazing structures associated in fair-weather or ambient suites (*e.g.*, Pemberton and Frey, 1984; Pemberton and MacEachern, 1997). In settings characterized by short-duration, high-frequency events (limiting colonization times) or high-energy events (leading to pronounced erosional amalgamation) the event beds may

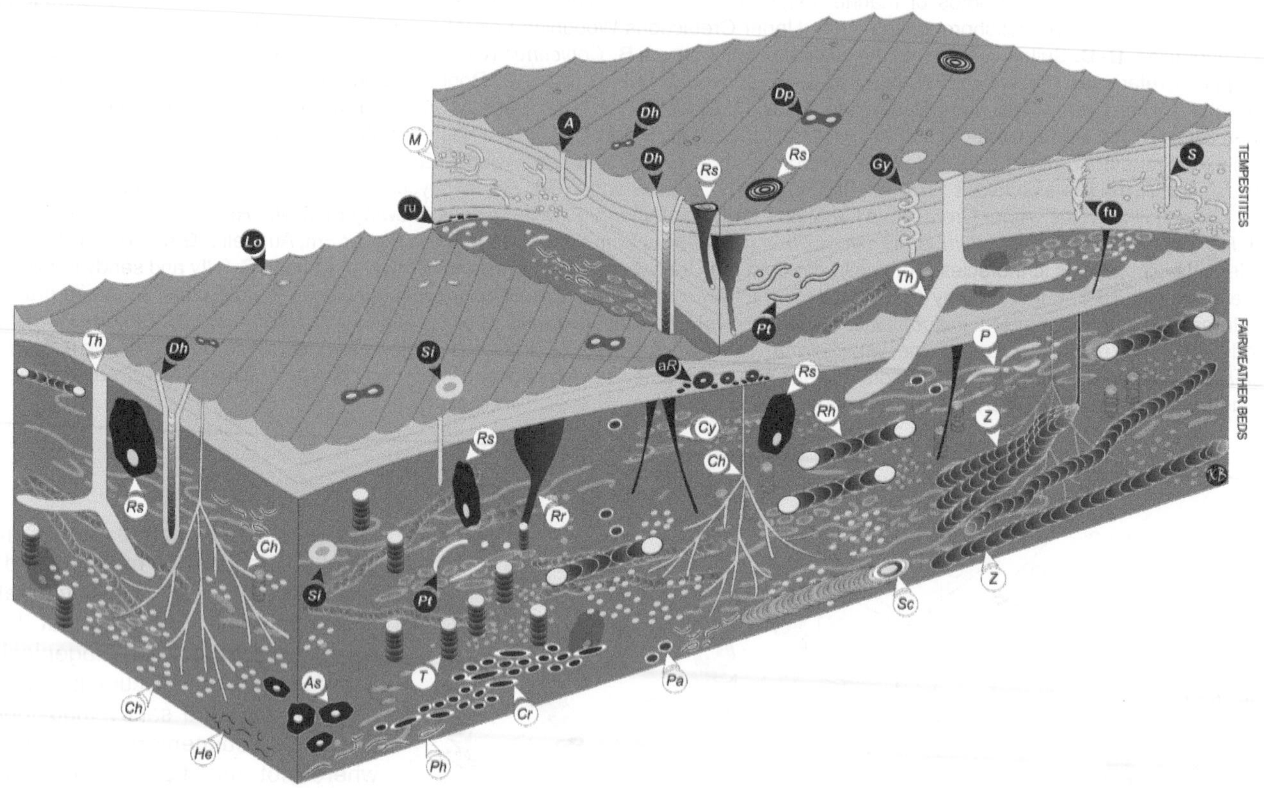

Figure 15. Schematic block diagram of the *Cruziana* Ichnofacies, with mixed association of *Skolithos* Ichnofacies elements related to tempestite emplacement. Trace-fossil abbreviations are as follows: *A: Arenicolites; As: Asterosoma; Ch: Chondrites; Cr: Cosmorhaphe; Cy: Cylindrichnus; Dh: Diplocraterion habichi; Dp: Diplocraterion parallelum;* fu: fugichnia; *Gy: Gyrolithes; He: Helminthopsis; Lo: Lockeia; M: Macaronichnus; P: Planolites; Pa: Palaeophycus heberti; Ph: Phycosiphon; Pt: Palaeophycus tubularis; Rh: Rhizocorallium; Rr: Rosselia rotatus; Rs: Rosselia socialis; S: Skolithos; Sc: Scolicia; Si: Siphonichnus; T: Teichichnus; Th: Thalassinoides*; and *Z: Zoophycos* . *Allochthonous Rosselia* mud balls (*aR*), and mudstone rip-up clasts (ru) are present locally, associated with event-beds. White labels record common ichnogenera of the *Cruziana* Ichnofacies. Black labels record ichnogenera associated with the tempestites.

be largely unburrowed, in contrast with the more pervasively bioturbated fairweather deposits; such scenarios lead to classical 'laminated-to-burrowed' bedding (*e.g.*, Pemberton and Frey, 1984). Owing to lower energy overall, burrows tend to be horizontal rather than vertical, although scattered vertical or steeply inclined burrows also occur. Large numbers of burrows can be present at stable, low-energy sites. Trails of epibenthic and endobenthic foragers also can be common, reflecting the abundance, diversity, and accessibility of food. Thus, the *Cruziana* Ichnofacies is characterized by;

- A mixed association of horizontal, inclined, and vertical structures, many of them corresponding to permanent to semi-permanent dwellings;
- Structures constructed by mobile organisms;
- Generally high diversity and abundance;
- Predominance of deposit-feeding structures with subordinate grazing structures; and
- Common overprinting of deep-tier over shallow-tier structures during continued burial and aggradation, locally leading to the preservation of composite and complex structures.

Zoophycos *Ichnofacies*

Of all recurrent marine ichnofacies, this ethological grouping (Figs. 13G, H and 16) is most debated and least understood. The ichnogenus *Zoophycos* has an extremely broad paleodepositional range; hence its designation as name-bearer for a supposedly depth-related ichnofacies has long been controversial. In popular bathymetric schemes, the *Zoophycos* Ichnofacies typically is portrayed as an intermediary between the *Cruziana* and *Nereites* ichnofacies, at a position corresponding approximately to the continental slope (Fig. 6; *cf.* Chapter 12). More specifically, the original designation placed it in flysch–molasse areas below wave base and free of turbidites, within a broad depositional gradient (Seilacher, 1967). Suites attributable to the ichnofacies are dominated by surface foraging and grazing structures, with associated complex deposit-feeding structures.

As re-evaluated (*e.g.*, Frey and Seilacher, 1980), one of the major environmental controls represented by the ichnofacies may be lowered oxygen levels associated with organic-rich sediments in calm-water settings. To the extent that these conditions actually occur on shelf-edge gradients, the popularized bathymetric placement of the ichnofacies is suitable. However, such generally dysoxic conditions, replete with a dominance of *Zoophycos*, are perhaps even better known in shallower water, epeiric deposits. Considering the above characteristics of the ichno-

Figure 16. Schematic block diagram of the *Zoophycos* Ichnofacies. Trace-fossil abbreviations are as follows: *As: Asterosoma; Ch: Chondrites; Cr: Cosmorhaphe; He: Helminthoida; P: Planolites; Ph: Phycosiphon; Sc: Scolicia; Sp: Spirophyton; Th: Thalassinoides*; and *Z: Zoophycos*.

facies, together with the widespread distribution of individual specimens of *Zoophycos* in both shallow- and deep-water deposits (Frey *et al.*, 1990), we speculate that the *Zoophycos*-making animal simply was broadly adapted in most ecologic respects: it tolerated not only a considerable range of water depths but also numerous substrate types, variable food resources, and different energy and oxygen levels. *Zoophycos* therefore appears in the *Cruziana* through *Nereites* ichnofacies simply because it reflects a useful way to feed along a surface and extends into deeper, progressively resource-poor sediments. Due to the singular prominence of *Zoophycos* in some settings, its predominance in recurring suites warrants its own ichnofacies designation. However, the less restrictive the depositional environment, the less distinctive is the ichnofacies (as a separate entity). In numerous places, the ichnofacies is hardly discernible in the broad transition from the *Skolithos* or *Cruziana* to the *Nereites* ichnofacies, especially on unstable ancient slopes that were subject to turbidity flows or swept by various shelf-edge contour currents (*see* Chapter 12) The role of the oxygen minimum zone (OMZ) on the slope to the distribution of the *Zoophycos* Ichnofacies has yet to be fully evaluated, but may also explain some of the paleogeographic distributions encountered in the ancient record.

Nereites *Ichnofacies*

In most respects, bathymetric implications of the *Nereites* Ichnofacies (Figs. 13I and 17) are less equivocal than those of any other recurrent ichnofacies. Although numerous trace fossils otherwise typical of shallow-water deposits locally range into deep-sea deposits, the reverse is not ordinarily true. The principal ethological groupings within this ichnofacies are surface grazing, farming and complex chemosymbiotic relationships with bacteria. In addition to water depth, turbidite deposition strongly influences organism behavior in the *Nereites* Ichnofacies. However, depth- and energy-related variables seem to be more important than turbidite deposition *per se* (Crimes and Fedonkin, 1994; Miller, 1993). For example, the trace assemblages persist today on abyssal plains beyond the reach of turbidity currents but are absent among well-developed, shallower water turbidite successions.

Nevertheless, most suites attributed to the *Nereites* Ichnofacies studied to date occur in turbidite-bearing successions, probably because the stratigraphic record is more complete, and because of taphonomic controls, wherein the soles of the event beds cast the predominantly horizontal structures that characterize the ichnofacies. The associated sediments thus may consist of virtually any lithology, except that the ratio of turbidite sand to hemipelagic or pelagic mud tends to diminish toward distal extremities of the deposit, and carbonates become increasingly scarce as the calcite compensation depth is approached. Partial or complete Bouma sequences are common locally, and physical sedimentary structures can include flute, groove, bounce, and load casts as well as prod marks, flame structures, and current ripples.

Animals exploiting lower bathyal to

Figure 17. Schematic block diagram of the *Nereites* Ichnofacies. Trace-fossil abbreviations are as follows: *Ch: Chondrites Cr: Cosmorhaphe; Fg: Fustiglyphus; He: Helminthoida; Lr: Lorenzinia; Ne: Nereites; P: Planolites; Pd: Paleodictyon; Ph: Phycosiphon; Sc: Scolicia; Sd: Spirodesmos; Sh: Spirophycus; Sp: Spirophyton; Sr: Spirorhaphe; Th: Thalassinoides; Ur: Urohelminthoida*; and *Z: Zoophycos* (figure after MacEachern *et al.*, 2007d).

abyssal environments have two major concerns: 1) scarcity of food, relative to more abundant supplies in shallower settings; and 2) periodic disruption by strong bottom currents or actual turbidity currents. In response to the latter, and over long spans of geologic time, the overall community ultimately developed two component parts: pre- and post-turbidite associations, as documented by their respective traces (*e.g.*, Miller, 1993). The pre-turbidite or ambient association is characteristic of quiet conditions and is dominant wherever the substrate is free of the influence of turbidity currents. Pre-turbidite animals constitute a stable, persistent community, well adapted to these quiet conditions. The prevailing ichnogenera are derived mainly from original early Paleozoic colonizers of the deep sea. The soft cohesive substrates typically contain low concentrations and patchy distributions of food, inducing trace makers to adopt complex geometric patterns to optimize exploitation of the substrate – such structures are referred to as 'graphoglyptids'. These pre-turbidite suites tend to be overwhelmed or eliminated by severe erosion or turbulence, however, and are replaced by the post-turbidite association after cessation of the turbidity current. Post-turbidite animals represent a more opportunistic, less stable community, better adapted to colonization of sandy turbidites, and are derived mainly from subsequent evolutionary immigrants from shallower water (Frey and Seilacher, 1980). These traces correspond to semi-permanent to permanent dwellings of deposit-feeding organisms and omnivores. Such suites may be entirely absent, however, where turbidites are buried beneath thick T_E units. As conditions revert to low-energy ambient settings, the pre-turbidite association gradually re-establishes itself.

In addition to pre- and post-turbidite associations, numerous turbidite-bearing submarine fans display ichnologic gradients along their depositional dips (*e.g.*, Crimes and Fedonkin, 1994). Where strong bottom currents occur in submarine canyons or fan channels, components of the *Skolithos* Ichnofacies may be present. Otherwise, proximal parts of submarine fans may be characterized by rosetted or radiating traces, and gently meandering forms of *Scolicia*. Medial parts of fans may be indicated by spiraled or tightly meandering traces, whereas patterned networks typify distal fan areas although other traces are generally present. *Zoophycos* is common locally in various settings, but it tends to be multilobed and, in places, is more complex than in the *Zoophycos* Ichnofacies.

The *Nereites* Ichnofacies *per se*, tends not to be readily identifiable along great expanses of abyssal plain, where sedimentation is exceedingly slow, and bioturbation is more or less constant rather than episodic. Without the casting ability of weakly erosive, sandy event beds, such as turbidites, recognition of surface graphoglyptids, upon which the *Nereites* Ichnofacies is based, is greatly inhibited – this taphonomic bias generally leads to these basinal facies being dominated by suites of *Planolites, Chondrites* and *Thalassinoides*.

Substrate-controlled Ichnofacies

Substrate-controlled ichnofacies encompass those suites that are not contemporaneous with the formation of the host media into which they are excavated. These suites represent the recolonization of the facies after it was deposited, buried and/or lithified, and as such are palimpsest and crosscut the original softground suites. Palimpsest suites do not correspond to the original depositional conditions of the facies, but rather, give insights into the conditions that operated during the period of recolonization. Many of these suites, therefore, correspond to stratigraphic discontinuities in the rock record (*see* below, and Chapter 2). There are three formally defined substrate-controlled ichnofacies.

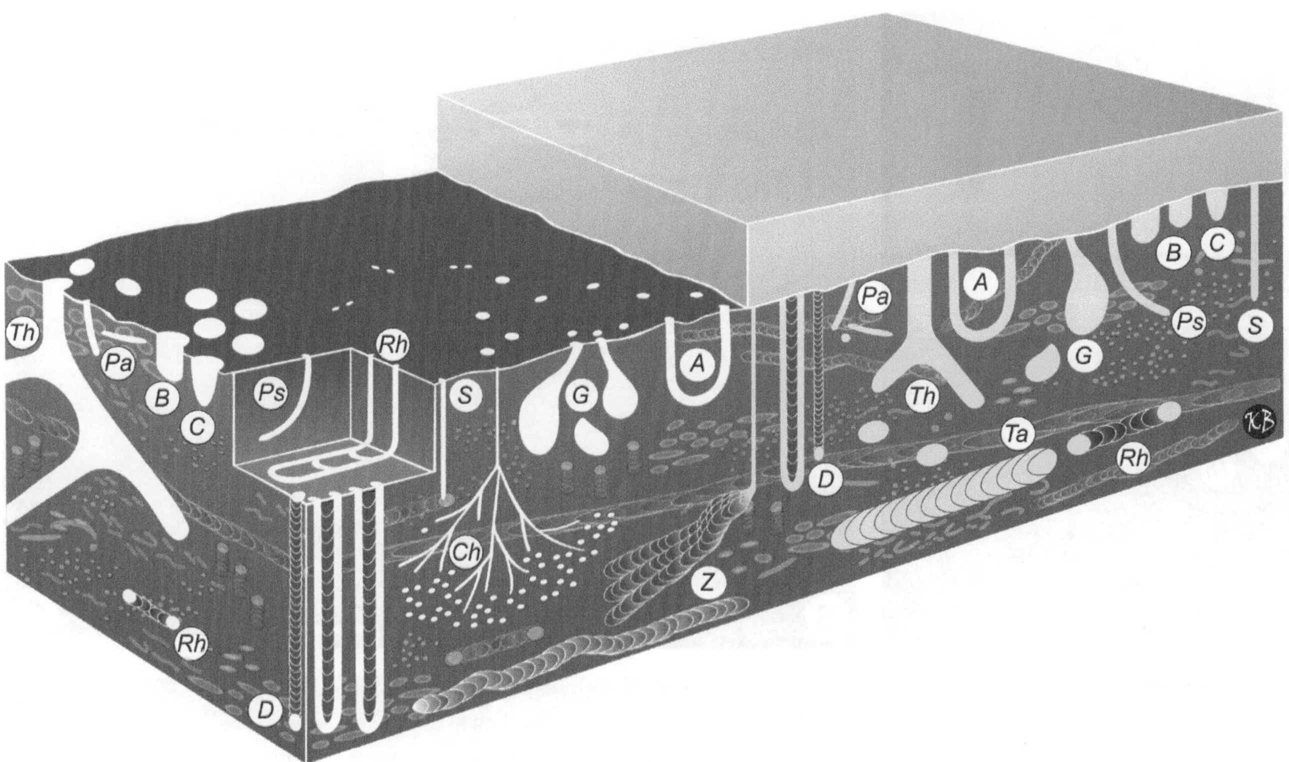

Figure 18. Schematic block diagram of the *Glossifungites* Ichnofacies. Trace-fossil abbreviations are as follows: *A: Arenicolites; B: Bergaueria; C: Conichnus; Ch: Chondrites; D: Diplocraterion; G: Gastrochaenolites; Pa: Palaeophycus; Ps: Psilonichnus; Rh: Rhizocorallium; S: Skolithos; Ta: Taenidium; Th: Thalassinoides*; and *Z: Zoophycos* (figure after MacEachern *et al.*, 2007d).

Glossifungites *Ichnofacies*

The *Glossifungites* Ichnofacies (Figs.18 and 19A, B) is characteristic of firm but unlithified substrates, such as dewatered muds, although incipiently cemented sands may also host these firmground ichnogenera. This ichnofacies demarcates a number of discontinuities, some of which have sequence-stratigraphic importance (*e.g.,* sequence boundaries, transgressive surfaces of erosion, and amalgamated sequence boundaries and flooding surfaces), but which can also be of autogenic derivation (*e.g.,* cut-bank margins of tidal channels, or periodically exposed intertidal flats. The sequence-stratigraphic significance of the *Glossifungites* Ichnofacies has been addressed by numerous workers (*cf.* Pemberton *et al.,* 2004).

Firmground ichnogenera are dominated by vertical to subvertical dwelling structures of suspension-feeding organisms. The most common structures correspond to *Diplocraterion, Skolithos, Psilonichnus, Arenicolites, Conichnus, Bergaueria*, and firmground expressions of *Gastrochaenolites*. Dwelling structures of inferred deposit-feeding organisms are also constituents of the ichnofacies, and include firmground *Thalassinoides, Spongeliomorpha, Taenidium, Palaeophycus, Chondrites, Rhizocorallium* and *Zoophycos*. Firmground elements commonly penetrate 20–100 cm below the discontinuity surface. Many shafts tend to be of large diameter (*e.g.,* 0.5–1.5 cm), particularly *Diplocraterion habichi* and *Arenicolites*. This scale of burrowing contrasts markedly with the predominantly horizontal and diminutive trace fossils typical of mudstone intervals. Firmground traces are also generally sharp-walled and unlined, reflecting the stable, cohesive nature of the substrate at the time of colonization and burrow excavation. Further evidence of substrate stability, which is atypical of soft muddy beds, is the passive nature of burrow fill (consisting of physically deposited media from the overlying bed, without trace-maker manipulation). This demonstrates that the burrow remained open after the trace maker vacated the domicile, thus allowing material from subsequent depositional events to be piped into the open tube. Preserved scratch-marked margins of some burrows also attest to the stiffness of the substrate during burrow excavation. The post-depositional origin of trace-fossil suites attributable to the *Glossifungites* Ichnofacies is clearly demonstrated by the ubiquitous crosscutting relationships with the previous softground assemblages. The final characteristic of the firmground suites is their tendency to demonstrate colonization in large numbers.

Teredolites *Ichnofacies*

The *Teredolites* Ichnofacies (Figs. 19C, D and 20) encompasses suites of borings excavated into *in situ* xylic (woody or coaly) substrates exposed to marine or marginal-marine conditions (Bromley *et al.,* 1984). Allochthonous xylic media (*e.g.,* transported logs in coastal or marine environments) pose problems in assigning the ichnofacies. Savrda *et al.* (1993) applied the term 'log-grounds' to high concentrations of allochthonous wood strewn across a depositional surface. Log-grounds may form useful mappable horizons

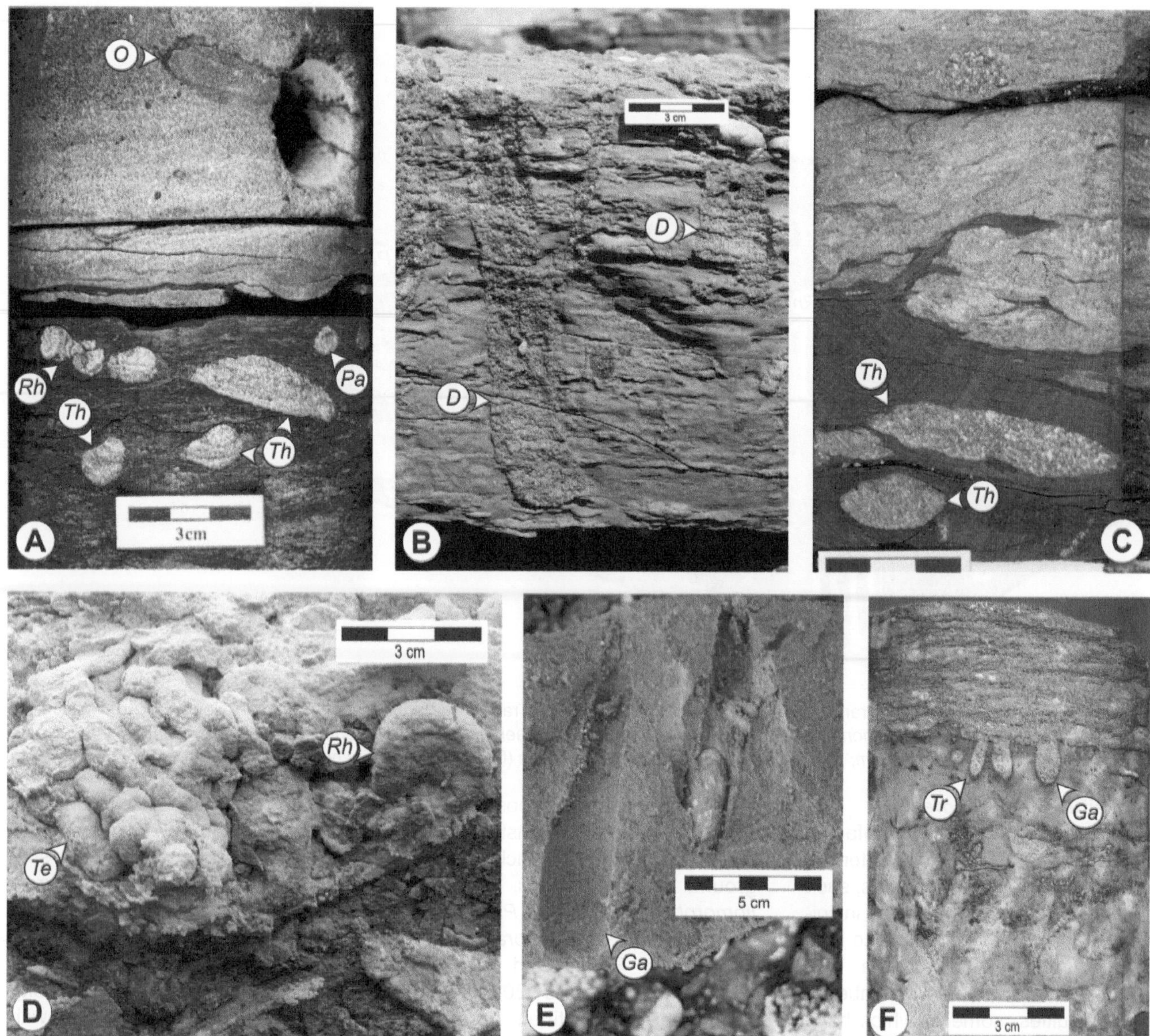

Figure 19. Photos of marine substrate-controlled trace-fossil suites and their ichnofacies designations. **A** and **B.** Suites attributed to the firmground *Glossifungites* Ichnofacies. **A.** Firmground *Thalassinoides (Th)*, *Rhizocorallium (Rh)*, and *Palaeophycus (Pa)* crosscutting a *Helminthopsis-Phycosiphon-Chondrites*-dominated suite in lower offshore mudstones. The overlying sandstone contains *Ophiomorpha (O)*, and occurs at the base of a tidal inlet deposit; Lower Cretaceous Viking Formation, Alberta. **B.** Firmground *Diplocraterion* (D) crosscutting offshore mudstones and subtending from a transgressive surface of erosion (TSE), Permian Pebbley Beach Formation, Australia. **C** and **D.** Suites attributed to the woodground *Teredolites* Ichnofacies. **C.** Woodground *Thalassinoides (Th)* developed in coal of the Lower Cretaceous Missisauga Formation, Scotian Shelf. **D.** *Teredolites (Te)* and *Rhizocorallium (Rh)* excavated into a coal layer in the Upper Cretaceous Ferron Formation, Utah. **E** and **F.** Suites attributed to the hardground *Trypanites* Ichnofacies. **E.** Bivalve-generated *Gastrochaenolites (Ga)*, bored into Late Triassic hematite-cemented continental sandstones of the Wolfville Formation along the margins of the Bay of Fundy, Nova Scotia. **F.** Crinoidal packstone overlying a carbonate hardground, marked by *Trypanites (Tr)* and *Gastrochaenolites (Ga)* bored into glauconitic and pyritic crinoidal wackestone, Big Valley Formation, Alberta. Photo is courtesy of Bill Martindale.

in transgressive successions, but do not constitute the ichnofacies if the wood was bored prior to wood emplacement, because the borings would not record contemporaneous colonization of a continuous substrate. Only woodground suites excavated into an *in situ* substrate, or bored into an allochthonous logground *after* log deposition constitute the *Teredolites* Ichnofacies. The colonization of a coal horizon is readily identified as *in situ* and is the archetypal expression of the *Teredolites* Ichnofacies. Recognizing that logground was bored post-depositionally is considerably more challenging. Isolated logs containing wood borings do not constitute the ichnofacies, because they are effectively clasts and the temporal relationship of the borings to the actual deposition of the log is uncertain. The sequence-stratigraphic applications of the *Teredolites* Ichnofacies have been addressed by Savrda *et al.* (1993) and Gingras *et al.* (2004).

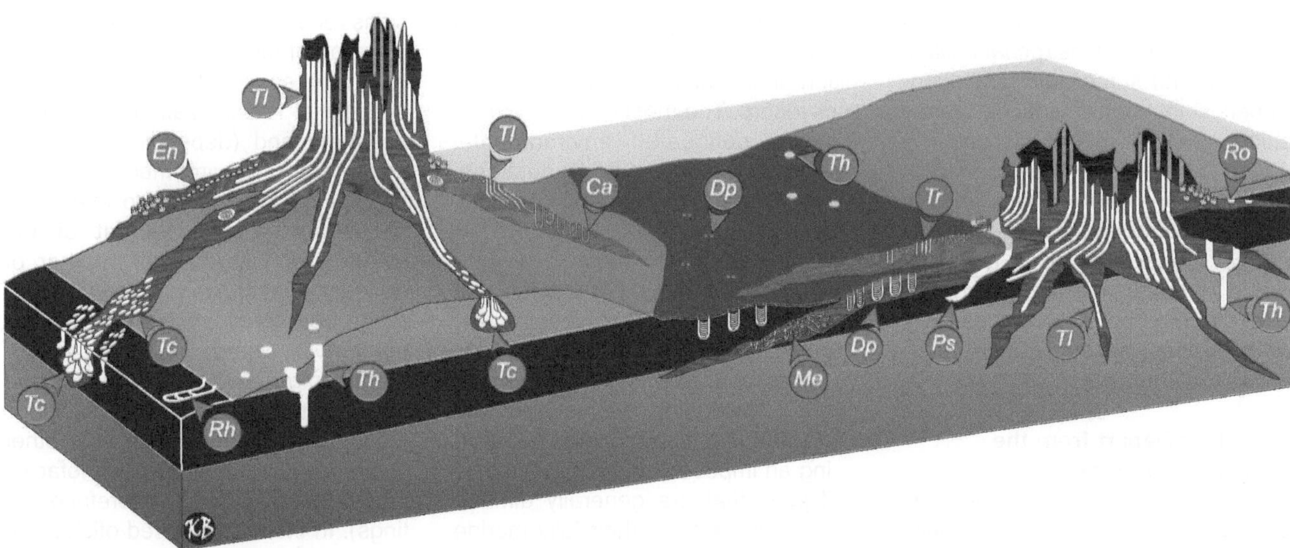

Figure 20. Schematic block diagram of the *Teredolites* Ichnofacies, showing a common intertidal expression. Trace-fossil abbreviations are as follows: *Ca: Caulostrepsis; Dp: Diplocraterion parallelum; En: Entobia; Me: Meandropolydora; Ps: Psilonichnus; Rh: Rhizocorallium; Ro: Rogerella; Tc: Teredolites clavatus; Th: Thalassinoides Tl: Teredolites longissimus*; and *Tr: Trypanites.* Red labels reflect genera recorded from ancient and modern occurrences. Green labels correspond to ichnogenera encountered only from modern occurrences (figure after MacEachern *et al.,* 2007b).

The *Teredolites* Ichnofacies, as described from the rock record, is confined to marine and marginal-marine settings. The principal namesake, *Teredolites*, reflects the dwellings of wood-boring bivalves, which in the present day only occur in fully marine to slightly salinity-reduced environments. The presence of the ichnogenera *Teredolites, Thalassinoides, Diplocraterion* and *Rhizocorallium*, excavated into xylic material is taken, therefore, to indicate largely marine conditions. Most rock-record occurrences of woodgrounds display a low diversity of trace fossils (*e.g.,* Bromley *et al.,* 1984; Savrda *et al.,* 1993). These low-diversity suites are dominated by penetrative borings attributable to *Teredolites longissimus* and *Teredolites clavatus*. Woodground trace suites commonly are monospecific; however, size-class variations have been observed in some of the boring assemblages, reflecting more than one spatfall. Modern wood-boring occurrences in the marine realm show higher diversities than are generally recorded from the rock record (Gingras *et al.,* 2004).

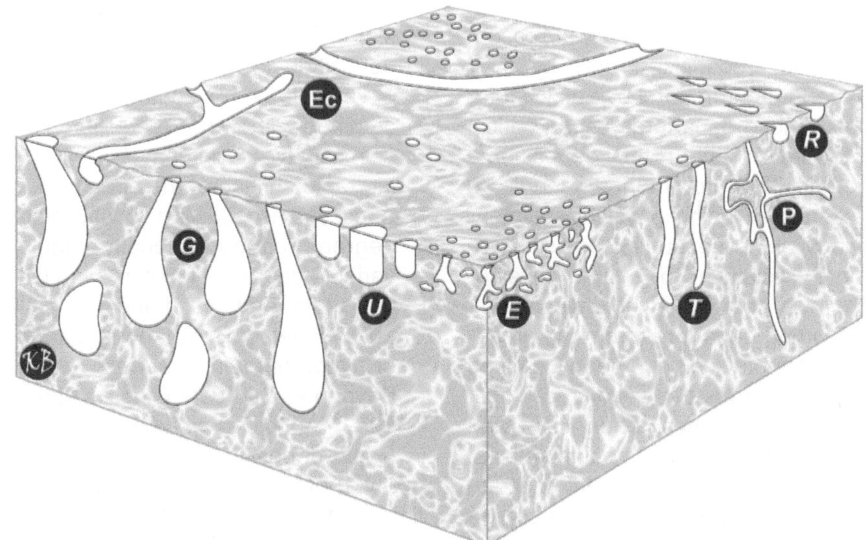

Figure 21. Schematic block diagram of the *Trypanites* Ichnofacies. Trace-fossil abbreviations are as follows: *E: Entobia;* Ec: echinoid grooves; *G: Gastrochaenolites*; P: polychaete borings; *R: Rogerella; T: Trypanites*; and *U: Uniglobites* (figure after MacEachern *et al.,* 2007b).

Trypanites *Ichnofacies*

The *Trypanites* Ichnofacies (Figs. 19E, F and 21) is characteristic of fully lithified marine substrates such as reefs (*see* Chapter 17), hardgrounds (*see* Chapter 14), rocky coasts, beach rock, unconformities, and other omission surfaces. As in the case of the *Teredolites* Ichnofacies, the concept does not apply to borings in individual shells, bones and clasts.

Suites attributable to the *Trypanites* Ichnofacies consist of sharp-walled, unlined, cylindrical, vase- and tear-shaped domiciles (*e.g., Trypanites* and *Gastrochaenolites*), irregular dwellings (*e.g., Entobia*), irregular pits or borings formed by barnacles (*e.g., Rogerella*), shallow anastomose systems excavated by sponges, bryozoans, suspension feeders or passive carnivores, and/or raspings and gnawings of algal grazers such as echinoids, chitons or limpets. Borings are distinctive, in that they cut through shells or grains, and are commonly oriented normal to the substrate, even when the surface is inclined or overhanging. Suites dominated by *Entobia* and

Trypanites record normal-marine conditions; however *Gastrochaenolites, Rogerella* and some other ichnogenera persist into some reduced salinity settings. Suites attributable to the *Trypanites* Ichnofacies are commonly intergradational with those of the *Glossifungites* Ichnofacies, and may crosscut former softground and firmground suites (Bromley, 1996). Suites commonly show moderately low diversities, although trace abundances may be high.

Suites that Depart from the Archetypal Ichnofacies

Many depositional settings are characterized by temporally and spatially variable physico-chemical stresses, leading to trace-fossil suites that depart from the archetypal Seilacherian ichnofacies suites (MacEachern *et al.,* 2007c). With increasing numbers of case studies, some of these may ultimately warrant erection of new ichnofacies; however, at present they are recognized only by their departures from archetypal ichnofacies. These departures impart critical information about the depositional setting that commonly cannot be ascertained using physical sedimentology alone because organisms are more sensitive to subtle physico-chemical changes in the environment. Recurring departures from the archetypal ichnofacies have been identified from numerous depositional settings, including estuaries, open bays and lagoons, storm-dominated shorelines, deltaic complexes, stagnant or stratified water bodies, and oxygen-depleted shelves and slopes.

Settings prone to physico-chemical stress are characterized by ichnological suites that are dominated by facies-crossing elements showing a high degree of infaunal opportunism (*i.e.,* the ability to exploit new ecological niches rapidly). Suites display reductions in diversity, and decreases in the range of ethological groupings. Stressed settings, as well, generally contain a greater proportion of simple structures generated by inferred trophic generalists (*i.e.,* omnivores that are capable of eating a variety of food types). Environmental stresses occur along a continuum from toxic to optimal; and, hence, familiarity with unstressed suites (*i.e.,* those attributable to the archetypal Seilacherian ichnofacies) is essential to the recognition of stressed assemblages in the rock record. The most common ichnologically delineated environmental stresses are salinity reductions, increased depositional rates, episodic deposition, heightened water turbidity, and reduced oxygenation.

The fundamental characteristics of the brackish-water (reduced salinity) ichnological model are:

1. Suites characterized by a persistent reduction in the number and diversity of ichnogenera, generating an impoverished marine suite;
2. Traces that are generally diminutive compared to their fully marine counterparts;
3. A predominance of simple structures (lined or unlined horizontal or vertical tubes) of inferred trophic generalists;
4. Suites comprising elements that record variations in substrate consistency, sediment caliber, and depositional rate (*e.g.,* alternations of vertical and horizontal dwellings, commonly showing re-equilibration or extensive spreitenation, recording rapid shifting of the structure);
5. Successions showing locally high degrees although locally highly variable degrees of bioturbation (BI 0-6) at the bed scale; and
6. Local development of monogeneric or monospecific suites (*e.g.,* Beynon *et al.,* 1988; Gingras *et al.,* 1999; Buatois *et al.,* 2005).

Rapid deposition rates are reflected by: i) overall decrease in bioturbation intensity; and ii) a paucity of elaborate, specialized-feeding structures (*e.g., Spirophyton* or *Zoophycos*) in favor of more mobile or temporary (sessile) deposit-feeding structures (*e.g., Planolites, Teichichnus* or *Cylindrichnus*; Nara, 1997; Gingras *et al.,* 2008). Dwellings that facilitate readjustment or re-equilibration (*e.g., Rosselia, Lingulichnus, Diplocraterion,* and *Conichnus*) are locally common. Where the settings record sporadic deposition, such as tempestites and turbidites, the partitioning of fair-weather infaunal communities and post-event communities becomes apparent (*e.g.,* Seilacher, 1991; Pemberton and Frey, 1984; Wetzel, 1991; Crimes and Fedonkin, 1994; Pemberton and MacEachern, 1997). Successions record partial to complete extermination of fair-weather communities, organism escape through the event bed, initial recolonization of the new substrate and (depending upon the magnitude of environmental contrast between the ambient and event-bed conditions) a replacement of the event suite with the fair-weather or resident suite. Juxtaposed suites may range from *Skolithos* Ichnofacies elements with *Nereites* Ichnofacies elements (*e.g.,* deep-sea turbidites), to suites attributable to the *Skolithos* Ichnofacies juxtaposed against other suites of the *Skolithos* Ichnofacies (*e.g.,* lower and middle shoreface settings). In storm-influenced offshore to distal lower shoreface settings, juxtaposition of suites typical of the *Cruziana* Ichnofacies with suites common to the *Skolithos* Ichnofacies is exceedingly common, and has, in the past, led to the term 'mixed *Skolithos–Cruziana* Ichnofacies'. This is not a true ichnofacies because such juxtapositions are the result of temporal variations in depositional energy and reflect a composite suite. In these scenarios, the alternation of such suites record in-place changes in energy and depositional rate. Mud turbidites, contourites, wave-assisted fluid-mud density flows, phytodetrital pulses, and freshet-induced hyperpycnal flood deposits are common to delta lobes, tidal shelves, continental slopes, and more rarely, large restricted bays or central basins adjacent to bayhead deltas in wave-dominated estuaries (*see* Chapters 8 to 12), but are less well-studied ichnologically. The depositional positions of these event beds, their substrate types, and the nature of food resources contained therein yield markedly different organism responses. Biogenic structures tend to be overwhelmingly those of facies-crossing deposit-feeders, but also include locomotion and resting structures, and structures of organisms capable of exploiting these anomalous layers after burial.

Increased water turbidity is commonly associated with persistent, suspended-sediment-laden distributary discharge on deltas and in brackish-water bays or mixing at the turbidity maximum zone of tidal-fluvial channels (*e.g.,* Allen *et al.,* 1980; Lesourd *et al.,* 2003). Ichnological suites are characterized by reduced bioturbation intensity, reduced ichnogenera diversity, and the limitation of ethological categories to those of locomotion, resting, deposit feeding and grazing (Gingras *et al.,* 1998; MacEachern *et al.,* 2005). Turbid water reduces primary productivity, clogs the feeding apparatus of endobenthic filter feeders, and increases the clastic content to be processed by suspension feeders and some carnivores (*e.g.,* anemones). Consequently, highly turbid settings display a pronounced impoverishment of ichnogenera normally attributed to the *Skolithos* Ichnofacies, and consist, instead, of facies-crossing elements attributable to the *Cruziana* Ichnofacies (*e.g., Rosselia, Planolites, Cylindrichnus,* and *Macaronichnus* isp.). The extension of suites attributable to the *Cruziana* Ichnofacies into sandstone-dominated marine facies is a common expression of the delta front (*e.g.,* Gingras *et al.,* 1998; Bann and Fielding, 2004; MacEachern *et al.,* 2005). Comparable impoverishment is typical of the turbidity-maximum zone of the tidal–fluvial transition (*e.g.,* Howard *et al.,* 1975).

Settings characterized by fluid mud include estuarine–channel systems prone to development of a turbidity maximum zone (*see* Chapter 9), or to delta distributaries and delta-front areas (*see* Chapter 10). Initial deposition of the flocculated clay yields soupground conditions. Ichnological characteristics include surface pascichnia, indistinct mottling, and rare, open, mucous-lined tubes that collapse readily during compaction (*e.g.,* Gingras *et al.,* 2007). Few of these structures are likely to survive into the rock record, unless the burrows possess some lithologic contrast with the enclosing sediment, or cross lithologically distinct layers (*e.g.,* mantle-and-swirl structures; Schieber, 2003). Biogenic structures generated after compaction are more commonly preserved, and include surface trails and resting structures, intrastratal deposit-feeding structures, and deep-tier dwelling and/or deposit-feeding structures from higher levels.

Reduced-oxygenation models are associated with 'oxygen-related ichnocoenoses' (ORI; Savrda and Bottjer, 1989). Reduced-oxygen settings range from the deep sea, slope, and shelf (*see* Chapters 8 and 12) to stratified lagoons, bays, estuaries, and abandoned tidal and distributary channels (*see* Chapters 9 to 11). Ichnological responses to dysoxic to anoxic conditions are reflected by:

- Reductions in ichnogenera diversity;
- Reduced trace-fossil abundance;
- Decreasing burrow diameter; and
- Decreasing depth of burrow penetration into the substrate (*e.g.,* Rhoads and Morse, 1971; Savrda and Bottjer, 1989; Wignall and Pickering, 1993; Martin, 2004).

More recent studies of present-day oxygen gradients indicate that some faunal responses to changing oxygenation at, and within, the bed are more complicated than hitherto proposed (Savrda, 2007). Some settings show remarkably high bioturbation intensity despite very reduced oxygen availability, and that neither burrow diameter nor tier depth necessarily track oxygen concentration. Levin *et al.* (2003) suggests that benthic food supply may locally be more important than oxygen, particularly where symbiont-bearing organisms may comprise the dominant infauna.

Savrda (2007) demonstrates the utility of expressing ORI in vertical successions, by evaluating transition-layer ichnofabrics and piped zones, as a means of assessing benthic oxygenation (both at the sediment–water interface and within the sediment itself). Earlier models concentrated on the identification of distinctive suites or of diagnostic ichnogenera (*e.g., Chondrites*), although recent work by Martin (2004) has called into question the recurrence of such expressions. Likewise, Schieber (2003) has questioned whether the reduced-oxygen association with black shales is borne out by more careful investigations – he provides a cautionary note by presenting a number of case studies of black shales that were previously regarded as anoxic, but which nevertheless contain features diagnostic of infaunal occupation. Ongoing investigations into oxygen-related ichnocoenoses are concerned with ORI profiles and high-resolution investigations using techniques such as SEM and thin sections. Recognition of ORI has important applications, including paleo-oceanographic reconstructions, characterization of paleoclimatic-mediated cycles (Milankovich-scale and El Niño-like effects), and the formation of reduced-oxygen *versus* oxygenated-condensed sections during maximum transgression (*cf.* Savrda, 2007), and potential relationships to oil and gas shales.

THE ROLE OF ICHNOLOGY IN GENETIC STRATIGRAPHY (ALLOSTRATIGRAPHY OR SEQUENCE STRATIGRAPHY)

Trace-fossil suites can be employed to aid in the recognition of various discontinuity types and to assist in their genetic interpretation. Ichnology can be employed to resolve surfaces of stratigraphic significance in two main ways:

1. Through the recognition of discontinuities using so-called 'omission suites', characterized by the presence of palimpsest softground ichnofacies and substrate-controlled ichnofacies (*i.e.,* the *Glossifungites, Trypanites,* and *Teredolites* ichnofacies), as well as the evidence of these ichnofacies overprinting the original softground suites; and
2. Through careful analysis of vertical softground (penecontemporaneous) ichnologic successions (analogous to facies successions; Pemberton *et al.,* 1992).

Integrating the data derived from omission suites with paleoecological data from vertically and laterally juxtaposed softground ichnological suites greatly enhances the recognition and interpretation of a wide variety of stratigraphically significant surfaces. Omission suites commonly record the colonization of an erosion surface wherein overlying strata have been removed and the underlying deposits have been compacted to varying degrees. Alternatively, these palimpsest suites may record the colonization of a surface of non-deposition, which has experienced auto-compaction and stiffening. Investigations of omission suites have shown that autogenically generated breaks are common to the terrestrial realm, inshore, intertidal settings, subtidal settings associated with channel migration, and to some slope settings. The tendency to produce autogenic erosion surfaces exposed to colonization is less of an issue in shelf and shallow-marine settings, where erosion of the substrate is typically followed closely by genetically associated deposition, closing the colonization window. In the case of cohesive substrates, autogenic breaks are associated with less indurated 'stiffgrounds' that demonstrate smaller, less penetrative and commonly somewhat compacted biogenic structures compared to true firmground counterparts that characterize allogenic discontinuities. Allogenic discontinuities vary in character spatially, depending upon the lithology of the exhumed substrate, the degree of coherence of the exhumed substrate, the energy regime at the time of colonization, and the paleoenvironment that prevailed during the colonization period. This leads to a wide range of possible trace-fossil suites. Studies of modern and ancient successions demonstrate that a single discontinuity may host omission suites that span the entire range from palimpsest softground, through firmground, woodground to hardground conditions. One should not hesitate, therefore, to consider correlating discontinuities demarcated by disparate trace-fossil suites.

Most discontinuities of sequence-stratigraphic importance (*see* Chapter 2) can be demarcated by trace-fossil omission suites. During falling relative sea level, regressive surfaces of erosion (RSE) and the subaqueous extensions of sequence boundaries (SB) are most prone to colonization. In particular, the bases of forced regressive and lowstand shorefaces and deltas (*see* Chapters 8 and 10) may contain palimpsest softground and firmground/stiffground omission suites. Such deposits pass seaward into correlative conformities (CC) that lack omission suites.

Flooding surfaces widely host omission suites, particularly where they are erosional (ravinement surfaces), and where they truncate or onlap earlier sequence boundaries. Some marine flooding surfaces and bay-margin flooding surfaces (BFS) are non-erosional or only weakly erosional, and do not lead to development of trace-fossil omission suites. In shelf and offshore settings, marine flooding surfaces dominate and commonly demarcate parasequence boundaries, and are locally associated with oxygen-related ichnocoenoses (ORI) and condensed sections. In lowered oxygen examples, trace fossils show reduced diversities, reduced sizes, smaller numbers, and shallower depths of penetration (Savrda and Bottjer, 1989; Martin, 2004); such scenarios may be characteristic of strongly oxygen-stratified basin settings. However, the presence of features such as oölites and shell accumulations associated with the marine flooding surfaces (including maximum flooding surfaces – MFS) support a shallow-water oxygenated setting.

Sediment starvation coupled with oceanic bottom currents, which may prevail in slope environments, may lead to well-oxygenated sea floors demarcated by stiffground, firmground, or even hardground omission suites that correlate with the MFS (*cf.* Savrda, 2007). Condensed sections typified by well-oxygenated conditions (*e.g.,* shallow-water passive margins, and settings with semi-permanent bottom currents) tend to display slow sedimentation with pronounced tiering, robust ichnogenera, elevated diversities, increased bioturbation intensities, and bed junctions characterized by numerous piped zones. Carbonate shelves also contain hardgrounds and borings of the *Trypanites* Ichnofacies that record rapid transgression (Coniglio and Dix, 1992; Domenech *et al.,* 2001), although there are several autogenic processes that are capable of creating hardgrounds (*see* Chapter 13).

In the inshore settings, embayments as well as estuarine incised valleys and other estuaries may contain low-energy BFS within their successions. In shallower water or higher-energy positions, the BFS may onlap an older sequence boundary during coastal retreat, and permit development of an omission suite. Higher-energy inshore conditions commonly lead to wave or tidal ravinement, which can exhume older deposits, and incise through and/or modify earlier discontinuities, particularly the sequence boundary, to produce a coplanar or amalgamated surface (*e.g.,* FS/SB). Coastal embayments and estuarine incised valleys are particularly prone to such discontinuity amalgamation. The continued transgression results in flooding over embayment and valley margins, and the transgressive ravinement of the interfluve areas.

The colonization of discontinuities yields critical information about the origin of the discontinuity and the paleoenvironmental conditions that followed. The close integration of ichnological characteristics with physical sedimentological facies analysis is necessary to achieve high-resolution reconstructions of paleogeography, paleoenvironment, and depositional architecture, permitting reliable interpretations of the genetic stratigraphy of the interval under study.

Integrating Ichnology with Facies Analysis

Here we will consider two situations – regressive and transgressive coasts – to show how the ichnological tools described above can be integrated with other sedimentological observations to refine facies analysis. The following descriptions, although typical of these regimes, are not intended to suggest that these are the only possible expressions of the environments represented. Regressive coasts (Fig. 22) showing strong wave influence and local development of deltas is a common scenario expressed by many Lower Cretaceous units of the Boreal Seaway in Alberta (*e.g.,* Falher Member cycles), and by Upper Cretaceous units such as the Eagle Sandstone of Montana, the Ferron Sandstone of Utah, and numerous cycles in the Book Cliffs of Utah (*e.g.,* the Blackhawk Formation; *see* Chapters 8 and 10 for more examples). Permian units of the south Sydney Basin in Australia (*e.g.,* Wasp Head and Snapper Point formations; Bann *et al.,* 2008) are comparable. In the second scenario reviewed here, the situation is transgressive, wherein the coastal regime is embayed and under erosive shoreface retreat (Fig. 23). Modern settings of the US east coast and the Dutch and German North Sea demonstrate comparable conditions,

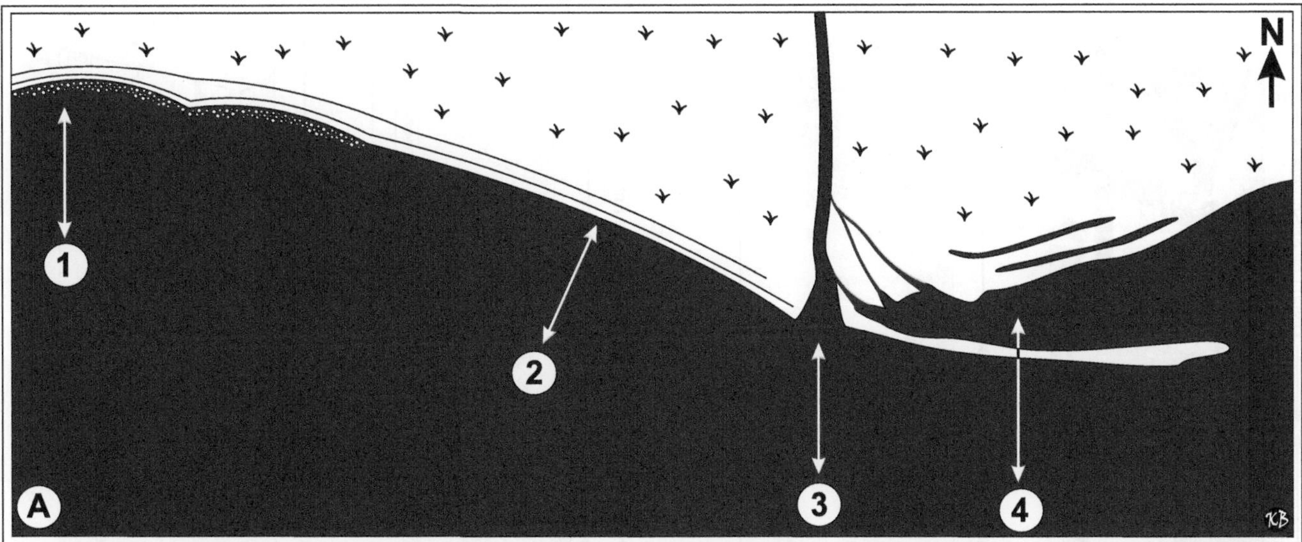

Figure 22. Paleogeographic map **A)** showing the hypothetical distribution of four facies successions (sections 1 to 4; Fig. 22**B–E)** along a highstand, regressive wave-storm-dominated coast. Trace-fossil abbreviations are: *Ar: Arenicolites; As: Asterosoma; Be: Bergaueria; C: Conichnus; Ch: Chondrites; Cr: Cosmorhaphe; Cy: Cylindrichnus; Dh: Diplocraterion habichi; Dp: Diplocraterion parallelum*; ea: equilibrium-adjustment structure; fu: fugichnia; *Gy: Gyrochorte; H: Helminthopsis; Ma: Macaronichnus* isp; *Ms: Macaronichnus segregatis; O: Ophiomorpha; P: Planolites; Pa: Palaeophycus heberti; Ph: Phycosiphon; Pt: Palaeophycus tubularis; Rh: Rhizocorallium; Rr: Rosselia rotatus; Rs: Rosselia socialis;* rt: rhizoliths; *S: Skolithos; Sf: Schaubcylindrichnus freyi; Si: Siphonichnus; T: Teichichnus; Ta: Taenidium; Th: Thalassinoides*; and *Z: Zoophycos*. Physical features: ss: soft-sediment deformation; and sy: syneresis cracks. See text for details.

but such scenarios are also apparent in certain cycles of the Viking Formation, as well as some cycles of the Bluesky Formation, and much of the McMurray Formation in Alberta. (*see* Chapter 11 for more examples). Within these constructs, integrating ichnology with the sedimentology can assist in differentiating the varying depositional environments lying along strike of one another, and in refining the paleogeographic reconstruction of the unit.

Progradational Wave-storm-dominated Coast

In this hypothetical shoreline (Fig. 22A), you, the facies analyst, have logged a number of cores to establish four recurring facies successions (Sections 1-4; Fig. 22B–E) that show predictable spatial distributions (*see* Chapters 2, 8 and 10). Careful stratigraphic assessment has led you to determine that these facies successions occur along depositional strike of one another within the same parasequence (a shoaling or progradational succession bounded by marine-flooding surfaces). You note the following characteristics – most sandstones show a strong storm overprint, manifest by oscillation ripple lamination and micro-hummocky cross-stratification in heterolithic zones, larger scale HCS and amalgamated HCS in more sand-dominated intervals, and erosionally amalgamated storm beds typified by SCS in the upper parts of most successions (*see* particularly sections 1 and 2; Fig. 22B and C). Capping these are multi-directional trough cross-stratified sandstones with current-ripple lamination, locally passing into low-angle planar-parallel stratification in wedge-shaped sets. In the case of sections 1 and 2 (Fig. 22B and C), the intervals show marked similarity. Both display reasonably high-diversity trace-fossil suites and high BI values, although the burrows are sporadically distributed and bioturbation generally decreases upwards (prominent in Section 2; Fig. 22C), owing, *in part*, to increasing tempestite emplacement and their erosionally amalgamated state, which serves to remove inter-storm deposits. Tempestites are typified by initial opportunistic colonization, progressively replaced by ambient fair-weather suites, particularly in the more heterolithic lower parts of the succession, locally with readjustment structures of *Rosselia, Teichichnus*, and *Diplocraterion*. Basal facies show higher diversities of grazing and deposit-feeding structures attributable to distal expressions of the *Cruziana* Ichnofacies passing into the archetypal *Cruziana* Ichnofacies. Upward, much of the section contains escape structures. Some beds are characterized by *in situ Rosselia* and *Ophiomorpha*. In the case of Section 1 (Fig. 22B), intervals tend to display the upward introduction of intrastratal deposit-feeding structures such as *Macaronichnus segregatis*. These lead to local increases in bioturbation intensity, and occur first in the zone of trough cross-stratification. The tops of these facies successions are locally crosscut by root structures.

Sections 3 and 4 (Fig. 22D, and E), however, show marked differences in the upper part of the interval. In the case of Section 3, the basal part is broadly similar to that of sections 1 and 2, but the upper interval shows a sharp, erosional transition to stacked bedsets of trough cross-stratified sandstone having mudstone rip-up clasts, thin mudstone interbeds, and rare current-ripple lamination marked by carbonaceous detritus. Above this, trough crossbeds give way to stacked successions of current ripples and climbing current ripples with soft-sediment deformation. This upper interval is distinctive in its paucity of bioturbation, most of which is associated with

Figure 22B. Caption page 47.

Figure 22C. Caption page 47.

structures that occur within the mudstone layers (*e.g., Planolites* and *Chondrites*), escape/equilibrium adjustment within the sandstone and *Ophiomorpha*, consistent with the predominance of elevated deposition rates. This is also supported by the presence of climbing ripples.

The facies succession in Section 4 (Fig. 22E), by contrast, shows an upward transition to more a markedly heterolithic interval that similarly coarsens upward. Sandstones show a predominance of wave-generated structures, although combined-flow ripple and current-ripple lamination are also intercalated. Upward, trough cross-stratification is present, with common mudstone rip-up clasts. Several beds show waning-flow features from horizontal planar-parallel lamina-

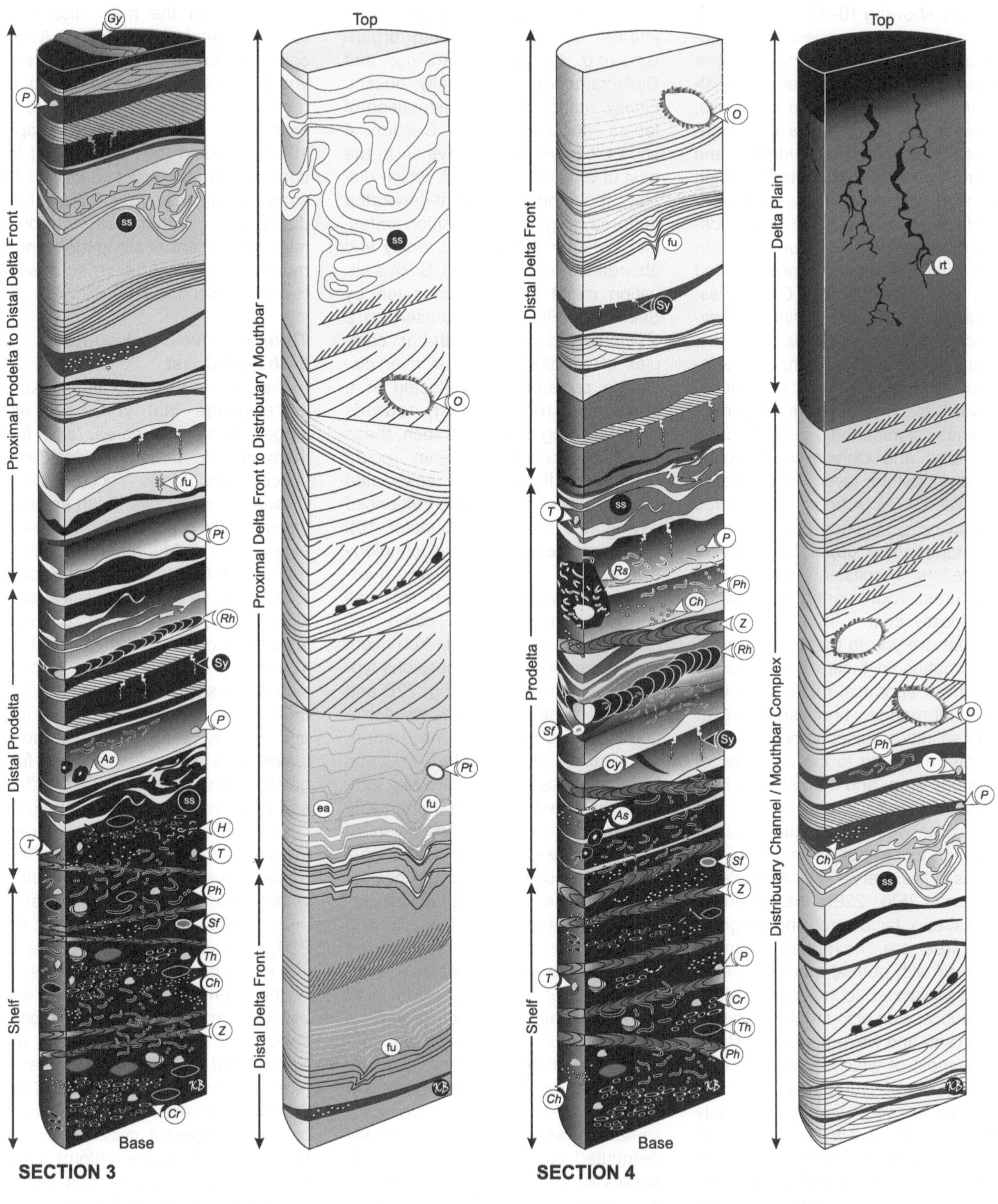

Figure 22D. Caption page 47.

Figure 22E. Caption page 47.

tion through current-ripple laminae. Sandstone beds are draped with carbonaceous mudstone, are locally silt-poor, and include small numbers of normally graded siltstone to claystone layers. Soft-sediment-deformation structures are common. You also note that syneresis cracks appear to be sporadically distributed through the unit. Ichnologically, the unit is also anomalous. Burrowing intensities are highly variable, ranging from BI 0-4. Most sandstone beds display BI 0-2, with mudstones showing variable silt contents and BI values ranging from 0-4. Claystone drapes on many sandstone beds have BI values of 0-1, with burrows mainly consisting of deep-tier structures. Diversity, however, is remarkably high, in the lower part of the succession, showing 10–12 ichnogenera, with many recording specialized feeding and grazing structures, indicative of periods of unstressed, normal-marine conditions. At the bed scale, however, numerous layers show highly reduced ichnological diversities and a predominance of facies-crossing elements, dominated by diminutive *Teichichnus, Planolites, Cylindrichnus, Rosselia*, and *Schaubcylindrichnus freyi*. These are interstratified with beds containing *Chondrites, Zoophycos, Rhizocorallium, Phycosiphon* and *Asterosoma*. The unit clearly shows marked temporal variations in depositional conditions. Upward, the diversity of ichnogenera declines significantly, with escape structures and *Ophiomorpha* typical of the sandstone layers, and isolated *Planolites, Teichichnus, Chondrites* and rare *Phycosiphon* present in the mudstones. The upper boundary of the facies succession shows root structures, root-bearing mudstones and, locally, coal.

What are you to make of this spatial variation in facies associations? The initial appraisal might be one of marginal-marine, brackish-water or other salinity stressed conditions (*see* Chapter 11) lying to the east side of the study area, and normal shoreface conditions (*see* Chapter 8) lying farther to the west. This interpretation, however, would not explain all of the details of the interval. For example, in Section 4 (Fig. 22E), the syneresis cracks and facies-crossing biogenic structures are, nevertheless, interstratified with normal-marine, diverse and specialized ichnological suites. The succession also shows a mixture of current- and wave-generated structures, most of which appear to have been draped by unburrowed, locally normally graded, and anomalously thick (> single slackwater period) mud layers of possible fluid-mud origin. Soft-sediment deformation in the heterolithic intervals and highly sporadic burrowing indicate marked variations in deposition rates. Much of the heightened burrowing is associated with mudstone deposition, yet at the bed scale there is marked variability in trace-fossil diversity, indicating temporal variations in the magnitude of paleoenvironmental stress. As an astute ichnological practitioner, you also note that vertical structures of inferred suspension-feeding organisms (*e.g., Conichnus, Skolithos*, and *Diplocraterion*) are virtually absent. Finally, mapping of the distribution of facies successions in Section 4 shows it to virtually always lie to the east of those of Section 3, which displays the strongest current-generated features, highest deposition rates, and lowest ichnological diversity and abundance (*see* Fig. 22A). Such a situation might be accommodated by Section 3 (Fig. 22D) representing a distributary-channel/mouth-bar complex of a delta, with most river discharge preferentially deflected in a downdrift direction (east) towards Section 4 (Fig. 22E). As such, these two facies successions may record heightened fluvial input associated with a mixed river-wave-influenced asymmetric delta (*see* Chapter 10). The characteristics of the ichnology are entirely consistent with river-induced physico-chemical stresses imposed on a deltaic coast (*cf.* MacEachern *et al.*, 2005). You might then consider facies successions typifying sections 1 and 2 (Fig. 22B and C) to represent a wave- to storm-dominated shoreline lying updrift, along-strike of the delta. Careful facies assessments, however, indicate that despite being remarkably similar to the facies succession of Section 1, the succession in Section 2 shows the common occurrence of current ripples, has higher proportions of claystone drapes passing upward from heterolithic basal facies into the more sandstone-prone proximal parts of the succession, and shows a general absence of vertical dwelling structures of inferred suspension-feeding organisms. The succession is also substantially thicker (not displayed in this schematic representation), shows more sporadically distributed burrowing, generally possesses lower bioturbation intensities, and possesses higher proportions of facies-crossing deposit-feeding structures (*e.g., Rosselia, Macaronichnus isp., Teichichnus*, and *Cylindrichnus*). From this, you could reasonably arrive at an interpretation of a wave- and storm-dominated shoreface succession (Section 1), grading along-strike into a wave- and storm-dominated delta front (Section 2), which lies updrift of the main distributary channel complex (Section 3). As you continue downdrift toward the east, you recognize that the delta becomes more river-influenced, owing to its asymmetry (Section 4; *see* Chapter 10). You would then predict that this condition would persist so long as the coast remained in the reach of the distributary system, whereupon it might grade farther along-strike into a more classical shoreface succession (similar to Section 1).

Transgressive Brackish-water Embayed Coast

In the second scenario (Fig. 23A), let us envisage that you have also logged numerous cores, and from them derived three recurring facies successions (facies successions B, C, and D), which overlie an erosional discontinuity cut into a fourth, areally extensive and older facies succession (herein referred to as Facies Succession A). These different facies successions are presented in schematic style in sections 1-3 (Fig. 23 B–D). You recognize that Facies Succession A consists of stacked cycles of thoroughly bioturbated mudstones and muddy sandstones. In some localities, a palimpsest suite of firmground *Thalassinoides, Skolithos*, and *Diplocraterion*, attributable to the *Glossifungites* Ichnofacies, separates the overlying facies successions from Facies Succession A, particularly where the discontinuity overlies muddy deposits (*e.g.*, Section 1; Fig. 23B). Where the discontinuity is cut into sandy parts of Facies Succession A, the overlying and underlying successions are separated by palimpsest softground *Diplocraterion* and *Skolithos* of the *Skolithos* Ichnofacies (*e.g.*, sections 2 and 3; Fig. 23C and D). In all localities, the underlying, coarsening-upward cycles of Facies Succession A display uniform bioturbation (BI 4-5), and contain diverse and robust trace-fossil suites corresponding to distal, archetypal and proximal expressions of the *Cruziana* Ichnofacies, from base to top.

At the western margin of the study area (Section 1; Fig. 23B), the overlying Facies Succession B characterizes well-developed fining-upward successions, marked by a basal scour surface with trough cross-strat-

Figure 23. Paleogeographic map **A)** of a hypothetical transgressively embayed coast, showing the distribution of three facies successions (Facies Successions **B, C** and **D)** erosionally overlying regionally extensive parasequences of Facies Succession A (Sections 1 to 3; B-D). Trace-fossil abbreviations are: *Ar: Arenicolites; As: Asterosoma; C: Conichnus; Ch: Chondrites; Cr: Cosmorhaphe; Cy: Cylindrichnus; Dh: Diplocraterion habichi; Dp: Diplocraterion parallelum*; ea: equilibrium-adjustment structure; fu: fugichnia; *G: Gyrolithes; Ma: Macaronichnus* isp; *Ms: Macaronichnus segregatis*; na: navichnia (mantle-and-swirl structures); *O: Ophiomorpha; P: Planolites; Ph: Phycosiphon; Pt: Palaeophycus tubularis; Rh: Rhizocorallium; Rs: Rosselia socialis;* rt: root rhizoliths; *S: Skolithos; Sch: Schaubcylindrichnus coronusSf: Schaubcylindrichnus freyi; Si: Siphonichnus; T: Teichichnus; Th: Thalassinoides*; and *Z: Zoophycos*. Physical features: sy: syneresis cracks. See text for details.

ified sandstones containing reactivation structures, mudstone rip-up clasts, and mudstone drapes (locally double drapes) on foresets. The thicknesses of crossbed sets decrease upwards and proportions of current-ripple lamination increase. Mudstone interbeds are thicker upward, imparting a marked heterolithic expression to the facies. Mudstones are dark, fissile and sporadically burrowed. Most burrowing is confined to the heterolithic part of the facies succession, and concentrated in the vicinity of the mudstone interbeds. The cores display a depositional inclination to the interbedded mudstones and sandstones. Toward the top of the interval, Facies Succession B is dominated by mudstones with local lenticular to wavy bedding containing small current ripples. Syneresis cracks are locally present in the mudstone beds. Burrowing tends to be reduced in diversity, variable in intensity, and sporadically distributed. Most burrowed zones show a BI of 1-3, although most units show BI 2 or less. The ichnological suite consists of sporadically distributed *Planolites, Cylindrichnus, Gyrolithes, Teichichnus, Palaeophycus*, and, in sandier zones, diminutive *Skolithos, Arenicolites*, and *Cylindrichnus*. Some mudstone beds display mantle-and-swirl structures (navichnia), characteristic of sediment-swimming organisms. This zone is erosionally truncated and capped by a gravel lag, passing into more pervasively burrowed (BI 5) silty mudstones with *Asterosoma, Zoophycos, Chondrites, Phycosiphon* and *Cosmorhaphe*.

Adjacent to this, toward the east (Fig. 23A), you encounter Facies Succession C overlying a sandier interval of Facies Succession A (Section 2; Fig. 23C). Palimpsest softground *Skolithos* separates the two successions in this locality. Facies Succession C shows a general coarsening-upward trend, consisting of moderately to thoroughly bioturbated sandy mudstones to muddy sandstones, capped by a thin fining-upward succession of sandy mudstones. The muddy units show rare, thin, low-angle undulatory parallel laminated sandstones (interpreted as micro-HCS) with oscillation ripple- and rare current-ripple laminated fine-grained sandstones, separated by pervasively burrowed sandy mudstones with BI values of 4-5. The muddy sandstones also contain thin zones of amalgamated parallel-laminated micro-HCS, having zones of silty to muddy sandstone showing BI 4-5. Thin mudstone interlaminae and interbeds are intercalated locally. The ichnological suite is fairly diverse and robust, and consists of fully marine suites attributable to the archetypal through to proximal expressions of the *Cruziana* Ichnofacies. You recognize that Facies Succession C is broadly similar to Facies

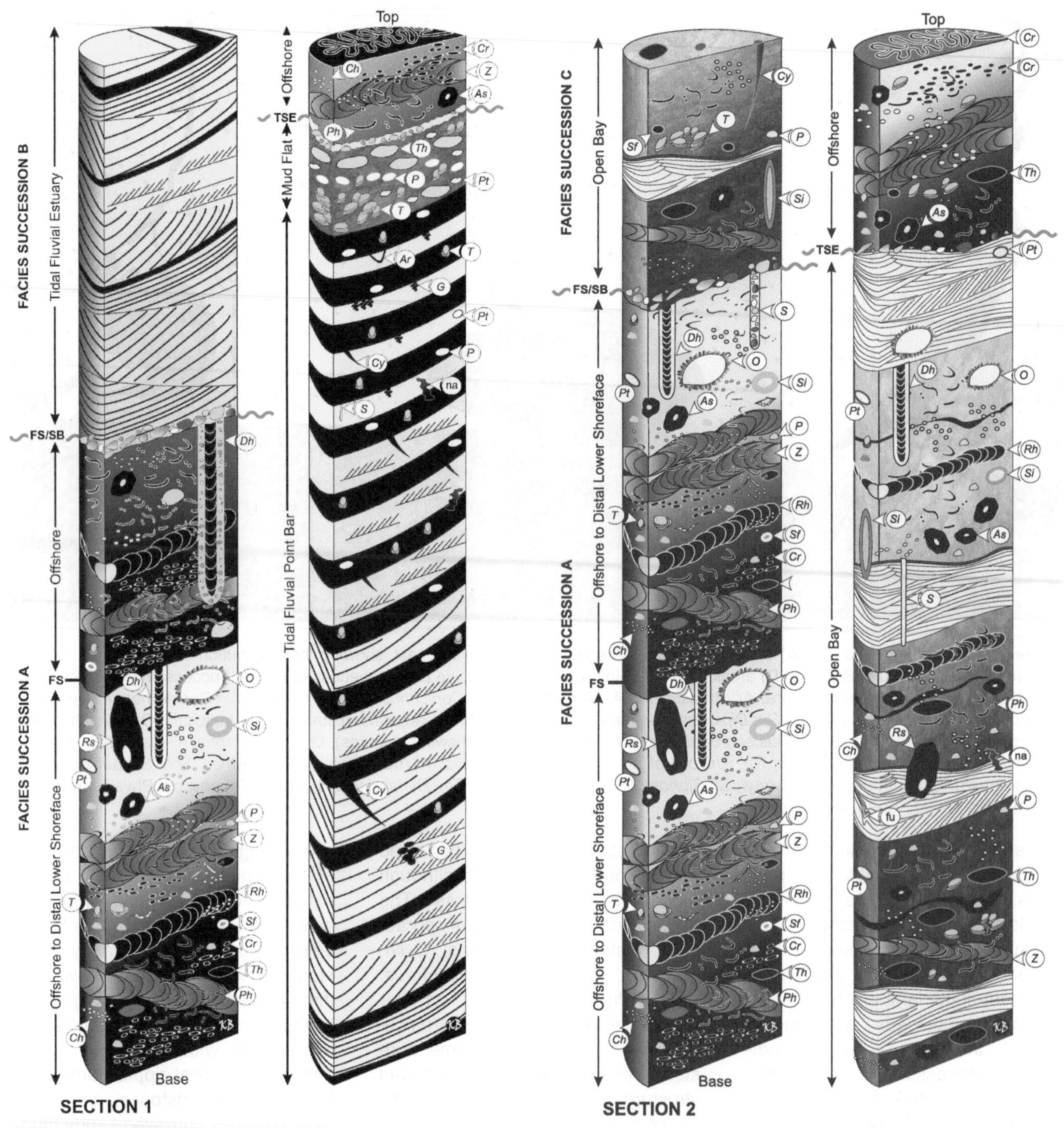

Figure 23B. Caption page 51.

Figure 23C. Caption page 51.

Succession A (Fig. 23C), although the trace-fossil suites in the overlying succession appear to reflect more sheltered and low-energy conditions for the same sediment calibers in Facies Succession A. Additionally, in contrast to the facies tracts of Facies Succession A, which show persistence along depositional strike measurable in 10s of kilometers, the facies tracts of Facies Succession C do not persist more than 5 kilometers along depositional strike and appear to become thinner and sandier, before onlapping and ultimately pinching out into a thin granule to pebble lag. The upper part of Facies Succession C is truncated and capped with a gravel lag, passing into pervasively burrowed (BI 5) silty and sandy mudstones containing abundant *Asterosoma, Thalassinoides, Zoophycos, Phycosiphon* and *Cosmorhaphe*.

Farther to the east, Facies Succession D is characterized by an anomalously thick heterolithic interval of laterally confined strata, overlying a discontinuity that incises deeply into units of Facies Succession A (Section 3; Fig. 23D). The palimpsest suite of traces that characterizes this surface (firmground *Thalassinoides*) is overlain by a zone of sanding-upward heterolithic facies, showing a complex association of current, oscillatory and combined-flow rippled thin sand-

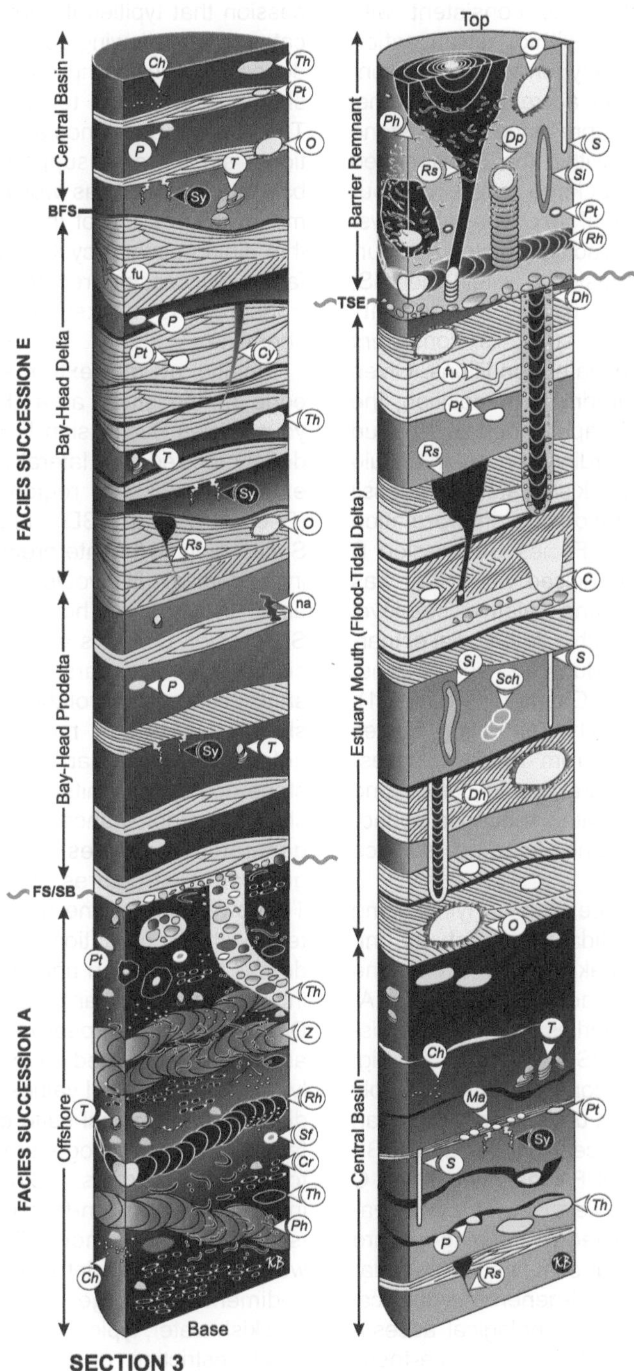

Figure 23D. Caption page 51.

stones, locally draped with dark, carbonaceous mudstone. Rare, thin, silty mudstone interbeds are intercalated. The facies displays variable bioturbation (BI 0-2), with a low-diversity suite consisting of facies-crossing ichnogenera (*e.g., Cylindrichnus, Ophiomorpha, Rosselia, Palaeophycus, Planolites*, and *Teichichnus*), as well as fugichnia.

This unit is sharply overlain by muddier heterolithic units comprising a composite bedset of oscillation-rippled and undulatory parallel-laminated sandstones, alternating with sporadically burrowed silty to sandy mudstones forming wavy to lenticular bedding. Dark, carbonaceous mudstone drapes locally mantle the sandstone layers, and some contain syneresis cracks. The trace-fossil suite is of low diversity, and bioturbation intensities vary from BI 1-4. Ichnogenera are diminutive, for the most part, and include a predominance of *Planolites, Teichichnus, Thalassinoides, Palaeophycus, Chondrites, Rosselia*, and *Macaronichnus* isp. Fugichnia, *Ophiomorpha, Skolithos* and *Diplocraterion* are locally associated with the sandstone beds.

The mudstone-prone heterolithics are overlain by sandstone-dominated heterolithics, which become sandier upward. Sandstones in this interval are mainly current-ripple laminated and horizontal planar-parallel laminated. Mudstone drapes are comparable to those of the underlying units, though generally thinner. Syneresis cracks are typically absent. Small amounts of aggradational current-ripple lamination are present, and thin burrowed muddy sandstones, locally with granules of chert, are intercalated. The facies typically shows sporadically distributed bioturbation (BI 0-3), although characterized by an increase in diversity and the introduction of more robust forms relative to underlying facies. Ichnogenera are typical of proximal expressions of the *Cruziana* Ichnofacies and of the *Skolithos* Ichnofacies. Ichnogenera are dominated by *Skolithos, Siphonichnus, Ophiomorpha, Diplocraterion, Schaubcylindrichnus, Palaeophycus, Rosselia, Conichnus* and rare *Bergaueria*, with some associated *Teichichnus, Cylindrichnus, Planolites*, fugichnia and *Chondrites*. Upward, Facies Succession D is erosionally overlain by pervasively bioturbated (BI 5) muddy sandstone comparable to units capping facies successions B and C in sections 1 and 2 (Fig. 23B, C–D). The facies displays ichnological characteristics typical of normal-marine conditions that nevertheless are discontinuous along depositional strike.

Although you have not worked the successions farther seaward, published articles show that these deposits are dominated by thinner and muddier parasequences (interpreted as shelf to upper offshore cycles), corresponding to distal expressions of your Facies Association A, truncated by an erosion surface locally containing a firmground suite of *Thalassinoides* and *Rhizocorallium* of the *Glossifungites* Ichnofacies. The discontinuity, which correlates to your lower discontinuity, is overlain by a lag of variable thickness and passes into thoroughly bioturbated (BI 5) silty and sandy mudstones of the offshore.

How could you interpret this spatial variation in facies successions? You should be struck immediately by the contrast in bioturbation style and facies uniformity of Facies Succession A, and the markedly heterolithic character of facies successions B–D. Facies Succession A can be interpreted as the result of slow, continuous deposition in a fully marine environment (*see* Chapter 8). The sanding-upward facies successions are reasonably regarded to reflect stacked parasequences deposited in a lower offshore to distal lower shoreface setting, comprising a progradational parasequence set that probably corresponds to a highstand systems tract. The overlying facies successions are separated from this parasequence set by a regionally extensive discontinuity, demarcated by palimpsest suites of trace fossils, indicating that the surface was colonized in a marine environment. The discontinuity displays considerable relief, at least locally, and cuts through different facies types of Facies Succession A. In several localities, the discontinuity is incised into distal (lower offshore) mudstones (*e.g.,* sections 1 and 3), indicating that initial exhumation of the discontinuity must have occurred during a relative sea-level fall. The character of facies successions B–D, however, indicates persistent, upward increasing marine influence on deposition, supporting some transgressive modification of the discontinuity.

Your analysis of Facies Succession B (Section 1; Fig. 23B) shows deep incision of the discontinuity into the underlying parasequence set. Overlying facies are recognized as current generated, and typically form fining upward intervals. Sand-bed thickness and bedform scale decrease upward, indicating progressive waning energy. The presence of reactivation structures, double mud drapes on foresets and rare syneresis cracks supports tidal overprint on deposition, probably in a setting with varying salinity. The upward increase in mudstone interbeds is consistent with inclined heterolithic stratification (IHS), most likely formed (in this context) by lateral accretion of channel margins. The presence of mudstone rip-up clasts within the crossbedded sandstone package supports your interpretation of a channelized system. The sporadically distributed burrowing in mudstones within the IHS, characterized by a low-diversity facies-crossing suite, is consistent with brackish-water conditions. Given the tidal overprint on deposition, the presence of rapidly deposited mud (probably recording deposition of fluid mud), and brackish-water trace-fossil suites, a reasonable interpretation would be that Facies Succession B lies within an incised valley that was excavated during relative sea-level fall and filled with a tidal-fluvial estuarine complex during ensuing transgression (*see* Chapters 9 and 11). The discontinuity separating Facies Succession A from Facies Succession B, therefore, would correspond to an amalgamated sequence boundary and marine flooding surface (FS/SB).

Facies Succession C, lying along strike of this tidal-fluvial estuary, initially appears akin to the fully marine deposits of Facies Succession A, although it overlies the regional discontinuity (FS/SB) (Section 2; Fig. 23C). The discontinuity is less deeply incised into the underlying facies than in Facies Succession B (Fig. 23B). The interval of Facies Succession C also shows a relatively thick, pervasively bioturbated coarsening-upward succession that contains both oscillatory- and current-generated physical structures. Your ichnological assessment shows that the trace-fossil suites are diverse and the ichnogenera robust, but they appear to record higher proportions of deposit-feeding and grazing structures than might be expected for the sediment caliber of the facies. Indeed, the paucity and thinness of storm beds, the presence of possible tide-generated current ripples, and the dominance of low-energy burrowing even in the muddy sandstones suggest that the setting was somewhat sheltered. The pervasive burrowing indicates that deposition rates were generally low. Nevertheless, the coarsening-upward succession that typifies the interval indicates that shallowing occurred during deposition, which implies a period of slow progradation to aggradation. The lack of persistence along depositional strike would support an open bay interpretation, as would the fully marine expression of the facies. The thin, fining-upward cycle capping the facies succession records renewed transgressive backstepping of the shoreline.

Finally, at the extreme eastern edge of the study area (Fig. 23A), your Facies Succession D overlies a deeply incised and laterally confined expression of the regional FS/SB (Section 3; Fig. 23D). Like Facies Succession B, the interpretation of an incised-valley fill would be reasonable. In contrast, however, Facies Succession D shows a complex transition from a coarsening-upward, sand-prone heterolithic interval, sharply overlain by mudstone-prone heterolithics and capped by sandstone-dominated units. Most facies show a predominance of oscillatory-generated structures with current-generated structures subordinate. The lower sand-prone unit is consistent with progradation. Mudstone drapes are common and suggest the interaction of fresh and marine water. Deposition rates appear to be generally high, as indicated by the generally low but variable BI values. The low-diversity ichnological suite comprises facies-crossing ichnogenera common to sandy substrates. Structures of inferred suspension-feeding organisms are uncommon. The facies would suggest the interplay of river-sediment discharge with standing brackish water, typical of a delta. The areal restriction of the succession within the valley profile would logically lead to the interpretation of a bayhead delta. The sharp contact separating this bayhead delta from the overlying mudstone-prone heterolithic facies would lead you to consider the surface to represent a bay flooding surface (BFS). This muddy heterolithic facies is dominated by sporadically burrowed mudstones, rapidly deposited mud drapes, and oscillation-rippled sandstone, strongly supporting the contention of a standing body of water adjacent to the bayhead delta. Generally reduced bioturbation intensities reflect high deposition rates, whereas the low-diversity ichnological suite is consistent with persistent brackish-water conditions. Rare syneresis cracks attest to fluctuating salinities. These mud-prone heterolithic deposits would be best interpreted as central-basin deposits. The overlying sand-prone heterolithics with greater proportions of current-generated structures are consistent with the flood-tidal delta accumulation in the estuary-mouth complex. This facies shows an absence of syneresis cracks, heightened ichnological diversity with more robust ichnogenera, and the introduction of some structures of inferred suspension-feeding organisms (*e.g., Conichnus* and *Bergaueria*). Such suites are typical of more uniform and generally normal-marine salinities. Deposition rates, however, appear to have remained elevated, as indicated by the low intensity of burrowing. The burrowed muddy sandstone unit capping Facies Association D is interpreted to record the erosional remnant of the back-stepping barrier complex at the estuary mouth during continued transgression (Section 3; Fig. 23D). As such, it represents a transgressive lag that mantles the ravinement surface (transgressive surface of erosion or TSE). This TSE would correlate to the erosion surfaces overlying facies successions B and C in sections 1 and 2 (Fig. 23B and C). The expression of Facies Association D is indicative of the tripartite zonation of facies types characteristic of a wave-dominated estuary (*see* Chapter 11).

Given this spatial variation of facies successions (Fig. 23A), you might reach the following interpretation. An older progradational parasequence set of a highstand systems tract (Facies Succession A), was exposed and locally incised during a relative fall in sea level. Sediment bypass appears to have taken place during development of the sequence boundary, as no fluvial or other lowstand deposits were recognized in the study area. Fluvial valleys were cut in the vicinity of Section 1 (Facies Succession B) and Section 3 (Facies Succession D). During the ensuing sea-level rise, the lowstand discontinuity was transgressively modified to form the FS/SB, and was colonized by infauna, producing palimpsest traces varying from suites attributable to softground omission to those of the *Glossifungites* Ichnofacies. In the western part of the transgressive shoreline, tidal energy was greater, and the valley in the area was filled with a tidal–fluvial estuarine complex (Facies Succession B) characterized by channel sandstones and IHS point-bar complexes within the zone of salt-water invasion (Section 1; Fig. 23B). Along strike of this tidal estuary, fully marine open-bay shorelines (Facies Succession C) slowly prograded and aggraded during the transgression, but ultimately sediment supply was low and these were gradually transgressed. Farther to the east, wave energy was greater, and the incised valley there was subjected to greater wave influence during transgression, allowing the mouth of the estuary to become barred (Section 3; Fig. 23D). The development of the barrier permitted a central basin or lagoon to form behind the barrier. The fluvial system draining the valley was obliged to deposit its sediment load into the central basin as a bayhead delta. As the transgression progressed, the central basin shifted landward across a bay-flooding surface, as did the more marine flood-tidal delta of the estuary-mouth complex. Continued transgression ultimately eroded the barrier leaving a transgressive lag on the ravinement surface capping the wave-dominated estuary deposits of Facies Succession D. This transgression also generated the TSE observed in sections 1 and 2. Your integrated ichnological–sedimentological interpretation of the interval not only accommodates the observations in the study area, but would also fit with the published expression of the interval in more basinward positions: older offshore marine parasequences cut by an amalgamated FS/SB, capped by a transgressive lag, and ultimately overlain by offshore mudstone.

Summary and Future Directions

The ichnofacies paradigm is based on temporally and spatially recurrent ichnological/sedimentological suites. Although the defined ichnofacies are especially linked to rock-record occurrences, the interpretations and understanding of most ichnofacies have been and will continue to be refined through studies of modern animals and their burrowing behaviors. When coupled with specific ichnological and sedimentological data, ichnofacies can be used not merely to discern the general character of the initial depositional environment, but to derive specific information regarding salinity, oxygenation, sedimentation rates, and the nature of observed cyclic accumulation (*e.g.,* rhythmic lamination/bedding). Ichnofacies also provide data that can be applied to sequence-stratigraphic concepts, including the identification of key surfaces and the interpretation of general transgressive–regressive cycles. Spatially, ichnofacies aid in delineating genetically related strata and help to identify facies boundaries. Trace-fossil suites aid in the recognition and genetic interpretation of various discontinuity types. Ichnology may be employed to resolve surfaces of stratigraphic significance in two main ways:

1. Through the identification of discontinuities using omission suites, comprising substrate-controlled ichnofacies and palimpsest softground suites; and
2. Through careful analysis of vertical softground ichnologic successions (analogous to facies successions).

Data derived from omission suites coupled with paleoecological data from vertically and laterally juxtaposed softground suites greatly enhance the recognition and interpretation of potentially significant stratigraphic surfaces.

Future work will see the increased integration of process-based neoichnological data with observations from the rock-record. Additionally, various schemes of numerical analysis and statistical evaluation are likely to interest certain workers. Owing to taphonomic considerations, the application of the ichnofacies concept to continental strata has proven difficult; for those who enjoy a challenge, this will be an area of research that will continue to expand. The development of additional ichnofacies is likely, but will need to be approached carefully. Workers should ensure that only the recurring aspects of suites derived from the analysis of case studies of broad geographic area and from numerous periods of geological time are employed, and lead to an ethologically distinct ichnofacies model. There are also a number of environmental parameters that have not been thoroughly tested – the role of temperature, for example, is poorly understood. Comparisons of suites from a variety of environments during greenhouse *versus* icehouse periods, and comparisons of low-latitude with high-latitude suites are challenges for the future.

More recently, we have seen the role of trace fossils expand into hydrocarbon production geology. 'Reservoir Ichnology' (*e.g.,* Pemberton and Gingras, 2005) assesses the influence of trace fossils on porosity and permeability trends in hydrocarbon-bearing strata. Although bioturbation routinely has been considered to be detrimental to reservoir quality, largely because biogenic churning of bedded sediment typically generates poorer sorting and thereby reduces permeability, high-resolution analyses are now showing that this is not exclusively the case. Intense bioturbation or cryptic bioturbation of originally laminated sandstones removes subtle heterogeneities, leading to more uniform reservoir characteristics. Additionally, bioturbation-enhanced bulk permeability has also been reported, ranging from biogenic modification of the primary depositional fabric through to diagenetic alteration – usually *recrystallization* – of the sedimentary matrix. Reservoir Ichnology is in its infancy and is a promising line of inquiry for the future.

Integrated ichnological-sedimentological facies models will continue to evolve, in order to explain the sedimentary characteristics of various depositional environments in ever greater detail and sophistication. Eventually the full integration of trace-fossil analysis with sedimentology will preclude the need for stand-alone chapters on ichnology. That, indeed, is a compelling challenge for future students of the science!

ACKNOWLEDGEMENTS

We would like to thank NSERC and the many oil companies who have supported our research into the applications of ichnology to facies analysis, genetic stratigraphy and reservoir characteristics. Dr. Stephen T. Hasiotis, Dr. Fiona J. MacEachern, and Dr. Bill Martindale kindly donated photos, and are warmly thanked for their generosity. Dr. Murray Gregory provided the stratigraphic information on and access to the New Zealand example of *Paleodictyon*. Floyd 'Bo' Henk provided the photo and stratigraphic information on the *Psilonichnus* specimen, and Woodside Petroleum Ltd. lent its permission to use the image in Fig. 13F. Also we apologize to the many colleagues whom we were not able to cite in the chapter, owing to the limitations placed on the length of the reference list. Finally, we dedicate this chapter to the memory of Dr. Robert (Bob) W. Frey (1938-1992), a truly gifted ichnological researcher who served as co-author for all previous versions of the ichnology chapter in *Facies Models*. The science as it stands today is underpinned by his insightful appraisal of animal–sediment relationships and its integration with the rock record. He is sorely missed.

REFERENCES

Basic sources of information

Basan, P.B., *ed.,* 1978, Trace Fossil Concepts, SEPM Short Course 5, Oklahoma City, 181 p.
First short course offered on ichnology. Useful papers on the sedimentological application of trace fossils in shallow-water shoreface environments.

Bromley, R.G., 1996, Trace Fossils: Biology, Taphonomy and Applications, Second edition: Chapman and Hall, London, 361 p.
First general textbook on the subject. An invaluable source of information that is profusely illustrated.

Bromley, R.G., Buatois, L.A., Mángano, M.G., Genise, J.F., and Melchor, R.N., *ed.,* 2007, Sediment-Organism Interactions: a Multifaceted Ichnology: SEPM Special Publication 89, 393 p.
Recent compilation of papers given at the first Ichnology Congress in Argentina, concerned more with paleontological aspects of Ichnology.

Curran, H.A., *ed.,* 1985, Biogenic Structures: Their Use in Interpreting Depositional Environments: SEPM Special Publication 35, 347 p.
Classic work on the application of trace fossils in paleoenvironmental interpretation.

Donovan, S.K., 1994, The Palaeobiology of Trace Fossils: Wiley and Sons, New York, 308 p.
A compendium volume with papers that are directed mainly towards the biological and paleontological aspects of ichnology.

Ekdale, A.A., Bromley, R.G., and Pemberton, S.G., 1984, Ichnology: Trace Fossils in Sedimentology and Stratigraphy: Society of Economic Paleontologists and Mineralogists, Short Course 15, 317 p.
First integrated short course on the subject of ichnology remains as a useful reference. The volume contains numerous diagrams and photographs and addresses most aspects of the science.

Frey, R.W., *ed.,* 1975, The Study of Trace Fossils: Springer-Verlag, New York, 562 p.
First integrated look at trace fossils using an applied approach, which remains an invaluable reference.

Häntzschel, W., 1975, Trace Fossils and Problematica, *in* Teichert, C., *ed.,* Treatise on Invertebrate Paleontology, Part W, Miscellanea, Supplement 1: Geological Society of America and University of Kansas Press, Boulder and Lawrence, 269 p.
Although outdated, it remains the most comprehensive compilation of ichnotaxonomy, badly in need of revision.

Hasiotis, S.T., 2002, Continental Trace Fossils: SEPM Short Course Notes No. 51, 131 p.
Excellent compilation of trace fossils associated with the continental realm. This volume is done as an atlas and illustrates a vast array of freshwater and continental ichnotaxa.

MacEachern, J.A., Bann, K.L., Gingras, M.K., and Pemberton, S.G., *eds.,* 2007a, Applied Ichnology, SEPM Short Course Notes 52, 380 p.
Most recent compilation of the applied aspects of ichnology, with 5 key conceptual papers and 13 case studies. Profusely illustrated with many high quality photographs and drawings.

McIlroy, D., *ed.,* 2004a, The Application of Ichnology to Palaeoenvironmental and Stratigraphic Analysis: Geological Society of London, Special Publication 228, 490 p.
Volume that resulted from a Lyell Meeting in London, which highlights a number of topical subjects using an ichnological point of view. The editor's introductory chapter lays out his "personal ethos" for the study of the science.

Miller, W. III, *ed.,* 2007, Trace Fossils: Concepts, Problems, Prospects: Elsevier, New York, 611 p.
Recent compilation volume that nicely summarizes the present state of ichnology, is akin to a recent update of the 1975 Frey volume.

Pemberton, S.G., *ed.,* 1992, Applications of

Ichnology to Petroleum Exploration, a core workshop: SEPM, Core Workshop 17, 429 p.
Product of the first ichnologically themed core conference, which emphasized the applications of ichnology to petroleum exploration. The volume consists of numerous case studies and contains many high-quality photographs of trace fossils in cores.

Reineck, H.-E., and Singh, I.B., 1980, Depositional Sedimentary Environments: With Reference to Terrigenous Clastics: Springer-Verlag, New York, second edition, 549 p.
A dated volume, which nevertheless continues to be one of the best integrations of sedimentology and ichnology particularly for present-day marine environments. The volume lays out the tenets of neoichnological work in the tradition of the German Senckenberg Institute. The section entitled 'Biological Parameters' is particularly insightful from the perspective of ichnology.

Seilacher, A.A., 2007, Trace Fossil Analysis: Springer, New York, 226 p.
A long-anticipated volume from the "Father of Modern Ichnology", it is a beautifully illustrated book with emphasis on the understanding of animal behavior from trace fossils.

Other references

Allen, G.P., Salomon, H.C., Bassoullet, P., du Penhoat, Y., and de Grandpre, C., 1980, Effects of tides on mixing and suspended sediment transport in macrotidal estuaries: Sedimentary Geology, v. 26, p. 69-90.

Bann, K.L., and Fielding, C.R., 2004, An integrated ichnological and sedimentological comparison of non-deltaic shoreface and subaqueous delta deposits in Permian reservoir units of Australia, *in* McIlroy, D., *ed.,* The Application of Ichnology to Palaeoenvironmental and Stratigraphic Analysis: Geological Society of London, Special Publicaiton 228, p. 273-307.

Bann, K.L., Tye, S.C., MacEachern, J.A., Fielding, C.R., and Jones, B.G., 2008, Ichnological and sedimentologic signatures of mixed wave- and storm-dominated deltaic deposits: Examples from the Early Permian Sydney Basin, Australia, *in* Hampson, G., Steel, R., Burgess, P., and Dalrymple, R., *eds.,* Recent Advances in Models of Siliciclastic Shallow-Marine Stratigraphy: SEPM Special Publication 90, p. 293-332.

Beynon, B.M., Pemberton, S.G., Bell, D.A., and Logan, C.A., 1988, Environmental implications of ichnofossils from the Lower Cretaceous Grand Rapids Formation, Cold Lake Oil Sands Deposit, *in* James, D.P., and Leckie, D.A., *eds.,* Sequences, Stratigraphy, Sedimentology: Surface and Subsurface: Canadian Society of Petroleum Geologists, Memoir 15, p. 275-290.

Bromley, R.G., Pemberton, S.G., and Rahmani, R.A., 1984, A Cretaceous woodground: the *Teredolites* Ichnofacies: Journal of Paleontology, v. 58, p. 488-498.

Buatois, L.A., and Mángano, M.G., 1995, The paleoenvironmental and paleoecological significance of the lacustrine *Mermia* ichnofacies: an achetypical subaqueous nonmarine trace fossil assemblage: Ichnos, v. 4, p. 151-161.

Buatois, L.A., and Mángano, M.G., 2004, Animal-substrate interactions in freshwater environments: applications of ichnology in facies and sequence stratigraphic analysis of fluvio-lacustrine successions, *in* McIlroy, D., *ed.,* The Application of Ichnology to Palaeoenvironmental and Stratigraphic Analysis: Geological Society of London, Special Publication 228, p. 311-333.

Buatois, L.A., Gingras, M., MacEachern, J.A., Mángano, M.G., Zonneveld, J.P., Pemberton, S.G., Netto, R.G., and Martin, A., 2005, Colonization of brackish-water systems through time: Evidence from the trace-fossil record: Palaios, v. 20, p. 321-347.

Chamberlain, C.K., 1978, Recognition of trace fossils in cores, *in* Basan, P.B., *ed.,* Trace Fossil Concepts: SEPM Short Course 5, Oklahoma City, p. 119-166.

Coniglio, M., and Dix, G.R., 1992, Carbonate slopes, *in* Walker, R.G. and James, N.P., *eds.,* Facies Models: Response to Sea Level Change: Geological Association of Canada, St. John's, Newfoundland, p. 349-373.

Crimes, T.P., and Fedonkin, M.A., 1994, Evolution and dispersal of deep-sea traces: Palaios, v. 9, p. 74-83.

Dalrymple, R.W., Makino, Y., and Zaitlin, B.A., 1991, Temporal and spatial patterns of rhythmite deposition on mud flats in the macrotidal Cobequid Bay-Salmon River Estuary, Bay of Fundy, Canada, *in* Smith, D.G., Reinson, G.E., Zaitlin, B.A., and Rahmani, R.A., eds., Clastic Tidal Sedimentology: Canadian Society of Petroleum Geologists Memoir 16, p. 136-160.

Domenech, R., De Gibert, J.M., and Martinell, J., 2001, Ichnological features of a marine transgression: Middle Miocene rocky shores of Tarragona, Spain: Geobios, v. 34, p. 99-107.

Droser, M.L., and Bottjer, D.J., 1986, A semiquantitative field classification of ichnofabric: Journal of Sedimentary Petrology, v. 56, p. 558-559.

Frey, R.W., and Pemberton, S.G., 1985, Biogenic structures in outcrops and cores. 1. Approaches to ichnology: Bulletin of Canadian Petroleum Geology, v. 33, p. 72-115.

Frey, R.W., and Pemberton, S.G., 1987, the *Psilonichnus* ichnocoenose, and its relationship to adjacent marine and nonmarine ichnocoenoses along the Georgia coast: Bulletin of Canadian Petroleum Geology, v. 35, p. 333-357.

Frey, R.W., and Seilacher, A., 1980, Uniformity in marine invertebrate ichnology: Lethaia, v. 13, p. 183-207.

Frey, R.W., Pemberton, S.G., and Saunders, T.D.A., 1990, Ichnofacies and bathymetry: a passive relationship: Journal of Paleontology, v. 64, p. 155-158.

Genise, J.F., Mángano, M.G., Buatois, L.A., Laza, J.H., and Verde, M., 2000, Insect trace fossil associations in paleosols: the *Coprinisphaera* ichnofacies: Palaios, v. 15, p. 49-64.

Gingras, M.K., MacEachern, J.A., and Pemberton, S.G., 1998, A comparative analysis of the ichnology of wave and river-dominated allomembers of the Upper Cretaceous Dunvegan Formation: Bulletin of Canadian Petroleum Geology, v. 46, p. 51-73.

Gingras, M.K., Pemberton, S.G., Saunders, T., and Clifton, H.E., 1999, The ichnology of brackish water Pleistocene deposits at Willapa Bay, Washington: variability in estuarine settings: Palaios, v. 14, p. 352-374.

Gingras, M.K., MacEachern, J.A., and Pickerill, R.K., 2004, Modern perspectives on the *Teredolites* Ichnofacies: observations from Willapa Bay, Washington: Palaios, v. 19, p. 79-88.

Gingras, M.K., Bann, K.L., MacEachern, J.A., and Pemberton, S.G., 2007, A conceptual framework for the application of trace fossils, *in* MacEachern, J.A., Bann, K.L., Gingras, M.K., and Pemberton, S.G., *eds.,* Applied Ichnology: SEPM Short Course Notes 52, p. 1-26.

Gingras, M.K., Pemberton, S.G., Dashtgard, S., and Dafoe, L., 2008, How fast do marine invertebrates burrow?: Palaeogeography, Palaeoclimatology, Palaeoecology, v. 270, p. 280-286.

Hasiotis, S.T., 2007, Continental ichnology: Fundamental processes and controls on trace fossil distribution, *in* Miller W III, *ed.,* Trace Fossils: Concepts, Problems, Prospects: Elsevier, New York, p. 268-284.

Howard, J.D., Elders, C.A., and Heinbokel, J.F., 1975, Animal-sediment relationships in estuarine point bar deposits, Ogeechee River-Ossabaw Sound, Estuaries of the Georgia Coast, U.S.A.: Sedimentology and Biology, V, Senckenbergiana Maritima, v. 7, p. 181-203.

Kotake, N., 1991, Non-selective surface deposit feeding by the *Zoophycos* producers: Lethaia, v. 24, p. 379-385.

Lesourd, S., Lesueur, P., Brun-Cottan, J.C., Garnaud, S., and Poupinet, N., 2003, Seasonal variations in the characteristics of superficial sediments in a macrotidal estuary (the Seine Inlet), France: Estuarine Coastal and Shelf

Science, v. 58, p. 3-16.

Levin, L.A., Rathburn, A.E., Gutiérrez, D., Muñoz, P., and Shankle, A., 2003, Bioturbation by symbiont-bearing annelids in near-anoxic sediments: implications for biofacies models and paleo-oxygen assessments: Palaeogeography, Palaeoclimatology, Palaeoecology, v. 199, p. 129-140.

MacEachern, J.A., Raychaudhuri, I., and Pemberton, S.G., 1992, Stratigraphic applications of the *Glossifungites* ichnofacies: delineating discontinuities in the rock record, *in* Pemberton, S.G., *ed.,* Applications of Ichnology to Petroleum Exploration, A Core Workshop: SEPM, Core Workshop 17, p. 169-198.

MacEachern, J.A., Bann, K.L., Bhattacharya, J.P., and Howell, C.D., 2005, Ichnology of deltas: organism responses to the dynamic interplay of rivers, waves, storms and tides, *in* Giosan, L. and Bhattacharya, J.P., *eds.,* River Deltas: Concepts, Models and Examples: SEPM Special Publication 83, p. 49-85.

MacEachern, J.A., Bann, K.L., Pemberton, S.G., and Gingras, M.K., 2007b, The ichnofacies paradigm: high resolution paleoenvironmental interpretation of the rock record, *in* MacEachern, J.A., Bann, K.L., Gingras, M.K., and Pemberton, S.G., *eds.,* Applied Ichnology: SEPM Short Course Notes 52, p. 27-64.

MacEachern, J.A., Pemberton, S.G., Bann, K.L., and Gingras, M.K., 2007c, Departures from the archetypal ichnofacies: effective recognition of environmental stress in the rock record, *in* MacEachern, J.A., Bann, K.L., Gingras, M.K., and Pemberton, S.G., *eds.,* Applied Ichnology: SEPM Short Course Notes 52, p. 65-93.

MacEachern, J.A., Pemberton, S.G., Gingras, M.K., and Bann, K.L., 2007d, The ichnofacies concept: a fifty-year retrospective, *in* Miller III, W., *ed.*, Trace Fossils: Concepts, Problems, Prospects, Elsevier, p. 50-75.

Martin, K.D., 2004, A re-evaluation of the relationship between trace fossils and dysoxia, *in* McIlroy, D., *ed.,* The Application of Ichnology to Palaeoenvironmental and Stratigraphic Analysis: Geological Society of London, Special Publication 228, p. 141-156.

McIlroy, D., 2004b, Some ichnological concepts, methodologies, applications and frontiers, *in* McIlroy, D., *ed.,* The Application of Ichnology to Palaeoenvironmental and Stratigraphic Analysis: Geological Society of London, Special Publication 228, p. 3-27.

Miller, W. III, 1993, Trace fossil zonation in Cretaceous turbidite facies, northern California: Ichnos, v. 3, p. 11-28.

Müller, A.H., 1962, Zur Ichnologie, Taxiologie und Ekologie, Fossiler Tiere, Teil 1: Freiberger Forschungsheft, C151, p. 5-30.

Nara, M., 1997, High-resolution analytical method for event sedimentation using *Rosselia socialis*: Palaios, v. 12, p. 489-494.

Pearson, N.J. and Gingras, M.K., 2006, An ichnological and sedimentological facies model for muddy point-bar deposits: Journal of Sedimentary Research, v. 76, p. 771-782.

Pemberton, S.G., and Frey, R.W., 1984, Ichnology of storm-influenced shallow marine sequence: Cardium Formation (Upper Cretaceous) at Seebe, Alberta, *in* Stott, D.F. and Glass, D.J., *eds.,* The Mesozoic of Middle North America: CSPG Memoir 9, p. 281-304.

Pemberton, S.G., and Gingras, M.K., 2005, Classification and characterizations of biogenically enhanced permeability: AAPG Bulletin, v. 89, p. 1493-1517.

Pemberton, S.G., and MacEachern, J.A., 1997, The ichnological signature of storm deposits: the use of trace fossils in event stratigraphy, *in* Brett, C.E., *ed.,* Paleontological Event Horizons: Ecological and Evolutionary Implications: Columbia University Press, p. 73-109.

Pemberton, S.G., MacEachern, J.A., and Frey, R.W., 1992, Trace fossil facies models: environmental and allostratigraphic significance, *in* Walker, R.G., and James, N.P., *eds.,* Facies Models: Response to Sea Level Change: Geological Association of Canada, St. John's Newfoundland, p. 47-72.

Pemberton, S.G., MacEachern, J.A., and Saunders, T., 2004, Stratigraphic applications of substrate-specific ichnofacies: delineating discontinuities in the rock record, *in* McIlroy, D., *ed.,* The Application of Ichnology to Palaeoenvironmental and Stratigraphic Analysis: Geological Society of London, Special Publication 228, p. 29-62.

Rhoads, D.C. and Morse, J.W., 1971, Evolutionary and ecologic significance of oxygen-deficient marine basins: Lethaia, v. 4, p. 413-428.

Savrda, C.E., 2007, Trace fossils and benthic oxygenation, *in* Miller III, W., *ed.,* Trace Fossils: Concepts, Problems, Prospects: Elsevier, New York, p. 149-158.

Savrda, C.E., and Bottjer, D.J., 1989, Trace-fossil model for reconstructing oxygenation histories of ancient marine bottom waters: application to Upper Cretaceous Niobrara Formation, Colorado: Palaeogeography, Palaeoclimatology, Palaeoecology, v. 74, p. 49-74.

Savrda, C.E., Ozalas, K., Demko, T.H., Hutchinson, R.A., and Scheiwe, T.D., 1993, Log grounds and the ichnofossil *Teredolites* in transgressive deposits of the Clayton Formation (Lower Paleocene), western Alabama: Palaios, v. 8, p. 311-324.

Schieber, J., 2003, Simple gifts and buried treasures – implications of finding bioturbation and erosion surfaces in black shales: The Sedimentary Record, p. 4-8.

Seilacher, A., 1967, The bathymetry of trace fossils: Marine Geology, v. 5, p. 413-428.

Seilacher, A., 1991, Events and their signatures - an overview, *in* Einsele, G., Ricken, W., and Seilacher, A., *eds.,* Cycles and Events in Stratigraphy: Springer-Verlag, New York, p. 222-226.

Taylor, A.M., and Goldring, R., 1993, Description and analysis of bioturbation and ichnofabric: Journal of the Geological Society of London, v. 150, p. 141–148.

Taylor, A., Goldring, R. and Gowland, S., 2003, Analysis and application of ichnofabrics: Earth-Science Reviews, v. 60, p. 227-259.

Walker, R.G., 1992, Facies, facies models and modern stratigraphic concepts, *in*: Walker, R.G., and James, N.P., *eds.,* Facies Models: Response to Sea Level Change: Geological Association of Canada, St. John's Newfoundland, p. 1-14.

Wetzel, A., 1991, Ecologic interpretation of deep-sea trace fossil communities: Palaeogeography, Palaeoclimatology, Palaeoecology, v. 85, p. 95-124.

Wignall, P.B., and Pickering, K.T., 1993, Paleoecology and sedimentology across a Jurassic fault scarp, northeast Scotland: Journal of the Geological Society of London, v. 150, p. 323-340.

4. INTRODUCTION TO SILICICLASTIC FACIES MODELS

Robert W. Dalrymple, Department of Geological Sciences and Geological Engineering, Queen's University, Kingston, ON, K7L 3N6, Canada

WHAT ARE SILICICLASTIC SEDIMENTS?

Siliciclastic sedimentary rocks consist of conglomerate, sandstone and shale that are derived by the erosion of pre-existing igneous, metamorphic and sedimentary rocks. They constitute more than 80% of all sedimentary rocks, with shale comprising over half of the total (Tucker, 2001; Potter *et al.*, 2005).

Siliciclastic sediments are composed of particles, or clasts (*i.e.*, they are 'clastic' deposits), that consist predominantly of silicate minerals (hence the prefix 'silici'). Unlike the biochemical sediments described in the second half of this book, these particles are made, not born. To borrow from the well-known analogy used for biochemical sediments (*see* Chapter 13), the 'factory' where the constituents of siliciclastic rocks are 'manufactured' is located in topographically high, commonly mountainous, source areas, where the creation of particles involves a spectrum of physical and chemical weathering processes, accompanied by erosion. Because these particles almost always originate on land, they are also referred to as 'terrigenous-clastic' sediments, even though most of them are deposited in marine environments.

A more appropriate analogy for siliciclastic sedimentation is that of a mail-delivery network. The main distribution routes are rivers, shelf currents and submarine-channel networks, whereas the transportation vehicles are the multitude of physical processes that move the sediment: winds, river currents, tidal currents, waves, oceanic currents and mass-transport processes such as debris flows and turbidity currents. Large grains moving slowly as bedload correspond to snail mail, whereas fine grains that move more rapidly in suspension are analogous to a priority courier service. A more comprehensive comparison of siliciclastic and biochemical sediments is provided in Table 1.

It is helpful to subdivide siliciclastic systems into three parts (Fig. 1): the source area where the particles are first created; the sediment-dispersal system that moves the particles from the source to the area where they are deposited; and the site of ultimate deposition (*cf.* Allen, 1997, p. 114). Such a subdivision is arbitrary because movement occurs within the source area, the particles continue to undergo weathering and abrasion during transport, and deposition, sometimes for relatively long periods, can occur during transport.

THE SEDIMENT SOURCE

In the case of biochemical sediments, the conditions within the area where the grains or crystals are produced

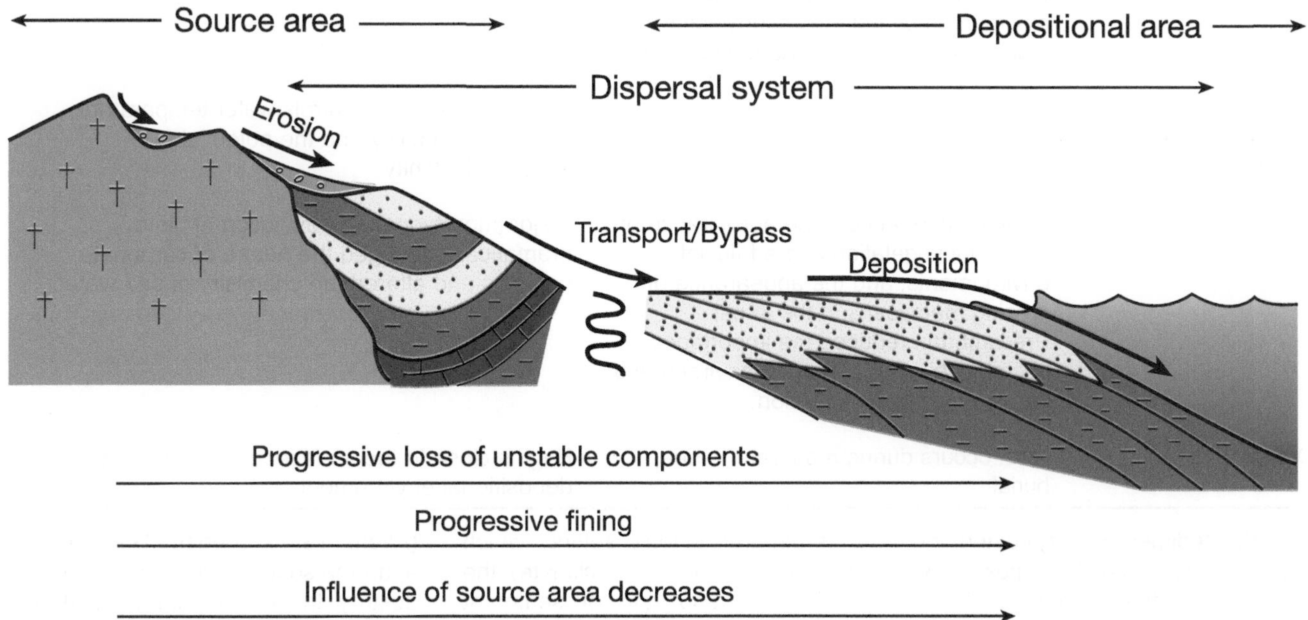

Figure 1. Simplified representation of the siliciclastic sedimentary system with its three overlapping components: source area (the 'sediment factory'), dispersal system and depositional area.

Table 1. Comparison of siliciclastic and biochemical sediments

Parameter	Siliciclastic	Biochemical
Source area	Older igneous, metamorphic and sedimentary rocks, generally located in topographically high areas at some distance from the site of deposition.	Essentially *in situ*.
Composition	Silicate minerals; controlled by composition of source area and by selective removal of weak grains by physical and chemical weathering, and abrasion during transport.	Carbonate, phosphate and evaporite minerals; controlled by chemical environment within depositional area.
Grain size	Controlled by relief and composition of the source area, hydraulic energy along the transport path and at the site of deposition, and selective deposition in areas with accommodation.	Controlled by type and size of organisms producing skeletons, biogenic fragmentation, dynamics of crystal growth; hydraulic sorting locally important.
Depositional environments	From terrestrial to deep marine.	Primarily shallow marine, with lesser amounts in deep-marine and lacustrine settings.
Tectonic settings	All tectonic environments although the largest sediment volumes are derived from tectonically active areas.	Areas distant from large influxes of fresh water and terrigenous-clastic sediment; most abundant on passive continental margins, isolated low-lying islands and intracratonic basins.
Importance of physical processes	Overwhelmingly important. Determines grain size and nature of physical sedimentary structures.	Moderate to minimal importance. Influences nature of biological community and causes some local transportation and sorting; may influence water chemistry as a result of mixing of water masses.
Influence of climate	Generally indirect. Influences nature of weathering, the terrestrial delivery system, and the prevalence of vegetation; determines the intensity and frequency of storms in coastal and shallow-marine environments, and the occurrence of glacial ice and the importance of winds.	Strong influence on water temperature and salinity, which controls the nature of the biological community and mineral chemistry of precipitates.
Influence of ocean chemistry and circulation	Minor importance.	Highly significant. Controls water temperature, nutrient availability and the potential for increased salinity.
Geologic age	Limited importance. The appearance of terrestrial vegetation altered fluvial environments, and the appearance of grazing metazoans in the latest Precambrian reduced the importance of microbial binding in marine environments and introduced bioturbation.	Major importance. The evolution of biota dramatically changed the nature of carbonate particles and altered the chemistry of sea water.
Diagenesis	Most occurs during moderate to deep burial.	Begins close to the surface, commonly within the depositional environment.

(*i.e.*, the sediment factory) determine almost all aspects of the deposits. By comparison, the influence of the source area is considerably less in the case of siliciclastic sediments. Perhaps its most important influence is its general tectonic setting, which determines (along with climate) the composition and rate of sediment supply, and the overall patterns of sediment dispersal. Physical- and chemical-weathering processes (Allen, 1997; Leeder, 1999) break apart the source rocks and selectively destroy those components that are physically weak (*i.e.*, those minerals and rocks that contain planes of weakness such as cleavage, foliation and fractures) and chemically unstable. Minerals containing large numbers of covalent bonds and few and/or weak cleavage planes (such as quartz) tend to survive and end up being more abundant in the deposits than in the source area. Rock types that are resistant to breakdown also become enriched in the deposits, relative to their abundance in the source.

The most direct influence of the source is, therefore, on the composition of the sediment. Although the mineralogy has a profound impact on the diagenesis of the sediment, sediment composition has relatively limited influence on how and where the grains are deposited. Differences in mineral density play a role in transportation and deposition, especially in the segregation of heavy minerals such as garnet and magnetite, but this has less influence than particle size on how the grains behave. Thus, mineralogy provides limited information about the environment in which the sediment accumulated, and is not incorporated into any facies model for a siliciclastic environment, in stark contrast to the case with biochemical facies models where sediment composition is an integral part of any model.

THE SEDIMENT-DISPERSAL SYSTEM

The sediment-dispersal system transfers grains from the source area to the site of permanent deposition (Fig. 1). A myriad of physical processes participate in the sediment transfer, and the development of robust interpretations of sedimentary successions requires a good understanding of the physics of these sediment-transport processes (*see* Chapter 2). A comprehensive treatment of this subject is beyond the scope of this chapter and readers are referred to such texts as Allen (1997), Leeder (1999) and Middleton and Southard (1984).

The overall driving force for the sediment-dispersal system is gravity, with source areas occurring in topographically elevated locations, and the depositional sites in topographically low areas. In settings with high relief and relatively short transport distances (*i.e.*, a few kilometers to several tens of kilometers), movement from the source to the site of deposition can be rapid, and particles spend little residence time in the source and dispersal system before being buried permanently. As a result, physical abrasion, breakage and chemical alteration is limited, yielding a sediment that is relatively coarse grained and compositionally immature, such that its mineralogy mimics closely the composition of the source area. At the other end of the spectrum, transport systems may reach thousands of kilometers in length. In such cases, particles may undergo long periods of temporary storage before being re-eroded and moved farther down the transport path. In such cases, particles experience greater amounts of physical and chemical weathering while they are *en route* to their final resting place, causing the deposit to be compositionally more mature and finer grained (Fig. 1).

Grain Transport Mechanisms

The most important sediment-transporting agents (air and water) have relatively low viscosity. In these media, grains are generally free to move relative to each other, with their exact behavior depending primarily on their size and, to a lesser extent, on their density and shape. This differential movement permits sediment sorting, which, at the finest scale, allows us to see lamination and bedding. In more viscous transport media such as glaciers (*see* Chapter 5) and debris flows (*see* Chapter 12), differential movement of grains is limited and the ability to sort them is greatly reduced. As a result, lamination and bedding are poorly developed.

In air and water, grain movement occurs either by bedload or suspension (Allen, 1997; Leeder, 1999; Middleton and Southard, 1984). *Suspended sediment* (the suspended load) consists of relatively fine-grained particles that have slow settling velocities. For these grains, the upward component of turbulence is able to offset their tendency to settle, allowing them to travel within the body of the flow. The maximum grain size that can move by this mechanism is determined by the intensity of the turbulence and increases as the flow speed increases. *Bedload* consists of particles that move in a thin zone, only a few grain diameters thick, directly above the stationary bed. Within this zone, grains move by several mechanisms, including rolling, sliding, creep as a result of impacts, and saltation (movement by a series of small hops up to five grain-diameters high). The grains that can be moved by the flow but are too big or dense to move by suspension (because their settling velocity is too large) are confined to the bedload layer. Grains that are small enough to be suspended may also be present in the bedload layer, either because the flow can carry no more material in suspension, or because such grains are on their way up into suspension or are falling out of suspension as a result of short-term fluctuations in flow strength. As a result, the sediment in the bedload layer is always more poorly sorted than the sediment in suspension, as is reflected in the nature of the deposits. In most circumstances, the amount of material moving as bedload is a small fraction (typically only a few percent) of the total amount of sediment in transport.

Causes of Down-system Fining

Because of its control on sediment transport behavior, particle size is one of the most fundamental properties of siliciclastic sediments. It also exerts a strong influence on the nature of the sedimentary structures present, as shown by the bedform phase diagrams for currents (Fig. 2; *see* also Chapter 6, Figure 6) and waves (Fig. 3; *see* Chapter 8). From an applied perspective, grain size and sorting are also strongly correlated with porosity and permeability. The larger scale dynamics of the sediment dispersal system exert the first-order control on the grain size of the sediment in a depositional environment, supplemented by local sorting. On average, over distances of tens to hundreds of kilometers, particles tend to become finer with distance along a sediment transport path. This phenomenon is well documented in fluvial systems (Parker, 1991; Frings, 2008) but occurs in all depositional environments. This fin-

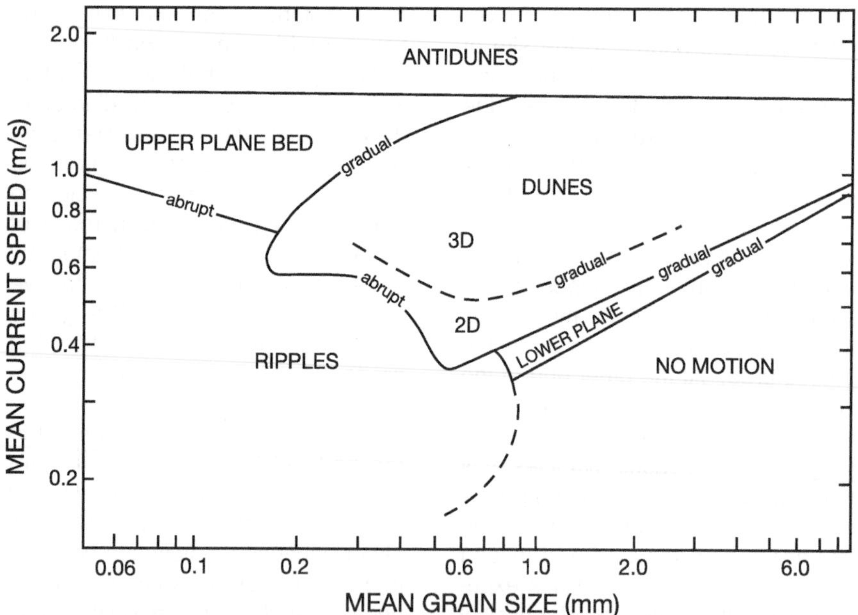

Figure 2. Current speed–grain size section through the three-dimensional phase diagram for the occurrence of current-generated bedforms. The third important variable, water depth, is less important than those shown here and the phase boundaries shift gently to higher current speeds as water depth increases. Modified after Southard and Boguchwal (1990). Reproduced with permission of SEPM (Society for Sedimentary Geology).

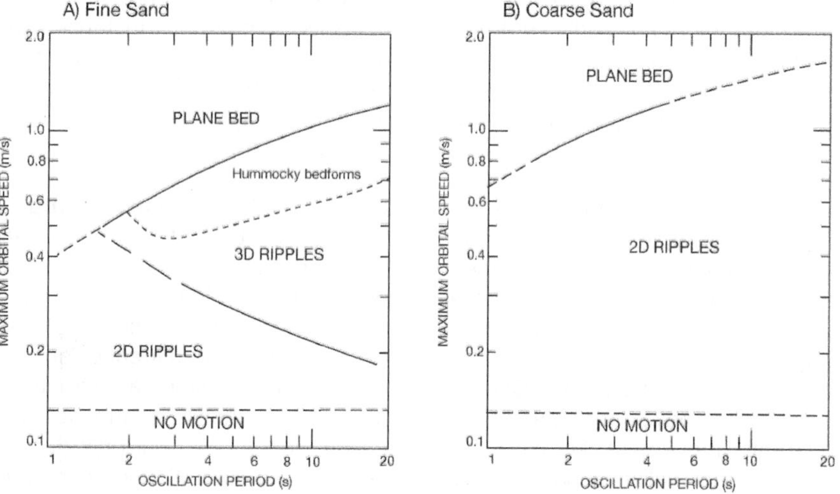

Figure 3. Simplified bedform phase diagrams for purely oscillatory flow in **(A)** fine sand and **(B)** coarse sand. Unlike the case with current-generated bedforms, the stability fields for wave-generated bedforms, and especially for the 'hummocky bedform' that is believed to generate hummocky cross stratification, are not well constrained, but it is known that the hummocky bedform does not occur in coarse sand (Cummings *et al.*, 2009). Modified after Southard (1991). Reproduced with permission.

ing trend is caused by two groups of processes: physical abrasion of grains, and selective deposition and preservation.

Abrasion

The physical breakage of particles, broadly termed 'abrasion' (Frings, 2008), takes many forms. Small-scale abrasion (*i.e.*, the breaking off of corners and the spalling of small chips in the silt- and clay-size range) is a relatively minor contributor to the fining trend. It is most important in the coarser size fractions that travel by bedload because they experience a higher frequency of more energetic collisions than do grains traveling relatively passively in suspension. Thus, bedload particles almost always display better rounding than finer grains (Pettijohn, 1957), and rounding generally increases as the distance of transport increases. Larger scale breakage of particles, such as splitting, is a more important contributor to the fining trend, but is also primarily limited to the bedload fraction, and to rock and mineral particles with inherent planes of weakness. The weakest particles are eliminated from the coarse-size fractions relatively rapidly (*e.g.*, Abbott and Peterson, 1978; Fig. 4), leading to an increase in the relative abundance of the more resistant rocks and minerals as distance from the source increases. The more easily broken minerals tend to become concentrated in the finer grain sizes; thus, feldspar is finer grained than quartz in most sandstone and shale.

Early workers thought that breakage was the major cause of down-system fining. The measured rates of particle abrasion are sufficient to explain the measured downstream fining in cases where soft and easily broken clasts are abundant, a situation that is most common very close to the source. The abrasion rates for the more resistant rock and mineral types that characterize more distal areas are, however, widely believed to be too low to account for the observed fining (Parker, 1991; Seal *et al.*, 1998). Thus, once the weak clasts have been removed and the sediment consists of resistant grains, or the grains are fine enough to spend most of their time in suspension, physical breakdown ceases to be an important factor (Parker, 1991; Frings, 2008).

Selective Deposition/Preservation

The second and most important contributor to the fining trend, in many circumstances, is the differential behavior of the various grain sizes during the cycles of erosion, transport and deposition that occur throughout their passage from the source to their final resting place. This is sometimes referred to as 'selective sorting' and has been attributed to the fact that the suspended finer grains move more often and more rapidly than coarser bedload particles. The resulting differences in the average transport speed cannot, however, generate

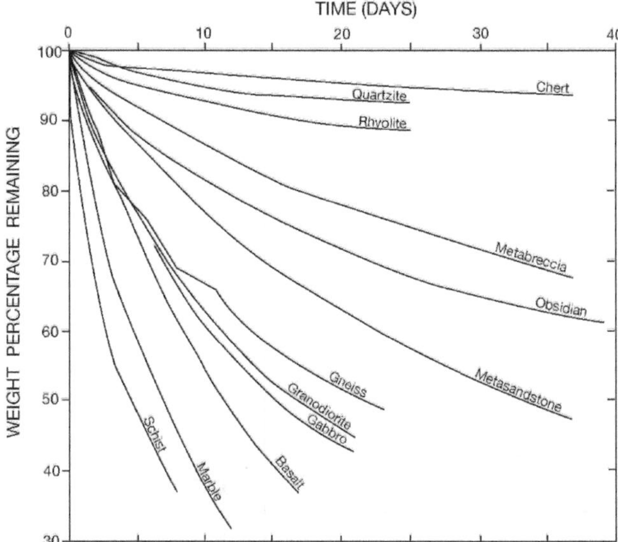

Figure 4. Relative resistance of different rock types to disintegration during transport, expressed as the weight lost as a function of time. The experimental apparatus consisted of a tumbling mill that simulated slow-speed river transport. Curves that descend steeply indicate weak rock types that are destroyed rapidly, whereas the more nearly horizontal curves indicate physically resistant rock types. Modified after Abbott and Peterson (1978). Reproduced with permission of SEPM (Society for Sedimentary Geology).

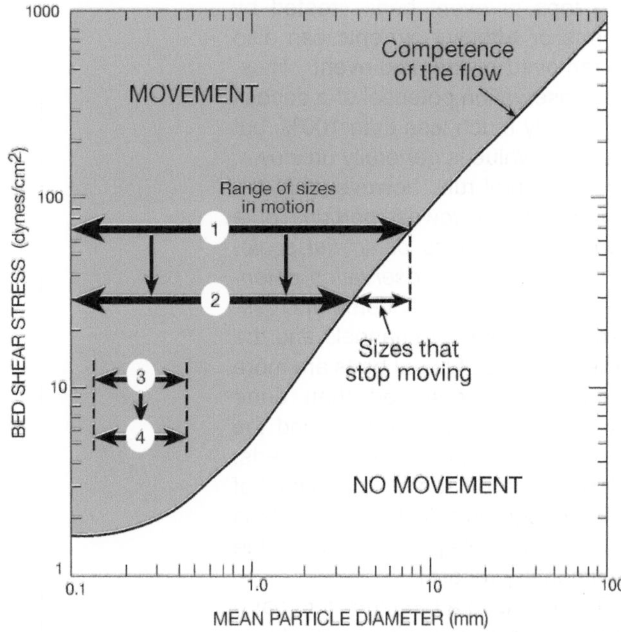

Figure 5. Plot of bed shear stress required for the initiation of motion by a unidirectional current as a function of grain size for well sorted sediment (the 'Shields curve'). The line indicates the coarsest grain size that a current of a given shear stress can carry (*i.e.*, the competence of that flow). As a current that is carrying a wide range of grain sizes slows (from 1 to 2), a range of coarse grain sizes falls below the threshold curve; these grain sizes stop moving. This is termed competence-driven deposition. By contrast, when a current that is carrying a narrow range of relatively fine grains slows (from 3 to 4), no sizes become too coarse to be deposited, but deposition will occur because of the decrease in capacity (termed capacity-driven deposition).

down-system fining unless there is net deposition and aggradation of the bed. Therefore, the down-system fining is more properly said to be caused by 'selective deposition' (Seal *et al.*, 1998) because the coarser particles are deposited closer to the source than the finer sediment. In simple terms, this is the result of two factors: decreasing hydraulic energy and the less well-documented concept of preservation potential.

Energy and Deposition The energy needed to initiate and maintain grain motion increases as particle size increases (Fig. 5). Thus, the energy level of the depositional medium determines its *competence*, which is indicated by the coarsest grain size that can be transported. It follows that a decrease in the energy of the transport system leads to the preferential deposition of (or failure to re-entrain) the coarsest fraction because of a decrease in competence (Fig. 5: decrease in energy from 1 to 2). Such deposition is referred to as competence-driven deposition (*cf.* Hiscott, 1994).

Deposition may also occur because of a decrease in the *capacity* of the flow (Hiscott, 1994), which refers to the amount of sediment that can be transported. The capacity (q_s) of a transport medium is a complex function of many variables, including the shear stress (or current speed and flow depth), grain size, grain density, and fluid density and viscosity (Yalin, 1977). The complex equations developed to predict capacity can be simplified to the general form of

$$q_s \propto U^n$$

where U is the depth-averaged current speed and n is an exponent with a value in the range of 3 to 5. Therefore, a small decrease in the current speed can lead to a large decrease in the capacity of the flow, which results in sediment deposition. While flow deceleration typically causes both competence-driven and capacity-driven deposition to occur at the same time, the two causes of deposition can occur independently. For example, if a flow is below capacity at all times, deceleration does not cause capacity-driven deposition, but may cause competence-driven deposition if coarse sediment is present. On the other hand, a flow that contains only fine-grained sediment and is at capacity must deposit sediment as the flow decelerates, even though all of the grains are fine enough to remain in transport (Fig. 5: energy decrease from 3 to 4). Overall, competence-driven sedimentation is more important than capacity-driven deposition as a cause of downstream fining.

Preservation Potential Not all deposition is permanent. For example, deposition on the lee side of a dune is followed almost immediately by erosion when the stoss side of the same bedform migrates past. Similarly, deposits formed on a point bar have a moderate to high probability of being eroded as the channel migrates back across the floodplain.

The tops of event beds created by storms or turbidity currents can also be removed by the next event. Thus, the preservation potential of a deposit is typically much less than 100%, but the exact value is generally unknown. It is a general rule, however, that *the topographically lowest portions of a deposit (regardless of spatial scale) have the highest preservation potential.* Therefore, the bottoms of cross-beds, the bases of channels and the lower portions of event beds are more likely to be preserved than dune crests, the top of point bars and the deposits capping event beds, although the preservation potential of such topographically high deposits is not zero (*i.e.*, they are present in the sedimentary record).

There is also a tendency for coarser grains to be deposited in the topographically lower part of a deposit. Thus, the coarsest grains tend to avalanche to the bottom of a dune slip face, or are concentrated at the base of channels (Fig. 6; *see* Chapter 6) and at the bottom of graded event beds in shelf and deep-water settings (*see* Chapters 8 and 12). Because of this, they are more likely to escape erosion than the overlying finer grained material. Thus, the coarser grains tend to remain in proximal areas while finer material moves farther along the transport path (Frings, 2008).

In areas with negative accommodation (*i.e.*, areas experiencing erosion), the current speed typically does not decrease in a down-system direction and there is little or no long-term preservation of deposits. Thus, downstream fining occurs only as a result of abrasion, with relatively rapid fining near the source as unstable rock and mineral species are eliminated, but with minimal fining after that (Fig. 7A; Frings, 2008). By contrast, the transition from an area with negative accommodation to an area with positive accommodation favors the development of downstream fining (Fig. 7B and C): the onset of sediment accumulation allows preferential deposition and preservation to remove the coarser sediment from the transport system. Because higher long-term rates of deposition produce more rapid down-system fining (Parker, 1991; Fig. 6), the rapidity of the lateral transition from negative to positive accommodation has an important influence on the rate of downstream fining: a spatially slow transition in accommodation produces gradual fining (Fig. 7B), whereas a spatially abrupt change in accommodation causes move rapid fining (Fig. 7C).

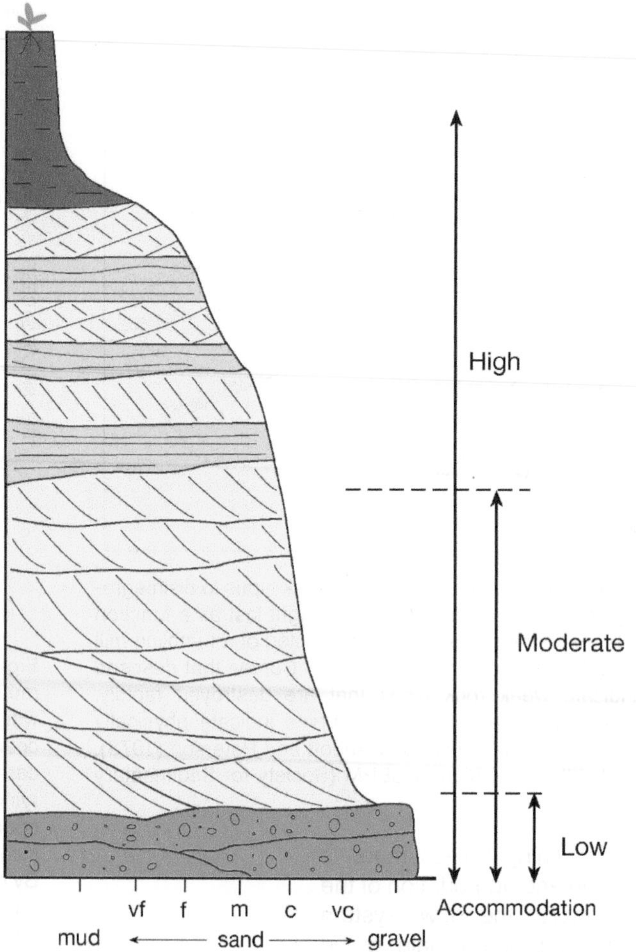

Figure 6. Upward-fining succession produced by a meandering-river point bar. As the accommodation increases, a progressively greater thickness of the succession is likely to be preserved, on average. In the example shown, low accommodation leads to preservation of only the basal gravel lag, allowing all sediment finer than gravel to bypass the area. If accommodation is high, the entire vertical succession including the muddy overbank deposits may be preserved. This causes a more rapid down-system fining than the low-accommodation case because a wider range of grain sizes is trapped in up-flow areas.

DEPOSITIONAL ENVIRONMENTS

Unlike the case with biochemical sediments that accumulate predominantly in a narrow range of shallow-marine environments (Table 2; *see* Chapter 13), siliciclastic sediments accumulate in a wide range of settings, from mountainous areas, through low-lying alluvial-plain and coastal environments, to shallow- and deep-marine settings. The nature of depositional environments and the deposits created within them is a function of many variables, which can be divided into two broad groups. The first of these are those processes that are intrinsic to the environment. They, acting on the available sediment, create the small-scale sedimentary structures and textures within the deposits, the larger scale geomorphology, and the dynamic behavior of the environment (*e.g.*, channel migration; delta-lobe switching) on relatively short time scales (*i.e.*, centuries to a few millennia). Such factors are termed *autogenic*. The second group of variables is those that are external to the local environment. They include the accommodation regime (both up-system and locally), climate, tectonic set-

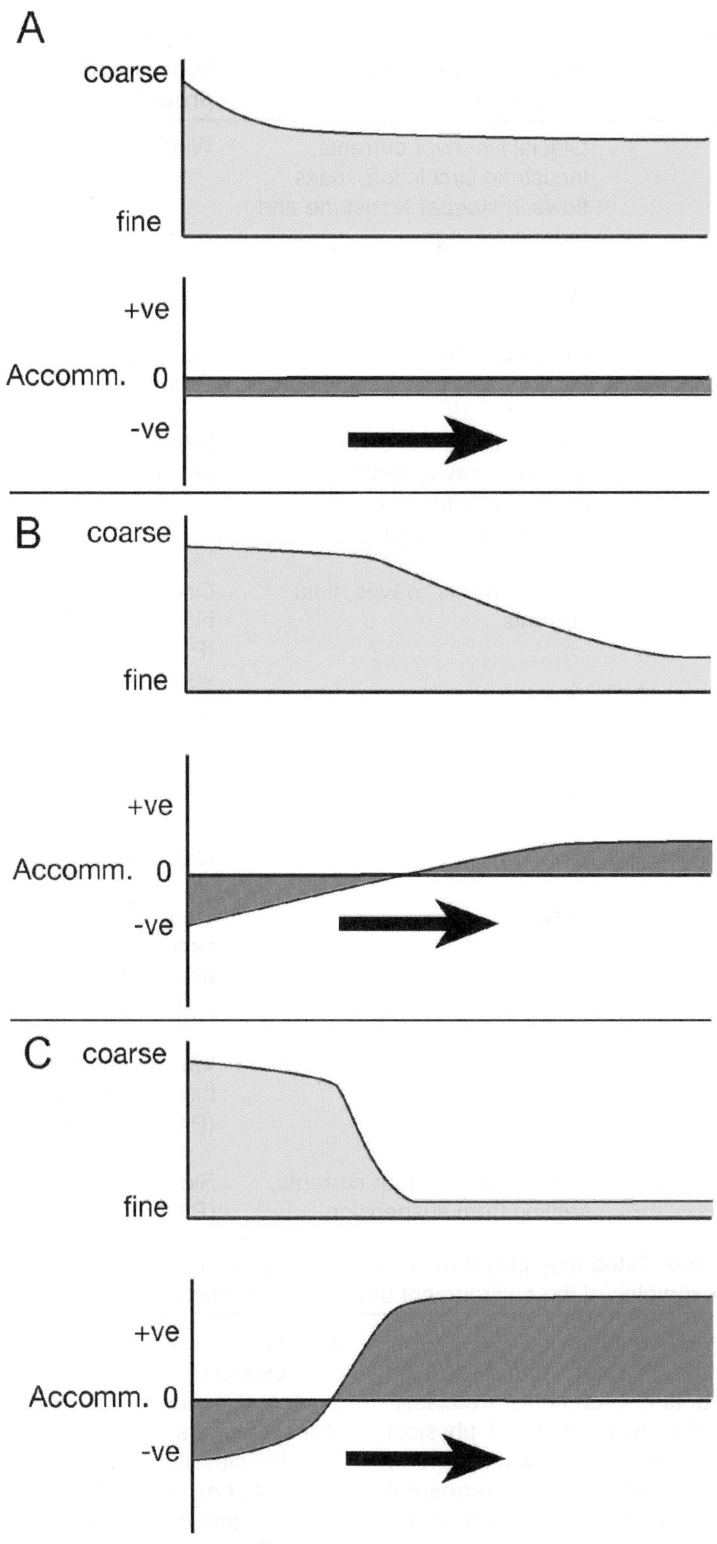

Figure 7. Relationship between downstream changes in accommodation and grain size for three situations. **A.** Continuous negative accommodation: grain size decreases slowly, as a result solely of abrasion and breakage. The rate of fining may be greatest in proximal areas because more of the sediment consists of weak rock types. This situation would occur on bypass surfaces during falling-stage and lowstand. **B.** Slow lateral change from negative to positive accommodation: the rate of grain-size fining increases in the area where accommodation becomes positive, but the rate of fining is not high. This situation would occur in low-gradient areas such as passive continental margins. **C.** Rapid lateral change from negative to highly positive accommodation (*e.g.*, a rift-basin margin), causing an abrupt decrease in grain size.

ting and biologic diversity. These *allogenic* factors create the broader context in which the environment sits and thereby have an influence on the grain size supplied to the environment, the precise nature of the processes that operate (*e.g.*, whether mass-flow processes are important, the relative influence of waves and tides, *etc.*), and the large-scale evolution of the environment (*e.g.*, whether it progrades or retrogrades).

It should be noted that the definitions of autogenic and allogenic used here differ from those employed in the emerging field of 'autostratigraphy' (Muto *et al.*, 2007). The concepts embodied in autostratigraphy blur the boundary between autogenic and allogenic as traditionally used, and generally apply to time scales that are longer than those that characterize autogenic processes as used in this volume. Their implications for facies models and sequence stratigraphy remain to be worked out fully.

Autogenic Controls on Sedimentation

Every environment possesses its own distinctive suite of physical processes (Reading, 1986; Allen, 1997; Posamentier and Walker, 2006). Thus, glacial environments (*see* Chapter 5) are dominated by the action of ice and the release of meltwater; fluvial environments (*see* Chapter 6) are dominated by unidirectional currents; most sediment deposition in eolian settings (*see* Chapter 7) is by wind; coasts (*see* Chapters 8–11) experience wave action and tidal currents in various combinations; and deep-marine environments (*see* Chapter 12) are dominated by mass-flow processes, including turbidity currents. Within each of these settings, the intrinsic processes create a distinctive geomorphology: ice produces a sculpted bed and moraines; wind generates dunes; currents, regardless of type, tend to create channels; and waves produce beaches and shorefaces. There is then a feedback between the geomorphology, the processes and the available sediment such that particular geomorphic settings are characterized by specific types of

Table 2. General classification of siliciclastic depositional environments and the processes operating in them

Major depositional environments	Major sub-environments	Major depositional processes*	Minor depositional processes
Terrestrial	Glacial and proglacial	Glacial ice, river currents, lacustrine circulation, mass flows in steeper lacustrine and marine settings	Wind
	Desert	Wind	River currents
	Fluvial	River currents	Vegetation (post-Ordovician), wind
	Lacustrine	Density circulation, river currents, waves, settling, organic productivity (Phanerozoic only)	Direct chemical precipitation
Coastal/shallow marine	Deltas	River currents, waves, tidal currents	Ocean circulation, biologic activity (Phanerozoic only), vegetation (post-Ordovician)
	Estuaries	Waves, tidal currents, river currents	Biologic activity (Phanerozoic only), vegetation (post-Ordovician)
	Wave-dominated settings	Waves	Tidal currents, ocean circulation, biologic activity (Phanerozoic only)
	Tide-dominated settings	Tidal currents	Waves, ocean circulation, biologic activity (Phanerozoic only)
Deep marine	Continental slopes, submarine fans, basin plains	Mass flows, contour currents, settling from suspension	Biologic activity (Phanerozoic only)

* In any given example, one or more of the processes listed may play a minor role. However, all of the processes listed in this column play a major role in at least some examples of the environment under consideration.

deposits. For example, the bases of channels contain the coarsest and highest energy deposits; the lower parts of shorefaces contain hummocky cross stratification; and deep-water slopes are characterized by fine-grained deposits that show evidence of sediment failure and slumping. The geomorphology and its associated deposits, in turn, migrate through time as a result of sediment deposition, creating vertical facies successions that are an expression of Walther's Law (*see* Chapter 2). It is these autogenic processes and relationships that are the basis of most of the facies models described in the subsequent chapters.

It follows, therefore, that siliciclastic depositional environments are classified by the assemblage of physical processes operating within them. The subdivisions of the terrestrial realm are well known and each of the main subdivisions is accorded its own chapter in this book (*e.g.*, glacial, eolian, fluvial and lacustrine). Similarly, the deep-marine environment, although physically large and complex, is also treated in one chapter. The coastal zone and its associated shallow-marine environments are, however, uniquely complex because of the interaction of both terrestrial and marine processes. This zone is also especially sensitive to changes in accommodation, as reflected in changes in relative sea level. Therefore, the classification of coastal environments has been the topic of much research and discussion (Boyd *et al.*, 1992; Orton and Reading, 1993; Harris *et al.*, 2002).

Coastal environments are most commonly subdivided on the basis of the dominant physical process, where 'dominance' is assigned to the process that transports and deposits the greatest volume of sediment (Table 2) because this process is responsible for producing the characteristics of the deposits, at scales ranging from the individual facies to their paleogeographic organization. Thus, wave-dominated environments have a simple, coast-parallel morphology; tide-dominated environments are characterized by coast-normal channels separated by elongate tidal bars; and river-dominated environments lack significant reworking of river-supplied sediment by waves or tidal currents (Fig. 8). Obviously, river-dominated settings are restricted to river mouths. Tide-dominated environments are also generally located at river mouths because such sites have sufficient coast-normal tidal flux to create strong tidal currents. Wave-dominated settings can also occur at river mouths, but predominate in the areas between rivers because river and tidal currents are generally less important there.

While the process-oriented approach to the classification of coastal environments is very useful, it is not the only important variable that influences coastal environments. They are especially sensitive to changes in the ratio between the rates of accommodation creation (A) and sedimentation (S) because they occur at the interface between the terrestrial and marine realms. As a result, the shoreline can change readily from regressive to transgressive, with a corresponding change in the sediment stacking pattern from progradational to retrogradational. Thus, it is important to distinguish between transgressive and regressive coasts (Boyd *et al*., 1992). This book takes both approaches, with separate chapters on wave-dominated (*see* Chapter 8) and tide-dominated (*see* Chapter 9) coasts, together with additional chapters on regressive (*i.e*., deltas; *see* Chapter 10) and transgressive (*i.e*., estuaries and barrier-lagoon complexes; *see* Chapter 11) settings.

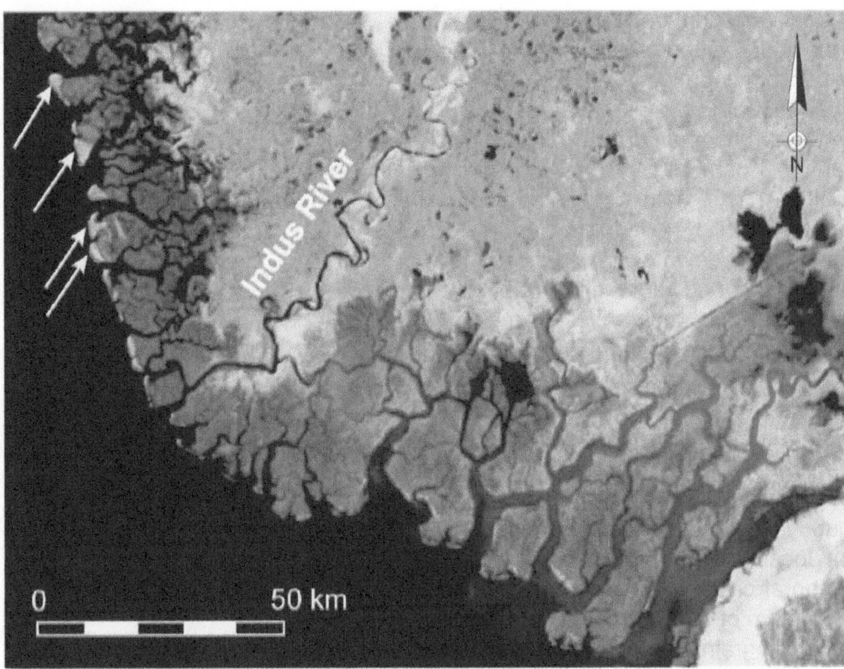

Figure 8. Landsat image of the Indus River delta, Pakistan. Karachi lies just off the image to the north along the coast. Most of the deltaic shoreline is tide dominated, as indicated by the network of coast-normal tidal channels, but the relative influence of waves increases to the northwest, as shown by the appearance of short, coast-parallel barrier islands (arrows). Image courtesy of NASA.

Allogenic Controls on Sedimentation

A variety of external factors influence how individual sedimentary environments behave and thereby influence the nature of their deposits. These include the available accommodation, the tectonic setting, climate and the evolving biologic community. Leeder (1999) contains a good review of the first three of these factors.

Accommodation

Accommodation (the space available for potential sediment accumulation) is the most fundamental control because the accommodation at any location determines whether sediment accumulates or not (Catuneanu, 2006). If the accommodation is zero (*i.e*., the equilibrium profile, either fluvial or marine, is coincident with the sediment surface), the site is an area of sediment bypass with no net deposition or erosion. If the accommodation is negative (*i.e*., the equilibrium profile lies below the sediment surface), net erosion occurs. Accumulation can only take place when and where there is positive accommodation, with deposition occurring until the sediment surface is raised back to the level of the equilibrium profile. (Of course, if all of the sediment has been deposited farther up the sediment-dispersal system, then no accumulation can take place, even if there is available accommodation.) Changes in accommodation as a result of eustatic sea-level change, tectonic movement or a change in energy level are incorporated in sequence-stratigraphic models as described in Chapter 2, and are considered in most modern facies models (*see* subsequent chapters, as well as Reading, 1996; Posamentier and Walker, 2006).

In cases where there is net sedimentation, a change in accommodation also influences the nature of the sedimentary record by changing the preservation potential of deposits formed at different elevations within a depositional environment (*cf*. Fig. 6). Successions that were formed at locations and times with low positive accommodation consist predominantly of topographically low facies (*e.g*., channel-base lags in fluvial environments; Shanley and McCabe, 1994; *see* also Chapter 6). Channels and sandy event beds (*e.g*., those formed by storms or turbidity currents) tend to be amalgamated with no intervening muddy deposits. By comparison, deposits formed at topographically high locations become volumetrically more important as accommodation increases, and the intervening muddy deposits are more abundant. Therefore, the same depositional environment can produce quite different successions as the accommodation changes.

As already discussed, the presence of accommodation along the transport path also exerts a strong control on the grain size of the sediment available for deposition at locations farther down system (Fig. 7). If there is no accommodation in more proximal areas, the sediment arriving is coarser grained than in the case where there is accommodation. If there is sufficient accommodation up-system to capture all of the bedload, then the deposits accumulating farther down system consist only of the muddy suspended load, regardless of the local energy level. If there are temporal changes in accommodation up-system of a depositional site, then the bulk grain size of the sediment accumulating must also vary temporally: the coarsest sediment accumulates when up-system accommodation is minimal (*i.e.*, during the falling stage; Posamentier and Morris, 2000; Yoshida *et al.*, 2007), and the finest sediment is deposited when accommodation is greatest (*e.g.*, during the transgressive systems tract).

The coastal zone is particularly sensitive to changes in accommodation. If the rate of creation of accommodation (A) is less than the rate of sediment accumulation (S), then the coast regresses, whereas the coast transgresses if A > S. This produces a fundamental change in the depositional environments (Boyd *et al.*, 1992): beach-ridge plains and deltas (*see* Chapters 8 and 10, respectively) occur in regressive settings, and barrier islands, lagoons and estuaries exist in transgressive situations (*see* Chapter 11).

Tectonic Setting

The tectonic setting of a depositional system exerts an important control on its nature, influencing such things as:

1. The composition of the source area and the length and steepness of the transport system, which together control the composition of the sediment supplied to the depositional environment (Fig. 9; Dickinson, 1988; Tucker, 2001) and the amount and style of sediment delivery (Milliman and Syvitski, 1992);
2. The spatial distribution of subsidence and accommodation (Fig. 10), which, as discussed above, influence the grain size of the sediment available at various points within the depositional basin (Fig. 7); and
3. The occurrence of certain depositional environments.

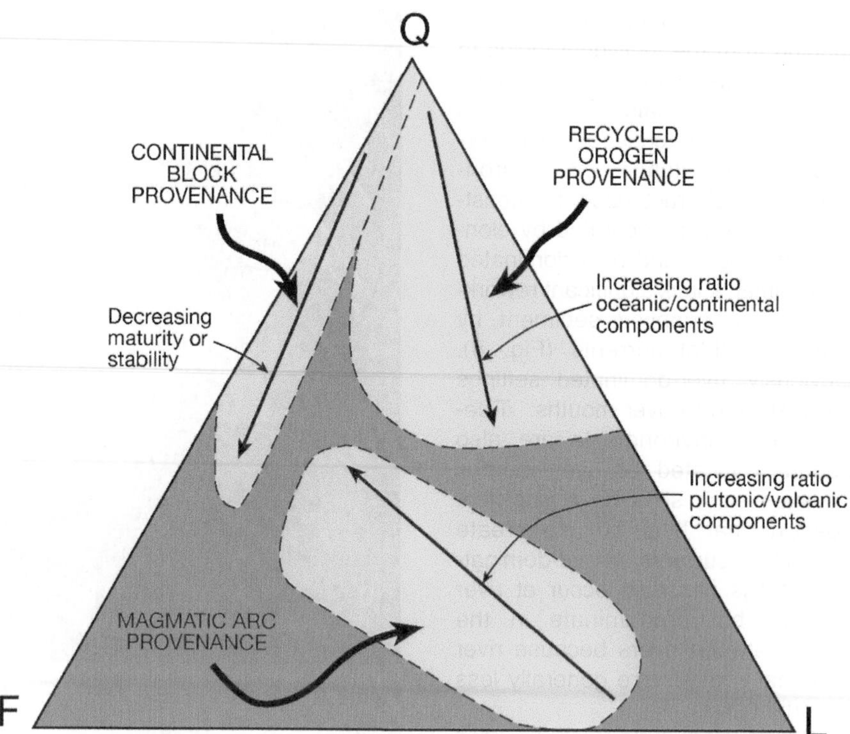

Figure 9. Triangular plot showing how tectonic setting influences the composition of sand-sized sediment. Q = quartz; F = feldspar; L = lithic (rock) fragments. Modified after Dickinson and Suczec (1979, Figure 1). AAPG©; reprinted by permission of AAPG whose permission is required for further use.

A full treatment of the diverse array of tectonic settings is beyond the scope of this chapter and readers are referred to Allen and Allen (2005) and Busby and Ingersoll (1995) for more detailed treatments.

Tectonically active areas such as collisional, rifted and strike-slip margins typically have short, steep fluvial catchments (Fig. 10A; Busby and Ingersoll, 1995; Leeder and Gawthorpe, 1987; Milliman and Syvitski, 1992), and the transfer of sediment to depositional sites tends to be rapid with little physical and chemical alteration. As a result, the sediment reflects the composition of the source area better, and is less compositionally mature, than that which accumulates in passive continental margins, intracratonic basins and, to a lesser extent, foreland basins (Dickinson and Suczec, 1979; Dickinson, 1988). Tectonically active areas supply large quantities of sediment relative to the size of the drainage basins (Milliman and Syvitski, 1992). Southeast Asia, in particular, with only 2% of the world's land surface, supplies 20–25% of all sediment to the global ocean. Thus, sedimentation is typically rapid, and extreme sedimentation events (*e.g.*, flash floods, volcanic eruptions and landslides) are much more common than in tectonically stable settings. Because of their small size and steep gradient, downstream fining is limited to the loss of the unstable components, and relatively coarse sediment is delivered to the depositional area. By contrast, there may be very rapid passage from negative to positive accommodation at the margin of the depositional area (this is most pronounced in rift and strike-slip basins) so that very rapid grain size and facies changes occur (*e.g.*, Larsen and Steel, 1978; Fig. 10A). Rapid subsidence in the depositional area commonly generates more complete successions than occur in areas with lower subsidence rates, and can cause the deposits of any depositional environment to form thick, vertically stacked successions because subsidence is able to keep pace with sedimentation (*e.g.*, Steel *et al.*, 1977; Clifton, 1981).

By comparison, passive continental margins and intracratonic basins (Fig. 10C; Busby and Ingersoll, 1995; Runkel *et al.*, 2007) typically have source areas with low relief, or, if there are areas of high relief in the headwaters of the drainage basins, they are far removed from the depositional area. Thus, the residence time of the sediment within the zone of weathering and transport is long, leading to the production of sediment that is compositionally mature (Dickinson and Suczec, 1979; Dickinson, 1988), something that is aided by the granitic/gneissic composition of the continental crust over which the rivers flow. Because there is usually little vertical movement within passive-margin and intracratonic basins, there may be only relatively little downstream fining because of the lack of accommodation (Fig. 10C). This is especially true because subsidence generally decreases toward the source area. This situation is in direct contrast with foreland basins, where the production of accommodation is greatest close to the source, which leads to more rapid down-system fining than is to be expected on passive margins (Fig. 10B).

Although almost any depositional environment can occur in any tectonic setting, the steep gradients of tectonically active areas make it more likely that alluvial fans and coarse-grained deltas are developed than in low-gradient settings (Gawthorpe and Colella, 1990). On collisional margins that face an open ocean (*e.g.*, the west coast of North and South America), coastal and shallow-marine environments tend to be wave dominated because shelves are narrow, a situation that allows oceanic waves to reach the coast with limited dissipation and mitigates against significant tidal amplification (Yoshida *et al.*, 2007). By contrast, rifted, strike-slip and foreland basins that are open to the ocean at one end (*e.g.*, the Bay of Fundy, the Gulf of California, the Persian Gulf and the Gulf of Papua) are favored sites for the development of tide-dominated environments (*see* Chapter 9) because of their funnel-shaped geometry. The narrowness of the shelf in all tectonically active settings makes it easy for sediment to be transferred to the basin-margin slope and beyond. Thus, deep-water deposits are abundant, and have a greater potential to occur at any time over a relative sea-level cycle than is the case in areas with lower gradients and broader shelves (Covault *et al.*, 2007). In the later situation, deep-water sedimentation is more likely to be confined to lowstands, as indicated in the standard sequence-stratigraphic model (*see* Chapters 2 and 12).

Sedimentary basins flanked by extensive, low-gradient continental areas (*e.g.,* passive continental margins, intracratonic basins and marginal seas) typically have most of their sediment delivered by a few large rivers (Milliman and Syvitski, 1992). Modern examples include the Amazon, Changjiang and Mississippi rivers. These large rivers tend to be fixed in location by structural features, and hence occupy one location for long periods of time. As a result, delta-related deposits are especially prominent in such areas. A mixture of wave- and tide-dominated coastal environments is present because an increase in the width of

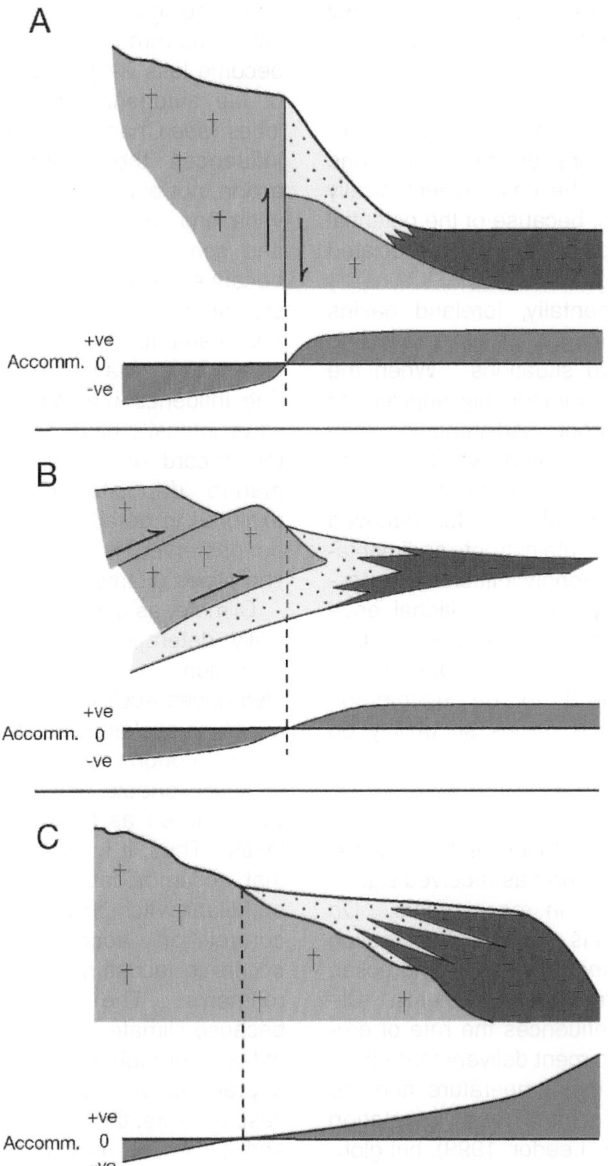

Figure 10. Schematic sections of **A)** a rifted or strike-slip basin, **(B)** a foreland basin, and **(C)** a passive continental margin, showing the spatial change in the development of accommodation. Sediment grain sizes will be influenced as shown in Figure 7.

the continental shelf enhances tidal action while damping waves because of frictional dissipation in shallow water (Yoshida *et al.*, 2007). Deep-water deposits are commonly organized into more discrete submarine fans than is the case in tectonically active areas, because of the potential for fewer point sources associated with the large rivers.

Environmentally, foreland basins may be thought of as alternating between two situations. When the shoreline is immediately adjacent to the thrust front, sedimentation may have more in common with tectonically active basins. On the other hand, when the shoreline is far removed from the mountain front, sedimentation is less strongly influenced by tectonic activity and depositional environments are more similar to other low-gradient basins. Deep-water deposits tend to be uncommon because such basins sit mainly on continental crust.

Climate

The influence of climate on siliciclastic sedimentation has received significant attention in recent years. On land, climate is crucial for the creation of glacial (*see* Chapter 5) and eolian (*see* Chapter 7) environments. Climate also influences the rate of erosion and sediment delivery through its control on the temperature and the amount of runoff and vegetation (Allen, 1997; Leeder, 1999), but globally the relief of the source is the single most important factor controlling sediment discharge (Milliman and Syvitski, 1992). Nevertheless, all else being equal, the sediment yield is lowest in areas with a moderate amount of precipitation and extensive vegetation coverage (Langbein and Schumm, 1958). Cecil and Dulong (2003) have also suggested that sediment yield is highest in areas that have well-developed seasonality with episodic but intense rainfall. The variations in sediment supply produced by climatic changes can cause rivers to incise or aggrade, and can produce an alternation in fluvial style between braided and meandering (Blum and Törnqvist, 2000; *see* Chapter 6). Climate also has a direct impact on the nature of paleosols.

In coastal and shallow-marine settings, the impact of sediment-supply variations remains important, but can become less easy to detect because of the autogenic switching of delta lobes (*see* Chapter 10). Climate also influences the intensity of wave action, not only because of latitudinal variations in the intensity, direction and seasonality of winds, but also because of temporal variations in storminess on time scales ranging from centuries to millennia (Billeaud *et al.*, 2007; Mayewski *et al.*, 2004). The influence that such changes in wave intensity have on the sedimentary record of coastal and shallow-marine deposits remains to be explored in detail. Their potential to be preserved is high in prograding shorelines (Hampson, 2000).

Climate, as governed by astronomically determined changes in solar insolation (*i.e.*, Milankovitch cyclicity), also drives eustatic sea-level change over time scales of less than a million years, through its control on the budget of ice sheets and the amount of water stored as groundwater and in lakes. Thus, it is widely appreciated that accommodation changes occur on Milankovitch time scales. Climatic control on accommodation also occurs in lacustrine and fluvial environments. The former is obvious because climate determines the level of lakes through variations in the ratio of precipitation to evaporation. In the case of rivers, the discharges of water and sediment may not change by comparable amounts as the precipitation-to-evaporation ratio changes. For example, an increase in water discharge might not be accompanied by a corresponding increase in sediment discharge, in which case the now more energetic river will be below capacity and will erode, whereas a decrease in water discharge without a comparable drop in the sediment discharge causes aggradation (Blum and Törnqvist, 2000). Thus, climate controls the location of the fluvial equilibrium profile: changes in climate could, therefore, create a stratigraphic expression that would mimic the results of a eustatic or tectonically caused rise or fall of the river mouth (*i.e.*, of base level). In the deep sea, climate influences sedimentation indirectly by means of the eustatic control on the location of the shoreline and, hence, on the delivery of sediment: deep-water accumulation is typically greatest during late falling-stage and early lowstand when river mouths are at the shelf edge although exceptions occur (*see* Chapter 12). Finally, climate also controls the intensity of the deep thermohaline circulation, which is reflected in the nature of contour-current deposits (Stow *et al.*, 1986)

Biological Evolution

Although the influence of plants and animals is not as important in siliciclastic sedimentation as it is for biochemical sediments, life does play a significant role at all stages in the erosion, transport and deposition of sediment. For example, it mediates weathering and the erosion of sediment as discussed above; it promotes deposition by the binding action of microbial mats and the trapping of particles by plants; and bioturbation overprints physical sedimentary structures and creates a wealth of new information about depositional conditions (*see* Chapter 3). Despite this, relatively little attention has been paid to how the major steps in the evolution of life have changed the nature of siliciclastic sediments. In general, the age of a deposit is taken to be of minimal importance in its interpretation, with facies models derived from the modern, or the Cretaceous, being applied routinely to rocks ranging from the Archean to the present (Donaldson *et al.*, 2002; Runkel *et al.*, 2007).

One of the two biological milestones that has received attention is the obvious impact that the appearance and evolution of crawling and burrowing organisms have had on sedimentary facies (*e.g.*, Sielacher *et al.*, 2005). In the Precambrian, the absence of bioturbation means that even the finest physical sedimentary structures are preserved intact. Less well understood, however, is the impact that the appearance of grazers had on the distribution of microbial mats: throughout the Precambrian, microbial mats must have covered most submarine surfaces, whereas during the Phanerozoic they were restricted to environments hostile to metazoans. The influence that the stabilizing influence of these mats had on siliciclastic sedimentation has not been studied in detail (*cf.* MacNaughton *et al.*, 1997).

The second evolutionary milestone of particular importance is the appearance of land plants (*see* review in Chapter 6). As plants slowly spread from low-lying areas into progressively higher and drier locations, they must have had a significant impact on the flashiness of fluvial discharge and the rate of sediment supply: vegetation tends to minimize variability by impeding runoff and binding the soil. It has been suggested that this has had a dramatic affect on the nature of fluvial sedimentation, with braided rivers predominating before the advent of land vegetation (Sønderholm and Tirsgaard, 1998). Terrestrial vegetation may have also caused a significant decrease in the effectiveness of wind as a sediment-transporting agent (Dalrymple *et al.*, 1985). Such ideas remain speculative and deserve additional study.

SUMMARY

The siliciclastic facies models presented in the subsequent chapters encapsulate, first and foremost, the autogenic physical processes that operate in each environment; a thorough understanding of these processes is fundamental to undertaking good paleo-environmental interpretations. Because these processes are governed by the laws of physics, they operate in broadly the same way regardless of the age of the deposit and its accommodation, tectonic or climatic setting. It is this 'universality' that allows facies models to be created and used. Our understanding of how the allogenic factors influence the way in which environments behave has increased dramatically over the last two decades. We have learned that the accommodation regime, tectonic setting, climate and the nature of the biota can have an important influence on which physical processes operate in a given environment and/or on the relative intensity of these processes. The allogenically determined context also exerts a strong influence on the grain size that is available for deposition, which, in turn, influences the nature of the physical sedimentary structures produced. The preservation potential of various sub-environments is also determined primarily by the allogenic factors such as accommodation and the tectonic regime.

As a result of this allogenic modulation of environmental behavior, no one facies model, or even a small set of them, can capture the full variability that is possible. The models provided in the following chapters represent the best possible generalizations that can be provided with our current understanding. They incorporate significant refinements in our knowledge of both autogenic and allogenic processes relative to what was presented in the previous edition of this book, and the models will continue to evolve as more research is done.

ACKNOWLEDGEMENTS

This chapter has been improved by the constructive comments provided by Noel James, Dale Leckie, Duncan MacKay, Guy Plint and Ron Steel.

REFERENCES

Basic sources of information

Allen, P.A., 1997, Earth Surface Processes: Oxford, Blackwell Science, 404 p.
The most comprehensive coverage available of the processes that operate on the Earth's surface, from weathering to orogenesis.

Allen, P.A. and Allen, J.R., 2005, Basin Analysis: Principles and Applications: Oxford, Blackwell Publishing, 549 p.
A thorough review of the dynamics of the Earth's crust and the origin of sedimentary basins.

Busby, C.J. and Ingersoll, R.V., 1995, Tectonics of Sedimentary Basins: Oxford, Blackwell Science, 579 p.
Excellent coverage of all major basin types; less mathematical than Allen and Allen (2005).

Catuneanu, O., 2006, Principles of Sequence Stratigraphy: Amsterdam, Elsevier, 375 p.
A state-of-the-art examination of sequence stratigraphy, both in general and as applied to all sedimentary environments.

Frings, R.M., 2008, Downstream fining in large sand-bed rivers: Earth-Science Reviews, v. 87, p. 39-60.
An excellent review of the processes responsible for one of the most overarching controls on siliciclastic sedimentary environments.

Leeder, M., 1999, Sedimentology and Sedimentary Basins–From Turbulence to Tectonics: Blackwell Science, Oxford, 592 p.
Comprehensive process-oriented coverage of weathering, sediment transport and deposition, and the dynamics of sedimentary environments, with a brief section on sedimentary basins.

Middleton, G.V. and Southard, J.B., 1984, Mechanics of Sediment Movement, 2nd edition: SEPM Short Course Notes, No. 3, variously paginated.
This remains the best overview of the detailed dynamics of sediment movement.

Posamentier, H.W. and Walker, R.G., 2006, Facies Models Revisited: SEPM Special Publication 84, 532 p. (CD only)
A more in-depth review of siliciclastic sedimentary environments than the present volume.

Southard, J.B., 1991, Experimental determination of bed-form stability: Annual Review of Earth and Planetary Science, v. 19, p. 423-455.
The best single review of the factors controlling the bedforms generated by currents and waves.

Tucker, M.E., 2001, Sedimentary Petrology: An Introduction to the Origin of Sedimentary Rocks, 3rd edition: Oxford, Blackwell Science, 262 p.
Very good coverage of all major sedimentary rock types.

Other references

Abbott, P.L. and Peterson, G.L., 1978, Effects of abrasion durability on conglomerate clast populations: Examples from Cretaceous and Eocene conglomerates of the San Diego area, California: Journal of Sedimentary Petrology, v. 48, p. 31-42.

Billeaud, I., Tessier, B., Lesueur, P. and Caline, B., 2007, Preservation potential of highstand coastal sedimentary bodies in a macrotidal basin: Example from the Bay of Mont-Saint-Michel, NW France: Sedimentary Geology, v. 202, p. 754-775.

Blum, M. and Tørnqvist, T.E., 2000, Fluvial responses to climate and sea-level change: a review and look forward: Sedimentology, v. 47, Suppl. 1, p. 2-48.

Boyd, R., Dalrymple, R.W. and Zaitlin, B.A., 1992, Classification of clastic coastal depositional environments: Sedimentary Geology, v. 80, p. 139-150.

Cecil, B. and Dulong, F.T., 2003, Precipitation models for sediment supply in warm climates, *in* Cecil, B. and Edgar, N.T., *eds.*, Climate Controls on Stratigraphy: SEPM, Special Publication 77, p. 21-27.

Clifton, H.E., 1981, Progradational sequences in the Miocene shoreline deposits, southeastern Caliente Range, California: Journal of Sedimentary Petrology, v. 51, p. 165-184.

Covault, J.A., Normark, W.R., Romans, B.W. and Graham, S.A., 2007, Highstand fans in the California borderland:

The overlooked deep-water depositional systems: Geology, v., 35, p. 783-786.

Cummings, D.I., Dumas, S. and Dalrymple, R.W., 2009, Fine-grained versus coarse-grained wave ripples generated experimentally under large-scale oscillatory flow: Journal of Sedimentary Research, v. 79, p. 83-93.

Dalrymple, R.W., Narbonne, G.M. and Smith, L., 1985, Eolian action and the distribution of Cambro-Ordovician shales in North America: Geology, v. 13, p. 607-610.

Dickinson, W.R., 1988, Provenance and sediment dispersal in relation to paleotectonics and paleogeography of sedimentary basins, *in* Kleinspehn, K.L. and Paola, C. *eds*., New Perspectives in Basin Analysis: New York, Springer-Verlag, p. 3-25.

Dickinson, W.R. and Suczec, C.A., 1979, Plate tectonics and sandstone compositions: American Association of Petroleum Geologists Bulletin, v. 63, p. 2164-2182.

Donaldson, J.A., Eriksson, P.G. and Altermann, W., 2002, Actualistic versus non-actualistic conditions in the Precambrian sedimentary record: reappraisal of an enduring discussion, *in* Altermann, W. and Corcoran, P.L., *eds*., Precambrian Sedimentary Environments: A Modern Approach to Ancient Depositional Systems: International Association of Sedimentologists, Special Publication 33, p. 3-13.

Gawthorpe, R.L. and Colella, A., 1990, Tectonic controls on coarse-grained delta depositional systems in rift basins, *in* Colella, A. and Prior, D.B., *eds*., Coarse-Grained Deltas: International Association of Sedimentologists, Special Publication 10, p. 113-128.

Hampson, G.J., 2000, Discontinuity surfaces, clinoforms, and facies architecture in a wave-dominated shoreface-shelf parasequence: Journal of Sedimentary Research, v. 70, p. 325-340.

Harris, P.T., Heap, A.D., Bryce, S.M., Porter-Smith, R., Ryan, D.A. and Heggie, D.T., 2002, Classification of Australian clastic coastal depositional environments based upon a quantitative analysis of wave, tidal, and river power: Journal of Sedimentary Research, v. 72, p. 858-870.

Hiscott, R.N., 1994, Loss of capacity, not competence, as the fundamental process governing deposition from turbidity currents: Journal of Sedimentary Research, v. 64A, p. 209-214.

Langbein, W.B. and Schuum, S.A., 1958, Yield of sediment in relation to mean annual precipitation: American Geophysical Union, Transactions, v. 39, p. 1076-1084.

Larsen, V. and Steel, R.J., 1978, The sedimentation history of a debris-flow dominated, Devonian alluvial fan—a study in textural inversion: Sedimentology, v. 25, p. 37-59.

Leeder, M.R. and Gawthorpe, R.L., 1987, Sedimentary models for extensional tilt-block/half-graven basins, *in* Coward, M.P., Dewey, J.F. and Hancock, P.L., *eds*., Continental Extensional Tectonics: Geological Society of London, Special Publication 28, p. 139-152.

MacNaughton, R.B., Dalrymple, R.W. and Narbonne, G.M., 1997, Early Cambrian braid delta deposits, Mackenzie Mountains, northwestern Canada: Sedimentology, v. 44, p. 587-609.

Mayewski, P.A., Rohling, E.E., Stager, J.C., Karlen, W., Maasch, K.A., Meeker, L.D., Meyerson, E.A., Gasse, F., van Kreveld, S., Holmgren, R., Lee-Thorp, J., Rosqvist, G., Rack, F., Staubwasser, M., Schneider, R.R. and Steig, E.G., 2004, Holocene climate variability: Quaternary Research, v. 62, p. 243-255.

Milliman, J.D. and Syvitski, J.P.M., 1992, Geomorphic/tectonic control of sediment discharge to the world ocean: The importance of small mountainous rivers: Journal of Geology, v. 100, p. 525-544.

Muto, T., Steel, R.J. and Swenson, J.B., 2007, Autostratigraphy: A framework norm for genetic stratigraphy: Journal of Sedimentary Research, v. 77, p. 1-12.

Orton, G.J. and Reading, H.G., 1993, Variability of deltaic processes in terms of sediment supply, with particular emphasis on grain size: Sedimentology, v. 40, p. 475-512.

Parker, G., 1991, Selective sorting and abrasion of river gravel. I: Theory: Journal of Hydraulic Engineering, v. 117, p. 131-149.

Pettijohn, F.J., 1957, Sedimentary Rocks, 2nd edition: New York, Harper and Brothers, 718 p.

Posamentier, H.W. and Morris, W.R., 2000, Aspects of the architecture of forced regressive deposits, *in* Hunt, D. and Gawthorpe, R.L., *eds*., Sedimentary Responses to Forced Regressions: Geological Society of London, Special Publication 172, p. 19-46.

Potter, P.E., Maynard, J.B. and Depetris, P.J., 2005, Mud and Mudstones, Introduction and Overview: New York, Springer, 297 p.

Reading, H.G., *ed*., 1996, Sedimentary Environments: Processes, Facies and Stratigraphy, 3rd edition: Oxford, Blackwell Science, 688 p.

Runkel, A.C., Miller, J.F., McKay, R.M., Palmer, A.R. and Taylor, J.F., 2007, High-resolution sequence stratigraphy of lower Paleozoic sheet sandstones in central North America: The role of special conditions of cratonic interiors in development of strata architecture: Geological Society of America Bulletin, v. 119, p. 860-881.

Seal, R., Toro-Escobar, C., Cui, Y., Paola, C., Parker, G., Southard, J.B. and Wilcock, P.R., 1998, Downstream fining by selective deposition: Theory, laboratory, and field observations, *in* Klingeman, P.C., Beschta, R.L., Komar, P.D. and Bradley, J.B., *eds*., Gravel-Bed Rivers in the Environment: Highlands Ranch, CO: Water Resources Publications, LLC, p. 61-84.

Seilacher, A., Buatois, L.A. and Mangano, M.G., 2005, Trace fossils in the Ediacaran-Cambrian transition; behavioural diversification, ecological burnover and environmental shift: Palaeogeography, Palaeoclimatology, Palaeoecology, v. 227, p.323-356.

Shanley, K. W., and McCabe, P. J., 1994, Perspectives on the sequence stratigraphy of continental strata: American Association of Petroleum Geologists Bulletin, v. 78, p. 544-568.

Sønderholm, M. and Tirsgaard, H., 1998, Proterozoic fluvial styles: response to changes in accommodation space (Rivieradal sandstones, eastern North Greenland): Sedimentary Geology, v. 120, p. 257-274.

Southard, J.B. and Boguchwal, L.A., 1990, Bed configurations in steady unidirectional water flows. Part 1: Synthesis of flume data: Journal of Sedimentary Research, v. 60, p. 658-679.

Steel, R.J., Mæhle, S., Nilson, H., Røe, S.L. and Spinnangr, Å., 1977, Coarsening-upward cycles in the alluvium of Hornelen Basin (Devonian) Norway: Sedimentary response to tectonic events: Geological Society of America Bulletin, v. 88, p. 1124-1134.

Stow, D.A.V., Faugères, J.-C., and Gonthier, E. 1986. Facies distribution and textural variation in Faro Drift contourites: velocity fluctuation and drift growth: Marine Geology, v. 72, p. 71-100.

Yalin, M.S., 1977, Mechanics of Sediment Transport, 2nd edition: New York, Pergamon Press, 298 p.

Yoshida, S., Steel, R. and Dalrymple, R.W., 2007, Depositional process changes – An ingredient in a new generation of sequence-stratigraphic models: Journal of Sedimentary Research, v. 77, p. 447-460.

5. GLACIAL DEPOSITS

Carolyn H. Eyles, School of Geography and Earth Sciences, McMaster University, Hamilton, ON, L8S 4K1, Canada

Nick Eyles, Department of Geology, University of Toronto, Toronto, ON, M5S 3B1, Canada

INTRODUCTION

The *cryosphere* is the name given to the approximately one fifth of the Earth's surface affected by the freezing of water. The cryosphere includes snowfields, valley glaciers, ice caps, ice sheets, floating ice such as icebergs, ice that forms on the surfaces of lakes, rivers and seas, and 'ground ice' that forms beneath landscapes frozen year-round (*permafrost*). The cryosphere has repeatedly expanded to cover *one third* of the global land area during the Pleistocene ice ages of the last two million years. During major glaciations, floating ice shelves and icebergs reached far onto continental shelves, where they influenced deep-marine environments, and changed ocean circulation by the release of huge volumes of meltwater (Benn and Evans, 1998; Dowdeswell and O'Cofaigh, 2002). Global sea level rose and fell as ice sheets waxed and waned, and influenced coastal evolution worldwide. In the more remote past, the Earth experienced six major intervals of glaciation when ice was present for tens of millions of years (*glacio-epochs;* Eyles, 2008). The earliest known glaciers formed about 2.8 billion years ago and some glaciations (those between roughly 750 and 600 million years ago) might have been so severe that they affected the entire planet (Fairchild and Kennedy, 2007; Hoffman, 2008)!

Understanding the formation and characteristics of glacial sediments has important and practical applications in northern regions such as Canada. These sediments underlie many large urban centers and contain aquifers that supply drinking water to millions of people. Groundwater exploration and management programs, investigations for waste-disposal sites, aggregate-resource mapping, and the cleanup of contaminated sites all require knowledge of the subsurface geology of glaciated terrains (Meriano and Eyles, 2009). The mineral-rich Precambrian shields of the northern landmasses are covered by extensive sheets of glacial sediment, and knowledge of ice dynamics and sedimentology is needed to locate economically valuable mineral resources, such as gold and diamonds, which lie, buried, beneath the cover of glacial deposits. The search for shallow gas, trapped in Pleistocene glacial sediments in Alberta, and for oil, coal and gas in older Paleozoic glacial strata in Brazil, Australia and India, has emphasized the importance of glacial sedimentology to energy exploration. Glacial sedimentology plays an important role in many other applications of environmental geology, such as in urban areas, in seismic-risk assessment and geological engineering, and is increasingly integrated with other disciplines such as geophysics.

Recent Developments

A major shift in focus has occurred since *Facies Models* was last published in 1992. Then, glacial facies modeling and knowledge of glacial processes was dominated by studies at modern glaciers flowing on hard rock (Fig. 1). In contrast, large Pleistocene ice sheets (and parts of today's Antarctic Ice Sheet) flowed across soft beds of wet sediment. Deformation and mixing of this sediment is now known to be an important process resulting in the formation of poorly sorted till (Boulton *et al.*, 2001; Evans *et al.*, 2006).

Also, there have been major advances in quantifying rates of glacial erosion (and thus landscape modification) as a consequence of analyses of cosmogenic isotopes and thermochronometry. The flux of glacial sediment from glaciated basins is understood better, and it is now known that significant chemical weathering can also take place in cold environments. Offshore, much has been learned about how ice sheets deposit sediment underwater on continental shelves and slopes (Boulton *et al.*, 1996; Domack *et al.*, 1999; Eyles *et al.*, 2001; Dowdeswell and O'Cofaigh, 2002; Heroy and Anderson, 2005). This knowledge has arisen as a consequence of oil and gas exploration, ocean drilling of deep-sea sediments to obtain climate records, and geophysical mapping of northern seafloors.

GLACIAL SEDIMENTARY ENVIRONMENTS

The glacial environment is one of the more difficult to summarize because glaciers can affect depositional processes both on land (*glacioterrestrial*) and offshore (*glaciomarine*), and there are many sub-environments within each of these settings (Figs. 2 and 3). In addition, the growth and decay of ice sheets gives rise to rapid time-transgressive deposition, commonly complicated by later reworking of deposits by marine and fluvial action. A broad *periglacial* zone surrounds ice sheets where it is too dry or slightly too warm for glaciers to grow. In this zone, freeze–thaw cycles and the deep-freezing of groundwater to form *ground ice* dominate sedimentary processes; there is also considerable potential for eolian processes to transport and deposit sediment such as loess (*see* Chapter 7).

Also, ice sheets expanded onto continental shelves when the sea level was lowered during glaciation. Much glacial sediment is ultimately preserved in deep-water continental-slope successions as debrites, muddy contourite drifts, and turbidites (Hooke and Elverhoi, 1996; Sejrup *et al.,* 2005; Tripsanas and Piper, 2008; *see* Chapter 12), and as horizons of ice-rafted debris in the deep ocean

Figure 1. A. Canada's most familiar glacier, the Athabasca Glacier, on the Icefields Parkway in Alberta. Most early glacial facies models were derived from study of easily accessible glaciers such as this one, which flows over bedrock. Pleistocene continental ice sheets behaved differently because they flowed over thick sediment. **B.** A geologist lying on bedrock looking up at the dirty base of a glacier flowing left to right (Glacier du Bosson, French Alps). The glacier is carrying debris within the basal ice (as englacial load). Observations at numerous glaciers show they transport very little englacial debris. The most effective means of moving sediment is where glaciers rest on soft beds composed of sediment that can be deformed and moved as the glacier flows (*see* Fig.5).

Figure 2. Much sediment in glaciated areas, such as the Copper River Valley, Alaska, shown here, is moved and deposited not just by glacial processes *per se*. A braided meltwater river leaving the ice front has reworked almost all primary glacial sediment such as tills and glacial landforms. Small lakes add further variability. In the lower part of the image braided and anastomosed rivers co-exist (*see* Chapter 6). Close to the ice front, debris is reworked by gravity and slumping off steep mountain slopes. Eolian activity is significant and small dunes are forming. Most of the sediment load is transported to the ocean downstream.

(Andrews, 1998). New information regarding glacially influenced deposition along continental margins is the key to understanding pre-Pleistocene glaciations, for which the sedimentary record is mainly preserved in marine basins (Eyles, 1993).

THE GLACIER SYSTEM

Glacial ice forms when snow accumulates and, at depth, undergoes repeated cycles of partial melting, refreezing and recrystallization. *Firn* is the material that forms at an intermediate stage between snow and ice, and has a density greater than 0.5 g/cm^3 (Fig. 4A). Glacial ice is formed with a density of 0.9 g/cm^3 with further burial and recrystallization. This process occurs within a few years in temperate areas, but takes many hundreds of years in the much colder and dryer Antarctic (Benn and Evans, 1998).

The formation of glacier ice takes place in the *accumulation zone* of a glacier or ice sheet (Fig. 4B). There, the mass of ice gained each year is greater than that lost by melting. At lower elevations and under warmer temperatures, glacier ice melts at greater rates than it is formed and the glacier loses mass. This area is called the *ablation zone*. The point on a glacier where there is neither gain nor loss of mass is termed the *equilibrium line* and its position can be approximated by the position of the snow line visible on a glacier at the end of the summer melt season.

Transfer of ice between the accumulation zone of a glacier and the ablation zone occurs through the process of *creep* or deformation. Glacier ice moves essentially under the influence of gravity in response to both vertical (compressive) and shear stresses. The rate of glacier movement is mostly dependent on the surface slope of the glacier, the thickness of the ice (shear stresses and rates of ice movement increase as ice thickness increases), and ice temperature ('warm' ice close to the melting point can deform and move much more rapidly than 'cold' ice). The *thermal regime* of a glacier is a description of the temperature of the ice, which affects not only the rate of movement but also the capacity of the ice to erode, transport and deposit sediment. *Cold-based* glaciers are typical of cold, high-latitude regions (*e.g.*, Antarctica), where the temperature at

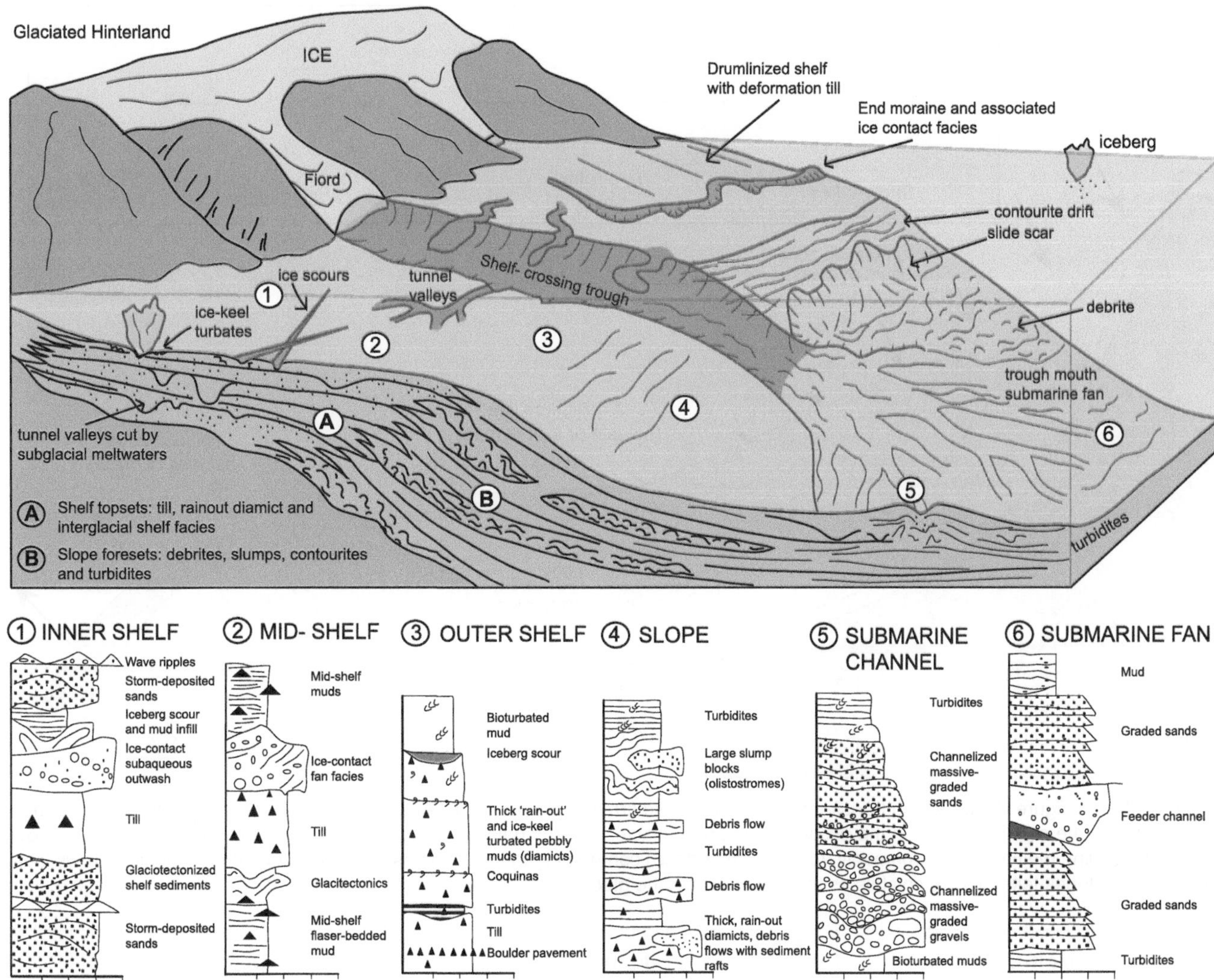

Figure 3. This figure illustrates a typical glaciated continental margin showing the principal glaciomarine environments and representative vertical profiles through sediment accumulating in these environments. Pleistocene glaciations have left a prominent glacial record on land but most sediment (perhaps as much as 90%) is deposited offshore on continental slopes, especially on trough-mouth fans. Glaciomarine deposits dominate the record of older glaciations in Earth history because the terrestrial record is easily eroded.

the base of the ice is well below the pressure melting point (*i.e.*, the temperature at which melting occurs, at the pressure present at the base of the glacier) and there is no water present. These glaciers typically move very slowly by internal deformation (*creep*) at rates of only a few meters per year and are ineffective in eroding bedrock. As a consequence, cold-based glaciers cannot create or move much sediment and are ineffective geomorphic agents. In warm and moist climates, such as those found in Alaska or the Canadian Rockies, ice is close to the pressure melting point (just below zero degrees centigrade) and moves by a combination of creep and by sliding over films of water at the ice base (Fig. 5); some refreezing of this water occurs (*regelation*) in the lee of bedrock irregularities creating an effective method for incorporating ('freezing on') debris into the ice base (Boulton, 1996). This debris is carried within a thin basal debris layer (usually less than 1m thick) that consists of irregular layers of ice and sediment (Fig. 1B). Armed with debris, rapidly flowing, warm-based ice (moving up to 250 m yr^{-1}) is highly abrasive and can readily carve into bedrock and transport large amounts of freshly broken glaciclastic sediment. This sediment may be carried away by subglacial rivers or may be transported within the ice as *englacial load* (Fig. 5A), or at the ice base as the *basal traction load* (Fig. 5B). Pre-existing or previously deposited sediment can also be transported below the ice base as a *subglacial deforming layer* (Fig. 5C). Observations at modern glaciers suggest that basal thermal conditions and sediment transport mechanisms beneath large ice masses such as continental-scale ice sheets are likely to be highly complex, with both spatial and temporal variability (Clarke, 2005).

Glacial Erosion: Processes and Products

Debris incorporated into the basal traction zone of a warm-based glacier can abrade underlying bedrock and produce smoothed substrates ornamented with features such as

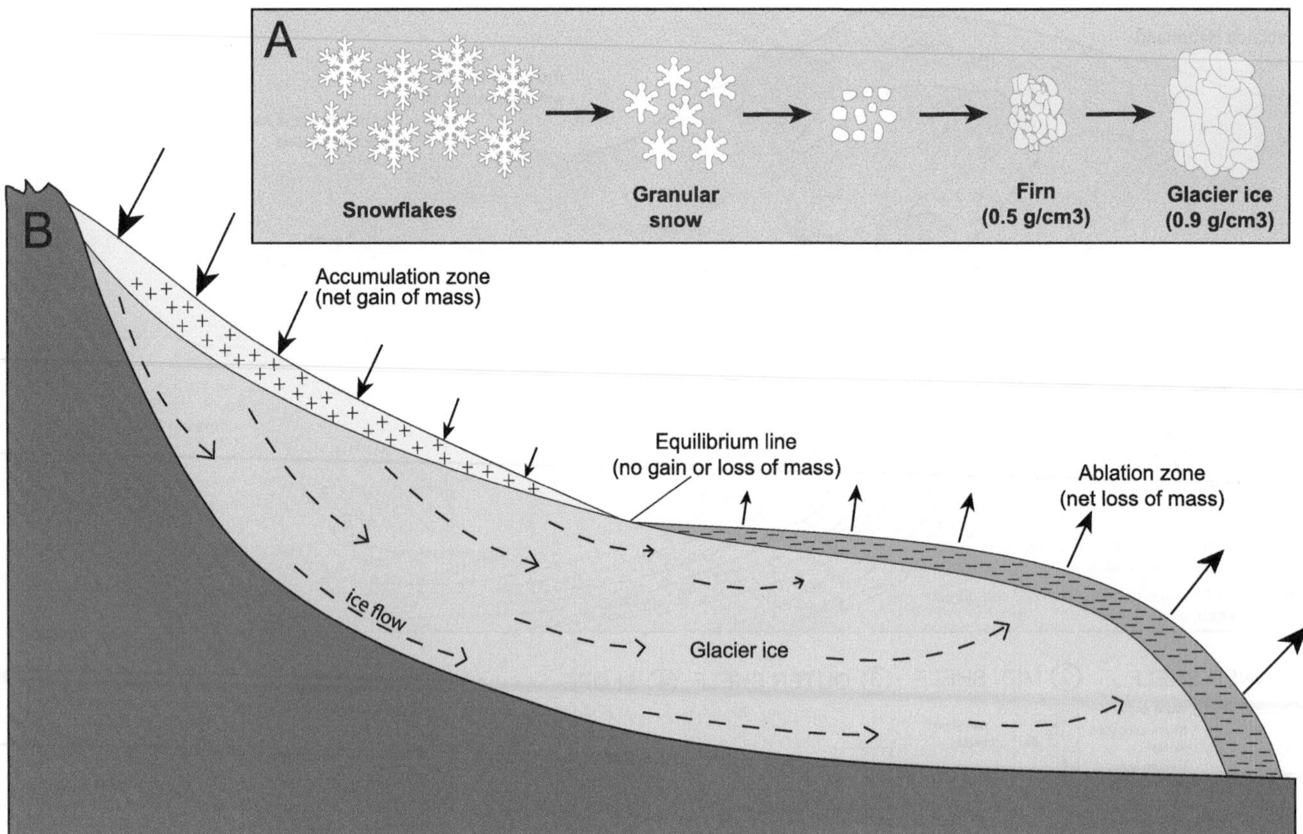

Figure 4. A. This figure illustrates the transformation of snow into firn and glacier ice. **B.** Anatomy of a valley glacier at the end of a melt season. Above the equilibrium line in the accumulation zone, glacier mass is gained through the addition of snow, firn and ice. Below the equilibrium line melting occurs (primarily from the ice surface) contributing to loss of mass in the ablation zone. Dashed arrows indicate flow trajectories taken by ice crystals as they move within the glacier.

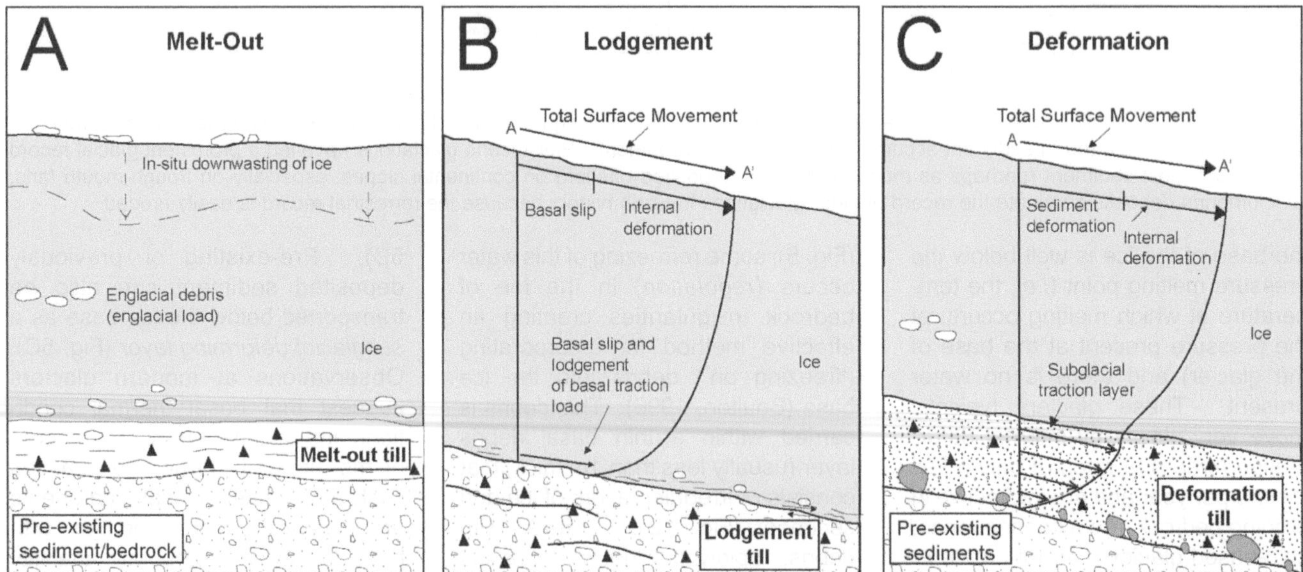

Figure 5. Till is produced in three principal ways. **A.** By basal melt-out under stagnant ice; **B.** By basal melt-out and lodgement onto a rigid substrate under flowing ice; and **C.** By the subglacial deformation of pre-existing sediment. The latter is the most effective in producing thick and extensive tills.

Figure 6. Features of glacial erosion. **A.** Striations and larger gouges on a glacially smoothed and abraded surface, northern Ontario. **B.** Series of crescentic chattermarks record former ice flow toward camera. Whitefish Falls, Ontario.

striations, gouges, grooves and chattermarks (Fig. 6; Benn and Evans, 1998; Hildes *et al.*, 2004). The process of freezing-on of subglacial meltwater in cavities at the ice base can also detach and incorporate loose blocks of sediment and bedrock into the basal traction layer, a process (called *plucking*) that creates a blunt or 'chopped off' down-ice end to otherwise smoothed bedrock features. The asymmetric low-relief landforms produced by the combined processes of abrasion and plucking are termed *roche moutonnée* (sheep rocks) and are used to indicate former ice-flow directions (Fig. 7A).

Debris carried within the basal traction layer of a glacier is also abraded, and shaped by abrasion and plucking processes during transport resulting in the streamlining and shaping of clasts (*e.g.*, *bullet-shaped boulders*; Fig. 7B) and the production of large quantities of silt (*glacial rock flour*).

Glacial Deposition: Processes and Products

A Note on Terminology: Tills and Diamicts

The most common sediment type ascribed to glacial processes is an extremely poorly sorted deposit composed of mud, sand, gravel and boulders (Fig. 8). However, the interpretation and even naming of such heterogeneous sediment is a challenge (Eyles *et al.*, 1983). Glaciers are able to transport and deposit poorly sorted admixtures of clasts (any particle larger than sand size) and matrix (predominantly mud) in a number of ways. Like concrete, this debris can be mechanically mixed as it is 'bulldozed' beneath or in front of the glacier; it can be carried *englacially* within the ice itself, or on its surface as *supraglacial* debris. This poorly sorted material was first identified in the early 1800s and given the name 'till' ('stony ground' in Scottish) but its origin was not immediately appreciated. The realization that till is a glacial deposit, extends over thousands of square kilometers of the mid-latitudes, and contains far-traveled boulders (*erratics*) was proposed in 1837 by Louis Agassiz. His proposal, that it had been deposited by large continental ice sheets that existed in the past, marked the dramatic beginning of paleoclimatology. Ancient *tillites* (lithified tills) bearing glacier-scratched (striated) clasts (Fig. 8C) were recognized soon afterwards.

Up to about 1950, any till-like rock or sediment was labeled 'glacial' because it was not appreciated that poorly sorted deposits form in a wide range of environments (Fig. 8). Landslide-derived debris flows deposited on land or under water (called debrites; Fig. 8D–F), pyroclastic flows, and volcanic mudflows (*i.e.* lahars; Fig. 8F) all deposit poorly sorted sediments. Meteorite impact and associated fallback processes are also associated with poorly sorted facies that, at first sight, can be confused with glacial deposits. With the sedimentological revolution of the 1960s several ancient deposits formerly interpreted as glacial, were

Figure 7. A. Roche moutonée, Whitefish Falls, Ontario. Former ice flow from left to right. Photo courtesy of S. Puckering. **B.** Bullet-shaped boulder surrounded by poorly sorted glaciofluvial sediment in front of the Saskatchewan Glacier, Alberta. The long axis of the boulder is oriented parallel to former ice-flow direction with the smooth, pointed end of the boulder facing up glacier and the blunt end facing down glacier (former ice flow was from right to left). Bullet-shaped boulders are distinctive features of glacial deposits and result from a combination of subglacial abrasion and plucking processes during transport.

re-evaluated and shown to be non-glacial debrites (Eyles, 1993; Eyles and Januszczak, 2004).

Today, the term *diamict* (diamictite when lithified) is used as a descriptive term for any poorly sorted deposit irrespective of origin. It carries no glacial connotation and is thus non-genetic. The term 'matrix-supported conglomerate' is an equivalent term favored by some. A till (tillite:rock) is a diamict(ite) formed as a result of the aggregation and direct deposition of debris transported by glacial ice. Tillites are most commonly confused with debrites (debris-flow deposits); the latter will be interbedded with other sediment gravity-flow facies such as turbidites (Fig. 8G and H; *see* also Chapter 12), whereas tillites will not. Poorly sorted facies left by meteorite impact, landsliding, or volcanic activity will be interbedded with other facies formed only in those settings. Diamict is a sediment type that can form in a variety of depositional settings and careful observation, recording and interpretation of its characteristics and those of associated facies are required to determine its origin. The construction of representative vertical profiles (*e.g.*, Fig. 3) and mapping of both vertical and lateral facies variability is an essential starting point in determining the broader depositional context of any diamict-hosting succession (Bennett *et al.*, 2002; Stow, 2007).

The three-dimensional geometry of the diamict unit also provides valuable information. Diamicts originating as tills are deposited directly by glaciers and are commonly associated with highly complex ice-contact facies deformed by ice melt, collapse, and bulldozing. Associated landforms provide additional clues as to depositional origin in Pleistocene terrains. In contrast, debrites deposited underwater in non-glacial settings are lensate, fill broad topographic lows and are intimately associated with other subaqueously deposited facies such as turbidites. Sediment is reworked downslope by mass flows in many glacial environments, especially in the glaciomarine environment (Fig. 3), but these deposits are not considered to be true tills as they were not deposited directly by the glacier.

Some workers have invested much effort in attempting to separate tills from non-tills by using simple criteria, such as the surface texture of sand grains, the orientation of clasts within the sediment (*clast-fabric analysis*) and grain size (Bennett *et al.*, 1999). These methods are not very effective, especially in ancient (lithified) successions and particularly when they are applied in isolation from field data showing the broader depositional context. Geochemical and mineralogical approaches are useful in cases where distinctive sediment sources can be identified, such as in the case of lahars in volcanic settings where the deposit consists predominantly of volcanic debris.

In the following sections, we shall examine the characteristics of

Figure 8. *(opposite page)* Poorly sorted admixtures of clasts and matrix are called by the non-genetic name *diamict* and form in a wide range of environments. The identification of diamicts deposited directly by glacier ice (*i.e.*, *till*) requires analysis of associated facies as well as facies characterisics. **A**. Subglacial (deformation) till overlying poorly sorted, ice-proximal glaciofluvial gravels and sands exposed in front of Saskatchewan Glacier, Alberta. This till deposit underlies a streamlined till plain. Outcrop is approximately 3 m high. **B**. 300-million-year-old glaciomarine diamictite exposed in the Carnarvon Basin of Western Australia **C.** Striated clast eroded from late Paleozoic diamictite, Carnarvon Basin, Australia. **D**. Exposure through an alluvial fan showing coarse-grained debris-flow diamict of Pleistocene age, British Columbia. Outcrop is approximately 5 m high. **E**. Diamict deposited by Pleistocene debris flows, Bow Valley, Alberta. **F**. Diamictite formed as a volcanic mudflow (lahar) contains clasts dominated by volcanic lithologies. St. Lucia. Largest clast is approximately 30 cm diameter. **G**. Geologist standing on a 3-m-thick diamictite (debrite) deposited by subaqueous debris-flow processes; turbidites occur below and above the diamictite providing important contextual information. Neoproterozoic Ghaub Formation, Namibia. **H**. Inverse grading (clasts become larger upwards) within turbidite associated with subaqueously deposited debrites; note well-defined imbrication of clasts indicating flow to right. Paleoproterozoic Gowganda Formation, Ontario. Outcrop is approximately 2 m high.

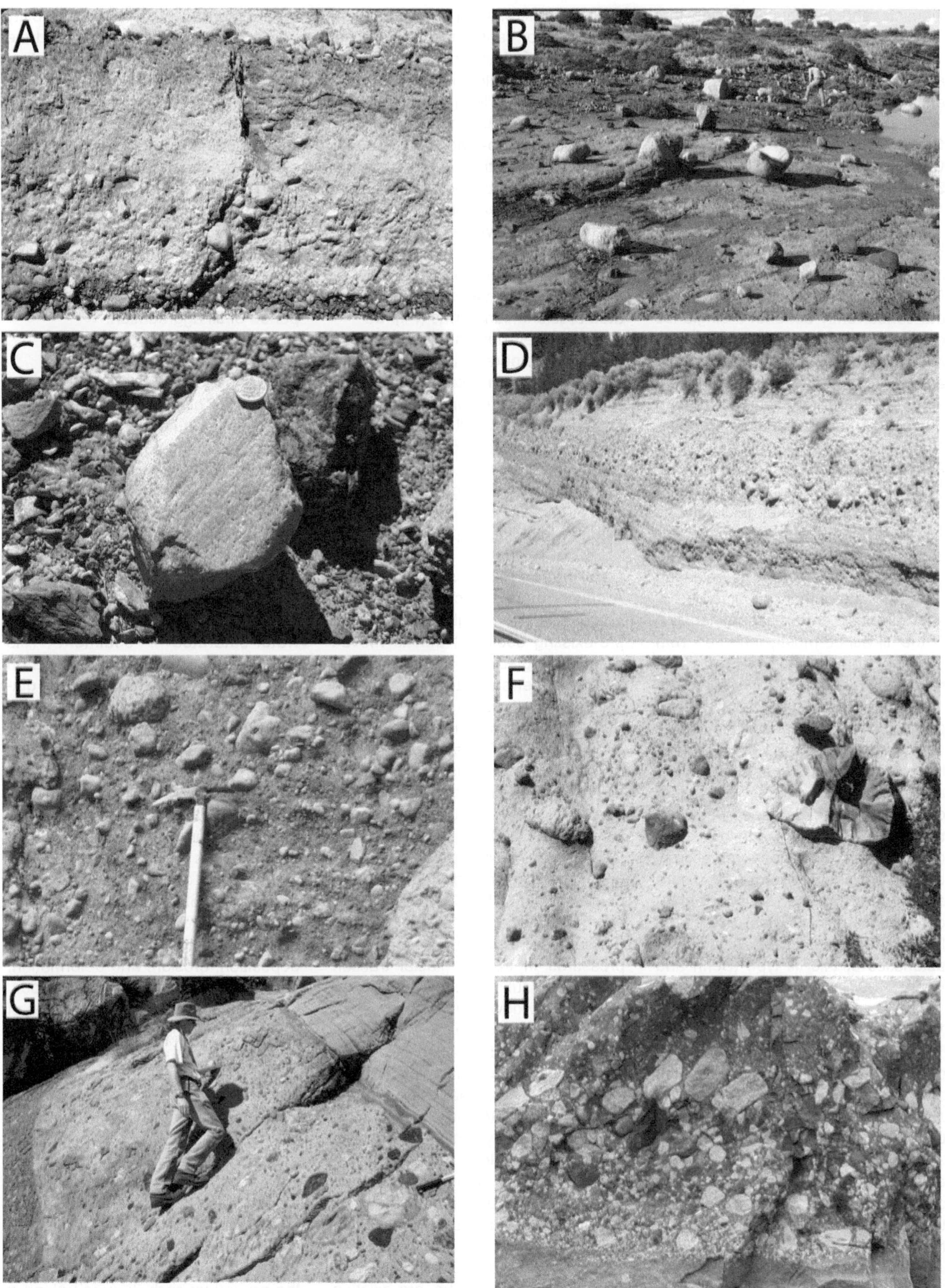

Figure 8. Caption on opposite page.

diamicts and associated facies that form in depositional systems found on land (glacioterrestrial and periglacial), and those that form offshore in marine environments (glaciomarine).

GLACIOTERRESTRIAL DEPOSITIONAL SYSTEM

Subglacial Settings: Till Formation

A number of processes are responsible for the deposition of subglacial tills beneath warm-based ice, including melt-out, lodgement and deformation (Fig. 5). These processes may operate concurrently beneath a single ice mass or sequentially through time. Many till successions or till complexes are therefore the result of not one, but several, superimposed till-forming processes (*see* Evans *et al.*, 2006).

Melt-out Processes

Till may be produced by the passive melt-out of basal and englacial debris under stagnant ice that is downwasting *in situ* (*melt-out till*; Fig. 5A). Melt-out till forms as debris is released from the ice either subglacially or supraglacially, and the characteristics of the till will be mostly inherited from the ice from which the debris is released. Melt-out tills may show foliation, or banding, that reflects textural and compositional variability of debris contained within the ice. The properties of the till can be substantially modified during or immediately after deposition, however, especially when till is released from ice with low debris content, and downslope remobilization of the sediment occurs. There are few descriptions of modern melt-out tills (Paul and Eyles, 1990) and few unambiguous distinguishing characteristics of this till type (Benn and Evans, 1998).

Lodgement Processes

Till can also form by the melt-out of debris from the base of moving ice and smearing of this debris onto the substrate (*lodgement till*; Fig. 5B). Lodgement of debris onto a rigid substrate (bedrock) produces lenticular beds of dense, over-consolidated diamict that may contain sub-horizontal shear planes and slickensided surfaces, in places where shear stresses within the accumulating till exceed the strength of the material and failure (slippage) occurs (Boulton, 1996). High shear stresses in the accumulating till also cause the preferential alignment of clast long axes parallel to ice flow. Bullet-shaped boulders are common in lodgement tills and are oriented with the streamlined, pointed end up-glacier (Fig. 7B). Measurement of the long-axis orientation of clasts embedded in subglacial tills can provide valuable data regarding former ice-flow directions. Lodgement tills lie on marked local and regional unconformities that may be ornamented with streamlined and striated erosional landforms (Fig. 6).

Both melt-out and lodgement processes operate beneath warm-based glaciers and can produce thin (< 20 m), but regionally extensive, sheets of subglacial deposits. The volume of englacial debris is strictly limited by continual melting at the base of warm-based glaciers and neither of these two processes are very effective in building up thick till sheets

Deformation Processes

A much more effective till-forming process is the subglacial mixing of pre-existing sediment that is moved within a subglacial traction layer comprising water-saturated debris with similar characteristics to wet concrete (Fig. 5C). This produces thick accumulations of *deformation till* created by subglacial shearing and mixing of pre-existing sediments and sediment melted out from the glacier (van der Meer *et al.*, 2003). This deforming layer allows sediment to move along under the glacier and also helps the glacier to flow (Figs. 9 and 10). In effect, the glacier almost floats across the bed as a result of high porewater pressures in the subglacial traction layer. The subsequent stiffening of this layer by dewatering leaves over-consolidated *deformation till*. A thick till deposit (up to 50 m) can be built up by repeated aggradation of till beds (Fig. 11). These deposits rest on deformed (glaciotectonized) substrate sediments, and typically include rafts of undigested older deposits (commonly outwash sediment) in their lower part, recording incomplete mixing of debris in the deforming layer. The characteristics of deformation tills vary widely according to the texture and composition of the pre-existing sediment incorporated into the deforming layer, their permeability and drainage characteristics, and the amount and type of strain the material has undergone. Deformation tills can range from structureless (well mixed and homogenized) to stratified, with distinct textural banding, and can show evidence of faulting or folding of incorporated sediment layers. *Boulder pavements*, distinctive horizons of clasts within the till, may also be indicative of subglacial deformation processes and form during episodic erosional phases within the overall period of till aggradation.

Subglacial Landforms

Deposition of subglacial till creates low-relief till plains (*ground moraine* of the older literature) decorated on their surface by elongate streamlined bedforms including *flutes* (long and thin; Fig. 10C) and *drumlins* (long and wide; Fig. 10E), oriented parallel to the direction of ice flow. Flutes are low relief, elongated ridges of sediment that commonly lie down-glacier of a large boulder. They form by subglacial sediment deformation as water-saturated sediment is squeezed into ice cavities formed on the lee side of obstructions on the glacier bed. Flutes have low preservation potential and although they are common in front of modern glaciers, they are rarely identified in older glacial landscapes. Drumlins are larger landforms and occur most commonly in swarms or fields. The origin of drumlins has prompted heated discussion and debate (Boulton, 1987; Menzies, 1989; Shaw *et al.*, 1989) and they have been variously interpreted as the product of subglacial deposition, erosion, and deformation, or of catastrophic meltwater floods. The most widely accepted theories of drumlin formation suggest that these bedforms are the product of either erosional streamlining of pre-existing sediment (where cores of older sediment are present below a thin drape of deformation till) or the selective deposition of thick units of deformation till. In both cases drumlins are formed under conditions of high shear stress. A streamlined and drumlinized till plain can be viewed as essentially a giant slickensided surface akin to the ground-up rock ('gouge') along a fault surface.

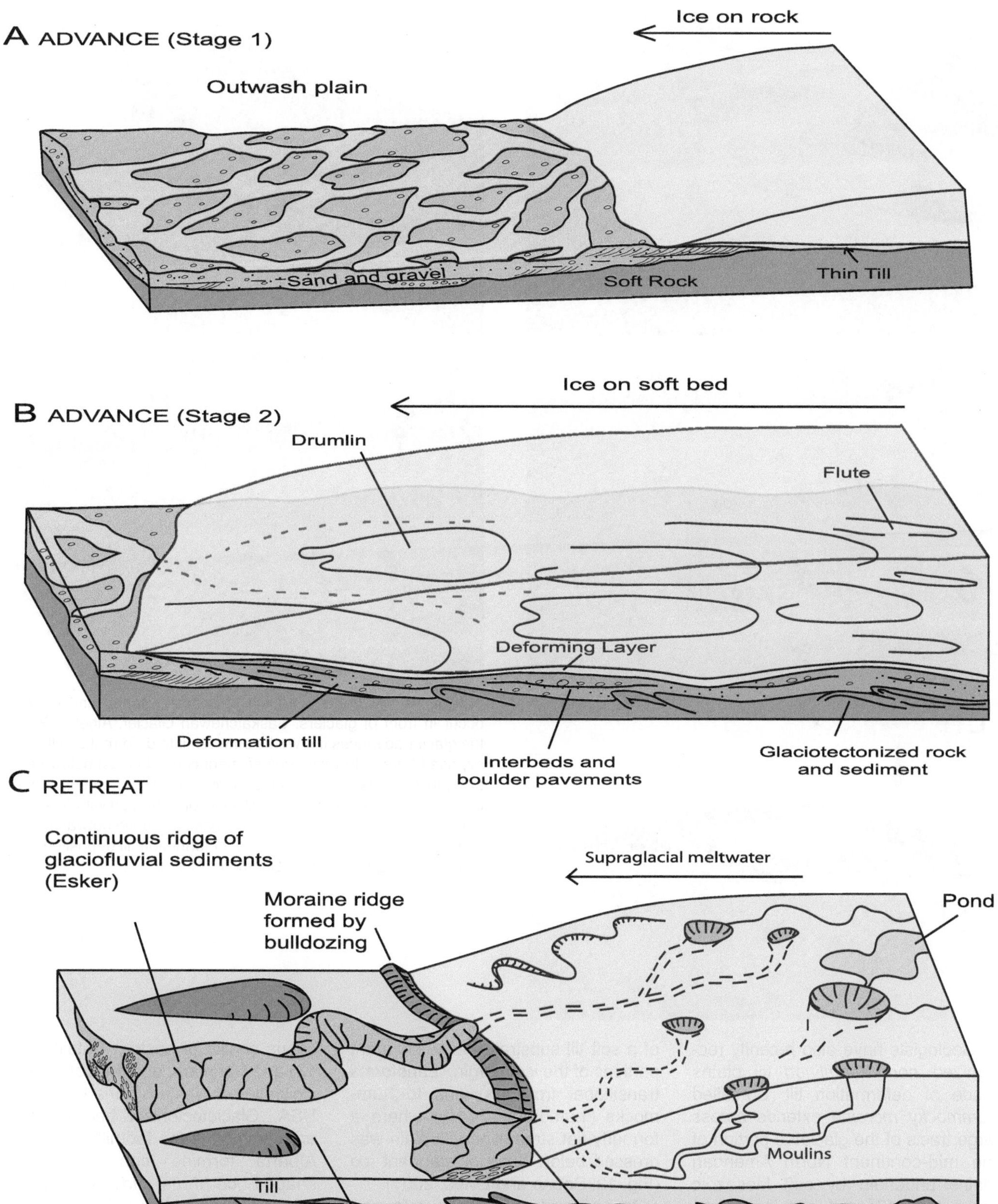

Figure 9. A. Ice sheets resting on bedrock do not deposit thick till. The generation and deposition of thick till occurs beneath the outer margins of glaciers resting on wet sediment. **B.** Till is produced by shearing of pre-existing sediment (typically outwash), cannibalized by the advancing ice front. At this stage, meltwater is drained through the bed as groundwater. **C.** During ice retreat, vast volumes of meltwater are produced by surface (supraglacial) melt. Waters are discharged to the ice-sheet bed through vertical shafts (moulins) and thence to the ice front. The plumbing system may become clogged with coarse-grained sediment leaving eskers and kames.

Figure 10. A. Large amounts of outwash sands and gravels occur in front of glaciers. Saskatchewan Glacier, Alberta. **B.** As the glacier advances it reworks outwash into deformation till seen exposed here below ice (ice/sediment contact is just below head of figure). **C.** Till is transported down glacier and smeared over underlying sediment to leave a till plain ornamented with flutes. **D.** Outcrop of deformation till (top of exposure) above outwash gravel. **E.** A 400-m-long drumlin built of deformation till deposited by a glacier moving left to right; Canmore, Alberta.

Geologists have also recently recognized *non-streamlined* till plains made of deformation till. So-called *hummocky moraine* extends across large tracts of the glaciated portion of the mid-continent North American plains underlain by soft Mesozoic rocks (Fig. 12) and has long been regarded as a supraglacial deposit formed on top of stagnant ice (see below). However, because much of this hummocky moraine is composed of the same clay-rich deformation till found in nearby streamlined landforms (drumlins) it is now recognized as the product of subglacial pressing of a soft till substrate below stagnant portions of the ice margin. Landforms transitional from drumlins to hummocks ('humdrums') occur where a formerly flat streamlined till plain was pressed below dead or stagnant ice (Fig. 12; Boone and Eyles, 2001).

In areas of soft, easily deformed bedrock, large bedrock rafts, sometimes hundreds of meters in length, are plucked from below the ice and shoved down-glacier, a process called *glaciotectonics*. This is analogous to thrust-belt tectonics typical of orogenic belts. Impressive glaciotectonic complexes occur on the Cretaceous chalks of eastern Britain (Fig. 13) and Denmark, and the soft Jurassic shales of Western Canada and the USA. Glaciotectonized bedrock rafts are common in the foothills region of Alberta forming concentric arc-shaped *ice thrust ridges.* These ridges formed below cold-based ice that was frozen to the underlying bedrock along the interior margin of the Laurentide Ice Sheet.

Associated Facies: The Role of Water

Subglacial till facies are most commonly associated with sediments

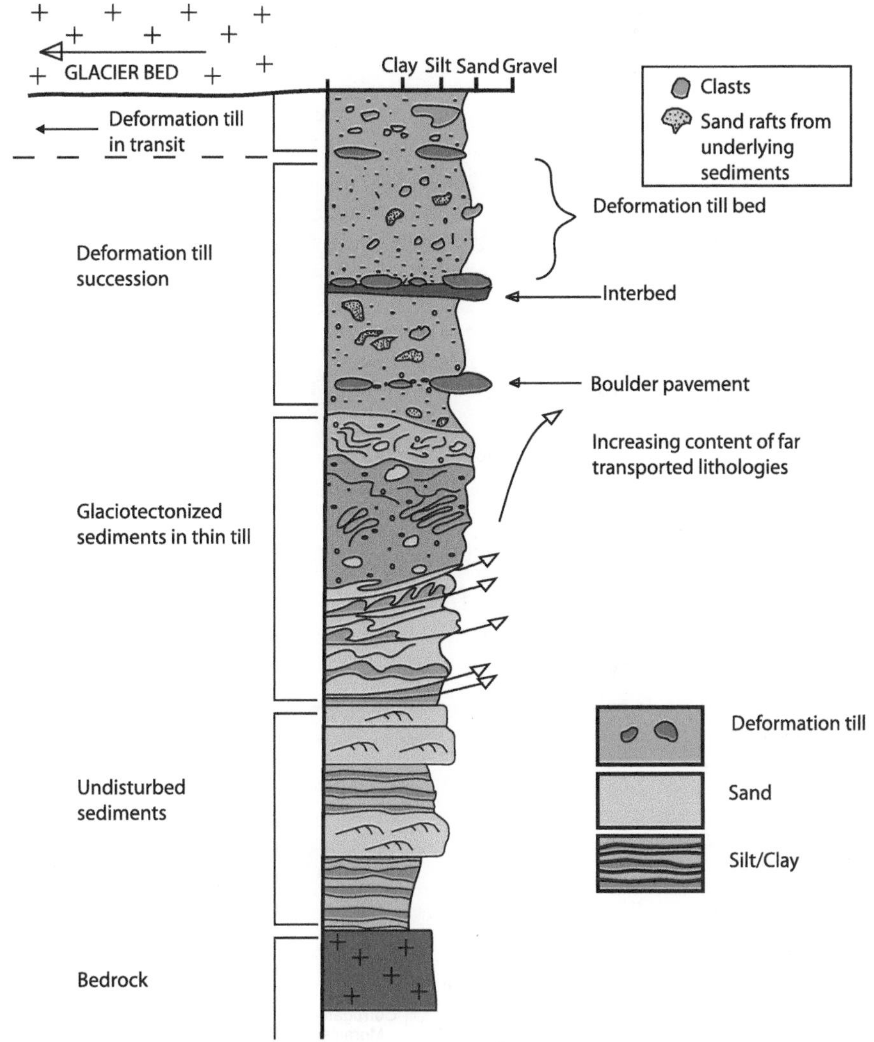

Figure 11. Idealized vertical profile through deformation till overlying deformed sediments.

deposited by running water (Sharp, 2005; Eyles, 2006). Melting of the surface of an ice sheet generates massive volumes of water. This drains to the ice base via shafts (called *moulins*; Fig. 9C) into subglacial tunnels or channels. Where ice overrides a hard rock substrate, channels are cut up into the ice ('R' channels; Fig. 14A), but these are cut down into the bed as Nye channels or 'tunnel valleys' in areas of soft rock or sediment (Benn and Evans, 1998). Fast flowing (up to 5 ms^{-1}) water moves through the tunnels carrying sediment as turbulent hyperconcentrated flows. These exit the ice front as efflux jets (similar to a garden hose) and rapidly deposit a variety of structureless and graded sand and gravel facies on ice-contact fan-deltas.

With ice retreat, the subglacial plumbing system decays and becomes choked with sediment. This leaves steep walled, sinuous-crested ridges of sand and gravel (*eskers*; Fig.14B) aligned sub parallel to ice flow. Eskers are most prevalent on areas of hard 'shield' rocks in northern Europe and Canada where they are commonly more than 100 km long. These ridges of rippled, cross-bedded and structureless gravel and sand were created in sinuous subglacial, englacial or supraglacial channels. Failure of the margins of the esker as the enclosing ice walls melt can cause extensive deformation and faulting of the relatively coarse-grained sediments. Eskers may take a variety of forms ranging from single elongate ridges to more complex *beaded* forms that record successive deposition of subaqueous fans in water ponded against the retreating ice margin.

Tunnel valleys are large, erosional channels cut into bedrock or sediment by subglacial meltwater under high hydrostatic pressure. These wide, flat-floored valleys can be infilled by a variety of glacial, glaciofluvial, glaciolacustrine or glaciomarine sediments and may not have any topographic expression on the ground surface. Tunnel valleys predominate on sedimentary lowlands, especially on continental shelves (*e.g.*, North Sea, Nova Scotian shelf; Fig. 3) and may result from sudden drainage of large subglacial or supraglacial lakes.

Catastrophic Subglacial Meltwater Floods. A number of subglacial landforms, including tunnel valleys, drumlins, and flutes have been attributed to formation by catastrophic meltwater floods (*e.g.*, Shaw *et al.*, 1989; Brennand and Shaw, 1994). The catastrophic meltwater-flood hypothesis requires that exceptionally large lakes, formed beneath, within or on top of, the Laurentide Ice Sheet, drained in a single event creating extensive subglacial sheet floods. These flood events are considered to be responsible for erosion of bedrock and sediment to form scoured bedrock surfaces and tunnel valleys, as well as for deposition of sediment in subglacial cavities to form drumlins. However, recent work argues strongly that the required floods are too large and extensive to be plausible and unlikely to have created these landforms (Clark *et al.*, 2005).

Glaciofluvial Deposits

Energetic braided rivers leaving the ice margin produce thick deposits of crudely bedded and very poorly sorted 'proximal' gravels on broad outwash fans (Miall, 1996). An absence of large-scale cross-stratified facies in such deposits (Fig. 15A) reflects a lack of deep channels and an oversupply of coarse debris (*see* Chapter 6). Portions of the ice margin are commonly buried under gravel where powerful meltstreams emerge and flow over low-standing portions of the glacier (Fig. 15B). The later melt of

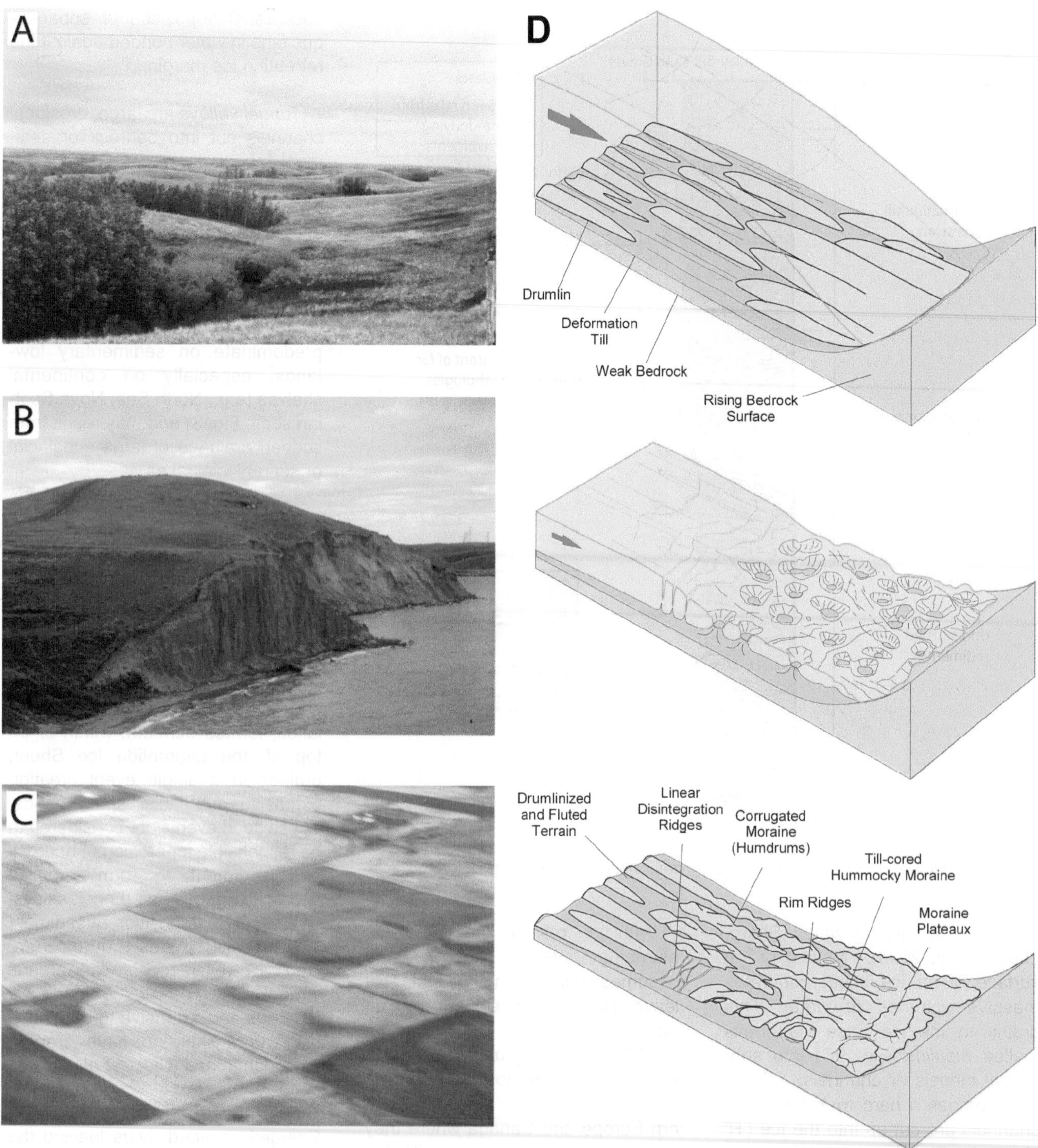

Figure 12. A. Hummocky moraine built of deformation till, such as this example from Alberta, Canada, occurs across large areas of the western glaciated prairies of North America. **B.** Deformation till exposed in outcrop through hummocks is no different to that exposed in adjacent drumlins. Height of outcrop is approximately 15 m. Alberta, Canada. **C.** From the air, a variety of hummock shapes include 'donuts' with rim ridges and inner depressions. Alberta, Canada. **D.** Schematic diagrams showing the formation of hummocky moraine as a result of stagnant ice pressing downward into soft deformation till. A down-glacier succession of landforms can be observed passing from drumlins and flutes in areas of relatively thin drift cover, to *humdrums* (hummocky drumlin forms), and hummocky moraine in areas where stagnant ice pressed into thicker deformation till. Modified from Boone and Eyles (2001).

Figure 13. Thrust and imbricated rafts of chalk moved below a Pleistocene ice sheet; Cromer, Eastern England.

dead ice and the collapse of overlying gravels leave prominent craters (*kettle holes*) and create *pitted outwash plains* (Fig. 15C; Benn and Evans, 1998).

In areas beyond the immediate ice terminus, multiple-channel (braided) rivers sweep across broad outwash plains or *sandar* (sing. *sandur* in Icelandic) depositing gravel and sand that become finer grained and increasingly organized with distance from the ice margin. Discharge characteristics of glaciofluvial streams are highly variable with diurnal, seasonal and annual fluctuations. Deposition on outwash plains is commonly dominated by large flood events (*jökulhlaups*) that may accomplish most of the annual sediment transport in a single period or event. The absence of vegetation along channel banks allows rapid rates of channel migration and exposed sediment is readily transported by eolian processes (*see* Chapter 7) resulting in deposition of wind-blown sand and silt (*loess*).

Glaciofluvial processes are important as they may completely rework sediment deposited by the glacier (Fig.10A), destroying any evidence that indicated the former presence of ice. This potential for reworking is a problem in the interpretation of ancient deposits as braided-river deposits occur in a wide range of depositional settings and a glacial connection may be difficult, if not impossible, to identify. Evidence must be sought from glacial clast shapes or striations, or from the presence or absence of features indicating cold climate or periglacial conditions (see below).

Supraglacial Settings

In glaciated mountains, glaciers act as conveyor belts moving large amounts of rock-fall debris in supraglacial positions (on the surface of the ice) and dumping it down valley. Valley glacial deposits are dominated by freshly broken, *supraglacial debris* (Figs. 16 and 17). The underlying subglacial till plain is buried under a cap of coarse bouldery debris dominated by local bedrock. A thick supraglacial debris cover can also develop in regions where compression of the ice margin, due to flow of ice into bedrock obstructions or ice-margin stagnation, results in complex folding and thickening of the basal debris layer. This thickened debris layer can be exposed at the ice surface during melting and downwasting to form a cover of supraglacial sediment with textural characteristics similar to those found in subglacial settings.

In areas dominated by deposition of sediment from supraglacial sources, a distinct hummocky topography evolves where ice melts slowly under an insulating cover of debris of variable thickness (Benn and Evans, 1998). Kettle holes form as ice down-wastes rapidly in poorly insulated areas and overlying sediment

Figure 14. A. Large 'R' channel exposed at the front of an Icelandic glacier (Kvíarjökull). Note figure (circled) standing to left of channel. **B**. If the channel becomes plugged with gravel, the melt of surrounding ice leaves a sinuous ridge called an esker. Esker is approximately 4 m high, Breidamerkurjökull, Iceland.

Figure 15. A. Glaciofluvial outwash; Ottawa Valley, Ontario. **B**. Portion of an Icelandic ice margin (Breidamerkurjökull) being buried by outwash. Melting of buried ice blocks creates craters several tens of meters in diameter, called kettle holes (**C**).

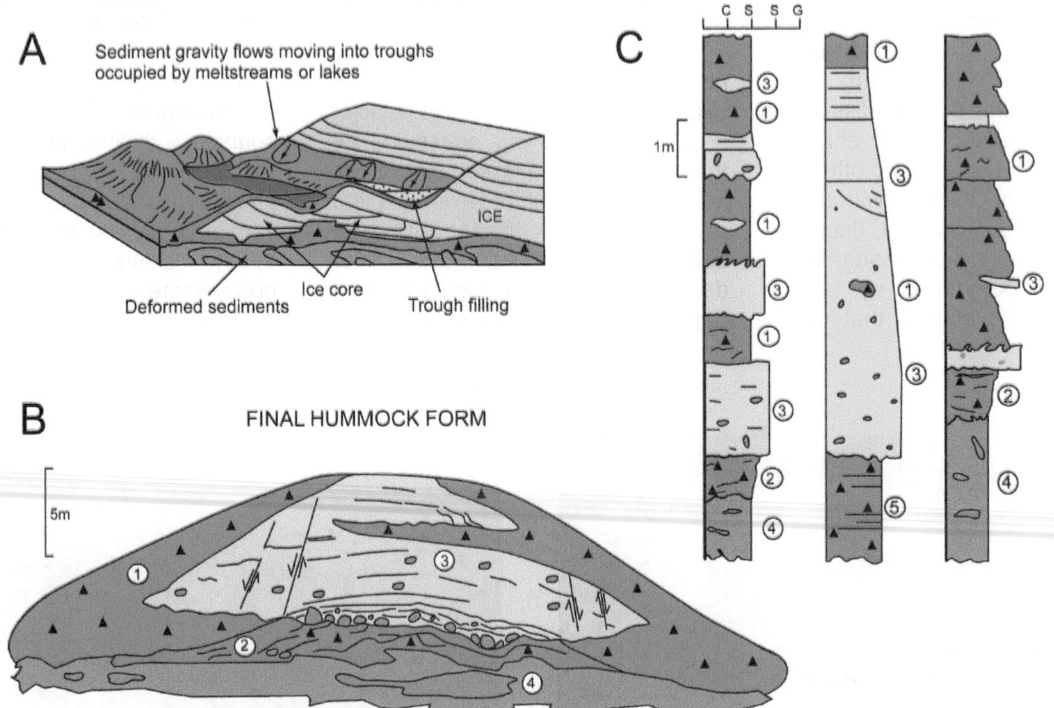

Figure 16. Facies model for supraglacial deposition where ice surface is covered by debris. **A**. Melt of buried ice results in widespread slumping and flow of sediment as it is lowered onto subglacial sediments below. Final ice melt produces a chaotic hummocky terrain. **B/C**. Typical hummock stratigraphy showing: 1: debris-flow diamict with rafts of other sediments, 2: diamict melted out from dead ice (melt-out till), 3: outwash gravels that accumulated in troughs, 4: glaciotectonically deformed subglacial sediment or rock, 5: subglacial till. Faults and slump structures are common throughout because of the loss of support by ice. Horizontal scale for logs shown in **C**: C= clay, S= silt, S=sand, G=gravel.

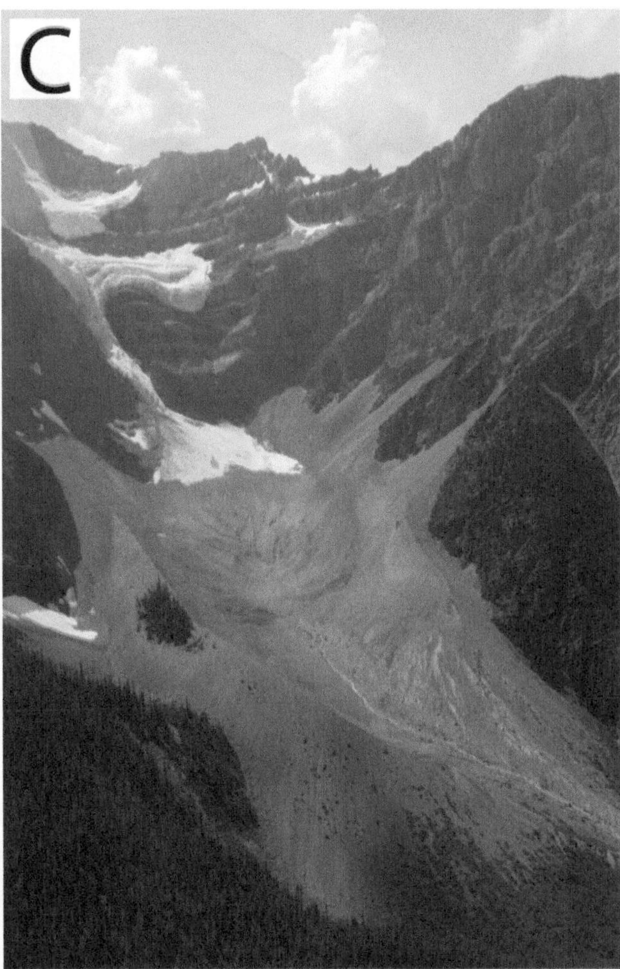

Figure 17. A. Most of the lower portion of the Sherman Glacier in southern Alaska is covered by rockfall debris, which is being conveyed downglacier and now insulates the underlying ice from melting. **B.** Typical rockfall debris being dumped at the ice margin. In many mountainous areas glaciers are much smaller than they were 100 years ago. The former extent of glaciers is often marked by prominent lateral-moraine ridges (**C**) that are composed of rockfall debris and can be cored by remnant masses of dead ice. These ridges are ultimately destroyed by mass wasting. Icefields Parkway, Alberta.

collapses. Mass flow of sediment into depressions on the ice surface creates localized thick debris piles, which are subsequently deposited as *hummocks* containing re-sedimented debris-flow diamicts (debrites), and slumped and deformed glaciofluvial and glaciolacustrine strata. Diamict facies may be structureless, graded and/or stratified, and commonly occur as stacked units that have channelized or lenticular, downslope-thickening geometry. These supraglacially deposited diamict facies commonly overlie subglacial tills and may be interbedded with glaciofluvial and glaciolacustrine facies (Fig. 16).

In glaciated valley settings, large volumes of sediment also accumulate between the glacier and the valley sidewalls as lateral-moraine ridges (Fig. 18). With ice retreat, these are destroyed quickly by mass wasting and accumulate as poorly sorted debrites along the valley floor, or are reworked by rivers. The term *paraglacial* has been used to refer to a short-lived phase occurring immediately after deglaciation when fluvial and mass wasting processes rework glacial sediment downslope.

Glaciolacustrine Settings

Lakes (*see* Chapter 21) are a common feature of terrestrial glacial settings because glaciers release large quantities of meltwater, create deep basins through erosion and isostatic depression, and block pre-existing drainage routes (Bennett *et al.*, 2002). Glacial lakes vary in form from narrow alpine types in areas of high relief, to those infilling large continental-scale basins. These large lakes are ponded in isostatically depressed continental interiors evacuated by ice sheets. Lake Agassiz is the most famous example, and extended over an area of about 1 000 000 km^2 of North America.

Although there is a broad range of glacial lake types, a simple distinction can be made between *ice-contact* and *non ice-contact* lakes (Fig. 19). A characteristic facies of *non ice-contact* lakes (Figs. 19B and 20; lakes fed by glacial meltwaters but lacking contact with an ice margin) in which seasonal variation in meltwater and sediment input occurs, consists of *varves*. Varves are annually produced couplets of relatively coarse- and fine-grained sediment (Fig. 21A). The coarse sediment layer (consisting of gravel, sand or silt) is deposited during spring and summer when significant supraglacial melting occurs and sed-

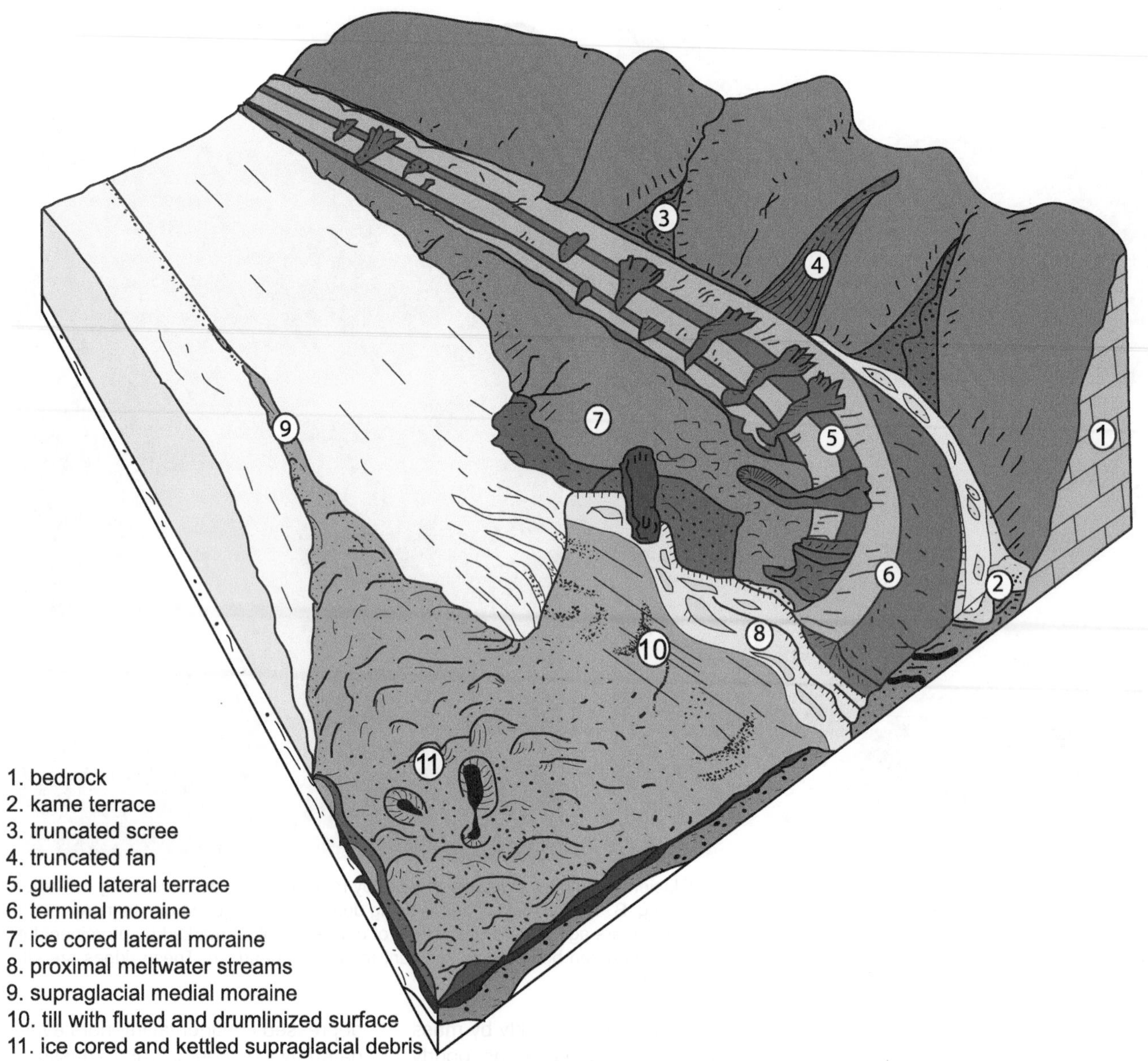

Figure 18. Facies model for sedimentation by valley glaciers.

iment-laden density underflows transport sediment across fan-delta lobes (Fig. 21B). A distinct succession of relatively coarse-grained sediment is deposited each summer and records the start, increase and ultimate decline of density-underflow activity (Ashley, 1975). Summer sediment layers thin and become finer grained distally into the basin. During winter, melting is suppressed and flow into the lake ceases, allowing fine-grained sediment suspended in the water column to settle to the lake floor as a layer of clay. This winter clay layer may show normal grading indicating deposition of suspended sediment beneath the ice cover of a closed lake. Clay-layer thickness is usually uniform across the lake basin. Winter clay layers may also contain units of coarser grained sediment transported into the basin by sediment gravity flows generated on unstable slopes along the basin margin (Fig. 20).

Successions of laminated silts and clays found in association with other glacial facies are commonly described as 'varves', regardless of demonstrated seasonal control on

Figure 20. *(opposite page)* Deposition in non ice-contact lakes. Coarse-grained sediment delivered by braided glacial meltstreams is deposited on extensive fan deltas; fine-grained sediment is transported into the basin by overflows, interflows and density underflows (Fig. 19). Sedimentation is strongly influenced by seasonal variations in the volume of meltwater entering the basin and is reflected in 'varved' sedimentation (1 to 3). Slumping on over-steepened delta fronts in winter may complicate this (I to V; *see* Shaw, 1977).

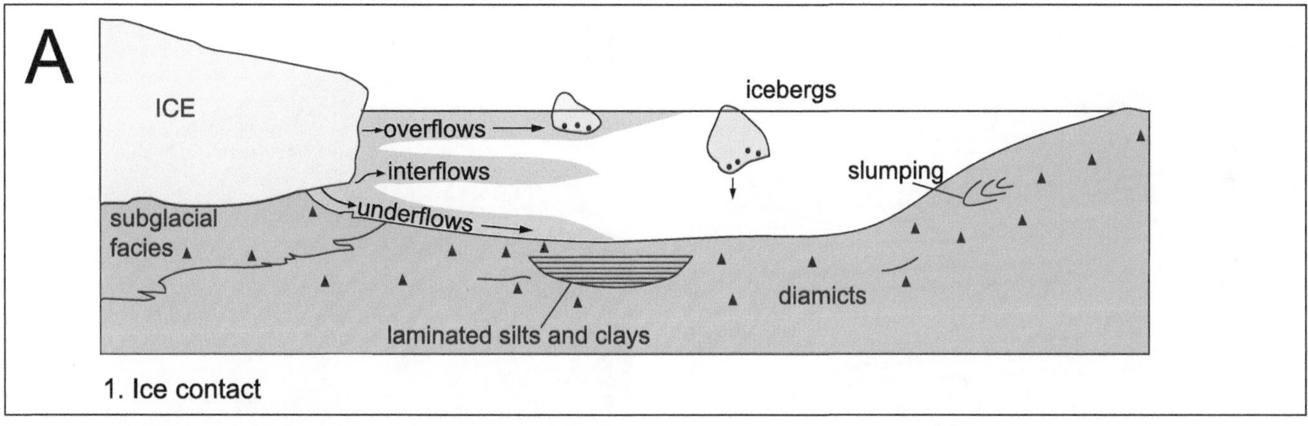

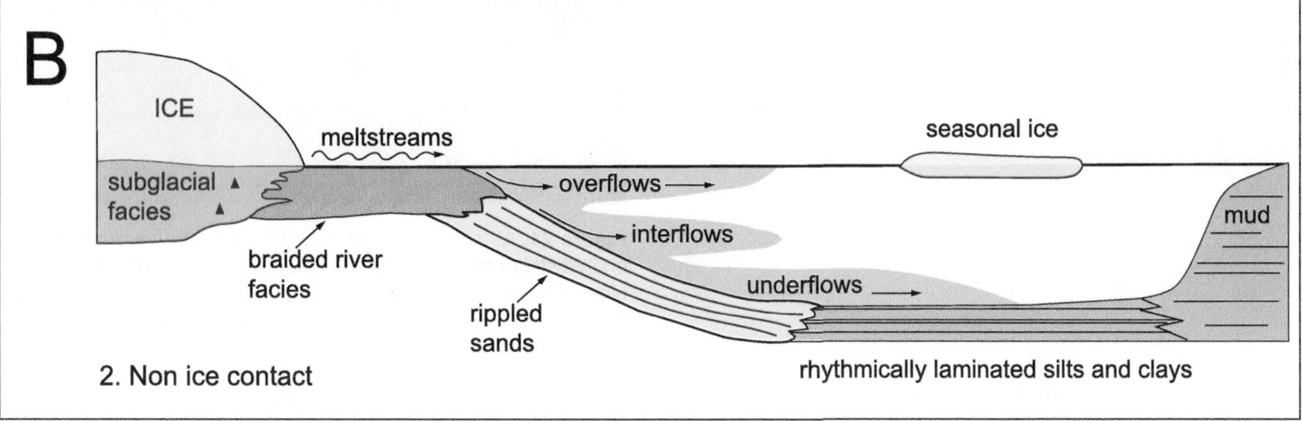

Figure 19. Facies models for the glaciolacustrine depositional system with ice-contact (**A**) and non ice-contact lakes (**B**).

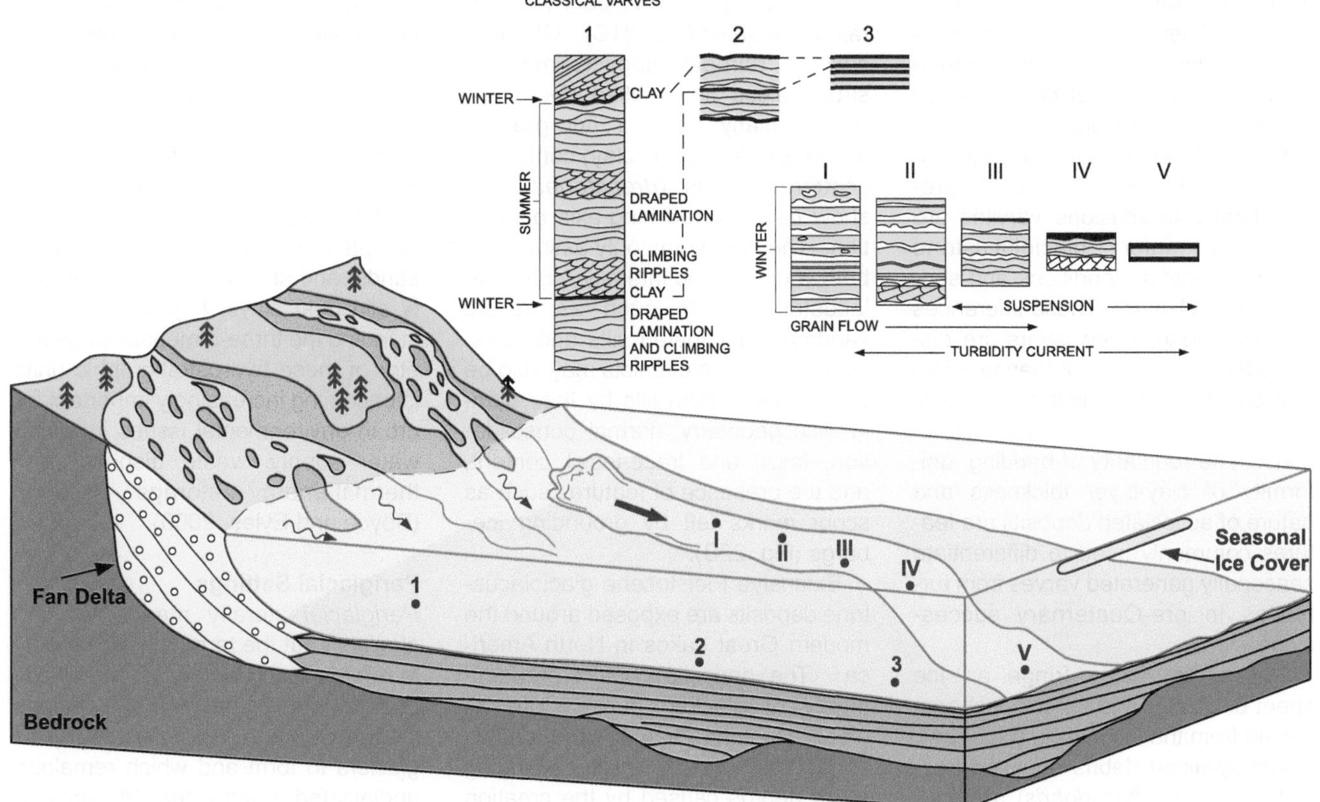

Figure 20. Caption on opposite page.

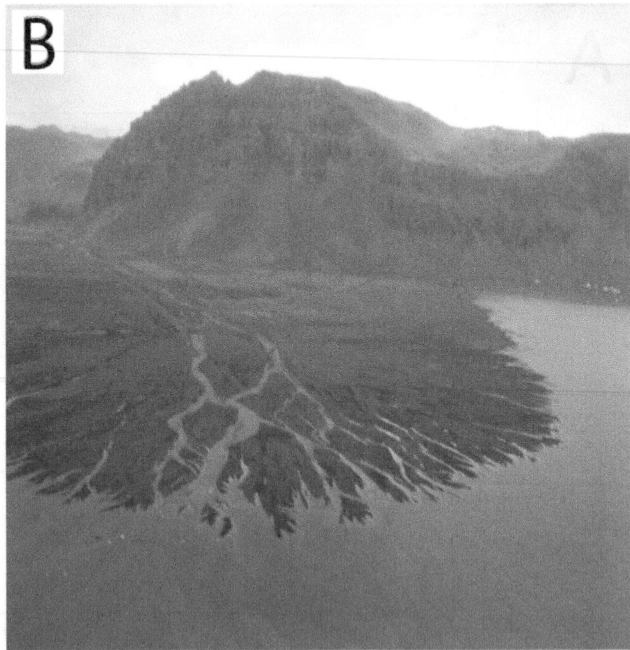

Figure 21. A. Laminated silt and clay couplets (varves) deposited in distal portions of a glacial lake basin. Numerous white granules of silt were brought in during summer underflows. **B.** Large fan delta fed by glacial melt-water streams in southeast Iceland. **C.** Fine-grained silty clay diamict with scattered clasts deposited by rain out processes in an ice-contact glacial lake; Scarborough Bluffs, Ontario. A cluster of ice-rafted debris lies above the ice pick.

their formation. Rhythmically laminated sediments can, however, accumulate in both lacustrine and marine settings as a result of deposition by discrete-event turbidity currents (turbidites) with no evident seasonal control (Fig. 20; Shaw, 1977). In pre-Quaternary successions 'varvites' are used to infer glaciolacustrine (glacioterrestrial) conditions and seasonality of climate. These inferences may be wrong, if the layers are discrete-event turbidites of marine origin and could severely distort paleogeographic and paleoclimatic reconstructions. The regularity of bedding, uniformity of clay-layer thickness and nature of associated deposits are features commonly used to differentiate seasonally generated varves from turbidites in pre-Quaternary successions.

Ice-contact lakes fringe an ice sheet or glacier and contain icebergs calved from the ice margin (Fig. 19A). Coarse-grained debris released from icebergs (*ice-rafted debris*) accumulates on the lake floor together with fine-grained sediment settling from suspension, producing poorly sorted *rain-out diamict* (Fig. 21C). Glaciolacustrine rain-out diamicts may be structureless or stratified (Fig. 22A) and typically have a fine-grained matrix (generally silty clay) containing scattered clasts (*dropstones*) that show no preferred long-axis orientation. They are commonly associated, both vertically and laterally, with other lacustrine facies such as deltaic sands and laminated silts and clays. Glaciolacustrine diamicts may also be discriminated from tills by their blanket-like geometry, normal consolidation, fossil and trace-fossil content, and the presence of features such as scour marks left by grounding icebergs (Fig. 22B).

Extensive Pleistocene glaciolacustrine deposits are exposed around the modern Great Lakes in North America. The deposits consist predominantly of stacked successions of glaciolacustrine diamict and deltaic sands that record rapidly changing water depths caused by the creation and removal of ice or sediment dams around the lake basins. Littoral and shallow-water sediments of such large lakes are commonly storm influenced and subject to deformation by floating ice masses (Eyles *et al.*, 2005). The interbedding of fine-grained diamicts and deltaic sands in these thick and extensive glaciolacustrine successions creates a stacked complex of aquifers (permeable sands) and aquitards (low permeability silts and clays; Fig. 22). Understanding the three-dimensional geometry of these hydrostratigraphic units is becoming increasingly important for urban environmental issues involving water supply, waste disposal and thermal-energy storage systems (Boyce and Eyles, 2000).

Periglacial Settings

Periglacial literally means 'around glaciers', but the term is used broadly to refer to both glacial and non-glacial cold-climate regions. These regions include cold areas that are too dry for glaciers to form and which remained unglaciated during the Pleistocene (*e.g.*, much of Yukon and Alaska), or lie in the sub-polar areas. In these

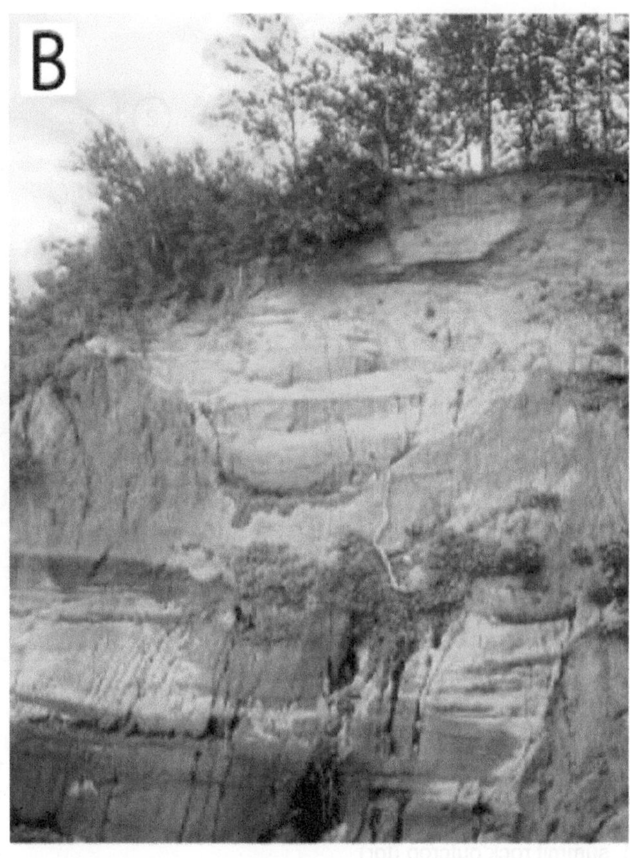

Figure 22. A. Stacked units of weakly stratified, fine-grained silty-clay diamict (mid part of section) and deltaic sands (upper part of section) exposed along the Scarborough Bluffs, Ontario. Middle (light-colored) diamict unit is approximately 6 m thick. **B.** Cross-sectional view through an iceberg scour in glaciolacustrine diamict made by a grounding iceberg and infilled with deltaic sands; Scarborough Bluffs, Ontario. Scour is approximately 4 m deep.

areas, short summers, frozen ground characterized by shallow depths of thawing in the summer months, the lack of precipitation and surface runoff, and strong winds limit erosion and deposition. As a result, sedimentological features created in periglacial settings are commonly associated with major unconformities separating stratigraphic successions. Mechanical weathering predominates as a result of freeze–thaw cycles. Thicker sediments may accumulate in valleys where slow down-slope movement of debris takes place under gravity in summer months, a process termed *solifluction* or *soil creep* (Fig. 23).

Where mean annual temperatures are less than −4°C the ground is permanently frozen producing *permafrost.* Permafrost is ground that stays frozen year round except for a shallow (< 2 m) surface thaw zone called the *active layer*. These conditions occur over 25% of the northern hemisphere (some 26 million km^2). Permafrost also occurs offshore (submarine permafrost) in areas flooded by the postglacial rise in sea level or where land is actively subsiding (*e.g.*, Mackenzie River delta area). In northern regions of Canada permafrost reaches a maximum reported depth of approximately 700 m.

Underground ice grows slowly and in forms that range from thin ice coatings on individual sediment grains to lenses, layers and masses of various shapes many meters to tens of meters in size. Considerable mechanical disturbance of surrounding sediment occurs as the larger ice masses grow. One of the most distinctive forms of ground ice are the large carrot-shaped *ice wedges* that grow in meters-deep cracks produced by ground contraction in severely cold climates that are too dry to have an insulating snow cover. Intersecting wedges create polygonal patterns on the ground surface (Fig. 23).

The growth of larger ground-ice masses often leaves pockets of unfrozen groundwater ('taliks'). In valleys where there are thick alluvial fills (and thus major aquifers that stay unfrozen) over-pressured groundwater is forcefully injected under artesian pressure toward the ground surface. Overlying sediment is bulged upward into small hills called *pingos* (Fig. 23).

During the summer the active layer is characterized by high water content as the upper part of the permafrost thaws but cannot drain due to the presence of impermeable frozen ground beneath. Seasonal refreezing of the thawed surface layer from the top down raises porewater pressures and creates widespread soft-sediment deformation structures and flame-like injections in the active layer. These are collectively referred to as *cryoturbation structures* (Fig. 24C). This process of sediment deformation is also expressed as honeycombed *patterned ground* (Fig. 24D).

Upon climatic warming, the thaw of buried ground ice results in subsidence of the land surface (*thermokarst)* and the creation of surface ponds (*thaw ponds;* Fig. 24B). The thaw of ice within pingos leaves crater-like depressions with a similar form to donuts, and the melt of ice wedges leaves open fissures that fill with windblown sand and silt, or slumped sediment thereby creating wedge shaped *ice-wedge casts* or *sand wedges*. The subsiding landscape may slowly become buried by

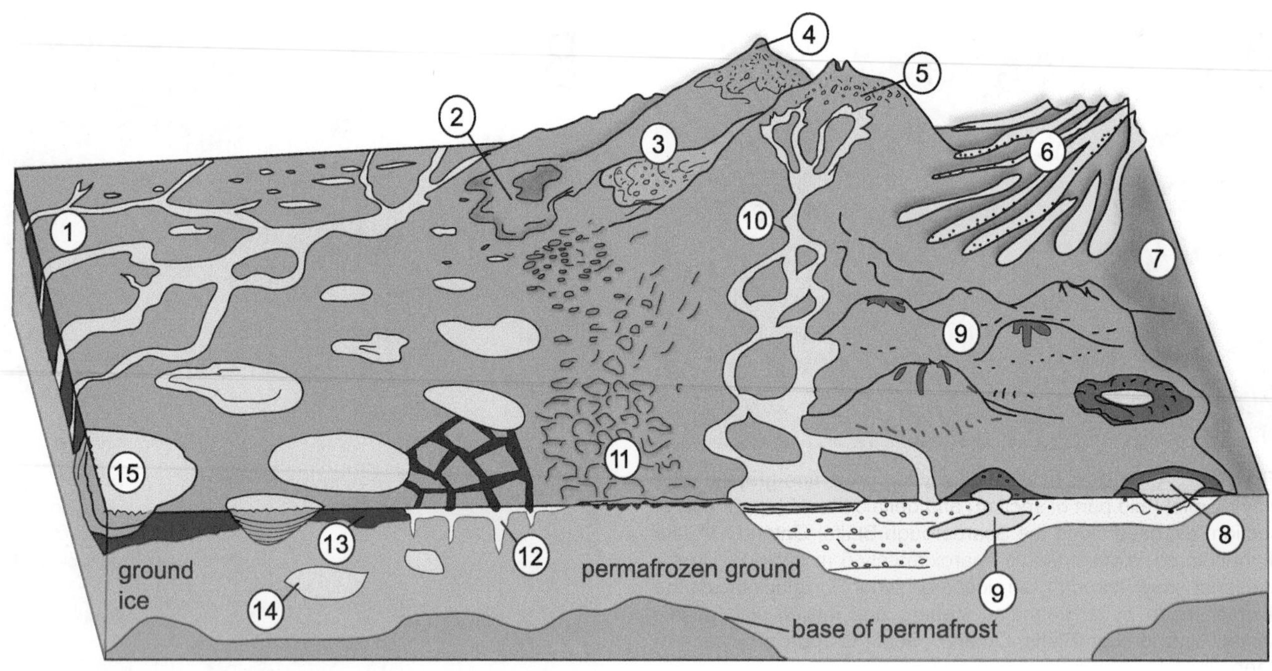

1. anastomosed river
2. creep of saturated sediment downslope (solifluction sheet)
3. creep of ice cemented debris (rock glacier)
4. summit rock outcrop (tor)
5. fresh-shattered rubble (blockfield)
6. eolian dunes
7. windblown sand and silt
8. degraded pingo
9. overpressured groundwater intruded into gravels creating a hydrolaccolith (pingo hummock)
10. snow melt fed braided stream
11. patterned ground (stone circles) and underlying cryoturbation structures
12. ice wedges forming polygonal network
13. active layer
14. unfrozen lens of water (talik)
15. thaw pond formed by melt of permafrost

Figure 23. Facies model for periglacial environments.

thick peat, which eventually strangles rivers into anastomosed types (*see* Chapter 6).

In periglacial regions, river-channel cross sections take on a distinct U-shape in response to the pushing of thick winter ice jams along the riverbed during spring break-up (Fig. 24A). Striated surfaces and boulder pavements can develop on the riverbed at the base of the jams. Similarly, tidal flats become ornamented with scrapes and scours made by winter ice floes that drift and shove large boulders around.

Eolian activity removes ('deflates') loose silty sediment left by meltwaters or exposed by the drainage of glacial lakes. Modern eolian dune fields occur in interior Alaska and parts of the Yukon where Pleistocene glaciations are recorded by thick deposits of wind-blown silt (loess; Fig. 23). In the Yukon and in China, loess deposits have yielded exceptional records of glacial and interglacial climates. The most extensive Pleistocene eolian-sand deposit in North America is the Nebraska Sand Hills and its coeval loess deposits of the Mississippi and Missouri river valleys.

The modern-day geographic extent of the periglacial environment is not matched by many examples from the ancient record, pointing to poor preservation of these deposits and/or lack of recognition of diagnostic features in lithified successions. Ancient sandstone wedges and cold-climate dune fields are known from the Neoproterozoic, and periglacial eolianites occur in late Paleozoic successions (*e.g.*, Cooper Basin) of Southern Australia where they host natural gas. The peat-covered Hudson Bay Lowlands is an excellent modern analog for the Carboniferous–Permian cold-climate coals found in the Karoo Basin of South Africa and Perth Basin of Western Australia.

GLACIOMARINE DEPOSITIONAL SYSTEMS

Glaciomarine environments are marine environments influenced by glacial processes. Regional climate is an important control on sedimentation style as it dictates the volume of meltwater reaching the marine environment. Temperate glaciomarine environments, for example, receive large volumes of meltwater and fine-grained sediment that are supplied directly to the shelf and result in high sedimentation rates. In contrast, glaciomarine environments fringing deeply frozen polar areas such as Antarctica are sediment starved as meltwater input is severely restricted (Anderson, 2002). In these settings chemical and biogenic sedimentation

Figure 24. A. Large masses of ice lie along the bed of a river during spring break-up in a periglacial region of northern Canada. **B.** Thaw ponds created by the local melt of permafrost during warm phases of the Holocene. Northwest Territories. Small pond in lower right of image is approximately 35 m long. **C.** Soft-sediment deformation structures (cryoturbation structures) caused by the overpressuring of porewater in silt-rich sediments during freezing. **D.** Patterned ground formed as coarse-grained sediment is preferentially sorted to the margins of intersecting polygons during cryoturbation, Spitzbergen.

is relatively important. Thick successions of glaciomarine deposits preserved in the ancient record are most likely to have formed in temperate settings.

Proximity to an ice margin is another important control on glaciomarine sedimentation because it determines whether the environment is dominated by glacial processes, as occurs in high-energy *ice proximal, ice-contact glaciomarine,* or *proglacial marine* settings, or by marine processes in *ice-distal* continental-shelf and slope settings.

Ice-proximal Glaciomarine Settings

In temperate regions, the ice-contact glaciomarine setting is a very dynamic environment dominated by powerful jets of meltwater exiting onto the seafloor from the mouths of subglacial tunnels (Lonne *et al.*, 2001). Deposition by underflows and sediment gravity flows results in a wide variety of massive and graded gravel and sand facies found in multi-storey, crosscutting channels on large subaqueous fans (Fig. 25). Lower density meltwater containing fine-grained suspended sediment rises to the surface, or to intermediate depths, in the water column forming plumes of suspended sediment. These plumes can extend many tens of kilometers from the ice front and deposit blankets of mud where the sediment settles out in quiet-water areas (*see* Chapters 8 and 10). Tides interact with plumes to produce laminated mud facies (*tidal rhythmites, see* Chapter 9) that may be difficult to distinguish from varves formed in glaciolacustrine settings. In ancient successions glaciomarine facies may be differentiated from glacioterrestrial facies by the presence of macro- and microfauna, although biological activity may be inhibited in the brackish water that characterizes ice-proximal glaciomarine settings (*cf.* Chapter 3).

Ice-proximal glaciomarine environments are very dynamic, and sedimentary successions formed in these settings consist of assemblages of heterogeneous facies types with rapid lateral and vertical facies variability, and irregular bed geometries (Fig. 25, Boulton *et al.*, 1996). In areas where ice-rafted debris is dropped to the seafloor from icebergs, *rain-out diamict* facies accumulate. These facies can be structureless or stratified and may have a wide range of textural characteristics depending on the nature and amount of ice-rafted debris, the supply of fines from meltwater plumes, and size sorting by wave and tidal currents on the seafloor. Sand-rich diamict facies commonly accumulate in areas subject to episodic traction currents and form part of a lithofacies continuum with pebbly sands and poorly sorted gravels.

Diamict facies are also generated

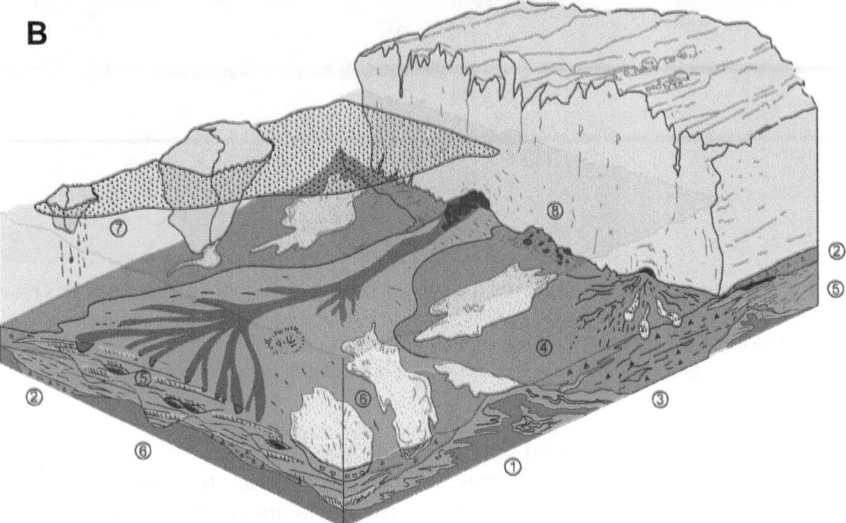

Figure 25. A. Ice-proximal glaciomarine environments, Columbia Glacier, Alaska. **B.** Facies model for glaciomarine deposits near the margin of a tidewater glacier. 1: glaciotectonized marine sediments or bedrock, 2: deformation till, 3: stratified diamict deposited by slumping of till and other debris, 4: mud with ice-rafted debris, 5: channeled gravel and sand of submarine fan, 6: slumps and sediment gravity-flow facies, 7: iceberg scour and plume of suspended sediment, 8: push ridge. The entire sediment package commonly takes the form of large ridges (morainal banks) reflecting short-lived pauses in the retreat of the ice margin.

by sediment gravity flows originating from the collapse of steep slopes. Mixing of heterogeneous sediment (gravels, sands, silts and clays) occurs during downslope transport to produce a range of diamict facies. Poorly sorted basal debris issuing from the ice front at the *grounding line* (where the ice begins to float) may also move downslope as debris flows and can contribute to the build-up of grounding-line fans.

Extremely high rates of sediment accumulation (up to 17 m year^{-1}) have been reported from ice-proximal glaciomarine settings in the fiords of Alaska. Seaward movement of the grounding line, which may occur on a seasonal basis due to the reduced rates of melting and iceberg calving in the winter, bulldozes ice-proximal sediment into an underwater ridge, or *morainal bank* (Fig. 25). These landforms commonly record the outer limit of grounded ice on continental shelves and are the marine equivalent of moraine systems formed in terrestrial settings. The bulldozed sediment contained within morainal banks is commonly deformed by folding and faulting (glaciotectonized). In cases where the ice advances and overrides a morainal bank, the sediment may be reworked into deformation till. Glacier retreat from a morainal bank produces a fining-upward succession passing from coarse-grained ice-proximal facies to laminated and massive mud and fine-grained rain-out diamicts. Isostatic uplift and emergence of morainal banks may bring these retreat successions above wave base, resulting in erosion and deposition of a cap of coarse-grained nearshore facies. In general, ice-contact glaciomarine deposits on the inner portion of continental shelves have a very low preservation potential. They are easily destroyed by shallow-marine erosion when the area experiences glacioisostatic uplift, by being flattened by grounding icebergs, and by erosion as a result of downslope collapse and mass wasting of sediment.

Ice-distal Glaciomarine Settings: Continental Shelves

Ice-distal glaciomarine environments can be found in a broad range of marine settings including those with relatively low relief, such as continental shelves and abyssal plains, and high-relief settings that include fiords and continental slopes. Sediment is supplied to ice-distal glaciomarine environments primarily by plumes of suspended fine-grained sediment, icebergs, mass-flow events and ocean currents (Eyles *et al.*, 1991).

The dominant sedimentary process on many temperate glacially influenced continental shelves is the production of extensive, blanket-like rain-out diamict facies formed by the mixing of fine-grained suspended sediment and coarser debris from icebergs (Fig. 26A). Rain-out diamicts can reach a thickness of many tens of meters and can be structureless or crudely stratified depending on textural characteristics of incoming sediment and the sorting capability of tide and wave-generated ocean currents. Rain-out diamicts formed in glaciomarine settings commonly contain abundant microfauna (foraminifera and radiolara) and shelly macrofauna (Fig. 26C). Rates of deposition can

Figure 26. Outer continental-shelf glaciomarine deposits (*see* Fig. 3) of Miocene to Pleistocene age (Yakataga Formation) are exceptionally well exposed on Middleton Island in the Gulf of Alaska. **A.** Thick successions of rain-out diamict produced by ice rafting and settling of suspended fines. These diamict facies are rich in marine micro-organisms (such as foraminifera) and shelly macrofauna. **B**. Large cluster of boulders within diamict produced by rainout from floating ice. **C**. Horizon of shelly fauna (coquina) within diamict form 'cold water carbonates' (*see* Chapter 15). Note ice-rafted clasts among the shells. **D.** Outcrop of glaciomarine boulder pavement within diamict (view along strike, beds dip to left of photo). The upper surfaces of clasts within the boulder pavement are planed off and striated indicating abrasion by a grounding ice shelf.

be as high as 1 m per thousand years. Ice-rafted debris can be transported many hundred kilometers from source areas. A 1.3-km-thick succession of ice-distal continental-shelf deposits dominated by rain-out diamicts is well exposed on Middleton Island in the Gulf of Alaska. There, thick planar-tabular units of rain-out diamict are interbedded with marine muds, boulder pavements, coquinas and storm-deposited sandstones (Eyles *et al.*, 1991; James *et al.*, 2009; Fig. 26).

Re-sedimentation processes can operate even on gentle slopes on glacially influenced continental shelves due to high porewater pressures generated in rapidly deposited fine-grained sediment. Sediment gravity flows may be triggered by seismic shock, storm events or iceberg grounding, and are recorded by units of structureless or stratified diamict (commonly containing flow-and-fold structures), which can be interbedded with turbidites. In the absence of associated turbidites or a displaced fauna, differentiation of unstratified diamicts, formed by rain-out processes from those formed as a result of downslope re-sedimentation processes, is difficult.

Many glaciated continental shelves show deep-water shelf-crossing troughs that were cut below extensive Pleistocene ice sheets (*e.g.*, the Laurentian Channel). These troughs are commonly the seaward extension of over-deepened valleys (fiords; Fig. 3) that extend inland, or of unfilled tunnel valleys. They funnel enormous volumes of glacial sediment, mostly by sediment gravity flow processes, to trough-mouth submarine fans at the base of the slope (Fig. 3). Deposits are dominated by coarse- and fine-grained turbidites, and debrites (debris-flow deposits). The latter may flow hundreds of kilometers beyond the base of slope onto abyssal plains.

The Effects of Changing Water Depths on Glaciated Continental Shelves

Sedimentation on glacially influenced continental shelves is very responsive to changes in water depth and energy regimes (Fig. 27). Many factors affect water depth and energy regimes on glaciated continental shelves including globally synchronous (eustatic) sea-level changes caused by the growth and melt of ice-sheets and localized glacio-isostatic changes caused by ice loading and unloading of the Earth's crust (see below). In situations where rapid advance of ice causes minimal isostatic depression of the shelf, glacio-eustatic lowering of sea level (or tectonic uplift of the shelf), can bring large areas of the continental shelf above storm wave base. This allows

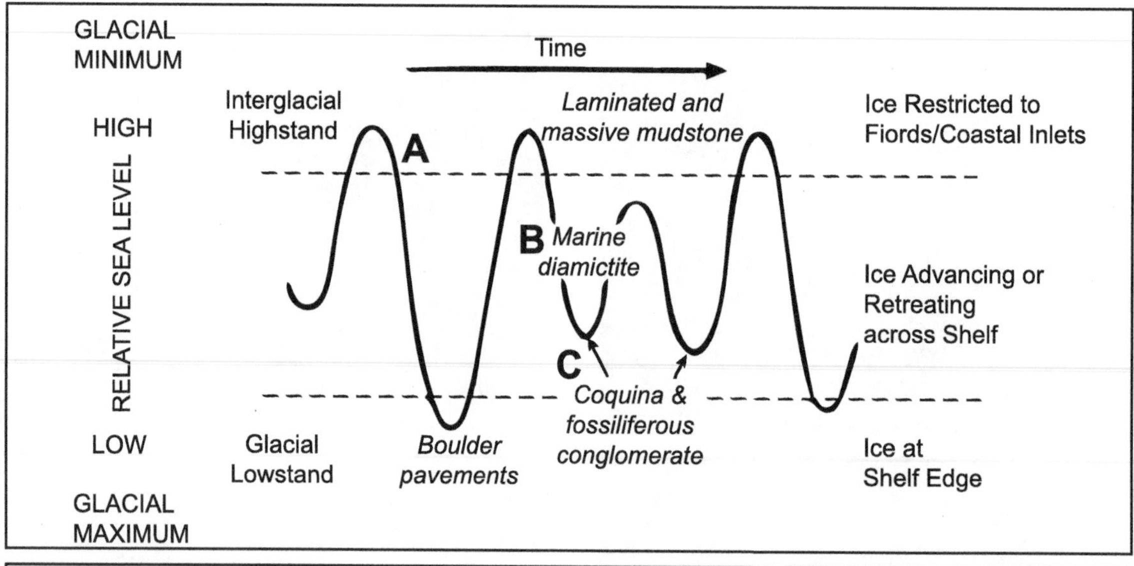

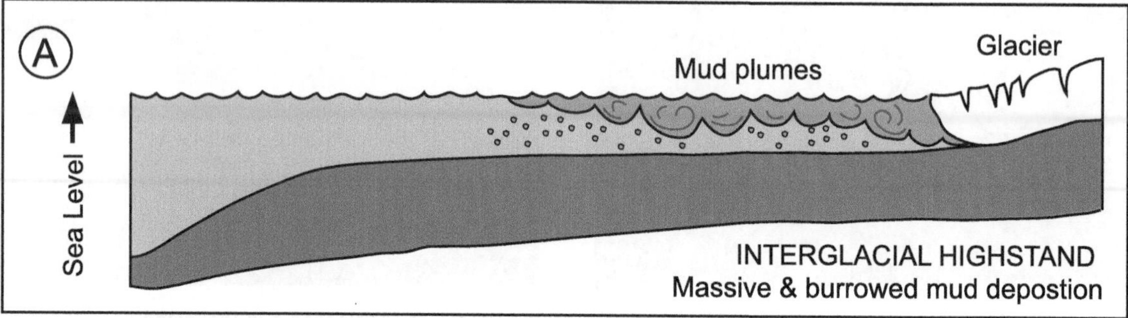

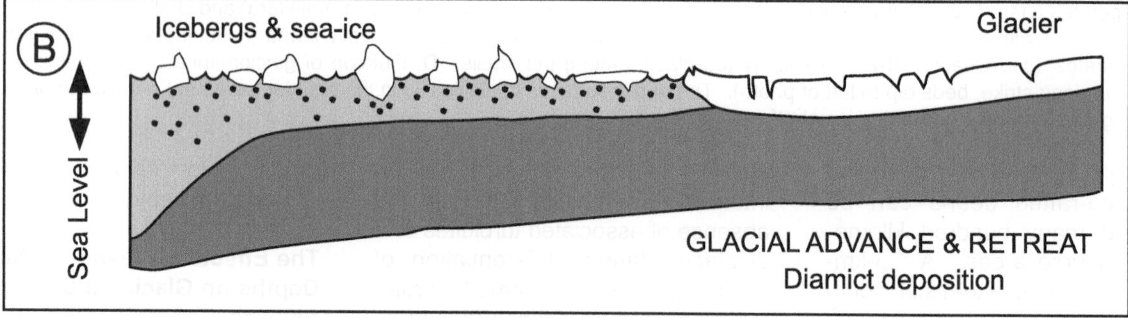

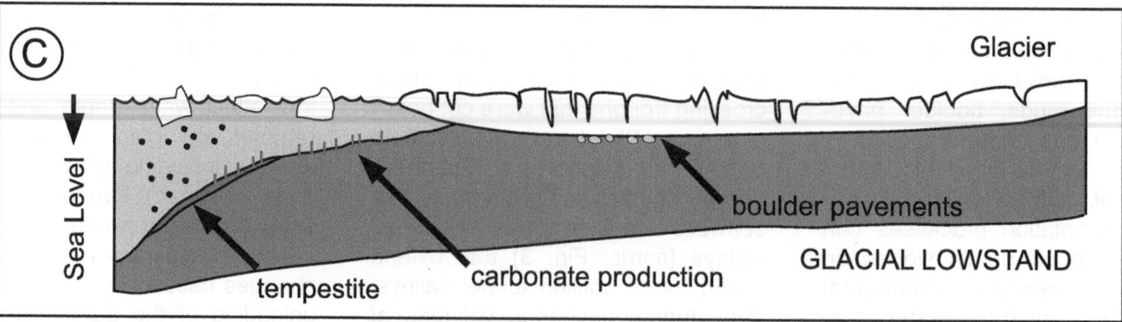

Figure 27. The effects of changes in water depth and proximity to the ice margin on sedimentation patterns on glaciated continental shelves (modified from James *et al.*, 2009). **A.** At times of relatively high sea level (*e.g.*, interglacial periods) low-energy mud deposition predominates. **B.** During ice advance or retreat across the shelf, water depths may be increasing or decreasing. Icebergs released from floating ice margins contribute coarse-grained sediment to form rain-out and re-sedimented diamicts. **C.** At times of relatively low sea level (*e.g.*, glacial lowstand) ice may advance across the shelf eroding and abrading previously deposited sediment to create marine boulder pavements and deposit tills. Basinward of the ice margin, epifaunal carbonate production is possible due to reduced fine-grained sediment input during cold conditions.

Figure 28. A. A 3-D multibeam sonar image of modern scours on the floor of the Beaufort Sea left by wind-driven bottom-dragging ice masses in water depths of 12 m (image courtesy Steve Blasco, GSC). **B**. Laminated shales buckled by grounding ice in the late Paleozoic of Brazil. Rock step in middle of photo is approximately 25 cm high.

erosion and winnowing of previously deposited diamict and the formation of boulder lags or disconformity surfaces. Boulder lags may be subsequently striated by overriding glacier ice, which may take the form of a fully or partially grounded ice shelf, to create marine boulder pavements (Fig. 26D). Deposition of fine-grained sediment is severely reduced during such sea-level lowstands due to persistent agitation of the water column by waves and currents. Previously deposited sediment is remobilized and sorted by storm-wave activity and may be redeposited as hummocky and swaley cross-stratified sands or as *tempestites* (*see* Chapter 8). Reduction of clastic sediment supply to a shelf under cold and/or high-energy conditions also allows the development of extensive colonies of calcareous invertebrates on boulder-lag surfaces forming *coquinas* (Fig. 26C; James *et al.*, 2009). Such carbonate-rich horizons found in glaciomarine successions may record episodes of reduced clastic sediment supply caused by reduced water depths and high-energy associated with glacial advance across the shelf. If water depth decreases sufficiently, ice margins fringing the marine environment may advance across the shelf causing erosion of the substrate or the deposition of till.

In this simple situation of minimal isostatic depression, water depths increase and energy levels decrease across the shelf during periods of ice retreat (or tectonic subsidence), allowing extensive horizons of bioturbated mud to accumulate. These deposits represent periods of enhanced clastic sediment supply to the marine environment and are commonly associated with highstand interglacial conditions (Fig. 27). However, this simple model of ice advance and retreat on glaciated continental shelves is complicated, in many instances, by the very complex interactions between glacio-eustatic, glacio-isostatic and tectonic influences on relative water depths (*see* section *Sequence Stratigraphy in Glaciated Basins* below). This makes paleoclimatic interpretation of disconformities, erosion surfaces and mud blankets in glacially influenced shelf successions extremely difficult.

Some glacially influenced continental margins (*e.g.*, Antarctica) are subsiding slowly and are characterized almost exclusively by deep-water environments. Such shelves can accommodate thick glacial and glaciomarine sediments, although sedimentation rates in cold glacial regimes are extremely slow. Antarctic shelf successions show a well-defined subsurface structure (evident on seismic records) of horizontal 'topsets' recording the buildup (aggradation) of the shelf through time. Till is an important component of these successions and records repeated deposition by shelf-crossing ice sheets (Fig. 3).

The Role of Floating Ice

There is now a new appreciation of the influence of floating ice masses on glaciomarine environments. These masses range from icebergs made of old glacier ice to younger seasonal and perennial ice masses (*pack ice*). Linear or curved furrows, produced by the grounding of iceberg and seasonal ice keels, are abundant on modern high-latitude shelves (Fig. 28A). Iceberg scours can be as wide as 50 m and several meters deep and may extend for several kilometers. In shallow water, continued disturbance (turbation) by ice keels re-suspends fine sediment producing coarse-grained *lag turbates* on the seafloor. The process of scouring may also trigger the sudden release of subsurface gas and water leaving pits in the seafloor called *pock marks*. Both iceberg scours and pock marks are easily destroyed by wave or re-sedimentation processes in shallow water and are rarely reported in the ancient glacial record (Fig. 28B).

Layers of Pleistocene ice-rafted debris have been discovered across large areas of the North Atlantic (*e.g.*, Heinrich layers; Andrews, 1998). Several times during the last 100 000 years, huge armadas of icebergs were released from rapidly calving margins of northern-hemisphere ice sheets leaving distinct ice-rafted horizons in deep-sea mud. These layers of ice-rafted debris form a rich repository of past environmental change related to relatively short-term climatic, glaciologic and/or paleo-oceanographic perturbations.

Figure 29. Glacially influenced continental-slope facies of the Yakataga Formation exposed in the coastal mountain ranges, Gulf of Alaska. **A.** Olistostrome unit (50 m thick) containing large contorted rafts of sandstone turbidites within diamictite facies records slumping of continental-slope deposits. **B.** Sandstone-filled sedimentary dikes injected downwards into diamictite record earthquake activity. Person for scale at lower left. **C.** Fine-grained turbidites with iceberg-rafted dropstones are typical deposits of glacially influenced continental slopes.

Ice-distal Glaciomarine Settings: Continental Slopes

Deep-water slopes are the largest global repositories of glacial sediments and the final resting place for much material transported by ice sheets from continental surfaces (Hjelsuen *et al.*, 2005). Sediments pushed across the shelf under ice sheets are moved into deeper water beyond the shelf edge by slumping and sediment gravity flows. Glaciated continental slopes are typically channelized with smooth interchannel areas (Fig. 3). In areas of smooth slopes, the oceanward growth (progradation) of the slope during successive glaciations is recorded on seismic profiles by large 'foresets' of downslope-thickening sediment wedges dominated by debris-flow facies (debrites) (Januszczak and Eyles, 2001). These are interbedded with interglacial contourite drifts where fine muddy sediment spilling from submarine fans is wafted along parallel to slope contours (*see* Chapter 12). Large-scale slumping of shelf-edge and upper-slope sediment is recorded by folded strata, rafts of displaced sediment and olistostromes (Fig. 29B), which are common features of ancient glaciomarine successions such as the Neoproterozoic Port Askaig Formation of Scotland (Eyles, 1993).

Active slope channels allow the transfer of large quantities of sediment from the outer shelf to deeper marine environments as sediment gravity flows including glaciogenic debris flows and turbidity flows. Massive and stratified diamicts and a variety of massive and graded gravel and sand facies are deposited in slope and base-of-slope channels and fans. Delivery of coarse-grained sediment directly to the shelf edge at times of ice advance across the shelf allows thick units of re-sedimented diamict, broadly channelized in cross section and thinning downslope, to accumulate in slope-cutting channels. Relatively well sorted, coarse-grained sediment transported by meltwater draining the ice provides a source of material for high-concentration turbidity currents (*see* Chapter 12). Stacked successions of unstratified and graded gravels and sands record such coarse-grained flows. Abandoned slope channels may be infilled with muds and rain-out diamicts.

Exceptional examples of relatively young continental-slope facies outcrop around the Gulf of Alaska coastline (Fig. 29; Eyles *et al.*, 1991). There, the subducting Pacific Plate has built an impressive accretionary complex now raised several hundred meters above sea level. Strata (Yakataga Formation) are thick (< 5 km), younger than 5 Ma, and dominated by broadly channelized successions of diamictites, graded conglomerates, sandstones and laminated mudstones (turbidites). Large-magnitude earthquakes are recorded by seismites consisting of olistostromes and dike networks that indicate sudden cracking of the seafloor and downward injection of sediment (Fig. 29). Glacially influenced continental-slope deposits of Pleistocene age are also found in passive-margin settings off the eastern coast of Canada and northeastern USA. Thick successions of glaciogenic debris-flow facies and turbidites are reported from large channels along this slope (*e.g.*, The

Gully, Laurentian Fan, and Trinity Trough).

Ichnology of Glaciated Continental Margins

Marine life occurs in abundance in the hostile (but nutrient-rich) environments around sea-going ice margins (Eyles and Vossler, 1992). Trace fossils are rare in rapidly deposited, coarse-grained ice-proximal marine deposits and increase in importance with distance from the ice margin. Most glaciated shelves and slopes show limited ichnological diversity and a low degree of bioturbation in relation to comparable non-glacial settings. Body-fossil evidence shows that polychaete worms, arthropods, molluscs and echinoderms are the dominant infaunal benthic organisms with deposit feeding being the predominant feeding strategy. The predominance of poorly consolidated muddy ice-rafted sediments in most glaciomarine environments excludes suspension feeders and limits development of the *Skolithos* Ichnofacies typical of non-glacial shelves. Rapid influxes of sediment in glaciomarine environments are recorded by escape structures (*fugichnia*), the presence of *Diplocraterion* and by the mass mortality of bivalves. Firmground burrows of the *Glossifungites* Ichnofacies are associated with erosional surfaces, boulder pavements and rare carbonate beds composed of mollusc valves (coquinas). Although gigantism is hypothesized to characterize cold-water faunas, diminutive feeding traces may reflect salinities lowered by glacial meltwater inputs (*cf.* Chapter 3). The downslope transport of food by turbidity currents and mass flows may explain the development of a diverse *Cruziana* Ichnofacies in water depths where in non-glacial settings the *Nereites* Ichnofacies is found.

SEQUENCE STRATIGRAPHY IN GLACIATED BASINS

Sequence stratigraphy involves subdivision of the sedimentary record into depositional sequences bounded by surfaces formed as a result of changes in accommodation and sedimentation (*see* Chapter 2). Recognition of boundaries resulting from changes in accommodation is extremely difficult in glaciated basins due to the number of processes affecting both accommodation space and sedimentation patterns. In glacioterrestrial environments, depositional sequences recording the advance and retreat of ice masses can take a variety of forms depending on factors such as local and regional topography, glacier dynamics, climate, substrate type, and drainage characteristics. Even within one glacial advance/retreat cycle the relationship between erosional and depositional processes may change both spatially and temporally. The identification of depositional sequences in such settings is not easy and must take into account all factors affecting the local and regional glaciological and depositional regime.

In glaciomarine basins, sequences may be identified on the basis of changing patterns of deposition resulting from relative sea-level change. Over the past 2.5 million years, high-frequency glacio-eustatic changes have profoundly affected coastal sedimentation patterns worldwide. The periodicity of these glacio-eustatic cycles is related to orbitally driven climate forcing (Milankovitch cycles) that for the past one million years has controlled the growth and decay of ice sheets on a 100 000 year cycle. Prior to 1 million years ago, glacial/interglacial cycles had a shorter 40 000 year periodicity. However, glacio-eustatic variations through a glacial cycle are not simple because Pleistocene ice sheets took many tens of thousands of years to grow but disappeared rapidly. In the early stages of the last glacial cycle, sea-level lowering was slow (< 5 m/1000 years) taking as much as 80 000 years to fall to its lowest position (~ 140 m below modern sea level) some 20 000 years ago. During deglaciation, sea level rose 20 m in less than 100 years and at times rose 4 m in 100 years. Some 8200 years ago, as much as 160 000 km^3 of freshwater from a huge glacial lake (Lake Agassiz in mid-continent North America) drained to the oceans in less than 12 months. This changed ocean thermohaline circulation and triggered abrupt changes in global climate. Sea-level recovery since the last glaciation, has slowed dramatically but is still underway. If the Greenland and Antarctic ice sheets were to melt, modern sea level would rise an additional 70 m. This complex, but profound, influence of ice-sheet growth and melting on eustatic sea-level changes indicates that glacial cycles should be considered as a primary control on the development of sedimentary sequences in marine basins during major glacial episodes in Earth history.

In addition to eustatic sea-level changes, the development of glaciomarine depositional sequences is also influenced by water-depth changes resulting from glacio-isostacy. Glacio-isostacy involves the loading (depression) and unloading (uplift) of the Earth's crust as large ice masses grow and decay. This isostatic depression or elevation of the crust creates relative changes in sea level in coastal areas of glaciated basins that are independent of eustatic sea-level change. The large Pleistocene ice sheets loaded the crust sufficiently to cause subsidence of their outer margins well below sea level and the ice-sheet margins began to float. This then allowed large-scale loss of ice mass through calving of icebergs and triggered rapid disintegration and retreat of the ice sheet. Variations in the magnitude of crustal loading across a glaciated basin, combined with the effects of eustatic sea-level changes can cause one part of the basin to experience a fall in relative sea level at the same time as water depths are increasing elsewhere in the basin. The combination of these effects in glaciated basins makes assessment of the sequence-stratigraphic significance of bounding surfaces extremely difficult.

A simple model for the formation of sequences on glaciated continental margins involves movement of ice out onto the shelf during times of glacio-eustatically lowered sea level (Fig. 27; *see* also above). This erosional event creates a sequence boundary. Sedimentary facies formed in ice-contact depositional environments against the advancing ice margin are unlikely to survive glacial advance but will be reworked

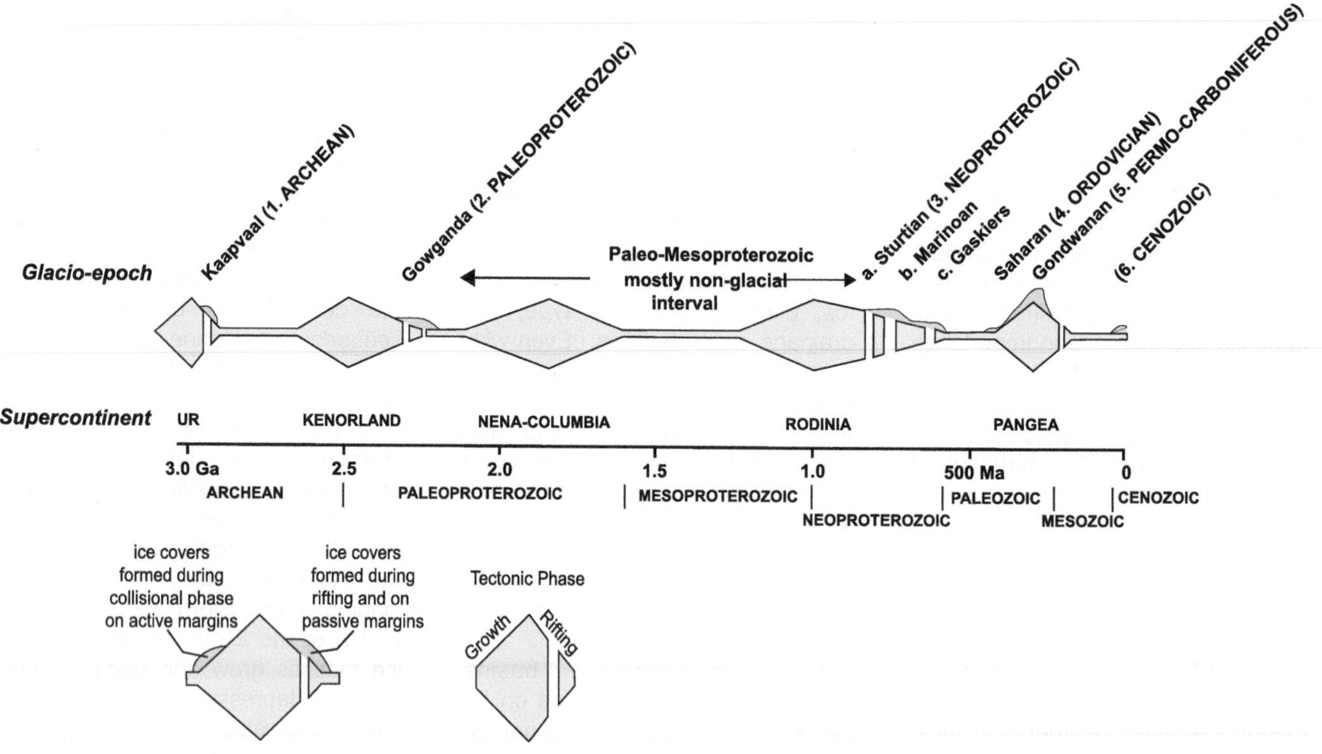

Figure 30. Timing of major episodes of multi-million-year-long glaciations (glacio-epochs) shown against supercontinent growth and rifting.

subglacially as deformation till or will be recycled to the ice front within morainal banks (Fig. 25). Ice advance will ultimately be halted by deeper water at the shelf break. Much of the sediment pushed across the continental shelf by advancing ice will be discharged down the slope. Glacio-isostatic depression of the continental-shelf edge may be sufficient also to initiate deglaciation by enhanced calving of icebergs along the ice margin. Most glaciomarine sedimentation occurs during glacier retreat when large volumes of meltwater are available. Thick, fining-upward successions, consisting of relatively coarse ice-contact deposits, overlain by rain-out and mass-flow diamicts and finally by laminated proglacial silts, are formed during glacier retreat. Soon after the glacier retreats onto land, crustal rebound results in rapid shallowing of the coastal margin. This causes uplift and erosion of glaciomarine deposits creating a bounding unconformity. Subsequent eustatic sea-level rise may flood coastal margins and mud deposition will resume across the shelf during interglacial conditions (Fig. 27).

THE PRE-PLEISTOCENE GLACIAL RECORD

A great expansion of interest in Earth's pre-Pleistocene glacial record has taken place in the last decade. The speculative snowball-Earth hypothesis (Evans, 2003; Etienne *et al.*, 2007; Hoffman 2008) for catastrophic late Proterozoic glaciations, the possible effects of cold climates on the biosphere, oil and gas exploration in late Paleozoic glacial successions, and the need to understand global-climate changes have driven this interest.

A Dominantly Marine Record

During Pleistocene glaciations less than 10% of all sediment produced by ice sheets was left on land. The bulk was delivered to marine environments by meltwater or by ice sheets bulldozing sediment across continental shelves. Most of the total flux ends up along the continental slope and in deep-water trough-mouth fans. Not surprisingly, Earth's pre-Pleistocene glacial record is overwhelmingly preserved in marine rocks.

Six lengthy episodes of cold climate occurred in Earth history when extensive ice cover existed for many millions of years (called *glacio-epochs*; Fig. 30; Eyles, 2008). Each epoch offers its own challenges to glacial sedimentologists and characteristic suites of facies are found in different tectonic settings and basins. The oldest epoch occured during the Archean around 2.8 billion (Ga) years ago (Young *et al.*, 1998), followed at 2.4 Ga in the Paleoproterozoic (Young *et al.*, 2001), a lengthy phase in the Neoproterozoic between 750 to about 600 million years ago (Ma) (Fairchild and Kennedy 2007), and briefly in the early Paleozoic around 440 Ma (Brenchly *et al.*, 2003). A long episode of glaciation occurred in the late Paleozoic (350 to 250 Ma; Eyles *et al.*, 1998) and the most recent, Cenozoic glaciation began some 40 million years ago culminating in the development of extensive continental ice sheets in the northern hemisphere after 2.5 Ma, but especially after 0.9 Ma during the Pleistocene.

The timing of pre-Pleistocene glaciations may be related to the breakup of successive supercontinents (Eyles and Januszczak, 2004). The bulk of ancient glacial deposits occurs in rift basins or in young passive margins formed by continental

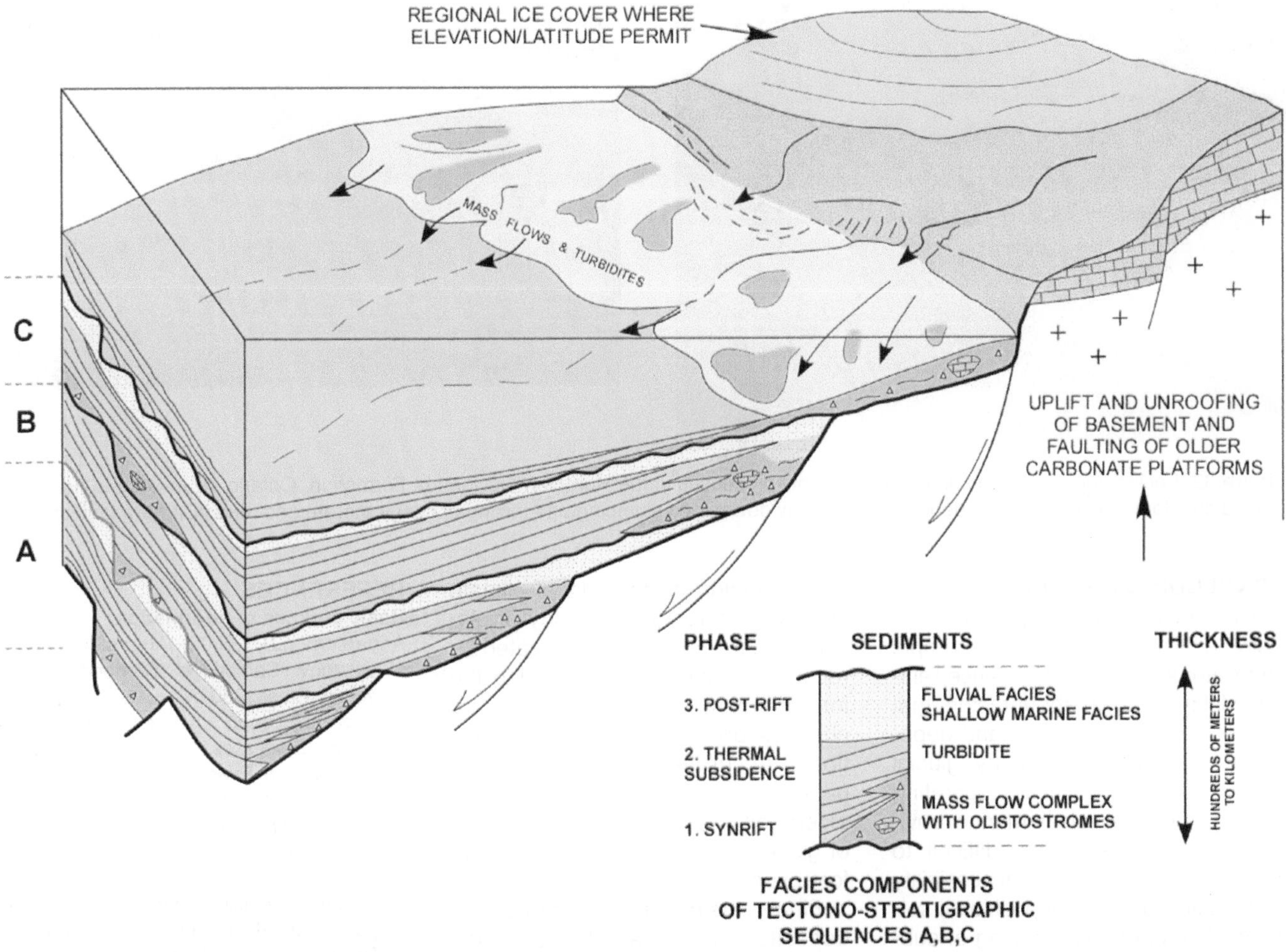

Figure 31. Tectonostratigraphic subdivision of the 2.4 billion year-old glacial Huronian Group in Northern Ontario. Facies can be grouped into distinct packages (tectonic sequences) recording episodic uplift of the basin margin. The lower part of each package consists of thick muddy glaciomarine diamictites succeeded by deeper water laminated shales as the offshore part of the basin subsided. Each package is capped by shallow-water facies as subsidence slowed and sandy deltaic facies prograde out into the basin.

breakup (Fig. 31). This depositional setting is well illustrated by the Paleoproterozoic Gowganda Formation of Ontario, Canada (~ 2.4 Ga), that records glaciation of a marine basin in a large rift basin along the ancient margin of the early North American continent.

Most of the glacial facies preserved within these ancient glaciated basins are the product of sediment gravity-flow processes and include thick successions of diamictite deposited by subaqueous debris flows (debrites) and turbidity currents (turbidites; Fig. 32). Terrestrial tillite is virtually non-existent. Basin tectonics controlled the long-term stratigraphic development of these basins, and facies can be grouped into distinct tectonostratigraphic successions (Fig. 31: Eyles, 2008). These record the tectonic history of the basin commencing with initial crustal stretching and the influx of coarse debris as debris flows from ice on uplifted rift shoulders, followed by rapid basin subsidence and the deposition of thick successions of deep-water turbidites, and an upper capping of shallow-water, sometimes fluvial, sandstone and conglomerate facies. These tectonostratigraphic successions identify the intimate relationship between climate and tectonics, a recurring theme in the study of ancient glaciations.

The Snowball Earth Debate

Some workers have argued for one or more *snowball Earth* events to have affected the Earth, particularly during the Neoproterozoic (between ~750 and ~600 Ma), when ice is hypothesized to have covered both polar and tropical regions, a kilometer-thick ice cover developed on oceans, and all hydrological and biological activity ceased for periods of up to 10 million years at a time (Hoffman, 2008). However, the argument for globally synchronous Neoproterozoic glaciation is not supported by detailed facies studies. Many of the so-called glacial till(ite)s are now recognized as sediment gravity-flow deposits formed in marine rift basins and associated facies indicate a fully functioning hydrologic system. Paleobiological records show that simple bacterial life prospered. Existing dating suggests a series of regional-scale diachronous glaciations tied to the progressive breakup and rifting of *Rodinia*. Emerging tectonic data indicate that ice covers waxed and waned as tectonic processes elevated crust along rift shoulders and created repositories for marine glaciomarine facies in rift basins offshore.

Figure 32. Facies deposited by sediment gravity-flow processes in the Huronian Gowganda Formation, Ontario. **A**. Inverse to normally graded conglomerates. **B.** Typical debrite produced by downslope slumping and mixing of conglomerate and mud. Camera lens cap for scale.

CONCLUDING REMARKS

Glacial deposits are complex and typically difficult to ascribe to simple depositional processes because relatively little is known about factors that control erosion, transport, and deposition of sediment in many glacial environments. Processes operating in subglacial and glaciomarine environments are particularly difficult to investigate due to the inaccessibility and hostility of these depositional settings. Current understanding of subglacial and glaciomarine processes is based, in part, on limited amounts of modern observational data and relies heavily on analysis and interpretation of previously deposited subglacial and glaciomarine deposits that range in age from Proterozoic to Recent. Glacial facies models require further development to provide better understanding of the range and interaction of processes in such complex settings.

Understanding glacial depositional systems requires not only an understanding of glaciers, their behavior and depositional products, but also of other depositional systems that may be influenced by glaciation. Fluvial, lacustrine, eolian, and marine systems can all be affected by glacial processes, either through direct proximity to glacial environments or though supply of sediment by glacial sources. Identifying a glacial signature in sedimentary successions is important for the interpretation of past climatic and paleoenvironmental conditions, and is particularly relevant today given the need for enhanced understanding of past global climate change. Interest in the study of glacial deposits has been invigorated by recent controversies surrounding the interpretation of extreme climate changes and snowball Earth conditions. Accurate reconstructions of past climate depend, to a large extent, on the accurate interpretation of glacial deposits and their mode of formation.

Finally, the understanding of glacial deposits, their characteristics and subsurface geometry is becoming increasingly important for resolution of environmental issues facing many communities in previously glaciated regions. Remediation of subsurface soil and water contamination, the search for groundwater resources, and the location of future waste-disposal sites all depend on glacial facies models to enable prediction of subsurface sediment types and their distribution. The application of glacial sedimentology to the resolution of environmental issues is likely to become increasingly important as urban centers expand.

ACKNOWLEDGEMENTS

This work was supported by NSERC through Discovery Grants awarded to C.H. Eyles and N. Eyles. Bob Dalrymple is thanked for his patience and careful review of the manuscript. Thanks also to Stacey Puckering, Jess Slomka, Riley Mulligan and Paul Durkin for drafting figures.

REFERENCES

Basic sources of information

Anderson, J.B., 2002, Antarctic Marine Geology: Cambridge, Cambridge University Press, 330 p.
Provides good overview of glacial marine processes.

Benn, D.I. and Evans, D.J.A., 1998, Glaciers and Glaciation: Arnold, 734 p.
Fundamental reading.

Bennett, M.R., Huddart, D. and Thomas, G.S.P., 2002, Facies architecture within a regional glaciolacustrine basin: Copper River, Alaska: Quaternary Science Reviews, v. 21, p. 2237-2279.
Excellent example of how to present a glacial facies study.

Boulton, G.S., Dobbie, K.E. and Zatsepin, S., 2001, Sediment deformation beneath glaciers and its coupling to the subglacial hydraulic system: Quaternary International, v. 86, p. 3-28.
The new paradigm in glacial geology; most sediment is moved underneath ice not within it.

Dowdeswell J. A. and O'Cofaigh, C., *eds.*, 2002, Glacier-Influenced Sedimentation on High-Latitude Continental Margins: Geological Society of London, Special Publication 203, 310 p.
Excellent source book.

Eyles, N., 2008, Glacioepochs and the supercontinent cycle after ~3.0 Ga: tectonic boundary conditions for glaciations: Palaeogeography, Palaeoecology, Palaeoclimatology, v. 258, p. 89-129.
Summary of the tectonic settings where ancient glacial facies were deposited and the challenges of recognising ancient cold climates.

Eyles, N., Eyles, C.H. and Miall, A.D., 1983, Lithofacies types and vertical profile models; an alternative approach to the description and environmental interpretation of glacial diamict sequences: Sedimentology, v. 30, p. 393-410.
Still highly relevant introduction to the

problem of diamict vs till.

Evans, D.J.A., Phillips, E.R., Hiemstra, J.F., and Auton, C.A., 2006, Subglacial till formation, sedimentary characteristics and classification: Earth-Science Reviews, v. 78, p. 115-176.
Comprehensive review of how till is formed.

Miall, A.D., 1996, The Geology of Fluvial Deposits: Sedimentary Facies, Basin Analysis and Petroleum Geology: Springer-Verlag Inc., 582 p.
Contains succinct overview of braided river deposits.

Sharp, M., 2005, Subglacial Drainage: Encyclopaedia of Hydrological Sciences, John Wiley and Sons, Ltd., p. 2587-2600.
Good introduction to the characteristics of the subglacial drainage system.

Stow, D.A.V.,2007, Sedimentary Rocks in the Field: A Colour Guide: London, Manson Publishing, 319 p.
Excellent 'how to' hands-on guide to facies descriptions for the student.

Other references

Andrews, J.T., 1998, Abrupt changes (Heinrich events) in late Quaternary North Atlantic marine environments: a history and review of data and concepts: Journal of Quaternary Science v.13, p. 3-16.

Ashley, G.M., 1975, Rhythmic sedimentation in glacial Lake Hitchcock, Massachusetts-Connecticut, *in* Jopling, A.V. and McDonald, B.C., *eds.*: Glaciofluvial and glaciolacustrine sedimentation: SEPM, Special Publication 23, p. 304-320.

Bennett, M.R., Waller, R.I., Glasser, N.F., Hambrey, M.J. and Huddart, D., 1999, Glacigenic clast fabric: genetic fingerprint or wishful thinking: Journal of Quaternary Science, v. 14, p. 125-135.

Boone, S. and Eyles, N., 2001, A geotechnical model for the formation of hummocky moraine by till deformation below stagnant ice: Geomorphology, v. 38, p. 109-124.

Boulton, G.S., 1987, A theory of drumlin formation by subglacial deformation, *in* Menzies, J. and Rose, J., *eds.*, Drumlin Symposium: Rotterdam, Balkema, p. 25-80.

Boulton, G. S., 1996, Theory of glacial erosion, transport and deposition as a consequence of subglacial sediment deformation: Journal of Glaciology, v. 42, p. 43-62.

Boulton, G.S., Van der Meer, J.J., Hart, J., Beets, D., Ruegg, G.H.J, Van der Watern, F.M. and Jarvis, J., 1996, Moraine emplacement in a deforming bed surge - an example from a marine environment: Quaternary Science Reviews, v. 15, p. 961-987.

Boyce, J. and Eyles, N., 2000, Architectural element analysis applied to glacial deposits; anatomy of a till sheet near Toronto, Canada: Geological Society of America Bulletin, v. 112, p. 98-118.

Brenchley, P.J., Carden, G.A.F., Hinta, L., Kaljo, D., Marshall, J.D., Martma, T., Meidla, T. and Nolvak, J., 2003, High-resolution stable isotopic stratigraphy of upper Ordovician sequences: constraints on the timing of bioevents and environmental changes associated with mass extinction and glaciations: Geological Society of America Bulletin, v.115, p. 89-104.

Brennand, T.A. and Shaw, J., 1994, Tunnel channels and associated landfroms, south-central Ontario: their implications for ice-sheet hydrology: Canadian Journal of Earth Sciences, v. 31, p. 505-522.

Clarke, G. K. C., 2005, Subglacial processes: Annual Review of Earth and Planetary Sciences, v. 33, p. 247-276.

Clarke, G.K.C., Leverington, D.W., Teller, J. T., Dyke, A. S. and Marshall S. J., 2005, Fresh arguments against the Shaw megaflood hypothesis: Quaternary Science Reviews, v. 24, p. 1533-1541.

Domack. E.W., Jacobsen, E.A., Shipp, S. and Anderson, J.B., 1999, Late Pleistocene–Holocene retreat of the West Antarctic Ice Sheet system in the Ross Sea: Part 2—Sedimentologic and stratigraphic signature: Geological Society of America Bulletin, v.111, p. 1517-1536.

Etienne, J.L., Allen, P.A., Rieu, R. and Le Guerroué, E., 2007, Neoproterozoic glaciated basins: a critical review of the Snowball Earth hypothesis by comparison with Phanerozoic basins, *in* Hambrey, M.J., Christoffersen, P., Glasser, N.F. and Hubbard, B., *eds.*, Glacial Sedimentary Processes and Products: International Association of Sedimentologists, Special Publication 39, p. 343-399.

Evans, D.A.D., 2003, A fundamental Precambrian-Phanerozoic shift in Earth's glacial style?: Tectonophysics, v. 375, p. 353-385.

Eyles, C.H., Eyles, N. and. Lagoe M.B., 1991, The Yakataga Formation: A six million year record of temperate glacial marine sedimentation in the Gulf of Alaska, *in* Anderson, J.B. and Ashley, G.M., *eds.*, Glacial Marine Sedimentation: Paleoclimatic Significance: Geological Society of America, Special Paper 261, p.159-180.

Eyles, C.H., Eyles, N. and Gostin, V., 1998, Facies and allostratigraphy of high latitude glacially influenced marine strata of the Early Permian southern Sydney Basin, Australia: Sedimentology, v. 45, p. 121-163.

Eyles, N., 1993, Earth's glacial record and its tectonic setting: Earth-Science Reviews, v. 35, p. 1-248.

Eyles, N., 2006, The role of meltwater in glacial processes: Sedimentary Geology, v.190, p. 257-268.

Eyles, N. and Vossler, S., 1992, Ichnology of a glacially-influenced continental shelf and slope; the Late Cenozoic Gulf of Alaska (Yakataga Formation): Palaeogeography, Palaeoecology, Palaeoclimatology, v. 94, p. 193-221.

Eyles, N. and Januszczak, N., 2004, Zipper Rift: Neoproterozoic glaciations and the diachronous break up of Rodinia between 740 and 620 Ma: Earth-Science Reviews, v. 65, p. 1-73.

Eyles, N., Daniels, J., Osterman, L. and Januszczak, N., 2001, Ocean Drilling Program Leg 178 (Antarctic Peninsula): Insights in to the sedimentology of continental shelf topsets and foresets: Marine Geology, v. 178, p. 135-156.

Eyles, N., Eyles, C.H., Woodsworth-Lynas, C. and Randall, T., 2005, The sedimentary record of drifting ice (early Wisconsin Sunnybrook deposit) in an ancestral ice-dammed Lake Ontario: Quaternary Research, v. 63, p. 171-181.

Fairchild, I. and Kennedy, M.J. 2007, Neoproterozoic glaciation in the Earth System: Journal of the Geological Society, v. 164, p. 895-921.

Heroy, D.C. and Anderson, J.B., 2005, Ice sheet extent on the Antarctic Peninsula region during the last glacial maximum; insights from glacial geomorphology: Bulletin of the Geological Society of America, v. 117, p. 1497-1512.

Hildes, D.H.D, Clarke, G. K.C., Flowers, G.E. and Marshall, S.J., 2004. Subglacial erosion and englacial sediment transport modeled for North American ice sheets: Quaternary Science Reviews, v. 23, p. 409-430.

Hjelsuen, B.O., Sejrup, H.P., Haflidason, H., Nygard, A., Ceramicola, S. and Bryn, P., 2005, Late Cenozoic glacial history and evolution of the Storegga Slide area and adjacent slide flank regions, Norwegian continental margin: Marine and Petroleum Geology, v. 22, p. 57-69.

Hoffman, P.F., 2008, Snowball Earth: status and new developments: GEO (IGC Special Climate Issue), v. 11, p. 44-46.

Hooke, R.L. and Elverhoi, A., 1996, Sediment flux from a fjord during glacial periods, Isfjorden, Spitsbergen: Global and Planetary Change, v.12, p. 237-249.

Januszczak, N., and Eyles, N., 2001, ODP drilling leads to a new model of shelf and slope sedimentation along the Antarctic continental margin. Geoscience Canada, v. 28 (4), p. 203-210.

James, N.P., Eyles, C.H., Eyles, N., Hyatt, E. and Kyser, K., 2009, Oceanographic significance of a cold-water carbonate environment: glaciomarine sediments of the Pleistocene Yakataga Formation, Middleton Island, Alaska: Sedimentology, v. 56, p. 367-398

Lønne, I., Nemec, W., Blikra, L.H. and Lauritsen, T., 2001, Sedimentary architecture and dynamic stratigraphy of a marine ice-contact system: Journal of Sedimentary Research, v. B71, p. 922-943.

Menzies, J., 1989, Drumlins - Products of controlled or uncontrolled glaciodynamic response?: Quaternary Science Reviews, v. 8, p. 151-158.

Meriano, M. and Eyles, N., 2009, Quantitative assessment of the hydraulic role of subglaciofluvial interbeds in promoting deposition of deformation till (Northern-Till, Ontario): Quaternary Science eviews, v. 28, p. 608-620.

Paul, M.A. and Eyles, N., 1990, Constraints on the preservation of diamict facies (melt-out tills) at the margins of stagnant glaciers: Quaternary Science Reviews, v. 9, p. 51-69.

Sejrup, H.P., Hjelstuen, B.O., Dahlgren, T., Haflidason, H., Kuijpers, A., Nygard, A., Praeg, D., Stoker, M., and Vorren, T.O., 2005, Pleistocene glacial history of the NW European continental margin: Marine and Petroleum Geology, v. 22, p. 1111-1129.

Shaw, J., 1977, Sedimentation in an alpine lake during deglaciation: Okanagan Valley, British Columbia, Canada: Geografiska Annaler, v. 59A, p. 221-240.

Shaw, J., Kvill, D. and Rains, B., 1989, Drumlins and catastrophic subglacial floods: Sedimentary Geology, v. 62, p. 177-202

Tripsanas E.K. and Piper D.J.W., 2008, Glaciogenic debris-flow deposits of Orphan Basin, Offshore Eastern Canada: Sedimentological and rheological properties, origin, and relationship to meltwater discharge: Journal of Sedimentary Research, v. 78, p. 724-744.

van der Meer, J.J.M., Menzies, J., and Rose, J., 2003, Subglacial till: the deforming glacier bed: Quaternary Science Reviews, v. 22, p. 1659-1685.

Young, G.M., Von Brunn, V., Gold, D.J.C. and Minter, W.E.L., 1998, Earth's oldest reported glaciation: physical and chemical evidence from the Archean Mozaan Group (~2.9 Ga) of South Africa: Journal of Geology, v. 106, p. 523-528.

Young, G.M., Long, D.G.F., Fedo, C.M. and Nesbitt, H.W., 2001, Paleoproterozoic Huronian basin: product of a Wilson cycle punctuated by glaciations and a meteorite impact: Sedimentary Geology, v. 141, p. 233-254.

6. ALLUVIAL DEPOSITS

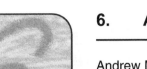

Andrew Miall, Department of Geology, University of Toronto, Toronto, ON, M5S 3B1, Canada

INTRODUCTION

Rivers are the major agent of sediment transfer from the continents into the world's seas and oceans. Surface drainage systems develop as soon as areas of continental crust are uplifted above sea level, and naturally evolve into systems of trunk rivers occupying the lowest-lying areas, with tributaries flowing into them from nearby highlands. Major rivers, such as the Mississippi–Missouri system in North America, and the Amazon system of South America can, with their tributaries, drain a significant proportion of a continental hinterland. For example, the Mississippi–Missouri system, the third-largest drainage system in the world, drains 41% of the 48 conterminous states of the United States, plus parts of Canada. Well-established river systems may persist while tectonic uplift takes place across them, resulting in deep incision of major channels. These are called *antecedent rivers*. The Colorado system and the Grand Canyon of the southwest United States, and the rivers draining southward into the Indian subcontinent from the Himalayan ranges are examples.

Major river systems may be extremely long-lasting because of the geological longevity of the structural controls that determine their position. Stratigraphic evidence indicates that the rivers draining into the present Gulf Coast, including the Mississippi, Brazos and Rio Grande, have occupied approximately their present courses across the coastal plain since the early Cenozoic. Such rivers are responsible for the removal and transfer of immense volumes of sediment. For example, some 25–30 km of surface erosion of the 1-Ga-old Grenville (tectonic) Province of eastern North America occurred prior to the Cambrian–Ordovician transgression, with much of the resulting sediment delivered by long-vanished rivers to the west coast of the continent, where it contributed to the formation of thick wedges of continental-margin sedimentary units of Proterozoic age. During the late Paleozoic, as much as 13 km of sediment was removed from the geologically young Appalachian system and transferred southwestward, and westward, into depositional basins on the southern and western margins of the continent. The Mississippi–Missouri system represents the latest giant river system to have been established across the continental interior. It evolved into its present configuration following the early Cenozoic Laramide uplifts of the Cordilleran ranges.

Alluvial rivers are defined as those flowing across their own deposits, in contrast to incised rivers that flow within eroded valleys. The terms *alluvial* and *fluvial* are used as synonyms. The distributary channels of deltas are alluvial, and so the contents of this chapter therefore also apply to that part of deltaic systems (*see* Chapter 10). From the headwaters of a river to its mouth, at sea level (base level), the volume of water in a river increases as it is joined by tributaries (unless aridity and infiltration cause excess water losses). The sediment load also increases in volume downstream, but becomes reduced in caliber as a result of sorting, abrasion, and selective deposition (*see* Chapter 4). The dynamic balance between water and sediment generates a *graded longitudinal profile* for the river, steeper at source, and flattening out to a fraction of a degree at the mouth. This profile may be continually changing in response to changes in the input of water and sediment, or tectonic movement along the river course. The deposits of alluvial rivers may be stratigraphically significant, especially in basins adjacent to areas of major uplift, where sediment thicknesses may reach 8–12 km. Such deposits serve as an invaluable record of the tectonic history of the structurally uplifted hinterland areas because the caliber of alluvial sediment, and its direction of transport, are a direct reflection of the steepness and direction of the slope imposed on drainage systems by tectonism.

Fluvial deposits are typically characterized by good porosity and permeability, and commonly constitute excellent aquifers. Many oil and gas reservoirs are hosted in fluvial units, and an understanding of the sedimentology and sequence stratigraphy of such deposits may comprise an important component of an exploration and production program in some petroleum provinces. Knowledge of fluvial architecture can become particularly useful during the enhanced recovery stage of exploitation, when the details of internal heterogeneity can be used to increase hydrocarbon recovery (Tyler, 1988; Martin, 1993). Some coal deposits are hosted in fluvial strata, and many placer deposits of gold, uranium, tin, and other heavy minerals are also located in non-marine units.

Fluvial deposits range from extremely coarse conglomerates to fine muds, reflecting the wide range of sub-environments within non-marine systems, from high-energy torrential streams in mountainous areas, to the broad, slow-moving channels of major trunk rivers. Chemical sediments constitute a minor component of most fluvial assemblages. They may include nodular carbonate deposits representing fossil soils, coal, and minor evaporites formed in non-marine sabkha settings (*see* Chapters 16, 20 and 21). Most fluvial sediment is transported by *traction currents*, the discharge characteristics of which may be deduced from the grain size and sedimentary structures of the resulting deposits

(Simons *et al.*, 1965), as discussed later.

The major sedimentary and landform elements of river systems are illustrated in Figure 1. The relative scales of the components of alluvial systems are described in Table 1 and illustrated in Figure 2. Studies of the sedimentary facies and sedimentary structures of fluvial deposits, and the geometry and relative arrangement of these components, constitute the basis of fluvial facies models, and are an essential part of the work to reconstruct the rivers that deposited them.

Following an introduction to the variability in fluvial styles in natural rivers, this chapter provides an introduction to the methods of facies and stratigraphic analysis of fluvial deposits, commencing with the recognition, classification and interpretation of individual lithofacies units. The next step is to demonstrate how these units are combined to form the major landscape elements of rivers, such as the component channels and bars. The preserved deposits of these features are termed *architectural elements*. Fluvial styles may be reconstructed from an analysis of the component lithofacies and architectural elements of a fluvial deposit. These yield insights into the *autogenic* controls on fluvial sedimentation, *i.e.,* the processes that naturally occur within a river system, such as the evolution of meanders, or the flooding of overbank areas during high-discharge events. The deposits of each fluvial style may be characterized by its own *facies model*, and several of the more common styles and their facies models are described here. On a basin-wide scale, river systems are controlled by *allogenic* processes, the processes of external forcing, that may include tectonism, climate change and, in coastal regions, sea-level change. Stratigraphic methods are utilized to study fluvial deposits at this scale, including methods of sequence stratigraphy. The chapter also includes a brief discussion of the tectonic settings of fluvial deposits.

Subsurface mapping methods are discussed briefly in the sections on architectural elements and cyclic sedimentation. Wireline logging has been a standard tool for many decades. This is now supplemented by several

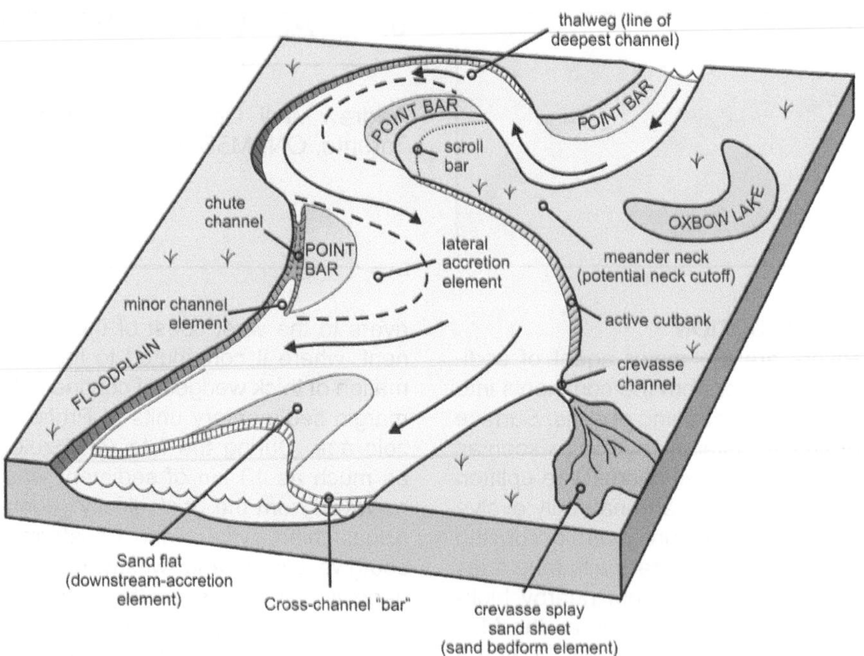

Figure 1. Block diagram of a river, illustrating some of the main terms used to describe the landscape elements of rivers.

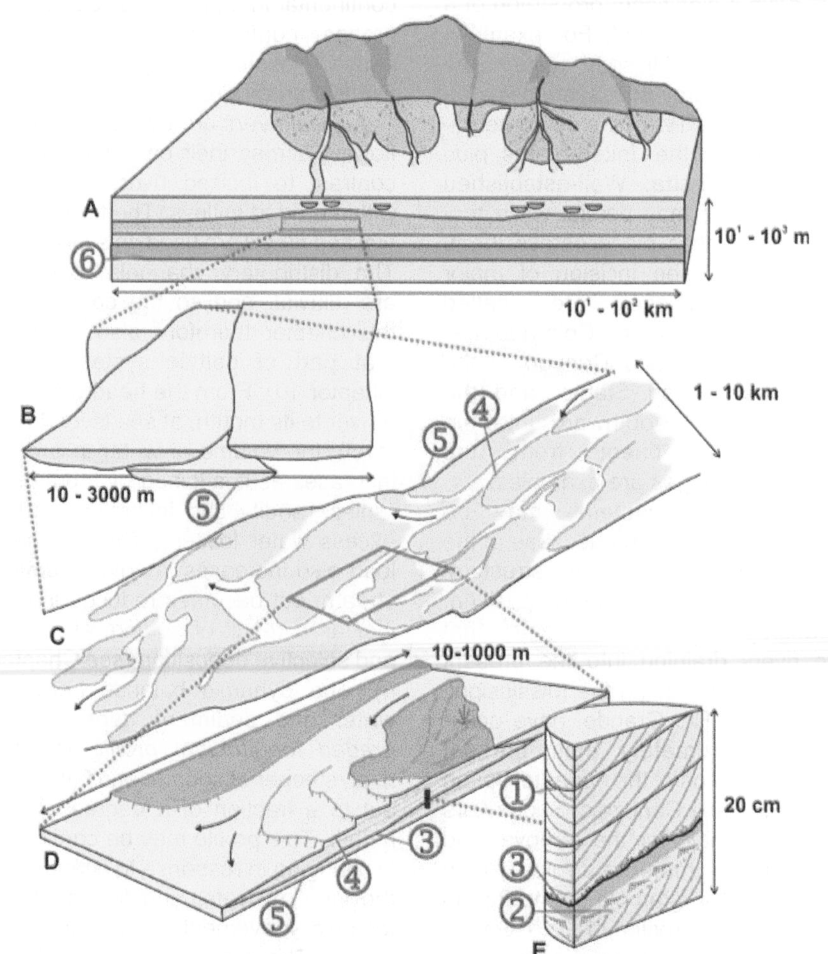

Figure 2. The hierarchy of lithologic units in a fluvial system. Each diagram represents an enlargement of the area shown by the dashed lines in the diagram above. The circled numbers refer to a system used for ranking the extent of the surfaces bounding units from the smallest to the largest scale.

Table 1. Hierarchy of depositional units in alluvial deposits

Group	Time scale of processes (yrs.)	Examples of processes	Instantaneous sedimentation rate (m/ka)	Fluvial, Deltaic depositional units	Rank and characteristics of bounding surfaces
1	10^{-6}	burst-sweep cycle		lamina	0th -order, lamination surface
2	10^{-5}-10^{-4}	bedform migration	10^{5}	ripple (microform)	1st -order, set bounding surface
3	10^{-3}	bedform migration	10^{5}	diurnal/seasonal dune increment, reactivation surface	1st -order, set bounding surface
4	10^{-2}-10^{-1}	bedform migration	10^{4}	dune (mesoform)	2nd -order, co-set bounding surface
5	100-10^{1}	seasonal events, 10-year flood	10^{2-3}	macroform growth increment	3rd -order, dipping 5–20° in direction of accretion.
6	10^{2}-10^{3}	100-year flood, channel and bar migration	10^{2-3}	macroform (*e.g.*, point bar, levee, splay), immature paleosol	4th -order, convex-up macroform top, minor channel scour, flat surface bounding floodplain elements
7	10^{3}-10^{4}	long term geomorphic processes (*e.g.*, channel avulsion)	10^{0}–10^{1}	channel, delta lobe, mature paleosol	5th -order, flat to concave-up major channel base
8	10^{4}-10^{5}	5th-order[1] Milankovitch[2] cycles, or response to fault pulse	10^{-1}	channel belt, alluvial fan, minor sequence	6th -order, sequence boundary; flat, regionally extensive or base of incised valley
9	10^{5}-10^{6}	4th-order[1] Milankovitch[2] cycles, or response to fault pulse	10^{-1}–10^{-2}	major dep. system, fan tract, sequence	7th -order, sequence boundary; flat, regionally extensive, or base of incised valley
10	10^{6}-10^{7}	3rd-order[1] cycles. Tectonic and eustatic processes	10^{-1}–10^{-2}	basin-fill complex	8th -order, regional disconformity

[1] The "order" terminology provided here is that of Vail *et al.* (1977)
[2] Milanovitch cycles: those generated by orbital forcing, which may cause eustatic (including glacioeustatic) sea-level change, or changes in fluvial discharge and sediment yield driven by climate change.

invaluable remote-sensing methods.

Quaternary fluvial sands and gravels underlie many cities, serve as shallow aquifers, as disposal sites for pollutants, and as sources of aggregates for construction and road building. A detailed knowledge of fluvial architecture may be very useful for construction, prospecting and exploitation. Ground-penetrating radar (GPR) has become an essential tool that can reveal internal sedimentological details with a resolution of a few decimeters, to depths of up to about 10 m. Stephens (1994), Beres *et al.* (1995), Bridge *et al.* (1998), Asprion and Aigner (1999) and Lunt and Bridge (2004) provide examples of the application of GPR surveys to modern and ancient fluvial systems.

High-quality reflection seismic data are increasingly used for the reconstruction of depositional systems. *Seismic geomorphology* refers to the technique for reconstructing ancient landscapes from three-dimensional seismic data (Davies *et al*., 2007). Fluvial channels and architectural elements such as meander bends, point bars and braid bars may be imaged on horizontal seismic sections. Good examples have been provided by Miall (2002), Sarzalejo and Hart (2006), and Ethridge and Schumm (2007). Pressure data may be used to test for fluid interconnectedness between sand bodies (Hardage *et al*., 1996), and can provide important supplementary information to architectural and stratigraphic reconstructions made from combinations of wireline-log and seismic data.

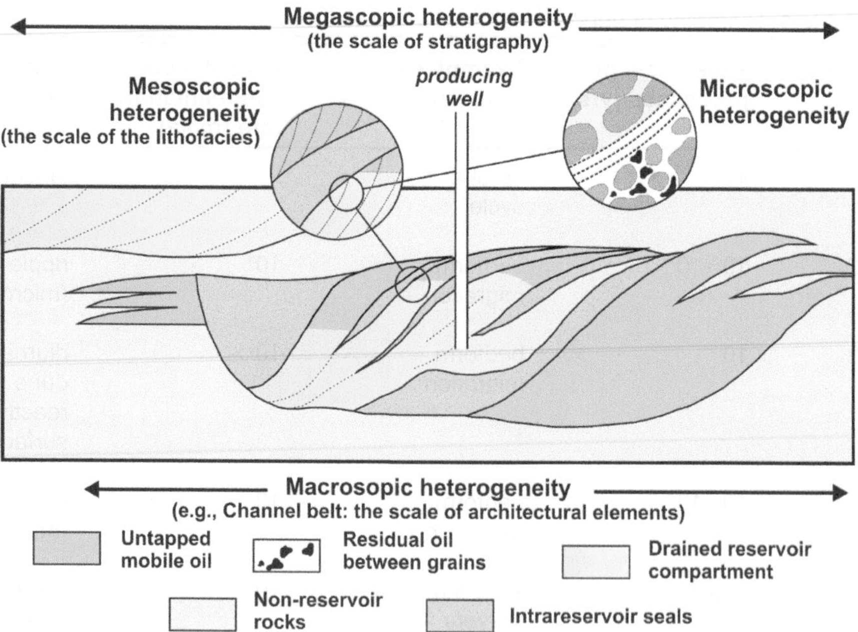

Figure 3. Schematic cross-section through a meander-belt complex, showing the various scales of heterogeneity, and the relationship of these heterogeneities to the distribution of reservoir fluids (Tyler, 1988).

THE HIERARCHY OF SCALES OF BOUNDING SURFACES AND DEPOSITIONAL UNITS

Geologists describe clastic deposits in terms of their natural depositional units, ranging from the largest, an entire basin-fill complex, to the smallest, an individual bed or lamina. Fluvial deposits can be naturally subdivided into a ten-fold hierarchy of depositional units (informally termed 'groups' in this chapter), based on their physical dimensions, their sedimentation rate, and the time scale they represent (Table 1). At each of these scales, rock bodies are enclosed by bounding discontinuities. These include the contacts between beds and sedimentary structures, channel scour surfaces, the base and top of stratigraphic units and the surfaces that define the major allogenic subdivisions of alluvial successions. Sequence boundaries, of 6th to 8th order in rank (in the terminology of Vail *et al*., 1977; Miall, 1996), range from structurally conformable discontinuities reflecting climatic or eustatic processes, to regional angular unconformities generated by tectonism.

A sedimentological classification of these surfaces (Table 1, Fig. 2) is a useful tool for field studies, facilitating field description and leading to more comprehensive interpretations. Petroleum-reservoir geologists recognize at least four scales of heterogeneity for purposes of calculating volumes and rates of production (Fig. 3). *Microscopic heterogeneity* is concerned with porosity–permeability variations at the scale of the individual sand grains. *Mesoscopic heterogeneity* is that made by bedding units and sedimentary structures. Bounding surfaces between individual bedding sets of the same lithofacies are classified as rank 1 bounding surfaces. Simple contacts between dissimilar lithofacies are classified as rank 2. *Macroscopic heterogeneity* includes the variability associated with the deposition of channels and bars. Bars constitute a category of architectural element in a channel and are bounded by surfaces of rank 4. Internal minor erosion surfaces and reactivation surfaces are classified as rank 3 (but may represent groups 3, 4 or 5 of Table 1, depending on their temporal significance). Major channels are rank 5. *Megascopic heterogeneity* deals with the variations across major sedimentary units and entire basins. This category encompasses lithostratigraphic units or stratigraphic sequences, bounded by surfaces of rank 6 and higher.

FLUVIAL STYLES

Rivers display a wide variety of channel and floodplain styles, reflecting the magnitude and caliber of the sediment load, the amount and variability of discharge, and bank stability. Four end-member fluvial styles are common (Fig. 4), and are the basis for the descriptions and analyses presented in this chapter. Fluvial style is governed mainly by the flow and sedimentary processes that operate during seasonal floods (Bridge, 2003). Such events are referred to as the *channel-forming discharge*. They may be equal to, or greater than, the *bankfull discharge*.

All rivers (and all fluid systems) generally meander as a result of turbulence, internal shear, and bank and bed friction. Straight channels, including canals and channels in flumes, commonly develop sinuous thalwegs because of these controls, with *alternate bars* on the insides of the bends (Figs. 4 and 5), so called because they develop on alternate banks of the channel. Fully developed meanders in sinuous rivers typically have areally extensive bars on the inside of the meander bend. In natural meandering rivers these are called *point bars*, and are amongst the most characteristic features of the meandering style (Figs. 1 and 4).

There is a natural transition between *braided* (low-sinuosity, multiple-channel rivers) and *meandering* (high-sinuosity, single-channel) styles, dependent upon channel slope and discharge. Thus, for a given slope, a river changes from meandering to braided as discharge is

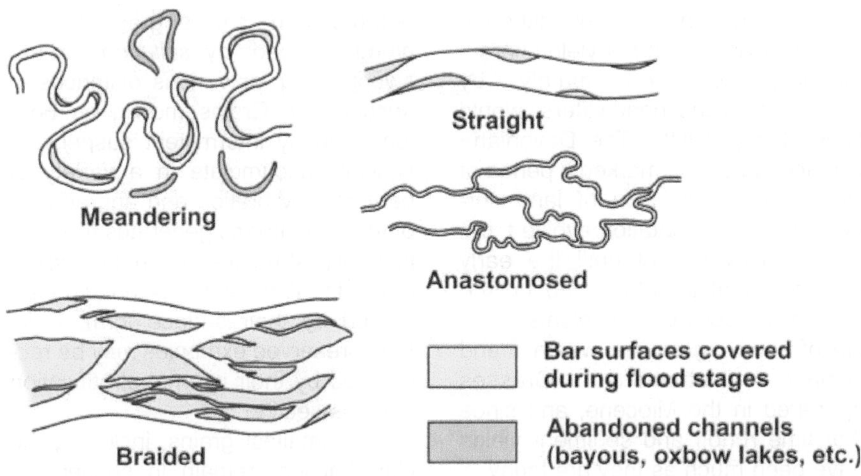

Figure 4. The four 'end-member' fluvial styles that represent some of the most common fluvial patterns occurring in nature.

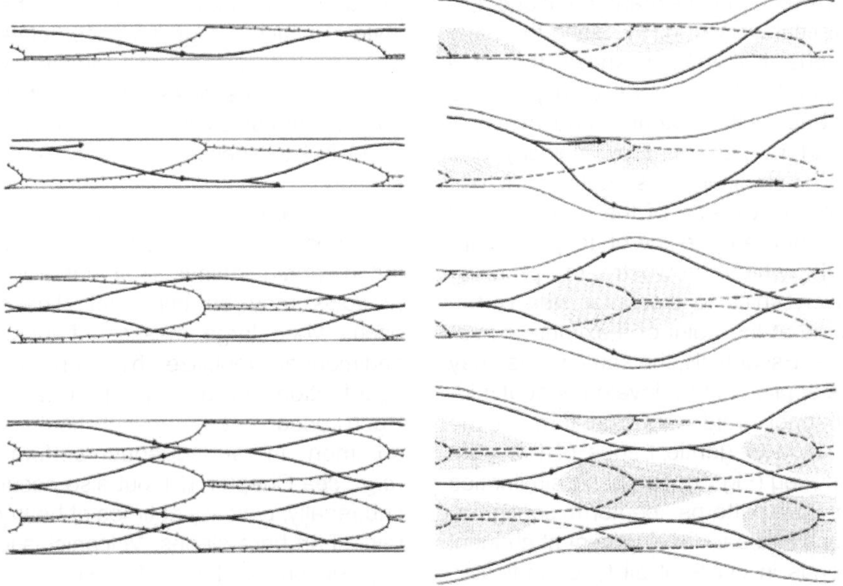

Figure 5. Idealized model for the early evolution of different channel patterns (right) from straight channels of different widths (left). Heavy curved lines show location of thalweg, lines with ticks show original position of alternate bars; remnants of these are shown by dashed lines in diagrams at right. Stippled areas are topographic highs (bars). Sinuous thalwegs lead to the development of rows of alternate bars. Narrow channels with single rows of bars (upper two rivers) evolve into meandering channels with point bars. Wider channels (lower two rivers) develop mid-channel bars and become braided (Bridge, 1985).

increased. For a few sets of rivers in specific climatic ranges, some of these relationships have been quantified. However, the data are inadequate to permit generalizations that can be applied to all geological conditions, a fact that considerably reduces the usefulness of quantitative geomorphology for geological reconstructions.

One of the primary prerequisites for braiding is a large quantity of coarse-grained sediment load (*bedload*). Leopold and Wolman (1957, p. 50) stated,

"*Braiding is developed by sorting as the stream leaves behind those sizes of the load which it is incompetent to handle ... if the stream is competent to move all sizes comprising the load but is unable to move the total quantity provided to it, then aggradation may take place without braiding.*"

The coarse, non-cohesive material comprising the bedload may be material transported along the river from upstream or introduced from tributaries, but may also include material eroded from non-cohesive banks, such as earlier braided-stream deposits (*e.g.*, terraces).

Braided rivers have several or many channels, separated by temporary bars and islands. They are commonly characterized by high discharge variability; *i.e.*, rivers in alpine and arctic areas that have seasonally high discharge variations, tend to be braided, as are ephemeral rivers in arid regions. In rivers of highly variable discharge, competency will be similarly variable, and there will be long periods of time during which the river will be unable to move at least the coarsest part of its bedload, which remains in temporary 'storage' as bar deposits. The flooding of channels, erosional diversion and initiation of new channels, migration of bars, and the movement of large quantities of bedload occur mainly during the occasional peak flood events (seasonal runoff, storm events).

Broad inland and coastal plains characterized by low slopes may be crossed by numerous *single-thread* or *anastomosed* (branching) rivers of low to moderate sinuosity. These rivers tend to be relatively stable in position and are therefore characterized by narrow channels, quite different from those of the braided and meandering styles. Anastomosed channels develop in areas of rapid aggradation, such as in confined, rapidly subsiding basins, or where rapid base-level rise is matched by an abundant sediment supply. The modern lower Rhine−Meuse system is a good example (*e.g.*, Törnqvist, 1993).

On the scale of an alluvial basin, the nature of the vegetation cover has an important effect on discharge characteristics, sediment load and fluvial style. In vegetated areas, runoff, following major rainfalls, is rarely catastrophic because the water is absorbed by the soil and plants and released slowly. Sediment is stabilized by roots, and vegetated banks are therefore highly resistant to erosion. This not only limits the supply of sediment to the river, but also reduces the bank widening that commonly accompanies braiding. In vegetated areas, because of the effects on discharge variability, bank stability and sediment supply, braiding is inhibited. The changes in river behavior when forests in upland catchment areas are destroyed (by fire or deforestation) are marked and have been well known for centuries. Peak discharge, the sediment yield and sediment calibre all increase, and a tendency may develop for debris flows to occur. Conditions were different in the geological past, prior to the evolution of land vegetation (see below).

Rivers are sensitive to longitudinal changes in the controlling variables and to temporal changes in these variables. For example, studies of rivers in the Canterbury Plains of New Zealand show that the transition from meandering to braided takes place at higher slopes as the caliber of the sediment load is increased. The development of the braided pattern is very sensitive to the local rate of supply of bedload to the river. A large volume of bedload, supplied, for example, from easily eroded banks, leads to channel shoaling and local flooding, bank incision and avulsion. Rivers may change in style as a result of triggering events. For example, a major flood, or a volcanic eruption, introducing unusual quantities of coarse bedload may change a river from a meandering to a braided style. Conversely, the damming of rivers, reducing their discharge variability, may lead to increased channel stability and the development of a meandering style.

The evolution of vegetation had a major effect on fluvial styles through geologic time because of the implications for bank stability and erodibility, and the rate of supply of sediment into river systems (Schumm, 1968; Davies and Gibliing, 2010). Prior to the Devonian, there was little or no land vegetation, and the land surface probably appeared much as arid areas do today, even where rainfall was high. Runoff would have been as flash floods and the sediment yield large. Braided fluvial styles predominated. From the Devonian to the end of the Paleozoic, vegetation was probably confined to nearshore and coastal plain areas, so that bank stabilization would have begun, although discharge and sediment-yield characteristics, controlled mainly by processes in the headwaters, would have changed little. The Devonian–Cretaceous period marked a period of increasing stabilization of land surfaces as land vegetation evolved, but it was probably not until the early Cenozoic that interfluve and upland areas were colonized by plants capable of surviving severe weather and climatic fluctuations. Grasses appeared in the Miocene, and since that time runoff and sediment yields have been much as they are today.

A useful idealized model for the evolution of different channel styles was developed by Bridge (1985) based on much experimental and theoretical work (Fig. 5). Since the early 1960s, it has been known that flow around a bend leads to a pressure set-up at the cutbank, and a helical overturn pattern. Flow velocity and shear stress increase outward across the convex side of the bend, toward the thalweg (the line of deepest channel), which, downstream, progressively shifts to the outer side of the bend at the point of maximum curvature. Essentially, the same model may be applied to the development of mid-channel braid bars that form at the outside of gentle bends of the local thalweg (Fig. 5). The basic similarities in flow patterns, bedform configurations and bar development at channel bends in rivers of all types indicates that studies of the geological results of these processes in small outcrops cannot hope to allow deductions about fluvial style, a point now made by many writers (*e.g.*, Miall, 1980; Bridge, 1985). Development of a complete model of a given ancient river depends on a thorough architectural analysis of the sediments, incorporating local and regional stratigraphic, facies, architectural-element, and paleocurrent data.

DEPOSITIONAL PROCESSES

Sediment is transported in rivers by two mechanisms, *traction currents* and *sediment gravity flows* (*see* Chapter 4). Traction currents are those that transport cohesionless sediment as dispersed grains, each moving individually. Large grains are moved by sliding or rolling along the bed (*bedload*); smaller grains bounce along the bed (by *saltation*) or are swept along for various distances in *suspension*. Grains moved as bedload and by intermittent suspension typically accumulate in a variety of bedforms. Migration and aggradation of these bedforms generates the sedimentary structures seen in ancient deposits. The bedforms are characterized by their surface form, while their preserved examples may be recognized by their internal stratification in cross-section.

The smallest grains, including the clay fraction, remain in suspension unless the water body comes to a complete stop, as in a floodplain pond or abandoned channel; at this time, the grains will slowly settle out. The style in which the grains accumulate to form stratification and sedimentary structures is the basis for the definition of lithofacies in traction-current deposits, as described in the next section.

Most in-channel alluvial sediments are deposited from traction currents, but in certain settings *sediment gravity flows* may be important. These occur when large masses of loose sediment are mobilized by failure and liquefaction on a sloping surface. Such occurrences are particularly common in subaqueous settings (*e.g., see* Chapter 12), but also occur subaerially, typically at times of heavy rainfall on bare slopes. Volcaniclastic deposits on the slopes of active volcanoes are particularly prone to this style of sediment transport and deposition. A flow may start as a landslide, and develop into a moving mass of wet, unsorted sediment termed a *debris flow*. Movement ceases when the flow looses momentum on a flat valley floor, and when the lubricating effect of pore waters has been lost by drainage into the substrate. The result is a largely chaotic deposit of poorly sorted debris, containing large pebbles, cobbles and even giant boulders, mixed together, and usually separated from each other by a finer grained matrix of sand, silt and clay. Close examination may reveal a subtle vertical size-segregation (grading) of the grains or the matrix, and a preferred orientation of the larger clasts that is imposed on the deposit by internal shear during the last few minutes before the flow 'freezes' on the basin floor.

Two terms, *competency* and *capacity*, are used to describe the ability of a river to transport sediment (*see* Chapter 4). Competence indicates the grain size of sediment that can be transported, and relates to the velocity and depth of flow. Capacity indicates the total volume of sediment that can be moved, and is a reflection of the magnitude of the discharge. In a later section, we discuss a suite of numerical models that explore the variations of these parameters and their effects on deposition.

Bankfull flow is the main channel-forming flow, and typically occurs during seasonal runoff, or during the rainy season in humid climates, or during rare catastrophic flow events in ephemeral environments. At this time, or at rare times when even this discharge is exceeded, flow may overtop the main channel and spread out onto the floodplain of the river. In braided systems, active channels typically occupy most of the valley floor, but in other fluvial environments, and especially in unconfined alluvial or coastal plains, the floodplain may be extensive. Sedimentary processes here include the settling of muds from the temporary ponds generated by overbank flow, sandy crevasse-splay deposits that spread directly from breaks (*crevasses*) in the channel bank and, in some settings, soils formed during drier phases. The type of soil may be a direct reflection of climatic setting, ranging from carbonized plant fragments, root beds and sideritic mudstones of reducing environments, to the calcretes that may form in semi-arid environments (*see* next section, Lithofacies).

LITHOFACIES

Observation and classification of lithofacies are now standard components of the facies-analysis methods for studying sedimentary rocks (*see* Chapter 2). Beds are classified on the basis of their primary depositional attributes, notably (in the case of fluvial clastic deposits) bedding, grain size, texture, and sedimentary structures. Biogenic structures and fossils may be important locally as additional descriptive attributes. Biochemical sediments, such as pedogenic calcretes, coal and evaporites, typically form minor components of most fluvial systems, but need their own careful characterization. The scale of an individual lithofacies unit depends on the level of detail incorporated in its definition. Facies may be defined very broadly, to encompass mapped stratigraphic units, or they may be defined finely, to accommodate the level of detail obtainable in the centimeter-by-centimeter logging typically carried out on core. For the purpose of architectural-element analysis, a relatively fine degree of description and subdivision is required.

It is good research practice to approach each new rock unit afresh, with the aim of making complete, unbiased observations of all important lithofacies attributes. However, sedimentological research has demonstrated that much of the apparent variability in sedimentary units disguises a limited range of basic lithofacies and biofacies types. The depositional processes that control the development of clastic fluvial lithofacies, such as traction-current transportation, and its accompanying fluid turbulence and the resulting effects on beds of clastic grains, are common to all rivers and obey the same physical laws everywhere, with the production of similar suites of lithofacies. For example, hydrodynamic structures, such as ripples and crossbedding, are formed by the migration of ripples and dunes. Considerable experimental work has shown that in the development of these bedforms there are consistent empirical relationships between bedform size and shape and a limited suite of physical parameters, of which the most important are the depth and velocity of the flow, and the grain-size of the sediment (Simons *et al.*, 1965). This has been termed the *flow-regime concept* (Fig. 6). Repetition of similar conditions leads to repetition of similar depositional products, which are then susceptible to a universal empirical classification in the field.

Miall (1977) reviewed braided-river deposits, and demonstrated a consistency in the lithofacies assemblages occurring in a wide range of modern and ancient sandy and gravelly sediments. A simple classification was proposed, making use of a two-letter code to facilitate quick field and laboratory identification and documentation. Use of this lithofacies scheme by a number of workers led to the recognition that the scheme could be applied to all fluvial deposits, not just those of braided type. In later work (Miall, 1978, 1996), the classification has been expanded with the definition of a number of additional minor but significant lithofacies types. Subsequently, the classification has been used by dozens of researchers and has become a standard field method for the examination of fluvial deposits.

The scheme is presented in Table 2. The capital letter in the facies code indicates the dominant grain size (G=gravel, S=sand, F=fine-grained facies, including very fine sand, silt and mud). The lower-case letter serves as a mnemonic for the characteristic texture or structure of the lithofacies (*e.g.*, p=planar crossbedding, ms=matrix-supported). Examples are illustrated in Figures 7, 8 and 9. Virtually all fluvial lithofacies are clastic in origin. The few exceptions are the nodular carbonate deposits (caliche) that develop in subsoils in semi-arid regions (Fig. 9 A, B), coal, and evaporite crystals, nodules or beds that develop in inland sabkhas (*see* Chapter 20).

It is recommended that the researcher use Table 2 as a basis for field research, while remaining open to the possibility that refinements are always possible, based on detailed observations of new units. Some of the lithofacies classes are gradational with others. For example, there may be no clear distinction in the field between lithofacies Sh and Sl; Sp and St are commonly difficult (or impossible) to distinguish in core; lithofacies Fl may contain minor coal streaks or carbonate nodules, leading to questions about a workable definition of lithofacies C and P for logging purposes. How thick should a sand bed be before it is separated out from Fl? Bed thickness cutoffs and accessory-percentage limits should be established at the outset of a project exercise to facilitate consis-

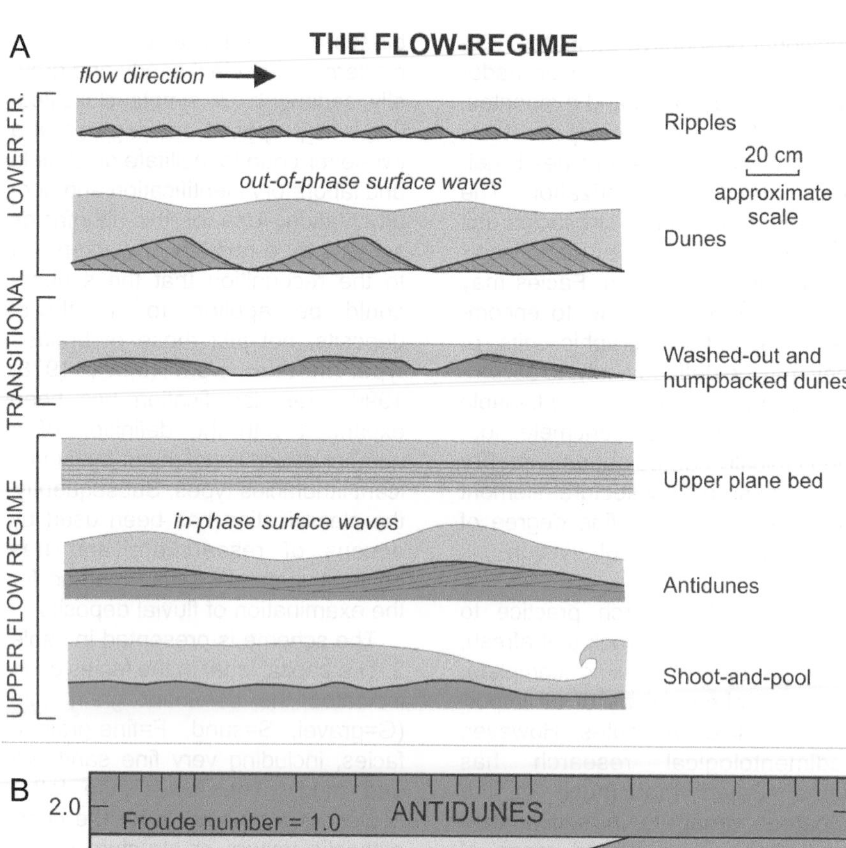

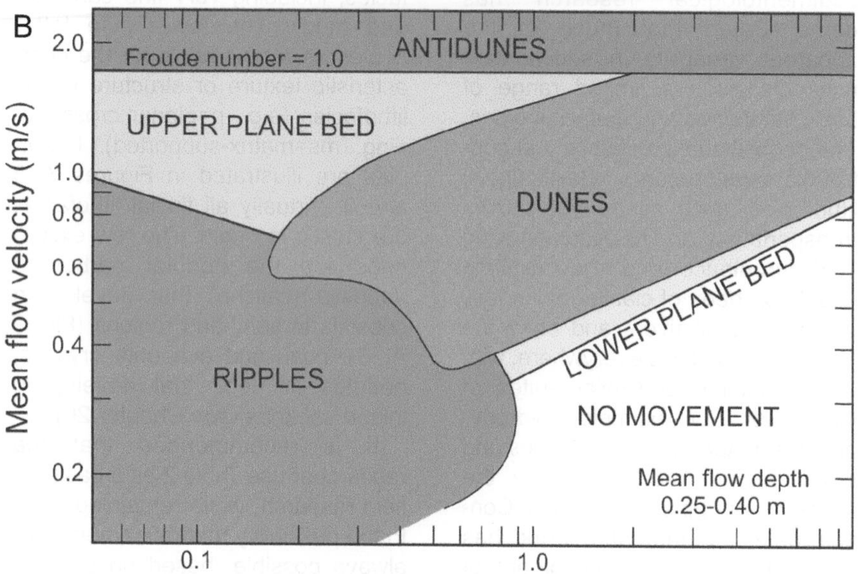

Figure 6. The flow-regime concept. **A.** The suite of bedforms that develops in sand of medium grain size at average depths of about 20 cm, under varying flow velocities. For example, if flow velocity is increased to about 0.4 m/s, ripples are replaced by dunes. These become progressively 'washed-out' at higher velocities, and at about 0.8 m/s a plane-bed condition develops, which is typically preserved in the rock record in the form of tabular sheets of sandstone with parting lineation on bedding planes (adapted from Simons *et al.*, 1965). **B.** Stability fields of bedforms developed over a grain-size range from coarse silt to pebbles, under constant flow depth and current speeds up to 2 m/s (adapted from Ashley, 1990).

tent logging practices.

The reader is also referred to a review of bedform classifications undertaken by Ashley (1990). This synthesis of modern work on bedforms of all types, in fluvial, tidal and other environments, resulted in a uniform approach to classification and a better understanding of the causes of morphological variation. Bedforms with heights exceeding 5 cm are now all classified as dunes. Those with straight crests give rise to planar-tabular crossbedding, and are referred to as *two-dimensional dunes*, whereas those exhibiting sinuous crests, called *3-D dunes*, are the source of trough crossbedding (Fig. 7). Measurement of crossbed orientation (the direction of dip of the foresets or the orientation of the axis of a trough) is the major basis for *paleocurrent analysis*, the technique for reconstructing the orientation of ancient river systems. Some aspects of paleocurrent analysis are also an important part of the analysis of architectural elements, as discussed in the following section.

ARCHITECTURAL ELEMENTS

When a river is viewed from the air, it can be seen to consist of various straight and curved channel reaches, and large areas of exposed gravel, sand, or mud, termed bars. In many rivers, the shape and distribution of these various features might seem chaotic, but most such units have distinctive surface forms, the terms for which, such as point bar, side bar, sand flat, chute channel, crevasse splay, *etc.*, constitute the familiar lexicography of fluvial geomorphology (some examples of this terminology are shown in Fig. 1). Their development follows certain relatively predictable patterns (Bridge, 1985, 1993, 2007; Miall, 1996) that leave their record in the resulting deposits. The channels and bars are the component depositional or geomorphic elements of the river, and the sediments that comprise them are termed architectural elements. An *architectural element* may be defined as a component of a depositional system equivalent in size to, or smaller than a channel fill, and larger than an individual lithofacies unit, characterized by a distinctive facies assemblage, internal geometry, external form and (in some instances) vertical profile. Some common examples are illustrated in Figures 10–17.

Architectural elements are amenable to descriptive and genetic classification, as are their component lithofacies. Miall (1985) proposed the following components of a descriptive classification, following the pioneering work of Allen (1983):

1. Nature of lower and upper bounding surface: erosional or gradation-

Table 2. Lithofacies classification

Facies code	Lithofacies	Sedimentary structures or textural organization	Interpretation
Gmm	matrix-supported, massive gravel	weak grading	plastic debris flow (high-strength, viscous)
Gmg	matrix-supported gravel	inverse to normal grading	pseudoplastic debris flow (low strength, viscous)
Gci	clast-supported gravel	inverse grading	clast-rich debris flow (high strength), or pseudoplastic debris flow (low strength)
Gcm	clast-supported massive gravel	-	pseudoplastic debris flow (inertial bedload, turbulent flow)
Gh	clast-supported, crudely bedded gravel	horizontal bedding, imbrication	longitudinal bedforms, lag deposits, sieve deposits
Gt	gravel, stratified	trough crossbeds	minor channel fills
Gp	gravel, stratified	planar crossbeds	transverse bedforms, deltaic growths from older bar remnants
St	sand, fine to v. coarse, may be pebbly	solitary or grouped trough crossbeds	sinuous-crested and linguoid (3-D) dunes
Sp	sand, fine to v. coarse, may be pebbly	solitary or grouped planar crossbeds	transverse and linguoid bedforms (2-D dunes)
Sr	sand, very fine to coarse	ripple crosslamination	ripples (lower flow regime)
Sh	sand, v. fine to coarse, may be pebbly	horizontal lamination parting or streaming lineation	plane-bed flow (super-critical flow)
Sl	sand, v. fine to coarse, may be pebbly	low-angle (< 15°) crossbeds	scour fills, humpback or washed-out dunes, antidunes
Ss	sand, fine to v. coarse, may be pebbly	broad, shallow scours	scour fill
Sm	sand, fine to coarse	massive, or faint lamination	sediment gravity-flow deposits
Fl	sand, silt, mud	fine lamination, small ripples	overbank, abandoned channel, or waning flood deposits
Fsm	silt, mud	massive	back-swamp or abandoned channel deposits
Fm	mud, silt	massive, desiccation cracks	overbank, abandoned channel, or drape deposits
Fr	mud, silt	massive, roots, bioturbation	root bed, incipient soil
C	coal, carbonaceous mud	plant, mud films	vegetated swamp deposits
P	paleosol carbonate (calcite, siderite)	pedogenic features: nodules, filaments	soil with chemical precipitation

modified from Miall (1978).

Two-dimensional dunes and planar crossbedding
Three-dimensional dunes and trough crossbedding
modern 2-D dune, Platte R., NB
modern 3-D dunes, Bay of Fundy
Sets of planar crossbedding, Cret., Banks Island, NWT
Sets of trough crossbedding, Cret., Athabasca Oil sands, AB
Planar crossbedding, Carb., Joggins, NS
Trough crossbedding, Permian, PEI

Figure 7. The two most common types of dunes and their preservation as crossbedded sand deposits.

Figure 8. Examples of fluvial lithofacies, illustrating the use of the lithofacies code. **A.** Modern debris-flow deposits, Death Valley, California; **B.** Devonian alluvial-fan deposits, Somerset Island, NWT; **C.** Fluvioglacial gravels, Banff, Alberta; **D.** Sets of trough crossbedding, Paleogene, Axel Heiberg Island, NWT; **E.** Low-angle crossbedding (Sl), possibly representing a transitional dune form (*see* Fig. 6A), Morrison Formation (Jurassic), near Gallup, New Mexico; **F.** Upper flow-regime plane-bed lamination (Sh), Whirlpool Sandstone (Silurian) near Hamilton, Ontario.

al; planar, irregular, curved (concave-up or convex-up);

2. External geometry: sheet, lens, wedge, scoop, U-shaped fill;
3. Scale: thickness, lateral extent parallel and perpendicular to flow direction;
4. Lithology: lithofacies assemblage and vertical succession;
5. Internal geometry: nature and disposition of internal bounding surfaces; relationship of bedding to external bounding surfaces (parallel, truncated, onlap, downlap); and
6. Paleocurrent patterns: orientation of flow indicators relative to internal bounding surfaces and to external form of element.

Miall (1985) originally suggested that there are eight basic architectural elements in fluvial deposits, based on a review of studies available up to the early 1980s, but subsequent work has provided a number of elaborations of this original classification and, as Fielding (1993) pointed out,

Figure 9. Examples of fine-grained deposits and chemical sediments. **A**, **B.** Caliche, Triassic, Spain; **C.** Interbedded floodplain mudstones and siltstones (Fl, Fm), Triassic, near Wingate, Arizona; **D.** Thin mudstones units form the recessive unit at the top of a succession of sheet sandstones formed by flash floods; Devonian, Somerset Island, NWT.

researchers may need to erect their own classification to reflect better the characteristics of the fluvial unit being investigated. A classification of architectural elements that occur within channels and those that occur on floodplains is provided in Tables 3 and 4, respectively.

In most fluvial systems, bars are developed over many seasons as a result of deposition along the sides of channels or in mid-channel positions. The most common form of bank-attached bar is the point bar (Figs. 10 and 11). The evolution of point bars on the inside of meandering channels is illustrated in Figure 5. Figure 11 illustrates how observations from either a modern or an ancient point bar may be used to construct a depositional model for a meandering fluvial channel.

One of the most diagnostic features of the point bar is that the dip of the accretionary units, which are separated by gently inclined master bedding planes, is oriented at a high angle to the main direction of flow. This is indicated by the current-rose diagram at the top right of Figure 10, where it is suggested that the orientation of these units is bimodal. Flow over the point-bar surface is more or less parallel to the strike of the accretionary units. Statistically, flow directions indicated by crossbed orientations tend to indicate the main direction of flow of the river, which is indicated by another rose diagram in Figure 10.

Bank-attached bars having the internal architecture of point bars are also common in braided systems but, as shown in Figures 2, 4, 5 and 10, bars in mid-channel positions are much more common. They may accrete downstream or laterally on either or both sides. Such lateral accretion forms an architectural element that is identical to a point bar, except for its mid-channel location. In places, the flow in a main braided channel may be oriented at a high angle to the overall channel orientation, as shown in Figure 10. In some of these, flow is at a high angle to the direction of accretion; in other cases, the accretion surfaces are oriented in a downstream direction. The documentation of these subtle differences is useful, because they reflect variations in fluvial style and can therefore be used in the construction of a depositional model for an ancient deposit. Miall (1985) proposed the definition of two types of architectural element to encompass these variations. Point bars were defined using the descriptive term lateral-accretion element, and those accreting in a downstream direction were termed downstream-accretion elements. An example of the latter is shown in Figure 12A. In this case, accretionary sets dip to the north (to the right in this view) and crossbed orientations indicate flow in the same direction. In a braided system, there may be little difference

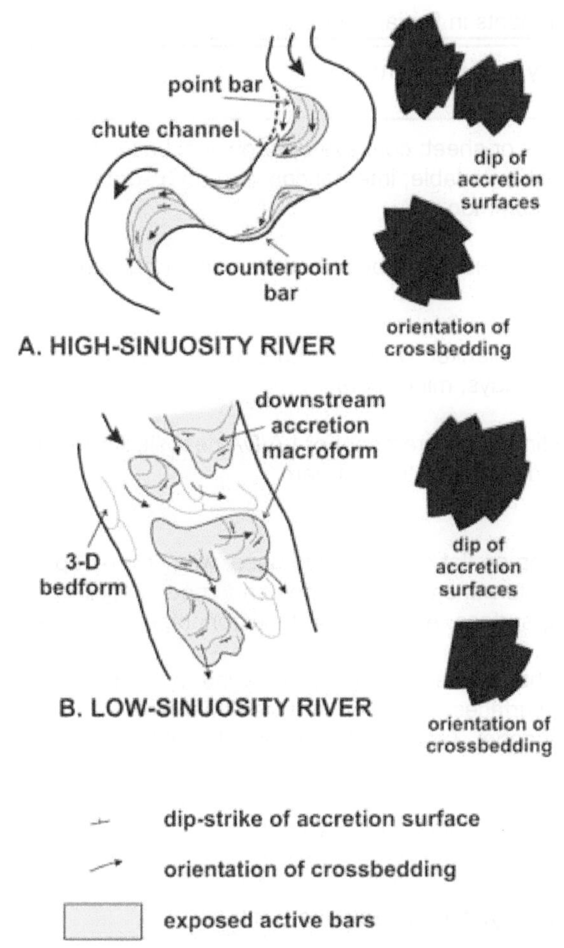

Figure 10. Examples of two fluvial styles, showing some typical architectural elements and indicating the range of orientations of dipping accretion surfaces and the variation in crossbed orientations. **A**. High-sinuosity river, such as a typical meandering river. Note that although there is overlap of crossbed and accretion-surface orientations regionally, locally the two are oriented nearly perpendicular to each other, as shown by the rose diagrams. A counterpoint bar is shown. Such bars can be distinguished from the more common point bars by the fact that the accretion surfaces exhibit curvature in plan view that is concave in the direction of dip of the master bedding, contrasting with the convex curvature of point-bar surfaces. Commonly such bars are characterized by finer grain sizes than other bars. **B.** Low sinuosity river - a typical sandy-braided river. Such rivers may exhibit relatively high local channel sinuosity, in which case bars accreting downstream may also be oriented at a high angle to the regional trend. In the example shown here, mean directions are skewed to the southeast, diverging from the overall channel orientation of south-southeast because of a major channel and bar complex (near the center of the reach) oriented in an easterly direction (Miall, 1994).

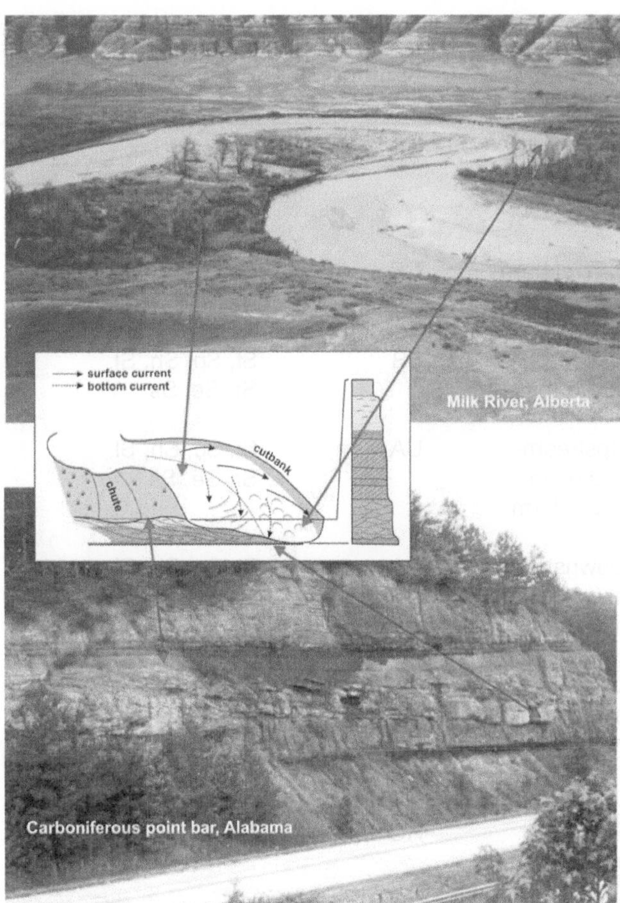

Figure 11. Observations from either a modern or an ancient point bar deposit may be used to construct a depositional model for a meandering fluvial channel, of which the point bar is typically the most important depositional component. The key characteristic of a point bar is the development of surfaces of lateral accretion that dip towards the apex of the meander bend, and upon which sediment is deposited as the meander enlarges or migrates, as erosion takes place on the outer bank. In an outcrop of an ancient deposit this accretionary geometry is visible in the form of large-scale, gently inclined master-bedding surfaces. In active, modern point bars the growth of the point bar can be deduced from the development of vegetation across the expanding bar surface, bushes or trees on the oldest part, scrub and grass on younger parts, and bare sand on the outermost edge of the bar, next to the water-filled channel. The block diagram model at the center shows the pattern of water flow through the bend. The surface flow entering the bend is directed toward the outer bank. There it turns downward and flows obliquely up across the point bar surface. Top photo courtesy D. G. Smith.

between the average orientation of crossbedding and the orientation of accretionary sets, as suggested by the rose diagrams for the low sinuosity river in Figure 10. Accretion may occur on the upstream flank of a mid-channel bar, but the deposits are rarely preserved.

Gradations between lateral and downstream accretion are to be expected, and a distinction may be made, such as that shown in Figure 13, where a pre-defined angle between channel flow direction and the orientation of accretionary dip is suggested as the cutoff for purposes of field definition. An angle of 60° has been used by this writer. This diagram also illustrates the system of documenting bounding surfaces (Table 1) that may be employed to provide some descriptive order to large outcrops. Major bounding surfaces (those of 5th or higher rank) are labeled with letters of the alphabet. Individual architectural elements, including accretionary bars and minor channels, are defined by 4th-order surfaces. Internal erosion surfaces are of 3rd order.

Table 3. A basic classification of the within-channel architectural elements in fluvial deposits

Element	Symbol	Principal facies assemblage	Geometry and relationships
Channels	CH	any combination	finger, lens or sheet; concave-up erosional base; scale and shape highly variable; internal concave-up 3rd-order erosion surfaces common.
Gravel bars and bedforms	GB	Gm, Gp, Gt	lens, blanket; usually tabular bodies; commonly interbedded with SB
Sandy bedforms	SB	St, Sp, Sh, Sl, Sr, Se, Ss	lens, sheet, blanket, wedge, occurs as channel-fills, crevasse splays, minor bars
Upstream-accretion macroform	UA	St, Sp, Sh, Sl, Sr, Se, Ss	lens, resting on bar remnant or LA/DA deposit. Accretion surfaces dipping gently upstream
Downstream-accretion macroform	DA	St, Sp, Sh, Sl, Sr, Se, Ss	lens resting on flat or channeled base, with convex-up 3rd-order internal erosion surfaces and upper 4th-order bounding surface. Accretion surfaces oriented downstream
Lateral-accretion macroform	LA	St, Sp, Sh, Sl, Se, Ss, less commonly Gm, Gt, Gp	wedge, sheet, lobe; characterized by internal lateral-accretion 3rd-order surfaces. Accretion surfaces oriented across channel. Typically downlaps onto flat basal erosion surface.
Scour hollows	HO	Gh, Gt, St, Sl	scoop-shaped hollow with asymmetric fill.
Sediment gravity flows	SG	Gmm, Gmg, Gci, Gcm	lobe, sheet, typically interbedded with GB
Laminated sand sheet	LS	Sh, Sl; minor Sp, Sr	sheet, blanket

Modified from Miall (1985). Facies classification from Miall (1996).

Table 4. Clastic architectural elements of the overbank environment

Element	Symbol	Lithology	Geometry	Interpretation
Levee	LV	Fl	Wedge up to 10 m thick, 3 km wide	Overbank flooding
Crevasse channel	CR	St, Sr, Ss	Ribbon up to a few hundred m wide, 5 m deep, 10 km long	Break in main channel margin
Crevasse splay	CS	St, Sr, Fl	Lens up to 10 by 10 km across, 0.1–6 m thick.	Delta-like progradation from crevasse channel into floodplain
Floodplain fines	FF	Fsm, Fl, Fm, Fr	Sheet, may be many km in lateral dimensions, up to 10s of m thick	Deposits of overbank sheet flow, floodplain ponds and swamps
Abandoned channel	CH(FF)	Fsm, Fl, Fm, Fr	Ribbon comparable in scale to active channel	Product of chute or neck cutoff

From Miall (1996).

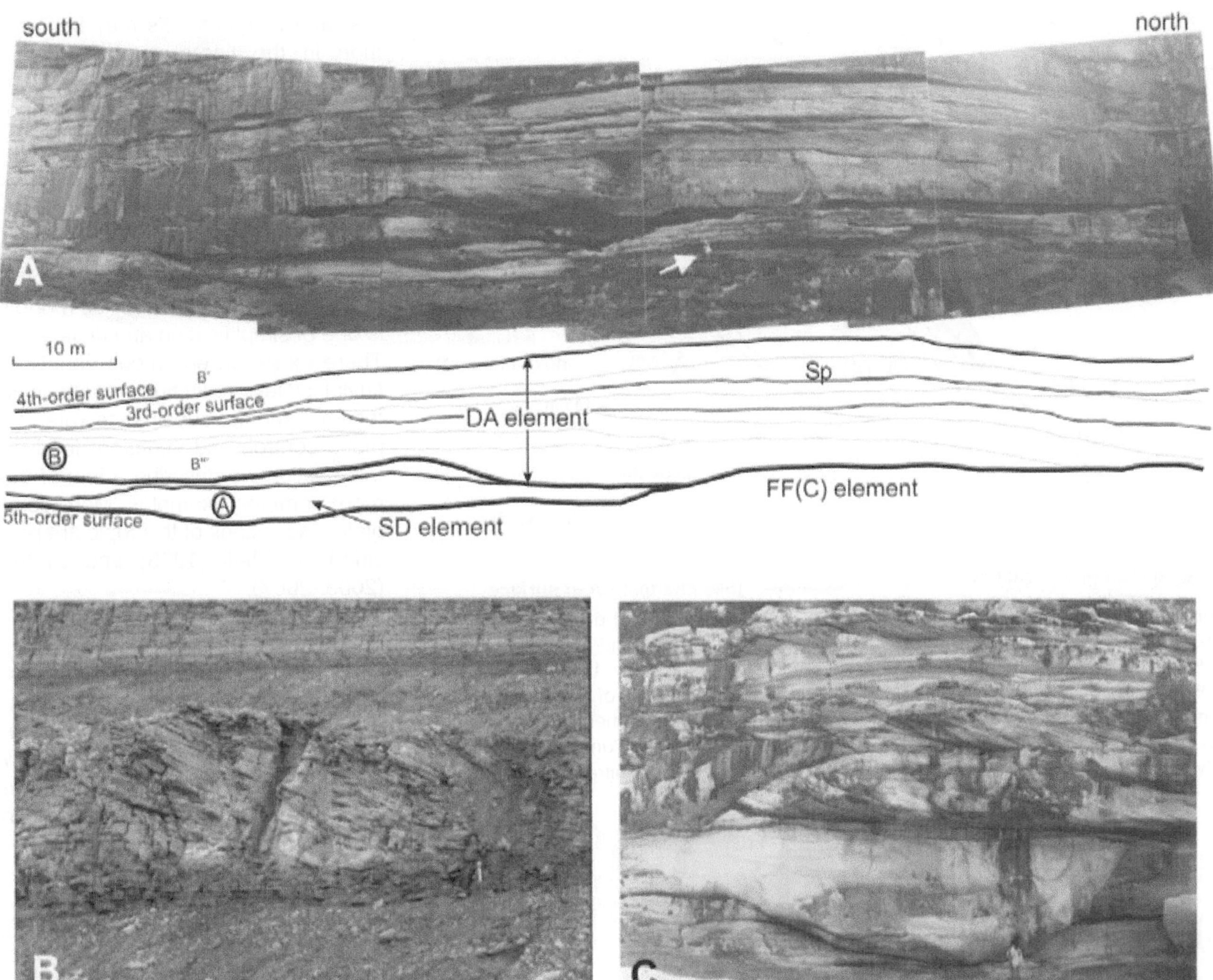

Figure 12. Examples of architectural elements. **A.** Downstream-accretion deposit (DA), resting on an element consisting of large individual dunes (SD) and an abandoned channel filled with fine-grained deposits (FF(C)). **B.** Lateral-accretion deposit; **C.** Hollow element; **A, C,** Hawkesbury Sandstone (Triassic), near Sydney, Australia; **B.** Tertiary Eureka Sound Group, Axel Heiberg Island, NWT.

Other examples of architectural elements are illustrated in Figure 12. Photograph C in Figure 12 is a hollow-element, the fill of an erosional hollow formed by turbulent scour at a confluence between tributaries or at a point where channels join downstream from a mid-channel bar, a process first described by Best (1987). The curved base and simple, homogeneous fill distinguish such structures from channels. Scour hollows may be up to five times deeper than typical channel depth (Best and Ashworth, 1997).

Other within-channel elements typically form laterally extensive tabular bodies that have essentially tabular internal architectures. For example, Figure 8B and C illustrate the gravel bar and bedform element, and Figure 8D is an example of the sandy bedform element.

Modern high-resolution remote-sensing techniques may yield invaluable information about the architecture of subsurface alluvial deposits. Figure 14 is an example of the mapping of point bars in cross-section and in plan view, using vertical and horizontal seismic sections together with wireline-log information. This type of data is becoming of increasing value in the search for stratigraphic oil and gas traps. Figure 15 illustrates the use of ground-penetrating radar to document the shallow subsurface, a method that has considerable value for the mapping of shallow aquifers and construction foundations in alluvial sands and gravels.

A classification of overbank architectural elements is provided in Table 4. These elements are distinguished mainly by their fine-grained nature, but architectural (shape) characteristics may also be diagnostic, with elements developing such geometrical forms as wedges, ribbons, lenses or sheets. The overbank is the only part of the fluvial system where chemical sediments may accumulate. In humid environments, plant growth generates soils and may lead to peat accumulations that can evolve, with burial, into coal. In more arid environments, evaporation and capillary action can lead to the formation of layers of calcrete nodules (Fig. 9A, B). Given lengthy exposure of an inactive floodplain (10^5 years) calcrete beds may become meters thick and extend for several kilometers, providing local stratigraphic markers. Bioturbation, invertebrate trace fos-

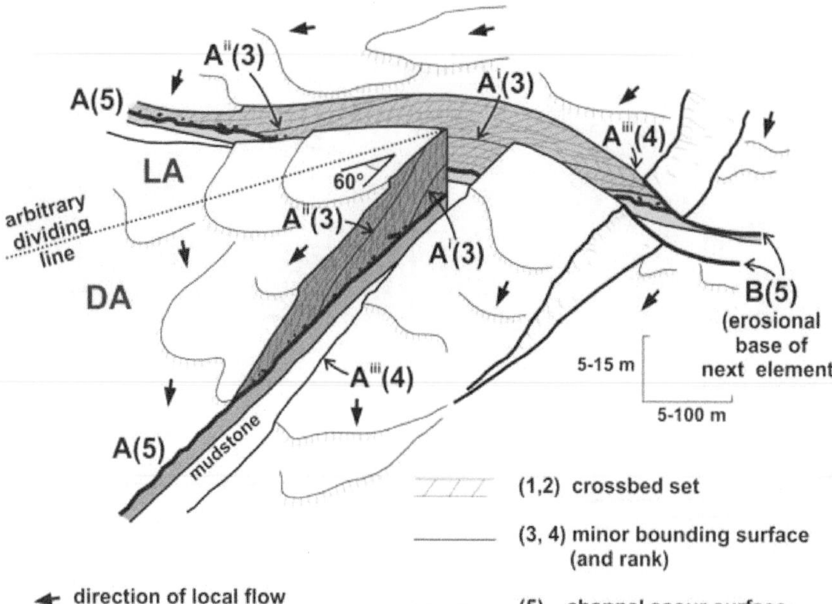

Figure 13. Dissection of a large braid bar showing both lateral and downstream accretion. Surfaces A and B are 5th-order surfaces defining the major sand bodies (*see* Table 1). Surfaces **A**ⁱ and **A**ⁱⁱ are minor (3rd-order) internal erosion surfaces caused by pauses in sedimentation or erosional episodes, and surface **A**ⁱⁱⁱ defines the top of the main element. Surface B is a 5th-order surface that defines the base of the next element. In the diagram, this is an erosion surface, and the next unit has yet to be deposited on it. The distinction between lateral- and downstream accretion may be made by an arbitrary cutoff, such as the 60° line shown at left.

sils, and animal tracks may be common in the fine-grained floodplain sediments (*see* Chapter 3 for a comprehensive review of terrestrial ichnofacies). Overbank elements are discussed further in the following section.

FACIES MODELS

As discussed in Chapter 2, the process of constructing *facies models* is one of simplification and distillation. There exists an enormous variety of fluvial styles, reflecting variations in tectonic setting, subsidence rate, climatic control, and sediment load. In this section we describe a few of the more common examples. More complete discussions of this topic are provided by Miall (1996) and Bridge (2003, 2007).

Meandering Rivers

The main process of formation of point bars by lateral channel migration has been described above. Meandering channels are typically sandy bedload rivers, but highly sinuous channels with lateral-accretion deposits may develop in coarse, grav-

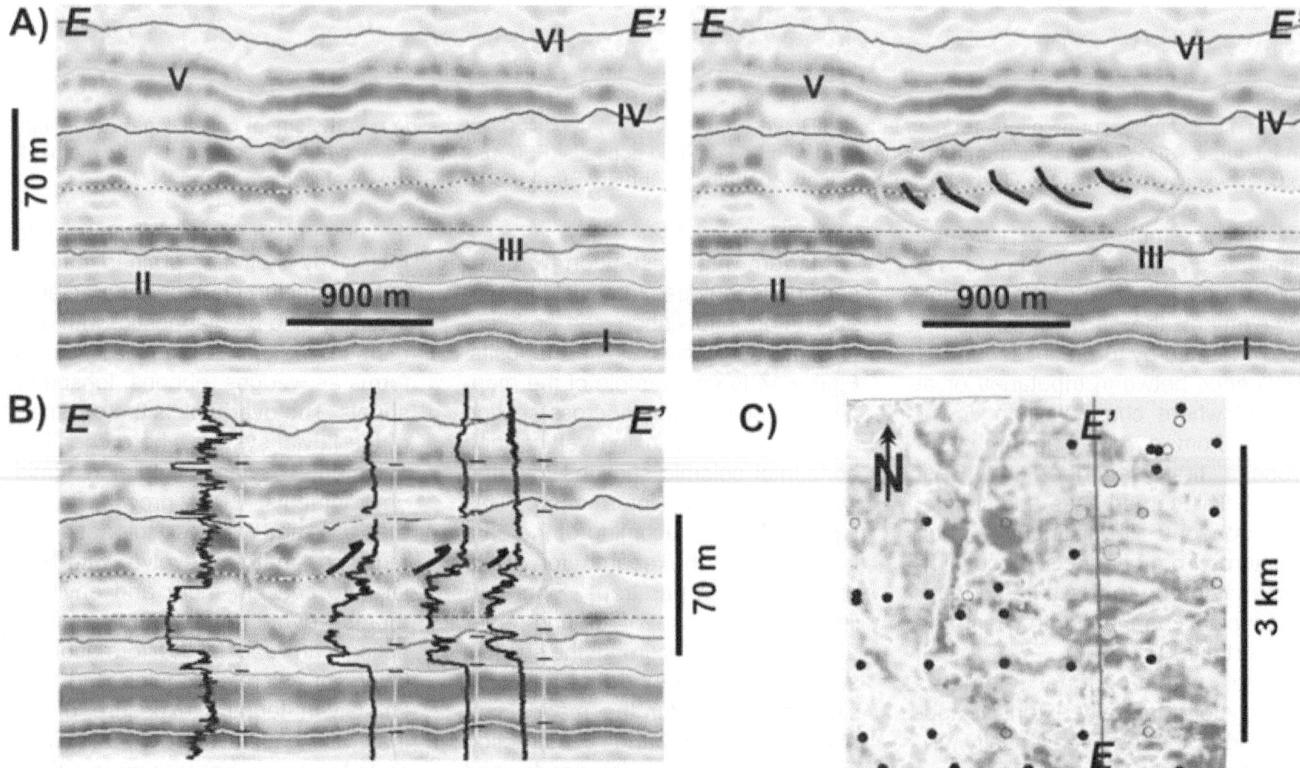

Figure 14. Use of core and 3-D seismic data to interpret fluvial architecture. An example from the Mannville Group of Alberta (Sarzalejo and Hart, 2006). Gamma-ray wireline-log data exhibit fining-upward cycles (inclined arrows). Lateral accretion sets are imaged in both horizontal **(C)** and vertical **(A, B)** seismic sections. The horizontal section **(C)** also shows the convex curvature of the point-bar surface.

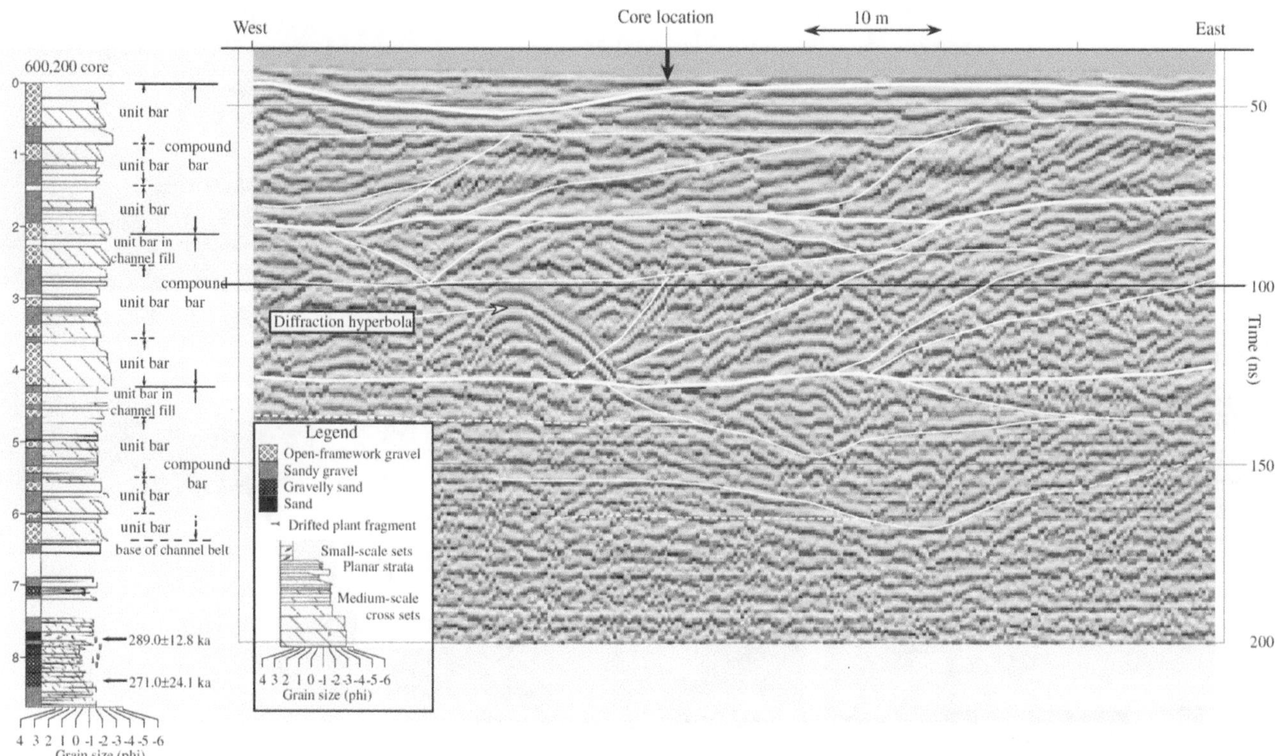

Figure 15. Ground-penetrating radar section through a modern gravelly braided river in Alaska and interpreted log of a core drilled along the survey line (Lunt and Bridge, 2004).

el-bed rivers (Fig. 17C) and in rivers dominated by a fine sand and silt sediment load (Fig. 17D).

Channel cutbanks are rarely preserved. Three examples from meandering-river systems are illustrated in Figure 16A, C and F. That shown in Figure 16C is an example of a meandering channel that was abandoned when it had only been partially filled with coarse bedload. A lateral-accretion deposit occupies the lower part of the channel, but the rest is filled with mudstone. Photo F illustrates an example of a cutbank where the evidence of bank failure has been preserved. The beds resting on the cutbank are nearly vertical, including a large block of sandstone that originally would have occupied the channel floor. The syndepositional origin of this deformation is indicated by the slump deposits, overlain by undisturbed, horizontal beds.

Other examples of meandering-channel deposits are illustrated in Figure 17. Photograph 17A in this figure illustrates a large sandy point-bar deposit that contains clasts of muddy floodplain sediment (Photograph 17B), probably derived by slumping from a nearby cutbank. Figure 17E illustrates one of the giant point-bar deposits that are the host for much of the heavy oil in the Athabasca Oil Sands of northern Alberta. The gently dipping accretion surfaces are clearly visible in this outcrop. This point bar is more than 25 m thick, indicating a channel depth of this scale. Most preserved channel-fill or point bar deposits range between 2 and 20 m in thickness. Giant rivers, such as the Mississippi, although of critical importance from the point of view of sediment flux, have rarely been located in the ancient record (Miall, 2006a).

Floodplain deposits are an integral part of most meandering-river successions. They consist mainly of fine-grained clastic units (an element classification is given in Table 4). A map of one of the Mississippi meanders (Fig. 18) illustrates the types of local depositional system that develop adjacent to meanders (although it needs to be borne in mind that the Mississippi is an exceptionally large river). Crevasse channels develop by breaching of a levee during high-runoff events, and may result in a permanent diversion of the main channel flow – a process termed *avulsion* – if this results in a slope advantage for the channel. That can occur if the river has been flowing within its own depositional channel (meander) belt for some time and has built up an *alluvial ridge*. Coarse bedload is diverted through the crevasse channel and spreads out into the overbank area as a *crevasse-splay*. Those in the Mississippi system may extend for up to 10 km from the meander bank. A smaller example is illustrated in Figure 17F. Here, a sandstone bed a few meters thick overlies a coal near the base of the outcrop. The convex-up shape of the deposit defines part of a low-amplitude cone, consistent with its origin as a delta-like splay extending from a crevasse channel, that would have been somewhere in the vicinity of the road at the right of the view. A small crevasse channel, filled with laminated silts and muds, is seen in Figure 9C. If crossbed directions can be measured in these deposits it would be expected that they would show orientations at high angles to the channel direction.

Wedge-shaped levee deposits formed of sediment deposited by unchannelized overbank floods

Figure 16. Examples of ancient fluvial deposits, highlighting channels and channel margins. **A**. Cenozoic fining-upward cycle, showing a cutbank, Bylot Island, Arctic Canada; **B**. A narrow, ribbon-like sandstone body, deposited in a single-thread river system, Permian, New Mexico. **C**. A point bar (lateral accretion shown by dashed lines) in a channel that was abandoned, with the remaining fill composed of mudstone (defined at base by dotted line), Jurassic, Yorkshire, UK; **D**, **E**. Channel and cutbank in an arid fluvial system, Jurassic, Arizona. Box in **E** indicates area shown in **D**. Note the truncated floodplain mudstone at left, in **D**, and block of mudstone resting on the cutbank; **F**. Cutbank with collapsed and rotated block of channel sandstone, Carboniferous, Kentucky; **G**. Sandstone on sandstone cutbank in braided system, Jurassic, New Mexico (person circled for scale); **H**, **I**. Box canyon formed by flash-flood erosion on an unconformity at the base of the Triassic succession, Arizona. Close-up of canyon wall shown in **I**.

occur adjacent to the channel in typical meandering systems. Levee deposits may be several meters thick and extend for several kilometers away from the cutbank. The backswamp in the Mississippi system (Fig. 18) was originally a saturated, boggy area where plant remains accumulated, although most such areas have been drained along the modern Mississippi. Some coal may develop in such environments, but, in many cases, the periodic influx of clastic sediment from the river prevents true coal from developing. The plant material is, instead, diluted with silt and mud, yielding carbonaceous shale or, at best, a coal high in 'ash' content – the noncombustible residue remaining after coal has been burned. In arid environments the overbank is the setting for the development of caliche, the nodular carbonate deposit formed by the concentration of groundwater solutes by evaporation (Fig. 9A, B).

Anastomosed Rivers

Two other important fluvial styles are illustrated in Figure 19. The anastomosed model describes systems of interconnected channels flowing through large backswamp areas. Channels are relatively stable in position, and therefore form narrow, ribbon-like sandstone bodies, encased in floodplain fines. Little lateral migration occurs because of the stability of the muddy banks, and elements such as point bars are therefore of minor importance. Channels are flanked by well-developed levees and by sheet-like crevasse-splay deposits. This fluvial style occurs in areas of rapid vertical aggradation. The Cumberland Marsh of northern Saskatchewan is a much-studied example (Smith *et al.*, 1989; Morozova and Smith, 1999, 2000). Other examples include the Magdalena River in Columbia, and parts of the Columbia River in British Columbia.

Figure 17. Examples of meandering river deposits. **A**, **B.** Large sandy point bar, Carboniferous, Kentucky; **B.** Close-up of mudstone clasts probably derived from cutbank erosion. Location shown by circle in **A; C.** Gravel lateral-accretion deposit, Triassic, Spain. The figures are standing opposite the horizontal base of the bar. Gently dipping surfaces higher in the outcrop are lateral accretion surfaces; **D.** Point bar composed mainly of very fine sandstone and siltstone, Tertiary, Ellesmere Island; **E.** Giant point-bar deposit, about 25 m high, Athabasca Oil Sands, near Fort McMurray, Alberta; **F.** Crevasse-splay deposit overlying a coal, Carboniferous, Kentucky.

A class of dryland *single-thread* to *anabranching* fluvial systems in arid areas, such as Australia, has been also grouped with this fluvial style in earlier studies as examples of anastomosed systems, but careful study of these rivers by North *et al.* (2007) has indicated that this is an inappropriate model. These river systems occur within broad alluvial plains rather than confined valleys. They are characterized by ribbon sandstones exhibiting low to moderate sinuosity, and which do not develop networks of interconnected sand bodies. Floodplain deposits contain indicators of aridity, such as calcrete-type paleosols. The ribbon sandstone shown in Figure 16B is interpreted as an example of a dryland system (North *et al.*, 2007).

Sand-bed Braided Rivers

There is a wide variety of fluvial styles of sand-bed braided rivers. Shallow, perennial rivers, such as much of the Platte River in Nebraska,

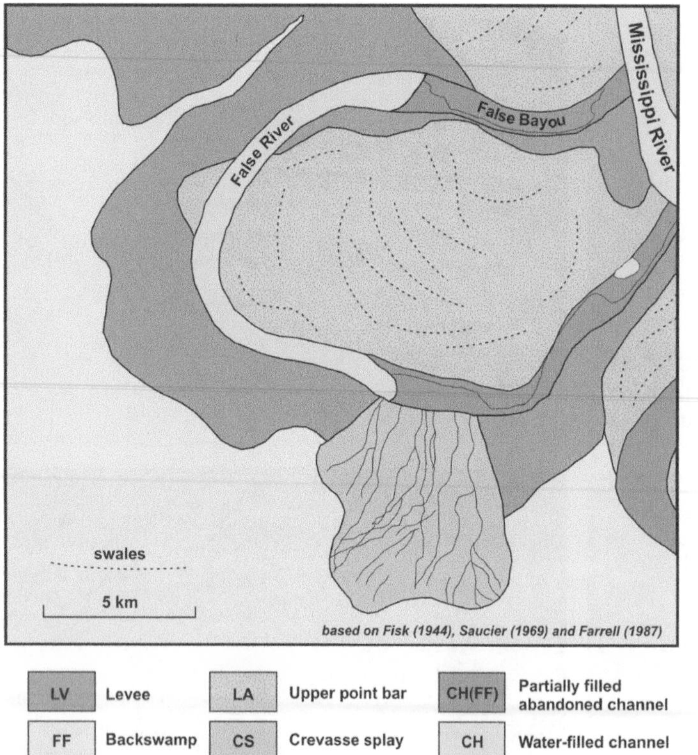

Figure 18. Landscape elements of a meandering-river floodplain; part of the modern Mississippi system (Farrell, 1987).

are characterized by the migration of large 3-D dunes, which deposit suites of planar-tabular crossbedding and associated minor structures (Fig. 19) of the sandy-bedform architectural element.

The most complex sandy braided deposits are those developed in large, perennial rivers, in which a wide variety of large bar forms develops, including both upstream-, lateral- and downstream-accretion deposits (Fig. 19). Lobate, planar-crossbedded 'unit bars' may coalesce to form wide, semi-permanent *sand flats*, which may become vegetated but could be removed in a major flood. The height of the lateral, downstream and upstream accretionary units formed on these flats provides an indication of channel depth, and is commonly in the 2–10 m range. The Saskatchewan River is an example of this fluvial style (Cant, 1978a, b; Sambrook-Smith *et al.*, 2006).

Braidplains in arid environments may be ephemeral and unchannelized, and are then characterized by tabular sandstone bodies several meters thick consisting of plane-laminated sandstone (element LS), or by flood sheets comprising thinning- and fining-upward assemblages of crossbedding and ripples (element SB; *e.g.*, Fig. 9D).

The Brahmaputra (Jamuna) River of Bangladesh is the largest active sandy-braided river system. It transports an enormous sediment load from the Himalayas to the Indian Ocean (Coleman, 1969; Ashworth *et al.*, 2000; Best *et al.*, 2003). Dunes there reach 6 m in height. A network of intersecting channels with actively

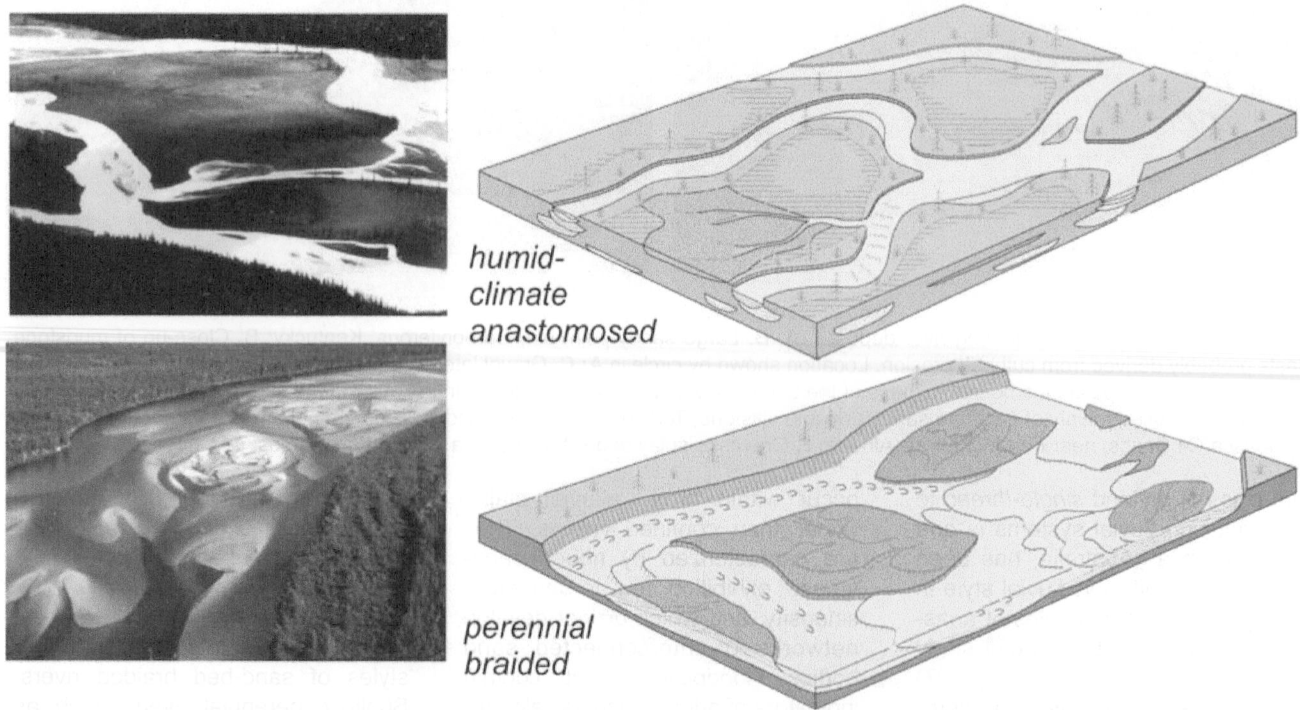

Figure 19. Two typical fluvial styles and the facies models developed to describe their deposits. Photos courtesy of D. G. Smith.

migrating bars and shifting channel margins constitutes a channel belt up to 8 km wide and 12 m deep, with local scour hollows reaching 50 m depths. No ancient deposit is known to have resulted from a river of this size, although the Hawkesbury Sandstone (Triassic) of the Sydney Basin, Australia (Rust and Jones, 1987; Miall and Jones, 2003), and Proterozoic deposits in Victoria Island, NWT (Rainbird, 1992) have both been identified as the deposits of giant braided rivers on the basis of the large size of the crossbed sets and interpreted channel depths.

Gravel-bed Braided Rivers

Proximal alluvial fans and fluvioglacial outwash streams are commonly gravel-bed braided rivers. Rivers flowing from areas of active uplift may also be dominated by gravel bedload (*e.g.*, rivers of the Canterbury Plains, South Island, New Zealand). Accretionary bars are deposited, just as in sand-bed rivers, including lateral- and downstream-accreting units. Where the river is shallow, it may be dominated by the gravel-bar and bedform element (Fig. 8C). Sand may accumulate in minor bar-top channels (Fig. 8B) and in protected areas on bar flanks during low flow.

In arid regions, where runoff is ephemeral and flashy, sedimentation may be dominated by rare, violent debris flows. The alluvial fans flanking Death Valley, California, are characterized by this process. The result is the deposition of coarse, poorly sorted sand, gravel and boulder deposits. Size sorting during transportation may lead to the development of graded bedding (Lithofacies Gmg; Fig. 8A).

CYCLES OF SEDIMENTATION

It has long been one of the common truisms of sedimentology that fluvial deposits exhibit *fining-upward successions*. The basis of this idea has been around since the 1940s (*see* Miall, 1996, Chapter 2), but became concrete with the work of Bernard *et al.* (1962) on modern depositional systems of the Gulf Coast of Texas, and Allen (1963) working with Devonian fluvial deposits in Britain.

Two separate but related autogenic processes lead to the development of these successions. Simple aggradation of a channel in any setting (*e.g.*, fluvial, tidal, submarine fan) results in decreasing flow depth and velocity and, consequently, in a decrease in the competency and capacity of the flow. The resulting facies successions may be identified in the subsurface by the recognition of *bell-shaped* wireline log profiles (SP or gamma-ray: Fig. 14). The development of a point bar and other accretionary deposits also leads to an upward decrease in grain size, as shown in Figure 11. It has long been argued, however, that the detailed composition of the fining-upward profile (*e.g.*, the amount of textural differentiation between the bottom and the top) is not diagnostic of fluvial style (Miall, 1980). Those in which significant thicknesses of floodplain fines are preserved have often been interpreted as the deposits of meandering systems, reflecting the large area of an alluvial valley or alluvial/coastal plain occupied by the floodplain in a meandering system. By contrast, braided systems tend to be wide, and may occupy most or all of an alluvial plain, or incised valley, leaving little room for a floodplain to develop. But such differences, however real in a modern setting, may be substantially distorted by the process of preservation in the rock record. Consider Figure 20. The column on the left, with its obvious upward fining and the presence of floodplain units, is typical of the type of deposit formed in a meandering fluvial environment, whereas that on the right, where cyclic facies changes are less pronounced, and floodplain units are not present, is considered to be characteristic of braided environments. However, the nature of the facies succession is also dependent on the rate of generation of accommodation, which may significantly alter the degree of preservation (the removal of the top of each facies succession; *see* Chapter 4), causing potential distortions in the resulting interpretation. Thus, the succession on the right in Figure 20 could be produced by a meandering river in a low-accommodation setting. This serves to emphasize the need to build interpretations based on the entire body of evidence, including, for example, a detailed description (wherever possible) of the assemblage of architectural elements and the paleocurrent patterns displayed by the unit under study.

As noted below, the changing balance between subsidence, sea-level change and sedimentation can also lead to the generation of allogenic cyclic successions. These may be described as stratigraphic sequences or tectonic cyclothems. Such cycles may extend across an entire basin margin, in contrast to cycles of autogenic origin, which are confined to individual channels or meanders. Mapping and correlation of cycles is therefore an important component of a basin analysis.

Godin (1991) provided a detailed study of the cycles present in an ancient sandy-braided succession and defined three types of cycle (Fig. 21). Cycles 1–6 m thick are developed by the generation of architectural elements, including sand sheets, downstream-accretion and lateral-accretion elements, and large scour hollows. They are within-channel cycles, of limited lateral extent. Channel-fill cycles may consist of stacked architectural elements and are typically 1–10 m thick, but may be much larger in the case of deposition by very large rivers. Major sandstone sheets 4–16 m thick may extend for many kilometers and, unlike the first two, may reflect allogenic causes, such as gentle basin tilting or base-level change (*see* next sections). Potentially, all large river systems could generate cycles showing these contrasting characteristics. Definition and mapping of the architectural elements and their bounding surfaces are essential for the correct identification of these types of cycle.

Ephemeral streams accumulate deposits as a result of flash floods and form successions consisting of stacked fining-upward sandstone sheets (Fig. 9D).

Wireline-log data are invaluable for subsurface mapping of facies and for the analysis of vertical profiles. Given the caveats discussed here regarding the incomplete preservation of cyclic fluvial deposits, log character may be used to character-

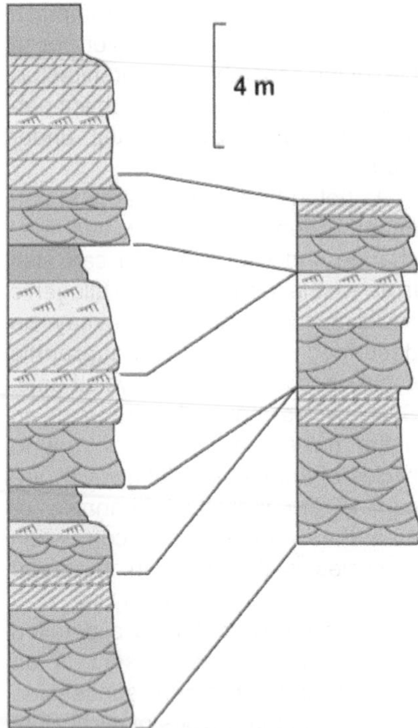

Figure 20. Contrasting degrees of preservation of fluvial fining-upward cycles. Each cycle in the column on the right is identical to the lower part of the cycle it is correlated with on the left. The difference between the two vertical profiles is, therefore, one of preservation, not of original depositional style. In slowly subsiding basins or other settings with low accommodation, channel systems may scour back and forth into their own earlier deposits, resulting in the stripping of the top of each cycle before burial and preservation. More rapid subsidence (i.e., times and places with higher accommodation) may result in a more complete preservation of a channel and floodplain succession (from Miall, 1980).

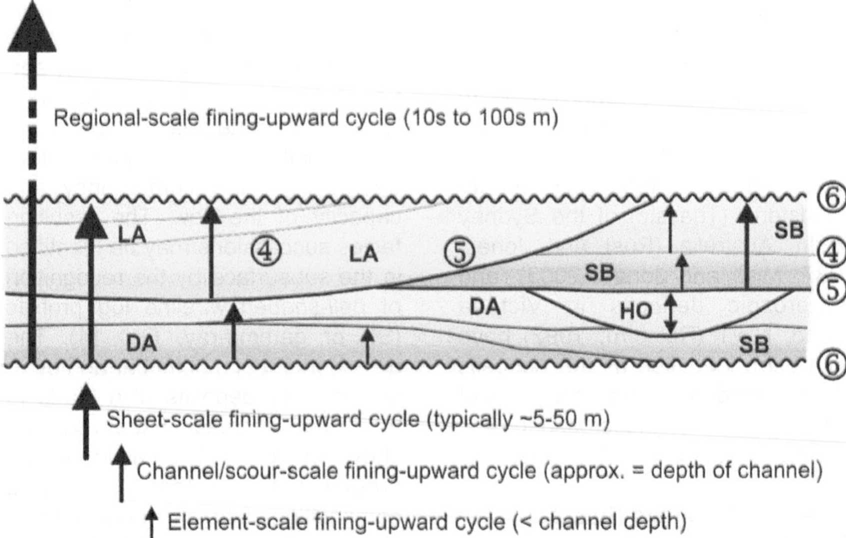

Figure 21. Four types of fining-upward cycles in fluvial deposits. Architectural elements are indicated by two-letter codes: LA: lateral-accretion deposit, DA: downstream-accretion deposit, HO: hollow element, SB: sand bars and bedforms. Numbers in circles refer to rank of bounding surface (*see* Table 1). Based on Godin (1991).

ize depositional style. This has been called *log facies analysis*. Figure 22 illustrates an example of the use of this method.

Many attempts have been made to develop quantitative architectural information from log data. Fielding and Crane (1987), Robinson and McCabe (1997) and Gibling (2006) compiled data relating channel width and depth for all fluvial styles. If the fluvial style can be deduced from log and core data, a set of quantitative relationships may be employed to estimate channel width, and, from this, some estimates of sandbody dimensions may be made – information of some use for reservoir-development purposes. However, interpretations of fluvial style are commonly ambiguous, and there are the problems of preservation mentioned earlier. Given this, there are significant limitations to the interpretations that can be constructed in this way (Miall, 2006b).

TECTONIC SETTING

Alluvial deposits are sensitive indicators of allogenic processes, such as tectonism and base-level change (Blair and Bilodeau, 1988; Paola *et al.*, 1992; Deramond *et al.*, 1993; Holbrook *et al.*, 2006). An examination of these controls is therefore an essential element of a basin analysis of a fluvial system. One of the most important allogenic controls is tectonic setting (Miall, 1996; Fig. 23). The thickest successions of alluvial deposits occur in rift basins (including 'failed' rifts and aulacogens), in forearc basins, retroarc and peripheral foreland basins, and those associated with hinterland deformation caused by plate collision, particularly foreland and strike-slip basins. Some of these basins are one sided, in that the alluvial apron forms a prismatic body flanking the sea or a major lake (Figs. 23 and 24). Drainage (paleocurrent direction) and the orientation of proximal-to-distal facies changes are essentially perpendicular to the basin margin. Ocean-margin basins on divergent (passive) plate margins are of this type, as are some foreland basins. In other cases the basin is two-sided, with alluvial aprons on both sides. The basin center may be occupied by a water body, or by a major trunk river that collects runoff and sediment, and disperses it along the length of the basin axis. The facies characteristics and orientation of this trunk river system are typically very different from those of the feeder system draining from the sides. Most rift basins, strike-slip fault-bounded basins, and many foreland basins are of this second type.

Fluvial deposits are not characteristic of basins floored by oceanic crust, such as oceanic trenches, subduction complexes and remnant ocean basins, where deep-water sediment gravity-flow deposits predominate. However, long-continued basin fill in areas of substantial sediment supply may generate accumulations that build above sea level, allowing preservation of nonmarine deposits (*e.g.*, the modern Mentawai Islands, which represent the subaerially exposed topmost part of the subduction complex adjacent to the Sumatra–Java trench).

Tectonic Control of Alluvial Sedimentation

Sloss (1962) used the term *clastic wedge* for the accumulations of clastic sediment derived from orogenic uplift that spread out across an adjacent foreland basin. These typically

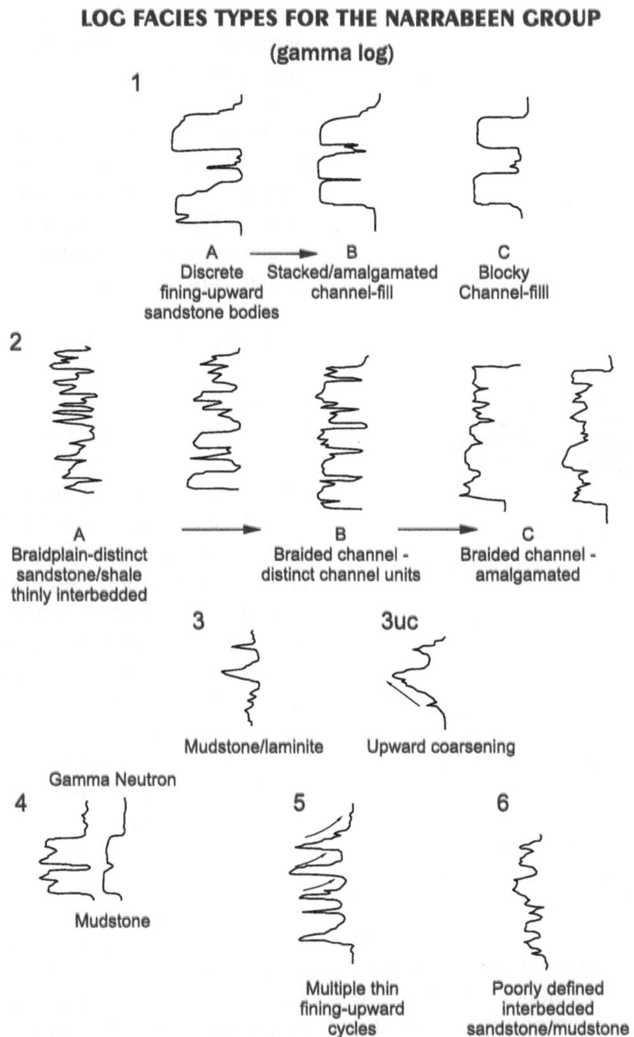

Figure 22. Application of log-facies analysis (numbers) to gamma-ray wireline logs through a fluvial deposit (Hamilton and Galloway, 1989). In such logs, a deflection to the left indicates a decreased gamma-ray value, which is usually indicative of coarser sediment in terrigenous-clastic successions.

approximate a wedge shape because they splay out from a source of limited areal extent (reflecting the localization of most tectonic episodes) and thin down dip as transport energy diminishes. The term clastic wedge has come to be more broadly applied to other tectonic settings, such as wedges of alluvial-fan deposits banked against a bounding fault in a rift basin. The original concept included the supposition that the geometry and timing of clastic deposits derived from the erosion of orogenic uplifts can be correlated to tectonic episodes in their source area (the term tectonic cyclothem highlights the same concept). For example, alluvial to shallow-marine Cenozoic clastic wedges of the Gulf Coast can be correlated with tectonism in the headwaters regions of the Gulf Coast rivers – the Cordilleran mountains of the western United States (Galloway, 1989). Other examples include the various pulses of nonmarine sediment that characterize the Alberta foreland basin (*e.g.*, Kootenay-Blairmore, Belly River-Paskapoo), which have been loosely correlated with terrane-accretion events along the western continental margin (Stockmal *et al.*, 1992).

Proximal (near source) deposits may reach enormous thicknesses (17 km of Tertiary fluvial and associated marine strata in the Western Trough of Burma, a forearc basin; 7 km of fluvial deposits in the Indus–Ganges trough of northern Pakistan, a foreland basin). Commonly they are structurally deformed and uplifted, indicating that the tectonism that uplifted the source area continued during deposition. Early basin-fill sediment may be cannibalized by uplift and erosion and fed back into the basin, where it is incorporated into the younger basin fill. It is common for proximal deposits to make up large-scale coarsening-upward cycles tens or hundreds of meters thick, recording increasing source-area relief and depositional slope during tectonism. These have been called *tectonic cyclothems* (Blair and Bilodeau, 1988). The sandstone-sheet cycles illustrated in Figure 21 include this type of allogenic deposit, but cyclothems resulting from regional uplift may be up to hundreds of meters thick and basin-wide in extent.

Distal deposits occupying the basin center, or those occurring along the coast, may interfinger with shallow-marine or lacustrine deposits (Fig. 24; *see* Chapters 10 and 21), the style of the interfingering providing a sensitive record of local fluctuations in base level.

The timing relationship between tectonism, subsidence and nonmarine sedimentation in foreland basins is variable, depending on the rigidity of the underlying crust and the magnitude and grain-size distribution of the sediment supply (including that introduced by axial drainage). Two contrasting models represent end-member conditions, the conventional *syntectonic* model of gravel progradation occurring contemporaneously with tectonism, and the *anti-tectonic* model, in which gravel progradation postdates tectonism during isostatic uplift and unroofing of the fold-thrust belt (Heller *et al.*, 1988). In the first case, a narrow belt of clastic sediments develops in the foredeep close to the proximal margin of the basin. In the second case, a widespread sheet of fluvial deposits may extend far across the basin, resting on the unconformity developed following the uplift (the Cretaceous Cadomin Formation of Alberta is an example of the latter). Climatic factors may also exert an important influence on the timing and extent of these clastic wedges, including the amount of rainfall that controls the rate of sedi-

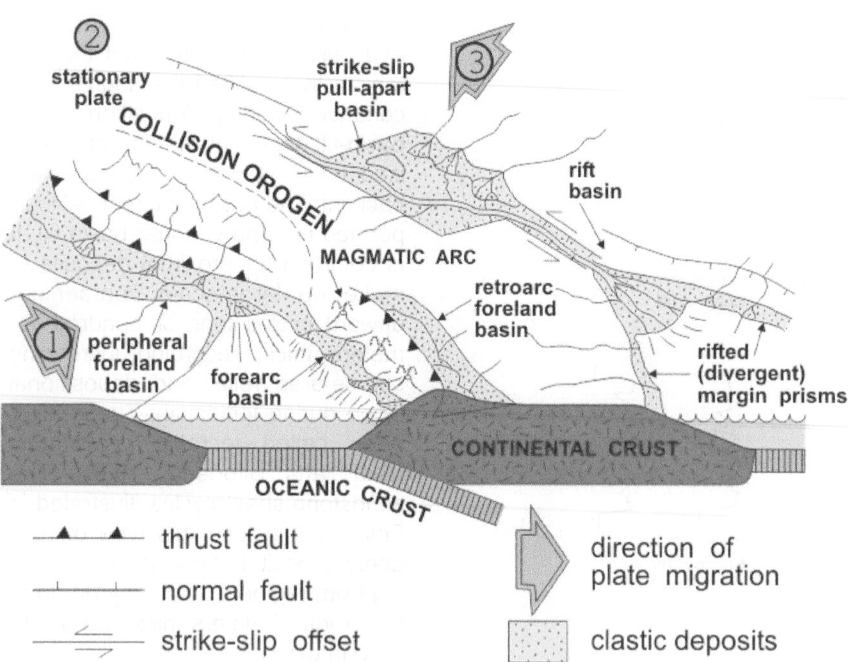

Figure 23. Tectonic settings in which fluvial deposits are common. Note the distinction between *transverse* drainage, which is oriented perpendicular to structural grain, and *axial* or *longitudinal* drainage, which typically consists of trunk streams flowing along the basin axis.

ment dispersal. Uplift of the fold-thrust belt increases orographic rainfall, which increases erosion and sediment yield. Where sediment supply outpaces subsidence, rivers draining transversely to the orogenic belt may develop an alluvial plain extending out across the entire basin. Jordan (1995) termed such basins *overfilled basins*. In basins with a more limited supply, alluvial drainage and sediment may be captured by a trunk stream flowing parallel and close to the structural axis of the basin — an *underfilled basin* in Jordan's terminology.

An important central concept in stratigraphy is *accommodation*, the space available for sedimentation and how this volume changes in response to allogenic forcing (sea-level change, tectonism, climate change, sediment supply, and the dependent variables fluvial competence and capacity). Paola *et al.* (1992) developed a fully quantitative numerical model for predicting the distribution of gravel across an asymmetrically subsiding basin, in which four major parameters could each be varied independently in turn, while keeping the others fixed. In the first case, slow increase of sediment flux leads to the development of a prograding pattern accompanied by increased sedimentation rate. This was termed a 'flux-driven' style of progradation. The second case shows that sediment is trapped in more proximal positions when subsidence rate is increased. This is the so-called 'subsidence-driven' model, in which progradation accompanies a diminished subsidence rate. The third case, termed the 'distribution-driven' model, shows the progradation that takes place when the gravel proportion is varied while the total sediment flux and subsidence rate are held constant. The fourth case shows that varying the amount of water available for transportation makes little difference to the gravel distribution in the basin.

Paola *et al.* (1992) also explored the relationship between basin scale and the rate of response to the various sedimentary controls. Each basin is characterized by a factor termed the 'basin equilibrium time', which is defined as the square of basin length divided by the water flux, or diffusivity (the rate of sediment transport). The equilibrium time is the time scale over which the basin is able to respond fully to periodic or cyclic changes in the controlling parameters (subsidence, sediment flux, *etc.*) by erosion or aggradation to generate an equilibrium longitudinal profile for the rivers in the basin. In large basins with long equilibrium times (millions of years), progradation is controlled primarily by changes in diffusivity or sediment supply. Where equilibrium times are relatively short, changes in subsidence rate appear to be the major control.

At the distal fringe of a clastic wedge, where it interfingers with marine deposits, the question of tectonic control versus control by sea-level change is more complex. Holbrook *et al.* (2006) introduced the useful concepts of *buttresses* and *buffers* to account for longitudinal changes in fluvial facies and architecture upstream from a coastline. A buttress is some fixed point that constitutes the downstream control on a fluvial graded profile. In marine basins this will be marine base level (sea level). In inland basins it will be lake level, or the lip or edge of a basin through which the trunk river flows out of the basin. The buffer is the space above and below the current graded profile; it represents the range of reactions that the profile may exhibit, given changes in upstream controls, such as tectonism or climate change that govern the discharge and sediment load of the river. For example, tectonic uplift may increase the sediment load, causing the river to aggrade towards its upper buffer limit. A drop in the buttress, as a result of a fall in sea (or lake) level, may result in incision of the river system, but if the continental shelf newly exposed by the fall in sea level has a similar slope to that of the river profile, there may be little change in the fluvial style of the river. In any of these cases, the response of the river system is to erode or aggrade towards a new dynamically maintained equilibrium profile that balances the water and sediment flux and the rate of change in *accommodation*. The zone between the upper and lower limits is the buffer zone and represents the available (potential) *preservation space* for the fluvial system.

Many basins contain regional unconformity surfaces composed of laterally amalgamated incised channel systems and the intervening

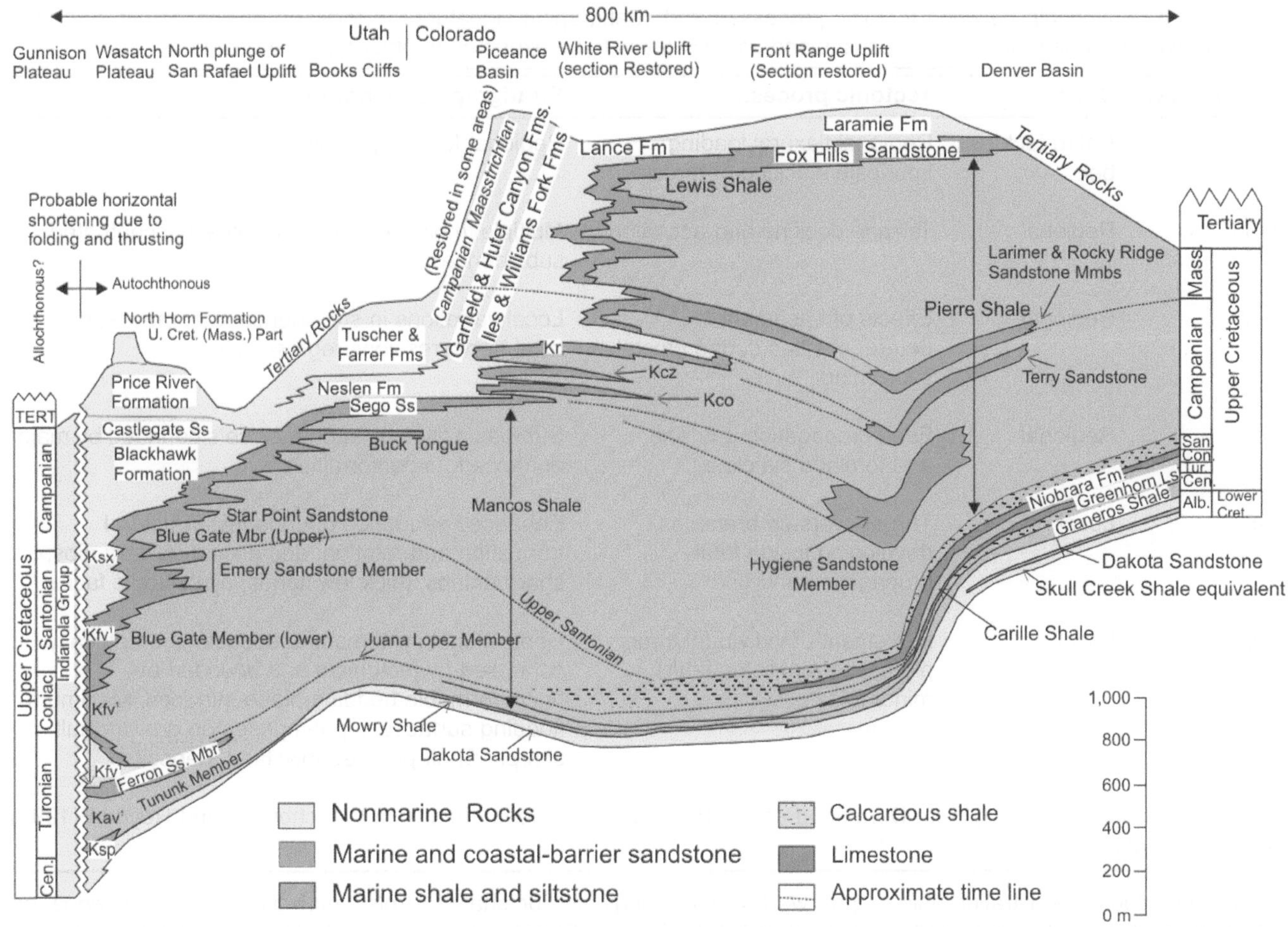

Figure 24. Stratigraphic cross section of Cretaceous–Tertiary rocks from central Utah to northeastern Colorado, across the Western Interior foreland basin. The prograding succession of marine and terrestrial deposits represents a clastic wedge. Clastic wedges of this type are very common in retroarc and peripheral foreland basins. Note the interfingering of marine and nonmarine strata (Molenaar and Rice, 1988).

upland, non-depositional or erosional interfluve areas (*e.g.*, Mannville Group of Alberta: Cant, 1998; J-Sandstone of the Denver Basin: Weimer, 1986). Cycles of sea-level change seem to be the most likely explanation for the extensive channel incision, although climate change may also modify the aggradational/degradational balance in a river system, as discussed below. Most large modern river systems have incised-valley fills beneath their present-day floodplain as a result of aggradation following the postglacial rise in sea level, or as a result of climate change associated with the end of the last glacial episode (*e.g.*, Blum, 1994).

Foreland basins are commonly characterized by clastic wedges of fluvial and coastal-plain strata that may be hundreds of meters thick, interbedded in complex stratigraphic relationships with marine shoreface and shelf deposits. Examples include many of the mid-Cretaceous deposits of the Alberta foreland basin (*e.g.*, Plint and Norris, 1991) and the Upper Cretaceous–Tertiary rocks of the Book Cliffs, Utah (Molenaar and Rice, 1988; Fig. 24). The overall basinward progradation of these clastic wedges can be attributed to the Sevier Orogeny, which caused basinward growth and cratonward displacement the fold-thrust belt. The individual progradational successions within the larger succession, such as the Ferron Sandstone and the Castlegate Sandstone, may have been generated either by tectonic mechanisms or eustatic changes in sea level (Ryer, 1984; Yoshida *et al*, 1996). The former seems more likely because control by eustatic sea-level change would lead to succession of cycles that could be correlated along the length of the basin. However, Krystinik and DeJarnett (1995) demonstrated that, at least in the Campanian to Maastrichtian record, detailed biostratigraphic correlation of seven sections between New Mexico and Alberta does not support the presence of through-going cycles on a millions- to tens-of-millions-of-years time scale, which would be expected if eustatic sea-level change was the predominant control.

Careful mapping in foreland basins and other tectonically active basins (mostly within the proximal, nonmarine strata of the basin fill) has generated the data in Table 5. Accommodation can be created and modified by tectonic processes on a time scale as short as a few tens of thousands of years. Major clastic wedges, such as those that prograde into the Alberta Basin from the Rocky Mountain uplift every few tens of millions of years, reflect terrane docking and accretion events (Stockmal *et al.*, 1992). Catuneanu *et al.* (1997, 1999) used bentonite horizons to

Table 5. The relationship between tectonic processes and stratigraphic signatures in foreland basins, at different time scales. A similar range of tectonic processes and time scales is to be expected in other types of basins

Duration m.y.	Scale	Tectonic process	Stratigraphic signature
> 50	Entire tectonic belt	Regional flexural loading, imbricate stacking of thrusts	Regional foredeep basin
10−50	Regional	Terrane docking and accretion	Multiple 'molasse' pulses; localized accentuated subsidence
10−50	Regional	Effects of basement heterogeneities during crustal shortening	Local variations in subsidence rate; may lead to local transgressions/regressions
> 5	Regional	Fault-propagation anticline and foreland syncline	Sub-basin filled by sequence sets bounded by major enhanced unconformities
0.5-5	Local	Thrust overstep branches developing inside fault-propagation anticline	Enhanced sequence boundaries; structural truncation and rotation; decreasing-upward dips; sharp onlaps; thick lowstands, syntectonic facies
< 0.5	Local	Movement of individual thrust plates, normal listric faults, minor folds	Depositional systems and bedsets geometrically controlled by tectonism and bounded by unconformable bedding-plane surfaces. Maximum flooding surfaces superimposed on growth-fault scarps. Shelf-perched lowstand deposits.

This table was adapted mainly from Deramond *et al.* (1993), with additional data from Waschbusch and Royden (1992) and Stockmal *et al.* (1992).

provide a detailed chronostratigraphic framework for foreland-basin deposits in Alberta, and demonstrated that the loading and unloading of the basin on a high-frequency time scale (10^5 years) generated repeated flexure around a hingeline separating the foredeep from the forebulge. Varban and Plint (2008) mapped stratigraphic relations in the Upper Cretaceous Kaskapau and Cardium formations of northwest Alberta and demonstrated the existence of high-frequency cycles having a 125 ka-cyclicity (of possible glacioeustatic origin) superimposed on onlap-offlap cycles controlled by movement of the foreland-basin forebulge having a periodicity of 0.5 to 0.7 m.y. The proximal parts of these units are composed of coastal-plain nonmarine facies. The alluvial fill of perched (wedge-top) basins in the Pyrenees contain unconformity surfaces that can be correlated to the development of imbricate faults in the underlying fold-thrust belt (Deramond *et al*., 1993; Fig. 25).

SEQUENCE MODELS FOR ALLUVIAL DEPOSITS

The concepts of sequence stratigraphy were defined originally for shallow-marine deposits (*see* Chapter 2), but are generally applicable to nonmarine deposits, even in entirely inland settings, where sea-level change has no influence on sedimentation. The concept of *accommodation*, defined earlier, is of central importance in the study of sequence stratigraphy. In basins that are partially or wholly marine, sea level is of fundamental importance, and in fluvial systems, sea level may operate as an important downstream control. The systems-tract terminology developed for marine sequences is applicable to this setting. But downstream base-level control (the *buttress* of Holbrook *et al.,* 2006) only influences the lower few tens of kilometers of a river system (Blum and Törnqvist, 2000). Fluvial sequences formed as a result of such a downstream control are discussed in the next section. Other allogenic controls must be taken into account for inland alluvial settings.

In the next section we discuss fluvial sequences generated under the primary influence of sea-level change, the major 'downstream' control. The influence of 'upstream' controls such as tectonic influence on regional slope and the tectonic and climatic control of the sediment supply, is discussed in the following section. In a coastal-plain alluvial basin, upstream from the influence of changing base level, and in the case of entirely nonmarine basins, upstream controls are of overriding importance, and may have a substantial effect on fluvial styles and depositional patterns. Sequence concepts can be useful in systematizing and interpreting these processes.

Two useful discussions of fluvial sequence stratigraphy were provided by Wright and Marriott (1993) and Shanley and McCabe (1994). A basic sequence model is illustrated in Figure 26.

Nonmarine Sequences Developed Under the Influence of Downstream Controls

Falling-stage Systems Tract (FSST) and Lowstand Systems Tract (LST)

Rivers naturally grade themselves to base level, typically developing a graded profile that decreases in slope toward the river mouth. A fall in base

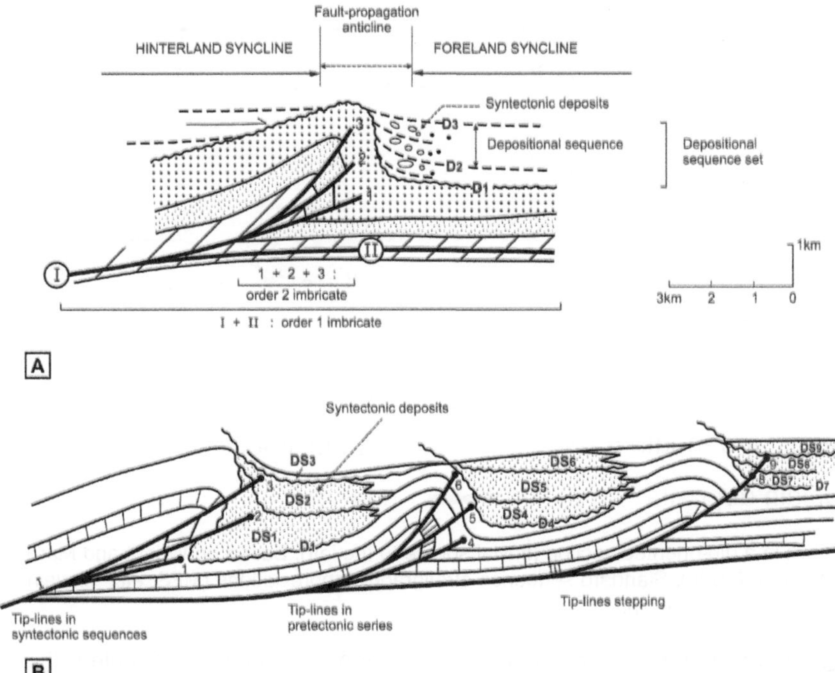

Figure 25. The relationship between the development of a fold-thrust belt and the stratigraphy of the adjacent foreland basin. Unconformities, numbered D1-D9, develop as imbricate thrust slices (1, 2, 3) develop in sequence. Uplift of each slice is recorded by a corresponding numbered unconformity in the basin (Deramond et al., 1993).

level may lead to several possible modes of adjustment (Schumm, 1993). If the fall in base level exposes a slope that is steeper than the river's graded profile, the river will erode its bed and an incised valley may develop. On more gentle slopes, the sinuosity of the river may increase, resulting in a constant channel slope. A river carrying a large sediment load may not undergo incision during base-level fall, but may extend its course with little change in its lower reaches, while continuing to prograde at the coastline (Blum and Törnqvist, 2000).

The late Cenozoic glaciation left most of the world's major rivers with incised valleys as a result of lower base levels. In many cases the drop in sea level took shorelines beyond the edge of the continental shelf and part way down the continental slope, causing widespread adjustments of graded profiles. Broad continental shelves, such as that of the Gulf Coast, are crossed by a complex of incised-valley fills and transgressive deposits formed during the Quaternary

Valleys, several kilometers wide and tens of meters deep, develop during the falling stage. Pauses in the base-level fall caused by periods of stable sea level may allow the valley to widen, and the evidence of such episodes may be preserved in the form of terrace remnants along the valley walls. A widespread erosion surface develops across the coastal plain. On the interfluves (the elevated areas between major river valleys) the rate of erosion may be slow, allowing for the development of widespread soils.

Sequence Boundary (SB)

In nonmarine systems, the sequence boundary represents the final position of the subaerial erosion surface immediately prior to the commencement of a new phase of base-level rise. The deep scour that occurs at the base of major rivers may cause this to be a very prominent surface within a fluvial succession (Best and Ashworth, 1997). The surface may cut down into a quite different facies succession, such as the transgressive or highstand shallow-marine or deltaic deposits of the preceding cycle, and the surface itself may be marked by a coarse lag deposit or evidence of extensive pedogenesis. However, in many systems this is not the case. Miall and Arush (2001) suggested the term *cryptic sequence boundary* for the erosion surfaces that develop on low-relief alluvial plains where, in outcrop, the sequence boundary appears identical to any other channel-scour surface. Such cryptic sequence boundaries in fluvial successions might be identified by sudden changes in detrital composition, major shifts in fluvial dispersal directions, and evidence of early cementation in the deposits immediately below the sequence boundary, all of which are indications of the extended period of time represented by this surface, during which the fluvial system continued to evolve.

Between the major river valleys, where downward erosion occurs at a slower rate, soils may be well developed, such as the caliche deposit illustrated in Figure 9. McCarthy and Plint (1998) explained how paleosoils may be used to define sequence boundaries and to establish a sequence stratigraphy within such deposits.

Transgressive Systems Tract (TST)

A rise in base level generates increased accommodation and a transgression of the lower course of the river is likely to occur. Incised valleys become estuaries, and have a range of depositional conditions ranging from fully marine at the mouth to fully nonmarine at the inland end of the estuary (*see* Chapters 9 and 11). There will be a decrease in slope of the lower course of the river, leading to a reduction in competency and, consequently, in the grain size of the sediment transported and deposited.

The fill of an incised valley may commence with a complex of amalgamated channel-fill deposits, regardless of the fluvial style, reflecting an extended phase of channel reworking during a period when little new accommodation is being added to the fluvial profile. With the commencement of a transgressive phase, the rate of accommodation generation increases and there is less channel scour into earlier deposits. Bar-top and floodplain deposits have a higher probability of preservation, so that the average

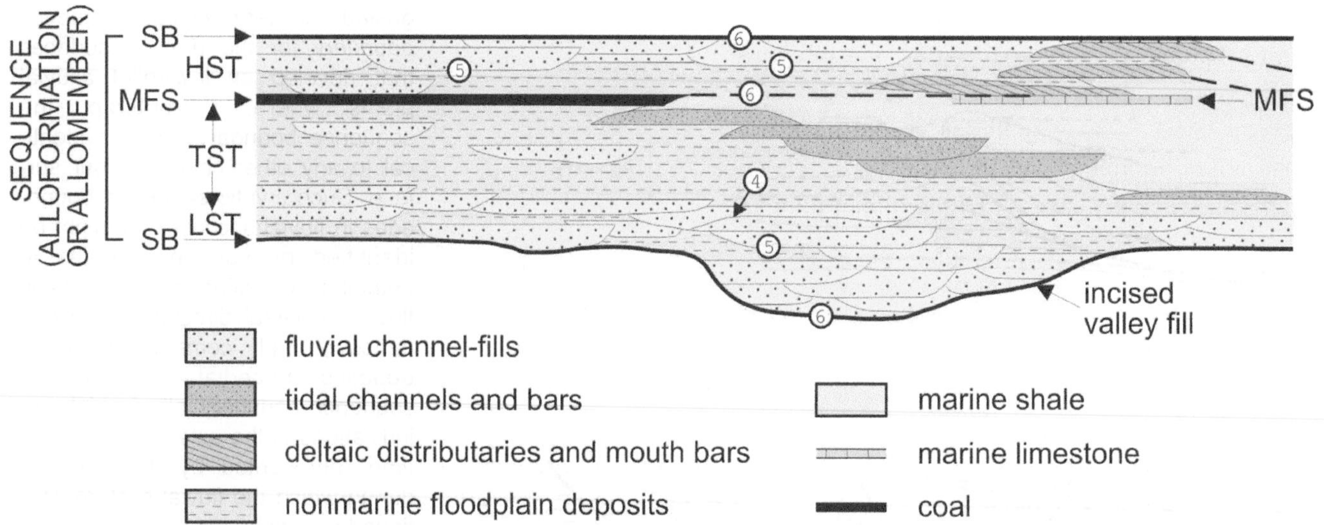

Figure 26. Sequence model for fluvial deposits shown in a dip-oriented section (fluvial source to the left). Based on Wright and Marriott (1993), Shanley and McCabe (1994) and Gibling and Bird (1994). Standard sequence-stratigraphic abbreviations (MFS, *etc.*, explained in text) and bounding-surface rankings (numerals in circles) are shown.

grain size of the deposits tends to decrease. A sheet-scale fining-upward cycle may be the result (Fig. 21).

Maximum-flooding Phase and Maximum-flooding Surface (MFS)

The limit of transgression is reached during the period when the rate of accommodation generation by rise in base level reaches its maximum. In some otherwise predominantly nonmarine fluvial systems this may be indicated by the appearance of marine influence in otherwise typical fluvial deposits (Fig. 26). Marine ichnofacies may be present (*see* Chapter 3), and tidal influence may occur in the form of tidal bedding, reversing crossbedding, sigmoidal bedding, and inclined heterolithic strata (IHS) (*see* Chapter 9). The extensive marine shales that mark the MFS in the marine realm may reach landward into an otherwise nonmarine succession.

Rapid accommodation generation may be reflected by high sedimentation rates, and provides ideal conditions for floodplain accumulation. The thickest and most laterally extensive coals typically occur during this phase of the base-level cycle (Bohacs and Suter, 1997). Channel complexes may be encased in thick floodplain fines and exhibit little lateral interconnectedness. Soils are likely to be immature.

Highstand Systems Tract (HST)

At the end of the phase of base-level rise, the rate of accommodation generation slows, leading to a reduction in sedimentation rate. Channel complexes become more laterally interconnected and soils will have more time to develop and become more mature. Those that develop on interfluves, well away from areas of active fluvial erosion, may survive to become the sequence boundary during the next phase of base-level fall.

Nonmarine Sequences Developed Under the Influence of Upstream Controls

Currie (1997) suggested an alternative terminology to replace the standard systems-tract terms used for marine basins because, obviously, terms that include such words as transgressive, highstand, etc., are inappropriate for basins that are entirely nonmarine. For falling-stage and lowstand deposits, Currie (1997) suggested the term *degradational systems tract*, for transgressive deposits, *transitional systems tract*, and for highstand deposits, *aggradational systems tract*. These terms provide analogous ideas regarding changes in accommodation and sediment supply and their consequences for depositional style.

As discussed above, *tectonic cyclothems* are alluvial deposits generated under the influence of upstream tectonic control. They may be tens to hundreds of meters thick. Coarsening-upward cycles are generated by increases in sediment delivery, which, in turn, reflect tectonic uplift, changes in subsidence rate, or climate change. Fining-upward cycles reflect the reverse processes. Heward (1978) described examples of ancient alluvial-fan successions that exhibit these stratigraphic variations.

Tectonic cyclothems constitute nonmarine analogues of stratigraphic sequences, and, just as the construction of a sequence nomenclature for marine basins has become an essential component of basin analysis, contributing significantly to the sophistication of regional interpretations, there is no reason why the same approach should not be used for nonmarine basins. However, within successions that are entirely nonmarine, especially within inland basins where correlation with marine units cannot be attempted, the definition of sequences and the recognition of sequence boundaries may be a challenging task. For example, a succession of nonmarine sequences showing considerable variation in thickness and fluvial style is described by López-Gómez and Arche (1993). These sequences occur in an inland rift basin, and were formed entirely under the influence of tectonic controls.

"Channels aggrade when sediment supply exceeds transport capacity of

the discharge regime, and degrade when the reverse is true" (Blum and Törnqvist, 2000, p. 12). Despite the truth of this statement, the response of fluvial systems to tectonic and climatic changes is complex and characterized by variable lag times (Shanley and McCabe, 1994), and the resulting sequences may not display the simple succession of depositional-systems tracts that are now becoming familiar and perhaps even standardized features of the coastal to shallow-marine sedimentary record. Detailed studies of modern river systems in areas such as the Gulf Coast (Blum, 1994; Blum and Price, 1994) and the Rhine-Meuse system of The Netherlands (Vandenberghe, 1993; Törnqvist, 1993, 1994; Törnqvist *et al.*, 1993, Vandenberghe *et al.*, 1994) are beginning to unravel the complex and out-of-phase responses of river systems to high-frequency external forcing, such as the glacioeustatic changes of the late Cenozoic. The following paragraphs discuss some of this work as illustrations of the general principal of complex response.

Blum (1994) demonstrated that nowhere within coastal fluvial systems is there a single erosion surface that can be related to lowstand erosion. Such surfaces are continually modified by channel scour, even during transgression, because episodes of channel incision may reflect climatically controlled times of low sediment load, which are not synchronous with changes in base level. This is particularly evident landward of the limit of base-level influence. Postglacial terraces within inland river valleys reveal a history of alternating aggradation and channel incision reflecting climate change, all of which occurred during the last postglacial rise in sea level. A major episode of valley incision occurred in Texas, not during the time of glacioeustatic sea-level lowstand, but at the beginning of the postglacial sea-level rise, which commenced at about 15 ka (Blum, 1994). The implications of this have yet to be resolved for inland basins where aggradation occurs (because of tectonic subsidence), rather than incision and terrace formation. However, it would seem to suggest that no simple relationship between major bounding surfaces and base-level change should be expected.

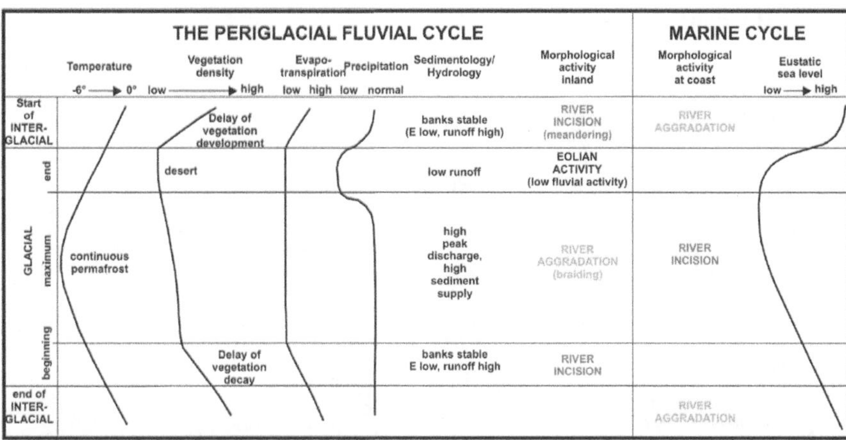

Figure 27. The relationship between temperature, vegetation density, evapotranspiration, precipitation, and sedimentary processes in river systems during glacial and interglacial phases. Adapted, in part, from work in the modern Rhine-Meuse system (Vandenberghe, 1993).

Figure 27 shows a model of fluvial processes in relationship to glacially controlled changes in climate and vegetation, based on the Dutch work. These studies, and those in Texas, deal with periglacial regions, where climate change and its effects were pronounced, even though the areas were not directly affected by glaciation. Vandenberghe (1993) and Vandenberghe *et al.* (1994) demonstrated that a major period of incision occurs during the transition from cold to warm phases because discharge increases while sediment yield remains low. Vegetation is quickly able to stabilize river banks, reducing sediment delivery, while evapotranspiration remains low, so that the runoff is high. Fluvial styles in aggrading valleys tend to change from braided during glacial phases to meandering during interglacials (Vandenberghe *et al.*, 1994). With increasing warmth, and, consequently, increasing vegetation density, banks become stabilized, and less sediment load is shed into the channel by bank erosion. Rivers of anastomosed or meandering style tend to develop, the former particularly in coastal areas where the rate of generation of new accommodation space is high during the period of rapidly rising base level (Törnqvist, 1993; Törnqvist *et al.*, 1993). Vandenberghe (1993) also demonstrated that valley incision tends to occur during the transition from warm to cold phases. Reduced evapotranspiration consequent upon the cooling temperatures occurs while the vegetation cover is still substantial. Therefore runoff increases, while sediment yield remains low. With the reduction in vegetation cover as the cold phase becomes established, sediment deliveries increase, and fluvial aggradation is re-established.

It is apparent that fluvial processes inland and those along the coast may be completely out of phase during the climatic and base-level changes accompanying glaciation. Within a few tens of kilometers of the sea, valley incision occurs at times of base-level lowstand, during cold phases, but the surface may be modified and deepened during the subsequent transgression until it finally becomes buried. Inland, major erosional bounding surfaces correlate to times of climatic transition, from cold to warm and from warm to cold, that is to say during times of rising and falling sea level, respectively.

The Dakota Group of northeast New Mexico and southeast Colorado provides a good example of an architecturally complex fluvial unit generated by a combination of upstream tectonic controls and downstream sea-level cycles (Fig. 28; Holbrook *et al.*, 2006). At the coastline, progradation and retrogradation creating three sequences was caused by sea-level cycles on a 10^5-year time scale. Each of these sequences can be traced up-dip towards the west, where they

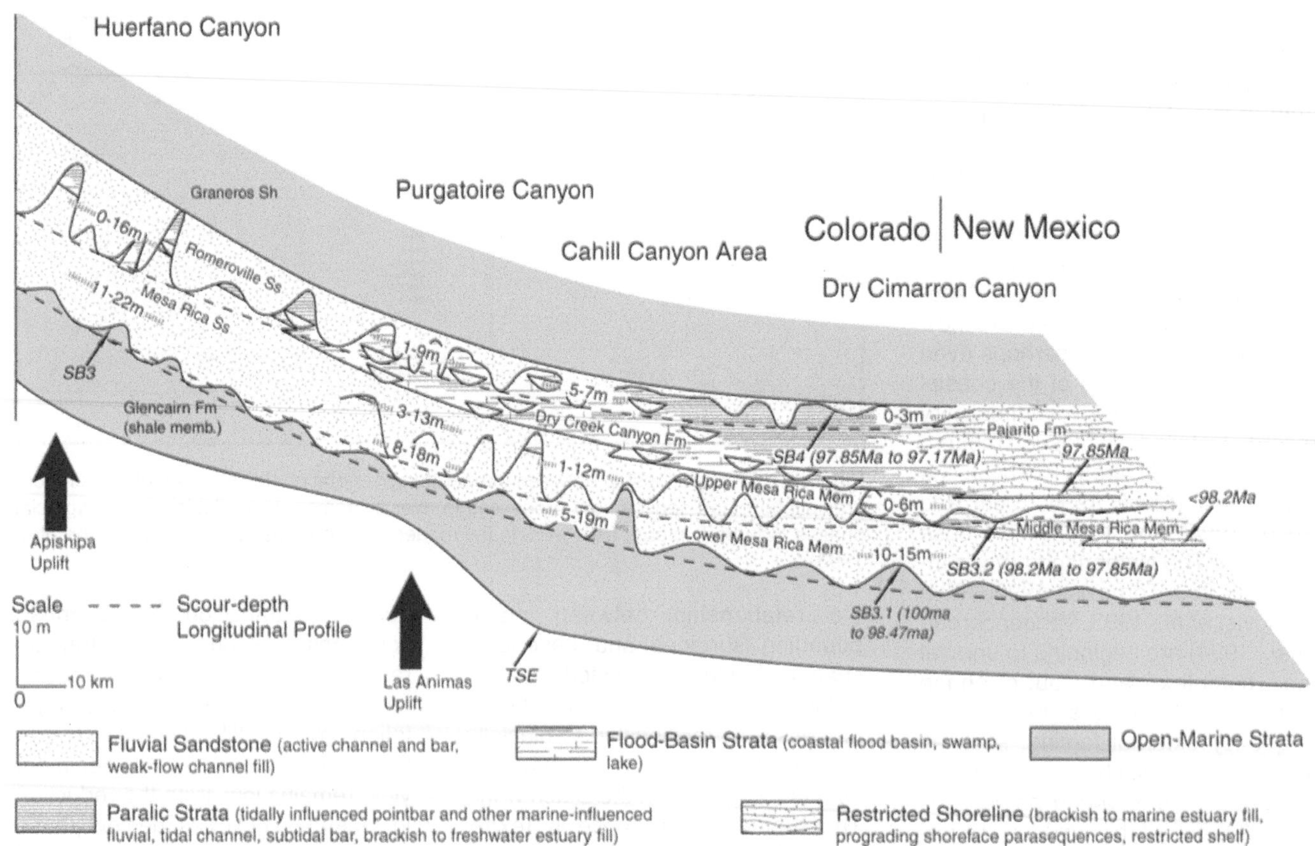

Figure 28. Longitudinal NW–SE section, approximately 250 km in length, through the mid-Cretaceous Dakota Group, from Colorado into the northeast corner of New Mexico. The internal architecture consists of a series of unconformity-bounded sandstone sheets that that reflect "deposition during repetitive valley-scale cycles of aggradation and incision." (Holbrook *et al.*, 2006, p. 164).

are composed of repeated cycles of aggradational valley-fill successions and mutually incised scour surfaces. These cycles reflect autogenic channel shifting within the limited preservation space available under conditions of modest, tectonically generated accommodation. This space is bounded by a lower buffer (in the Holbrook *et al.* (2006) terminology) set by the maximum depth of local channel scour, and an upper buffer set by the ability of the river to aggrade under the prevailing conditions of discharge and sediment load.

CONCLUSIONS

Application of the principle of uniformitarianism, 'the present is the key to the past', has enabled sedimentary geologists to develop a deep understanding of the sedimentary processes responsible for ancient fluvial deposits. In the case of alluvial systems, the measurement and monitoring of active channels, and the trenching and coring of surface deposits have provided a wealth of information, from which, together with comparative data from the ancient record, the foundation of the modern subject of facies analysis has been constructed. However, one of the problems with this approach is that the product of modern processes may be ephemeral. Surface deposits may be removed by the next flood, meandering channels scour into earlier channel and floodplain deposits, and so on. Modern deposits have widely varying preservation potential. For example, late-season bar-top ripples have a very low preservation potential, whereas deep scour hollows formed at tributary confluences during peak seasonal runoff have a much higher preservation potential. This is particularly critical where sedimentological information is to be used to assist in the analysis of subsurface reservoirs and aquifers, where it is the architecture of the selectively preserved deposits that is of interest. This is why improvements in geophysical techniques, especially GPR and 3-D seismic, have been so important in our ability to interpret ancient fluvial successions. In many cases they can now provide much essential information about the three-dimensional architecture of the preserved alluvial deposits of interest. Detailed architectural study of large two-dimensional outcrops of ancient deposits also may provide much important information.

Knowledge of sedimentology at the level of individual facies and facies successions is as important as ever for the exploitation of heterogeneous fluvial oil and gas reservoirs because every deposit is unique, and commonly the details of facies, architecture and stratigraphy are not readily amenable to simple analysis. The review of the sedimentology of alluvial deposits provided here is but an introduction to a broad and complex subject. Even our knowledge of modern sedimentary systems is incomplete. For example, very few studies have examined in detail the sedimentology of modern arid-region rivers and rivers in tropical rainforest settings.

Similarly, relatively little is known about how river systems in inland settings not connected to the sea respond to external forcing. The scope for further study of fluvial systems and their deposits, therefore, remains considerable.

REFERENCES

Basic sources of information

Blum, M. D., and Törnqvist, T. E., 2000, Fluvial responses to climate and sea-level change: a review and look forward: Sedimentology, v. 47, p. 2-48.
A useful review of allogenic processes.

Bridge, J. S., 2003, Rivers and Floodplains: Forms, Processes and Sedimentary Record: Oxford, Blackwell, 491 p.
A comprehensive textbook on fluvial processes and deposits. Particularly strong on modern processes, fluvial hydraulics and bedform generation, and on numerical models of channel form and evolution.

Miall, A. D., 1996, The Geology of Fluvial Deposits: Sedimentary Facies, Basin Analysis and Petroleum Geology: Springer-Verlag Inc., Heidelberg, 582 p.
Comprehensive treatment of ancient fluvial deposits. Detailed chapters on mapping methods (surface and subsurface), sequence stratigraphy, and the stratigraphy and facies of fluvial oil and gas reservoirs.

Shanley, K. W., and McCabe, P. J., 1994, Perspectives on the sequence stratigraphy of continental strata: American Association of Petroleum Geologists Bulletin, v. 78, p. 544-568.
This remains the best overall review of the sequence stratigraphy of fluvial deposits.

Other references

Allen, J. R. L., 1963, Henry Clifton Sorby and the sedimentary structures of sands and sandstones in relation to flow conditions: Geologie en Mijnbouw, v. 42, p. 223-228.

Allen, J. R. L., 1983, Studies in fluviatile sedimentation: bars, bar complexes and sandstone sheets (low-sinuosity braided streams) in the Brownstones (L. Devonian), Welsh Borders: Sedimentary Geology, v. 33, p. 237-293.

Ashley, G. M., 1990, Classification of large-scale subaqueous bedforms: a new look at an old problem: Journal of Sedimentary Petrology, v. 60, p. 160-172.

Ashworth, P. J., Best, J. L., Roden, J. E., Bristow, C. S., and Klaassen, G. J., 2000, Morphological evolution and dynamics of a large, sand braid-bar, Jamuna river, Bangladesh: Sedimentology, v. 47, p. 533-555.

Asprion, U., and Aigner, T., 1999, Towards realistic aquifer models: three-dimensional georadar surveys of Quaternary gravel deltas (Singen Basin, SW Germany): Sedimentary Geology, v. 129, p. 281-297.

Beres, M., Green, A., Huggenberger, P., and Horstmeyer, H., 1995, Mapping the architecture of glaciofluvial sediments with three-dimensional georadar: Geology, v. 23, p. 1087-1090.

Bernard, H. A., Leblanc. R. J., and Major, C. J., 1962, Recent and Pleistocene geology of southeast Texas, *in* Rainwater, E. H., and Zingula, R. P., *eds.*, Geology of the Gulf Coast and central Texas: Geological Society of America Guidebook for 1962 Annual Meeting, p. 175-224.

Best, J. L., 1987, Flow dynamics at river channel confluences: implications for sediment transport and bed morphology, *in* Ethridge, F. G., and Flores, R. M., *eds,* Recent and Ancient Nonmarine Depositional Environments: SEPM Special Publication 31, p. 27-35.

Best, J. L., and Ashworth, P. J., 1997, Scour in large braided rivers and the recognition of sequence stratigraphic boundaries: Nature, v. 387, p. 275-277.

Best, J. L., Ashworth, P. J., Bristow, C. S., and Roden, J., 2003, Three-dimensional sedimentary architecture of a large mid-channel sand braid bar, Jamuna River, Bangladesh: Journal of Sedimentary Research, v. 73, p. 516-530.

Blair, T. C., and Bilodeau, W. L., 1988, Development of tectonic cyclothems in rift, pull-apart, and foreland basins: sedimentary response to episodic tectonism: Geology, v. 16, p. 517-520.

Blum, M. D., 1994, Genesis and architecture of incised valley fill sequences: a late Quaternary example from the Colorado River, Gulf Coastal Plain of Texas, *in* Weimer, P., and Posamentier, H. W., *eds.*, Siliciclastic Sequence Stratigraphy: Recent Developments and Applications: American Association of Petroleum Geologists Memoir 58, p. 259-283.

Blum, M. D., and Price, D. M., 1994, Glacio-eustatic and climatic controls on Quaternary alluvial plain deposition, Texas coastal plain: Gulf Coast Association of Geological Societies, Transactions, v. 44, p. 1-9.

Bohacs, K., and Suter, J., 1997, Sequence stratigraphic distribution of coaly rocks: fundamental controls and paralic examples: American Association of Petroleum Geologists Bulletin, v. 81, p. 1612-1639.

Bridge, J. S., 1985, Paleochannel patterns inferred from alluvial deposits: a critical evaluation: Journal of Sedimentary Petrology, v. 55, p. 579-589.

Bridge, J. S., 1993, The interaction between channel geometry, water flow, sediment transport and deposition in braided rivers, *in* Best, J. L., and Bristow, C. S., *eds.*, Braided Rivers: Geological Society of London, Special Publication 75, p. 13-71.

Bridge, J. S., 2007, Fluvial facies models: recent developments, *in* Posamentier, H. W., and Walker, R. G., *eds.*, Facies Models Revisited: SEPM Special Publication 84, p. 85-170.

Bridge, J. S., Collier, R., and Alexander, J., 1998, Large-scale structure of Calamus River deposits (Nebraska, USA) revealed using ground penetrating radar: Sedimentology, v. 45, p. 977-986.

Cant, D. J., 1978a, Bed forms and bar types in the South Saskatchewan River: Journal of Sedimentary Petrology, v. 48, p. 1321-1330.

Cant, D. J., 1978b, Development of a facies model for sandy braided river sedimentation: comparison of the South Saskatchewan River and Battery Point Formation, *in* Miall, A. D., *ed.*, Fluvial Sedimentology: Canadian Society of Petroleum Geologists Memoir 5, p. 627-640.

Cant, D. J., 1998, Sequence stratigraphy, subsidence rates, and alluvial facies, Mannville Group, Alberta foreland basin, *in* Shanley, K. W., and McCabe, P. J., *eds.*, Relative Role of Eustasy, Climate, and Tectonism in Continental Rocks: SEPM Special Publication 59, p. 49-63.

Catuneanu, O., Sweet, A. R., and Miall, A. D., 1997, Reciprocal architecture of Bearpaw T-R sequences, uppermost Cretaceous, Western Canada Sedimentary Basin: Bulletin of Canadian Petroleum Geology, v. 45, p. 75-94.

Catuneanu, O., Sweet, A., and Miall, A. D., 1999, Concept and styles of reciprocal stratigraphies: Western Canada foreland system: Terra Nova, v. 11, p. 1-8.

Coleman, J. M., 1969, Brahmaputra River: channel processes and sedimentation: Sedimentary Geology, v. 3, p. 129-239.

Currie, B. S., 1997, Sequence stratigraphy of nonmarine Jurassic-Cretaceous rocks, central Cordilleran foreland-basin system: Geological Society of America Bulletin, v. 109, p. 1206-1222.

Davies, N. S. and Gibling, M. R., 2010, Cambrian to Devonian evolution of alluvial systems: The sedimentological impact of the earliest land plants: Earth-Science Reviews, v. 98, p. 171–200

Davies, R. J., Posamentier, H. W., Wood, L. J., and Cartwright, J. A., *eds.*, 2007, Seismic Geomorphology: Applications to Hydrocarbon Exploration and Production: Geological Society of London, Special Publication 277, 274 p.

Deramond, J., Souquet, P., Fondecave-Wallez, M.-J., and Specht, M., 1993, Relationships between thrust tectonics and sequence stratigraphy surfaces in foredeeps: model and examples from the Pyrenees (Cretaceous-Eocene, France, Spain), *in* Williams, G. D., and

Dobb, A., *eds.*, Tectonics and Seismic Sequence Stratigraphy: Geological Society of London, Special Publication 71, p. 193-219. .

Ethridge, F. G., and Schumm, S. A., 2007, Fluvial seismic geomorphology: a view from the surface, *in* Davies, R. J., Posamentier, H. W., Wood, L. J., and Cartwright, J. A., *eds.*, Seismic Geomorphology: Applications to Hydrocarbon Exploration and Production: Geological Society of London, Special Publication 277, p. 205-222.

Farrell, K. M., 1987, Sedimentology and facies architecture of overbank deposits of the Mississippi River, False River region, Louisiana, *in* Ethridge, F. G., Flores, R. M., and Harvey, M. D., *eds.*, Recent Developments in Fluvial SEPM Special Publication 39, p. 111-120.

Fielding, C. R., 1993, A review of recent research in fluvial sedimentology: Sedimentary Geology, v. 85, p. 3-14.

Fielding, C. R., and Crane, R. C., 1987, An application of statistical modelling to the prediction of hydrocarbon recovery factors in fluvial reservoir sequences, *in* Ethridge, F. G., Flores, R. M., and Harvey, M. D., *eds.*, Recent Developments in Fluvial Sedimentology: SEPM Special Publication 39, p. 321-327.

Fisk, H.N., 1944, Geological investigation of the alluvial valley of the lower Mississippi River: Mississippi River Commission, Vicksburg, Mississippi, 78 p.

Galloway, W. E., 1989, Genetic stratigraphic sequences in basin analysis II: Application to northwest Gulf of Mexico Cenozoic basin: American Association of Petroleum Geologists Bulletin, v. 73, p. 143-154.

Gibling, M. R., 2006, Width and thickness of fluvial channel bodies and valley fills in the geological record: a literature compilation and classification: Journal of Sedimentary Research, v. 76, p. 731-770.

Gibling, M. R., and Bird, D. J., 1994, Late Carboniferous cyclothems and alluvial paleovalleys in the Sydney Basin, Nova Scotia: Geological Society of America Bulletin, v. 106, p. 105-117.

Godin, P., 1991, Fining-upward cycles in the sandy-braided river deposits of the Westwater Canyon Member (Upper Jurassic), Morrison Formation, New Mexico: Sedimentary Geology, v. 70, p. 61-82.

Hamilton, D. S., and Galloway, W. E., 1989, New exploration techniques in the analysis of diagenetically complex reservoir sandstones, Sydney Basin, NSW: The Australian Petroleum Exploration Association Journal, v. 29, p. 235-257.

Hardage, B. A., Levey, R. A., Pendleton, V., Simmons, J., and Edson, R., 1996, 3-D Seismic imaging and interpretation of fluvially deposited thin-bed reservoirs, *in* Weimer, P., and Davis, T. L., *eds.*, Applications of 3-D Seismic Data to Exploration and Production: AAPG Studies in Geology 42, p. 27-34.

Heller, P. L., Angevine, C. L., Winslow, N. S., and Paola, C., 1988, Two-phase stratigraphic model of foreland-basin sequences: Geology, v. 16, p. 501-504.

Heward, A. P., 1978, Alluvial fan sequence and megasequence models: with examples from Westphalian D - Stephanian B coalfields, northern Spain, *in* Miall, A. D., *ed.*, Fluvial Sedimentology: Canadian Society of Petroleum Geologists Memoir 5, p. 669-702.

Holbrook, J., Scott, R. W., and Oboh-Ikuenobe, F. E., 2006, Base-level buffers and buttresses: a model for upstream versus downstream control on fluvial geometry and architecture within sequences: Journal of Sedimentary Research, v. 76, p. 162-174.

Jordan, T. E., 1995, Retroarc foreland and related basins, *in* Busby, C. J., and Ingersoll, R. V., *eds.*, Tectonics of Sedimentary Basins: Blackwell Scientific Publications, Oxford, p. 331-362.

Krystinik, L. F., DeJarnett, B. B., 1995, Lateral variability of sequence stratigraphic framework in the Campanian and Lower Maastrichtian of the Western Interior Seaway, *in* Van Wagoner, J. C., and Bertram, G. T., *eds.*, Sequence Stratigraphy of Foreland Basins: American Association of Petroleum Geologists Memoir 64, p. 11-25.

Leopold, L. B., and Wolman, M. G., 1957, River channel patterns; braided, meandering, and straight: U. S. Geological Survey Professional Paper 282-B.

López-Gómez, J., and Arche, A., 1993, Architecture of the Cañizar fluvial sheet sandstones, Early Triassic, Iberian Ranges, eastern Spain, *in* Marzo, M., and Puigdefábregas, C., *eds.*, Alluvial Sedimentation: International Association of Sedimentologists Special Publication 17, p. 363-381.

Lunt, I. A., and Bridge, J. S., 2004, Evolution and deposits of a gravelly braid bar, Sagavanirktok River, Alaska: Sedimentology, v. 51, p. 415-432.

McCarthy, P. J., and Plint, A. G., 1998, Recognition of interfluve sequence boundaries: integrating paleopedology and sequence stratigraphy: Geology, v. 26, p. 387-390.

Martin, J. H., 1993, A review of braided fluvial hydrocarbon reservoirs: the petroleum engineer's perspective, *in* Best, J. L., and Bristow, C. S., *eds.*, Braided Rivers: Geological Society of London, Special Publication 75, p. 333-367.

Miall, A. D., 1977, A review of the braided river depositional environment: Earth-Science Reviews, v. 13, p. 1-62.

Miall, A. D., 1978, Lithofacies types and vertical profile models in braided river deposits: a summary, *in* Miall, A. D., *ed.*, Fluvial Sedimentology: Canadian Society of Petroleum Geologists Memoir 5, p. 597-604.

Miall, A. D., 1980, Cyclicity and the facies model concept in geology: Bulletin of Canadian Petroleum Geology, v. 28, p. 59-80.

Miall, A. D., 1985, Architectural-element analysis: A new method of facies analysis applied to fluvial deposits: Earth-Science Reviews, v. 22, p. 261-308.

Miall, A. D., 1994, Reconstructing fluvial macroform architecture from two-dimensional outcrops: examples from the Castlegate Sandstone, Book Cliffs, Utah: Journal of Sedimentary Research, v. B64, p. 146-158.

Miall, A. D., 2002, Architecture and sequence stratigraphy of Pleistocene fluvial systems in the Malay Basin, based on seismic time-slice analysis: American Association of Petroleum Geologists Bulletin, v. 86, p. 1201-1216.

Miall, A. D., 2006a, How do we identify big rivers, and how big is big?: Sedimentary Geology, v. 186, p. 39-50.

Miall, A. D., 2006b, Reconstructing the architecture and sequence stratigraphy of the preserved fluvial record as a tool for reservoir development: a reality check: American Association of Petroleum Geologists Bulletin v. 90, p. 989-1002.

Miall, A. D, and Arush, M., 2001, Cryptic sequence boundaries in braided fluvial successions: Sedimentology, v. 48, p. 971-985

Miall, A. D., and Jones, B., 2003, Fluvial architecture of the Hawkesbury Sandstone (Triassic), near Sydney, Australia: Journal of Sedimentary Research, v. 73, p. 531-545.

Molenaar, C. M., and Rice, D. D., 1988, Cretaceous rocks of the Western Interior Basin, *in* Sloss, L. L., *ed.*, Sedimentary cover–North American Craton: U.S., The Geology of North America: Boulder, Colorado, Geological Society of America, v. D-2, p. 77-82.

Morozova, G. S., and Smith, N. D., 1999, Holocene avulsion history of the lower Saskatchewan fluvial system, Cumberland Marshes, Saskatchewan-Manitoba, Canada, *in* Smith, N. D., and Rogers, J., *eds.*, Alluvial Sedimentology VI: International Association of Sedimentologists Special Publication 28, p. 231-249.

Morozova, G. S., and Smith, N. D., 2000, Holocene avulsion styles and sedimentation patterns of the Saskatchewan river, Cumberland Marshes, Canada: Sedimentary Geology, v. 130, p. 81-105.

North, C. P., Nanson, G. C., and Fagan, S. D., 2007, Recognition of the sedimentary architecture of dryland anabranching (anastomosed) rivers: Journal of Sedimentary Research, v. 77, p. 925-938.

Paola, C., Heller, P. L., and Angevine, C. L., 1992, The large-scale dynamics of grain-size variation in alluvial basins, 1: theory: Basin Research, v. 4, p. 73-90.

Plint, A. G., and Norris, B., 1991, Anatomy of a ramp margin sequence: facies successions, paleogeography and sediment dispersal patterns in the Muskiki and Marshybank formations, Alberta Foreland Basin: Bulletin of Canadian Petroleum Geology, v. 39, p. 18-42.

Rainbird, R. H., 1992, Anatomy of a large-scale braid-plain quartzarenite from the Neoproterozoic Shaler Group, Victoria Island, Northwest Territories, Canada: Canadian Journal of Earth Sciences, v. 29, p. 2537-2550.

Robinson, J. W., and McCabe, P. J., 1997, Sandstone-body and shale-body dimensions in a braided fluvial system: Salt Wash Sandstone Member (Morrison Formation), Garfield County, Utah: American Association of Petroleum Geologists Bulletin, v. 81, p. 1267-1291.

Rust, B. R., and Jones, B. G., 1987, The Hawkesbury Sandstone south of Sydney, Australia: Triassic analogue for the deposit of a large braided river: Journal of Sedimentary Petrology, v. 57, p. 222-233.

Ryer, T. A., 1984, Transgressive-regressive cycles and the occurrence of coal in some Upper Cretaceous strata of Utah, U. S. A., *in* Rahmani, R. A., and Flores, R. M., *eds*., Sedimentology of coal and coal-bearing sequences: International Association of Sedimentologists Special Publication 7, p. 217-227.

Sambrook-Smith, G. H., Ashworth, P. J., Best, J. L., Woodward, J., and Simpson, C. J., 2006, The sedimentology and alluvial architecture of the sandy braided South Saskatchewan River, Canada: Sedimentology, v. 53, p. 413-434.

Sarzalejo, S., and Hart, B. S., 2006, Stratigraphy and lithologic heterogeneity in the Mannville Group (southeast Saskatchewan) defined by integrating 3-D seismic and log data: Bulletin of Canadian Petroleum Geology, v. 54, p. 138-151.

Saucier, R.T., 1969, Geological Investigation of the Mississippi River area, Artonish to Donaldsonville, LA: US Army Corps of Engineers, Waterways Experiment Station, Technical Reports, S-69-4, Vicksburg, Mississippi, 20 plates

Schumm, S. A., 1968, Speculations concerning paleohydrologic controls of terrestrial sedimentation: Geological Society of America Bulletin, v. 79, p. 1573-1588.

Schumm, S. A., 1993, River response to baselevel change: implications for sequence stratigraphy: Journal of Geology, v. 101, p. 279-294.

Simons, D. B., Richardson, E. V., and Nordin, C. F., 1965, Sedimentary structures generated by flow in alluvial channels, *in* Middleton, G. V., *ed*., Primary Sedimentary Structures and their Hydrodynamic Interpretation: SEPM Special Publication 12, p. 34-52.

Sloss, L. L., 1962, Stratigraphic models in exploration: American Association of Petroleum Geologists Bulletin, v. 46, p. 1050-1057.

Smith, N. D., Cross, T. A., Dufficy, J. P., and Clough, S. R., 1989, Anatomy of an avulsion: Sedimentology, v. 36, p. 1-23.

Stephens, M., 1994, Architectural element analysis within the Kayenta Formation (Lower Jurassic) using ground-probing radar and sedimentological profiling, southwestern Colorado: Sedimentary Geology, v. 90, p. 179-211.

Stockmal, G. S., Cant, D. J., and Bell, J. S., 1992, Relationship of the stratigraphy of the Western Canada foreland basin to Cordilleran tectonics: insights from geodynamic models, *in* Macqueen, R. W., and Leckie, D. A., *eds*., Foreland Basin and Fold Belts: American Association of Petroleum Geologists Memoir 55, p. 107-124.

Törnqvist, T. E., 1993, Holocene alternation of meandering and anastomosing fluvial systems in the Rhine-Meuse delta (central Netherlands) controlled by sea-level rise and subsoil erodibility: Journal of Sedimentary Petrology, v. 63, p. 683-693.

Törnqvist, T. E., 1994, Middle and late Holocene avulsion history of the River Rhine (Rhine-Meuse delta, Netherlands): Geology, v. 22, p. 711-714.

Törnqvist, T. E., van Ree, M. H. M., and Faessen, E. L. J. H., 1993, Longitudinal facies architectural changes of a Middle Holocene anastomosing distributary system (Rhine-Meuse delta, central Netherlands): Sedimentary Geology, v. 85, p. 203-219.

Tyler, N., 1988: New oil from old fields: Geotimes, v. 33, No. 7, p. 8-10.

Vail, P. R., Mitchum, R. M., Jr., Todd, R. G., Widmier, J. M., Thompson, S., III, Sangree, J. B., Bubb, J. N., and Hatlelid, W. G., 1977, Seismic stratigraphy and global changes of sea-level, *in* Payton, C. E., *ed*., Seismic stratigraphy - applications to hydrocarbon exploration: American Association of Petroleum Geologists Memoir 26, p. 49-212.

Vandenberghe, J., 1993, Changing fluvial processes under changing periglacial conditions: Z. Geomorph, N.F., v. 88, p. 17-28.

Vandenberghe, J., Kasse, C., Bohnke, S., and Kozarski, S., 1994, Climate-related river activity at the Weichselian-Holocene transition: a comparative study of the Warta and Maas rivers: Terra Nova, v. 6, p. 476-485.

Varban, B. L., and Plint, A. G., 2008 Sequence stacking patterns in the Western Canada foredeep: influence of tectonics, sediment loading and eustasy on deposition of the Upper Cretaceous Kaskapau and Cardium Formations: Sedimentology, v. 55, p.395-421.

Waschbusch, P. J., and Royden, L. H., 1992, Episodicity in foredeep basins: Geology, v. 20, p. 915-918.

Weimer, R. J., 1986, Relationship of unconformities, tectonics, and sea level change in the Cretaceous of the Western Interior, United States, *in* Peterson, J. A., *ed*., Paleotectonics and Sedimentation in the Rocky Mountain Region, United States: American Association of Petroleum Geologists, Memoir 41, p. 397-422.

Wright, V. P., and Marriott, S. B., 1993, The sequence stratigraphy of fluvial depositional systems: the role of floodplain sediment storage: Sedimentary Geology, v. 86, p. 203-210.

Yoshida, Shuji, Willis, A., and Miall, A. D., 1996, Tectonic control of nested sequence architecture in the Castlegate Sandstone (Upper Cretaceous), Book Cliffs, Utah: Journal of Sedimentary Research, v. 66, p. 737-748.

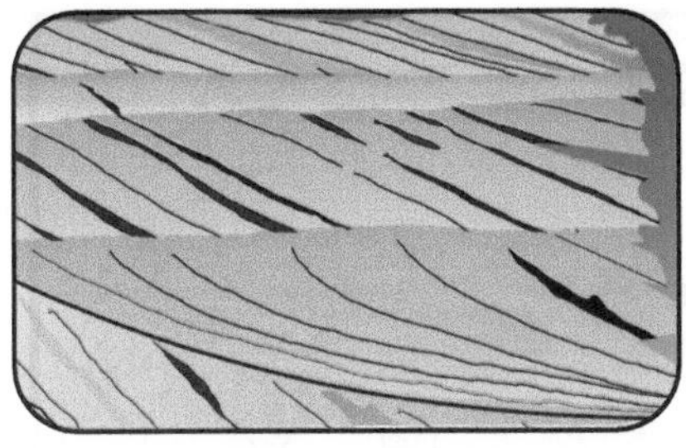

7. EOLIAN SYSTEMS

Michael E. Brookfield, Institute of Earth Sciences, Academia Sinica, 128 Academia Road, Sec. 2, Nankang, Taipei 115, Taiwan, ROC

Simone Silvestro, International Research School of Planetary Sciences, Universita d'Annunzio, Viale Pindaro, 42 65127 Pescara, Italy

INTRODUCTION

Ancient desert sandstones are amongst the most beautiful of sedimentary rocks (Fig. 1) and have been recognized in rocks as old as the early Proterozoic. Interpreting such deposits involves the study of existing eolian systems and their constituent features, relating these to processes of formation, and developing criteria for recognizing their development in space and through time. This chapter concentrates on those features directly related to wind action, even though, in many deserts, water action is equally or more important than wind in determining their characteristics (Brookfield, 2008). It does not go into the details of sediment mobilization, transport and deposition that are covered well in many other publications (Lancaster, 1995; *see* Chapter 2). We first look at the processes which form identifiable features in modern deserts and examine how to identify them in ancient rocks. Then, we use these features as the basis for constructing models for individual bedforms and develop larger scale facies models by combining individual bedform models, with examples of their application to the deposits of ancient deserts. Last, we consider some of the external controls on desert development that are relevant to the stratigraphic interpretation of eolian deposits. As this chapter deals with entire eolian systems, we include consideration of erosional features, which may be preserved under contemporary or younger sediments; some extraterrestrial examples, particularly from Mars, wind-blow silt and clay (loess), and those paleosols that are commonly developed in desert environments.

Eolian features develop where a protective cover of vegetation is sparse or absent and winds can pick up sand and dust – usually this means in dry climates. Semi-arid to extremely arid (hyperarid) conditions affect about one third of the Earth's surface (Cook *et al.*, 1993) and are planet wide on other bodies in the solar system (Carr, 2006). Extremely arid conditions (with no rainfall for periods of one year or longer) cover only about 4% of the Earth (Goudie, 2002), but, in the past, could have extended over much greater areas within supercontinents. The largest modern deserts occur around 30°N and S of the equator (Fig. 2A), where air in the equatorial Hadley circulation begins to descend, becoming denser and drier. Despite this, there are many desert areas extending to the poles on Earth (Seppala, 2003), and Mars and Venus are both entirely desert (Fig. 2B and C). On Earth, there are three main desert sedimentary environments: alluvial fans and ephemeral streams, inland sabkhas or playas, and sandy deserts - also called 'sand seas' or ergs (Goudie *et al.*, 2000). The first two are covered in other chapters (*see* Chapters 6 and 21); here, this chapter concentrates on sandy deserts and their associated environments. Sandy deserts form only about twenty percent of the area of modern terrestrial deserts. The rest is composed of eroding mountains (40%), stony areas or serirs (10–20%), desert flats (10–20%) and smaller areas of dry washes, volcanic cones and badlands, where erosion rather than deposition takes place (Goudie, 2002). The amount of eolian sand in the world's deserts varies greatly, being partly dependent on climate, wind regime, basin area and the nature of the source rocks. In North America, alluvial fans are much more important (30%) and sandy deserts much less important (less than 1%) than elsewhere, the one exception being the Gran Desierto, Sonora, Mexico. On Mars and Venus, wind is the dominant surface process. On Mars, as the atmospheric pressure is only 1% of Earth's, the thresh-

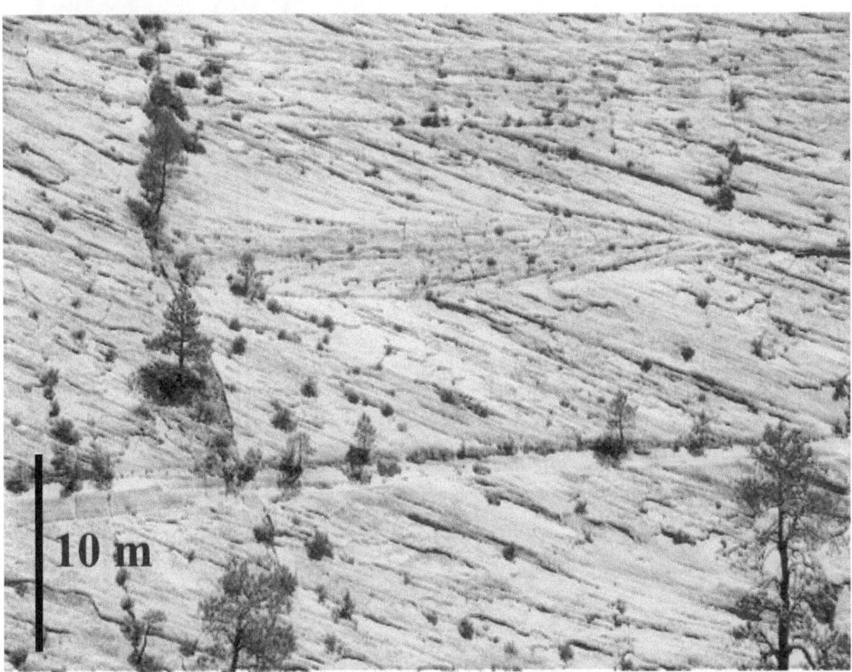

Figure 1. Beautiful large-scale crossbeds in the Navajo Sandstone, Zion, Utah.

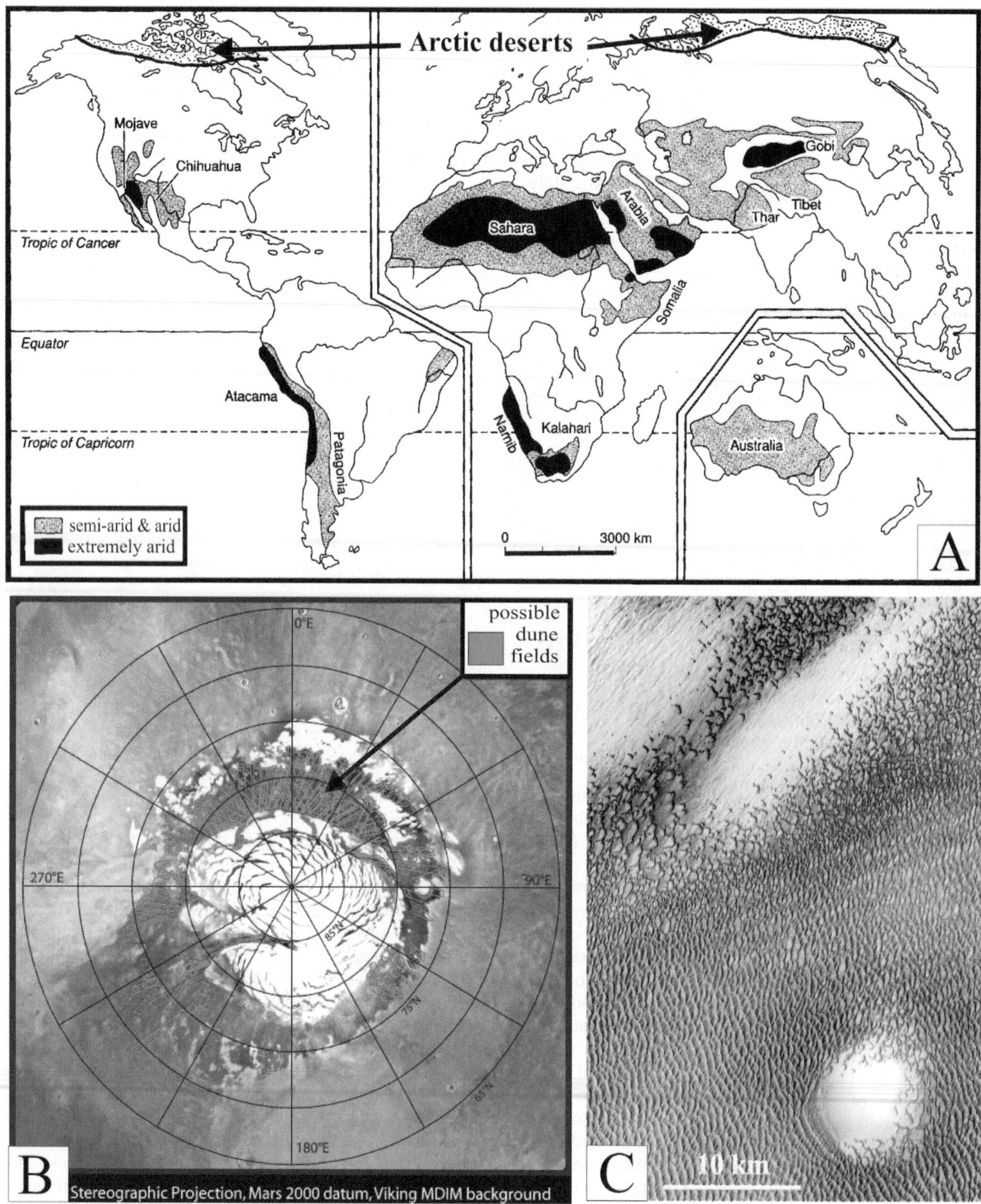

Figure 2. A. Arid zones on Earth (after Goudie, 2002, Fig 1.1). **B.** Mars north polar ergs (courtesy NASA). **C.** Detail of north polar erg, Mars (courtesy NASA).

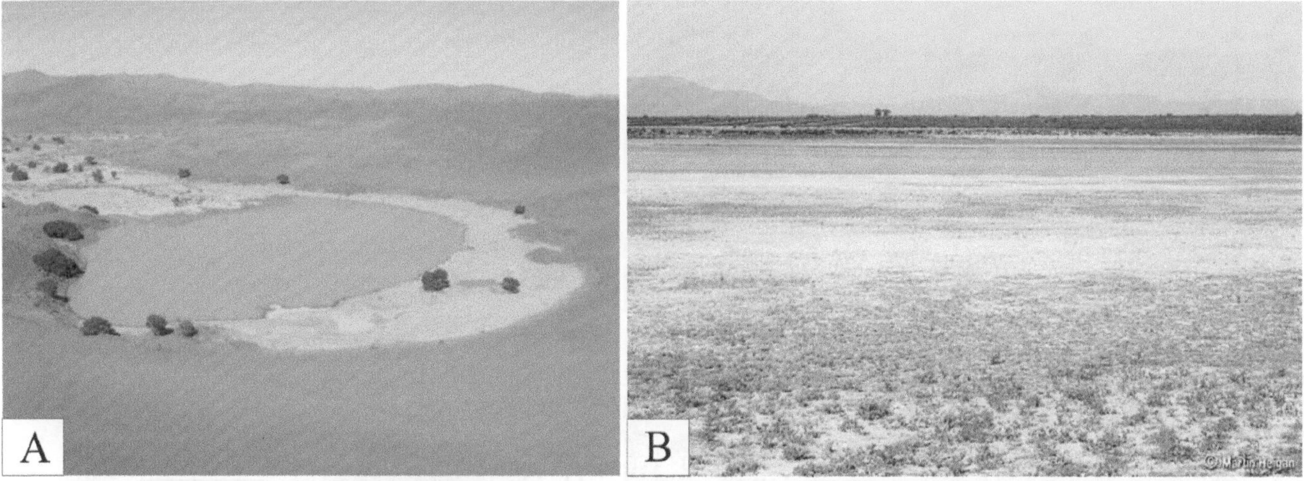

Figure 3. Pans. **A**. Flooded pan, Namib Desert. **B.** Dry pan and evaporites, Kersfontein, South Africa.

Figure 4. Reg. **A**. General view of surface of Gilf Kebir, Libyan Desert, Egypt. **B**. Close-up view of **A**.

old frictional speed required to move sand in saltation is 10 times greater than on Earth. Despite this, the higher wind speeds and lesser gravity of Mars lead to higher rates of sand transport (Greeley and Iversen, 1985). Wind transports sand as a creeping carpet, moved forward by grain impacts, and in saltation, whereas dust (silt and clay) is normally carried in suspension. Both sand and dust may erode underlying materials. In the following two sections, we compare the modern erosional and depositional features and soils in deserts against ancient examples.

EROSIONAL FEATURES

Erosional features are the result of abrasion by sand- and dust-laden winds. They tend to occur in higher upland areas where airflow is increased, at the margins of deserts where the change from erosion to deposition occurs, and in areas where climate changes have increased wind erosion as a result of a loss of vegetation and/or an increase of wind speeds. Ancient erosional features may be preserved by any younger sediment including eolian deposits (common) and river and lake deposits.

Pans are shallow depressions with smooth rounded outlines that normally develop in finer grained, (commonly lake), sediment. Water collects in them during rains and they may develop evaporitic crusts (Fig. 3) as well as lunette dunes on their downwind margins. Identifying ancient pans requires detailed analysis of the relationships of interbedded facies.

Regs (or serirs) are stone lags that form by removal of sand and dust. They may be as vast as sand seas, and can be mantled by intricately interlocking stone pavements that armor the surface and protect underlying sediment from erosion (Fig. 4). Ancient regs should be found between bedrock/fluvial sediment and overlying eolian deposits, but few have been described (Zavalla *et al.*, 2005).

Fluted surfaces occur on rocks hard enough to be polished by sandblasting. They consist of scalloped interlocking surfaces with patterns indicating the average direction of sand-moving winds (Fig 5). They commonly occur on materials that are more resistant than the surrounding areas. Fluting is found on some bedrock surfaces underlying ancient eolian sandstones and is also common on Mars.

Ventifacts are pebbles facetted by sand-laden winds and may show two (zweikanter), three (dreikanter) or more facets (Fig. 6). Like fluted surfaces and yardangs (*see* below) they tend to preserve one predominant

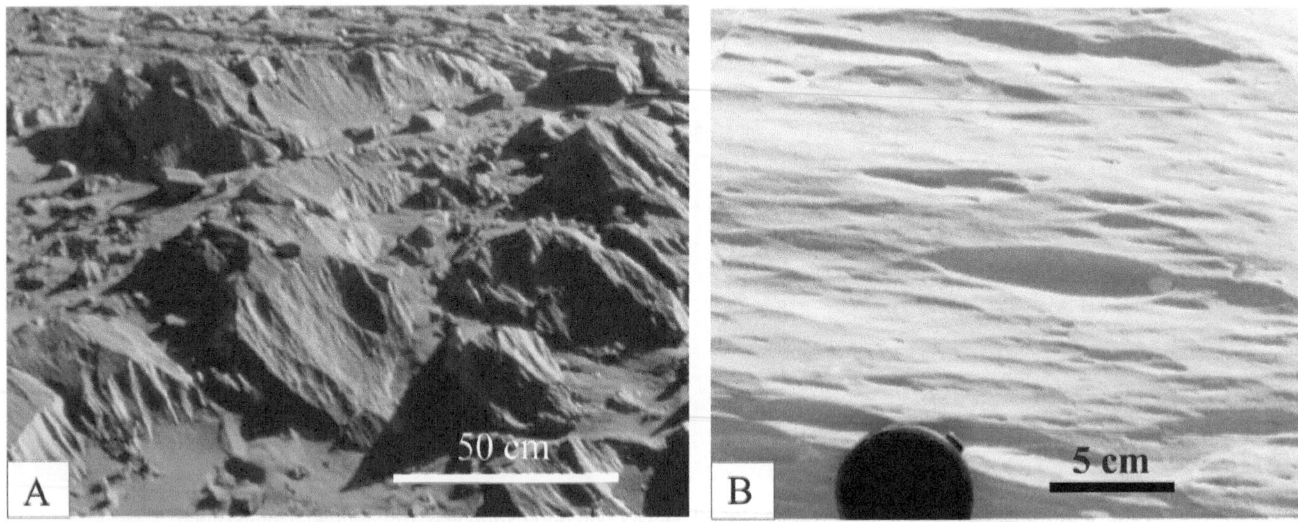

Figure 5. Fluting. **A**. Eocene limestone, south of Farafara, Egypt. **B**. Detail of **A** showing preferentially eroded gypsum micronodules showing the wind direction.

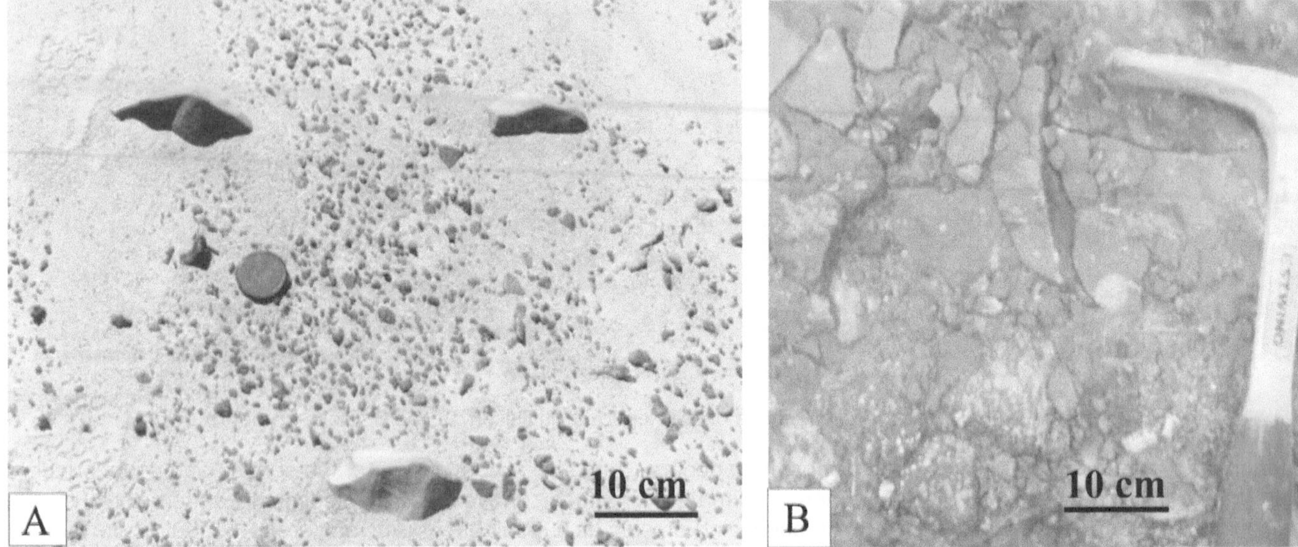

Figure 6. Ventifacts. **A**. Aligned group and clasts' long axes perpendicular to the dominant wind, Libyan Desert, Egypt. **B**. Reworked ventifact in Lower Permian breccia, Thornhill, Scotland.

wind direction, presumably the strongest. Where well-preserved, they can be used to determine ancient wind directions, for example on proglacial Quaternary outwash plains.

Yardangs are wind-abraded ridges formed of cohesive material which may be either indurated sediment or hard rock. The classic form is a streamlined teardrop shape (like drumlins) with a steep upwind slope, a gentler downwind slope, and smooth sides running up to a sharp ridge crest. They vary in size from one to thousands of meters in length and also vary greatly in morphology (Fig. 7). Yardangs, especially those in softer lake sediments, can show the average wind direction during their formation as well as the directions of local airflow around them. Their orientation, relative to sand streaks (and their orientation on hard and soft materials), may differ indicating changing wind patterns with time, those in the softer sediments usually being younger than those in hard rocks. Ancient yardangs, with coarse lags and granule ripples between them, occur on penecontemporaneous cemented dune sandstones at the Permo-Triassic unconformity in Utah (Tewes and Loope, 1992).

Ridges and swales are large features marking zones of sand transport. They occur between the large sand seas of the Sahara and Mars and occupy vast areas. On Earth, discontinuous rock ridges that have a dark desert patina alternate with lighter swales occupied by variable amounts of quartz sand. The ridges and swales are oriented parallel to the prevailing wind and their deflection around mountainous obstacles (Fig. 8). No fossil examples are yet known.

DEPOSITIONAL FEATURES

Modern depositional features consist of all grain sizes from gravel to clay. Gravel deposits are primarily lags (*see* above) although very strong winds can also pile small pebbles into

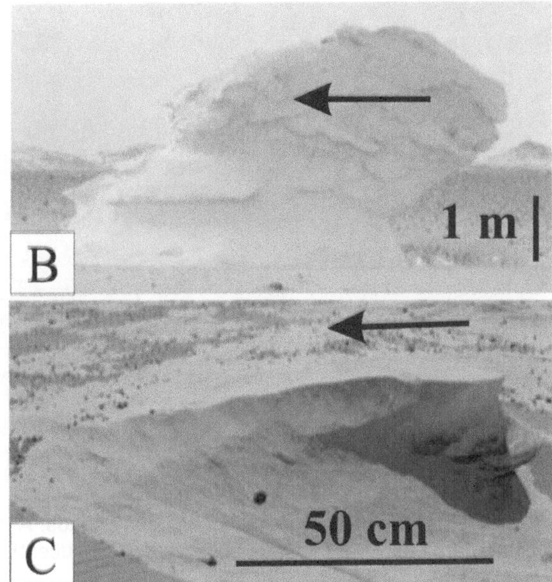

Figure 7. Yardangs of different sizes. **A**. In horizontal sedimentary rocks, Lut Desert, Iran. **B**. In lake silts, Farafara, Egypt. **C**. Mini-yardang in Eocene limestone, south of Farafara, Egypt. Arrows indicate dominant wind direction.

flow-transverse ridges (or megaripples), as in the Antarctic dry valleys. Sand deposits accumulate behind obstacles and in topographic lows to form accumulations of various thicknesses and extents. Silt and clay deposits accumulate in topographic lows and where wind velocities decrease: they form extensive deposits marginal to Quaternary glaciers. Eolian sand deposits are the most impressive and occur in three main settings: sandy deserts, coastal dunes and proglacial outwash plains; the transport and depositional processes are the same.

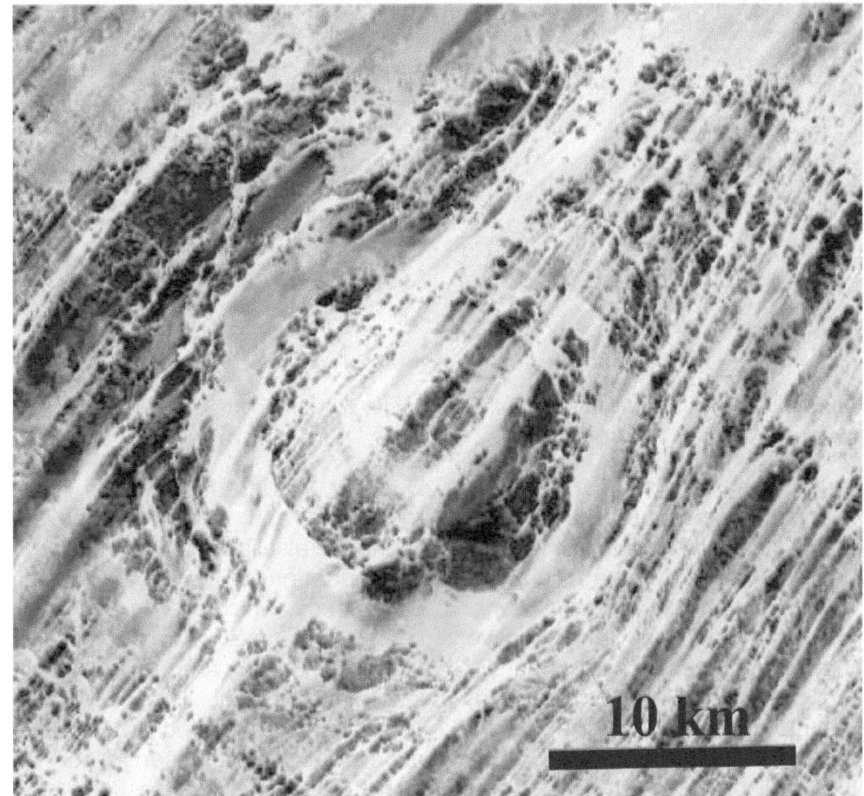

Figure 8. Ridges and swales (radar image) around Aorounga crater on Tibesti massif, northeast Africa, showing deflection of winds around high-standing erosional remnant of an impact crater.

Eolian Sand Composition and Texture

Most eolian sand is enriched in quartz because the physical processes rapidly break down minerals with cleavages into fine dust, although the composition also reflects available source materials, climate and transportation distance. For example, sand dunes on Mars are dark because they consist of coarse basalt grains (about 0.5 mm diameter) rather than quartz sand. The texture of eolian sands on Earth falls into three sub-groups: well-sorted to very well-sorted fine coastal dune sands; moderately sorted to well-sorted fine to medium grained inland dune sands; and poorly sorted interdune and reg sands (Ahlbrandt, 1979). Dune sands vary in mean grain size from 0.68ø (1.6 mm) to 3.4ø (0.1 mm). Though planar-laminated and rippled interdune sand-grains may consist of coarse well-rounded sands, most interdune and reg sands are bimodal in the sand fraction and have a higher silt and clay content than the adjacent dune sand, either due to infiltration or protection from erosion by the larger grains. In inland dune fields with predominantly unidirectional winds, there is progressive sorting and finer mean grain size as one moves down-

Figure 9. Fine- to medium-grained eolian dune sands. **A**. Well-rounded, mature quartz sands. **B**. Sub-rounded, submature sands, with ilmenite (dark), feldspar (cloudy), and quartz (translucent) grains.

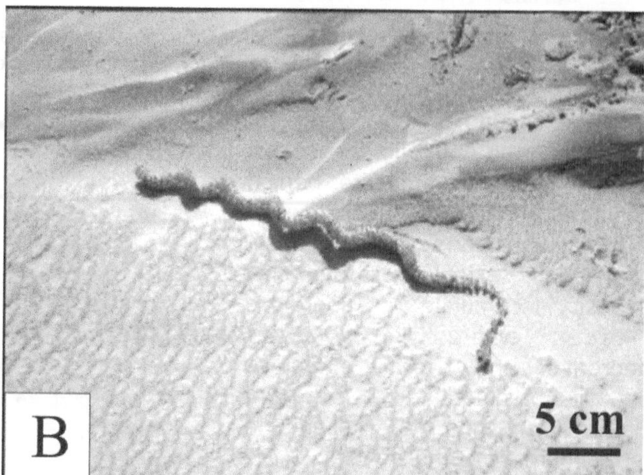

Figure 10. Adhesion structures. **A**. Adhesion warts. **B**. Adhesion ripples (foreground), Sapelo Island, Georgia, with whelk egg case in center. (**B,** courtesy of Paul Howell)

wind from the sand source. In reversing and multidirectional wind regimes, sand accumulates in dunes that have very slow net rates of migration. Because of the fluctuating conditions, dune crest and base tend to have more divergent mean sizes than in unidirectional wind regimes, but attempts to relate grain-size properties to bedform hierarchies and dune types (Wilson, 1973) have not been confirmed. Sources of sand can often be determined from their composition and modifications. For example, alluvial sand often has iron-oxide coatings (grain cutans) due to precipitation from pore waters, whereas beach sands are clean due to chemical and physical removal of such coatings. Separate alluvial and beach sources can be inferred from such differences in coastal sand seas such as the Gran Desierto and Namib deserts. The roundness and maturity of eolian sands can also indicate the amount of transportation (Fig. 9). Proglacial dune sands, which are derived from nearby immature outwash deposits (*see* Chapter 5), are usually much less mature and less rounded than desert sands deposited in warm climates. In many deserts, coarse quartz sand and granules get very well-rounded by the wind over time since, unlike water, there is little cushioning of impacts during transportation.

Eolian Surface Structures

A variety of surface structures are common on eolian bedforms and include adhesion structures (Fig. 10), animal tracks and trails (Fig. 11), and marks made by vegetation, rain and hail. All may be preserved by renewed deposition of sand and are fairly common in ancient eolian sandstones (McKee, 1982).

Eolian Sand Laminae

The detailed structures of eolian laminae differ from subaqueous laminae formed in similar ways because of the different characteristics of the fluid. Hunter (1977) proposed four main types of eolian laminae based on a study of small coastal dunes.

Planebed lamination is produced by wind velocities too high for ripple formation and is analogous to upper-regime planebed in aqueous deposits. It occurs only in coarse sands and granules and can be used to recognize sand-sheet deposits (Fig. 12A). The grains are commonly very well rounded quartz as extreme

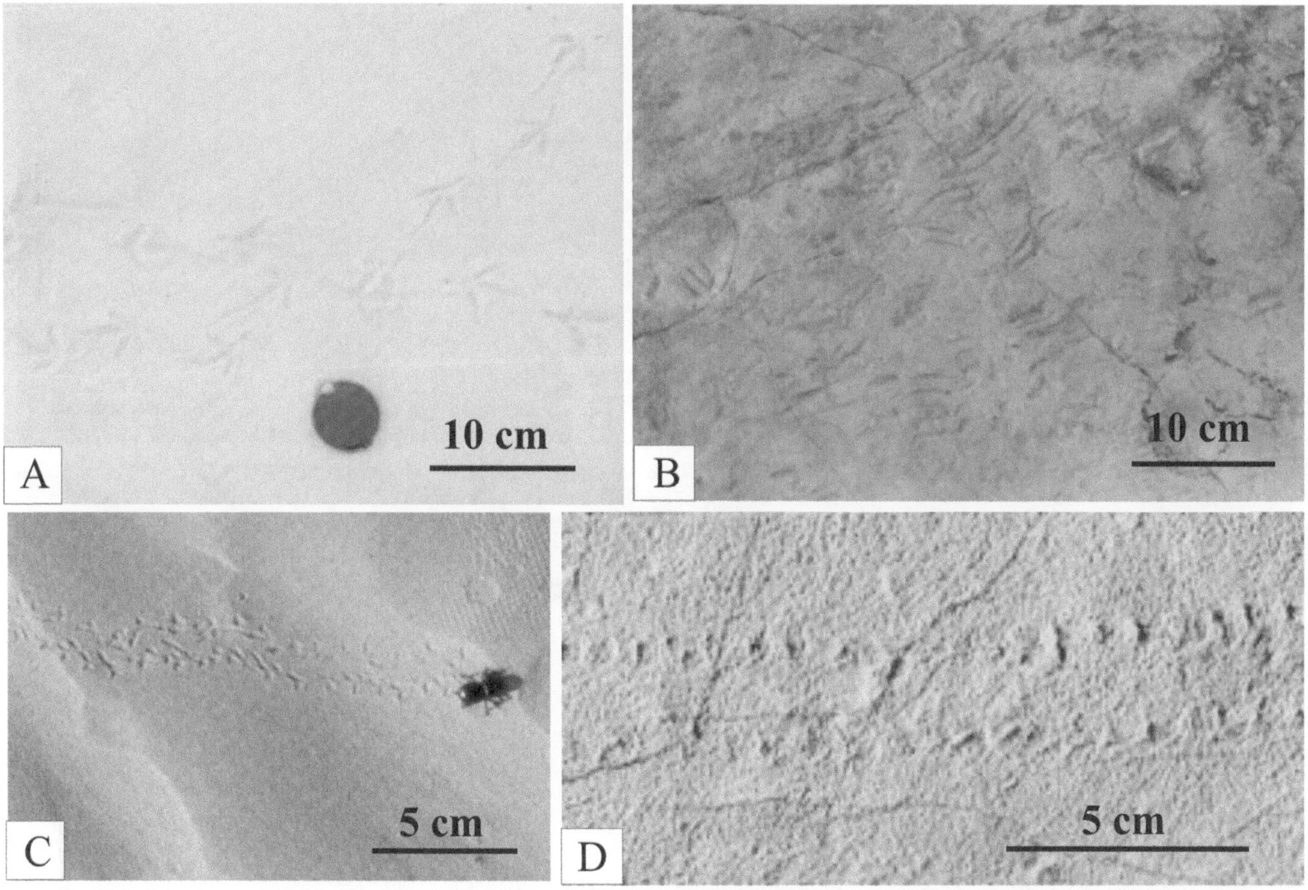

Figure 11. Bird tracks: **A**. Recent grainfall sands, **B**. Ancient grainfall sandstone; Insect tracks: **C**. Recent rippled sands. **D**. Permian sandstone (Antarctica).

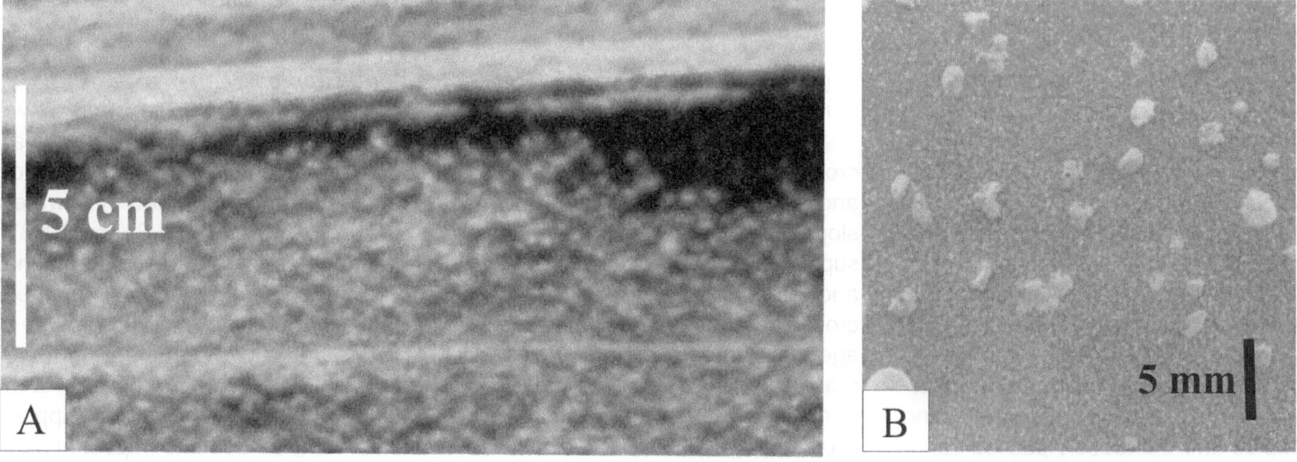

Figure 12. A. Planebed lamination in coarse sand (Permian, Arran, Scotland). **B**. Planebed lamination (forming) in coarse sand and granules (Opportunity landing site, Mars).

abrasion occurs in the moving grain carpets (Fig. 12B). Unlike water, there is no lower-regime planebed sand deposit in wind, for obscure reasons which may be related to the thickness of the viscous boundary layer. Upper plane bedding is caused by planing or flattening of bedforms as the dominant mode of sediment transport shifts from creep to saltation and suspension.

Climbing-ripple lamination closely resembles aqueous varieties, but the laminae coarsen upwards and the ripple foresets are difficult to recognize because of the low relief of the moving ripples (Fig. 13A). Hunter (1977) distinguished two main types: *rippleform strata* occur when the ripples foresets can be identified (Fig. 13B), while *translatent strata* occur when only the bounding surfaces between migrating ripples are visible (Fig. 13C and D). Sets in both types are inversely graded and relatively closely packed (average porosity 39%). Rippleform strata indicate a

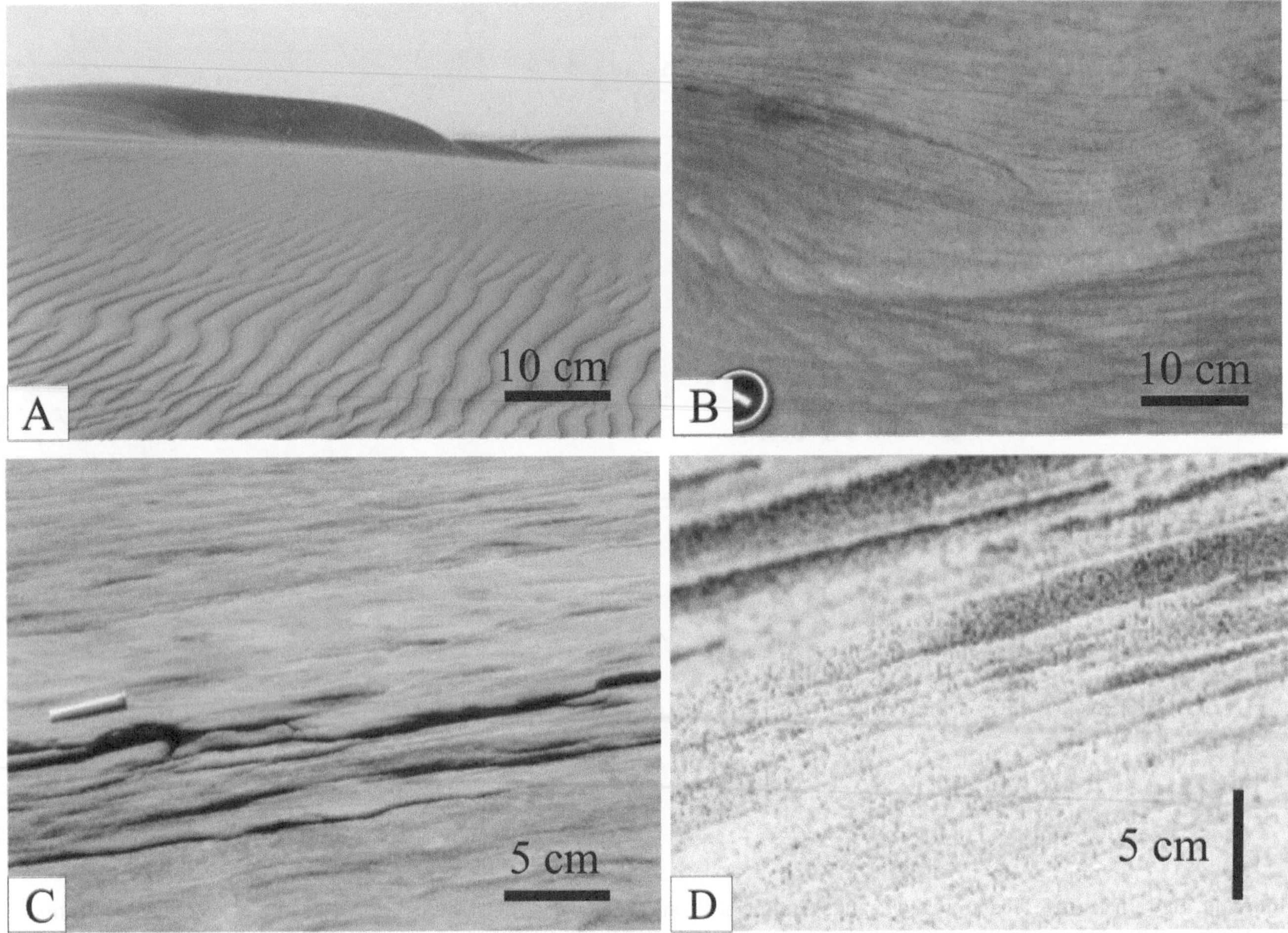

Figure 13. Climbing-ripple lamination. **A**. Climbing ripples moving down an interdune saddle forming rippleform strata. **B**. Oblique section through rippleform strata (Permian, Arran, Scotland). **C.** Translatent strata, Permian, Arran, Scotland. **D**. Detail of translatent strata showing upward-coarsening laminae.

very high rate of sedimentation relative to the rate of migration, producing steep angles of climb.

Grainfall lamination is produced by deposition from suspension, typically in the lee of obstacles such as dunes (Fig. 14A). Grain segregation is relatively poor and laminae are difficult to see (Fig. 14B). Packing has a porosity (average 40%) that is intermediate between that of closely packed traction laminae of planebed and climbing ripples and the loosely packed sandflow strata described below. Graded strata are produced by changes in wind velocity which controls the size of sand in transport and flight distances of individual grains. Distinctive features of grainfall-produced strata are:

- Gradual thinning or tapering downwind (*e.g.*, down a slipface and across an interdune area); and
- Extreme variability of thickness, ranging from less than 1 mm (wind gusts of a few seconds) to 10 cm or more (sustained gusts).

Sandflow lamination (avalanche crossbedding) is caused by slumping and consequent grainflow down a slope, typically a dune slipface, that is supplied with sand by creep, saltation and grainfall (Fig. 15A). Sandflow cross-strata are loosely packed (average porosity 45%), interfinger with grainfall laminae near their base, and form lenses parallel to the strike of the slipface (Fig. 15B, C)

Eolian Bedforms

Wilson (1972) proposed three main scales of eolian bedforms, ripples, dunes and draas, from his studies in the Sahara, to which we can add sand sheets. The dune–draa distinction, however, is not universally accepted, and a descriptive classification based on form and complexity should probably be used (Table 1).

Wind Ripples

Ripples formed by the wind have wavelength/ripple-height ratios that are much greater than those characterizing ripples formed in water (eolian ~ 15; water ~ 9; Sharp, 1963). Wind ripples vary from small and fine grained to large and coarse grained (Fig. 16), reaching a height of over 2 meters and a wavelength of up to 43 meters on the Argentine Puna Plateau (Milana, 2009). Eolian ripples have been divided into impact ripples and aerodynamic ripples (*see* discussion in Pye and Tsoar, 2009), but there is no break between their properties and the division is controversial (*see* discussion in Milana, 2009). Measurements from Mars suggest a wavelength/height ratio of ~6.7, twice as steep as terrestrial wind ripples, but the reason for this is not known.

Dunes

Dunes may be small or large, simple

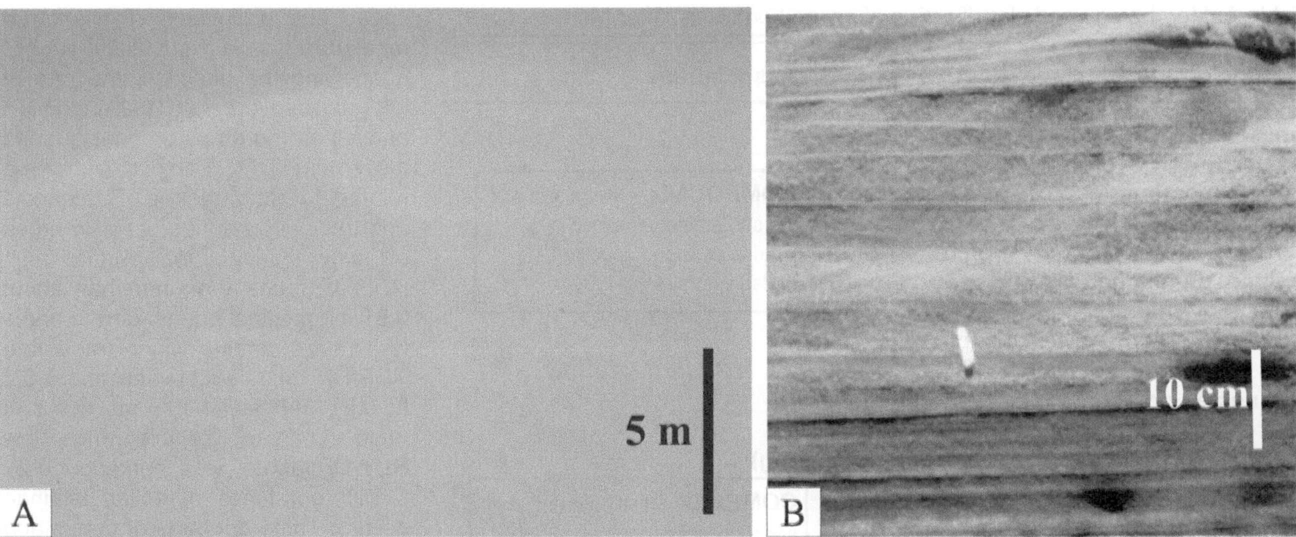

Figure 14. Grainfall. **A**. Fine sand 'smoking' over a dune slipface (Gran Desierto, Mexico). **B**. Oblique section through grainfall laminae (Permian, Arran, Scotland).

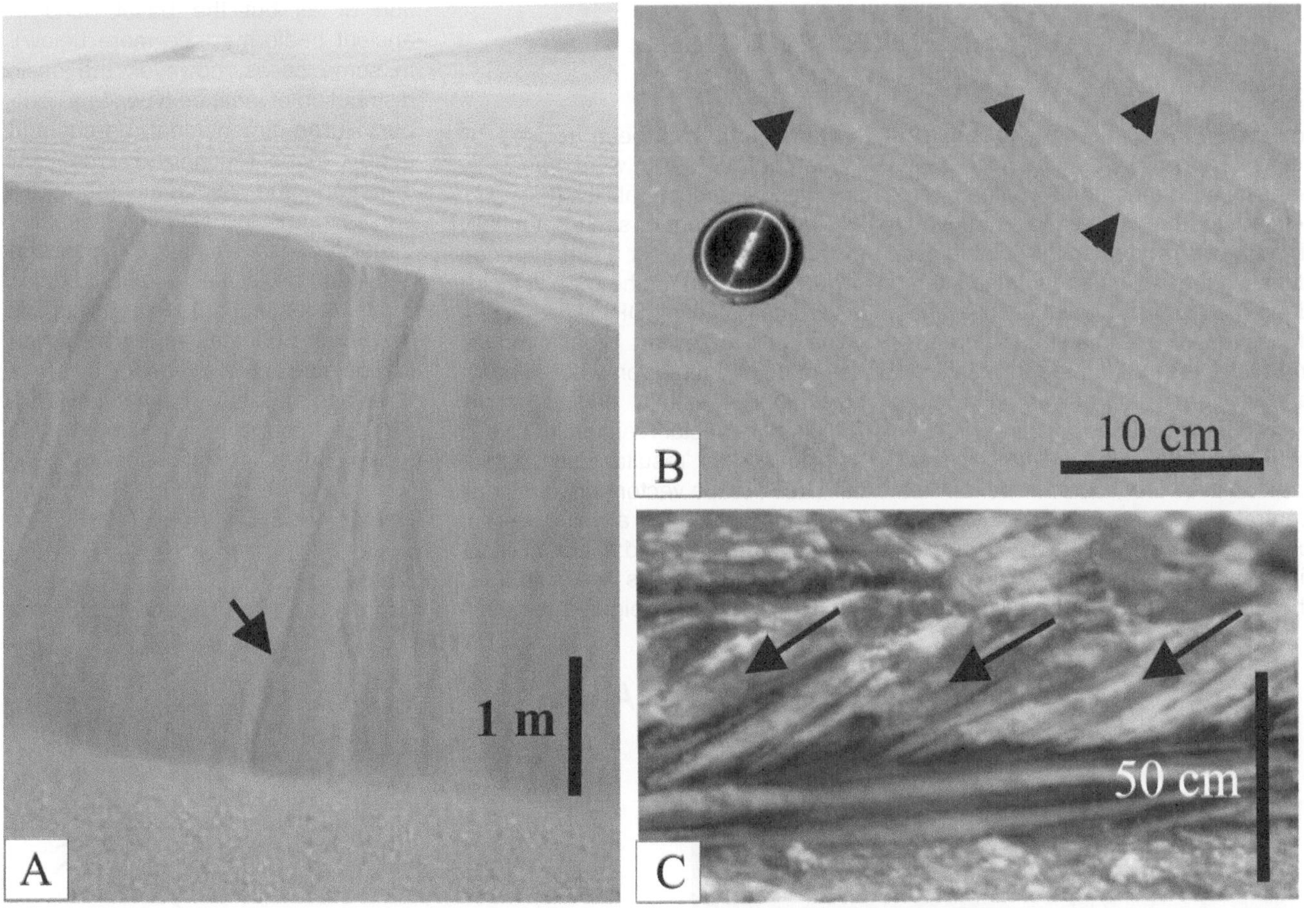

Figure 15. Sandflow cross-strata. **A**. Lobate sandflow down modern slipface (Gran Desierto, Mexico). **B**. Strike section of sandflow strata alternating with grainfall (Permian, Arran, Scotland). **C.** Dip section of sandflow strata, alternating with grainfall (Permian, Arran, Scotland). Arrows indicate sandflows and sandflow strata.

or complex, and linear (transverse or longitudinal) or three-dimensional, and vary from 0.1 to 100 meters in height. The most important difference between ripples and dunes is that *dunes scale with the flow that forms them*; in other words, deep flows make large dunes – provided, of course, that there is enough time, space and sand for the dunes to develop fully. Dunes may be classified into barchan, transverse, longitudinal and star varieties, though the transverse and longitudinal types are commonly grouped together as lin-

Table 1. Morphology and classification of eolian bedforms. After McKee (1979)

Morphology	Name	Associations
Sheet-like	**Sheet sands**	
Thin elongate strips	**Streaks**	**COMPOUND** – two or more of the same type combined by overlap or superimposition (Wilson's draa)
Circular to elliptical mound, dome-shaped	**Dome**	
Crescent in plan view	**Barchan**	
Connected crescents	**Barchanoid (aklé)**	
Asymmetrical ridges	**Transverse (reversing)**	
Symmetrical ridge	**Linear (seif)**	**COMPLEX** – two different basic types occuring together, either superimposed (Wilson's draa), or adjacent
Central peak with arms	**Star (pyramidal)**	
U-shaped	**Parabolic**	

ear dunes with a variation from flow transverse through oblique to longitudinal (Table 1). Longitudinal and transverse dunes show differences related to the way sand is moved through the systems. In longitudinal dunes, sand moves through the interdune corridors and obliquely over the dunes resulting in coarse sand creeping along the corridors and mantling and armoring the lower windward parts of the dunes. In transverse dunes the sand is moved directly over the dunes so that coarse sand is concentrated on the windward slope, crest and upper slipface, whereas finer sand saltates to accumulate on the lower slipface and adjacent lee-side interdunes.

Dune form has been related by Fryberger (1979) to wind variability and sand transport ability (Fig. 17). The gross amount of sand potentially moved by the wind for a stated period of time, weighted for the wind velocity, is the drift potential (DP), which has a value but no direction. What is moved depends, of course, on what is available. The 'resultant drift direction' (RDD) is the vector resultant of all drift directions. The 'resultant drift potential' (RDP) is the vector sum of all drift potentials and gives a value for the net movement of sand: if actual wind-speed measurements are available, the RDP can be weighted for both speed and grain size. These three measures give a very useful way of quantifying the direction and rate of net sand movement (RDD and RDP) as well as the gross amount of sand movement (DP). These can then be related to wind regime. Barchanoid and transverse dunes occur in areas of fairly unidirectional winds (high RDP/DP ratios – greater than about 0.8); longitudinal dunes form in areas with more variable winds (moderate RDP/DP ratios – between about 0.2 to 0.8); and star bedforms occur in regions of very variable winds (low RDP/DP ratios – less than about 0.2).

Different types of eolian laminae occur on different parts of dunes and can be used to recognize their different components (Fig. 18A), though there is a problem of non-preservation of all but the basal parts in ancient bedforms (see more below). In some cases, however, the basic distribution of laminae types in a complex three-dimensional barchanoid dune can be recognized even after deflation: the slipfaces (sandflow cross-strata), the saddles (climbing translatent strata) and the passage between the slipfaces and saddles (grainfall lamination) (Fig. 18B). This dune was truncated at the height on the lee-side actually reached by sandflows and must have been fairly small. Further information on stratification types and their relationship to dune morphology is given by Rubin and Hunter (1982) and Rubin (1987).

Draas

Draas are large sand bedforms

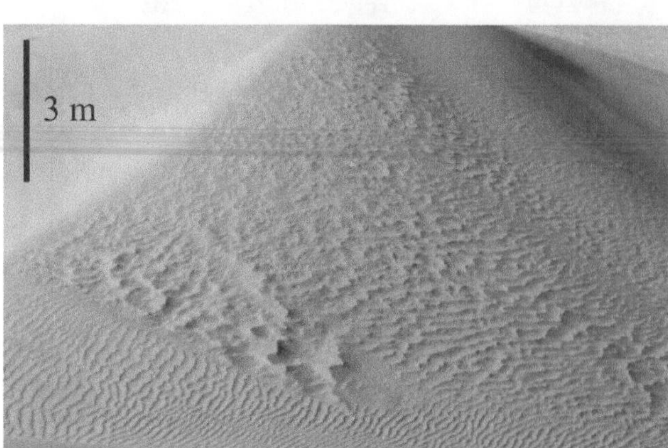

Figure 16. Wind ripples on the lee slope of an Algerian dune: regular fine-sand ripples in foreground, and larger coarse-sand ripples behind.

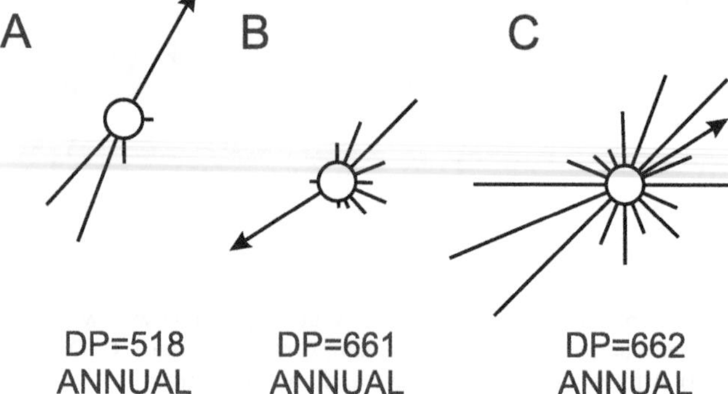

Figure 17. Characteristic relationship between wind regimes and dune types: **A**. narrow unimodal winds and barchanoid dunes, Peru; **B**. bimodal winds and linear dunes, Mauritania; and **C**. complex winds and star dunes, Libya. DP = drift potential; arrows indicate resultant drift direction (RDD) (*see* text). After Fryberger (1979).

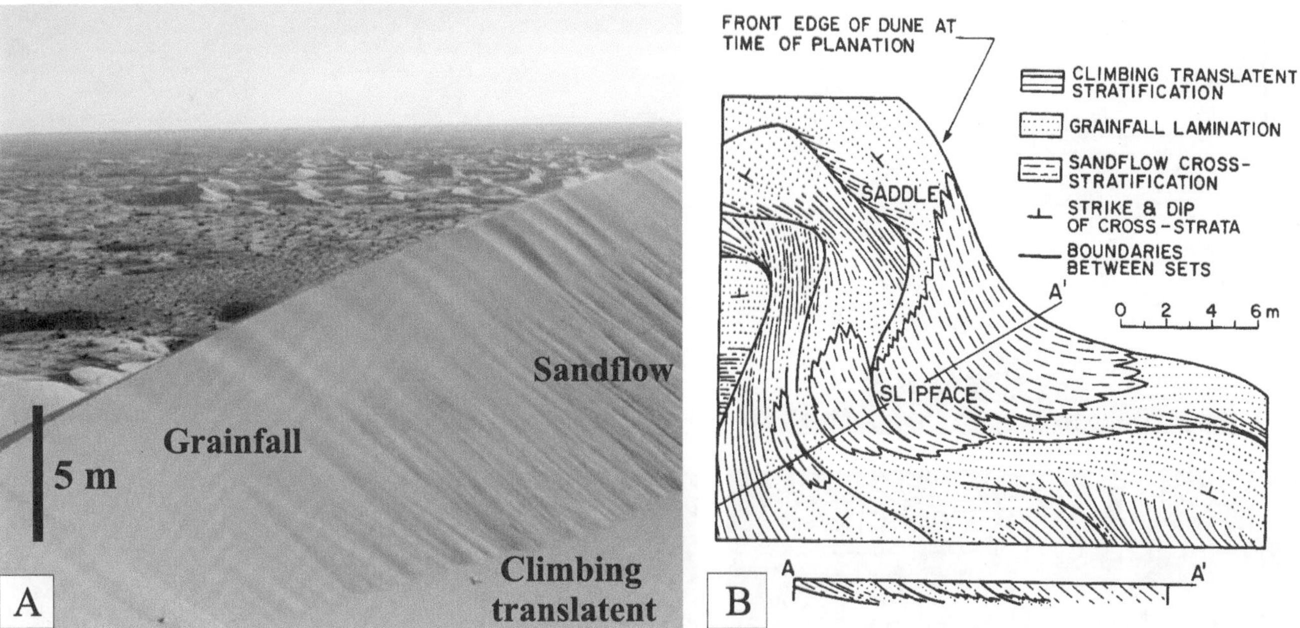

Figure 18. Distribution of strata types on dunes. **A**. Sandflow, climbing translatent and grainfall on arm of a star draa (NW Gran Desierto, Mexico). **B**. Truncated barchan dunes showing plan view distribution of strata (Padre Island, Texas; from Hunter, 1977).

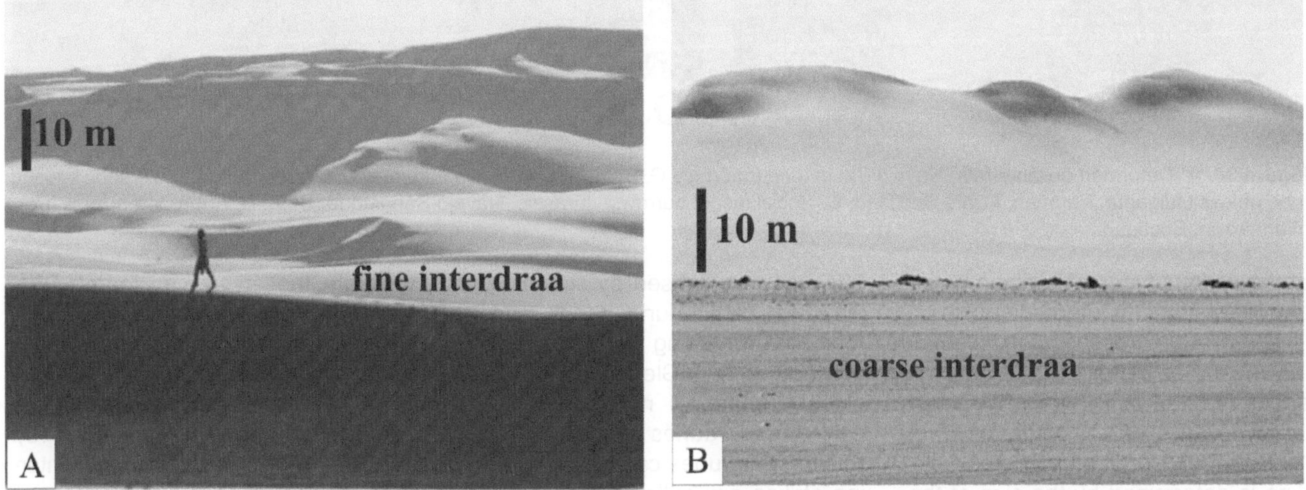

Figure 19. Draa. **A**. Transverse draa, Ubari Sand Sea, Libya. **B**. Longitudinal draa, Libyan Desert, Egypt.

between 20 and 450 meters high, characterized by the superimposition of smaller dunes on them. Like dunes they scale with depth of flow, the thickness of the atmospheric boundary layer (ABL), though the dune–draa relationship is also dependent on the structure of the ABL. Both transverse and longitudinal draa occur in sand seas (Fig 19). Modern draa are rarely in equilibrium with the present-day wind regime since they take hundreds to thousands of years to readjust to changing conditions (Wilson, 1973). Identification of ancient draa depends on the recognition of hierarchies of bounding surfaces (see below).

Internal Structures

Internal structures of modern dunes and draa can rarely be observed and direct knowledge of their larger primary internal structures is still limited to older trenching studies (*e.g.*, McKee, 1966), although, since the last edition of this book, ground-penetrating radar studies are increasingly being used to gain indirect knowledge of bedding geometry (*e.g.*, Bristow *et al.*, 1996). Radar has good penetration in dry sand, though it picks up hydrological boundaries as well (Overmeeren, 1998). Together with thermoluminescence dating, ground-penetrating radar can identify different stages of bedform growth and destruction in Quaternary ergs (*see* Supersurfaces below). We can not be sure, however, that the deeper preserved structures relate to the modern wind regime, so such studies need to be supplemented with computer modeling of the structures produced by the diversity of migrating bedforms (Rubin, 1987; Schwammie and Herrmann, 2004). Ancient eolian sandstones are now also being analyzed with ground-penetrating radar (Jol *et al.*, 2003). The internal structures of large sand bedforms consist

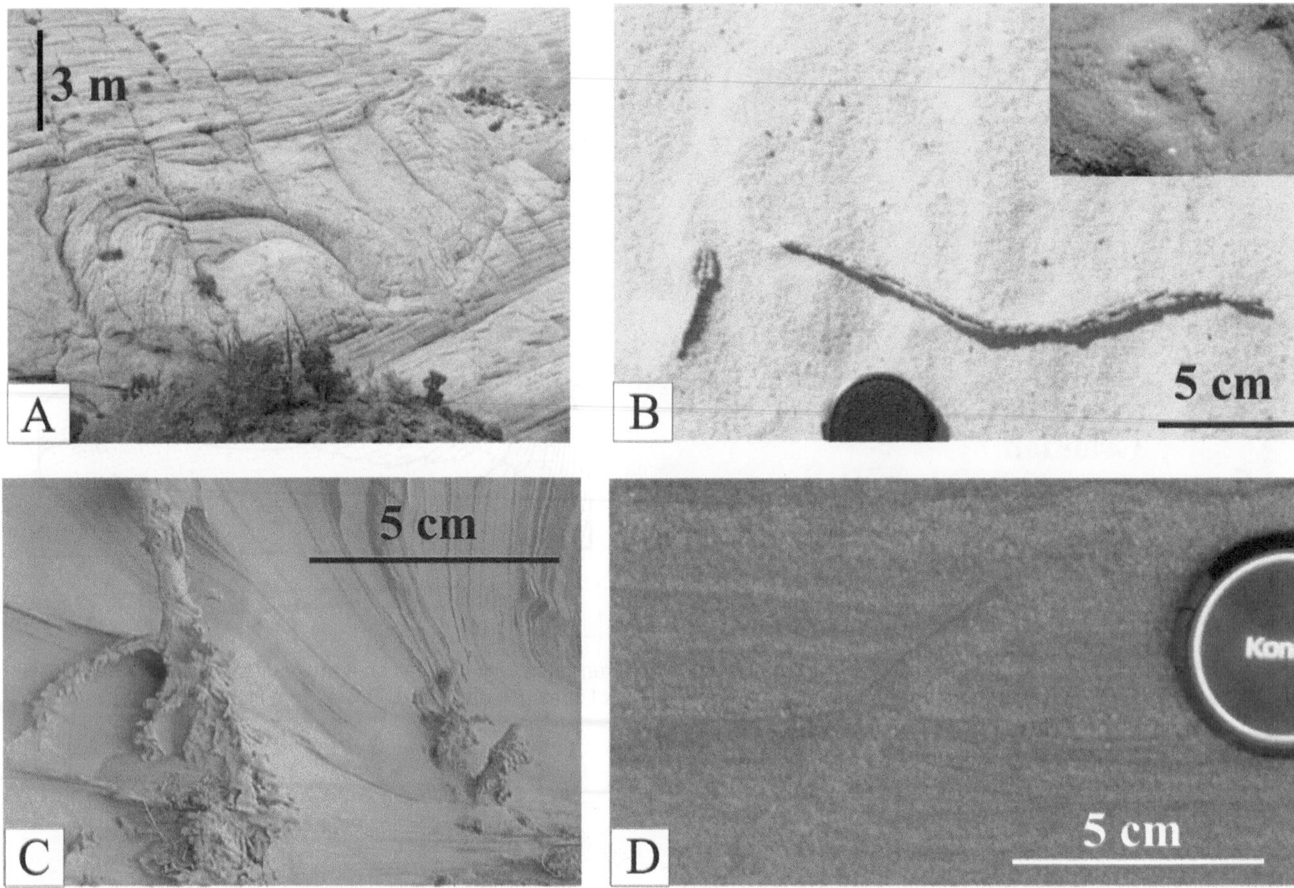

Figure 20. A. Deformed eolian strata, Navajo Sandstone (courtesy G. Bryant). **B**. Modern fulgarites, Egyptian desert. Inset shows fused tube of fossil fulgarite, Permian, Arran, Scotland. **C.** Social insect burrows, Jurassic, Navajo Sandstone, Utah. **D**. Arthropod burrow, Permian, Arran, Scotland.

of the various types of laminae noted above arranged in interbedded sets.

Large-scale crossbedding, consisting mostly of sandflow and grainfall strata, is often considered characteristics of large eolian bedforms (Fig. 1). But, the size of internal crossbedding is determined by the size of the bedform which forms it and also by the rate of climb of the bedform, and both eolian and subaqueous bedforms in shelf seas and deep rivers form similar structures (Allen, 1982; Rubin and Hunter, 1982; *see* also Chapters 6 and 9). Only the most rapidly climbing eolian bedforms preserve more than a small part of the lower lee-slopes and bedform climbing has to be considered before any interpretation can be made of ancient eolian sandstones (*see* Bounding Surfaces below). Secondary internal structures give information on post-depositional conditions.

A number of other sedimentary structures may also be present. *Deformed eolian crossbedding* (Fig. 20A) may be caused by slumping of water-saturated dune sands and is found in ergs being inundated by the sea (Hurst and Glennie, 2008) and also by seasonal melting of ice in cold-climate dunes (Koster, 1988). Fused silica tubes caused by lightening strikes are called *fulgarites* (Fig. 20B). The *burrows and traces* of various organisms, including animal burrows and tracks, and rootlets, can be common at some horizons and indicate less arid conditions (Fig. 20C and D) (Ekdale *et al*., 2007).

Interdunes and Interdraa

Interdunes and interdraa are an integral part of eolian bedform systems (Kocurek, 1981). In deserts that have a limited sand supply, interdunes consist of lag deposits, coarse sand sheets and small isolated dunes, and episodically flooded sabkhas. Here, the water table may reach the surface and control the amount and type of sediment deposition. Because of the way in which the sand is transported through the systems, longitudinal bedforms tend to have coarse lags, sand sheets and dunes in the interdune areas, whereas transverse bedforms tend to have sabkhas and fine sand dunes between them (Fig. 19). The size of the interdune (and interdraa) areas is also dependent on sand supply and on the stage of development of the erg in which they occur. Many modern deserts have extensive interdunes because of their relatively recent development. Ancient interdune deposits may either be coarse lags or fine-grained stream and lake deposits showing evidence of desiccation (*see* Chapters 6 and 21).

Sand Sheets

Sand sheets consist of flat areas of plane-bedded coarse sand usually overlying finer material and are basically lag deposits in a zone of transport and bypassing. Although sand sheets tend to be small, some are very extensive and may have subdued superimposed bedforms. For example, the Selima sand sheet in southern Egypt consists of a vast plain of coarse sand (120 000 km^2) overlying pedogenically altered sand (Fig. 21A and B). The only bedforms present, in places, are low sand ridges and small dunes of coarse sand (zibar; Maxwell and Haynes, 2001). Alternating light and dark chevron-shaped sand patterns of long wavelength (130 to 1200 m) and low amplitude (10 to 30 cm) occur downwind from low scarps and major dune fields, and move at up to 500

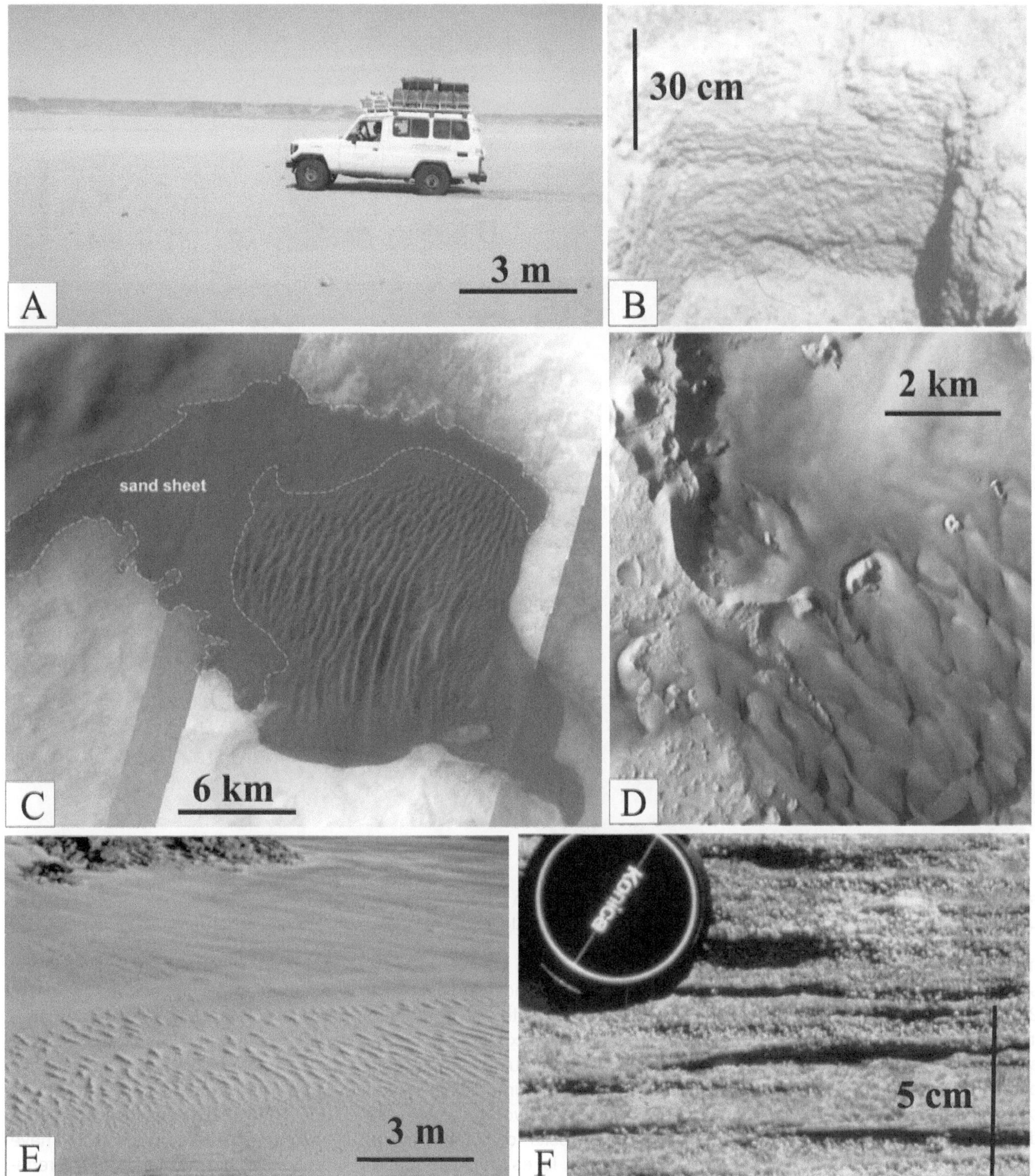

Figure 21. A. Selima sand sheet, southern Egypt, with Gilf Kebir in background. **B**. Pit in Selima sand sheet showing planar coarse sand undergoing pedogenesis, including accumulation of clay cutans around grains and dust infiltration. **C**. Sand sheet surrounding a Martian erg composed of transverse dunes (dominant wind to the upper left in the image). **D**. Sand sheet within a Martian crater passing through breached crater walls into large linguoid bedforms (**C**, **D** courtesy NASA). **E**. Zibar on sand sheet, Mauritania. **F**. Planar coarse sand and granule laminae possibly due to migrating zibar, Permian, Arran, Scotland.

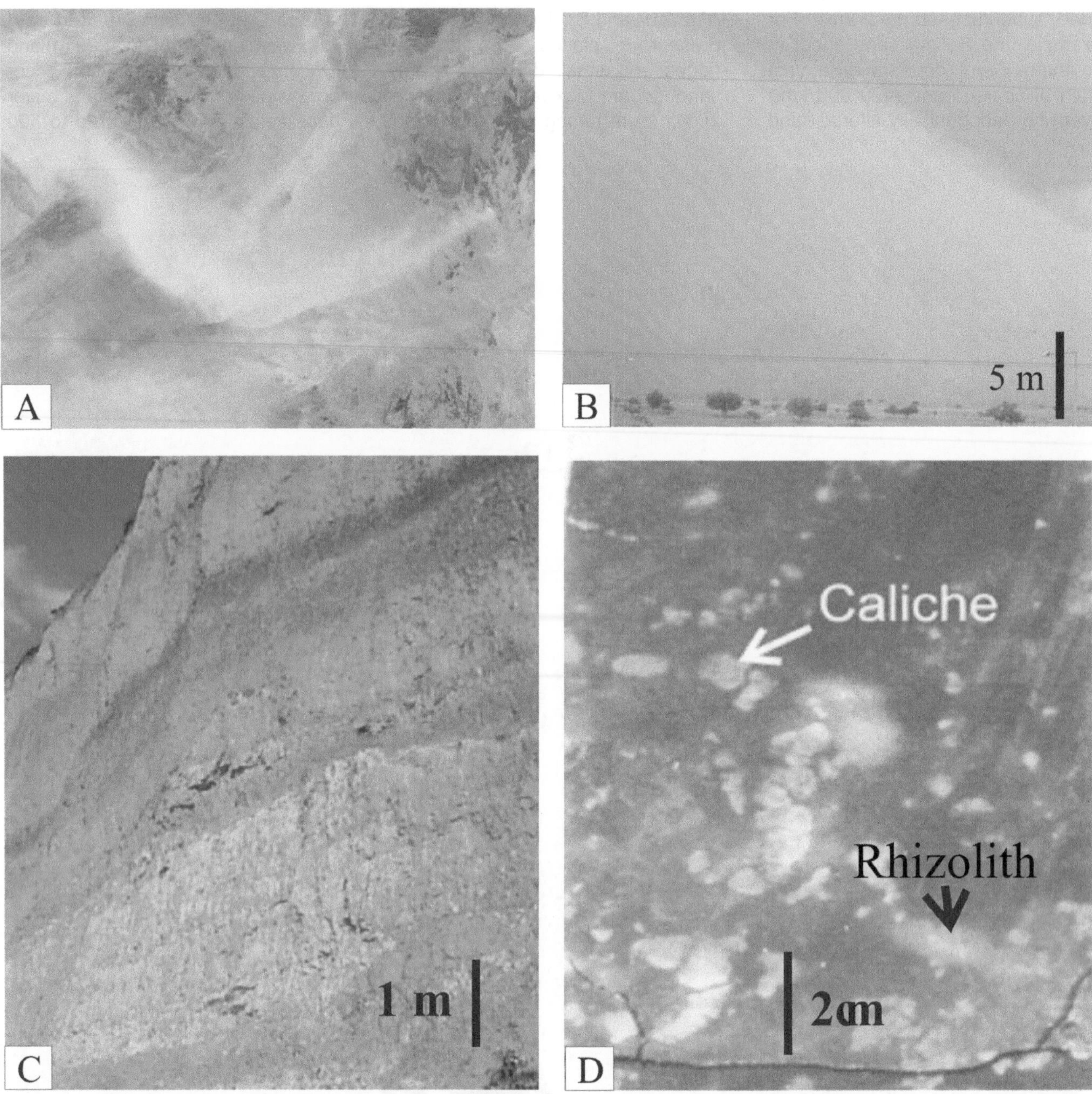

Figure 22. A. Dust plume over Erg Chech, Sahara (courtesy NASA). **B**. Dust storm, Sudan. **C**. Quaternary loess and soil horizons, Llanzhou, central China. **D**. Loessite with calcrete and rhizoliths (rootlet traces), Permian, USA (courtesy Kansas Geological Survey).

meters per year. On Mars, sand sheets cover enormous areas (Fig. 21C and D). Sand sheets are an important part of some ancient sands seas and are recognized by plane bedding and low-angle lamination in coarse sands; they may contain deposits formed by coarse-sand megaripples called zibar (Fig. 21E) and these can be recognized in ancient eolian sandstones (Fig. 21F) (Biswas, 2004). The low sand ridges are probably undetectable in ancient rocks.

Eolian Dust

Eolian dust (silt and clay) is blown vast distances both from modern deserts and from exposed proglacial deposits to accumulate in adjacent areas. Large amounts of fine sediment can be reworked from the exposed deposits of former inland and proglacial lakes, such as mega-Lake Chad, and transported long distances (Fig. 22A and B). Though the precise origin of eolian silts (*loess*) is vigorously debated, Quaternary loess deposits form some of the most fertile soils in the northern hemisphere and accumulated adjacent to the proglacial deposits of former Quaternary ice sheets (Fig. 22C). Some ancient loess deposits (*loessite*) have been recognized (Soreghan *et al.*, 2007; Fig. 22D), but they should be commoner, even though many have probably been reworked and redeposited by rivers considering their erodability, or deposited in the sea where they have been modified by marine processes. The main criteria for recognition is their fine grain size

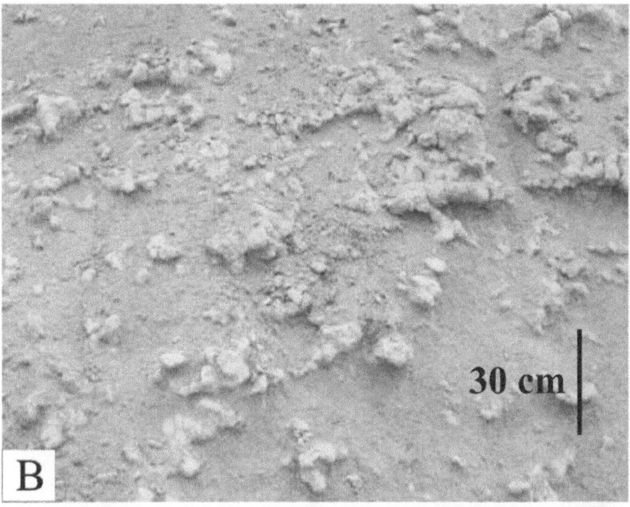

Figure 23. A. Recent structureless calcrete, Australia. **B**. Ancient nodular caliche, Cretaceous, Jabalpur, India, formerly interpreted as a marine limestone.

(silt to very fine sand) and structureless nature. Paleosols develop during non-depositional phases that may be climatically determined.

SOILS AND PALAEOSOLS

Soils are alteration horizons on exposed land surfaces, either rock or sediment, and mark periods of non-deposition (diastems) in a succession. Their development and characteristics depend on substrate type, relief, climate, biota and duration of exposure. Because most soils are controlled by the biota, soils have changed with the evolution of land biotas through time, and only post-Miocene paleosols (with the evolution of grasses) can be compared directly with Recent soils. In semi-arid climates, seasonal rains cause alternate wetting and drying with the build-up of carbonate horizons (*calcretes*, or *caliche*; *see* Chapter 20) which are good indicators of semi-arid conditions, although well-developed calcretes have been confused with marine limestones in the past (Fig. 23). In such climates, carbonate retention increases with decreasing rainfall due to decreased leaching. The carbonate-rich soils developed in Quaternary wind-blown silts and clays (loess) mark diastems in accumulation and may be useful in regional correlation. Conversely, in arid to hyperarid climates, usually only incipient alteration occurs and only weakly developed soil horizonation is present. The current arid–hyperarid transition on Earth marks a shift from biotic to largely abiotic conditions. Under hyperarid conditions, biological effects, silicate transformations and the loss of less stable constituents by leaching are no longer the dominant forces in pedogenesis: in hyperarid soils, carbonate retention decreases with decreasing rainfall and nitrate concentration increases. The near absence of both biological and dissolution processes in hyperarid soils make them fundamentally different from other soils on Earth (Ewing *et al.*, 2006). The interpretation of paleosols requires, however, not only knowledge of modern soils and the processes which form them, but also of how their biological make-up and formative processes have evolved through time. Readers are referred to Retallack (2001) for more information on the recognition and interpretation of paleosols.

SAND SEAS (ERGS)

Eolian sand accumulates in vast sand seas, called ergs. In the largest desert on Earth, the Sahara (7 million km^2), there are several major ergs, which individually can cover areas as large as 500 000 km^2 (twice the area of Nevada; Fig. 24A). They are located in physiographic or structural basins with long histories of sediment accumulation, including extensive Tertiary and Pleistocene fluvial sediments (Breed *et al.*, 1979). On Earth, modern eolian deposits are rarely more than 100 meters thick due to extreme Quaternary climatic fluctuations. On Mars, the major ergs are marginal to the northern ice cap and have deposits up to 200 m thick (Figs. 2B, C and 24B), while on Titan the ergs are equatorial (Fig. 24C). The main reason for the accumulation of sand in an erg seems to be the presence of a topographic depression or wind shadow (Wilson, 1973). As in aqueous environments, accumulation is frequently the result of increasing depth of flow or diversion around or over an obstacle and the consequent drop in fluid velocity and sediment transport capacity (Mainguet and Remini, 2004; *see* Chapter 4). In ergs the flow depth is the height of the atmospheric boundary layer, commonly between 1 and 2 km on Earth and around 5 km on Mars. Two ergs can be used to illustrate linear-dominated and transverse-dominated bedform systems.

The Fachi–Bilma erg is situated in the southern Sahara and is dominated by longitudinal bedforms that are generally orientated parallel to the mean wind direction (Mainguet and Callot, 1974; Fig. 25A). This erg is partly developed in the wind shadow of the Tibesti massif which deflects the winds around it (Fig. 24A). Within the erg there is a definite spatial zonation of dune types. Barchans occur on all sides marking zones of intermittent deposition. Inwards, the barchans coalesce into larger sinuous longitudinal dunes and then into larger compound longitudinal draa. In the upwind part of the erg a zone of large star draa over 100 meters

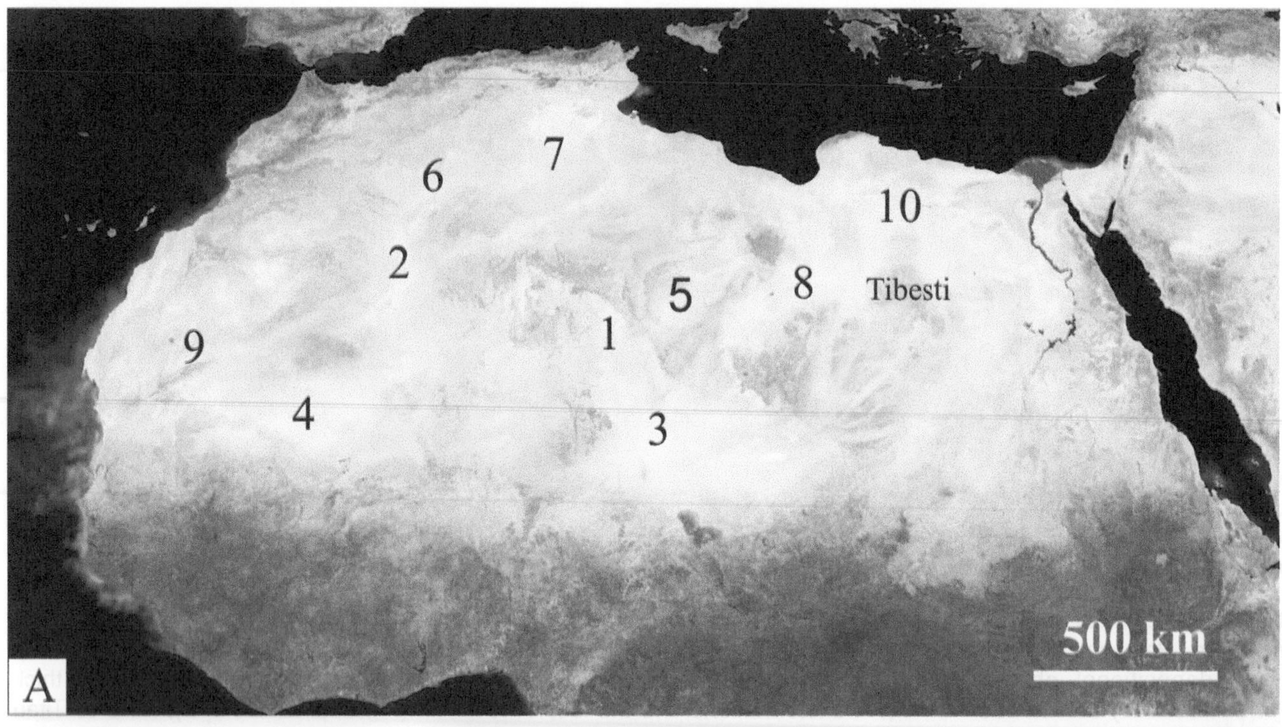

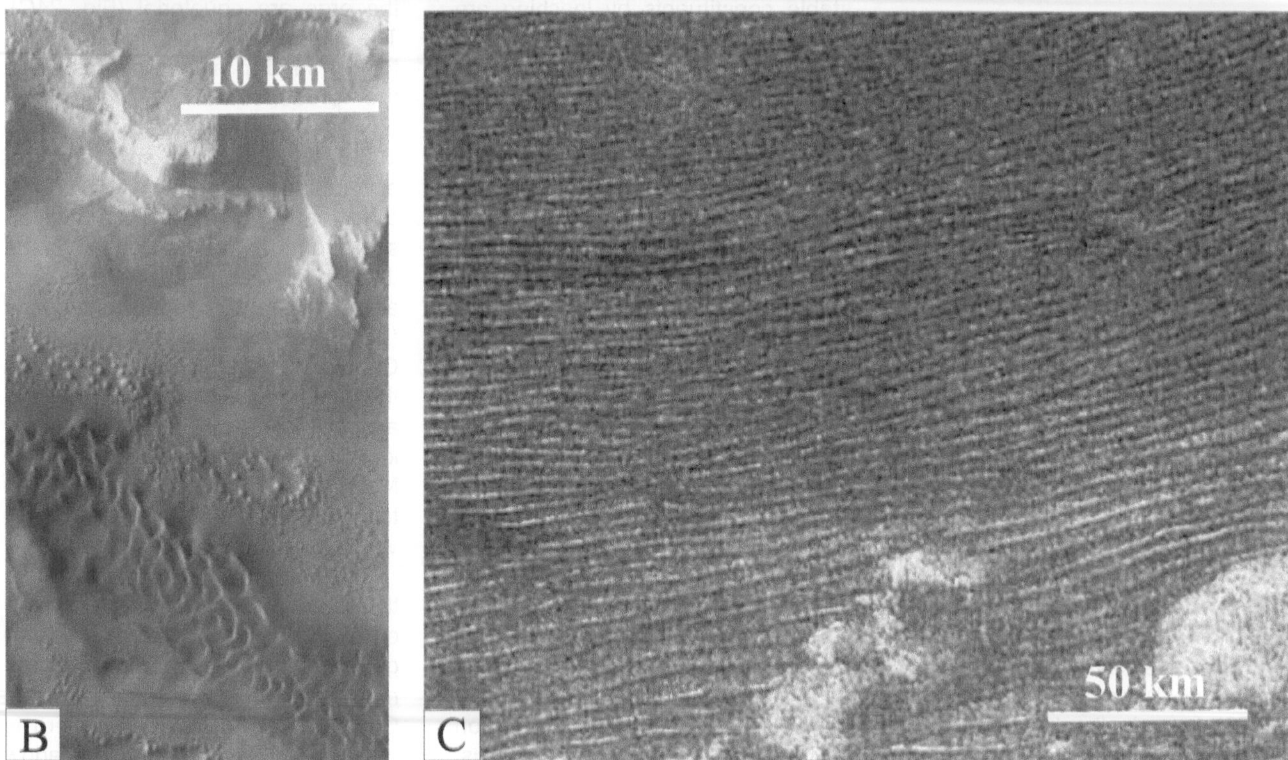

Figure 24. A. Ergs in the Sahara: 1– Iferuoane, 2– Chech, 3– Fachi-Bilma, 4– Adrar, 5– Murzuk , 6– Great western, 7– Great eastern, 8– Rebiana, 9– Iguedi, and 10– Libyan. **B**. Mars erg. **C**. Titan erg.

high occur in the zone of turbulence and fluctuating wind direction immediately downwind of the Tibesti massif. Longitudinal dunes are common in many ergs on Earth and also cover as much as 20% of Titan's surface (Lorenz *et al*., 2007)

The southeastern part of the Gran Desierto of Mexico is dominated by transverse bedforms (Fig. 25B) as are many ergs on Earth and the North Polar erg on Mars. The rest of the Gran Desierto and many other ergs are dominated by mixed bedforms, such as the lines of star draa on linear draa of the Great Eastern erg in Algeria (Fig. 26), and the Aglaonice erg on Venus. Breed *et al*. (1979) show bedform maps based on satellite photography of many of the major ergs in the world in which similar regular patterns can be seen.

Thus, as a basis for eolian facies models we have static architectural

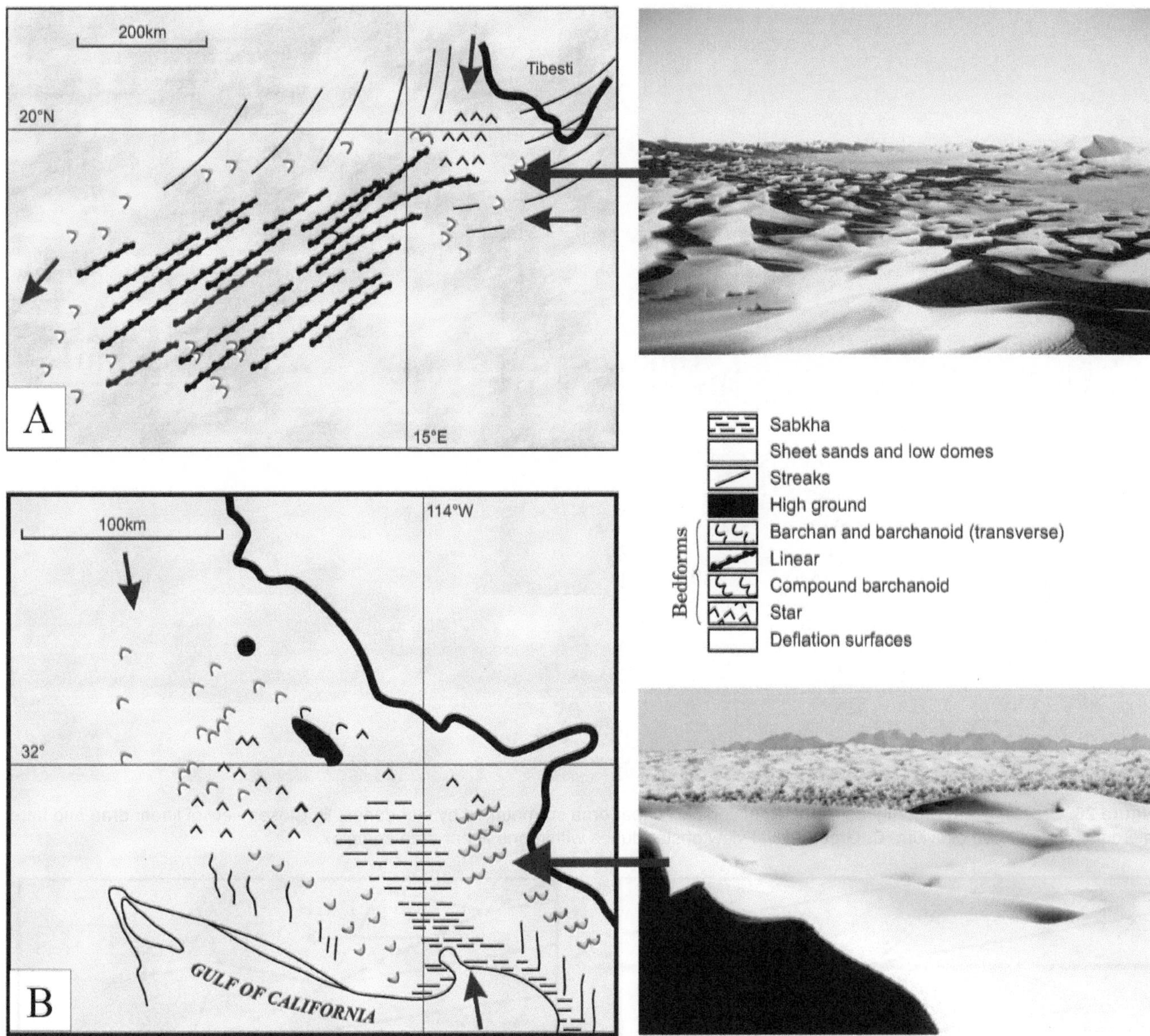

Figure 25. A. Bedform distribution in the Fachi–Bilma erg, south-central Sahara, and view of complex linear draa at edge of star-draa area. **B**. Bedform distribution in the Sonoran Desert, Mexico, and view of compound barchanoid draa in the SE.

elements (stratification types on different parts and types of bedforms) and architectural units (bedforms and interdunes in modern deserts). We now need a dynamic dimension to see how various combinations of migrating bedforms with their characteristic strata can form thick, extensive ancient eolian sandstones.

STRATIFICATION AND BOUNDING-SURFACE MODELS

Stratification and its large-scale organization result from the migration of bedforms of different types as they climb at various angles and directions over one another. But bedforms generated by moving sand are rarely that simple and are also affected by topography, including existing bedforms. Simple stratification models involve simple migrating systems and form simple architectural assemblages. For example, simple transverse dunes migrating across an erg may give simple tabular crossbeds separated by planar bounding surfaces. Simple transverse dunes migrating over and down the lee-side of a larger transverse draa generate slightly more complicated patterns including down-wind dipping surfaces that reflect the lee-side slope of the draa. Changing winds modify existing bedforms and these bedforms also interact with the wind to produce a variety of disequilibrium forms.

Rubin (1987) showed using computer modeling the enormous variation and complexity of stratification patterns that are possible. Bedforms climbing at angles gentler than their windward (stoss-side) slopes erode the preceding bedforms, producing erosion surfaces bounding sets of cross-strata (Brookfield, 1977; Fig. 27). These bounding surfaces occur at all scales: even translatent strata are really only surfaces bounding wind ripples. The only limitation is that a train of bedforms must migrate. In addition to these autogenic surfaces that are of limited lat-

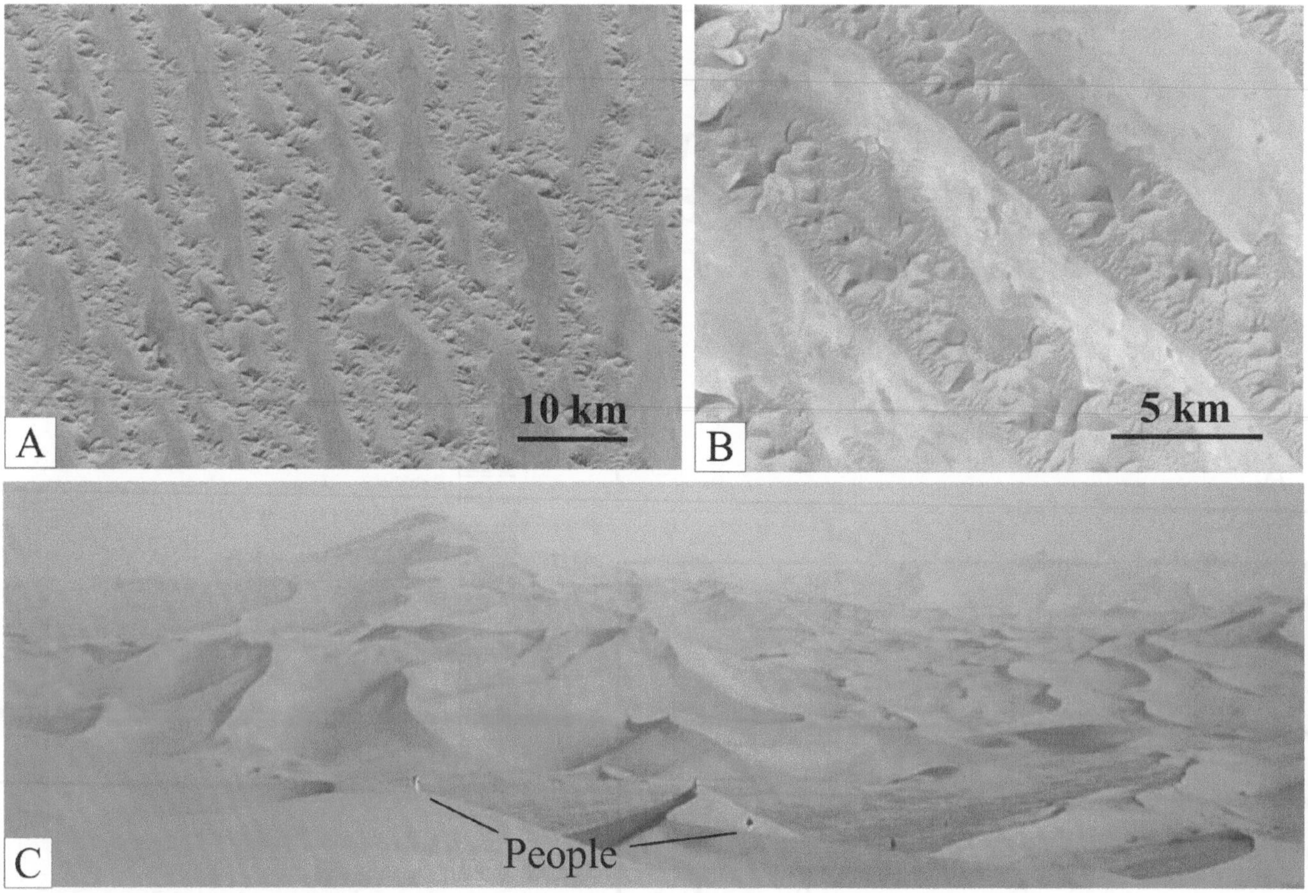

Figure 26. Great eastern erg, Algeria. **A**. Lines of complex linear draa surmounted by star dunes. **B**. Closer view of linear draa and bare interdune corridors on alluvium. **C**. Ground-level view of star dunes within one linear draa complex.

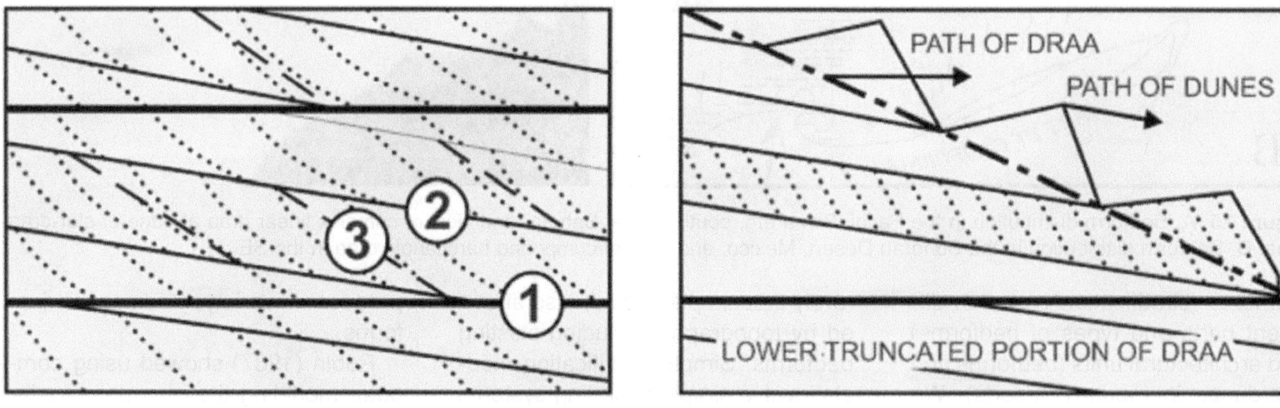

Figure 27. Eolian bounding surfaces and their origin (from Brookfield, 1977).

eral extent (probably a few kilometers at most), other more extensive surfaces cut across ancient eolian sandstones truncating all underlying surfaces. These *supersurfaces* are due to external (allogenic) controls on the erg and are discussed in a later section.

A hierarchy of autogenic surfaces may be recognized (Brookfield, 1977; *see* also Chapter 6). *First-order* surfaces are flat-lying bedding planes cutting across all underlying eolian structures and are attributed to the migration of draa or dunes (Figs. 27 and 28). *Second-order* surfaces often lie between first-order surfaces and usually dip downwind with variable inclinations. These surfaces are attributed to transverse dunes climbing down the lee slopes of draa or to the lateral migration of linear dunes across the draa lee slope (Fig. 29). *Third-order* surfaces bound bundles of laminae within cosets of cross-laminae. They are reactivation surfaces and are attributed to erosion of the dune's lee face followed by renewed deposition due to local fluctuations in wind direction and velocity (Fig. 29).

Simple dune and draa systems should lack the second-order surfaces, but in fact many do show them because few systems are in the dynamic equilibrium assumed in climbing bedform theories. Second-

Figure 28. Horizontal first-order surfaces bounding simple cross-strata caused by migration of simple bedforms (dunes; Navajo Sandstone, Utah). Bleaching of sandstone below each first-order surface may be due to sporadic rains percolating downwards in interdune areas.

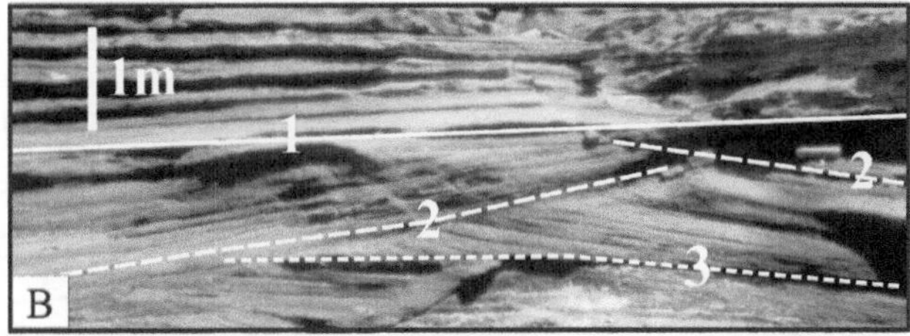

Figure 29. Second-order (2) and third-order (3) surfaces between subhorizontal first-order surfaces (1) caused by migration of a complex or compound bedform (draa). **A**. Dip section, Permian, Dumfries, Scotland, **B**. Strike section, Permian, Arran, Scotland.

order surfaces are especially common in seasonally reversing dune systems and can be formed by dunes periodically overtaking one another and coalescing. All dune-draa systems must show a resultant sand drift, even longitudinal systems. Net unidirectional climbing is thus a near certainty, though some star draa systems may climb only very slowly, if at all, and preserve only a very thin lowermost part of the system.

These stratification models are for uniform assemblages of one type of dune-interdune or draa–dune-interdune system. In ergs many different systems are present, so it is necessary to consider the possible stacking of different systems during the evolution of a desert. This requires making use of the typical spatial distribution of bedform types in a desert, and the vertical and lateral variations in stratification formed as they migrate and as the desert evolves.

FACIES MODELS

The facies models presented here are based on the two fundamental concepts discussed above.

1. Stratification types vary in proportion and location in different types of bedforms in modern deserts. These can be used to recognize the processes which form laminae in ancient eolian sandstones and, in principle, to reconstruct the various bedforms from the distribution of the laminae (Hunter, 1977).
2. Strata are generated by hierarchies of migrating bedforms of different sizes and shape which climb over one another at different angles and in differing directions. The resulting bounding surfaces can be used to recognize the nature of the bedform systems in ancient eolian sandstones (Brookfield, 1977; Rubin and Hunter, 1982; Rubin, 1987).

Synthetic models are the best hope for interpreting ancient eolian sandstones (Rubin, 1987) because there are not enough well-documented studies of stratification in recent eolian bedforms on which to base facies models (and we object to using ancient examples for this purpose as this leads to circular reasoning). Also, what can be seen, even in Quaternary eolian bedforms, is unlikely to be anything like an equilibrium response to Recent or even last glacial wind conditions, because large dunes and draa have enormous reconstitution times of thousands of years and could not possibly have adjusted to the rapid alternations of climate and wind regimes during and since the Quaternary (Wilson, 1973). Ancient ergs might have formed under more uniform conditions, which may explain why ancient eolian sandstone systems appear simpler than modern ones (*e.g.*, Kocurek, 1988). Direct comparison between Recent and ancient eolian deposits also suffers from the fact that only the lowest parts of eolian bedforms are usually preserved. Rates of bedform migration are normally at least one order of magnitude greater than the rate of vertical build-up (Brookfield, 1977). Thus, only the lowest parts (of the order of one tenth or less) are preserved. Thus, we have to reconstruct large bedforms almost entirely from structures formed on their basal lee slopes.

Facies models should show the relationships among contemporary facies in a specific system. They should not initially take into account disruption of facies relationships due

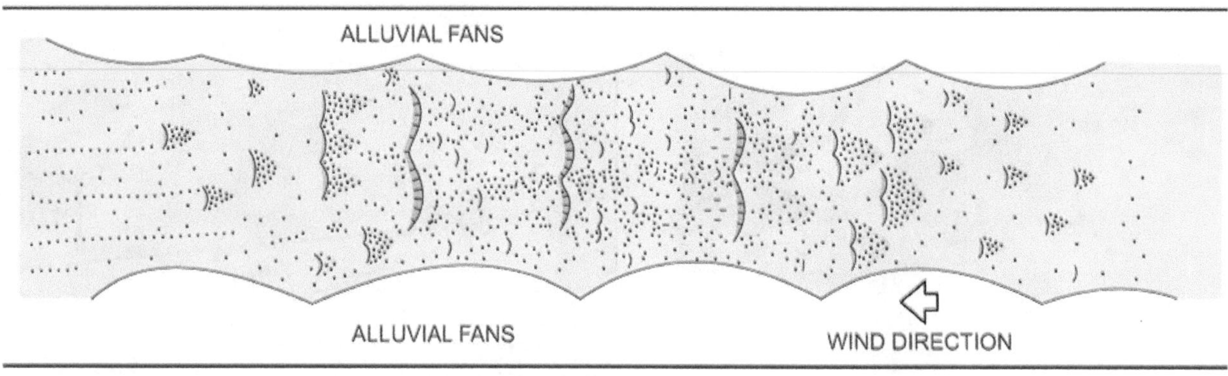

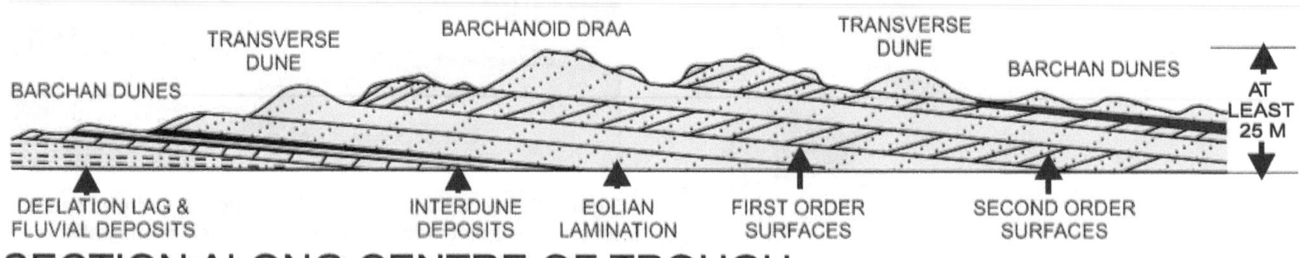

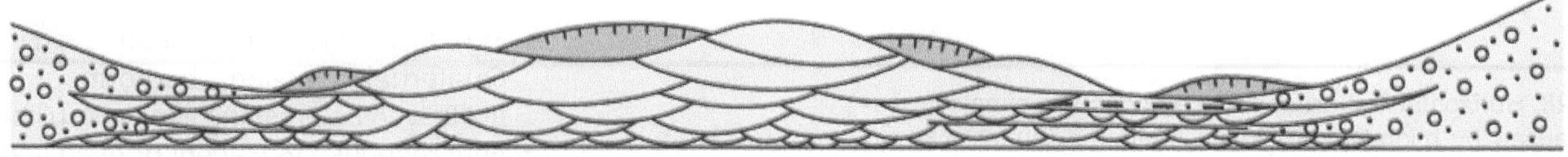

Figure 30. Synthetic facies model for transverse bedforms in an enclosed basin oriented parallel to the dominant wind direction. Vertical scale is exaggerated such that the bedforms appear to be climbing too steeply, and the troughs shown in the transverse section (perpendicular to wind) are too concave. Only first-order surfaces are shown on the transverse section; some second-order surfaces are shown on the longitudinal (wind-parallel) section. Basin is at least 5 km across.

to major external controls. For this reason, the two different models shown in Figures 30 and 31 are based on the steady migration of predominantly transverse and predominantly longitudinal bedforms during steady subsidence. These may be considered simplifications of the eastern part of the Gran Desierto (Fig. 25B) and of the Fachi–Bilma erg (Fig. 25A), respectively.

The transverse-bedform model (Fig. 30) is based on unidirectional winds and upward and lateral build-up of sand within an enclosed basin with marginal alluvial fans. The fans allow us to keep the first-order bounding surfaces horizontal and provide a continuous supply of sand: in a wide extensive plain, these bounding surfaces might be convex upward on a large scale and resemble successive stacked shells, or they might show interfingering of eolian sands with finer grained fluvial and lake deposits. The model involves the initial development of sand patches and barchan dunes with the onset of aridity, followed by the successive development of transverse dunes and compound transverse draa at the climax of aridity. Climatic fluctuations during erg development lead to complex interfingering of fluvial and eolian deposits.

The longitudinal-bedform model (Fig. 31) follows the same pattern, except that in this case the early sand patches and barchans are followed by linear dunes and linear complex draa and, eventually, in more complex wind regimes, by star draa. These longitudinal patterns have been observed rarely in ancient eolian sandstones, perhaps because such bedforms are characteristic of deserts with net through-flow of sand and little net deposition, or because such bedforms migrate laterally, laying down deposits which mimic those of transverse bedforms (Rubin *et al.*, 2008). The bedforms, as in the Fachi–Bilma erg, rest on alluvium and lag deposits. Note that the internal structures of the individual sets cannot be shown at the scale of these models. In both models the thickest units are in the centre of the erg, where the largest bedforms occur. Erg migration can lead to systematic changes in the nature and thickness of eolian strata in specific areas. In an ideal desert, we would expect that thick compound superimposed cross-strata would characterize the interior parts. Toward the margins there would be increasing numbers of thicker interdune deposits with smaller simpler cross-strata and possibly interbeds of fluvial deposits (Fig. 32).

Depositional models have been developed for ancient ergs (Karpeta, 1990; Kocurek, 1981; Mounteney, 2006). Karpeta (1990) used an almost identical procedure to the one advocated above to interpret the Permian

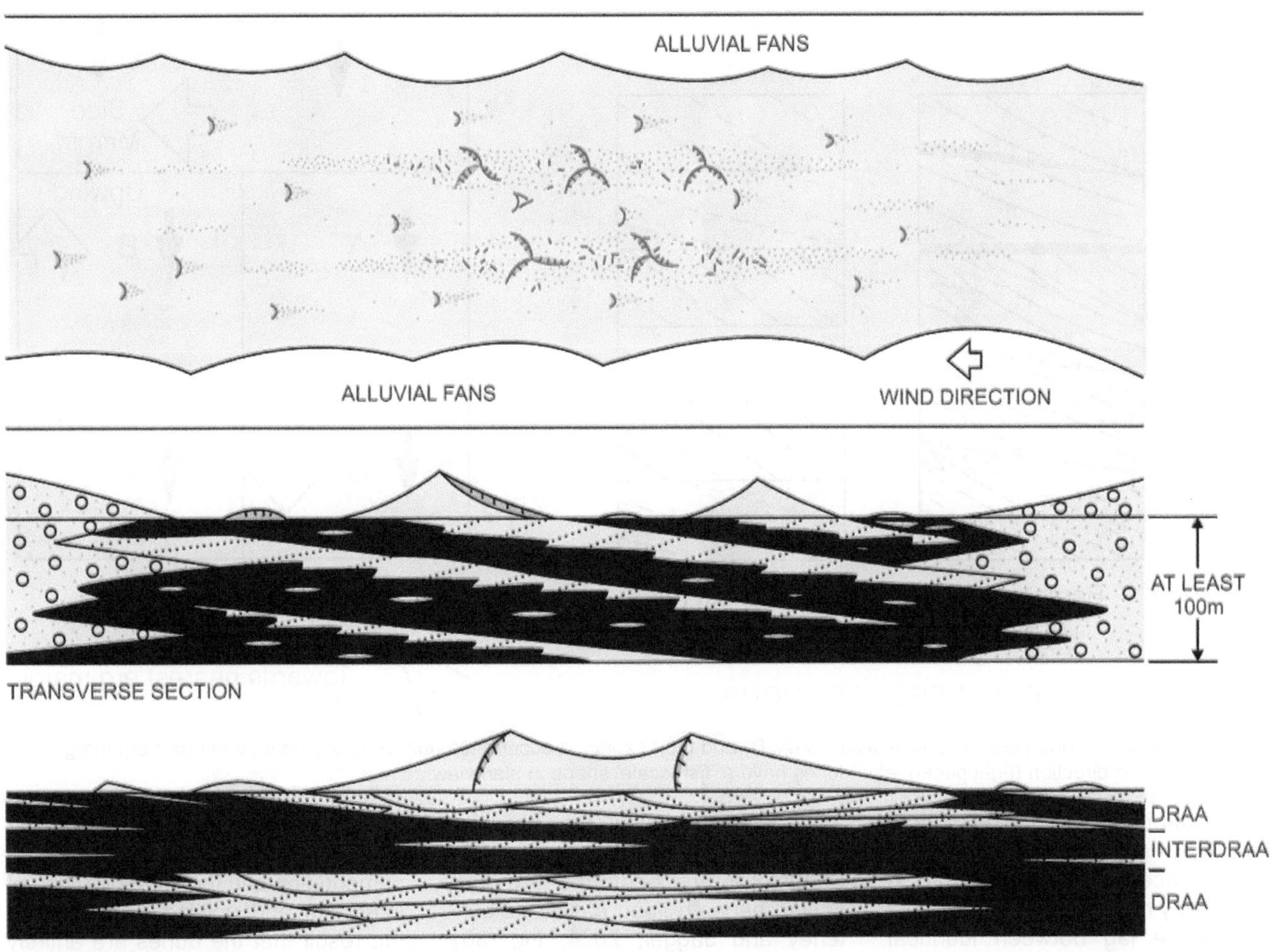

Figure 31. Synthetic facies model for longitudinal bedforms in the same enclosed basin shown in Figure 30. Note that some lateral migration of wind-parallel linear draa is permitted – without migration, vertical stacks of linear and star draa deposits would be separated by thick intedraa lags, barchan dunes and fluvial deposits (lags, dunes and fluvial deposits are shown in black). Basin is at least 5 km across.

Bridgnorth Sandstone in Britain and is a good example of how to use the facies models. He first defined facies based on the nature of the stratification (the laminae types), the orientation of foresets, and the nature of the bounding surfaces (Fig. 33). Thus, his facies A and B are sandflow strata alternating with laminated grainfall strata between first-order surfaces, with poles to crossbedding giving the shape of the slipfaces: the presence or absence of second-order bounding surfaces permits differentiation between simple and compound bedforms. His facies C to F are strata between second-order surfaces, interpreted as smaller crescentic dunes (C), dome-shaped dunes (D), linear dunes (F), and small oblique scour pits in the troughs between the larger dunes/draa (E). These facies were then grouped into facies associations (above) which he interpreted as different types of dune–interdune systems (Fig. 34). There are now many examples of such interpretations including those that show interactions with fluvial systems (*e.g.*, Tanner and Lucas, 2007; Mountenay and Jagger, 2004).

On Mars, the sedimentary rocks of the Burns Formation (exposed in all craters visited by the Opportunity rover; Fig. 35) consist of grains of evaporate minerals (0.3 to 0.8 mm in diameter). The lower unit has rippleform, grainfall and grainflow lamination cut by a third-order bounding surface attributed to a change in wind direction (Grotzinger *et al.*, 2005). The change from the lower to the middle unit is marked by a scoured surface caused by a change in wind direction and erosion of partially cemented sandstone while the contact between the middle and the upper unit is interpreted as a diagenetic boundary. The upper two units have several sets of planar-laminated (translatent) strata with the uppermost unit also showing signs of water action. Thus, the eolian complexities on Mars seem as real as those on Earth.

SUPERSURFACES

Kocurek (1981) noted synchronous and regionally extensive horizons in the Navajo Sandstone that cut all earlier surfaces and which are now known as supersurfaces. These horizons separate stratigraphic packages and record major interruptions in basin history. Such supersurfaces, which represent unconformities, disconformities, and/or flooding surfaces, are now an essential part of allostratigraphy and sequence stratigraphy. Although often consid-

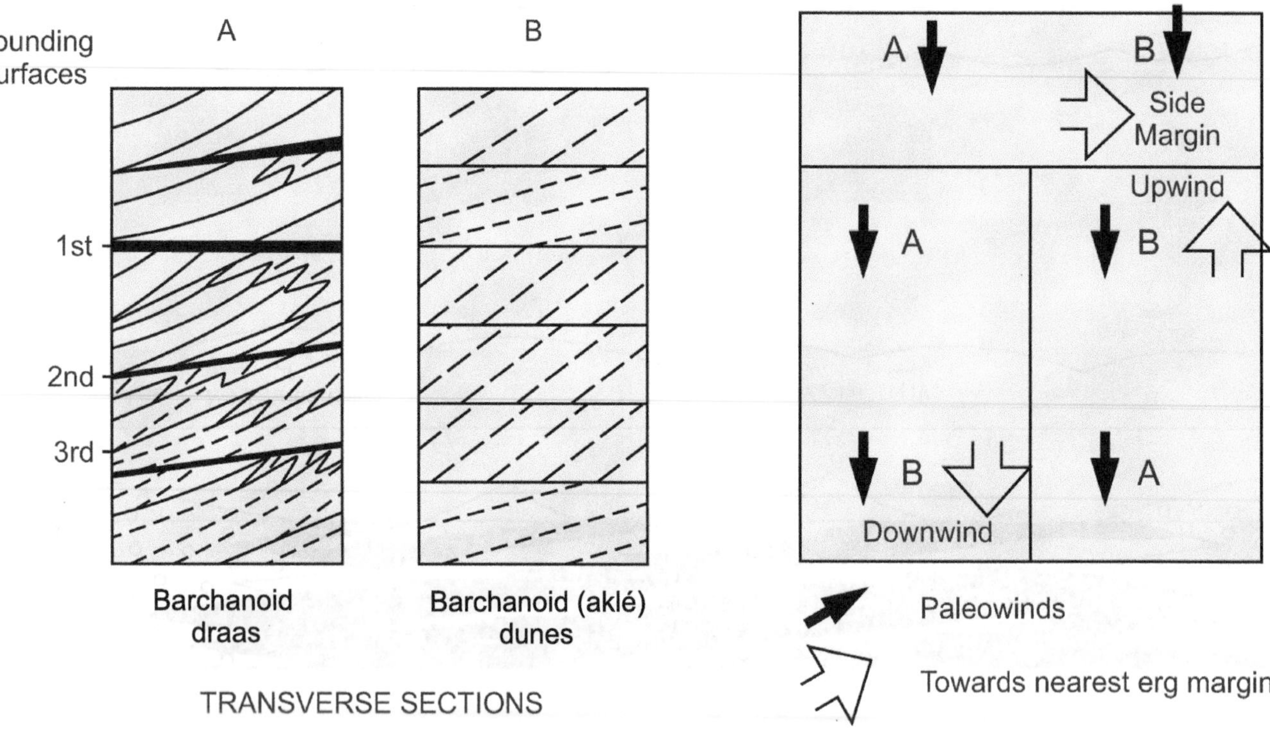

Figure 32. Synthetic wind-parallel vertical sections (**A**, **B**) and their expected occurrence relative to the nearest inferred erg margin and the resultant wind direction (right panel). Aklé dunes have a 'fish-scale' shape in plan view.

ered isochronous on a geological time scale, they need not be in detail. Such supersurfaces may be difficult to detect: they may be marked by only a thin coarse lag between identical eolian sands, even though they record a complex history (Jones and Blakey, 1993). Kocurek (1988) considered that supersurfaces fall into three broad categories which are discussed below with examples. Of course, the three types are not mutually exclusive: for example, sea-level changes may cause changes in climate and vice versa.

Surfaces Formed During Regional Termination of an Erg because of Climatic Change

The vegetation-covered Nebraska sand hills were once a large early Quaternary erg, now preserved beneath vegetation. The resulting supersurface is essentially the developing complex rolling land surface that is being smoothed gradually by erosion of the hills and deposition in the depressions. An unusual surface of this type surmounts the Permian Yellow Sands of northeastern England where linear draa were buried rapidly by the flooding of the Zechstein Sea so that rounded profiles of individual draa up to 50 meters high are preserved below the overlying Marl Slate. The Cretaceous Etjo Sandstone erg bedforms are preserved beneath flood basalts (Mountney and Jagger, 2004; Fig. 36). Desert-wide alternations of wet and dry climatic conditions, and hence of eolian and fluvial/lacustrine strata, such as those of the Quaternary of the Sahara, have now been inferred for some ancient sand seas, including the Jurassic Navajo Sandstone (Fig. 37). Most ancient ergs have, however, undergone much greater erosion before being preserved.

Surfaces Formed by the Contraction of Erg Boundaries because of Changes of Sea Level or Tectonic Setting

The Gran Desierto of northern Mexico (Fig. 13B) shows deflating bedforms around its western and southern margins, due to a combination of climate, sea-level and tectonic causes (Lancaster, 1993); such surfaces can be of local or desert-wide extent. Blakey and Middleton (1983) noted many such surfaces in the Mesozoic of the western USA. (Fig. 38). These surfaces are especially common in coastal deserts where sea-level changes directly affect the desert, although sea-level changes can also affect groundwater levels farther inland. This may cause chemical cementation and the immobilization of interdune and upwind sources, with the result that the dunes are entirely deflated, leaving a planar scoured or Stokes surface. Sand sheets may form on these surfaces under specific conditions.

Surfaces Formed by the Migration of Sand Seas

The migration of an entire sand sea over time may form a basal leading and an upper trailing supersurface. These would, however, simply form the lower and upper bounding surfaces to the eolian deposits. Within an extensive inland basin, changes in climate, sea level, tectonics or source could trigger the migration of points and areas of sand accumulation (Wilson's (1971) "sand nodes"). Thus, an erg could migrate around within a basin leading to multiple, closely spaced and intersecting supersurfaces. The best example of this that we know of is from the Permian Cedar Mesa Sandstone, Utah (Mountney, 2006).

Sequence-stratigraphic concepts have been applied to the interpretation of eolian deposits, particularly to coastal accumulations which are

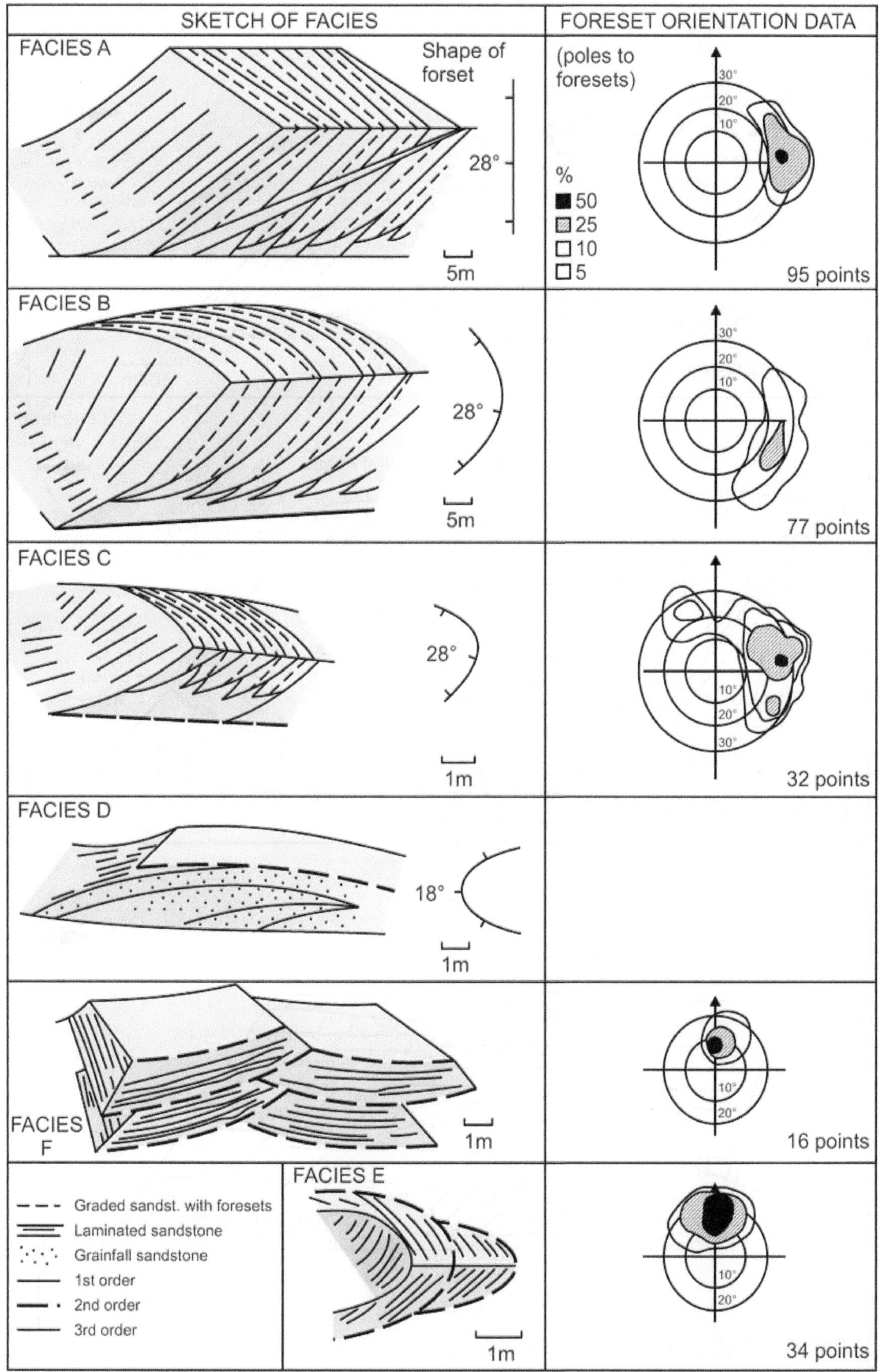

Figure 33. Three-dimensional sketches of facies based on foreset orientation and shape, Bridgnorth Sandstone, Permian, UK. Modified from Karpeta (1990).

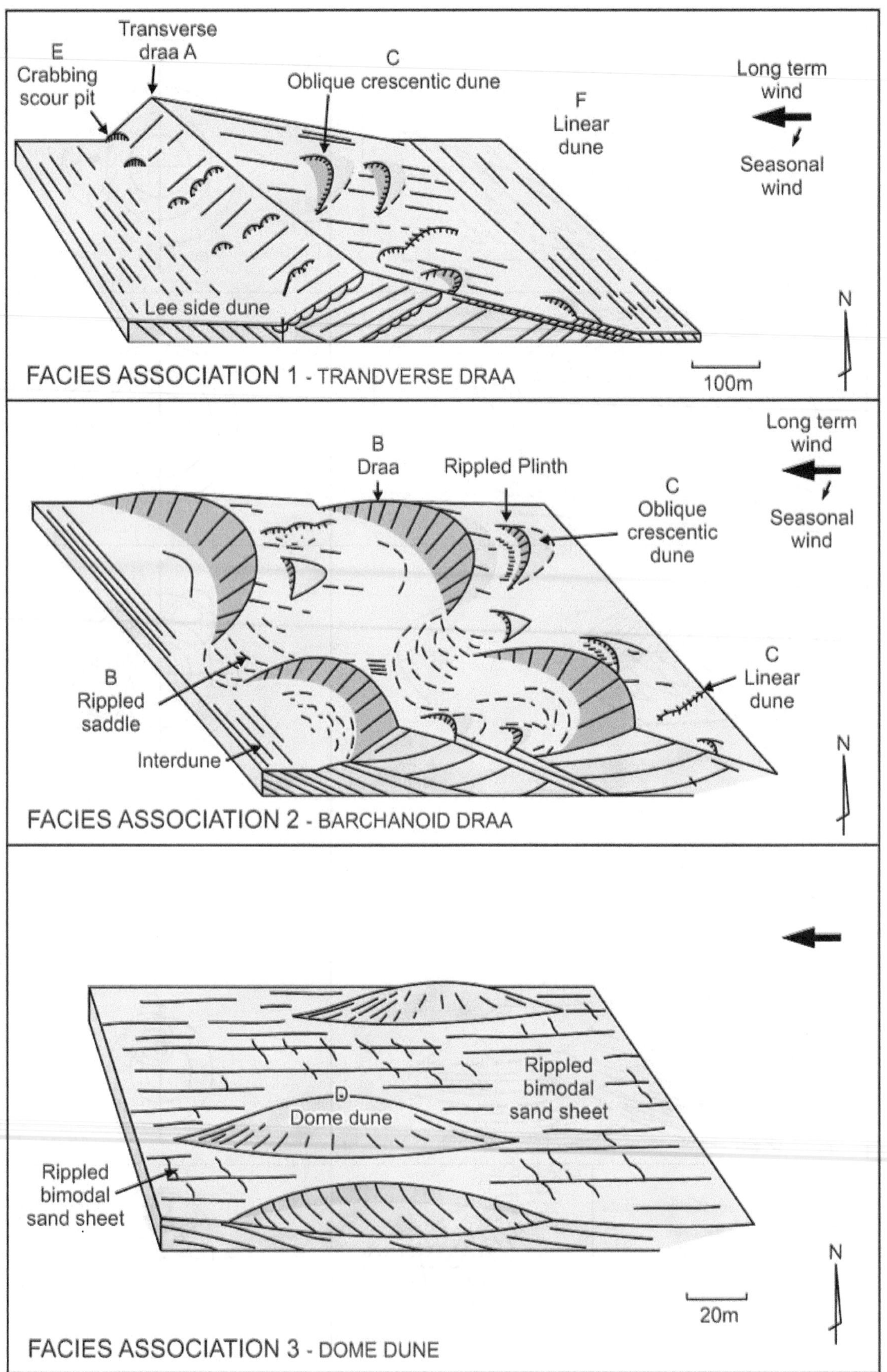

Figure 34. Facies associations in the Bridgnorth Sandstone, Permian, UK. Modified from Karpeta (1990). Crabbing scour pit is a scour moving sideways across the slipface. Rippled plinth is a flat upwind surface with ripples.

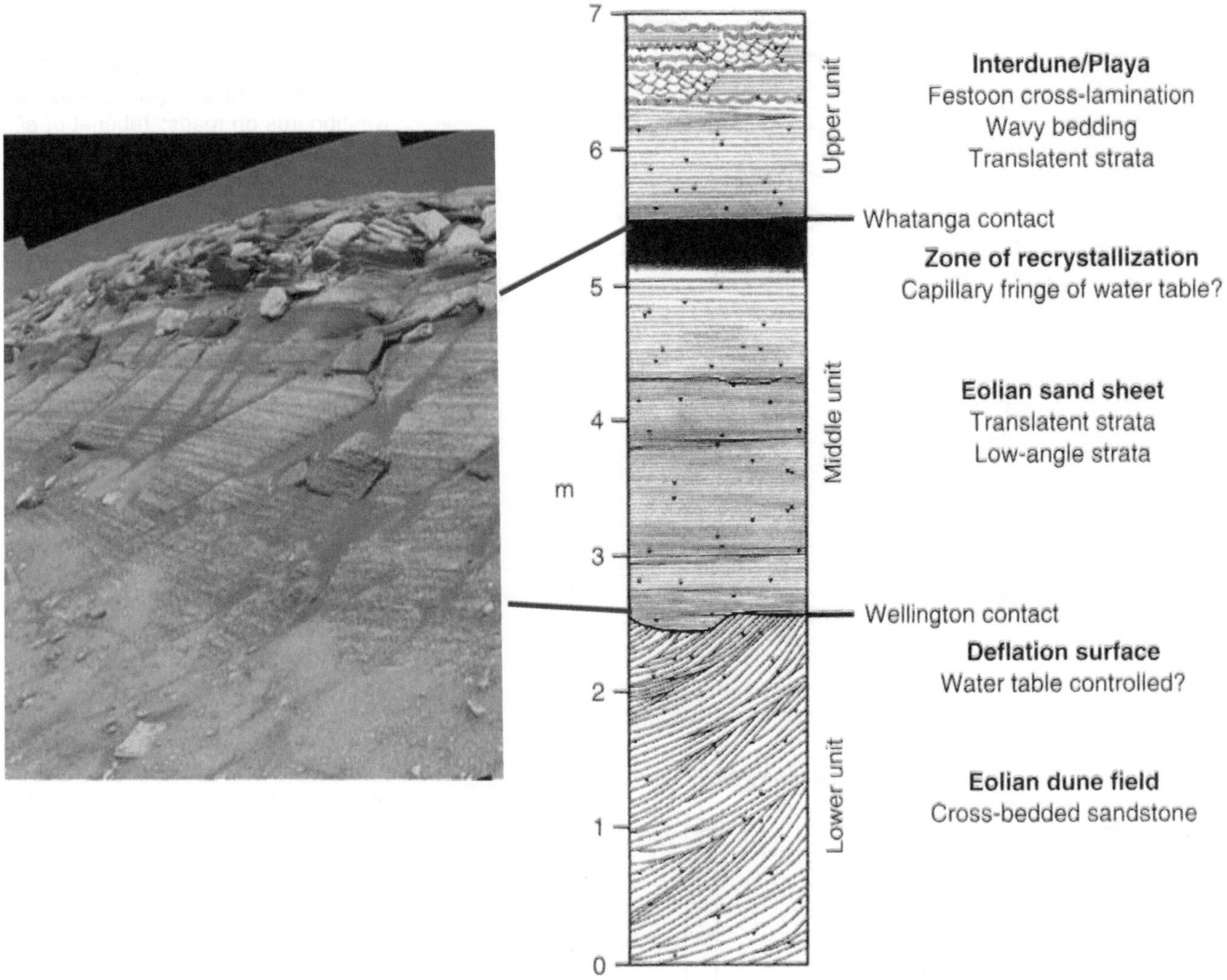

Figure 35. Strata and interpreted section at Meridianum Planum, Mars (modified Grotzinger *et al.*, 2005).

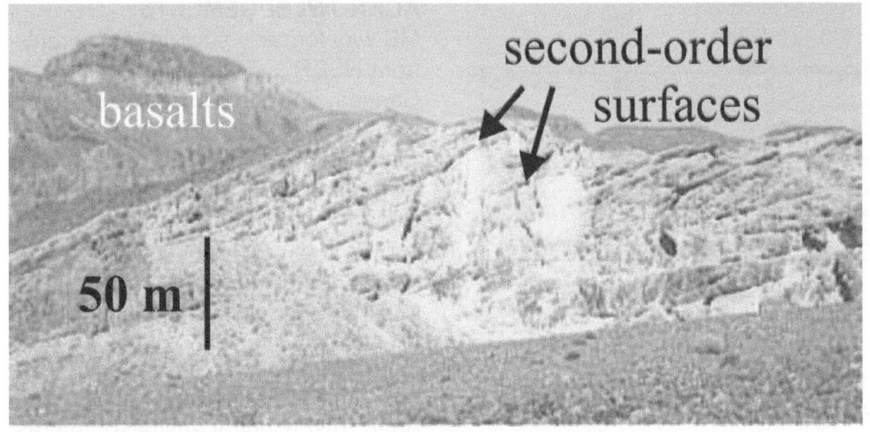

Figure 36. Downwind dipping second-order bounding surfaces in transverse draa of the Cretaceous Etjo Formation, Namibia, buried by flood basalts. Cliff face is 90 m high. (Courtesy N.P. Mounteney).

directly affected by the relative changes of sea level that form the basis for sequence stratigraphy (Veiga *et al.*, 2002). Interior basin ergs with shallow water tables can also have eolian successions determined by base-level changes caused by fluctuating water tables and the extent of lakes (themselves controlled by climate) (Carr-Crabaugh and Kocurek, 1998). Furthermore, the climatically controlled enormous areal variations of interior basin lakes such as Lake Chad and Lake Eyre cause synchronous flooding of very large areas and deposition of lake facies alternating with accumulation of eolian facies (Brookfield, 2008).

TOWARD A DYNAMIC INTERPRETATION OF EOLIAN DEPOSITS

Dynamic interpretations of both modern and ancient sand seas, based on autogenic bedform migration, and allogenic climatic change and tectonic subsidence variations and their associated factors, are discussed in Kocurek (1999), Lancaster (1995, Chapter 8), and Mounteney *et al.* (1999). Many such interpretations are controversial, as there is no consensus about what causes the form, shape and distribution of bedforms in

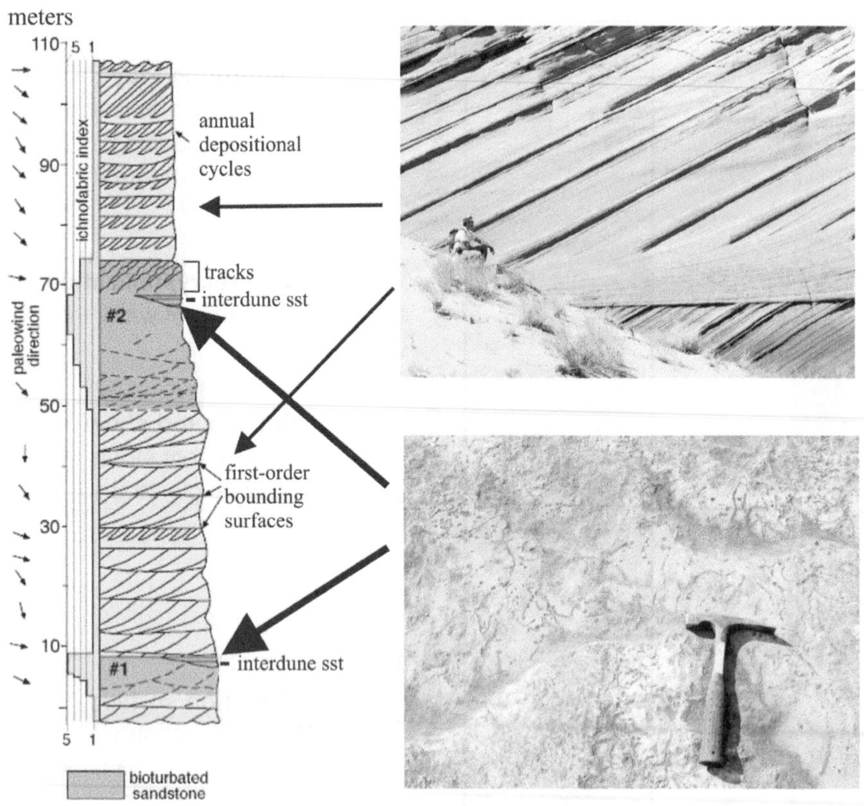

Figure 37. Alternating wet and dry cycles in the Navajo Sandstone (courtesy D.B. Loope).

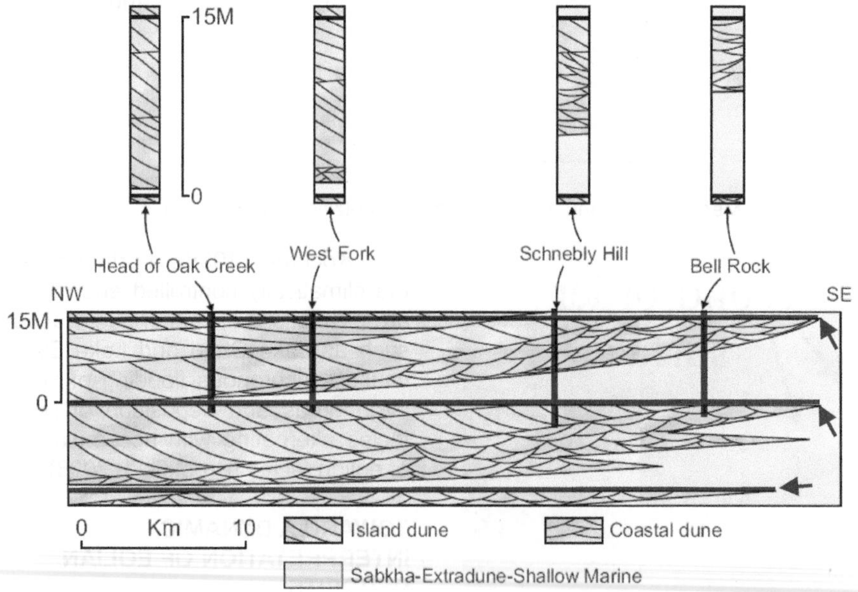

Figure 38. Correlation of sea-level controlled supersurfaces (heavy horizontal lines, arrowed) in part of the Coconino Sandstone, western USA. From Blakey and Middleton (1983).

modern deserts (Kocurek and Ewing, 2005; Besler, 2008). Experimentation is difficult above ripple sizes, and wind-tunnel experiments do not resemble natural flows, because the former are basically pipe flow whereas natural flows have a free surface (the top of the atmospheric boundary layer). Large-scale dynamic studies in modern deserts are limited because dunes and draa are not adjusted to the current wind regime because of their enormous lag time (Allen, 1974). The development of large transverse and longitudinal bedforms has been related to structures in the atmospheric boundary layer such as longitudinal vortexes (visualized as cloud streets; Etling and Brown, 1993), bouncing vortices (analogous to washboards on roads; Taberlet *et al.*, 2007) controlled by kinematic waves at the atmospheric boundary layer/troposphere transition, and internal gravity waves (Scorer, 1978; Sorbjan, 1989), but such explanations remain largely untested. Models of autogenic bedform migration need to consider bedform/airflow interactions and the resulting sand-transport patterns, and we need better theoretical models relating bedform development to atmospheric boundary-layer thickness and structure (Besler, 2008). Using only surface winds to interpret the flow structures molding eolian bedforms is like ignoring the turbulence above about 10 cm in deep-water flows. There is currently little effort being made to integrate research on planetary atmospheric structure and dynamics (worked on by meteorologists) with research on eolian-bedform development (worked on by sedimentologists), either on Earth or elsewhere. Yet, without knowing how modern eolian bedforms interact with the atmosphere and respond to climatic change, it is impossible to interpret ancient ones. Life-long research opportunities exist in this field.

ACKNOWLEDGEMENTS

MB was formerly supported by grants from NSERC (Solid Earth Sciences), Canada, and is now supported by the Institute of Earth Sciences, Academia Sinica, Taiwan (via its director Borming Jahn) and by the Postgraduate Department of Geology, Jammu University, India (via Ghulam Bhat). For his studies in modern deserts he is indebted, for field organization and discussions, to Gary Kocurek, Nick Lancaster, and Grady Blount (Mexico), Rod Wells, Roger Callen and Brian Rust (Central Australia), András Zboray and Dabuka Expeditions (Egypt, Sudan), Suzanne Leroy and Pedro Costa (Mauritania), Pointe Afrique (Algeria, Niger, Chad), and to the local inhabitants (especially in Algeria, Rajasthan and Tarim) for food and accommodation (where available). SS is supported by grants from the Agenzia Spaziale Italiana. We thank Bob Dalrymple for his careful editing and constructive suggestions.

REFERENCES

Basic sources of information

Allen, J.R.L., 1982, Sedimentary structures; their character and physical basis: Amsterdam, Elsevier, v. 1, 594 p.; v. 2, 644 p.
The finest, most comprehensive and thoughtful text ever written on sedimentary structures. Essential reading for all sedimentologists.

Besler, H., 2008, The Great Sand Sea in Egypt: Elsevier, Amsterdam, 250 p.
A study of one very large modern sand sea.

Carr, M.H., 2006, The Surface of Mars: Cambridge University Press, Cambridge, 307 p.
Essential reading for students of eolian features.

Goudie, A., Livingstone, I. and Stokes, S., *eds.*, 2000, Aeolian Environments, Sediments, and Landforms: J.Wiley & Sons, New York, 336 p.
Comprehensive reviews of aspects of both modern and ancient deserts.

Greeley, R. and Iversen, J.D., 1985, Wind as a Geological Process on Earth, Mars, Venus and Titan: Cambridge University Press, New York, 333 p.
An interesting comparison between eolian processes on different large solar-system bodies with atmospheres

Lancaster, N., 1995, Geomorphology of Desert Dunes: Routledge, London, UK., 290 p.
A great introduction to the subject, with alternatives thoughtfully considered.

Pye, K. and Tsoar, H., *eds.*, 2009, Aeolian Sand and Sand Dunes: Springer, Berlin, 475 p.
A good summary of eolian bedforms.

Rubin, D.M., 1987, Cross-bedding, Bedforms and Paleocurrents: SEPM, Concepts in Sedimentology and Paleontology, v. 1, 187 p.
An essential guide to the relationships of bedforms, bedding and bounding surfaces.

McKee, E.D., *ed.*, 1979. A Study of Global Sand Seas: United States Geological Survey Professional Paper 1052, 429 p.
Many essential satellite and other studies.

Scorer, R.S., 1978, Environmental Aerodynamics : Ellis Horwood Ltd., Chichester, 488 p.
One of the most thoughtful summaries of atmospheric dynamics with many insights into boundary layer structure and development.

Seppala, M., 2003, Wind as a Geomorphic Agent in Cold Climates: Cambridge University Press, Cambridge, 368 p.
It does not have to be hot – just dry – for eolian processes to dominate a landscape.

Sorbjan, Z., 1989, Structure of the Atmospheric Boundary Layer: Prentice Hall, Englewood Cliffs, New Jersey, USA, 317 p.
You need to know the characteristics of any 'fluid' before you can interpret the bedforms formed by it (including mass flows, water, air, and ice).

Wilson, I.G., 1973, Ergs: Sedimentary Geology, v. 10, p. 77-106.
The impetus for modern studies on sand seas.

Other references

Ahlbrandt, T.S., 1979, Textural parameters of eolian deposits: United States Geological Survey Professional Paper 1052, p. 21-51.

Allen, J. R L., 1974, Reaction, relaxation and lag in natural sedimentary systems: general principles, examples and lessons: Earth Science Reviews, v. 10, p. 263-342

Biswas, A., 2004, Coarse aeolianites: sand sheets and zibar-interzibar facies from the Mesoproterozoic Cuddapah Basin: Sedimentary Geology, v. 174, p. 149-160.

Blakey, R.C. and Middleton, L.T., 1983, Permian shoreline aeolian complexes in central Arizona; dune changes in response to cyclic sealevel changes, *in* Brookfield, M.E. and Ahlbrandt, T.S., *eds.*, Eolian Sediments and Processes: Amsterdam, Elsevier, p. 551-581.

Breed, C.S., Fryberger, S.G., Andrews, A., McCauley, C., Lennartz, F., Gebel, D. and Horstman, K., 1979, Regional studies of sand seas using Landsat (ERTS) imagery: United States Geological Survey Professional Paper 1052, p. 305-397.

Bristow, C.S., Pugh, J. and Goodall, T., 1996, Internal structure of aeolian dunes in Abu Dhabi determined using ground-penetrating radar: Sedimentology, v. 43, p. 995-1003.

Brookfield, M.E., 1977, The origin of bounding surfaces in ancient eolian sandstones: Sedimentology, v. 24, p. 303-332.

Brookfield, M.E., 2008, Palaeoenvironments and palaeotectonics of the arid to hyperarid intracontinental latest Permian – late Triassic Solway basin (U.K.): Sedimentary Geology v. 210, p. 27-47.

Carr-Crabaugh, M., and Kocurek, G., 1998, Continental sequence stratigraphy of a wet eolian system: a key to relative sea-level change: SEPM Special Publication 59, p. 213-228.

Ekdale, A.A., Bromley, R.G., and Loope, D.B. 2007, Ichnofacies of an ancient erg: A climatically influenced trace fossil association in the Jurassic Navajo Sandstone, southern Utah, USA, *in* Miller, W. III., *ed.*, Trace Fossils: Concepts, Problems, Prospects: Elsevier, Amsterdam, p. 562-574.

Etling, D. and Brown, R.A., 1993, Roll vortices in the planetary boundary layer: a review: Boundary-Layer Meteorology, v. 65, p. 215-248.

Ewing, S.A., Sutter, B., Amundson, R., Owen, J., Nishiizumi, K., Sharp, W., Cliff, S.S., Perry, K., Dietrich, W.E. and McKay, C.P., 2006, A threshold in soil formation at Earth's arid-hyperarid transition: Geochemica et Cosmochemica Acta, v. 70, p. 5293-5322.

Fryberger, S.G., 1979, Dune form and wind regime: United States Geological Survey Professional Paper 1052, p. 137-169.

Goudie, A.S., 2002, Great Warm Deserts of the World: Oxford University Press, New York, 444 p.

Grotzinger, J.P., Arvidson, R.E., Bell, J.F., Calvin, W., Clark, B.C., Fike, D.A., Golombek, M., Greeley, R., Haldemann, A., Herkenhoff, K.E., Jolliff, B.L., Knoll, A.H., Malin, M., McLennan, S.M., Parker, T., Soderbolom, L., Sohl-Dickstein, J.N., Squyres, S.W., Tosca, N.J. and Watters, W.A., 2005, Stratigraphy and sedimentology of a dry to wet eolian depositional system, Burns formation, Meridiani Planum, Mars: Earth and Planetary Science Letters, v. 240, Issue 1, p. 11-72.

Hunter, R.E., 1977, Basic types of stratification in small eolian dunes: Sedimentology, v. 24, p. 361-387.

Hurst, A. and Glennie, K.W., 2008, Mass-wasting of ancient dunes and sand fluidization during a period of global warming and inferred brief high precipitation: the Hopeman Sandstone (Late Permian), Scotland: Terra Nova, v. 20, p. 274-279.

Jol, H.M., Bristow, C.S., Smith, D.G., Junck, M.B. and Putnam, P., 2003, Stratigraphic imagining of the Navajo Sandstone using ground-penetrating radar: The Leading Edge, v. 22, p. 882-887.

Jones, L.S. and Blakey, R.C., 1993, Erosional remnants and adjacent unconformities along an eolian-marine boundary of the Page Sandstone and Carmel Formation, Middle Jurassic, South-Central Utah: Journal of Sedimentary Petrology, v. 63, p. 825-859.

Karpeta, W.P., 1990, The morphology of Permian paleodunes – a reinterpretation of the Bridgnorth Sandstone around Bridgnorth, England in the light of modern dune studies: Sedimentary Geology, v. 69, p. 59-75.

Kocurek, G., 1981, Significance of interdune deposits and bounding surfaces in aeolian dune sands: Sedimentology, v. 28, p. 753-780.

Kocurek, G., 1988, First-order and super bounding surfaces in eolian sequences - bounding surfaces revisited: Sedimentary Geology, v. 56, p. 193-206.

Kocurek, G., 1999, The aeolian rock

record: *in* Goudie, A.S., Livingstone, I. and Stokes, S., *eds.*, Aeolian Environments, Sediments and Landforms: J. Wiley, Chichester. p. 239-259.

Kocurek, G. and Ewing, R., 2005, Aeolian dune field self-organization – implications for the formation of simple versus complex dune field patterns: Geomorphology, v. 72, p. 94-105.

Koster, E.A., 1988, Ancient and modern cold-climate aeolian sand deposition; a review: Journal of Quaternary Science, v. 3, p. 69-83.

Lancaster, N., 1993, Origins and sedimentary features of supersurfaces in the northwestern Gran Desierto Sand Sea: International Association of Sedimentologists, Special Publication 16, p. 71-83.

Lorenz, R, Radebaugh, J., Paillou, Ph. and Cassini RADAR Team, 2007, Radar imaging of sand dunes on Titan and Earth: Geophysical Abstracts, v. 9, p. 4604.

Mainguet, M. and Callot, Y., 1974, Air photo study of typology and interrelations between the texture and structure of dune patterns in the Fachi-Bilma Erg, Sahara: Zeitschrift fur Geomophologie, v. 20, p. 62-69.

Mainguet, M. and Remini, B., 2004, Le role des mega-obstacles dans la formation et le faconnement des ergs: quelques exemples du Sahara: Larhyss Journal, v. 3, p. 13-23.

Maxwell, T.A. and Haynes, C.V. Jr., 2001, Sand sheet dynamics and Quaternary landscape evolution of the Selima Sand Sheet, southern Egypt: Quaternary Science Reviews, v. 20, p. 1623-1647.

McKee, E.D., 1966, Structure of dunes at White Sands National Monument, New Mexico (and a comparison with structures of dunes from other selected areas): Sedimentology, v. 7, p. 1-70.

McKee, E.D., 1982, Sedimentary Structures in Dunes of the Namib Desert, Southwest Africa: Geological Society of America Special Paper 108, 64 p.

Milana, J.P., 2009, Largest wind ripples on Earth? Geology, v. 37, p. 343-346.

Mountney, N.P., 2006, Periodic accumulation and destruction of aeolian erg sequences in the Permian Cedar Mesa Sandstone, White Canyon, southern Utah, U.S.A: Sedimentology, v. 53, p. 789-823.

Mountney, N.P., Howell, J.A., Flint, S.S. and Jerram, D.A., 1999, Climate, sediment supply and tectonics as controls on the deposition and preservation of the aeolian-fluvial Etjo Sandstone Formation, Namibia: Journal of the Geological Society, London, v. 156, p. 771-778.

Mounteney, N.P. and Jagger, A., 2004, Stratigraphic evolution of an aeolian erg margin system: the Permian Cedar Mesa Sandstone, SE Utah USA: Sedimentology, v. 51, p. 713-743.

Overmeeren, R.A. van, 1998, Radar facies of unconsolidated sediments in The Netherlands: a radar stratigraphy interpretation: Journal of Applied Geophysics, v. 40, Nos.1-3, p. 1-18.

Retallack, G.J., 2001, Soils of the Past: 2nd edition. Blackwell, Oxford, 404 p.

Rubin, D.M. and Hunter, R.E., 1982, Bedform climbing in theory and nature: Sedimentology, v. 29, p. 121-138.

Rubin, D.M., Tsoar, H. and Blumberg, D.G., 2008, A second look at western Sinai seif dunes and their lateral migration: Geomorphology, v. 93, p. 335-342.

Schwammie, V. and Herrmann, H., 2004, Modelling transverse dunes: Earth Surface Processes and Landforms, v. 29, p. 769-784.

Sharp, R.P., 1963, Wind ripples: Journal of Geology, v. 71, p. 617-636.

Soreghan, G.S., Moses, A.M., Soreghan, M.J., Hamilton, M.A., Fanning, C.M. and Link, P.K., 2007, Palaeoclimatic inferences from upper Palaeozoic siltstone of the Earp Formation and equivalents, Arizona-New Mexico (USA): Sedimentology, v. 54, p. 707-719.

Taberlet, N., Morris, S.W. and McElwaine, J.N., 2007, Washboard road: the dynamics of granular ripples formed by rolling wheels: Physical Review Letters, v. 99, 068004 (4 pages).

Tanner, L.H. and Lucas, S.G., 2007, The Moneave Formation: sedimentologic and stratigraphic context of the Triassic-Jurassic boundary in the Four Corners area, southwestern U.S.A: Palaeogeography, Palaeoclimatology, Palaeoecology, v. 244, p. 111-125.

Tewes, D.W. and Loope, D.B., 1992, Palaeo-yardangs: wind-scoured landforms at the Permo-Triassic unconformity, Sedimentology, v. 39, p. 251-261.

Veiga, G.D., Spalletti, L.A., and Flint, S., 2002, Aeolian/fluvial interactions and high-resolution sequence stratigraphy of a non-marine lowstand wedge: the Avile Member of the Agrio Formation) Lower Cretaceous), central Neuquen Basin, Argentina: Sedimentology, v. 49, p. 1001-1019.

Wilson, I.G., 1971, Desert sandflow basins and a model for the origin or ergs: Geographical Journal, v. 137, p. 180-199.

Wilson, I.G., 1972, Aeolian bedforms - their development and origin: Sedimentology, v. 19, p. 173-210.

Zavala, C., Maretto, H. and Di Meglio, M., 2005, Hierarchy of bounding surfaces in aeolian sandstones of the Jurassic Tordillo Formation (Nequen Basin, Argentina): Geologica Acta, v. 3, p. 133-145.

8. WAVE- AND STORM-DOMINATED SHORELINE AND SHALLOW-MARINE SYSTEMS

A. Guy Plint, Department of Earth Sciences, The University of Western Ontario, London, ON, N6A 5B7, Canada

INTRODUCTION

About 80% of the world's coasts and shelf seas are dominated by hydraulic and sedimentary processes related to storms. These processes are reflected in an abundance of sedimentary rocks that preserve evidence of wave processes, and episodic erosional and depositional events. Shelf seas are, however, inherently ephemeral features that exist when sea levels are relatively high, reverting to dry land when continental margins and epeiric seas are drained at lowstand. A shelf is considered to be wave-dominated when wave-related processes dominate over tidal processes; however, it is not the *absolute* magnitude, but the *relative* magnitude of each process that determines the sedimentary response (Davis and Hayes, 1984).

Shelf seas include a continuum of depositional environments from beach and shoreface through inner to outer shelf settings. Deposition commonly takes place in relatively shallow water (typically < 100 m) and on surfaces with a low topographic gradient. As a result, shorelines have the potential to move laterally for many tens of kilometers, both as a result of sediment accumulation along the coast, and also as a result of relative rises and falls in sea level. For example, submerged glaciofluvial sediments and the presence of deltas near many shelf margins record a eustatic fall to about 120 m below present sea level by about 26 ka ago during the peak of the Wisconsinan glaciation, that left most shelves subaerially exposed. This was followed by rapid sea-level rise during deglaciation, reaching the present level about 5–6 ka ago (Peltier and Fairbanks, 2006). A new pulse of sea-level rise began in the middle of the 19th century, and continues at a rate of about 2 mm/yr (or more), mainly as a result of the melting of Greenland ice, with lesser contributions from ocean warming and expansion (Miller and Douglas, 2006). The geological record shows that transgressive and regressive events are a ubiquitous feature of shelf deposits throughout Earth history, although they generally have involved less dramatic sea-level changes than those of the Pleistocene.

The sediments that accumulate on shelves have very complex stratigraphy, punctuated by numerous, commonly closely spaced erosion surfaces, that record both wave scour in transgressive shorelines and also subaerial erosion, as shelf deposits were exposed to fluvial processes during regression and relative sea-level fall. Stratigraphic and facies complexity is further increased by potentially rapid changes in the intensity of wave and tidal processes linked to changes in the width and depth of shelves, and by changes in the caliber of sediment delivered to the shoreline (Yoshida *et al.*, 2007).

Evolving Ideas

Previous editions of *Facies Models* provided a series of snapshots of the evolution of thinking about shelf deposits. In the first edition (Walker, 1979), the principal issue was the question of how sand was transported to the outer part of shelves, and the degree to which it was reworked by tidal, storm and oceanographic currents. Coarsening-upward successions, some containing hummocky cross-stratification (HCS) and capped by conglomerate, were seen as particularly difficult to explain in terms of known shelf processes, and no coherent facies model could be presented. The second edition (Walker, 1984) continued to focus on the hydrodynamic interpretation of HCS and the attendant problem of sand delivery to shelves. HCS was widely viewed as a type of large storm-wave-formed ripple, but which defied re-creation in a laboratory, and which would be suicidal to try to observe first-hand during a storm! Storm-driven shelf flows were acknowledged to be potentially important agents of along- and across-shelf sand transport, although, perversely, geological evidence seemed to indicate sand transport by shore-perpendicular currents (compare Leckie and Krystinik, 1989 with Duke, 1990). Ancient examples of 'shelf sand ridges' were thought to be gradationally rooted in shelf mud, raising the problem of how sand was transported and concentrated far out on the shelf.

By the third edition (Walker and Plint, 1992), the question of HCS genesis had been largely solved by innovative flume studies (Southard *et al.*,1990), subsequently elaborated by Dumas *et al.* (2005) and Dumas and Arnott (2006). Many offshore sand ridges had also been recognized for what they were, namely detached shoreline sediments (*i.e.*, no longer in physical continuity with highstand shoreline deposits closer to shore), that were deposited during falling stage, lowstand or early transgressive systems-tract time. The detached shoreline sandstone bodies were bounded above and below by erosion surfaces, and had become 'stranded' on the shelf by subsequent marine transgression (*e.g.*, Bergman and Walker, 1987; Plint, 1988; Posamentier *et al.*, 1992). This theme was subsequently elaborated by many others (*e.g.*, Walker and Wiseman, 1995; Mellere and Steel, 1995; Bergman and Snedden, 1999). Stacked, coarsening-upward successions were interpreted as the product of repeated, relative sea-level changes, a topic scarcely mentioned in previous editions.

It is important to appreciate that the progress in understanding shelf

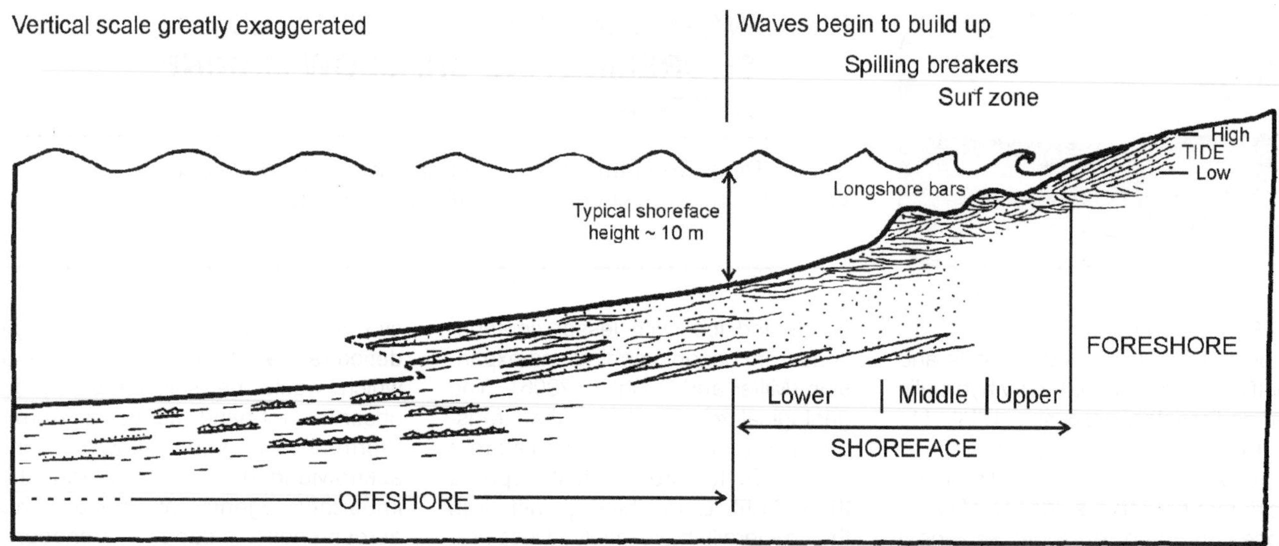

Figure 1. Generic shoreline to shelf profile showing the main morphological elements. Waves become asymmetric as they enter the shoreface zone, and water moves landward under spilling breakers, building longshore bars and driving the longshore circulation system (*see* Figure 5). The shoreface has a steeper gradient than the shelf, and typically ranges up to about 10 m in height, depending on grain size and wave climate. (*See* Figure 11 for an alternative shoreface-shelf profile).

deposits during the late 1980s and early 1990s was driven primarily by *stratigraphic* studies, rather than detailed facies analysis, which had been the focus of sedimentology since the 1960s. In particular, it became apparent, initially from subsurface studies in Alberta (where drilling data are exceptionally abundant), that seemingly minor erosional contacts and thin pebble beds separating disparate sedimentary facies were in fact the record of vertical movements of sea level, and dramatic lateral shifts in shoreline position (*e.g.*, Plint *et al.*, 1986, 1988; Boreen and Walker, 1991; Pattison and Walker, 1992). Despite the renewed emphasis on *stratigraphy*, detailed sedimentological analysis was, of course, an essential prerequisite to recognition of the sometimes subtle facies offsets that marked sequence boundaries and flooding surfaces.

The release of SEPM Special Publication 42 (Wilgus *et al.*, 1988), containing the Exxon sequence-stratigraphic models, precipitated a revolution in thinking about all depositional systems, but particularly those in marginal-marine, shelf, and slope to basin settings. Storm-dominated shelf deposits provided the 'nursery' for sequence-stratigraphic concepts because the effects of relative sea-level changes could be clearly observed, in terms of the creation and removal of space ('accommodation') between sea surface and seabed, and corresponding lateral excursions of the shoreline which were, in turn, reflected in vertical changes, both gradual and abrupt, in the nature of the facies.

Now, twenty-two years after SEPM 42 was published, interest in sequence stratigraphy is unabated, and the focus has shifted to looking at shelves in a more holistic way. The holistic approach is not only concerned with the transport and deposition of sand (with its obvious interest to the petroleum industry), but also with mud, because mud forms the bulk of shelf sediments, and it is much more difficult to understand, both in terms of transport and depositional processes (*e.g.*, Nittrouer *et al.*, 2007). On the technical front, continued improvements in the resolution of seismic imaging, particularly in 3-D, has had a dramatic impact on our ability to visualize sediment bodies beneath the surface, and hence understand their origins better.

MORPHOLOGICAL ELEMENTS

The main morphological elements of a modern shoreline- to shelf-depositional system are shown in Figure 1. The foreshore, which is synonymous with beach, is that portion of the shelf that lies above the low-tide line, and which is dominated by the swash and backwash of breaking waves. The shoreface lies below the low-tide level and is characterized by sandy sediment, a relatively steep gradient (approximately 1:200 or about 0.3°), and frequent transport of sand by shoaling waves. Modern progradational shorefaces are typically 5–10 m high, depending on the wave climate. In contrast, modern transgressive shorefaces fronting barrier islands tend to be taller, up to 25 m, but they are sand-starved, eroding systems that are generating an erosion surface and are unlikely to be represented by progradational shoreface sandbodies; the latter will typically not exceed 10 m thick in the rock record (Clifton, 2006). Seaward, the shoreface grades into the offshore zone where the gradient is much lower (about 1:1000 to 1:2000, or less), and sediment tends to be dominated by mud. On high energy progradational coasts, however, sandy sediment may extend tens of kilometers out onto the shelf in several tens of meters of water. In consequence, it may not be possible to recognize, in vertical sections of rocks, the topographic break at the toe of the shoreface on the basis of a distinct facies change. Geologists sometimes choose the base of mud-free sandstone (swaley cross-stratification; SCS) and amalgamated HCS as an operational 'base of shoreface' (*e.g.*,

Leckie and Walker, 1982; McCrory and Walker, 1986; Plint and Walker, 1987), whereas others interpret underlying HCS interstratified with bioturbated mud to represent the lower part of the morphological shoreface (*e.g.*, Hampson and Storms, 2003). In other settings where little sand is available, the shoreface may consist mostly of mud (Rine and Ginsberg, 1985).

The gradient of the beach and upper shoreface is strongly influenced by the grain size of the sediment. Coarse-grained gravelly beaches are steep, and waves break close to shore, with much of the energy reflected back to sea, whereas sandy beaches slope more gently, and waves dissipate their energy across a wide zone, sometimes occupied by breaker bars that may extend seaward for hundreds of meters (Clifton, 2006). Breaker bars, troughs and rip channels have a relief of up to a few meters.

On modern shelves, for example the US Middle Atlantic Bight, the lower part of the shoreface may be molded into large shoreface ridges that trend oblique to the beach (Swift and Field, 1981; Swift *et al.*, 1981; Snedden *et al.*, 1999). These shoreface ridges, composed of fine- to coarse-grained sand, are mantled by various types of subaqueous dunes and ripples (Fig. 2). The ridges originate in the middle shoreface, in places associated with ebb-tidal deltas, and, because of transgression, may be stranded on the shelf as shelf ridges with heights of 6–9 m, lengths of 8–14 km and widths of 1–2 km (Rine *et al.*, 1991). In the relatively sand-starved Gulf of Mexico, approximately shore-parallel shelf sand ridges are present in water depths of 20–35 m and are 30–120 km long, < 6 km wide and ~4 m thick (McBride *et al.*, 1999). In the Gulf of Lions (Mediterranean Sea), which has relatively low wave energy, Berné *et al.* (1998) described late glacial lowstand delta-shoreface sandstones, erosively overlain by an elongate transgressive sandbody about 1 km wide. The transgressive sandbody lies in about 90 m of water and comprises dunes (now moribund) up to 10 m high. In all these examples, and others, the shelf sand ridges lie immediately above a marine transgressive surface cut by landward translation of the shoreface during the Holocene transgression. The sand forming the ridges was originally supplied by underlying lowstand shoreface deposits, ebb-tidal deltas, or paleo-valley-fills, but has been more-or-less completely remolded by both storm- and tide-driven currents (Snedden and Dalrymple, 1999; *see* Chapters 9 and 11). Although storms repeatedly move sand over shelf ridges, and their internal structure and external morphology continues to evolve, there is no clear evidence that shelf ridges have nucleated and grown in an original offshore position.

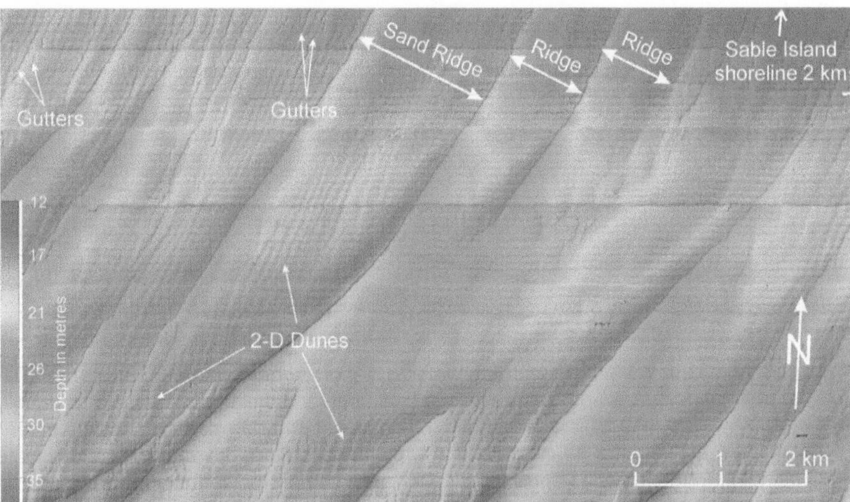

Figure 2. Shaded-relief multi-beam image off the SW coast of Sable Island (Nova Scotia shelf) showing sand ridges (heights 1–13 m, mean about 5 m), and superimposed 2-D dunes (heights 10 cm – 4 m, mean about 50 cm). Large wave ripples with a wavelength of ~1 m (similar to those shown in Figure 13), are mostly developed in coarser sand and gravel in dune troughs, but these smaller features are not imaged at this scale. Cores show that storms have moulded the crossbedding in the top 30–50 cm of dunes into low-angle lamination interpreted as HCS and SCS. Steep-sided, shore-normal erosional gutters 20–60 cm deep are present mainly in water < 20 m deep (range 10–40 m). After Li and King, 2007; image courtesy of Michael Li.

PHYSICAL PROCESSES ON THE BEACH AND UPPER SHOREFACE

Sediment transport on the beach and upper shoreface is driven by waves, which generate both oscillatory currents through the orbital motion of water under the wave, and alongshore and offshore-directed rip currents generated as waves shoal and break. Good reviews of these processes are given by Komar (1976) and Clifton (2006). As waves enter shallow water, the circular, deep-water motion of water particles is distorted to an elliptical pattern that becomes a back-and-forth motion at the seabed (Fig. 3). During fair weather, the landward stroke of waves as they cross the shoreface tends to be stronger than the seaward stroke, resulting in a net landward drift of coarser sediment toward the beach.

As waves enter water that is ~1.3 times the wave height, the frictional drag of the seabed causes the wave crest to topple forward as a breaker in the surf zone, and some of the energy of the wave is translated into a landward rush of water up the beach - the swash. The swash is capable of moving large pebbles that tend to accumulate at the top of the beach, and is fast enough to produce plane lamination in sand (Fig. 4). The landward translation of water in the surf zone drives two main current systems: longshore currents driven by waves that approach the beach obliquely, and a cell circulation of rip- and longshore currents (Komar, 1976; Fig. 5). Rip currents, which can be very powerful, flow seaward through narrow channels (typically 20–40 m wide, 1–2 m deep) cut in the upper shoreface, which is the zone in which waves are breaking (Fig. 3). Rip currents are capable of transporting sand and gravel (Hunter *et al.*, 1979), but they die out sea-

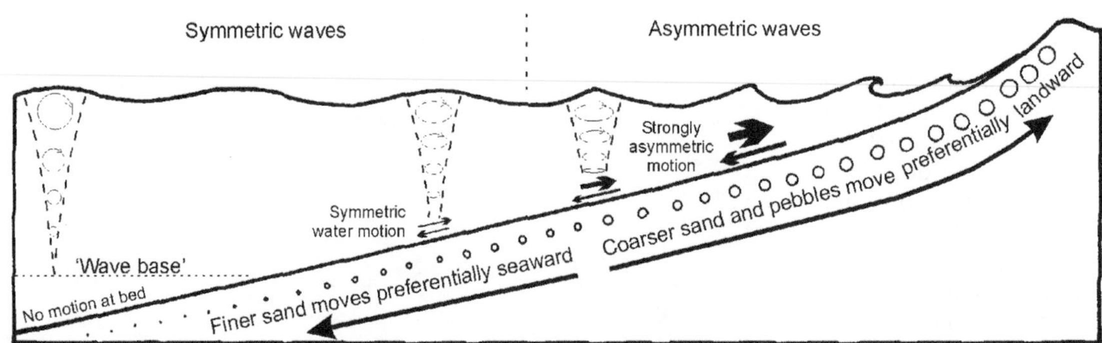

Figure 3. Behavior of waves in shoaling water. Offshore, where water depth is greater than approximately half the surface wavelength, waves are not capable of moving the seafloor sediment, and the seafloor is said to lie below 'wave base'. (Note that 'wave base' is highly variable, dependent on surface wavelength and sediment grain size!) As waves enter shallower water above wave base, water particles at the seabed move back and forth in a symmetrical pattern driven by the orbital water motion under the wave. Closer to shore, the oscillation pattern becomes strongly asymmetric, with a short, powerful landward surge under the wave crest, and a longer, weaker seaward flow under the wave trough. This flow asymmetry preferentially drives coarser sediment landward and finer sediment seaward, resulting in an upward-coarsening facies succession. Modified from Clifton (2006).

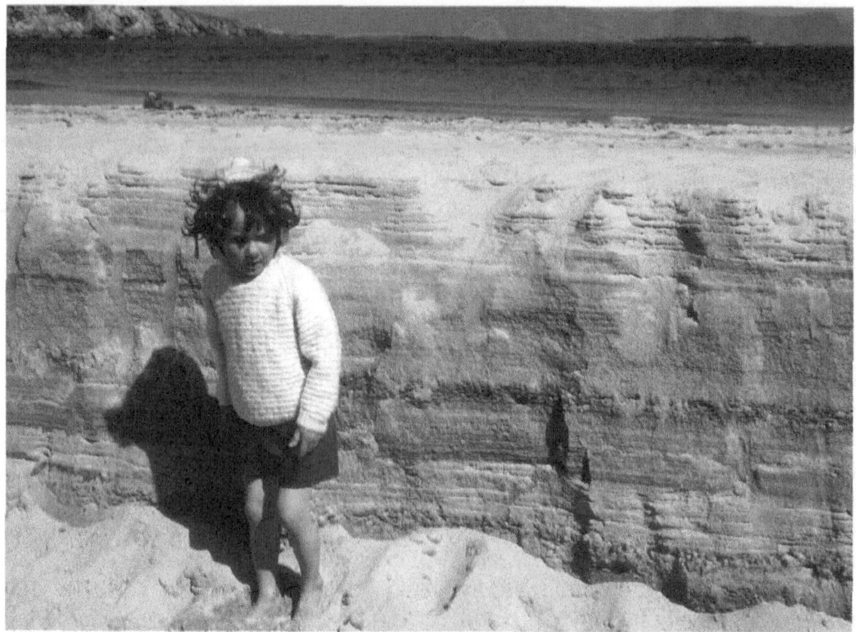

Figure 4. Strike view of planar lamination, picked out by heavy mineral laminae in fine sand, created under upper flow regime conditions produced by wave swash on the foreshore. This lamination dips gently seaward, at the inclination of the beach face. Keem Bay, Co. Mayo, Ireland.

ward of the surf zone and do not transport sand or gravel more than a few hundred meters offshore.

STORM-INDUCED CURRENTS ON THE SHELF

Although storm conditions exist for only a small portion of the year at any particular point on a shelf, their effects in terms of erosion and sediment transport are disproportionately large. Storm winds are responsible for two main types of current on the shelf: relatively slow-moving unidirectional, coast-parallel to coast-oblique geostrophic flows that result from wind stress on the sea surface, and fast-moving, oscillatory flows due to wave motion propagating to the seabed (Swift *et al.*, 1986).

During a storm, frictional coupling between the wind and the sea surface causes water to move downwind. The moving surface water entrains successively deeper layers, each of which is deflected to the right (in the northern hemisphere; left in the southern hemisphere) by Coriolis force. The net effect of Coriolis deflection is to cause the bulk of the shelf water mass to move at as much as 90° to the right of the wind direction. Water thus driven toward the shore causes the sea surface to rise, sometimes by as much as several meters, forming a coastal set-up (also known as a 'storm surge'). The elevated sea surface in turn generates a hydraulic head which drives a bottom current that flows seaward down the pressure gradient (Fig. 6). This bottom current is also deflected to the right by the Coriolis force, resulting in a geostrophic flow that moves obliquely offshore. The geostrophic flow becomes more shore-perpendicular if the hydraulic gradient steepens (because wind strength increases), but if the hydraulic gradient diminishes, Coriolis force exerts a proportionately greater influence and the flow swings more shore-parallel. A similar geostrophic flow, but in the opposite direction, will develop if water is blown offshore. Geostrophic flows persist as long as there is a wind-driven pressure gradient, typically lasting for a few days in the case of an Atlantic winter storm (Swift *et al.*, 1986). It is important to appreciate that geostrophic flows are a continuous response to a pressure gradient and may last for many hours or even days, and are not the result of a short-lived 'storm-surge ebb' related to the end of storm winds, or to rapid drainage of water from coastal lagoons after the storm has passed.

Observations on Modern Shelves

Storm-dominated shelves are influenced by various types of current, but the most important, in terms of sediment transport, are geostrophic flows and wave-induced oscillatory flows; both types of current typically operate together during storms (*e.g.*, Swift *et al.*, 1986; Duke, 1990; Nittrouer and Wright, 1994). The resulting combined flows are very effective in moving bedload because the oscillatory currents, which may reach several meters/second in the ~10-cm thick wave boundary layer just above the seabed (Fig. 6), very effectively throw sediment into suspension, allowing it to be carried down-current by the lower velocity, unidirectional geostrophic flow.

Storm-generated flows have now been measured on many shelves. For example, geostrophic flows resulting from northeasterly winter storms on the middle Atlantic continental shelf of the United States generate near-bottom velocities of up to about 0.7 m/s in depths of 10–20 m, with flow directed both along-shelf and obliquely offshore (Swift *et al.*, 1981; Swift and Field, 1981). In the Gulf of Mexico, tropical storms and hurricanes generate flow velocities close to the seabed of up to about 3 m/s (Snedden *et al.*, 1988; Snedden and Nummedal, 1991). Figure 7 shows an example of sediment transport on the Texas shelf as a result of a moderate storm, which generated waves of up to 9 second period and 2 m height; the resulting combined flows were capable of moving fine sand in depths of up to about 70 m. On the Bering Shelf, Nelson (1982) found that storm-driven combined flows had moved fine

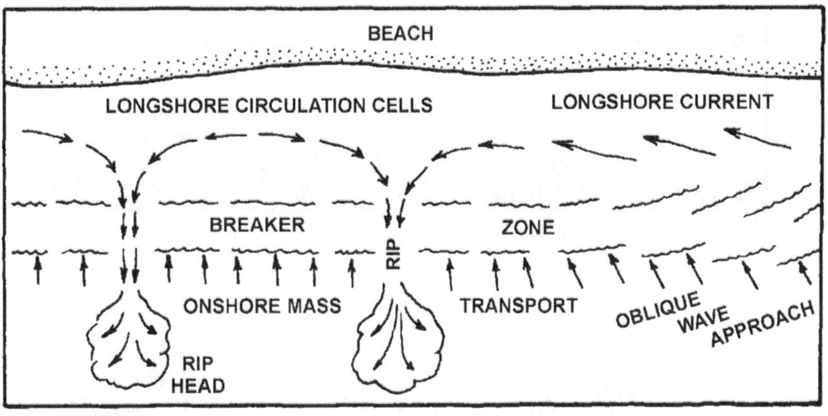

Figure 5. Water circulation system in the nearshore area. As waves cross the shoreface and enter the breaker zone, water moves landward, piling up against the beach. Circulation cells (left side of diagram) carry water and sand seaward in narrow rip-current systems that carve shallow channels across the upper part of the shoreface. Rip currents rapidly dissipate, and offshore sand transport ceases seaward of the breaker zone. If waves approach the beach obliquely (right side of diagram), the longshore current system tends to flow in one direction and is capable of moving huge volumes of sand along the coast. Based on Komar, (1976).

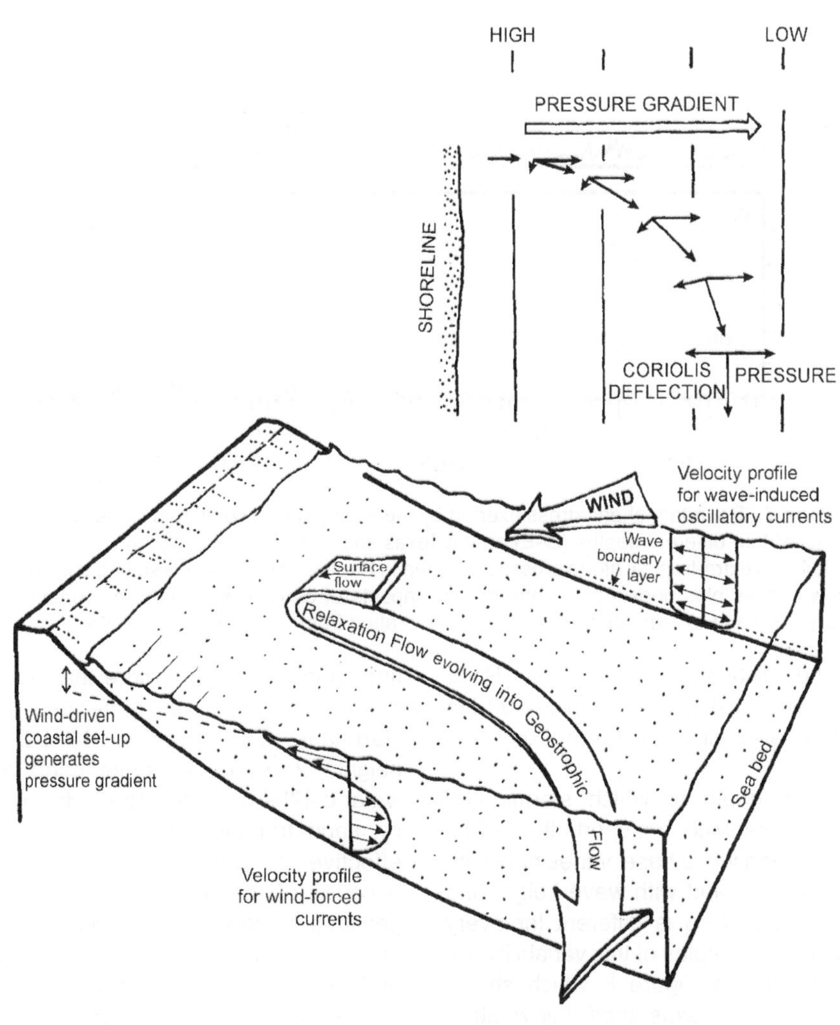

Figure 6. The storm-driven shelf current system. Onshore winds drive surface water landward, creating a coastal set-up ('storm surge'). The resulting hydraulic pressure gradient (upper diagram) drives lower layers of water seaward as a relaxation flow. Coriolis force deflects the relaxation flow to the right (in northern hemisphere, left in southern) resulting in a coast-parallel to coast-oblique geostrophic flow. Wave-driven oscillatory currents suspend sandy sediment in the thin wave boundary layer, and, in combination with the weaker but unidirectional geostrophic current, tend to cause sand grains to migrate gradually along- and offshore through the process of combined flow. Based on Swift *et al.* (1986) and Duke (1990).

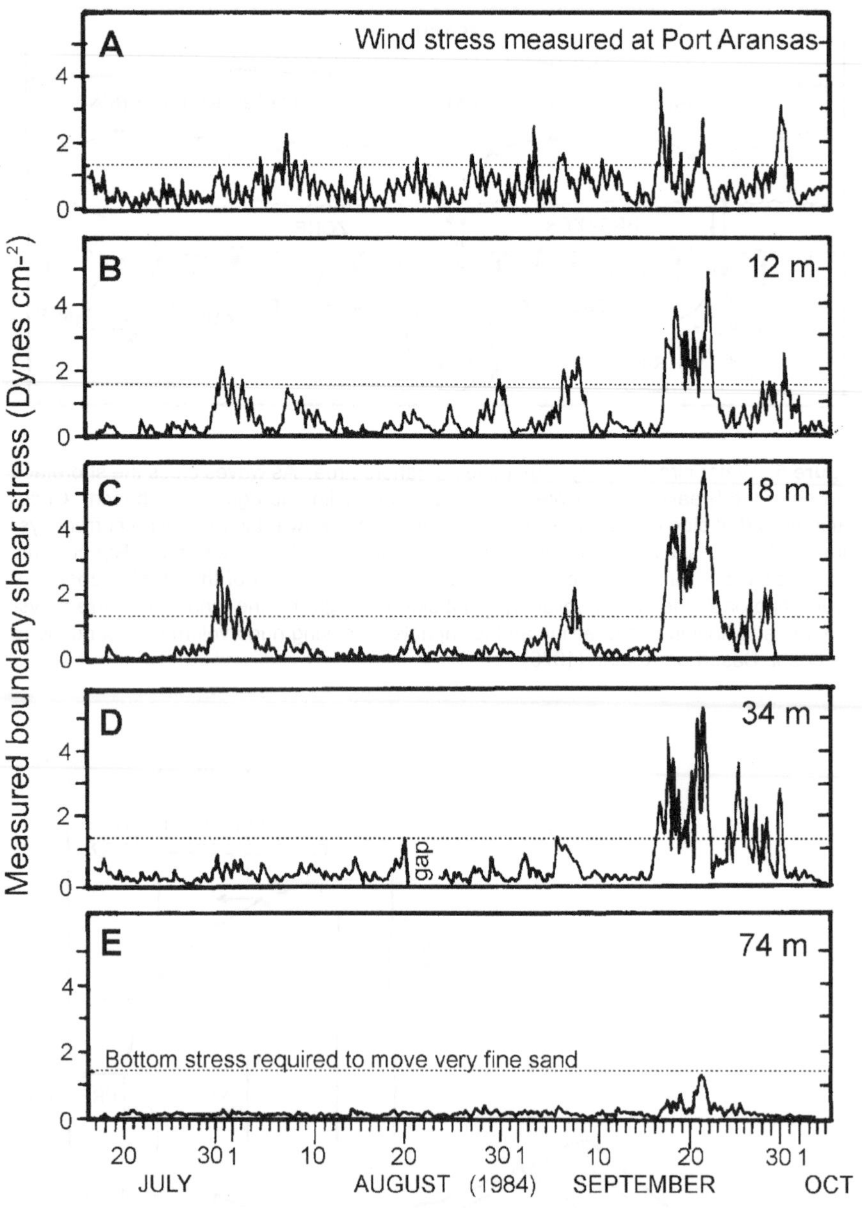

Figure 7. Boundary shear stress exerted by the wind **(A)** and combined waves and currents at four different depths **(B–E)** on the Texas shelf between 16th July and 4th October, 1984. An extra-tropical storm crossed the area in September, and the resulting combined flow (waves plus geostrophic current) was capable of causing significant movement of very fine sand to a depth of nearly 70 m. Re-drafted from Snedden *et al.* (1988).

sand offshore from the Yukon River delta for a distance of up to 100 km over a shallow platform 20–30 m deep.

The depth to which storms can cause bedload transport, the 'effective storm wave base' varies both with grain size, and with wave height and period, and so is different for every storm. Examples of this variability are illustrated in Figure 8 which shows that, on the Texas shelf, fair-weather waves (period 6 sec., height 0.5 m), are calculated to be able to move very fine sand in water no more than 7 m deep, whereas waves generated by Hurricane Carla (period 14 sec., height 12 m) could move sand at water depths down to 115 m. In reality, sand transport during storms is effective at even greater depths because of the added affect of the geostrophic flow. On the west coast of the United States, the shelf is steep and narrow, and the huge fetch afforded by the Pacific Ocean allows very large storm waves to develop. In consequence, shelf sediments are very strongly influenced by combined flows, which have been measured at > 1 m/s close to the seabed at a depth of 50 m (only ~6 km offshore), where the sediment consists of well-winnowed, fine to very fine sand (Caccione *et al.*, 1999).

Hurricane Carla, one of the most powerful recorded storms in the Gulf of Mexico, crossed the Texas shelf in September 1961 and generated a storm bed that subsequently (1984–5) was sampled in 23 box cores (Snedden and Nummedal, 1991). These cores revealed a bed of sand distributed along at least 200 km of the shelf to the SW of the hurricane track (Fig. 9). This storm-deposited sand bed was thickest (7 cm) and coarsest (very fine sand) on the inner shelf and thinned and fined seaward to coarse silt. The base of the bed was sharp and slightly erosive, and typically showed an upward succession from plane-parallel lamination or massive sand to gently inclined lamination having a gradational, bioturbated upper transition to overlying mud (Snedden and Nummedal, 1991). Core cross-sections of the Carla storm bed show that it maintained a fairly uniform thickness parallel to shore, but thinned rapidly in water more than about 30 m deep, with an eventual pinch out somewhere seaward of the 75 m isobath about 50 km from shore (Fig. 10). Although bottom currents were not measured during Hurricane Carla, comparison with other storms, and model calculations, indicate that the Carla storm bed was generated by a southwesterly directed, along-shelf and obliquely offshore-directed combined flow with near-bottom current velocities that approached 3 m/s. On the Bering shelf, Nelson (1982) documented storm beds that consisted of fine to very fine sand that thinned from ~20 cm inshore to 1 cm ~100 km offshore. Other examples of storm-generated, along- and across-shelf flows are summarized by Forristal *et al.* (1977) and Morton (1988). It is important to bear in mind that burrowing organisms may greatly modify storm deposits. For example, the sharp-based sand deposited by Hurricane Alicia (1983) in the Gulf of Mexico, was completely homogenized by burrowing organisms within a few months (Morton,1988).

Although combined flows are recognized as the principal sediment-transporting processes operating on storm-dominated shelves, Myrow and Southard (1996) pointed out that gravity can potentially also play a significant role in transporting sediment downslope, even across shelves. Although shelves slope too gently to initiate or support classical turbidity currents through the process of autosuspension, Myrow and Southard (1996) argued, from geological observations, that to explain the thickness of some ancient storm-deposited sandstone beds, it was necessary to infer that 'conventional' combined flows had been supplemented to some degree by density-driven offshore flow. Subsequent observations on modern shelves (*e.g.*, Traykovski *et al.*, 2000; Wright *et al.*, 2001; Wright and Friedrichs, 2006; Friedrichs and Scully, 2007), showed that waves are capable of generating dense suspensions of mud in the thin wave boundary layer, which, under the influence of gravity, will flow downslope. Observations by Wright *et al.* (2002) suggested that fine sand could also move seaward as a wave-supported density flow, but *only* over the steep inner shelf within a few kilometers of shore. As emphasized by Pattison *et al.* (2007), there is as yet, no evidence of wave-supported density flows moving significant volumes of even very fine sand across modern shelves, although various Cretaceous shelf sandstone beds have recently been interpreted as 'shelf turbidites' (Edwards *et al.*, 2005a; Pattison *et al.*, 2007).

Despite the very significant progress that has been made in our

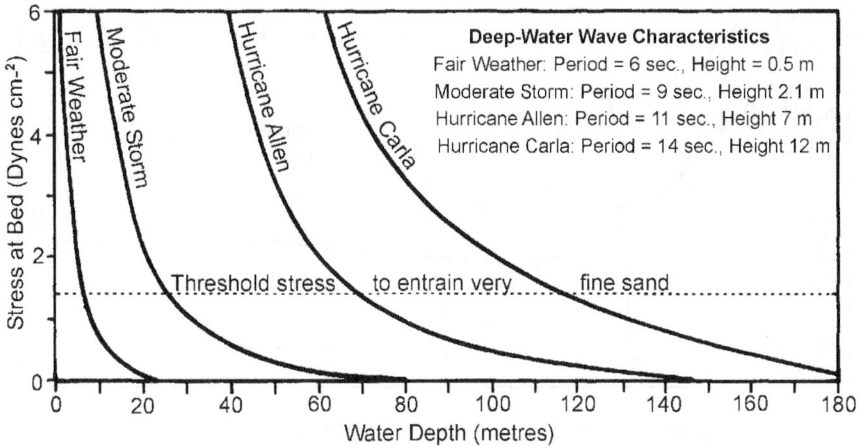

Figure 8. Calculated effectiveness of waves in causing entrainment of very fine sand. This diagram shows that 'wave base' depends on the period (wavelength) and height of waves. Thus 'fair weather' waves can entrain very fine sand down to about 7 m depth, whereas the 14 second, 12-m-high waves generated by Hurricane Carla (September 1961) were capable of moving very fine sand to a depth of over 115 m. In reality, sand transport operates to a deeper depth than indicated by these curves because of the added stress caused by the geostrophic component of flow. For example, the 'Moderate Storm' curve represents the *calculated* effectiveness of waves in the September 1984 storm (Fig. 7), predicting sand movement to about 26 m. However, actual measurements (Fig. 7, panels C, D) showed that combined flows could cause vigorous sand movement at 34 m and incipient movement down to about 70 m depth. Effective wave base would be shallower for coarser sediment. Redrafted from Snedden *et al.* (1988).

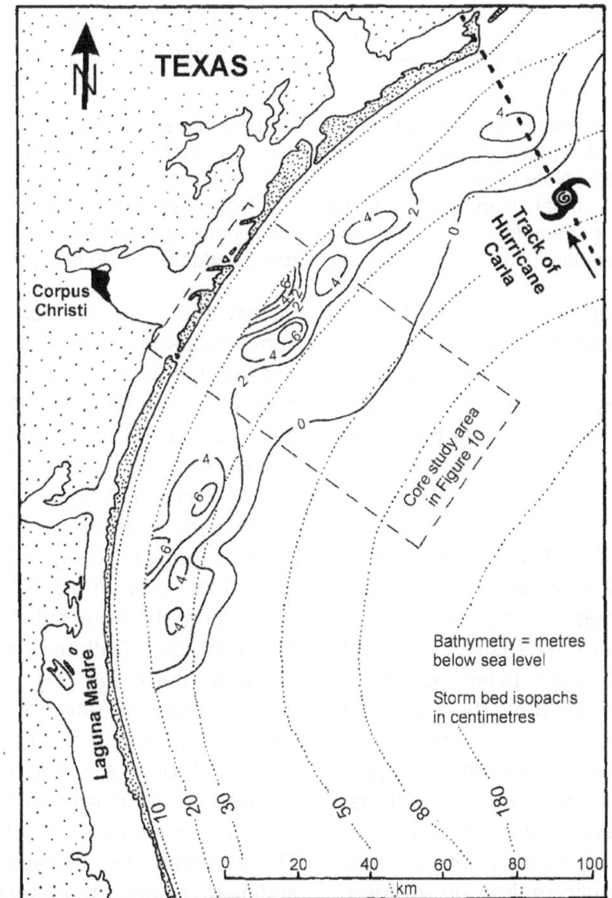

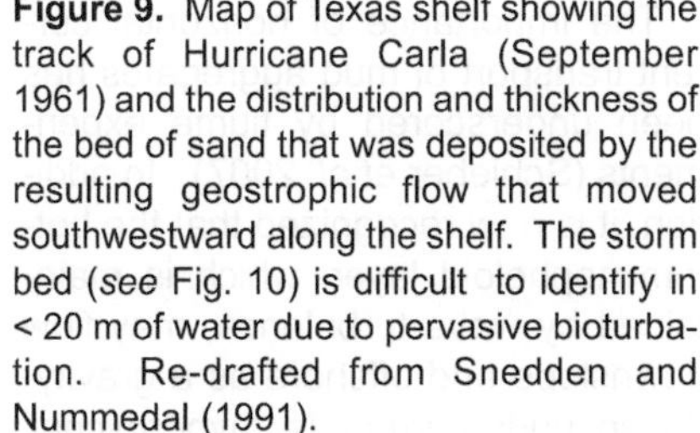

Figure 9. Map of Texas shelf showing the track of Hurricane Carla (September 1961) and the distribution and thickness of the bed of sand that was deposited by the resulting geostrophic flow that moved southwestward along the shelf. The storm bed (*see* Fig. 10) is difficult to identify in < 20 m of water due to pervasive bioturbation. Re-drafted from Snedden and Nummedal (1991).

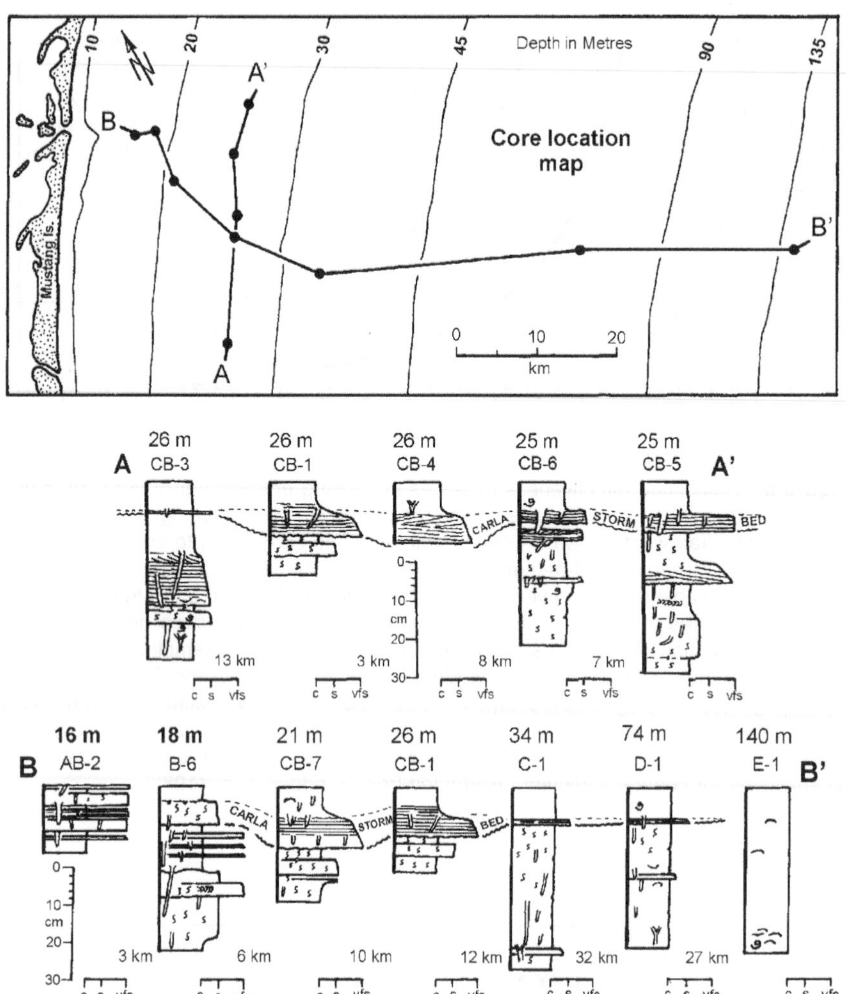

Figure 10. Representative strike (A) and dip (B) core cross-sections (located in sketch map above, and in Figure 9) showing the thickness and character of the Hurricane Carla storm sand bed. Note relative continuity of storm bed along isobaths (A), and seaward thinning and eventual disappearance beyond about 75 m water depth (B). The base of the sand bed is sharp, locally scoured, and the dominant sedimentary structure is parallel to gently undulating lamination. Re-drawn from Snedden and Nummedal (1991).

understanding of shelf sediment transport, there still appears to be some disparity between the processes and deposits observed on modern shelves, and those inferred to have operated on ancient shelves; evidently, more research is necessary!

TRANSPORT AND DEPOSITION OF MUD ON SHELVES

Modern shelves, having been recently flooded, are extensively mantled with coarse-grained 'palimpsest' sediments deposited during the preceding glacioeustatic lowstand, and hence the shelf surface has not yet equilibrated with present conditions. Where modern sediments are accumulating in equilibrium with a high-energy hydraulic regime on a steep, narrow shelf (*e.g.*, California), mud is concentrated on the mid shelf, and passes landward into sandy sediments (Ogston *et al.*, 2000; Dunbar and Barrett, 2005). Where wave energy is lower (*e.g.*, western Adriatic coast; western margin of the East China Sea), mud may accumulate in a shore-parallel body hundreds of kilometers long, molded by permanent along-shelf currents (tidal, thermohaline, *etc.*; Cattaneo *et al.*, 2007; Liu *et al.*, 2007; Fig. 11). Thus, shelf seas can be very efficient at trapping terrigenous mud, in a variety of ways, such that only a small percentage escapes to the deep sea (unless the shoreline is close to the shelf edge). The evolution of ideas regarding shelf sediment transport, especially mud, are reviewed by Hill *et al.* (2007), on which the following brief summary is based.

Numerous case studies show that mud entering shelf seas is rapidly lost from buoyant ('hypopycnal') riverine plumes of freshwater, which, because of Coriolis deflection and coastal currents, rarely extend seaward more than 10–20 km before they are deflected coast-parallel. Suspended silt and clay particles settle rapidly to the sea bed, typically < 10 km from the river mouth because the grains are 'repackaged' into larger aggregates through three different processes: saline water promotes particle aggregation by electrochemical coagulation; biogenic aggregation occurs when filter-feeding organisms concentrate mud particles into fecal pellets; and flocculation results in the bonding of mineral particles by dissolved organic molecules. The relative importance of these three processes varies with environmental setting, water chemistry, climate, *etc.*, and the overall relative importance of these three processes is not known. Nevertheless, in many environments, aggregate particles settle at about 1 mm/s and accumulate at the seafloor as a turbid bottom nepheloid layer (*i.e.*, a near-bed layer of increased suspended-sediment concentration).

Although mud aggregates initially settle within ~10 km of river mouths, mapping shows that mud belts may extend very much farther along- and across-shelf. This is explained in terms of the resuspension of mud by waves, particularly during storms. The resuspended mud may then be carried horizontally ('advected') farther along-shelf and across-shelf by geostrophic flows, and by long-term coastal currents, building a large prismatic mud wedge characterized by seaward-dipping clinoforms inclined at up to about 1° (*e.g.*, Nittrouer *et al.*, 1996; Cattaneo *et al.*, 2003, 2007; Liu *et al.*, 2007; Fig. 11).

The importance of horizontal current transport of mud aggregates has been underscored by flume experiments (Schieber *et al.* 2007). In addition, it is now recognized that the bottom nepheloid layer, which is maintained by wave turbulence, may flow downslope and offshore as a gravity-driven undercurrent (a 'hyperpycnal'

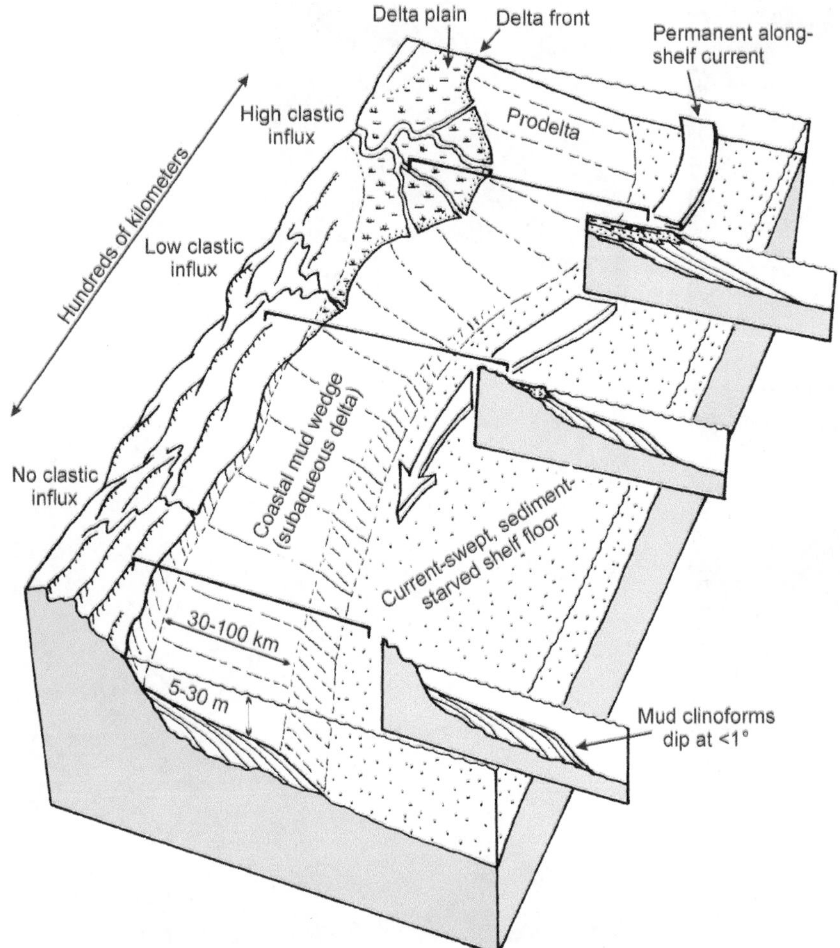

Figure 11. Generic model for the accumulation of a shore-parallel mud wedge down-drift from a major river mouth. Mud is repeatedly re-suspended by storms and carried along the shelf by geostrophic flows, and, importantly, by permanent along-shelf currents. The resulting mud wedge may be hundreds of kilometers long and extend seaward for over 100 km. The top of the mud wedge is limited by ambient wave energy and typically lies in 20–30 m of water. The seaward margin of the wedge accretes laterally forming large-scale mud clinoforms that downlap onto a sediment-starved shelf mantled with a veneer of transgressive sediment, commonly shelly sand and gravel. Diagram based mainly on Cattaneo *et al.* (2003, 2007, western Adriatic) and Liu *et al.* (2007, East China Sea) but applicable to many other mud-rich shelves (*e.g.*, Amazon, Fly).

flow). For example, Ogston *et al.* (2000) and Traykovski *et al.* (2000) found that when the Eel River (California) was in flood at the same time that the shelf was stirred by storm waves, a 50-cm-thick layer of fluid mud was observed in 60 m of water, flowing offshore at up to 30 cm/s. The hyperpycnal flow finally died out in water too deep for waves to maintain the bottom nepheloid layer.

Even more extreme concentrations of fluid mud having sediment concentrations of 10–400 g/l have been observed in some estuaries and off the mouths of major rivers such as the Amazon, particularly during times of high river discharge. In the Amazon, fluid mud 1–2 m thick forms in the river mouth through tidally enhanced mixing of fresh and salt water (Kineke *et al.*, 1996). About 20% of this mud is carried alongshore by a NW-flowing coastal current where it accumulates in seaward-prograding mud banks and tidal flats. The concentration of mud, and hence the viscosity of the water, are so high that waves from the South Atlantic are almost completely damped out (Wells and Coleman, 1981; Rine and Ginsburg, 1985; Nittrouer *et al.*, 1996). The bulk of the Amazon mud is transported obliquely northward across the shelf to build a subaqueous mud delta in 15–70 m of water. The top surface of the delta is defined by the depth of wave resuspension and the clinoform dips seaward at up to 1:200, downlapping onto a relict sand deposit (Nittrouer *et al.*, 1996; Fig. 11).

Muddy Sediments

Mud-dominated coasts and shelves, as exemplified by the NE coast of Brazil and Suriname, are characterized primarily by thick (meter-scale) units of homogeneous mud, commonly deposited during river floods, punctuated by faint parallel, wavy and lenticular laminations of silt or very fine sand. Fluid mud inhibits colonization by infauna and hence burrows are rare. However, seasonal episodes of wave or current scour expose more consolidated mud which is then colonized by burrowing fauna, producing a succession of massive muds punctuated by scoured and burrowed surfaces (Rine and Ginsburg, 1985; Kuehl *et al.*, 1996). Sand is rare and confined to storm-related laminae on the inner shelf, and to thin, narrow chenier ridges at the shoreline that form during times of mud starvation and coastal erosion. In deeper water, on the subaqueous delta clinoform, bioturbation becomes more abundant at the expense of laminated mud as the sedimentation rate diminishes. Good examples of some ancient mud-dominated shallow-marine successions are given in Schieber *et al.* (1998).

Recent experimental studies of mud deposition (Schieber *et al.*, 2007), coupled with observations of ancient muddy-shelf deposits (*e.g.*, MacQuaker and Bohacs, 2007; MacQuaker *et al.*, 2007; Varban and Plint, 2008a; Plint *et al.*, 2009), are leading to the realization that much mud deposition does not take place passively through vertical settling of particles, but instead is a more dynamic process, involving long-distance horizontal transport of mud aggregates, and a complex history of depositional, erosional and biotic colonization events, commonly in water much shallower than traditionally envisioned. It appears that mudrocks are undergoing their own 'facies revolution'!

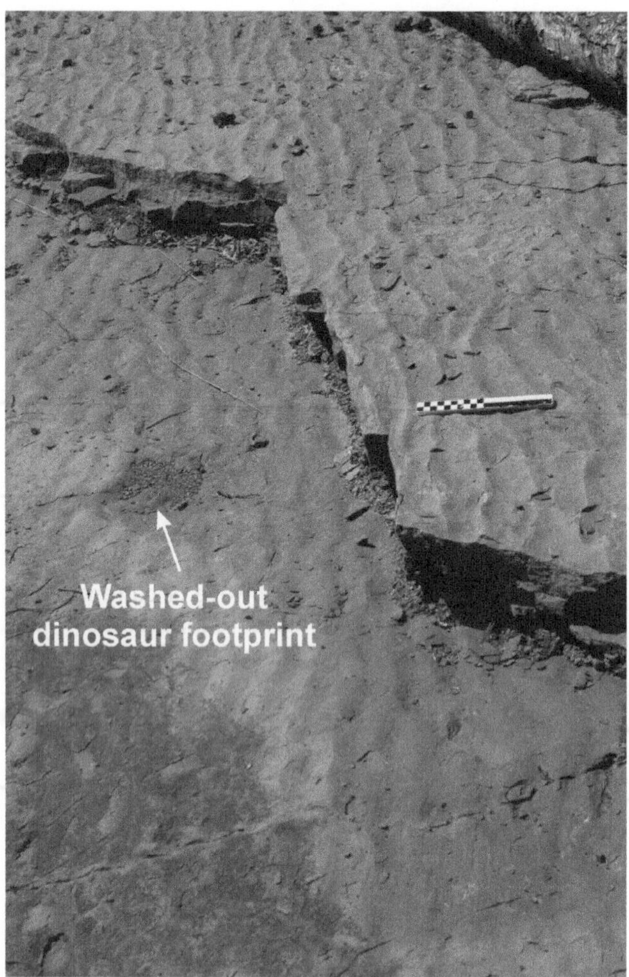

Figure 12. Examples of symmetrical, straight-crested small oscillation ripples from an Upper Cretaceous beach deposit. Doe Creek unit (Cenomanian), Flatbed Creek, BC. Scale bar = 20 cm.

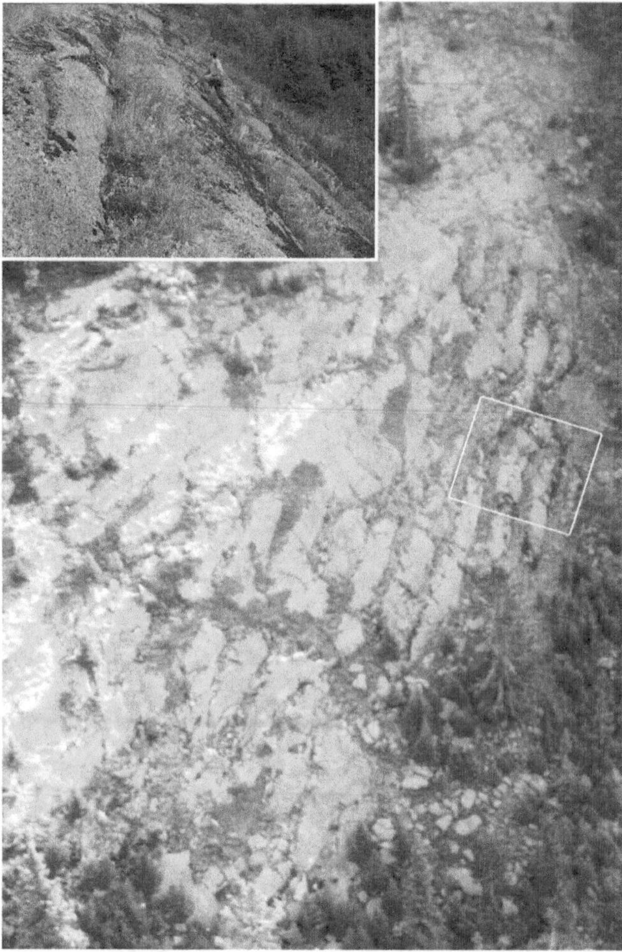

Figure 13. Oblique aerial view of large oscillation ripples in fine gravel mantling a marine ravinement surface at the base of the Shaftesbury Formation (Albian) at Deadhorse Meadows, Alberta. Wavelength of ripples in lower right is ~ 2.5 m and height 30–40 cm. Inset shows ripples seen at ground level.

SEDIMENTARY STRUCTURES IN STORM-DEPOSITED SANDSTONES

The physical sedimentary structures that form in sandy shoreface to shelf sediments are generated primarily by combined flows, embodying both longer term (hours–days) unidirectional currents, and short-term (seconds) oscillatory currents. The nature of these flow types is very different. The boundary layer beneath a unidirectional flow is relatively thick (meters to tens of meters) with a gentle velocity gradient, whereas an oscillatory boundary layer is very thin (< 10 cm) because it is formed and destroyed on a timescale of seconds with each back-and-forth stroke of the waves (Swift *et al.*, 1986; Fig. 6). Thus, storm waves, with their large orbital diameters, can generate high oscillatory velocities (> 50 cm/s) only a few centimeters above the bed; in fine sand, such velocities are capable of generating plane-bed conditions at the expense of dunes or large ripples (Clifton, 1976).

The most important sedimentary structures found in shelf deposits are various types of small wave (oscillatory) ripples (Fig. 12), large wave ripples (Fig. 13), hummocky cross-stratification (Fig. 14), swaley cross-stratification (SCS), and planar lamination (Fig. 15). In addition to these primarily wave-formed structures, subaqueous dunes (Fig. 2), forming dm-scale crossbedding are generated by combined flows with a strong unidirectional component (Dumas *et al.*, 2005; *see* also Aigner (1985) for good illustrations of storm-formed sedimentary structures, and Cheel and Leckie (1993) for a good review of HCS).

Influence of Grain Size

It is important to appreciate that the types of sedimentary structure that develop on a clastic shelf are dependant not only on the hydraulic conditions but also on the sediment grain sizes available (Clifton, 2006). For example, under unidirectional flows, it is not possible for subaqueous dunes to form if the grain size is less than ~0.15 mm (upper fine sand; Harms *et al.*, 1982). Similarly, under strong oscillatory flows, HCS is generally formed in coarse silt to fine sand (and rarely medium sand; *e.g.*, Amos *et al.*, 1996; Li and Amos, 1999; Li and King, 2007), whereas the same flows will produce large symmetrical oscillatory ripples in coarser sand and gravel (Cheel and Leckie, 1993; Dumas *et al.*, 2005; *see* Chapter 4).

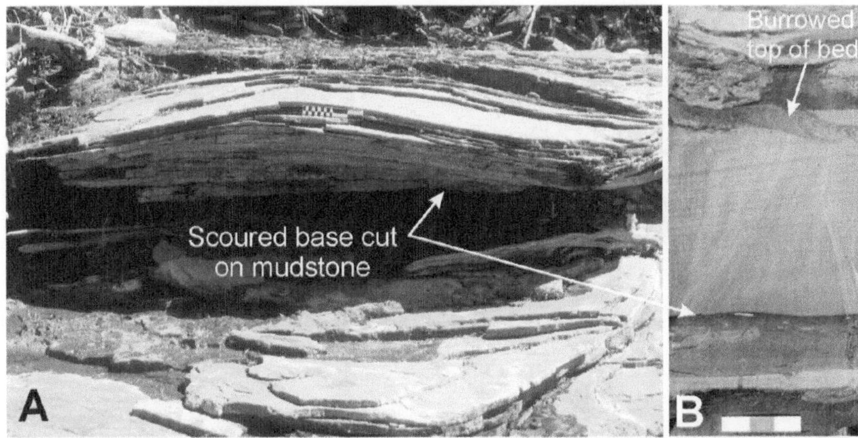

Figure 14. **A**. Transverse section through hummocky cross-stratification showing upward-arching laminae and subtle truncation on left side. This example is slightly asymmetrical, suggesting net migration toward the right. Marshybank Formation (Coniacian), Belcourt Creek, BC. Scale bar = 20 cm. **B**. HCS in core showing scoured base, subtly diverging laminae and burrowed top. From Cardium Formation (Turonian) in well 6-22-54-19W5, 1970 m, Alberta. Scale bar = 3 cm. **C.** Plan view of HCS bed shown in **A**, showing exhumed hummock and swale topography. Upper surface of hummocks is ornamented with strongly 3-D combined-flow ripples that record diminishing wave and current strength as the storm waned (*see* also Fig. 19). Marshybank Formation (Coniacian), Belcourt Creek, BC. **D**. Amalgamated hummocky cross-stratification in lower shoreface deposits. Note local preservation of cm-scale mud beds that record several episodes of post-storm mud deposition, alternating with erosion and HCS formation during storms. Chungo Member, Wapiabi Formation (Santonian), Thistle Creek, Alberta; scale bar = 20 cm.

Ripples and Hummocky Cross-Stratification

One of the best descriptions and illustration of ancient wave-rippled sediments is provided by de Raaf *et al.* (1977), on which Figure 16 is based. Experimental studies by Dumas *et al.* (2005) and Dumas and Arnott (2006) have greatly clarified the hydraulic conditions under which ripples and HCS form in very fine sand. In these experiments, the orbital diameter and velocity of the oscillatory current (U_o) and the velocity of the superimposed unidirectional current (U_u), were varied.

When oscillatory currents (U_o) were between ~20 to ~50 cm/s and the unidirectional component of flow (U_u) was small (< 10 cm/s), the stable bedforms were small symmetric ripples. They have a wavelength of about 10 cm, height about 1 cm, sharp continuous crest-lines and a symmetrical profile (Fig. 12). As U_u was increased to 10−25 cm/s, ripple wavelength and height increased somewhat, and the ripple crests became more rounded and broke up into short curved segments creating small asymmetric ripples that have a

Figure 15. **A**. Swaley cross-stratification in shoreface sandstone, showing multiple low-angle scours and upward-flattening lamina sets. Permian Pebbley Beach Formation, New South Wales. Australia; scale bar = 20 cm. **B**. Swaley cross-stratification in fine sandstone in core showing characteristic upward-flattening lamina sets. Cardium Formation (Turonian), in well 15-30-64-9W6, 1974 m, Alberta. **C**. Planar lamination in a storm bed consisting of upper very fine sandstone. Note lag of small mud chips at the base of the bed, marking a phase of erosion, and burrows descending from the top, marking a pause in deposition. Ben Nevis Formation (Aptian–Albian) in well North Amethyst K-15, 2354 m, offshore Newfoundland. Photo with permission of Husky Energy Ltd.

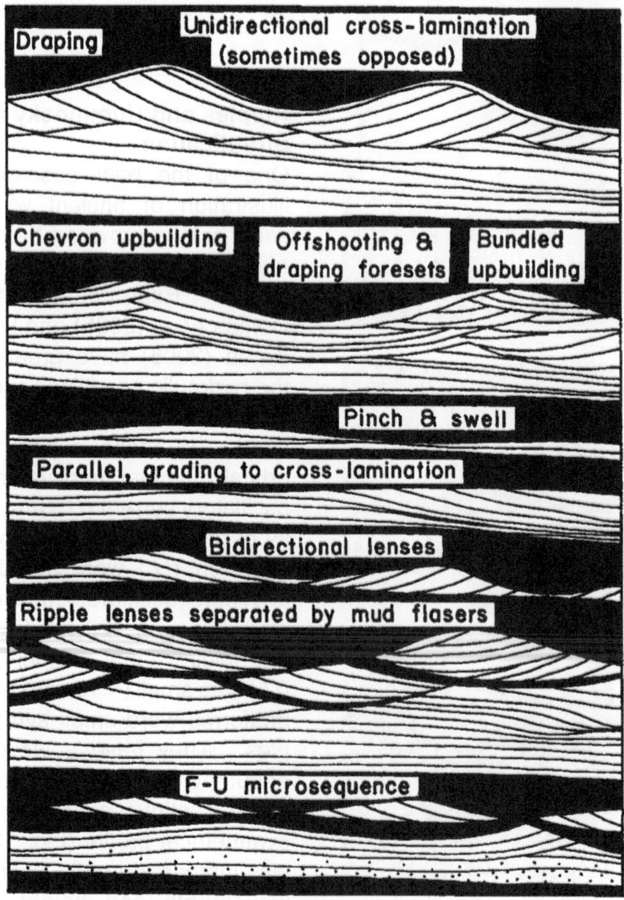

Figure 16. Sketch summarizing characteristic features of cross-lamination formed by small wave ripples. Modified from de Raaf *et al.* (1977).

3-D geometry; they are combined-flow ripples (Figs. 14C and 17).

When the oscillatory current U_o was increased to ~50–100 cm/s (to simulate large storm waves), but U_u was only 0–5 cm/s, large symmetric ripples with a wavelength of ~1–2 m and sharp discontinuous crests were the stable bedform. However, if U_u was increased slightly to 5–10 cm/s, the large ripples co-existed with rounded, dome-shaped symmetrical forms with a wavelength of 1.5–3 m and height of 10–25 cm that Dumas *et al.* (2005) termed 'hummocky bed forms'. A further slight increase in U_u to ~> 10 cm/s produced asymmetric large ripples which were quite strongly 3-D, with wavelengths of 1–> 4 m and an average height of ~20 cm. When oscillatory velocities were increased to ~90–120 cm/s, small and large ripple bedforms were washed out and replaced by plane bed.

The experimental results of Southard *et al.* (1990), Dumas *et al.* (2005) and Dumas and Arnott (2006) confirm what has long been surmised, that HCS is the product of large waves, commonly accompanied by weak currents (*e.g.*, Harms *et al.*, 1975; Duke 1985). However, HCS does not appear to be the product of a distinct bed phase but instead

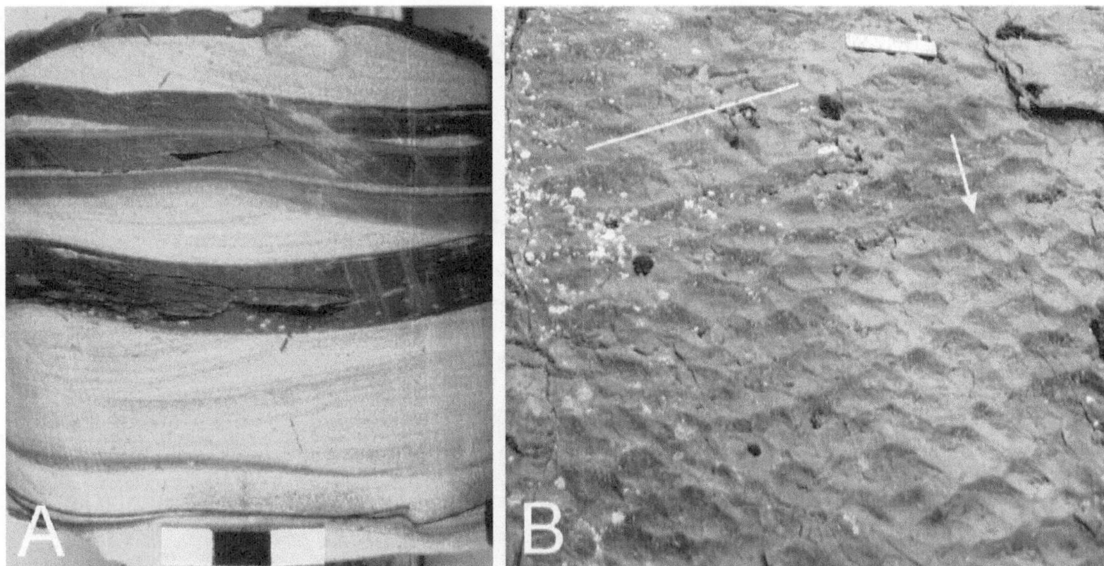

Figure 17. **A**. Combined-flow ripples seen in core. Note symmetrical external form of ripple, reflecting short-term oscillatory wave motion, but unidirectional dip of internal lamination indicates the net transport direction under the influence of a unidirectional flow. Cardium Formation (Turonian) in well 6-13-52-16W5, 1875 m, Alberta. Scale bar = 3 cm. **B**. Plan view of 3-D combined-flow ripples showing vaguely linear, wave-formed crest-lines (white line), but asymmetrical crests indicate preferential migration towards the viewer (arrow). Dowling Member of Wapiabi Formation (Santonian), Blackstone River, Alberta. Scale bar = 20 cm.

records transitional bed states between symmetric and asymmetric large ripples, and between asymmetric large ripples and plane bed. Non-migrating (isotropic) hummocky bedforms develop when the oscillatory current is > ~50 cm/s and the unidirectional component is < 5 cm/s. With an increase in U_u to 5–10 cm/s, hummocky bedforms migrate down current, producing anisotropic HCS which, with a further increase in U_u, can evolve into stratification resembling trough crossbedding. Thus HCS is most likely to form above storm wave base under combined flows having a strong oscillatory component but a weak unidirectional component, and where sediment aggradation rates are sufficient to preserve the hummocky bedforms. Isotropic HCS is predicted to evolve into anisotropic HCS closer to shore where unidirectional currents are stronger (Dumas and Arnott, 2006). Yang *et al.* (2006) have shown that the wavelength of HCS is related to wave orbital diameter, controlled by water depth, with hummock wavelength becoming shorter with decreasing water depth in the surf zone.

Swaley Cross-stratification

In shoreface deposits, it is commonly observed that interbedded mudstone and sandstone with HCS grades up into mud-free sandstone characterized by gently undulating fine lamination with local, shallow, broadly circular scours draped and filled with upward-flattening lamina sets (Fig. 15A, B). This structure was called swaley cross-stratification by Leckie and Walker (1982), who inferred from its geometric similarity to HCS, and its consistent stratigraphic position above HCS and below beach deposits, that it probably represented a storm wave-formed structure of the middle shoreface. The experimental results of Dumas and Arnott (2006) support the geologically based interpretation of SCS, indicating that this structure forms during storms under oscillatory-dominant combined-flow conditions. The preponderance of erosional swales over constructional hummocks is attributed to deposition in shallower water where the aggradation rate is lower and scouring more frequent. Dumas and Arnott (2006) predict that the greater strength of unidirectional currents in the upper shoreface would result in laterally migrating swales that would be recorded in the rock record as angle-of-repose cross-stratification, a structure conventionally attributed to migrating 3-D dunes.

Planar Lamination

Planar lamination (Fig. 15C) develops readily in fine to very fine sand under combined flows where the unidirectional component may only be a small fraction of the oscillatory component (Arnott and Southard, 1990). Planar lamination is typically developed immediately above the erosional base of a storm bed, and may grade up into HCS or wave ripples. This organization suggests that planar lamination develops during the early stage of sediment settling around the peak of the storm, when both the oscillatory and unidirectional components of flow were strongest (Cheel, 1991).

Dunes

Two- and three-dimensional dunes (Ashley, 1990; 'megaripples' and 'sandwaves' in older literature) are very common on modern shorefaces and sandy shelves where they develop in upper fine- to coarse-grained sand under fair-weather longshore and onshore currents (Clifton *et al.*, 1971; Hunter *et al.*, 1979). Migration of dunes produces trough and tabular crossbedding, typically a few dm in thickness. In shorefaces composed of fine sand, dune crossbedding is typically preserved only in the upper shoreface where longshore bars provide protection from waves.

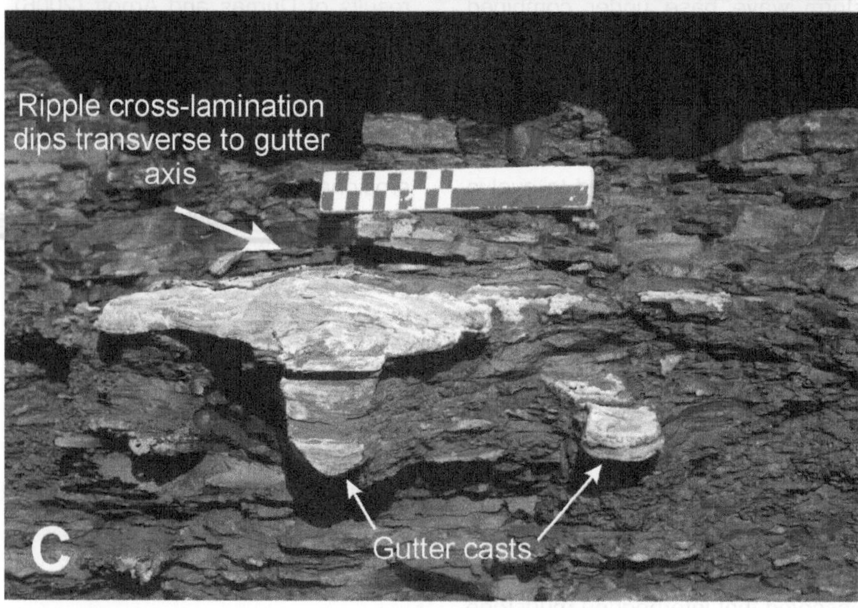

Figure 18. **A**. Transverse view of gutter cast in lower shoreface facies. Note steep to overhanging walls cut in mud and HCS in the gutter fill. Marshybank Formation (Coniacian), Muskeg River, Alberta. Scale bar = 20 cm. **B**. Longitudinal view of a gutter cast showing small flute-like scours on the walls, and undulating hummocky lamination in the fill. Vertical structures above scale bar (20 cm) are sand-filled dikes of uncertain origin. Doe Creek unit (Cenomanian), Flatbed Creek, BC. **C**. Small gutter cast in offshore mudstone facies. Gutter hangs below an isolated combined-flow wave ripple. Wave-ripple crest is parallel to the gutter axis, and ripple crosslamination dips to right. This structure might record seafloor erosion by an along-shelf geostrophic flow. Doe Creek unit (Cenomanian), Flatbed Creek, BC. Scale bar = 20 cm.

However, where medium- or coarser sand is available, crossbedding, commonly with complex orientations, will be widely developed across the shoreface (Wignall *et al.*, 1996; Clifton, 1981, 2006).

Large 2-D and 3-D dunes that have spacings of tens to hundreds of meters and heights of a few decimeters to a few meters are also widely developed on modern shelves where they indicate broadly along-shelf flows (Swift *et al.* 1979; Li and King, 2007; Fig. 2). These bedforms are active only during storms and tend to become mantled with mud and degraded by burrowing organisms during fair weather.

Gutter Casts

Gutter casts, which are elongate scour structures usually cut in mud and filled with sand, are typically 2–50 cm wide, 2–30 cm deep (although larger examples are not uncommon), and straight or gently sinuous. The cross-section is variable, with U or V shapes common. Walls are steep to undercut, and the ends may be steep or gently flared (Fig. 18A). The upper surface of the gutter cast is commonly ornamented with wave ripples oriented transverse to the long axis of the gutter cast, although in other cases, gutter and ripples are near parallel (*e.g.*, Kerr and Eyles, 1991). Excellent descriptions of gutters with plane laminated to massive fills are given by Myrow (1992) who interpret-

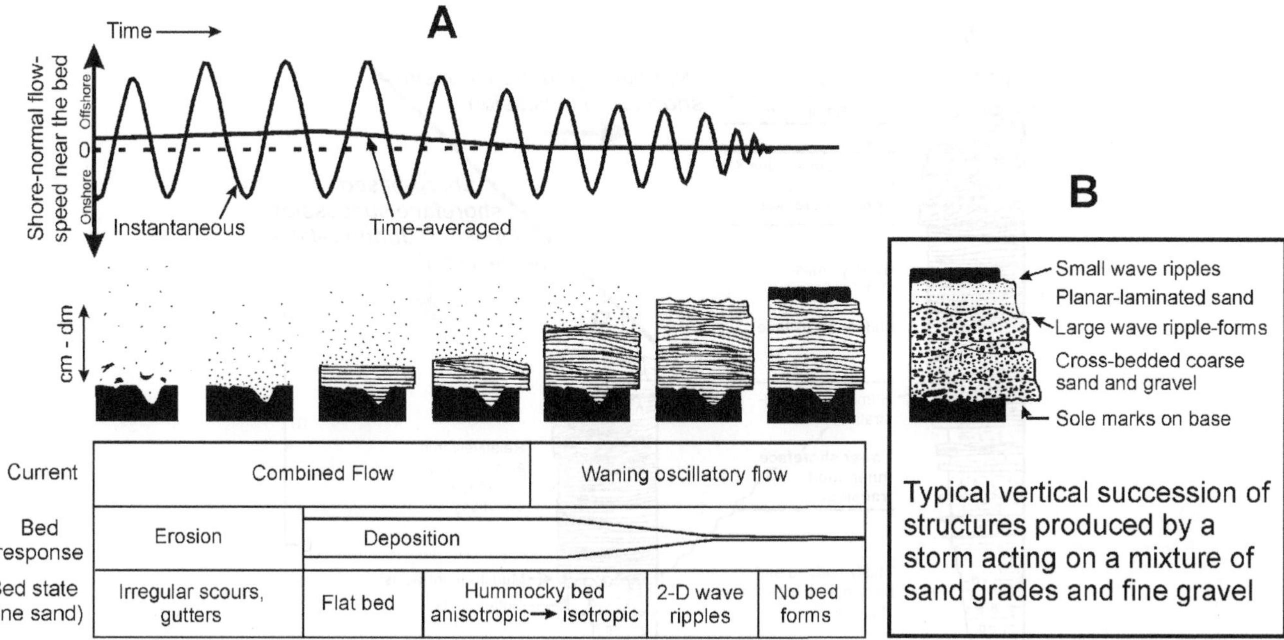

Figure 19. A. The development of an idealized event bed in fine sandstone as a result of storm-generated combined flow. High-frequency, wave-generated oscillatory currents are dominant, but unidirectional geostrophic flow during the peak of the storm provides a net offshore component. During the rising phase of the storm, sediment is suspended and the muddy bed is eroded, forming a variety of sole marks and gutters. As the storm starts to wane, initially planar-laminated sand is deposited under powerful combined flow, but this evolves to HCS, which initially may be anisotropic due to the influence of the unidirectional flow component. As the storm wanes, continued sediment settling under largely oscillatory flow produces isotropic HCS, eventually mantled by small wave ripples. **B.** The results of the same storm conditions when the substrate is a mixture of sand and fine gravel. Dunes, both 2-D and 3-D, migrate during strong combined flow to produce crossbedding, which may be of unusually low inclination due to the affect of superimposed wave motion. Large symmetrical wave ripples and low-angle to planar-laminated sand record the transition to dominantly oscillatory flow as the storm wanes. The whole bed is mantled by small wave ripples and a mud drape (Based on Cheel, 1991; Cheel and Leckie, 1993).

ed these structures to have been cut and filled by offshore-directed storm flows in a nearshore environment. Other examples from Cretaceous rocks of western Canada are similar in many respects to those of Myrow (1992), but differ in having an ubiquitous fill of fine-grained HCS and rippled sandstone (Plint, 1996; Plint and Nummedal, 2000; Fig. 18B), suggestive of filling (but not necessarily cutting) during strong wave action. Amos *et al.* (2003) observed shore-normal gutter casts forming in 10–40 m of water on the shoreface off Sable Island (Nova Scotia shelf; Fig. 2). Gutter formation and filling took a few hours and occurred only during strong coastal downwelling due to onshore directed storm winds, supporting the interpretation of Myrow (1992).

Gutter casts are common in shoreface successions, associated with HCS, but they also occur in thinly interbedded wave-rippled sandstone and mudstone typical of more offshore environments. Where shorelines can be mapped regionally, gutter casts in nearshore HCS facies tend to be oriented shore-perpendicular (Leckie and Krystinik, 1989; Myrow, 1992; Plint, 1996; Plint and Nummedal, 2000; Fig. 2), whereas smaller gutters typical of thinner bedded, more offshore facies tend to be shore-oblique to shore-parallel (Aigner, 1985; Hart *et al.*, 1990; Hay *et al.*, 2003; Varban and Plint, 2008a; Fig. 18C); the latter may record geostrophic flows in areas too deep to have experienced strong wave action.

Storm Beds

The development of an idealized storm bed is summarized in Figure 19A, which portrays the response of fine sand on the seafloor to various flow states during the rising and waning stages of a storm. Figure 19B summarizes the sedimentary structures that develop under similar hydraulic conditions but where the bed consists of a mixture of sand and fine gravel.

FACIES SUCCESSIONS ON STORM-DOMINATED SHELVES

Storm-dominated shelves and coasts tend to be linear, and hence produce relatively simple, tabular rock units compared to deltaic coasts, which have an irregular shoreline and corresponding lateral facies complexity (*see* Chapter 10). Coastal progradation will produce an essentially tabular body in which the basic stratigraphic motif is a sandier upward succession that records a progressive upward increase in the influence of waves and currents as the shoreline progrades. The succession may culminate in subaerial beach deposits and even alluvial sediments if the accommodation is entirely filled and the top of the succession has not subsequently been removed by transgressive erosion (Fig. 20). The details of the succession will vary depending on variables such as available grain sizes, proportion of sand/gravel to mud, wave and tidal energy, biological activity, shelf slope, subsidence rate and rate of sediment supply (*see* additional dis-

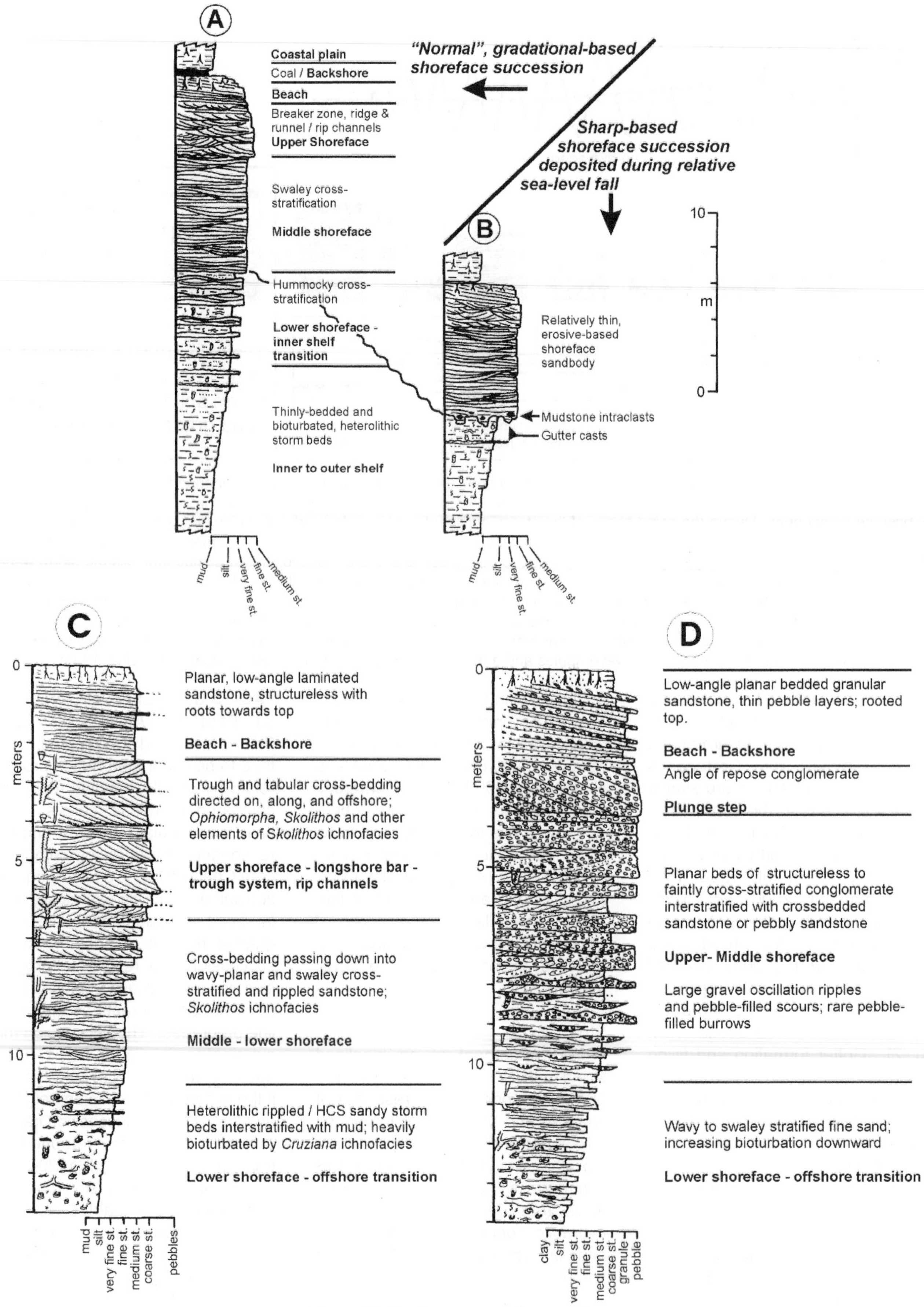

Figure 20. Caption on opposite page, top.

Figure 20. *(opposite page)* **A**. Facies succession representative of a prograding, wave-dominated shoreface to shelf setting dominated by fine-grained sand deposited during stable or slowly rising sea level. Muddy and highly bioturbated shelf deposits pass gradationally upward through thinly bedded rippled and thicker bedded HCS sandstone into mud-free SCS and crossbedded sandstone of the middle and upper shoreface. **B**. A sharp-based shoreface sandstone deposited during a period of relative sea-level fall (a forced regression – *see* Figure 24 for a modern analog and Figure 30 for an ancient example). At such times, reduced accommodation in the nearshore zone eliminates space for the shelf-to-shoreface transitional zone, and shoreface sandstones rest erosively on a wave-scoured surface typified by abundant gutter casts. This surface is referred to as a regressive surface of marine erosion. Both columns based on numerous examples from the Cretaceous Western Interior. **C**. Summary facies succession for a prograding wave-dominated shoreface dominated by medium to coarse-grained sand. Much of the middle and upper shoreface is dominated by crossbedding deposited by dunes migrating up the shoreface under the influence of shoaling waves, or along- and offshore under the influence of longshore and rip-circulation systems, or geostrophic currents. Finer grained sands of the lower shoreface have wavy-planar lamination, SCS or HCS, grading down into heavily bioturbated sandy silts of the inner shelf (Based on Clifton, 2006, and other sources). **D**. Summary facies succession for a shoreface dominated by conglomerate. Much of the shoreface is composed of tabular beds of conglomerate with little or no internal structure, separated by minor sandy interbeds (*e.g.*, Fig. 27). Locally, large-scale oscillation ripples in gravel may be preserved. The top of the shoreface may be marked by a thick, seaward-dipping avalanche set representing the plunge step at the toe of the foreshore. Foreshore sediments are planar-bedded sandstone and granule-stone with imbricated pebbles, penetrated by roots at the top. Trace fossils tend to be rare. Based on various Cretaceous shoreface successions in Western Canada, Hart and Plint (1995) and Clifton (2006).

---

cussion in Chapter 10). The effects of sea level and other allogenic changes are discussed further below.

Shoreline Trajectory

A storm-dominated shoreline and shelf succession can prograde when the rate of sediment supply exceeds accommodation rate ('normal regression'), and/or during relative sea-level fall ('forced regression') when accommodation is removed (see below; Plint, 1991; Helland-Hansen and Martinsen, 1996; Naish and Kamp, 1997). The stratigraphic consequences of different shoreline trajectories were discussed by Helland-Hansen and Martinsen (1996) who recognized 'accretionary' and 'non-accretionary' shorelines. Accretionary regression takes place at least in part due to the addition of sediment, either during slow sea-level rise ('normal regression') or slow fall ('forced regression'), whereas non-accretionary regression is simply the result of sea-level fall, with negligible addition of sediment. Accretionary transgression takes place when the shoreline climbs up over contemporaneous coastal-plain deposits, allowing the accumulation of a nonmarine 'back-barrier wedge' that is coeval with transgressive sediments on the shelf. Non-accretionary transgression occurs when the shoreline moves parallel to the land surface and no space is generated to accommodate terrestrial sediment landward of the beach. A logical consequence of this approach is the need to recognize four systems tracts (*see* Chapter 2), rather than the three conventionally recognized in the Exxon scheme.

Late Pleistocene and Quaternary Prograding Shoreface Systems

Modern shelves were flooded by post-glacial sea-level rise over the last 15 ky and hence most sediments represent a transgressive systems tract. As a result, few prograding linear shoreline systems are available as analogs for ancient highstand and falling-stage systems tracts (following systems-tract terminology of Plint and Nummedal, 2000). One of the best examples is the broadly arcuate, 200 km long microtidal shoreline of Nayarit on the Pacific coast of Mexico (Fig. 21A and B). The shoreface has prograded about 15 km in the last 3.6 ky at an average rate of about 3 m/year (Curray *et al.*, 1969). Fine to medium sand is supplied both by small coastal rivers as well as by marine erosion of older transgressive deposits. The powerful wave-driven longshore drift system has built a sandbody about 7 m thick that grades seaward into mud (Fig. 21C).

The coast of Brazil also provides good analogues for prograding wave-dominated coasts. On the semi-arid, wave-dominated NE Brazilian coast of Rio Grande do Norte, relative sea level peaked about 1.4 m above present at ~5.6 ka and has since fallen (Caldas *et al.*, 2006), leading to strandplain progrdation. The area lacks river mouths, and, instead, sand is supplied by west-directed longshore currents, and by eolian dunes blown in the same direction. Foreshore deposits are about 3.5 m thick and step seaward and downward, with the toe of the upper shoreface sands resting on the eroded surface of the underlying, transgressive lagoonal deposit (Fig. 22; *see* also Chapter 11). This shoreface provides an example of a falling-stage ('forced regressive') facies succession that lacks muddy heterolithic lower shoreface deposits due to lack of accommodation (contrast Figure 20A and B). Similar falling-stage strandplains from the Paranaguá coastal plain in southern Brazil are of late Pleistocene (105–85 ka) and Holocene (5.1–0 ka) ages, and both are inclined seaward, indicating deposition during two separate phases of relative sea-level fall (Lessa *et al.*, 2000). Other examples of prograding Brazilian strandplains were described by Dominguez *et al.* (1987) who showed that coastal progradation could be localized, not only around river mouths, but also where longshore-drift systems converged, where there was no direct fluvial input.

On the Rio Grande do Sul strandplain of southern Brazil, wave-dominated late Quaternary (125 ka) strandplain deposits are exposed on

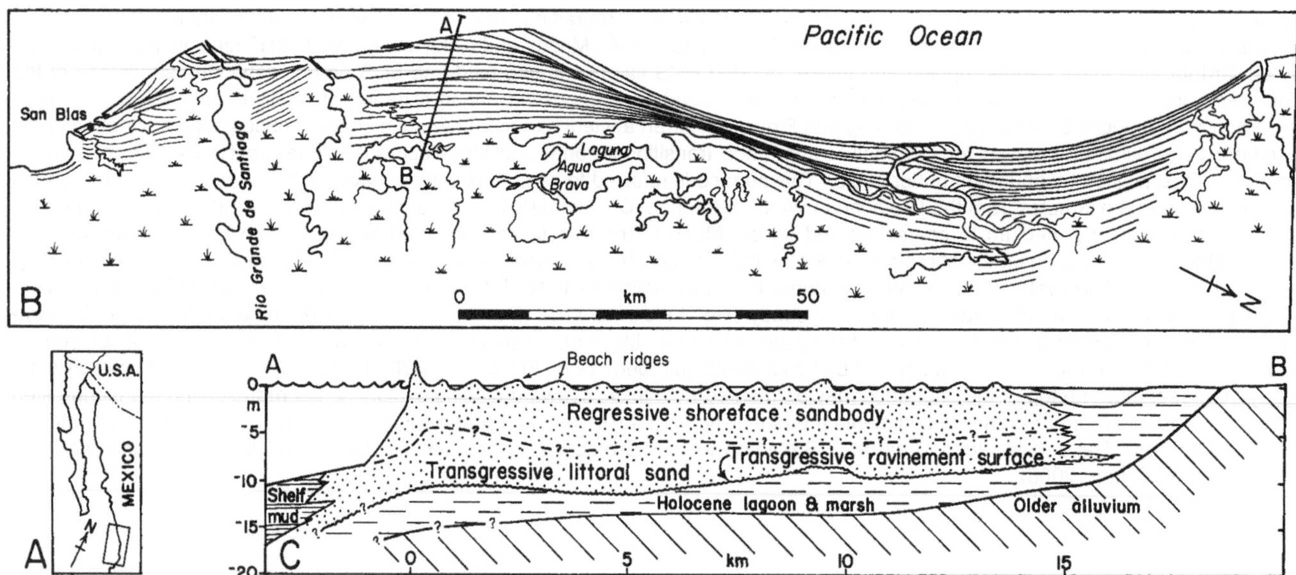

Figure 21. A. Coast of Nayarit, Mexico, showing (**B**) an arcuate wave-dominated strandplain, with up to 280 sub-parallel beach ridges, that have accreted in about 4500 years of coastal progradation. Changes in wind patterns and in the position of river mouths have resulted in localized erosion and reorientation of the shoreline. **C**. Cross-section showing Holocene transgressive ravinement surface overlain by transgressive sand, in turn overlain by the prograding shoreface–beach–ridge complex. Modified from Curray *et al.* (1969).

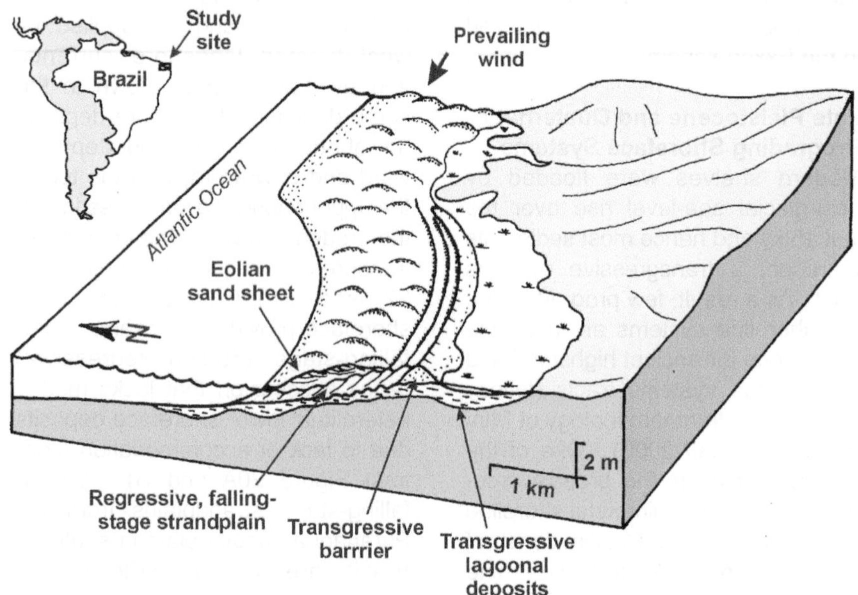

Figure 22. Example of a transgressive barrier and lagoon system that formed on the semi-arid NE coast of Rio Grande do Norte, Brazil, during the later stages of the Holocene transgression at about 5.6 ka. Subsequent sea-level fall of about 1.4 m has caused progradation of a falling-stage strandplain sandbody that downlaps directly onto the underlying marine ravinement surface cut across lagoonal sediments. The ravinement surface may have experienced further erosion during sea-level fall. Strong alongshore prevailing winds have blanketed the strandplain with eolian dune deposits. Re-drawn from Caldas *et al.* (2006).

land (Tomazelli and Dillenburg, 2007). Very fine to fine grained, slightly silty glauconitic sand with HCS represents the lower to middle shoreface, and grades up into fine sand with abundant crossbedding and *Ophiomorpha*, indicative of nearshore bars in the upper shoreface. Low-angle laminated beach facies are overlain by several meters of eolian dune sandstone (Fig. 23). Quaternary highstand and falling-stage deposits from wave-dominated shelves in New Zealand and Australia were described by Bradshaw and Nelson (2004). Seismic and GPR profiles, drilling, and ^{14}C and thermoluminescent dating of fine- to medium-grained shoreface sands beneath the Tuncurry coast in SE Australia provide a very clear link between transgressive and regressive deposition and glacioeustatic sea-level changes of about 80 m (Fig. 24).

Facies Succession in Ancient Prograding Shoreface Sandbodies

There are many ancient examples of sheet-like sandbodies characterized by HCS and SCS that are interpreted to have been deposited in linear, wave-dominated shorefaces (*see* Reading and Collinson, 1996 for review). The literature is strongly biased towards the Cretaceous of the Western Interior of North America, both because of excellent exposure in the Rocky Mountain Foothills, and because these sandstones form important hydrocarbon reservoirs and hence are extensively drilled (Midtgaard, 1996; Olsen *et al.*, 1999; Fitzsimmons and Johnson, 2000; Mellere and Steel, 2000; Hampson and Storms, 2003; Edwards *et al.*, 2005b; Varban and Plint, 2008a, b).

Wave-dominated shelf to shoreface successions that were deposited under conditions of stable or rising sea level typical of the highstand systems tract show a gradually

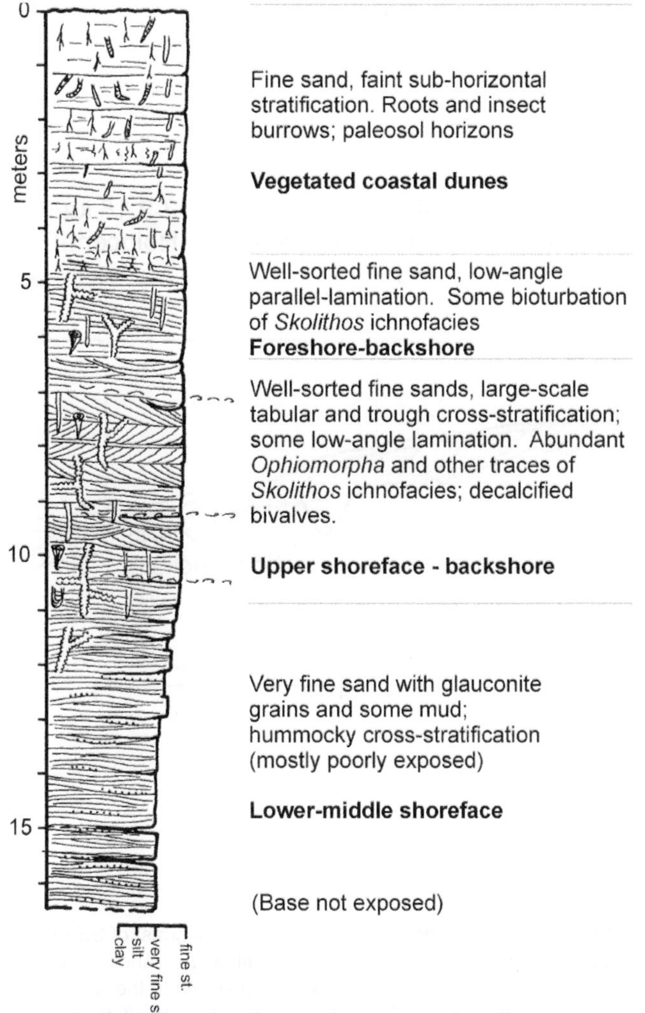

Figure 23. Graphic log showing facies succession in a strandplain shoreface sandbody exposed in sand pits located 6–8 m above sea level on the coast of Rio Grande do Sul (southern Brazil). Deposition took place in a wave-dominated microtidal shoreface during late Pleistocene interglacial highstand and early eustatic fall at about 125 ka. This succession bears close similarity to ancient strandplain shoreface deposits (*e.g.*, Fig. 20 A and B), the notable difference being the importance of eolian deposits atop the Pleistocene succession. Compiled from description by Tomazelli and Dillenburg (2007).

thickening-upward succession of sandstone beds inter-stratified with mud (Fig. 20A). Where the available sediment is primarily fine sand and mud, the base of a complete succession begins with bioturbated (*Cruziana* Ichnofacies) silty mud or silty sand with little preserved primary stratification, representative of the shelf (Fig. 25A; details of trace-fossil assemblages are not discussed here, *see* Chapter 3). Upward, cm-scale very fine sandstone beds appear. The bases of beds are sharp, and bed tops, although burrowed, may preserve wave- or combined-flow ripples (Fig. 25B and C). Higher in the succession, sandstone beds are dm-scale and sharp-based with HCS, separated by bioturbated mudstone (Fig. 14A and B); hummocky sands become more amalgamated upward as fair-weather mud layers are progressively eroded away by storms (Fig. 14D). The rippled and HCS sandstone facies probably represent inner shelf to lower shoreface environments. The HCS sandstone beds progressively thicken and amalgamate up-section, giving way to mud-free SCS sandstone, commonly accompanied by a change from very fine to fine-grained sand and a change from the *Cruziana* to the *Skolithos* trace-fossil assemblage. The upper few meters of the succession are typically upper fine to medium-grained sandstone with cross-bedding representing dunes migrating in longshore troughs and rip channels on the upper shoreface (Fig. 20A). Gently seaward-dipping planar lamination represents the beach swash zone (Figs. 4 and 26), the top of which may grade into massive, root-bioturbated sand of the backshore zone.

Coasts supplied with medium- or coarse-grained sand will show a shoreface succession dominated by crossbedding (potentially oriented along-, on- and offshore; Hunter *et al.* 1979; Clifton, 1981), and possibly large-scale oscillation ripples, these being the stable bedforms in these coarser sediments (*e.g.*, Wignall *et al.* 1996; Clifton, 2006; Fig 20C). Close to river mouths, conglomerate may be abundant, forming large wave ripples (Fig. 13) and crudely to well-stratified gravels and sandstones (Figs. 20D, 27 and 28; Clifton, 1981, 2003; Leckie, 2003; Leckie and Walker, 1982; Varban and Plint, 2008a; and review in Hart and Plint, 1995). Mapping in various conglomeratic Cretaceous formations suggests that gravel typically migrates alongshore from river mouths by no more than about 5–10 km.

Sand Dispersal Across the Shelf

Although many modern shelves trap mud within ~ 50 km of shore, and sand even closer to the coast (*see* for example Cattaneo *et al.* 2007 and discussion therein; Fig. 11), strongly wave-influenced shelves such as off Texas (Snedden *et al.*, 1988) and the Bering shelf (Nelson, 1982), show that in some settings, it is possible for sand to be transported many tens of kilometers offshore during major storms, with sand transport being particularly effective if the water is only 20–30 m deep.

Where paleo-shorelines in ancient shelf deposits can be mapped with confidence, such as along the western margin of the North American Western Interior Seaway, it is evident that fine sand was transported offshore for many tens of kilometers. For example, Varban and Plint (2008a, b) using outcrop and log control showed that on a low-gradient Turonian ramp, heterolithic strata with HCS extended ~30 km seaward

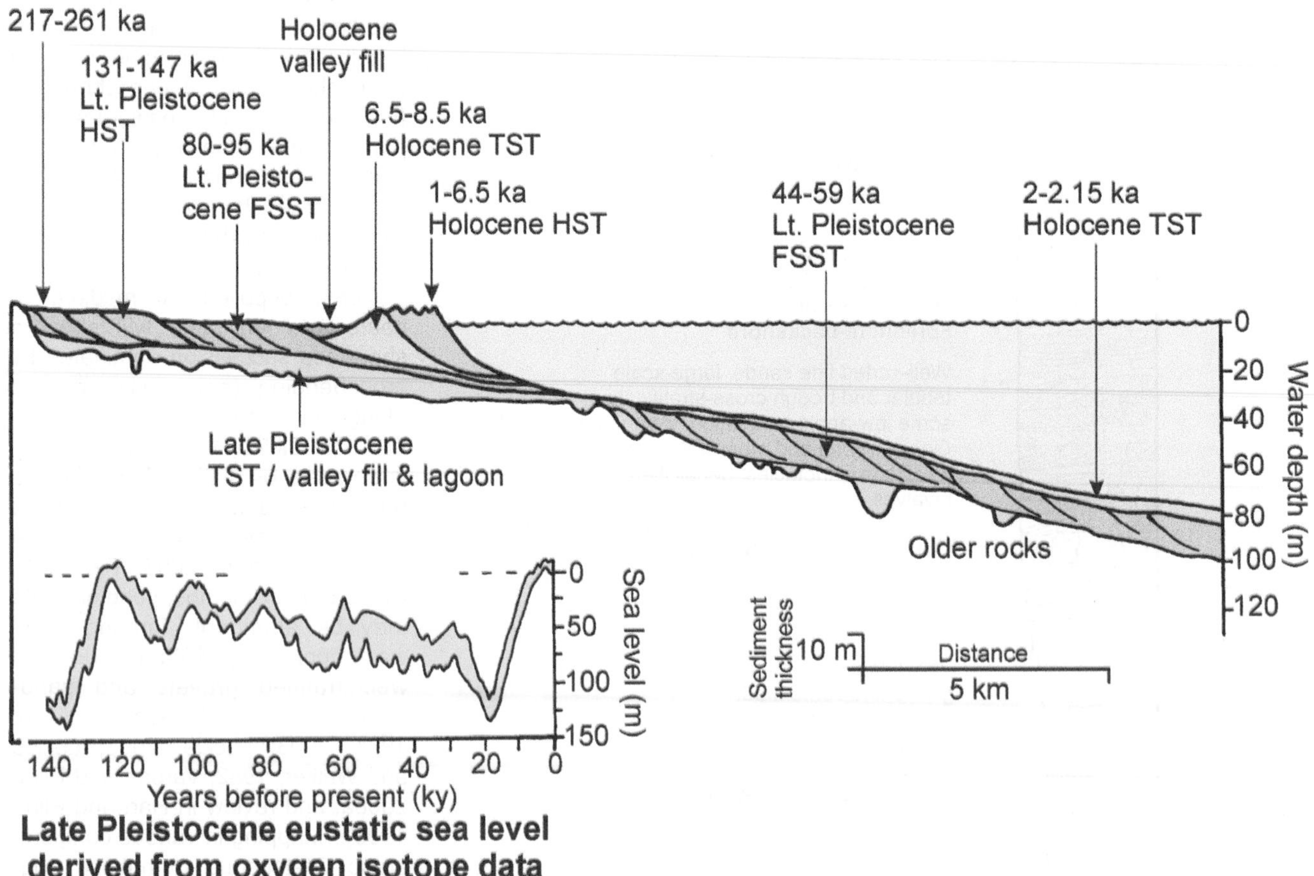

Figure 24. Regressive and transgressive barrier and strandplain deposits on, and beneath, the Tuncurry coast of SE Australia. Isotopic and thermo-luminescent dating of the various sediment packages, coupled with seismic imaging, allow the various sedimentary units to be assigned to systems tracts that can be directly linked to glacio-eustatic changes over the last 150 ky. Note the well-developed falling-stage systems tract (FSST – termed a 'regressive system tract' by Bradshaw and Nelson, 2004), extending far offshore beneath the present sea floor. Re-drawn from Bradshaw and Nelson (2004) from original data in Roy *et al.* (1997); isotopic and thermo-luminescent data from multiple sources referenced in Bradshaw and Nelson (2004) (*See* also Chapter 11, Figure 34.)

of SCS shoreface sandstone, before grading into cm-bedded very fine sandstone and mudstone, which extended another ~70 km farther seaward (Fig. 29). The water was inferred to have been only a few tens of meters deep across this ramp, and no clinoforms could be recognized.

There is no evidence that shorelines migrated seaward of the SCS zone, and hence it appears that fine and very fine sand were transported by combined flows for up to about 100 km from the contemporaneous shoreline; silty sediments extended a farther ~100 km offshore before grading into clay.

Middle Turonian to Lower Coniacian rocks of the Cardium Formation in Alberta also include extensive heterolithic deposits that extend tens of kilometers beyond the shoreline, and which form major hydrocarbon reservoirs (Walker and Eyles, 1988; Leggitt *et al.*, 1990). On the lower Paleozoic epeiric platform of Minnesota and Wisconsin, HCS sandstones are interpreted to extend about 200 km offshore; such extensive sand dispersal is attributed to shallow water, a low shelf gradient and an extremely low subsidence rate (Runkel *et al.*, 2007).

RESPONSES OF SHELF SYSTEMS TO ALLOGENIC FORCING

All sedimentary systems are influenced by external, or *allogenic* processes, the most important of which are rates of change of eustatic sea level, subsidence, and sediment supply (*see* Chapters 2 and 4). It is convenient to think of changes in sea level and subsidence (both of which can go up or down) as simply change in *accommodation rate*, which is the rate at which space is either made available for sediment accumulation (*i.e.*, due to tectonic subsidence and/or eustatic rise), or the rate at which accommodation is removed (*i.e.*, due to tectonic uplift and/or eustatic fall). Unfortunately, it is very difficult for geologists interpreting the rock record to separate eustatic and tectonic effects.

Most marine shelf sediments accumulate when the shoreline is prograding and relative sea level is either rising slowly or falling. Because modern shelves were very recently (~15–5 ka) flooded by the post-glacial transgression, sedimentary geologists are presented with innumerable examples of transgressive barrier–lagoon– estuary systems (*see* Chapter 11), but very few prograding strandplains. By contrast, the rock record of shelf seas is dominated by prograding deltas (*see* Chapter 10) and strandplain deposits, and barriers

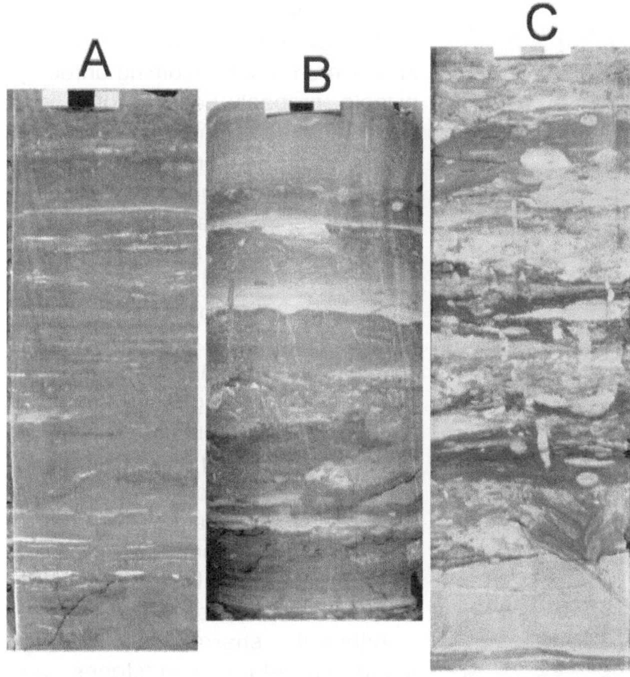

Figure 25. **A**. Heavily bioturbated silty mudstone with remnant, less bioturbated silty storm beds, deposited many tens of kilometers offshore. Cardium Formation (Turonian) in well 12-18-54-21W5, 1304 m, Alberta. Scale bar = 3 cm. **B**. Graded storm-deposited beds of very fine sand and silt, inter-stratified with dark mudstone, deposited tens of kilometers offshore. Note sharp and locally scoured bases to sandstone beds, and pervasive bioturbation that, in places, has completely destroyed the original stratification (compare with Figure 10). Cardium Formation (Turonian) in well 16-5-55-18W5, 1875 m, Alberta. Scale bar = 3 cm. **C**. Intensely bioturbated sandy storm beds with remnants of HCS and ripples, probably deposited in the lower shoreface-inner shelf transition zone. Cardium Formation (Turonian), well 6-22-54-19W5, 1971 m, Alberta. Scale bar = 3 cm.

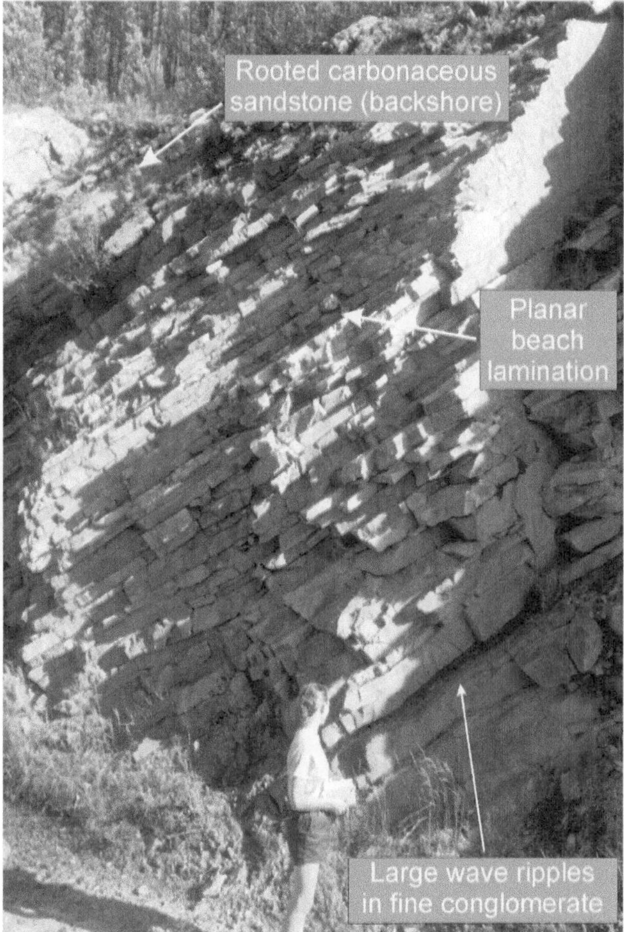

Figure 26. Planar-laminated fine sandstone deposited in a foreshore swash zone (compare with Figure 4), capped by rooted carbonaceous sandstone of the backshore zone. Large oscillation ripples in fine conglomerate were deposited in 2–3 m of water in the upper shoreface. Cardium Formation, Grande Cache, Alberta.

are rare, most being preserved as the most landward part of strandplains (*e.g.*, Figs. 21, 22 and 24).

Ancient Examples of Forced Regressive Deposits

Subsurface study of the Cardium Formation in Alberta revealed lateral transitions between gradationally based and sharp-based shoreface sandstones (Plint, 1988). In the latter, SCS sandstone rests erosively on muddy shelf facies, and the transitional succession of rippled and HCS sandstone beds is absent (Fig. 20B). The Cretaceous rocks of western Canada provide numerous other examples of sharp-based shoreface sandbodies, commonly with large, shore-normal gutter casts on the basal surface (*e.g.*, Plint, 1991, 1996; Plint and Norris, 1991; Plint and Kreitner, 2007). The compressed or absent facies transition between muddy shelf and sandy shoreface deposits indicated that shoreface progradation had taken place during relative sea-level fall and accommodation loss, a process called 'forced regression' by Plint (1991) to emphasize its dependence on relative sea-level fall, rather than on sediment supply. Subsequent studies (*e.g.*, Posamentier *et al.*, 1992; Nummedal *et al.* 1993; Ainsworth 1994; Nummedal and Molenaar, 1995) confirmed the widespread occurrence of sharp-based, forced-regressive shoreface sandbodies, further documented in Hunt and Gawthorpe (2000).

The regressive surface of marine erosion (RSME) below the shoreface sandbody is the result of wave scour of the inner shelf, a process that is simply 'transgressive ravinement in reverse'. Sharp-based shoreface sandstone bodies are typically thin (3−7 m); particularly thin examples of about 1−2 m have been described by Plint and Kreitner (2007). The presence of rooted sediment at the top of the shoreface indicates that the sand body is thin because progradation took place into water of comparable depth (*i.e.*, as little as 1−2 m), rather than because of subsequent transgressive erosion. Although a RSME has been widely documented at the base of SCS sandstone facies, Tamura *et al.* (2007) found that, in cores through a Japanese Holocene strandplain, the most prominent erosion surface to form during forced regression was at the base of the longshore trough system in about 5−6 m of water. In this example, an erosion surface at the base of SCS/HCS facies in the lower

Figure 27. A. Well-segregated and stratified fine conglomerate and medium-grained sandstone interbeds from a gravelly shoreface. Conglomerate beds appear mostly structureless, although low-angle cross-stratification is visible locally. The conglomerate–sandstone couplets probably represent storm event beds. **B**. Tabular crossbedded medium-grained pebbly sandstone inter-stratified with apparently structureless conglomerate; crossbedding indicates along-shore migration of dunes. Cardium Formation, Bay Tree, British Columbia. Scale bar in both photos = 15 cm; photographs courtesy of Bruce Hart.

shoreface was not recognizable, possibly because of the limited core view, or because formation of such an RSME is favored by a particular combination of wave energy and bathymetric profile.

Detailed study of very well-exposed Cretaceous strata in Utah (Hampson and Storms, 2003) has shown that within an individual, upward-coarsening progradational succession (also termed a parasequence), minor regressive surfaces of marine erosion and minor transgressive surfaces can alternate on a dip scale of a few kilometers. Transgressive events erode and flatten the shoreface profile, which steepens again during progradation. Each RSME records the response of the shoreface equilibrium profile to either a minor relative sea-level fall or an increase in storminess, whereas the transgressive surfaces may represent a decrease in storminess or a minor sea-level rise; these changes are interpreted to operate on time-scales of 10^2 to 10^5 years. The RSME documented by Hampson and Storms (2003) extend seaward only a few kilometers before becoming unrecognizable, probably because of a relatively high accommodation rate and steep shelf gradient. In contrast, many RSME mapped in Western Canada can be traced downdip for tens of kilometers as an apparently single surface, probably reflecting a low accommodation rate and low shelf slope.

Plint and Nummedal (2000) proposed a falling-stage systems tract (FSST) for sediments deposited on clastic ramps during relative sea-level fall. This was a logical complement to the three systems tracts of the original Exxon sequence, and supported the conceptual models of Helland-Hansen and Martinsen (1996).

Although sharp-based, falling-stage shoreface sandstones are widely documented from low-gradient ramps of the Western Interior, Pekar *et al.* (2003) describe a contrasting situation from wave-dominated Oligocene shelf deposits beneath the New Jersey shelf. These rocks show evidence for deposition during relative sea-level fall in the form of offlap, truncation, and downward shifts in facies, yet it is not possible to recognize a basinward shift in onlap, nor a distinct falling-stage or a lowstand systems tract. As a result, highstand, falling stage, and lowstand systems tracts merge into a single, offlapping package. This is explained in terms of low subsidence rate, low sediment supply and high wave energy, which generated a by-pass zone across the shelf in 10−20 m of water. As sea level fell, submarine wave scour formed an erosion surface that was a submarine extension of the subaerial unconformity (the sequence boundary) to landward.

Naish and Kamp (1997) and Browne and Naish (2003) were able to establish detailed links between Plio-Pleistocene systems tracts and eustatic sea-level changes, showing that highstand deposits were sometimes *conformably* overlain by falling sea-level deposits (*see* Fig. 31A, stages A-D) (their 'regressive systems tract') making it impossible to precisely differentiate highstand and falling-stage systems tracts on the basis of facies succession and

Figure 28. Well-stratified upper shoreface and foreshore deposits consisting of pebbly, medium to coarse-grained sandstone. Pleistocene of San Francisco area, California. Scale bar (circled) = 20 cm.

bounding surfaces. The lack of erosional contact between highstand and falling-stage facies was attributed to a high subsidence rate, which reduced the erosive effects of shallowing. The studies of Naish and Kamp (1997) and Pekar *et al.* (2003) are a reminder that some stratigraphic successions are not easily divided into systems tracts.

Erosional Dissection of Sheet-like Shelf Sandbodies due to Sea-level Fall

The Cardium Formation of Alberta hosts oilfields that have an elongate, broadly shore-parallel shape, and a reservoir composed of heterolithic, rippled and HCS-bearing fine sandstone. The lenticular shape of the producing field, and location tens of kilometers seaward of broadly contemporaneous sandy shoreface deposits suggested interpretation of the Cardium pools, and many other 'offshore' sandbodies in the Western Interior, as shelf sand ridges (*e.g.*, Walker, 1984). Later studies (*e.g.*, Plint *et al.*, 1986, Bergman and Walker, 1987; Plint, 1988; Wadsworth and Walker, 1991) showed that muddy and heterolithic shelf deposits of the Cardium Formation had suffered major *erosional dissection* as a result of repeated relative sea-level falls and rises. Regional mapping of one of the most spectacular of these erosion surfaces (the E5 surface in the Cardium Formation; Walker and Eyles, 1991) using a database of > 6000 wells showed that the producing fields of HCS sandstone owed their linear shape not to the primary depositional morphology of the sandstone, but to erosional dissection, by up to 20 m, of an original sheet-like body. Hydraulic considerations indicated that the regional blanket of conglomerate draping the E5 erosion surface must have been derived from an original fluvial deposit, laid down at sea-level lowstand. Morphologically, the E5 surface resembles a staircase in which the 'steps' represent a succession of wave-cut platforms, and the 'risers' mark the locations of successive erosional shorefaces carved between pulses of sea-level rise. Although the sediments below E5 preserve no direct evidence of subaerial emergence, the *morphology* of the erosion surface strongly suggests formation as a succession of marine wave-cut terraces. The overlying conglomerate, of fluvial origin, has been completely reworked in a marine setting.

Detached Shoreface Deposits: 'Offshore Bars'

Sandy or conglomeratic sediments concentrated in long (commonly tens

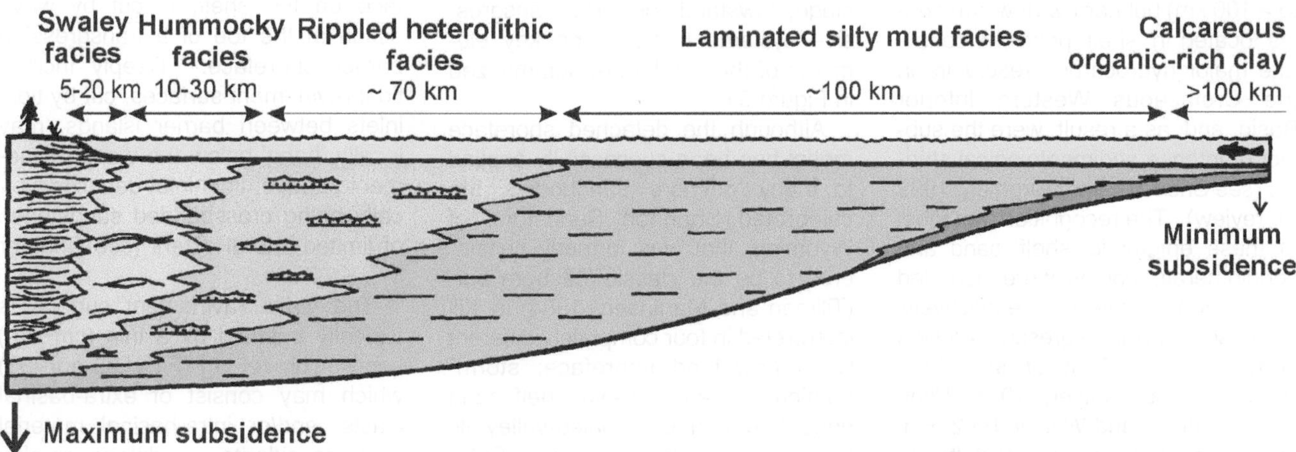

Figure 29. Summary of facies distributions across a low-gradient foreland-basin ramp represented by the Kaskapau Formation (Turonian) in Alberta and BC. Facies were mapped in outcrop and core, and from well-log responses. In this example, fine and very fine sand are interpreted to have been transported up to about 100 km offshore by combined flows moving along the shelf to the SE; silty muds accumulated for about 100 km farther seaward before grading into claystone. Re-drawn from Varban and Plint (2008a).

Figure 30. Sharp-based shoreface sandstone resting erosively on offshore black silty mudstone, the basal regressive surface of marine erosion is ornamented with large gutter casts oriented shore-perpendicular. The top of the shoreface is marked by a pebble-veneered ravinement surface, overlain by dark, laminated offshore marine mudstone. Dunvegan Formation (Cenomanian), Grande Cache area, Alberta.

to > 100 km) but narrow (few km) bodies located in 'shelf' positions constitute major hydrocarbon reservoirs in the Cretaceous Western Interior Basin, and, as a result, were the subject of intense sedimentological studies (*see* Snedden and Bergman, 1999 for review). The recognition that most of these enigmatic 'shelf' sand and conglomerate bodies were bounded below and above by, respectively, regressive and transgressive erosion surfaces (*e.g.*, Plint *et al.*, 1986, Bergman and Walker, 1987; Plint, 1988; Pattison and Walker 1992; Fig. 30), led to a major paradigm shift: 'offshore bars' became 'detached shoreface deposits' (e.g., Bergman and Snedden, 1999) that may have been deposited during the falling-stage, lowstand, or early transgressive systems tracts. The key elements of this model are summarized in Figure 31.

Although the detached shoreface model has been successfully applied to many 'offshore' sandbodies, the celebrated Shannon Sandstone of Wyoming, that was formerly considered to be the classic 'offshore bar' (Tillman and Martinsen, 1984) is still interpreted in four completely different ways (lowstand shoreface; storm-dominated transgressive shelf sand ridge; tide-dominated paleo-valley-fill; transgressive tidal sand ridge; Suter and Clifton, 1999). Despite intense effort to find consensus, there seems to be no single model that fits all the facts, perhaps necessitating a hybrid interpretation that blends, for example, a truncated regressive 'lowstand shoreface' erosively overlain by a 'transgressive shelf ridge'. It appears to the writer that it is critical to establish in more detail the regional sequence-stratigraphic relationships within the mudstones that separate the Shannon from coeval nearshore sandstones to the West. Clearly, not all 'offshore sandstones' have neatly-packaged explanations!

Transgressive Deposits

Although modern shelves provide a spectacular example of transgressive deposition, caution is necessary when using the high-amplitude, high-rate, high-frequency glacioeustatic changes of the Pleistocene as a basis for interpreting older sea-level changes, most of which were probably of both lower amplitude and rate (Cattaneo and Steel, 2003; Suter, 2006).

During relative sea-level rise across a storm-dominated clastic shelf, terrigenous sediment is trapped in aggrading estuarine and coastal-plain environments (*see* Chapters 9 and 11), which greatly reduces the volume and caliber of new sediment delivered to the shoreline and shelf. This trapping mechanism appears to be the principal reason for the abrupt decrease in both sediment grain size, and sedimentation rate above marine transgressive surfaces, rather than simply the increase in water depth.

The wave ravinement surface, the most important and widespread surface on the shelf, is cut by wave action at the toe of a transgressing barrier shoreface. Deeply incised tidal ravinement surfaces, cut by tidal inlets between barrier islands, may locally 'hang' below the wave ravinement surface, tidal-inlet deposits typically being crossbedded sandstones of limited lateral extent (*see* Chapter 11).

The wave ravinement surface is typically mantled by a thin (cm–dm) sand or gravel lag (Figs. 32 and 33) which may consist of extra-basinal clasts, and/or intra-basinal material such as siderite, sandstone or clay pebbles derived from immediately underlying rocks, and may also include marine shells and shark teeth concentrated by wave-winnowing of

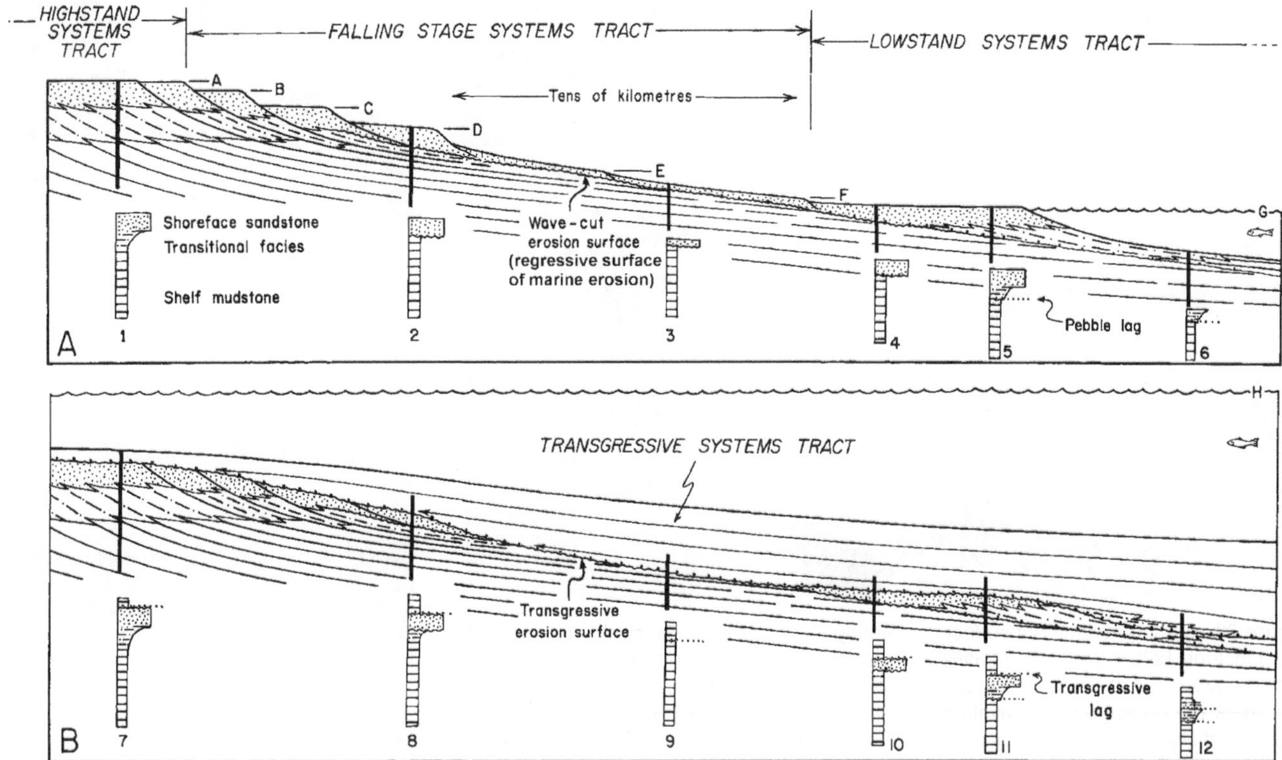

Figure 31. Generalized model for development of progradational shoreface sandstone bodies and detached 'offshore bars' on a low-gradient, wave-dominated shelf. In panel **A**, a wave-dominated strandplain (highstand systems tract) progrades as sea level slowly rises (level A), producing a gradational-based succession (column 1). As relative sea level falls through levels B-F, accommodation is removed through the process of forced regression which may lead to development of an anomalously thin sheet sandstone body (falling-stage systems tract) with a thin or absent lower transition to more offshore facies (columns 2, 3; *e.g.*, Fig. 30). Sea level reaches lowstand at level G, where wave erosion carves a new lowstand shoreface profile into muddy shelf deposits. Continued deposition fills this space with a lowstand shoreface sandbody (lowstand systems tract) that is erosive-based along its landward margin (column 4), but which may become gradational-based as it progrades into deeper water (columns 5, 6). In panel **B,** lowstand deposition is terminated when the rate of sea-level rise exceeds the capacity of sediment supply to fill the new space, leading to marine transgression, eventually reaching level H. The shoreline is driven landward, and wave erosion in the toe of the transgressing barrier shoreface, which is completely eroded in this example, strips off a variable thickness (decimeters to meters) of underlying sediment forming a wave ravinement surface. This is typically mantled by a gravel lag (*e.g.,* Fig. 32) but otherwise may be more cryptic (Fig. 35), and all evidence of subaerial exposure is typically removed (*e.g.*, Fig. 33). Ravinement erosion strips off falling-stage shoreface sandstones (columns 8, 9) and the entire regressive package is blanketed by transgressive marine mudstone. Lowstand sandstones are left isolated and enclosed in offshore marine mudstone (columns 10–12), producing a top-truncated lowstand shoreface masquerading as an 'offshore bar'. Diagram based on Plint (1988).

the sea floor in a sediment-starved setting. If the sediments below the ravinement surface are sandy, this material may be reworked into transgressive, crossbedded and bioturbated sand sheets or ridges (Rine *et al.*, 1991), although convincing examples of ancient ridges (*e.g.*, Posamentier, 2002) are rare. This might be because ridges tend to become moribund as water deepens, after which they are slowly degraded by storm process to form a more tabular 'transgressive sand sheet' (Snedden and Dalrymple, 1999; Suter, 2006; *see* Chapters 9 and 11). The most typical and widespread transgressive shelf sediments are, however, sheets of bioturbated mudstone, sometimes with a significant glaucony content, which rest directly on the transgressive lag. Sheets of shelf mudstone extending for hundreds of kilometers are documented in Cretaceous shelf deposits of the Western Interior basin (*e.g.,* Varban and Plint, 2005, 2008a). Detailed correlation of a variety of shelf systems has shown that subtle siltier upward successions in mudstones deposited > 100 km offshore (Fig. 34), can be correlated with nearshore facies that provide evidence of coastal progradation and relative sea-level fall; the subtle distal shelf successions are attributed to increased wave reworking and winnowing of the muddy shelf surface during sea-level fall (Williams *et al.* 2001; Macquaker *et al.*, 2007; Tyagi *et al.*, 2007).

PERSPECTIVE ON THE MODEL

In the previous iteration of *Facies Models* (Walker and Plint, 1992), a well-established facies model for a prograding shelf-shoreline system was presented, based heavily on Cretaceous Western Interior case studies (Figs. 20A and 31). Although valid for environments dominated by fine sand, this model was justly criticized (*e.g.*, Clifton, 2006) for not considering coarser grained systems where the influence of grain size on bedform stability would result in a succession dominated by various types of crossbedding rather than

Figure 32. Marine ravinement surface separating sideritic paleosols, coals and channel-fill sandstone below from well-stratified offshore marine silty storm beds above. Inset photo shows detail of chert cobbles and pebbles resting on the ravinement surface, where they mark the passage of a gravel-bearing transgressive barrier beach (scale bar = 15 cm). Cardium Formation (Turonian–Coniacian) overlain by Muskiki Formation (Coniacian), Mistanusk Creek, BC.

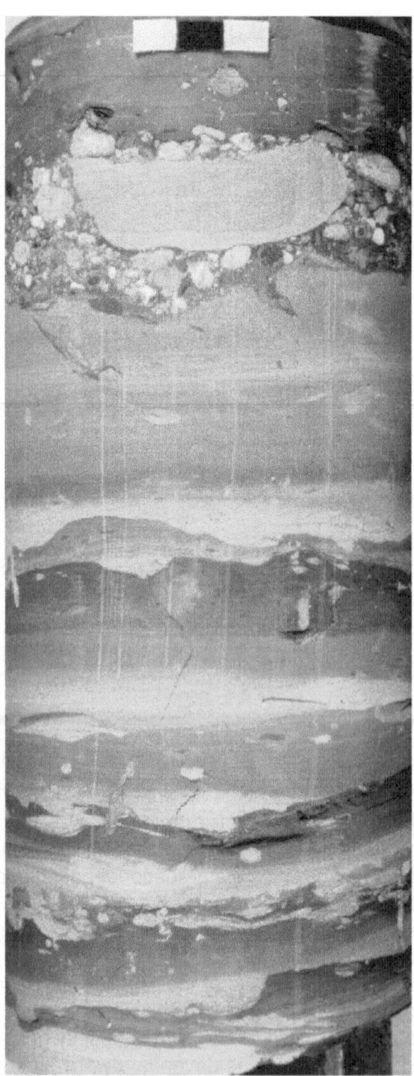

Figure 33. Marine ravinement surface (E5 surface of Cardium Formation), with lag of extrabasinal chert pebbles reworked from falling-stage fluvial and lowstand shoreface deposits, including a lithified sandstone pebble eroded from the immediately underlying (and formerly subaerially exposed) heterolithic storm-bed facies. In this example, evidence for subaerial exposure has been completely removed by wave erosion in the transgressive shoreface. Well 6-13-52-16W5, 1872 m, Alberta. Scale bar = 3 cm.

HCS and SCS. Figure 20C and D now illustrate synoptic facies successions for coarser grained environments, and additional detail is given in Clifton (2006).

The importance of deposition during the falling limb of the relative sea-level curve is reflected in the fact that many examples of shorelines that have prograded far out onto the shelf are now assigned to the falling-stage (alternatively termed 'forced regressive' or 'regressive') systems tract, rather than to the 'highstand' systems tract as was more common 15–20 years ago.

The major obstacle to the erection of an actualistic facies model is that modern storm-dominated shelves are not in equilibrium with highstand conditions, and many are still in transgressive mode. Modern shelf sediments are largely palimpsest sands and gravels reworked from pre-Holocene lowstand deposits. As such, they provide potential analogs for the thin transgressive lags and bioturbated 'sheet sands' at the base of many ancient shelf successions. However, the extensive blankets of regressive mudstone that dominate the ancient record of shelf deposition are at present poorly developed to non-existent. If orbitally driven glacioeustatic cyclicity (and the human race) persist to some 50 000 years into the future, sedimentologists will be much better placed to observe shelf-mud deposition during falling sea-level as the next 100 ky glacial cycle progresses!

The transport of mud across shelves raises intriguing, but until recently, largely ignored questions. Although mud delivered by rivers is initially swept along-shelf and deposited in a band only a few tens of kilometers wide, it is clear that mud is transported much farther offshore if storm waves are able to stir the seabed and re-suspend the mud, permitting additional steps of movement both along and across the shelf, driven by geostrophic flow and, potentially, wave-supported density flow. The great extent of some ancient shelf mudstone blankets (*e.g.*, in the Cretaceous Western Interior) implies that the seafloor was probably significantly shallower and flatter than currently envisaged, across which storm waves repeatedly re-suspended mud. Mud moved primarily as aggregates carried horizontally in a turbid layer at the sea bed. On steep, narrow, tectonically active shelves, and to a lesser extent on divergent continental margins, the rapid offshore increase in water depth would limit the breadth of the wave re-suspension zone, although conversely, steeper slopes would enhance the effectiveness of density-driven offshore flows. On all shelves, episodes of relative sea-level fall would promote seaward extension of the mud blanket (Ridente and Trincardi, 2005), and simultaneously promote rapid progradation of sandy shoreface deposits as nearshore accommodation was

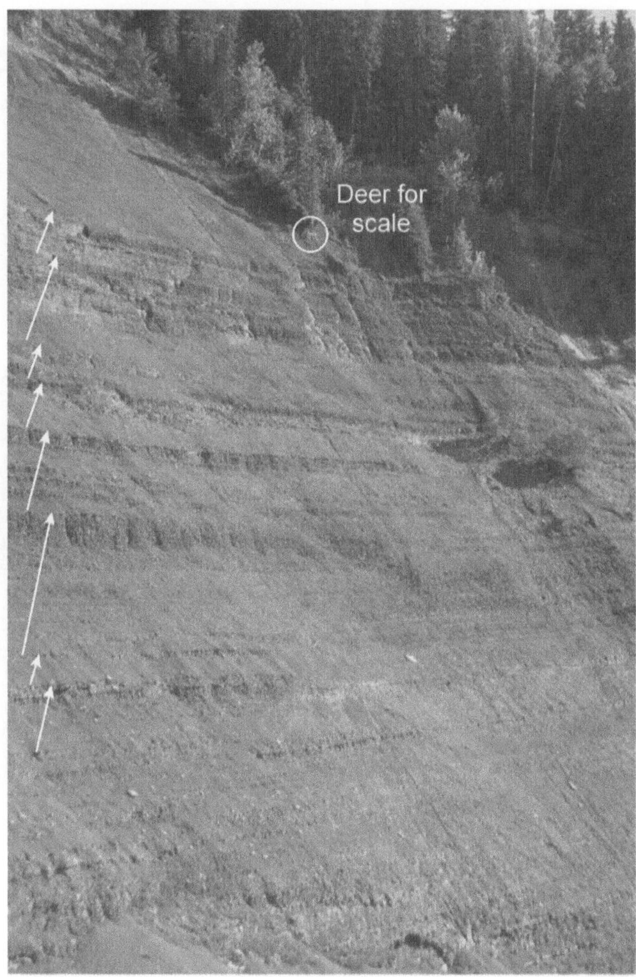

Figure 34. Subtle siltier upward parasequences in sandy siltstone shelf facies deposited many tens of kilometers offshore. Although subtle, many of these parasequences can be correlated to nearby well logs (*e.g.,* Varban and Plint, 2005), and then traced for 100–200 km in subsurface, suggesting that they indicate regional changes in sedimentation rate, probably related to allogenic transgressive-regressive events caused by relative sea-level changes, rather than autogenic processes such as delta-switching. Kaskapau Formation (Turonian), Murray River, BC. Photo courtesy of Bogdan Varban.

removed. Depending on such factors as shelf gradient, sediment supply, wave energy and subsidence rate, the resulting shoreface succession might be either sharp-based or gradational-based but nevertheless is likely to show some evidence of an accelerated seaward shift of facies in response to accommodation loss.

A possible consequence of relative sea-level fall is the development of one or more detached, falling-stage, lowstand, or transgressive shoreface sandbodies as suggested by the embryonic 'lowstand shoreface model' of Plint and Walker (1992). Evidence from Plio-Pleistocene shelf deposits, and the more ancient record, indicates that these detached shoreface models are robust, but need to be adapted to local conditions of bathymetry, topography, sediment texture, hydraulic regime and rates of sea-level change and sediment supply. The degree to which falling-stage and lowstand deposits are subsequently reworked on a transgressive shelf appears to be highly variable, which probably accounts for the variability displayed by transgressive shelf sandbodies (*e.g.*, Snedden and Dalrymple, 1999). .

Any interpretation of shelf deposits should be open to the possibility that the rocks preserve a record (sometimes decidedly cryptic!), of a migrating shoreline, subaerial emergence and deposition of fluvial sediment. Such consideration is necessary, despite the fact that obvious sandy shoreline deposits may be absent, and evidence of emergence and fluvial deposition may be represented by nothing more than a scoured and perhaps biologically bored surface, mantled by a few pebbles and bioclasts, or by totally reworked fluvial sands (Fig. 35). To this potential complexity should be added the possibility of short-term changes in the relative influence of waves and tides related to changing basin physiography, and also subtle syn-depositional tectonic deformation of the seafloor. The resulting sediment bodies will preserve the imprint of these diverse influences, both in the pattern of original *deposition*, and also in the pattern of post-transgressive *preservation* (*e.g.*, Martinsen, 2003). Efforts to decipher these diverse influences will no doubt continue to perplex and stimulate many future generations of sedimentary geologists!

ACKNOWLEDGEMENTS

My work on Cretaceous shelf sediments has been supported by the Natural Sciences and Engineering Research Council of Canada, by the Department of Mines, Energy and Resources Canada, by the University of Western Ontario, and by numerous oil companies including Amoco, Anadarko, B.P., Canadian Hunter, Devon, EnCana, Exxon-Mobil, Home, Husky, Imperial, Petro-Canada, Talisman, Texaco, Union-Pacific, Unocal, and Wascana. Results from western Canada owe much to the efforts of generations of my graduate students who have toiled over innumerable well logs, as well as up and down steep Rocky Mountain hillsides, to unravel the details of Cretaceous shelf deposits. Additional information and commentary on various aspects of shelf sedimentation were provided by Carl Friedrichs, Michael Li, Joe Macquaker and Dag Nummedal To all these institutions, companies and individuals, I express my sincere appreciation. I am grateful to Xavier Roca, Bogdan Varban, David Uličný and, especially, Bob Dalrymple for their perceptive and constructive reviews; responsibility for errors and omissions nevertheless remains with the author.

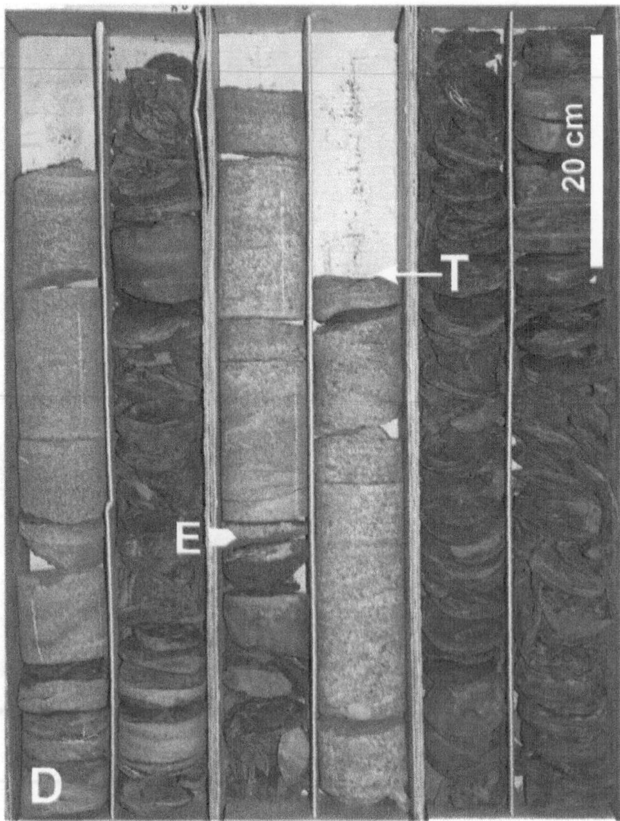

Figure 35. Examples of subtle ravinement surfaces, sequence boundaries and reworked lowstand deposits. **A**. Subtly pedogenically modified, rubbly marine mudstone with carbonaceous tree roots sharply overlain by grey laminated marine mudstone. This combined ravinement surface and sequence boundary is marked by a subtle color and textural change, and by a few shell fragments and fish bones. Scale bar = 20 cm. Doe Creek unit of Kaskapau Formation (late Cenomanian), Flatbed Creek, BC. **B**. Black chert pebbles (circled) capping light brown, sideritized pyritic silty mudstone marking an extensive transgressive surface at the top of a subtle siltier upward succession. The chert pebbles are interpreted to have been fluvially emplaced during subaerial emergence at lowstand. Physical evidence of subaerial emergence was removed during subsequent transgression, when the reworked pebbles were blanketed by offshore mudstone (*see* column 9 in Fig. 31). Westgate Formation (Albian) in well 16-32-40-1W5, 1669 m, Alberta; photo courtesy of Xavier Roca. **C**. Two units of marine mudstone with marine bivalves, separated by a 10 cm coal underlain by a 10 cm rooted rubbly pedogenic zone. There is no shallow marine sandstone associated with the shallowing event that led to emergence. Coal accumulation was a response to early TST deposition and the base of the coal represents a sequence boundary. Doe Creek unit of Kaskapau Formation (late Cenomanian), Flatbed Creek, BC. **D**. Medium- to coarse-grained, unbioturbated pebbly crossbedded sandstone sharply overlying a regionally extensive erosion surface 'E' cut on black, offshore mudstone. The sandstone unit (which thins laterally to a thin veneer (*e.g.*, Fig. 35B) over a few km) is interpreted to have originally been deposited by a river when the underlying mudstone was emergent at lowstand. The fluvial sand was then completely reworked by marine processes following marine transgression, also at surface 'E'. Continued deepening halted sandstone reworking, after which the transgressive sandstone sheet was blanketed (at 'T') with black offshore mudstone. Westgate Formation (late Albian) in well 10-33-34-5W5, 2479 m, Alberta; photo courtesy of Xavier Roca.

REFERENCES

Key sources of information

Bergman, K.M. and Snedden, J.W., *eds.*, 1999, Isolated Shallow Marine Sand Bodies: Sequence Stratigraphic Analysis and Sedimentologic Interpretation: SEPM Special Publication 64, 362 p.
Collection of papers summarizing state of the art thinking on 'offshore' sandstone bodies, with particular emphasis on the enigmatic Shannon Sandstone, the genesis of which still eludes a consensus interpretation.

Clifton, H.E., 2006, A re-examination of facies models for clastic shorelines, *in* Posamentier, H.W. and Walker, R.G., *eds.*, Facies Models Revisited: SEPM Special Publication 84, p. 293-337.
Summary of nearshore deposits and processes, based on decades of practical experience, on land and in the sea!

Hunt, D. and Gawthorpe, R.L., *eds.*, 2000, Sedimentary Responses to Forced Regressions: Geological Society of London, Special Publication 172, 383 p.
Collection of papers describing many examples of sedimentary successions deposited during relative sea-level fall.

Nittrouer, C.A. and Wright, L.D., 1994, Transport of particles across continental shelves: Reviews of Geophysics, v. 32, p. 85-113.
Brief summary of major processes responsible for transporting sediment across shallow-marine shelves.

Nittrouer, C.A., Austin, J.A., Field, M.E.,

Kravitz, J.H., Syvitski, J.P.M. and Wiberg, P.L. *eds.*, 2007, Continental Margin Sedimentation: From Sediment Transport to Sequence Stratigraphy: International Association of Sedimentologists, Special Publication 37, Blackwell Publishing, Oxford, 549 p.
Compilation of comprehensive, state-of-the-art studies of sediment transport and stratigraphy on contrasting Atlantic and Pacific shelves of the USA.

Reading, H.G. and Collinson, J.D., 1996, Clastic coasts, *in* Reading, H.G., *ed.*, Sedimentary Environments: Processes, Facies and Stratigraphy: Blackwell Science, Oxford, p. 154-231.
Comprehensive summary of all major sedimentary environments; enormous reference list.

Schieber, J., Zimmerle, W. and Sethi, P.S., 1998, Shales and mudstones, 1. Basin studies, sedimentology, and paleontology: E. Schweizerbart'sche Verlagsbuchhandlung, Stuttgart, 384 p.
Collection of review papers and case studies highlighting the sedimentology and stratigraphy of mudrocks.

Suter, J.R., 2006, Clastic Shelves, *in Posamentier*, H.W. and Walker, R.G., *eds.*, Facies Models Revisited: SEPM Special Publication 84, p. 339-397.
Beautifully illustrated summary of current ideas on shelf processes and deposits, including both wave and tidal-dominated systems.

Wilgus, C.K., Hastings, B.S., Posamentier, H.W., van Wagoner, J.C., Ross, C.A. and Kendall, C.G. St.C., *eds.*, 1988. Sea Level Changes: An Integrated Approach: SEPM Special Publication 42, 407 p.
Now dated but still important collection of papers outlining theory and application of sequence stratigraphy.

Other references

Ainsworth, R.B., 1994, Marginal marine sedimentology and high-resolution sequence analysis; Bearpaw-Horseshoe Canyon transition, Drumheller, Alberta: Bulletin of Canadian Petroleum Geology, v. 42, p. 26-54.

Aigner, T., 1985, Storm Depositional Systems. Lecture Notes in Earth Sciences: Springer-Verlag, Berlin, 174 p.

Amos, C.L., Li, M.Z. and Choung, K-S., 1996, Storm-generated hummocky stratification on the outer Scotian Shelf: Geo-Marine Letters, v. 16, p. 85-94.

Amos, C.L., Li, M.Z., Chiocci, F.L., LaMonica, G.B., Cappucci, S., King, E.H. and Corbani, F., 2003, Origin of shore-normal channels from the shoreface of Sable Island, Canada: Journal of Geophysical Research, v. 108C, p. 3094, doi:10.1029/2001JC001259, 2003.

Arnott, R.W. and Southard, J.B., 1990, Exploratory flow-duct experiments on combined-flow bed configurations, and some implications for interpreting storm-event stratification: Journal of Sedimentary Petrology, v. 60, p. 211-219.

Ashley, G.M., 1990, Classification of large-scale subaqueous bedforms; a new look at an old problem: Journal of Sedimentary Research, v. 60, p. 161-172.

Bergman, K.M. and Walker, R.G., 1987, The importance of sea-level fluctuations in the formation of linear conglomerate bodies; Carrot Creek Member of Cardium Formation, Cretaceous Western Interior Seaway, Alberta, Canada: Journal of Sedimentary Petrology, v. 57, p. 651-665.

Berné, S., Lericolais, G., Marsset, T., Bourillet, J.F. and de Batist, M., 1998, Erosional offshore sand ridges and lowstand shorefaces: examples from tide- and wave-dominated environments of France: Journal of Sedimentary Research. v. 68, p. 540-555.

Boreen, T. and Walker, R.G., 1991, Definition of allomembers and their facies assemblages in the Viking Formation, Willesden Green area, Alberta: Bulletin of Canadian Petroleum Geology, v. 39, p. 123-144.

Bradshaw, B.E. and Nelson, C.S., 2004, Anatomy and origin of autochthonous late Pleistocene forced regression deposits, east Coromandel inner shelf, New Zealand: implications for the development and definition of the regressive systems tract: New Zealand Journal of Geology and Geophysics, v. 47, p. 81-99.

Browne, G.H. and Naish, T.R., 2003, Facies development and sequence architecture of a late Quaternary fluvial-marine transition, Canterbury Plains and shelf, New Zealand: implications for forced regressive deposits: Sedimentary Geology, v. 158, p. 57-86.

Caccione, D.A., Wiberg, P.L., Lynch, J., Irish, J. and Traykovski, P., 1999, Estimates of suspended-sediment flux and bedform activity on the inner portion of the Eel continental shelf: Marine Geology, v. 14, p. 83-97.

Caldas, L.H.O., Oliveira, J.G., Medeiros, W.E., Stattegger, K., and Vital, H., 2006, Geometry and evolution of Holocene transgressive and regressive barriers on the semi-arid coast of NE Brazil: Geo-Marine Letters, v. 26, p. 249-263.

Cattaneo, A. and Steel, R.J., 2003, Transgressive deposits: a review of their variability: Earth-Science Reviews, v. 62, p. 187-228.

Cattaneo, A., Correggiari, A., Langone, L. and Tricardi, F., 2003, The late-Holocene Gargano subaqueous delta, Adriatic shelf: Sediment pathways and supply fluctuations: Marine Geology, v. 193, p. 61-91.

Cattaneo, A., Trincardi, F., Asioli, A. and Correggiari, A. 2007, The western Adriatic shelf clinoform: energy-limited bottomset: Continental Shelf Research, v. 27, p. 506-525.

Cheel, R.J., 1991, Grain fabric in hummocky cross-stratified storm beds: genetic implications: Journal of Sedimentary Petrology, v. 61, p. 102-110.

Cheel, R.J. and Leckie, D.A., 1993, Hummocky cross-stratification: Sedimentology Review No. 1. Oxford, UK, Blackwell Scientific Publications, p. 103-122.

Clifton, H.E., 1976, Wave-formed sedimentary structures - a conceptual model, *in Davies*, R.A. and Ethington, R.L., *eds.*, Beach and Nearshore Sedimentation: SEPM Special Publication 24, p. 126-148.

Clifton, H.E., 1981, Progradational sequences in Miocene shoreline deposits, southeastern Caliente Range, California: Journal of Sedimentary Petrology, v. 51, p. 165-184.

Clifton, H.E., 2003, Supply, segregation, successions, and significance of shallow marine conglomeratic deposits: Bulletin of Canadian Petroleum Geology, v. 51, p. 370-388.

Clifton, H.E., Hunter, R.E. and Phillips, R.L., 1971, Depositional structures and processes in the non-barred, high-energy nearshore: Journal of Sedimentary Petrology, v. 41, p. 651-670.

Curray, J.R., Emmel, F.J. and Crampton, P.J.S., 1969, Holocene history of a strandplain, lagoonal coast, Nayarit, Mexico*:* *in* Castanares, A.A. and Phleger, F.B., *eds.*, Coastal Lagoons - A Symposium. Mexico City, Universidad Nacional Autonoma, p. 63-100.

Davis, R.A. and Hayes, M.O., 1984, What is a wave-dominated coast?: Marine Geology, v. 60, p. 313-329.

deRaaf, J.F.M., Boersma, J.R. and van Gelder, A., 1977, Wave-generated structures and sequences from a shallow marine deposit, Lower Carboniferous, County Cork, Ireland: Sedimentology, v. 24, p. 451-483.

Dominguez, J.M.L., Martin, L. and Bittencourt, A.C.S.P., 1987, Sea-level history and Quaternary evolution of river mouth associated beach-ridge plains along the south-east Brazilian coast: a summary, *in* Nummedal, D., Pilkey, O.H. and Howard, J.D., *eds.*, Sea-Level Fluctuations and Coastal Evolution. SEPM Special Publication 41, p. 115-127.

Duke, W.L., 1985, Hummocky cross-stratification, tropical hurricanes, and intense winter storms: Sedimentology, v. 32, p. 167-194.

Duke, W.L., 1990, Geostrophic circulation or shallow marine turbidity currents? The dilemma of paleoflow patterns in storm-influenced prograding shoreline systems: Journal of Sedimentary Petrology, v. 60, p. 870-883.

Dumas, S. and Arnott, R.W.C., 2006, Origin of hummocky and swaley cross-stratification - The controlling influence of unidirectional current strength and aggradation rate: Geology, v. 34, p. 1073-1076.

Dumas, S., Arnott, R.W.C. and Southard, J.B. 2005, Experiments on oscillatory-flow and combined-flow bed forms: implications for interpreting parts of the shallow-marine sedimentary record: Journal of Sedimentary Research, v. 75, p. 501-513.

Dunbar, G.B. and Barrett, P.J., 2005, Estimating palaeobathymetry of wave-graded continental shelves from sediment texture: Sedimentology, v. 52, p. 253-269.

Edwards, C.M., Hodgson, D.M., Flint, S.S. and Howell, J.A., 2005a, Contrasting styles of shelf sediment transport and deposition in a ramp margin setting related to relative sea-level change and basin floor topography, Turonian (Cretaceous) Western Interior of central Utah, USA: Sedimentary Geology, v. 179, p. 117-152.

Edwards, C.M., Howell, J.A. and Flint, S.S., 2005b, Depositional and stratigraphic architecture of the Santonian Emery Sandstone of the Mancos Shale: Implications for Late Cretaceous evolution of the Western Interior foreland basin of Central Utah, USA: Journal of Sedimentary Research, v. 75, p. 280-299.

Fitzsimmons, R. and Johnson, S., 2000, Forced regressions: recognition, architecture and genesis in the Campanian of the Bighorn Basin, Wyoming, *in* Hunt, D and Gawthorpe, R.L.G., *eds*., Sedimentary Responses to Forced Regressions: Geological Society of London, Special Publication 172, p. 113-139.

Forristal, G.Z., Hamilton, R.C. and Cardone, V.J., 1977, Continental shelf currents in tropical storm Delia: observations and theory: Journal of Physical Oceanography, v. 87, p. 532-546.

Friedrichs, C.T. and Scully, M.E., 2007, Modeling deposition by wave-supported gravity flows on the Po River prodelta: From seasonal floods to prograding clinoforms: Continental Shelf Research, v. 27, p. 322-337.

Hampson, G.J. and Storms, J.E.A., 2003, Geomorphological and sequence stratigraphic variability in wave-dominated, shoreface-shelf parasequences: Sedimentology, v. 50, p. 667-701.

Harms, J.C., Southard, J.B., Spearing, D.R. and Walker, R.G., 1975, Depositional Environments as Interpreted from Primary Sedimentary Structures and Stratification Sequences: SEPM Short Course 2, 161 p.

Harms, J.C., Southard, J.B. and Walker, R.G., 1982, Structures and Sequences in Clastic Rocks: SEPM Short Course 9, 249 p.

Hart, B.S. and Plint, A.G., 1995, Gravelly shoreface and beachface deposits, *in* Plint, A.G., *ed*., Clastic Facies Analysis - a Tribute to the Research and Teaching of Harold G. Reading: International Association of Sedimentologists, Special Publication 22, Blackwell Science, Oxford, p. 75-99.

Hart, B.S., Vantfoort, R.J. and Plint, A.G., 1990, Discussion of: Is there evidence for geostrophic currents preserved in inner to middle-shelf deposits?: Journal of Sedimentary Petrology, v. 60, p. 633-635.

Hay, M.J., Brettle, M.J., Tyagi, A., Varban, B.L., Kreitner, M.A., and Plint, A.G., 2003, Sediment dispersal patterns on a huge muddy shelf: Middle Cretaceous Shaftesbury to Cardium interval, Alberta and British Columbia: Canadian Society of Petroleum Geologists, Annual Convention, Calgary June 2-6, 2003. Abstract on CD. [http://www.cspg.org/conventions/abstracts/2003abstracts_author.htm]

Helland-Hansen, W. and Martinsen, O.J., 1996, Shoreline trajectories and sequences: description of variable depositional-dip scenarios: Journal of Sedimentary Research, v. 66, p. 670-688.

Hill, P.S. and 11 others, 2007, Sediment delivery to the seabed on continental margins, *in* Nittrouer, C.A. *et al., eds*., Continental Margin Sedimentation, from Sediment Transport to Sequence Stratigraphy: International Association of Sedimentologists, Special Publication 37, Blackwell Publishing, Oxford, p. 49-99.

Hunter, R.E., Clifton, H.E. and Phillips, R.L., 1979, Depositional processes, sedimentary structures and predicted vertical sequences in barred nearshore systems, southern Oregon coast: Journal of Sedimentary Petrology, v. 49, p. 711-726.

Kerr, M. and Eyles, N., 1991, Storm-deposited sandstones (tempestites) and related ichnofossils of the Late Ordovician Georgian Bay Formation, southern Ontario, Canada: Canadian Journal of Earth Sciences, v. 28, p. 266-282.

Kineke, G.C., Sternberg, R.W., Trowbridge, J.H. and Geyer, W.R., 1996, Fluid-mud processes on the Amazon continental shelf: Continental Shelf Research, v. 16, p. 667-696.

Komar, P.D., 1976, Beach Processes and Sedimentation: Englewood Cliffs, New Jersey, Prentice Hall, 429 p.

Kuehl, S.A., Nittrouer, C.A., Allison, M.A., Faria, L.E.C., Dukat, D.A., Jaeger, J.M., Pacioni, T.D., Figueiredo, A.G. and Underkoffler, E.C., 1996, Sediment deposition, accumulation, and seabed dynamics in an energetic fine-grained coastal environment: Continental Shelf Research, v. 16, p. 787-815.

Leckie, D.A., 2003, Modern environments of the Canterbury Plains and adjacent offshore areas, New Zealand - an analog for ancient conglomeratic depositional systems in nonmarine and coastal zone settings: Bulletin of Canadian Petroleum Geology, v. 51, p. 389-425.

Leckie, D.A. and Krystinik, L.F., 1989, Is there evidence for geostrophic currents preserved in the sedimentary record of inner to middle-shelf deposits?: Journal of Sedimentary Petrology, v. 59, p. 862-870.

Leckie, D.A. and Walker, R.G., 1982, Storm- and tide-dominated shoreline in Cretaceous Moosebar- Lower Gates interval - outcrop equivalents of Deep Basin gas trap in western Canada: American Association of Petroleum Geologists Bulletin, v. 66, p. 138-157.

Leggitt, S.M., Walker, R.G. and Eyles, C.M.,1990, Control of reservoir geometry and stratigraphic trapping by erosion surface E5 in the Pembina-Carrot Creek area; Upper Cretaceous Cardium Formation, Alberta, Canada: American Association of Petroleum Geologists Bulletin, v. 74, p. 1265-1182.

Lessa, G.C., Angulo, R.J., Giannini, P.C. and Araújo, A.D., 2000, Stratigraphy and Holocene evolution of a regressive barrier in south Brazil: Marine Geology, v. 165, p. 87-108.

Li, M. Z. and Amos, C.L., 1999, Sheet flow and large wave ripples under combined waves and currents: field observations, model predictions and effects on boundary layer dynamics: Continental Shelf Research, v. 19, p. 637-663.

Li, M. Z. and King, E.L., 2007, Multibeam bathymetric investigations of the morphology of sand ridges and associated bedforms and their relation to storm processes, Sable Island Bank, Scotian Shelf: Marine Geology, v. 243, p. 200-228.

Liu, J.P., Xu, K.H., Li, A.C., Milliman, J.D., Velozzi, D.M., Xiao, S.B. and Yang, Z.S., 2007, Flux and fate of Yangtze River sediment delivered to the East China Sea: Geomorphology, v. 85, p. 208-224.

Macquaker, J.H.S. and Bohacs, K.M., 2007, On the accumulation of mud: Science, v. 318, p. 1734-1735.

Macquaker, J.H.S., Taylor, K.G. and Gawthorpe, R.L., 2007, High-resolution facies analysis of mudstones: Implications for paleoenvironmental and sequence stratigraphic interpretations of offshore ancient mud-dominated successions: Journal of Sedimentary Research, v. 77, p. 324-339.

Martinsen, R.S., 2003, Depositional remnants, part 1: Common components of the stratigraphic record with important implications for hydrocarbon exploration: American Association of Petroleum Geologists Bulletin, v. 87, p. 1869-1882.

McBride, R.A., Anderson, L.C., Tudoran, A., Roberts, H.H., Sen Gupta, B.K. and Byrne, M.R., 1999, Holocene stratigraphic architecture of a sand-rich shelf and origin of linear shoals: northeastern Gulf of Mexico, *in* Bergman, K.M. and

Snedden, J.W., *eds.*, Isolated Shallow Marine Sand Bodies: Sequence Stratigraphic Analysis and Sedimentologic Interpretation. SEPM Special Publication 64, p. 95-126.

McCrory, V.L.C. and Walker, R.G., 1986, A storm- and tidally-influenced prograding shoreline - Upper Cretaceous Milk River Formation of southern Alberta, Canada: Sedimentology, v. 33, p. 47-60.

Mellere, D. and Steel, R.J., 1995, Variability of lowstand wedges and their distinction from forced-regressive wedges in the Mesaverde Group, southeast Wyoming: Geology, v. 23, p. 803-806.

Mellere, D. and Steel, R., 2000, Style contrast between forced regressive and lowstand/transgressive wedges in the Campanian of south-central Wyoming (Hatfield Member of the Haystack Mountains Formation*), in* Hunt, D. and Gawthorpe, R.L., *eds.*, Sedimentary Responses to Forced Regressions: Geological Society of London, Special Publication 172, p. 141-162.

Midtgaard, H., 1996, Inner-shelf to lower shoreface hummocky sandstone bodies with evidence for geostrophic influenced combined flow, Lower Cretaceous, West Greenland: Journal of Sedimentary Research, v. 66, p. 343-353.

Miller, L. and Douglas, B.C., 2006, On the rate and causes of twentieth century sea-level rise; Philosophical Transactions of the Royal Society, Series A. v. 364, p. 805-820.

Morton, R.A., 1988, Nearshore responses to great storms, *in* Clifton, H.E., *ed.*, Sedimentologic consequences of convulsive geologic events: Geological Society of America, Special Paper 229, p. 7-22.

Myrow, P.M., 1992, Pot and gutter casts from the Chapel Island Formation, southeast Newfoundland: Journal of Sedimentary Petrology, v. 62, p. 992-1007.

Myrow, P.M. and Southard, J.B., 1996, Tempestite deposition: Journal of Sedimentary Research, v. 66, p. 857-887.

Naish, T. and Kamp, P.J.J., 1997, Sequence stratigraphy of sixth-order (41 k.y.) Pliocene-Pleistocene cyclothems, Wanganui Basin, New Zealand: A case for the regressive systems tract: Geological Society of America Bulletin, v. 109, p. 978-999.

Nelson, C.H., 1982, Modern shallow-water graded sand layers from storm surges, Bering Shelf: a mimic of Bouma sequences and turbidite systems: Journal of Sedimentary Petrology, v. 52, p. 537-545.

Nittrouer, C.A., Kuehl, S.A., Figueiredo, A.G., Allison, M.A., Sommerfield, C.K., Rine, J.M., Faria, L.E.C. and Silveira, O.M., 1996, The geological record preserved by Amazon shelf sedimentation: Continental Shelf Research, v. 15, p. 817-841.

Nummedal, D. and Molenaar, C.M., 1995, Sequence stratigraphy of the Gallup Sandstone*, in* Van Wagoner, J.C. and Bertram, G.T., *eds.*, Sequence Stratigraphy of Foreland Basin Deposits: American Association of Petroleum Geologists Memoir 64, p. 277-310.

Nummedal, D., Riley, G.W. and Templet, P.L., 1993, High-resolution sequence architecture: a chronostratigraphic model based on equilibrium profile studies*, in* Posamentier, H.W., Summerhayes, C.P., Haq, B.U. and Allen, G.P., *eds.*, Sequence Stratigraphy and Facies Associations: International Association of Sedimentologists, Special Publication 18, Blackwell Science, Oxford, p. 55-68.

Ogston, A.S., Caccione, D.A., Sternberg, R.W. and Kineke, G.C., 2000, Observations of storm and river flood-driven sediment transport on the northern California continental shelf: Continental Shelf Research, v. 20, p. 2141-2162.

Olsen, T., Mellere, D. and Olsen, T., 1999, Facies architecture and geometry of landward-stepping shoreface tongues: the Upper Cretaceous Cliff House Sandstone (Mancos Canyon, southwest Colorado): Sedimentology, v. 46, p. 603-625.

Pattison, S.A.J. and Walker, R.G., 1992, Deposition and interpretation of long, narrow sandbodies underlain by a basinwide erosion surface: Cardium alloformation, Cretaceous Western Interior Seaway, Alberta, Canada: Journal of Sedimentary Petrology, v. 62, p. 292-309.

Pattison, S.A.J., Ainsworth, R.B. and Hoffman, T.A., 2007, Evidence of across-shelf transport of fine-grained sediments: turbidite-filled shelf channels in the Campanian Aberdeen Member, Book Cliffs, Utah, USA: Sedimentology, v. 54, p. 1033-1063.

Pekar, S.F., Christie-Blick, N., Miller, K.G. and Kominz, M.A., 2003, Quantitative constraints on the origin of stratigraphic architecture at passive continental margins: Oligocene sedimentation in New Jersey, U.S.A.: Journal of Sedimentary Research, v. 73, p. 227-245.

Peltier, W.R. and Fairbanks, R.G., 2006, Global glacial ice volume and last glacial maximum duration from an extended Barbados sea level record: Quaternary Science Reviews, v. 25, p. 3322-3337.

Plint, A.G., 1988, Sharp-based shoreface sequences and "offshore bars" in the Cardium Formation: Their relationship to relative changes in sea level*, in* Wilgus, C.K. *et al., eds.,* Sea Level Changes: An Integrated Approach. SEPM Special Publication 42, p. 357-370.

Plint, A.G., 1991, High-frequency relative sea level oscillations in Upper Cretaceous shelf clastics of the Alberta Foreland Basin: Evidence for a Milankovitch- scale glacio-eustatic control*? in* Macdonald, D.I.M., *ed.*, Sedimentation, Tectonics and Eustacy: International Association of Sedimentologists, Special Publication 12, Blackwell Science, Oxford, p. 409-428.

Plint, A.G., 1996, Marine and nonmarine systems tracts in fourth order sequences in the Early-Middle-Cenomanian, Dunvegan Alloformation, northeastern British Columbia, Canada*, in* J. Howell and J.D. Aitken, *eds.*, High Resolution Sequence Stratigraphy: Innovations and Applications: Geological Society of London, Special Publication 104, p. 159-191.

Plint, A.G. and Kreitner, M.A., 2007, Extensive, thin sequences spanning Cretaceous foredeep suggest high-frequency eustatic control: Late Cenomanian, Western Canada foreland basin, Geology, v. 35, p. 735-738.

Plint, A.G. and Norris, B., 1991, Anatomy of a ramp margin sequence: Facies successions, paleogeography and sediment dispersal patterns in the Muskiki and Marshybank formations, Alberta Foreland Basin. Bulletin of Canadian Petroleum Geology, v.39, p. 18-42.

Plint, A.G. and Nummedal, D., 2000, The falling stage systems tract: Recognition and importance in sequence stratigraphic analysis*, in* Hunt, D. and Gawthorpe, R.L., *eds.*, Sedimentary Responses to Forced Regressions: Geological Society of London, Special Publication 172, p. 1-17.

Plint, A.G. and Walker, R.G., 1987, Cardium Formation 8. Facies and environments of the Cardium shoreline and coastal plain in the Kakwa field and adjacent areas, northwestern Alberta: Bulletin of Canadian Petroleum Geology, v. 35, p. 48-64.

Plint, A.G., Walker, R.G. and Bergman, K.M., 1986, Cardium Formation 6. Stratigraphic framework of the Cardium in subsurface: Bulletin of Canadian Petroleum Geology, v. 34, p. 213-225.

Plint, A.G., Walker, R.G. and Duke, W.L., 1988, An outcrop to subsurface correlation of the Cardium Formation in Alberta*, in* James, D.P. and Leckie, D.A., *eds.*, Sequences, Stratigraphy, Sedimentology: Surface and Subsurface, Canadian Society of Petroleum Geologists, Memoir 15, p. 167-184.

Plint, A.G., Tyagi, A., Hay, M.J., Varban, B.L., Zhang, H. and Roca, X., 2009, Clinoforms, paleobathymetry, and mud dispersal across the Western Canada Cretaceous foreland basin: evidence from the Cenomanian Dunvegan Formation and contiguous strata. Journal of Sedimentary Research, v. 79, p. 144-161.

Posamentier, H.W., 2002, Ancient shelf

ridges - A potentially significant component of the transgressive systems tract: Case study from offshore northwest Java. American Association of Petroleum Geologists, Bulletin, v. 86, p. 75-106.

Posamentier, H.W., Allen, G.P., James, D.P. and Tesson, M., 1992, Forced regressions in a sequence stratigraphic framework: concepts, examples and exploration significance: American Association of Petroleum Geologists, Bulletin, v. 76, p. 1687-1709.

Ridente, D. and Trincardi, F., 2005, Pleistocene 'muddy' forced regression deposits on the Adriatic shelf: A comparison with prodelta deposits of the late Holocene highstand mud wedge: Marine Geology, v. 222, p. 213-233.

Rine, J.M. and Ginsburg, R.N., 1985, Depositional facies of a mud shoreface in Suriname, South America - a mud analogue to sandy, shallow-marine deposits: Journal of Sedimentary Petrology, v. 55, p. 633-652.

Rine, J.M., Tillman, R.W., Culver, S.J. and Swift, D.J.P., 1991, Generation of late Holocene sand ridges on the middle continental shelf of New Jersey, USA - evidence for formation in a mid-shelf setting based on comparisons with a nearshore ridge: *in* Swift, D.J.P., Oertel, G.F., Tillman, R.W. and Thorne, J.A., *eds*., Shelf Sands and Sandstone Bodies. International Association of Sedimentologists, Special Publication 14, Blackwell Science, Oxford, p. 395-423.

Roy, P.S., Zhuang, W-Y., Birch, G.F., Cowell, P.J. and Li, C., 1997, Quaternary geology of the Forster-Tuncurry coast and shelf, southeast Australia. Geological Survey of New South Wales, Report GS 1992/201, 405 p.

Runkel, A.J., Miller, J.F., McKay, R.M. and Palmer, A.R., 2007, High-resolution sequence stratigraphy of lower Paleozoic sheet sandstones in central North America: The role of special conditions of cratonic interiors in the development of stratal architecture: Geological Society of America, Bulletin, v. 119, p. 860-881.

Schieber, J., Southard, J. and Thaisen, K., 2007, Accretion of mudstone beds from migrating floccule ripples, Science, v. 318, p. 1760-1763.

Snedden, J.W. and Nummedal, D., 1991, Origin and geometry of storm-deposited sand beds in modern sediments of the Texas continental shelf, *in* Swift, D.J.P., Oertel, G.F., Tillman, R.W. and Thorne, J.A., *eds*., Shelf Sands and Sandstone Bodies: International Association of Sedimentologists, Special Publication 14, Blackwell Science, Oxford, p. 283-308.

Snedden, J.W. and Bergman, K.M., 1999, Isolated shallow marine sandbodies: Deposits for all interpretations, *in* Bergman, K.M. and Snedden, J.W., *eds*., Isolated Shallow Marine Sand Bodies: Sequence Stratigraphic Analysis and Sedimentologic Interpretation: SEPM Special Publication 64, p. 1-11.

Snedden, J.W. and Dalrymple, R.W., 1999, Modern shelf sand ridges: from historical perspective to a unified hydrodynamic and evolutionary model, *in* Bergman, K.M. and Snedden, J.W., *eds*., Isolated Shallow Marine Sand Bodies: Sequence Stratigraphic Analysis and Sedimentologic Interpretation: SEPM Special Publication 64, p. 13-28.

Snedden, J.W., Nummedal, D. and Amos, A.F., 1988, Storm- and fair-weather combined flow on the central Texas continental shelf: Journal of Sedimentary Petrology, v. 58, p. 580-595.

Snedden, J.W., Kreisa, R.D., Tillman, R.W., Culver, S.J. and Schweller, W.J., 1999, An expanded model for modern shelf sand ridge genesis and evolution on the New Jersey Atlantic shelf, *In* Bergman, K.M. and Snedden, J.W., *eds*., Isolated Shallow Marine Sand Bodies: Sequence Stratigraphic Analysis and Sedimentologic Interpretation: SEPM Special Publication 64, p. 147-163.

Southard, J.B., Lambie, J.M., Federico, D.C., Pile, H.T. and Weidman, C.R., 1990, Experiments on bed configurations under bidirectional purely oscillatory flow, and the origin of hummocky cross-stratification: Journal of Sedimentary Petrology, v. 60. p. 1-17.

Suter, J.R. and Clifton, H.E., 1999, The Shannon Sandstone and isolated linear sand bodies: Interpretations and realizations, *in* Bergman, K.M. and Snedden, J.W., *eds*., Isolated Shallow Marine Sand Bodies: Sequence Stratigraphic Analysis and Sedimentologic Interpretation. SEPM Special Publication 64, p. 321-356.

Swift, D.J.P. and Field, M.E., 1981, Evolution of a classic sand ridge field: Maryland sector, North American inner shelf: Sedimentology, v. 28, p. 461-482.

Swift, D.J.P., Freeland, G.L. and Young, R.A., 1979, Time and space distribution of megaripples and associated bedforms, Middle Atlantic Bight, North American Atlantic Shelf: Sedimentology, v. 26, p. 389-406.

Swift, D.J.P., Young, R.A., Clarke, T.L., Vincent, C.E., Niedoroda, A. and Lesht, B., 1981, Sediment transport in the Middle Atlantic Bight of North America: synopsis of recent observations, *in* Nio, S-D., Shüttenhelm, R.T.E. and van Weering, T.C.E., *eds*., Holocene Marine Sedimentation in the North Sea Basin: International Association of Sedimentologists, Special Publication 5, Blackwell Science, Oxford, p. 361-383.

Swift, D.J.P., Han, G. and Vincent, C.E., 1986, Fluid processes and sea-floor response on a modern storm-dominated shelf: Middle Atlantic shelf of North America. Part I: the storm-current regime, *in*: Knight, R.J. and McLean, J.R., *eds*., Shelf Sands and Sandstones: Canadian Society of Petroleum Geologists, Memoir 11, p. 99-119.

Tamura, T., Nanayama, F., Saito, Y., Murakami, F., Nakashima, R. and Watanabe, K., 2007, Intra-shoreface erosion in response to rapid sea-level fall: depositional record of a tectonically uplifted strand plain, Pacific coast of Japan: Sedimentology, v. 54, p. 1149-1162.

Tillman, R.W. and Martinsen, R.S., 1984, The Shannon shelf-ridge sandstone complex, Salt Creek anticline area, Powder River Basin, Wyoming, *in* Tillman, R.W. and Siemers, C.T., *eds*., Siliciclastic Shelf Sediments: SEPM Special Publication 34, p. 85-142.

Tomazelli, L.J. and Dillenburg, S.R., 2007, Sedimentary facies of a last interglacial coastal barrier in south Brazil: Marine Geology, v. 33, p. 33-45.

Traykovski, P., Geyer, W.R., Irish, J.D. and Lynch, J.F., 2000, The role of wave-induced density-driven fluid mud flows for cross-shelf transport on the Eel River continental shelf: Continental Shelf Research, v. 20, p. 2113-2140.

Tyagi, A., Plint, A.G. and McNeil, D.H., 2007, Correlation of physical surfaces, bentonites, and biozones in the Colorado Group from the Alberta Foothills to south-western Saskatchewan, and a revision of the Belle Fourche / Second White Specks formational boundary: Canadian Journal of Earth Sciences, v. 44, p. 871-888.

Varban, B.L. and Plint, A.G., 2005, Allostratigraphy of the Kaskapau Formation (Cenomanian-Turonian) in subsurface and outcrop: NE British Columbia and NW Alberta, western Canada foreland basin, Bulletin of Canadian Petroleum Geology, v. 53, p. 357-389.

Varban, B.L. and Plint, A.G., 2008a, Palaeoenvironments, palaeogeography, and physiography of a large, shallow, muddy ramp: Late Cenomanian-Turonian Kaskapau Formation, western Canada foreland basin: Sedimentology, v. 55, p. 201-233.

Varban, B.L. and Plint, A.G. 2008b, Sequence stacking patterns in the western Canada foredeep: Influence of tectonics, sediment loading and eustasy on deposition of the Upper Cretaceous Kaskapau and Cardium formations: Sedimentology, v. 55, p. 395-421.

Wadsworth, J.A. and Walker, R.G., 1991, Morphology and origin of erosion surfaces in the Cardium Formation (Upper Cretaceous, Western Interior Seaway, Alberta) and their implications for rapid sea level fluctuations: Canadian Journal of Earth Sciences, v. 28, p. 1507-1520.

Walker, R.G., 1979, Shallow marine sands, *in* Walker, R.G., *ed*., Facies Models: Geoscience Canada, Reprint Series 1, p. 75-89.

Walker, R.G., 1984, Shelf and shallow

marine sands, *in* Walker, R.G., *ed.*, Facies Models, 2nd Edition., Geoscience Canada Reprint Series 1, p. 141-170.

Walker, R.G. and Plint, A.G., 1992, Wave and storm dominated shallow marine systems, *in* Walker, R.G. and James, N.P., *eds.*, Facies Models, Responses to sea-level change: Geological Association of Canada, Waterloo, Ontario, p. 219-238.

Walker, R.G. and Eyles, C.H., 1988, Geometry and facies of stacked shallow-marine sandier-upward sequences dissected by erosion surface, Cardium Formation, Willesden Green, Alberta: American Association of Petroleum Geologists Bulletin, v. 72, p. 1469-1494.

Walker, R.G. and Eyles, C.H., 1991, Topography and significance of a basin-wide sequence-bounding erosion surface in the Cretaceous Cardium Formation, Alberta, Canada: Journal of Sedimentary Petrology, v. 61, p. 473-496.

Walker, R.G. and Wiseman, T.R., 1995, Lowstand shorefaces, transgressive incised shorefaces, and forced regressions: examples from the Viking Formation, Joarcam area, Alberta: Journal of Sedimentary Research, v. 65, p. 132-141.

Wells, J.T. and Coleman, J.M., 1981, Physical processes and fine-grained sediment dynamics, coast of Suriname, South America: Journal of Sedimentary Petrology, v. 51, p. 1053-1068.

Wignall, P.B., Sutcliffe, O.E., Clemson, J. and Young, E., 1996, Unusual shoreface sedimentology in the Upper Jurassic of the Boulonnais, northern France: Journal of Sedimentary Research, v. 66, p. 577-586.

Williams, C.J., Hesselbo, S.P., Jenkyns, H.C. and Morgans-Bell, H.S., 2001, Quartz silt in mudrocks as a key to sequence stratigraphy (Kimmeridge Clay Formation, Late Jurassic, Wessex Basin, UK): Terra Nova, v. 13, p. 449-455.

Wright, L.D. and Friedrichs, C.T., 2006, Gravity-driven sediment transport on continental shelves: A status report: Continental Shelf Research, v. 26, p. 2092-2107.

Wright, L.D., Friedrichs, C.T. and Scully, M.E., 2002, Pulsational gravity-driven sediment transport on two energetic shelves: Continental Shelf Research, v. 22, p. 2443-2460.

Wright, L.D., Friedrichs, C.T., Kim, S.C. and Scully, M.E., 2001, Effects of ambient currents and waves on gravity-driven sediment transport on continental shelves: Marine Geology, v. 175, p. 25-45.

Yang, B., Dalrymple, R.W. and Chun, S., 2006, The significance of hummocky cross-stratification (HCS) wavelengths: evidence from an open-coast tidal flat, South Korea: Journal of Sedimentary Research, v. 76, p. 2-8.

Yoshida, S., Steel, R.J. and Dalrymple, R.W., 2007, Changes in depositional processes - an ingredient in a new generation of sequence-stratigraphic models: Journal of Sedimentary Research, v. 77, p. 447-460.

9. TIDAL DEPOSITIONAL SYSTEMS

Robert W. Dalrymple, Department of Geological Sciences and Geological Engineering, Queen's University, Kingston, ON, K7L 3N6, Canada

TIDES AND TIDAL CURRENTS

Origin of Tides

The term *tide* is most commonly used to refer to any *periodic* (regularly repeating) fluctuation in the water level (Fig. 1) that is generated by the deformation of the ocean surface by the gravitational attraction of the moon and sun. More precisely, such a tide is an *astronomical tide*, and is distinct from a *wind tide*, which is a water-level fluctuation caused by changes in the strength and direction of the wind. Wind tides are small (usually only a few decimeters of water-level change) and are generally not important sedimentologically, except in cases where astronomic tides are negligible, such as in microtidal areas. During major storms, however, they may reach several meters in height and are called *storm surges* (*see* Chapter 8). Unlike astronomic tides, wind tides are generally not periodic; they are not considered further here.

Both the sun and moon play a role in generating astronomic tides; however, the moon, although a smaller mass than the sun, exerts a tide-generating force that is slightly more than twice as large as that of the sun because it is much closer to the earth. One of the best non-mathematical discussions of the origin of tides is given by the Open University Course Team (1989). Allen (1997) and Kvale (2006) also provide comprehensive discussions of the origin and nature of tides.

Lunar tides represent the vector sum of two forces: 1) the gravitational attraction of the moon, which is strongest on the side of the earth facing the moon, and 2) the centrifugal force caused by the revolution of the earth–moon system about its common center of mass (Fig. 2). This centrifugal force is of uniform intensity and direction all over the earth. On the side of the earth facing the moon,

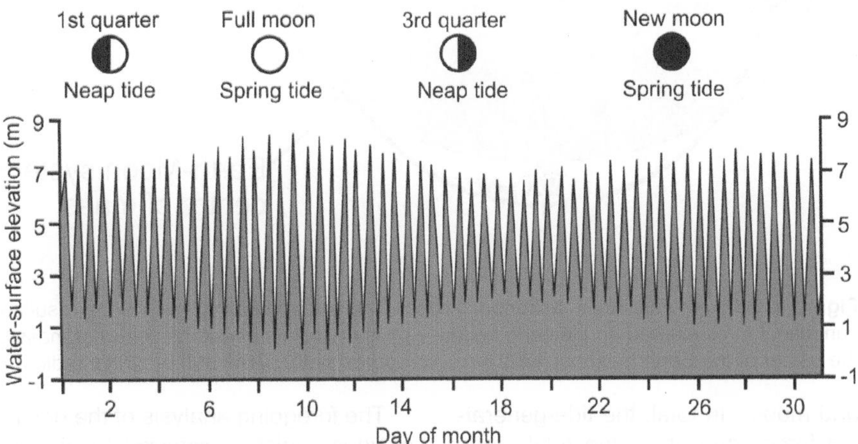

Figure 1. Predicted tidal variations in water level at Saint John, New Brunswick. Following standard practice, the datum is approximately the spring low-tide elevation. The tides are semi-diurnal with a pronounced variation in range between neap and spring tides. A diurnal inequality is evident during days 8–12 and 22–28.

the gravitational attraction of the moon is greater than the oppositely directed centrifugal force, while the reverse is true on the other side of the earth. The water in the oceans therefore piles up in two bulges, one underneath the moon, and the other on the opposite side of the earth. As a result of the earth's rotation, the bulges appear to travel around the earth as two 'tidal waves', causing water levels to rise and fall regularly. Rising water levels are known as the *flood tide*, whereas the fall is the *ebb tide*. The period for one complete *tidal cycle* produced by the moon is just slightly longer than one-half of a day (12.42 hours) and is therefore said to be *semi-diurnal*. (This is longer than 12 hours because the moon orbits in the same direction as the earth rotates, and any point on the surface of the earth requires a little more than one revolution to return to the peak of the tidal bulge). Because the earth's axis is inclined with respect to the plane of the moon's orbit most of the time, any given point on the earth's surface passes closer to the crest of one tidal bulge than the other, thereby adding a *diurnal* (once a day) component to the tidal spectrum, such that the two daily high and low tides are of unequal elevation (Fig. 2). This *diurnal inequality* is absent when the moon is on the equator, something that happens with a period of 13.66 days. The semi-diurnal tide predominates in most ocean basins today, but may be damped out in some circumstances, leaving the diurnal tide as the dominant tidal period. At the present time, this is most common in the Pacific Ocean, where diurnal tides are relatively more important. Situations that have a semi-diurnal tide with a significant diurnal signal are said to have *mixed tides*. Semi-diurnal, mixed and diurnal tides have all been reported from ancient sedimentary successions.

The magnitudes of the tide-generating forces exerted by the moon and sun vary over a number of different periods because of such astronomical factors as the changing declination and eccentricity of the orbits of the moon relative to the earth, and of the earth relative to the sun. Additional periods of variation exist because of the interaction between the sun

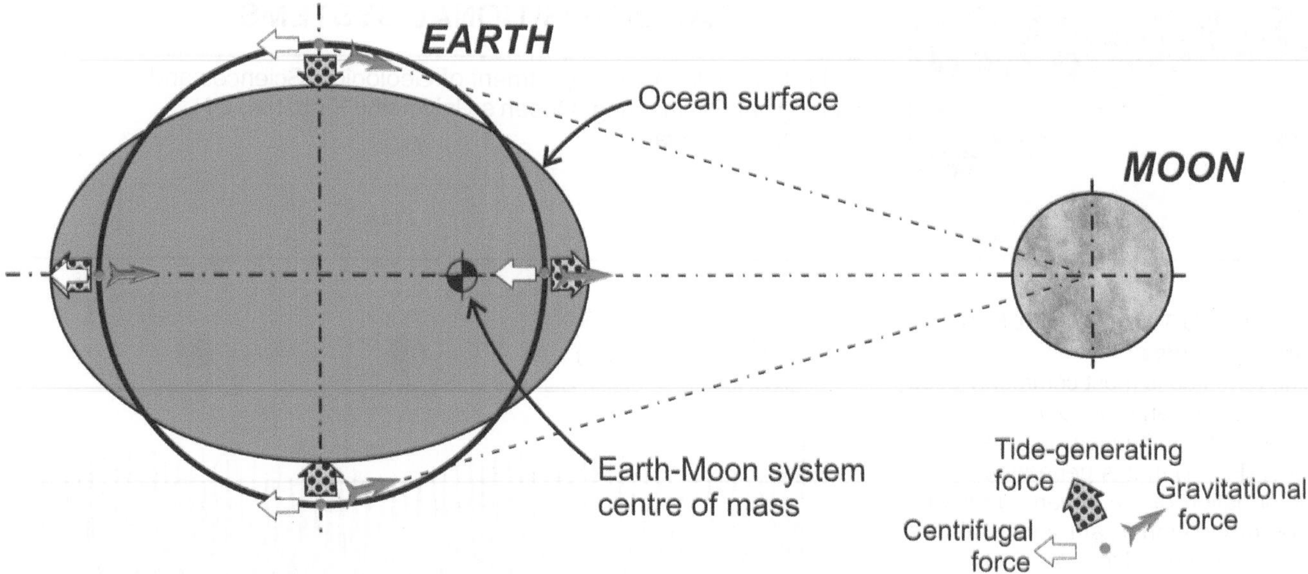

Figure 2. Origin of equilibrium astronomic tides. The tide-generating force is the resultant of the moon's gravitational attraction and the centrifugal force exerted on the earth because of its revolution about the center of mass of the earth–moon system, which is offset from the center of the Earth by about 4700 km (*i.e.*, approximately 75% of the earth's radius).

and moon. In total, the tide-generating forces and, thus, the *tidal range* (the difference in water level between successive high-tide and low-tide levels), vary over more than 25 different periods, the shortest being the semi-diurnal period mentioned above, whereas the longer ones are several decades in length. Most of the longer periods are generally thought to be of limited significance, geologically, except in the analysis of data obtained from tidal rhythmites (*see* below).

One of the more important shorter periods is a biweekly variation in tidal range caused by the positions of the moon and sun relative to the earth. When the sun and moon lie in a straight line relative to the earth (*i.e.*, at the times of full and new moon), their effects add to produce greater than average tidal ranges which are referred to as *spring tides*. By contrast, when the sun and moon are at right angles relative to the Earth (*i.e.*, at the first and third quarters of the moon) their forces counteract each other and the tidal range is smaller than average. Such smaller tides are *neap tides*. For semi-diurnal tides, this neap–spring variation has a period of 14.77 days, and contains 28 tidal cycles. Such semi-monthly variations in tidal range are also generated by other astronomic factors, including the varying declination of the moon (Kvale, 2006).

The foregoing analysis of the origin of tides, which is referred to as the 'equilibrium theory', provides a basic understanding of tidal phenomena, but it ignores several important factors such as the presence of continents and the fact that the rotating earth generates a Coriolis 'force' (actually an acceleration) that influences all water motions. A fuller understanding requires a 'dynamic theory' of tidal-wave motion. The continents provide a first-order control over the path followed by each tidal wave as it migrates around the earth. These paths can be seen on maps of *cotidal lines* that show the location of the crest of the tidal wave as a function of time (Fig. 3). (For the global map of co-tidal lines, *see* Figure 8.8 of Allen, 1997.) The motion of the tidal wave is also controlled by the Coriolis effect, which deflects the tidal wave to the right-hand side (relative to the forward motion of the tidal wave) of the ocean or sea in the northern hemisphere (Fig. 3) and to the left in the southern hemisphere. Some of the tidal energy entering an ocean basin or sea is also reflected back out. The interaction of the outgoing wave with the incoming tidal wave, under the influence of the Coriolis acceleration, creates *amphidromic points*, which are fixed locations about which the tidal wave rotates, generally in a counter-clockwise direction in the northern hemisphere (Fig. 3). These amphidromic points are displaced to the left side of a water body looking in the direction that the tidal wave is moving in the northern hemisphere, and to the right in the southern hemisphere.

Tidal Range

The tidal range varies dramatically from place to place on the earth's surface, from nearly zero to a present-day maximum of 16.3 m. A useful categorization of tidal ranges subdivides them into three classifications: *microtidal* (0–2 m range), *mesotidal* (2–4 m range) and *macrotidal* (> 4 m range). Some workers also recognize megatidal conditions, where the tidal range exceeds 10 m. Only 6 locations in the modern world have such enormous tides (Archer and Hubbard, 2003), but they may have been more common in the distant geological past because the moon was closer to the earth.

Because the tide-generating forces are small, only the open oceans develop significant tides, and even they typically have a range of less than 1 meter; such conditions prevail at most islands in the open ocean (*e.g.*, Hawaii). Smaller bodies of water, including enclosed seas such as the Mediterranean Sea and the water on continental shelves, cannot develop an appreciable tide of their own. The tides that are observed on continental shelves are due to the 'forcing' action of the oceanic tide. As the tidal wave moves from the open ocean onto a shelf, shoaling and shoreline convergence associated with coastal embayments concentrates the energy within the tidal wave into a progressively smaller cross-sectional area, and the tidal range increases. The influence of shoreline convergence and shallowing on the tidal range is illustrated in Figure 4, which shows how the tidal range increases into the Bay of Fundy. Such a landward increase in tidal range, followed by a sharp decrease in tidal range to the limit of tidal action, characterizes many coastal embayments and river mouths and is termed *hypersynchronous* behavior.

Particularly large ranges occur if the *natural period* of the water body (*i.e.*, the period for the water to 'slosh' back and forth in the absence of any forcing action) is close to one of the astronomically determined tidal periods, in which case *resonant amplification* takes place. This situation is shown dramatically by the Bay of Fundy–Gulf of Maine system, where the tidal range reaches 16.3 meters. The natural period of this system is *ca.* 13.3 hours, which is close to, but not in perfect agreement with, the semi-diurnal tide (12.42 hr). Even larger tides would occur if there were a perfect match. The natural period of a modern or ancient water body cannot be estimated by simple means because of the complex way that the tidal wave interacts with the coastal geometry and submarine bathymetry; only numerical modeling is able to determine this period. On straight continental shelves such as the United States east coast, the tidal range increases as the width of the shelf increases. Theoretically, resonance and the largest tidal ranges occur when the shelf width is of the order of 200–400 km, the precise width depending on the water depth (Howarth, 1982). Tidal range is also controlled by amphidromic systems: the tidal range increases outward from an amphidromic point where the range is zero (Fig. 3).

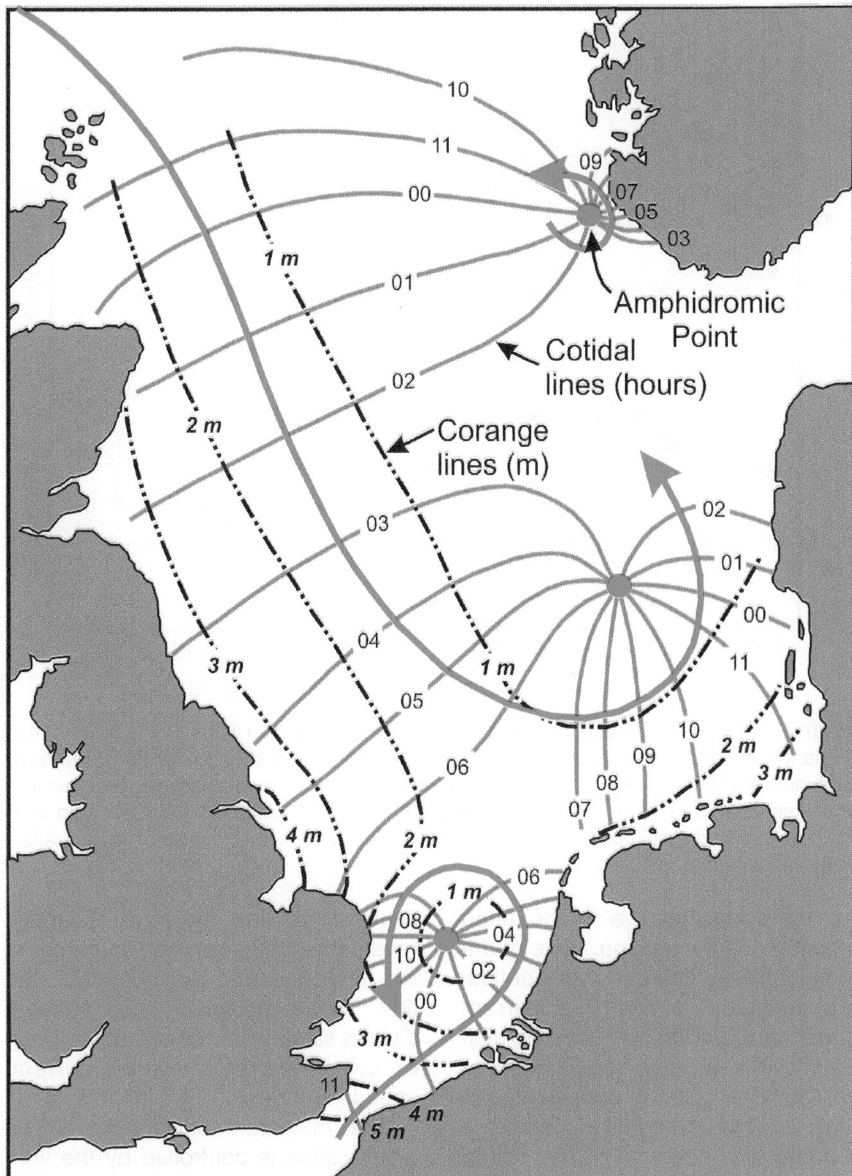

Figure 3. Amphidromic tidal system in the North Sea. The blue co-tidal lines show the times of high water in lunar hours, and indicate the direction of propagation (red arrows) of the tidal wave (anticlockwise) around each amphidromic point (red circles). The tidal range (black dash-dot lines) increases outward from each of the three amphidromic points, with the highest ranges occurring in coastal embayments such as the Wash (4 m) and German Bight (3 m). In general, the ranges are much larger on the English coast than on the Danish and Norwegian coast, because Great Britain lies to the right of the incoming tidal wave, which is banked up against this coast because of the Coriolis effect. Modified after Houbolt (1968).

When all of these factors are taken into consideration, the largest tidal ranges occur within funnel-shaped embayments on passive continental margins (Archer and Hubbard, 2003). Rift basins, such as the Bay of Fundy, Canada, and the Gulf of Cambay, India, are especially prone to having large tides, which might recur over many sea-level cycles. Foreland basins are also likely to attain a geometry that favors amplification of the tide, especially when they are partially flooded to produce a headward-tapering embayment such as occurs today in the Persian Gulf. By comparison, tectonically active continental margins with their narrow shelves are much less likely to experience large tidal ranges.

Tidal Currents

The tidal range is an important and widely cited parameter, but, by itself, it generally has only an indirect influence on the nature of the sedimentary deposits. The tidal currents created by the changes in water elevation

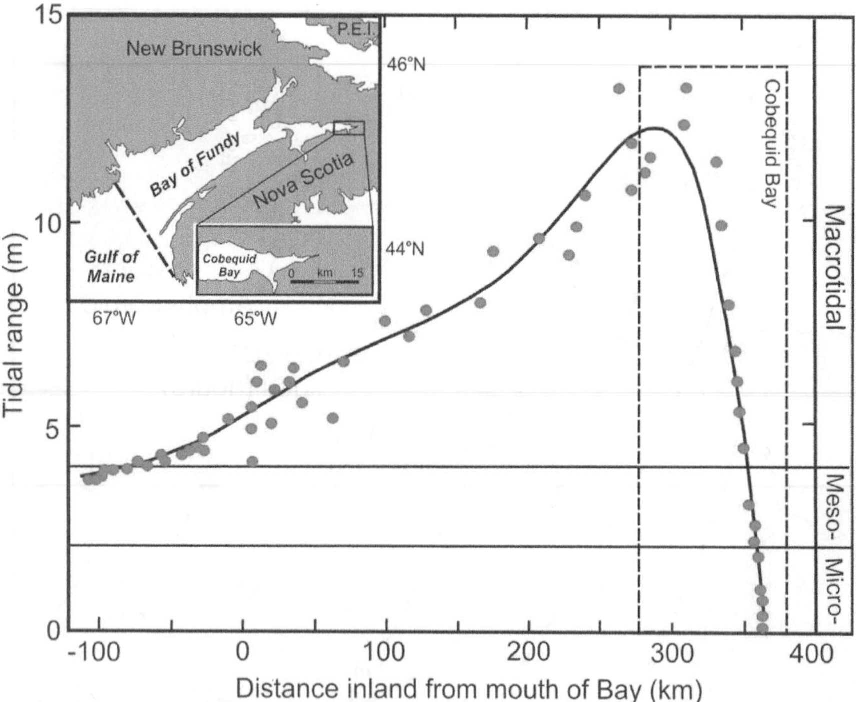

Figure 4. Variation in average tidal range as a function of location within the Bay of Fundy. The landward increase in tidal range is a result of the funnel-shaped nature of the bay, combined with shallowing, which concentrates the tidal energy into an ever-decreasing cross-sectional area. The landward decrease in ranges landward of the maximum is a result of frictional dissipation and the rise of the bed above the low-water level. After Archer and Hubbard (2003).

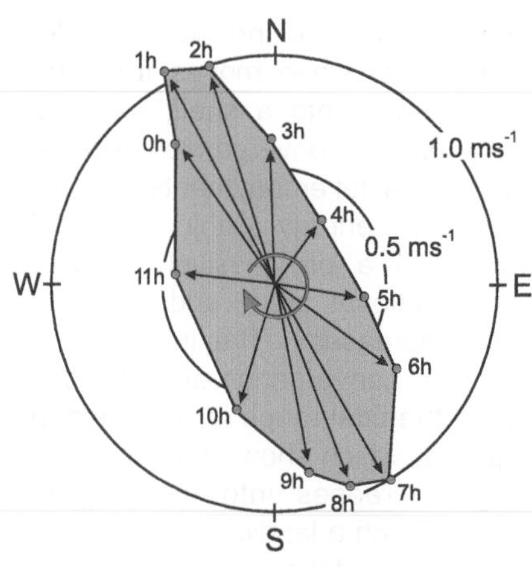

Figure 5. Rotary tidal currents and the tidal ellipse. The tidal ellipse is constructed by joining the tips of the successive current-velocity vectors that have a length scaled to the current speed. Successive vectors rotate clockwise (red arrow). Note that current speed never falls below 0.3 ms^{-1} so there is no slack-water period.

do, however, have a pronounced impact on sedimentation. One of the most important characteristics of tidal sedimentation is that the current speed and flow direction change systematically over each semi-diurnal, or diurnal, tidal cycle. In open-shelf settings, far from the confining influence of a coastline, the tidal current speed and direction change in such a way that the tips of sequential current-velocity vectors trace out a path termed a *tidal ellipse* over a single tidal cycle (Fig. 5). These tides, which are termed *rotary tides*, have no *slack-water* period when the current speed decreases to zero. In the ideal case, the tidal ellipse is a circle with equal speeds in all directions, but in nearshore areas, the presence of the shoreline causes current speed to be higher parallel to the coast, and weaker in an onshore–offshore direction, producing elongated tidal ellipses (Fig. 6). The currents flow into and out of the mouth of rivers, with flow directions that are at a high angle to the general trend of the shoreline. In such settings, as well as in narrow seaways and straits such as the English Channel where the shorelines constrain the flow, the currents simply reverse by 180° between the incoming flood tide and the outgoing ebb tide. These *rectilinear tides* have a distinct slack-water period at the time of current reversal, when current speeds are zero.

The maximum current speed at any location is controlled by the volume of water that must pass that point in each half tidal cycle, a quantity that is termed the *tidal prism*. The volume of water is a function of the tidal range *and* the size of the area being drained and flooded. Large tidal ranges are commonly associated with strong tidal currents, and sedimentation in most macrotidal areas such as the Bay of Fundy (Fig. 4) is tide dominated because of these strong currents. Globally, however, the tidal range varies by less than two orders of magnitude, whereas the area being flooded and drained can vary by many orders of magnitude, from small lagoons to very large bodies of water. Therefore, the tidal prism and the strength of the tidal currents are more strongly influenced by area than by tidal range, and there is no one-to-one correspondence between tidal range and the strength of the tidal currents. For example, a large tidal range does not generate significant currents if the tidal prism is small. Thus, straight macrotidal coasts experience only weak tidal currents (Yang *et al.*, 2005), whereas the neighboring river mouths and embayments with the same tidal range have stronger tidal currents because they have a larger tidal prism. Conversely, fast currents can occur in areas with a small tidal range if the estuary or embayment is large relative to the cross-sectional area of the mouth through which the flow passes. This is the case, for example, in Chesapeake Bay: it is microtidal but has a very small mouth relative to its size and maximum speeds at the entrance are typically greater than 0.5 ms^{-1}, and locally exceed 1 ms^{-1}. Within the broad inner estuary, current speeds are much less. Many seaways and straights also have strong tidal currents, even in cases where the tidal range is small, because they link large bodies of water. The passages between the islands of Japan and the English Channel (Fig. 7) are examples of this.

The hypersynchronous behavior that characterizes many coastal embayments and river mouths (Fig. 4) is reflected in the currents, with the speeds increasing inward to a 'tidal maximum' where the tidal currents are fastest, beyond which the speeds

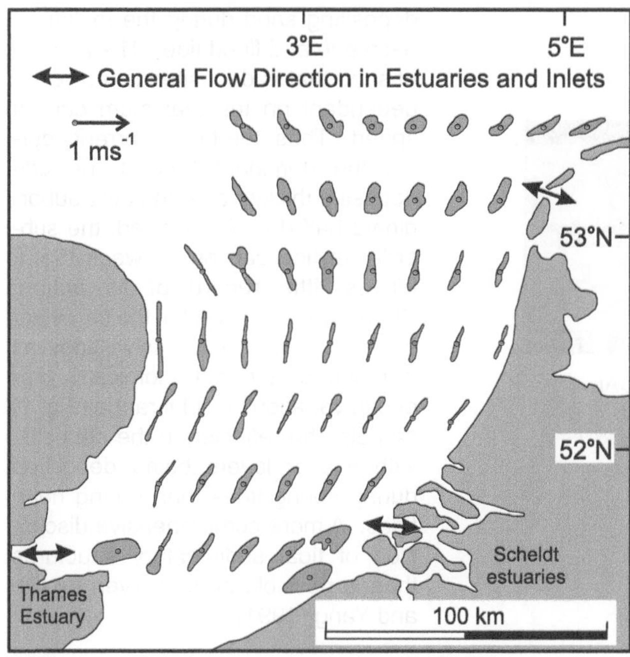

Figure 6. Tidal-current ellipses for currents 1 m above the bed in the southern North Sea. Note the extreme elongation of ellipses in the south due to the coastal constriction, and the more circular shapes in the north. Current directions in the mouth of the Thames River and Scheldt estuaries are approximately perpendicular to those on the shelf. Red dots indicate the measurement station and the zero-speed point at the centre of the ellipse. Modified after Houbolt (1968, Figure 4) and McCave (1971, Figure 9).

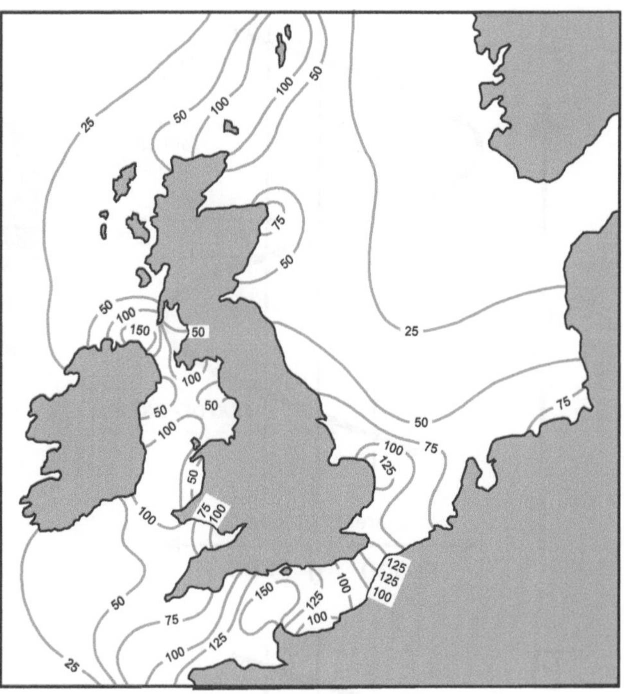

Figure 7. Maximum near-surface current speeds (cm s^{-1}) during spring tides on the continental shelf around the British Isles. Modified after Howarth (1982).

decrease toward the tidal limit (Dalrymple and Choi, 2007). This location is typically some distance landward of the coast. Strong tidal currents may also occur near the outer edge of the continental shelf if a large volume of water must move onto and off of the shelf during each tidal cycle. A good example is given by Georges Bank, at the outer edge of the shelf between Nova Scotia and Cape Cod, where tidal-current speeds exceed 0.5 ms^{-1} in all areas shallower than 60 m, with peak values up to 1.2 ms^{-1}. Current speeds are typically less than 0.2 ms^{-1} farther inboard in the Gulf of Maine.

In the open ocean, the tidal wave is symmetrical and, consequently, the speed and duration of the ebb and flood currents are equal. In many coastal areas, however, deformation of the tidal wave and the interaction of the flow with the bottom topography produce inequalities between the flood and ebb currents. As a result, most areas experience a *net* or *residual* transport of sediment in the direction of the stronger (*dominant*) current. The weaker current that flows in the opposite direction is termed the *subordinate* current. Just as wind waves become steeper on their front side as they approach a beach, the tidal wave also becomes asymmetric as it enters shallow water, with the flood tide being of shorter duration than the ebb tide. (The tidal equivalent of a breaking wave is called a *tidal bore*.) This, in turn, creates the tendency for flood currents to be faster than ebb currents in coastal areas. Because the sediment-transport capacity of a current increases exponentially as the speed increases, the shorter duration of the flood tide is more than compensated by the higher speed, in most cases, leading to a tendency for '*flood dominance*' and the net onshore movement of sediment. Fluvial discharge may accentuate the ebbing tide in rivers to the point that areas of *ebb dominance* and net seaward movement of sediment are created. Topographic irregularities on the sediment surface (*e.g.*, tidal bars and channel bends) and along the coast (*e.g.*, headlands) also complicate the patterns of residual sediment transport. The exposed side of such an irregularity experiences stronger currents than the sheltered side, but the areas of stronger and weaker currents switch locations when the current reverses. This process generates oppositely directed zones of ebb- and flood-dominance on either side of the irregularity. Typically, these areas of ebb and flood dominance alternate spatially in a system of mutually evasive, flood-dominated and ebb-dominated channels, producing a complicated pattern of sediment transport, as has been documented in Bay of Fundy estuaries (Dalrymple *et al.*, 1990).

TIDAL SEDIMENTARY STRUCTURES

The *periodic* (*i.e.*, regular) changes in current speed and direction that characterize tidal currents produce several diagnostic sedimentary structures. The current speed is close to zero at both high and low tide in cases where the tidal currents are rectilinear. These *slack-water* periods allow fine-grained suspended sediment to settle to the bed, producing a *mud drape*. Current speeds reach a maximum and are commonly capable of transporting and

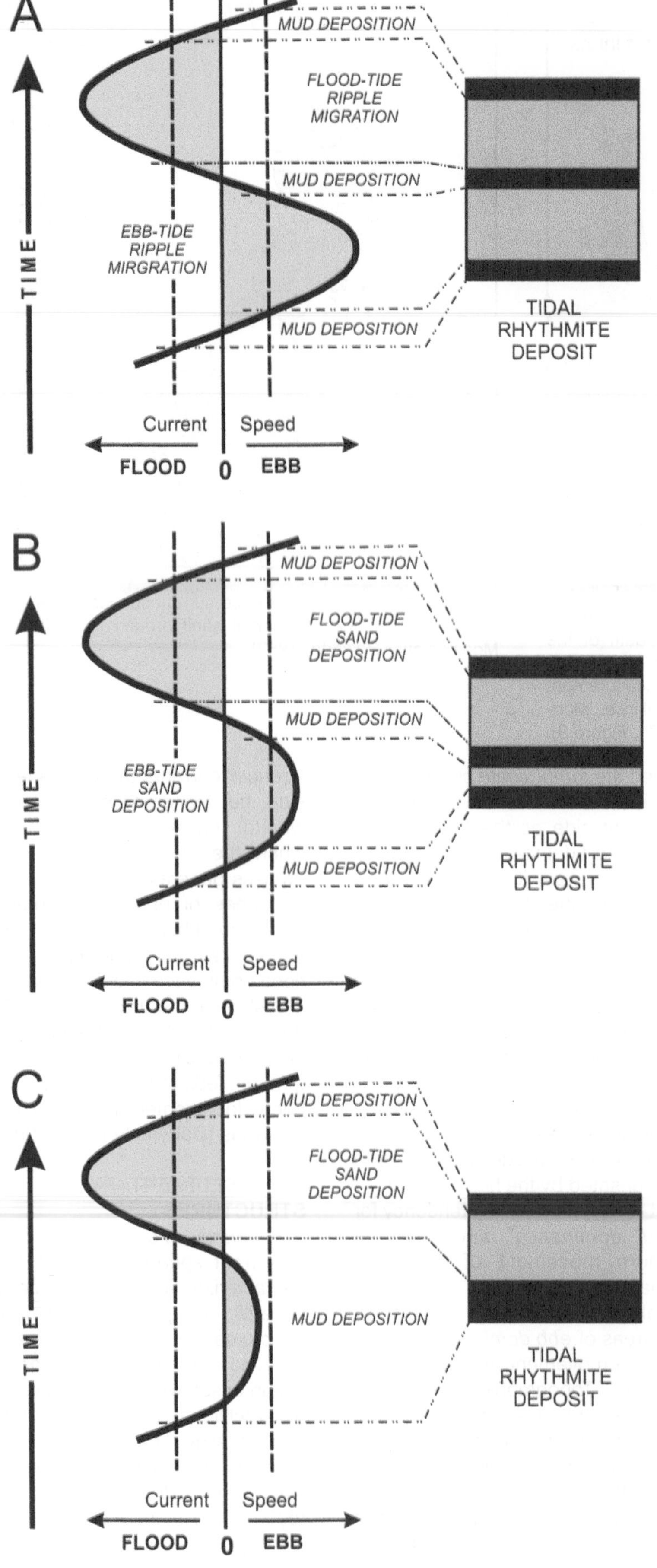

depositing sand during the middle of each ebb and flood tide. The amount of sand that is deposited is, however, dependant on the maximum current speed. Thus, the faster currents during the dominant half of the tide deposit a thicker layer than the subordinate half (Fig. 8). Indeed, the subordinate tide can be so weak that it leaves little record of its action. Changes in the speed of the dominant current between successive tides as a result of the astronomically produced variations in tidal range (Fig. 1) can also be reflected in the deposits, with thicker layers being deposited during spring tides than during neap tides. A more comprehensive discussion of tidal sedimentary structures than is possible here is given by Nio and Yang (1991).

Tidal Rhythmites

Horizontal laminations consisting of alternating sandy/silty and muddy material that show cyclic changes in layer thickness due to diurnal and neap–spring variations in tidal-current speed (Fig. 9) are termed *tidal rhythmites* (Dalrymple *et al.*, 1991; Williams, 1991; Tessier *et al.*, 1995). The subordinate tide is typically too weak to create a deposit, so that each pair of coarse and fine laminae represents a complete tidal cycle (Fig. 8C). The coarse layers commonly contain no internal structure because the sediment is deposited from suspension, but in areas or at times with higher peak current speeds, ripple crosslam-

Figure 8. Variation in tidal-current speed over a single tidal cycle and the corresponding tidal-rhythmite deposit. **A.** A symmetrical tide, with equal duration and magnitude of the flood and ebb currents, generates two sand layers of equal thickness. **B.** A flood-dominant situation, but with an ebb current that is capable of depositing a thin sand layer. The resulting alternation of thicker and thinner sand layers would mimic the pattern created by the diurnal inequality in a system such as (**C**). **C.** A highly asymmetric tide with an ebb current that is never capable of transporting sand. This situation is quite common in the tidal-rhythmite occurrences documented in modern and ancient successions. Based on Dalrymple *et al.* (1991).

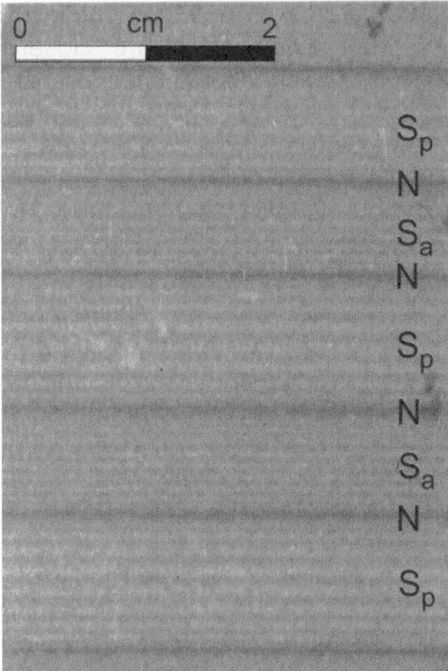

Figure 9. Tidal rhythmites from the Elatina Formation (Neoproterozoic) of South Australia. Note how the thickness of the lighter colored coarser (silty) layers varies cyclically, from thicker during spring tides ('S') to thinner during neap tides ('N'). The subscript 'p' refers to perigee when the Moon is at its closest to Earth, whereas the subscript 'a' refers to apogee (the Earth–Moon separation is greatest). Note how the thickness of the spring-tide laminae is greater during perigee and thinner at apogee. See Williams (1991) for more details on these deposits.

ination may be present. In the ideal semi-diurnal case, there will be 28 sand–mud lamina pairs in a neap–spring cycle, but more than 28 can occur if some of the subordinate tides also deposit a sand layer. Fewer than the expected number occurs if the currents become so weak during neap tides that recognizable laminae are not generated, or if the current speed becomes so high during spring tides that there is erosion. Tidal-rhythmite successions that depart significantly from the regularity seen in the ideal cases (Fig. 9) are common, and have been termed *non-cyclic rhythmites*, to differentiate them from those that show good tidal cyclicity (*cyclic rhythmites*). Long-term records with complete tidal-cycle information are rare, but can be used to deduce such things as:

1. The nature of the tides (diurnal *versus* semidiurnal);
2. Seasonal variations in river discharge;

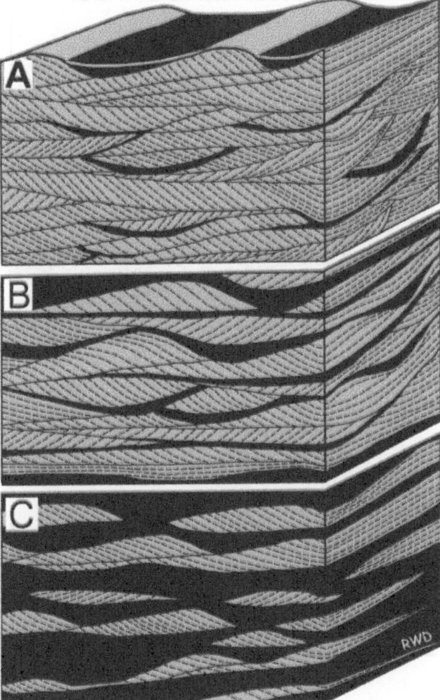

Figure 10. Flaser bedding (**A**), wavy ('tidal') bedding (**B**) and lenticular bedding (**C**). After Reineck and Singh (1980).

3. The elevation of deposition within the intertidal zone; and even
4. The Earth–Moon distance.

The development of cyclic tidal rhythmites requires relatively special circumstances. Sediment supply must be high to permit at least a millimeter of deposition in every 12-hour period. This condition is satisfied only along the margins of channels and in delta-front and prodeltaic settings. Strong tidal currents are necessary to suspend the sediment that ultimately settles to create the cyclic rhythmites. For this reason, it has been suggested that good cyclic rhythmites are formed most commonly in macrotidal environments, although they have been reported from areas with a tidal range as small as 2 m. Sedimentation must also occur in a protected setting so that extraneous events such as storm waves or wind tides do not disturb the regularity of tidal deposition. Thus, most modern cyclic rhythmites are found in the middle and inner parts of estuaries and deltas. In fact, many cyclic tidal rhythmites might have formed in the freshwater-tidal zone close to the landward limit of tidal action.

Flaser, Wavy and Lenticular Bedding

Flaser, wavy and lenticular bedding (Fig. 10) are a gradational spectrum of heterolithic deposits consisting of alternating rippled sand and mud. They differ from cyclic tidal rhythmites by the absence of clear ordering of layer thickness and can be considered as non-cyclic rhythmites. The variation from flaser to lenticular bedding represents an increase in the deposition and preservation of mud. In the deposit, the amount of mud is a direct reflection of the amount of mud in suspension: areas with high suspended-sediment concentrations are more likely to have wavy and lenticular bedding, whereas areas with low suspended-sediment concentrations (*i.e.*, cleaner water) will more often produce flaser bedding. Flaser bedding is also more likely to occur in areas with stronger currents because these currents can erode freshly deposited mud from the ripple crests. It should be noted, however, that flaser, wavy and lenticular bedding can be generated by processes other than tidal currents. For example, episodic flooding of overbank areas in a fluvial system or weak storm-wave action in deeper water can produce such heterolithic deposits. One must look for hints of cyclicity such as the diurnal inequality or neap–spring cycles, or for the presence of bipolar current-ripple orientations, before ascribing a tidal origin to a given example.

Tidal Bundles and Reactivation Surfaces

In areas where tidal currents are strong enough to generate subaqueous dunes, the crossbedding produced by them contains regularly spaced internal discontinuities that are produced by the periodic flow reversals. *Reactivation surfaces* will be produced if the subordinate current is capable of eroding the lee face of the dunes formed by the preceding dominant current (Fig. 11). The amount of truncation on these surfaces depends on the intensity of the subordinate current: stronger currents lead to greater erosion and more prominent reactivation surfaces, whereas weak subordinate currents produce minimal truncation.

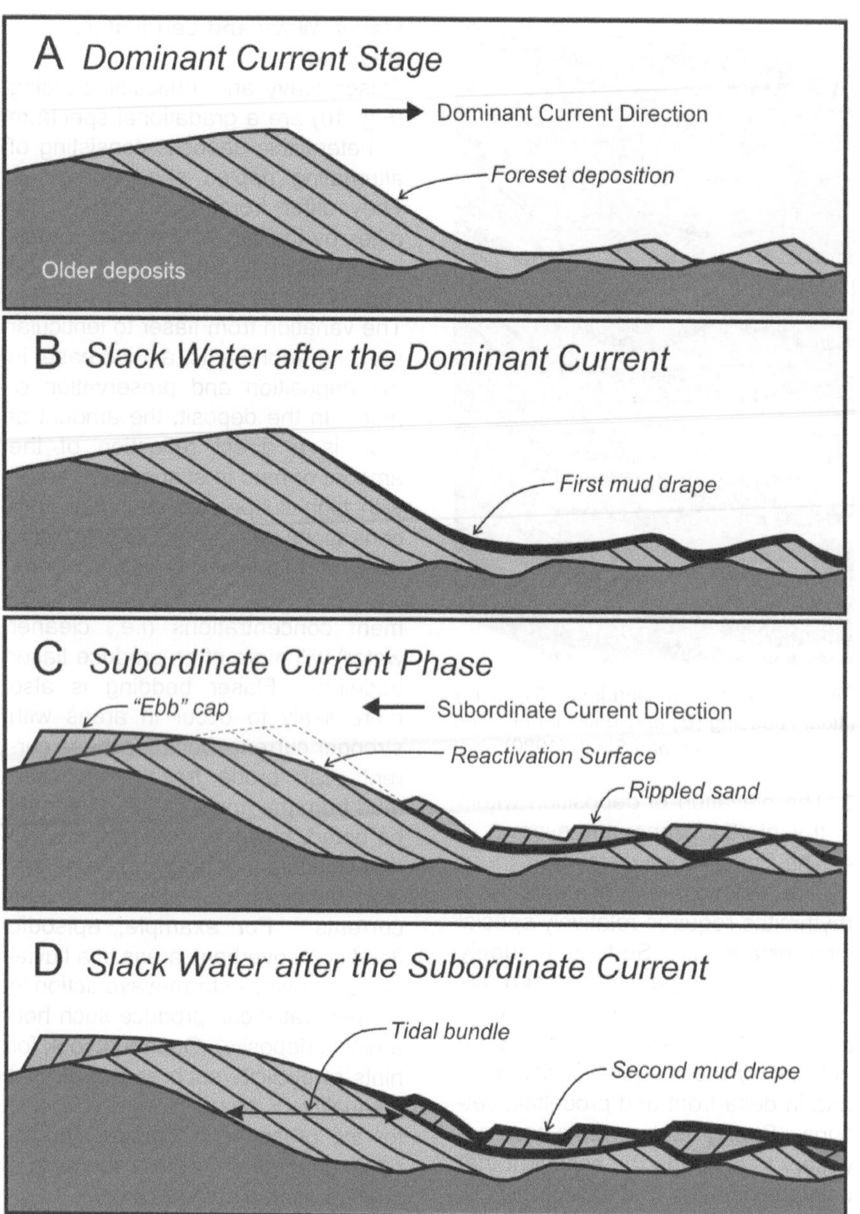

Figure 11. Structures produced in a dune during a tidal cycle. In this example, the currents exhibit a pronounced asymmetry and suspended-sediment concentrations are moderately high, allowing the deposition of pronounced mud drapes. The presence of two drapes is most common in subtidal settings. A tidal bundle is the deposit of the dominant portion of the tidal cycle. Based on Visser (1980, Figure 3).

(Note that a reactivation surface occurs *within* an otherwise continuous set of crossbedding produced by a single dune; this term should not be used to refer to the erosion surface between sets formed by two different dunes.) Ripples generated by the subordinate current can mantle the erosion surface (Fig. 12A; Kohsiek and Terwindt, 1981; Dalrymple and Rhodes, 1995). These ripples are an excellent indicator of tidal action, but care must be taken to distinguish them from reverse-flow ripples produced by circulation in the flow-separation eddy. The latter do not extend above the dune-toeset region, whereas the former can be present far up the lee face, if the reactivation surface has a gentle inclination.

Mud drapes can also be deposited on the lee face during one or both slack-water periods, if the suspended-sediment concentration in the water column is high enough (Figs. 11 and 12). The amount of sand deposited by the subordinate current is typically small, and hence the mud drapes deposited after the dominant and subordinate tides are closely spaced. Such double mud drapes can be expected in subtidal settings, whereas only a single mud drape (the one formed at high slack water) is common in intertidal settings because water does not stand over the surface at low slack water. Exceptions occur (Fenies *et al.*, 1999), so caution must be exercised in using double drapes as an indication of a subtidal origin. Mud drapes might not be deposited if the water contains no suspended sediment, there is wave action, or there are rotary tides that have no slack-water period.

The deposit of a single, dominant tide is called a *tidal bundle*, whether bounded by reactivation surfaces or mud drapes (Visser, 1980; Figs. 11 and 12A). Cyclic variations in the thickness of these bundles (Fig. 13) are produced by the change in current strength over the neap–spring cycle. Bundled crossbedding, therefore, commonly shows a periodic change from sandy, spring-tide intervals with widely spaced mud drapes, to muddier, neap-tide deposits in which ripple cross-lamination is more abundant. The base of the crossbed set can undulate up and down: the set base will be lower (*i.e.*, there will be deeper trough scour) during spring tides, and the set base will be higher during neap tides when the height of the dune is less (Dalrymple and Rhodes, 1995). The spacing of these undulations will equal the distance of dune migration over the neap–spring cycle. The angle of the reactivation surfaces can also change, from minimal truncation during neap tides to more prominent, lower angle truncation during spring tides. As with tidal rhythmites, there should be 28 tidal bundles within a complete neap–spring cycle if the tides were semi-diurnal, but there will be fewer than 28 if the current speeds drop below the threshold for sand movement during neap tides. An excellent ancient example of tidal bundles has been described by Allen and Homewood (1984).

Compound Dunes

Compound dunes (*i.e.*, large to very large dunes upon which there are smaller, *simple dunes*; Fig. 14A; Dal-

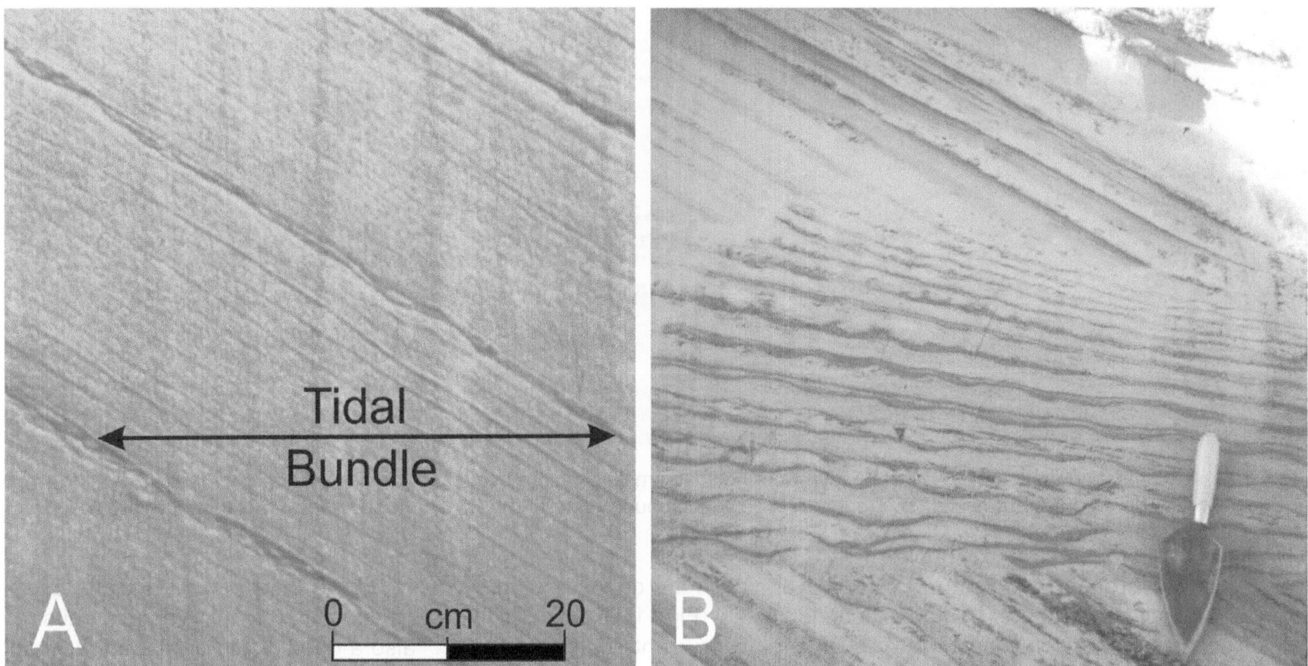

Figure 12. Bundled crossbedding in subtidal sands of the Oosterschelde estuary, the Netherlands. **A**) Close-up of dune foresets showing several tidal bundles that are bounded by thin mud couplets that enclose subordinate-current ripples. **B**) Outcrop photo showing repetitive mud drapes in the foresets and bottomsets of a large crossbed. Note how the mud drapes are concentrated in the bottomsets, causing them to be muddier than the foresets. Trowel 24 cm long.

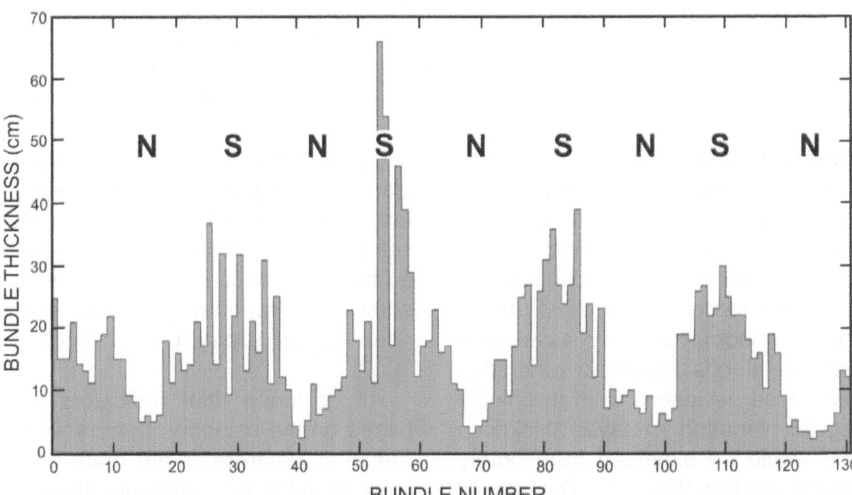

Figure 13. Variation of bundle thickness within a crossbed from the Oosterschelde estuary, the Netherlands (*cf.* Fig. 12). 'N' = neap tides; 'S' = spring tides. The average number of bundles between successive neap tides is 27. The tendency for the thickness of successive bundles to be alternately larger and smaller is due to the diurnal inequality. After Visser (1980, Figure 4).

rymple and Rhodes, 1995) are a common feature in most sandy inshore and continental-shelf tidal environments. These bedforms, which were previously called tidal *sand waves*, range in height from less than 1 m to more than 20 m, and have wavelengths of 10 m to several hundred meters (Belderson *et al.*, 1982; Dalrymple and Rhodes, 1995). Larger bedforms generally occur in deeper water. These large dunes are typically asymmetrical (Fig. 14A), with their steeper (lee) face inclined in the direction of net sediment transport and bedform migration. Lee-face inclinations vary from the angle of repose (32–35°) to nearly zero, but are typically about 5° (Belderson *et al.*, 1982; Dalrymple, 1984; Dalrymple and Rhodes, 1995).

The deposits of compound dunes are characterized by internal discontinuities created by the troughs of the smaller and faster-moving superimposed dunes as they migrate down the lee face of the more slowly migrating larger dune (Figs. 14B and 15). If the superimposed dunes are small relative to the larger dune, then the depth of erosion of the brink will be minimal and the larger dune will generate simple, large-scale crossbedding with reactivation surfaces in its upper part (Fig. 15A). If, however, the superimposed dunes are large, then more gently dipping *compound crossbedding* will be produced (Fig. 15B and C), in which each inclined crossbed was created by one superimposed dune. Allen (1980) hypothesized that the inclined erosion surfaces within compound dunes, which are termed *master bedding planes* (E2 surfaces in Fig. 15), are formed by the subordinate tide, but field observations (Dalrymple, 1984) have shown that the subordinate currents are generally too weak to move enough sediment to produce such surfaces. Instead, reactivation surfaces created by the subordinate tide (E1 surfaces in Fig. 15) will be present within the smaller crossbeds.

The geometry shown by compound crossbedding is equivalent to

Figure 14. A. Compound dune in the Bay of Fundy. Note how the small, simple dunes are migrating up and over the crest of the larger dune. Shovel (foreground) for scale. **B.** Trench through a compound dune in the Bay of Fundy showing compound crossbedding. Meter stick for scale.

the forward-accretion architectural element of Miall (*see* Chapter 6), and the paleocurrent direction measured from the small crossbeds should fall in the same quadrant as the dip direction of the master bedding planes. Typically, the deposits of compound dunes coarsen upward because the current speed is higher near the crest of the larger dune than in the sheltered trough. The trough and bottomset region can be the site of mud deposition (Figs. 12B, 15A and C), and the bottomset deposits can also be intensely bioturbated. Figure 16 shows a vertical succession through such a deposit. Previously, this succession was interpreted as representing a tidal bar (Mutti *et al.*, 1985; Dalrymple, 1992; Willis, 2005), but recent work has shown that the bodies on which this succession was based contain the forward-accretion deposits expected within compound dunes rather than the lateral-accretion deposits found within tidal bars (Olariu *et al.*, 2008; *see* below).

Herringbone cross-stratification, which refers to adjacent crossbeds with opposed (bipolar) dip directions, is widely used as an indication of tidal deposition. The origin of bipolar crossbedding has been somewhat of a mystery, because the development of mutually evasive channel systems commonly produces a unidirectional paleocurrent pattern at any one location, even though, when summed over many outcrops, the overall paleocurrent distribution may be bipolar. Herringbone cross-stratification appears to be particularly common, however, within the deposits of compound dunes (Fig. 15B and C): the subordinate current can produce dunes migrating up the lee face that can then be buried and preserved beneath sediment deposited during the subsequent dominant tide (Dalrymple, 1984). In 2-D outcrops of trough crossbedding, care must be taken to avoid misinterpreting an end-on view of troughs as herringbone cross-stratification.

Recognition of Tidal Deposits

Complete neap–spring successions of tidal rhythmites or tidal bundles are diagnostic of tidal deposition; however, rhythmite or tidal-bundle successions are rarely complete. Furthermore, random variability associated with river floods, storms or wind tides can cause deviations from the highly regular variation in layer thickness that would be expected from purely tidal deposition (Fig. 13). De Boer *et al.* (1989) have developed a statistical test that can be used to determine whether a suspected rhythmite or tidal-bundle succession is sufficiently regular to be interpreted as tidal, with confidence; however, even known tidal deposits can fail this test (Dalrymple *et al.*, 2003). Therefore, a tidal origin must be based on less than definitive evidence in many cases.

Deposits that contain an abundance of heterolithic bedding or lamination, which is not intensely bioturbated because of rapid sedimentation in a brackish-water setting (*see* Chapter 3), is likely to be tidal in origin. The presence of a moderate abundance of double mud drapes and the alternation of thick–thin lamina pairs is also a good indicator that the deposit is tidal. Landward-directed paleocurrents, other than those created by flow separation on the downstream side of tight meander bends, are also strongly indicative of tidal action. An abundance of crossbedding is suggestive of tidal sedimentation in shallow-marine deposits, because few other processes can create sufficiently strong currents there. It has been suggested that tidal crossbedding is also more regular (*i.e.*, that there is less variation between successive sets) than the crossbedding generated by rivers, because tidal processes show less random variability than river floods (Dalrymple and Choi, 2007).

All of these tidal indicators are based on the balance of probabilities (these characteristics are more common in tidal environments than in other settings); however, any one of these features could potentially be formed by other processes. Thus, widely scattered occurrences of double mud drapes, thick–thin lamina alternations, herringbone cross-stratification and reactivation surfaces are all possible in almost any environment and should not be over-interpreted as being of tidal origin.

TIDE-DOMINATED ENVIRONMENTS

Introduction

There is a diverse array of tidal sedimentary environments, stretching from the limit of tidal action in rivers, through deltaic and estuarine settings, to open-coast tidal flats and continental shelves. Three major morphological sub-environments are widespread in this spectrum of tidal environments: *tidal channels*, *tidal flats* and *tidal bars* (also called *tidal-current ridges* in shelf settings). These sub-environments are, in fact, intimately related in many settings, even though there has been a tendency to consider them as distinct entities. For example, within tide-dominated estuaries and deltas (*see* more below), tidal channels contain, or are flanked by, tidal bars, and the upper part of channel margins consist of tidal flats. Tidal flats occur in isolation from channels only along open, unbarred coasts, and tidal bars are independent of channels only on the continental shelf.

Most tidal environments occupy the transition zone between purely fluvial processes and fully marine conditions. As a result, there is a longitudinal gradient of depositional processes that is reflected in the morphology and facies of tidal deposits. It is beyond the scope of this chapter to discuss these trends in detail and only the major aspects are considered below. See Dalrymple and Choi (2007) for an extended discussion.

In tide-dominated estuaries and deltas, tidal-current speeds typically increase inland because of the hypersynchronous nature of the tide, reaching a maximum some distance landward of the coast, and decreasing from there to the tidal limit, which

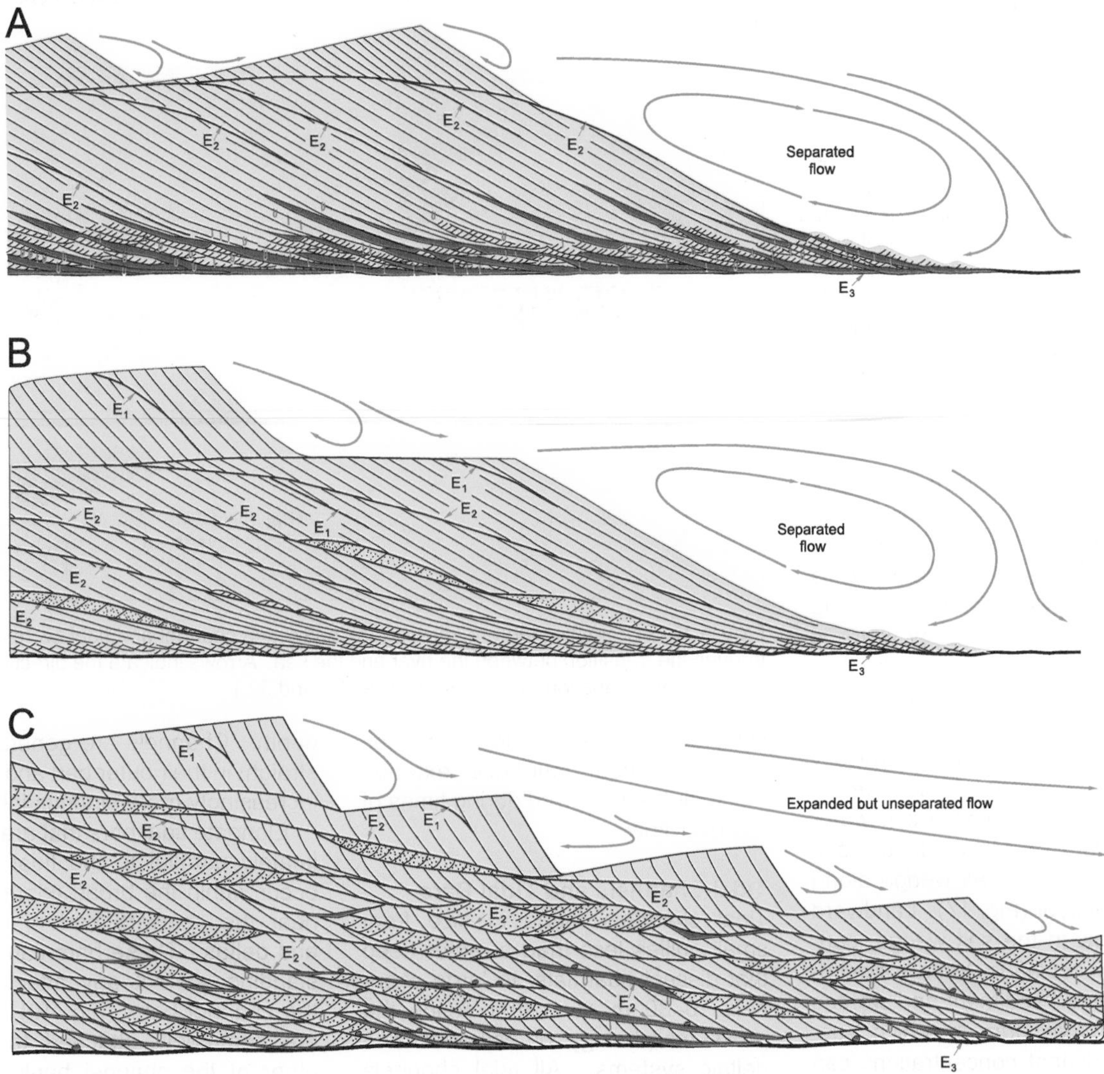

Figure 15. Schematic sections through compound dunes showing the range of possible structures. **A.** Large-scale simple crossbedding with randomly spaced reactivation surfaces (E_2 surfaces), each of which is generated as the trough of a small superimposed dune passes over the brink of the larger dune. Mud drapes deposited during slack-water periods are shown in the toeset region of the larger dune. **B.** Compound crossbedding with low-angle master bedding planes (E_2 surfaces) separating the individual simple crossbeds, each of which is formed by a medium to large superimposed dune as it migrates down the larger dune's lee face (*cf.* Fig. 14). Note how the bedforms produced by the subordinate tide extend far up the master-bedding surfaces, a feature that distinguishes them from ripples formed by the flow-separation eddy. Reactivation surfaces produced by the subordinate tide (E_1 surfaces) occur within the crossbeds. **C.** Compound crossbedding similar to **(B)** but with smaller superimposed dunes and lower angle master-bedding planes (E_2 surfaces). Note how the crossbeds become thicker upward. Bioturbated muddy deposits accumulate in the low-energy trough producing a pronounced upward-coarsening succession (*see* Fig. 16). The vertical scale of the sections can range from *ca.* 50 cm to several meters. **(A)** and **(B)** modified after Allen (1980, Figure 8, Class IIA and IVA, respectively).

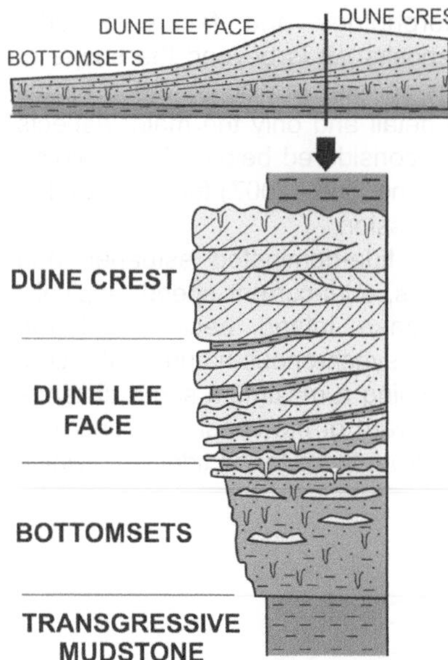

Figure 16. Schematic vertical succession of the deposit created by a compound dune, based on the Eocene of the Ager Basin, northern Spain. The vertical succession is typically about 5 m thick. The sediment body shown here was formerly interpreted as a 'tidal bar', but subsequent work has shown that it was formed by a large compound dune (Olariu *et al.*, 2008). After Mutti *et al.* (1985; reproduced with permission of Wiley-Blackwell).

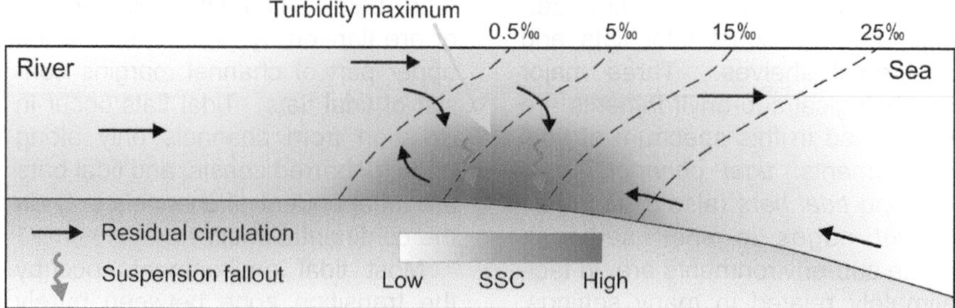

Figure 17. Development of a salt wedge in a partially mixed fluvial–marine transition zone, with residual seaward flow in the surface layer and landward flow near the bed, a pattern that is termed 'estuarine circulation'; it occurs in both estuaries and deltas. Flocculation and this residual circulation traps fine-grained sediment near the landward tip of the salt wedge, producing a turbidity maximum. 'SSC' = suspended-sediment concentration.

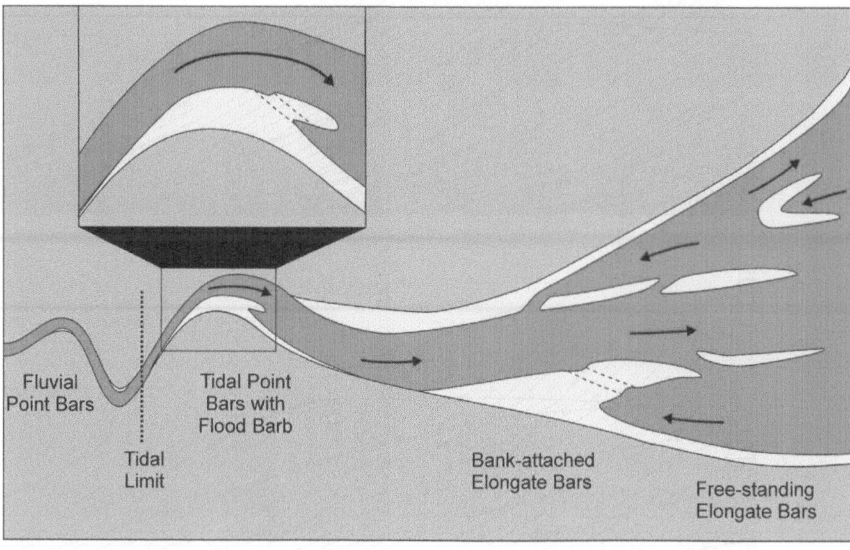

Figure 18. Schematic diagram showing the change in channel and tidal-bar morphology through the transition between the river and the sea. Arrows indicate the direction of local residual sand transport. (*See* also Figures 30 and 32.)

may be located tens to hundreds of kilometers landward of the coast. The water displays a landward decrease in salinity in cases where there is appreciable river discharge. The most landward limit of salty water is invariably located seaward of the tidal limit and a fresh-water tidal reach is always present. The transition between fresh and salt water typically exhibits some degree of *estuarine circulation*, in which lighter fresh water flows outward on top of denser salt water that forms a *salt wedge* along the channel floor (Fig. 17). Landward residual flow in the salt wedge, coupled with *flocculation*, traps suspended fine-grained sediment and creates a *turbidity maximum* in which suspended-sediment concentrations can reach very high values (Dalrymple and Choi, 2003). Settling of this suspended sediment during slack-water periods can generate *fluid mud*, which, by definition, has a suspended-sediment concentration greater than 10 gl^{-1}. Seasonal variations in river discharge cause these features to migrate up- and down-river by 10s of kilometers. River discharge also influences the distance to which tides penetrate: tidal influence extends much farther landward during times of low river flow.

Tidal Channels and Tidal Bars

Tidal channels range in size from small gullies less than a meter deep that dissect tidal flats, to large channels 10–30 m deep that define the gross morphology of estuarine and deltaic systems. All tidal channels display an exponential seaward widening that is referred to as 'funnel-shaped' (Fig. 18), a pattern that exists because tidal discharge decreases in a landward direction, reaching zero at the tidal limit, beyond which the channel characteristics are determined by fluvial discharge and slope. It is not known whether there are corresponding longitudinal changes in water depth. Channels are generally fairly straight in their outer part and become increasingly more sinuous inland.

Elongate tidal bars that are approximately parallel to the tidal currents occur within the broad tidal channels near the seaward part of estuaries and deltas. These bars can be free-standing without any attachment to the channel bank (Fig. 19), but they are commonly attached to one or other of the channel banks at their landward end, producing a headward-terminating, flood-dominant channel termed a *flood barb*, which is flanked by a seaward-tapering elongate bar. The channel on the other side of the bar is usually ebb-dominated, such that there is oppositely directed net sediment transport on either side of the bar crest. Successive bars along the channel typically attach to oppo-

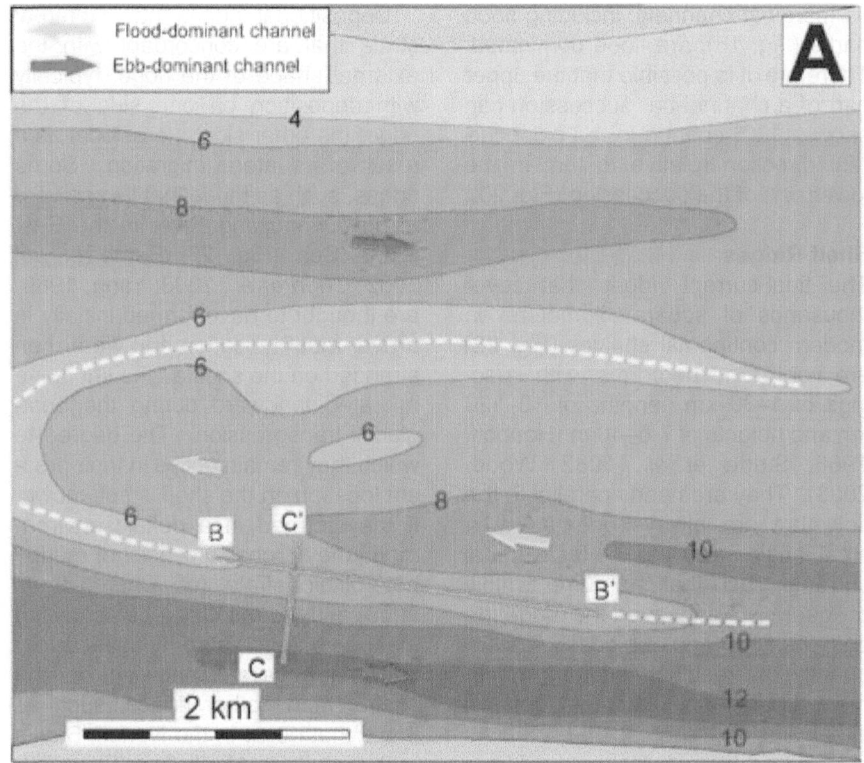

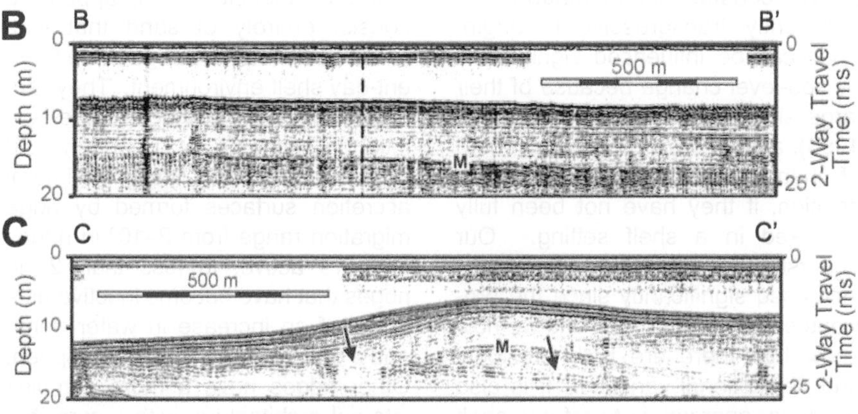

Figure 19. Bathymetry (**A**) and seismic sections (**B**, **C**) through an elongate tidal bar in a distributary channel of the Fly River delta, Papua New Guinea, showing horizontal stratification in a strike-parallel direction (B-B') and dipping strata in the transverse section (C-C'). This lateral-accretion geometry is produced by deposition on the flank of the bar as a result of southerly migration of the tidal channel on the south side of the bar. After Dalrymple *et al.* (2003, Figures 5C, 15C, D).

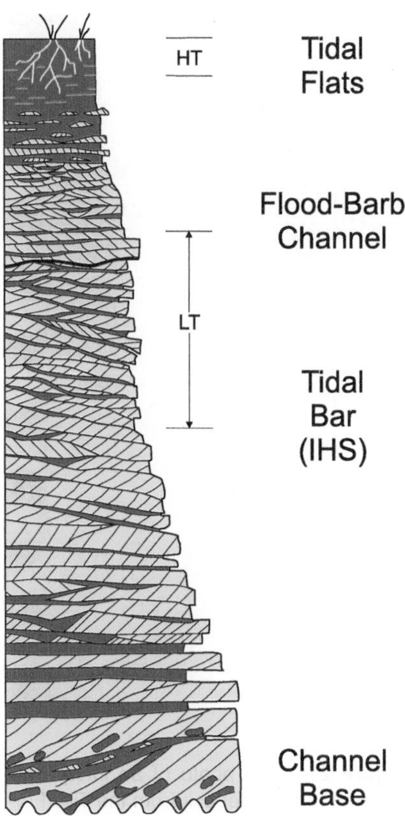

Figure 20. Schematic vertical profile through a tidal-channel–tidal-bar–tidal-flat succession, such as might be generated in the sandy part of the tidal-fluvial transition zone of an estuary or delta, in the area occupied by the turbidity maximum. Muddier versions of this succession also occur. The high-tide (HT) level is situated near the top of the succession. The low-tide (LT) level is located within the sandy sediments. Near the tidal limit where the tidal range is small, it will lie within the deposits of the flood-barb channel, whereas in more seaward areas, it might be located one-half to two-thirds of the way down from the top.

site banks, because of the gentle sinuosity of the main channel. Farther landward where the channel is more sinuous, the elongate bars become shorter and attach to the bank on the seaward side of each tidal point bar (Fig. 18). The available data indicate that all tidal bars, regardless of whether they are bank-attached or free standing, migrate sideways because of erosion on the outside or cut-bank side of the channel bend and deposition on the bar that occupies the adjacent point-bar location. As a result, tidal bars generate lateral-accretion deposits (Fig. 19) that comprise most of the upward-fining channel succession (Fig. 20).

Channel-bottom (thalweg) deposits contain the coarsest sand and/or gravel available at that location within the tidal-channel system (Barwis, 1978; Smith, 1987). Shell debris may be abundant in the more seaward part of such systems. Mud clasts can be abundant. Crossbedded sandstone is common, but the sedimentary structures depend on the prevailing combination of grain size, current speed and water depth. The channel-bottom deposits are very muddy if fluid muds are present, with anomalously thick (> 5–10 mm), homogenous mud layers interbedded with the coarsest sand or gravel in the local succession (Dalrymple *et al.*, 2003; Ichaso and Dalrymple, 2009). This situation can produce a distinctive deposit that has very bimodal grain sizes (Fig. 21). The deposits become less muddy upward because the fluid muds were concentrated at the channel bottom (Fig. 20). The sand fines upward, howev-

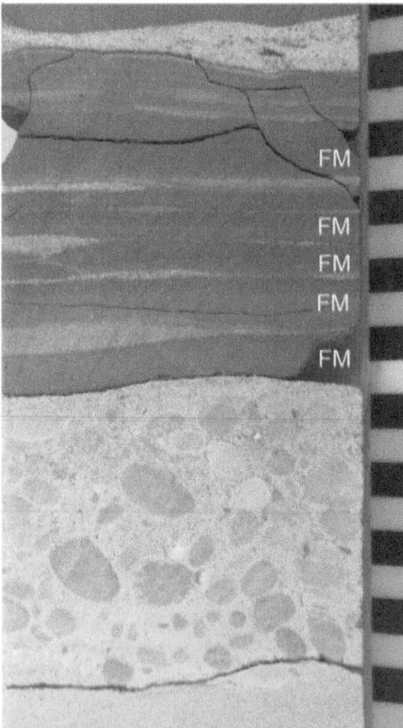

Figure 21. Core photo of an interpreted channel-bottom deposit from the Tilje Formation (Jurassic), offshore Norway, showing thick, homogeneous fluid-mud (FM) layers, interbedded with coarse-grained deposits. *See* Ichaso and Dalrymple (2009) for details. Scale divisions (right) one centimeter each.

er, as the amount of mud decreases. The upper part of the succession, which is generated by the flanking tidal flats, becomes muddier again (*see* more about tidal flats below).

Inclined heterolithic stratification (IHS; Thomas *et al.*, 1987; *see* also Smith, 1987) is a common constituent of tidal-channel–tidal-bar successions (Fig. 20). The proportion of sand generally decreases upward and can vary substantially from location to location, ranging from almost entirely cross-bedded and rippled sand to almost entirely mud (Allison *et al.*, 2003; Pearson and Gingras, 2006). The detailed structures present may include tidal rhythmites, but they are not ubiquituous: cyclic rhythmites are likely to be most abundant in the more sheltered, landward part of tidal-channel systems. Seasonal cyclicity produced by variations in river discharge may be an important cause of the alternation between sandy and muddy beds that comprise the IHS.

The main channels connected to the rivers are ebb dominated, whereas marginal channels, including flood barbs (Fig. 18), are flood dominated. Therefore, it is possible that the upper part of a channel-bar succession can show a 180° change in the paleocurrent direction relative to that in the lower part of the succession (Fig. 20).

Shelf Ridges

The tidal-current ridges that cover thousands of square kilometers on modern continental shelves (Fig. 22) are very large bedforms, with spacings of 1–30 km, lengths of 10–120 km and heights of 7.5–40 m (Houbolt, 1968; Stride *et al.*, 1982; Wood, 2003). They are nearly parallel to the flow, but are always oriented at a small angle (< 20°) to the fastest tidal currents, such that the currents flow obliquely up and over the ridge on both the flood and ebb tides. Consequently, the residual sediment movement is in opposite directions on either side of the crest (Fig. 23). All shelf ridges that are far removed from active coastal sedimentation are apparently transgressive in origin. They can be influenced significantly by sea-level change because of their large size (Snedden and Dalrymple, 1998), and they can contain a record of changing sea level and shoreline position, if they have not been fully reworked in a shelf setting. Our understanding of these ridges has increased significantly since the previous edition of this book and we now know that there are both erosional and depositional varieties. It is also no longer appropriate to refer to shelf ridges as 'sand ridges' as has been done, because some consist of muddy deposits.

Erosional ridges have an external form that is discordant relative to the internal structure of the ridge, with truncation of strata on the flanks of the ridge (Fig. 24). The deposits within the ridge can be of any composition. Those that have been documented in the Yellow Sea (Jin and Chough, 2002) are composed of muddy heterolithic strata that were deposited in coastal environments earlier during the transgression. Tidal currents on the shelf were responsible for eroding material from the troughs between the ridges, and for depositing a thin sandy lag that mantles one or both sides of these ridges.

Depositional ridges have internal strata that are concordant with the external shape of the ridge, typically with deposition on one side of the ridge; the other side is erosional as a result of its lateral migration. Some ridges, such as those that lie seaward of the Changjiang River in the East China Sea (Fig. 25; Berné *et al.*, 2002; Chen *et al.*, 2003; Yang, 1989) are thought to have formed initially in a river-mouth location and were then stranded on the shelf as the shoreline migrated landward during the post-glacial transgression. The degree to which they remain active in their present location on the shelf is not known. If they formed in a deltaic environment, they probably consist of muddy heterolithic strata (*cf.* Chen *et al.*, 2003) and are mantled by a sandy lag that has been created by tidal-current and storm-wave winnowing on the shelf. Many other ridges, such as those that occur in the Norfolk Banks in the southern North Sea (Figs. 22B and 26; Houbolt, 1968), appear to consist entirely of sand that was reworked to form the ridge in the present-day shelf environment. They represent 'fully evolved' ridges in the evolutionary scheme of Snedden and Dalrymple (1998). Dips of the lateral accretion surfaces formed by ridge migration range from 2–10° in ridges that are active, to less than 2° in ridges that have become inactive as a result of an increase in water depth and a decrease in tidal-current speed. Other ridges have a more complex internal architecture, with a core that probably consists of coastal sediment that is overlain by sand that was deposited in the present-day shelf environment (Berné *et al.*, 1994).

Active ridges composed of medium to coarse sand are covered by simple and compound dunes; consequently, they are predominantly crossbedded (Houbolt, 1968; Berné *et al.*, 1994). Compound crossbedding may be common within such ridges; in the situation described by Reynaud *et al.* (1999), this cross-stratification is up to 10 m thick because the ridge was formed in a deep-water setting near the shelf edge, in the high-energy zone that occurs where the tidal wave moves onto the shelf. At the other end of the energy spectrum, it is possible to get ridges that consist entirely

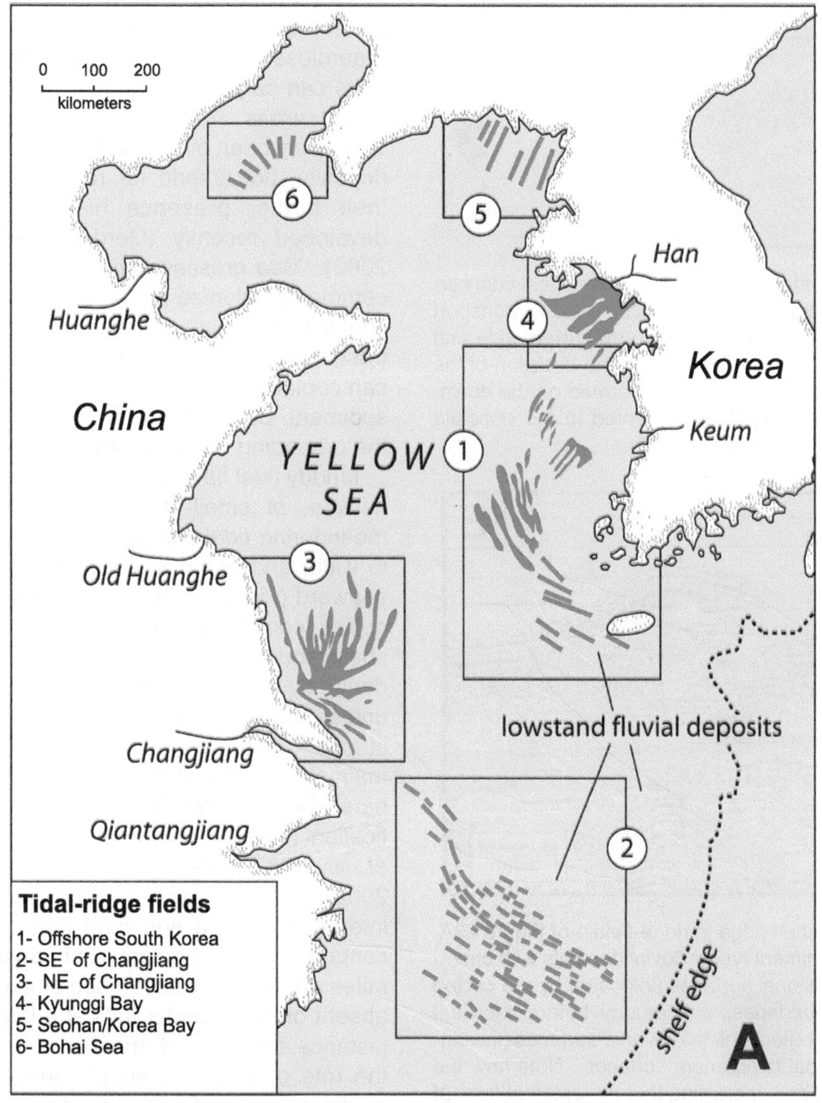

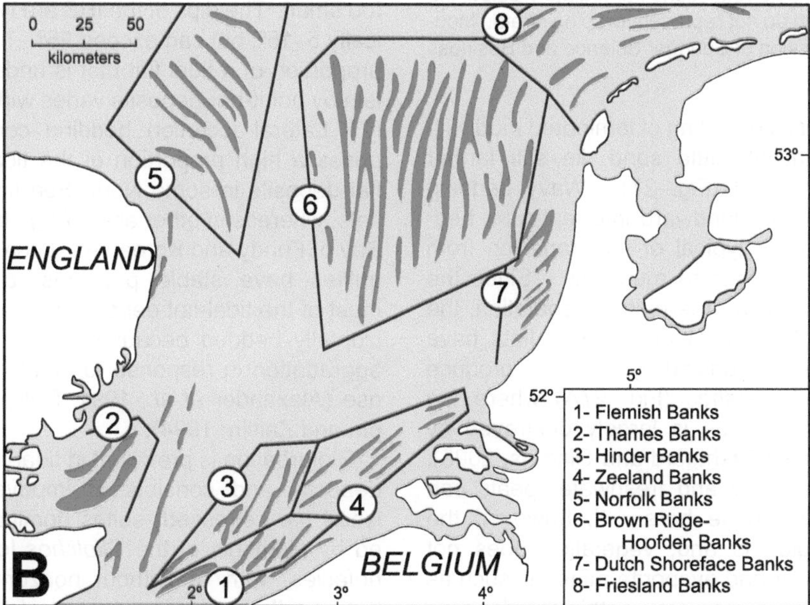

Figure 22. Distribution of tidal-current ridges in **A**) the Yellow and East China seas and **B**) the southern North Sea.

of bioturbated fine sand. Shelf ridges that have become moribund lack large-scale bedforms. They can become draped by strata that thicken away from the ridge crest (Reynaud *et al.*, 1999) and contain ripple cross-lamination and/or storm deposits with a greater degree of burrowing. The vertical succession of grain sizes in sandy shelf ridges is thought to coarsen upward because current and wave action is most intense in the shallow water on the ridge crest.

Tidal Flats

Tidal flats are widespread along shorelines that have a large tidal range (Amos, 1995). They extend from the *supratidal zone*, through the *intertidal zone*, and into the shallow portion of the *subtidal zone* (Fig. 27). Two distinct types of tidal flat can be recognized, those in channelized areas sheltered from significant wave action, and those that occur along open marine coasts with no protection from waves.

Sheltered, Channel-related Tidal Flats

Such tidal flats border the tidal channels within deltas and estuaries, as well as the tidal channels that occur in back-barrier lagoonal settings (*see* Chapter 11). These tidal flats become wider in a seaward direction through the fluvial-marine transition. The outer edge of these flats is bordered by a tidal channel (Fig. 27); thus, the tidal flat represents the uppermost part of the channel–tidal-bar succession (Fig. 20). The currents in the channel flow sub-parallel to the local shoreline, and the speeds of both the flood and ebb currents decrease away from the channel thalweg toward the flanking supratidal area. Consequently, sediment in and adjacent to the channel is typically sandy, and passes gradually into mud near the high-tide line. Sand flats are absent in cases where a small channel lies within the mudflat associated with a larger nearby channel. The deposition of mud on the upper part of the tidal flats is also facilitated by the processes of *settling lag* and *scour lag* that operate in these settings (Dalrymple and Choi, 2003).

The *sand flats* that occupy the

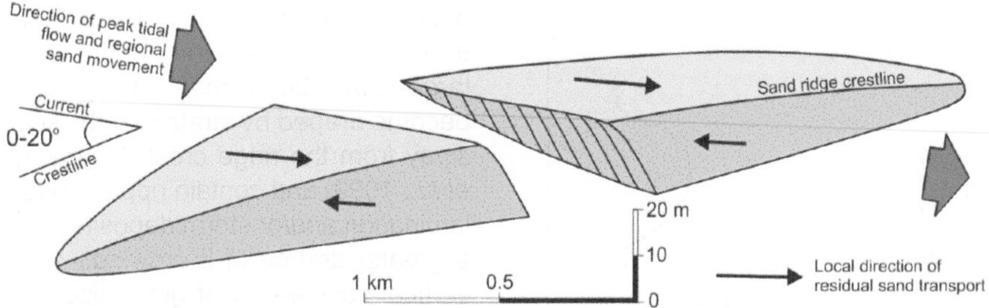

Figure 23. Model of sediment transport around a tidal sand ridge. The strongest tidal currents flow at a slightly oblique angle to the ridge crest so that the local, residual transport is in opposite directions on either side of the crest. Erosion on the up-current side and net deposition on the down-current flank causes the ridge to move in the direction of the regionally stronger current. The lateral-accretion deposits that are formed on the down-current flank should contain small-scale crossbedding that is oriented in the opposite direction to the regional transport. Modified after Stride *et al.* (1982).

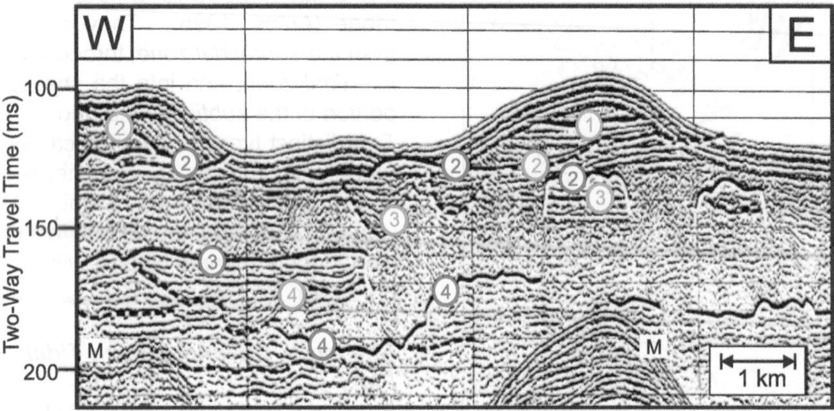

Figure 24. Seismic section through an erosional shelf ridge in ridge-field 1 of Figure 22A. The succession is interpreted to consist of two sediment types: fluvial deposits with chaotic, inchoherent relections overlying high-relief erosional surfaces (inidcated by red circled numbers) that are inferred to be sequence boundaries; and muddy heterolithic tidal deposits with high-amplitude, laterally continuous reflections that overlie surfaces (indicated by blue numbers) that are interpreted to be tidal ravinement surfaces. Note how the reflectors within the ridge are truncated at the seabed, indicating that the external form of the ridge was produced by erosion. The conformable reflections immediately below the sediment-water interface are a seismic artefact and do not represent draping strata. Modified after Jin and Chough (2002), with kind permission of Springer Science and Business Media.

lower portion of most tidal flats can contain dune crossbedding in areas with high current speeds, and ripple cross-lamination in places where the current speed is lower and/or the grain size is too fine to allow the formation of dunes. Thickness of the crossbeds should decrease toward the shore because the dunes get smaller in shallower water. Parallel lamination can also occur in the sand flats located in the area with the highest tidal-current speeds in macrotidal estuaries (Dalrymple *et al.*, 1990). Mud flasers (Fig. 10A) can be present in the deposits of rippled sand flats. *Mixed flats*, which contain mud layers formed by slack-water settling, lie shoreward of the sand flats. *Mud flats,* consisting of laminated mud with relatively little sand, lie still farther landward (Fig. 27). Wavy bedding passing landward into lenticular bedding is typical of the transition from mixed flats to mud flats. Since the evolution of terrestrial vegetation, the highest parts of the tidal flats have been colonized by plants to produce *salt marshes* (Fig. 27A) where the stratification is largely destroyed by roots (Dashtgard and Gingras, 2005). Saltwater and freshwater peats can accumulate here. If, however, the region is arid, vegetation does not occur and evaporite minerals such as gypsum can grow within the deposits of the upper intertidal zone (*see* Chapters 16 and 20). Desiccation cracks are most abundant in the upper intertidal and supratidal zones, regardless of climate, and microbial mats can carpet the surface, even in humid areas. Their presence has commonly been overlooked in ancient deposits, but criteria for recognizing their former presence have been developed recently (Gerdes *et al.*, 2000). Sea grasses and mangroves commonly colonize large parts of the tidal flats in tropical climates. By contrast, tidal flats in mid to high latitudes can contain ice-rafted debris and soft-sediment deformation produced by the grounding of ice blocks.

Muddy tidal flats are dissected by a network of small- to medium-sized meandering *tidal gullies* that increase in width and depth as they coalesce seaward (Fig. 27). The bottom of the gullies commonly contains a lag of shells and mud clasts. Sand is abundant only in the more seaward areas, unless the tidal creek connects to a stream with terrestrial drainage. The main site of deposition is on the point bars where inclined heterolithic stratification (IHS) is developed (Thomas *et al.*, 1987; Pearson and Gingras, 2006). These deposits consist of interbedded sand and mud that can contain tidal rhythmites; such rhythmites are less well developed or even absent on the broader flats at a large distance from the channel, because the rate of sediment supply and the space available for aggradation are too small. The dips of the IHS are typically 5–15°, but can exceed 25°. The proportion of a tidal flat that is underlain by point-bar deposits varies widely. Lateral-accretion bedding comprises a high proportion of the tidal-flat deposits in some North Sea tidal flats, whereas in other areas (*e.g.*, the Bay of Fundy and Korean west coast) gullies have stable positions, and most of the tidal-flat deposits are horizontally bedded because of vertical aggradation in response to sea-level rise (Alexander *et al.*, 1991; Dalrymple and Zaitlin, 1994).

Bioturbation is prevalent in tidal-flat deposits and consists of impoverished (*i.e.*, stressed) suites dominated by elements of the *Skolithos* Ichnofacies, with or without horizontal burrows that are more typical of the *Cruziana* Ichnofacies (*see* Chapter 3). The degree of stress reflected in

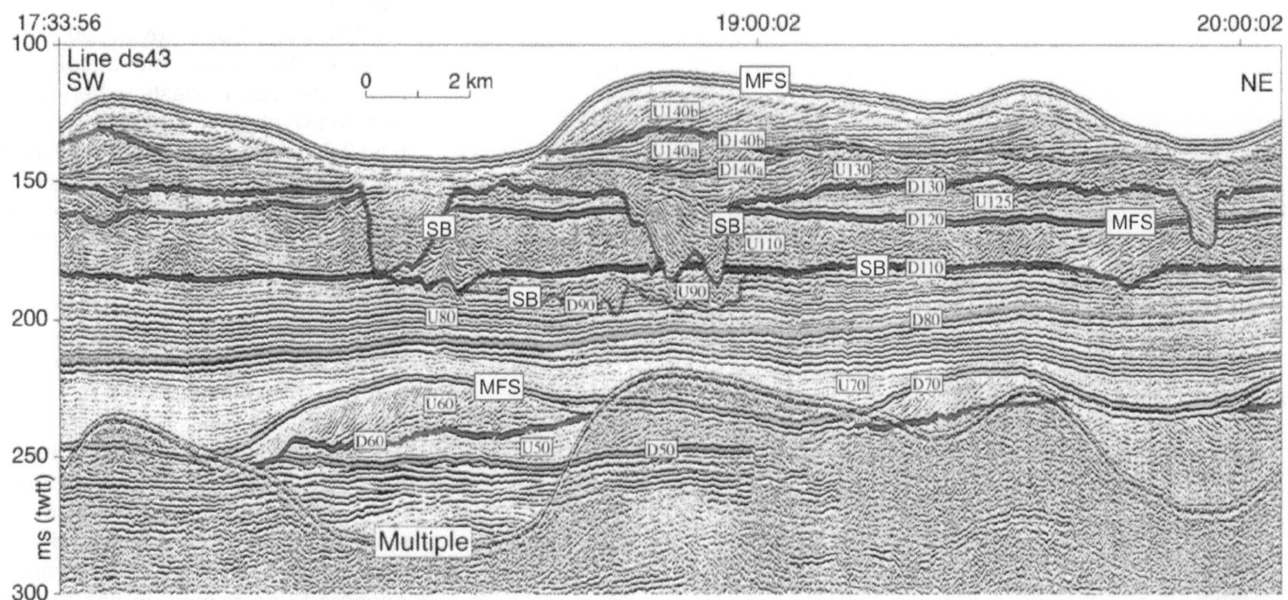

Figure 25. Seismic section from the East China Sea, seaward of the Changjiang Delta. Modern shelf tidal ridges (U140a, b) that contain lateral-accretion bedding are present on the surface. Note that the ridges migrate over each other, and locally have an erosional basal surface. A number of units with a deeply erosional base (U90, U110 and U130) and consisting of sandy fluvial and tidal-fluvial deposits alternate with gently seaward-dipping muddy strata (U50, U70, U80 and U125) that were deposited by prograding interglacial highstand and falling-stage deltas. The locations of inferred sequence boundaries ('SB') and maximum flooding surfaces ('MFS') are indicated. An older set of transgressive shelf ridges is preserved in U60. After Berné *et al.* (2002). Reprinted with permission from Elsevier.

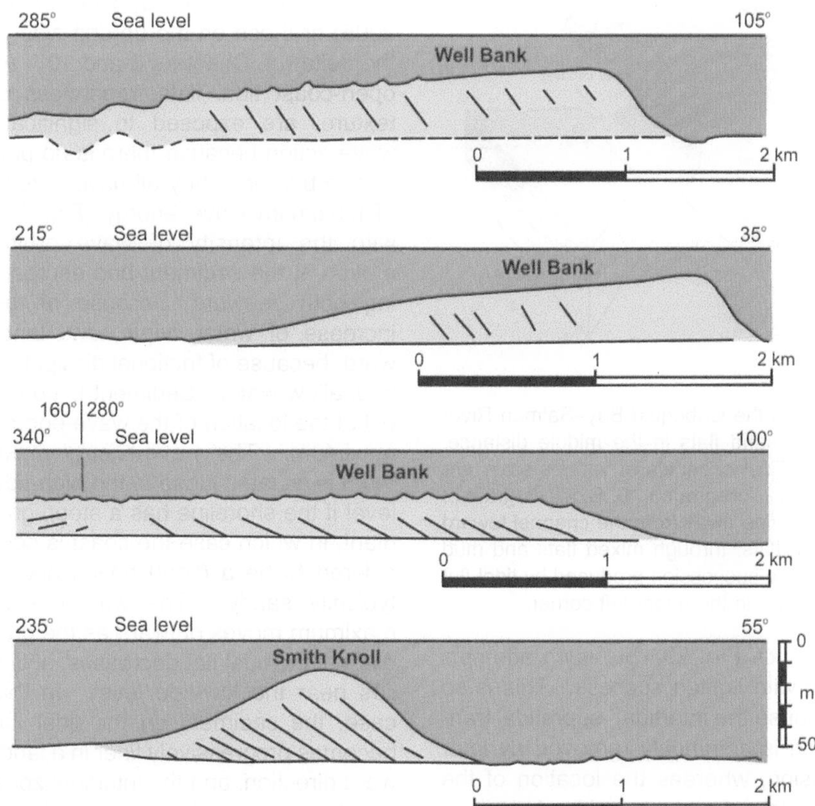

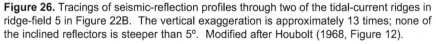

Figure 26. Tracings of seismic-reflection profiles through two of the tidal-current ridges in ridge-field 5 in Figure 22B. The vertical exaggeration is approximately 13 times; none of the inclined reflectors is steeper than 5°. Modified after Houbolt (1968, Figure 12).

the trace-fossil assemblage (*i.e.*, by a reduction in diversity and trace-fossil size) increases toward the high-tide level as well as landward into areas with reduced salinity (*e.g.*, Yang *et al.*, 2007). The degree to which the physical sedimentary structures are over-printed by bioturbation depends on the salinity and the rate of sedimentation and/or reworking. Thus, bioturbation is typically less in areas of lower salinity; and the deposits that accumulate in close proximity to a channel margin, where physical reworking and the rate of sediment deposition are higher, are less intensely bioturbated than the deposits that accumulate on broad tidal flats that are at a distance from a channel.

Progradation of a channel-associated tidal flat generates an upward-fining succession (Figs. 20 and 27B; Weimer *et al.*, 1982; Terwindt, 1988; Dalrymple, 1992). The succession begins with an erosion surface that is created by the thalweg of the adjacent tidal channel. There is a gradual upward decrease in the grain size and thickness of sand beds, and an increase in the proportion of mud. The thickness of such successions

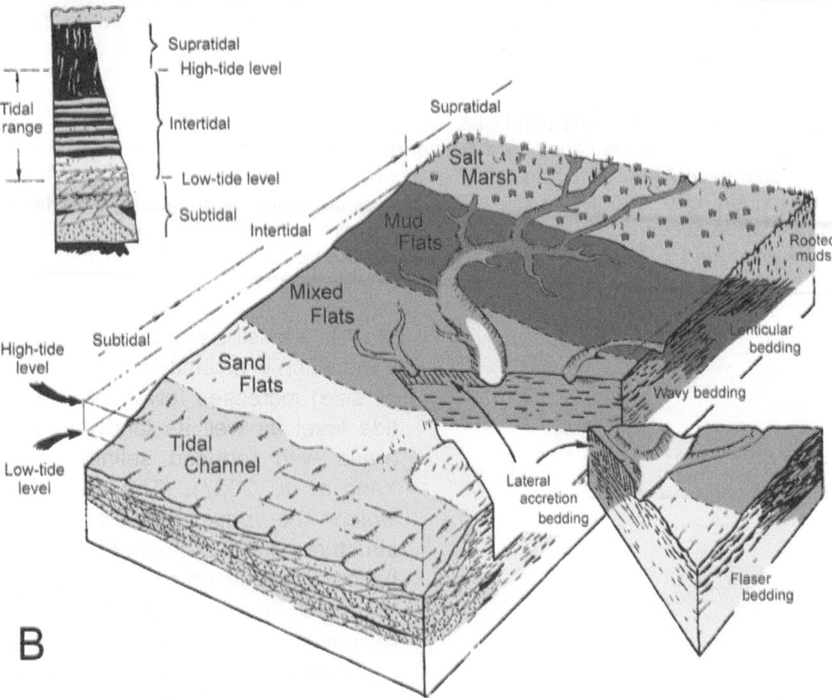

Figure 27. A. Oblique airphoto of a broad tidal flat in the Cobequid Bay–Salmon River estuary, Bay of Fundy, showing the transition from sand flats in the middle distance, through mudflats to salt marsh in the foreground. Tidal channels of various sizes are present. Field of view approximately 1 km across in the foreground. **B.** Block diagram of a typical, channel-associated tidal flat. The tidal flats fine away from the channel toward the high-tide level, passing gradationally from sand flats, through mixed flats and mud flats, to salt marshes. An example of the upward-fining succession produced by tidal-flat progradation, as a result of channel migration, is shown in the upper left corner.

ranges from about 1 m to more than 10 m and approximates the depth of the channel before compaction. Numerous descriptions of such successions were published in the 1970s and 1980s (see selected references in Dalrymple, 1992). Some workers have attempted to estimate the tidal range by determining the sediment thickness between the interpreted levels of high and low tide (Terwindt, 1988; *cf.* Fig. 20), but such attempts met with limited success. This is so because the intertidal–supratidal transition is commonly removed by later erosion, whereas the location of the subtidal–intertidal transition is typically cryptic because it occurs within sandy sediments in which exposure indicators are rare and difficult to recognize.

Open-coast Tidal Flats

Tidal flats are widely developed along low-gradient upper mesotidal and macrotidal open coasts. Very little was known about the deposits of such tidal flats when the previous edition of this volume was written. This situation has changed markedly over the last 5 years, because of significant new work along the Chinese and Korean coasts. This work indicates that several useful generalizations can be made, even though our knowledge remains less than that for channel-associated flats.

Open-coast tidal flats vary widely in their sedimentary character, ranging from those that are dominated by sand, to those that are composed predominantly of mud. This difference appears to be related, in part, to the available sediment: tidal flats adjacent to and down-drift from a major river, such as those close to the mouth of the Changjiang River (Fan *et al.*, 2004), tend to be much muddier than those that lack a nearby source of mud (*e.g.*, the tidal flats along the western coat of Korea; Yang *et al.*, 2005) or those on the up-drift side of the delta (*cf.* Chapters 8 and 10). All open-coast tidal flats, regardless of texture, are exposed to significant wave action because there is no protective barrier. They all have a zone of maximum wave energy (Fig. 28), with the intensity of wave orbital motion at the sediment bed decreasing both seaward, because of the increase of water depth, and landward, because of frictional dissipation in shallow water. Sediment is coarsest at the location of the wave-energy maximum. The wave-energy maximum is situated close to the high-tide level if the shoreline has a steep gradient, in which case the coast is considered to be a (tidal) beach and is typically sandy. The wave-energy maximum moves outward as the gradient of the tidal flat decreases, until it lies near the low-tide level. In this case, the sediment on the tidal flat becomes progressively finer in a landward direction, and the intertidal zone can be composed predominantly of mud (*cf.* Fan *et al.*, 2004). In intermediate cases such as those described by Yang *et al.* (2005), the zone of maximum wave energy and coarsest sediment lie part way across the inter-

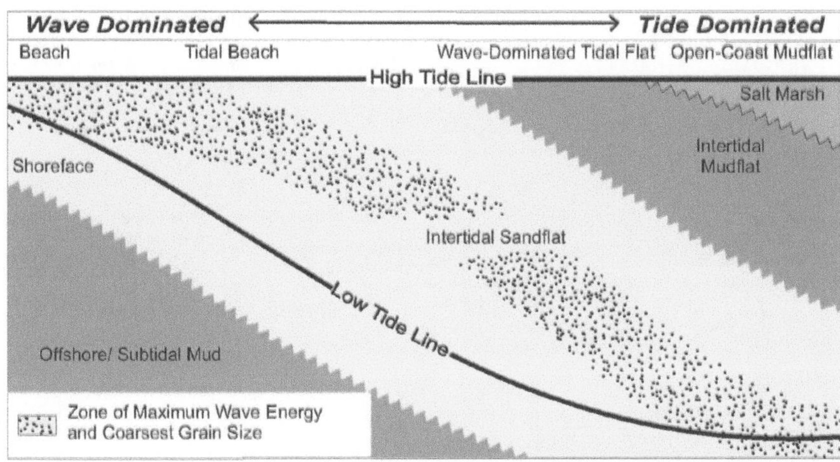

Figure 28. Schematic representation of the transition from a normal wave-dominated beach and shoreface (at left) to a tide-dominated open-coast tidal flat (right side) such as those adjacent to the Changjiang River delta (Fan *et al.*, 2004). The wave-dominated tidal flats described by Yang *et al.* (2005) occupy an intermediate position.

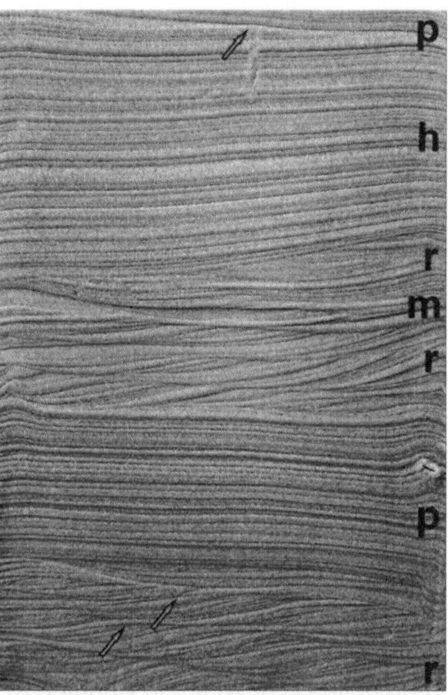

Figure 29. Epoxy peel from the middle intertidal zone of a sandy, wave-dominated open-coast tidal flat, western Korea, showing the typical structures produced by storm waves. The upward transition from parallel lamination ('p') to wave ripples ('r') is interpreted to represent decreasing energy as water depth decreases during the ebbing tide, followed by the formation of a thin mud drape ('m') immediately before low-tide emergence. The subsequent transition from ripples ('r') to hummocky cross stratification ('h') represents increasing energy as the water deepens during the flood tide. Rare vertical burrows indicated by arrows. Peel 25 cm wide. Photo courtesy of B.C. Yang.

tidal zone and there can be a narrow muddy tidal flat in the upper intertidal and supratidal zone.

Sedimentation on sandy open-coast tidal flats is typically wave dominated despite the existence of a large tidal range (Yang *et al.*, 2005). This potentially counter-intuitive situation occurs because the tidal currents are weak as a result of the small shore-normal tidal prism. By comparison, nearby river mouths and embayments can be tide dominated (Yang *et al.*, 2007). Sedimentary structures on these tidal flats consist of storm-generated graded beds that contain HCS and wave ripples (Fig. 29). Distinction from true shorefaces is difficult, but Yang *et al.* (2008) have suggested that the storm beds can contain evidence of tidal modulation of wave energy: wave energy is highest at high tide, but non-existent at low tide when the tidal flat is exposed. This can lead to cyclic variation in the nature of the sedimentary structures.

The nature of muddy open-coast tidal flats has been explored in areas close to the mouth of the Changjiang River where sedimentation rates are very high (Fan *et al.*, 2004). An onshore fining trend predominates. Tide-generated sand–mud lamination is prevalent in the lower and middle intertidal zone; erosion surfaces mantled by a sandy and mud–pebble lag record storm-wave activity. Salt-marsh deposits occur in the upper intertidal and supratidal zones. Bioturbation in this situation is minimal because of rapid aggradation. Areas more distant from a river mouth might be more intensely bioturbated. Tidal lamination in such areas might be preserved only within rapidly deposited storm beds.

Variations in the supply of mud to open-coast tidal flats, as a result of delta-lobe switching, can cause the surface sediment to alternate between muddy, during episodes of active progradation, and sandy, during periods of sediment starvation and erosion. Decadal and longer variations in storminess might have a similar affect. Such erosive episodes can lead to the generation of a beach ridge, termed a *chenier,* composed of sand with variable amounts of shell debris (Thompson, 1968; *see* also Chapter 8).

The fringing tidal flats in some back-barrier lagoons (*see* Chapter 11) also pass outward into subtidal muds. Such tidal flats will not be significantly coarser in the lower intertidal zone unless there is appreciable wave action in the lagoon. McIlroy *et al.* (2005) have interpreted successions of this type in the Nequen Formation (Jurassic) in Argentina.

TIDE-DOMINATED DEPOSITIONAL SYSTEMS

The morphological elements described above are assembled in various ways in larger depositional systems. The characteristics of these systems, including the particular elements present and their spatial and stratigraphic distribution, are controlled by (paleo-)geographic location and the trajectory of the shoreline (transgressive or regressive). Tide-dominated deltas are the primary regressive tidal system, whereas transgressive situations are characterized by tide-dominated estuaries and continental shelves. Seaways and straits – narrow constrictions that connect two larger water bodies – are also the site of tide-dominated sedimentation, and can exist in both regressive and transgressive situations.

Tide-dominated Deltas

Deltas are regressive coastal environments at the mouths of rivers that prograde because of the input of sediment by a river. A considerable amount of work has been done on tide-dominated deltas over the last decade, but they remain poorly understood because of the large range of variability that they exhibit and the complex array of sub-environments that are present. Consequently, it is too early to claim that a solidly based facies model exists for these systems. The following paragraphs highlight the main attributes of tide-dominated deltas as they are known currently. Readers are referred to Chapter 10 for a more comprehensive discussion of deltas, including the principles of their classification, the dispersal of sediment from river mouths, and the broad subdivision of deltaic environments (*e.g.*, subaerial and subaqueous delta plains, mouth-bar area, delta front and prodelta). An abbreviated treatment of tide-dominated deltas is provided in Chapter 10 and a more extensive review is given in Willis (2005).

Tide-dominated deltas occur in mesotidal and macrotidal settings, and experience extensive reworking of river-supplied sediment by tidal currents. Many of the largest rivers in the world have deltas that experience strong tidal reworking, examples including the Amazon, Changjiang, Ganges–Brahmaputra and Fly rivers. Such large rivers might be especially prone to strong tidal influence because of the large tidal prism caused by the low-gradient terrain over which they flow; small, steep-gradient rivers, by contrast, are less likely to have tide-dominated deltas.

Strong currents entering the river mouth create a series of straight to gently sinuous coast-normal tidal channels that are separated by elongate tidal bars (Fig. 30). These channels become narrower and more sinuous inland (Fig. 18), but lack the high-sinuosity reach that characterizes estuaries (see below). The major channels have a quasi-regular spacing that is of the order of 15–40 km in the large deltas cited above. Such separations are required in order that each channel has a sufficiently large tidal catchment area that the flow can maintain the channel; channels that are closer together are likely to become filled with sediment. Channel depths are commonly only 10–15 m, but reach up to 35 m in the Han River delta in Korea. Relief between the channel thalweg and the crest of the adjacent bar becomes less in a seaward direction and the channels merge into the flat seafloor in the outer delta-front or prodelta area. The subaqueous delta plain, which extends from the seaward limit of exposure to the top of the prodelta slope and is generally less than about 15–20 m deep, is particularly wide in tide-dominated deltas, reaching up to 100 km in width. The tidal bars that separate the main channels in the subaqueous delta plain display broad flat tops because they have aggraded as close to sea level as the dynamic conditions permit. Their surface rises in a landward direction and they eventually become emergent, forming the islands that separate the distributary channels (Fig. 30). These distributaries converge in a landward direction, forming a distinct funnel-shaped geometry. Tidal flats comprise only a narrow fringe along the sides of the distributary channels, but can reach several kilometers in width on the fronts of the islands between these channels. Coastal areas adjacent to the main river mouth display one of two morphologies. In cases such as the Ganges–Brahmaputra and Indus deltas, the areas flanking the delta are occupied by a complex network of sinuous tidal channels (Allison *et al.*, 2003). The Amazon and Changjiang deltas, by comparison, are flanked by broad, open-coast tidal flats, particularly on their down-drift side, which have few or no tidal channels.

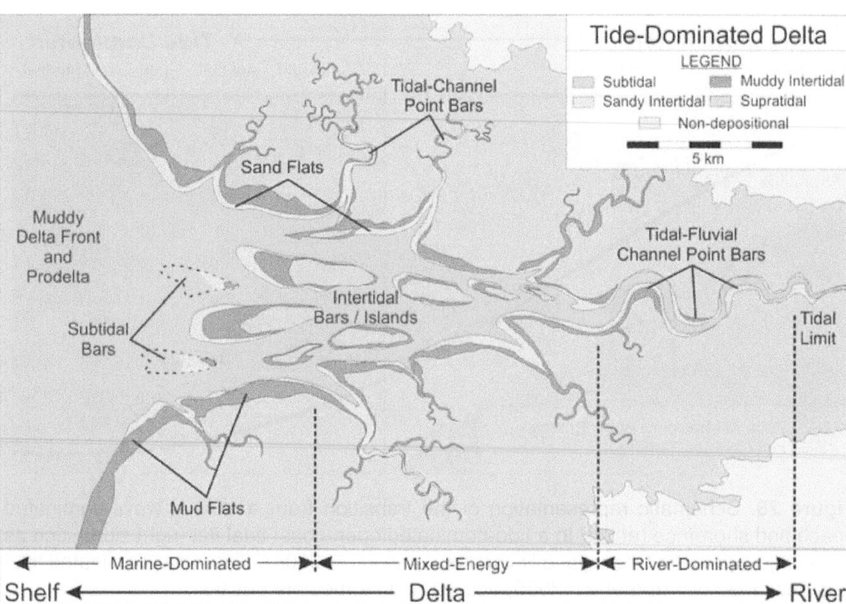

Figure 30. Distribution of environments within a tide-dominated delta. Image courtesy of D. MacKay.

The progradation of tide-dominated deltas generates the upward-coarsening succession that typifies all deltas (Fig. 31; Dalrymple *et al.*, 2003; Davies *et al.*, 2003; Willis, 2005). The base of the succession consists of prodeltaic mud that shows diverse levels of bioturbation. High levels of bioturbation characterize up-drift areas and locations distal to active distributary channels (*cf.* Chapter 3). Deposits immediately seaward of active distributaries are heterolithic and sparsely bioturbated because of rapid deposition. They contain thick, rapidly deposited fluid-mud layers that have flowed offshore from the delta-front area in response to wave action in shallow water (Kuehl *et al.*, 1996; Dalrymple *et al.*, 2003). Earlier suggestions that tidal rhythmites might be common in prodelta deposits have not been borne out by observations of modern deltas: depositional conditions are too variable because of waves to permit uninterrupted deposition of tidal layering. Two notable exceptions to this have been reported from the ancient record: the Elatina Formation (Neoproterozoic) of Southern Australia (Fig. 9; Williams, 1991), and the Mississippian Pride Shale of West Virginia (Miller and Eriksson, 1997), both of which contain exceptionally long and complete tidal-rhythmite successions in prodelta deposits. The conditions under which these

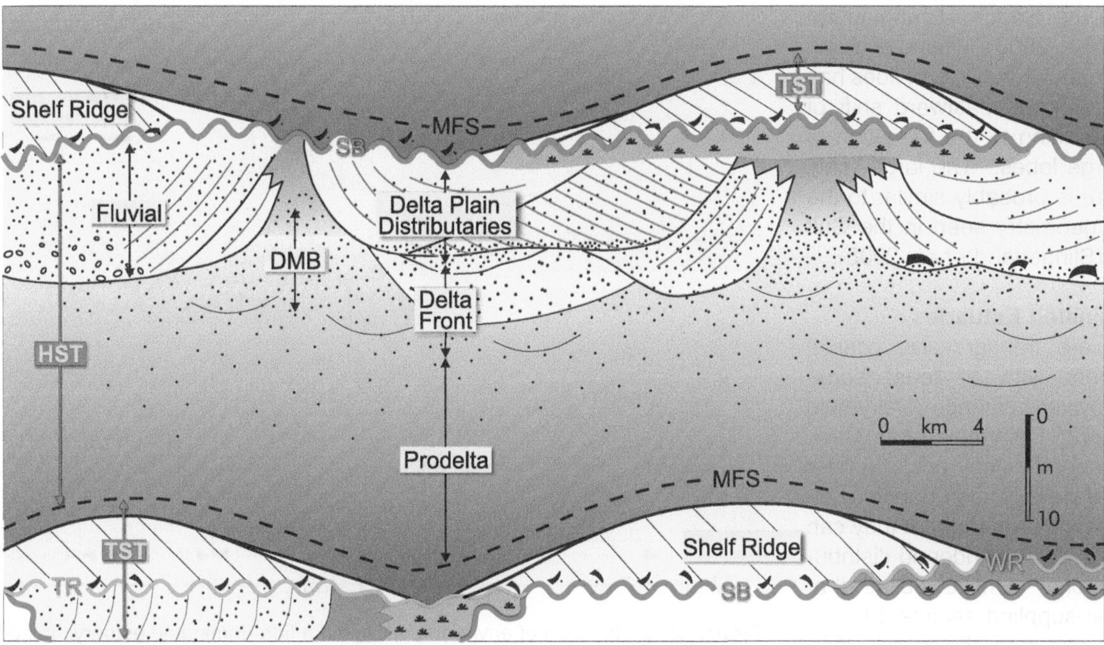

Figure 31. Hypothetical coast-parallel section through a tide-dominated delta based on the Fly River and Changjiang River deltas. The upward-coarsening succession from prodeltaic to delta-front/distributary-mouth-bar (DMB) deposits is extensively channelized in its upper part. Sandy deposits of the mouth-bar region are commonly erosively based. Shelf ridges created during transgressions overlie shelly lags and are buried by prodelta mud with their morphology largely intact (cf. Fig. 25). Based on Figure 18 of Dalrymple *et al.* (2003).

deposits formed appear to be unusual in their ability to exclude wave action. The Elatina Formation is associated with the 'Snowball Earth' glaciation, which is thought to have been the most extensive in Earth history: perhaps extensive sea ice inhibited wave action. By contrast, the Pride Shale might have accumulated at the head of an embayment in which wave action was minimal. A possible modern analog is provided by the deltas at the head of fiords where tidal rhythmites have been reported.

Shallow-water, delta-front and mouth-bar deposits overlie the prodelta mud. Unlike river- and wave-dominated deltas, there is no well-developed mouth-bar at the end of each distributary (Fig. 30). Instead, tidal currents dissect the mouth bar, creating a series of shore-normal, elongate tidal bars. The deposits of this area should, therefore, contain multiple channels (Figs. 30 and 31). There is an overall landward and upward coarsening, but gradational upward-coarsening successions might be uncommon because of lateral migration of the channels. The deposits contain a mixture of wave- and tide-generated sedimentary structures (Dalrymple *et al.*, 2003), unless the area is sheltered from wave action. In deltas where the rivers supply mainly fine to very fine sand (*e.g.*, the Fly River; Dalrymple *et al.*, 2003), the tidal bars at the river mouth contain hummocky cross stratification formed by storm waves, and medium- to thick-bedded heterolithic stratification that contains wave-generated fluid-mud deposits is also present in exposed delta-front settings (Kuehl *et al.*, 1996; Dalrymple *et al.*, 2003). Short tidal-rhythmite successions can be developed in more sheltered areas such as low in channels (Jaeger and Nittrouer, 1995) or in more landward locations. Tidally generated fluid-mud layers can also be common in channel-bottom areas. In deltas supplied with medium to coarse sand, the tidal-bar deposits consist predominantly of crossbedding produced by dunes (Davies *et al.*, 2003). A superbly described ancient example is provided by the Frewens sandstone (Frontier Formation, Cretaceous) of Wyoming (Willis *et al.*, 1999; *see* extended description in Chapter 10). Mud drapes are abundant in the crossbeds, which have pervasive seaward-directed paleocurrents because of river dominance.

The subaerial delta plain is occupied by a series of headward-tapering distributary channels separated by vegetated islands (Fig. 30). One of these channels is typically the active distributary, while the others carry lesser amounts of river discharge. Those with minimal fluvial influence can be considered as 'abandoned' and are actively reworked by tidal currents. These channels may be overall coarser grained than the active distributaries (*cf.* Dalrymple *et al.*, 2003). Bhattacharya (Chapter 10) has suggested that these channels may be longer-lived than the distributaries of other deltas because the channels are kept open by tidal currents during abandonment, rather than becoming filled. Periodic reoccupation might, therefore, lead to complex stacking of channel deposits if relative sea level is rising during delta progradation (Fig. 31). These channel successions will show an upward fining of the sand fraction, but the basal part can contain abundant fluid-mud layers (Fig. 20) that will be especially abundant in the bottomsets of any compound dunes (Fig. 15) that are present. Inclined heterolithic stratification should be common.

The only delta in which the history of delta-lobe switching is well documented is the Ganges–Brahmaputra (Allison *et al.*, 2003). Here, avulsion of the river above the limit of tides has caused a stepped eastward shift of the delta, leading to the formation of several large lobes. The large-scale architecture is probably similar to the mounded geometry seen in the Gulf of Papua (Slingerland *et al.*, 2008).

Tide-dominated Estuaries

Estuaries are transgressive coastal embayments with at least some amount of river influence (Dalrymple *et al.*, 1992; Dalrymple and Choi, 2007). Generally, they occur in river valleys that were incised during a relative sea-level lowstand, but they can also occur within abandoned distributary channels. Unlike deltas that export river-supplied sediment to the sea, estuaries trap the river's load landward of the main coast and import sediment from the sea. Tide-dominated estuaries that have a significant amount of sand supplied by transgressive coastal erosion tend to be sand dominated, as is the case in the Cobequid Bay–Salmon River system in the Bay of Fundy (Dalrymple *et al.*, 1990, 1991). Most of the following discussion focuses on this type of estuary. In some cases, however, the estuary may be much muddier, either because the river supplies large quantities of mud, or because mud is moved into the estuary from the sea. This mud is typically supplied by a nearby large river that has established a delta: the mud exported from the large river is advected along the coast, causing down-drift estuaries to be mud dominated in their outer part, as is the case with the Qiantangjiang estuary, which lies immediately south of the Changjiang River delta (Li *et al.*, 2006; Fig. 22A). By comparison, estuaries up-drift of the delta are likely to be sandy.

The axial portion of coarse-grained tide-dominated estuaries (Fig. 32) is occupied by channels that contain sandy sediment. The sand is coarsest at both the mouth and head of the estuary because these areas are closest to the two sediment sources. The channel deposits are finest in the tidal–fluvial transition zone in the middle of the estuary, but, unlike wave-dominated estuaries, no lagoonal deposits are present (Dalrymple *et al.*, 1992). Consequently, the distinct tripartite facies distribution (sandy barrier–muddy lagoon–sandy bay-head delta; *see* Chapter 11) of wave-dominated estuaries does not occur in tide-dominated systems. Instead, tide-dominated estuaries are bordered by muddy tidal flats and marshes that occupy the space between the axial channels and the valley walls.

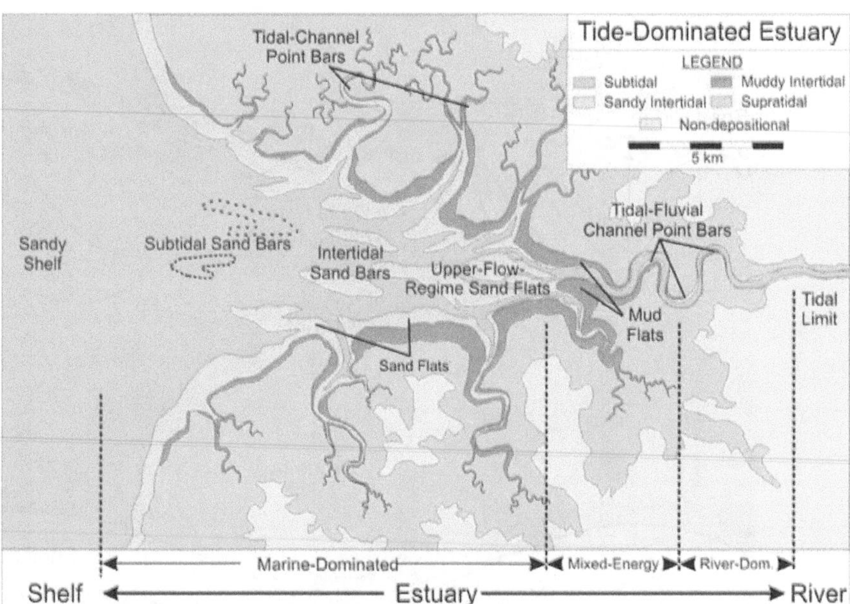

Figure 32. Distribution of environments within a tide-dominated estuary. Image courtesy of D. MacKay.

The most seaward facies consists of elongate sand bars (Figs. 32 and 33) that are separated by mutually evasive ebb-dominant and flood-dominant channels (Dalrymple *et al.*, 1990). The bars are typically covered by dunes, and the bar sediments are composed of crossbedded, medium to coarse sand in which the deposits of compound dunes (Figs. 14 and 15) are likely to be common. Lateral shifting of the channels and their associated bars produces an erosionally based upward-fining trend because current speeds are greatest in the channels and decrease toward the bar crests. On average, the tidal-current speed increases into the estuary, so that the area headward of the elongate bars is characterized by extensive sand flats with parallel lamination. Still farther headward, these flats grade into the tidally influenced portion of the river channel (Figs. 32 and 33). Parallel lamination is present here too, but crossbedding containing mud drapes may become abundant farther headward as the sand size increases into the river. Tidal rhythmites occur in the bordering mud flats (Dalrymple *et al.*, 1991), and IHS can be present in the tidal point bars that occur in this region. The muddy channel sands of the inner estuary pass gradationally into coarser fluvial sediments at the tidal limit.

The facies within a tide-dominated estuary shift headward during active transgression. In the process, migrating tidal channels coupled with wave action erode all, or part, of the more landward facies, producing one or more *tidal ravinement surfaces*. The resulting stratigraphic succession depends on the shoreline trajectory (*cf.* Plink-Björklund and Steel, 2006). If the shoreline migrates landward with a nearly flat trajectory, because of a slow rise of relative sea level, most of the estuarine succession is removed. Figure 34 shows an ideal case of essentially complete preservation, due to a steeply rising shoreline trajectory. The succession consists of a stack of erosionally based channels, each of which is more distal in its character than those beneath: the basal deposits in the valley axis will be fluvial in character, whereas those at the top, immediately below the overlying offshore shales of the next progradation, will consist of estuary-mouth or transgressive-shelf sand-ridge deposits (*see* below). Channel deposits within the valley

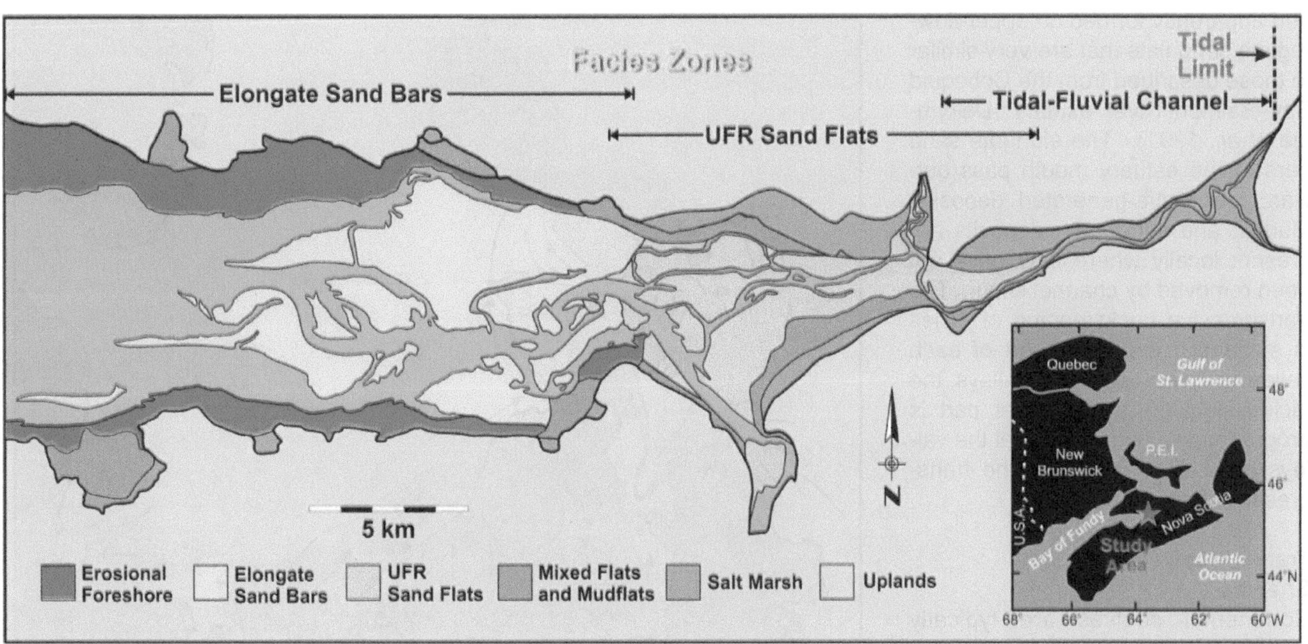

Figure 33. Facies distribution in the Cobequid Bay–Salmon River macrotidal estuary, Bay of Fundy. Trangressive erosion of the foreshores occurs in the outer part of the system even though the salt marshes and mudflats in the inner part have begun to prograde. 'UFR' = upper flow regime. After Dalrymple *et al.* (1990, Figure 2).

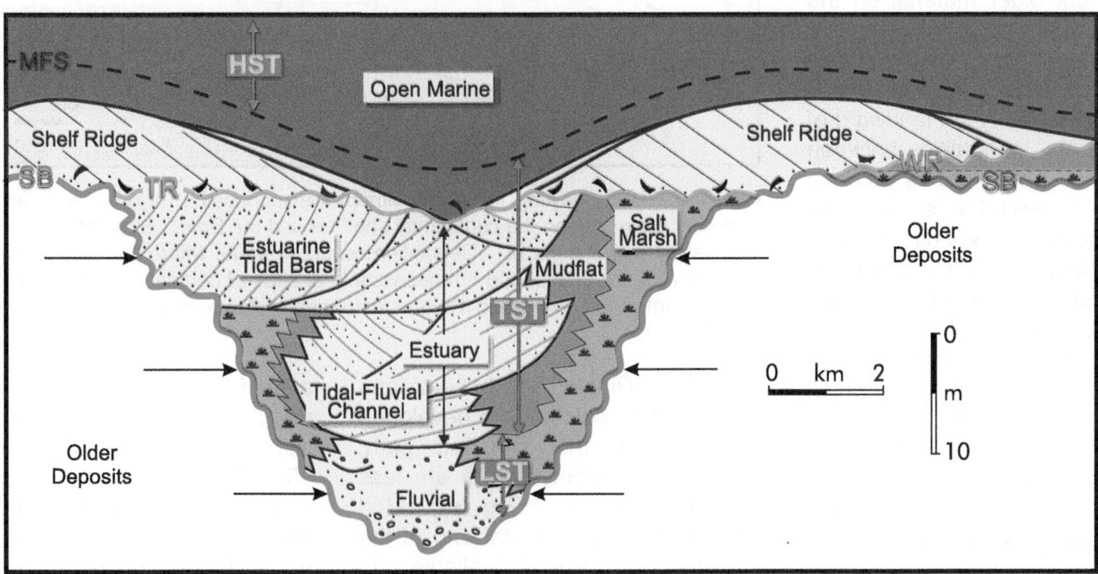

Figure 34. Hypothetical, coast-parallel section showing the stratigraphy of a tide-dominated estuary situated in an incised valley, overlain by transgressive shelf ridges that are, in turn, buried by prodelta mud of the next highstand. The estuarine succession is shown with complete preservation of the estuary-mouth sand bars (*see* Figs. 32 and 33). Less than complete preservation is more likely. The estuarine portion of this figure is based on the Cobequid Bay–Salmon River and Severn River estuaries, while the marine part is based on the southern North Sea and the area seaward of the Changjiang River delta. This section represents the transgressive counterpart of Figure 31.

can be flanked by muddy tidal-flat and salt-marsh deposits along the valley margins. The lower portion of this succession has the highest preservation potential, but it is possible for the deep channels between the estuary-mouth sand bars to scour completely to the base of the valley, remove any fluvial and tidal-fluvial deposits, and modify the geometry of the sequence boundary.

Exceptionally well preserved tide-dominated estuary deposits have been described from three incised valleys in the Eocene of Spitzbergen (Plink-Björklund and Steel, 2006). Each consists of a stacked succession of channel deposits that show the coarse → fine → coarse grain-size trend as they are walked out from the river to the estuary mouth. Paleocurrents are ebb dominated in the inner, fluvially influenced part of the estuary, and flood dominated in the outer, elongate sand-bar portion, just as has been documented in modern tide-dominated estuaries (Dalrymple *et al.*, 1991). Parallel-laminated fine sandstone characterizes the middle portion of the system and apparently formed on upper-flow-regime sand flats that are very similar to those described from the Cobequid Bay–Salmon River estuary (Dalrymple *et al.*, 1991). The elongate sand bars at the estuary mouth pass outward into wave-generated deposits. Mudflat and salt-marsh deposits are present locally where they have not been removed by channel scour. The transgressive backstepping of facies is evident in the lower part of each valley fill, but in two of the valleys, the facies stacking in the upper part is progradational, indicating that the valleys have filled *in situ* at the transgressive limit.

Transgressive Continental Shelves

Continental shelves are typically starved of new sediment during transgression, because river-borne material is trapped in estuaries. As a result, tidal currents (*e.g.*, Fig. 7) and storm waves rework older material on the shelf (*cf.* Stride, 1982), producing a *transgressive surface of marine erosion* (also called a *ravinement surface* in areas that formerly sat landward of the shoreline). Inequalities between the flood and ebb currents produce large areas in which the residual sediment transport is in one direction. In each of these regionally extensive *transport paths*, sediment is eroded from the sea floor in the up-current part of the path. Coarse material will remain as a shelly gravel lag in which large wave ripples can be present (Fig. 36). These erosional areas can be sculpted into erosional shelf ridges if the sea floor is underlain by easily eroded sediments (Fig. 24). The sediment removed is transported down current, and deposited where tidal-current speeds decrease, either at the head of embayments or in deeper water areas farther offshore (Fig. 35). Flow-parallel *sand ribbons* and isolated simple and compound dunes occur in the zone of bypassing. Depositional areas on shelves are covered by sand sheets that can cover thousands of square kilometers, as is the case in the areas surrounding the British Isles. Compound dunes are extensively developed, and sand ridges occur if there is sufficient sediment and the maximum current speed exceeds 100 cm s^{-1} (Stride *et al.*, 1982). Rippled sand sheets and isolated sand patches are developed farther down the transport path, where tidal current speeds are less than about 50 cm s^{-1} (Fig. 36). Mud accumulates at the distal end of the transport path, if it is available. In general, however, transgressive sand sheets are composed of clean sand that contains little mud.

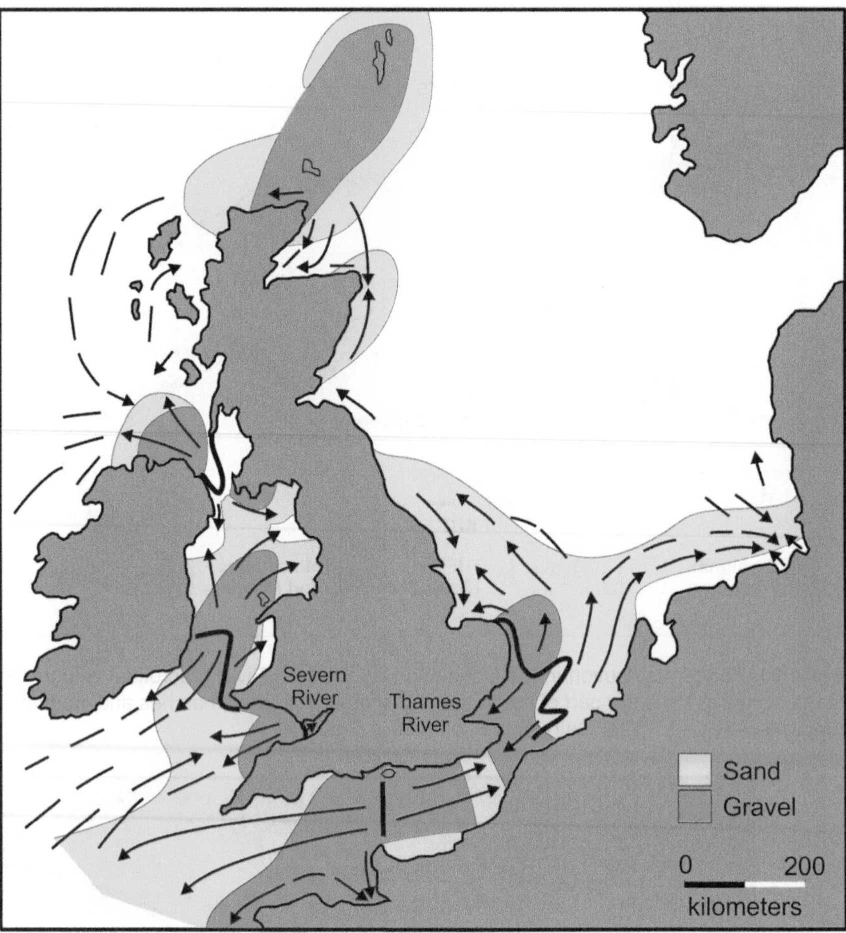

Figure 35. Map showing the net sediment-transport paths (arrows) and distribution of sediment grain size on the continental shelf around the British Isles (brown = gravel; yellow = sand; white = mud). Deposition occurs in areas where sediment moves into an area with lower speeds, and not necessarily at sites of bedload convergence. Note the transport into the Severn and Thames estuaries. After Johnson *et al.* (1982) and Stride *et al.* (1982).

Transgressive tidal sand sheets rest on a flooding surface and are typically only a few meters thick: the one in the southern North Sea *averages* less than 5 m. They will be much thinner than this (perhaps only a few decimeters thick) in areas without abundant sand or in the troughs between any shelf sand ridges. At ridge crests, this transgressive 'lag' may be 10–20 m thick; as a rule of thumb, any transgressive lag that exceeds 3–5 m thick might represent a shelf ridge. Structures in the proximal portion of tidal sand sheets consist of compound crossbedding (Fig. 15). Due to the dominance of one current, crossbed orientations are primarily unidirectional. Within sand-ridge fields, however, deposition occurs preferentially on the side that is dominated by the regionally weaker current (Fig. 23). These sand-ridge deposits can, therefore, have a paleocurrent direction that is the opposite of that in other parts of the sand sheet. Crossbed set thickness decreases in the down-transport direction, due to the decreasing current speed (Fig. 36), and pass distally into rippled and burrowed fine sand and mud. The relative influence of storms increases as the tidal current speeds decrease, and the distal part of the sand sheet might contain storm-generated structures. Within

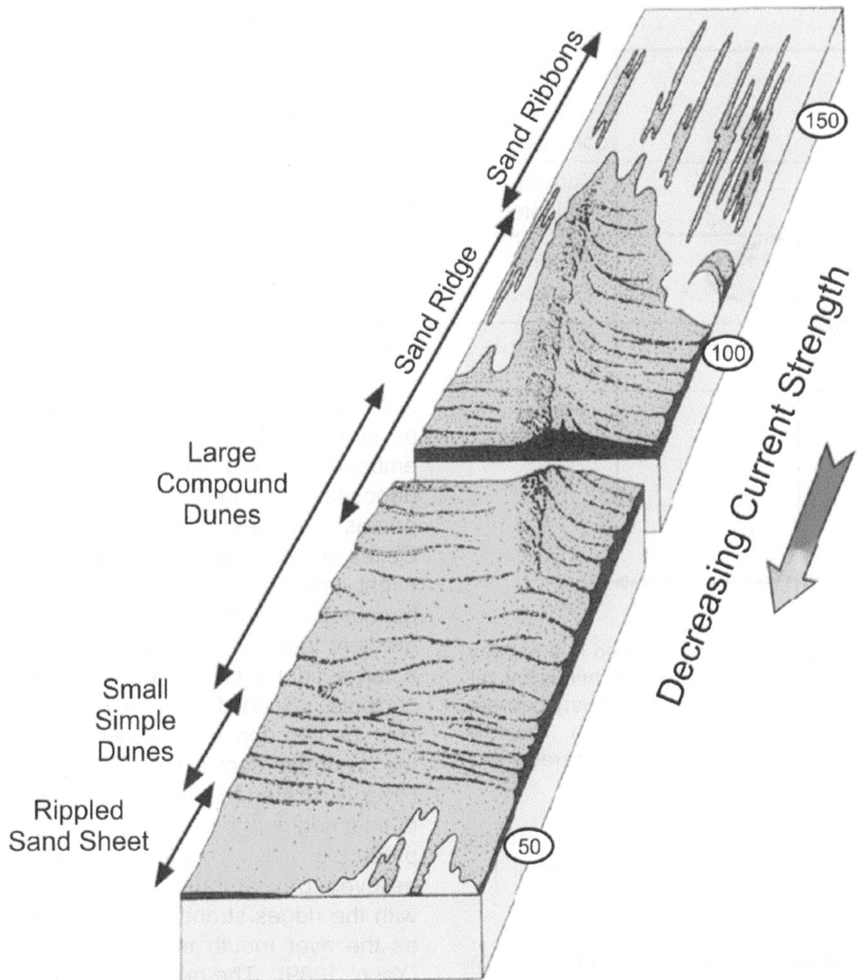

Figure 36. Idealized distribution of bedforms along a shelf sediment-transport path, with the typical maximum (spring-tide), near-surface current speeds (cm s^{-1}; circled) associated with each bedform type. Modified from Belderson *et al.* (1982) and Suter (2006). Reproduced with kind permission of Springer Science and Business Media, and SEPM (Society for Sedimentary Geology).

active sand ridges, it is believed that the sediments coarsen upward because wave action is most intense on the ridge crests. During a transgression, however, the progressive increase in water depth causes current speeds at any location to decrease through time. The sand ridges can become moribund and sediment is then eroded from the ridge crests by storm waves, and deposited as a drape on the ridge flanks (Reynaud *et al.*, 1999). It is highly likely that some of the ridge topography will remain despite the crestal erosion, because the volume of the ridge is so large. The subdued ridge is eventually mantled by mud when coastal/deltaic progradation resumes (Figs. 25, 31 and 34).

In some circumstances, transgression leads to the abandonment of estuary-mouth ridges on the shelf. For example, Yang (1989) proposed that the sand ridges that lie seaward of the Changjiang River (Fig. 22A) originated at the mouth of the river and were stranded on the shelf as the coast moved westward during the Holocene transgression. The ridges closest to the modern river mouth are being buried by prodeltaic muds of the present-day delta (Fig. 25).

Seaways and Straits

The narrow seaways and straits that link two larger water bodies are a unique depositional setting that has only recently been recognized as a distinct sedimentary environment (Anastas *et al.*, 2006). They represent a constriction and any water motion is accentuated; even small differences in water elevation at either end of a strait may generate substantial currents. One of the most common causes of currents in seaways and straits are tides, although unidirectional oceanic currents may also play a role. These currents cause such areas to be current-dominated, even though the adjacent larger bodies of water are wave dominated. The English Channel (Fig. 7) and Torres Strait between Australia and Papua New Guinea are two examples.

Relatively little is known about sedimentation in seaways and straits. Inequalities between the tidal currents in either direction generate sediment transport pathways with net transport in one direction, as is seen in the English Channel (Fig. 35). The superposition of a unidirectional oceanic current on the tidal motion would have the same affect. As a result, dunes are widespread wherever there is mobile medium to coarse sand, and generate cross-bedding with a unidirectional paleocurrent orientation (Anastas *et al.*, 2006). Depending on the proximity of rivers entering the strait from the sides, the deposits can range from being either relatively clean sand (perhaps with a significant biogenic component in cases where there is little river influence, to being quite muddy. Deposition is likely to be greatest in areas of flow expansion on the down-current side of constrictions, leading to the development of a subaqueous delta-like deposit.

The response of seaways and straits to variations in sea level might be complex. They will be occupied by rivers if they are exposed during times of low sea level. When they are inundated, changes in water depth might cause either an increase or decrease in current speed. Anastas *et al.* (2006) have hypothesized that each seaway has a particular water depth at which the current speed is greatest (Fig. 37). Both shallowing and deepening of the water from this optimal depth leads to a decrease in current speed, the former because friction becomes important, and the latter because the cross-sectional area of the seaway becomes too great to act as a constriction. Almost no data exist to test

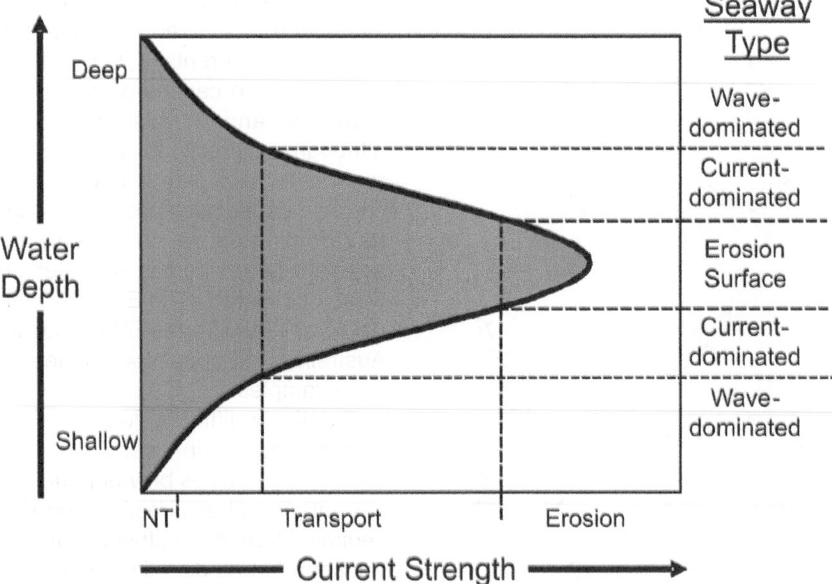

Figure 37. Proposed relationship between water depth and current speed in a seaway, and the nature of the deposits formed within it (right side). There are no numerical values on the axes because the depth at which the peak speed occurs might vary between examples, depending on such factors as bottom roughness and seaway width. The peak speed might not reach into the erosion field in all cases. 'NT' = no current-generated sediment transport. After Anastas *et al.* (2006).

this idea; more work is warranted because straits and seaways are moderately common in the geological record. Anastas *et al.* (2006) provided an example dominated by carbonate deposits, and the succession interpreted by Mutti *et al.* (1985) as deltaic has recently been reinterpreted by Olariu *et al.* (2008) as having accumulated in a seaway. The succession described by Mellere and Steel (1996) might also be a seaway deposit. All three examples are characterized by large-scale crossbedding, which exists because of the large water depth in which the dunes formed. Unlike estuarine or deltaic settings where crossbedded sandstone can also form, mud drapes are essentially absent, and the trace-fossil assemblage records open-marine salinity.

STRATIGRAPHIC ORGANIZATION OF TIDAL SUCCESSIONS

The stratigraphic organization of tidal deposits, including the systems tract in which tidal deposits are most likely to occur, is highly variable. Two factors have a particularly strong influence on this organization: the temporal variability of the strength of tidal currents in response to changing coastal morphology, and the general accommodation regime (overall low *versus* high accommodation).

Timing of Tidal Dominance

As discussed earlier in this chapter, the occurrence of strong tidal currents, and hence the potential for the development of tidal dominance, depends first and foremost on the existence of a large tidal prism and secondarily on the presence of a large tidal range. Topographic constrictions are also particularly conducive to the development of strong currents. The conditions that favor strong tidal influence are not restricted to any particular portion of a relative sea-level cycle; consequently, tidally dominated deposits can occur in any systems tract, although there is a general tendency for them to be most common in transgressive successions.

This tendency exists because numerous embayments and estuaries are created as the rising water level inundates the irregular topography created by terrestrial erosion during the preceding lowstand. Relatively large tidal prisms exist in these locations, and tidal currents are accentuated because of the topographic constrictions provided by valley walls. Furthermore, estuaries and embayments occupy topographic lows and the tidal deposits within them have a higher chance of preservation than is the case elsewhere along the coast. Thus, transgressive deposits, and especially those occupying incised valleys, tend to contain an abundance of tidal facies (*e.g.*, Plink-Björklund and Steel, 2006; Plink-Björklund, 2008). By comparison, regressive coasts of the highstand, falling-stage and lowstand systems tracts tend to be straighter because estuaries and embayments are filled with deltaic deposits, leading to reduced tidal prisms and less flow constriction. Deltas at the mouth of large rivers might provide an exception to this because, as noted earlier, their low gradient leads to the development of a large tidal prism and strong tidal currents. Such deltas might, therefore, be tide dominated regardless of the systems tract; the Changjiang River illustrates this point. The chain of tidal ridges that lies seaward of the present-day river mouth (Fig. 22A) is believed to have formed sequentially, with the ridges stranded on the shelf as the river mouth moved landward (Yang, 1989). The modern highstand delta is also tide dominated, suggesting that this system has been tide dominated throughout a significant part of the latest sea-level cycle. Furthermore, this system appears to have retained the same depositional character over many preceding sea-level cycles (Berné *et al.*, 2002; Fig. 25), with the development of tidal-current ridges during each transgression. In the ancient, the deltas that comprise the Sego Sandstone in the Book Cliffs of Utah are interpreted to have been tide-dominated during both regressive and transgressive phases (Willis and Gabel, 2001). Such long-term persistence of tidal action may also occur in structurally controlled basins, where the conditions necessary for the enhancement of the tidal current occurs again and again over many sea-level cycles: rift basins are especially prone to tidal domination at their head (*e.g.*, the Bay of Fundy and Gulf of California) because of their tapered geometry. The Nequen backarc rift basin in Argentina (Jurassic), for example, appears also to have experienced strong tidal currents throughout several complete sea-level cycles, as all systems tracts are tidally influenced to tidally dominated (McIlroy *et al.*, 2005). This is also the case in the Yellow Sea, where numerical modeling and facies observations indicate that almost identical tidal deposits are present in the modern and previous highstand deltas of the Han River (Choi and Dalrymple, 2004).

The relationship between relative sea level and the strength of tidal currents does, however, differ from basin to basin, depending on how the local paleo-geography changes as a function of sea level. Passive continental margins are more likely to experience strong tidal action during highstands when the shelf is wide (Yoshida *et al.*, 2007) because the tidal range at the coast increases as shelf width increases (Howarth, 1982), until the shelf width reaches one quarter of the wavelength of the tidal wave (*i.e.*, approximately 250 km for a water depth of 50 m), at which point resonance occurs with the lunar semi-diurnal tide. The tidal range then decreases if the shelf width increases still more.

The Cretaceous western interior foreland basin of North America, by comparison, appears to have been most strongly tidally influenced during lowstands (Mellere and Steel, 2000; Yoshida *et al.*, 2007). Examples of such lowstand tidal deposits are provided by the Frewens sandstone (Willis *et al.*, 1999) and the Sego Sandstone (Willis and Gabel, 2001). Two factors might have contributed to this:

1. The seaway was narrow at this time so tidal currents may have been enhanced by flow constriction; and
2. Fault-generated topography on the sea floor had a greater influence on the currents than at highstands, perhaps creating localized constrictions that enhanced tidal flow.

By contrast, highstand shorelines in the basin are almost universally wave dominated.

Influence of Accommodation

Tidal depositional systems are controlled by accommodation in the same way as other environments. Tidal channels tend to become amalgamated in low-accommodation settings, in the same manner that fluvial channels do (*see* Chapter 6), generating complex stacks of laterally and vertically juxtaposed channel deposits. As accommodation increases, there is progressively greater preservation of muddy sediments that accumulated either in prodeltaic-shelf or muddy tidal-flat environments. The Sego Sandstone (Cretaceous; Fig. 38) and the Quaternary of the East China Sea (Fig. 25) illustrate the range of possible stratigraphic expressions of tidal systems. The situation shown in Figure 24 from the Yellow Sea represents a degree of amalgamation that is similar to that seen in the Sego Sandstone.

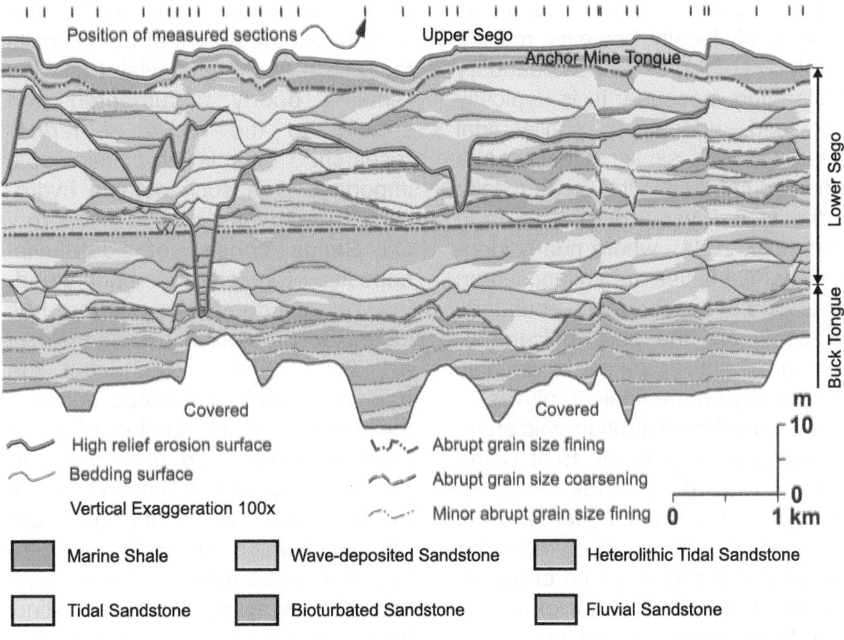

Figure 38. East–west (nearly shoreline parallel) cross section of the Sego Sandstone in the Book Cliffs of Utah, showing a nested succession of tidally influenced channel deposits, interspersed with minor amounts of wave-dominated prodeltaic mudstone (*e.g.*, the Anchor Mine Tongue). Image courtesy of B. Willis (after Figure 3b of Willis and Gabel, 2001). Reproduced with permission of Wiley-Blackwell.

The Sego Sandstone provides a well-documented example of a succession that accumulated in a low-accommodation setting (Willis and Gabel, 2001; Fig. 38). Over the 1.4 Ma during which this unit accumulated, only about 40–50 m of accommodation was created. The succession consists of three broadly tabular sandstone bodies, each of which can be correlated for tens of kilometers along strike, that consist of amalgamated, sandy, tide-dominated channels. These composite sandstone bodies are separated by muddier intervals that record offshore and prodeltaic sedimentation in a wave-dominated regime (Fig. 38). This juxtaposition of tide- and wave-dominated sedimentation does not require system-wide changes in the strength of the tidal currents or waves, but is more likely the product of a mixed-energy regime in which wave energy predominated in areas seaward of the coast, whereas tidal currents dominated sedimentation in the channels that occupied the region landward of the coast (*cf.* Dalrymple and Choi, 2007; Yoshida *et al.*, 2007; Plink-Björklund, 2008). Gradational upward-coarsening deltaic successions are present only locally; much more commonly, the tidal sandstones are bounded at their base by an erosion surface created by scour in tidal channels. Picking the sequence boundaries can be extremely difficult in this complex network of anastomosing erosion surfaces. Willis and Gabel (2001) identified the sequence boundary as the surface with the highest erosional relief. This surface is overlain at its lowest points by coarse-grained fluvial sandstone that lacks tidal indicators such as mud-draped foresets. This sandstone is present only locally, however, making correlation of the sequence boundary challenging because it is typically over- and underlain by similar tidal facies. A very similar situation has been documented in the modern delta of the Han River, Korea (Choi and Dalrymple, 2004), where nearly identical tidal-flat facies are juxtaposed on either side of the sequence boundary. Paleosols are the best indicator of exposure, but they are preserved only sporadically on the tidal interfluves.

Quite a different stratigraphic architecture is present in the East China Sea, seaward of the modern Changjiang River delta (Berné *et al.*, 2002; Fig. 25). Here, accommodation has been created at a rate of 300 m per million years, an order of magnitude faster than that characterizing the Sego Sandstone. As a result, there is little or no amalgamation of channels. Instead, sandy intervals, which are concentrated in valley fills and transgressive shelf ridges, are separated by thick successions of prodeltaic mud. The external form of the tidal ridges is well preserved because of passive burial by prodeltaic deposits in the manner suggested by Snedden and Dalrymple (1998), unlike the situation in the Sego Sandstone (Fig. 38) where extensive dissection by later channels has destroyed the external geometry of the tidal ridges.

PRESENT STATUS AND FUTURE DIRECTIONS

Since the early 1990s, when the previous edition of this book was written, attention has shifted away from the tidal-flat environment and the simple demonstration of tidal influence in a succession toward a more evenly balanced, holisitic view of tidal depositional systems. Our knowledge of a wide range of tidal environments has increased dramatically, with the publication of studies that have expanded our understanding of the open-coast tidal flats, tidal-current ridges, tide-dominated deltas and seaways/straits. The recent attempt to synthesize the criteria by which one can tell where a specific tidal deposit formed in the transition between a purely fluvial setting and open-marine conditions (Dalrymple and Choi, 2007) pulls together a wealth of information, but the nature of the deposits formed in the tidal-fluvial transition remains poorly documented and should be the focus of further research. This is of considerable importance, as such notable hydrocarbon reservoirs as the McMurray Oil Sands consist of tidal-fluvial deposits. Similarly, our understanding of tide-dominated deltas has increased dramatically (Dalrymple *et al.*, 2003; Willis, 2005), but the amount of variability noted between the relatively small number of documented modern and ancient examples is so great that it remains difficult to determine which attributes are noise and which are worthy of retention in a facies model. The recognition of seaways/straits as a distinct tidal environment is so recent that few examples exist and it is difficult to provide a synthesis for this setting. Notable improvements have occurred in our knowledge of the deposits of tidal shelf ridges, as a result of studies in the area around the British Isles and the East China and Yellow seas. The recognition that both depositional and erosional ridges exist complicates the interpretation of ancient deposits, but provides new ways to explain features that had previously seemed enigmatic. Modern, open-coast tidal flats have also received considerable attention (Fan *et al.*, 2004; Yang *et al.*, 2005), but this new model has not been widely applied to the ancient record yet.

Our view of when tide-dominated sediments are likely to occur has also matured considerably, with the recognition that tidal dominance can occur at any point in a relative sea-level cycle. The recent appreciation that individual sedimentary basins might preferentially experience strong tidal action at a particular portion of a relative sea-level cycle (Yoshida *et al.*, 2007) could provide a degree of predictability that had not existed before. Despite these dramatic improvements in our understanding, tidal depositional systems remain less well documented than many other environments. We are in urgent need of a greater number of case studies in both modern and ancient successions; without them, the distillation process that must occur in order to build facies models cannot proceed.

ACKNOWLEDGEMENTS

I thank Noel James and Skip Davis for their helpful comments on an earlier version of the manuscript, the Natural Sciences and Engineering Research Council of Canada for its continued support of my research on tidal systems, and the many students and colleagues who have helped me to understand tidal depositional systems better. I also thank Geoff Reith for drafting many of the figures.

REFERENCES

Basic sources of information

Allen, P.A., 1997, Earth Surface Processes: Oxford, Blackwell Science, 404 p.
Chapter 8 of this book contains an excellent discussion of the origin and nature of tides, with more detail than is provided here.

Dalrymple, R.W. and Choi, K.S., 2003, Sediment transport by tides, *in* Middleton, G.V., *ed.*, Encyclopedia of Sediments and Sedimentary Rocks: Dordrech, Kluwer Scientific, p. 606-609.
A review of the fundamentals of sediment transport by tidal currents.

Dalrymple, R.W. and Choi, K.S., 2007, Morphologic and facies trends through the fluvial-marine transition in tide-dominated depositional systems: A systematic framework for environmental and sequence-stratigraphic interpretation: Earth-Science Reviews, v. 81, p. 135-174.
A review of the variation in processes, morphology and facies through the fluvial–marine transition in tide-dominated environments.

Dalrymple, R.W. and Rhodes, R.N., 1995, Estuarine dunes and barforms, *in* Perillo, G.M.E., *ed.*, Geomorphology and Sedimentology of Estuaries: Developments in Sedimentology, v. 53, p. 359-422.
A comprehensive synthesis of information on the morphology, dynamics and deposits formed by the bedforms and bars that are an integral part of most tidal systems.

Nio, S.-D. and Yang, C.-S., 1991, Diagnostic attributes of clastic tidal deposits: a review, *in* Smith, D.G., Reinson, G.E., Zaitlin, B.A. and Rahmani, R.A., *eds.*, Clastic Tidal Sedimentology: Canadian Society of Petroleum Geologists Memoir 16, p. 3-28.
This is the place to start reading about the origin of tidal bundles, rhythmites and neap-spring cycles.

Stride, A.H., *ed.*, 1982, Offshore tidal sands: processes and deposits: New York, Chapman and Hall, 222 p.
Although older, this book provides a

good summary of transgressive shelf sedimentation around the British Isles. Excellent chapters on tidal processes, bedforms and facies distribution.

Willis, B.J., 2005, Deposits of tide-influenced river deltas, *in* Giosan, L. and Bhattacharya, J.P., *eds.*, River Deltas – Concepts, Models, and Examples: SEPM Special Publication 83, p. 87-129.

The most comprehensive review available of tide-influenced deltas.

Other references

Allen, J.R.L., 1980, Sand waves; a model of origin and internal structure: Sedimentary Geology, v. 26, p. 281-328.

Allen, P.A. and Homewood, P., 1984, Evolution and mechanics of a Miocene tidal sandwave: Sedimentology, v. 31, p. 63-81.

Allison, M.A., Khan, S.R., Goodbred, S.L., Jr. and Kuehl, S.A., 2003, Stratigraphic evolution of the late Holocene Ganges-Brahamputra lower delta plain: Sedimentary Geology, v. 155, p. 317-342.

Alexander, C.R., Nittrouer, C.A., Demaster, D.J., Park, Y.-A. and Park, S.-C., 1991, Macrotidal mudflats of the southwestern Korean coast: A model for interpretation of intertidal deposits: Journal of Sedimentary Petrology, v. 61, p. 805-824.

Amos, C.L., 1995, Siliciclastic tidal flats, *in* Perillo, G.M.E., *ed.*, Geomorphology and Sedimentology of Estuaries: Developments in Sedimentology, v. 53, p. 273-306.

Anastas, A.S., Dalrymple, R.W., James, N.P. and Nelson, C.S., 2006, Lithofacies and dynamics of a cool-water carbonate seaway: Mid-Tertiary, Te Kuiti Group, New Zealand, *in* Pedley, M. and Tucker, M., *eds.*, Cool-Water Carbonates: Depositional Systems and Paleoenvironmental Controls: Geological Society of London, Special Paper 255, p. 245-268.

Archer, A.W. and Hubbard, M.S., 2003, Highest tides of the world, *in* Chan, M.A. and Archer, A.W., *eds.*, Extreme Depositional Environments: Mega End Members in Geologic Time: Geological Society of America, Special Paper 370, p. 151-173.

Barwis, J.H., 1978, Sedimentology of some South Carolina tidal-creek point bars, and a comparison with their fluvial counterparts, *in* Miall, A.D., *ed.*, Fluvial Sedimentology: Canadian Society of Petroleum Geologists, Memoir 5, p. 129-160.

Belderson, R.H., Johnson, M.A. and Kenyon, N.H., 1982, Bedforms, *in* Stride, A.H., *ed.*, Offshore Tidal Sands: Processes and Deposits: New York, Chapman and Hall, p. 27-57.

Berné, S., Trentesaux, A., Stolk, A., Missiaen, T., de Batist, M., 1994, Architecture and long term evolution of a tidal sandbank: The Middelkerke Bank (southern North Sea): Marine Geology, v. 121, p. 57-72.

Berné, S., Vagner, P., Guighard, F., Lericolais, G., Liu, Z., Trentesaux, A., Yin, P. and Yi, H.I., 2002, Pleistocene forced regressions and tidal sand ridges in the East China Sea. Marine Geology, v. 188, p. 293-315.

Chen, A., Saito, Y., Hori, K., Zhao, Y. and Kitamura, A., 2003, Early Holocene mud-ridge formation in the Yangtze, offshore China: A tidal-controlled estuarine pattern and sea-level implications: Marine Geology, v. 198, p. 245-257.

Choi, K.S. and Dalrymple, R.W., 2004, Recurring tide-dominated sedimentation in Kyonggi Bay, west coast of Korea in Holocene and late Pleistocene sequences: Marine Geology, v. 212, p. 81-96.

Dalrymple, R.W., 1984, Morphology and internal structure of sand waves in the Bay of Fundy: Sedimentology, v. 31, p. 365-382.

Dalrymple, R.W., 1992, Tidal depositional systems, *in* Walker, R.G. and James, N.P., *eds.,* Facies Models- Response to Sea Level Change: Geological Association of Canada, p. 195-218.

Dalrymple, R.W., Baker, E.K., Harris, P.T. and Hughes, M., 2003, Sedimentology and stratigraphy of a tide-dominated, foreland-basin delta (Fly River, Papua New Guinea), *in* Sidi, F.H., Nummedal, D., Imbert, P., Darman, H. and Posamentier, H.W., *eds.*, Tropical Deltas of Southeast Asia - Sedimentology, Stratigraphy, and Petroleum Geology: SEPM Special Publication 76, p. 147-173.

Dalrymple, R.W., Knight, R.J., Zaitlin, B.A. and Middleton, G.V., 1990, Dynamics and facies model of a macrotidal sandbar complex, Cobequid Bay-Salmon River estuary (Bay of Fundy): Sedimentology, v. 37, p. 577-612.

Dalrymple, R.W., Makino, Y. and Zaitlin, B.A., 1991, Temporal and spatial patterns of rhythmite deposition on mudflats in the macrotidal, Cobequid Bay-Salmon River estuary, Bay of Fundy, Canada, *in* Smith, D.G., Reinson, G.E., Zaitlin, B.A. and Rahmani, R.A., *eds.*, Clastic Tidal Sedimentology: Canadian Society of Petroleum Geologists, Memoir 16, p. 137-160.

Dalrymple, R.W. and Zaitlin, B.A., 1994, High-resolution sequence stratigraphy of a complex, incised valley succession, the Cobequid Bay- Salmon River estuary, Bay of Fundy, Canada: Sedimentology, v. 41, 1069-1091.

Dalrymple, R.W., Zaitlin, B.A. and Boyd, R., 1992, Estuarine facies models: conceptual basis and stratigraphic implications: Journal of Sedimentary Petrology, v. 62, p. 1130-1146.

Dashtgard, S.E. and Gingras,M.K., 2005, Facies architecture and ichnology of recent salt-marsh deposits: Waterside Marsh, New Brunswick, Canada: Journal of Sedimentary Research, v. 75, p. 596-607.

Davies, C., Best, J. and Collier, R., 2003, Sedimentology of the Bengal shelf, Bangladesh: comparison of late Miocene sediments, Sitakund anticline, with the modern, tidally dominated shelf: Sedimentary Geology, v. 155, p. 271-300.

De Boer, P.L., Oost, A.P. and Visser, M.J., 1989, The diurnal inequality of the tide as a parameter for recognizing tidal influences: Journal of Sedimentary Petrology, v. 59, p. 912-921.

Fan, D., Li, C., Wang, D., Wang, P., Archer, A.W. and Greb, S.F., 2004, Morphology and sedimentation on open-coast intertidal flats of the Changjiang Delta, China: Journal of Coastal Research, v. 81, p. 23-35.

Fenies, H., de Ressequier, A. and Tastet, J.-P., 1999, Intertidal clay-drape couplets (Gironde Estuary, France: Sedimentology, v. 46, p. 1-15.

Gerdes, G., Klenke, T. and Noffke, N., 2000, Microbial signatures in peritidal siliciclastic sediments: a catalogue: Sedimentology, v. 47, p. 279-308.

Houbolt, J.J.H.C., 1968, Recent sediments in the southern bight of the North Sea: Geologie en Mijnbouw, v. 47, p. 245-273.

Howarth, M.J., 1982, Tidal currents of the continental shelf, *in* Stride, A.H., *ed.*, Offshore Tidal Sands: Processes and Deposits: New York, Chapman and Hall, p. 10-26.

Ichaso, A.A. and Dalrymple, R.W., 2009, Tidal and wave-generated fluid-mud deposits in the Tilje Formation (Jurassic), offshore Norway: Geology, v. 37, p. 539-542.

Jaeger, J.M. and Nittrouer, C.A., 1995, Tidal controls on the formation of fine-scale sedimentary strata near the Amazon River mouth: Marine Geology, v. 125, p. 259-281.

Jin, J.H. and Chough, S.K., 2002, Erosional shelf ridges in the mid-eastern Yellow Sea: Geo-Marine Letters, v. 21, p. 219-225.

Johnson, M.A., Kenyon, N.H., Belderson, R.H. and Stride, A.H., 1982, Sand transport, *in* Stride, A.H., *ed.*, Offshore Tidal Sands: Processes and Deposits: New York, Chapman and Hall, p. 58-94.

Kohsiek, L.H.M. and Terwindt, J.H.J., 1981, Characteristics of foreset and topset bedding in megaripples related to hydrodynamic conditions on an intertidal shoal, *in* Nio, S.-D., Shüttenhelm, R.T.E.and van Weering, Tj.C.E., *eds.*, Holocene Marine Sedimentation in the North Sea Basin: International Association of Sedimentologists, Special Publication 5, p. 27-37.

Kuehl, S.A., Nittrouer, C.A., Allison, M.A., Fraia, L.E.C., Dakut, D.A., Maeger, J.M.

Pacioni, T.D., Figgueiredo, A.G. and Underkoffler, E.C., 1996, Sediment deposition, accumulation, and seabed dynamics in an energetic fine-grained coastal environment: Continental Shelf Research, v. 16, p. 787-815.

Kvale, E.P., 2006, The origin of neap–spring cycles: Marine Geology, v. 235, p. 5-18.

Li, C., Wang, P., Fan, D. and Yang, S., 2006, Characteristics and formation of late Quaternary incised-valley-fill sequences in sediment-rich deltas and estuaries: case studies from China, *in* Dalrymple, R.W., Leckie, D.A. and Tillman, R.W., *eds.*, Incised Valleys in Time and Space: SEPM Special Publication 85, p. 141-160.

McCave, I.N., 1971, Sand waves in the North Sea off the coast of Holland: Marine Geology, v. 10, p. 199-225.

McIlroy, D., Flint, S., Howell, J.A. and Timms, N., 2005, Sedimentology of the tide-dominated Jurassic Lajas Formation, Neuquen Basin, Argentina, *in* Veiga, G.D., Spelletti, L.A., Howell, J.A. and Schwarz, E., *eds.*, The Neuquen Basin, Argentina: A Case Study in Sequence Stratigraphy and Basin Dynamics: Geological Society of London, Special Publication 252, p. 83-107.

Mellere, D. and Steel, R.J., 1996, Tidal sedimentation in Inner Hebrides half grabens, Scotland: the Mid-Jurassic Baerreraig Sandstone Formation, *in* De Batist, M. and Jacobs, P., *eds.*, Geology of Siliciclastic Shelf Seas: Geological Society of London, Special Publication 117, p. 49-79.

Mellere, D, and Steel, R.J., 2000, Style contrast between forced regressive and lowstand/transgressive wedges in the Campanian of south-central Wyoming, *in* Hunt, D., and Gawthorpe, R.L., *eds.*, Sedimentary Responses to Forced Regressions: Geological Society of London, Special Publication 172, p. 51-75.

Miller, D.J. and Eriksson, K.A., 1997, Late Mississippian prodeltaic rhythmites in the Appalachian Basin: A heirarchical record of tidal and climatic periodicities: Journal of Sedimentary Research, v. 67, p. 653-660.

Mutti, E., Rosell, J., Allen, G.P., Fonnesu, F. and Sgavetti, M., 1985, The Eocene Baronia tide dominated delta-shelf system in the Ager Basin, *in* Mila, M.D.and Rosell, J., *eds.*, Excursion Guidebook: International Association of Sedimentologists, 6th European Regional Meeting, Lleida, Spain, p. 579-600.

Olariu, C., Steel, R., Dalrymple, R.W., Gingras, M.K. and Rubino, J.-L., 2008, Tidal dunes of the Eocene Baronia Sandstone, Ager Basin, Spain: Distinguishing tidal dunes from tidal bars; Why Bother? Abstract CD, 2008 American Association of Petroleum Geologists Annual Convention, April 20-23, San Antonio, TX.

Open University Course Team, 1989, Waves, Tides and Shallow-Water Processes: New York, Pergamon Press, 187 p.

Pearson, N.J. and Gingras, M.K., 2006, An ichnological and sedimentological facies model for muddy point-bar deposits: Journal of Sedimentary Research, v. 76, p. 771–782.

Plink-Björklund, P., 2008, Wave-to-tide facies change in a Campanian shoreline complex, Chimney Rock Tongue, Wyoming-Utah, U.S.A., *in* Hampson, G.J., Steel, R.J., Burgess, P.M. and Dalrymple, R.W., *eds.*, Recent Advances in Models of Siliciclastic Shallow-Marine Stratigraphy: SEPM Special Publication 90, p. 265-291.

Plink-Björklund, P. and Steel, R., 2006, Incised valleys on an Eocene coastal plain and shelf, Spitsbergen – Part of a linked shelf-slope system, *in* Dalrymple, R.W. Leckie, D.A. and Tillman, R.W., *eds.*, Incised Valleys in Space and Time: SEPM Special Publication 85, p. 281-307.

Reineck, H.-E. and Singh, I.B., 1980, Depositional Sedimentary Environments, 2nd edition: New York, Springer-Verlag, 549 p.

Reynaud, J.-Y., Tessier, B., Dalrymple, R.W., Marsset, T., de Batist, M., Proust, J.-N., Bourillet, J.F. and Lericolais, G., 1999, Eustatic and hydrodynamic controls on the architecture of a deep shelf sand bank (Celtic Sea): Sedimentology, v. 46, p. 703-721.

Slingerland, R., Driscoll, N.W., Milliman, J.D. and Miller, S.R., 2008, Anatomy and growth of a Holocene clinothem in the Gulf of Papua: Journal of Geophysical Research, v. 113, F01S13, doi:10.1029/2006JF000628.

Smith, D.G., 1987, Meandering river point bar lithofacies models: Modern and ancient examples compared, *in* Ethington, F.G., Flores, R.M. and Harvey, M.D., *eds.*, Recent Developments in Fluvial Sedimentology: SEPM Special Publication 39, p. 83-91.

Snedden, J.W. and Dalrymple, R.W., 1998, Modern shelf sand bodies: an integrated hydrodynamic and evolutionary model, *in* Bergman, K. and Snedden, J.W., *eds.*, Isolated Shallow Marine Sandbodies: Sequence Stratigraphic Analysis and Sedimentologic Interpretation: SEPM Special Publication 64, p. 13-28.

Stride, A.H., Belderson, R.H., Kenyon, N.H. and Johnson, M.A., 1982, Offshore tidal deposits: sand sheet and sand bank facies, *in* Stride, A.H., *ed.*, Offshore Tidal Sands: Processes and Deposits: New York, Chapman and Hall, p. 95-125.

Suter, J.R., 2006, Facies models revisited: clastic shelves, *in* Posamentier, H.W. and Walker, R.G., *eds.*, Facies Models Revisited: SEPM Special Publication 84, p. 339-397.

Tessier, B., Archer, A.W., Lanier, W.P. and Feldman, H.R., 1995, Comparison of ancient tidal rhythmites (Carboniferous of Kansas and Indiana, USA) with modern analogues (the Bay of Mont-Saint-Michel, France): International Association of Sedimentologists, Special Publication 24, p. 259-271.

Terwindt, J.H.J., 1988, Palaeo-tidal reconstructions of inshore tidal depositional environments, *in* de Boer, P.L., van Gelder, A. and Nio, S.-D., *eds.*, Tide-Influenced Sedimentary Environments and Facies: Boston, D. Reidel Publishing Company, p. 233-263.

Thomas, R.G., Smith, D.G., Wood, J.M., Visser, J., Calverley-Range, E.A. and Koster, E.H., 1987, Inclined heterolithic stratification - terminology, description, interpretation and significance: Sedimentary Geology, v. 53, p. 123-179.

Thompson, R.W., 1968, Tidal Flat Sedimentation on the Colorado River Delta, Northwestern Gulf of California: Geological Society of America, Memoir 107, 413 p.

Visser, M.J., 1980, Neap-spring cycles reflected in Holocene subtidal large-scale bedform deposits: a preliminary note: Geology, v. 8, p. 543-546.

Weimer, R.J., Howard, J.D. and Lindsay, D.R., 1982, Tidal flats and associated tidal channels, *in* Scholle, P.A. and Spearing, D., *eds.*, Sandstone Depositional Environments: Tulsa, America Association of Petroleum Geologists, p. 191-245.

Williams, G.E., 1991, Upper Proterozoic tidal rhythmites, South Australia: sedimentary features, deposition, and implications for the earth's paleorotation, *in* Smith, D.G., Reinson, G.E., Zaitlin, B.A. and Rahmani, R.A., *eds.*, Clastic Tidal Sedimentology: Canadian Society of Petroleum Geologists, Memoir 16, p. 161-178.

Willis, B.J. and Gabel, S., 2001, Sharp-based, tide-dominated deltas of the Sego Sandstone, Book Cliffs, Utah, USA: Sedimentology, v. 48, p. 479-506.

Willis, B.J., Bhattacharya, J.P., Gabel, S.L. and White, C.D., 1999, Architecture of a tide-influenced river delta in the Frontier Formation of central Wyoming, USA: Sedimentology, v. 46, p. 667-688.

Wood, L.J., 2003, Predicting tidal sand reservoir architecture using data from modern and ancient depositional systems, *in* Grammer, G.M., Harris, P.M. and Eberli, G.P., *eds.*, Integration of Outcrop and Modern Analogs in Reservoir Modeling: American Association of Petroleum Geologists Memoir 80, p. 1-22.

Yang, B.C., Dalrymple, R.W. and Chun, S.S., 2005, Sedimentation on a wave-dominated, open-coast tidal flat, southwestern Korea: summer tidal flat – winter shoreface: Sedimentology, v. 52, p. 235-252.

Yang, B.C., Dalrymple, R.W., Chun, S.S., Johnson, M.K. and Lee, H.J., 2008,

Tidally modulated storm sedimentation on open-coast tidal flats, southwestern Korea: Distinguishing tidal-flat from shoreface storm deposits, *in* Hampson, G.J., Steel, R.J., Burgess, P.R. and Dalrymple, R.W. *eds.,* Recent Advances in Models of Siliciclastic Shallow-marine Sedimentation: SEPM Special Publication 90, p. 161-176.

Yang, B.C., Dalrymple, R.W., Gingras, M.K., Chun, S.S. and Lee, H.J., 2007, Up-estuary variation of sedimentary facies and ichnocoenoses in an open-mouthed, macrotidal, mixed-energy estuary, Gomso Bay, Korea: Journal of Sedimentary Research, v. 77, p. 757-771.

Yang, C.-S., 1989, Active, moribund and buried tidal sand ridges in the East China Sea and the southern Yellow Sea: Marine Geology, v. 88, p. 97-116.

Yoshida, S., Steel, R. and Dalrymple, R.W., 2007, Depositional process changes – An ingredient in a new generation of sequence-stratigraphic models: Journal of Sedimentary Research, v. 77, p. 447-460.

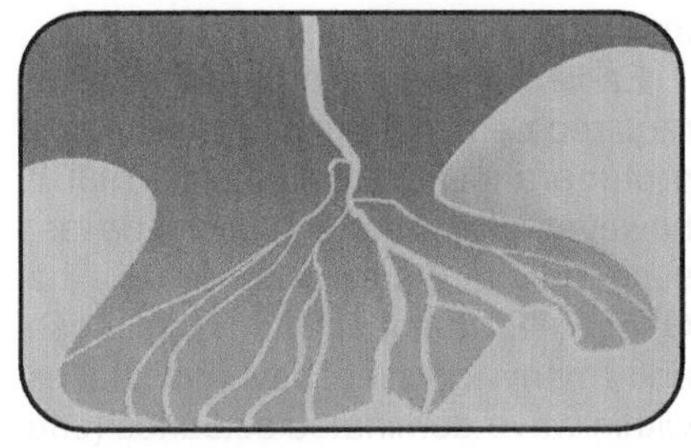

10. DELTAS

Janok P. Bhattacharya, Department of Earth and Atmospheric Sciences, SR1 Rm 312, University of Houston, 4800 Calhoun Road, Houston, Texas 77204-5007, USA

INTRODUCTION

The shoreline is one of the most critical boundaries in the transfer of sediment from land to sea. Much of this sediment is carried by rivers and deposited in the form of deltas (Fig. 1). Prediction of the growth and decay of modern deltas is critical in coastal wetland areas, such as in Louisiana and much of Asia, where rampant dam building has caused an immense decrease in river discharge into the world's oceans. This has resulted in enormous stress to these coastal ecosystems as they experience subsidence and land loss, combined with the devastating affects of coastal storms such as hurricanes and cyclones. From an economic perspective, deltas have been estimated to host close to 30% of all of the world's oil, coal, and gas deposits (Tyler and Finley, 1991). Significant fresh-water resources also occur in delta deposits. Exploitation of deltaic aquifers or hydrocarbon reservoirs requires a robust predictive facies model. This chapter reviews basic definitions, describes river-mouth processes and deltaic sub-environments and their facies characteristics, and ends by considering the regional-scale allogenic controls on deltaic successions, which may be elucidated by sequence-stratigraphic analysis.

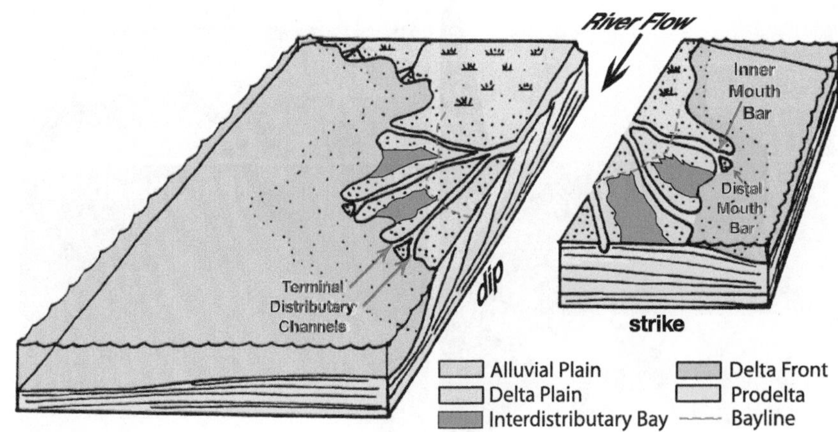

Figure 1. A. Block diagram of a river delta showing main environments and facies.

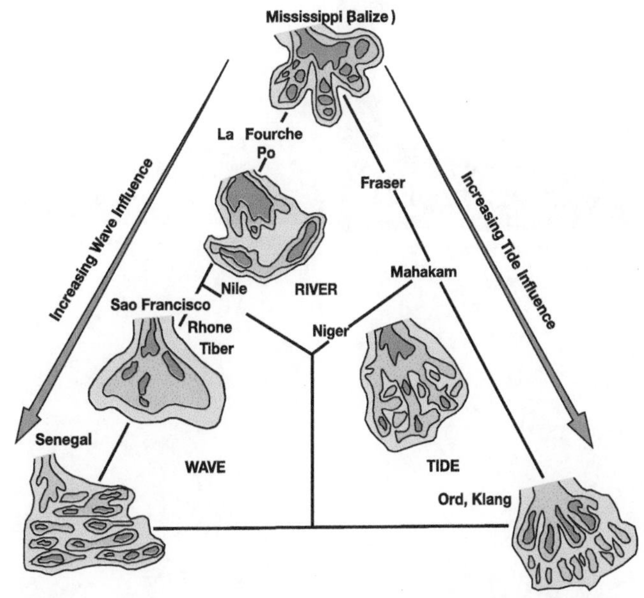

Figure 2. Sandbody geometries (as shown by isopach patterns) of the six delta types of Coleman and Wright (1975) plotted on the river-, wave- and tide-dominated tripartite classification of Galloway (1975; from Bhattacharya and Walker, 1992). Note that all sandbodies narrow toward a point (fluvial) source. Also note similarity of the tide-dominated delta to the river-dominated end-member.

DELTA DEFINITIONS AND FUNDAMENTAL CONCEPTS

Deltas are defined as subaerial landforms and their submarine extensions built directly by, or associated with, rivers at the point where they feed into a standing body of water. Deltas are fundamentally regressive and all deltas must be, to some degree, river-influenced. The regressive nature of delta deposits yields two other fundamental characteristics, the development of upward-shallowing vertical facies successions as well as seaward-dipping clinoform bedding patterns (Fig. 1). The fact that deltas are sourced by rivers also results in the deposition of lobate or elongate sediment bodies (Figs. 2 and 3). The point-sourced nature of the deposit may be masked by alongshore dispersal of sediment: sand may be reworked into a coast-parallel shoreface by wave action, whereas mud can be transported many tens of kilometers alongshore, to form a chenier plain that is linked to the updrift delta. The coalescence of multiple lobes along depositional strike may also blur the simplified shapes shown

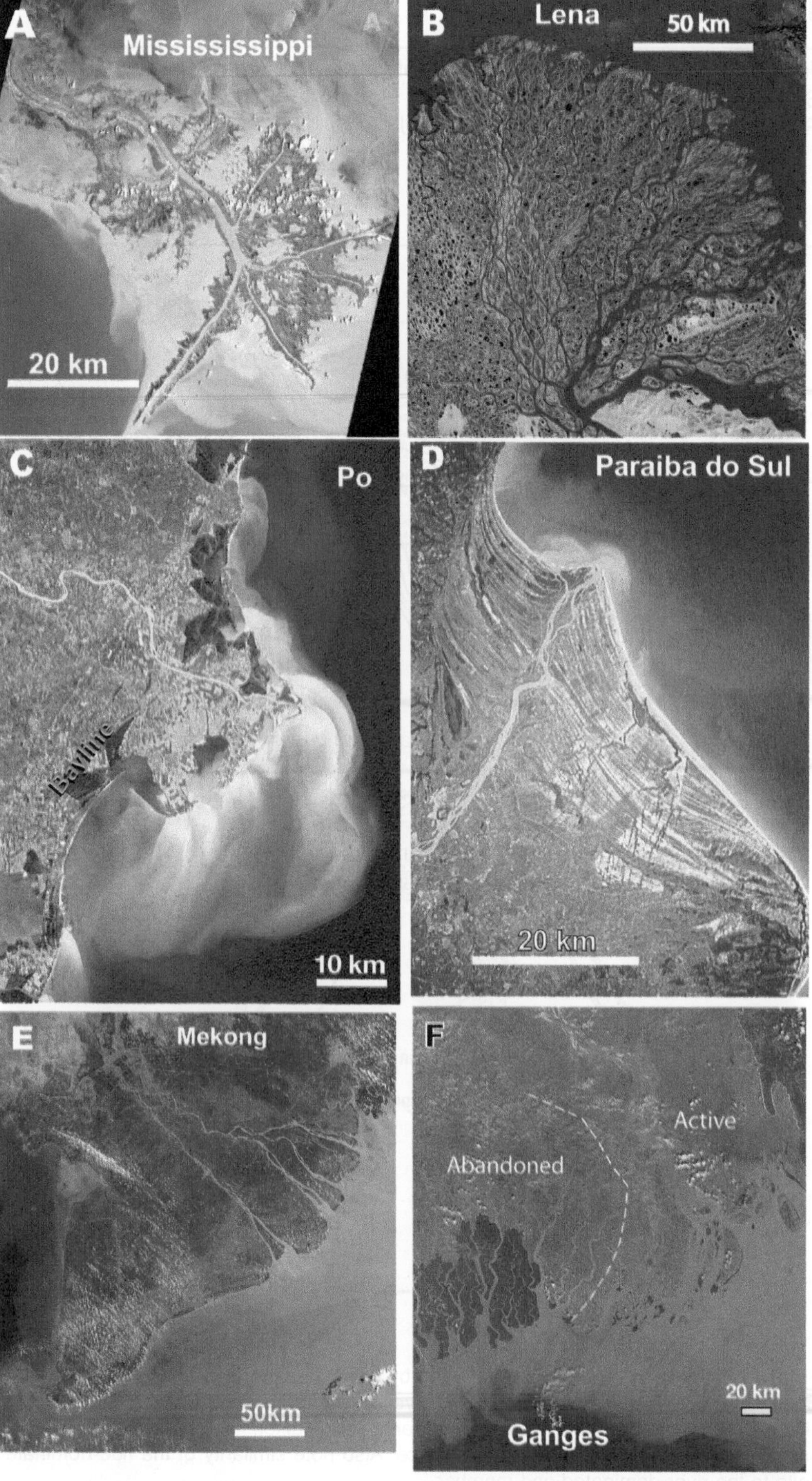

Figure 3. Satellite and shuttle photos of various modern deltas. **A**. River-dominated Mississippi birdfoot delta. **B**. River-dominated Lena River Delta (Russian Arctic) showing numerous orders of branching with many tens of terminal distributary channels. **C**. Po Delta, Italy; dashed line shows the bayline indicating the landward limit of salt water. This also marks the division between the upper and lower delta plain. Note also south-directed plumes of sediment. **D**. Wave-dominated coastline composed of multiple beach ridges associated with the Paraiba do Sul, Brazilian coast. **E**. Mixed-energy, wave- and tide-influenced Mekong Delta, Vietnam. Note muddy shallow coastal area, with prominent muddy peninsula (the Camau) in the south. **F**. Tide-dominated Ganges–Brahmaputra Delta shows complex network of tidal channels. Dashed line separates abandoned from active part of delta. North is to the top in each delta. Photos courtesy of NASA.

in Figures 2 and 3.

Earlier definitions of the term delta required demonstration of a shoreline protuberance (e.g., Elliott, 1986); however, a number of modern deltas, such as the Amazon in Brazil, the Ganges–Brahmaputra in Bangladesh and many smaller bayhead deltas within wave-dominated estuaries (*see* Chapter 11), lie within incompletely filled embayments. Therefore, the presence of a protuberance is secondary to an overall regressive character of fluvial origin in defining a delta. This contrasts with 'estuarine depositional systems', which are defined geologically as being transgressive and receiving sediment derived both from the marine and fluvial realms (Dalrymple *et al.,* 1992; *see* Chapters 9 and 11). However, the term estuary has also traditionally been defined on an oceanographic basis as the area where marine and fresh-waters mix to produce brackish water (Pritchard, 1967). Using this definition, all marine deltas will have an estuarine component near the river mouth. However, during abandonment, embayed parts of delta lobes, such as distributary channels or interdistributary bays, may be transformed into transgressive estuaries, which satisfy both the geologic and oceanographic definitions.

The term delta has also been applied broadly to many ancient progradational successions or clastic wedges that show a marine to non-marine transition, or which contain a marine-fluvial or lacustrine-fluvial interface (Alexander, 1989). Although these deposits are coastal deposits (*sensu lato*), they may have been formed of sediment that was dispersed broadly along the shoreline (*see* below) rather than being a more localized and point-sourced sediment body at a river mouth. Thus, the identification of a shoreline deposit as being specifically deltaic requires information showing that the deposits were linked to a feeding river. This can be done by detailed correlations that allow specific channel deposits to be linked to lobate delta deposits, or by mapping of sediment-body geometries that suggest a point-source for the sediment supply (*e.g.*, Fig. 2). Where a deltaic origin cannot be demonstrated in shoreline deposits,

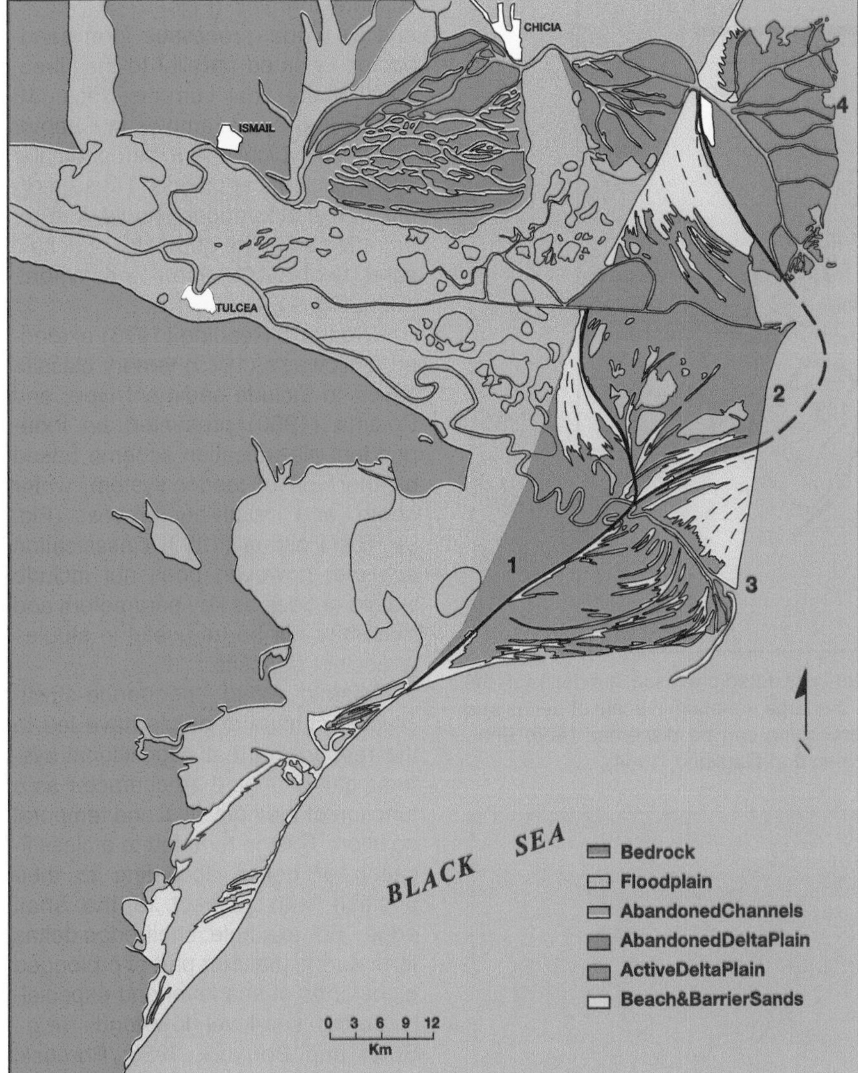

Figure 4. Depositional environments and history of the Danube River delta plain. The highest discharge northern branch feeds the river-dominated delta lobe 4, comprising numerous bifurcating distributary channels with only minor wave reworking. The southern-most branch feeds the distinctly asymmetric wave-influenced lobe 3. The updrift side of lobe 3 comprises amalgamated beach ridges whereas the downdrift side comprises river-dominated bayhead deltas building behind a wave-formed barrier island. The asymmetry is preserved in the older lobes 1 and 2. The central lobe 2 is largely inactive and is presently being destroyed. Sands from lobe 2 are carried south by longshore drift to accumulate in the vicinity of lobe 3. From Bhattacharya and Giosan (2003); after Panin *et al.* (1983).

more general environmental terms such as *paralic* (defined as pertaining to deposits laid down on the landward side of a coast), *strandplain* or *coastal plain* may be preferable in order to avoid inappropriate geometric implications.

Deltas range in scale, from continental-scale systems, such as the modern Mississippi Delta in the Gulf of Mexico (Fig. 3A) that have an area of about 28 500 km^2, to smaller components within other depositional systems, such as the bayhead deltas that lie at the landward end of many wave-dominated estuarine or lagoonal systems. Many continental-scale deltas, such as the Danube in the Black Sea (Fig. 4) and the Mississippi (Fig. 5), may contain smaller scale crevasse deltas within larger scale delta lobes, resulting in a complex, hierarchical facies architecture.

Most deltas produce inherently cyclic deposits, as they pass through phases of progradation and abandonment (Fig. 6). Delta cycles are typically driven by river avulsion, which causes old lobes to be abandoned and new lobes to grow. Some types of wave- or tide-influenced deltas may show a lower propensity for avulsion and lobe switching than more river-dominated deltas, like the Mississippi, which has experienced seven major avulsions over the last 9000 years (Boyd *et al.*, 1989; Fig. 7). Different scales of avulsions may occur, including wholesale regional avulsion of trunk channels, in which an entire delta lobe may be abandoned, versus local avulsions of terminal distributary channels, which may affect only a small portion of the delta (Olariu and Bhattacharya; 2006; Figs. 5 and 7).

Barrier-shoreface deposits may form during the transgressive abandonment phase of a delta. The classic example is the Chandeleur Islands in the Gulf of Mexico, which are the remnants of a now abandoned older lobe of the Mississippi delta (Boyd and Penland, 1989; Figs. 6 and 7). During abandonment, waves and tides may also cause erosion of the upper parts of a delta deposit forming top-truncated delta lobes. These abandonment surfaces and associated transgressive facies, historically referred to as the 'destructive phase', form key units and bounding surfaces in sequence-stratigraphic studies (*e.g.*, flooding surfaces) that have long been used to correlate and map individual delta lobes within paralic clastic wedges (*e.g.*, Scruton, 1960).

CLASSIFICATION OF DELTAS

The commonly used tripartite classification of deltas (Fig. 2; Galloway, 1975) is based on the idea that variations in the ratio of fluvial, wave and tidal processes result in different and identifiable plan-view morphology of the resulting deposit, as well as distinctive internal facies successions. Of course, there is a natural tendency for workers to force-fit their particular example into one of the end-member categories, despite the fact that most deltas are likely to be mixed-influence and plot somewhere within the triangle. Many modern deltas, such as the Danube in the Black Sea (Fig. 4), show a mixture of delta types both between and within discrete lobes. This can create prob-

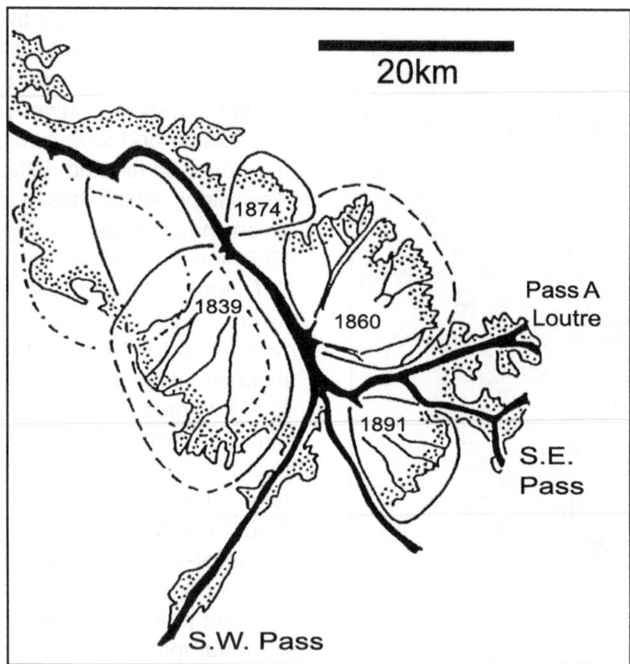

Figure 5. Infilling of interdistributary bays by historically dated crevasse 'subdeltas' in the modern Mississippi birdfoot delta (Fig. 3A). Note the large variation in scale of deltas and distributary channels. At least three orders of branching can be discerned (from Bhattacharya and Walker, 1992; simplified from Coleman and Gagliano, 1964).

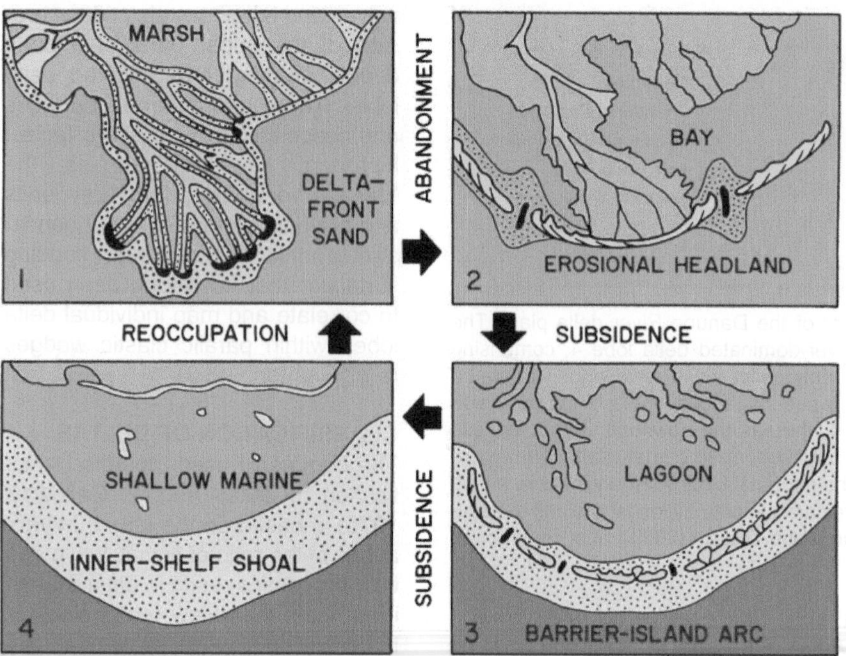

Figure 6. Evolution of Mississippi Delta lobes from progradation to abandonment (simplified from Boyd *et al.*, 1989). Each delta lobe goes through an initial phase of progradation, during which it shows a river-dominated character. As it is abandoned, it transforms into a wave-dominated barrier-lagoon system as a result of subsidence and transgression. The barrier is ultimately drowned to form a relict shelf shoal.

lems for interpreters, especially in subsurface studies where the nature of a depositional system is typically determined on the basis of sparse well or core data that may not represent the whole system.

Long-held models suggest that river-dominated deltas display an overall digitate or lobate morphology (Figs. 2, 3A, B, 5, 6, and 7) whereas wave-influenced deltas show smooth-fronted lobes that have arcuate to cuspate margins (Figs. 2, 3C, D, 4, and 8). Tidal processes form sandbodies oriented parallel to the directions of the tidal currents (Figs. 3E and F; and *see* examples in Chapter 9), which typically flow perpendicular to the regional shoreline. These morphological differences should, in turn, be reflected in the geometry of deltaic sand bodies (Coleman and Wright, 1975; Fig. 2).

Orton and Reading (1993) extended Galloway's (1975) ternary classification to include sediment type, and Postma (1990) presented an independent classification scheme based on the type of feeder system, water depth, and mouth-bar process (Fig. 9). The Postma (1990) classification scheme, however, does not include waves or tides as key parameters and thus may not be as useful in studies of ancient deposits.

Seismic- and sequence-stratigraphic studies of deltas have led to the recognition that depositional systems can change their character as a function of their physical and temporal position. This, in turn, led to a classification of deltas according to their position with respect to the shelf edge. For example, shelf-edge deltas form during the later part of prolonged highstands of sea level and especially during sea-level lowstands (*e.g.*, Suter and Berryhill, 1985; Porebski and Steel, 2003, 2006; Fig. 10). Deltas deposited at the shelf edge are commonly unstable and develop impressive growth faults with up to kilometers of throw. Sandbodies in these settings commonly are aligned parallel to strike (Fig. 10), but this elongation is controlled by subsidence along the growth faults rather than by wave processes. They typically form much thicker facies successions than shelf-phase deltas, which lie in a more landward position, are deposited in shallower water, and may show less complete preservation of the initial deposits (*see* sequence-stratigraphy section below).

DELTAIC PROCESSES

Sediment Dispersal at River Mouths

A delta forms when a river of sediment-laden fresh water enters a standing body of water, loses its competence to carry the sediment, and

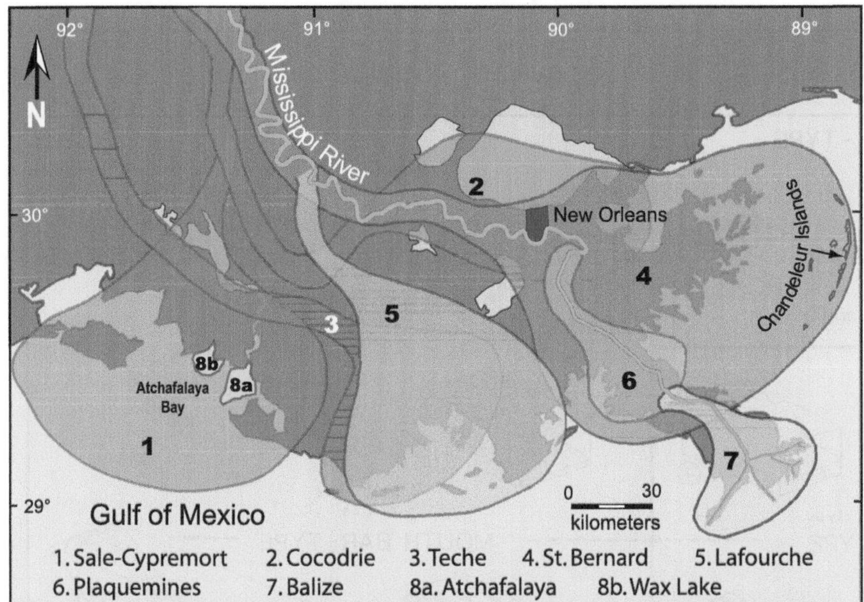

Figure 7. Evolution of Mississippi Delta lobes over past 9000 years (after Kolb and Van Lopik, 1966). Lobe switching occurs about once every 1000 years. Note that the Chandeleur Islands are a transgressive barrier system formed following abandonment of the St. Bernard lobe (*see* also Fig. 6 and Chapter 11, Figs. 8 and 9). The natural propensity of the over-lengthened Mississippi River is to avulse into Atchafalaya Bay. Artificial diversion of Mississippi water into the Atchafalaya River has built a new delta (*see* Fig. 13).

thus deposits it, typically near the river mouth (Fig. 11). Jet theory has been widely applied to explain the dynamics of how river plumes interact with the body of standing water (*e.g.*, Wright, 1977). The internal facies distribution and external morphology of a deltaic deposit depends upon:

1. Whether the river outflow is denser (hyperpycnal), equally dense (homopycnal) or less dense (hypopycnal) than the standing body of water;
2. The interaction of the river plume with marine processes, which can include waves, tides, storms, and ocean currents, and biogenic reworking;
3. The physical position of the delta in the basin, such as location relative to the shelf edge;
4. The caliber of sediment fed to the shoreline, as well as the caliber of sediment moving along shore or along shelf (*i.e.*, gravel, sand or mud);
5. Depositional slopes;
6. The degree to which river-derived sediment is reworked by marine processes during delta progradation; and
7. The degree of post-depositional erosion, such as the top truncation of a delta by waves that is common during lobe abandonment and subsequent transgression.

Historically, most marine deltas have been assumed to be hypopycnal because fresh water is less dense that seawater. In hypopycnal flows, the fresh river water rides on top of

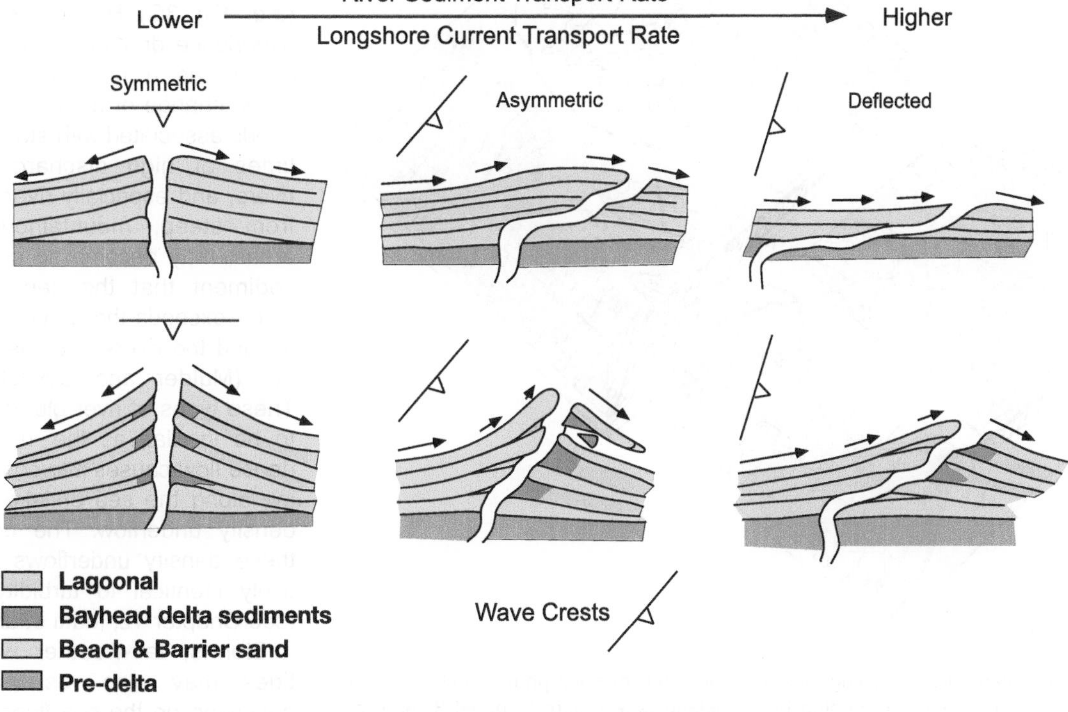

Figure 8. Morphology of wave-influenced deltas. Top row represents lower river discharge compared to bottom row. The river discharge plume acts as a groyne that traps sediment updrift. (Modified after Bhattacharya and Giosan, 2003).

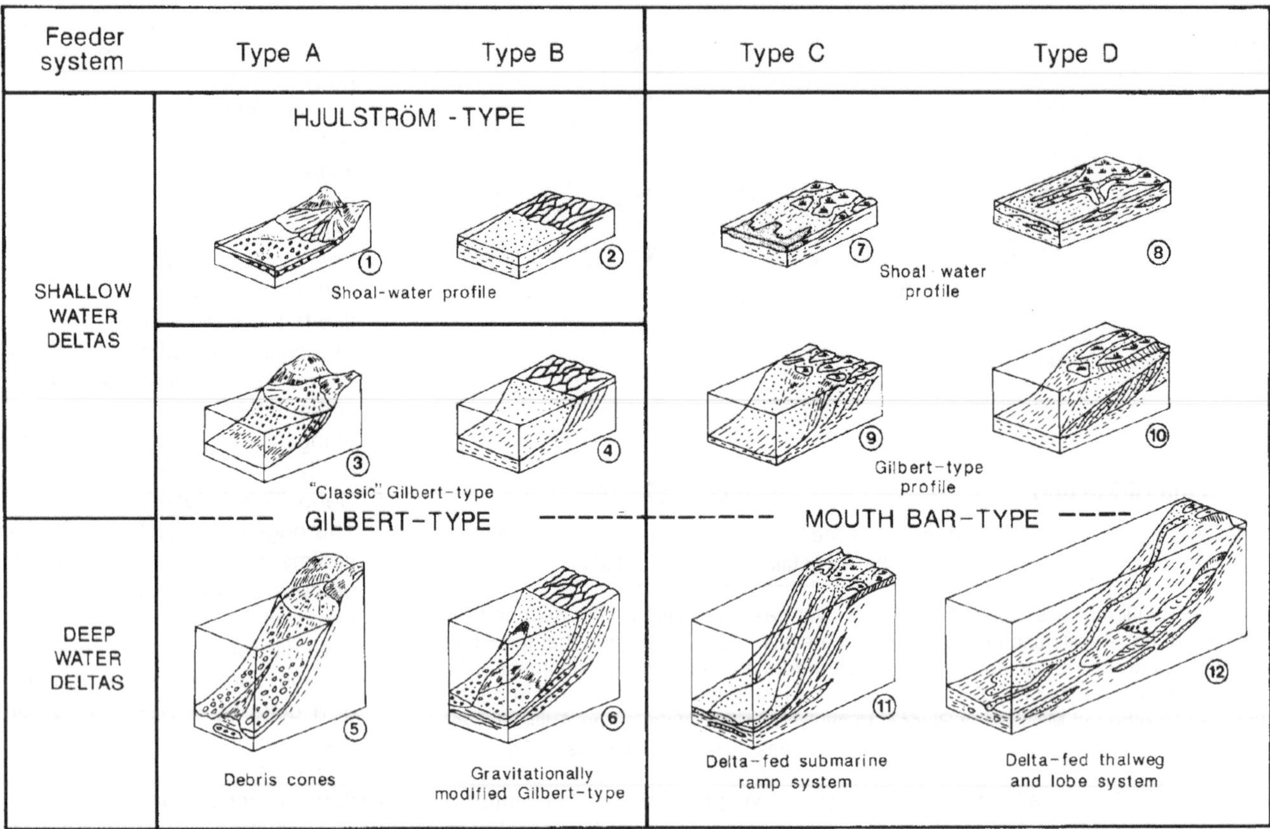

Figure 9. Classification of coarse-grained deltas incorporating type of feeder system, water depth, and type of mouth-bar process. (From Reading and Collinson, 1996; after Postma, 1990). (Reproduced by permission of Blackwell Publishing).

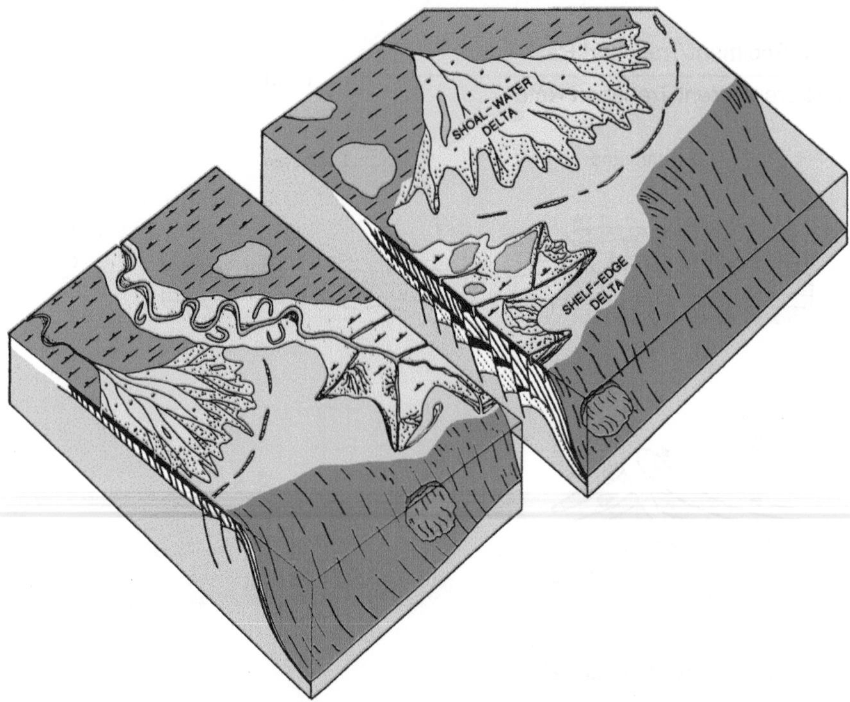

Figure 10. Block diagram contrasting lobate shoal-water (or shelf-phase) deltas, shown in the abandonment phase with fringing barrier island (*see* Figure 6, panel 3) and the coevel active shelf-edge delta. Note thickening of facies across growth faults in the shelf-edge delta. (From Bhattacharya and Walker, 1992; after Edwards, 1981).

the seawater as a buoyant plume (Fig. 11B) visible in satellite images (*e.g.*, Fig. 3C). However many rivers experience dramatic changes in discharge as a function of seasonal runoff change or as a result of major floods associated with storms. During times of high discharge, smaller rivers, and especially rivers that flow from steep, mountainous source areas, may become so choked with sediment that the sediment-laden water exceeds the density of seawater and the rivers become hyperpycnal (Mulder and Syvitski, 1995). These types of river plumes are said to be inertial, as the inertia of the dense flow causes it to continue flowing along the sea or lake bed as a density underflow. The deposits of these density underflows are effectively identical to turbidity currents (*see* Chapter 12; Mutti *et al.*, 2003).

Storms, fair-weather waves and tides may also resuspend mud deposited on the sea floor during an earlier river flood. This mud may subsequently migrate along the shelf forming a dilute, hyperpycnal

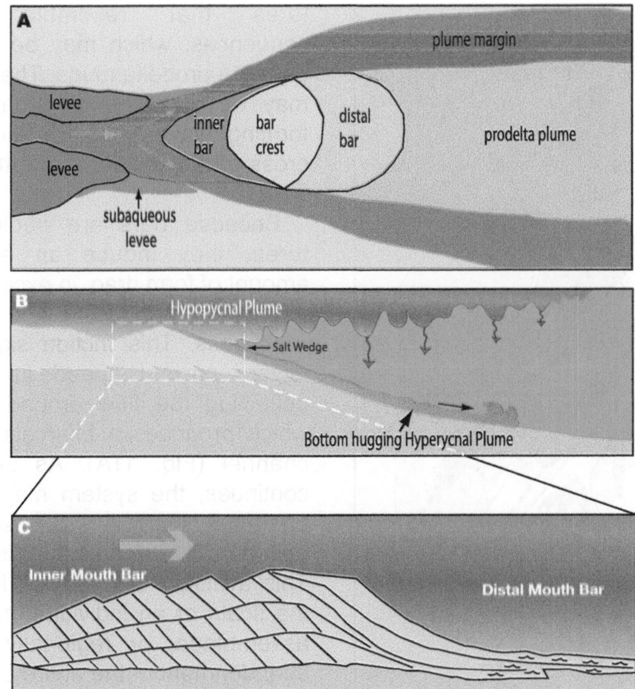

Figure 11. River-plume processes in river-dominated deltas. **A**. Plan view of expanding plume. **B**. Illustration of hypopycnal buoyant plume and hyperpycnal 'inertial' plume. **C**. Detail of mouth bar showing climbing dunes on back side of bar feeding single large foresets (avalanching grain flows) on the bar front. Based on Bates (1953), Wright (1977) and Nemec (1995).

geostrophic fluid-mud belt (Fig. 12; Bentley, 2003). Many river deltas alternate between hypopycnal and hyperpycnal conditions, even in fully marine settings. Many rivers may also produce both hyper- and hypopycnal plumes at the same time (Nemec, 1995; Kineke *et al.*, 2000; Fig. 11B). Homopycnal conditions are the least common, as only small density differences are required for a flow to become either hypopycnal or hyperpycnal. In lacustrine settings, hyperpycnal plumes are more common.

In addition to buoyancy and inertia, interaction of the river plume with the ambient water and the sea floor causes frictional deceleration. Much of the active sand or gravel deposition occurs, at least initially, in a *distributary mouth bar* (also referred to as the stream-mouth or middle-ground bar; Figs. 1 and 11). This process operates in its purest form in river-dominated deltas and the classic mouth bar is a fundamental architectural element of such systems. Mouth bars may coalesce to form bar assemblages that in turn build larger scale delta lobes (Fig. 13). Individual mouth bars can be on the order of

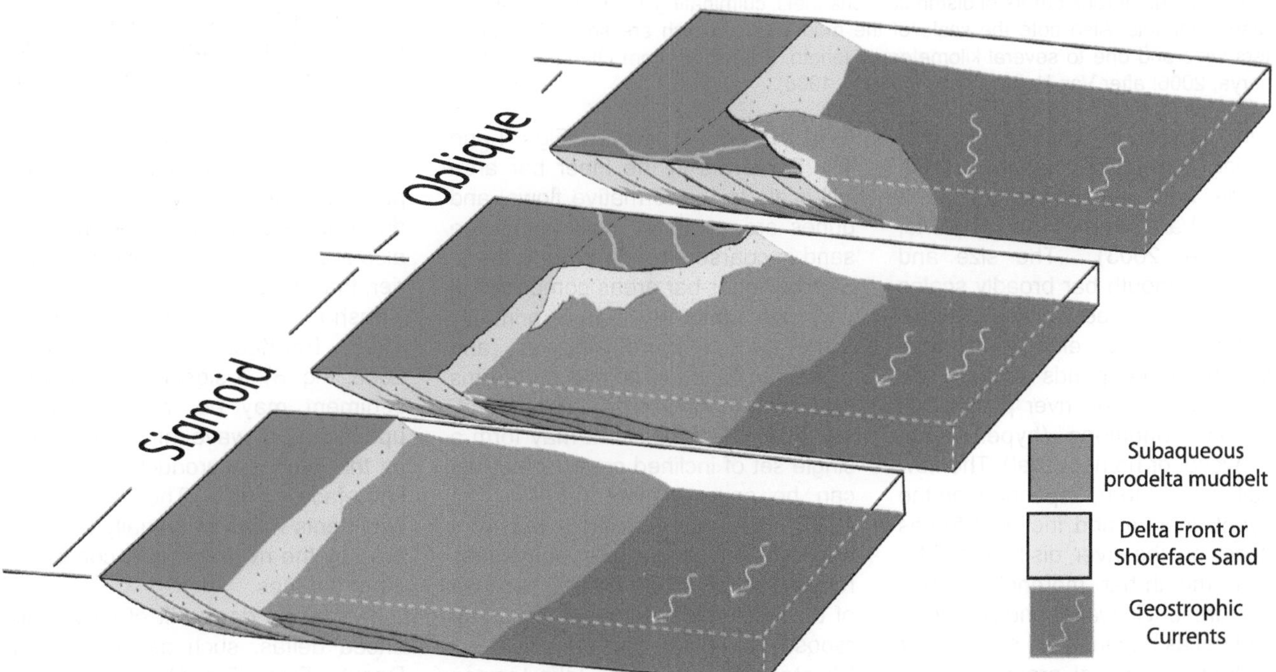

Figure 12. Block diagram of subaqueous prodelta mudbelt, showing along-shelf elongation. Oblique clinoforms are associated with the sandy delta front (*see* also Fig. 15A), whereas sigmoidal clinoforms are associated with the muddy subaqueous prodelta (*see* also Fig. 15B). Note that the two down-drift blocks show a distinctive 'double clinoform' that merges into a single clinoform in the top panel. Although no scale is shown, the subaqueous mudbelt may extend for several tens of kilometers in an offshore direction and several hundred kilometers alongshelf. Coriolis forces and wind regime are important causes of the down-drift deflection of mud plumes (*cf.* Fig. 3C). Such mudbelts may be a few to several tens of meters in thickness. Modified after Vakarelov (2006). *See* also Chapter 8, Figure 11.

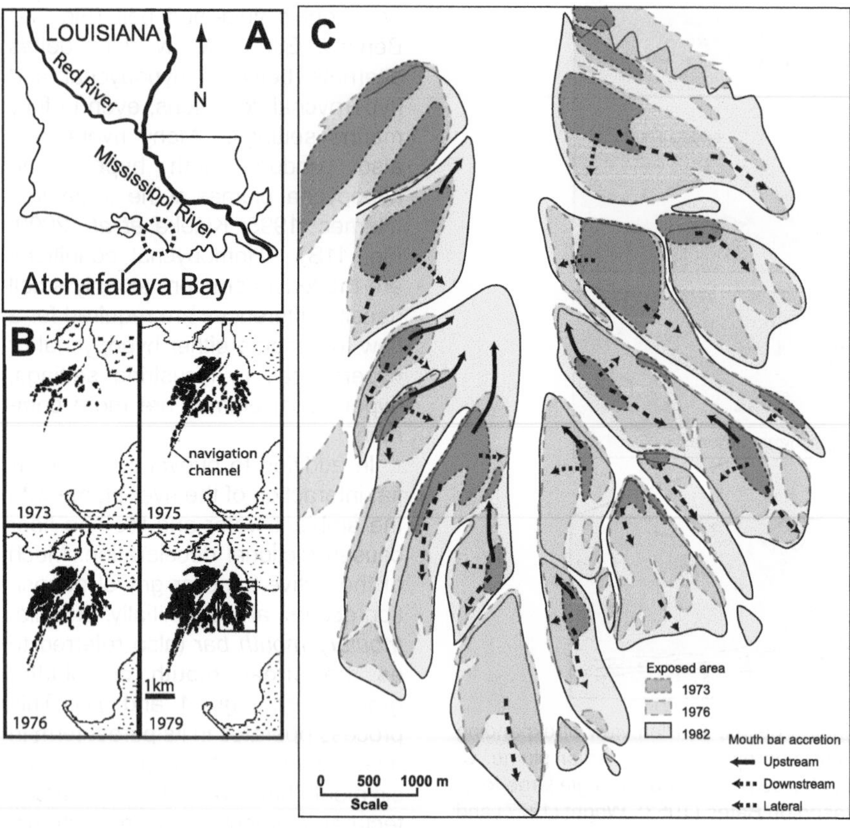

Figure 13. Development of a shallow-water delta in Atchafalaya Bay, Mississippi Delta. **A**. Location Map, **B**. River-dominated lobe forms by the coalescing of distributary mouth bars (black), suggesting friction dominance. **C**. As the delta grows, the mouth bars accrete upstream, downstream and laterally (compare 1976 and 1982 shorelines). Note that there are numerous orders of distributary channels, culminating in small terminal distributary channels. Also note the scale of the mouth bars, which are several hundred meters wide and one to several kilometers in length. (Simplified from Olariu and Bhattacharya, 2006; after Van Heerden and Roberts, 1988).

several kilometers long and a kilometer wide in relatively large systems like the modern Atchafalaya Delta (Fig. 13; Van Heerden and Roberts, 1988; Tye, 2003). The size and shape of a mouth bar broadly scales to the size of the feeding river, whereas the shape (*e.g.*, elongation, width to length ratio) depends on the angle of dispersion of the river plume and the flow conditions (hyperpycnal, hypopycnal or homopycnal). The flow condition is, in turn, dependant on the buoyant, inertial and frictional forces that act on the river discharge. The classic mouth-bar morphology may not be present in wave- and tide-dominated deltas because the sediment is reworked by basinal processes and transformed into other sandy geomorphic elements such as tidal bars, barrier bars, or shorefaces (Fig. 8).

Mouth bars can be divided into a proximal, *inner bar*, where flow is constricted, and a distal *outer bar* characterized by flow expansion (Figs. 1 and 11). Bedforms in the inner bar area will scale to the formative flows, and dunes are common, so long as the sand is coarse enough to form them. Shallow inner-bar areas composed of very fine sand may show primarily plane bed or current ripples, and more rarely supercritical bedforms forming humpback cross strata (Fielding, 2006). The bar-front may form a single set of inclined cross strata that can be several meters thick (Fig. 11C). Coarse-grained bar-front deposits like these are sometimes referred to as Gilbert deltas, because of their resemblance to the angle-of-repose, gravelly foresets described in Pleistocene deltaic lake-terrace deposits in Utah (Gilbert, 1885).

Bar-front deposits may grade into thinner sheet-like frontal splays that form distal-bar deposits. Frontal splays form extensive sandy beds containing classic waning-flow features that resemble Bouma sequences, which may be interbedded with prodelta muds. These sands may be reworked by storm waves forming wave-rippled or hummocky cross-stratified beds (Mutti *et al.*, 2003).

Because bars are bedload features, they induce an enormous amount of form drag, in excess of that associated with grain and bedform roughness. This friction significantly lowers current speed as well as deflecting the flow around the bar, which produces a bifurcation of the channel (Fig. 11A). As bifurcation continues, the system may become unstable initiating an autogenic upstream avulsion of the distributary. This avulsion can happen locally, at the scale of an individual bar or bar assemblage, or regionally causing abandonment of the entire delta lobe.

Wave Reworking

Waves smooth out and elongate mouth bars in a shore-parallel direction. The ability of waves to extend a bar downdrift (Fig. 8) will depend on the ratio of sediment deposition by river floods to the transport capacity of the longshore drift, as well as the obliquity of wave approach (Bhattacharya and Giosan, 2003). In deltas with high wave energy or very infrequent floods (*e.g.,* centennial floods), mouth bars may be extended for many kilometers or more alongshore, passing gradationally into shoreface deposits (*sensu stricto*). A river can also act as a groyne, or hydraulic barrier, that traps sediment carried in the longshore drift system (Dominguez, 1996; Bhattacharya and Giosan, 2003; Fig. 8). A significant amount of sediment may be trapped on the updrift side of wave-influenced deltas by this process, producing an updrift shoreface 'wing'. These shoreface sediments may not actually be delivered by the river and may come from other sources.

In regions downdrift of wave-influenced deltas, such as seen in the Danube Delta (Fig. 4), alternations of barrier-island lagoon systems may form, with river-fed bay or lagoon-head deltas at their landward margin. Farther downdrift, sediments may be very muddy (Fig. 12), and may form progradational muddy shoreline complexes, capped by thin and narrow beach deposits, consisting of reworked shells or sand, that are termed *cheniers*. These are found in association with the Atchafalaya/Mississippi deltas (Augustinius, 1989), the muddy Camau Peninsula that lies downdrift of the Mekong River in Vietnam (Ta *et al.* 2005; Fig. 3E), and the muddy Suriname coast, which is fed by the Amazon Delta (Rine and Ginsburg, 1985).

Tidal Reworking

Where tidal currents are sufficiently strong, as occurs within embayed parts of deltas or in deltas deposited within larger scale embayments (*see* Chapter 9), mouth bars may be dissected or elongated in a shore-normal direction, such as is seen in deltas of the Mekong River in Vietnam (Fig. 3E), the Mahakam in Kalimantan, and the Ganges–Brahmaputra in the Bay of Bengal, Bangladesh (Fig. 3F). Tides may also cut deep tidal channels or scours along the shoreline which may be filled with mud or sand.

DELTA ENVIRONMENTS

Most deltas (*i.e.*, those with large amounts of muddy sediment) comprise five main geomorphic environments of deposition (Figs. 14 and 15):

1. A nearly flat subaerial delta plain (where river processes dominate);
2. A steeply dipping sand-dominated shoreface or bar-front (dips typically greater than 1°);
3. A broader and relatively flat, sandy subaqueous delta plain (also termed the subtidal platform), which extends into about 10–15 meters water depth, which corresponds to fair-weather wave-base;
4. A shallow-dipping (less than 0.1°) distal foreset region that is typically silty or muddy; and
5. A relatively flat, clay-rich bottomset region.

The nomenclature of the subaqueous regions of deltas is not widely agreed upon. The term delta front commonly is used to refer to the steeply dipping sandy shoreface or bar-front and its extension into the subtidal platform (*e.g.*, Bhattacharya and Walker, 1992), but the term delta front has also been applied to the

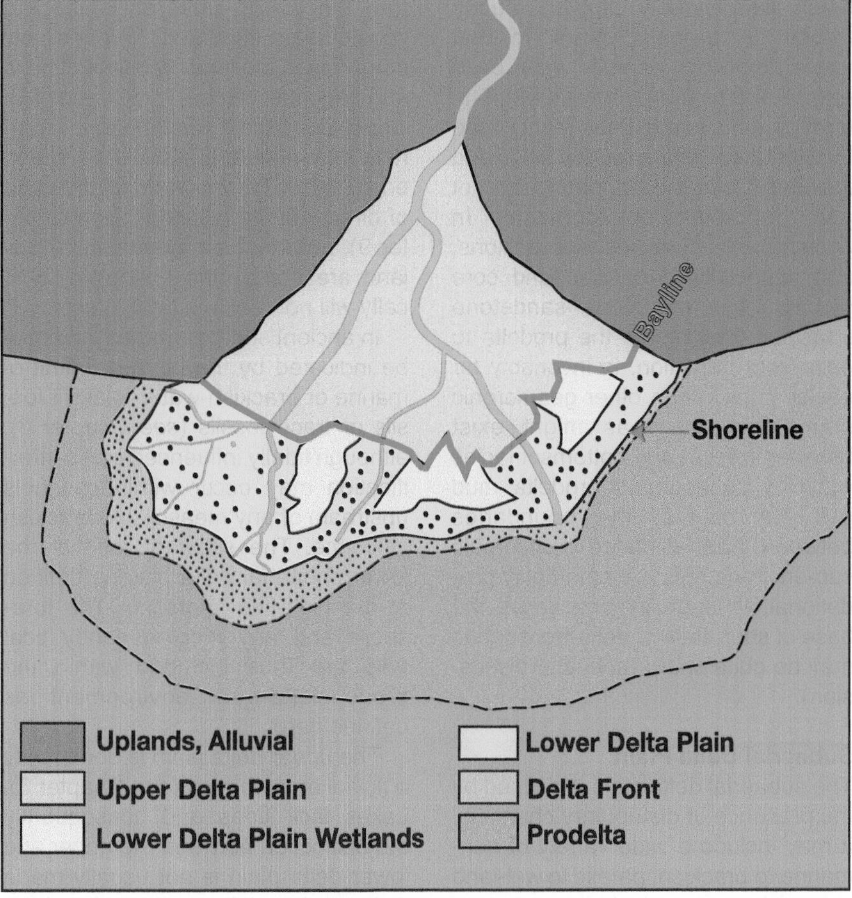

Figure 14. Major plan-view subdivisions of a delta. The upper delta plain is essentially non-marine and is characterized by distributive river systems.

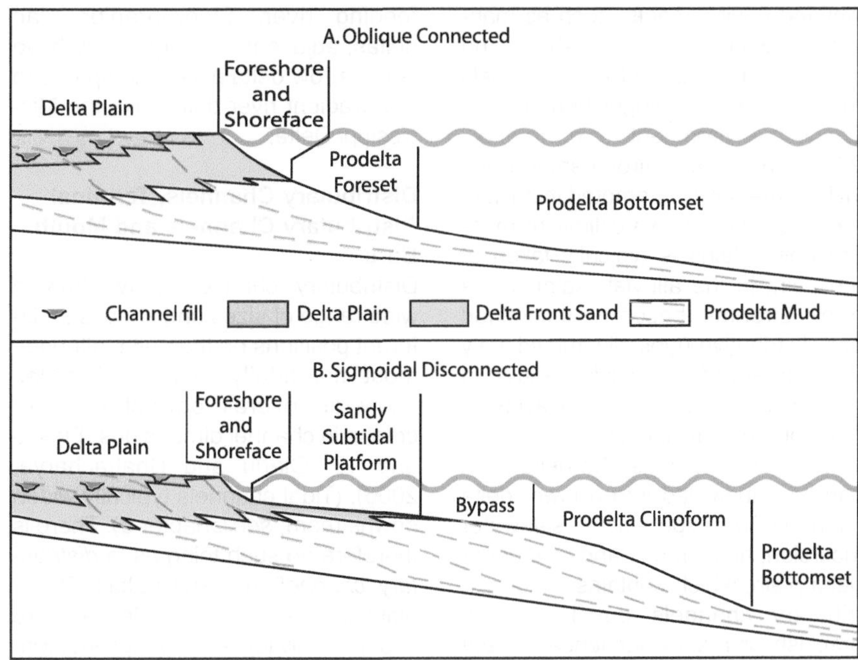

Figure 15. Prograding delta clinoforms. **A**. Oblique sandy shoreface connected to prodelta muds. **B**. Sigmoidal, prodelta mud plume is disconnected from the sandy subtidal platform and delta-front/shoreface, producing a double clinoform. *See* also Figure 12.

much less steeply dipping muddy foreset or prodelta clinoform that forms below fair-weather wave base (*e.g.*, Roberts and Sydow, 2003). The term prodelta is most commonly used to refer to this entire muddy area, and includes both the clinoform foreset and bottomset environments. In ancient vertical facies successions, and especially in outcrop and core studies, the mudstone–sandstone transition, that marks the prodelta to delta-front transition, is invariably far easier to pick than other geomorphic boundaries, such as might exist between foreset and bottomset facies within a subaqueous prodelta mud belt. Of course, the boundaries between many of these geomorphic sub-environments are commonly gradational, although, in some cases, the base of shoreface or delta-front sands may be quite sharp (*see* later discussion).

Subaerial Delta Plain

The subaerial delta plain is defined by the presence of distributary channels. It may include a wide variety of non-marine to brackish, paralic to wet-land sub-environments, including swamps, marshes, tidal flats, lagoons, and interdistributary bays. Subtly different wetland environments, such as mangrove swamps, salt marshes, and intertidal mudflats and tidal channels may be readily distinguished in modern delta environments (*e.g.*, Ta *et al.*, 2005), but these environmental subdivisions are rarely attempted in ancient settings. The landward limit of modern delta plains is typically taken at the point in the alluvial realm where trunk streams become unconfined and distributive (typically immediately downstream of the alluvial valley). In many cases, this is the nodal avulsion point on an alluvial plain.

In modern settings, the delta plain can be subdivided into a lower delta plain, which experiences salt-water incursion, and a more landward upper delta plain, which contains major distributary channels but lacks any brackish or marine influence, except for limited tidal influence. The demarcation between these areas is referred to as the bayline, originally defined by Posamentier *et al.* (1988) as the break in slope of the river profile, and envisioned to coincide with the landward limit of onlapping coastal plain sediments. It effectively demarcates the landward limit of bays and lagoons (Figs. 1, 3C and 14). Upper-delta-plain distributary channels may nevertheless still be affected by tides, far landward of the area of direct marine incursion (*see* Chapter 9), but adjacent interchannel lowland areas (*i.e.*, the floodplain) typically will not feel any tidal affects.

In ancient settings, the bayline may be indicated by the landward limit of marine or brackish-water-tolerant fossils or trace fossils (*see* Chapter 3), although tidally influenced cross-stratification may occur within channels upstream of any measurable brackish influence. The seaward limit of the lower subaerial delta plain is defined at the high-tide shoreline. The foreshore and any fringing sandy tidal flats are thus included within the broad delta-front environment as defined here.

The upper delta plain is dominantly a fluvial environment (*see* Chapter 6). Lakes lack tides and consequently the distinction between the upper and lower delta plain is not usually made in lacustrine deltas, although a bayline may still be defined, if there is a slope change associated with the feeding river. Steep-gradient fan deltas, adjacent to scarps, may have very narrow delta plains, compared to low-gradient river deltas like the Mississippi Delta.

Distributary Channels, Terminal Distributary Channels and Mouth Bars

Distributary channels may show a wide range of sizes and shapes in different positions on the delta, although most non-tidally influenced deltas show an overall downstream decrease in channel dimensions (Figs. 3 and 13; Olariu and Bhattacharya, 2006). (Tidal channels typically widen downstream; *see* Chapter 9). There is therefore no such thing as '*a distributary channel*' in many deltas. Delta-plain channels tend to be few and are separated by lowland floodplains, with their associated levee and crevasse-splay deposits (*see* Chapter 6). These interdistributary areas can be small in area, especially in deltas fed by braid plains, such as is common in glacially fed systems (*see* Chapter 5). Replacement of the floodplain by channels will depend on the avulsion frequency and the relative rates of channel migration and basin subsidence or floodplain aggradation (Bristow and Best, 1993).

Channels within the upper delta plain are essentially a variety of fluvial channel (*see* Chapter 6 for facies criteria) but it may be tricky to determine whether such channels are distributive in an ancient example. To determine this, one should look for progressive changes in dimensions of channel-sandstone bodies commensurate with that expected in a distributive system (*i.e.*, downstream decrease in channel width and depth).

The smallest channels are referred to as 'terminal distributary channels' and are intimately associated with mouth bars (Figs. 1, 13 and 16). Terminal distributary channels form at the boundary between the delta plain and delta front (Olariu and Bhattacharya, 2006). The channels may be formed depositionally, by coalescence of adjacent mouth bars, in which case there may not be basal erosion, or they may form by avulsion and scour, in which case they will have a basal erosion surface, but typically with only a few meters of relief. Terminal distributary channels may show a greater proportion of upstream accretion than fluvial channels and they can also extend several kilometers offshore as subaqueous chutes, forming a channelized to scoured facies within the delta front (Olariu and Bhattacharya, 2006).

In river-dominated muddy, hypopycnal settings, especially those with minimal tides or waves, the positions of distributary channels may be fixed for long periods. This leads to the formation of elongate bar fingers, such as in the deeper water Mississippi 'birdfoot' delta (Fig. 3A and 5), as well as in very shallow-water deltas, such as the Volga Delta, in the Caspian Sea, and many bayhead and lagoonal deltas including the Colorado River Delta in Matagorda Bay, Texas. In sandier systems deposited in shallower water, or not stabilized by human interference, distributaries switch more rapidly and coalesce to form more lobate deltas, as in the Atchafalaya (Fig. 13) and Lafourche (Fig. 7) lobes in the Gulf of Mexico,

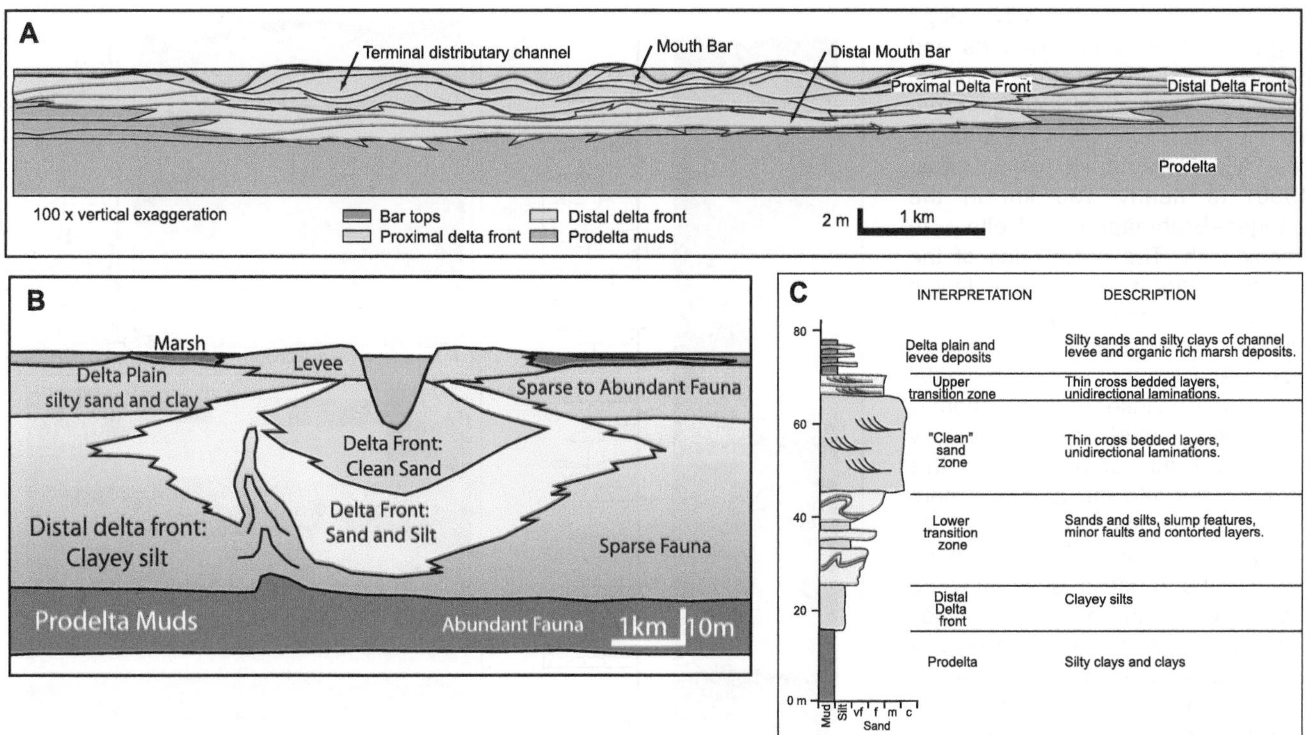

Figure 16. Cross sections and measured vertical section of terminal distributary channels. **A**. Shallow-water, Atchafalaya-type delta forms a broad delta-front sand sheet containing a complex association of terminal distributary channels and mouth bars (*see* also Figure 13). **B**. Deeper water Mississippi-type delta forms a narrower bar-finger sand (*see* Figure 3A). In both cases, the terminal distributary-channel deposits are fully contained within the 'clean' delta-front sand and comprise less than about 30% of the total upward-coarsening facies succession, as shown in **C**. (Simplified from Olariu and Bhattacharya, 2006).

and the Lena Delta in the Russian Arctic (Fig. 3B) (Olariu and Bhattacharya, 2006).

In extreme cases in some wave-influenced deltas, the distributary channel may be 'deflected' to flow parallel to the shoreline (Fig. 8). Thus, relative to river-dominated deltas, the progradation rate of wave-influenced deltas is slowed because sand is being exported alongshore, to feed progradation elsewhere along the coast. This allows rivers feeding wave-influenced coasts to maintain a higher slope that inhibits avulsion (Swenson, 2005). As a consequence, wave-influenced deltas typically have only a few active distributary channels (Fig. 3C and D), whereas river-dominated deltas can have tens to hundreds of active terminal distributary channels in wave-dominated deltas (Figs. 3B and 13). Terminal distributary channels may be significantly larger, wider, and deeper than in river-dominated deltas because bifurcation in wave-influenced deltas is inhibited. If there are small sea-level fluctuations that expose the shoreface, such terminal distributary channels may become entrenched, forming small-scale incised distributary channels that resemble incised valleys, albeit with rather local incision (Barton *et al.*, 2004).

Many tidally influenced deltas show distributary channels that are stable for hundreds to thousands of years, such as in the Mekong Delta, Vietnam (Ta *et al.*, 2005; Fig. 3E) and the Ganges–Brahmaputra Delta in Bangladesh (Kuehl *et al.*, 2005; Fig. 3F). This increased channel stability, relative to river-dominated deltas, may occur because tidal action keeps channels flushed of sediment and allow channels and bars to remain stable for longer time periods. Increased channel stability results in far more elongate sand bodies, with higher length-to-width ratios than are typically found in river-dominated delta fronts (ratio of 10 versus 2 respectively; Reynolds, 1999).

The previous version of this chapter (Bhattacharya and Walker, 1992) showed several examples of so-called river-dominated distributary channels in ancient birdfoot deltas. These examples showed thick sandy distributary channels eroding into marine prodelta shales, despite the fact that this is rare in modern deltas. Some of these sandstones are over 30 m thick and entirely cut out delta-front sandstones that are 10 m thick. All of the previous examples of deeply incised distributary channels illustrated in Bhattacharya and Walker (1992), such as the Pennsylvanian Booch Delta in Oklahoma, and the incised 'feeder' channels in the Cretaceous Dunvegan Formation in Western Canada, are now interpreted as incised-valley fills rather than distributary channels (*e.g.,* Bowen and Weimer, 2003; Plint and Wadsworth, 2003).

Delta Front: Shoreface/Bar-front and Subtidal Platform

The *delta front* is defined as the intertidal shoreline and adjacent gently dipping (*i.e.*, nearly flat) subtidal platform. It comprises the area dominated by sand or gravel, although mud may be present (Figs. 14 and 15). The width of this subtidal platform can be up to several tens of kilometers depending on the tidal range (*e.g.*, Ta *et al.*, 2005; Roberts and Sydow, 2003). The width is higher in meso- and macrotidal settings, ranging from a few kilometers in the mesotidal Alta Delta in Norway (Korner, 1990) to nearly 100 km in the Ganges–Brahmaputra Delta in Bangladesh. The outer edge of the subtidal platform is typically at, to just below, fair-weather wave-base and much of the subtidal platform may be highly affected by waves and tidal currents. Because of the generally high energy, mud does not usually settle over this area, although this area grades into the adjacent mud-dominated prodeltaic shelf (Fig. 12). During large river floods or major storms, the prodelta shelf may be blanketed by shallow-marine sandy turbidites or storm-beds forming distinctly interbedded deposits that are variably referred to as transitional offshore, 'toe-of-shoreface' or distal delta-front facies, depending on whether they are thought to be linked to a shoreface, to a mouth bar, or both (*e.g.*, Mutti *et al.*, 2003).

River-dominated proximal delta fronts typically consist of coalesced mouth bars separated by terminal distributary channels (Fig. 16). Wave-dominated delta-fronts may comprise shorefaces and their subtidal extensions, but may contain a greater proportion of mud and may typically show lower levels of bioturbation than non-deltaic shorefaces (*see* also Chapter 8). Tidally influenced delta fronts are less well studied, but may consist of tidally reworked sandy mouth-bars that accrete laterally and down-stream (Willis, 2005). In tidal systems, a greater proportion of thin mud layers may be deposited within the delta front as tidal mud drapes or as eroded mud-chip layers.

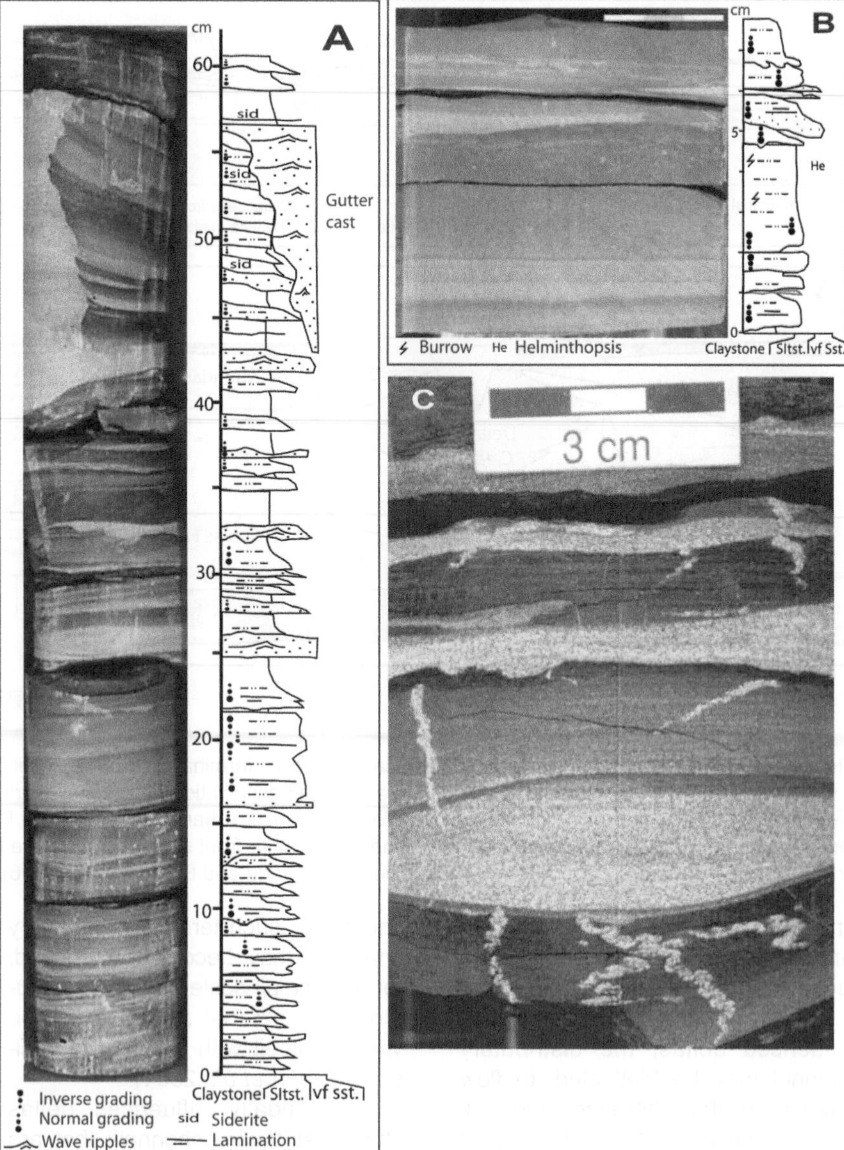

Figure 17. Photos and measured section of cores through river-dominated prodelta facies of the Dunvegan Formation, Alberta. **A**. Interbedded claystones, siltstones and very fine-grained sandstones with storm-produced, sand-filled gutter cast near the top. **B**. Interbedded normally and inversely graded very fine-grained sandstones, siltstones and sparsely burrowed mudstones from river-dominated prodelta of the Dunvegan Formation. **C**. Dewatering (syneresis) cracks in interbedded graded sandstones and mudstones of the Dunvegan Formation. (Photo courtesy of James MacEachern). Scale bars in **B** and **C** are 3 cm long.

Prodelta

The *prodelta* has historically been interpreted as the area where fine mud and silt settles slowly out of suspension, although it is becoming increasingly recognized that many prodelta mud belts contain rapidly deposited hyperpycnal fluid muds formed by wave-assisted density currents (*see* Chapters 8 and 9), or by rapidly collapsing hypopycnal plumes, suggesting that suspension settling may be more rapid than previously thought. Prodelta deposits may show highly variable levels of bioturbation, depending on sedimentation rates and the influence of brackish water associated with hyperpycnal flows (Bhattacharya and MacEachern, *2009*; *see* Chapter 3). Prodelta muds may merge seaward with fine-grained hemipelagic and commonly calcareous sediment of the distal basin floor and grade landward into the sandy delta-front facies of the subtidal platform. The preservation of silty or sandy laminae or beds within the prodelta, especially showing normal or inverse grading, is generally taken to indicate deposition by river floods and, hence, mark proximity to the river mouth (Fig. 17). By contrast, deposits that are totally bioturbated form in areas away from the active river (MacEachern *et al.*, 2005). Where the sediments are rhythmically laminated, a tidal influence may be inferred (Willis, 2005; Fig. 18C).

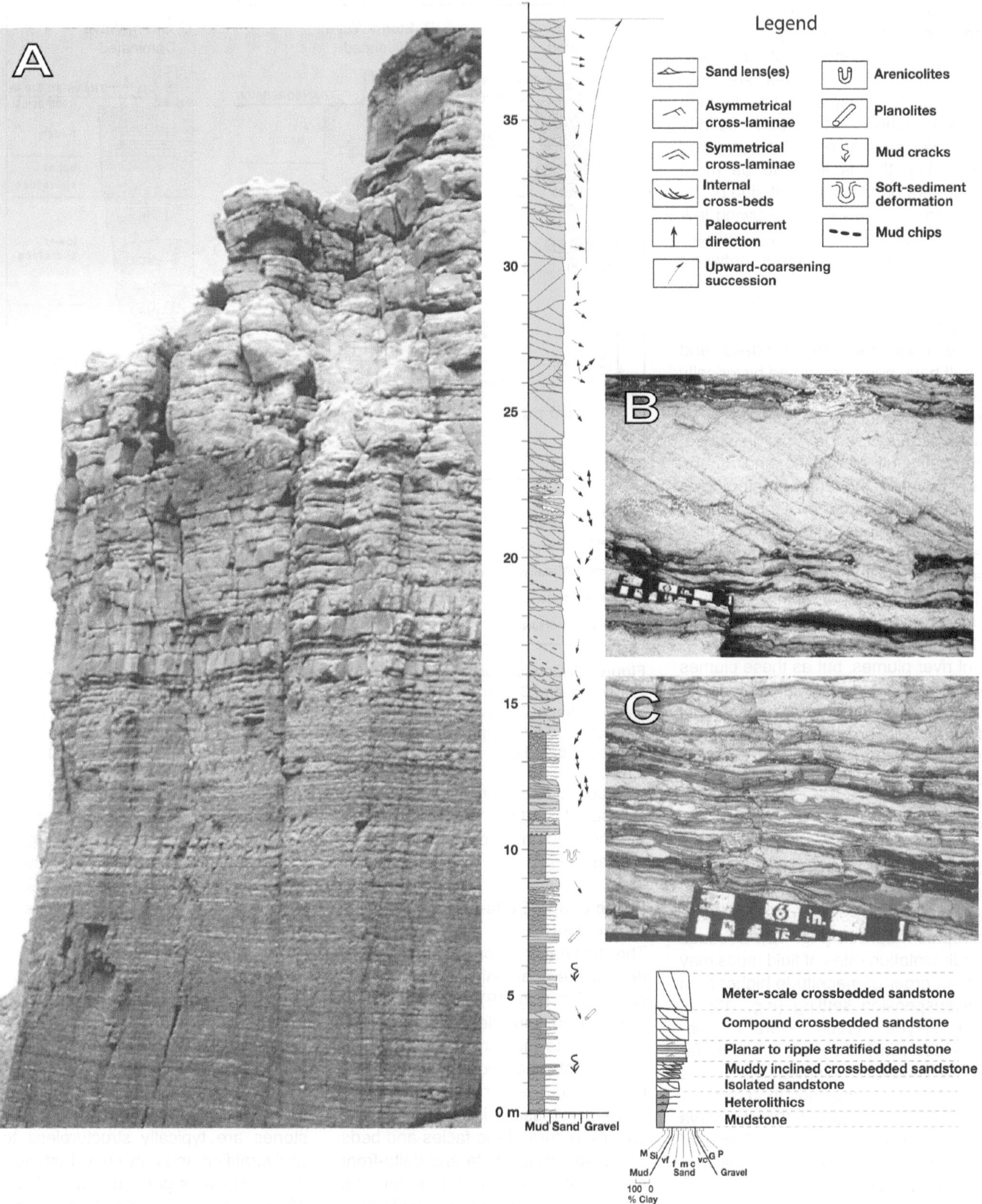

Figure 18. Facies succession through tide-dominated delta front of the Cretaceous Frewens Sandstone, Wyoming. **A**. Upward-coarsening facies succession, and adjacent measured section. **B**. Double mud-drapes on forests of seaward-directed, dune-scale cross-strata, indicative of tidal modulation of river outflow. **C**. Heterolithic, lightly burrowed subtidal prodelta facies at the base of the succession. The presence of mudstone laminae at every scale is diagnostic of tidal influence.

Because of the abundant suspended sediment, filter-feeding organisms that produce open vertical burrows of the *Skolithos* Ichnofacies are suppressed (*see* Chapter 3). Locally, high sedimentation rates and salinity changes may also inhibit burrowing (Fig. 17); this is likely to be more important on the downdrift side of the river mouth. However, burrowing levels may be very high along bed surfaces representing times between river floods when normal salinity and slow sedimentation conditions resume.

Although the term prodelta and shelf have been presented historically as mutually exclusive environments, many of the world's muddy inner shelves, such as the Amazon, Bay of Bengal, Gulf of Mexico and others, are covered by tens-of-meters-thick mud belts that can be thought of as the subaqueous extension of their associated deltas (*e.g.*, Nittrouer *et al.*, 1986; Neill and Allison, 2005; Fig. 12; *see* also Chapter 8, Fig. 11). Much of this muddy sediment was originally deposited by suspension out of buoyant river plumes, but as these plumes collapse the sediment concentrates at the seabed forming a hyperpycnal fluid-mud layer that may be kept in suspension by waves or moved by storm-generated currents (*e.g.,* Hill *et al.*, 2007). Mud may also be introduced directly onto the seafloor by hyperpycnal river flows (Mulder and Syvitsky, 1995). Much of this mud is deflected by the Coriolis force producing an along-shelf-trending geostrophic mud belt (Fig. 12). Because sedimentation rates of fluid muds may be an order of magnitude higher than by suspension settling (*e.g.*, 10s of cm per year versus several millimeters per year), they are probably far more important volumetrically in the construction of the shelf than has historically been realized (Bhattacharya and MacEachern, 2009). Some of this mud can be reworked onshore, forming the chenier plains referred to above.

VERTICAL FACIES SUCCESSIONS

Although they provide an incomplete representation of the 3-D complexity of deltaic deposits, idealized vertical facies successions provide essential points of reference for outcrop or subsurface studies. Cores, well logs and measured vertical sections remain the most common data type in most studies.

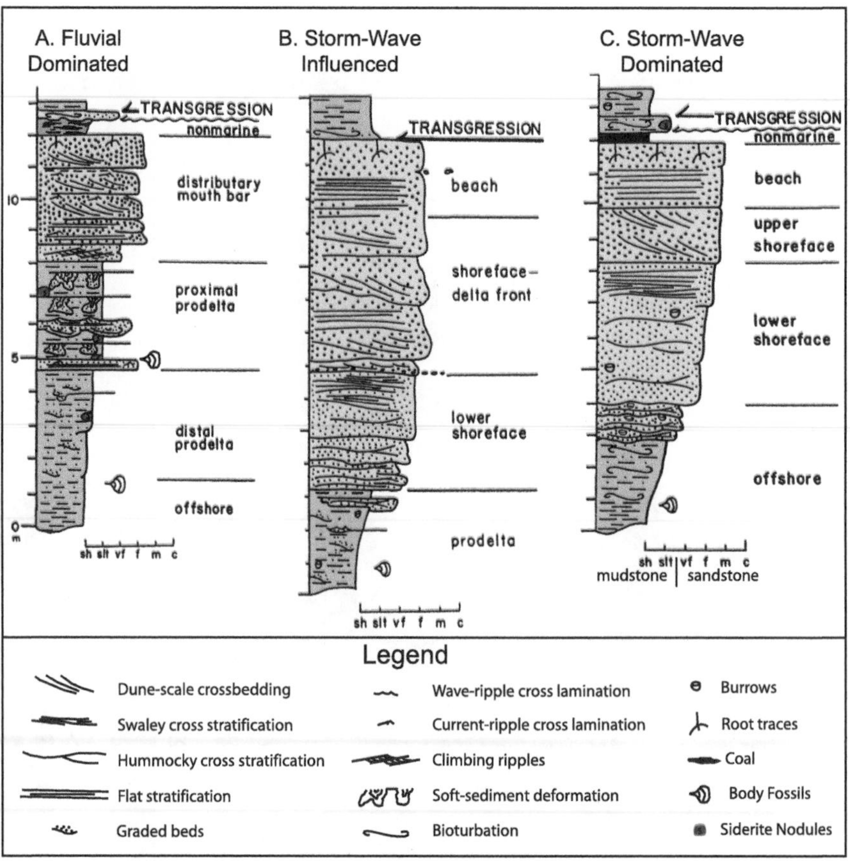

Figure 19. Comparison of delta-front successions in **A**. River-dominated, **B**. Storm-wave-influenced, and **C**. Storm-wave-dominated deltas in the Upper Cretaceous Dunvegan Formation, Alberta. The river-dominated succession is the most irregular texturally, but all are characterized by an upward-coarsening grain-size trend. Basal mudstones are increasingly bioturbated with decreasing river influence. After Bhattacharya and Walker (1992).

Prodelta and Delta-front Facies Successions

The hallmark of progradation of a delta is the coarsening-upward facies succession (Fig. 19), showing a transition from muddier facies of the prodelta into the sandier facies of the delta front and mouth-bar environments, and finally into the delta plain if the succession is complete. The specific nature of the facies and beds in prograding prodelta and delta-front successions will depend on the processes influencing sediment transport, deposition and reworking. Upward-coarsening facies successions can also be produced by the progradation of wave-formed deltaic shorefaces, such as occur on both the updrift and downdrift margin of wave-influenced deltas (Fig. 8). As a delta progrades, the accumulation and preservation of the delta-front to delta-plain succession is highly dependant on shoreline trajectory (Fig. 20; Helland-Hansen and Gjelberg, 1994), regardless of the type of delta.

River-dominated Delta-front and Prodelta Successions

In river-dominated deltas, prodelta sediments are typically heterolithic, laminated to thin-bedded mudstones, with or without sandstones (Figs. 17, 19A and 21). Siltstones and sandstones are typically structureless to well stratified and may show both normal and inverse grading (Figs. 17 and 19) reflecting deposition from hyperpycnal density underflows generated at the river mouth during high discharge floods and storms. Inverse grading is particularly diagnostic of the acceleration of sediment-laden hyperpycnal plumes that have reached maximum capacity during a

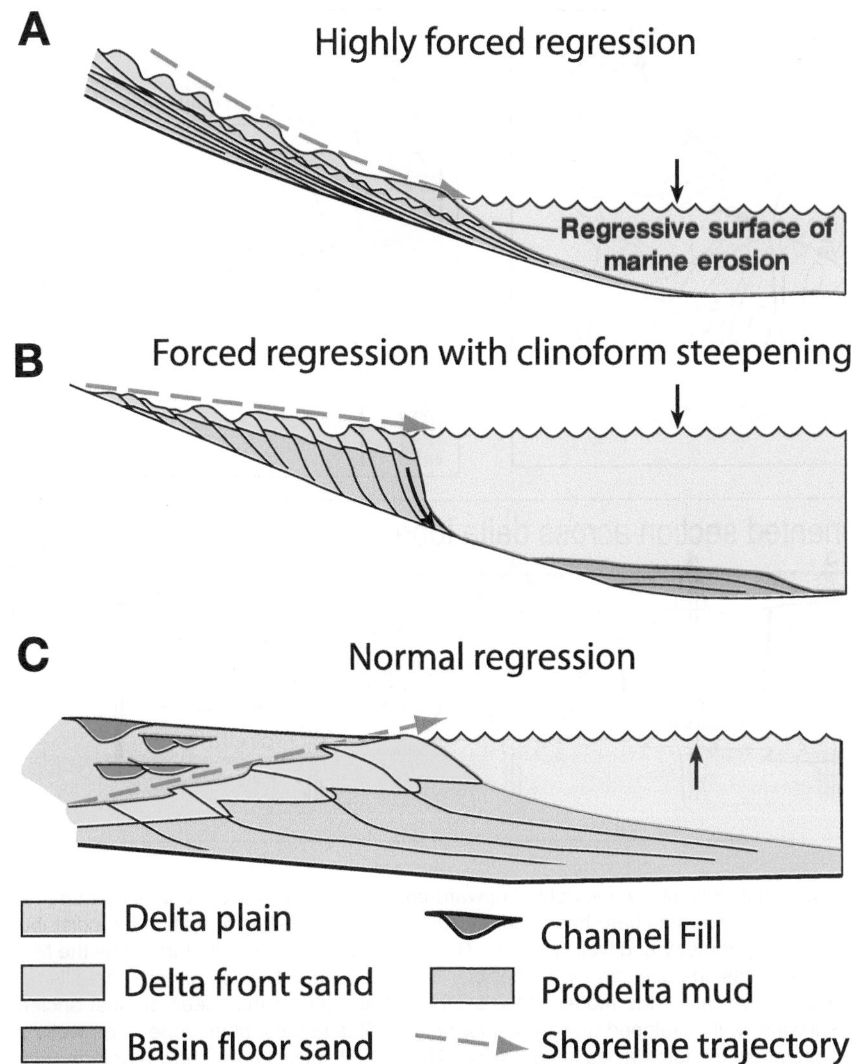

Figure 20. Examples of 'forced' and 'normal' regression (simplified after Helland-Hansen and Gjellberg, 1994). **A**. Sharp-based shoreline deposits are produced when the trajectory of a falling shoreline is greater than seafloor slope (*see* Chapter 8). **B**. Gradational-based deposits are predicted to form when falling shoreline trajectory is equal to or less than the seafloor slope. Oversteepening can cause sediment gravity flows that are deposited on the basin floor. In both cases of forced regression (**A** and **B**) there is no subaerial accommodation and delta-topset facies are essentially absent, as they are prone to erosion during subaerial exposure. Thin topset facies may also be reworked or eroded during the subsequent transgression. This contrasts with normal regression, **C**, where the shoreline trajectory rises. As a consequence, subaerial accommodation is positive and significant accumulation of delta-topset facies (*i.e.*, river channels, interdistributary bays, *etc.*) can occur. Thick paralic and non-marine facies accumulate and are more likely to be preserved under such conditions (*Cf.* Chapter 8, Fig. 31).

waxing river flood (Mulder *et al.*, 2003). The amount of bioturbation can be variable. Wave-formed structures may occur at the tops of graded sandstone beds, but will be less abundant than in a more wave-influenced setting. If floods occur during major coastal storms, sets of highly aggrading wave-rippled sandstone may occur (Fig. 22A) and hummocky cross stratification may be abundant, especially if the storms affect both the river and the coastal areas (so-called storm floods of Wheatcroft, 2000; *see* also Mutti *et al.*, 2003). Soft-sediment deformation features are also common in river-dominated deltas because of the high sedimentation rates (Figs. 19A and 22B). The formation of over-pressured prodelta muds, caused by deposition of overlying heavier sand, may inhibit expulsion of pore fluids in underlying muds leading to reduced strength. This may cause remobilization of the overlying sand resulting in abundant soft-sediment deformation, such as load casts, mud diapirs, dewatering structures and development of growth faults (*e.g.*, Bhattacharya and Davies, 2004; Coleman *et al.*, 1983), as well as slumps and slides (Fig. 10). The scale of deformation usually relates to the thickness of overpressured prodelta mud, and can be tens of meters in scale, but, where such deltas are closer to the shelf-edge, or where deltas overlie salt deposits, the entire basin slope may become involved producing regional-scale faulting, with throws of several kilometers, such as occur in the Niger Delta (Evamy *et al.*, 1978) and parts of the Gulf of Mexico.

Sandy delta-front facies predominantly reflect deposition from rapidly decelerating unidirectional flows in distributary mouth-bar environments. Structures may include unidirectional current ripples and crossbedding, flat-stratified sandstones, massive graded beds and Bouma sequences (Fig. 22C). High rates of deposition result in rapid burial and preservation of structures formed by unidirectional or oscillatory flows. Sorting, especially in gravelly systems, may be poor to moderate. Variations in discharge of the river may produce an irregular coarsening-upward succession, with interbedded mudstones and sandstones throughout (Figs. 19A and 21).

Abundant dewatering cracks (Fig. 17C) may reflect syneresis, produced by the flocculation and contraction of clays as a result of salinity changes, or they may reflect mechanical dewatering of sediment with high initial porosity (*e.g.*, Pratt, 1998). Abundant siderite nodules may be produced by early diagenesis, which commonly requires fresh water influx to reduce sulfate activity, although this may characterize any area where freshwater and saltwater mixing occurs, such as in bays, lagoons and estuaries (Bhattacharya, 2006).

Wave-influenced Delta Fronts

Updrift margins of asymmetric wave-influenced deltas (Fig. 8) are usually characterized by a coarsening-upward shoreface succession, such

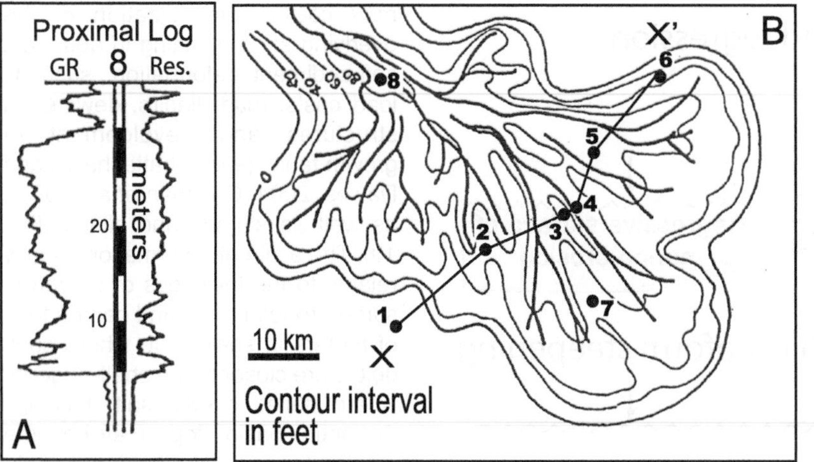

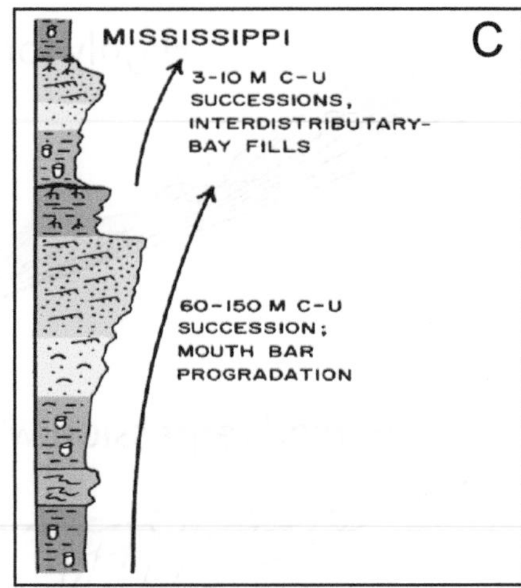

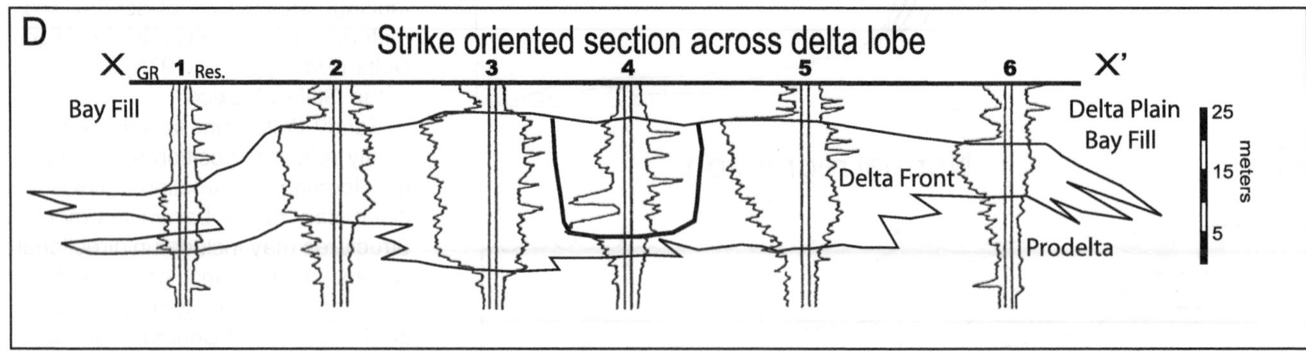

Figure 21. Cores and well logs from the modern Mississippi Delta show a variety of upward-coarsening facies successions (gamma ray (GR) and resistivity (Res.) logs are shown). **A**. Proximal log shows a sharp-based sandy succession suggesting formation by a distributary channel. **B**. Sandstone ispoach map shows the lobate geometry and well locations. **C**. Vertical facies succession formed by the Mississippi Delta, showing thicker prodelta to delta-front successions overlain by thinner bay-fill successions. Distributary channel fills are expected, but are not shown here. Thickness values are typical for the modern Mississippi, but are thicker than seen in most ancient delta successions. **D**. Cross section of delta lobe showing well developed upward-coarsening profiles in most of the lobe (*e.g.*, wells 2, 3, 5 and 7; see B for locations). The fringes of the lobe show muddier facies successions. The bay-fill successions show more irregular, fining and coarsening-upward successions (*e.g.*, compare coarsening-upward bay-fill succession in wells 3 and 6 versus the fining to irregular successions in wells 1, 2, 4, and 5). Modified from Coleman and Prior (1982).

as is shown in Figure 19C. The proportion of fair-weather wave-produced structures (such as wave ripples and dune-scale cross-stratification) is greater updrift. Sandy sediment may be texturally more mature and better sorted in up-drift areas where river influence is less (Dominguez *et al.*, 1987). Updrift prodelta mudstones, if present at all, may be more bioturbated, thinner, and sandier than in river-dominated settings, whereas downdrift regions may be very muddy (*cf.* Fig. 12).

In the geological record, a single vertical facies succession of this type indicates a prograding wave-dominated shoreface. In the previous version of this chapter (Bhattacharya and Walker, 1992) it was emphasized that three-dimensional control is necessary before such a shoreface can be positively ascribed to a delta. However, recent studies of many so-called 'classic' shoreface successions are showing locally high ichnological stress in intervening mudstones, which may be a direct indicator of a river plume nearby that induces a brackish-water or high-sedimentation stress (*e.g.*, Hampson and Howell, 2005). However, I emphasize that it is difficult to produce a thick, upward-coarsening facies succession in the absence of the relatively high sediment supply usually associated with rivers. However, if there is no evidence of high sediment turbidity and low salinity, as might be suggested by trace fossils, this may indicate deposition occurred far enough from a river mouth, either in an updrift or downdrift direction, that the term shoreface or strandplain should be used rather than delta in the strict sense.

Downdrift regions typically consist of sandy barrier islands alternating with muddy lagoonal or bay-fill deposits, such as are seen in the Danube and many other wave-influenced deltas (Fig. 4, Bhattacharya and Giosan, 2003). These may resemble the storm-flood deltaic-shoreface deposits shown in Figure 19B or the irregular river-dominated examples shown in Figure 19A. Facies may show indications of high sedimentation rates and fresh-water influence (*e.g.*, soft sediment deformation, climbing current ripples, brackish-water fauna). The downdrift lagoonal facies may also contain heterolithic tidal strata, depending on the tidal range. The lagoon or bay may also be filled partly by coarsening- or thickening-upward facies successions that represent progradation of lagoon-head or bayhead deltas. These successions will look similar to the fluvial

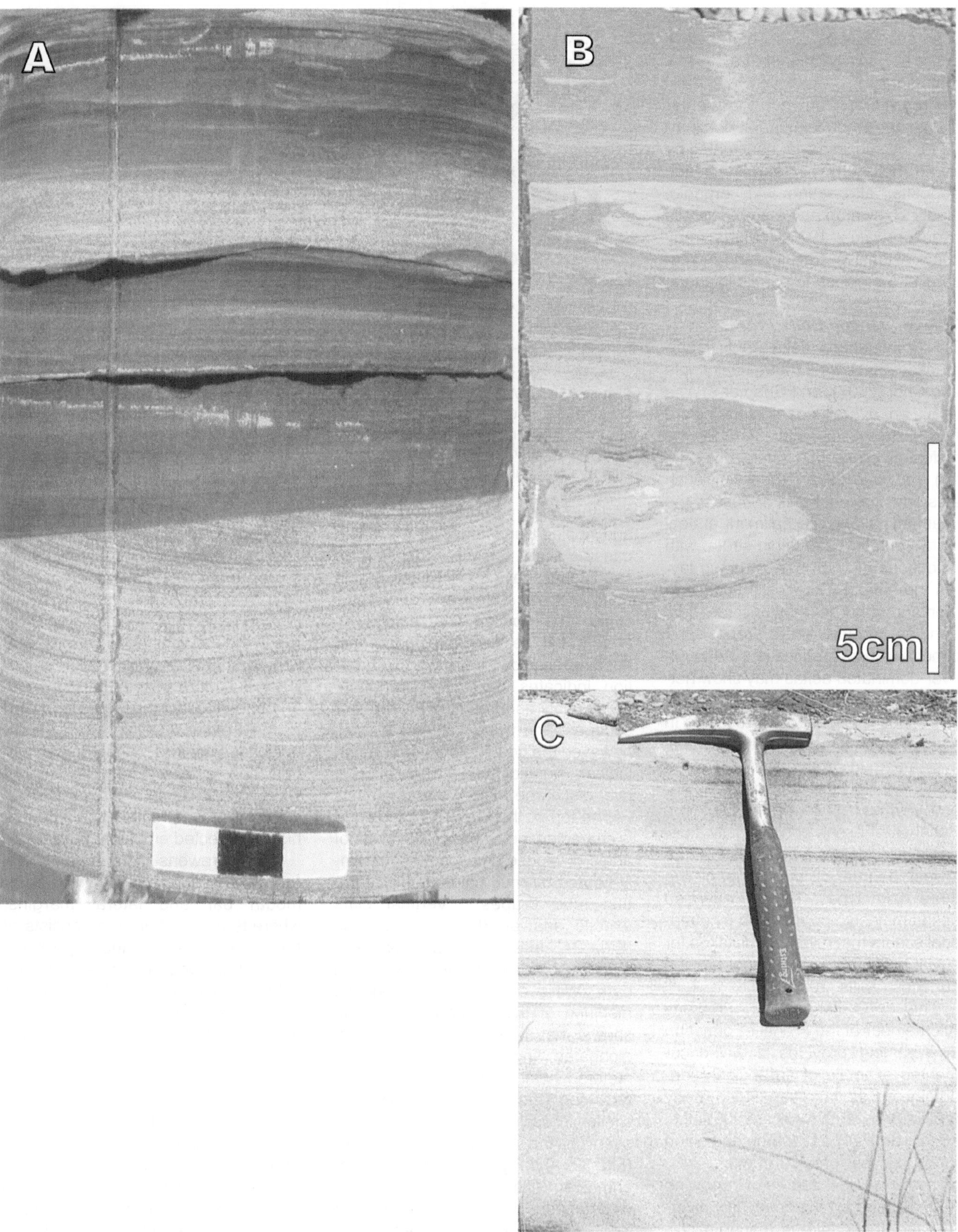

Figure 22. A. Aggrading wave ripple in prodelta facies of the Dunvegan Formation. **B**. Prodelta mudstones of the Kavik Formation, Prudhoe Bay Field, Alaska, containing soft-sediment deformation features (ball and pillow structure). **C**. Amalgamated graded beds in the Cretaceous Panther Tongue Delta, Utah, which may have been formed by sustained sediment gravity flows. (*See* Olariu and Bhattacharya, 2006; In **A**, scale is 3 cm).

successions described above or the tidal successions described below, but are typically thinner (i.e., a few meters thick).

Prograding downdrift chenier plains may produce mud-flat-dominated facies successions, punctuated by thin shell-rich or sandy beach deposits, alternating with organic, plant-rich swamp and marsh deposits (Kaczorowski, 1978). Muds may show extensive cracking and indications of a brackish to paralic infauna.

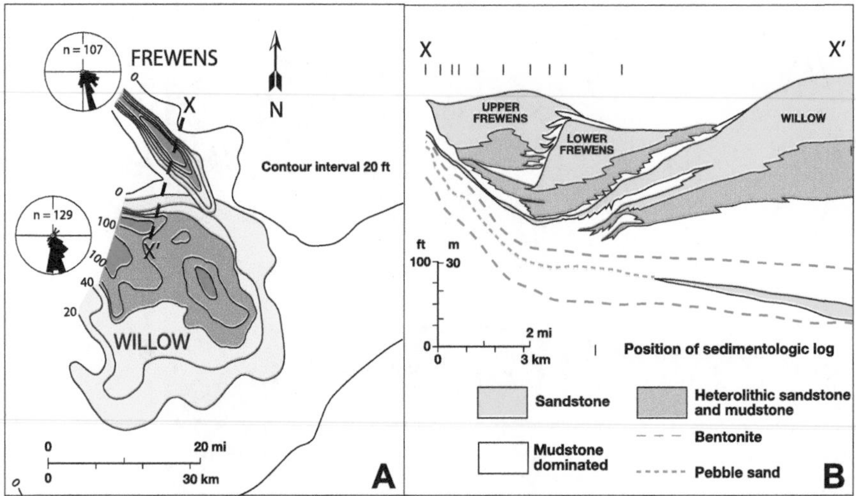

Figure 23. A. Map and **B.** cross section of delta lobes and fingers in the Cretaceous Lower Belle Fourche Member, Frontier Formation, Wyoming. **B**. Strike cross section shows overlapping lens-shaped delta bodies. Facies details of the Frewens tide-dominated delta sandstone are shown in Figures 18 and 24. (After Willis *et al.*, 1999; Bhattacharya and Willis, 2001).

Tide-influenced Delta Fronts

Tidally influenced delta fronts (summarized in Willis (2005), Dalrymple and Choi (2007), and Chapter 9) also show an overall coarsening-upward facies succession, but internally the facies may reflect tidal influence (Fig. 18), provided other processes, such as storms, do not mask or destroy the tidal lamination. The hallmark of tidal delta fronts is heterolithic strata, with ubiquitous mud drapes and mud layers throughout. Diagnostic tidal sedimentary features may include tidal bundles, double mud drapes and bimodal cross-stratification. Although river-dominated deltas may show heterolithic strata, mud layers typically are deposited seasonally, resulting in a more sporadic distribution. In tidal systems, mud commonly is deposited daily, resulting in a far more regular and rhythmic style of interbedding. Because stresses change daily, tide-dominated deltas typically show the lowest degree of burrowing of the three delta types. Tidally influenced deltas may also be modified by deep-tidal scours that might be mistaken for fluvial or distributary channel erosion surfaces (Willis, 2005).

The Cretaceous Frewens Allomember of the Frontier Formation in Wyoming provides a well-documented example of an ancient, tide-dominated deltaic sandstone body (Willis *et al.*, 1999; Willis, 2005; Figs. 18, 23 and 24). Upward-coarsening facies successions, 30 m thick, show an extremely low degree of burrowing (Fig. 18) and an extremely low diversity and abundance of marine microfossils indicating a high stressed, probably brackish-marine setting, similar to that observed by Dalrymple *et al.* (2003), in the modern Fly Delta. Tidal features abound, including heterolithic, wavy-bedded mudstones and rippled sandstones at the base, passing into thicker crossbedded sandstones having abundant double mud drapes and reactivation surfaces. The upward coarsening, the preferential seaward-directed cross-bedding and low degree of bioturbation indicates deltaic progradation, presumably as a result of the input of sediment and fresh water by a river; however, the tidal features throughout indicate significant tidal modulation of river discharge.

The top of the Frewens sandstone is characterized by meter-thick sets of angle-of-repose crossbeds, commonly floored by mud chips derived from thin mud drapes. Vertical cliffs expose seaward-dipping clinoforms (Fig. 24) interpreted as seaward-migrating, tidally influenced mouth bars (Willis *et al.*, 1999). Regional mapping of the Frewens sandstone shows that it forms two extremely elongated 'fingers', each a few kilometers in width and about 20 kilometers long (Fig. 23). These are very similar in scale to the elongate bars seen in the Ganges–Brahmaputra (Fig. 3F) and Fly deltas. The adjacent and older Willow Allomember forms a well-defined depositional lobe and lies seaward of the Frewens delta tongue (Fig. 23). It shows a greater proportion of wave-dominated shoreface deposits in the downdrift, southern region, but these pass northward into tidally influenced crossbedded sandstones, with well-developed double mud drapes and tidal bundles. This increase in tidal influence is attributed to the presence of a structural embayment in which tidal currents were enhanced and into which the Frewens Delta subsequently prograded. The Willow Allomember is thus a more mixed-energy, wave- and tide-influenced delta as opposed to the Frewens Delta, which shows ubiquitous tidally dominated stratification and a near-complete lack of wave-formed sedimentary structures, which reflects progradation into the highly protected embayment.

In the Frewens Delta, mud is preferentially deposited on the outer mouth bar and lateral margins, whereas the central area consists of tidally modified, sandy proximal mouth bars and bedforms (Fig. 24), although abundant mud lamina and mud chips are preserved even within the otherwise primarily sandy facies. The partitioning of mud along the lateral margins shows similarities to the tide-dominated estuary model of Dalrymple *et al.* (1992), but, unlike the estuarine facies model that depicts a transgressive system, thick offshore prodelta mudstones as well as the pronounced coarsening-upward record the regressive and thus deltaic nature of the Frewens Allomember.

Delta-plain Successions

Distributary Channels

Distributary channels within the non-

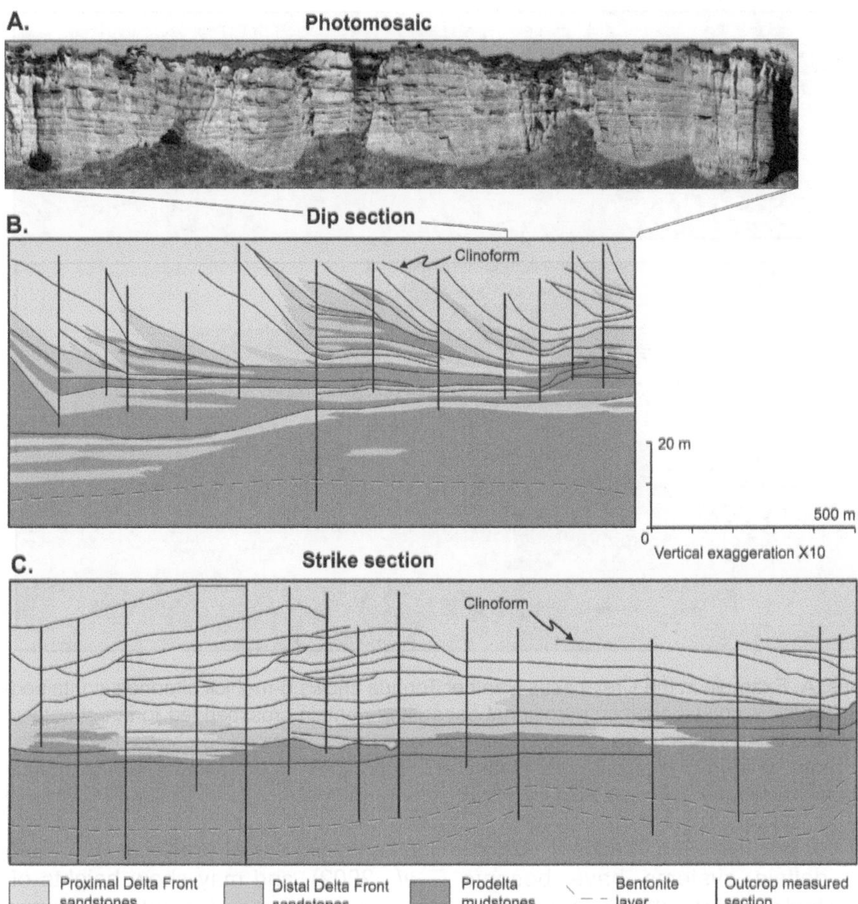

Figure 24. Outcrop showing complex internal architecture in the Cenomanian (Upper Cretaceous) tide-dominated river delta of the Frewens Allomember, Frontier Formation, central Wyoming. **A**. Photomosaic of part of the outcrop emphasizing the steepness of the clinoforms. **B**. Dip-oriented section of the prograding delta, showing the seaward-dipping clinoforms. **C**. Strike section showing bi-directional downlap, forming a classical lens-shaped geometry. In both cases, muddy bottomset facies interfinger with the sandy foreset facies forming a shazam-type facies boundary. Clinoform dip varies from 5–15°. (Note that **B** and **C** contain significant vertical exaggeration). These dips are considerably higher than described in most modern tide-influenced deltas, perhaps because of the very sandy nature of this example. Detailed facies shots and measured sections are shown in Figure 18. (Modified from Willis *et al*., 1999).

marine upper delta plain will look largely fluvial, and typically comprise erosionally based, fining-upward single-storey deposits (*see* Chapter 6), although they may show tidal influence. These should contrast with confined tributive or trunk incised-valley systems, which may comprise multi-storey, multilateral valley fills (*see* Boyd *et al*., 2006). In the lower delta plain, distributary channels may show brackish-water or tidal indications, such as tidal bundles, herringbone crossbedding, and sparse restricted marine trace and body fossils, depending on the degree of river dominance, the seasonal and longer term discharge variability. Although the salt-water wedge will migrate no farther landward than the bayline, tidal effects may be felt farther upstream. As a consequence, rhythmic alternations of mud and sand may be seen in wholly freshwater channels (Gastaldo *et al*., 1995; Dalrymple and Choi, 2007; van den Berg *et al*., 2007; *see* Chapter 9).

Small, shallow terminal distributary channels may be wholly contained within broader upward-coarsening delta-front deposits (Figs. 16 and 25; Olariu and Bhattacharya, 2006). Even in deeper water systems like the Mississippi, the active distributary channel comprises less than 30% of the total thickness of the overall coarsening-upward facies succession (Fig. 16B and C). Fining upwards is common in these channel deposits. Channels may be actively filled by sandy bar deposits or passively filled by mud, reflecting abandonment. Sandstones may be relatively structureless and poorly stratified, indicating rapid deposition and infilling (Gani and Bhattacharya, 2007; Olariu and Bhattacharya, 2006), or they may show well-developed sedimentary structures, and evidence of upstream accretion reflecting confined flow on the upstream margin of the mouth bar.

Interdistributary Areas

Interdistributary and interlobe areas commonly contain a series of relatively thin, stacked coarsening- and fining-upward muddy facies successions that may be capped by coal, carbonaceous shale, paleosols, or intertidal-flat deposits (Fig. 21). These successions are usually a few meters thick, and may not show coarsening-upward facies successions that are as pronounced as those found in prograding deltaic lobes (Elliott, 1974; Tye and Coleman, 1989). In the upper delta plain, these interdistributary areas are essentially alluvial floodplains (*see* Chapter 6) and, depending on the climate, may contain coals or carbonaceous shales (reflecting swamp and marsh environments), lacustrine deposits (containing lacustrine deltas and crevasse splays) and paleosols. In the lower delta plain, interdistributary bays will have a brackish to marine character, typically containing heterolithic strata with a reduced diversity and size of ichnofauna that will merge distally with the prodelta. Depending on the relative influence of tides, the heterolithic strata may or may not show rhythmic bedding. In some cases, as the delta progrades, these brackish-bay deposits may become completely isolated and closed off to become delta-plain lakes, swamps or marshes. In cores or outcrops, bay and proximal prodelta deposits may be indistinguishable, as they reflect intergradational environments.

The proportion of sandy lobe versus muddy inter-lobe facies successions will depend on the nature and type of delta system, the stability of distributary channels, and the

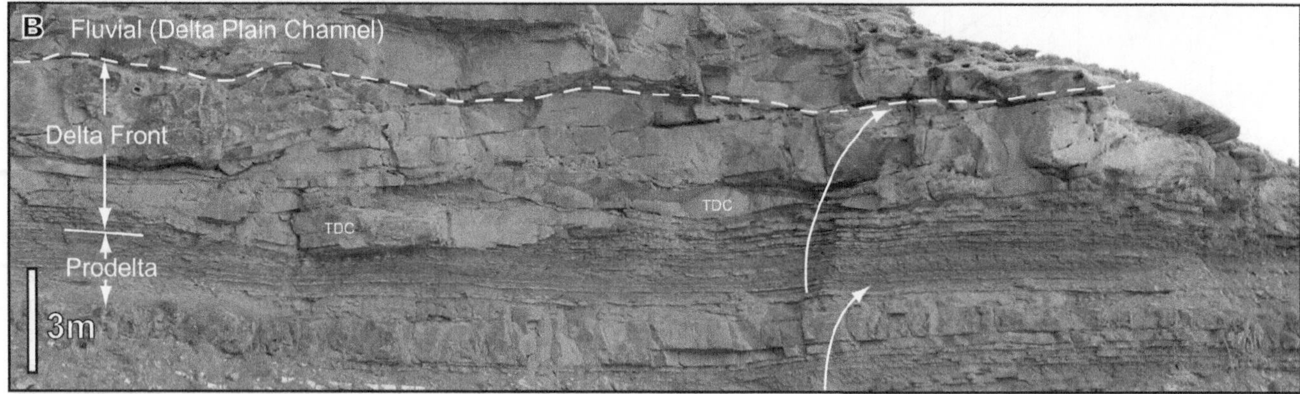

Figure 25. Photos of terminal distributary channels. **A**. Example in the Cretaceous Panther Tongue shows 6-m-thick channels contained within upward coarsening delta-front sandstone. Channels show undulating erosional bases and contain crossbedded to structureless sandstones and abundant mud-chips (*see* also Olariu and Bhattacharya, 2006). **B**. Cretaceous Ferron Sandstone shows 1–2-m-thick scoured channels within the upward coarsening sandy delta-front facies. The terminal distributary channels are primarily trough cross-bedded and pass upward into thicker distributary and trunk channels. *See* also Figure 16.

amount of non-marine accommodation. If the shoreline trajectory is upward (Fig. 20C), thick accumulations of mud-prone paralic and non-marine facies can accumulate behind an aggrading shoreface or delta front.

Wave-influenced systems, like the Po (Fig. 3C) and the Danube (Fig. 4), may contain significant lagoonal and bay mudstones in regions downdrift of the river mouth. River-dominated deltas may contain extensive interdistributary bays, especially in deltas like the Volga and the Mississippi where the delta lobes are very elongate. Tide-dominated deltas do not typically have interdistributary bays and are more likely to be flanked by abandoned distributary channels or by extensive tidal flats (Fig. 3F), dissected by tidal channels (Dalrymple and Choi, 2007).

FACIES ARCHITECTURE OF DELTAS

Bedding geometry and lateral facies variability can be addressed by the use of seismic data, ground-penetrating radar, continuous outcrop data, and correlation and interpolation of well data. Facies architectural studies of deltas significantly lag behind those of fluvial, deep-water and aeolian systems, but more architectural studies of deltaic systems have become available recently (Willis *et al.*, 1999; Mellere *et al.*, 2002; Chidsey *et al.*, 2004; Olariu and Bhattacharya, 2006; Gani and Bhattacharya, 2007).

Dip Variability

In cross-sectional dip view, deltas have historically been divided into topset, foreset and bottomset regions (Fig. 15), which are sometimes referred to by the term 'clinoform'. As discussed above, some deltas contain two foresets, a sandy foreset of the delta front or shoreface, and a less steeply dipping muddy foreset of the prodelta clinoform (Figs. 12 and 15).

Sandy delta-front foresets show dips that can range from a few degrees up to the angle of repose in Gilbert-type deltas (Fig. 26). These sandy foresets typically have a concave-upward profile (Figs. 24, 27 and 28). Shoreface elements (*see* Chapter 8) show dip angles typically less than 1°, whereas mouth bars typically form more steeply dipping clinoforms (Figs. 24, 25 and 26). By contrast, prodelta foresets, regardless of whether they are attached or disconnected (Fig. 15), slope at < 1°, have a more sigmoidal profile (*e.g.*, Liu *et al.*, 2002), and may show heights of up to a couple of hundred meters, although lesser heights are common. Topset facies show either undulating beds, associated with channel deposits or flat-lying deposits associated with coastal and delta-plain facies. (Fig. 28B).

The overall prograding clinoform geometry of deltas may consist of simple seaward-dipping, gently curved shingles or a more complex S-shaped sigmoidal pattern (Figs. 9, 15, 24 and 26–29) (Bhattacharya, 2006). This geometry can be seen in down-dip seismic profiles of modern and ancient deltas (Figs. 27 and 28). Many superb Quaternary examples have been documented (*e.g.*, Sidi *et al.*, 2003; Anderson and Fillon, 2004). This clinoform geometry can also be reconstructed in core and well-log cross sections (Figs. 29 and 30), and can be seen in many large outcrops (Figs. 24 and 26; Mellere *et al.*, 2002; Chidsey *et al.*, 2004; Gani and Bhattacharya, 2005).

In shelf-edge systems, clinoforms commonly steepen toward the shelf edge (Porebski and Steel, 2003, 2006; *e.g.*, Figs. 12, 27 and 28). This steepening reflects the fact that, at

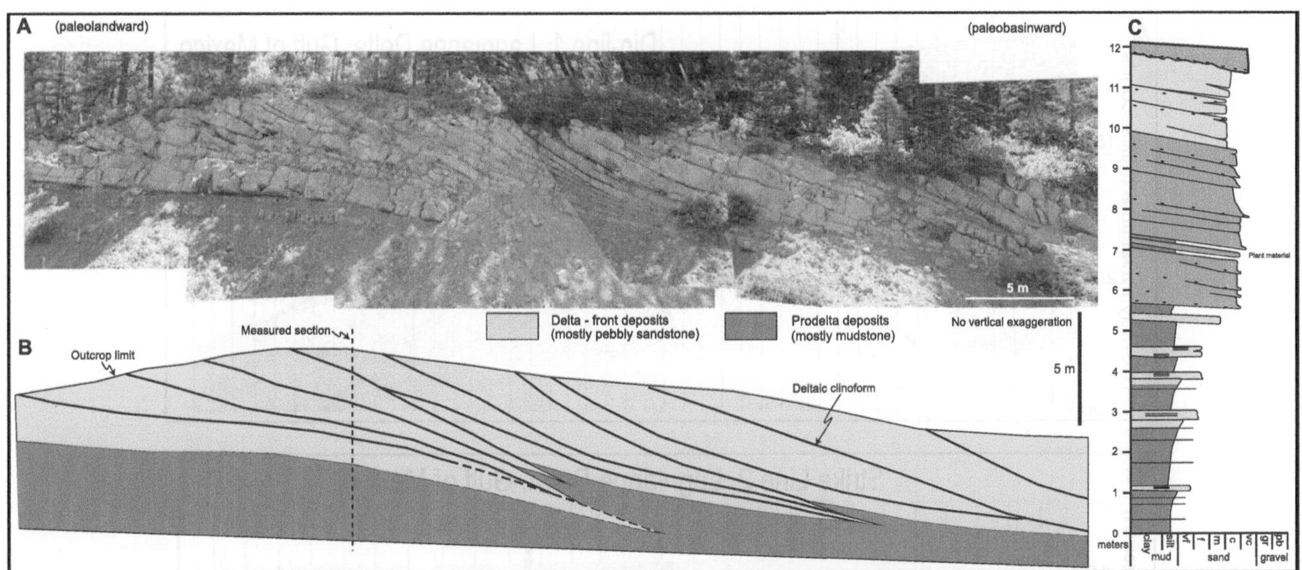

Figure 26. Details of facies interfingering at the base of a small-scale outcrop example of a gravelly, Pennsylvanian 'Gilbert' delta, Taos Trough, New Mexico. **A**. Outcrop photomosaic. **B**. Line drawing of bedding with facies interpretation. Note that clinoforms dip steeply seaward (average 13°). **C**. Lithologic column of this coarse-grained delta (position of the measured section is shown in **B**). Note that despite the apparent sharp base in the measured section, lateral interfingering between delta front and prodelta is clear, which indicates the danger of inferring a forced regression on the basis of a single measured section. From Gani and Bhattacharya (2005).

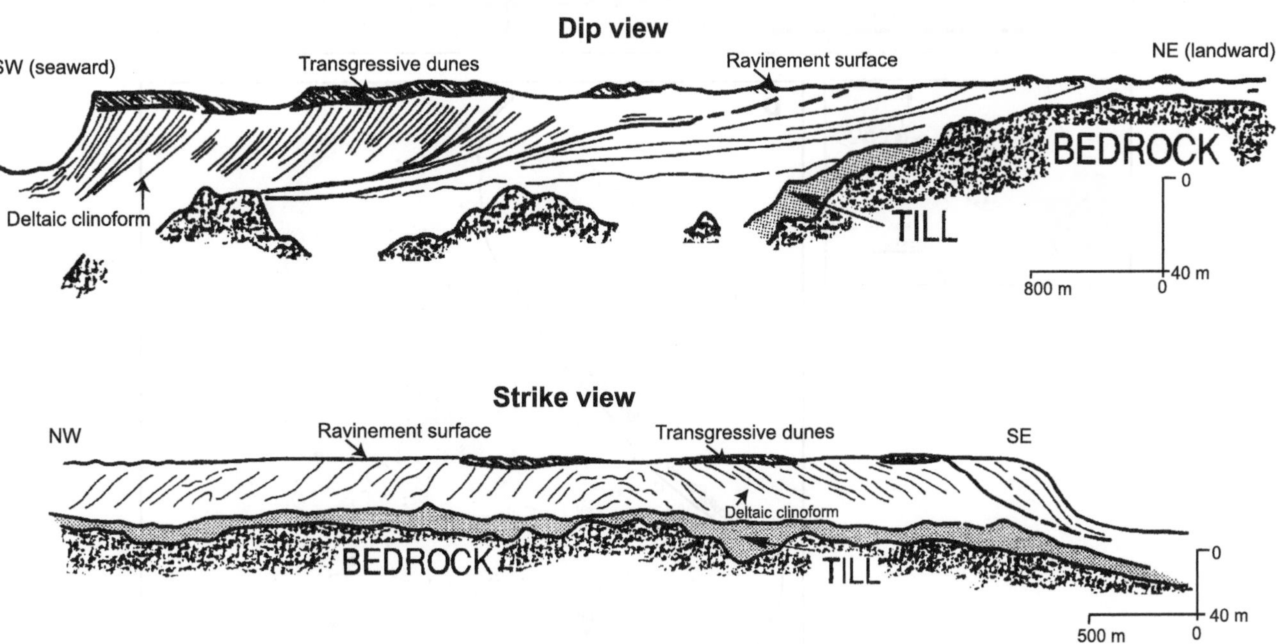

Figure 27. Dip- and strike-oriented sections showing bedding geometry of top-truncated, lowstand deltas, based on shallow seismic profiles collected seaward of the Natashquan River, Gulf of St. Lawrence, Canada. The reworked transgressive shelf sediments on top of the deltas, which presently lie in 20 m of water, represent a small fraction of the overall sediment body. Note the progressive increase in clinoform dip seaward. (Modified after Hart and Long, 1996).

the shelf edge, the delta-front slope merges with the continental slope. As accommodation exceeds the ability of the delta to fill the space, its front first steepens and then fails, supplying sediment to the deeper slope and basinal environment (*e.g.*, Figs. 10 and 20B; *see* Chapter 12). Clinoform gradients of shelf edge-deltas average between 1° and 8°, resulting in significant instability of shelf-edge sediments, where large-scale synsedimentary deformation features are common (Fig. 10). Slopes are typically much lower in deltas deposited on the mid- to inner-shelf, or in basins that lack a well-developed shelf edge. Soft-sediment deformation features may still be important in such low-slope basins, but are correspondingly smaller scale.

Strike Variability

Although the older seismic and sequence-stratigraphic literature is rife with dip-oriented cross sections,

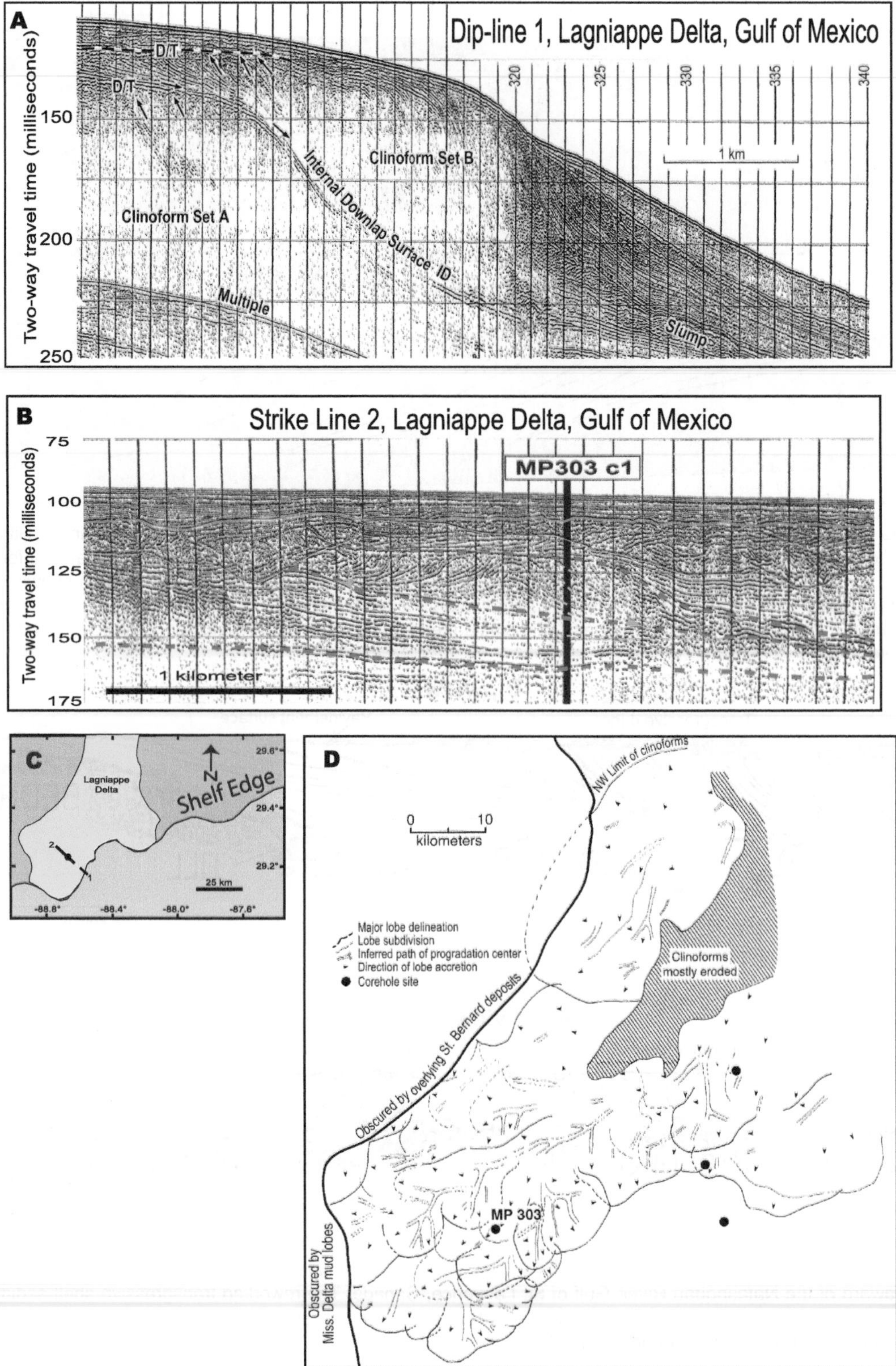

Figure 28. Seismic geometry of the Lagniappe Delta, Gulf of Mexico. **A**. Dip line showing seaward-dipping clinoforms. Internal downlap surface (ID) reflects downlap of successive delta lobes during progradation of the delta. Some of the downlap surfaces also show truncation (D/T surfaces) suggesting that they are longer-term transgressive surfaces separating successive deltas. **B**. Strike line showing lens-shaped ('mounded') delta lobes with bi-directional downlap and concave-upward bottomsets (pink lines). Mouth-bar assemblages are shown by green dashed lines. Convex-upward seismic facies (red) above the red erosion surface are fluvial-bar and distributary-channel deposits. Relatively flat-lying uppermost units, above the blue transgressive surface, are coastal and delta-plain muds. **C**. Map showing outline of lobe. **D**. Detailed mapping of seismic facies shows sub-lobes, bar-assemblages and associated terminal distributary channels. Core MP303 c1 shows a predominantly upward-coarsening facies succession. After Roberts *et al.* (2004).

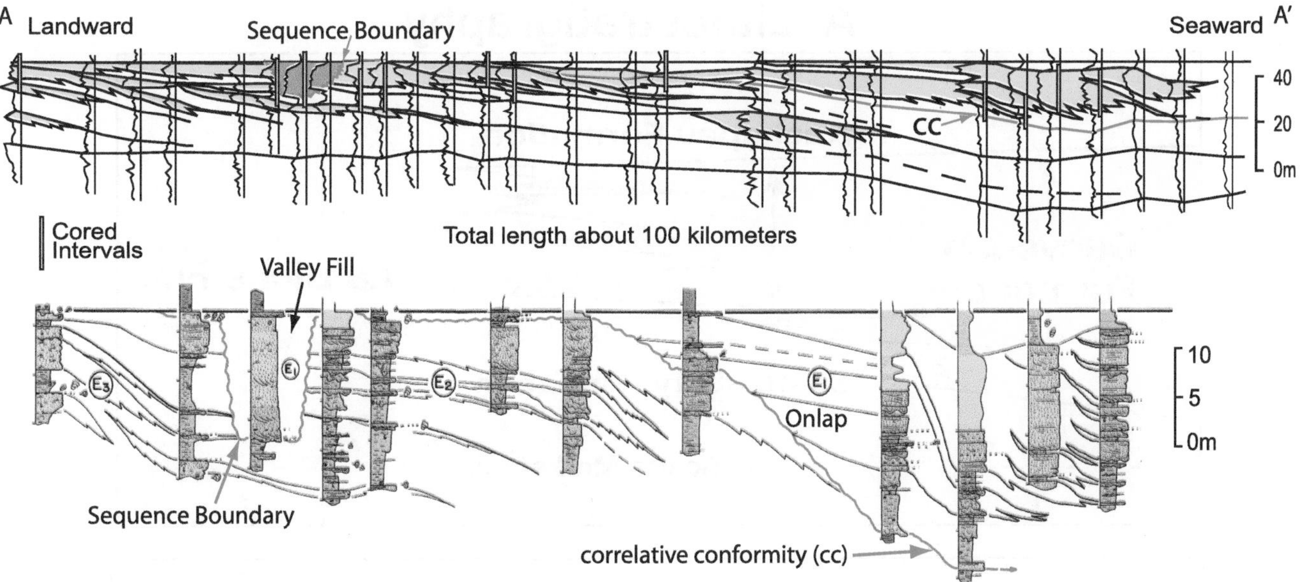

Figure 29. Dip-oriented well-log and core cross sections within Allomember E of the Upper Cretaceous Dunvegan Formation, showing offlapping clinoforms and sequence boundaries associated with the valley fill. The valley overlies a sequence boundary and is interpreted to feed the onlapping lowstand delta. This is a classic example of an attached lowstand (Ainsworth and Pattison, 1994). Distally, the sequence boundary passes into its correlative conformity underneath the lowstand delta. Location of cross section shown in Figure 33. Modified from Bhattacharya (1991).

newer studies emphasize strike-oriented variability (see examples in Anderson and Fillon, 2004). Strike variability (*i.e.*, the timing and spacing of overlapping lenses or lobes) is dependant on the number, spacing, and avulsion frequency of distributary channels. It also depends on seafloor bathymetry, especially if there is differential subsidence or uplift, whether related to tectonics or simple differential compaction, as deltas commonly fill low areas on the sea floor.

Along strike, sandy shingled delta-front foresets may pass into muddier sigmoidal clinoforms downdrift (Fig. 12). Overlapping delta lobes result in lens-shaped, bidirectionally downlapping units that exhibit a mounded appearance (Figs. 21D, 23B, 24C, 27 and 28B). Along-strike spacing may be a predictable outcome of 'compensation-style' deposition of successive lobes whereby a younger lobe tends to lie in the low area between two older lobes (*cf.* Fig. 7). Lobes in river-dominated deltas may be more pronounced than in other types of deltas because along-strike reworking is not significant and abrupt facies transitions can occur between distributaries and interdistributary areas (Bhattacharya, 1991).

SEQUENCE STRATIGRAPHY OF DELTAS

Allostratigraphy and sequence stratigraphy involve the correlation of bounding discontinuities through different rock types that yields a fundamentally different picture of genetic stratal relationships than older lithostratigraphic approaches (Fig. 30). Sequence-stratigraphic analysis allows far more accurate paleogeographic maps to be constructed (compare Figs. 31 and 32) because of the ability to subdivide the succession into smaller temporal increments. The inherently cyclic nature of deltaic sedimentation, wherein regressive phases are abruptly topped by transgressive units or flooding surfaces, produces upward-coarsening facies successions termed *parasequences* (Van Wagoner *et al.*, 1990; *see* Chapter 2) that are the dominant motif of deltaic systems (Fig. 19). When these short-term autogenic cycles (typically less than a few thousand years) are convolved with a longer term allogenic signal, such as Milankovitch-scale glacio-eustatic rises and falls of global sea level (10s of thousand of years), distinctive *parasequence sets* form that define *systems tracts* (*see* Chapter 2). The stacking pattern and facies development of the parasequences in the systems tract depends on the shoreline trajectory (Fig. 20; Catuneanu, 2006).

These concepts are best illustrated by example. Historically, the term delta has been generally applied to many clastic wedges, such as the Devonian–Carboniferous-age Catskill Delta wedge in the northeastern USA, and the Cretaceous Dunvegan, Ferron and Frontier formations in Western North America (Bhattacharya, 2006).

The Dunvegan Delta

The Dunvegan Formation of Alberta represents a heterolithic wedge of mudstones, sandstones and conglomerates up to 300 m thick, deposited from the actively rising Western Cordillera into the adjacent Cretaceous Interior Seaway. The term 'Dunvegan Delta' has been applied to this entire undifferentiated sedimentary package (Figs. 30A and 31). A previous map of all sandstones within the lithostratigraphically defined wedge (Fig. 31) shows several broadly lobate bodies with greater thickness, but it was practically impossible to determine which lobe belonged to which channel, without more detailed sequence-stratigraphic correlations.

In a study area of about 100 000

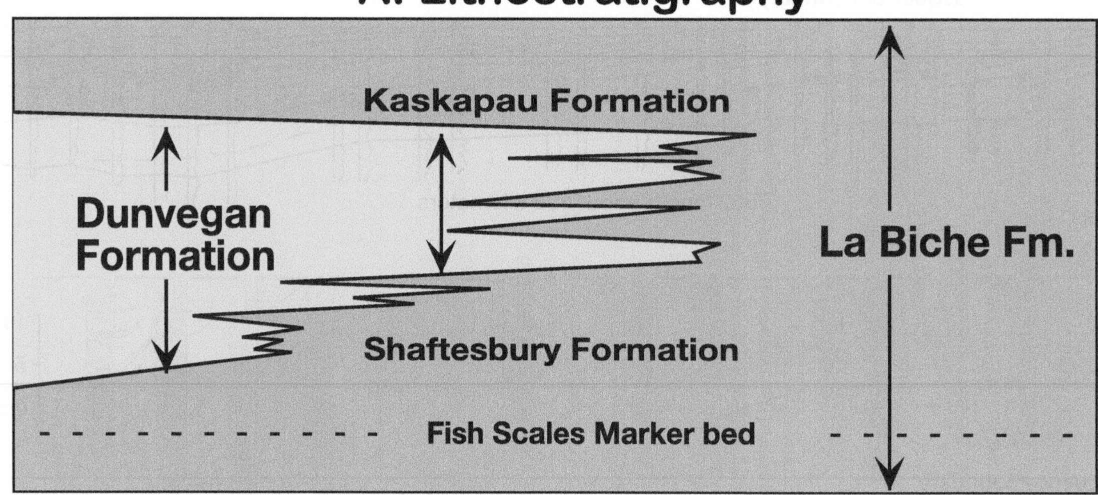

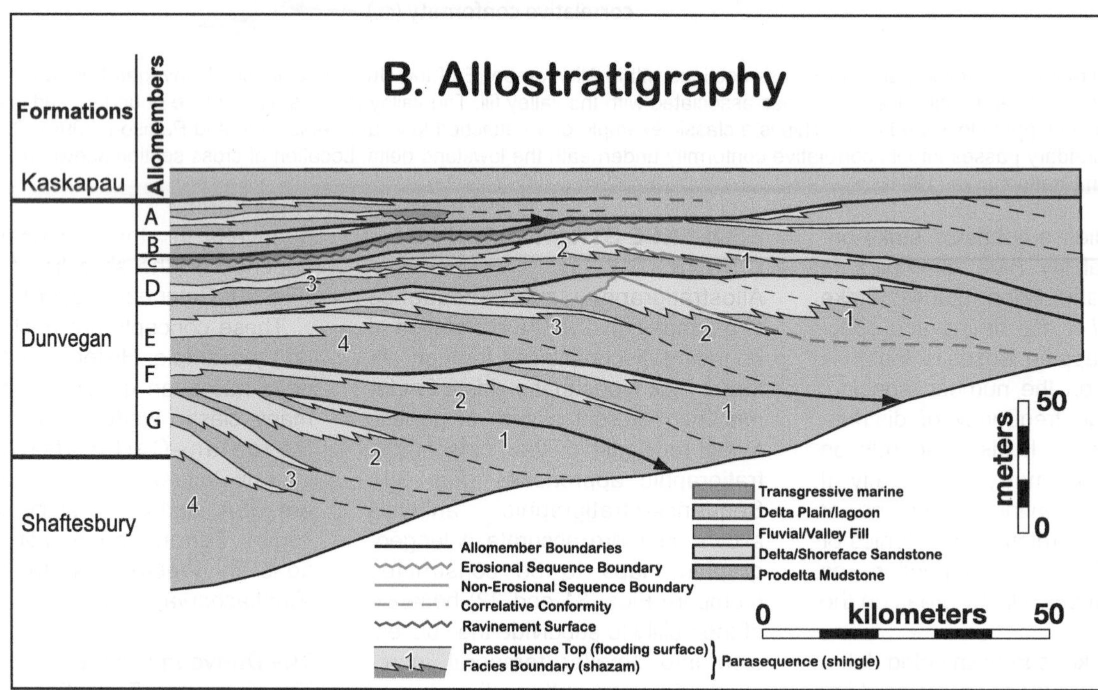

Figure 30. Cross section across the Alberta Foreland Basin illustrating the difference between a lithostratigraphic (**A**) and an allostratigraphic (**B**) interpretation of the Upper Cretaceous Dunvegan Formation (Bhattacharya, 1994). The lithostratigraphic representation depicts a homogenous, wedge-shaped sandstone body that tapers to the east. Some inter-fingering of the distal end of the Dunvegan Formation into the La Biche Formation shales is shown. The allostratigraphic interpretation shows that the Dunvegan comprises several stacked, disconformity-bounded allomembers (A to G). Each allomember consists of several smaller scale, offlapping, shingled units that map as delta lobes (*e.g.*, Fig. 32). Each allomember is bounded by a regional transgressive flooding surface. These regional flooding surfaces and smaller scale 'shingle' boundaries show that the Dunvegan consists of numerous sandy-compartments, bounded by mudstones. Some of these sandstones, such as in G1 and F are apparently completely surrounded by shale and thus may form separate reservoir compartments. Cross section located in Figure 33.

km^2 in the subsurface of Alberta, Bhattacharya (1994) and Plint (2000) recognized several through-going transgressive surfaces (Fig. 30B). These bounding discontinuities were used to subdivide the Dunvegan wedge into several allomembers (A-G; Fig. 30B), each of which could be subdivided further into shingled offlapping parasequences (Figs. 29 and 30B). Several incised valleys, previously thought to be mere distributary channels, were recognized within the allomembers. They rest on sequence boundaries and define depositional sequences and systems tracts (Bhattacharya, 1994; Plint and Wadsworth, 2003; Figs. 29, 30, 32 and 33). Cross sections (Figs. 29 and 30) show that the lowstand deltas are attached to the rest of the highstand wedge (Ainsworth and Pattison, 1994).

The new sequence-stratigraphic interpretation allowed mapping of paleoenvironments within individual parasequences (Figs. 32 and 33) as well as linking of specific delta systems to their fluvial feeder systems. The more river-dominated deltaic sys-

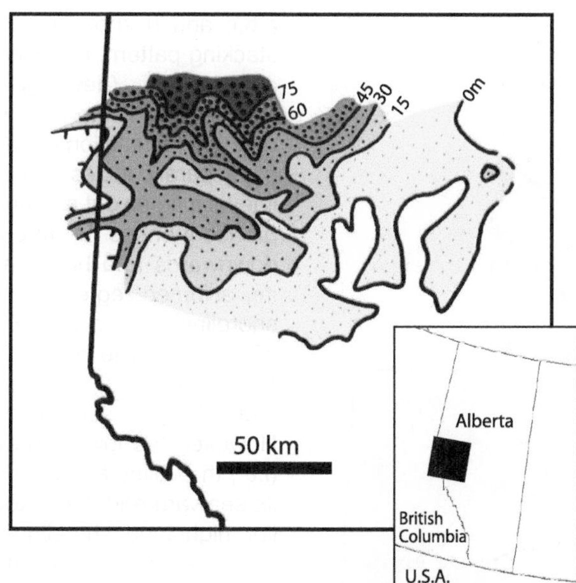

Figure 31. Sandstone isolith map of the undifferentiated Dunvegan clastic wedge (after Burk, 1963).

tems in the lower allomembers (E, F, and G, Fig. 30B) represent deposition during an overall late highstand to lowstand of relative sea level, whereas the barrier-shoreface systems in the younger allomembers (A-D, Fig. 30B) indicate deposition during a time of overall transgression and highstand of sea level. Each allomember records an early history of progradation with a positive shoreline trajectory, followed by downstepping indicating a negative trajectory, or forced regression, with the consequent formation of an incised valley and deposition of a lowstand delta (Fig. 34).

More recent work (Plint and Wadsworth, 2003) in the landward realm has demonstrated landward thickening between the younger allomember boundaries, suggesting

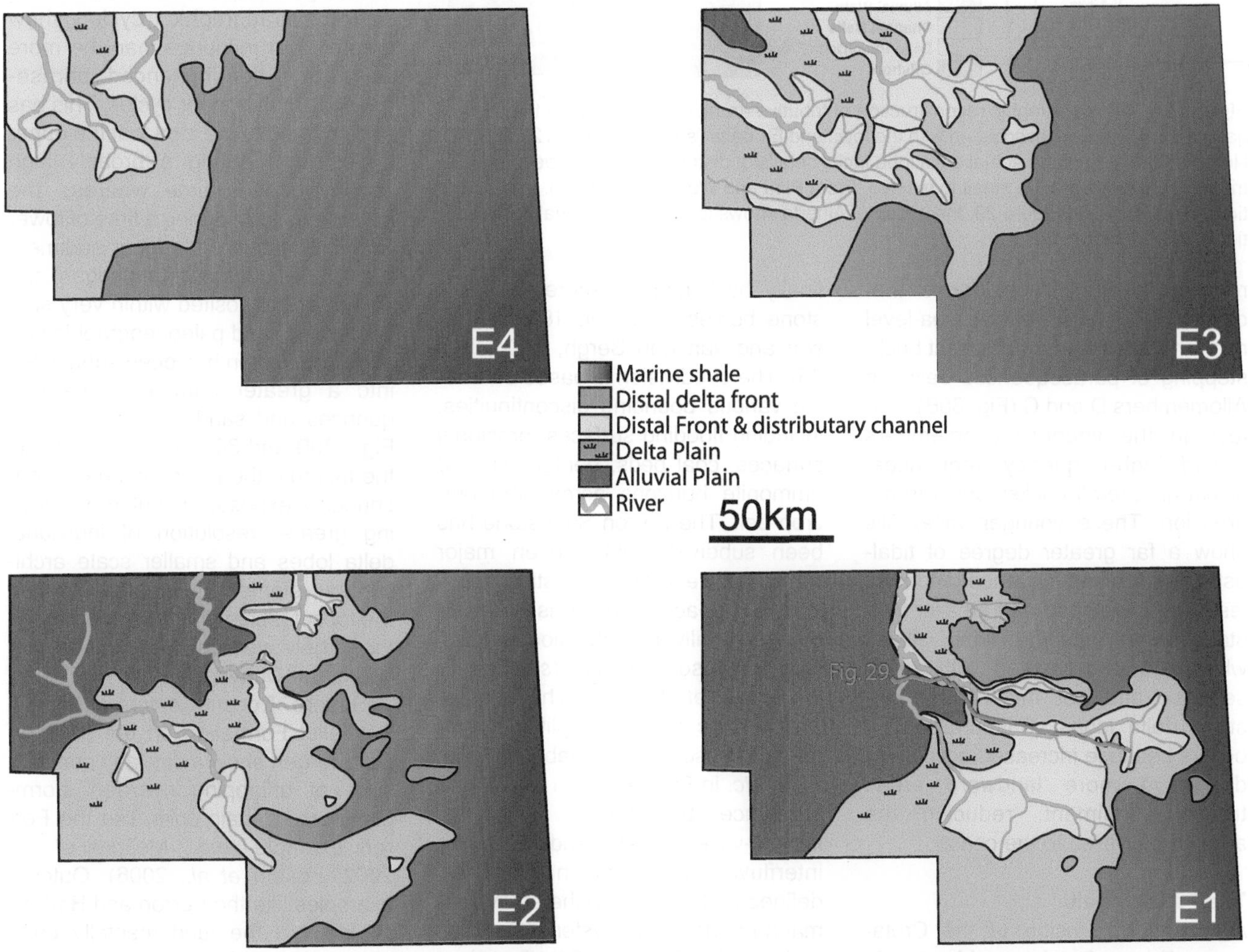

Figure 32. Paleogeographic maps of successive offlapping parasequences within Allomember E of the Dunvegan Formation (compare with Fig. 31). Note numerous small deltas associated with the highstand parasequences E4 to E2. The youngest parasequence, E1, is accompanied by fluvial entrenchment, development of an incised valley, and formation of a single, large delta lobe. Parasequence E1 is thus interpreted as a lowstand delta. The E1 valley system has been mapped by Plint and Wadsworth (2003) and is shown in Figure 33. From Bhattacharya (1991).

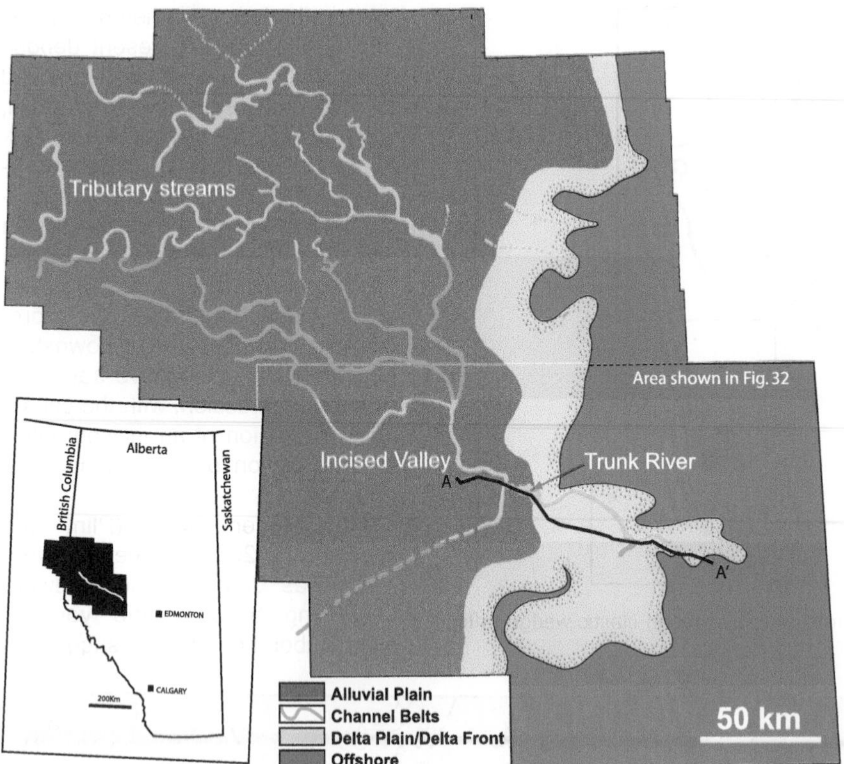

Figure 33. Paleogeographic reconstruction of tributive–trunk fluvial system in the Dunvegan drainage network associated with the lowstand E1 delta (*see* also Figures 29 and 32). Note that the map of the actual delta lobe lacks details of distributive pattern because distributary channels are too small to resolve. From Plint and Wadsworth (2003). Cross section A-A' is shown in Figure 29. Inset map of Alberta shows location of regional cross section shown in Figure 30.

renewed tectonic subsidence that caused the overall relative sea-level rise. This results in a significant backstepping of parasequences between Allomembers D and C (Fig. 30B). Valleys in the younger allomembers record high-frequency sequences, within an overall backstepping transgression. These younger valley fills show a far greater degree of tidal-estuarine, marine influence, and feed less well-developed and smaller lowstand deltas than the older valleys, which are largely filled with fluvial sediment and feed much larger lowstand delta lobes. This presumably occurs because increased accommodation in more landward areas trapped sediment, reducing the amount delivered to the coast.

The Ferron 'Delta'

A recent subdivision of the Cretaceous Ferron Sandstone Member in central Utah, based on nearly continuous outcrop exposures, shows a complicated series of seaward-stepping, offlapping, to aggrading and finally backstepping shoreline sandstone bodies (Gardner, 1995; Garrison and van den Bergh, 2004; Fig. 35). The subdivision is based on tracing various bounding discontinuities, including flooding surfaces, erosional surfaces, coal beds, bentonites, and ammonite horizons across the outcrop belt. The Ferron Sandstone has been subdivided into seven major transgressive–regressive 'stratigraphic cycles', each of which is bounded by a regionally traceable flooding surface and associated coal (*see* papers in Chidsey *et al.*, 2004). These larger stratigraphic cycles are split into numbered parasequences (labeled 1a,b, 2a,b, *etc.* in Figure 35). In addition, 5 sequence boundaries, linked to incised-valley systems and correlative interfluve paleosols have been defined and traced into the equivalent marine parts of the system (Garrison and van den Bergh, 2004). Examination of the parasequences shows that they can be organized into sets that indicate changing shoreline trajectory. For example, parasequences 1z, a,b,c and d show a progradational stacking pattern, but also show some aggradation (defining a positive shoreline trajectory), with consequent accumulation of delta-plain and coaly paralic facies behind the shoreline as it steps out and up. In contrast, parasequences 2f through 2j step downwards and basinwards, indicating a forced regressive, or negative, shoreline trajectory (*see* also Figures 20 and 34). The base of 2f also passes landward into an incised valley, indicating a sequence boundary. However, the lowstand systems tract (*i.e.*, the valley and the delta lobes at its seaward end) lie on top of the earlier highstand shorelines of parasequences 2a to 2d, indicating an attached lowstand (Ainsworth and Pattison, 1994).

The strongly seaward-stepping deltaic parasequences in the lowest Ferron stratigraphic cycle show greater river influence than the more wave and tide-influenced parasequences in the upper cycles. This has been related to the idea that the upper cycles form during a broadly high accommodation time whereas the lower units form during a time of lower accommodation and higher sediment supply. Although the Dunvegan and Ferron are deposited within very similar tectonic and paleogeographic settings, the Ferron has been subdivided into a greater number of parasequences and sand bodies (compare Figs. 30B and 36). This likely reflects the fact that the Ferron is nearly continuously exposed in outcrop, allowing greater resolution of individual delta lobes and smaller scale architectural elements, especially within the topset facies, than is possible with the largely well-log based cross-section of the Dunvegan.

Superb Eocene outcrops of the Battfjellet Formation on the coast of Spitzbergen show a similarly complex suite of offlapping river-fed storm-dominated deltaic units, like the Ferron example (*e.g.*, Mellere *et al.*, 2002; Løseth *et al.*, 2006). Outcrop examples like the Ferron and Battfjellet indicate the fundamentally complex nature of facies compartmentalization within fluvio-deltaic reservoirs.

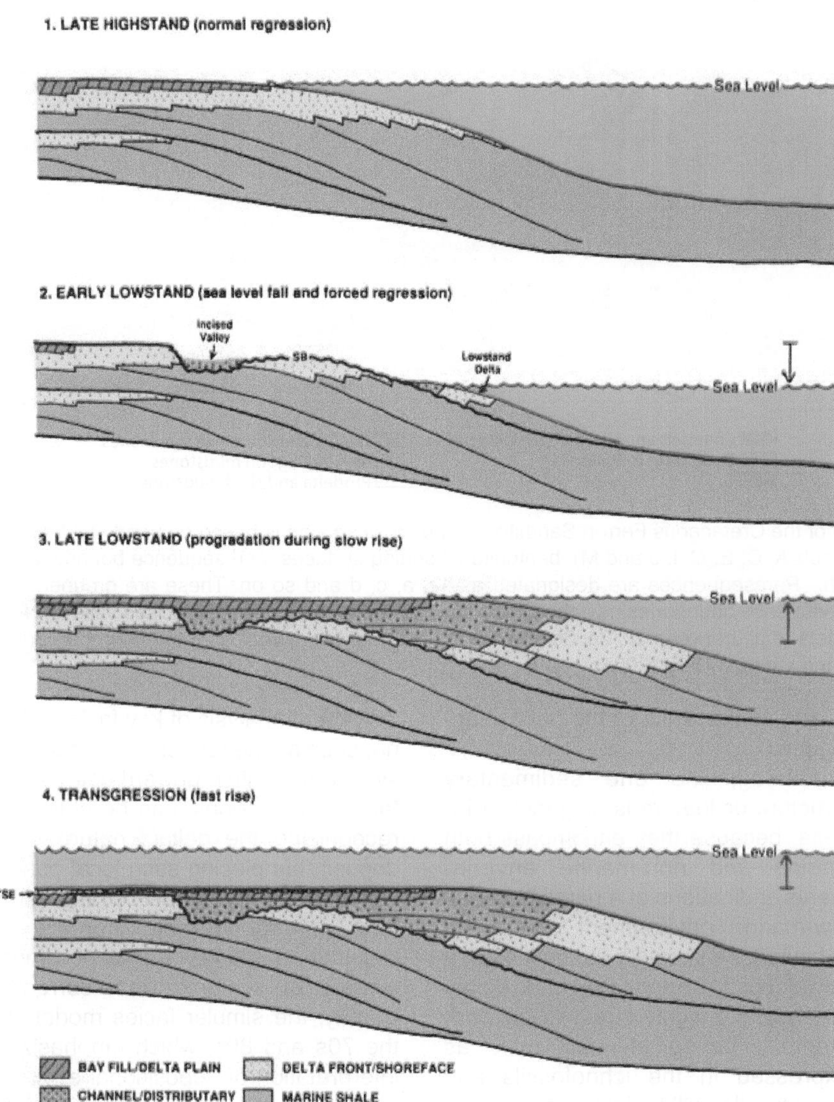

Figure 34. Cross-sectional evolution of a Dunvegan Allomember, such as shown in Figure 29. Deposition begins with late highstand regression (1), with an upward shoreline trajectory and deposition of delta-plain facies. This is followed by the early lowstand (2), characterized by development of a forced regression with fluvial incision and a downstepping trajectory. During late lowstand aggradation (3) there is a return to a positive trajectory, with infilling of the incised valley and deposition of more coastal plain facies. Note that this represents an 'attached lowstand'. This is followed by transgression (4), with a transgressive surface of erosion (TSE) that removes some of the delta-plain topset facies. Transgressive erosion, in this case, is not sufficient to produce a detached lowstand.

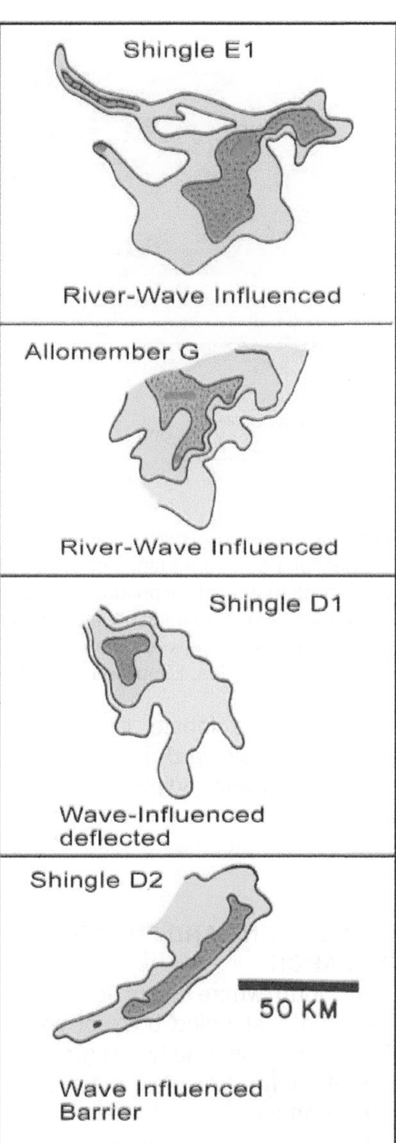

Figure 35. Spectrum of sandbody geometries shown by sandstone isolith maps of shingles (i.e., parasequences) within various allomembers in the Upper Cretaceous Dunvegan Formation (Alberta). Lower parasequences (G and E1) show greater fluvial influence, whereas upper parasequences (D2 and D1) are more wave influenced. From Bhattacharya and Walker (1992) based on data presented by Bhattacharya and Walker (1991). Mapping of isoliths within entire allomember does not reveal deltaic shape as clearly.

Top-truncated Deltas and the Shelf-sand Problem

Sequence stratigraphy has also been useful in the re-interpretation of seemingly isolated 'shelf' sandstones or 'offshore bars' as depositional remnants of top-truncated lowstand deltas and shoreface deposits (Snedden and Bergman, 1999; Bhattacharya and Willis, 2001; Martinsen, 2003; *see* also Chapter 8). These deposits comprise meters to tens-of-meters thick, upward-coarsening facies successions that are surrounded by 'shelf' mudstone. Broadly coeval paralic facies, such as typify the high-accommodation and unequivocally deltaic Dunvegan and Ferron sandstones described above, are commonly found 10s to 100s of kilometers landward.

The basinally isolated nature of these sediment bodies shows that they are detached lowstand deposits (Ainsworth and Pattison, 1994; *see* Chapter 8), with a high-degree of top-truncation. Although delta-plain, paralic and non-marine facies are eroded from the top of every succession by shoreline ravinement during transgression (*see* Chapter 11), sediment bodies nevertheless show internal evidence of a deltaic origin. The coarsening-upward facies successions resemble those described above, commonly showing a low to moderate degree of shallow-marine

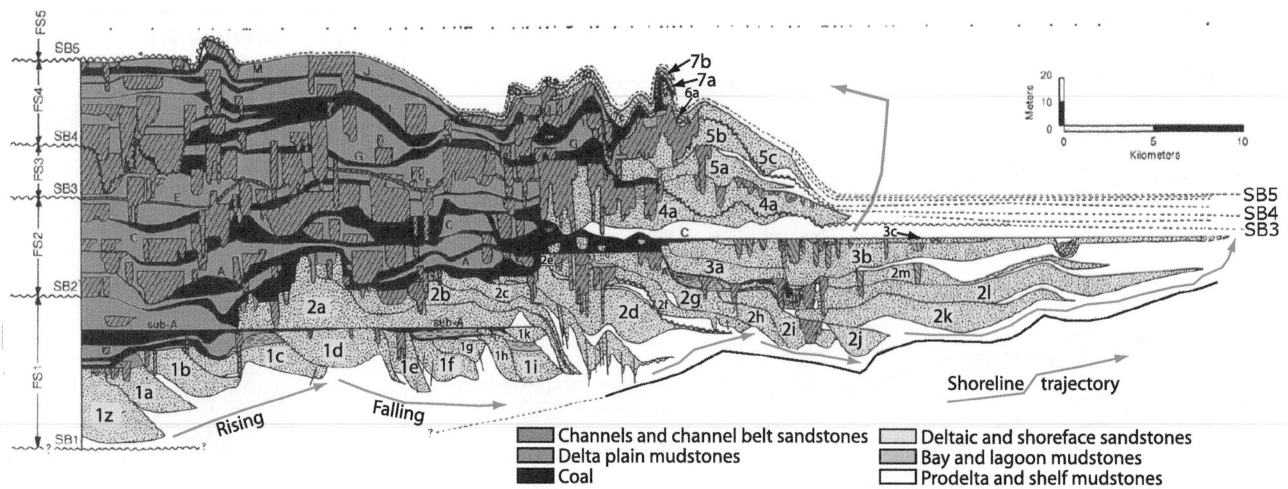

Figure 36. Sequence-stratigraphic cross section of the Cretaceous Ferron Sandstone. The cross section is based on nearly 100 measured sections and correlations of coals (lettered sub-A, C, E, G, I, J and M), bentonites, flooding surfaces, and sequence boundaries in nearly continuous cliff exposures in central Utah. Parasequences are designated as 1z, a, c, d and so on. These are grouped into parasequence sets. In addition, more recently, 5 sequence boundaries have been identified based on recognition of incised valleys. Note that the boundaries between the parasequence sets (*e.g.*, between 1k, 2a, 3c and 4a) are taken at major coal seams, rather than at the sequence boundaries. Simplified after Garrison and van den Berg (2004).

burrowing. The lobate to elongate geometry, basinward-dipping clinoform bedding and radiating paleocurrents, are also consistent with a deltaic origin (Bhattacharya and Willis, 2001).

CONCLUSIONS AND FUTURE RESEARCH

Deltas form where rivers feed sediment into a standing body of water. These fundamentally regressive deposits form broadly lobate sediment bodies that internally show coarsening upward facies successions as delta-front sands prograde over more distal prodelta muds. If such a progradation occurs during a drop of sea level, non-marine and paralic delta-plain facies are seldom preserved, forming top-truncated prodelta and delta-front deposits (Fig. 20). Where progradation occurs during rising sea level, delta-plain facies, including paralic wetland muds and distributary channels, may be preserved. River-supplied sediment can be reworked or remobilized by waves, which commonly carry sediments alongshore and can also cause significant delta-top erosion, especially during transgression. Tides typically dissect delta deposits into elongate tidal bars or ridges. In dip view, deltas form distinct seaward-dipping clinoform deposits, whereas in strike view they form bidirectional downlapping lens-shaped deposits, whose width relates to the width of the depositional lobe.

Although no one sedimentary structure or feature is diagnostic of a delta, because they encompass both marine and non-marine environments, indications of a paralic to shallow-marine environment should be exhibited, as should a connection to a fluvial point source. Because deltas are typically highly stressed environments, biological activity, as expressed in the ichnofossils and body fossils, will typically show a distinct diminution in size, diversity, and abundance compared to more stable marine environments.

Although river deltas may form depositional systems in their own right, this chapter emphasizes that, spatially, many deltas include significant components of other depositional systems, such as the barrier-lagoon and shoreface elements that are found in many wave-dominated deltas. River-fed deltas may also form as smaller components of other depositional systems, such as the bayhead deltas within estuaries. Prograding muddy chenier plains may form downdrift as a result of coast-parallel advection of river-derived mud plumes. Analysis of temporal variability also shows that, as deltas become abandoned, they may evolve from a regressive delta into a transgressive estuary, barrier-lagoon, shoreface or shelf deposit.

Careful analysis of key facies criteria, such as the reduction in trace-fossil diversity within upward coarsening facies successions, can be critical in recognizing the deltaic nature of a deposit, but placing such local observations into a broader stratigraphic and paleogeographic context is required to identify the depositional system or systems tract correctly. Clearly, the simpler facies models of the 70s and 80s, which emphasized interpretation of depositional environments using vertical profiles and codified facies schemes, are no longer adequate to characterize a complex, 3-D depositional system, such as a delta. Although basic descriptions of the rocks and construction of measured vertical sections are still an essential starting point in interpreting ancient sedimentary systems, a much more 3-D view is needed to encapsulate the known variability of deltaic deposits. Such facies analysis should incorporate concepts of facies preservation, bounding discontinuities and sequence stratigraphy, and facies architectural analysis. As with all other facies models, the important point is the understanding of how environments change as the controlling parameters change, rather than assigning arbitrary names.

The previous edition of *Facies Models* suggested that tide-dominated deltas did not exist and were better characterized as estuaries. Since then numerous studies of modern tide-dominated deltas, as well as several ancient examples, have allowed more general models to be proposed for this type of delta (*see* also Chapter 9). Also, virtually every example of a distributary channel presented in previous editions has been re-interpreted as an incised valley, forcing a re-evaluation of what distributary channels actually look like (Olariu and Bhattacharya, 2006).

Facies architectural analysis has proved to be a robust approach to the analysis of fluvial, deep-water, and eolian systems, but explication of coastal and shallow-marine systems remains nascent because of extreme facies complexity (*e.g.*, Reynolds, 1999; Gani and Bhattacharya, 2007). 3-D seismic geomorphological studies are revolutionizing our understanding of deep-water facies architecture, especially in channelized systems in which depositional elements have sharp boundaries, but many deltaic elements have gradational boundaries and remain poorly imaged in such data. Sequence-stratigraphic analysis provides a way to interpret time-stratigraphic evolution, but most studies emphasize dip-variability (*e.g.*, shoreline-trajectory concepts) and strike-variability is still poorly understood.

Scientific attention is now being drawn to the bounding mudstone facies, previously referred to as 'offshore shelf' deposits and otherwise largely ignored, but which now are increasingly recognized to be intimately linked to deltaic processes. Although muddy chenier plains lie downdrift of many modern deltas, there remain very few ancient examples of muddy deltaic coastlines or chenier plains (Walker and Harms, 1971; Hovikoski *et al.*, 2008).

Lastly, numerical and experimental flume modeling of deltas is an important and burgeoning field of research (*e.g.*, Paola, 200; Overeem *et al.*, 2005; Swenson, 2005; Syvitsky *et al.*, 2007, Muto *et al.*, 2007). These models attempt to predict coastal growth and decay of modern systems, as well as predict reservoir heterogeneity for subsurface applications. Integration of these models with data from field examples will be a fruitful approach to predicting the behaviour and complexity of deltaic depositional systems.

ACKNOWLEDGEMENTS

The author would like to acknowledge Robert Dalrymple for his thorough review and careful editing. My research on deltaic systems was made possible by funding to the Quantitative Sedimentology Consortium at the University of Houston by Anadarko, BP, Chevron, Nexen and Shell.

REFERENCES

Basic sources of information

Anderson J.B. and Fillon, R.H., *eds.*, 2004, Late Quaternary stratigraphic evolution of the Northern Gulf of Mexico Margin: SEPM Special Publication 79, 311 p.
Excellent compendium of papers, that outline the recent sequence-stratigraphic history and facies architecture of shelf-edge deltas in the Gulf of Mexico.

Bhattacharya, J.P., 2006, Deltas, *in* Walker, R.G., and Posamentier, H., *eds.*, Facies Models revisited: SEPM Special Publication 84, p. 237-292.
Recent lengthy summary of deltaic facies models, with an emphasis on facies architecture, and on which this paper is partly based.

Broussard, M.L., *ed.*, 1975, Deltas, Models for Exploration: Houston Geological Society, 555 p.
Dated but still key reference on deltas, both modern and ancient. Includes important synthesis papers by Galloway, and Coleman and Wright (see other references).

Chidsey, T.C., Adams, R.D. and Morris T.H. *eds.*, 2004, The fluvial-deltaic Ferron Sandstone: Regional-to-Wellbore-scale Outcrop Analog Studies and Applications to Reservoir Modeling: American Association of Petroleum Geologists, Studies in Geology, v. 50, 568 p.
Valuable case history of one of the best-studied ancient delta systems in the Cretaceous Interior of North America.

Colella, A. and Prior, D.B. *eds.*, 1990, Coarse-grained deltas: International Association of Sedimentologists, Special Publication 10, 357 p.
Key reference on coarse-grained deltas and fan-deltas.

Coleman, J.M. and Prior, D.B., 1982, Deltaic environments, *in* Scholle, P.A. and Spearing, D.R. *eds.*, Sandstone Depositional Environments: American Association of Petroleum Geologists Memoir 31, p. 139-178.
Mississippi-centric but still very useful reference on deltaic facies. Color illustrations of cores through the modern Mississippi are especially illustrative.

Fisher, W.L., Brown, L.F., Scott, A.J. and McGowen, J.H., 1969, Delta systems in the exploration for oil and gas, a research colloquium: Texas Bureau of Economic Geology, Austin, TX. 204 p.
Dated but still key reference on deltas that outlines the concept of depositional systems.

Giosan, L. and Bhattacharya, J.P., *eds.*, 2005, River Deltas: Concepts, Models and Examples: SEPM Special Publication 83, 502 p.
Recent compendium of papers on deltas including summary papers on tide-influenced deltas by Willis, numerical modeling by Overeem et al., *and ichnology of deltas by MacEachern* et al.

Helland-Hansen, W. and Gjelberg, J. G., 1994, Conceptual basis and variability in sequence stratigraphy: a different perspective: Sedimentary Geology, v. 92, p.31-52.
This is a fundamental and largely theoretical paper that outlines the basic concepts of shoreline trajectory and facies preservation as a function of sea-level change.

Mulder, T., Syvitski, J.P.M., Migeon, S., Faugeres, J.C. and Savoye, B., 2003, Marine hyperpycnal flows: initiation, behavior and related deposits: a review: Marine and Petroleum Geology, v. 20, p. 861-882.
The latest paper by Mulder and colleagues that synthesizes their thinking on how to recognize the deposits of river-fed hyperpycnal flows.

Nemec, W., 1995, The dynamics of deltaic suspension plumes, *in* Oti, M.N and Postma, G., *eds.*, Geology of Deltas: Balkema, Rotterdam, p. 31-93.
Thorough synthesis of delta-plume processes.

Nittrouer C. A., Kuehl, S. A., DeMaster D. J. and Kowsmann, R. O., 1986, The deltaic nature of Amazon shelf sedimentation: Geological Society of America Bulletin, v. 97, p. 444-458.
Important early paper documenting the extent of delta-linked muddy shelf deposits.

Olariu, C. and Bhattacharya, J.P., 2006, Terminal distributary channels and delta front architecture of river-dominated delta systems: Journal of Sedimentary Research, v. 76, p. 212-233, Perspectives, DOI: 10.2110/jsr.2006.026.
Synthesis of processes and facies of terminal distributary channels.

Orton, G. and Reading, H.G., 1993, Variabilty of deltaic processes in terms of sediment supply, with particular emphasis on grain size: Sedimentology, v. 40, p. 475-512.
Major synthesis of deltaic systems that examines and classifies deltas according to the role of sediment caliber in

controlling morphology and processes.

Porebski, S.J. and Steel, R.J., 2003, Shelf-margin deltas: their stratigraphic significance and relationship to deepwater sands: Earth Science Reviews, v. 62, p. 283-326.
Synthesis and review of shelf-margin deltas and their linkage to deep-water systems.

Reading, H.G. and Collinson, J.D., 1996, Clastic Coasts *in* Reading, H.G., *ed.*, Sedimentary Environments; Processes, Facies and Stratigraphy, 3rd Edition: Blackwell Science, Oxford, p. 154-231.
Advanced synthesis of clastic coasts, including deltas, by the originators of the facies-models concept. Still a valuable resource of information.

Reynolds, A.D., 1999, Dimensions of paralic sandstone bodies: American Association of Petroleum Geologists Bulletin, v. 83, p. 211-229.
Important attempt to compile data on the dimensions of deltaic and related sandstone bodies.

Sidi, F.H., Nummedal, D., Imbert, P., Darman, H. and Posamantier, H.W., *eds.*, 2003, Tropical Deltas of Southeast Asia – Sedimentology, Stratigraphy, and Petroleum geology: SEPM Special Publication 76, 269 p.
Much-needed compendium of papers on previously under-documented Asian delta systems. Many good examples based on 2-D seismic data and several good examples of tide-influenced deltas.

Whateley, M.K.G. and Pickering, K.T., *eds.*, 1989, Deltas: Sites and Traps for Fossil Fuels: Geological Society of London, Special Publication 41, 360 p.
Compendium of papers, including summary of the Nile delta by Sestini.

Wright, L.D., 1977, Sediment transport and deposition at river mouths: a synthesis: Geological Society of America Bulletin, v. 88, p. 857-868.
Important synthesis of river mouth processes and depositional features.

Other references

Ainsworth R.B. and Pattison, S.A.J., 1994, Where have all the lowstands gone? Evidence for attached lowstand systems tracts in the Western Interior of North America: Geology, v. 22, p. 415-418

Alexander, J., 1989, Deltas or coastal plain? With an example of the controversy from the Middle Jurassic of Yorkshire, *in* Whateley, M.K.G. and Pickering, K.T., *eds.*, Deltas: Sites and Traps for Fossil Fuels: Geological Society of London, Special Publication 41, p. 11-19.

Augustinius, P.G.E.F., 1989, Cheniers and chenier plains: a general introduction: Marine Geology, v. 90, p. 219-230.

Bates, C.D., 1953, Rational theory of delta formation: American Association of Petroleum Geologists Bulletin, v. 37, p. 2119-2162.

Bhattacharya, J.P., 1991, Regional to sub-regional facies architecture of river-dominated deltas in the Alberta subsurface, Upper Cretaceous Dunvegan Formation, *in* Miall, A.D. and Tyler, N., *eds.*, The Three-Dimensional Facies Architecture of Terrigenous Clastic Sediments, and its Implications for Hydrocarbon Discovery and Recovery: SEPM, Concepts and Models in Sedimentology and Paleontology, v. 3, p. 189-206.

Bhattacharya, J.P. 1994, Cretaceous Dunvegan Formation strata of the Western Canada Sedimentary Basin, *in* Mossop, G.D. and Shetsen, I., *eds.*, Geological Atlas of the Western Canada Sedimentary Basin: Canadian Society of Petroleum Geologists, and Alberta Research Council, p. 365-373.

Bhattacharya, J.P. and Davies, R.K., 2004, Sedimentology and structure of growth faults at the base of the Ferron Member along Muddy Creek, Utah, *in* Chidsey, T.C., Adams, R.D. and Morris, T.H., *eds.*. The Fluvial-Deltaic Ferron Sandstone: Regional-to-Wellbore-scale Outcrop Analog Studies and Applications to Reservoir Modeling: American Association of Petroleum Geologists, Studies in Geology 50, p. 279-304.

Bhattacharya, J.P. and Giosan, L., 2003, Wave-influenced deltas: geomorphological implications for facies reconstruction: Sedimentology, v. 50, p. 187-210.

Bhattacharya, J.P., and MacEachern, J.A., 2009, Hyperpycnal rivers and prodeltaic shelves in the Cretaceous seaway of North America: Journal of Sedimentary Research, v. 79. p. 184-209.

Bhattacharya, J. P. and Walker, R. G., 1992, Deltas, *in* Walker, R. G. and James, N. P., *eds.*, Facies Models: Response to Sea-level Change: Geological Association of Canada, p. 157-177.

Bhattacharya, J.P. and Willis, B.J., 2001, Lowstand deltas in the Frontier Formation, Powder River Basin, Wyoming; Implications for sequence stratigraphic models, U.S.A.: American Association of Petroleum Geologists Bulletin, v. 85, p. 261-294.

Boyd, R. and Penland, S., 1989, A geomorphic model for Mississippi delta evolution: Gulf Coast Association of Geological Societies, Transactions, v. 38, p. 443-452.

Boyd, R., Suter, J. and Penland, S., 1989, Sequence stratigraphy of the Mississippi delta: Gulf Coast Association of Geological Societies, Transactions, v. 39, p. 331-340.

Boyd, R., Dalrymple, R.W., and Zaitlin, B.A., 2006, Estuarine and Incised Valley Facies Models *in* Walker, R.G., and Posamentier, H., *eds.*, Facies Models revisited: SEPM Special Publication 84, p. 171-235.

Bowen, D.W. and Weimer, P., 2003, Regional sequence stratigraphic setting and reservoir geology of Morrow incised-valley sandstones (Lower Pennsylvanian): Eastern Colorado and western Kansas: American Association of Petroleum Geologists Bulletin, v. 87, p. 781-815.

Burk, C.E, Jr. 1963, Structure, isopach and facies maps of Upper Cretaceous marine successions, west-central Alberta and adjacent British Columbia: Geological Survey of Canada, Paper 62-31, 10 p.

Catuneanu, O., 2006, Principles of Sequence Stratigraphy: Elsevier, 336 p.

Coleman, J.M. and Gagliano, S.M., 1964, Cyclic sedimentation in the Mississippi River deltaic plain: Gulf Coast Association of Geological Societies, Transactions, v. 14, p. 67-80.

Coleman, J.M., Prior, D.B. and Lindsay, J.F., 1983, Deltaic influences on shelf edge instabilities, *in* Stanley, D.J. and Moore, G.T., *eds.*, The Shelf Break: Critical Interface on Continental Margins: SEPM Special Publication 33, p. 121-137.

Coleman, J.M. and Wright, L.D., 1975, Modern river deltas: variability of processes and sand bodies, *in* Broussard, M.L., *ed.*, Deltas, Models for Exploration: Houston Geological Society, Houston, TX, p. 99-149

Dalrymple, R.W., Zaitlin, B.A. and Boyd, R., 1992, Estuarine facies models: conceptual basis and stratigraphic implications: Journal of Sedimentary Petrology, v. 62, p.1130-1146.

Dalrymple, R. W., Baker, E. K., Harris, P. T. and Hughes, M. G., 2003, Sedimentology and stratigraphy of a tide-dominated foreland-basin delta (Fly River, Papua New Guinea), *in* Sidi, F.H., Nummedal, D., Imbert, P., Darman, H., and Posamantier, H.W., *eds.*, Tropical Deltas of Southeast Asia – Sedimentology, Stratigraphy, and Petroleum Geology: SEPM Special Publication 76, p. 147-174

Dalrymple, R.W. and Choi, K., 2007, Morphologic and facies trends through the fluvial–marine transition in tide-dominated depositional systems; A schematic framework for environmental and sequence-stratigraphic interpretation: Earth-Science Reviews, v. 81, p. 135-174.

Dominguez, J.M.L., 1996, The São Francisco strandplain: a paradigm for wave-dominated deltas? *in* De Baptist, M. and Jacobs, P., *eds.*, Geology of Siliciclastic Shelf Seas: Geological Society of London, Special Publication 117, p. 217-231.

Dominguez, J.M.L., Martin, L. and Bittencourt, A.C.S.P., 1987, Sea-level history and Quaternary evolution of river mouth-associated beach-ridge plains along the east-southeast Brazilian Coast: a summary, *in* Nummedal, D., Pilkey, O.H. and Howard, J.D., *eds.*, Sea-level Fluctuations and Coastal Evolution: SEPM Special Publication 41, p. 115-127.

Edwards, M.B., 1981, Upper Wilcox Rosita delta system of South Texas: growth-faulted shelf-edge deltas: American Association of Petroleum Geologists Bulletin, v. 65, p. 54-73.

Elliott, T., 1974, Interdistributary bay sequences and their genesis: Sedimentology, v. 21, p. 611-622.

Elliott, T., 1986, Deltas, *in* H.G. Reading, *ed.*, Sedimentary Environments and Facies: Blackwell Scientific Publications, Oxford, p. 113-154.

Evamy, D.D., Haremboure, J., Kamerling, P., Knapp, W.A., Molloy, F.A. and Rowlands, P.H., 1978, Hydrocarbon habitat of Tertiary Niger delta: American Association of Petroleum Geologists Bulletin, v. 62, p. 1-39.

Fielding, C.R., 2006, Upper flow regime sheets, lenses and scour fills: Extending the range of architectural elements for fluvial sediment bodies: Sedimentary Geology. v. 190, p. 227-240

Galloway, W.E., 1975, Process framework for describing the morphologic and stratigraphic evolution of deltaic depositional systems, *in* Broussard, M.L., *ed.*, Deltas, Models for Exploration: Houston Geological Society, Houston, TX, p. 87-98.

Gani, M.R. and Bhattacharya, J.P., 2005, Bedding correlation vs. facies correlation in deltas: Lessons for Quaternary stratigraphy, *in* Giosan, L. and Bhattacharya, J.P., *eds.*, River Deltas: Concepts, Models and Examples: SEPM Special Publication 83, p. 31-48.

Gani, M.R. and Bhattacharya, J.P., 2007, Basic building blocks and process variability of a Cretaceous delta: Internal facies architecture reveals a more dynamic interaction of river, wave, and tidal processes than is indicated by external shape: Journal of Sedimentary Research, v. 77, p. 284-302.

Garrison, J.R., Jr. and van den Bergh, T.C.V., 2004, The high-resolution depositional sequence stratigraphy of the Upper Ferron Sandstone Last Chance Delta: an application of coal zone stratigraphy, *in* Chidsey, T.C., Adams, R.D. and Morris, T.H., *eds.*, The Fluvial-Deltaic Ferron Sandstone: Regional to Wellbore Scale Outcrop Analog Studies and Applications to Reservoir Modeling: American Association of Petroleum Geologists, Studies in Geology 50, p. 125-192.

Gastaldo, R.A., Allen, G.P. and Huc, A.-Y., 1995, The tidal character of fluvial sediments of the modern Mahakam river delta, Kalimantan, Indonesia, *in* Flemming, B.W. and Bartholoma, *eds.*, Tidal Signatures in Modern and Ancient Sediments: International Association of Sedimentologists, Special Publication 24, p. 171-181.

Gilbert, G.K., 1885, The topographic features of lake shores: United States Geological Survey, Annual Report, 5th (1883 -1884), p. 69-123.

Hampson, G.J., and Howell, J.A., 2005, Sedimentologic and geomorphic characterization of ancient wave-dominated deltaic shorelines: Upper Cretaceous Blackhawk Formation, Book Cliffs, Utah, U.S.A., *in* Giosan, L., and Bhattacharya, J.P., *eds.*, River Deltas: Concepts, Models and Examples: SEPM Special Publication 83, p. 133-154.

Hart, B.S. and Long, B.F., 1996, Forced regressions and lowstand deltas: Holocene Canadian example: Journal of Sedimentary Research, v. 66, p. 820-829.

Hovikoski, J., Lemiski, R., Gingras, M., Pemberton,G. and MacEachern, J.A., 2007, Preliminary Notes on the Depositional Setting of the Late Cretaceous Alderson Member (Lea Park Formation), *in* Summary of Investigations 2007, V. 1, Saskatchewan Geological Survey, Saskatchewan Industry Resources, Miscellaneous Report, 2007-4.1, CD-ROM, Paper A-10, 11 p.

Hovikoski, J., Lemiski, R., Gingras, M., Pemberton, G., and MacEachern, J.A., 2008, Ichnology and Sedimentology of a Mud-Dominated Deltaic Coast: Upper Cretaceous Alderson Member (Lea Park Fm), Western Canada, Journal of Sedimentary Research; v. 78; p. 803-824.

Kaczorowski, R.T., 1978, The chenier plain and modern coastal environments, southwestern Louisiana: Houston Geological Society, Special Volume, p. 1-45.

Kineke, G.C., Woolfe, K.J., Kuehl, S.A., Milliman, J.D., Dellapenna, T.M. and Purdon, R.G. 2000, Sediment export from the Sepik River, Papua New Guinea: Evidence for a divergent dispersal system: Continental Shelf Research, v. 20, p. 2239-2266.

Kolb, C.R. and Van Lopik, J.R. 1966, Depositional environments of the Mississippi River deltaic plain, southeastern Louisiana, *in* Shirley, M.L. and J.A. Ragsdale, *eds.* Deltas: Houston Geological Society, Houston, TX, p.17-62.

Kuehl, S.A., Allison, M.A., Goodbred, S.L. and Kudrass, H., 2005, The Ganges-Brahmaputra delta. *in* Giosan, L. and Bhattacharya, J.P., *eds.*, River Deltas: Concepts, Models and Examples: SEPM Special Publication 83, p. 413-434.

Liu, J.P., Milliman, J.D. and Gao, S., 2002, The Shandong mud wedge and post-glacial sediment accumulation in the Yellow Sea: Geo-Marine Letters, v. 21, p. 212-218.

Løseth, T.M., Steel, R.J., Crabaugh, J.P. and Schellpeper, M., 2006, Interplay between shoreline migration paths, architecture and pinchout distance for siliciclastic shoreline tongues: evidence from the rock record: Sedimentology, v. 53, p.735-767.

Martinsen, R. S., 2003, Depositional remnants, Part 1. Common components of the stratigraphic record with important implications for hydrocarbon exploration and production: American Association of Petroleum Geologists Bulletin, v. 87, p. 1869-1882.

Mellere, D., Plink-Björklund, P. and Steel, R., 2002, Anatomy of shelf deltas at the edge of a prograding Eocene shelf margin, Spitsbergen: Sedimentology v. 49, p. 1181-1206.

Mulder, T. and Syvitsky, J.P.M., 1995, Turbidity currents generated at river mouths during exceptional discharge to the world's oceans: Journal of Geology, v. 103, p. 285-298

Muto, T., Steel, R. J. and Swenson, J. B., 2007, Autostratigraphy: A framework norm for genetic stratigraphy: Journal of Sedimentary Research, v. 77, p. 2-12.

Mutti, E., Tinterri, R., Benevelli, G., di Biase, D. and Cavanna, G., 2003, Deltaic, mixed and turbidite sedimentation of ancient foreland basins: Marine and Petroleum Geology, v. 20, p. 733-755

Neill, C.F., and Allison, M.A., 2005, Subaqueous deltaic formation on the Atchafalaya Shelf, Louisiana: Marine Geology, v. 214, p. 411-430.

Overeem, I., Syvitski, J.P.M. and Hutton, E.W.H., 2005, Three-dimensional numerical modeling of deltas, *in* Giosan, L. and Bhattacharya, J.P., *eds.*, River Deltas: Concepts, Models and Examples, SEPM Special Publication 83, p. 13-30.

Paola, C., 2000, Quantitative models of sedimentary basin filling: Sedimentology, v. 47, p. 121-178.

Panin, N., Panin, S., Herz, N. and Noakes, J.E. 1983, Radiocarbon dating of Danube Delta deposits: Quaternary Research, v.19, p.249-255.

Plint, A.G., 2000, Sequence stratigraphy and paleogeography of a Cenomanian deltaic complex: the Dunvegan and lower Kaskapau formations in subsurface and outcrop, Alberta and British Columbia, Canada: Bulletin of Canadian Petroleum Geology, v. 47, p. 43-79.

Plint, A.G., and Wadsworth, J.A., 2003, Sedimentology and palaeogeomorphology of four large valley systems incising delta plains, Western Canada foreland basin; implications for Mid-Cretaceous sea-level changes: Sedimentology, v. 50, p. 1147-1186.

Porebski, S.J. and Steel, R.J., 2006, Deltas and sea-level change: Journal of Sedimentary Research, v. 76, p. 390-403.

Postma, G., 1990. Depositional architecture and facies of river and fan deltas: a synthesis, *in* Colella, A. and Prior, D.B., *eds.*, Coarse-grained Deltas: International Association of Sedimentologists, Special Publication 10, p. 13 – 27.

Pratt, B.R., 1998, Syneresis cracks: subaqueous shrinkage in argillaceous sediments caused by earthquake-induced dewatering: Sedimentary Geology, v. 117. p. 1-10.

Pritchard, D. W., 1967, What is an estuary, physical viewpoint, *in* Lauf, G. H., *ed.*, Estuaries: American Association for the Advancement of Science, Washington D.C., Publication Number 83, p. 3-5.

Rine, J.M. and Ginsburg, R.N., 1985, Depositional facies of a mud shoreface in Suriname, South America. A mud analogue to sandy, shallow-marine deposits: Journal of Sedimentary Research, v. 55, p. 633-652.

Roberts, H.H., Fillon, R.H., Kohl, B., Robalin, J.M. and Sydow, J.C., 2004, Depositional architecture of the Lagniappe Delta; sediment characteristics, timing of depositional events, and temporal relationship with adjacent shelf-edge deltas, *in* Anderson, J.B. and Fillon, R.H., *eds.*, Late Quaternary Stratigraphic Evolution of the Northern Gulf of Mexico Margin: SEPM Special Publication 79, p. 143-188

Roberts, H.H. and Sydow, J., 2003, Late Quaternary stratigraphy and sedimentology of the offshore Mahakham Delta, East Kalimantan (Indonesia), *in* Hasan Sidi, F., Nummedal, D., Imbert, P., Darman, H. and Posamantier, H.W., *eds.*, Tropical Deltas of Southeast Asia – Sedimentology, Stratigraphy, and Petroleum geology: SEPM Special Publication 76, p. 125-145.

Scruton, P.C., 1960, Delta building and the deltaic sequence, *in* Shepard, F.P., Phleger, F.B. and van Andel, T.H., *eds.*, Recent Sediments Northwest Gulf of Mexico: American Association of Petroleum Geologists, Tulsa, OK, p. 82-102.

Snedden, J. W. and Bergman, K. M., 1999, Isolated shallow marine sand bodies: Deposits for all interpretations, *in* Bergman, K. M. and Snedden, J. W., *eds.*, Isolated shallow marine sand bodies: Sequence stratigraphic analysis and sedimentologic interpretation: SEPM Special Publication 64, p. 1-12.

Swenson, J.B., 2005, Relative importance of fluvial input and wave energy in controlling the timescale for distributary-channel avulsion, Geophysical Research Letters, 32, L23404, doi:10.1029/2005GL024758.

Syvitski, J.P.M., Pratson, L.F., Wiberg, P.L., Steckler, M.S., Garcia, M.H., Geyer, W.R., Harris, C.K., Hutton, E.W.H., Imran, J., Lee, H.J., Morehead, M.D. and Parker, G., 2007, Prediction of margin stratigraphy, *in* Nittrouer, C.A. *et al.*, *eds.*, Continental-Margin Sedimentation; From Sediment Transport to Sequence Stratigraphy: International Association of Sedimentologists, Special Publication 37, p. 459-530.

Ta, T.K.O, Nguyen, V.L., Tateishi, M., Kobayashi, I. and Saito, Y., 2005, Holocene delta evolution and depositional models of the Mekong River Delta, southern Vietnam, *in* Giosan, L. and Bhattacharya, J.P., *eds.*, River Deltas: Concepts, Models and Examples: SEPM Special Publication 83, p. 453-466.

Tye, R.S., 2003, Geomorphology; An approach to determining subsurface reservoir dimensions: American Association of Petroleum Geologists Bulletin, v. 88, p. 1123-1147.

Tye, R. S. and Coleman, J. M., 1989, Depositional processes and stratigraphy of fluvially dominated lacustrine deltas: Mississippi Delta Plain: Journal of Sedimentary Petrology, v. 59, p. 973-996.

Tyler, N. and Finley, R. J., 1991, Architectural controls on the recovery of hydrocarbons from sandstone reservoirs, *in* Miall, A. D. and Tyler, N., *eds.*, The Three-dimensional Facies Architecture of Terrigenous Clastic Sediments, and Its Implications for Hydrocarbon Discovery and Recovery: SEPM, Concepts and Models in Sedimentology and Paleontology 3, p.1-5.

Vakarelov, B.K., 2006, Controls on Stratigraphic Architecture in Shallow Marine Environments: Tectonic Forcing And Mud Geometries, University of Texas at Dallas, Ph.D. Dissertation, 122p.

van den Berg, J.H., Boersma, J.R. and van Gelder, A., 2007, Diagnostic sedimentary structures of the fluvial-tidal transition zone - Evidence from deposits of the Rhine and Meuse: Netherlands Journal of Geosciences, v. 86, p. 287-306

Van Heerden, I.L. and Roberts, H.H., 1988, Facies development of Atchafalaya delta, Louisiana: a modern bayhead delta: American Association of Petroleum Geologists Bulletin, v. 72, p. 439-453.

Van Wagoner, J.C., Mitchum, R.M., Campion, K.M. and Rahmanian, V.D., 1990, Siliciclastic sequence stratigraphy in well logs, cores, and outcrops: Tulsa, OK: American Association of Petroleum Geologists, Methods in Exploration Series 7, 55 p.

Walker, R.G. and Harms, J.C., 1971, The "Catskill Delta": a prograding muddy shoreline in central Pennsylvania: Journal of Geology, v. 79, p. 381-399.

Willis, B.J., 2005, Tide-influenced river delta deposits, *in* Giosan, L., and Bhattacharya, J.P., *eds.*, River Deltas: Concepts, Models and Examples: SEPM Special Publication 83, p. 87-129.

Willis, B.J., Bhattacharya, J.B., Gabel., S.L. and White, C.D, 1999, Architecture of a tide-influenced delta in the Frontier Formation of Central Wyoming, USA: Sedimentology, v. 46, p. 667-688.

11. TRANSGRESSIVE WAVE-DOMINATED COASTS

Ron Boyd, ConocoPhillips Company, Houston, Texas, 77079, USA
and University of Newcastle, Callaghan, 2308, NSW, Australia

INTRODUCTION

Transgressive wave-dominated coasts occur at the critical interface between land and sea and include such important components as estuaries, lagoons and barrier islands. This interface has special human significance as it is where a large proportion of the earth's population lives, including many of its major cities and ports such as New York, London, Tokyo, Sydney, Houston, Shanghai and Mumbai. Hence, it is also particularly sensitive to the potential impact of global warming and sea-level rise. In addition, significant economic resources occur in deposits formed in this setting, making them important targets for petroleum exploration and production. Such deposits also serve as aquifers for urban water supply.

Modern ocean coastlines experienced a sea-level rise of over 100 m from *ca.* 18 ka BP (kiloannum before present) to around 6 ka BP. As a result, most of the world's coastlines, away from major sediment sources, retain a transgressive form and much modern research has been focused on these coasts (*e.g.*, Dalrymple *et al.*, 1992, 1994, 2006; Boyd *et al.*, 1992, 2006). In contrast, fewer investigations have documented ancient transgressive coastlines (*e.g.*, Cattaneo and Steel, 2003; Coe, 2003; Davis, 1985).

CONCEPTS OF COASTAL ORGANIZATION AND CLASSIFICATION

The coastal interface between terrestrial and marine environments is constantly changing at both the human and geological time scales. This change is expressed conceptually as the interplay between coastal transgression and regression (Fig. 1). These terms are often confused, and here we define transgression as the relative landward motion of the shoreline and regression as its relative seaward motion. Transgression and regression are in turn the result of the interaction between coastal erosion/deposition and sea-level rise/fall (*e.g.*, Curray, 1964; *see* discussion on relationship between accommodation and sedimentation in Chapter 2). Note that transgressions and regressions do not result solely from sea-level changes, but are a composite response to the interaction of coastal sediment flux and changes in relative sea level resulting from the sum of tectonic and eustatic components. Hence, regression can occur under rising relative sea level if sufficient sediment volume is supplied to the coast, a condition termed 'normal regression' (*see* Chapter 2).

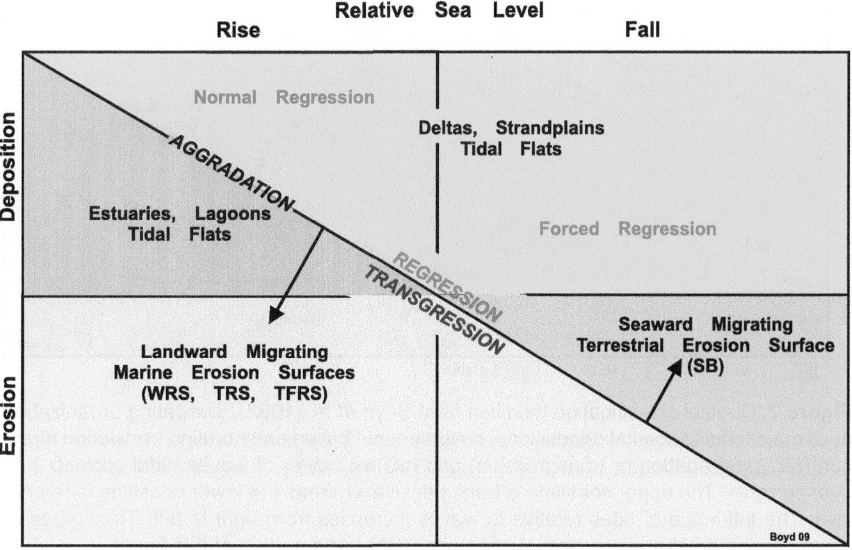

Figure 1. Relative sea level (a proxy for accommodation–A) plotted against erosion/deposition (a proxy for sediment flux–S). Diagonal line from bottom right to top left indicates a balance between relative sea level (RSL) and sediment flux such that the shoreline remains in the same location (A = S); aggradation occurs when deposition matches RSL rise. Conditions above this line result in regression while conditions below this line produce transgression. Blue zone indicates the situations under which estuaries, lagoons and transgressive tidal flats form, whereas the yellow zone indicates conditions where deltas, strandplains and regressive tidal flats occur. Erosional surfaces formed during transgression (*see* Figure 29) include the wave ravinement surface (WRS), tidal ravinement surface (TRS) and tidal/fluvial ravinement surface (TFRS). The main erosional surface formed during regression is the sequence boundary (SB). Modified after Curray (1964) and Boyd *et al.* (1992).

Classification of coastal geomorphology and sediments is often also accomplished by reference to the dominant physical process controlling the coastal zone, the two most important being waves and tidal currents. Davies (1964) was one of the first to recognize the significance of these two processes and used them to divide the world's coastlines into wave-dominated and tide-dominated categories. Subsequent studies confirmed Davies' approach (*e.g.*, Hayes, 1975) and also recognized that the key to understanding these coastal processes was to consider the ratio of wave to tidal power (*e.g.*, Harris *et al.*, 2002). Wave-dominated coasts are commonly found in areas of strong persistent winds and relatively low tidal power, such as between lati-

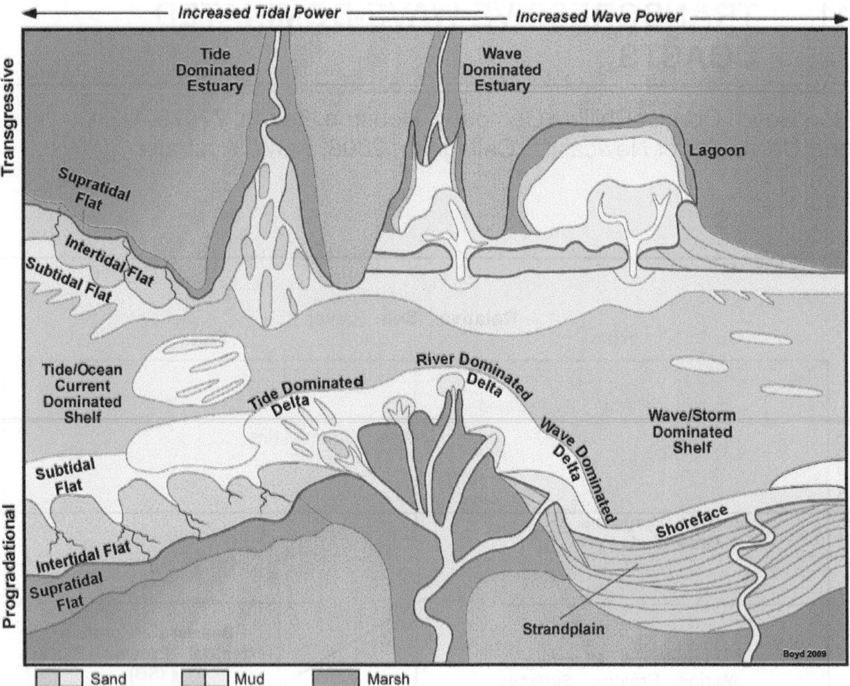

Figure 2. Coastal classification modified from Boyd *et al.* (1992), illustrating organization of all major clastic coastal depositional environments based on shoreline translation direction (*i.e.*, progradation or transgression) and relative power of waves, tidal currents and river currents. The upper coastline is transgressive whereas the lower coastline is regressive. The influence of tides relative to waves increases from right to left. Transgressive wave-dominated shorelines occupy the upper right-hand portion of this figure.

tudes 35–60° N and S, and particularly on west-facing shorelines in these temperate regions.

A sedimentologic and stratigraphic classification of coasts can be constructed by combining the concept of transgression and regression with the relative importance of waves and tides, as well as the influence of rivers (Fig. 2). On transgressing coasts, the wave-dominated region consists of eroding linear coastlines, barrier-lagoon shorelines occupying embayments with little fluvial input, and wave-dominated estuaries at the mouths of river valleys. Shelf shoals and ridges consisting of reworked shoreline deposits lie seaward of transgressing coastlines (Fig. 2). Landward of transgressing coastlines, major clastic deposition is typically constrained to river valleys.

TRANSGRESSIVE COASTAL GEOMORPHOLOGY

Transgressing coasts receive sediment from two sources: the rivers that reach the coast, often by way of the incised valleys that are created during periods of relative sea-level lowstand, and wave processes that erode pre-existing deposits as the coast moves landward.

The rivers that supply sediment to the coast create a variety of fluvial-channel and floodplain geomorphological elements in the area upstream of the tidal limit and contain channel, bar/sand flat, levee, crevasse splay and floodplain elements (*see* Figure 1, Chapter 6). Outside of the river valleys prior to transgression, most of the land surface is composed of an interfluve exposure surface (Fig. 3) that is characterized by dry, oxidized soils and vegetation. The first indication of the incoming transgression in rivers and valley-fill deposits is marked by the appearance of freshwater tidal conditions in the river channels, a situation where the flow reverses during each tidal cycle, but no saline water penetrates the region. Farther downstream in the saline tidal reach, river channels change their form from straighter to sinuous and back to straighter as they approach the estuary (*e.g.*, Dalrymple *et al.*, 1992; *see* Chapter 9). Mud can be deposited during slack-water periods, and fluvial sediment is deposited at the upstream end of estuaries in bayhead deltas (Figs. 4 and 5). These deltas are similar to the marine deltas described in Chapter 10, but are commonly more river-dominated and smaller. Like their marine counterparts, they contain a subaerial delta plain, distributary channels, mouth bars and a delta front. Fine-grained sediment, escaping the bayhead delta, is transported farther seaward into the estuarine central basin that occupies the low-energy region between the bayhead delta and the marine shoreline (Fig. 6). This low-energy zone results from the intersection in the central-basin region of a seaward decrease of river energy and a landward decrease of marine energy. The zone, where marine and fresh waters mix, is often an area of increased mud deposition coinciding with a turbidity maximum in the overlying estuarine water column (Figs. 3 and 6). Areas of brackish to saline water separated from the ocean by a sandy barrier but not occupying the site of former river valleys and deltas are termed lagoons (Figs. 2 and 7). Low-lying estuary margins are commonly occupied by tidal flats, marshes and beaches (Figs. 3 and 7).

At the seaward end of the system, marine sources of energy (*i.e.*, waves and tidal currents) and sediment control the coastal geomorphology, with

Figure 4. *(opposite page)* Example of a wave-dominated estuary, Port Stephens, New South Wales, Australia, showing a merged landscape/seascape DEM. This view illustrates the tripartite division into an outer marine-dominated flood-tidal delta and barriers (right; pink and yellow), a deeper central basin (middle; blues) and an inner fluvially dominated river valley and bayhead delta (Karuah River upper left; pink and yellow). Depth color bar at right in meters below sea level. Original bathymetry constructed from 65 000 digitized leadline soundings. Note the fan-shaped sandy flood-tidal delta channels (*cf.* Figure 11) and the muddy, flat, bayhead prodelta. The mounded features in the central basin are oyster reefs.

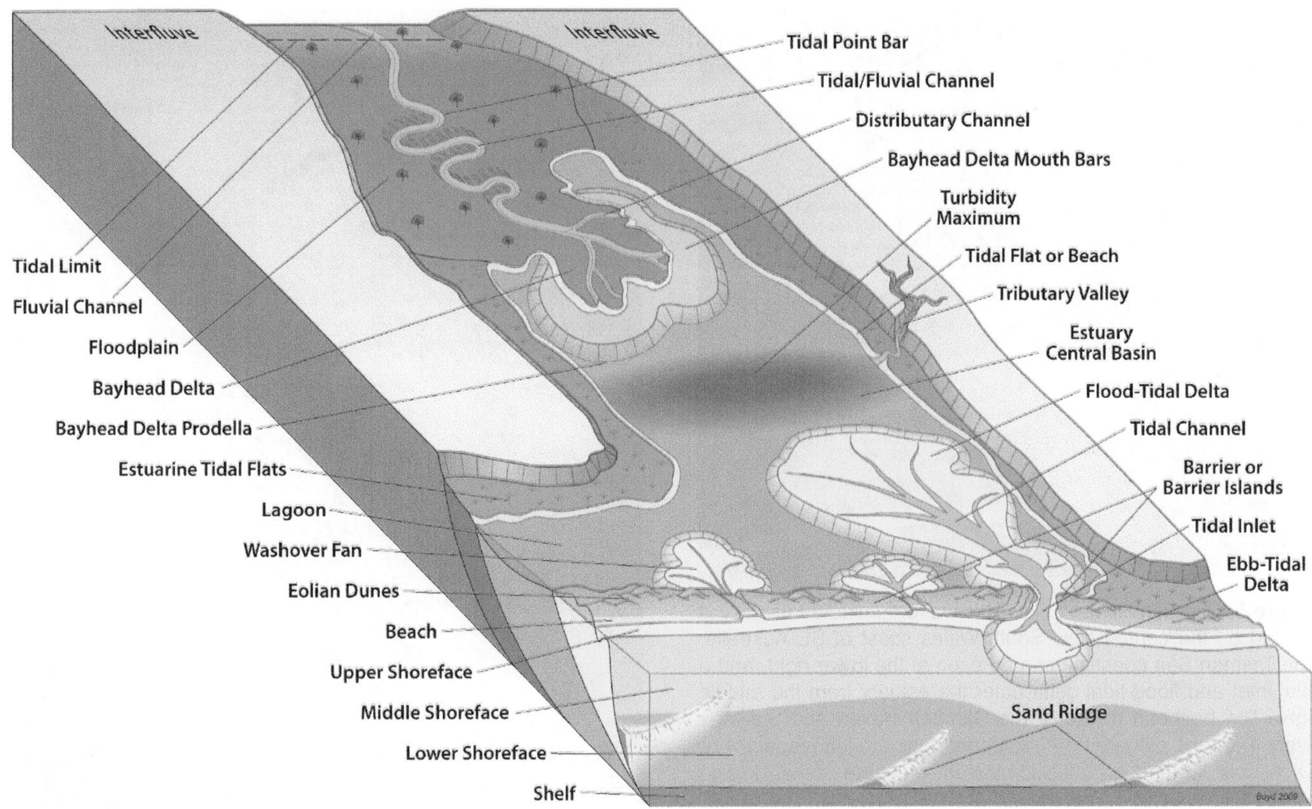

Figure 3. Geomorphology of transgressive wave-dominated coasts. This figure shows a shore-perpendicular incised valley containing a river, a downstream transition to a tidal-fluvial channel and a linked bayhead delta. Farther seaward is a muddy estuarine central basin, separated from the ocean by a transgressive barrier cut by tidal inlets and washovers (*cf.* Fig. 8). The shoreline of the estuary is occupied by beaches and/or tidal flats. A lagoon stretches parallel to the shore on the left and has no fluvial source (*cf.* Figs. 6 and 7). The barriers are fronted by a landward-migrating erosional shoreface. The small grass symbol identifies brackish marsh facies whereas the tree symbol indicates fresh-water swamps.

Figure 4. Caption on opposite page.

Figure 5. Camden Haven Lake. This wave-dominated estuary is located on the central New South Wales coast of SE Australia. The Tasman Sea coastline can be seen at the lower right, and a tidal inlet and flood-tidal delta enter the estuary from the middle right. The bayhead deltas of the Camden Haven and Stewarts rivers enter the estuary from the upper right and left respectively. Each has formed a river-dominated elongate delta. Image sourced from Google Earth.

Figure 6. The North Carolina coast at Cape Hatteras exhibits a range of transgressive wave-dominated coastal features. A narrow barrier-island chain broken by tidal inlets makes up the Outer Banks to the east (right). A series of three incised valleys occupied by estuaries (darker blues and browns) occur in the center of this satellite image. The zone of turbidity maximum is marked by the brown water in the estuaries. The flooded areas on the interfluves between the valleys are occupied by lagoons.

sediment being derived mainly from longshore and offshore sources. Transport of sand across the mouths of estuaries and embayments by wave action produces barriers that separate estuarine and lagoonal water bodies from the open ocean (Fig. 8). Transgressive barriers frequently have a low profile and are the site of washover channels and fans where shoreline sands are transferred from the beach to the landward side of the barrier (Fig. 9). Higher profile barriers may be capped by eolian dunes, extensive vegetation and barrier lakes. During transgression, rising sea level inundates low-lying areas behind the barrier, raising the water table and producing extensive paralic swamps and wet, boggy gley soils (Figs. 7 and 8).

Barriers are frequently cut by channels termed tidal inlets that allow the tides to flow back and forth between the open ocean and back-barrier water bodies. These tidal inlets frequently migrate in the direction of longshore transport, and sediments deposited in the channels are subsequently transported seaward into ebb-tidal deltas (Figs. 3 and 10) that project out into the open ocean, and landward into flood-tidal deltas (Figs. 4 and 11) that prograde into estuaries and lagoons. Each tidal delta (*e.g.*, Fig. 12) has a characteristic morphology of tidal channels and bars (*e.g.*, Hayes, 1975), with ebb-tidal deltas and inlets tending to be larger and more numerous (and hence barrier islands shorter) on more tide-influenced coasts, and flood-tidal deltas being better developed and tidal inlets less common (and hence barriers longer) on more wave-influenced coasts (Fig. 13).

The nearshore zone on transgressive coastal barriers has all the geomorphological features that are found on regressive shorelines (*see* Chapter 8), including beaches, nearshore bars, troughs and rip channels, and a shoreface stretching seaward from the barrier to the inner shelf (Fig. 3). On more mixed-energy coasts, barriers tend to become lower and merge into tidal flats where extensive tidal channel, marsh or evaporite flats develop landward of the shoreline. Transgressive shorelines have erosional, not depositional, shorefaces and their landward movement results in the production of an erosional surface and a transgressive sand sheet seaward of the base of the shoreface (Fig. 14).

PROCESSES CONTROLLING TRANSGRESSIVE WAVE-DOMINATED COASTS

Wave processes influence coastal sedimentation by effecting both shore-normal and shore-parallel sediment transport (Fig. 15). Shoaling waves tend to develop an onshore asymmetry that transports water and sand particles towards the shoreline. Wave impact on the seabed is a function of depth, resulting in a concave-

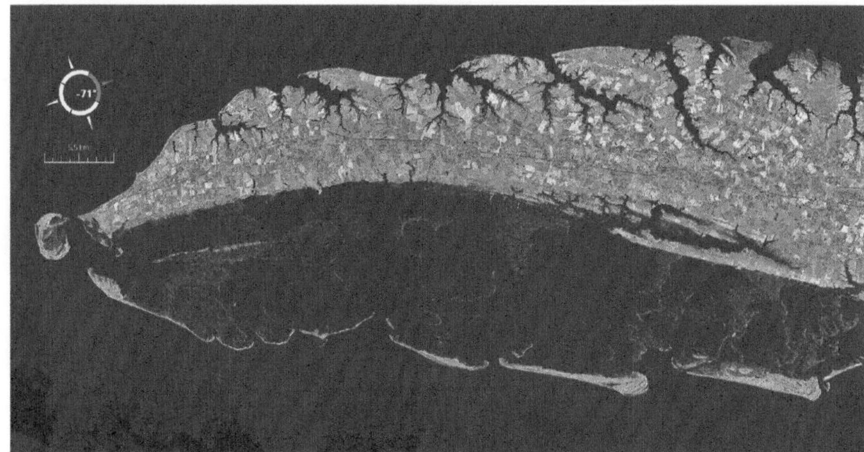

Figure 7. This satellite image of the Virginia, USA coast, shows a transgressive wave-dominated shoreline composed of barrier islands and tidal inlets immediately to the north of the entrance to Chesapeake Bay (at lower left). This coastline has minimal fluvial input as evidenced by the small streams and drainage areas on the Delmarva Peninsular mainland. Hence, the water bodies enclosed by these barriers are best described as lagoons.

upward shoreface profile and a seaward-fining grain-size trend. The shoreface profile represents an equilibrium surface and its translation through time and space is a critical process in forming coastal stratigraphy. Wave motion at the seabed is a function of wave length and, in major storms, can extend across the entire continental shelf. The term 'wave base' has often been applied to the depth to which wave motion causes sediment motion on the seabed, and a value of approximately half the wavelength of the surface waves is widely used to determine this depth. However, the depth of the shoreface base during transgression is more important in controlling stratigraphy, and this is determined more by the modal wave size, which is a product of wave power and its frequency of occurrence, with the base of the shoreface typically occurring in water depths closer to 10–15% of the surface wavelength. For example, along the SE coast of Australia, the modal wave is 1.5 m high with a 10 second period (Short, 1979), which equates to a modal wavelength of around 150 m, and the shoreface base occurs in water depth of 15–25 m (Boyd *et al.*, 2004). Coasts with larger modal waves such as along the southern coast of Australia (Short, 1984) have correspondingly deeper shoreface bases (up to 30 m deep), whereas coasts with smaller modal waves such as the Gulf of Mexico have shallower shorefaces (around 5–10 m deep – *see* comparison in Boyd and Penland, 1984)).

Excess water accumulating at the shoreline as a result of wind and oblique wave approach is transformed into a longshore current that

Figure 8. The southern Chandeleur Islands, located east of the Mississippi Delta in Louisiana, are a transgressive barrier-island arc that consists of a series of vegetated washover fans separated by overwash channels and ephemeral tidal inlets. This infra-red image shows the marsh in red with numerous erosional scours from the impact of hurricanes, fronted by a thin beach. Seaward of the beach, transgressed marsh can be seen cropping out in the surf zone. Since this image was taken in 1982, repeated hurricane impact has reduced these barriers to remnant marsh islands and submerged shoals. The subaerial portion will soon disappear leaving only a submerged sandbody.

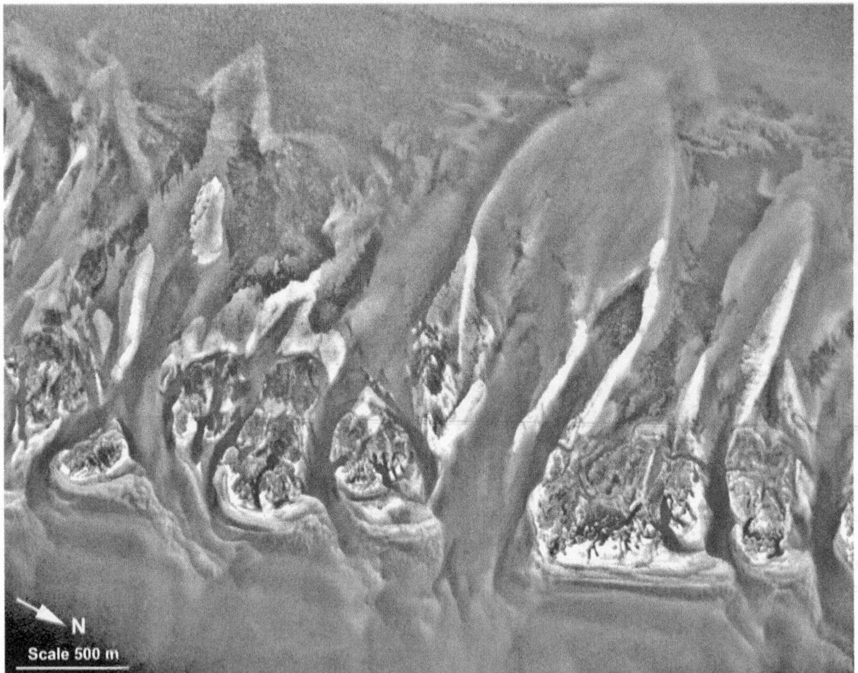

Figure 9. Vertical airphoto of the Chandeleur Islands, Louisiana, with north to the right. Washover deposits approximately 1 km wide and up to 2 km long can be seen extending to the NW from the Gulf of Mexico shoreline at the bottom. These washover deposits were formed by force 5 Hurricane Camille in 1972 and consist of central washover channels between remnants of eolian dunes and marsh, and distal fan deposits that prograde into Chandeleur Sound (top).

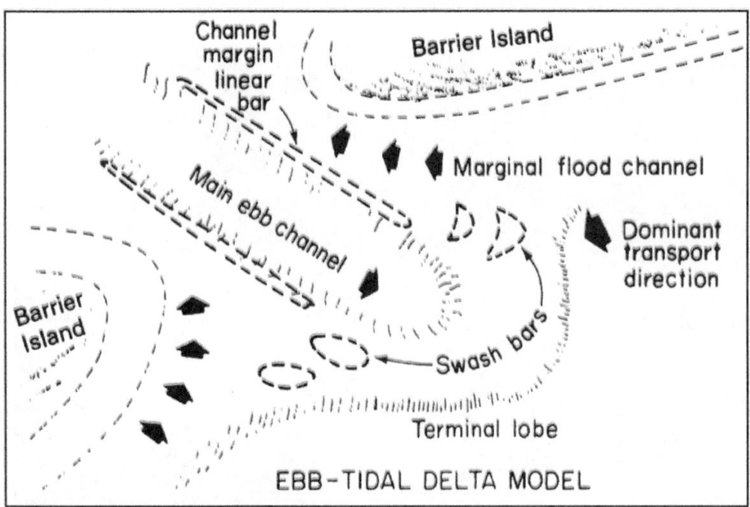

Figure 10. Ebb-tidal deltas consist of a main ebb channel that is the seaward continuation of the tidal inlet, flanked laterally by channel-margin linear bars and by swash bars on the terminal lobe. Marginal flood-dominant channels are developed on one or both sides of the main ebb channel. The main channel is often asymmetric due to downdrift migration, with a deeper thalweg downdrift, and an inclined shallower prograding spit on the updrift side (from Hayes, 1975).

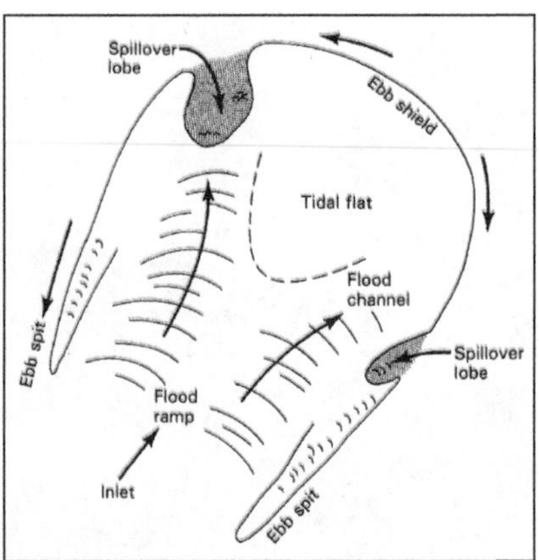

Figure 11. Flood-tidal deltas develop at the landward end of tidal inlets, and consist of an inclined flood ramp that shoals landward and bifurcates into multiple flood-dominant channels around one or more middle-ground tidal-flat shoals or bars. Return ebb flow is deflected around an ebb shield forming lateral spillover lobes and spits (from Hayes 1975).

moves sediment parallel to the coast (Fig. 15), providing the raw material for elongating barriers across coastal embayments, and for tidal processes to carry into estuaries. During storms, overwash of barriers results from a combination of wind stress, low atmospheric pressure, wave setup and wave run-up, and can cause the water surface to be elevated over 10 m above normal sea level (*e.g.*, Boyd and Penland, 1981). Excess water that cannot be removed by offshore or alongshore flow results in landward movement of water and sediment across the barrier and the creation of washover deposits (*e.g.*, Schwartz, 1975). Eolian processes acting alone may also contribute to landward transport when onshore wind blows sand from the beach to the backshore and on into coastal dune fields and wind-tidal flats (Fig. 12).

Estuaries and open-water bodies landward of the shoreline that experience tidal action exchange water with the open ocean on a semi-diurnal or diurnal cycle (Fig. 16). The volume of water that must be exchanged is called the *tidal prism* (*see* Chapter 9), and is the product of tidal range and water-body area. The influence of tidal processes is most pronounced at tidal inlets and channels that cross the shoreline. At these locations, flood- (landward-oriented) dominant currents take shoreline sand and remove it through tidal channels to flood-tidal deltas and tidal flats in the back-barrier area. In this way, tidal, washover and eolian processes recycle sediment from the front to the back of the barrier as the transgression proceeds. The depth of tidal inlets and tidal channels is a direct response to the tidal prism and the number of tidal inlets or channels available to exchange the tidal volume. However, tidal inlets are likely to be deeper where vigorous wave-driven longshore currents deposit sand on the updrift side of the inlet, causing it to narrow, thereby accentuating cur-

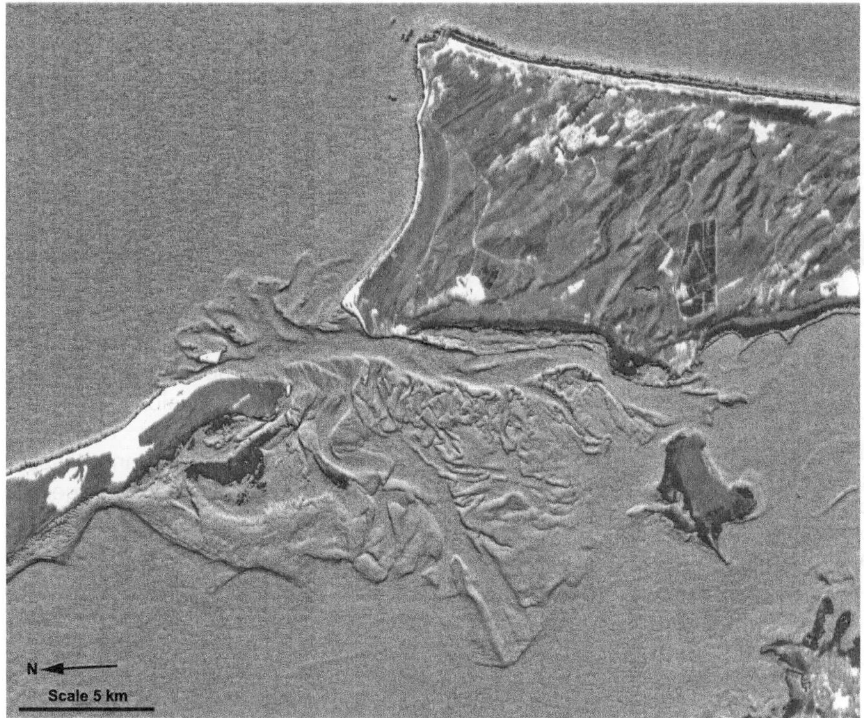

Figure 12. South Passage tidal inlet between Stradbroke Island and Moreton Island barriers enclosing Moreton Bay (at lower right), SE Queensland, Australia (north is to the left). This high wave-energy coast has a well-developed flood-tidal delta with an area of approximately 130 km². A small ebb-tidal delta exists, despite the high wave energy in the Tasman Sea (upper left) because of the large tidal prism of Moreton Bay (area over 1000 km²; maximum tidal range of 2 m). Compare with Figures 10 and 11. Extensive development of eolian dunes can be seen on both Stradbroke and Moreton islands. (Source NASA).

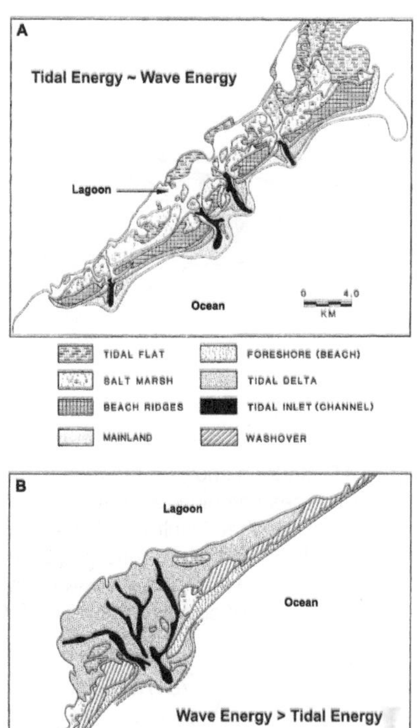

Figure 13. Differences between coastal sandbodies in areas with high wave energy (**B**) versus areas with more nearly equal wave and tidal energy (**A**). In areas with a relatively large tidal prism, barriers are short, often drumstick-shaped (wide at one end and tapered at the other; *see* Fig. 7) and are separated by an inlet with well-developed ebb-tidal deltas. In areas with higher wave energy, barriers are longer (*see* Fig. 6), there are fewer tidal inlets, and flood-tidal deltas are better developed than ebb-tidal deltas (from Tye and Moslow, 1982).

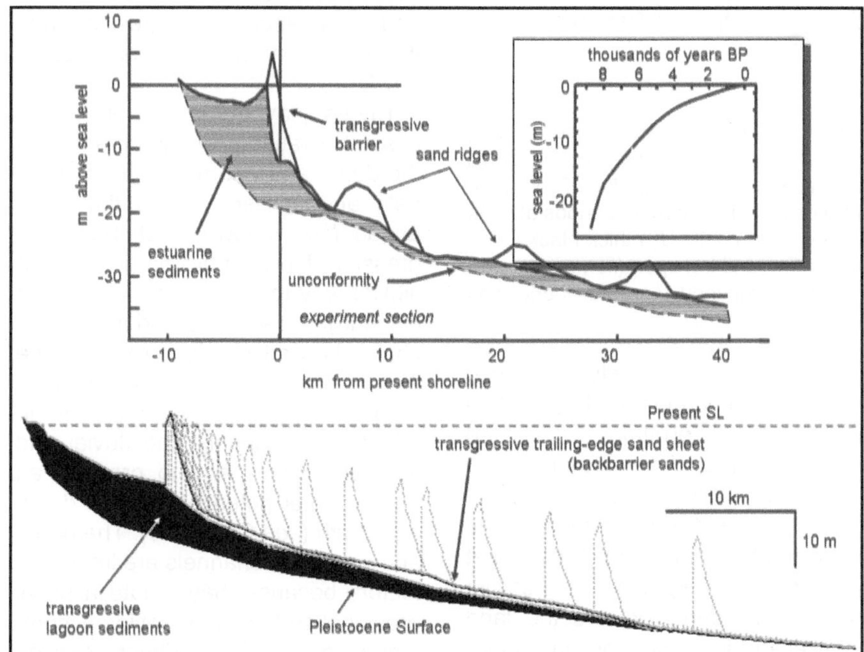

Figure 14. The upper figure shows the stratigraphy of the coast near Duck, NC, while the lower figure shows the result of a quantitative geometrical model showing the successive positions of the transgressing shoreline and barrier (dotted lines) and the predicted preservation of transgressive lagoonal sediments, overlain by reworked shelf sands after transgression of the shoreface. The Holocene relative sea-level curve used in the model is shown in the inset. Modified from Cowell *et al.* (1999).

rent scour. Tidal asymmetry is common in tidal flows, favoring net sediment transport in the direction of the faster current (Fig. 16). The reduction in speed at slack water during tidal reversals allows for periods of mud deposition (Fig . 16).

Less is known about processes and sediment transport on mixed-energy coasts where tidal and wave action is more nearly equal, but indications are that in such places the coastal zone is widened, tidal-flat environments expand, extensive development of tidal channels may occur, and barriers become progressively shorter and tidal inlets more numerous as the tidal prism increases. Nevertheless, storm reworking by waves remains an effective process that maintains a shoreface, contributes to its landward translation

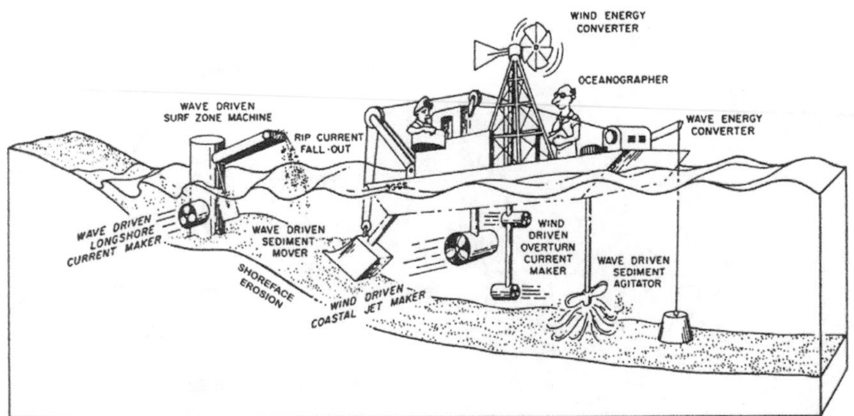

Figure 15. Cartoon showing the marine forces impacting transgressive coasts. Wind and wave processes combine to create nearshore circulation that drives alongshore transport and rip currents that result in offshore transport. Wave erosion (wave-driven sediment mover) results in shoreface retreat. Elevation of water levels against the shore is effected by the onshore wave motion, and onshore and alongshore wind motion, causing barrier overwash. Modified from Niedoroda (1980).

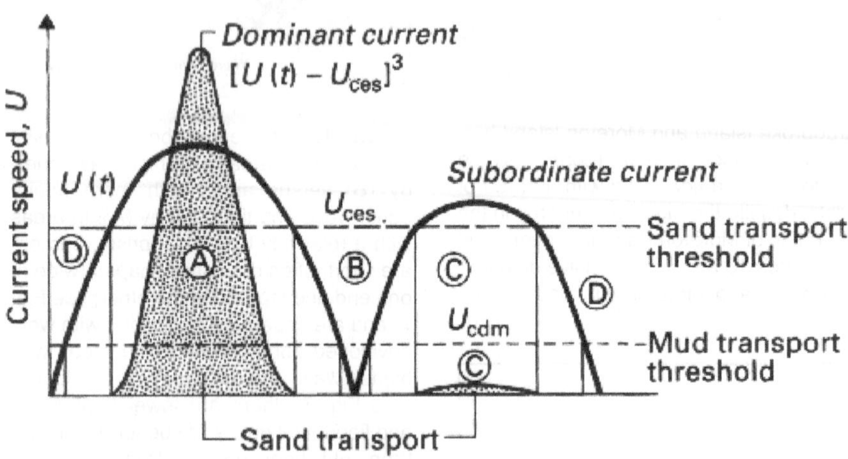

Figure 16. An asymmetric tidal-velocity curve for an ebb–flood cycle showing differing thresholds for sand and mud transport (horizontal dashed lines) and times of sand (**A**, **C**) and mud (**B**, **D**) deposition. This type of asymmetric tidal motion can produce flood-dominant transport through tidal inlets forming flood-tidal deltas, and alternate periods of sand and mud deposition resulting in IHS beds in tidal/fluvial channels. U(t) thick black line = variation of velocity with time, U_{ces}= threshold velocity for sand transport (*i.e.*, sand transport above this line), U_{cdm} = threshold velocity for mud transport (*i.e.*, mud deposition below this line). From Allen (1982).

and generates a range of wave-dominated sedimentary structures and facies (*e.g.*, Yoshida *et al.*, 2007).

The rise of relative sea level that causes shoreline transgression results from a eustatic (global sea-level) rise and/or the lowering of the land surface by tectonic and sediment-loading processes. *Eustasy* refers to global changes in sea level due to water mass added to, or removed from, the oceans and has both short and long-term components, many of which are cyclical and driven by tectonic and climatic forcing. Quaternary Earth history, in particular, has been strongly influenced by climatic changes and glaciation operating at orbitally controlled Milankovitch cyclicity, resulting in major eustatic fluctuations of 10s of meters to over 100 m on time scales of 20–400 ka and a corresponding set of shoreline transgressive/regressive cycles. Subsidence lowers the land surface and is a common and ongoing process in most sedimentary basins. Compaction results from a reduction in sediment volume due to loading. It also lowers the land surface, especially in thick, rapidly deposited sedimentary sections such as those beneath major deltas (*see* Chapter 10).

Clastic sediment is primarily derived from rivers and deltas and is moved laterally into adjacent coastal systems by along-shore transport. Sediment flux from hinterland drainage basins is primarily a function of drainage basin relief and climate (*see* Chapter 4) and depends on topography, water discharge, sediment size and transport capacity (*see* review in Blum and Tornqvist, 2000). Shoreline transgression at river mouths will commonly result in drowned river valleys or transgressed deltas, where coastal submergence produces estuaries. In these locations, river-supplied sediment is deposited inland from the coastline and accumulates farther up-valley in fluvial, bayhead delta and estuarine central basin settings. A rise in relative sea level flattens the equilibrium profile of rivers, leading to upstream storage of sediment, contributing further to transgression. During high rates of relative sea-level rise, such as those experienced during the latest Pleistocene and early Holocene times, even large continental drainages such as the Mississippi were unable to maintain their deltas and experienced transgression and the formation of estuaries (Frazier, 1974).

Rivers play an important role in coastal stratigraphy by determining the depth and volume of fluvial channels and incised valleys that form ahead (*i.e.*, landward) of the transgression. The river discharge acts in a similar way to tidal prism, with higher discharges requiring wider, deeper channels, and incision occurs if the flow is able to carry more sediment than is supplied from upstream. In tidal-fluvial channels, the fluvial and tidal discharges add to produce an even greater potential channel depth and width (*see* Chapter 9). The depth and volume of channels are important factors because they create a basal erosional surface and provide accommodation in which sediment can be stored ahead of the transgression (Figs. 17 and 18). Away from major river mouths and deltas, transgression results in inundation of coastal embayments, forming bays, lagoons, and estuaries in adjacent smaller val-

Figure 17. Oblique airphoto of an incised valley (Red Deer River, SE Alberta, Canada). Such valleys are eroded by fluvial processes, with an exposure surface on the interfluves outside the valley, a continuation of this sequence-boundary unconformity at the base of the valley, and a fill of fluvial deposits that is at least one fluvial-channel thick. This view shows a typical terrestrial valley prior to transgression. During and after transgression, the remaining space in the valley will be filled by estuarine and perhaps even marine deposits (source Google Earth).

Figure 18. 3-D seismic imagery of incised-valley topography on the sub-Cretaceous unconformity in the Western Canada Sedimentary Basin. This image shows the extensive interfluve area between valleys, the incised sinuous valley course and potential point-bar fluvial infill in the base. Like Figure 17, this image gives an understanding of the topography that is present prior to a transgression (image courtesy of H. Posamentier).

leys, or erosion of linear coasts. As these embayments become larger during the transgression, the tidal prism increases and the landward transport of sand is enhanced.

TRANSGRESSIVE COASTAL FACIES AND FACIES ASSOCIATIONS

Facies formed in transgressive coastal settings can be grouped into estuarine and barrier facies associations. The first of these associations is formed in a more landward location and consists of sediments deposited in:

1. Tidal-fluvial channels;
2. Bayhead deltas; and
3. Estuarine central basins, lagoons, and their margins.

The second, more seaward association is formed near the shoreline and consists of sediment deposited in;

1. barrier; and
2. inlet-associated locations.

Estuarine Facies Association

In the most landward locations experiencing transgression, valley formation and fluvial deposition will precede the transgression in river valleys (*see* Chapter 6); elsewhere (*i.e.*, on interfluves), most of the land surface will be occupied by erosional or non-depositional landscapes (Fig. 18). Following Walther's Law (*see* Chapter 2), the next deposits formed during transgression in upstream estuarine locations will be tidal-fluvial channel, marsh, tidal flat and bayhead-delta facies (*e.g.*, Fig. 5 and Fig. 19 vertical profile 7). Tidal-fluvial channels are similar to fluvial channels in that they have erosional bases with channel lags and are filled with coarser grained sand and gravels (from both mud clast and lithic sources) that contain dune and ripple crossbedding and planar-stratified beds (*see* vertical profiles 7 and 8, Fig. 19; *see* also Chapter 9). However, a major difference from the preceding fluvial deposits is that, here, tidal indicators such as bioturbation and muddy drapes on bedforms are present (Fig. 16; *see* Chapter 9 for a full discussion of tidal indicators), as well as a shift from fresh-water fauna and flora with swamps and trees, to brackish-water organisms and traces with salt-marsh and mangrove vegetation. Tidal-fluvial channels usually occur in relatively low-gradient situations, producing sinuous channel forms (Figs. 3 and 5). These sinuous channels typically deposit muddy layers on tidal point bars, generating *inclined heterolithic stratification* (IHS; Fig. 20). The tidal channels typically generate fining-upward successions, resulting from the coarser grained channel deposits being overlain by tidal point bars and fine-grained tidal-flat and floodplain deposits including coaly beds (*e.g.*,

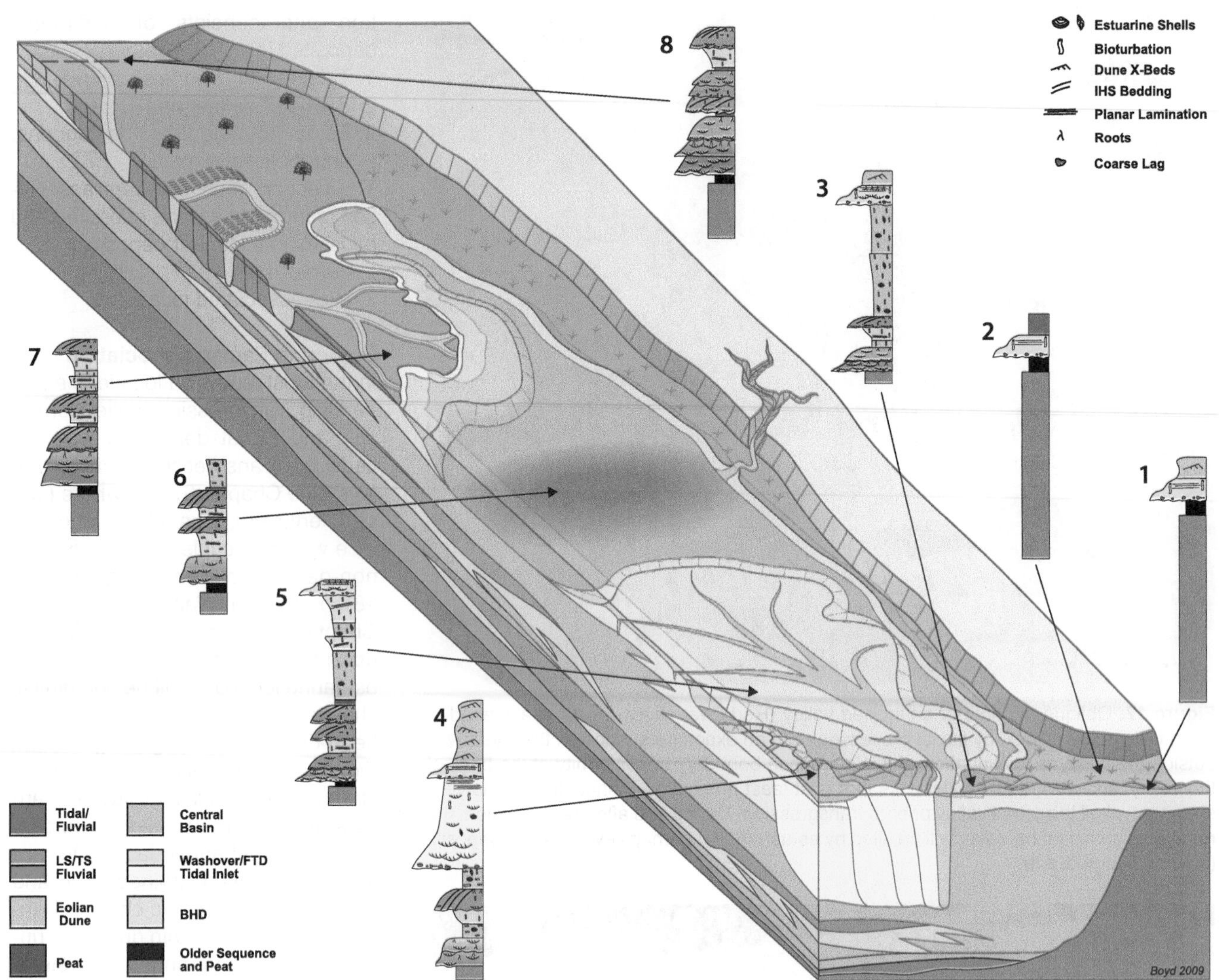

Figure 19. Facies model for transgressive wave-dominated coasts. This cut-away block diagram, derived from Figure 3, shows shore-parallel and shore-perpendicular sections. The shore-parallel section is located immediately behind the shoreline and passes through the barrier and migrating tidal inlet. The shore-perpendicular section extends from the shoreline to beyond the tidal limit in an incised-valley setting. A number of retrogressively stacked fluvial to bayhead-delta parasequences are preserved in the incised-valley fill, and overlie lowstand fluvial deposits (reddish color) and the sequence boundary at the base of the valley. An estuarine central basin, washovers, a flood-tidal delta and marsh deposits lie behind the barrier. Eight representative vertical profiles are shown; profile width is proportional to grain size (or a proxy such as the gamma-ray log response).

Rahmani, 1988). As well as the sequence-bounding unconformity at the base of the valley, lateral migration of tidal channels produces channel-base erosional surfaces in these upstream estuarine facies (Fig. 19). Such transgressive erosional surfaces are often termed *ravinement surfaces.*

The bayhead delta itself usually takes a river-dominated form due to its protected location (Figs. 3 and 5), but can also have extensive tidal influence and even be tidally dominated as is the case in the Gironde estuary (*e.g.*, Allen and Posamentier, 1993); it is rarely wave-dominated. Prodelta facies are usually muddy, laminated to bioturbated (Fig. 20), and are overlain by delta-front and distributary-mouth-bar sands and silts, and finally by muddy and organic-rich delta-plain deposits (vertical profile 7, Fig. 19). Although formed during an overall transgression, bayhead deltas are progradational in nature, and can generate multiple back-stepping packages (*i.e.*, parasequences; *see* Chapter 2) within the estuarine fill (side panel of Fig. 19). The unfilled muddy central estuarine basin is a lower energy zone relative to the landward fluvial zone and the marine coastal area farther seaward (vertical profile 6, Fig. 19). When first deposited, these muds are laminated, but over time the slow rate of deposition and high biological activity in the central basin result in the accumulation of mostly organic-rich, bioturbated muds (Fig. 20), typically with shelly body fossils. The deposition of mud is aided by the interaction of fresh and saline water within the turbidity-maximum zone (*e.g.*, Nichols and Biggs, 1985; *see* Chapter 9) that often occupies the central basin of wave-dominated estuaries. Along the margins of the estuary, tidal mud flats and channels occur along with sandy beaches. These deposits typically coarsen upward as the shoreline of the estuary progrades, and can contain tidal rhythmites in more strongly tide-influ-

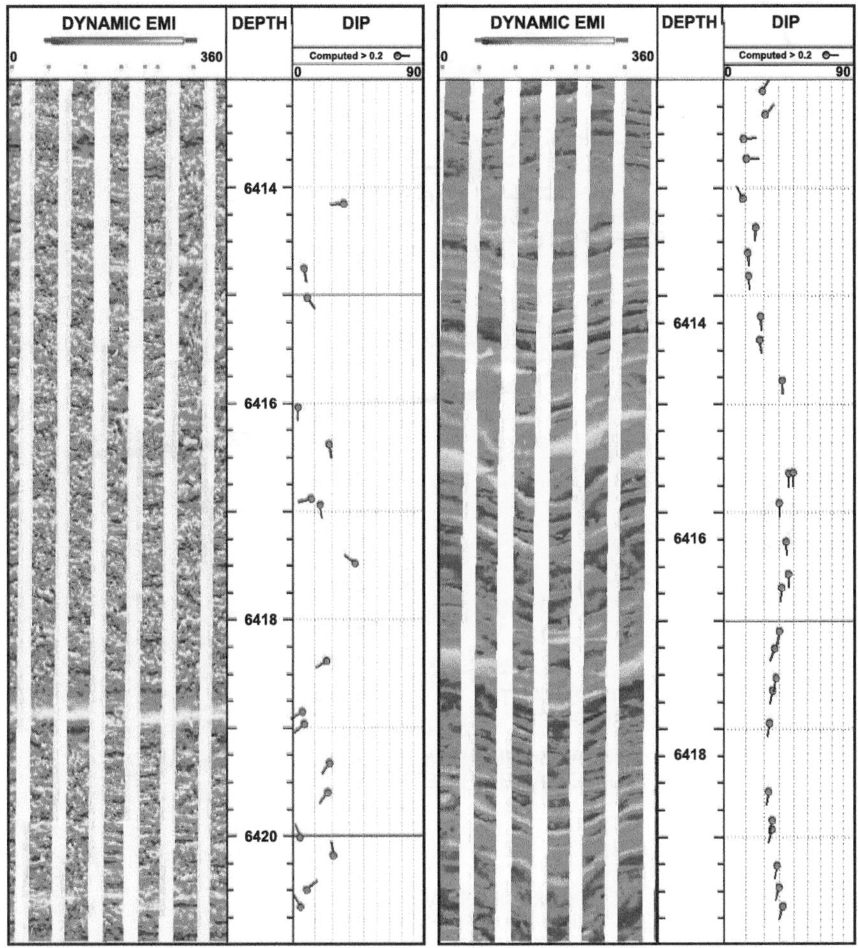

Figure 20. EMI electrical image logs showing 6 vertical strips making up an unrolled circular view of the inside of a borehole through the Morrow Sandstone, and illustrating the ability of EMI data to distinguish different estuarine depositional settings. The image on the left shows mainly structureless to parallel-laminated sandstone and shale from a central-basin to bayhead–prodelta setting, whereas the image on the right shows dipping IHS beds from a tidal-fluvial channel setting. The 'tails' of the green 'tadpoles' at the right side show stratal dip directions: these are relatively low and random in the left image, whereas, in the lower right image, the IHS beds dip uniformly to the SSW at 20-35°. Numbers in the depth column are in feet. High resistivity values are yellow and low values are brown.

enced settings, as well as organic-rich marsh muds and tidal-channel sands containing mud intraclasts. The thickness of bayhead-delta and estuary-margin deposits is determined by the water depth in the estuary, together with the additional space added by the rising sea level, but is usually only a few meters (Figure 19); by comparison, marine deltas may be tens to hundreds of meters thick (*see* Chapter 10).

Although fluvial valley-fill, bayhead delta and estuarine central-basin deposits form the basal part of transgressive wave-dominated coastal successions, they are most frequently deposited in regressive parasequences that are retrogressively stacked. A typical vertical succession within one of these parasequences consists of a tidal-fluvial ravinement surface overlain first by estuarine basin muds that pass upward into bayhead delta and tidal-fluvial channel deposits; continuing step-wise, relative sea-level rise results in a series of flooding surfaces separating a series of retrogradational-stacked upper-estuarine parasequences. However, within each parasequence, progradation occurs and results in an overall coarsening upward internal trend (Fig. 19). Bioturbation style in this upper estuarine association (*see* Chapter 3) is distinctive (MacEachern and Pemberton, 1994), in that it generally consists of an assemblage with limited species diversity, characterized by forms such as *Gyrolithes, Teichichnus* and *Planolites* created by organisms that are tolerant of fluctuating salinity. In addition, the stressful upper estuarine conditions often result in impoverished, diminutive faunas. Microfauna in these conditions are also affected by the stressful salinity conditions resulting in a similar reduction in species diversity (Boyd and Honig, 1992) and development of distinctive marsh-foraminifera assemblages dominated by agglutinated foraminifera. Salt-marsh grasses, such as *Spartina* spp and larger mangrove tree species and their derived pollen, form distinctive floral associations in estuarine settings. Coastal peat formed in both freshwater swamps and brackish marshes is a distinctive facies that is commonly found in transgressive settings. Peat is best formed and preserved when the rate of water-table rise (and linked relative sea-level rise) is equivalent to the rate of peat production. These two rates have been comparable during many transgressions, leading to the development of extensive coal beds. Conversely, because of dropping groundwater tables, peat is rare during falls in relative sea level.

Barrier Facies Association

Most transgressing barriers have a low profile and consist of a washover terrace with ephemeral eolian dunes, fronted by a beach and shoreface (Figs. 6–9). More stable barriers that have greater sediment supply, more effective eolian transport or slower rates of relative sea-level rise can develop higher dunes (Fig. 12) and be longer lasting. This type of barrier can also experience stepwise transgression/regression cycles (Fig. 21), during which the barriers can grow vertically and even temporarily regress, forming progradational parasequences within the overall transgression (*e.g.*, Swift, 1975; Sixsmith *et al.*, 2008). Overwash processes (Schwartz, 1975) typically transport marine sand from the seaward side of the barrier, in sheets or channels, for distances of a few meters to many kilometers (Fig. 9) landward into estuaries or lagoons (*e.g.*, Boyd and Penland, 1981). These overwash deposits contain clean marine sand that typically has an erosional base with a coarse lag (*e.g.*, vertical profiles 1–4, Fig. 19). They include shallow channel deposits and have a sheet to radial fan geometry that thins in a landward direction. The contained fauna can consist of mixtures of open-marine and estuarine/lagoonal organisms. Washover thickness is typically only several centimeters to a few meters and is controlled by the depth of water on the landward side of the barrier.

The other major facies deposited landward of the barrier during transgression are those formed in flood-tidal deltas and their associated tidal inlets and channels (Figs. 6–12, vertical profile 4, Fig. 19). Many washover deposits originate from breaches in the barrier that begin life as tidal inlets that do not have sufficient tidal prism to remain open (*e.g.*, Fig. 8). For larger tidal prisms in situations where wave processes are not energetic enough to close the breach in the barrier, a permanent tidal inlet is established that acts to transfer marine sediment into estuaries and lagoons. Sand moving alongshore under the influence of waves is deposited on an elongating spit platform on the updrift side of the inlet (Fig. 22). The tidal flow is diverted to the downdrift side where the thalweg erodes into the channel bank. In this way, the inlet may migrate laterally tens of kilometers during transgression (Fig. 19) and rework significant amounts of barrier and estuarine sediment (*e.g.*, Hoyt and Henry, 1967; Ashley and Sheridan, 1994; FitzGerald *et al.*, 2004). As the depth (and width) of the tidal inlet is determined by the tidal prism, estuaries and lagoons with large areas and tidal ranges require large channel cross sections to cope with the flow. Such inlets can be tens of meters deep and may be similar in size to, or even larger than, coeval fluvial channels. For example, the tidal channel of Barataria Pass on the old Lafourche complex of the Mississippi River delta is approximately 50 m deep, whereas the main channel of the Mississippi River reaches its greatest depth of about 60 m between

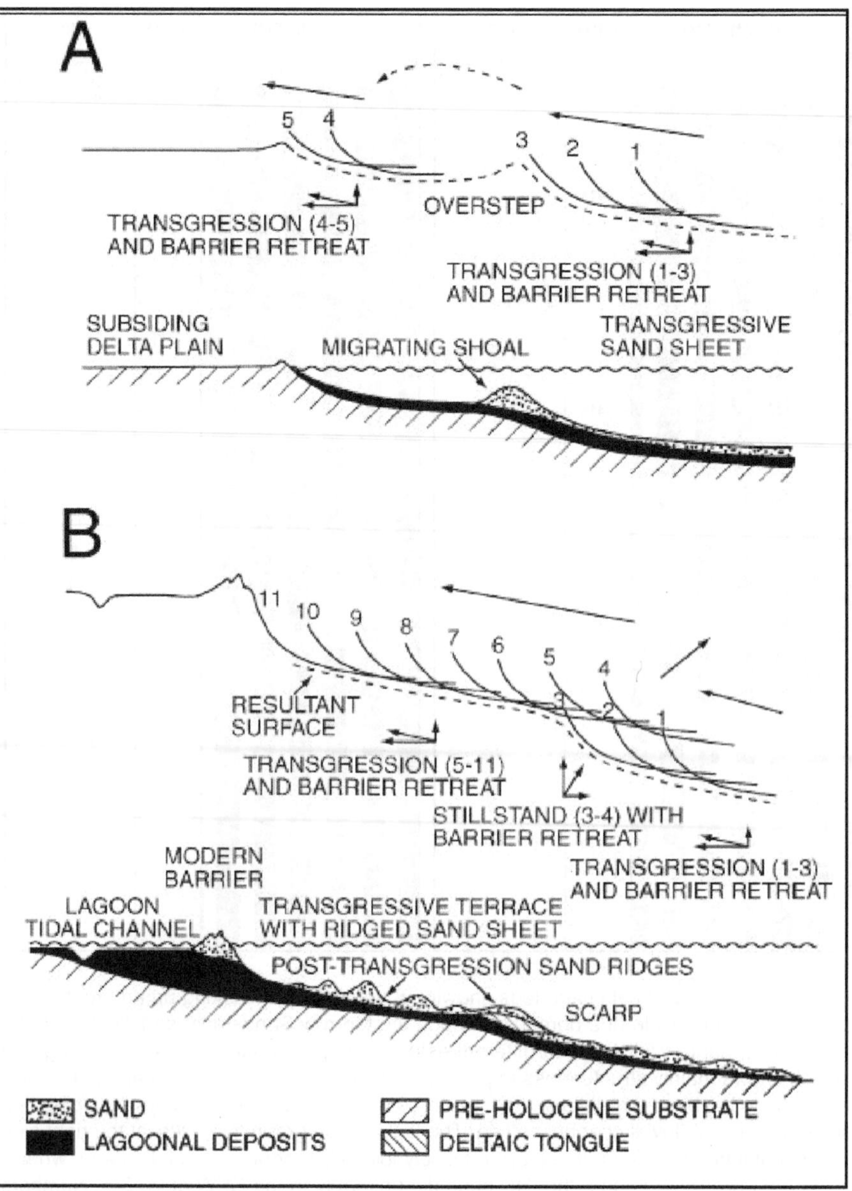

Figure 21. A. Continuous transgression and barrier retreat in which the barrier is overstepped and continues to be reworked as a shelf shoal. This is based on the fate of barriers such as the Chandeleur Islands (Figs. 8 and 9) in the Mississippi delta. **B.** Punctuated transgression where the transgression is interrupted by a period of stillstand, progradation and then renewed transgression, during which sand left behind on the shelf is reworked into shelf ridges. Arrows show the trajectory of the shoreface. The dashed line in the upper part of each section is the wave ravinement surface. After Swift (1975).

Baton Rouge and New Orleans. Sedimentary successions built by tidal inlets are similar to those formed by rivers in that they have an erosional channel base (vertical profile 4, Fig. 19; and Fig. 23), a lower fill consisting of crossbedded sand, and an upper fill consisting of low to high-angle prograding spit clinoforms overlain by beach and eolian dune deposits (*e.g.*, Kumar and Sanders, 1974).

The flood-tidal delta consists of sand that is transported through the tidal inlet by flood-dominant currents. Its surface is dissected by a network of tidal channels that separate tidal sand bars (Figs. 4 and 12), and its landward margin consists of prograding tidal-delta foresets. Tidal channels are often tightly meandering in plan view, and may exhibit IHS bedding like their more landward tidal-fluvial counterparts. Sediment composition and maturity of the sand and mud moved through the tidal delta system reflect their marine source, which may

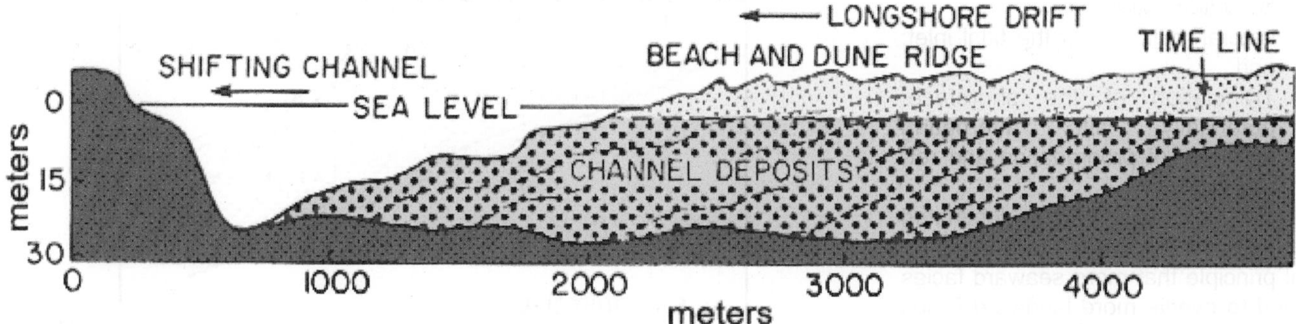

Figure 22. Generalized cross section parallel to the shoreline illustrating the lateral migration of a tidal inlet to the left, erosion of the downdrift side of the inlet and development of a barrier inlet cut-and-fill succession composed of channel deposits overlain by beach and eolian dune facies. The details of the channel fill are shown in Figure 23.

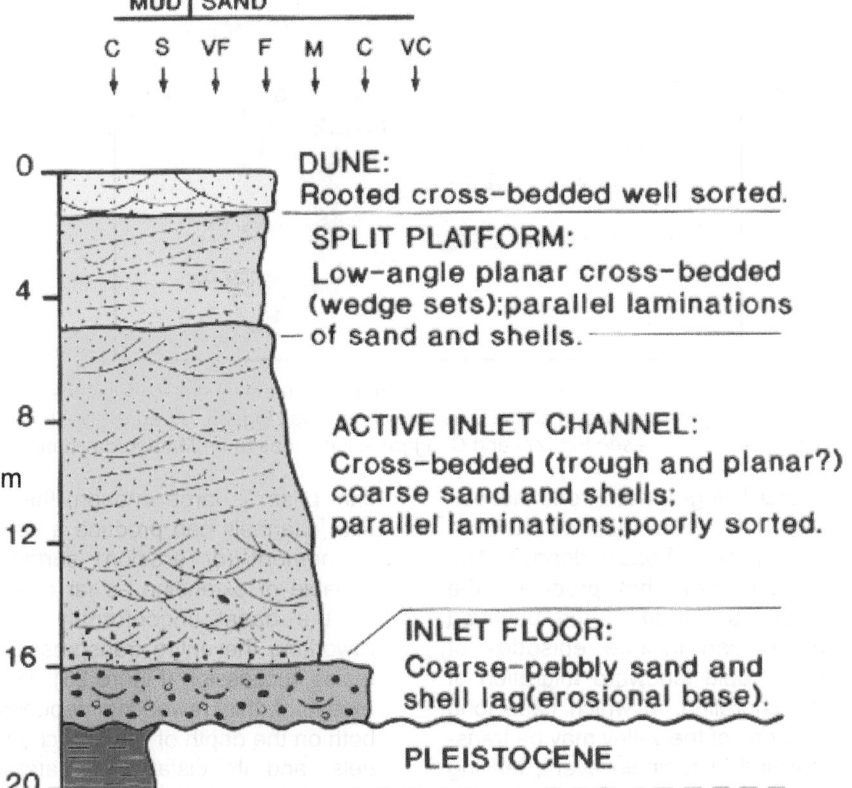

Figure 23. Tidal-inlet channel fill illustrating the erosional base, coarse lag and fining-upward grain size trend in the inlet-channel fill. The channel succession is overlain by a finer-grained, lower-angle spit platform, beach and eolian dune facies. Modified from Moslow and Tye (1985).

differ significantly from the composition and maturity of the fluvially supplied sediment. Sediment in the flood-tidal delta decreases in grain size from its source at the marine shoreline toward the landward margin of the tidal-delta front. Because of its proximity to the open ocean and the daily supply of water with an open-marine salinity, the flora and fauna of the flood-tidal delta is much more marine-dominated than its bayhead delta counterpart, resulting in assemblages of trace and body fossils that have higher diversity and more marine attributes (*see* Chapter 3). Over time, tidal channels extend into the estuary and erode into the previously deposited tidal-delta sand, and can scour deeply enough to erode into the underlying central-basin mud. As the shoreline transgresses, the landward migration of the flood-tidal delta complex produces a coarsening-upward sandy unit that overlies estuarine central-basin mud (vertical profiles 4 and 5, Fig. 19), and is dissected by erosionally based, upward-fining tidal-channel deposits (Boyd and Honig, 1992).

The shoreface lies seaward of the transgressing barrier and has an overall morphology and sediment character that is similar to that described for a regressive shoreface (*see* Chapter 8), with subenvironments that include eolian dunes, beach, ebb-tidal delta, surfzone/upper shoreface, and middle and lower shoreface. Transgressive shorefaces experience net erosion due to insufficient input of sand combined with the landward loss of sediment into washovers, eolian dunes and tidal inlets. Consequently, the shoreface migrates landward, eroding the sediment that makes up the barrier superstructure (Fig. 14). Because of its lack of preservation during transgression, and its treatment in Chapter 8, the shoreface is not discussed further here.

FACIES MODELS

Galloway and Hobday (1996) identify three main mechanisms for creation of stratigraphy; progradation, lateral migration and aggradation. In the transgressive coastal setting, these mechanisms operate in six main ways:

1. Lateral migration of fluvial and tidal-fluvial channels and vertical aggradation of the floodplains forming the lower part of the valley fill;
2. Episodic seaward progradation of the bayhead delta and estuarine shoreline;
3. Aggradation of the estuarine central basin;
4. Landward progradation of the

flood-tidal delta;
5. Lateral migration of the tidal inlet; and
6. Translation of the shoreface, potentially with alternating episodes of progradation and retrogradation.

Facies models for wave-dominated transgressive coasts follow the general principle that more seaward facies tend to overlie more landward facies (Fig. 24A), although there may be local variations in this due to short-lived progradational episodes. Since transgression commonly involves lower depositional volumes than its regressive counterpart and a number of effective erosional processes, erosion surfaces created by wave and tidal processes (*i.e.*, wave and tidal ravinement surfaces) will also be an important component of the resulting transgressive stratigraphy. Hence, an appropriate facies model for a setting in which a single complete facies succession is preserved will consist of the following components (Fig. 19).

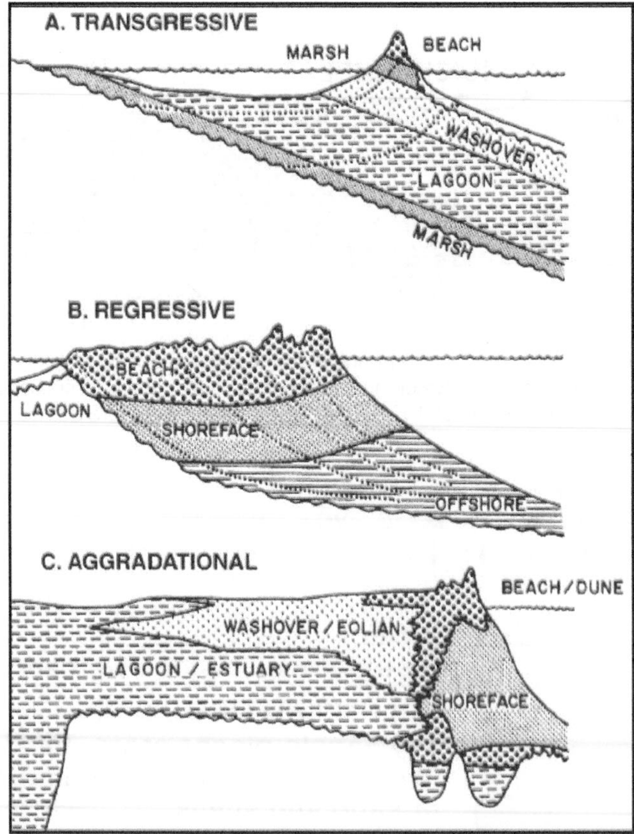

Figure 24. Galloway and Hobday (1996) presented three main models of coastal stratigraphy: **A.** transgressive, based on the Delaware coast – *see* Fig. 33, **B.** regressive, based on Galveston Island – *see* Fig. 27) and **C.** aggradational, based on Matagorda Island.

- Fluvial deposits are usually present at the base of the succession, especially in incised-valley settings where they overlie a basal unconformity (*e.g.,* Figs. 17 and 18). They will be overlain by tidal-fluvial channel deposits in which fluvially deposited sand becomes increasingly interbedded with mud and shows evidence of bioturbation as the transgression progresses. In the case of abandoned delta lobes, deltaic deposits underlie transgressive estuarine facies. Part of the fluvial sands may be lowstand in age, but the upper part is commonly a component of the overall transgressive succession.
- Flooding surfaces typically separate tidal-fluvial channel units from one or more overlying progradational bayhead-delta units. These parasequences are themselves separated by a flooding surface in most cases, but can also be bounded by scour surfaces produced by distributary or tidal channels. Such surfaces can extend down to the basal unconformity. The successive bayhead-delta coarsening-upward successions gradually become more distal and muddy as the fluvial system retreats landward, producing an overall fining-upward trend that culminates in the deposition of a muddy central-basin deposit. The transgression that produces the landward retreat of the bayhead deltas can include episodes of more rapid landward migration of the shoreline, in which case long sections of the valley may be transgressed in a single event, leaving bayhead-delta deposits as coarser-grained ‘beads' in the valley, separated by intervals of estuarine muddy facies (Bowen and Weimer, 2004). This episodic retreat is likely when the valley gradient alternates between steeper and flatter segments, and/or the rate of sea level rise varies.
- With ongoing transgression, the next unit deposited will be the prograding flood-tidal deltas and washovers that interfinger with the central-basin muds as the shoreline migrates landward (vertical profile 5, Figure 19). Continued transgression results in reworking of the flood-tidal delta deposits by tidal channels and tidal inlets. If the tidal prism is large enough, these tidal channels can produce a tidal ravinement (or erosion) surface capable of extending as far down as the basal unconformity and reworking the entire transgressive facies succession (Figure 25). The extent of this reworking depends both on the depth of the tidal channels, and its distance of lateral migration. Reinson (1992) recognized the importance of this process and made the laterally migrating tidal-inlet succession one of his three main models for coastal sedimentation. The culmination of the coarsening-upward succession resulting from landward shoreline migration is the development of barrier deposits including eolian dunes, beaches and shoreface units, unless these deposits are removed by later shoreface erosion, which is a common occurrence. Although this entire barrier succession is not frequently preserved in the deposits of a single transgression, the coastline can experience several

episodes of transgression and regression during the overall transgression (*e.g.*, Swift, 1975; Hampson *et al.*, 2008). In the process, barrier facies can be fully or partially preserved (Fig. 21).

The extent to which the preceding idealized vertical succession (Fig. 19) is actually realized depends on both positioning within the transgressive landscape (*e.g.*, in a valley versus on an interfluve), and the relationship of the rate of relative sea-level rise to the rate of sedimentation. Thus, the full succession is only possible in the axis of a paleo-incised valley where both fluvial channels and tidal inlets were developed (*e.g.*, vertical profiles 3, 4, Fig. 19). Away from this axis, transgressive deposits will be thinner, and composed more frequently of floodplain, marginal bayhead delta, estuarine central basin and basin-margin deposits, or washover/tidal-flat facies (*e.g.*, vertical profiles 1, 2, Fig. 19).

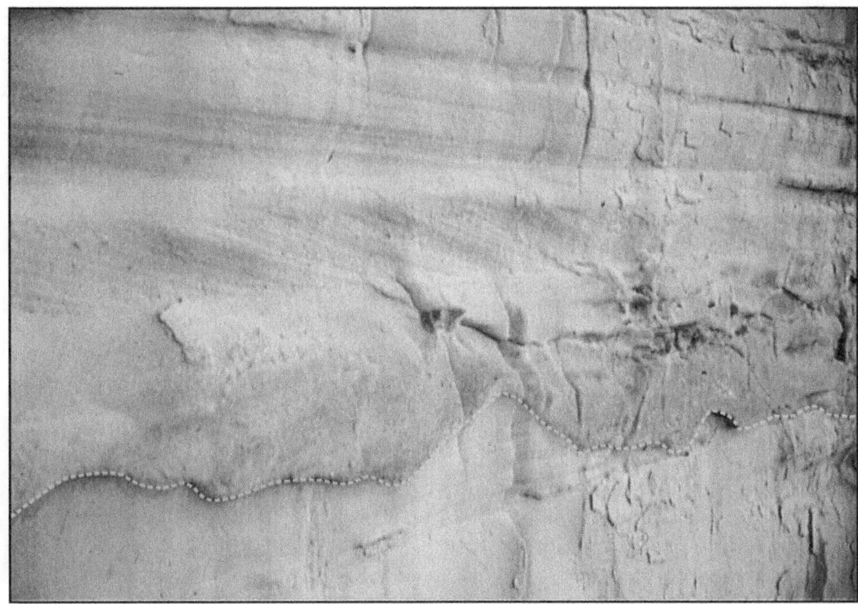

Figure 25. Outcrop photo of a tidal-ravinement surface (TRS; yellow dashed line) at the base of a tidal channel in the Waratah Sandstone (Permian) from the northern Sydney Basin, Australia. Horizontally bedded, regional deposits occur below the TRS, whereas herring-bone, cross-stratified coarse sandstones at the base of the channel fill overlie the TRS. The TRS has over 20 cm of relief and is mantled by a lag deposit.

Facies-model Variability

Walker's original formulation of the facies-models concept (*e.g.*, Walker 1984; *see* also Chapter 2) identified the need to synthesize information from numerous ancient and modern examples whose variability was 'distilled' away to produce a model that would serve as a norm for comparison. The features discussed below represent significant examples of this variability. As models evolve and acquire additional complexity, what was once regarded as local variability or 'noise' that was distilled away can now be incorporated as an important component of the 'signal'.

Transgressive and Regressive Limits

Coastal transgression is defined by the landward migration of the shoreline, which produces a stratigraphic succession in which more landward facies underlie more seaward facies (Fig. 24A). This distinctive characteristic can become blurred, however, at two critical points in an accommodation/sediment-flux cycle, namely the transgressive and regressive limits that are also termed the reversal or turnaround positions. At these locations, the balance between accommodation and sediment flux is equal and then reverses. In sequence-stratigraphic terms (see below) these positions are represented by the *maximum flooding surface* and the *transgressive* or *maximum-regressive surface* that mark the transition from retrogradation to progradational parasequence stacking and *vice versa*, respectively.

On many of the world's modern coastlines, the major Holocene eustatic rise terminated around 3–6.5 ka BP, although there have been numerous minor fluctuations and local variations since. In coastal locations that receive large amounts of river-supplied sediment, such as major deltas including the well-described Mississippi, Rhine–Meuse, Rhone, Niger and Danube, the regional shoreline migration direction reversed soon after the relative stillstand began (*e.g.,* Frazier, 1974), and these locations have become progradational deltaic shorelines. Elsewhere, the sediment flux has not been sufficient to fill the transgressive-limit estuaries and these retain the inherited transgressive geomorphology (Fig. 26). However, during a relative stillstand, the barrier can receive sediment that is moved onshore from the shelf or alongshore from adjacent eroding headland or deltaic sources. In this case, the shoreline at the transgressive limit begins to prograde, and, instead of a simple transgressive or regressive stratigraphy, this type of shoreline exhibits a transitional situation that retains an overall transgressive geomorphology but a stratigraphy the exhibits the beginning of progradation or aggradation (Figs. 24B, C, 26 and 27).

The case of a transition from transgressive to regressive stratigraphy is well illustrated by the textbook example of Galveston Island, Texas (Figs. 26 and 27). Here, the regional coastal geomorphology has the classic appearance of a barrier island, with a long thin coastal sand body cut by tidal inlets, separating a large estuarine system (Galveston Bay) from the Gulf of Mexico. However, widely referenced initial investigations by Bernard *et al.* (1959), and subsequent investigations by Rodriguez *et al.* (2004), have identified a transgressive basal surface, overlain by a regressive strandplain/shoreface system (Fig. 27). Most of Galveston Island was deposited since the sea-level stillstand, beginning around 5.5 ka BP, from material supplied alongshore from the Brazos River Delta. A similar composite stratigraphy was identified for many coastal sand bodies in SE Australia by Roy (1994), where progradation was caused by the addition of sand that moved landward across the inner shelf and shoreface after stillstand (see below). Again, the Australian examples appear paradoxical as they have the appearance of barriers that have undergone recent transgression, but their internal stratigraphy exhibits a composite transgressive-regressive character, indicating that they preserve the reversal position and contain the maximum flooding surface. Even the classic progradational geomorphology of the Dutch coast (van Straaten, 1965; Fig. 28) and the Nayarit coast, Mexico (Curray *et al.*, 1969; *see* Chapter 8, Fig. 21), is underlain by transgressive deposits, indicating that these examples also formed at the transgressive limit and are composite features, albeit ones that have progressed farther into the regressive highstand than Galveston Island.

Figure 26. Satellite image illustrating the geomorphology of Galveston Island, Texas. Galveston Island is separated from the Texas mainland to the west by the approximately 5 km wide West Bay estuary, and from coastal headlands to the NE and SW by the Bolivar Roads and San Luis Pass tidal inlets respectively. Although Galveston Island has the general appearance of a barrier island, the presence of beach ridges and its internal stratigraphy (*see* Fig. 27 whose location is shown by a yellow line) indicates that it is a composite transgressive/regressive feature.

Aggradational Barrier at Transgressive Limit

Another variation on this theme of preservation at the transgressive limit is the potential for a gradual reversal in which the supply of sediment is sufficient to balance the later stages of sea-level rise that immediately precede the stillstand, causing the shoreline to aggrade prior to progradation (Fig. 24C). This situation was proposed initially by Dickinson et al. (1972) for modern coastlines and has become a component of many parasequence-stacking patterns in sequence-stratigraphic textbooks (*e.g.*, Van Wagoner *et al.*, 1991; Coe, 2002). However, it is not well documented in modern or ancient studies. Recently Simms *et al.* (2006) have shown that Mustang Island, Texas, began to accumulate in its present position as early as 9.5 ka BP while sea level was still rising rapidly, and was supplied by sufficiently large amounts of sediment derived from adjacent eroding deltas and coastal headlands that it aggraded in place, producing a succession over 20 m thick consisting mainly of beach, dune and barrier washover sand with little mud. Since at least 7.5 ka BP, substantial parts of Mustang Island have been reworked by tidal inlets, with up to 1–2 km of the island's length being reworked to a depth of 15 m in less than 1000 years (Morton and McGowen, 1980). As the expected

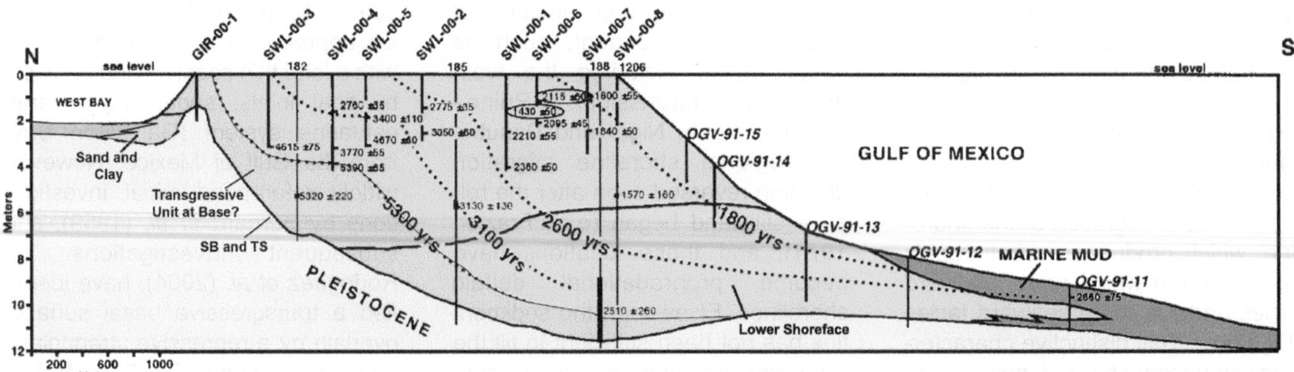

Figure 27. Shore-normal cross section through Galveston Island, Texas, with data derived from Bernard *et al.* (1970), Sirinigan and Anderson (1994) and Rodriguez *et al.* (2004). See Figure 26 for location. Galveston Island is traditionally regarded as a barrier island, and is underlain by a transgressive ravinement surface and separates the West Bay and Galveston Bay estuary from the Gulf of Mexico. However, the chronology of the deposits within the barrier indicates that it is a composite transgressive/regressive feature formed mostly through progradation since 5.3 ka BP. It thus contains the reversal or turn around position and the maximum flooding surface. Based on the reported chronology, the MFS could lie either at the top of the Pleistocene, or along the 5300 year isochron (making the basal unit transgressive; *cf.* Figs. 19 and 27).

thickness of barrier/strandplain deposits is approximately equal to the depth of the shoreface base (here 10–15 m), any greater thickness indicates unusual circumstances, in this case, the occurrence of aggradation during the later stages of the sea-level rise.

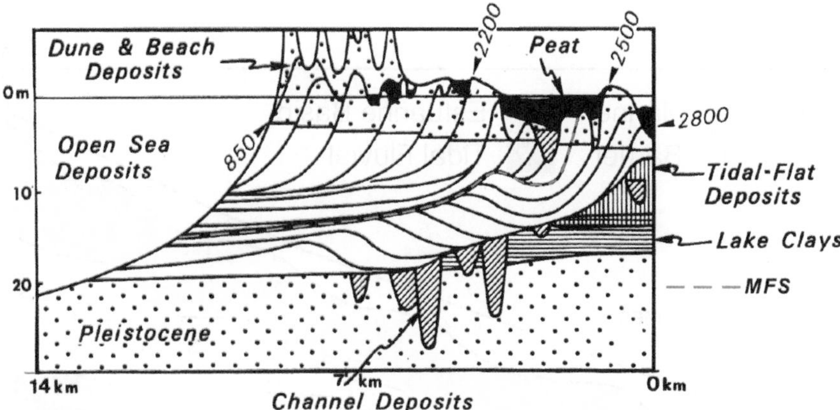

Figure 28. Shore-normal cross section through the Dutch coast (van Straaten, 1965) consists of an underlying transgressive barrier-estuary deposit with mainly landward-dipping strata deposited prior to 2500 years BP, overlain by a regressive strandplain unit with seaward-dipping shoreface strata deposited since 2500 years BP. Estuarine tidal flats and lake deposits are present behind and below the transgressive unit; extensive peat beds overlie it. This stratigraphy is similar to that beneath Galveston Island, in that it preserves the reversal position and MFS.

Variation in Transgressive Preservation

Several early workers recognized that the degree of preservation of transgressive deposits varied from place to place (*e.g.*, Fischer, 1961; Curray, 1964; Swift, 1968). Swift (1968) provided the first comprehensive review of the processes of transgressive coastal erosion (*i.e.*, ravinement), identifying both wave- and tidal-ravinement surfaces and, in subsequent papers, showed that the trajectory of the ravinement surfaces controlled the resulting stratigraphy. The concept of ravinement-surface trajectory was formalized by Helland-Hansen and Martinsen (1996) and extended by Cattaneo and Steel (2003). The advantage of the trajectory approach is that it recognizes that transgressive preservation is the vector sum of sea-level rise and the landward translation of the shoreface (and any other erosional surfaces). It also allows for a rigorous categorization of transgressive stratigraphy by noting the relative order and stratigraphic level of the four main erosional or ravinement surfaces that can be present in coastal successions (Fig. 29): fluvial incision at the base of the incised valley (sequence boundary), tidal-fluvial ravinement surface(s), tidal-ravinement surface(s) formed in tidal inlets and related channels, and the wave-ravinement surface formed at the shoreface. In a transgressive setting, the fluvial erosion surface precedes formation of the tidal-ravinement surfaces which, in turn, precede the wave-ravinement surface. This order controls the potential influence of each surface, with later surfaces having the potential to rework earlier surfaces and deposits lower in the stratigraphic succession. During transgression, the shoreface is the last erosional surface to migrate landward. In this respect, the shoreface acts like a regional snowblower or bulldozer blade, scraping off elements of the landward stratigraphy and redistributing the sediment onshore, alongshore and farther offshore. In contrast, the effect of other erosional surfaces is more local and restricted to a channel belt. Using these concepts, the following four main categories of transgressive stratigraphy can be identified (Fig. 29):

- No preservation below the wave-ravinement surface (WRS)

If no relative sea-level rise occurs during transgression, the wave-ravinement surface will carve a horizontal trajectory that removes all material above the base of the shoreface. If the shoreface base is the deepest erosional surface, no transgressive stratigraphy will be preserved. This situation is most common on interfluves or where the underlying sequence boundary (SB) occurs at a shallow depth, such as in lagoons and off-axis valley locations.

- Preservation only between the SB and WRS

In this situation, the WRS trajectory is relatively flat and erodes down below most other surfaces, leaving only partial preservation of fluvial facies, and possibly tidal-channel and tidal-inlet deposits. In this situation, the transgressive surface can exist within the fluvial fill of the incised valley or it may correspond to the tidal/fluvial ravinement surface (TFRS).

- Preservation between the SB and TFRS and below the WRS

Here the fluvial deposits, at the base of the valley, will be preserved below the TFRS. Deposits formed by tidal-fluvial channels, bayhead-delta facies and central-basin facies, and possibly also some tidal channel and tidal inlet fill facies, can also be present, depending on the depth of erosion by the WRS.

- Preservation between the SB and TFRS, between the TFRS and the tidal ravinement surface (TRS), and below the WRS

This situation occurs where the sequence boundary is deeper and hence a thicker succession accumulates, or where the three erosion surfaces have a steeper trajectory. These conditions result in retention of a more complete succession, including fluvial deposits, part to all of the tidal-fluvial and central-basin facies, and part to all, of the tidal-channel, tidal-inlet, flood-tidal delta and other barrier facies. In this situation, an important factor in the generation of stratigraphy is the role of the tidal inlet as its trajectory migrates both landward and alongshore, providing the potential to rework and replace much of the original barrier superstructure, and even much of the central-basin deposits, in the process.

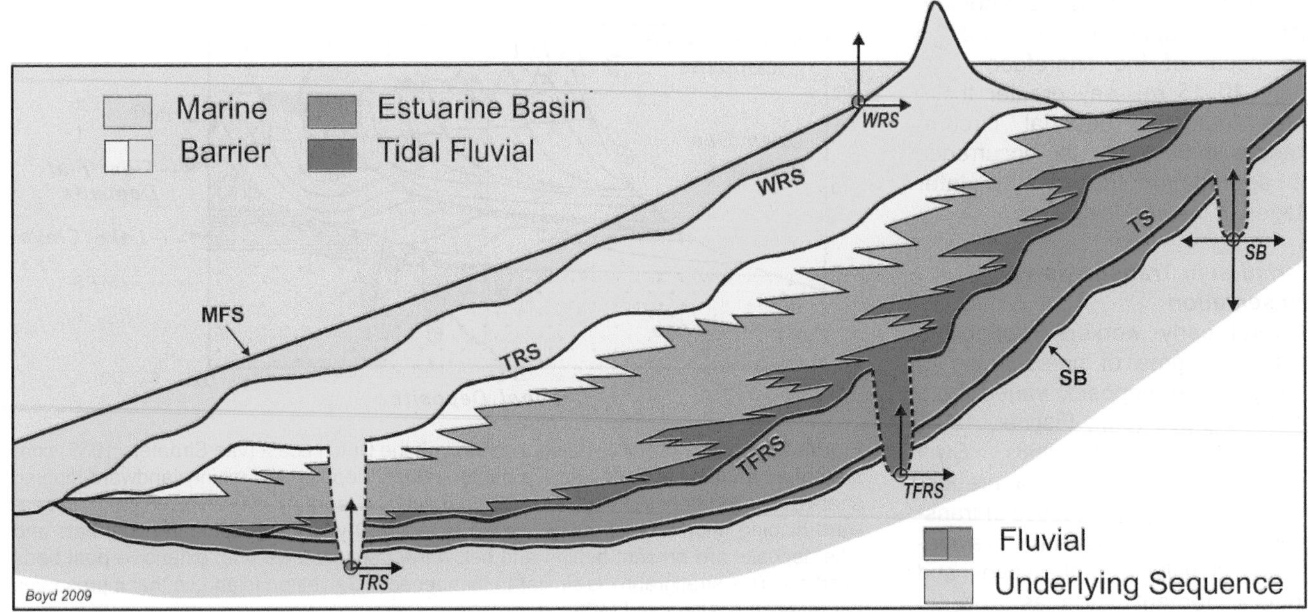

Figure 29. Schematic shore-normal section showing the relative order and stratigraphic level of the four main erosional or ravinement surfaces found in incised valleys and transgressive coastal settings: SB = sequence boundary; TFRS = tidal/fluvial ravinement surface; TRS = tidal ravinement surface; and WRS = wave ravinement surface. The transgressive surface (TS) marks the transition from lowstand regression to transgression, whereas the maximum flooding surface (MFS) marks the reversal from transgression to highstand regression. Red circles and arrows indicate vector translation components of each erosional surface. See text for additional discussion.

In any of the scenarios described here, transgressive shelf sand and mud can overlie the WRS.

As a result of erosion, coastal barrier deposits and others high in the stratigraphic succession are not preserved in most transgressive settings, and occur typically only at transgressive/regressive reversal points, and locations where aggradation takes place. Conversely, incised-valley deposits and facies, such as fluvial deposits that occur low in the stratigraphic succession, are common components of transgressive stratigraphy, particularly in basins with low accommodation (*e.g.,* Zaitlin *et al.,* 2002). The high degree of variability in transgressive preservation is a function of the numerous independent processes that control deposition and erosion: the depth of fluvial erosion is controlled by base level, river discharge and sediment flux; the depth of tidal and tidal-fluvial erosion is controlled primarily by the tidal prism; and the depth of wave erosion is determined mainly by wave energy.

The previous discussion considered only the simple situation of continuous linear transgression where the rates of erosion, sediment input and relative sea-level rise remain relatively constant. However, these rates commonly vary and cause changes in the trajectory or the shoreline and, in turn, produce changes in the degree of preservation along the path of the transgressing shoreline. In very low gradient settings that experience high rates of subsidence and compaction such as the US Gulf Coast, it is possible for the retreating shoreline to be submerged, with the mainland shoreline jumping landward (*see* Fig. 21), 'overstepping' the barrier (*i.e.,* 'transgressive submergence'; Penland *et al.,* 1988). Under these conditions, the mainland shoreline continues to migrate landward after submergence of the barrier, leaving the original shoreline-generated sand body to be reworked into an inner shelf shoal (Fig. 30).

In other situations, the shorter term trajectory may alternate between transgression and regression within an overall longer term transgression. This situation, identified by Swift (1968, 1975; Fig. 21) and others on the US Atlantic coast, and by Frazier (1974) in the Gulf of Mexico, produces composite transgressive and regressive episodes of deposition termed stepped or punctuated transgression. A much greater volume of transgressive sediment can be preserved in this situation.

Other Sources of Stratigraphic Variability

Grain Size and Sediment Composition Considerable differences in transgressive stratigraphy should be expected from gravelly and muddy shorelines as well as from mixed clastic and carbonate settings. Gravelly shorelines are likely to be much steeper, and more inclined to leave coarse lags and shoals behind on the inner shelf after transgression, than their finer grained counterparts. Muddy shorelines, on the other hand, are flatter and tend to be cohesive unless recently deposited. In many instances, sandy deposits are swept landward into estuaries, leaving older, muddy back-barrier deposits exposed on the beachface and shoreface (Fig. 8).

In general, carbonate sedimentation occurs only in the absence of significant clastic input and is well suited to conditions of shoreline transgression where clastics are sequestered in up-dip fluvial and estuarine deposits. Transgressive successions can be capped by oolitic–bioclastic grainstone, peritidal fenestral microbial lime mudstone or coralline algal bindstone/ boundstone (*e.g.,* Mitchell *et al.,* 2001) The mixture of carbonates and clastics in transgressive settings can strongly influence diagenetic pathways and control the development of future porosity and permeability.

Mixed-Energy Shorelines On wide embayed modern shorelines such as the Georgia Bight, Gulf of Carpentaria and German Bight, shorelines exhibit a transition from higher wave energy at the margins of the bight to higher tidal energy in the axis. This transition is accompanied by a progressive shortening of barrier islands (Fig. 13), more closely spaced tidal inlets and more extensive development of ebb-tidal deltas and estuarine tidal flats (Barwis and Hayes, 1979). Estuaries may also experience a range of wave/tidal energy regimes extending from purely wave dominated, such as have been described above, to tide-dominated settings such as the Bay of Fundy (Dalrymple *et al.*, 1992; *see* Chapter 9), with intermediate mixed-energy estuaries such as the Gironde Estuary (Allen and Posamentier, 1993) occurring in transitional settings. In the Gironde, a wave-dominated barrier is present at the mouth of the system, but the bayhead delta is tide-dominated and the central basin includes tidal sand bars as well as muddy basinal deposits.

Wind assists the landward migration of barrier deposits by removing sand from the beach face and transferring it into eolian dunes behind, and on top of, the barrier. In this way, eolian processes can contribute to erosion, enhancing the landward translation of the shoreline. This effect may be enhanced in arid and high-latitude regions where vegetation is less effective at binding the sand.

Climatic Variability Considerable stratigraphic differences occur because of differences in climate. High-latitude coasts (*e.g.*, FitzGerald and Rosen, 1987) are often icebound for many months of the year and experience long periods of low energy, interspersed with high-energy storms. They also experience complex relative sea-level changes because of ice loading and unloading, as well as movement of the associated glacial forebulge. In addition to normal sediment sources, extensive glacial deposits form local sources that contain a wide range of sediment sizes, including coarse gravel. Coastlines are commonly irregular and coastal sediment transport is characterized by short-distance transport from local sources into adjacent embayments. Ice-free high latitude areas, especially between 40-60 degrees N and S and on the western sides of continents, experience high wave energy and hence deeper shorefaces and coarser grainsizes. By comparison, low-latitude areas tend to experience lower wave energy and relatively higher tidal energy, and rare but intense storms, such as tropical cyclones. Increased tropical weathering produces abundant fine-grained sediment leading to a potentially higher proportion of muddy coasts.

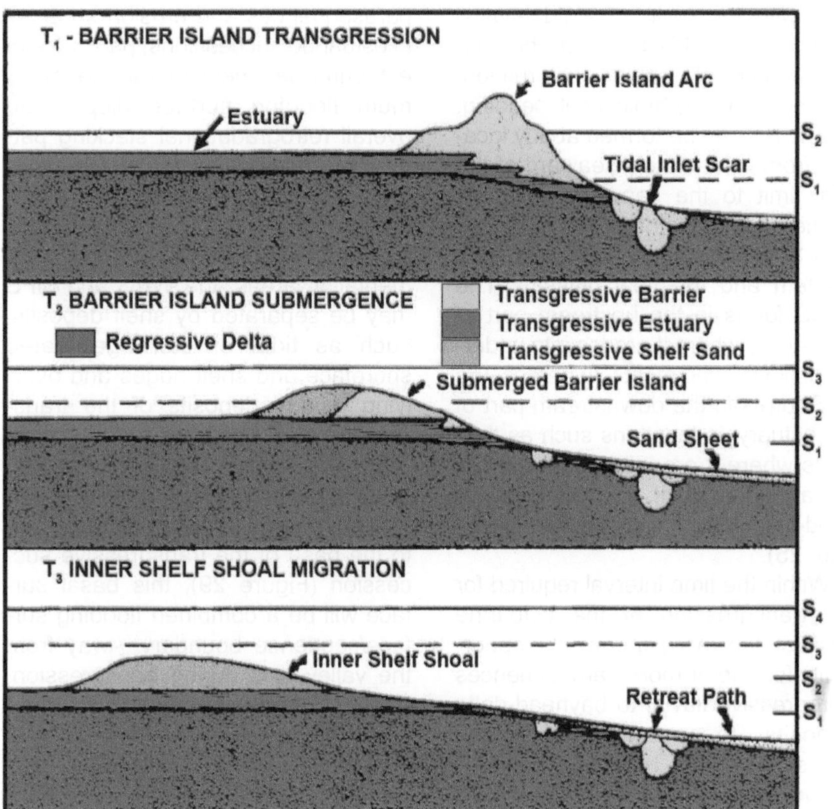

Figure 30. Three stages in the evolution of an inner-shelf shoal by the process of transgressive submergence of a barrier island, based on the abandoned delta lobes of the Mississippi delta (*see* Figs. 8 and 9). Note that the product of this process is a reworked marine sand body, not a preserved barrier. From Penland *et al.* (1988).

SEQUENCE STRATIGRAPHY OF TRANSGRESSIVE WAVE-DOMINATED COASTS

Sediments deposited in transgressive wave-dominated coastal settings are a component of the transgressive systems tract in high-frequency sequences (Catuneanu, 2006), but they can occur in any systems tract in long-duration (*e.g.*, 3rd-order) sequences and sequence sets (*see* Chapter 2). Transgressive successions are typically preserved best in valley-fills (*see* Fig. 17 for an example of a fluvial valley prior to transgression), where they overlie a basal sequence boundary and any lowstand fluvial deposits that might be present at the base of the valley (*cf.* Van Wagoner *et al.*, 1990). The transgression is expressed first by a change in fluvial depositional style (*e.g.*, less amalgamation of channel deposits, and the appearance of peat, wetter soils and lacustrine facies) associated with a transition from forward to backward-stepping parasequences. This change represents the transgressive surface (Fig. 29). If depositional rates are low, this change in depositional style may coincide with the onset of tidal influence and the formation of an erosional flooding surface that is coincident with the base of a tidal-fluvial channel (*i.e.*, TFRS), and the first appearance of marine bioturbation, IHS beds or rhythmic tidal bedding. The TFRS can be formed at any location from immediately seaward of the tidal limit to the seaward extent of bayhead-delta distributaries, and has also been termed the bayhead-delta diastem and the tidal diastem. The TFRS forms in the upstream part of an estuary where it erodes into underlying fluvial sediments, whereas the TRS forms in the downstream part of the estuary, in locations such as tidal inlets where it erodes into underlying fluvial, tidal-fluvial, estuarine-basin, flood-tidal delta and barrier facies (Fig. 25).

Within the time interval required for the transgression of the estuarine basin up the valley, there is the opportunity for one or more parasequences of regressive fluvial to bayhead-delta deposits to accumulate. These parasequences have a wedge-shaped geometry that thins landward to where it onlaps, and seaward to where it downlaps (Fig. 19). At the seaward end of the estuary, a similar set of parasequences may be formed by the flood-tidal delta prograding landward into the estuary. The marine end of the estuary can also develop erosional surfaces resulting from lateral migration of tidal channels in the flood-tidal delta and tidal inlet (*e.g.*, Clifton 1994); these surfaces are the true TRS. The WRS that caps the estuarine succession can frequently be a composite erosion surface formed by the passage of several major storms that erode the underlying transgressive systems-tract deposits to varying depths. It can also be a stepped erosional surface formed by punctuated transgression.

At the end of formation of the transgressive systems tract, the shoreline reaches the transgressive limit and parasequences begin to aggrade and/or prograde. The final flooding surface where this transition occurs identifies the maximum flooding surface (Fig. 29). This surface has special significance for coastal deposits as it represents the end of shoreline transgression, and signifies future estuarine infilling and the end of estuarine sedimentation. Despite the potentially complex arrangement of parasequences and erosion surfaces in estuarine successions, packages of estuarine sediment below the maximum flooding surface display an overall retrogradational stacking pattern. In situations where there is continual reworking and little or no deposition on the shelf, the WRS and MFS coincide. Alternatively, in high sedimentation areas, the WRS and MFS may be separated by shelf deposits, such as tidal or storm-generated shoreface and shelf ridges and overlying muddy deposits of the transgressive systems tract (*see* Chapter 8).

In situations where the tidal- or wave-ravinement surfaces penetrate to the base of the transgressive succession (Figure 29), this basal surface will be a combined flooding surface/sequence boundary. Away from the valley axis during transgression, the basal sequence boundary can be onlapped by a range of peripheral deposits such as marshes and swamps, distal bayhead and flood-tidal deltas, organic deposits such as peat or oyster reefs, or eolian deposits, followed eventually by one or more ravinement surfaces and open-marine deposits.

CASE STUDIES FROM TRANSGRESSIVE WAVE-DOMINATED COASTS

Quaternary Example: West Netherlands Basin

One region that has been the subject of extensive modern investigations is the West Netherlands Basin on the North Sea coast of western Europe (Figs. 28 and 31). Here, the stratigraphic succession produced during the rapid sea-level rise in the early Holocene and subsequent more stable period has been documented by Tornqvist (1998) and Busschers *et al.* (2007) among many others. Hijma *et al.* (2009) provide a particularly detailed set of cross sections of transgressive early–mid–Holocene valley-fill and estuarine deposits of the Rhine–Meuse Delta that accumulated when rates of RSL rise were 8–10 mm a^{-1} and tidal ranges were 1.5–2 m (*i.e.*, microtidal). A complex suite of fluvial, fluvial-tidal and bayhead-delta facies is present (Fig. 31). The late Pleistocene evolution of the Rhine–Meuse area began with valley incision followed by fluvial deposition, coupled with eolian activity. Holocene RSL rise and associated flooding resulted in widespread peat formation and fluvial aggradation, followed by the establishment of an estuarine tidal (central) basin into which the Rhine River deposited a bayhead delta and widespread tidally influenced silty clays (Fig. 31). The seaward side of this system was occupied by sandy barriers and tidal inlets. When the rate of RSL rise slowed after 6 ka BP, the tidal basin quickly filled leading to widespread peat formation. Tidal inlets closed and regressive coastal strandplains formed (Fig. 28).

Quaternary Example: Delaware Coast, USA

A well-documented example of a modern transgressive coastline was provided by Kraft and co-workers (*e.g.*, Demerest and Kraft 1987) from the Delaware coast of the eastern USA (Figs. 32 and 33). This coastline stretches for 40 km alongshore from Maryland to Cape Henlopen and takes in the Rehoboth and Bethany Beach headlands and the estuaries of Delaware, Indian River and Rehoboth bays. Barriers have built parallel-to-shore spit systems up to 25 m thick into Delaware Bay, and sandbodies up to 10 m thick have accumulated at the mouths of the smaller bays. These sand accumulations consist mainly of tidal inlet, flood-tidal delta and washover facies. Extensive shore-parallel and shore-normal drilling transects identified the major contribution of eroding headlands to the coastal sediment budget and the variability in accommodation behind the barriers that results from the presence of incised valleys. In the deepest incised valleys, transgressive successions over 25 m thick accumulated prior to passage of the shoreface and formation of the WRS. At the present shoreline, these sediments are composed of approximately 50% each of sandy barrier and muddy estuarine/lagoonal deposits. However, as shown in Figure 33, up to 8 m of erosion is occurring on the modern shoreface, resulting in the removal of most of the barrier deposits and preservation primarily of the basal flood- and ebb-tidal deltas, some estuarine and lagoonal sediment, and

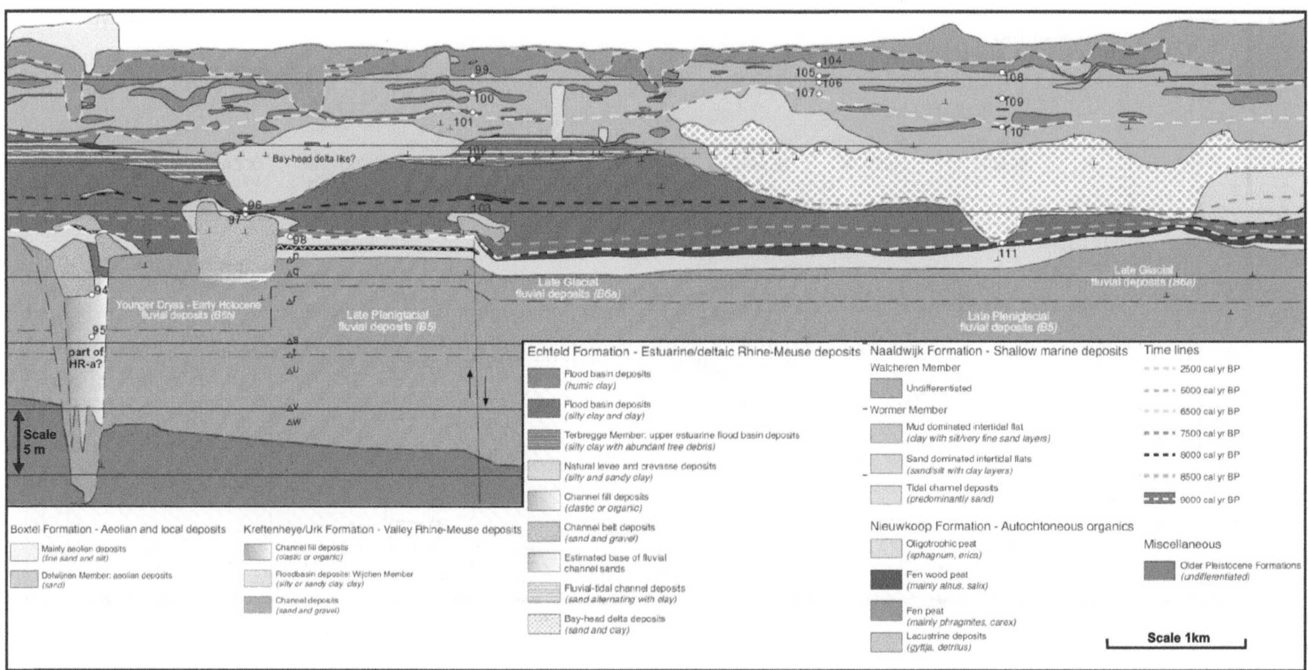

Figure 31. Transgressive stratigraphy of the Rhine–Meuse delta from Hijma *et al.* (2009). This is part of several long, shore-parallel cross sections showing a complex distribution of transgressive facies including interpreted bayhead deltas and flood-tidal deltas.

the basal fluvial deposits. On the headland segments of the coast, shoreface erosion is generating a combined SB/WRS where it truncates the valley interfluves.

Quaternary Example: Central NSW Coast, Australia

The wave-dominated central New South Wales (NSW) coast of Australia has accumulated multiple transgressive/regressive stratigraphic cycles during the Quaternary. These cycles have been well documented, for example in the Tuncurry area (Roy *et al.*, 1995; Fig. 34; *see* also Chapter 8, Fig. 24). Results show a series of Quaternary fine to coarse, mature quartzose shoreline sandbodies, separated by thinner, muddier estuarine deposits. The Holocene Tuncurry barrier/ strandplain shows a composite transgressive/regressive architecture. The basal and more landward transgressive unit (S1(tr)) is wedge-shaped and up to 22 m thick, contains estuarine macrofauna and was deposited during the Holocene sea-level rise from 8.5 to 6.2 ka BP. The overlying and more seaward regressive unit (S1(rg)) is also about 20 m thick, contains mainly marine macrofauna and was deposited between 6 ka BP and the present, after sea level stabilized. Progradational time lines within the system approximate the modern shoreface profile but the transgressive barrier has a lower clinoform slope (0.007 or 0.38° versus 0.011 or 0.63°). Estuarine deposits of the Wallamba River and Wallis Lake lie behind the modern barrier and contain a tidal channel and an extensive flood-tidal delta that extends 8.5 km into the estuary. It was mostly deposited between 2–4 ka BP.

Adjacent to the Holocene barrier/ strandplain at Tuncurry and elsewhere on the NSW coast, there are a series of older barrier/strandplain features, although those at Tuncurry are the best described and dated. Here, composite coastal sandbodies 12–14 m thick occur landward of the Holocene sandbody and form a seaward-descending and younging series of highstand barrier/strandplain sandbodies (S7a-S5c in Fig. 34). On the inner to mid shelf at a depth of -30 to -60 m there is another composite sandbody dating from 43–59 ka BP. This submerged barrier/strandplain is also composed of multiple sandbodies separated by transgressive estuarine deposits, and represents deposition during an overall falling sea level, resulting in a forced-regressive succession punctuated by transgressive events. Each of the Tuncurry shoreline sandbodies thus occupies the turnaround position from transgression to regression of one of 5 separate, high-frequency sea-level cycles (Fig. 34 inset). There has been little tectonic or compactional subsidence and hence the shoreline sandbodies are located today at the approximate sea level that existed at their time of deposition. Three sandbodies were deposited at, or slightly higher than, present-day sea level and are landward of the current shoreline. One sandbody was deposited at a lower sea level than present and is located seaward of the modern shoreline. This stratigraphic architecture reflects lateral stacking of shoreline deposits resulting from a lack of accommodation because there is no tectonic subsidence or compaction. It contrasts markedly with the Cretaceous Hosta Sandstone example (*see* below) where rapid creation of accommodation resulted in vertical stacking of transgressive/regressive shoreline sandbodies in an overall retrogradational architecture. However, the New South Wales example is similar to laterally stacked transgressive Pleistocene shorelines in Delaware and transgressive/regressive raised Pleistocene shorelines in Georgia and Texas (*e.g.*, Demarest and Kraft,

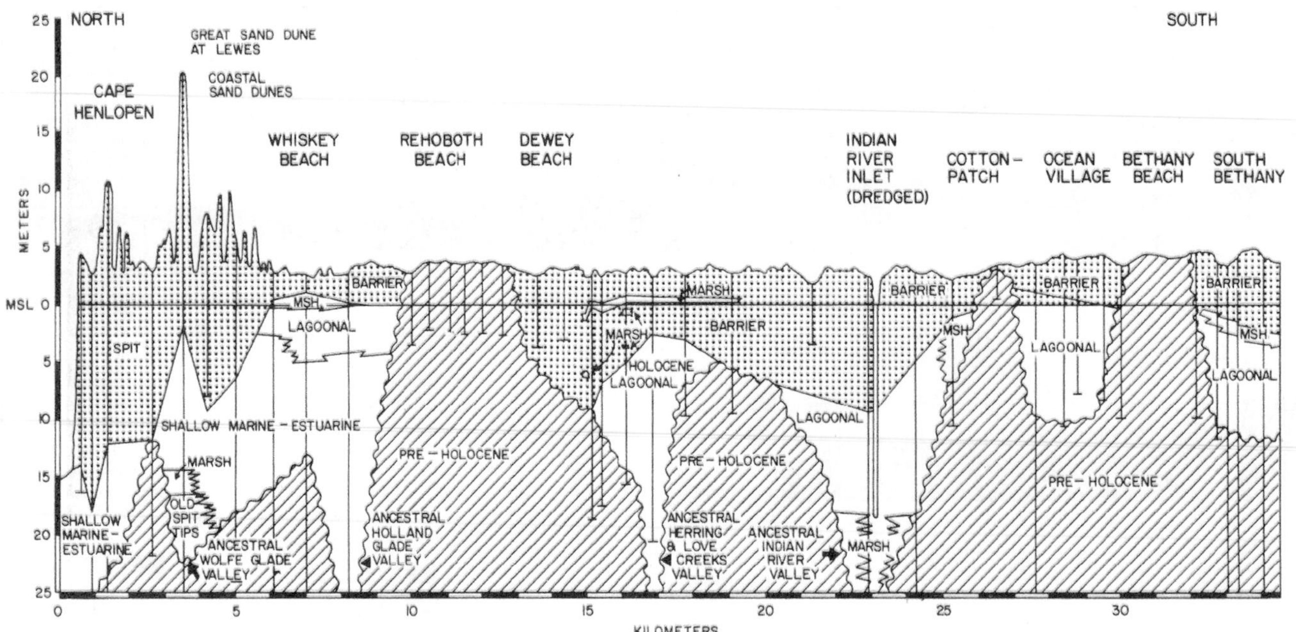

Figure 32. Transgressive stratigraphy of the Delaware coast, USA (from Kraft *et al.*, 1987). This shore-parallel section, obtained on the present-day barrier, shows valley incision to a depth of over 25 m and a transgressive fill of estuarine facies at the base, an upward transition to lagoonal and marsh facies, followed by an overlying barrier and spit unit. Shoreface erosion will remove the upper 7–10 m of this succession.

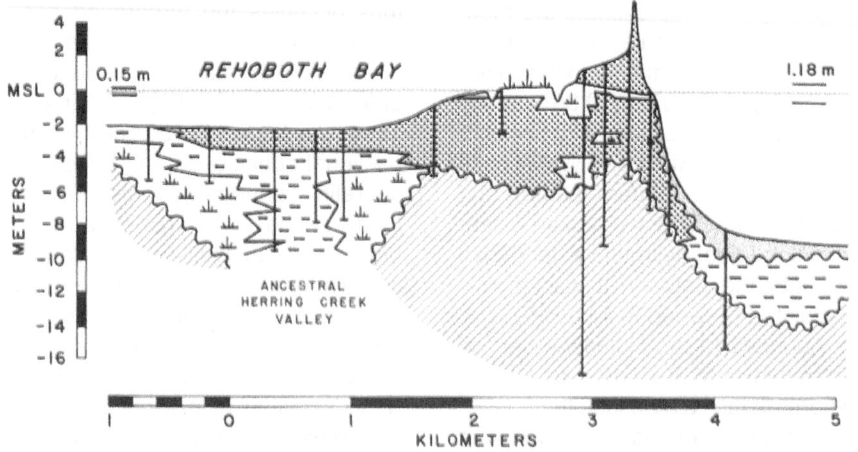

Figure 33. Shore-normal cross section of Rehoboth Bay, Delaware, USA (from Kraft *et al.*, 1987). This section shows that scour in a tidal inlet has modified the sequence boundary. The barrier superstructure is composed mainly of flood-tidal delta and washover facies, with the modern beach and dunes making up only a thin cap. Rehoboth Bay is a former valley filled with lagoonal and estuarine facies behind the barrier, whereas the shoreface is currently eroding to a depth of about 7–10 m below present sea level. Comparison with Figure 32 suggests that, if transgression continues, the WRS will remove most of the barrier; only the deeper parts of the valley fill will be preserved. The short horizontal lines around mean sea level (MSL) indicate tidal range in the ocean and the estuary.

1987).

Ancient Example: Cretaceous Point Lookout Sandstone, San Juan Basin, USA

A series of investigations of the well-exposed Cretaceous Cliff House and Point Lookout sandstones in the San Juan Basin of New Mexico (*e.g.*, Devine, 1991; Olsen *et al.*, 1999) has shown that they are part of a series of transgressive-regressive cycles that characterize deposition in the North American Cretaceous foreland basin (Fig. 35). Sixsmith *et al.* (2008) describe a detailed example of an overall transgressive sandstone from the Cretaceous Hosta Tongue of the Point Lookout Sandstone (Figs. 36 and 37). Sixsmith *et al.* (2008) interpret the Hosta Tongue to consist of three major depositional environments:

1. A broad (2–6-km wide), sandy back-barrier platform containing blind tidal-channel networks separated by dunes and barforms;
2. Barrier islands punctuated by tidal-inlet channels; and
3. A wave-dominated shoreface and inner shelf that is 3–6-km wide.

The deposits of these environments are stacked transgressively in response to a net shoreline retreat of 7.5 km during a relative sea-level rise of 50 m.

This net-transgressive succession is punctuated, however, by five transgressive/regressive cycles that form a retrogradationally stacked parasequence set (*sensu* Van Wagoner *et al.*, 1990) with WRS separating the parasequences. The landward part of each parasequence (Figs. 36 and 37A) consists of tide-dominated back-barrier deposits (tidal-channel fills, tidal flats, washover fans and swash-platform facies) and a seaward portion (Figs. 36 and 37B) consisting of wave-dominated shallow-marine facies (lower to upper shoreface and offshore-transition to inner-shelf deposits of the Mancos Shale).

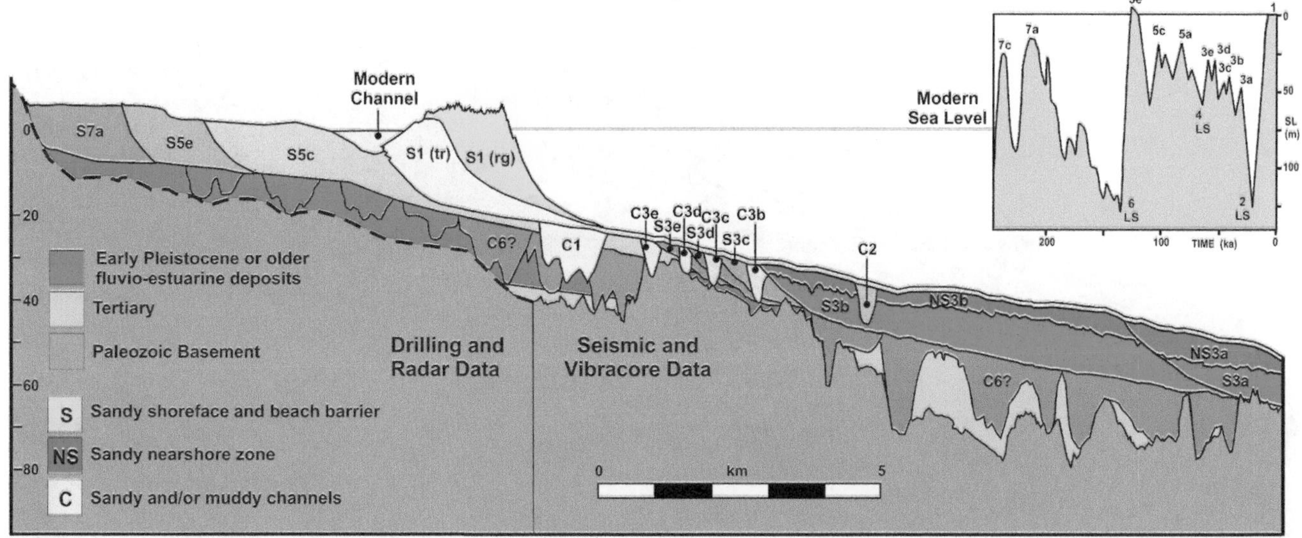

Figure 34. Shore-normal cross section through the Quaternary coastal stratigraphy at Tuncurry on the central New South Wales coast of SE Australia. This complex stratigraphy illustrates a situation in which little vertical accommodation has been added and sediment has accumulated by lateral stacking of successive high-frequency sequences, which can be correlated to the eustatic sea-level curve (inset). The present-day barrier and strandplain unit consists of a composite transgressive (S1tr) and regressive (S1rg) unit in front of a modern estuary and channel fill. This style is mimicked by numerous transgressive/regressive cycles on the inner shelf, more of which were deposited during the falling stage of oxygen-isotope stage 3 (3a–3e), and also by those further landward (S5 and S7). Overall, each package consists of a transgressive estuarine and barrier unit fronted by a regressive strandplain unit at the RSL reversal position, with each of these positions corresponding to a sea-level highstand. Compiled from Roy *et al.* (1995), with additional interpretations of J. Wadsworth and R. Boyd. See text for more discussion.

Ancient Example: Carboniferous Morrow Formation, Anadarko Basin, USA

The Carboniferous (Lower Pennsylvanian) Anadarko Basin of Oklahoma, Texas, Kansas and Colorado occupied a low-gradient intra-cratonic depression that experienced repeated transgressive-regressive cycles through Morrowan time (approximately 7 cycles over the 6 Ma from 318–312 Ma). Shoreline movements covered distances of 400 km between the highstand and lowstand positions (Sonnenberg *et al.*, 1990). These shorelines occupied an approximately equatorial position and experienced both wave and tidal influences. Upper Morrow transgressive deposits consist mainly of incised-valley fills (Fig. 38), whereas the Lower Morrow also includes transgressive barrier and shoreface deposits. In the Upper Morrow of Colorado, Kansas and the Panhandle region of Texas and Oklahoma, conditions alternated between exposed marine deposits at lowstand, traversed by incised valleys that filled with fluvial and estuarine deposits during transgression, and highstand black shales and limestones. Krystinik and Blakeney (1990) and Bowen and Weimer (2003) described transgressive deposits in the Morrow as occurring in long linear valleys with up to seven major cycles of fill (Fig. 38), commonly in the same valley trend whose position is controlled by underlying faults. Individual valleys are over 250 km long, between 0.8 and 6.4 km wide and up to 30 m deep.

Morrow valleys are typically incised into underlying marine limestone and shale of the preceding highstand, and commonly have a basal fluvial-channel and floodplain fill. Fluvial channels of a single cycle are generally relatively thin (2–5 m, but may be up to 10 m), and are composed of coarse-grained sandstone and conglomerate, pedogenically altered floodplain shale and crevasse-splay deposits. The SB and basal fluvial deposits are then overlain by finer grained estuarine sandstone and shale (Fig. 38), followed by a marine ravinement/flooding surface. Preserved shoreline deposits outside the valleys are rare, and within valleys the most common estuarine facies are those lowest in the stratigraphic succession including tidal/fluvial channels, bayhead delta and estuarine central-basin deposits. The preserved Morrow stratigraphy reflects a combination of low gradients and large glacio-eustatic sea-level oscillations that led to rapid transgressions along flat shoreline trajectories.

Bowen and Weimer's (2003) studies of petroleum fields along the length of the Morrow valleys found that amalgamated fluvial deposits predominate in proximal locations in Colorado, mixed tidal–fluvial facies are abundant in intermediate positions, and tidal facies are most abundant in distal locations in Kansas. Detailed facies studies by Wheeler *et al.* (1990) and Buatois *et al.* (2002) showed, however, that the shoreline trajectory was locally steep, resulting in the preservation of estuary-mouth facies and interbedding of transgressive estuarine-valley fills and regressive open-marine offshore to shoreface successions. This architecture reflects a punctuated transgression and the development of a 'beaded' transgressive valley fill with isolated occurrences of sandier estuary-mouth deposits.

Figure 35. Outcrop photo of the transgressive Cliff House Sandstone, SW Colorado, showing the transition from a thick barrier sandbody on the right (red arrow approximately 5 m high for scale) to a thin, interbedded sand and shale on the left (yellow arrow). The red arrow also identifies the sharp ravinement contact between the gray-colored organic-rich coastal-plain shales of the Menefee Formation below and the medium-grained Cliff House barrier sands above. Photo courtesy of Bruce Hart.

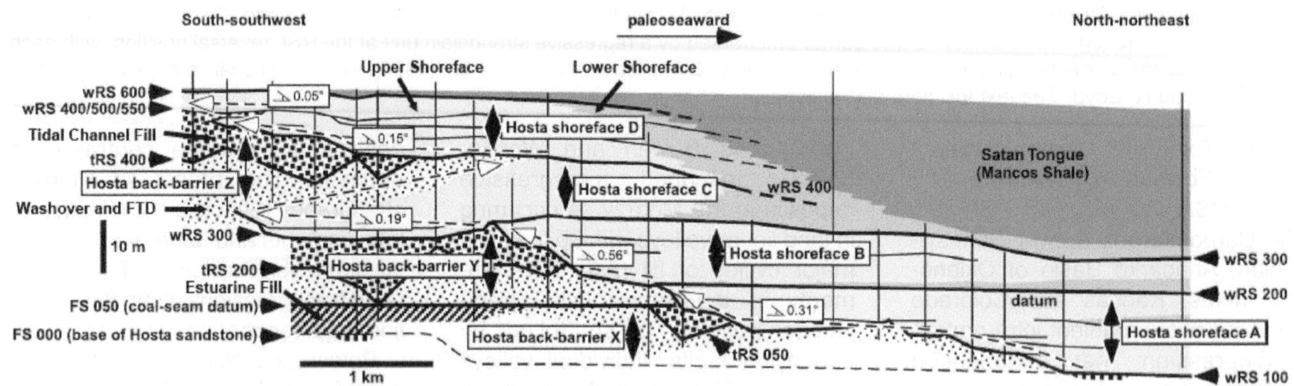

Figure 36. Stratigraphy of the Hosta Sandstone along depositional dip, showing four cycles, each of which contains transgressive back-barrier tidal-delta and washover facies (X–Z) at its landward end, and regressive shoreface facies (A–D) to the NNE. The overall shoreline trajectory exhibits a zig-zag path (net transgressive, with short regressive episodes) in response to variations in sediment supply and relative sea level, and illustrates an example of a step-wise or punctuated transgression From Sixsmith *et al.* (2008). .

RECENT AND FUTURE DEVELOPMENTS

Facies models for coastal systems were developed relatively early, during the 1960s to 1970s, and have been augmented over the following three decades with the addition of models for estuary and incised-valley deposits, and the integration of sequence-stratigraphic concepts. Hence, facies models for transgressive wave-dominated coasts are relatively mature, with a firmly established summary model (Figs. 19 and 29) and an appreciation of the main types and causes of variability. Nevertheless, there are still aspects of this depositional setting that require further investigation. In addition, many of the recent advances in this field, like those for other environments, result from the application of new technology that allows the environment and its deposits to be viewed in new ways, thereby providing new insight and understanding.

Recent Advances and New Technologies

In modern environments, improved sidescan sonar (Thieler *et al.*, 2001) and multibeam techniques (Goff *et al.*, 2005; Mayer et al., 2007) permit visualization of large areas of the sea or estuary floor with previously unavailable sub-decimeter resolution, documenting processes such as shoreface retreat and estuary bedform migration. Ground-penetrating radar allows the acquisition of vertical cross sections to a depth of up to 60 m, with a vertical resolution of less than a decimeter. This technique has been used to visualize the internal stratigraphy of barrier deposits in a wide variety of marine (Fitzgerald *et al.*, 1992; Moller and Anthony, 2003) and lacustrine settings (Smith *et al.*, 2003).

In ancient deposits, LIDAR ((LIght Detection And Ranging) allows researchers to create very high-resolution, digital outcrop models that can be visualized in three dimensions and from any selected perspective (Bellian *et al.*, 2005; Fabuel-Perez *et al.*, 2009). The intensity and character of

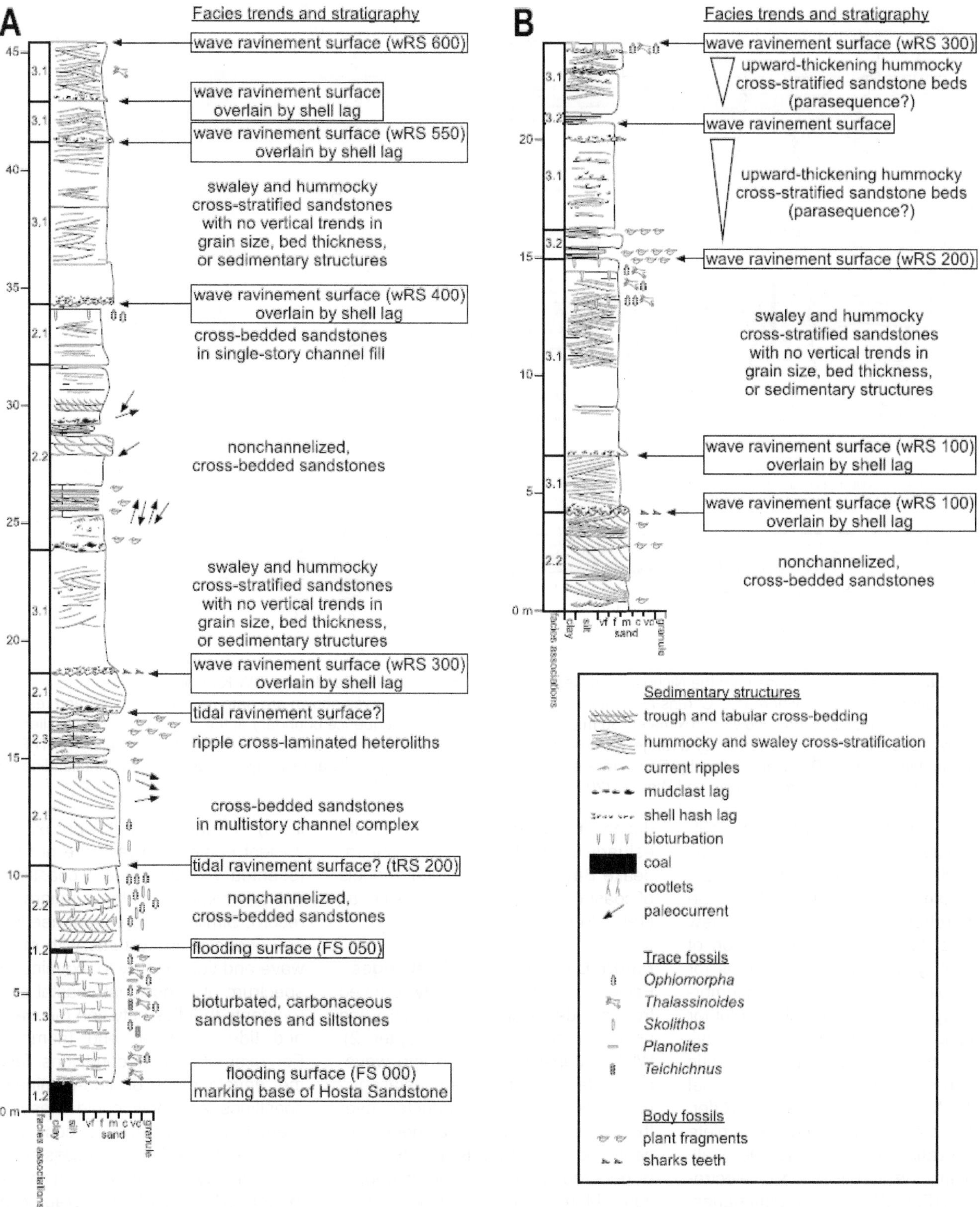

Figure 37. Logged sections through the Hosta Sandstone illustrating facies successions and key stratigraphic surfaces in (**A**) proximal and (**B**) distal settings, as might be seen at the left (**A**) and right (**B**) sides, respectively, of Figure 36. From Sixsmith *et al.* (2008).

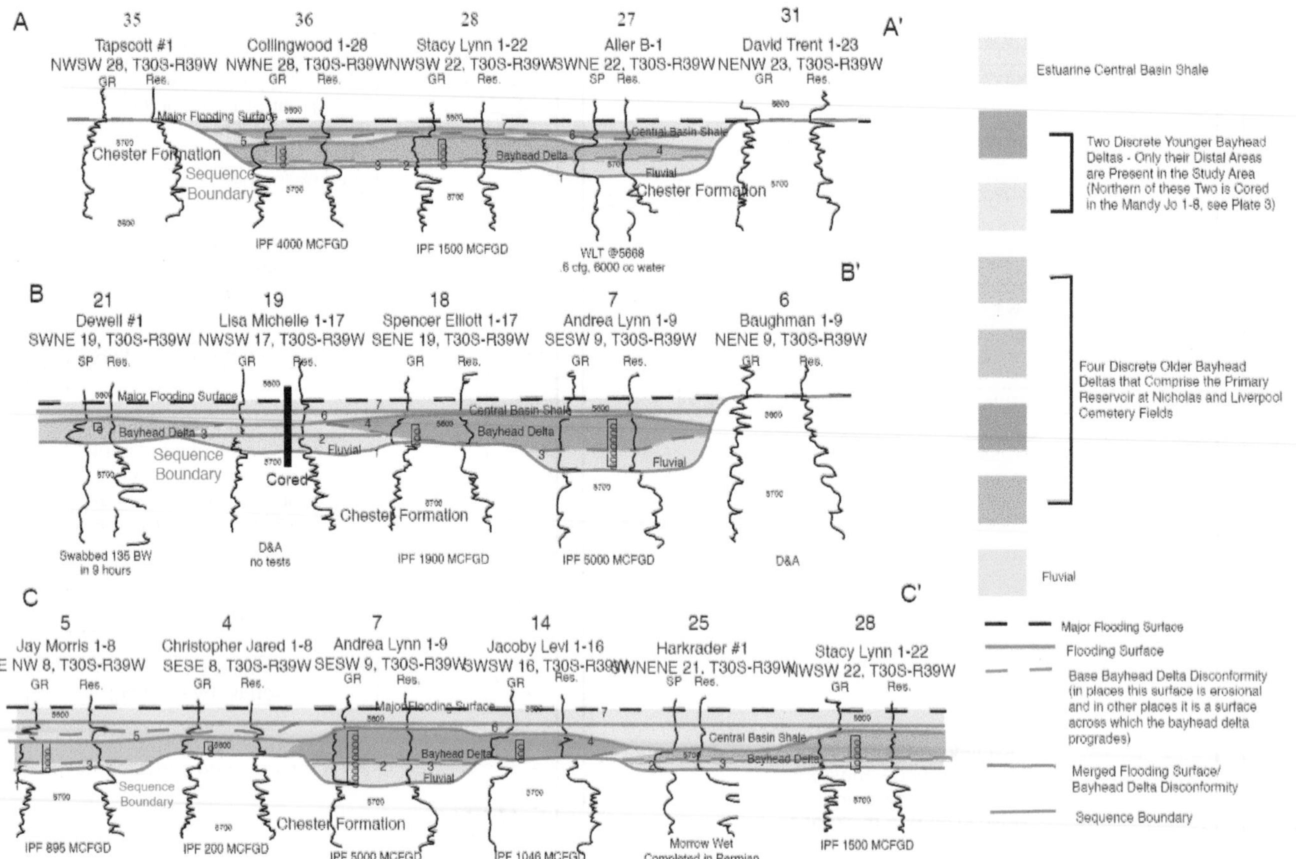

Figure 38. Three cross sections from the Nicholas and Cemetery fields in the Morrow Sandstone, SW Kansas. Cross sections A-A' and B-B' run across the valley, whereas C-C' runs along the length of the valley. Log curves shown are gamma ray on the left and resistivity on the right. These sections show approximately 30 m of valley incision with a sequence boundary at the base. The valley is filled first with fluvial sediment, which is overlain by a range of transgressive bayhead-delta facies and estuarine central-basin shale. The TFRS is shown as the Bayhead Delta Disconformity. Each cross section is about 5 km long, wells are evenly spaced along the section line and depths shown are in feet. From Bowen and Weimer (2004)

the laser light reflected from the outcrop holds promise as a tool for characterizing facies. In the subsurface, 3D-seismic techniques allow the researcher to obtain plan-view images that show the distribution of deposit types that rival airphotos for their ability to visualize paleo-geomorphology, and are especially useful for imaging the shape of incised valleys (Fig. 18). High-resolution acoustic (borehole televiewer) or electrical (FMI/EMI) borehole images (Rider, 1996) provide an almost complete visualization of the borehole wall, enabling the identification of electrofacies (Fig. 20), facies boundaries, fractures, sedimentary structures and even bioturbation without the need for core.

Quantitative techniques also hold significant potential to increase our understanding of transgressive coastal systems. A wide range of numerical models have been produced, including those dealing with estuarine circulation (*e.g.*, Chesapeake Bay), and geometrical models of coastal evolution (*e.g.*, Cowell *et al.* (1999; Fig. 14)

Future Directions and Challenges

Facies models are ideally derived from a wide database of modern and ancient examples (*see* Chapter 2). While extensive research into wave-dominated transgressive coastal environments has been conducted over the past six decades, it has been concentrated in a relatively narrow range of geographic and stratigraphic locations. More future work needs to be undertaken in situations such as polar and tropical settings that are not dominated by mid-latitude climate. For example, it should be possible to use climate models to predict the geographic distribution of past facies associations (*e.g.*, higher wave energy on west-facing high-latitude coastlines). Situations that approximate ancient epeiric seaways are not represented well by modern coastlines, but are common in the stratigraphic record. Similarly, more work needs to be conducted in systems with mixed wave and tidal energy to examine the spectrum of transitional coastal environments that lies between the wave- and tide-dominated end members. For example, what conditions favor the development of tide-dominated coastlines and how could these be predicted in ancient successions? What would the facies successions of more tidally dominated but wave-influenced coastal stratigraphy be like?

Although work has been done on stratigraphic successions where carbonate and clastic deposition alternates, there have been few studies in transgressive environments of contemporaneous carbonate and clastic deposition, even though this is relatively common in tropical locations. In modern settings, much work has been conducted on present-day highstand shorelines, but more needs to be done on the preserved submarine depositional record on the continental shelf (*e.g.*, Nordfjord *et al.;* 2006), and on the style of deposition at the lowstand turnaround position where regression changes to transgression (*e.g.*, Ponten and Plink-Bjorklund, 2009). The detailed nature of the relative sea-level record is not well known as it applies to the history of transgression. Recent work suggests that the melting of polar ice caps and marine ice sheets was episodic, and that the subsequent global sea-level changes are periodically rapid and step-wise. An important problem that needs to be addressed is what the coastal response is to this type of sea-level rise, and whether the parasequence stacking patterns and flooding surfaces observed in the ancient record are a response to rapid episodic changes in eustasy or if they are autocyclic and a response mainly to variations in sediment supply. Larger scale issues that need further work include the influence that spatial differences in subsidence have on the variability of transgressive deposits. For example, how do transgressive deposits vary between the lowstand and highstand shorelines on old divergent margins versus young convergent margins? In the latter situation, subsidence pattern can vary greatly in both magnitude and direction over both time and space. Also, how do transgressive deposits vary in response to very different rates of RSL rise that probably existed during more stable greenhouse periods and highly variable icehouse conditions? During transgression, the rate of RSL rise controls the vertical vector of shoreface translation but what controls the horizontal vector? Does the amount of time available or the substrate resistance produce variations in this horizontal vector? Can the result be predicted in advance?

In general, although there have been numerous studies of modern transgressive coastal environments, there have been relatively few detailed studies of their ancient counterparts. Modern environments cannot reproduce the range of sediment supply and RSL combinations that occurred throughout earth history and additional studies of the spectrum of ancient transgressive coastal environments are needed. This, combined with the expanded use of the range of new visualization technologies such as digital outcrop construction, 3-D seismic and image logs, together with the ability to simulate the entire range of stratigraphic variability in quantitative models, represents the future path toward developing a more comprehensive suite of facies models for transgressive coasts.

ACKNOWLEDGEMENTS

This chapter benefited from many external contributions, and the author wishes to thank the following in particular: ConocoPhillips for providing time and resources to complete the manuscript; R.W. Dalrymple for professional editing; D. Swift and A. Niederoda for assistance with Figure 15; S. Lang for Figure 12; P. Cowell for Figure 14; H. Posamentier for Figure 18; K. Maier and R. Reid for Figure 20; and B. Hart for Figure 35.

REFERENCES

Basic sources of information

Boyd, R., Dalrymple, R. and Zaitlin, B.A., 1992, Classification of clastic coastal depositional-environments: Sedimentary Geology, v. 80, p. 139-150.
Classifies transgressive wave-dominated environments in the overall context of coastal depositional settings and provides useful definitions of most systems.

Boyd, R., Dalrymple, R. and Zaitlin, B.A., 2006, Estuarine and incised-valley facies models, *in* Posementier, H. W. and Walker, R. G., *eds*., Facies Models Revisited: SEPM Special Publication 84, p. 171-235.
Recent and well-illustrated review of estuarine and incised-valley sedimentology and stratigraphy in CD format.

Cattaneo, A. and Steel, R. J., 2003, Transgressive deposits, a review of their variability: Earth Science Reviews, v. 62, p. 187-223.
Comprehensive review of transgressive clastic deposits with excellent bibliography and advanced treatment of sequence-stratigraphic concepts.

Catuneanu, O., 2006, Principles of Sequence Stratigraphy: Elsevier, Amsterdam, 375 p.
This recent text provides an authoritative view of sequence stratigraphy that includes many concepts of transgressive coastal stratigraphy.

Coe, A.L., *ed*., 2003, The Sedimentary Record of Sea Level Change: Open University: Cambridge University Press, Cambridge, 287 p.
Well illustrated overview of sea-level changes and resulting sedimentation with especially useful treatment of shoreline sedimentation in Part 3.

Dalrymple, R. W., Zaitlin, B. A. and Boyd, R., 1992, Estuarine facies models: Conceptual basis and stratigraphic implications: Journal of Sedimentary Petrology, v. 62, p. 1130-1146.
Early perspectives paper that presented one of the first comprehensive facies models for estuaries.

Dalrymple, R.W., Boyd, R. and Zaitlin, B.A., 1994, Incised-Valley Systems: Origin and Sedimentary Sequences: SEPM Special Publication 51, 391 p.
One of the first volumes to provide a systematic study of incised valleys; includes conceptual framework and numerous modern and ancient case studies.

Dalrymple, R.W., Leckie, D.A. and Tillman, R.W., 2006, Incised Valleys in Space and Time: SEPM Special Publication 85, 348 p.
Most recent summary volume on incised valleys, enabling comparison of valleys of different ages and settings.

Davis, R.A., 1985, Coastal Sedimentary Environments: Springer, New York, 716 p.
An older but detailed treatment of most coastal environments, including chapters on estuaries, tidal inlets, shorefaces, tidal flats and barriers.

Other references

Allen, G. P. and Posamentier, H. W., 1993, Sequence stratigraphy and facies model of an incised valley fill: The Gironde Estuary, France: Journal of Sedimentary Petrology, v. 63, p. 378-391.

Allen, J.R.L., 1982, Sedimentary Structures, their Character and Physical Basis. Developments in Physical Sedimentology, 30A and B: Elsevier, Amsterdam, 592 p.

Ashley, G.M. and Sheridan, R.E., 1994, Depositional model for valley fills on a passive continental margin, *in* Dalrymple, R.W., Boyd, R. and Zaitlin, B.A., *eds*., Incised-Valley Systems: Origin and Sedimentary Sequences: SEPM Special Publication 51, p. 285-302.

Bellian, J. A., Kerans, C. and Jennette, D. C., 2005, Digital outcrop models: Applications of terrestrial scanning LiDAR technology in stratigraphic modeling: Journal of Sedimentary Research, v. 75, p. 166-176.

Barwis, J.H. and Hayes, M.O., 1979, Regional patterns of modern barrier island and tidal inlet deposits as applied to paleoenvironmental studies, *in* Ferm, J.C. and Horn, J.C., *eds.*, Carboniferous Depositional Environments in the Appalachian Region: University of South Carolina, Carolina Coal Group, p. 472-498.

Bernard, H.A., Major, C.F., Jr. and Parrot, B.S., 1959, The Galveston barrier island and environs: a model for predicting reservoir occurrence and trend: Gulf Coast Association of Geological Societies Transactions, v. 9, p. 221-224.

Bernard, H.A., Major, C.F., Jr., Parrot, B.S. and LeBlanc, R., 1970. Recent sediments of southeast Texas – A field guide to the Brazos alluvial and deltaic plains and the Galveston barrier island complex. The University of Texas at Austin, Bureau of Economic Geology, Guidebook 11, 132 p.

Blum, M.D. and Törnqvist, T.E., 2000, Fluvial responses to climate and sea-level change: a review and a look forward: Sedimentology, v. 47, Supplement 1, p. 2-48.

Bowen D.W. and Weimer, P., 2003, Regional sequence stratigraphic setting and reservoir geology of Morrow incised-valley sandstones (Lower Pennsylvanian), eastern Colorado and western Kansas: American Association of Petroleum Geologists Bulletin, v. 87, p. 781-815.

Bowen D.W. and Weimer, P., 2004, Reservoir geology of Nicholas and Liverpool Cemetery fields (Lower Pennsylvanian), Stanton County, Kansas, and their significance to the regional interpretation of the Morrow Formation incised-valley-fill systems in eastern Colorado and western Kansas: American Association of Petroleum Geologists Bulletin, v. 88, p. 47-70.

Boyd, R. and Penland, S., 1981, Washover of deltaic barriers on the Louisiana Coast: Transactions of the Gulf Coast Association of Geological Societies, v. XXI, p. 243-248.

Boyd, R. and Penland, S., 1984, Shoreface translation and the Holocene stratigraphic record: Examples from Nova Scotia, the Mississippi Delta and Eastern Australia: Marine Geology, v. 60, p. 391-412.

Boyd, R. and Honig, C., 1992, Estuarine sedimentation on the Eastern Shore of Nova Scotia: Journal of Sedimentary Petrology, v. 62, p. 569-583.

Boyd, R., Ruming, K. and Roberts, J.J., 2004, Geomorphology and surficial sediments of the SE Australian continental margin: Australian Journal of Earth Sciences, v. 51, p. 743-764.

Buatois, L.A., Mangano, M.G., Alissa, A. and Carr, T.R., 2002, Sequence stratigraphic and sedimentologic significance of biogenic structures from a late Paleozoic marginal- to open-marine reservoir, Morrow Sandstone, subsurface of southwest Kansas, USA: Sedimentary Geology, v. 152, p. 99-132.

Busschers, F.S., Kasse, C., Van Balen, R.T., Vandenberghe, J., Cohen, K.M., Weerts, H.J.T., Wallinga, J., Johns, C., Cleveringa, P. and Bunnik, F.P.M., 2007, Late Pleistocene evolution of the Rhine-Meuse system in the southern North Sea basin: imprints of climate change, sea-level oscillation and glacioisostacy: Quaternary Science Reviews, v. 26, p. 3216-3248.

Cowell, P.J., Roy, P.S., Cleveringa, J. and de Boer, P.L., 1999, Simulating coastal systems tracts using the shoreface translation model: SEPM Special Publication 62, p. 165-175.

Curray, J.R., 1964. Transgressions and regressions, *in* Miller, R.L., *ed.*, Papers in Marine Geology: New York, Macmillan, p. 175-203.

Curray, J.R., Emmel, F.J. and Crampton, P.J.S., 1969, Holocene history of a strand plain lagoonal coast, Nayarit, Mexico, *in* Castanares, A.A. and Pfleger, F.B., *eds.*, Lagunas Costeras: Simposium UNAM-UNESCO, p. 63-100.

Davies, J.L., 1964, A morphogenic approach to world shorelines: Zeitschrift fur Geomorphology, v. 8, p. 27-42.

Demarest, J.M., II and Kraft, J.C., 1987, Stratigraphic record of Quaternary sea levels: implications for more ancient strata, *in* Nummedal, D., Pilkey, O.H. and Howard, J.D., *eds.*, Sea-level Fluctuations and Coastal Evolution: SEPM Special Publication 41, p. 223-239.

Devine, P.E., 1991, Transgressive origin of channeled estuarine deposits in the Point Lookout Sandstone, Northwestern New Mexico: a model for Upper Cretaceous, cyclic regressive parasequences of the U.S. Western Interior: American Association of Petroleum Geologists Bulletin, v. 75, p.1039-1063.

Dickinson, K.A., Berryhill Jr., H.L. and Holmes, C.W., 1972, Criteria for recognizing ancient barrier coastlines, *in* Rigby, J.K. and Hamblin, W.K. *eds.*, Recognition of Ancient Sedimentary Environments: SEPM Special Publication 16, p. 192–214.

Fabuel-Perez, I., Hodgetts, D. and Redfern, J., 2009, A new approach for outcrop characterization and geostatistical analysis of a low-sinuosity fluvial-dominated succession using digital outcrop models: Upper Triassic Oukaimeden Sandstone Formation, central High Atlas, Morocco: American Association of Petroleum Geologists Bulletin, v. 93, p. 795-827.

Fischer, A.G., 1961, Stratigraphic record of transgressive seas in light of sedimentation on Atlantic coast of New Jersey: American Association of Petroleum Geologists Bulletin, v. 45, p. 1656-1666.

FitzGerald, D.M. and Rosen, P.S., *eds.*, 1987, Glaciated Coasts: New York, Academic Press, 364 p.

FitzGerald, D. M., Baldwin, C. T., Ibrahim, N. A. and Humphries, S. M. 1992, Sedimentologic and morphologic evolution of a beach-ridge barrier along an indented coast: Buzzards Bay, Massachusetts, *in* Fletcher, C. H., III and Wehmiller, J. F., *eds.*, Quaternary Coasts of the United States: Marine and Lacustrine Systems: SEPM Special Publication 48, p. 65-75.

FitzGerald, D.M., Kulp, M., Penland, P., Flocks, J. and Kindinger, J., 2004, Morphologic and Stratigraphic Evolution of Ebb-Tidal Deltas along a Subsiding Coast: Barataria Bay, Mississippi River Delta: Sedimentology, v. 15, p. 1125-1148.

Frazier, D.E., 1974, Depositional episodes: their relationship to the Quaternary stratigraphic framework in the northwestern portion of the Gulf Basin: University of Texas at Austin, Bureau of Economic Geology, Geological Circular, v. 4, 28 p.

Galloway, W.E. and Hobday, D.K, 1996, Terrigenous Clastic Depositional Systems 2nd edition: New York, Springer-Verlag, 489 p.

Goff, J. A., Mayer, L. A., Traykovski, P., Buynevich, I., Wilkens, R., Raymond, R., Glang, G., Evans, R. L., Olson, H. and Jenkins, C., 2005, Detailed investigation of sorted bedforms. or rippled scour depressions, within the Martha's Vineyard Coastal Observatory, Massachusetts: Continental Shelf Research, v. 25, p. 461-84.

Griffiths, C.M., Salles, T, Li, F. and Dyt, C.P., 2009, Using SEDSIM to predict the response of coastal sediments to climate change in Australia: Proceedings of the International Conference on Climate Change, The Environmental and Socio-economic Response in the Southern Baltic Region, Szczecin, Poland, May 2009, p. 104-106.

Harris, P. T., Heap, A. D., Bryce, S. M., Porter-Smith, R., Ryan, D. A. and Heggie, D., T., 2002, Classification of Australian clastic coastal depositional environments based on a quantitative analysis of wave, tide and river power: Journal of Sedimentary Research, v. 72, p. 858-870.

Hayes, M.O., 1975, Morphology of sand accumulations in estuaries: an introduction to the symposium, *in* Cronin, L.E., *ed.*, Estuarine Research, Vol. II: New York, Academic Press, p. 3-22.

Helland-Hansen, W. and Martinsen, O.J., 1996, Shoreline trajectories and sequences: description of variable depositional-dip scenarios: Journal of Sedimentary Research, v. 66, p. 670-688.

Hijma, M.P., Cohen, K.M., Hoffmann, G., Van der Spek, A.J.F. and Stouthamer, E., 2009, From river valley to estuary: the evolution of the Rhine mouth in the early to middle Holocene (western Netherlands, Rhine-Meuse delta), Netherlands: Journal of Geosciences — Geologie en

Mijnbouw, v. 88, p. 13-53.

Hoyt, J.H. and Henry, V.J., 1965, Significance of inlet sedimentation in the recognition of ancient barrier islands: Wyoming Geological Association, 19th Field Conference Guidebook, p. 190-194.

Kraft, J.C., Chrzastowski, M.J., Belknap, D.F., Toscano, M.A., and Fletcher, C.H., III, 1987, The transgressive barrier-lagoon coast of Delaware: Morphostratigraphy, sedimentary sequences and responses to relative sea level rise, *in* Nummedal, D., Pilkey, O.H., and Howard, J.D., *eds.*,Sea-level Fluctuations and Coastal Evolution: SEPM Special Publication 41, p. 129-143.

Krystinik, L.F. and Blakeney, B.A., 1990, Sedimentology of the upper Morrow Formation in eastern Colorado and western Kansas, *in* Sonnenberg, S. A., Shannon, L. T., Rader, K., von Drehle, W. F. and Marting, G. W., *eds.*, Morrow Sandstones of Southeast Colorado and Adjacent Areas: Rocky Mountain Association of Geologists, p. 37-50.

Kumar, N. and Sanders, J.E., 1974, Inlet sequence: a vertical sequence of sedimentary structures and textures created by the lateral migration of tidal inlets: Sedimentology, v. 21, p. 491-532.

MacEachern, J.A. and Pemberton, S.G., 1994, Ichnological aspects of incised valley fill systems from the Viking Formation of the Western Canada Sedimentary Basin, Alberta, Canada, *in* Dalrymple, R.W., Boyd, R. and Zaitlin, B.A., *eds.*, Incised Valley Systems–Origin and Sedimentary Sequences: SEPM Special Publication 51, p. 129-158.

Mayer, L.A., Raymond, R., Glang, G., Richardson, M.D., Traykovski, P. and Trembanis, A.C., 2007, High-resolution mapping of mines and ripples at the Martha's Vineyard Coastal Observatory: IEEE Journal of Oceanic Engineering, v. 32, p. 133-149.

Moller, I. and Anthony, D., 2003, A GPR study of sedimentary structures within a transgressive coastal barrier along the Danish North Sea coast. *in* Bristow, C. S. snf Jol, H. M., *eds.*, Ground Penetrating Radar in Sediments: Geological Society of London, Special Publication 211, p. 55-65.

Morton, R.A. and McGowen, J.H., 1980, Modern depositional environments of the Texas coast: University of Texas at Austin. Bureau of Economic Geology 20, 167 p.

Moslow, T.F., and Tye, R.S., 1985, Recognition and characterization of Holocene tidal inlet sequences, Marine Geology, 63, 129-151.

Nichols, M.M. and Biggs, R.B., 1985, Estuaries, *in* Davis, R.A., Jr.,*ed.*, Coastal Sedimentary Environments, 2nd edition: New York, Springer-Verlag, p. 77-186.

Niedoroda, A. W., 1980, Shoreface-Surf Zone Sediment Exchange Processes and Shoreface Dynamics: NOAA Marine Ecosystems Analysis Program, Boulder, Colorado, 79 p.

Nordfjord, S., Goff, J.A., Austin, J.A. Jr. and Gulick, S.P.S., 2006, Seismic facies of incised-valley fills, New Jersey continental shelf: implications for erosion and preservation processes acting during latest Pleistocene–Holocene transgression: Journal of Sedimentary Research, v. 76, p. 1284-1303

Olsen, T.R., Mellere, D. and Olsen, T., 1999, Facies architecture and geometry of landward-stepping shoreface tongues: the Upper Cretaceous Cliff House Sandstone (Mancos Canyon, south-west Colorado): Sedimentology, v. 46, p. 603- 625.

Penland, S., Boyd, R. and Suter, J.R., 1988, Transgressive depositional systems of the Mississippi delta plain: a model for barrier shoreline and shelf sand development: Journal of Sedimentary Petrology, v. 58, p. 932-949.

Ponten, A. and Plink-Bjorklund, P., 2009, Process regime changes across a regressive to transgressive turnaround in a shelf-slope basin, Eocene central basin of Spitsbergen: Journal of Sedimentary Research, v. 79, p. 2-23

Pritchard, D.W., 1967, What is an estuary? Physical viewpoint, *in* Lauff, G.H., *ed.*, Estuaries: American Association for the Advancement of Science, Publication 83, p. 3-5.

Rahmani, R.A., 1988, Estuarine tidal channel and nearshore sedimentation of a Late Cretaceous epicontinental sea, Drumheller, Alberta, Canada, *in* de Boer, P.L., van Gelder, A. and Nio, S.D., *eds.*, Tide-Influenced Sedimentary Environments and Facies: Boston, D. Reidel Publishing Company, p. 433-481.

Reinson, G.E., 1992. Transgressive barrier island and estuarine systems, *in* Walker, R.G. and James, N.P., *eds.*, Facies Models–Response to Sea Level Change: Geological Association of Canada, p. 179-194.

Reynaud, J.Y., Tessier, B., Proust, J.N., Dalrymple, R., Bourillet, J.F., DeBatist, M., Lericolais, G., Berné, S. and Marsset, T., 1999, Architecture and sequence stratigraphy of a late Neogene incised valley at the shelf margin, southern Celtic Sea: Journal of Sedimentary Research, v. 69, p. 351-364.

Rider, M.H., 1996, The Geological Interpretation of Well Logs, 2nd edition: Caithness, Whittles Publishing, 256 p.

Rodriguez, A.B., Anderson, J.B., Siringan, F.P. and Taviani, M., 2004, Holocene evolution of the east Texas coast and inner continental shelf: along-strike variability in coastal retreat rates: Journal of Sedimentary Research, v. 74, p. 405-421

Roy, P. S., 1994, Holocene Estuary Evolution - Stratigraphic Studies from Southeastern Australia, *in* Dalrymple, R.W., Boyd, R. and Zaitlin, B.A., *eds.*, Incised Valley Systems–Origin and Sedimentary Sequences: SEPM Special Publication 51, p. 241-263.

Roy, P., Cowell, P.J., Ferland, M.A. and Thom, B.G., 1995, Wave dominated coasts, *in* Carter, R.W.G. and Woodroffe, C.D., *eds.*, Coastal Evolution: Cambridge, Cambridge University Press, p. 121-186.

Schwartz, R.K., 1975, Nature and genesis of some washover deposits. US Army Corps of Engineers, Coastal Engineering Research Center, Technical Memoir 61, 69 p.

Short, A.D., 1979, Three dimensional beach stage model: Journal of Geology, v. 87, p. 553-571.

Short, A.D., 1984, Beach and nearshore facies, southeast Australia: Marine Geology, v. 60, p. 261-282.

Simms, A.R., Anderson, J.B. and Blum, M., 2006, Barrier-island aggradation via inlet migration: Mustang Island, Texas: Sedimentary Geology, v. 187, p. 105-125.

Siringan, F.P., and Anderson, J.B., 1993, Seismic facies, architecture and evolution of the Bolivar Roads tidal inlet/delta complex, east Texas Gulf Coast: Journal of Sedimentary Petrology, v. 63, p. 794-808.

Sixsmith, P. J., Hampson, G. J., Gupta, S., Johnson, H. D. and Fofana, J. F., 2008, Facies architecture of a net transgressive sandstone reservoir analog: The Cretaceous Hosta Tongue, New Mexico: American Association of Petroleum Geologists Bulletin, v. 92, p. 513-547.

Sonnenberg, S. A., Shannon, L. T., Rader, K. and Von Drehle, W. F., 1990, Regional structure and stratigraphy of the Morrow Series, southeast Colorado and adjacent areas, *in* Sonnenberg, S. A., Shannon, L. T., Rader, K., von Drehle, W. F. and Martin, G. W., *eds.*, Morrow Sandstones of Southeast Colorado and Adjacent Areas: Rocky Mountain Association of Geologists, p. 1-8.

Smith, D. G., Simpson, C. J., Jol, H. M., Meyers, R. A. and Currey, D. R., 2003, GPR stratigraphy used to infer transgressive deposition of spits and a barrier, Lake Bonneville, Stockton, Utah, USA, *in* Bristow, C. S. and Jol, H. M., *eds.*, Ground Penetrating Radar in Sediments: Geological Society of London, Special Publication 211, p. 79-86.

Swift, D.J.P., 1968, Coastal erosion and transgressive stratigraphy: Journal of Geology, v. 76, p. 444-456.

Swift, D.J.P., 1975, Barrier – island genesis: evidence from the central Atlantic shelf, eastern U.S.A.: Sedimentary Geology, v. 14, p. 1-43.

Thieler, E.R., Pilkey, O.H., Cleary, W.J. and Schwab, W.C., 2001, Modern sedimentation on the shoreface and inner continental shelf at Wrightsville Beach, North Carolina, USA: Journal of Sedimentary Research, v. 71, p. 958-970.

Törnqvist, T.E., 1998, Longitudinal profile evolution of the Rhine-Meuse system during the last deglaciation: interplay of climate change and glacio-eustasy?: Terra Nova, v. 10, p. 11-15.

Van Straaten, L.M.J.U., 1965, Coastal barrier deposits in South and North Holland in particular in the areas around Scheveningen and Ijmuiden: Mededel. Geol. Stichting, v. 17, p. 41-75.

Van Wagoner, J.C., Mitchum, R.M., Campion, K.M. and Rahmanian, V.D., 1990, Siliciclastic Sequence Stratigraphy in Well Logs, Cores, and Outcrops: Concepts for High-Resolution Correlation of Time and Facies: Tulsa, American Association of Petroleum Geologists Methods in Exploration Series, No. 7, 55 p.

Walker, R.G., 1984, General introduction: Facies, facies sequences and facies models, *in* Walker, R.G., *ed*., Facies Models, 2nd edition: Geological Association of Canada, Geoscience Canada, Reprint Series 1, p. 1-9.

Yoshida, S., Steel, R.J. and Dalrymple, R.W., 2007, Changes in depositional processes – an ingredient in a new generation of sequence-statigraphic models: Journal of Sedimentary Research, v. 77, p. 447-460.

Zaitlin, B.A., Warren, M.J., Potocki, D., Rosenthal, L. and Boyd, R., 2002, Depositional styles in a low accommodation foreland basin setting: an example from the Basal Quartz (Lower Cretaceous), southern Alberta: Bulletin of Canadian Petroleum Geology, v. 50, p. 31-72

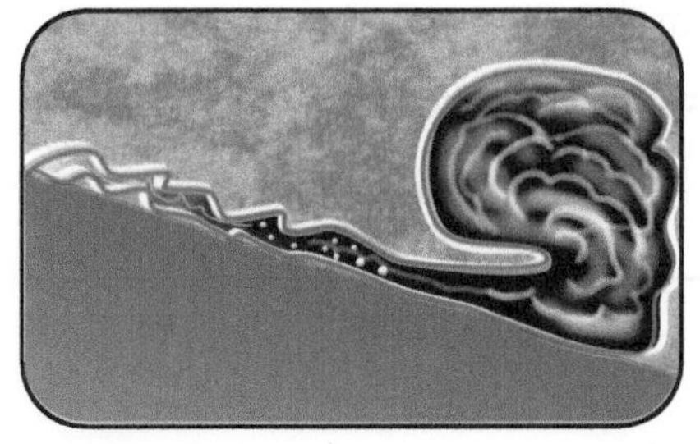

12. DEEP-MARINE SEDIMENTS AND SEDIMENTARY SYSTEMS

R. William C. Arnott, Department of Earth Sciences, University of Ottawa, Ottawa, ON, K1N 6N5, Canada

INTRODUCTION

The deep marine is a unique sedimentary environment compared to all others because of its inaccessibility and the enormous spatial scale of many of its constituent depositional systems. For, example, the modern Bengal Fan, which has been accumulating sediment for only about 55 m.y., is 3000 km long, over 1400 km wide, > 5000 m thick, and contains an estimated sediment volume of $4x10^6$ km^3 (Table 1). (For a global inventory of fan systems, see the wall chart of Barnes and Normark, 1985). Accordingly, the principal investigative tool for modern deep-marine environments is seismic, including an array of techniques that range from high-frequency, high-resolution but shallow-penetrating surveys, to lower frequency, more deeply penetrating but lower resolution surveys. Three-dimensional seismic with its ability to image features in both plan and cross-sectional views has proven to be especially useful. Over the past decade, this work has resulted in the publication of many stunning seismic images that have improved greatly our understanding of the deep-marine sedimentary system (*e.g.*, Weimer *et al.*, 1991; Posamentier and Kolla, 2003; Posamentier and Walker, 2006). Nevertheless, the seismic method generally suffers from limited vertical resolution – minimum vertical resolution of industry seismic data is commonly of the order of 10 meters. To fill the gap, the geological community must also utilize outcrop studies, but here too a number of shortcomings are recognized (Fig. 1). Perhaps most importantly, the outcrop record is inherently 2-dimensional with, at best, local 3-D perspectives (*i.e.*, valley cuts). Also, in most cases, the horizontal scale of most outcrops is small compared to the spatial scale of most deep-marine architectural elements, and, more profoundly, their parent sedimentary system. Nevertheless, over the past decade or so, much research, including work on a number of seismic-scale outcrops, is helping to bridge the gap between the ancient outcrop record and modern seismic images, to merge these two independent datasets into a single, coherent package (*see*, for example, Nilsen *et al.*, 2007). Nevertheless, the emphasis here, as in other chapters in this volume, is on the lithological characteristics of the strata that make up the deep-marine sedimentary record, and also on how these strata are distributed spatially (space) and vertically (time) in the geological record. Excellent reviews of the seismic-scale attributes can be found elsewhere (*e.g.*, Posamentier and Kolla, 2003; Posamentier and Walker, 2006).

BRIEF HISTORY OF DEEP-MARINE SEDIMENTOLOGY

Turbidites are a ubiquitous feature in the deep-marine sedimentary record, and are deposited from subaqueous turbidity currents. However, because the formative process is largely hidden from direct observation, the connection between process and deposit remained speculative and anecdotal since as early as the late 1800s. Knowledge of deep-marine processes took a major step forward following the laying of the first successful Atlantic seafloor telegraph cable in 1866. In 1899, Benest (*in* Heezen and Ewing, 1952) reported that

> *"accidents to cables have already been valuable ... in directing attention to hitherto unsuspected forces constantly in action altering the features of the sea bottom".*

The most famous cable disruption occurred on November 18, 1929, on the southern margin of the Grand Banks off east coast Canada. On that day, an estimated 7.2 magnitude earthquake occurred and disrupted telegraph communication between North America and Europe. Several submarine cables were apparently broken at the time of the earthquake, presumably as a result of seabed movement. More puzzling, however, was the fact that a number of other cables became deactivated successively after the earthquake, some as much as 13 hours later (Heezen and Ewing, 1952; Piper *et al.*, 1999). Furthermore, the cables broke progressively in a single (offshore) direction. It was hypothesized that a turbidity current moving at considerable speed was responsible (Heezen and Ewing, 1952). At about the same time, Kuenen and Migliorini (1950) began experimenting with sediment dispersions released into a basin of still-standing water. Upon release the dispersion formed a bottom-hugging turbidity current that produced a deposit with a characteristic upward decrease in grain size. Based on these results, these authors suggested that the common occurrence of upward-fining beds in the deep-marine geological record might reflect deposition from turbidity currents. Beds showing this upward-fining character were observed in flysch deposits of the European Alps by Bouma (1962). In addition to the upward fining, he also noted a characteristic vertical succession of sedimentary structures, the origin of which was then unknown. Shortly thereafter, experimentalists illustrated the variety of bedforms that formed under unidirectional currents of various speeds (Simons *et al.* 1965; and now many others). Based on those results, the characteristic suite of sedimentary structures observed by Bouma, the so-called Bouma turbidite sequence, reflects deposition from a decelerating turbidity current. At about the same time, Middleton (1966a, b, 1967) published a series of important papers based on flume experiments, describing the

Table 1. Dimensional characteristics of five deep-marine turbidite systems. Data from wall chart of Barnes and Normark (1984).

Name	Location	Age	Length, Width (km)	Maximum Thickness (m)	Volume (km^3)	Dominant Grain Size	Grain-size Range
Amazon	Brazilian margin	Middle Miocene	700, 700 (max)	4200	7x10^5	mud	pebbles to mud
Bengal	Bay of Bengal	Eocene	2800 (min), 1400 (max)	> 5000	4x10^6	mud	mud to medium sand
Laurentian	coast eastern Canada	Quaternary	1500 (max), 400	2000	1x10^5	mud to med-fine sand	mud to gravel
Mississippi	Gulf of Mexico	Pleistocene	540, 570	4000	2.9x10^5	silty mud	mud to gravel
Navy	coast southern California	Late Pleistocene		900	75	sandy silt	mud to gravel

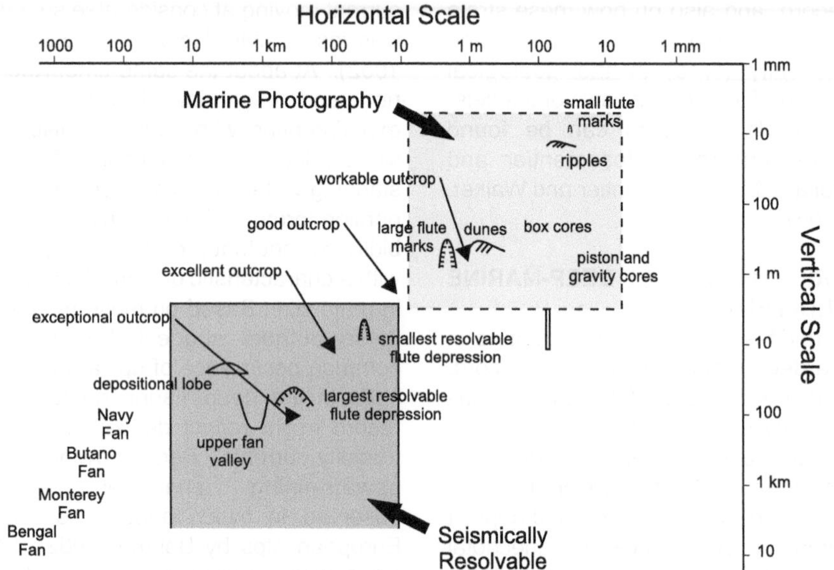

Figure 1. Vertical and horizontal scales of modern deep-marine systems (Bengal, Monterey and Navy fans), ancient turbidite systems (Butano), some deep-marine sediment features/elements, and outcrop and core (adapted from Barnes and Normark, 1985). Note how the scale of most outcrops is dwarfed by the scale of seismic data sources, but more importantly, the size of deep marine fans and their constituent depositional elements.

various parts of a turbidity current and the influence of sediment concentration on depositional patterns and characteristics.

Today, it has been well established that sediment-gravity flows, principally turbidity currents and debris flows, in addition to mass-movement processes, are the major formative agents of the deep-marine sedimentary record. A superb illustration is the 1929 Grand Banks event, which is now known to have been initiated by widespread failure of a surficial layer about 20–25 m thick (Piper *et al.*, 1999). The slide, which comprised numerous overlapping slumps formed by progressive retrogressive failure, moved downslope, and in areas of high slope transformed into a debris flow and ultimately into a sustained, fast-moving (~70 km/hr) turbidity current. This current eroded on the continental slope and eventually deposited a turbidite up to 1 m thick on the Laurentian submarine fan. The turbidite contains about 185 km^3 of sediment – using railway boxcars to conceptualize volume, 185 km^3 equates to 1.29 billion boxcars, or a train almost 22 million kilometers long, and one that would wrap around the world over 558 times!

SEDIMENT TRANSPORT MECHANISMS AND DEPOSITS

Unlike continental and shallow-marine sedimentary systems, sand and gravel transport in the deep marine is dominated by sediment-gravity flows and mass-movement processes. Finer grained sediments, principally silt and clay, although present in sediment-gravity flows and their related deposits, are transported mostly in suspension.

Mass-movement Deposits – Slides and Slumps

The gravity-driven downslope movement of coherent to semi-coherent masses of sediment along discrete failure planes is termed mass movement or mass wasting. Such movements, which can range up to hundreds of kilometers in distance, occur when the driving force, gravity,

Figure 2. Mass-movement deposits. **A.** Slide deposit (strata are vertically dipping due to later (Cordilleran) tectonic deformation and stratigraphic top is toward the left). Base of slide is indicated by the solid black line. Note the extensively deformed and brecciated strata in the lower part of the slide, including a large rotated block (Neoproterozoic Isaac Formation, British Columbia). **B.** Slump deposit with extensive internal ductile deformation (base of slide indicated by solid white line; Carboniferous Gull Island Formation, Ireland).

exceeds the tensile strength of the parent sediment pile (which depends upon a wide range of internal conditions, including pore-fluid pressure, sediment composition and consolidation, and the occurrence of mechanically weak layers, among others). The mechanisms responsible for triggering the initial instability include oversteepening, seismic loading, cyclic storm-wave loading, rapid accumulation and underconsolidation, gas charging, gas-hydrate dissociation, seepage, *etc.* (*e.g.*, Locat and Lee, 2002). Once initiated, movement continues until the resisting forces, principally friction along the basal failure plane, exceed the gravitational driving force and *en masse* deposition takes place. Following earlier workers, two end-member kinds of mass-movement deposits are recognized: slides and slumps, where the principal difference between the two is the intensity and nature of internal deformation. In the case of slides, deformation is comparatively minor and is dominated by brittle deformation, mostly in the form of bedding-parallel surfaces of detachment, although shear deformation may be intense in a thin zone near the base of the unit (Fig. 2A). In slumps, deformation shows an element of rotation and typically is more intense and ductile in character (Fig. 2B). Soft-sediment 'slump' folds are abundant, and are commonly tightly folded with the axial plane sub-parallel to planes of internal shear.

Sediment-gravity Flows

Gravity currents occur when a more-dense fluid moves through and displaces a less-dense fluid. Sediment-gravity currents are a type of gravity current, wherein the density contrast is produced by the presence of suspended sediment. Early classification schemes for sediment-gravity flows recognized four types of end-members: debris flows, grain flows, fluidized/liquefied flows, and turbidity currents (Middleton and Hampton, 1976), each differentiated by the primary mechanism for suspending sediment and maintaining the density contrast: matrix strength, grain collision, escaping pore fluid and fluid turbulence, respectively. More recently, an alternative scheme by Mulder and Alexander (2001) proposed a classification based on flow properties and sediment-support mechanisms, and recognized two types of end-member flows: cohesive flows and frictional flows. Here, a modified version of this classification will be adopted.

Cohesive Flows — Debris Flows

Cohesive flows, which are more commonly termed debris flows and mud flows, are sediment-gravity flows where the volume concentration of the solid and fluid phases are of the same order of magnitude, and the occurrence of a cohesive matrix imparts a pseudoplastic rheology to the flow (*e.g.*, Mohrig *et al.*, 1998; Mulder and Alexander, 2001). Hereafter the term debris flow, which is entrenched in the geological literature, will be used to refer to all cohesive flows. Particles are principally suspended by cohesive forces provided by a matrix of fluid and fine-grained sediment (generally a silt–clay mixture. Note that in both subaerial and subaqueous debris flows (*see* also Chapter 6), the amount of clay-size particles needed to generate sufficient yield strength to suspend larger particles can be surprising low, possibly of the order of 2–4% by volume. In addition to matrix strength, buoyancy effects, particle-particle interaction (collisions and near misses), hindered settling, elevated pore pressure, and in some cases fluid turbulence may provide additional support for suspended particles. Deposition occurs when one or more of the following exceed the driving gravitational force: intrinsic shear strength of the sediment-water mixture, grain-contact friction and friction along the flow boundaries, which then causes the flow to freeze inward, either *en masse*, or more gradually from areas of lower shear near the flow's surface toward those of higher shear at the base.

Debris-flow deposits, or debrites, form sheetlike to lobe-shaped masses that have steep margins as a result of the strength of the moving

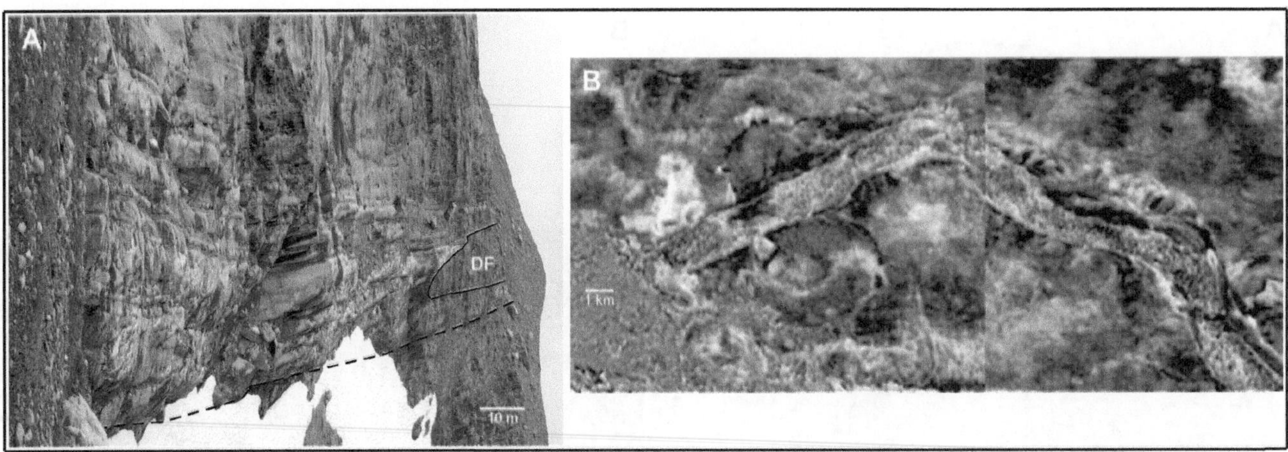

Figure 3. A. Ancient example of a partly debris-flow-filled channel. Base of channel is indicated by dashed black line. After accumulating several meters of sandstone, the channel became partly plugged by a debris-flow deposit (DF; Neoproterozoic Isaac Formation, British Columbia). Thereafter, channel-filling sandstones onlap and then overlap the debris-flow deposit. **B.** Seismic image of the chaotic reflections of a debris-flow deposit that has exploited a pre-existing seafloor channel (photo courtesy of Henry Posamentier).

and deposited sediment mass. Deposits range widely in scale, but can be up to tens of kilometers wide and over 100 m thick, although much thinner beds are more common. At their downflow terminus toe-thrusts are common due to a rapid down-flow reduction in flow speed. Bases of debris-flow deposits are commonly planar and non-erosional, because of the strength of the moving mass and the damping of large-scale fluid turbulence. Nevertheless, deeply scoured bases are observed. In some cases, these erosive basal contacts represent the opportunistic occupation of a pre-existing seafloor channel (Fig. 3), whereas others are thought to be formed by a rigid part of the flow 'ploughing' through the underlying seafloor sediment (*see* Posamentier and Walker, 2006, their Figures 155 and 157) or being dragged along the surface forming linear grooves up to 40 m deep, several hundred meters wide and extending longitudinally for more than 20 km (Posamentier and Kolla, 2003). The upper surface can also be uneven and ranges between flat to highly rugose. Internally, debris-flow deposits range from mud- to sand-rich, typically with a disorganized, poorly sorted character (*e.g.*, Nemec and Steel, 1984; Fig. 4A). On seismic images debris-flow deposits, especially those interpreted to be mud rich, exhibit a distinctive chaotic or reflection-free character. Where present, clasts ranging from sand grains to enormous blocks are generally dispersed throughout a fine-grained matrix. In some deposits, clasts of incorporated soft sediment are contorted and commonly show a subtle to well-developed orientation with their longest dimension oriented subparallel to the base of the deposit (Fig. 4B). Preferential particle alignment and particle deformation is likely the result of shearing within the sediment mass during transport, which then may be accentuated by post-depositional compaction and consolidation.

Although cohesive, some debris flows are capable of movement over long distances. For example, Gee *et al.* (1999) reported a modern debris flow off the west coast of Africa with a run-out distance of about 700 km. Such large distances may be attributed to hydroplaning and the attendant reduction of friction between the bed and the overlying flow. Elevated hydrodynamic pressure exerted on the forward part of the flow causes a wedge of ambient fluid (*i.e.*, seawater) to penetrate beneath the flow and separate it from the bed (Mohrig *et al.*, 1998). As a result, frictional resistance at the base of the flow is significantly reduced and the weight of the overlying debris flow becomes borne by the fluid. Importantly, the overriding debris flow must have sufficiently low permeability (*i.e.*, be sufficiently muddy) so that the basal fluid layer does not dissipate.

Debris flows are also capable of spawning turbidity currents. Owing to mixing along its leading edge, sediment is eroded from the debris flow and cast into a developing turbulent suspension above the flow. However, owing to the low permeability of most debris flows, water infiltration is minimal and the amount of sediment eroded and transferred into the turbidity current is very small (< 1%). Also, so long as the debris flow and overlying turbulent flow are moving at a similar velocity, the amount of erosion along the interface between them is negligible. But, if the debris flow stops, then the turbulent suspension can rework the top of the debris-flow deposit as it becomes detached from the parent debris flow and continues farther downslope (Fig. 4C). Conversely, it has been reported that, at an abrupt reduction in slope, turbidity currents can be partially transformed into a debris flow. As the turbulent flow decelerates rapidly and turbulence is reduced, suspended particles fall rapidly toward the bed where the ubiquitous occurrence of cohesive mud particles eventually increases yield strength to the point where the particles become supported by matrix strength.

Mass transport deposits (MTDs) are observed often in shallow- and deep-penetrating seismic images. The MTDs are typically erosively based deposits that occur on a range of scales, including enormous features that cover areas up to several 1000 km^2 and are several 100s of meters thick. The dimensions of MTDs, including their length:width ratio, are controlled by the geomorphology and position of the sediment

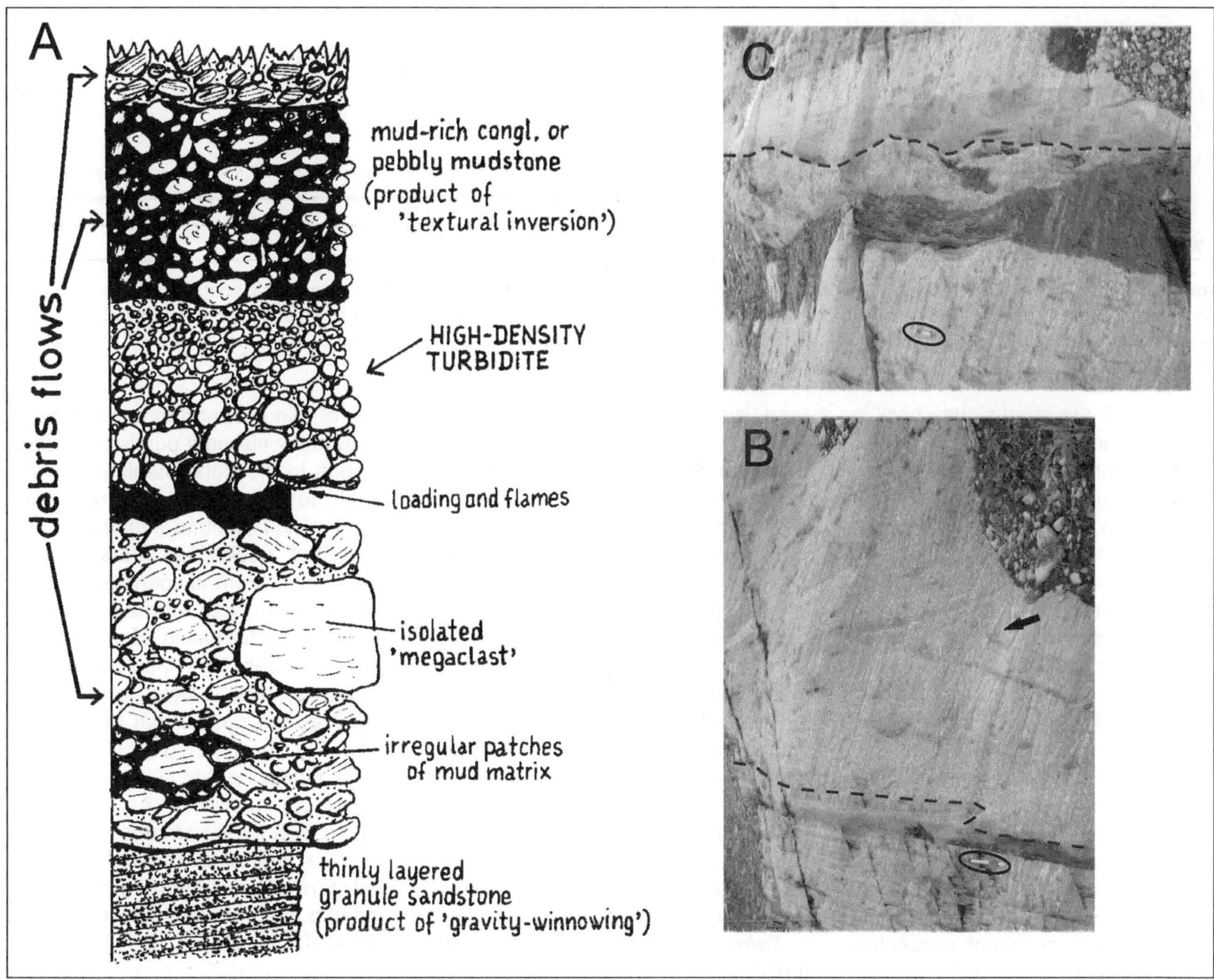

Figure 4. Debris-flow deposits (debrite). **A.** Schematic diagram showing typical characteristics of subaqueous debris-flow deposits (from Nemec and Steel, 1984). **B.** Sharp-based debris-flow deposit overlying a succession of thin- and medium-bedded turbidites (below the dashed line). Note the dispersed quartz pebbles and mudstone clasts, some of which have been folded (arrow) due to shearing within the moving cohesive mass (pencil, circled, for scale). **C.** Pebbly debris-flow deposit overlain by a T_{bc} turbidite (pencil, circled, for scale). Black layer is a large mudstone clast. The anomalous concentration of quartz pebbles in the upper part of the debrite is attributed to reworking by a genetically related turbidity current that may have deposited the overlying T_{bc} turbidite. The base of this bed (dashed line) has loaded passively into the underlying debrite, which at the time must have been water saturated and poorly compacted (Neoproterozoic Isaac Formation, British Columbia).

source, wherein large, expansive deposits are related to instability initiated along the upper part of the basin margin, at or near the shelf edge, whereas smaller deposits form in the basin by failure along local slopes (detached sediment sources; Moscardelli and Wood, 2008). Internally MTDs consist predominantly of a complex, typically disorganized assemblage of slump, slide and debris-flow deposits that can locally be interstratified with (organized) channel and overbank strata. The occurrence of large MTDs in the stratigraphic column represents a major change in sedimentation regime within the basin. In many cases these changes have been interpreted to be related to allocyclic processes, and therefore to have sequence-stratigraphic significance (*see* sequence-stratigraphy section below).

Frictional Flows

According to Mulder and Alexander (2001), frictional flows form a continuum from mass movements (slumps/slides – *see* above) to a variety of different kinds of sediment suspensions subdivided on the basis of the dominant mechanisms of sediment support. The most common the frictional flows, however, are turbidity currents.

Turbidity Currents Based on flow and sediment characteristics, turbidity currents consist of three distinct but not sharply bounded parts: head, body and tail (*see* review by Kneller and Buckee, 2000; Fig. 5). The head of turbidity currents is the sediment-rich part of the flow and the site where most of the mixing with the ambient fluid occurs. It is characterized by a sharp, overhanging nose,

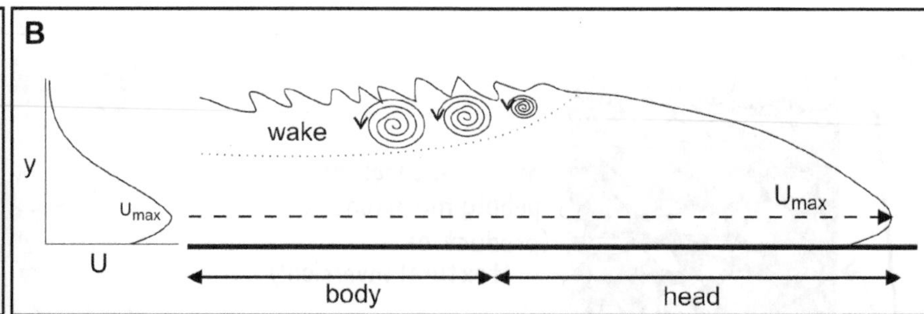

Figure 5. A. Well-developed head of an experimental turbidity current. **B.** Line diagram illustrating the typical shape and velocity profile of a turbidity current (modified after Kneller and Buckee, 2000). Note that, unlike an open-channel flow such as a river, the velocity maximum occurs in the lower part of the flow. Note also the extensive mixing (due to interfacial instability) that occurs along the upper part of the current.

above which the head slopes back in the upstream direction due to the resistance of the stationary overlying fluid. This generates a strong shear that sweeps sediment-rich fluid from the head backward toward the body of the current. To sustain the current, the head must be continually provided with new sediment supplied from the body, which moves forward faster than the head. Because of differences in settling velocity, coarser sediment tends to accumulate in the lower part of the head whereas finer sediment is moved upward and backward into the body of the current, the consequence being that, with time, the flow becomes longitudinally differentiated in terms of grain size. At the tail of the flow, sediment concentration is low and, as a consequence, flow speed is slower and eventually decreases to zero.

Increasingly it is being recognized that most natural turbidity currents are moderately to highly density stratified. Moreover, most natural turbidity currents are typically of higher sediment concentration and made up of sediment significantly more poorly sorted and coarser grained than that contained in laboratory currents. Unfortunately, the effect of sediment concentration, especially high concentrations, on the nature of the current and how it deposits sediment, remain poorly understood. In the early 1980s, Lowe (1982) published a theoretical classification for turbidity-current deposition, wherein he recognized two kinds of turbidity current and their related deposits: low-density and high-density. Classical turbidites, as described originally by Bouma (1962), are interpreted to be deposited by low-density turbidity currents (*see* also Mulder and Alexander, 2001). The adjective, low-density, refers to the concentration of suspended sediment in the flow, which, based on the earlier work of Bagnold (1954, *in* Mulder and Alexander, 2001), is thought to be approximately 9% sediment volume, or less, in low-density flows (Fig. 6A). Above that value, the closeness of adjacent grains begins to damp fluid turbulence, and hence turbulence alone is insufficient to suspend sediment fully, especially coarse sediment. Accordingly, additional support mechanisms like dispersive pressure, hindered settling, and buoyancy are needed and become increasingly more effective with high sediment concentration.

Turbidites deposited by low-density turbidity currents consist of all, or part, of the idealized succession described by Bouma (1962), and reflect decelerating flow speed (Fig. 7). However, if characteristics of the deposit were formed simply by a decelerating unidirectional shear flow, then for sediment coarser than lower fine sand (> 0.15 mm), upper stage plane bed (b-division) should not be succeeded by current ripples (c-division), but instead by dunes (medium-scale cross-stratification). Only in rare instances does dune cross-stratified sandstone occur where it should – sandwiched between upper-stage plane bed and current-ripple cross-stratification. Therefore, what intuitively should be the norm is in fact the exception – but why? One explanation that has been advanced is that most of the decelerating turbidity currents passed too quickly through the dune stability field. Although appealing, it has been argued that dunes can be formed from a flat bed in a matter of a few tens of minutes, and that many natural turbidity currents persisted at flow speeds in the dune stability field for much longer periods (Arnott and Hand, 1989). Another suggestion has been that, under high rates of sediment fallout from suspension, upper-stage plane bed remains stable because the formation of dunes and, in some cases, also ripples, is inhibited (Lowe, 1988). However, in many turbidites the ripples that formed the c-division show a negligible angle of climb, indicating that fallout rates are, in fact, commonly low. An alternative explanation for the absence of dunes might be the effect of high near-bed sediment concentration on the inception of dunes. Under such conditions, bed-surface defects are prevented from being amplified into dunes and, as consequence, plane bed persists, even at flow speeds that in a clear-water flow would form dunes. Finally, when near-bed sediment concentrations have been sufficiently reduced that bed defects can grow into bedforms, the flow is moving too slowly and/or the sediment is too fine to form dunes, and ripples form instead. In the case of a T_{bd} turbidite, where neither dune nor ripple cross-stratification is present, it is argued that sediment concentration remains sufficiently high for long enough that neither dunes or current ripples form and plane bed remained stable until the end of traction transport.

A significant part of the sand- and gravel-rich deep-marine sedimentary record consists not of classical Bouma turbidites, but instead of structureless, normally graded and, less commonly, ungraded or inversely graded sandstone and conglomerate, which equate to the T_a division of

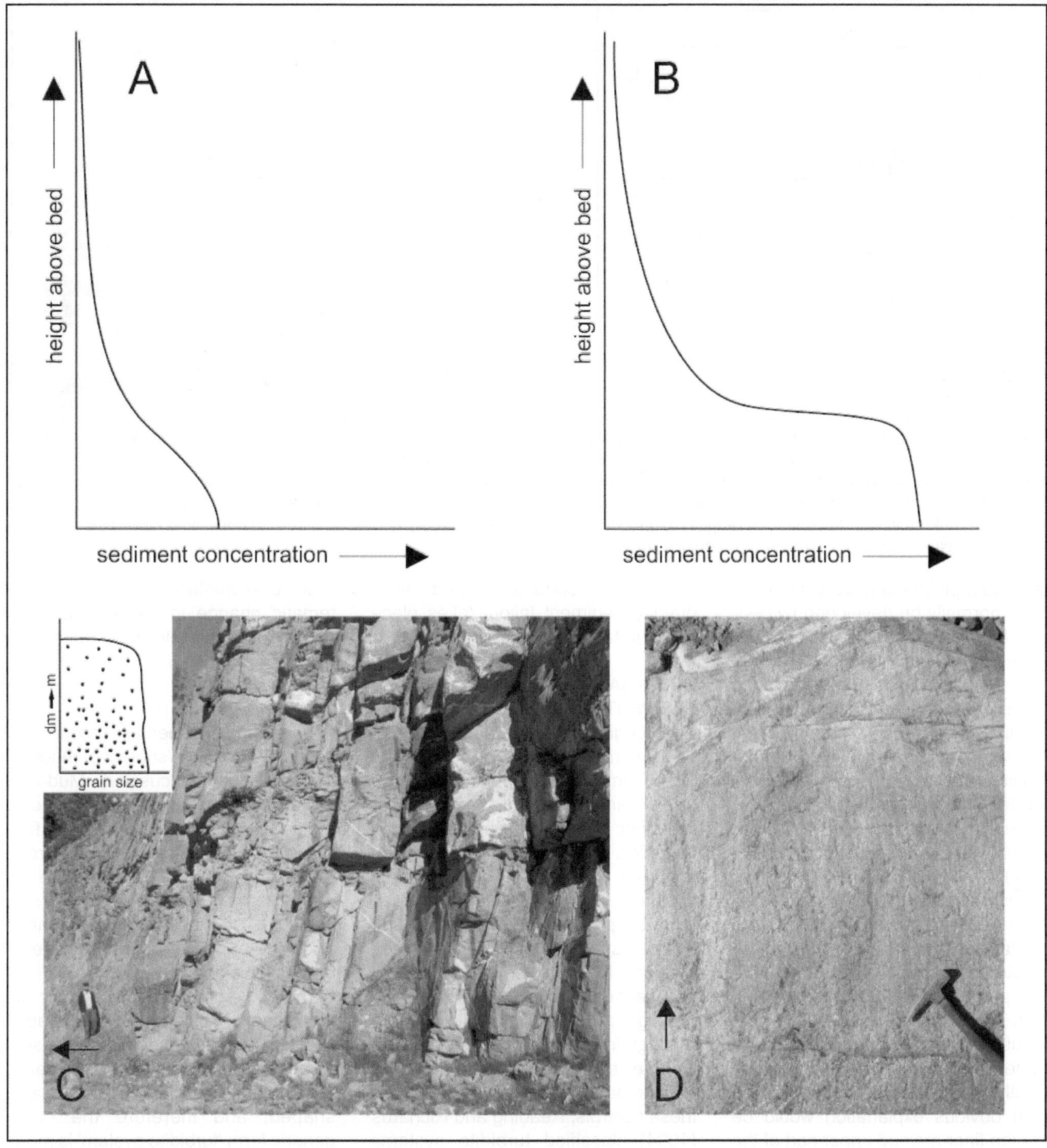

Figure 6. Idealized sediment-concentration profile in a low concentration **A.** and high concentration **B.** turbidity current. **C.** Succession of stacked decimeter- to meter-thick T_a/S3 beds. Inset depicts a typical structureless, coarse-tail graded T_a/S3 bed. **D.** Coarse-tail graded, structureless T_a/S3 bed. Arrows in **C**, **D** indicate stratigraphic tops.

Bouma (1962), or the S3/R3 division of Lowe (1982). In general, such beds are several decimeters to a few meters thick (Fig. 6C) and are either amalgamated or separated by a thin finer grained interval. Also, beds are characteristically poorly sorted and coarse-tail graded, wherein only the coarsest part of the grain-size distribution fines upward, typically with an upward decrease in the abundance of the coarsest grains (*e.g.*, Sylvester and Lowe, 2004; Fig. 6D). Mudstone intraclasts are common and, in many cases, are concentrated near the top of the bed. In addition, water-escape features, including pillar and dish structures, are common in some beds. However, in spite of their abundance, the origin of structureless beds remains a major source of debate in the geological literature (*see*, *e.g.*, Stow and Johansson, 2000), although there is growing consensus that they are deposited rapidly from high-concentration (high-density) suspensions – sedimentation

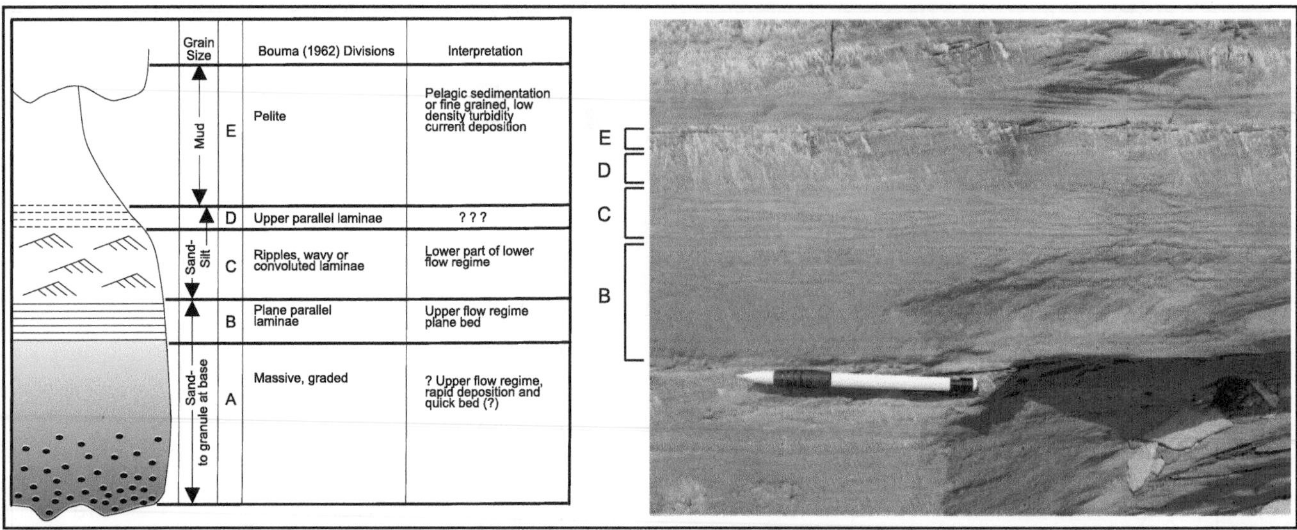

Figure 7. Line diagram illustrating a complete idealized turbidite (after Blatt *et al.*, 1984). The lowermost a-division consists of normally graded sandstone or conglomerate deposited from suspension, which then is overlain by the planar stratified b-division (upper plane bed). This, in turn, is overlain by small-scale cross-stratified sandstone (formed by current ripples) of the c-division, overlain by the subtly to well-interlaminated sandstone/siltstone and mudstone d-division capped finally by mudstone of the e-division. Photo on right is an outcrop example of a T_{bcde} turbidite that ranges from medium-grained sandstone at its base to silty mudstone at the top. The a-division is missing because deposition started with bedload transport.

being so rapid that any lamination that might normally be developed by bedload transport is not visible. Under these conditions sediment 'raining' from suspension (*i.e.*, capacity-driven sedimentation; *see* Chapter 4) entraps fine-grained sediment that otherwise could be maintained in suspension, but nevertheless gets incorporated into the accumulating bed, producing a relatively poorly sorted deposit (*e.g.*, Sylvester and Lowe, 2004). With time and a reduction in sedimentation rate, beds become better sorted and, in some cases, are overlain by tractional sedimentary structures, especially planar-lamination, producing T_{ab}-type successions. This, then, raises the question: what about the origin of beds that do not grade upward into planar lamination or other tractional structures? The most obvious explanation would be that the finer grained upper part of the bed was eroded by a later transport event. Although appealing, it cannot explain beds overlain by a fine-grained, typically mud-rich, layer. Here, the lack of tractional structures at the top of the bed must be related to highly efficient sediment bypass following the earlier episode of rapid sediment fallout. During bypass, sediment fallout all but ceases and dunes and ripples are prevented from forming by the maintenance of high sediment concentration, which lasts until low-energy conditions and fine-grained sediment fallout takes place from the tail of the flow.

DEEP-MARINE ARCHITECTURAL ELEMENTS

The origin and characteristics of deep-marine clastic systems depend on a complex assemblage of autogenic and allogenic processes, including: changes of global sea level, tectonics, sediment flux and composition, and the nature of the sediment supply system. For example, sediment supply controls the volume and internal stratigraphy of the system, whereas the number and nature of sediment entry points controls its morphology and sediment distribution. Also, grain size, which is a function of climate and provenance, controls sedimentation patterns. Based on these controls, Reading and Richards (1994) classified turbidite systems based on morphology, recognizing point-source fans, multiple-source ramps, and line-source aprons. Further, based on the dominant grain size, they also distinguished between gravel-rich, sand-rich, mixed sand-mud, and mud-rich systems (Fig. 8). Gravel- and sand-rich systems tend to be small (radius of a few to a few tens of kilometers) and grade rapidly to fine-grained basin-floor deposits. Sand-mud systems, however, are much larger (radius up to a few hundreds of kilometers) and exhibit a systematic change in depositional elements and their internal stratigraphy down the transport pathway. It is these systems that make up much of the sandstone-rich part of the ancient deep-marine sedimentary record, and accordingly are the subject of much of the subsequent discussion. Mud-rich systems are the largest, and range in radius from several tens to a few thousand kilometers, and today represent the most voluminous deep-marine sedimentary systems (*see* Table 1). Collectively, mud-, sand- and gravel-rich deep-marine sedimentary systems have generally been termed 'deep-sea fans' because of their common semi-conical shape. However, many modern deep-sea 'fans', and by extension ancient fans, are in fact elongate or irregularly shaped, and therefore the more generic term 'turbidite system' is more appropriate (Bouma *et al.*, 1985). In addition, confusion exists because the various parts of a turbidite system, which in a downflow direction consist of the upper, middle and lower fan, have been variously defined by different authors. The recent classification by Pirmez *et al.* (2000), which is based on the modern Amazon Channel, subdivides the system based on the spatial patterns of sediment erosion and deposition (Fig. 9). From proximal to distal, the fan subdivisions

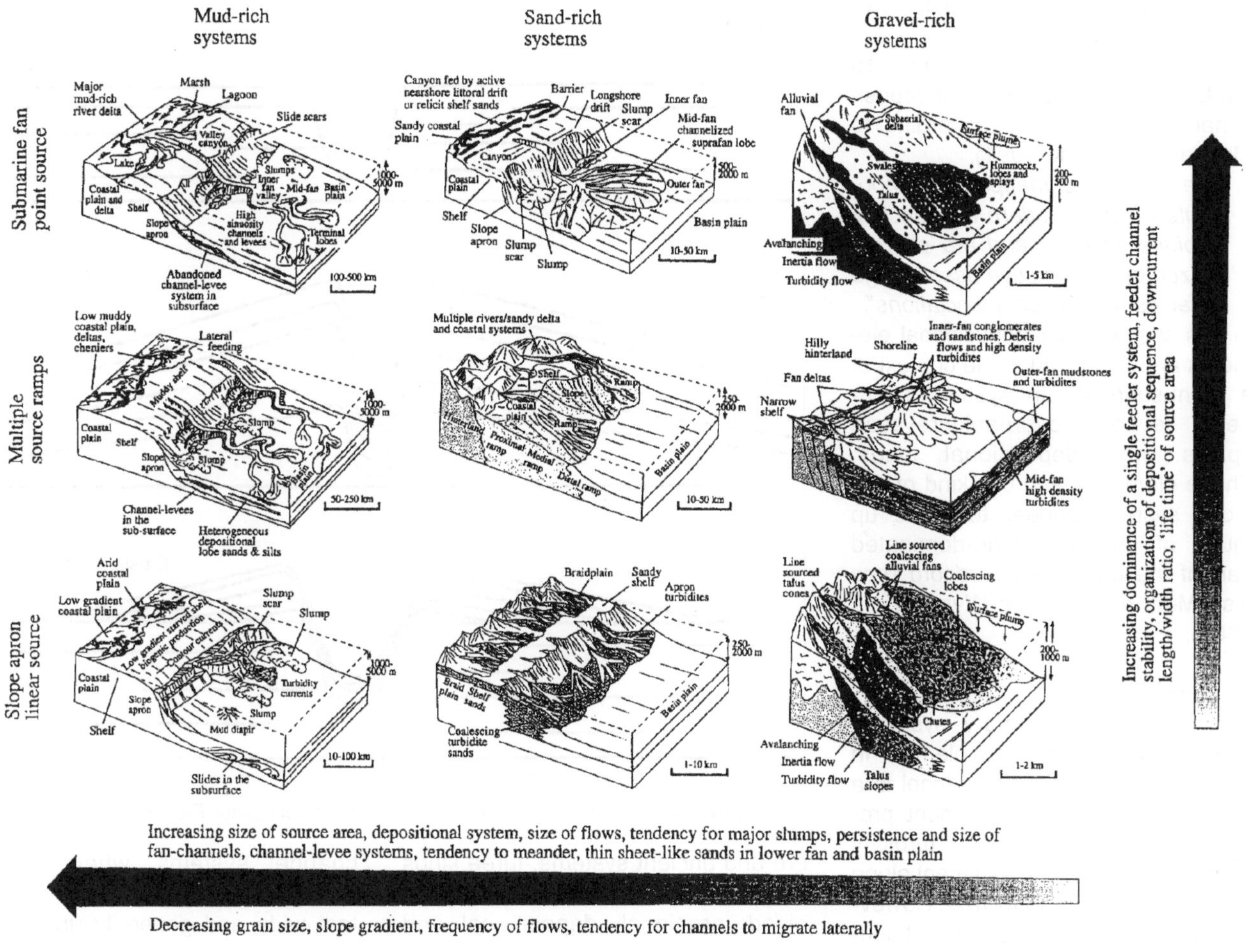

Figure 8. Idealized models for deep-marine sedimentary systems based on volume, caliber and nature of sediment input (from Stow and Mayall, 2000, based on Reading and Richards, 1994).

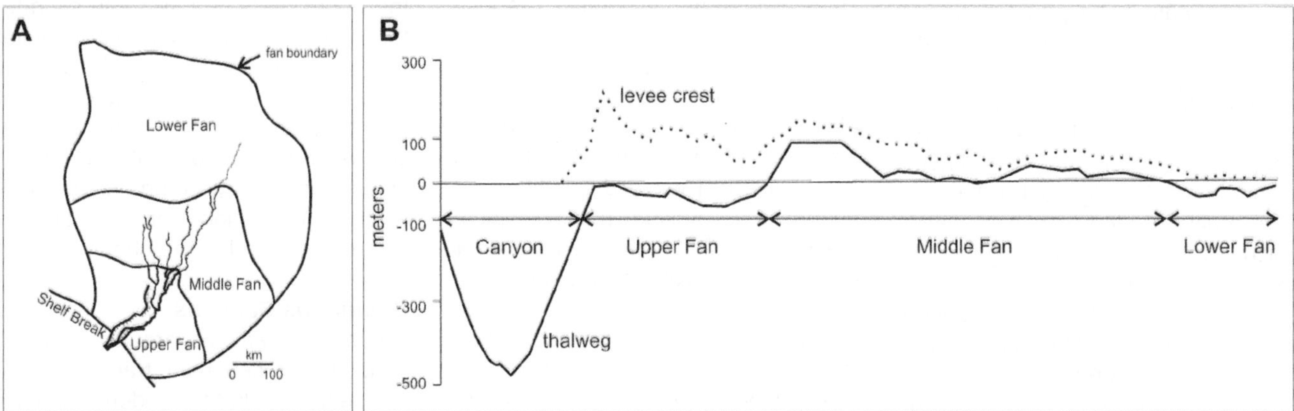

Figure 9. A. Plan view of the Amazon Fan (modified after Flood *et al.*, 1991). Note the basinward change from a single-thread channel to a distributive network, the longest system being the modern Amazon Channel, which extends 900 km beyond the present shelf-slope break). **B.** Basinward transect along the modern Amazon Channel (from Pirmez *et al.*, 2000). Subdivision of the system is based on the relationship between the elevation of the channel thalweg and levee crest relative to the adjacent seafloor surface. Datum (zero line) is the general level of the seafloor outside of the channel.

are: submarine canyon – zone of net erosion; upper fan – a zone of net sediment bypass, wherein the channel thalweg (the deepest part of the channel), which is bounded on both sides by constructional levees, lies at about the same elevation as the surrounding seafloor (*i.e.*, the area external to the channel); middle fan – zone of net sediment deposition caused by flow expansion related to loss of flow confinement, and accordingly, where the thalweg lies generally above the surrounding seafloor; and lower fan – the area lying downflow from the middle fan where the rate of net deposition decreases and the thalweg elevation more closely approximates the elevation of the surrounding seafloor.

Internally turbidite systems are made up of an assemblage of depositional elements, which according to Mutti and Normark (1991),

> *"are the basic mappable components of both modern and ancient turbidite systems and are characterized by a distinctive assemblage of facies and facies associations".*

In this chapter, four depositional elements are discussed in the context of a point-sourced fan system: channels, levees, overbank/crevasse splays, and depositional lobes. These elements, and this kind of turbidite system, appear to make up much of the sandstone-dominated part of the deep-marine record (*see* also Mutti *et al.*, 2003; Wynn *et al.*, 2007).

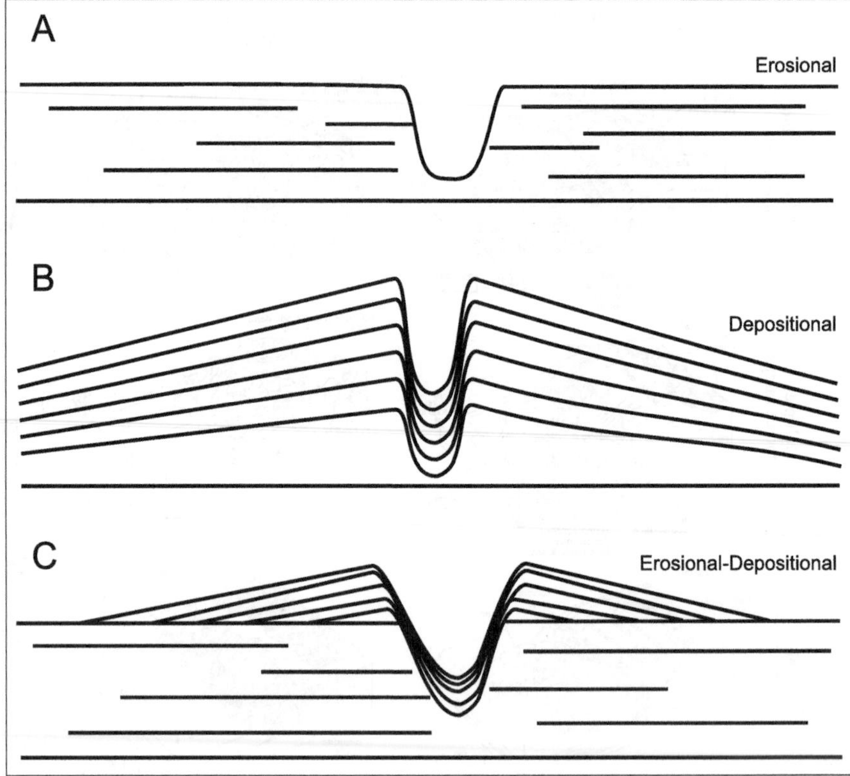

Figure 10. Nomenclature for deep-marine channels (after Pickering *et al.*, 1995).

Channels

As in many sedimentary systems, channels are a common element in deep-marine settings. A channel is a negative topographical element produced mostly by confined turbidity currents that transport sediment along a major, long-term pathway. However, channels can also be sites of sediment deposition or erosion. Like channels in the continental realm, the condition of erosion, bypass or deposition is controlled by the sediment characteristics and boundary conditions of the system – changes in one or both of these parameters will effect a change in the channel system. Deep-marine channels, like fluvial channels, continually seek a longitudinal profile graded to a base level, which in the deep marine is generally taken to be gravity base, but, more practically, is defined as the position where the flow becomes unconfined at the upcurrent end of the terminal lobe (*e.g.*, Pirmez *et al.*, 2000). Where the channel gradient and sediment-transporting flows are in equilibrium, channels migrate laterally along a plane parallel to the equilibrium profile and most of the sediment is bypassed to areas farther downflow. Erosion and deposition, on the other hand, represent conditions where the channel profile lies above and below the graded profile, respectively (*see* Chapters 2 and 6).

Based on observations from modern and ancient systems, three kinds of deep-marine channels are recognized: erosional, depositional and mixed erosional-depositional (Fig. 10). Erosional channels are bounded by a scour surface that clearly truncates older strata. In contrast, depositional channels are bounded by well-developed channel-margin levees that, with time, become progressively elevated above the surrounding seafloor. Mixed channels show a combination of levee deposition and channel-axis erosion, and, in which, the channel floor may lie above, or below, the level of the adjacent seafloor.

Channels can also be classified on the degree of confinement of the channel, which typically changes systematically downflow (*e.g.*, Posamentier and Kolla, 2003). Highly confined channels are those in which flow is contained mostly within the channel. Generally, these channels occur in the upflow parts of a turbidite system and include submarine canyons and erosional channels, the dimensions of which are commonly of the order of kilometers to several kilometers wide and > 100 m deep. Farther downflow, these channels change into leveed-channel systems, where channel dimensions range up to a few kilometers wide and generally up to about 100 m deep (*i.e.*, the relief between the thalweg and the adjacent levee crests). Here, flows and, in particular, the upper parts of flows, can escape the confines of the channel to build channel-bounding levees. These systems, in turn, are succeeded downflow by a complex of poorly confined channels associated with a lobate sedimentary body, commonly termed a depositional lobe or fan, at the downstream end of the channel.

Submarine Canyons

Submarine canyons are the primary conduits for sediment transport into the deep sea. In the modern oceans, submarine canyons range up to more than 2.5 km deep and 100 km wide (*see* Normark and Carlson, 2003), whereas the examples of ancient canyons described so far are typically much smaller, reaching only slightly more than 1 km deep and 10 km wide. The fill of a submarine canyon is typically stratigraphically complex and lithologically variable. In part, this relates to temporal and spatial differences in sediment source, which

Figure 11. A, **B.** Submarine-canyon strata consisting of contorted mudstone-rich slump deposits most probably sourced from local canyon-wall collapse (Pennsylvanian Jackfork Formation, Arkansas). **C.** 135-m-deep canyon incised into thin-bedded and very thin-bedded continental-slope turbidites. Canyon fill consists mostly of graded, structureless coarse-grained sandstone and conglomerate (**D**). (Hector Formation, Lake Louise, Alberta).

varies from local canyon-wall collapse, to an up-dip feeder system sampling a shelf and/or continental sediment source. Fills dominated by local wall collapse are typically fine-grained with common mud-rich mass-movement (slump and slide) and debris-flow deposits (Fig. 11A, B); an example is provided by the subsurface late Paleocene Yoakum and Lavaca canyon fills (Galloway *et al.*, 1991). Fine-grained sediment deposited from suspension may also occur as a drape infilling part or all of the canyon. Coarse sandstone and conglomerate occur as isolated elements in mud-rich fills, but also dominate some fills, especially those formed in tectonically active areas (Fig. 11C). In most cases, coarse sediment occurs as thick-bedded, structureless beds deposited by high-concentration turbidity currents (Fig. 11D).

Erosional Channels

Although the smaller scale erosional channels down-system from submarine canyons also owe their existence to erosion, their fill is different, and typically consists of a significant proportion of sand- and gravel-rich strata deposited by turbidity currents and other frictional flows. Although the geological literature is replete with erosional channel-fill models (*e.g.,* Beaubouef *et al.*, 1999; Mayall and Stewart, 2000; Samuel *et al.*, 2003), the stages common to most models include: channel inception, sediment bypass, channel filling and channel abandonment. *Channel inception* is marked by a period of successive flows with high transport efficiency that scour-out a throughgoing topographical feature that serves as the conduit for later flows. Sediment transported during this stage is carried further basinward and deposited in more distal areas. This stage is then succeeded by the *channel-bypass stage* wherein flows range between complete bypass (with no further erosion) to incomplete bypass. Incomplete bypass is commonly indicated by the deposition of laterally discontinuous beds (due to erosion by subsequent flows), intercalation of coarse- and fine-grained deposits, the common patchy occurrence of tractional sedimentary structures, especially dune cross-stratification and coarse-grained lags, and thin drapes of fine-grained sediment, a heterogeneous assemblage of lithofacies herein termed the 'bypass facies'. The next stage, *channel fill*, is characterized by a change toward flows that have a lower transport efficiency that initiate sediment deposition within the channel, eventually filling part, or all, of it. Later, as a result of a diversion of flow at a point upstream (avulsion), or diminution of flow for other reasons (*e.g.*, sea-level rise), the channel system is abandoned and becomes a site of mostly fine-grained deposition that drapes any residual topography. Superimposed on this idealized succession of events are episodes of reactivation, particularly during the channel-filling stage, which serve to temporarily rejuvenate the system but not reverse its long-term filling trend (*e.g.*, Samuel *et al.*, 2003).

Erosional channels are commonly reported from seismic images, and much less commonly from the ancient outcrop record. This disparity may be the result of two factors:

1. Many seismically resolved examples of erosional channels are very large, commonly measuring more than 100 m deep (thick) and a kilometer to many kilometers wide (*e.g.*, Mayall and Stewart, 2000; Deptuck *et al.*, 2003; Abreu *et al.*, 2003) and therefore are on a scale significantly larger than most outcrops; and
2. If strata inside and outside the channel margin are of similar lithology, or are poorly exposed, recognizing the channel-bounding surface in outcrop may be difficult, even though the channel fill and the surrounding deposits may have distinctly different acoustical properties and therefore are easily differentiated on seismic.

Seismic images also show that erosional channels are commonly bordered by well developed levees. Deptuck *et al.* (2003) recognized two end-member kinds of erosional channels and their related levee deposits. Large-scale channels and their related 'outer' levees represent the master erosional channel that is typically several kilometers wide and > 100 m deep (Fig. 12). These features were infilled, partly or completely, by smaller scale channels with their associated 'inner' levees, and are described in the next section.

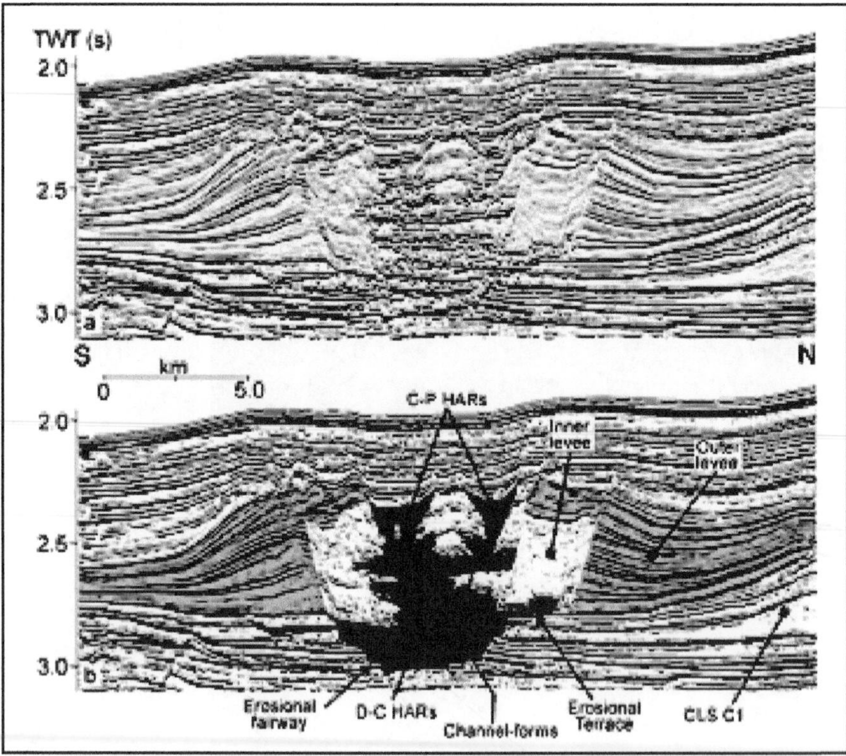

Figure 12. Uninterpreted and interpreted seismic profile of a channel-levee system in the Indus Fan, Bay of Bengal (Deptuck *et al.*, 2003). Outer levees bound master channels that are several kilometers wide, which are filled with the deposits of smaller channels (HARs) and their associated inner levees.

Heterogeneous strata deposited by incompletely bypassed flows form the basal unit of the channel fill (Fig. 13). This is commonly succeeded by contorted seismic reflectors interpreted to represent slump/slide (mass-movement) and debris-flow deposits produced by collapse of the channel margins. These strata are overlain by a thick succession of sandstone or, less commonly, conglomerate that forms a tabular or sheetlike unit comprising amalgamated small-scale channel fills. During this early stage of channel fill, these smaller scale channels tend to be poorly confined with low to moderate sinuosity and a high width-to-depth ratio (*see* leveed channels below). These channels are succeeded by channels with higher sinuosity and a lower width-to-depth ratio (*see* later: confined sinuous channels). Furthermore, these channels commonly show an upward increase in angle of climb, reflecting an increased rate of channel aggradation relative to the rate of lateral migration (Peakall *et al.*, 2000). Eventually, the entire channel system, which at this point has been either partly or completely filled, is abandoned and blanketed by a layer of thin-bedded turbidites and hemipelagic suspension deposits.

Leveed Channels

Leveed channels are more commonly reported from outcrops because of their smaller size compared to large erosional channels,. Although smaller, leveed-channel fills still range up to a few kilometers wide and 100 m thick (Fig. 14). In addition to occurring as an independent channel element, leveed channels are also important stratal components in larger scale erosional channel fills as described above.

Leveed channels typically have a sinuous planform (Fig. 14A, B) with levees of varying development along their margins. Based on the degree of confinement, two end-member types of leveed channels are recognized: poorly confined and highly confined. In both types, channel-fill strata terminate abruptly along an erosion surface defining the outer-bend margin of the channel. Strata cut by the erosion surface are either genetically related levee deposits or strata related to an older channel. Along the inner-bend side, however, channel-fill strata either grade continuously into levee deposits in poorly confined channels, or onlap them in highly confined channels. Also, owing to the effects of (fluid) inertia and the resulting tendency of a current to continue along a straight line while the channel floor bends beneath it, levees and levee deposits are always best developed along the outer bend of all channels.

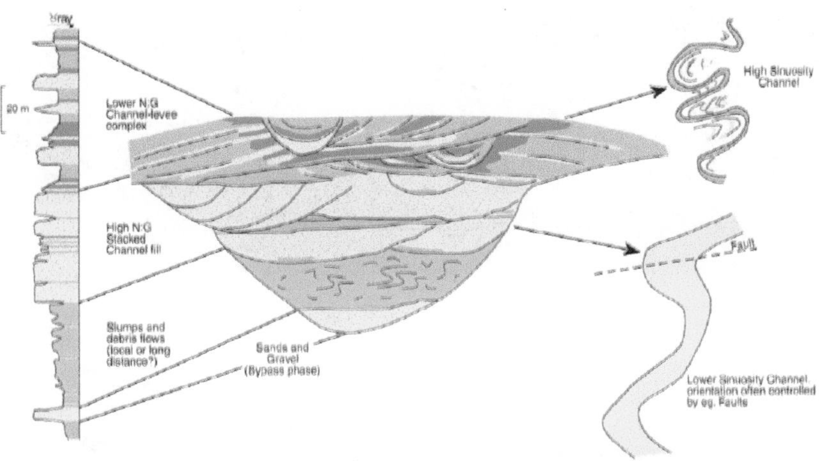

Figure 13. Idealized model for the fill of an erosional channel (Mayall and Stewart, 2000).

Poorly Confined Leveed Channels

Channel Deposits: The base of poorly confined leveed channels is commonly asymmetric, with a steeper margin along one side (analogous to the cut bank of a sinuous fluvial channel; Fig. 15). In addition, the channel base is often characterized by a step-flat morphology, indicating the episodic but systematic step-like lateral migration of the entire channel system (*e.g.*, Eschard *et al.*, 2003; Navarro *et al.*, 2007). This asymmetry is present also in the nature of the relationship between the channel-fill and adjacent levee strata. Along the steep margin, levee strata are either in erosional contact with channel strata (*see* 'outer-bend' levee deposits below) or are separated by a thin, fine-grained bypass unit (Beaubouef *et al.*, 1999). In contrast, on the opposite side of the channel (analogous to the point bar of a sinuous fluvial channel; *see* Chapter 6), channel-fill strata either onlap or grade laterally into levee deposits (*see* 'inner-bend' levee deposits below). The fill of leveed channels is, in fact, composed of the fill of myriad smaller channels, which, because of extensive amalgamation, are difficult to trace individually in outcrop. Nevertheless, the fill of an individual channel is of the order of a few to several meters thick and tens to, at most, a few hundred meters wide. The 3-D amalgamation of these

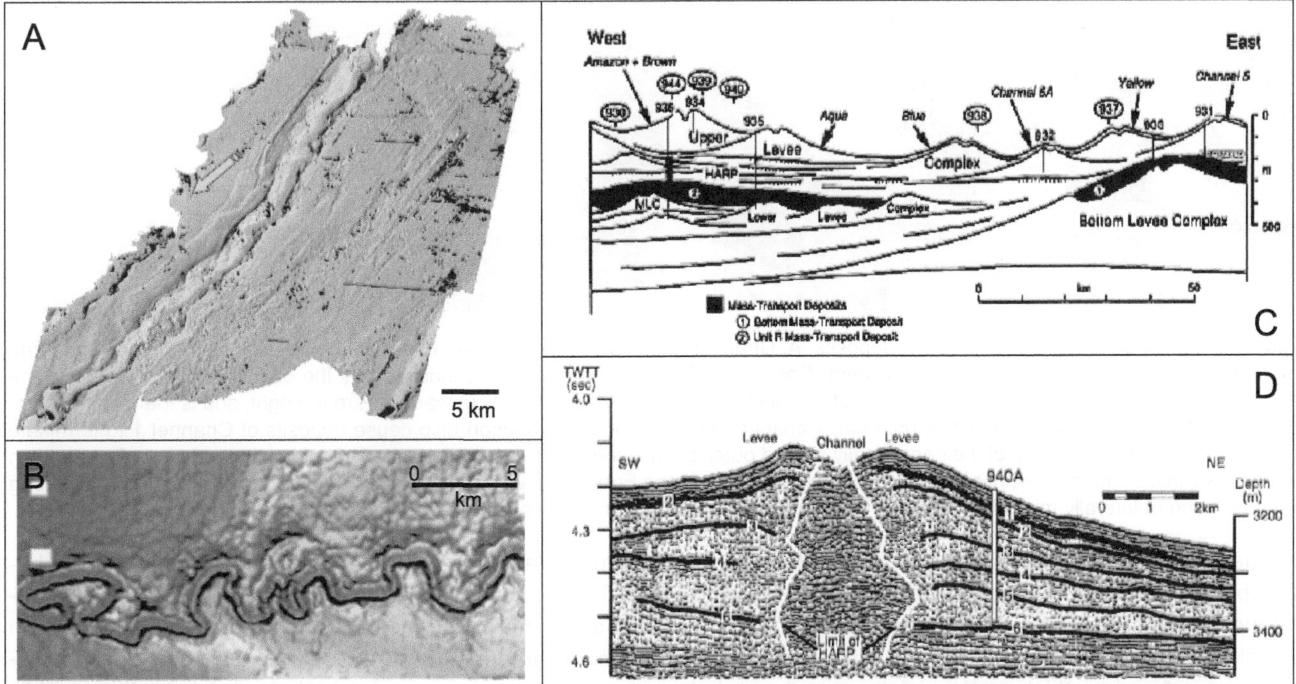

Figure 14. Sinuous leveed channels. **A.** Low sinuosity, large-scale (width > 1 km) (Pleistocene) Einstein Channel in the Gulf of Mexico (modified after Posamentier and Walker, 2006). **B.** High sinuosity, small-scale (100s m wide) modern Amazon channel-levee system in about 3500 m water depth (from Pirmez *et al.*, 2000). **C, D.** Seismic profiles across multiple **(C)** and single **(D)** channel-levee complexes in the Amazon Fan (from Hiscott *et al.* (1998) and Piper and Normark (2001) respectively). In **C,** note the organized offset stacking, termed compensational stacking, of successive complexes (*i.e.*, younger channels are located preferentially in the topographic lows between two older systems). Note also the extensive blanket of mass-transport deposits that underlies the uppermost stack of channel-levee complexes.

Figure 15. Poorly confined leveed-channel deposits. **A.** Channel fill 1 is a sharp-based, up to about 80 m-thick sandstone/conglomerate unit sharply overlain by a second channel system (Channel fill 2), the base of which is indicated by the dashed orange line. Note the sharp, terraced channel base (solid orange line) of Channel fill 1 that ascends obliquely upward toward the right, and is the result of combined vertical and lateral channel migration. Lateral channel migration and aggradation also cause deposits of Channel 1 to terminate abruptly as they overstep strata of their genetically related outer-bend levee. **B.** Along the opposite (left) side, channel strata fine and thin continuously into strata of the inner-bend levee (Neoproterozoic Isaac Formation, western Canada). **C, D.** Sharp, terraced margin along the outer bend of laterally migrating, vertically aggrading channels (Upper Cretaceous Tres Pasos, southern Chile).

channels forms a channel unit, which is probably the most readily identified channel succession in outcrop. Channel units range from several meters to a few tens of meters thick and commonly show a well-developed fining- and thinning-upward trend. As noted above, their base is typically marked by the presence of coarse sediment and abundant mudstone intraclasts. Also, in many cases, coarse, amalgamated strata grade abruptly laterally into finer, more stratified deposits. Strata in the axial part are generally thick- to very thick-bedded, coarse-tail normally graded or structureless conglomerate or sandstone. Beds are typically amalgamated with variable lateral continuity (ranging from tens to hundreds of meters laterally). These coarse-grained deposits are T_a turbidites deposited by gravel- and sand-rich, high-concentration turbulent flows. Toward the margin of channel units, especially those higher in the leveed-channel fill, coarse-grained strata tend to thin rapidly (generally over < 100 m), and grade from amalgamated sandstone to less amalgamated, more thin- to thick-bedded sandstone with intervening mudstone. Sandstone beds consist of T_a and T_{ab} turbidites and are interca-

Figure 16. A. Proximal outer-bend levee deposits composed of medium- and thin-bedded turbidites. Thicker beds typically consist of T_{bcde} turbidites that thin rapidly away from the channel margin, whereas thinner beds consist of T_{cde} turbidites that change little in thickness laterally. **B.** Thin to very thin-bedded T_{cde} and T_{de} turbidites of the distal levee (field notebook for scale). As in the proximal levee, thin beds in the distal levee show little thickness change laterally. Photos **A** and **B** are from the Neoproterozoic Isaac Formation, western Canada.

lated with thin-bedded T_{cd} and T_{cde} Bouma turbidites. The thinning and fining of strata toward the margin of channel units suggest that the higher-energy, axial part of the channel is flanked by lower-energy conditions.

Levee Deposits: Subaqueous levees build upward by the addition of sediment when flows overtop the margins of the channel. Flow overspill and related levee aggradation occur in three ways: flow stripping, inertial overspill, and continuous overspill. Flow stripping and inertial overspill occur at channel bends, and preferentially along the outer bank. In the case of flow stripping, the upper fine-grained part of the flow becomes separated from the lower, coarse-grained part that remains confined to the channel. Inertial overspill occurs when an energetic flow is unable to follow the sinuous thalweg and 'runs-up' the channel margin, causing even the lower parts of the flow to escape the channel. Continuous overspill takes place where the thickness of the flow exceeds the depth of the channel, leading to loss of the flow above the height of the levee along both sides of the channel. Levee growth along the inner-bend side of the channel and the straight segments between channel bends comes about by continuous overspill. Upon escaping the channel, the flow expands rapidly and collapses, resulting in elevated rates of sedimentation immediately adjacent to the channel, with rapidly decreasing rates of deposition farther from the channel. This lateral variation in average sedimentation rate gives levee deposits their distinctive 'gull wing' geometry on seismic images (Fig. 14C). As a levee aggrades, the relief between the channel floor and levee crest increases, eventually allowing only the upper, more dilute portion of the through-going flows to overspill. This process is responsible for the upward thinning and fining trends associated with many levee deposits.

Owing to their fine-grained nature, levee deposits are typically not well exposed, although notable exceptions exist. Along the outer-bend side of the channel, levee deposits tend to be thick and sand-rich, whereas those on the inner-bend side are significantly finer and thinner. Paleocurrents, typically measured from the c-division of turbidites, are generally oriented very oblique to paleoflow in the main channel. Closest to the channel and along the outer bend, (the proximal levee facies) strata consist mostly of thin- to medium-bedded, fine to medium-grained sandstone T_{bc} turbidites interstratified with thin-bedded, very fine- to fine-grained sandstone/siltstone T_{cde} turbidites and common thick T_a sandstones (Fig. 16A). In many modern and interpreted ancient proximal-levee deposits, small-scale (ripple) cross-lamination commonly shows evidence of vertical aggradation (*i.e.*, climbing) because of sediment fallout from suspension as the flow expands, with the angle of climb generally decreasing away from the channel margin. However, the presence of climbing ripples is not a universal feature of levee deposits as shown by the Windermere Supergroup (Navarro *et al.*, 2007). This may be due to a somewhat coarser grain size and/or poorer sorting, but the cause remains uncertain.

With increasing distance from the channel, the strata become thinner. Typically, thicker beds thin rapidly over 100s of meters whereas thinner beds show little lateral change. As a consequence, strata occurring a few to several 100s of meters from the channel in the distal levee are composed predominantly of thin-bedded, very fine to fine-grained sandstone/siltstone T_{cde} turbidites (Fig. 16B). The c-division consists of one to at most a few sets of non-climbing ripple cross-stratification. Locally, distal levee strata are intercalated with overbank or crevasse-splay deposits (*see* below).

Levee strata on the inner side of channel bends are distinctly thinner and generally finer grained than those on the outer side of the bend. Deposits consist predominantly of fine sandstone and siltstone T_{cde} turbidites, which, locally, are interbed-

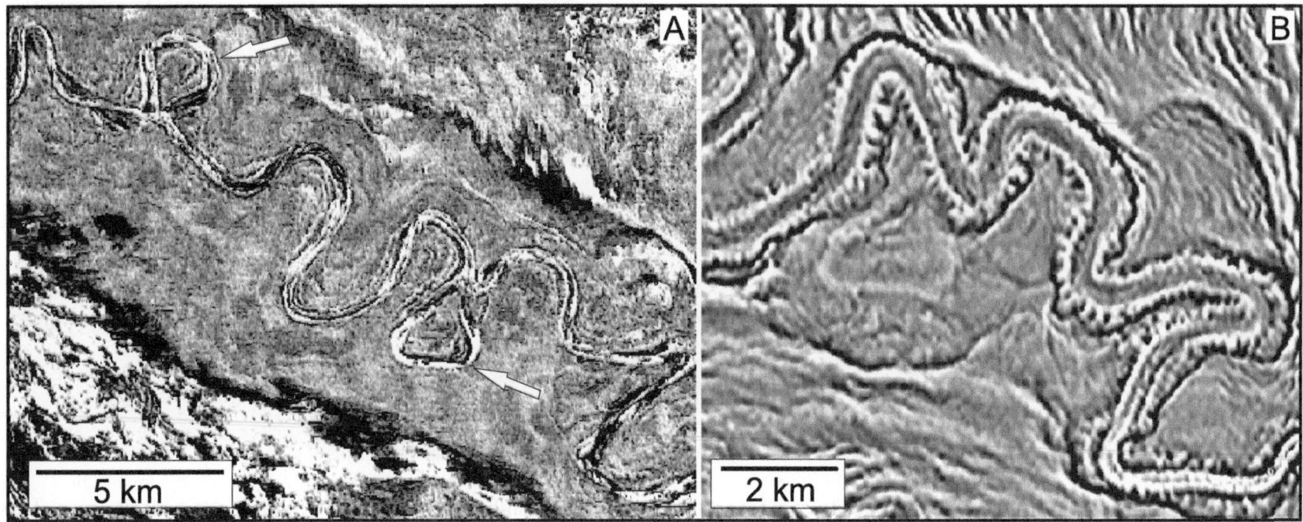

Figure 17. Highly sinuous small-scale (< 1 km wide) leveed channel system, the Joshua Channel (Pleistocene), Gulf of Mexico (Posamentier and Kolla, 2003; Posamentier and Walker, 2006). **A.** Seismic horizon slice illustrating the highly sinuous nature of the channel. Note also the meander cut-offs (oxbows) indicated by arrows. **B.** Seismic curvature map showing the well-developed levees that bound the channel. Note the common slump scars along the levees on both sides of the channel.

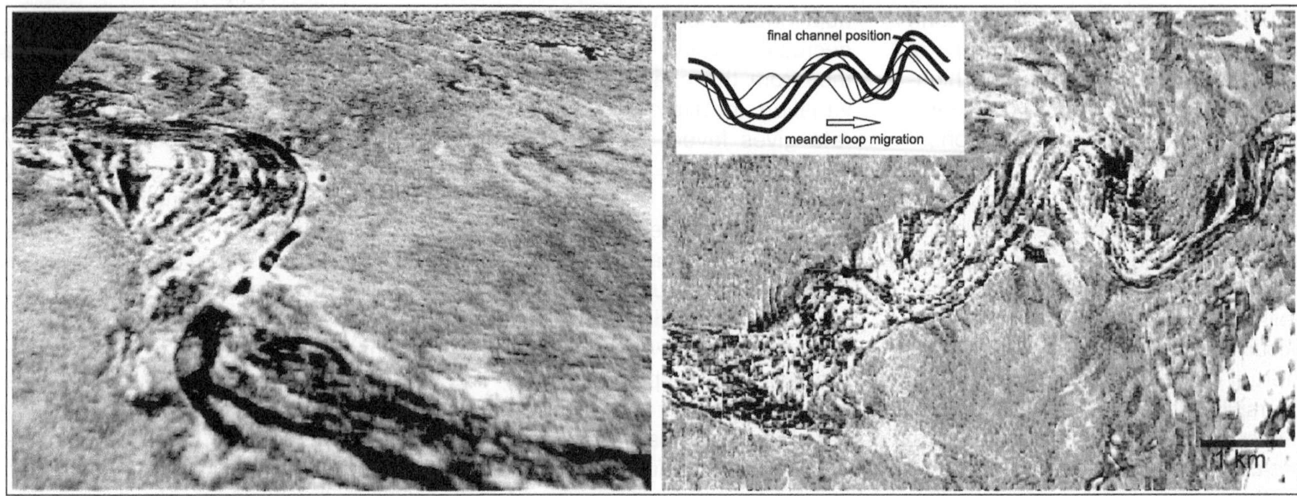

Figure 18. Organized channel pattern created by the lateral offset pattern of successive channel-fill elements (images courtesy of Henry Posamentier). Inset sketch from Posamentier and Kolla (2003).

ded with thicker sandstone beds. These latter beds consist typically of medium- to thick-bedded, fine to medium sandstone T_c turbidites composed of multiple (3 or 4) ripple cross-stratified sets. Like on the opposite side of the channel, strata thin and fine away from the channel, but the rate of change is much greater and reflects the lower-energy nature of overspilling flows on the inner-bend side of the channel. An important difference between inner- and outer-bend levee strata is that, in many places, inner-bend strata are continuous with and grade laterally into channel-fill strata, indicating a continuum between channel and levee deposition.

Highly Confined Leveed Channels: Channels of this variety are significantly smaller than those described above – channel width and depth are of the order of tens to a few hundreds of meters and a few tens of meters, respectively; sinuosity is generally higher too (Fig. 17). In outcrop, highly confined channels occur mostly as disconnected channels fills that locally are clustered laterally and/or vertically. Clustering is attributed to younger channels exploiting remnant seafloor topography formed by incompletely filled older channels (Fig. 18). In addition, lateral-accretion deposits are well developed, indicating systematic deposition on the inner bend of laterally migrating sinuous channels (Arnott, 2007a; Fig. 19A, B).

Channel bases are generally planar and horizontal with only local small-scale (few cm) scours. In places, however, the surface shows a step-like geometry, rising abruptly upward by a few to several meters. The top of the sandy channel fill interfingers with thinly bedded turbidites (Fig. 19A and B). The strata within the channel fill show a distinctive and consistent dip of up to about 7 to 12°, and, like similar features observed in meandering fluvial systems, are interpreted to be lateral-

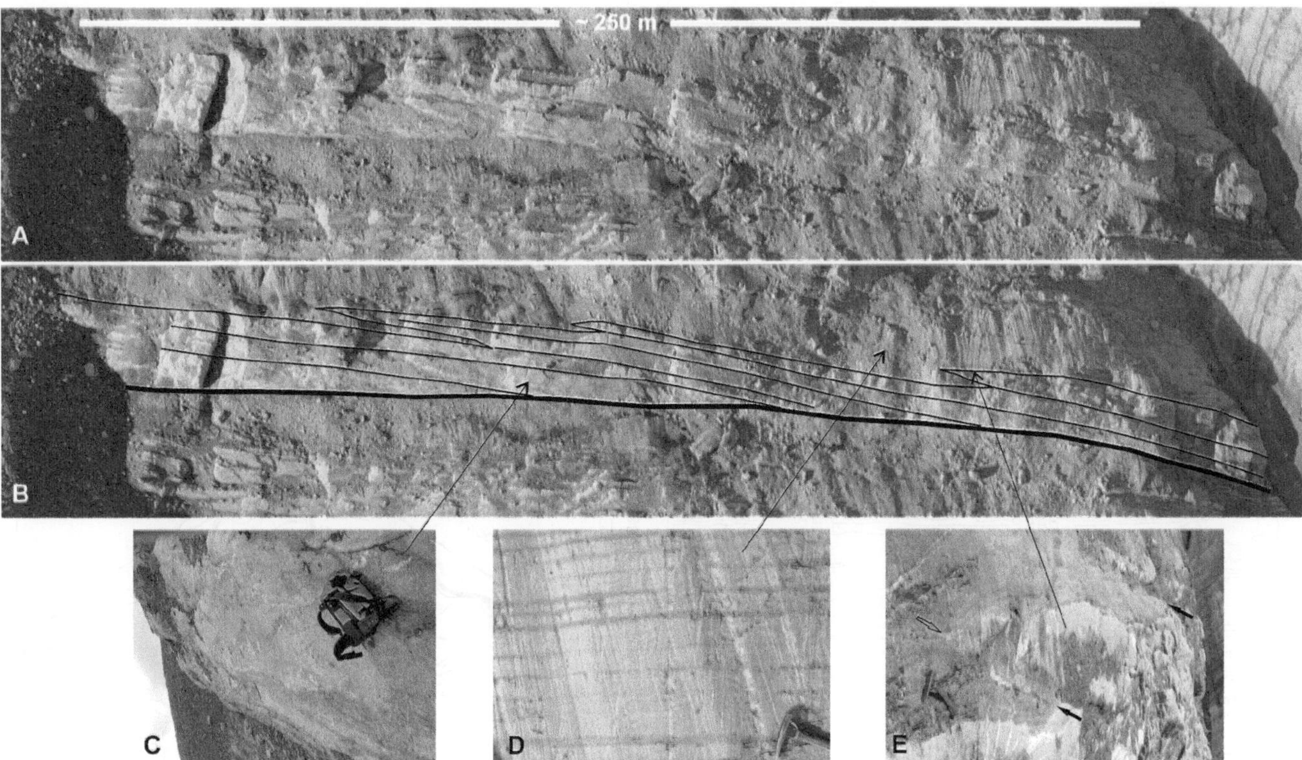

Figure 19. A, **B.** Lateral-accretion deposits (LADs) in a highly confined leveed channel from the Neoproterozoic Isaac Formation, western Canada. Note the dipping LADs, which are inclined at an angle of 7-12°, and are interpreted to have accumulated on the inner-bend margin (point bar) of a laterally accreting sinuous channel (from Arnott, 2007a). Also, note the sharp, planar basal contact, but the interfingering of the sandy beds with thin, mud-rich deposits at the top. **C.** Structureless, graded sandstone near the base of LADs. At their upper end, such coarse beds pinch-out abruptly upward (open arrow in **E**) into thin-bedded T_{cde} and T_{de} turbidites (**D**), which in turn become truncated downward by the coarse beds (solid arrows in **E**).

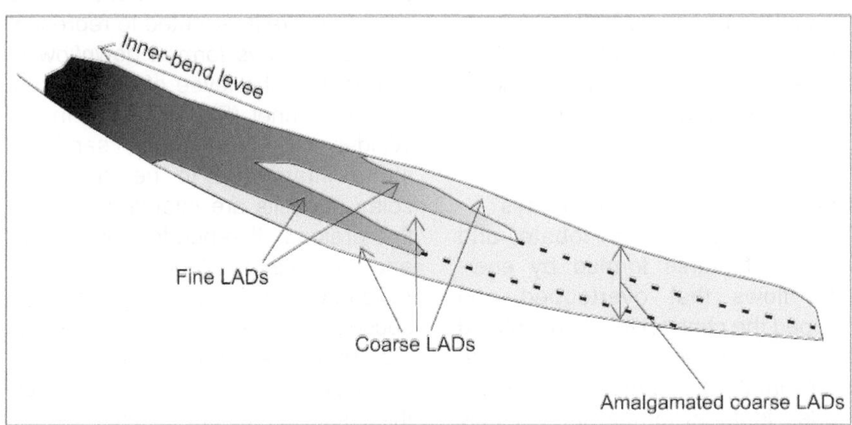

Figure 20. Idealized model of lateral-accretion deposits (LADs) formed by a laterally migrating deep-marine sinuous channel. Each coarse and fine LAD consists of several beds and indicates that there were longer term repetitive alternations in the nature of the flows.

accretion deposits (LADs) formed on the inner bend of a laterally migrating sinuous channel. The fill of the channel consists of lower and upper parts. Beds in the lower part are typically thick- and very thick-bedded, and generally amalgamated (Fig. 20). They consist mostly of sharp-based, graded sandstone and less common granule or fine pebble conglomerate (Fig. 19C). Bases of beds are commonly scoured and, in places, completely erode underlying beds. Mudstone generally occurs as localized patches of intraclasts. Stratigraphically upward, sand-/gravel-rich strata change little in grain size but thin, typically becoming medium bedded; mudstone intraclasts are absent. In the upper part of the channel fill, fine-grained strata (mudstone and thinly bedded turbidites) become interstratified with the coarse-grained deposits (Fig. 19A, B). It is noteworthy that coarse-grained beds terminate abruptly up-slope (Figs. 19D, E and 20), which contrasts markedly with the gradual lateral trend observed in poorly confined leveed channels. Also, very near the terminus of each bed, coarse strata consist of a small number of graded, poorly sorted beds capped by planar laminated or dune cross-stratified sandstone. Down-dip from the terminus of each coarse bed, fine-grained strata are truncated as the coarse beds amalgamate (Figs. 19E and 20). Near their termination, fine-grained strata consist of almost complete Bouma-division turbidites (Fig. 19E) that, up dip, thin, fine and become dominated by upper-division turbidites (T_{cde}).

The coarse-grained beds in the LADs represent coarse sediment deposition on the lower part of the

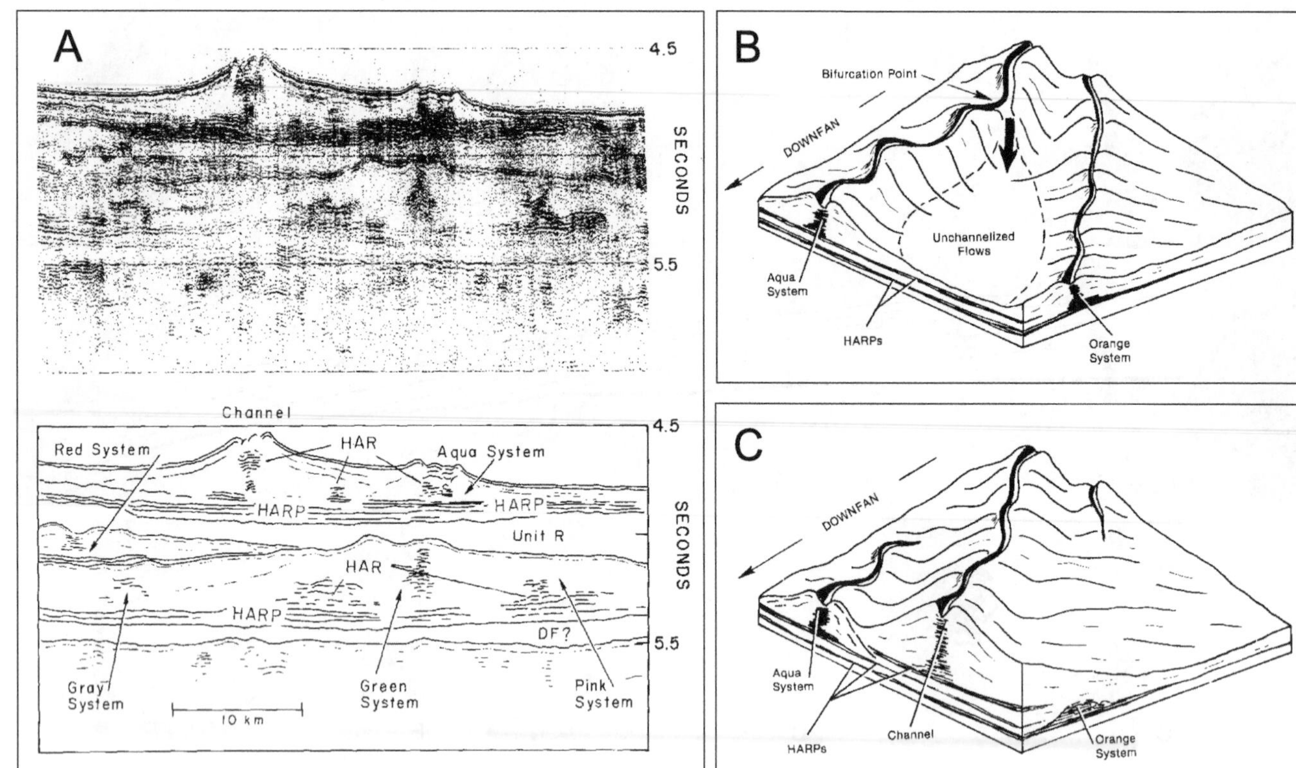

Figure 21. A. Seismic profile and interpretive sketch of channel-levee complexes in the Amazon Fan (Flood *et al.*, 1991). High amplitude, sheet-like packages, termed HARPs, commonly underlie the high-amplitude reflectors (HARs) that are interpreted to be channel deposits. HARPs are interpreted to be crevasse-splay deposits (**B**) formed during the initial stages of an avulsion, which in many cases are overlain by their genetically related channel (**C**).

channel margin. In the LADs fine-grained beds fine and thin abruptly beyond the termini of the coarse LADs, and represent the fine-grained part of the inner-bend levee onto which the coarse-grained LADs onlap. On the opposite, or outer-bend, margin of the channel, coarse-grained LADs typically terminate abruptly against fine, thin-bedded turbidites of a slightly older levee system.

The rhythmic interstratification of coarse- and fine-grained beds, which is especially well developed and preserved in the upper part of channel fills, reflects recurring changes in sediment transport through the channel system. Deposition of the fine beds most likely represents periods of highly efficient turbidity currents that bypass the channel bend and transport much of their coarse sediment load farther downdip. Periodically, these conditions are interrupted by episodes of less efficient turbidity currents that result in deposition of a small number of beds that make up each coarse interval. Currently, the cause for the rhythmic alternation of fine and coarse beds in the LADs is poorly understood, but most likely relates to repetitive changes in local or regional flow and/or channel conditions (Arnott, 2007a).

Overbank and Crevasse Splays

Overbank splays are lobate and sheetlike features formed by energetic flows that overtopped and escaped the channel in an unconfined manner. Crevasse splays, on the other hand, are larger scale lobe-shaped features formed immediately downflow of a crevasse channel that is incised into the channel margin and proximal levee (*cf.* Chapter 6). To date, few unequivocal examples of crevasse- or overbank-splay deposits have been reported from the geological record. Work on the Amazon Fan (*e.g.*, Flood *et al.*, 1991) identified a distinctive seismic facies that commonly underlies channels (Fig. 21A). Termed a HARP (high-amplitude reflection package) because of its high acoustic impedance, these strata have a sheet-like geometry and are assumed to be sand-rich. These packages are interpreted to represent crevasse splays formed downflow of a breach in the levee of an adjacent active channel (Fig. 21B, C). In the Windermere Supergroup, sand-rich strata interpreted to be crevasse-splay deposits are interstratified with fine-grained, thin-bedded basin-floor and distal-levee deposits (Fig. 22; Arnott 2007b). These crevasse-splay successions, which are up to several meters to a few tens of meters thick, consist of decimeter to several meter-thick units of medium- to thick-bedded structureless sandstone containing common mudstone intraclasts. In places, these strata are interbedded with units composed of upper division (T_{cde}) sandstone turbidites. Structureless sandstones are poorly sorted and coarse-tail graded with a matrix of fine sandstone, siltstone and mudstone, and are interpreted to have been deposited rapidly by capacity-driven deposition immediately downflow of an area of rapid flow expansion. The thinner graded beds are deposited on the periphery of the collapsing sediment cloud. The intercalation of the two types of beds is related to the lateral wandering of the zone of flow expansion.

At its headward end, a crevasse splay is joined to the breach in the adjacent parent channel by a crevasse channel, which forms a sharp-based, distinctly coarser grained unit within a background of fine-grained levee deposits. Crevasse-channel deposits tend to be thin (up to a few meters thick) and consist of amalgamated thick-bedded, massive or normally graded sandstone or (less commonly) conglomerate. Discontinuous beds of single-set-thick dune cross-stratified sandstone are common also.

Overbank splays, on the other hand, form from large magnitude, high concentration, coarse-grained turbidity currents that overspill the adjacent channel without confinement (Fig. 22). Their deposits consist of single to multiple beds forming units up to 2–5 m thick. Internally, units consist of amalgamated thick-bedded, medium-grained sandstone turbidites that commonly comprise complete Bouma turbidites.

Basin-floor Deposits

In the proximal part of the basin floor, leveed channels terminate downflow in a thick, laterally extensive sediment body variously termed a depositional lobe, distributary-channel complex, sand-sheet deposits, or frontal-splay complex, for it is here that highly confined flows emanating from the leveed channels become unconfined and depositional (Fig. 23). The depositional lobes range up to several tens of meters thick and 100 km wide. Farther basinward, they become finer and thinner, and eventually are replaced by hemipelagic and pelagic deposits. Loss of confinement and the lateral spreading of the flow can be the result of a reduction in slope and/or loss of sufficient fine-grained sediment to build channel-margin levees (Posamentier and Kolla, 2003). Characteristics of the transition from channels to depositional lobes are principally controlled by grain size. In coarse-grained systems, the depositional lobe connects directly with the leveed channel, but in mud-rich systems, the lobe is separated by a transitional zone marked by large scours, sediment waves and sediment mounds. Interpreted transition-zone deposits in the ancient stratigraphic record consist also of an array of lithofacies suggesting both sediment bypass and sediment deposition. Bypass features include shallow scour surfaces draped by fine- and/or coarse-grained (lag) sediment that commonly is planar laminated or dune cross-stratified, and units up to about 5 m thick composed of compensationally stacked scour-based lenticular sandstone. Depositional features include sheetlike sandstones that are similar to sheetlike splay deposits described below.

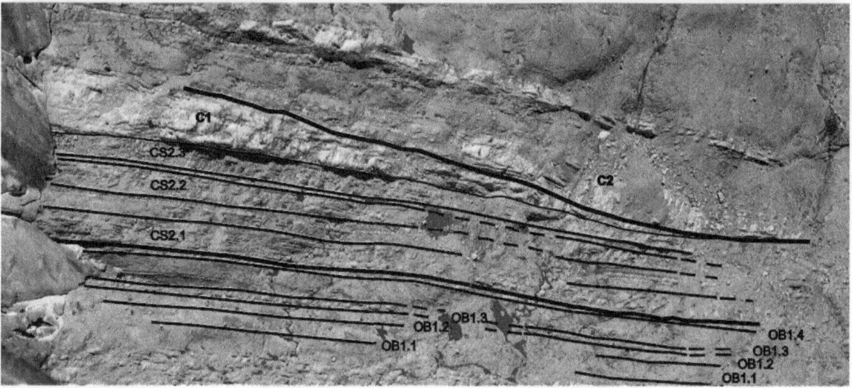

Figure 22. Aerial photo of strata from the Neoproterozoic Isaac Formation, western Canada showing overbank-splay deposits (OB1.1-OB1.4), and crevasse-splay deposits (CS2.1-CS2-3) and their genetically related crevasse-channel fills (C1, C2) – note, strata are vertically dipping (see Arnott, 2007b). Overbank-splay deposits occur as units, one to a few beds thick, comprising medium- to thick-bedded, coarse-grained, more complete turbidites (T_{bcde}) interbedded with thin-bedded, upper-division turbidites (T_{cde}, T_{de}). This succession is then overlain abruptly by a thick crevasse-splay deposit consisting of three several-meter-thick packages of matrix-rich structureless sandstone (CS2.1–CS2-3) intercalated with few-meter-thick packages composed of classical turbidites. Crevasse-splay deposits are then sharply overlain by a lateral-accreting channel deposit that comprises two separate channel fills (C1, C2).

Depositional Lobes

The planform geometry and size of a depositional lobe depends principally on the sand-mud ratio of the sediment supply. In sand- and gravel-rich systems, these features tend to be areally restricted because of rapid deposition, and form discrete elements on the order of a few kilometers wide and a few meters to several tens of meters thick that decrease in thickness and grain size rapidly downflow.

Recent work by Deptuck *et al.* (2007) showed that such late Pleistocene features off the coast of Corsica are of the order of 2–19 km² in area, 9–20 m thick, have a length-to-width ratio of < 1–2, and show no evidence of progradation but instead backstep upslope. (A similar backstepping pattern is noted in proximal mouth-bar deposits; *see* Chapter 10). Based on seismic expression and shallow piston cores, strata are dominated by amalgamated sand

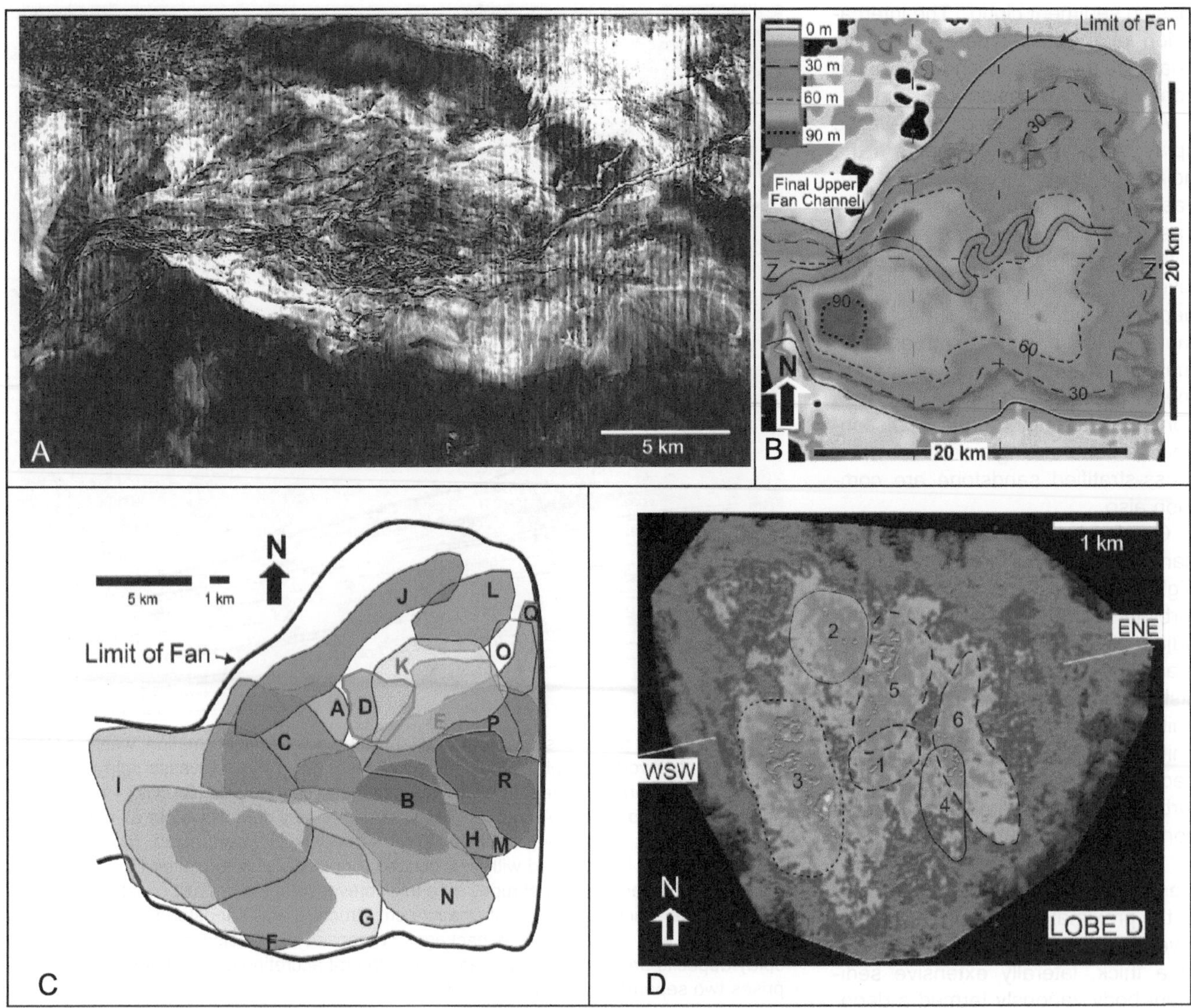

Figure 23. A. Seismic image of a Pleistocene depositional-lobe complex developed at the downflow terminus of a leveed channel, Gulf of Mexico (Posamentier and Kolla, 2003). **B.** Isopach map of a Pleistocene depositional-lobe complex, Indonesia (Saller et al., 2008). **C.** Map illustrating the internal stratigraphic complexity of the lobe complex in **B**, which consists of 18 (A-R) discrete lobe deposits, which formed during the lowstand systems tract. During the late lowstand to early transgressive systems tract, however, a change to more mud-rich flows caused the leveed channel at the headward end of the lobe complex to extend basinward ('final upper fan channel'), which caused lobe deposition to shift basinward too. **D.** Detail of a single depositional lobe (Lobe D – outlined in red in **C**) showing that it too is composed of several even smaller, discrete splay elements (labeled 1-6). It is these smaller splay elements that are generally observed in the ancient record and here are termed sheet-like splay elements.

that represents 50% of the planform area and 75% of the total volume of the depositional lobe. In contrast, more mud-rich systems form broader, more sheetlike features that are up to about 10 km wide and become finer and thinner more gradually downflow. Also, their stratigraphic composition is significantly more complex than those in sand-rich systems.

Based on seismic interpretation and observations in the rock record, such lobe systems consist commonly of three recurring architectural elements: deep channels, shallow channels, and sheetlike splay deposits (Figs. 23B, C, D and 24). Deep channels show up to a few tens of meters of incision, with channel margins that are locally steep (few to several meters of relief over a lateral distance of only a few tens of meters; *e.g.*, Meyer and Ross, 2007). These channels are filled with a variety of lithofacies, including heterogeneous assemblages of coarse- and fine-grained strata related to incomplete bypass, and also thick-bedded, amalgamated sandstone (*e.g.*, Johnson *et al.*, 2001; Fig. 24C). Deep channels are interpreted to represent the principal conduit that supplies sediment to the depositional-lobe complex. Shallow channels, which occur downflow of the deep channels, show minor relief along their base (scour only a few meters deep over a few hundred meters laterally), and exhibit a consistent internal stratigraphy that mostly is aggradational. In the channel axis, strata range up to about 5–10 m thick and consist of thick- to very thick-bedded amalgamated sandstone (T_a beds), which, when traced laterally over a few hundred meters, show a

Figure 24. A. Ancient basin-floor deposits (Neoproterozic upper Kaza Group, western Canada) comprising 3 principal depositional elements: shallow channels, deep channels, and sheetlike depositional lobes — note strata are dipping vertically. Yellow arrows indicate location of figures **B**, **C** and **D**. In all photos stratigraphic top is to the left. **B.** Shallow, erosionally based channel with only subtle relief along its base (black dashed line). From the margin toward the axis (direction indicated by red arrow), strata show a rapid increase in sandstone:mudstone ratio as thin-bedded turbidites pass laterally into amalgamated sandstone in the channel axis. **C.** Deep channel with prominently scoured base (solid black line), overlain by a heterogeneous bypass facies. **D.** Sheetlike depositional lobes of the order of 15 m thick that form laterally extensive bodies composed of amalgamated sandstone (hammer (circled) for scale). Base and top of lobe indicated by double-headed arrow.

gradual but systematic change to thick-/medium-bedded complete turbidites that become progressively thinner, finer and less complete turbidites, and eventually pass into single-set, thin-bedded, fine-sandstone T_{cde} turbidites (Fig. 24B). Infilling of a shallow channel is probably a result of reduced efficiency caused by deposition farther downflow in the sheetlike splay element, which then is followed by, or is coeval with, the initiation of a new channel and lobe element elsewhere.

Outboard of the terminus of the shallow channels, flows become unconfined and highly depositional, forming the sand-rich sheetlike splay element that farther downflow forms a distal apron of thin- and very thin-bedded fine-grained turbidites intercalated with hemipelagic and pelagic mudstone. Currently, details of this transition are poorly understood. Nevertheless, in the more proximal sand-rich part of the depositional lobe, strata consist of laterally extensive, tabular units that range from a few meters to a few tens of meters thick, but typically are of the order of 15 m thick (*e.g.*, Meyer and Ross, 2007; Fig. 24D). Typically, the base of each unit is marked by a sharp increase in grain size compared to the underlying strata, which, in many cases, consists of a few meter-thick succession of structureless, mudstone-intraclast-rich, coarse-tail-graded sandstone beds that resemble overbank and crevasse-splay deposits. The sheetlike splay element consists almost entirely of amalgamated sandstone that can be mapped over several kilometers. Although difficult to discern because of amalgamation, beds are generally thick to very thick bedded and seemingly form a random pattern of bed-scale cut-and-fill with no organized internal architecture such as compensational stacking of beds.

CONTOURITES

Although first theorized in the mid-1930s to exist, the occurrence of contour-following deep-sea currents, or contour currents, was first demon-

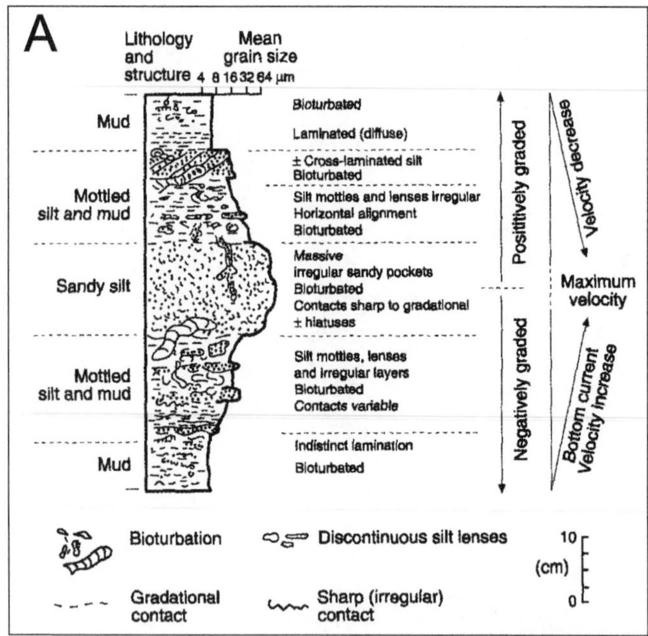

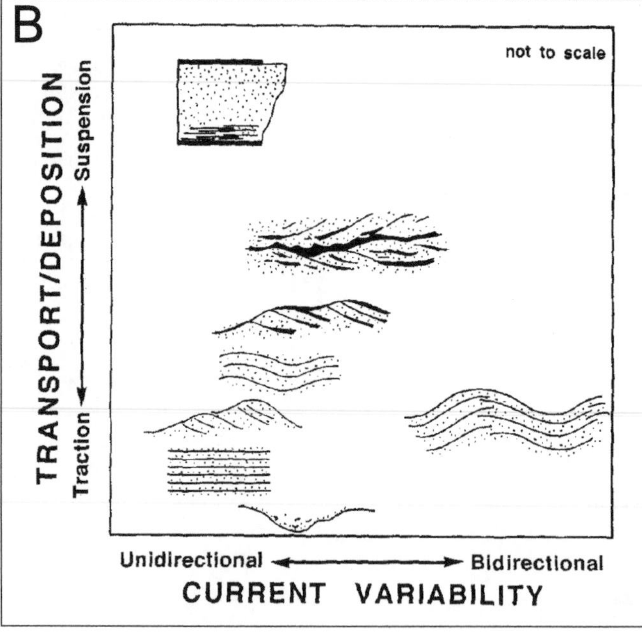

Figure 25. A. Characteristics of an idealized coutourite deposit (after Stow *et al.*, 2002). **B.** Small-scale (core and outcrop) characteristics of contourite deposits (Ito, 1996). Temporal variations in the speed and/or direction of the contour current cause many of the contourite structures (*e.g.*, mud flasers, reactivation surfaces) to resemble features that occur in tidal deposits (*see* Chapter 9), although complete reversals are generally absent in contourite deposits.

strated some 30 years later on the continental rise off eastern North America. Deposits of these currents, which are known as contourites, were shown to be characterized by features that distinguish them from better known turbidites, and were later discovered to be the principal constituent in large (tens to hundreds of km long, few tens of kilometers wide, and up to over 1 km high), elongate, slope-parallel sediment bodies termed sediment drifts. These deposits owe their existence to deep-water bottom currents that form part of the global thermohaline or wind-driven circulation system. These semi-permanent currents generally flow parallel to the slope, but locally, especially because of topographical effects, can be diverted obliquely up or down the slope.

In polar regions, cold surface water, and also more saline water formed by surface-water freezing, descends to the basin floor initiating a large-scale (global) flow system. As a consequence of the rotation of the Earth and the Coriolis effect, these currents, which have speeds of about 1–2 cm/s, become deflected toward the western side of ocean basins. There, they are constrained by the continental slope and their speed increases to about 10–20 cm/s and, where locally constricted, can reach speeds of over 2 m/s. However, even at their generally lower speeds, contour currents are approximately at the threshold velocity for very fine and fine-grained sand, and hence are an important sediment-transporting agent in the deep sea.

The characteristics of contourites are controlled largely by sediment supply (Stow *et al.*, 2002). They can be composed of terrigeneous clastic, volcaniclastic, or carbonate sediment, and grain size can range from mud to sand and admixtures of both, although mud and silt are most common. In addition, gravel particles occur, but are restricted to local areas of high energy and attendant seafloor reworking and winnowing. (Note that in most cases gravel is brought to the area by glacial ice rafting). Sorting is generally moderate to good, except in areas with low current speeds and slow sedimentation where bioturbation has mixed the deposit. Moreover, bioturbation, which generally is dominated by forms of the *Nereites* ichnofacies (*see* Chapter 3), can be moderate to intense, and can destroy primary sediment layering. Traction structures are common and, depending on flow speed and grain size, include small-scale (current ripple) and large-scale (dune) cross-stratification, and also scour marks. Paleocurrents are commonly, but not strictly, parallel to the slope. Contourites often occur as composite units 20–30 cm thick (Fig. 25A). Most successions consist of a basal upward-coarsening interval consisting of muddy to silty to sandy contourites overlain by a unit that fines upward to muddy contourites (Gao *et al.*, 1998; Stow *et al.*, 2002). This upward change indicates a systematic temporal change in flow speed and/or sediment supply, and which recent evidence suggests occurs on time scales that closely parallel Milankovitch periodicities, suggesting a relationship between orbital forcing of climate and changes in bottom-current velocity (Stow *et al.*, 2002). Where turbidites and contourites coexist, they may be difficult to differentiate. However in an interpreted Plio-Pleistocene turbidite–contourite succession, Ito (1996) suggested that contourites can be identified based on minor inverse grading within the ripple cross-stratified unit, in addition to intercalated layers or drapes of mud within ripple cross-stratified sandstone/siltstone units (Fig. 25B). Moreover, contourites commonly contain internal erosion surfaces, typically lack an ordered vertical succession of features like a classical turbidite, and, where interstratified with turbidites, are bounded sharply on their base and top. Collectively these features indicate fluctuating bottom-current speed, and, in many cases, the oscillation between traction and suspension sedimentation.

SEQUENCE STRATIGRAPHY OF DEEP-WATER DEPOSITS

Early sequence-stratigraphic models for deep-marine siliciclastic deposits were based on wide passive continental margins with a well-developed shelf–slope break. Along such margins, the supply of continent-derived sediment into deep water depends on the state of the continental shelf, which, in large part, is controlled by the position of the shoreline. During highstand when the shelf is wide, clastic sediment is sequestered in marginal-marine and continental settings and sediment flux into the deep sea is much reduced. Lowstand, on the other hand, and especially when rivers reach the shelf–slope break, is a time of voluminous sediment supply and active deposition in deep water. It has been pointed out, however, that in situations where the continental shelf is narrow, for example along the coast of California, or where submarine canyons have incised into the coastal rivers, for example in the modern Congo Delta, sediment, especially sand, is almost continuously supplied to deep water irrespective of the position of relative sea level. Notwithstanding this important variant, much of the following discussion will follow the lead of the early models and consider a point-sourced sediment system where characteristics of the sediment supply, namely the sediment flux, sediment caliber and sand:mud ratio, change in a systematic manner over a relative sea-level cycle (Fig. 26).

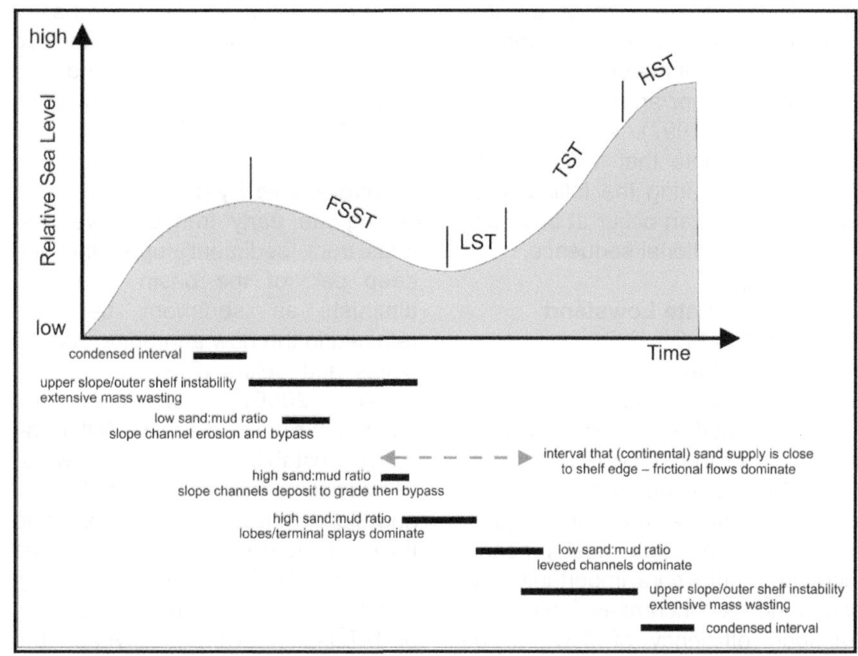

Figure 26. Idealized model relating characteristics of sediment transport and deposition to relative sea level (RSL) on a passive continental margin with a single point-source sediment supply (modified after Posamentier and Walker, 2006). With falling RSL widespread gravitational instability on the outer shelf/upper slope results in extensive mass wasting and incision of canyons and slope channels, and bypass of sediment to the base of slope and basin floor. As the shoreline approaches the shelf-slope break and flows become more enriched in sand, a period of deposition occurs in the slope channels to bring them into grade, whereafter flows bypass the slope and feed basin-floor depositional lobes. Later, the sand:mud ratio begins to decrease, which enhances levee growth and the development of leveed channel systems. Eventually as sediment flux decreases and the deep-marine part of the system becomes progressively more starved of sediment, ongoing slope instability may cap the succession with mass-transport deposits. Eventually a condensed interval forms at the next highstand.

Falling-stage and Early Lowstand Systems Tracts

During falling relative sea level (RSL), as the influence of the river-supplied sediment begins to be felt at the shelf edge, the deep-marine system becomes reactivated with thick, mud-rich (but also sand transporting), high-density flows (*i.e.*, high transport-efficiency flows). Accordingly, the character of the sediment-gravity flows change, causing them to be out of grade with the existing slope gradient, which, in this case, triggers erosion, with the depth of incision decreasing downflow. On the upper and mid slope, submarine canyons and large erosional channels become excavated and the eroded sediment is moved downflow. At the base of slope, the depth of incision is minor, but nevertheless this area too is generally an area of bypass. In contrast, the basin floor becomes an active depocenter, the deposits consisting mostly of fine sediment eroded from the slope. Any coarse sediment is sourced from slope erosion, but can also be derived from relict and palimpsest sediment on the shelf and, as time passes, increasingly from the advancing shelf deltas.

The falling stage is characterized also by widespread gravitational instability along the upper slope and outer shelf, caused by factors such as hydro-eustatic uplift (due to the reduced thickness of the overlying water column), stratal overpressures, seafloor fluid seepage, gas-hydrate decomposition and sublimation, oversteepened slope gradient (as the system progrades into rapidly deepening water near the relict shelf edge; *see* Chapter 10), and erosion along margins of canyons and channels. The consequence is extensive erosive mass wasting in the form of slumps, slides and cohesive debris flows, which singly, or in combination with turbidity-current erosion, excavate major canyon systems. The eroded sediment is moved downslope producing mass-transport deposits (MTDs) on the slope and basin floor. Observations from the modern indicate that MTDs occur on a range of scales, and include those of enormous scale such as the late Pleistocene MTDs on the Amazon Fan, which are of the order of 50–100 m thick and cover an area of 10^4 km^2 (Piper *et al.*, 1997). However, it is important to note that MTDs, which are common during the falling-stage systems tract, can occur at any position in a depositional sequence.

Lowstand to Late Lowstand Systems Tracts

Shelf deltas have now migrated far out onto the shelf, and in some cases may form shelf-edge deltas perched at the top of the slope (see Chapter 10). As a consequence, sediment supply into the basin changes significantly: sediment flux most probably increases, but, more importantly, the sand:mud ratio increases causing the transport efficiency of flows to be reduced. In the submarine canyons flows still generally bypass and canyon-wall collapse continues. Nevertheless, coarse-grained canyon fills indicate out-of grade systems, which, upon re-establishing equilibrium, return to being a sediment transfer conduit. On the middle and lower part of the slope, channels are mostly low to moderate sinuosity and poorly confined. They are initially at grade and neither erode nor aggrade. With time and increasing sand:mud ratio, channel gradients become less than that of the (equilibrium) graded profile promoting extensive in-channel deposition and aggradation. The net result of these conditions is the formation of a sheetlike sandstone body consisting of amalgamated channel sandstones that upward becomes increasingly more interstratified with fine-grained strata (*i.e.*, channel fills become separated by out-of-channel deposits). At the same time, but downflow in the depositional lobe, sediment supply has been significantly reduced because of deposition in the upslope channels, and because the increased sand:mud ratio causes this part of the system to retrograde (*i.e.*, backstep upslope). As the lowstand progresses, slope channel-levee systems eventually reach an equilibrium profile, and as a consequence, sediment flux into more distal parts of the turbidite system increases. Accordingly, depositional-lobe complexes become the loci of sediment deposition and prograde basinward, although, because of extensive lateral compensational stacking, the rate of progradation may be low (*see, e.g.*,, Saller *et al.*, 2008).

Transgressive Systems Tract

During the early transgressive systems tract, sediment supply into the deep part of the basin begins to diminish as sediment becomes retained in the now aggrading coastal plains and deltas (Posamentier and Walker, 2006). Moreover, the sand:mud ratio decreases. Within the canyon instability along canyon walls, in addition to parts of the shelf edge and slope that are re-equilibrating to the newly developing basin conditions (Posamentier and Kolla, 2003), results in the common emplacement of MTDs. Flows that bypass the canyon, however, form highly sinuous, confined channel systems in the middle and lower parts of the slope. Compared to the lowstand, flows now are highly stratified with a coarse basal layer supporting a thick fine-grained upper part. In addition, channel sinuosity is typically higher, which, in turn, favors the development of channels having a lower width:depth ratio and greater channel relief (Pirmez *et al.*, 2000). Collectively these conditions result in the formation of well-developed fine-grained levees bordering erosively based, laterally-migrating confined channels. Stratigraphically upward, these channels become significantly more aggradational, suggesting that rates of deposition have increased both in the channel and on the levees. Farther downflow, sediment supply has diminshed, and so too the sand:mud ratio. Enhanced levee development causes the leveed-channel to depositional lobe transition to prograde basinward, but, because of comparatively low sediment flux, at a much reduced rate (*e.g.*, Saller *et al.*, 2008). Eventually sediment flux to the basin floor becomes sufficiently reduced that depositional-lobe complexes begin to retrograde.

Highstand Systems Tract

Rising relative sea level eventually floods the ever-widening continental shelf. As a consequence, sediment supply (especially of sand and gravel) into the deep part of the basin is effectively shut off as continental sediment becomes trapped in up-dip fluvial and coastal estuarine systems. Farther basinward, sedimentation rates are low and dominated by deposition of fine-grained siliciclastic and biogenic material from suspension, forming a condensed horizon rich in pelagic sediment. Where shelf conditions are appropriate (water depth, salinity, temperature, nutrients, *etc.*), however, carbonate production may be initiated, interrupting for the first time the succession of purely siliciclastic deposits, but then only to become deactivated during the subsequent fall of relative sea level.

SUMMARY AND CONCLUSIONS

The inaccessibility and immense scale of the deep-marine sedimentary environments makes it especially challenging for sedimentological research. Our current state of knowledge has benefited significantly from recent improvements made in seismic acquisition and resolution. These data suggest that much of the modern deep-marine sedimentary record can be subdivided into 2 end-member categories: an expansive unit consisting of simple, concordant reflectors indicating mostly fine-grained suspension fallout deposition that blankets the seafloor, and a significantly more complex unit, both locally and regionally, which locally interrupts the fine-grained blanket. It consists of sand and, less commonly, gravel-size sediment that was transported and deposited by a variety of sediment gravity flow and mass-movement processes. It is these latter kinds of deposits that are most commonly described in the sedimentological literature, and, based mostly on seismic data, include: channels, levees, overbank/crevasse splays, and depositional lobes.

Channels are diverse in kind, size and origin. Submarine canyons and other kinds of erosional channels owe their existence to erosion by moving currents, mass wasting, or a combination of both. Fills of these channels are similarly diverse and range between locally derived, typically fine-grained sediment, to sandstone and conglomerate composed of particles derived from an upslope sediment source. Ideally, these erosive channels pass downflow, into leveed channels and then farther basinward, and with an associated loss of flow confinement, into depositional lobes. Depending on the degree of lateral confinement, leveed channels can be subdivided into two kinds: poorly and highly confined channels. Poorly confined channels tend to have lower sinuosity, be larger in width, and form much thicker channel deposits. Along the outer margin of channel bends, channel-fill strata terminate abruptly. Along the opposite side, or inner bend, channel-fill strata show two kinds of termination. In poorly confined channels, coarse channel strata grade laterally into fine, thin-bedded turbidites of the inner-bend levee, suggesting significant flow overspill along that margin. In contrast, (coarse) channel strata of highly confined channels onlap abruptly and become interfingered with fine, thin-bedded inner-bend levee strata, suggesting containment of at least the basal, coarse-grained part of the flow below the top of the inner-bend levee. In addition, channel strata on the inner side of bends show negligible upward or lateral change in grain size, and lateral-accretion deposits are well developed

Leveed channels are bound on their sides by levees, which are best developed at channel bends. Outer-bend levees of poorly confined channels show a systematic but rapid lateral thinning and fining of moderately thick beds away from the channel margin. This trend suggests a genetic relationship between channel and levee, and relates to lateral loss of flow competence and/or capacity as flows, or the upper part of flows, escape the channel and travel across the adjacent levee. Intercalated with the medium beds are thinner, finer beds that show little lateral change, and, after a few hundred meters away from the channel margin, become indistinguishable from thicker beds that had previously thinned and fined laterally. In highly confined channels, outer-bend levees consist typically of fine, thin-bedded turbidites that change little laterally. In this case, the channel margin represents an erosion surface that has incised vertically and laterally into levee strata deposited by an older highly confined channel. Crevasse splay and overbank splay deposits are also associated with outer bends. These features, which appear to be best developed in poorly confined channels, are the result of flow overspill, possibly aided by inertial run-up. Overbank splays tend to be smaller and thinner, and commonly consist of a single or small number of often complete Bouma turbidites. Crevasse splays are larger, longer-lived features that comprise an array of lithofacies, including strata indicating rapid flow expansion and high rates of deposition (structureless, matrix-rich sandstones).

Depositional lobes are major sediment bodies that occur downflow of leveed channels. These units comprise three subunits: deep channels, shallow channels and sheetlike splay deposits. Deep channels, which are the principal sediment-input conduit for the lobe, are deeply incised and filled, at least in part, with heterogeneous strata related to incomplete bypass, and in some cases also by thick amalgamated sandstone. Shallow channels form a distributive network immediately downflow of the deep channel. Although channel bases show meters of erosion, it is distributed over hundreds of meters laterally. Typically, these channels are filled with sand-rich strata that show a systematic axis-to-margin fining and thinning trend. Unconfined flow at the downflow terminus of the shallow channels builds up the sheetlike splay unit that consists almost exclusively of amalgamated sandstone. Eventually these strata will pass gradually outward into fine-grained (suspension) deposits of the seafloor (possibly interrupted locally by contourite deposits), however at present the details of that change are poorly understood.

Like in most other sedimentary systems, sedimentation in the deep-marine environment is influenced significantly by changes of relative sea level, especially where the continental shelf is wide and terrestrial sediment is point sourced. During stages of falling relative sea level, widespread gravitational instability causes extensive mass wasting on the slope and the formation of extensive mass-transport deposits on the lower slope and basin floor. During the lowstand, slope canyons and erosional channels pass downflow into leveed channels, depositiional lobes and eventually suspension deposits on the distal basin floor. Spatial and temporal characteristics of deposition along this transport/deposition pathway depend on the complex interaction of myriad sedimentological variables, including sediment supply, sediment caliber, sediment mineralogy, channel gradient, and flow density and stratification, in addition to possible high-frequency changes of relative sea level. Nevertheless, as relative sea level rises, sediment supply becomes progressively reduced as river-borne sediment becomes increasingly sequestered in shelf and coastal-plain environments. Eventually a condensed horizon consisting of mostly suspension deposits is formed, and marks the minimum level of sediment supply into the deeper parts of the basin.

In the last version of this book, Walker (1992) noted that

> *"Submarine fan studies are presently in a state of flux. The seismically derived information from modern fans ... cannot be applied very easily in the geological record where the scale of observation is commonly very much smaller. Conversely, features in the geological record cannot be related to parts of modern fans".*

Over the past several years, however, the minimum resolution of seismic images has improved enormously and the number of seismic-scale outcrops examined has increased significantly. Together these developments have narrowed the gap considerably, but still more needs to be done. Of particular note is the need to groundtruth the origin of a number of seismic-reflection patterns. Do they faithfully mimic the character of the geology, or might some part of the pattern be an artifact of geology and wave interference or diffraction characteristics, for example, in places where bed thickness and facies change rapidly? Such details might have dramatic impact on how our models predict stratal architecture and connectivity. Future areas of research must also stress the fine-grained part of the deep-marine sedimentary record. Although typically poorly exposed, it is these strata that make up most of the deep-marine sedimentary record. Also, because of their abundance and ubiquity, these strata probably add significantly to hydrocarbon reservoirs. Lastly, and maybe most profoundly, a concerted effort is needed to understand better the mechanisms that transport and deposit sediment in the deep marine (*e.g.*, McCaffrey *et al.*, 2001). For example, the effect of sediment concentration on turbulence structure and intensity in turbidity currents, or how different levels of density stratification in the flow manifest themselves in the geological record are not understood well. These and many other important topics are now ripe for investigation as the seemingly endless advances in analytical instrumentation make what was formerly impossible, possible.

REFERENCES

Basic sources of information

Bouma, A.H., Normark, W.R. and Barnes, N.E., *eds.*, 1985, Submarine Fans and Related Turbidite Systems. Frontiers in Sedimentary Geology: Springer-Verlag, New York, 351 p.
This volume provides an excellent summary of the seismic attributes of many modern submarine-fan systems. A smaller number of ancient examples is also presented.

Kuenen, P.H. and Migliorini, C.I., 1950, Turbidity currents as a cause of graded bedding: Journal of Geology, v. 58, p. 91-127.
Seminal paper that established the modern turbidite concept.

McCaffrey, W.D., Kneller, B.C. and Peakall, J., *eds.*, 2001, Particulate Gravity Currents: International Association of Sedimentologists, Special Publication 31, 302 p.
A mostly theoretical and experimental approach to the study of particulate gravity currents. Papers provide current ideas on the mechanisms of sediment transport and deposition in these flows.

Mulder, T. and Alexander, J., 2001, The physical character of subaqueous sedimentary density flows and their deposits: Sedimentology, v. 48, p. 269-299.
This paper provides a thorough review of sediment gravity flows and their deposits, and proposes a classification scheme based on sediment concentration.

Mutti, E., Steffens, G.S., Pirmez, C. and Orlando, M., *eds.*, 2003, Turbidite: models and problems: Marine and Petroleum Geology, v. 20, p. 523-933.
A collection of 24 papers discussing modern and ancient slope-to-basin-floor sedimentary systems as well as transport mechanisms and depositional processes.

Nilsen, T., Shew, R., Steffens, G. and Studlick, J., *eds.*, 2007, Atlas of Deep-Water Outcrops: American Association of Petroleum Geologists, Studies in Geology 56, 504 p. + CD-ROM papers.
An extensive, well-illustrated compilation of case examples of deep-marine sedimentary rocks from around the world, including important statistical data.

Posamentier, H.W. and Walker, R.G., 2006, Deep Water Turbidites and Submarine Fans, *in* Posamentier, H.W. and Walker, R.G., *eds.*, Facies Models Revisited: SEPM Special Publication 84, p. 399-520.
Recent review of deep-marine sedimentation with a comprehensive compilation of seismic images that illustrate a wide array of depositional elements.

Piper, D. J. W. and W. R. Normark, 2001, Sandy fans; from Amazon to Hueneme and beyond: American Association of Petroleum Geologists Bulletin, v. 85, p. 1407-1438.
A review of submarine fans of varying scale, tectonic setting, sedimentology, stratigraphy, and sediment-supply sources.

Reading H.G. and Richards, M., 1994, Turbidite systems in deep-water basin margins classified by grain size and feeder system: American Association of Petroleum Geologists Bulletin, v. 78, p. 792-822.
This important contribution sheds light onto a generally under-appreciated control on the distribution and pattern of deep-marine sedimentation, namely the nature of the sediment feeder system.

Stow, D.A.V., Pudsey, C.J., Howe, J.A., Faugères, J-C. and Viana, A.R., *eds.*, 2002, Deep-Water Contourite Systems: Modern Drifts and Ancient Series, Seismic and Sedimentary Characteristics: Geological Society of London, Memoirs 22, 464 p.
Major compilation of papers devoted to the study of sedimentation patterns, processes and deposits in contourite systems.

Weimer, P. and Link, M.H., *eds.*, 1991, Seismic Facies and Sedimentary Processes of Submarine Fans and Turbidite Systems: Springer-Verlag, New York, 449 p.
Collection of papers that illustrate the seismic character of a variety of ancient and modern turbidite systems.

Wynn, R.B., Cronin, B.T. and Peakall, J., 2007, Sinuous deep-water channels: Genesis, geometry and architecture - sinuous deep-water channels: Marine and Petroleum Geology, v. 24, p. 341-564.
This collection of 11 papers is an excellent reference on the most recent concepts related to deep-marine sedimentology and stratigraphy.

Other references

Abreu, V., Sullivan, M., Pirmez, C. and Mohrig, D., 2003, Lateral accretion packages (LAPs): an important reservoir element in deep water sinuous channels: Marine and Petroleum Geology, v. 20, p. 631-648.

Arnott, R.W.C. 2007a, Stratal architecture and origin of lateral accretion deposits (LADs) and conterminuous inner-bank levee deposits in a base-of-slope sinuous channel, Lower Isaac Formation (Neoproterozoic), east-central British Columbia, Canada: Marine and Petroleum Geology, v. 24, p. 515-528.

Arnott, R.W.C. 2007b, Stratigraphic architecture and depositional processes of a proximal crevasse splay and genetically related, sinuous channel fill, Isaac Formation, British Columbia, Canada, *in* Nilsen, T., Shew, R., Steffens, G. and Studlick, J., *eds.*, Atlas of Deep-Water Outcrops: American Association of Petroleum Geologists, Studies in Geology 56, CD-ROM, Chapter 126, 12 p.

Arnott, R.W.C. and Hand, B.M. 1989, Bedforms, primary structures and grain fabric in the presence of sediment rain: Journal of Sedimentary Research, v. 59, p. 1062-1069.

Barnes, N.E. and Normark, W.R., 1985, Diagnostic parameters for comparing modern submarine fans and ancient turbidite systems, *in* Bouma, A.H., Normark, W.R. and Barnes, N.E., *eds.*, Submarine Fans and Related Turbidite Systems. Frontiers in Sedimentary Geology: Springer-Verlag, New York, p. 13-14 and wall chart.

Beaubouef, R.T., Rossen, C., Zelt, F.B., Sullivan, M.D., Mohrig, M.D. and Jennette, D.C., 1999, Deep-water sandstones, Brushy Canyon Formation, west Texas: American Association of Petroleum Geologists Continuing Education Course Note Series 40, 62 p.

Blatt, H., Middleton, G. and Murray, R., 1984, Origin of Sedimentary Rocks, 2nd Edition: Prentice Hall, Englewood Cliff, New Jersey, Chapter 5, p. 127-205.

Bouma, A.H., 1962, Sedimentology of Some Flysch Deposits: Elsevier, Amsterdam, 168 p.

Deptuck, M.E., Steffens, G.S., Barton, M. and Pirmez, C., 2003, Architecture and evolution of upper fan channel-belts on the Niger Delta slope and in the Arabian Sea: Marine and Petroleum Geology, v. 20, p. 649-676.

Deptuck, M. E ., Piper, D.J.W., Savoye, B. and Gervais, A., 2007, Dimension and architecture of late Pleistocene submarine lobes off the northern margin of East Corsica: Sedimentology, v. 54 p.1-31.

Eschard, R., Albouy, E., Deschamps, R., Euzen, T. and Ayub, A., 2003, Downstream evolution of turbiditic channel complexes in the Pab Range outcrops (Maastrichtian, Pakistan): Marine and Petroleum Geology, v. 20, p. 691-710.

Flood, R.D., Manley, P.L., Kowsmann, R.O., Appi, C.J. and Pirmez, C., 1991, Seismic facies and Late Quaternary growth of Amazon Submarine Fan, *in* Weimer, P. and Link, M.H., *eds*., Seismic Facies and Sedimentary Processes of Submarine Fans and Turbidite Systems: Springer-Verlag, New York, p. 415-433.

Galloway, W.E., Dingus, W.F. and Paige, R.E., 1991, Seismic and depositional facies of Paleocene-Eocene Wilcox Group submarine canyon fills, northwest coast Gulf Coast, U.S.A, *in* Weimer, P. and Link, M.H., *eds*., Seismic Facies and Sedimentary Processes of Submarine Fans and Turbidite Systems: Springer-Verlag, New York, p. 247-271.

Gee, M.J.R., Masson, D.G., Watts, A.B. and Allen, P.A., 1999, The Saharan debris flow: an insight into the mechanics of long runout submarine debris flows: Sedimentology, v. 46, p. 317-335.

Gao, Z., Eriksson, K.A., Youbin, H., Shunshe, L. and Jianjua, G., 1998, Deep-Water Traction Current Deposits – a Study of Internal Tides, Internal Waves, Contour Currents and Their Deposits: Beijing, V.P.S Science Press, 128 p.

Heezen, B.C. and Ewing, M., 1952, Turbidity currents and submarine slumps, and the 1929 Grand Banks earthquake: American Journal of Science, v. 250, p. 849-873.

Hiscott, R.N., Pirmez, C. and Flood, R.D., 1998, Amazon submarine fan drilling: a big step forward for deep-sea fan models: Geoscience Canada, v. 24, p. 13-24.

Ito, M., 1996, Sandy contourites of the Lower Kazusa Group in the Boso Peninsula, Japan: Kuroshio current-influenced deep-sea sedimentation in a Plio-Pleistocene forearc basin: Journal of Sedimentary Research, v. 66, p. 587-598.

Johnson, S.D., Flint, S., Hinds, D. and Wickens, H.DV., 2001, Anatomy, geometry and sequence stratigraphy of basin floor to slope turbidite systems, Tanqua Karoo, South Africa: Sedimentology, v. 48, p. 987-1023.

Kneller, B. and Buckee, C., 2000, The structure and fluid mechanics of turbidity currents: a review of some recent studies and their geological implications: Sedimentology, v. 47, p. 62-94.

Locat, J. and Lee, H.J., 2002, Submarine landslides: advances and challenges: Canadian Geotechnical Journal, v. 39, p. 193-212.

Lowe, D.R., 1982, Sediment gravity flows: II. Depositional models with special reference to the deposits of high-density turbidity currents: Journal of Sedimentary Petrology, v. 52, p. 279-297.

Lowe, D.R., 1988, Suspended-load fallout rate as an independent variable in the analysis of current structures: Sedimentology, v. 35, p. 765-776.

Mayall, M. and Stewart, I., 2000, The architecture of turbidite slope channels, *in* Weimer, P., Slatt, R.M., Coleman, J., and others, *eds*., Deep-Water Reservoirs of the World: Gulf Coast Section SEPM, p. 578-586.

Meyer, L. and Ross, G.M., 2007, Channelized lobe and sheet sandstones of the Upper Kaza Group basin-floor turbidite system, Castle Creek South, Windermere Supergroup, B.C., Canada, *in* Nilsen, T., Shew, R., Steffens, G. and Studlick, J., *eds*., Atlas of Deep-Water Outcrops: American Association of Petroleum Geologists, Studies in Geology 56, CD-ROM, Chapter 126, 22 p.

Middleton, G.V., 1966a, Experiments on density and turbidity currents I: Canadian Journal of Earth Sciences, v. 3, p. 523-546.

Middleton, G.V., 1966b, Experiments on density and turbidity currents II: Canadian Journal of Earth Sciences, v.3, p. 627-637.

Middleton, G.V., 1967, Experiments on density and turbidity currents III: Canadian Journal of Earth Sciences, v. 4, p. 475-505.

Middleton, G.V. and Hampton, M.A., 1976, Subaqueous sediment transport and deposition by sediment gravity flows, *in* Stanley, D.J. and Swift, D.J.P., *eds*., Marine Sediment Transport and Environmental Management: John Wiley, New York, p. 197-218.

Mohrig, D., Whipple, K.X., Hondzo, M., Ellis, C. and Parker, G., 1998, Hydroplaning of subaqueous debris flows: Geological Society of America Bulletin, v. 110, p. 387-394.

Moscardelli, L. and Wood, L. J., 2008, New classification system for mass transport complexes in offshore Trinidad: Basin Research, v. 20, p. 73-98.

Mulder, T. and Alexander, J., 2001, The physical character of subaqueous sedimentary density flows and their deposits: Sedimentology, v. 48, p. 269-299.

Mutti, E. and Normark, W.R. 1991, An integrated approach to the study of turbidite systems, *in* Weimer, P. and Link, M.H., *eds*., Seismic Facies and Sedimentary Processes of Submarine Fans and Turbidite Systems: Springer-Verlag, New York, p. 75-106.

Navarro, L., Khan, Z. and Arnott, R.W.C., 2007, Depositional architecture and evolution of a deep-marine channel-levee complex: Channel 3, Castle Creek south, Isaac Formation, Windermere Supergroup, B.C., Canada, *in* Nilsen, T., Shew, R., Steffens, G. and Studlick, J., *eds*., Atlas of Deep-Water Outcrops: American Association of Petroleum Geologists, Studies in Geology 56, CD-ROM, Chapter 127, 22 p.

Nemec, W. and Steel, R.J., 1984, Alluvial and coastal conglomerates: their significant features and some comments on gravelly mass-flow deposits, *in* Koster, E.H. and Steel, R.J. *eds*., Sedimentology of Gravels and Conglomerates: Canadian Society of Petroleum Geologists Memoir 10, p. 1-31.

Normark, W.R. and Carlson, P.R., 2003, Giant submarine canyons: is size any clue to the importance in the rock record, *in* Chan, M.A. and Archer, A.W. *eds*., Extreme Depositional Environments: Mega end Members in Geologic Time: Geological Society of America Special Paper 370, p. 1-15.

Peakall, J., McCaffrey, W.D., Kneller, B., Stelting, C.E., McHargue, T. and Schweller, W.J., 2000, A process model for the evolution of submarine channels: implication for sedimentary architecture, *in* Bouma A.H., and Stone, C.G., *eds*., Fine-grained Turbidite Systems: AAPG Memoir 72/SEPMSpecial Publication 68, p. 73-88.

Pickering, K.T., Hiscott, R.N. and Hein, F.J., 1995, Deep Marine Environments: Chapman and Hall, London, 416 p.

Piper, D.J.W., Pirmez, C., Manley, P.I., Long, D., Flood, R.D., Normark, W.R. and Showers, W., 1997, Mass-transport deposits of the Amazon Fan, *in* Flood, R.D., Piper, D.J.W., Klaus, A. and Peterson, L.C. *eds*., Proceedings of the Ocean Drilling Program, Scientific Results, v. 155, p.109-146.

Piper, D.J.W., Cochonat, P. and Morrison, M.L., 1999, The sequence of events around the epicentre of the 1929 Grand Banks earthquake: initiation of debris flow and turbidity current inferred from sidescan sonar: Sedimentology, v. 46, p. 79-97.

Pirmez, C., Beauboeuf, R.T., Friedmann, S.J. and Mohrig, D.C., 2000, Equilibrium profile and baselevel in submarine channels: examples from Late Pleistocene systems and implications for the architecture of deepwater reservoir, *in* Weimer, P., Slatt, R.M., Coleman, J. and others, *eds*., Deep-Water Reservoirs of the World: Gulf Coast Section SEPM, p. 782-805.

Posamentier, H.W. and Kolla, V., 2003, Seismic geomorphology and stratigraphy of depositional elements in deepwater settings: Journal of Sedimentary Research, v. 73, p. 367-388.

Saller, A., Werner, K., Sugiaman, F., Cebastianti, A., May, R., Glenn, D. and

Barker, C., 2008, Characteristics of Pleistocene deep-water fan lobes and their application to an upper Miocene reservoir model, offshore East Kalimantan, Indonesia: American Association of Petroleum Geologists Bulletin, v. 92, p. 919-949.

Samuel, A., Kneller, B., Raslan, S., Sharp, A. and Parsons, C., 2003, Prolific deep-marine slope channels of the Nile Delta, Egypt: American Association of Petroleum Geologists Bulletin, v. 87, p. 541-560.

Simons, D.B., Richardson, E.V. and Nordin, C.F. Jr., 1965, Sedimentary structures generated by flow in alluvial channels, *in* Middleton, G.V., *ed.*, Primary Sedimentary Structures and their Hydrodynamic Interpretation: SEPM Special Publication 12, p. 34-52.

Stow, D.A.V. and Johansson, M., 2000, Deep-water massive sands: nature, origin and hydrocarbon implications: Marine and Petroleum Geology, v. 17, p. 145-174.

Stow, D.A.V. and Mayall, M., 2000, Deep-water sedimentary systems: new models for the 21st century: Marine and Petroleum Geology, v. 17, 125-135.

Sylvester, Z. and Lowe, D.R., 2004, Textural trends in turbidites and slurry beds from the Oligocene flysch of the east Carpathians, Romania: Sedimentology, v. 51, p. 945-974.

Walker, R.G., 1992, Turbidites and submarine fans: *in* Walker, R.G. and James, N.P., *eds.*, Facies Models: Response to Sea Level Change: Geological Association of Canada, Chapter 13, p. 239-275.

13. Introduction to Biological and Chemical Sedimentary Facies Models

Noel P. James, Department of Geological Sciences, Queen's University, Kingston, ON, K7L 3N6, Canada

Alan C. Kendall, School of Environmental Sciences, University of East Anglia, Norwich, UK, NR4 7TJ, England

Peir K. Pufahl, Department of Earth and Environmental Science, Acadia University, Wolfville, NS, B4P 2R6, Canada

WHAT ARE BIOLOGICAL AND CHEMICAL SEDIMENTS?

The sediments described in the following chapters comprise a wide range of deposits that have the same heritage; they are chemical precipitates that can involve the participation of organisms from microbes to metazoans. They are mostly marine sediments, but are also important components of lake deposits throughout geologic history. For ease of discussion they are herein called biochemical. Organic-rich sediments (oil shales, coals) are also biochemical sediments but, for the most part, constitute elements of other models. Of those sediments considered here, carbonates are the most widespread and most obviously biogenic. Evaporites are the least biological and most chemical and form in the most extreme environments. Bioelemental deposits, a new term that encompasses phosphorite, chert, and iron formation, are the most spatially and temporally discrete.

THE SEDIMENTS ARE BORN, NOT MADE

This deceptively simple phrase encapsulates the main theme of biochemical sedimentation. It also highlights the differences between such deposits and the terrigenous clastic facies described in the preceding chapters. Terrigenous clastic sediment is principally a disintegration product of parent rock wherein resultant particles are transported to the depositional environment. Once there, patterns of texture and fabric are impressed upon them by the local hydraulic regime. The signatures of such facies are in their sedimentary structures and grain-size variations (Table 1). Biochemical sediments are 'born' as precipitates or skeletons within the depositional environment (Fig. 1). For carbonate sediments, this attribute of originating largely in place has profound consequences, namely:

1. Sediment composition is fundamental in characterizing the depositional environment;
2. Grain-size variations need not signal changes in hydraulic regime;
3. Large structures such as platforms are produced entirely by sediments formed in place, they are self-generating and self-sustaining; and
4. The temporal and spatial style of accumulation depends upon the nature of the sediments themselves;

Implications for evaporites include:

1. Mineralogy reflects the type of water being evaporated and the salinity of bottom brines;
2. Many sediments are never transported but grow in place, at, or below the sediment surface; and
3. Most are shallow-water (<10 m) or are pelagic deposits.

Even more importantly for bioelemental deposits, the sediments reflect a combination of hydraulic regime, seawater chemistry, and authigenic processes.

Important attributes are:

1. Sediments are mostly precipitates formed directly from seawater as well as *in situ* authigenic minerals precipitated within the sediment;
2. Unlike carbonates, grain-size variations generally signal changes in hydraulic regime; and
3. Lateral facies trends are usually the result of large-scale ocean circulation patterns.

As carbonate sediments and facies have become better understood (Bathurst, 1975; Wilson, 1975; Scholle *et al.*, 1983; Tucker and Wright, 1999), attention has shifted to

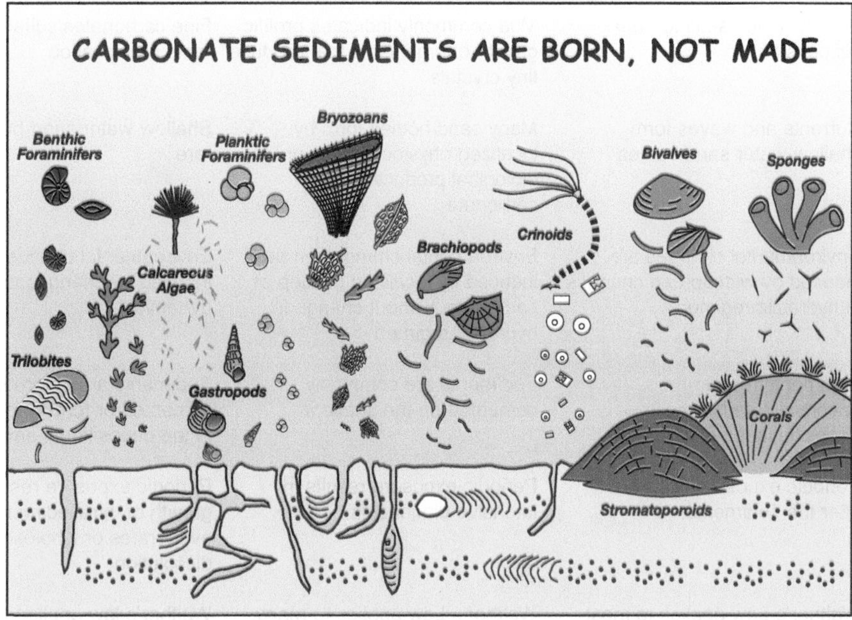

Figure 1. A sketch of some important carbonate and siliceous sediment-producing organisms depicting their disintegration to form sediment in the environment where they lived.

Table 1. Differences between terrigenous clastic and biochemical sediments

Terrigenous Clastic	Carbonate	Evaporite	Bioelemental
Climate is no constraint, sediments occur worldwide	Most sediments occur in shallow marine environments	Most sediments occur in shallow-water or mud-flat environments	Most sediments occur in middle to distal shelf environments
Sediments are both terrestrial and marine	Sediments are mostly marine	Sediments occur only in restricted terrestrial and marine environments	Sediments are mostly marine
Grain-size reflects hydraulic energy of the environment	Grain size reflects the size of skeletons and precipitated grains	Crystal size reflects nucleation and growth rate, or diagenetic alteration	Grain size reflects nucleation and growth rate within the water column or during authigenesis as well as hydraulic energy
Mud indicates settling from suspension	Mud commonly indicates prolific growth of organisms that produce tiny crystals	Fine carbonates/sulfates indicate rapid precipitation	Fine sediment indicates numerous nucleation sites and suspension settling
Currents and waves form shallow-water sand bodies	Many sand bodies form by localized physiochemical or biological production of carbonate	Shallow-water sand bodies are rare	Currents and waves form shallow sand bodies by reworking and winnowing precipitated sediments
Environmental changes are induced by widespread changes in hydraulic regimen	Environmental change can be induced by localized buildup of carbonate, without change in hydraulic regimen	Environmental change is induced by changes in basin dynamics	Environmental change can be induced by changes in physical oceanography and hydraulic regime
Sediments remain unconsolidated in the depositional environment	Sediments are commonly cemented on the seafloor	Sediments are commonly cemented or form crystal crusts in the depositional environment	Sediments can be cemented or unconsolidated on the seafloor
Periodic exposure does not alter the sediments	Periodic exposure results in intensive diagenesis	Periodic exposure results in growth of intrasediment evaporates or wholescale dissolution	Periodic exposure depends on deposit type and may or may not result in diagenesis
Walther's Law applies to most deposits	Walther's Law applies to many, but not all, deposits	Walther's law applies to few deposits	Walther's Law applies to most deposits

the spatial and temporal dynamics of sediment accumulation (Loucks and Sarg, 1993; Harris *et al.*, 1999, Schlager, 2005; Lukasik and Simo, 2008). At the same time a wealth of new information has arisen about carbonate deposition outside the tropics (Nelson, 1988; James and Clarke, 1997) and in non-actualistic, especially Precambrian, environments (Grotzinger and James, 2000). Over the last 30 years, compilations for evaporites have appeared (Dean and Schreiber, 1978; Sonnenfeld, 1984; Schreiber, 1988; Melvin, 1991; Warren, 2006). These syntheses have evolved from strictly chemical facies models to progressively more sedimentological treatments. Understanding of phosphorite, chert and iron formation depositional processes (Gross, 1983; Burnett and Froelich, 1988; Fralick and Barrett, 1995; Glenn *et al.*, 2000; Simonson, 2003; Trendall and Blockley, 2004) has provided new information about the temporal changes in seawater chemistry and ocean circulation required to produce bioelemental sediments (Föllmi *et al.*, 1994; Maliva *et al.*, 1989, 2005). Over the past few years, much has also been learned about the importance of bacteria in precipitating certain Precambrian and Phanerozoic bioelemental facies (Konhauser *et al.*, 2002).

The library of modern and ancient examples has grown to the point where true comparative sedimentology between modern and many ancient deposits can now be practiced, revealing previously unrecognized and universal themes in sedimentation. Nevertheless, as illustrated in the following chapters, consensus has yet to be reached on many fundamentals of sediment dynamics, and, in the case of some bioelemental deposits, the processes that drive precipitation.

THE DEPOSITIONAL EDIFICE

Words used to describe biochemical deposits vary greatly, as do their meanings (Ginsburg and James, 1975, Tucker and Wright, 1990, Glenn *et al.*, 2000; Warren, 2006). Although they can form across the environmental spectrum, the thickest carbonate deposits (Fig. 2) accrete as platforms, whereas the thickest evaporite deposits form within basins. Bioelemental sediments accumulate in an array of depositional settings, but the thickest and most aerially extensive are associated with upwelling along the margins of epeiric platforms and unrimmed shelves.

Platforms and Ramps

A *platform* (Fig. 3) is a large edifice formed by the accumulation of sediment in an area of subsidence. Most such structures have flat tops, steep sides, can be several kilometers thick, and can extend over many hundreds of square kilometers. A *shelf* is a platform tied to an adjacent land mass. The adjacent hinterland is a potential source of terrigenous clastic sediment, fresh water, terrigenous organic matter, and nutrients. A flooded craton many hundreds to thousands of kilometers across is called an *epeiric platform* when covered by shallow seawater (an epeiric sea). A *bank* is an isolated platform surrounded by deep ocean water and cut off from terrigenous clastic sediments and terrestrial runoff. An *atoll* is a specific type of bank commonly developed on a subsiding volcano. Such structures can contain *insular phosphorites*, deposits that are derived, in part, from the guano of nesting seabirds. Carbonate atolls and banks may be so dominated by reefs that their geological expressions are termed *reef complexes*.

The platform edge is a critical interface. A *rimmed platform* (Fig. 3) has a segmented to continuous rampart of reefs or sand shoals along the margin that absorbs ocean waves. Modern warm-water carbonate platforms are generally rimmed because corals are prolific and construct large reefs along the edges of shelves and banks. By absorbing waves and swells and dissipating storm energy, the rim allows a variety of lower energy leeward environments to form in back-reef settings. It can also restrict water circulation, increasing the possibility of evaporite formation. Accumulation space (accommodation) is limited by sea level and facies are strongly differentiated.

An *unrimmed* or *open platform* (Fig. 3) is one in which there is no shelf-edge barrier. Unrimmed platforms occur today on the leeward side of large tropical banks and are the norm in all cool-water settings. A *ramp* (Fig. 3) is an unrimmed shelf that slopes gently basinward at angles of less than 1 degree. Nearshore, wave-agitated facies grade into deeper water, low-energy deposits, and there is generally no discernable shallow-water break in slope, although some ramps can be distally steepened. Because oceanic waves and swells sweep directly onto, and across, the shallow seafloor of open shelves and ramps:

1. The energy level of many shallow-water environments is high;
2. Nearshore facies are complex;
3. Sediment can easily be transported into deeper water;
4. Only in the nearshore zone can sedimentation rate keep pace with rate of sea-level rise; and
5. Subtidal accumulation space is controlled as much by the depth of wave abrasion as by sea level (*see* Chapter 15).

Because unrimmed platforms and ramps are affected by the same physical processes as terrigenous clastic shelves (*see* Chapters 8 and 9), their facies have similar physical attributes. What sets them apart is the continual in-place production of biochemical sediment across the environmental spectrum and early cementation, the latter especially so in warm-water settings.

Lagoons and widespread shallow platforms can become restricted and

Figure 2. Massive cliff-forming Upper Devonian limestones (at right) overlain by well-bedded Mississippian carbonates (snow covered) in the front ranges, Rocky Mountains, Alberta, Canada. Sediments that form these massive platforms were mostly deposited in shallow water, the Devonian rocks are commonly reefal whereas the Mississippian strata formed on ancient ramps.

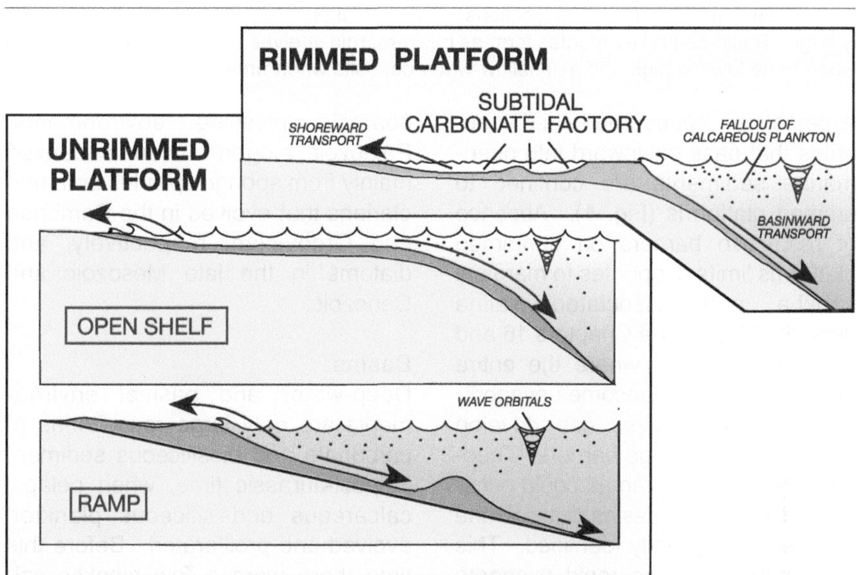

Figure 3. A sketch of rimmed and unrimmed platforms illustrating the position of the carbonate factory and the main directions of sediment transport. Note that open ocean waves and swells impinge on open shelves and ramps but are absorbed at the edge of the rimmed platform. Offshore banks are the same as these shelves and can have rimmed (generally windward) and unrimmed (generally leeward) margins.

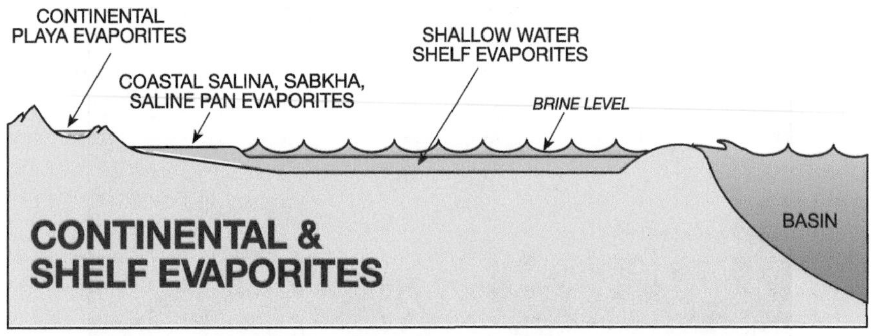

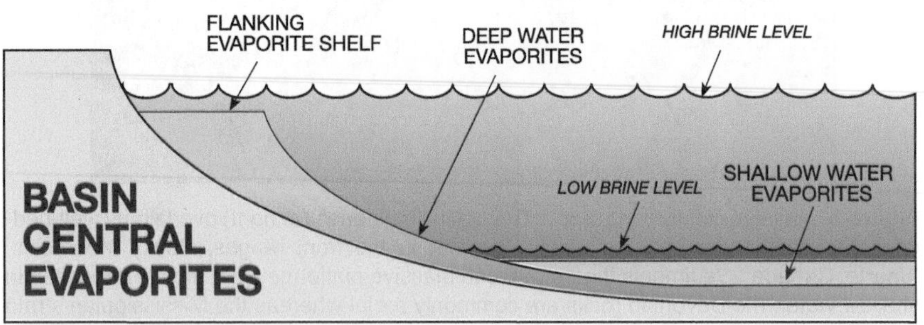

Figure 4. A sketch illustrating locations of the evaporite factory and resulting facies in continental, shelf and basin-central locations. Continental playa evaporites can be extremely large. Basin-central evaporites form as basin-margin shelves and deep-water deposits when brine level is high and as shallow-water deposits when brine level is low.

hypersaline. Widespread shelf evaporites that pass basinward into open-marine sediments are confined to rimmed platforms (Fig. 4). Absence of restrictive barriers on unrimmed platforms limits evaporites to marginal sabkha and associated salina deposits (Fig. 5; *see* Chapters 16 and 20). In situations where the entire depositional basin becomes evaporitic, *evaporitic shelves* can develop without platform edge barriers. Theoretically, evaporite ramps could occur around evaporite basins but not one has been confidently identified. This is probably because rapid evaporite accumulation is restricted to shallow brines. Evaporites are either confined to marginal shelves built on the shallowest parts of older ramps, or they form in the deepest parts, when the adjacent basin desiccates.

Widespread bioelemental deposits only form on *unrimmed shelves, ramps and unrestricted epeiric platforms* because they need to be connected to the open ocean to ensure a sustained source of upwelled seawater rich in Fe, Si, and P (Fig. 6; *see* Chapter 19). Continental margin iron formation and abiogenic chert dominate such settings in the Precambrian, whereas phosphorite and biogenic chert characterize Phanerozoic coastal upwelling environments. Paleozoic biogenic chert is derived mainly from sponge spicules and radiolarians that evolved in the Cambrian and Ordovician, respectively, and diatoms in the late Mesozoic and Cenozoic.

Basins

Deep-water and basinal environments are only large repositories of carbonate and biosiliceous sediment in post-Jurassic time, when pelagic calcareous and siliceous plankton evolved and proliferated. Before this time there were a few planktic calcareous (*e.g.*, stylolinids) and siliceous (*e.g.*, radiolarians) organisms but they did not form extensive deposits. Devonian to Triassic basinal sediments are principally siliciclastic deposits containing minor pelagic siliceous and carbonate skeletal components, with carbonate contributions coming mainly from shallow platforms via sediment gravity flows. Pre-Devonian basinal sediments are wholly siliciclastic with only minor biosiliceous components.

All large and thick evaporites occupy depositional basins that undergo partial or complete desiccation during isolation from the sea. Such *evaporite basins* can be flanked by *evaporitic shelves*, or alternatively, earlier carbonate shelves that become emergent during evaporitic episodes (Fig. 4). Evaporites cease to form during periods of flooding and the same basins then become open- or restricted-marine environments.

Phosphorite does not form in basins because the authigenic processes that concentrate P within sediment cannot operate there. Basinal chert is either biogenic (sponges and plankton) or, especially in the Precambrian, associated with exhalative iron formation around hydrothermal vents at spreading centres and in volcanic arcs (Fig. 6).

THE SEDIMENT FACTORY

Carbonates

The *carbonate factory* (Figs. 1 and 3) is generally the shallow, illuminated seafloor, water column, or lake floor. Particles of all grain sizes are born here (Fig. 7A and B); they crystallize as skeletons or precipitate out of seawater, either directly or via biochemical mediation. Sediments mostly remain in place forming widespread neritic or 'subtidal' deposits (*see* Chapter 14 and 15) or reefs and mounds (*see* Chapter 17). Much of the abundant fine fraction is resuspended periodically and piles up as muddy peritidal flats against highs on the platform and along the shoreline (*see* Chapter 16). Fine sediment is also swept seaward where, together with sediment gravity flows originating at the margin, it accumulates on the slope and on the adjacent basin floor (*see* Chapter 17). Regardless of the ultimate site of accumulation, it is the factory that is at the core of carbonate sedimentation. All carbonate facies and carbonate stratigraphy depend on the 'health' of this production unit, benthic and pelagic.

Evaporites

The *evaporite factory* (*see* Chapter 20) can operate in terrigenous clastic and carbonate settings (Fig. 4) and may have been associated with Precambrian iron formations. Evaporites (Fig. 7C) similarly accumulate rapidly in shallow-water conditions, both marine and terrestrial (Fig. 5). Export of sediment does occur but it is substantially less than in carbonate settings because so much of the sedi-

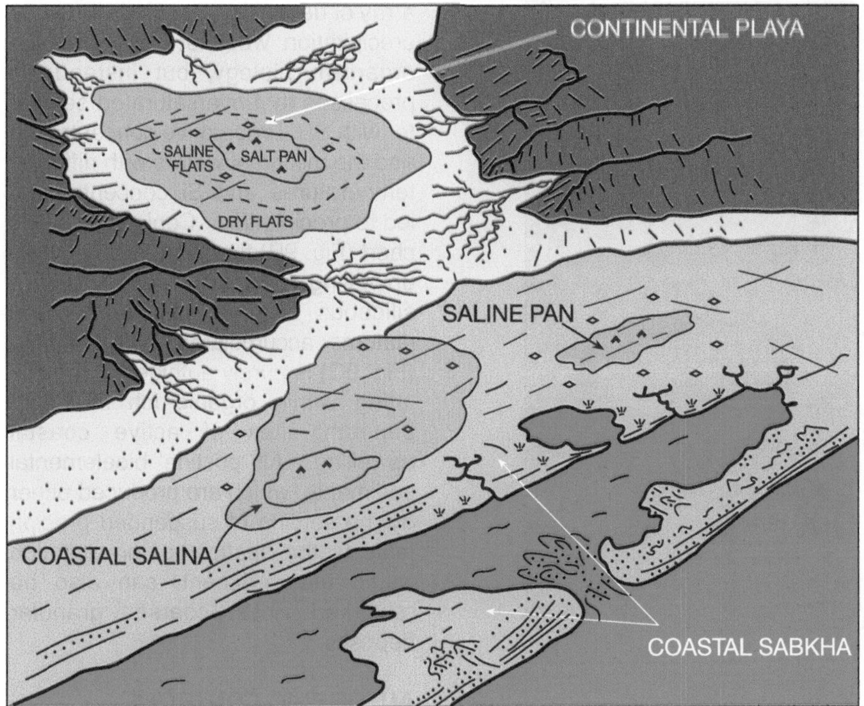

Figure 5. Modern evaporite depositional environments (original diagram courtesy C.R. Handford). Evaporites today form in non-marine interior basins with playa lakes, salt pans, and mud flats (continental sabkhas), coastal supratidal mud flats (marine sabkhas) and marine-fed, coastal lagoons and salt pans (salinas).

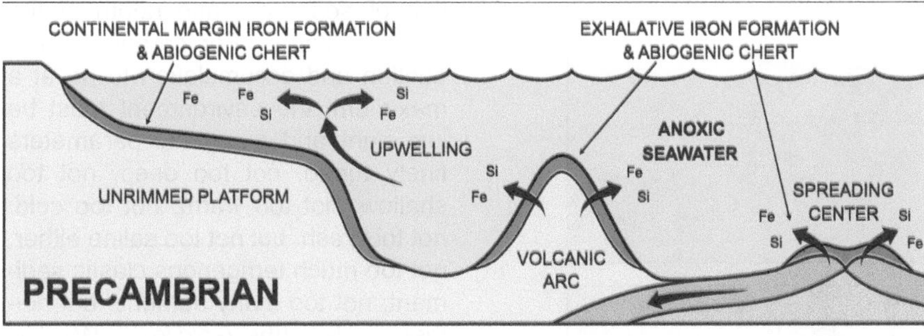

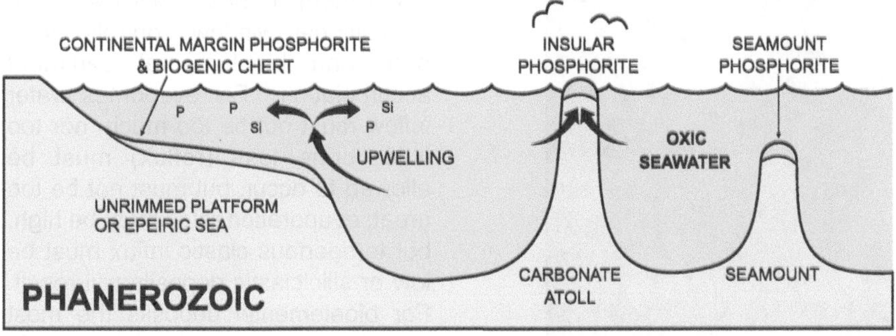

Figure 6. A sketch showing where Precambrian and Phanerozoic bioelemental sediments accumulate. Large deposits form on passive continental margins with coastal upwelling. Basinal iron formation and chert form around hydrothermal vents that provide a constant supply of Fe and Si. Insular phosphorite is derived from seabird guano on carbonate atolls. Convective pumping of P-rich ocean water through the atoll may cause further enrichment. Seamount phosphorite is likely insular in nature, but occurs at depth because of subsidence beneath the ocean surface.

ment is lithified initially or quickly becomes cemented. Environments (*see* Table 2 in Chapter 20) range from continental lakes to coastal salinas and sabkhas to shelf-wide complexes. Evaporites can also develop in shallow- and deep-water basin centres.

Facies models can be devised for evaporite systems because most deposition occurs via physiochemical precipitation. The original overall composition and mineralogy of an evaporite provides information about the type of water being evaporated, the brine temperature and salinity, and the degree of basin isolation. They are, however, difficult to model because:

1. Early facies descriptions were entirely chemical in nature and so difficult to integrate into a modern sedimentary facies model;
2. Evaporites exposed in outcrop (uncommon) are much altered, and subsurface information is limited;
3. No areas of present-day evaporite deposition compare in size to many ancient basins (Fig. 8);
4. Changes in depositional conditions are rapid, profound and commonly result in superimposed facies, making the deposits non-Waltherian in character;
5. New environments typically destroy and replace older ones, rather than moving laterally, so that associated facies are unreliable when interpreting poorly preserved deposits; and
6. Evaporites are susceptible to wholesale, post-depositional change that can remove all primary depositional features.

Bioelemental Deposits

The *bioelemental factory* produces sediment within the full spectrum of marine environments, wherever there is a sustained supply of Fe, Si, and P, (Fig. 6; *see* Chapter 19). Deposit types reflect temporal changes in seawater chemistry as well as differences in the physiochemical and biologic processes promoting precipitation. Iron formation deposition (Fig. 9A) likely reflected a combination of redox sensitive, physiochemical and biologic precipitation processes within the water column and the sediment. Continental margin iron formation accumulated in relation to upwelling, whereas exhalative iron formation relied directly on hydrothermal Fe. Precambrian chert formed in a broad

Figure 7. A. A bedding plane outcrop of Upper Ordovician limestone with numerous brachiopods; Anticosti Island, Québec, Canada; **B.** A ~30 m-high cliff of Cretaceous chalk with numerous chert bands, Sussex, southern England (image courtesy C. Reid); **C.** Holocene bedded gypsum that crystallized on the floor of Marion Lake, Yorke Peninsula, South Australia.

array of depositional settings because precipitation was not dependent on organism biology, but instead on processes that oversaturated seawater with Si. Evaporative concentration and the mixing of waters with different temperatures and Si concentrations led to precipitation. Copious biogenic chert (Fig. 9B) formed in Phanerozoic upwelling environments where siliceous sponge spicules, and diatoms accumulated. Phosphorite (Fig. 9C) is purely authigenic, precipitating within organic-rich sediment beneath sites of active coastal upwelling. All 'pristine' bioelemental sediments, which are produced either by the settling of suspended precipitates to the seafloor or precipitation within the sediment, can also be reworked into coarse granular deposits.

AUTOGENIC CONTROLS

Sediment production and accumulation in the biochemical factory is determined by the rates of subsidence, terrigenous clastic input, position of sea level, and depth of the euphotic zone. For carbonate production and accumulation to be at a maximum, the environment must be just right and autogenic parameters finely tuned; not too deep, not too shallow; not too warm, not too cold; not too fresh, but not too saline either; not too much terrigenous clastic sediment; not too many nutrients but neither too few (the *Goldilocks Window* of Goldhammer *et al.*, 1990). A similar depositional 'window' controls evaporite and bioelemental sediment accumulation. For evaporites water inflow must not be too much, nor too little: brine loss (reflux) must be allowed to occur, but must not be too great; evaporation rates must be high, but terrigenous clastic influx must be low or siliciclastic deposits will result. For bioelemental deposits the most important factors are the delivery of P, Si, and Fe and the ability of seawater-porewater to maintain precipitation. As in carbonates and evaporites, high terrigenous clastic input is detrimental or the precipitates will be swamped. Principal autogenic controls are climate and oceanography (including basin or shelf restriction) but organism biology can be a major factor.

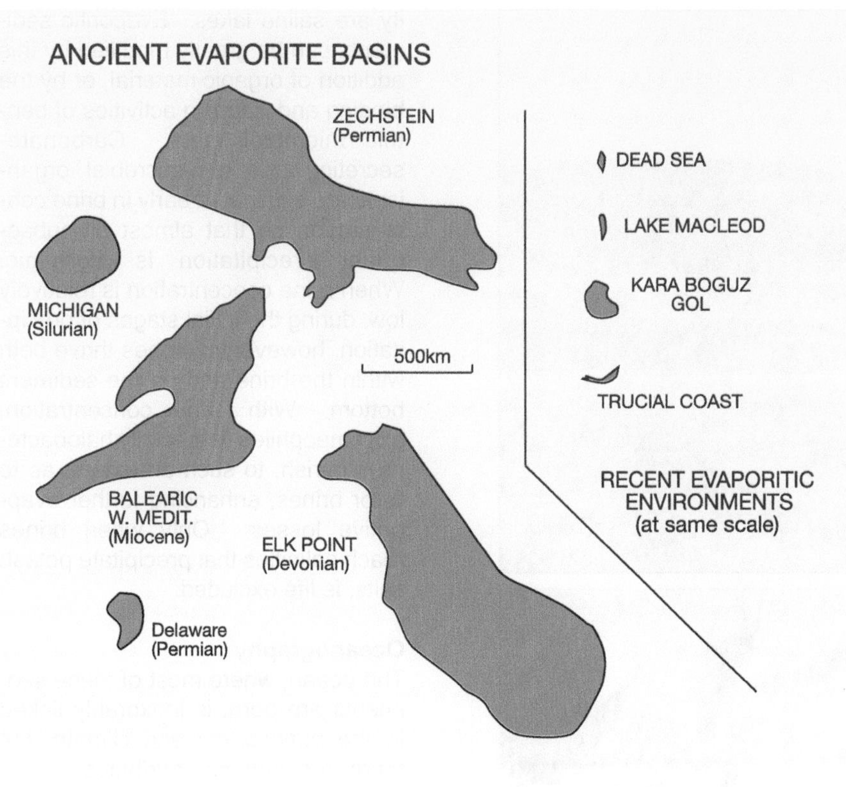

Figure 8. Comparison of size (aerial extent) of Recent (right) and ancient (left) evaporite settings.

Organism Biology

Most carbonate and biosiliceous sediments today are produced biologically or by biochemical mediation (Flugel, 2004). There are four sorts of particles:

1. *Precipitates* - those grains formed by direct or biologically mediated chemical precipitation, *e.g.,* ooids and some muds.
2. *Biofragments* - the calcareous and siliceous sheaths, shells, tests, and spicules of the myriad of sessile, burrowing, drifting or swimming invertebrates, microbes and algae that live in shallow and deep water across the environmental spectrum.
3. *Peloids* - grains of microcrystalline carbonate, generally agglutinated or cemented feces or diagenetically altered grains, and.
4. *Intraclasts* - fragments of consolidated, hardened or lithified sediment (*see* Scholle and Ulmer-Scholle (2003) for examples).

Phototrophic organisms are particularly important because they can rapidly produce enormous amounts of sediment. Such organisms are mostly photosynthetic microbes and calcareous algae, or mixotrophs, invertebrates such as living stony corals, some sponges, giant clams, and big foraminifers that have symbiotic phototrophs (light-dependent micro-organisms) in their tissues (Fig. 1; Table 2). These symbionts, by taking over some metabolic functions, allow otherwise small- or modest-sized invertebrates to precipitate large mineralized skeletons but do confine such organisms to the photic zone. This bargain, large size for light dependence, restricts rapid carbonate or biosiliceous deposition to very specific settings. A caution here - it is far from certain which calcareous invertebrates contained symbionts in the past, especially in the Paleozoic.

Carbonate-producing organisms can be partitioned into a *Photozoan Association* of plants and animals many of which are light-dependent and a *Heterozoan Association* of animals that are mostly light-independent and grow across the neritic environmental spectrum into deep-water environments (James, 1997).

Microbes, the scum of the earth, are profoundly important in biochemical sedimentation. They drill into particles, trap grains, and induce precipitation both in the water column and in the sediment. Although dominated by photosynthetic cyanobacteria on the illuminated seafloor, heterotrophic bacteria thrive in lightless, deep-water, hemipelagic and pelagic environments. These microbes, most of whom are bacteria, together with their organic matrix, form mm-thick *microbial mats* that bind carbonate grains together and with induced carbonate precipitation produce stromatolites, structures that dominate Proterozoic and Archean carbonate and iron-formation facies. Just beneath the seafloor, heterotrophic bacterial communities and *biofilms*, submillimeter veneers of bacterial microcolonies, produce a range of authigenic carbonate cement types and grains. Such communities promote *phosphogenesis* within the sediment, creating *in situ* phosphatic peloids and crusts (Föllmi *et al.*, 1991). Microcrystalline carbonate particles that precipitate in the slime are generally called *automicrite* because the material develops in place, as opposed to *allomicrite* that forms as sediment in the water column or from the disintegration of benthic algae.

But organisms are not just sediment producers. Grasses of various sorts (all post-Jurassic) are also efficient sediment trappers, binders and stabilizers (Demicco and Hardie, 1994). Ichnofauna burrow through the sediment, move and sort particles, ingest, process and excrete sediment, form grains (some peloids), and diagenetically alter tiny particles. Boring macro and micro-organisms excavate into grains and hard substrates, and in the process break down the carbonate host and produce sediment grains. Large skeletons form elevations above the seafloor and create new environments. The biochemical platform is truly a living thing and, as the biosphere has evolved, so the character of such structures has changed dramatically through geologic time. Evaporitic environments are rarely completely lifeless, although hypersalinity does reduce organism diversity to low levels. In fact, the world's regions of highest organic productiv-

Figure 9. A. Massive 40 m high spoil piles of Paleoproterozoic iron formation, Schefferville, Labrador, Canada; **B.** Hummocky cross-stratified, spiculitic chert, Upper Permian, Sverdrup Basin, Arctic Canada (finger scale); **C.** Nodular Miocene phosphate, Victoria, southern Australia.

ity are saline lakes. Evaporitic sediments can be greatly modified by this addition of organic material, or by the binding and trapping activities of benthic microbial mats. Carbonate-secreting and non-microbial organisms are eliminated early in brine concentration so that almost all subsequent precipitation is inorganic. When brine concentration is relatively low, during the initial stages of precipitation, however, microbes thrive both within the brine and on the sediment bottom. With further concentration, extremeophiles (especially halobacteria) flourish, to such an extent, as to color brines, enhancing further evaporitive losses. Only when brines reach salinities that precipitate potash salts, is life excluded.

Oceanography

The ocean, where most of these sediments are born, is inexorably linked to the atmosphere and climate, but there are inherent attributes of the marine environment that directly affect the generation of biochemical sediment.

1. *Water Temperature*

A water temperature of about 20°C partitions carbonates in modern shallow seas into a warm-water, low-latitude realm and a cool- to cold-water mid- to high-latitude realm (Nelson, 1988; James and Clarke, 1997; Fig. 10). Carbonates of all types, both photozoan and heterozoan, are produced on *warm-water platforms* (*see* Chapter 14) but such settings are dominated by direct precipitates and particles from phototrophic organisms. On modern *cool-water platforms* (*see* Chapter 15), the photic zone is of less importance because carbonates are generated almost entirely by non-phototrophic heterozoan organisms; also, sedimentation rates seem to be much lower. Whereas warm and cool water platforms and ramps can be differentiated in Cenozoic and Mesozoic sequences, their recognition in Paleozoic and Precambrian carbonates, although claimed, are less certain.

There is clear temperature stratification in open-ocean situations. Warm to cool surface waters extend down to depths of -50 or -100 m, below which there is a rapid decrease

Table 2. Sedimentary aspect of modern warm- and cool-water carbonate components and their ancient counterparts

Component	Modern, Warm Water	Modern, Cool Water	Ancient Counterpart
Large elements of reefs and and biogenic mounds	Corals (with symbionts)	ABSENT	Corals, Stromatoporoids, Stromatolites, Coralline Sponges, Rudist Bivalves, Archaeocyathans
Whole organisms form granule-sized particles	Large Benthic Foraminifers (with symbionts)	ABSENT	Large Benthic Foraminifers (*e.g.*, Fusilinids)
Remain whole or break apart into several pieces to form sand- and gravel-size particles	Bivalves, Red Algae	Bivalves, Red Algae, Brachiopods, Barnacles	Red Algae, Brachiopods, Cephalopods, Trilobites
Whole skeletons that form sand- and gravel-size particles	Gastropods, Small Benthic Foraminifers	Gastropods, Small Benthic Foraminifers	Gastropods, Small Benthic Foraminifers
Spontaneously disintegrate upon death to form many sand-sized particles	Green (Codiacean) and Red Algae	Red Algae, Bryozoans, Echinoderms	Phylloid Algae, Pelmatozoans and other echinoderms, Bryozoans
Concentrically laminated or micritic sand-sized grains	Ooids, Peloids	ABSENT	Ooids, Peloids
Medium sand sized and smaller particles in basinal deposits	Planktic Foraminifers, Coccoliths, Pteropods	Planktic Foraminifers, Coccoliths, Pteropods	Planktic Foraminifers and Coccoliths (Post-Jurassic), Stylolinids
Encrust on or inside hard substrates, to build up thick deposits or fall off upon death to form sand grains	Encrusting Foraminifers, Red Algae, Bryozoans	Encrusting Foraminifers, Red Algae, Bryozoans	Red Algae, Renalcids, Encrusting Foraminifers, Bryozoans
Spontaneously disintegrate upon death to form lime mud	Green Algae (Dasyclaceans)	Red Algae, Bryozoans, Serpulid Worms	Dasyclad Green Algae
Trap, bind, and facilitate precipitation of fine-grained sediment to form mats, stromatolites, and thrombolites	Bacteria and Other Microbes	Bacteria and Other Microbes	Calcimicrobes and microbes (especially pre-Ordovician)

in temperature (the seasonal thermocline) with depth. The deep ocean below about 1000 m is <4°C. A similar stratification can be present in enclosed basins and shallow seas, but the temperature in many such basins, especially if estuarine-type circulation (fresher surface water flowing seaward over sea water) is present, can be roughly the same from surface to seafloor. Finally, if the water is deep enough, the combination of cold temperatures and high pressures leads to complete carbonate dissolution (beneath the Carbonate Compensation Depth) and so the deep seafloor is either clay or biosiliceous ooze.

Most evaporite basins and deep saline lakes are thermally stratified. In some bottom waters are warm, whereas in others it is the surface brines that preferentially heat up. Colored surface brines are particularly efficient at absorbing light and heating up the water accordingly.

2. *Water Circulation*

Unrimmed shelves and ramps (Fig. 3) have free circulation, and open-marine conditions are the norm. Fine-grained carbonates, suspended by storms and by organism activities, can be carried from where they form to quieter or deeper environments. If unrimmed shelves are located in regions of upwelling bioelemental sediments dominate. Like carbonates, such sediments also show a basinward decrease in grain size. Rimmed shelves commonly have inferior circulation and fine-grained sediment transport is curtailed. Carbonates are typically muddy, even in shallow environments subject to frequent high-energy cyclonic storms. Furthermore, poor water circulation results in departures from oceanic water composition. Freshwater inflow into restricted shelves, if only seasonal, causes brackish conditions nearshore, or even bottom anoxia where persistent density stratification results. In arid climates, restricted shelves become hypersaline. For evaporites to form, however, the degree of restriction required is extreme to the extent that large evaporite deposits are confined to basins that have become isolated or nearly isolated from the open sea. Evaporative losses in such basins are not offset by inflow, and the water level drops below sea level; a process termed *evaporative drawdown*. Most, but not all Phanerozoic evaporitic cherts are found in lakes within volcanic terranes, the silica being derived from rapid chemical weathering. By contrast, Precambrian evaporitic cherts precipitated in coastal marine environments. Cherts with pseudomorphs of evaporite crystals are sometimes the only preserved evidence of evaporite deposition – the evaporite having been completely dissolved away.

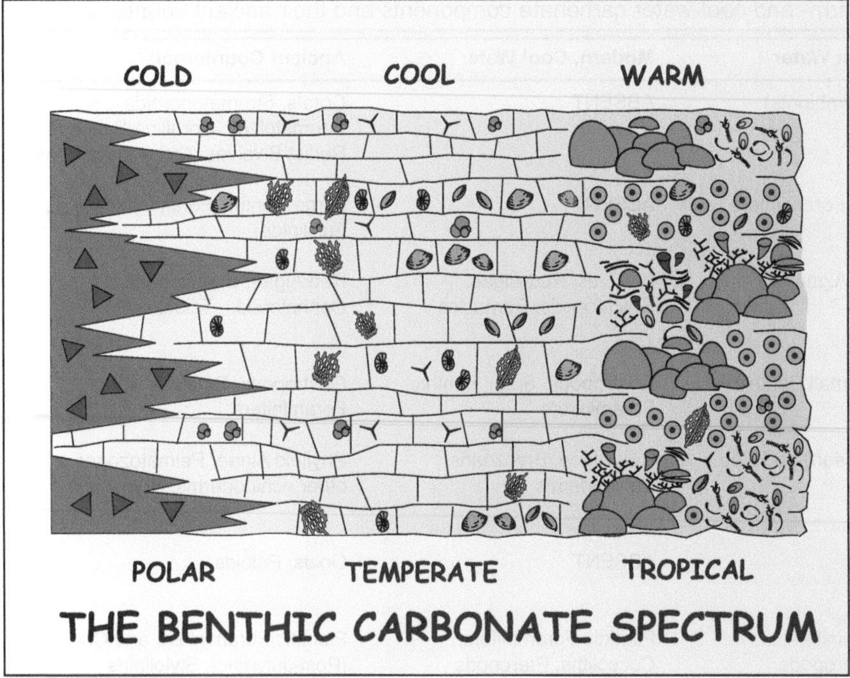

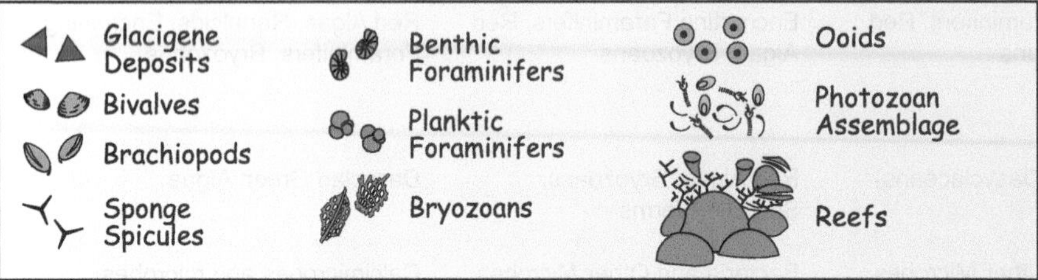

Figure 10. A sketch depicting the range of carbonate sediments produced in tropical to polar depositional environments.

3. *Salinity*

Increasing salinity reduces biotic diversity (*see* Chapter 20 - although this is commonly difficult, in early stages, to separate from the effects of fluctuating salinity) and above 40‰ most invertebrates disappear; calcareous algae, however, continue to be sediment producers for a short time until they too eventually vanish with increasing salinity. Occasionally salt-tolerant skeleton-producing invertebrates thrive in salinities too high for calcareous mud-producing algae. These shells may accumulate in enormous numbers, as in Shark Bay (Western Australia). This serves as a reminder that not all hypersaline carbonates are muddy. The amount of carbonate generated by direct precipitation or microbially induced precipitation, however, increases with increasing salinity and the sediments grade insensibly into fine-grained evaporitic carbonates.

Different evaporites are products of changing brine salinity, but what is less evident is the quantity of salts that have been removed from the world ocean at certain times in the past. These removals significantly reduced ocean salinity to the extent that thermohaline ocean currents must have been affected and marine life would have had to adapt to less saline conditions – perhaps up to a 10% drop in salinity.

4. *Nutrients*

Nutrients seem to be a mixed blessing! All living things need nutrient elements, particularly P and N and, in some settings, Fe and Si. Depending on the nutrient, bioessential elements may enter the ocean from land *via* weathering and runoff, diffusion from the atmosphere, or through emission at hydrothermal vents. Once in seawater, these elements are recycled to the surface ocean via upwelling of bottom water containing nutrients liberated at depth by the bacterial degradation of dead plankton.

Tropical environments away from land are typically nutrient-deficient because organisms remove nutrients and the near-surface waters are strongly temperature stratified, and so upwelling of nutrient-rich waters is generally prevented. Such regions are called *oligotrophic* because all available bioessential elements are rapidly reused by organisms and so few nutrients remain in the water column. It is paradoxical that these nutrient deserts are regions of extensive carbonate sedimentation. This is because $CaCO_3$ precipitates more easily in warm seawater and because many of the organisms (*e.g.*, corals, sponges) have symbionts that aid in calcification or silicification. Whether similar organisms had symbionts in the past is a topic of active debate (*see* Chapter 17).

The down side is that because such benthic organisms in clear-water, tropical environments are so adapted to low nutrient levels, when nutrients increase they act as fertilizers promoting the prolific growth of phytoplankton, soft algae, and filter feeders; corals and similar organisms are simply overwhelmed (Mutti and Hallock, 2003) and the photozoan assemblage is replaced by a heterozoan one. Furthermore, such nutrients promote the growth of phytoplankton to such an extent that sheer numbers limit light penetration. Human introduction of nutrient elements into shallow tropical systems has, sadly, in this way resulted in the demise of many modern reefs.

Mesotrophic environments, locations where nutrients are in good supply, are therefore dominated by filter-feeding invertebrates such as molluscs, bryozoans, sponges, brachiopods, and echinoderms. Environments where nutrients are overwhelming, especially regions of coastal upwelling and river mouths, are sites of prolific marine life and concomitant deposition of bioelemental sediment. These *eutrophic* or *hypertrophic* environments generally lack a healthy benthic biota because of the almost complete lack of oxygen at depth. Instead the seafloor is colonized by bacterial communities that thrive in the accumulating organic-rich sediment.

Seawater evaporation initially concentrates nutrients as well as major elements, and this leads to other types of hypertrophic environments. Increased salinities eliminate most scavengers, make bottom brines anoxic, and promote organic matter preservation. This accounts for the common association of organic-rich sediments with ancient evaporites and the link between evaporites and hydrocarbon source rocks.

5. *Light Penetration*

Light penetration, which controls the phototrophs, varies with water depth, latitude (controls light refraction into the water), and water clarity. The photic zone is about 70 m deep in most places and it is in this zone that carbonate production is highest, being greatest in the upper 10–20 m. Clarity is, however, markedly reduced where turbidity is high as a result of siliciclastic runoff, particularly on west sides of tropical/subtropical oceans, or upwelling and high surface productivity above the outer parts of platforms on the eastern sides of oceans. It can be as shallow as a few tens of meters in such situations. Whereas carbonates have restricted distributions at these sites, bioelemental sediments are abundant.

6. *Oxygenation*

Because well-oxygenated waters are essential for the growth of skeletal invertebrates, any decrease in dissolved oxygen reduces diversity and subsequently the abundance of such organisms in a predictable way. Partial to complete anoxia can be induced in intermediate and bottom waters by:

- Stratification of the water column via pronounced temperature or salinity layering, reducing or arresting vertical mixing;
- Significantly increasing salinity; or
- Dramatic increase in nutrient supply to the surface ocean.

These commonly cause intense primary productivity resulting in a rain of copious dead and decaying phytoplankton to the seafloor (phytodetritus or pelagic snow). As this organic matter cascades through the water column its bacterial decay gradually depletes dissolved oxygen.

Oxygenation is also necessary for precipitation in some bioelemental systems. Nowhere was this more important than in large Paleoproterozoic iron formations, where production of oxygen by photosynthesis led to vast hematite and magnetite precipitates across the shelves.

Climate

Evaporites are more sensitive to climate than any other sediment type. They provide unequivocal evidence of arid, but not necessarily warm, conditions (they occur in Antarctic playa lakes). Nevertheless, almost all ancient evaporites formed in warm climates because, when the atmosphere is undersaturated with water, more evaporation occurs at higher temperatures.

Aridity is also prevalent in:

1. *Mountain-bounded basins at any latitude,* because the mountains act as rainfall-snowfall traps and barriers, resulting in basin-floor, rain-shadow deserts;
2. *Continental areas in low to middle latitudes isolated from the ocean,* where temperatures are high and humidity low, creating high evaporation rates capable of generating brines that can even precipitate potash salts;
3. *Regions adjacent to cool, upwelling oceanic currents* which generate cool onshore winds with high humidity but low moisture content that warm as they move inland, lowering their humidities and promoting evaporation; and
4. *Mid-latitude high-pressure atmospheric belts* where cool, dry air descends;

The general climatic setting also partly controls the rate at which terrigenous clastic sediments are delivered to a shelf or basin by fluvial and aeolian processes. This fresh water and siliciclastic influx can partially or completely suppress the production of carbonate by:

- Lowering the salinity;
- Reducing water transparency;
- Clogging the feeding and/or respiratory apparatus of sessile benthic organisms; and/or,
- Increasing nutrient and particulate organic content of the water.

Climate also controls the amount and rate of evaporation or rainfall, and thus the composition of seawater, in shallow, partially restricted basins on or alongside continents. Furthermore, climate controls the 'storminess' of the ocean and whether high energy is constant or episodic.

Evaporite mineralogy can be controlled by temperature and atmospheric composition. Such restrictions allow primary – early diagenetic evaporites to be used as paleoclimate indicators. Anhydrite appears to require temperatures above 35°C to form and most potash salts are temperature sensitive. Brine temperatures at the time of halite crystal growth can be determined from fluid inclusions – sometimes with surprising results; depositional temperatures during the formation of some Silurian salts from the Michigan Basin were between 5–20°C. Nahcolite ($NaHCO_3$) in Eocene lake deposits from Colorado requires the atmosphere to have had a minimum of 1125 ppm of CO_2 (today = ~375 ppm).

Autostratigraphy

This term refers to the propensity of some biochemical factories, once they start producing sediment, to outpace sea-level rise and so generate a shallowing-upward stratigraphic succession. *Vertical accretion* or *aggradation* typifies carbonate and evaporite systems, in that sediments are produced in place and so, with time, inexorably build up towards sea level, (or the local brine level in a desiccated evaporite basin) even though they may never reach it (*see* Chapter 20). *Lateral accretion* of carbonates occurs because the factory is too shallow to retain the sediments or because they are piled up against a shoreline, and so forming a prograding sedimentary wedge. Bioelemental deposits do not usually produce a shallowing-upward stratigraphic motif because they typically accumulate more slowly and never fill available accommodation space.

Whereas both lateral and vertical accretion will occur on a carbonate shelf, vertical accretion is probably more typical on carbonate banks (because there are few highs except for reefs around which strandline facies can nucleate) and in large evaporite basins. Walther's Law is clearly applicable to lateral accretion but may not always apply during vertical accretion, because new environments can be created simply by shallowing or restricting water movement. Carbonate and evaporite platforms, if exposed, will be subject to diagenesis in the meteoric realm, the nature and extent of which depends upon both intrinsic (mineralogy, previous history, porosity) and extrinsic (climate, time) factors.

As carbonate and evaporite production and accumulation are predominantly shallow-water phenomena and the controls are dynamic, the depositional systems are always in imminent danger of being shut down. For carbonates, this occurs by intersecting, or being intersected by a critical interface that limits deposition. Such interfaces are:

1. Air–water (exposure and shutdown of all carbonate production);
2. Base of photic zone (the demise of phototrophic organisms);
3. Base of the zone of wave abrasion (above which bottom currents and wave action may lead to erosion and cementation);
4. Strong O_2 minimum zone (extinction of all higher plants and animals);
5. Thermocline (water is too cold for most carbonate-producing organisms); and
6. Pycnocline (the salinity is too great for most organisms).

These interfaces cause the *bounding discontinuities* and are responsible for the conspicuous packaging of carbonate facies successions. Changes in the water balance within evaporitic settings are commonly fast, and result in equally abrupt vertical changes in evaporite mineralogy and brine depth (and consequent evaporite facies). This may also lead to abrupt cessation of evaporite deposition or to its sudden appearance.

ALLOGENIC FORCING

Since the biochemical factory is generally the ocean (although *see* Chapter 21), any change in the ocean affects the factory and so alters the character and stratigraphy of the deposit. In terms of biochemical sediments, three principal allogenic drivers, tectonics, celestial mechanics, and biological evolution affect the nature of the ocean.

Tectonics

Tectonics is arguably the most important allogenic driver, even though it generally operates on a long-time scale. Global tectonics determines the patterns of marine chemistry, sea level, and ocean circulation, which in turn affect the health of the biochemical factory. Local tectonics control the rate and style of subsidence and the nature of the foundation upon which, a carbonate platform nucleates, or, on which, evaporites accumulate.

Global Tectonics

Relative tectonic activity in the ocean basin, *i.e.,* mid-ocean ridge spreading rates and attendant volcanism, profoundly affects sea level and, in so doing, is important for both siliciclastic and biochemical sedimentation. Increased mid-ocean spreading leading to flooding of the cratons vastly increases the area for carbonate sedimentation, *e.g.,* Ordovician and Cretaceous. It also seems to coincide with greenhouse climatic conditions and to departures from present-day ocean-water compositions brought about by increased mid-ocean ridge spreading rates. These changes are thought to determine the mineralogy of invertebrate calcareous skeletons and ooids (Calcite and Aragonite seas; Stanley and Hardie, 1998). Such variations are also coupled with changes in the type of potash evaporites with $MgSO_4$-impoverished evaporites forming at time of greater spreading rates. Calcitic particles are usually preserved whereas aragonitic skeletons can be lost via dissolution just below the seafloor or later during meteoric diagenesis. Thus the apparent rates of accumulation may be appreciably greater during calcite sea times (middle Paleozoic and middle to late Mesozoic) than during aragonite sea times.

Because platforms and basins are on shifting tectonic plates, they can drift out of latitudes where carbonate, evaporite, or bioelemental deposition is favored. Carbonate platforms can, for example, move equatorward out of a mid-latitude, cool-water belt and the sediments change to those dominated by tropical, warm-water invertebrates, and higher accumulation rates. A good example is the northward movement of the Australian Plate into the tropics during the Cenozoic, leading to formation of the Great Barrier Reef. Alternatively, carbonate deposition may be swamped by terrigenous clastics as the platform moves into a more humid climate belt. Similarly, evaporitic basins may be transported away from areas of negative water deficits so that evaporite accumulation ceases and either terrigenous clastics or carbonates take their place.

Another important effect is the modification of global ocean circulation patterns by the opening or closing of ocean gateways. This, in turn, affects water temperature and nutrient flow. The most recent example is the opening of the passageway between Australia and Antarctica south of Tasmania in the Southern Ocean during the Oligocene, isolation of Antarctica, and formation of our modern icehouse world.

Local Tectonics

The effect of antecedent topography is most obvious during the early stages of rifting. Carbonate platforms nucleate on horsts and lake deposits including evaporites accumulate in grabens. On newly rifted, passive continental margins the rate of subsidence is predictable. Any changes in these rates occur slowly. Intracratonic basins have the same subsidence style and rate as mature passive margins. Thus, the space for sediment production and accumulation has been similar in these settings throughout geologic time. Platforms and ramps on the inboard side of foreland basins, towards the craton, have similar subsidence rates, but the style is more irregular and less predictable. Both subsidence and uplift occur as the basin responds to lithospheric flexure. Carbonate structures and evaporite accumulations in strike-slip basins and/or on thrust complexes are affected by unpredictable episodic and dramatic subsidence or uplift.

Carbonate platforms are most frequent along low and mid-latitude passive margins as:

1. Small banks atop horsts on newly rifted crust;
2. Huge shelves along mature mar-

gins. This is also where large bioelemental deposits generally form, on passive continental margins in mid-latitude settings where coastal winds promote upwelling;

3. Giant offshore banks, which continue to grow and coalesce as antecedent rift-blocks subside and are buried;
4. Atolls around oceanic volcanoes; and
5. Carbonate platforms may also form as shelves around the margins of, and as banks within, shallow intracratonic basins.

Although not as common on convergent margins, they do form irregular banks and shelves encircling fragmented and obducted tectonic slivers. Ramps are prevalent as the initial or foundation stages of rimmed platforms or as structures in their own right during geological periods when there were few large skeletal organisms to build reefs and form a rim. They occur most frequently in intracratonic basins or along the cratonward side of foreland basins.

Apart from climatic control, large evaporite deposits only form in:

1. Hydrologically closed sedimentary basins in which outflow is less than inflow;
2. Originally deep depressions or areas that subside rapidly during evaporite accumulation; and
3. Settings starved of terrigenous clastic sediment.

Three types of basins are particularly susceptible to restriction and therefore potential sites of evaporite precipitation (Kinsman, 1974).

1. *Intracratonic basins* are long-lived, and have active subsidence histories lasting tens of millions of years. Evaporite deposition tends to be episodic so that evaporites are sandwiched between other sediment types. Although aerially extensive, the evaporites are typically thin (tens to hundreds of meters). If the evaporites are marine-derived, calcium sulfates are over-represented relative to halite. Some basins accumulate non-marine evaporites. Facies consist of shallow-water and mudflat successions (with both shelf and basin-central distributions) but episodes of basin starvation, together with subsidence, can generate depositional basins with several hundred meters of relief, in which 'deep water' evaporites can accumulate.
2. *Divergent plate-margin basins* occur within continental rift valleys, juvenile actively spreading oceans such as the modern Red Sea, and failed rifts (aulacogens). Evaporites can be thick, exceeding 5 km in some instances. They are generally composed of halite but also may contain $MgSO_4$-deficient potash salts. Such compositions suggest the influence of hydrothermal fluids, either as the source of the evaporites or substantially modifying a seawater source. Where continued spreading has generated wide oceans, evaporite deposition essentially ceases and the older evaporites now occur as linear belts along, and beneath, the newly formed passive continental margin sequences.
3. *Remnant-ocean basins* at convergent-plate boundaries, under appropriate conditions of restriction and climate, can change into enormous and deep evaporite basins. Desiccation of these basins, under conditions of intense evaporative loss, may create shallow water environments potentially several kilometers below world ocean level. The best known of these was the Mediterranean during the Late Miocene. These are situations that have no parallel today.

Celestial Mechanics

Milankovitch forcing of sea-level change, on the meter to decameter scale, via celestially driven changes in solar insolation and resulting continental glaciation, is equally applicable to all sedimentary deposits because it determines accommodation. Such small-scale sea-level changes are arguably the principal cause, along with autogenic shallowing-upward, of much cyclicity in carbonates, with or without evaporites. Large-scale changes, on the 20- to 100-m scale, however, have more dramatic consequences. In siliciclastic systems this results in movement of the source area to the shelf edge, transformation of the shelf into a terrestrial environment, and sediment delivery directly to the slope and basin. In carbonate platforms, however, sea-level fall is catastrophic (Fig. 11). It leads to exposure of the whole platform, the carbonate factory is shut down, only a puny platform exists on the upper slope; the former platform is exposed to meteoric diagenesis and the basin is sediment starved - the system is shut down. In ramps, the facies belts simply move downslope with falling sea level and the system carries on. In evaporite systems, the basin margin or platform edge is a critical interface whose elevation determines the degree to which the depositional environment is connected to the open ocean (Fig. 4). If the crest of the edge is below sea level, carbonates are deposited but once sea level falls to or below the edge then, with the flow from the open ocean impeded and under appropriate climatic conditions, evaporites begin to precipitate in the restricted neritic or basinal environments. In bioelemental sedimentary systems such changes in sea level can either shut down or move the locus of upwelling. What a difference geometry makes!

Biological Evolution

We rely heavily upon observations from modern depositional environments to interpret ancient sedimentary successions and to construct facies models. This approach works, and is seen to work because the composition of most sedimentary particles has remained the same through time; a quartz sand grain or an ooid is the same in the Pleistocene, Permian or Precambrian. Because organisms have changed with time, however, (*see* Flugel, 2004 for examples) it is difficult, at first glance, to compare modern and specific ancient biochemical sedimentary facies. This tends to intimidate those who are beginning the study of such sedimentary rocks, but it should not. Carbonate- and silica-secreting organisms in the rock record, when viewed as sediment producers, do have living counterparts, although they may not even belong to the same phylum. Thus 'carbonates are like Shakespeare; the plays remain the same, only the actors change' is

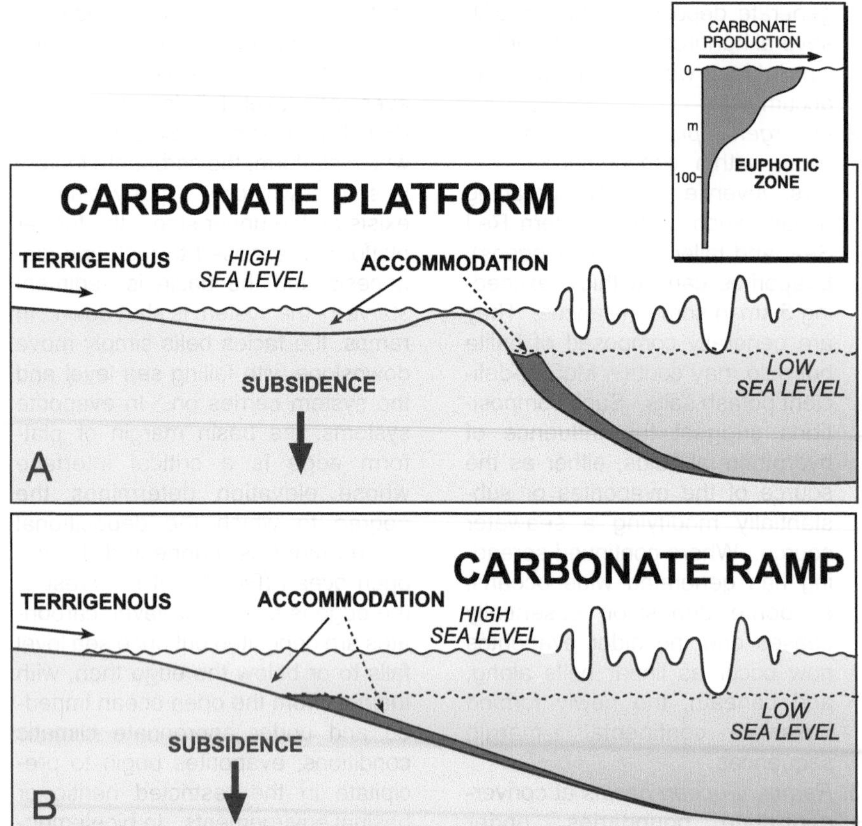

Figure 11. A sketch illustrating the main controls (subsidence, terrigenous clastic input, change of carbonate production with depth, and sea-level fluctuation) that affect sedimentation across a carbonate platform (**A**) and a carbonate ramp (**B**). Deposition occurs in all environments during sea-level highstand. When sea level falls platform deposition ceases except for a narrow shelf whereas ramp deposition merely shifts downslope.

true to a first degree but is more complex in detail. This is true because, despite the numerous groups of organisms with hard parts, there are only two ways in which these hard parts are arranged, either as whole, rigid skeletons (*e.g.*, foraminifers, snails, corals) or as numerous individual segments held together in life by organic matter (*e.g.*, trilobites, sponges, clams, crinoids). Table 2 lists the more important modern components and their potential fossil equivalents.

The sedimentology of organism remains should not be casually equated to their paleoecology. Evolution of the 'modern' carbonate-producing community was gradual and took place in the Mesozoic. The shallow-water Paleozoic brachiopod–pelmatozoan–tabulate and rugose coral–stromatoporoid–trilobite assemblages were replaced by a green and red calcareous algae–scleractinian coral–mollusc–foraminiferal consortium. Many survivors from the Paleozoic era, such as the crinoids, stromatoporoids (coralline sponges), brachiopods, and bryozoans, are still with us, but are somewhat different, and abundant only in deep, cool, or less saline settings. Interestingly, these refugees (together with barnacles) produce much of the sediment on many modern cool-water platforms.

Extending such thinking to the Archean and Proterozoic (Grotzinger and James, 2000) is a stretch indeed! The lack of metazoans, dominance of microbes, and astonishingly active 'abiotic' precipitation makes these deposits truly non-actualistic; their only modern analogues probably lie in unusual extreme modern environments, such as hot springs and alkaline lakes. This is nowhere more apparent than with iron formation where the physiochemical and biological processes of precipitation on the young earth are still not completely understood.

Finally, reef-building metazoans appear and disappear through geological time (*see* Chapter 17). Flat-topped carbonate platforms can only form if there is an elevated rim. Such a rim must be formed by reef-building elements (stromatolites included) and so such platforms are only present at those times in geological history when reef-building organisms thrived (*see* Chapter 17); ramps are the norm at all other times.

Global Evolution

Chemical and biological deposits, because they were inexorably tied to tectonic, biological, oceanographic, and climatic change, were prisoners of time; they changed as the world changed. A generalized and approximate plot of the character and abundance of these deposits through geologic history illustrates such dramatic variations (Fig. 12).

Carbonate deposits are profoundly different through geologic history although the facies motifs are surprisingly similar. Microbial carbonates (*e.g.*, stromatolites) characterize the Archean and Proterozoic, calcareous invertebrates and algae dominate the Paleozoic but pelagic carbonates become overwhelming in the Cretaceous. This latter change is mainly felt in the deep sea but it did also affect neritic deposition in the form of prolific chalks (Fig. 7B). It also seems to have led to the reduction in microbial carbonates and the presence of fewer synsedimentary cements.

It is difficult to produce a temporal record of evaporites because accumulation rates are determined by the frequency of saline giants, and these are, in turn, largely controlled by global tectonics, which influence the frequency of sedimentary basins that can be restricted or isolated from the world ocean (*see* Tectonics above). Nevertheless, on a global scale, there have been times when evaporites were few (*e.g.*, present day), other times when there were massive evaporites but they were geographically restricted (*e.g.*, Silurian of North America and probably the most voluminous deposit in the world, the early Jurassic of the Gulf of Mexico), and yet still other times when there were definite coeval evaporitic peaks with deposits present on several continents (*e.g.*, Devonian and Permo-Tri-

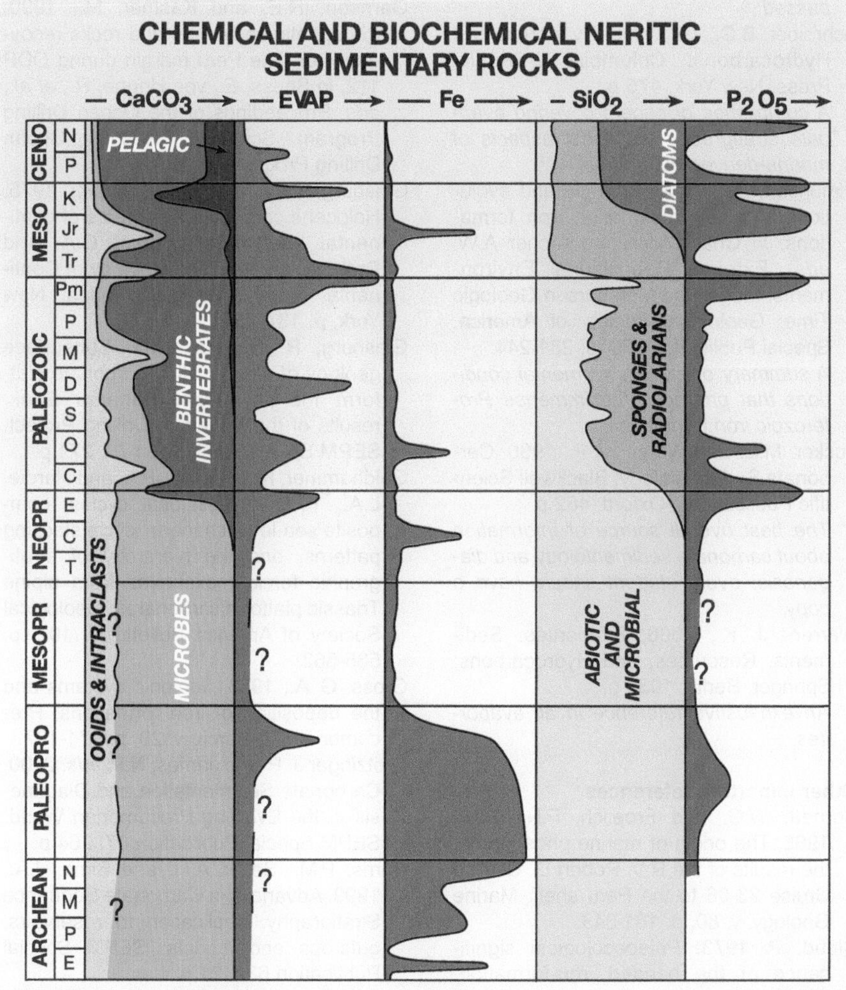

Figure 12. A generalized sketch illustrating the broad changes of marine biological and chemical sedimentary rocks through geological time.

assic). Many evaporites formed within extensional basins (*e.g.*, Jurassic Louanne Salt of the Gulf of Mexico; Permian Zechstein of the North Sea) and their present-day geographical extent is a stretched and more extensive version of their originally smaller depositional basins. These changes make assessment of salt volumes difficult or impossible to determine. Other salt basins (particularly older ones) may have entirely lost their evaporites through later subsurface dissolution by groundwater. Evidence of Paleoproterozoic evaporites, for example, is scarce, and mainly in the form of breccias with halite casts.

The Archean was distinguished by exhalative iron formation. Paleoproterozoic continental margin iron formation was due to the advent of an oxygen-stratified ocean and a shift to upwelling-driven, neritic sedimentation. Disappearance of iron formation at the beginning of the end of the Paleoproterozoic probably reflects a change in ocean chemistry to sulphidic conditions. Its Lazurus-like reappearance during the Neoproterozoic 'snowball' icehouse glaciations is poorly understood. The largest Phanerozoic iron deposits are ironstones, which accumulate as transgressive deposits on continental shelves during sea-level rise.

Whereas Precambrian chert was largely abiotic or microbially mediated Phanerozoic chert is predominantly biogenic. Phosphorite is a Phanerozoic phenomenon, except for a period just after the Paleoproterozoic icehouse. Phosphorite peaks are related to periods of enhanced accumulation through various ocean–climate feedbacks that either intensify upwelling or increase P delivery to the world ocean.

Sadly, with biological and chemical sedimentary deposits, the present is not a straightforward key to interpreting the past, but is just an important guide! Each student of the rock record must be cognizant of the time in which they are working before making any intelligent facies model interpretation of these fascinating rocks!

SUMMARY

The characteristics of all sediments or sedimentary rocks reflect the environment in which they accumulated and the fundamental difference between biochemical and siliciclastic sediments is that the grains (or crystals) of biochemical deposits record specific attributes of the depositional environment. This extends the scope of the environmental interpretation beyond what is possible in siliciclastic sediments (source area reconstructions aside). Thus, paleoceanographic studies are achievable in a way that they are not in terrigenous clastic sediments. The important caveat is, however, that because of biological evolution, atmospheric and paleoceanographic change, these sediments have varied dramatically through geologic time. The Lyellian maxim that the present is the key to the past is true in terms of process, but not necessarily true in terms of product.

ACKNOWLEDGEMENTS

This overview has benefited greatly from careful reviews by Brian Pratt and Bob Dalrymple.

REFERENCES

Basic sources of information

Bathurst, R.G.C., 1975, Carbonate Sediments and their Diagenesis: Elsevier, Amsterdam, 658 p.
A classic that contains good summaries of carbonate components and their modern environments of deposition.

Demicco, R. V., and Hardie, L.A., 1994, Sedimentary Structures and Early Diagenetic Features of Shallow Marine Carbonates: Atlas Series No. 1, 265 p.
A guide to the sedimentary attributes of carbonates at the macroscale with prolific illustrations; a must for every sedimentologist.

Flugel, E., 2004, Microfacies of Carbonate Rocks: Springer-Verlag, 976 p.
An exhaustive synthesis of all aspects of carbonate deposition: an invaluable reference.

Föllmi, K. B. *ed.*, 1994, Concepts and Controversies in phosphogenesis: Geologicae Helveticae, v. 87, p. 639-788.
Includes excellent overviews of phosphorite chemistry and deposition.

Fralick, P.W. and Barrett, T.J., 1995, Depositional controls on iron formation associations in Canada, *in* Plint, A.G., *ed.*, Sedimentary Facies Analysis: International Association of Sedimentologists, Special Publication, v. 22, p. 137-156.
A fundamental contribution to understanding the depositional context of iron formation.

Glenn, C. R., Prevot-Lucas, L., and Lucas, J., *eds.*, 2000, Marine Authigenesis: From Global to Microbial: SEPM Special Publication, v. 66, 536 p.
A must have for anyone wishing to understand the synsedimentary precipitation of bioelemental sediments.

Kendall, A. C., 1988, Aspects of evaporite basin stratigraphy, *in* Schreiber, B. C., *ed.*, Evaporites and Hydrocarbons, Columbia University Press, New York, p. 11-65.
Contains sections dealing with basinal models, the inapplicability of Walther's law, and relations of evaporites to sea-level changes.

Maliva, R. G., Knoll, A. H., Siever, R., 1989, Secular change in chert distribution: a reflection of evolving biological participation in the silica cycle: Palaios, v. 4, p. 519-532.
A succinct overview of the environments of chert deposition and the global silica cycle.

Maliva, R. G., Knoll, A.H., and Simonson, B.M., 2005, Secular change in the Precambrian silica cycle: insights from chert petrology: Geological Society of America Bulletin, v. 117, p. 835-845.
An important contribution to understanding the accumulation of Precambrian chert.

Melvin, J. L., *ed.*, 1991, Evaporites, Petroleum and Mineral Resources: Elsevier, Amsterdam, 556 p.
Includes several papers on non-marine, sabkha and shallow-water evaporites, some with excellent bibliographies.

Schlager, W., 2005, Carbonate Sedimentology and Sequence Stratigraphy: Society of Economic Paleontologists and Mineralogists, Concepts in Sedimentology and Paleontology, no. 8, 200 p.
A succinct and readable summary of carbonate sedimentology and how it can be applied to sequence stratigraphy.

Scholle, P. A., Bebout, D.G. and Moore, C.H., *eds.*, 1983, Carbonate Depositional Environments: American Association of Petroleum Geologists, Memoir 33, 708 p.
A synthesis of all carbonate (and some evaporitic) depositional environments, illustrated with colour photographs and diagrams - a critical source book for any student, the images will never be surpassed.

Schreiber, B.C., *ed.*, 1988, Evaporites and Hydrocarbons: Columbia University Press, New York, 475 p.
A compilation of papers covering evaporite stratigraphy and most aspects of marine-derived evaporites.

Simonson, B.M., 2003, Origin and evolution of large Precambrian iron formations, *in* Chan, A.W. and Archer A.W. *eds.*, Extreme Depositional Environments: Mega End Members in Geologic Time: Geological Society of America, Special Publication 370, p. 231-244.
A summary of the environmental conditions that produced the immense Proterozoic iron formations.

Tucker, M.E. and Wright, V. P., 1990, Carbonate Sedimentology: Blackwell Scientific Publications, Oxford, 482 p.
The best overall source of information about carbonate sedimentology and diagenesis; every student should have a copy.

Warren, J. K., 2006, Evaporites: Sediments, Resources, and Hydrocarbons: Springer, Berlin, 1035 p.
An exhaustive reference in all evaporites.

Other important references

Burnett, W.C. and Froelich, P.N., *eds.*, 1988, The origin of marine phosphorite; the results of the R.V. Robert D. Conrad Cruise 23-06 to the Peru shelf: Marine Geology, v. 80, p. 181-343.

Cloud, P. 1973, Paleoecological significance of the banded iron-formation: Economic Geology, v. 68, p.1135-1143.

Crevello, P.D., Wilson, J.L., Sarg, J.F., and Read, J.F., *eds.*, 1989, Controls on Carbonate Platform and Basin Development: SEPM Special Publication 44, 405 p.

Dean, W.E, and Schreiber, B.C., *eds.*, 1978, Notes for a short course on marine evaporites: Society of Economic Paleontologists and Mineralogists, Short Course no.4, 395 p.

Filippelli, G. M., 1997, Controls on phosphorous concentration and accumulation in oceanic sediments: Marine Geology, v. 139, p. 231-240.

Föllmi, K.B., Garrison, R.E. and Grimm, K.A., 1991, Stratification in phosphatic sediments: illustrations from the Neogene of California, *in* Einsele, G., Ricken, W. and Seilacher, A., *eds.,* Cycles and Events in Stratigraphy, Springer, p. 492-507.

Föllmi, K.B., Weissert, H., and Lini, A., 1994 1993, Nonlinearities in phosphogenesis and phosphorous-carbon coupling and their implications for global change, *in* Wollast R., Mackenzie, F.T., and Chou, L., *eds.* Interactions of C, N, P, and S Biogeochemical Cycles and Global Change. NATO ASI Series. Series I: Global Environmental Change, Berlin, Springer-Verlag, v. 4, p. 447-474.

Garrison, R.E, and Kastner, M., 1990, Phosphatic sediments and rocks recovered from the Peru margin during ODP 112, *in* Seuss, E., von Heune, R., *et al., eds.* Proceedings of the Ocean Drilling Program, Scientific Results; Ocean Drilling Program, p. 111-134.

Ginsburg, R.N. and James, N.P., 1975, Holocene carbonate sediments of continental shelves, *in* Burke, C.A. and Drake, C.L., *eds.*, The Geology of Continental Margins: Springer-Verlag, New York, p. 137-155.

Ginsburg, R. N., *ed.* 2001, Subsurface geology of a prograding carbonate platform margin, Great Bahama Bank: results of the Bahamas Drilling Project: SEPM Special Publication 70, 271 p.

Goldhammer, R.K., Dunn , P.A. and Hardie, L.A., 1990, Depositional cycles, composite sea-level changes, cycle stacking patterns, and the hierarchy of stratigraphic forcing: examples from alpine Triassic platform carbonates: Geological Society of America Bulletin, v. 102, p. 535-562.

Gross, G. A., 1983 Tectonic systems and the deposition of iron formations: Precambrian Research, v. 20, p. 171-187.

Grotzinger J. P. and James, N.P. *eds*, 2000, Carbonate Sedimentation and Diagenesis in the Evolving Precambrian World: SEPM Special Publication 67, 364 p.

Harris, P.M., Saller, A.H. and Simo, J.A., 1999, Advances in Carbonate Sequence Stratigraphy: application to reservoirs, outcrops and models: SEPM Special Publication 63, 421 p.

James, N.P., 1997, The cool-water carbonate depositional realm, *in* James, N.P. and Clarke, J.A.D., *eds.*, Cool-water Carbonates, SEPM Special Publication No. 56, p. 1-22.

James, N. P., and Clarke, J.L., *eds.*, 1997, Cool-Water Carbonates: SEPM Special Publication 56, 440 p.

Kinsman, D.J.J., 1974, Evaporite deposits on continental margins: 4th Symposium on Salt, Cleveland, Ohio, Northern Ohio Geological Society, v. 1, p. 255-259.

Klein, C., 2005, Some Precambrian banded iron formations (BIFS) from around the world: Their age, geologic setting, mineralogy, metamorphism, geochemistry, and origin: American Mineralogist, v. 90, p. 1473-1499.

Konhauser, K. O., Hamade, T., Raiswell, R., Morris, R.C., Ferris, F.G., Southam, G., Canfield, D.E., 2002, Could bacteria have formed the Precambrian banded iron formations? Geology, v. 30, p. 1079-1082.

Loucks, R.G. and Sarg, J.F. *eds.*, 1993, Carbonate Sequence Stratigaphy: American Association of Petroleum Geologists Memoir 57, 545 p.

Lukasik, J., and Simo, A.E., 2008, Controls on Carbonate Platform and Reef Development: SEPM Special Publication 89, 431 p.

Mutti, M., and Hallock, P., 2003, Carbonate systems along nutrient and temperature gradients: some sedimentological and geochemical constraints: International Journal of Earth Sciences, v. 92, p. 465-475.

Nelson, C.S., *ed.,* 1988, Cool Water Carbonate Sediments: Sedimentary Geology, v. 60, 367 p.

Ohmoto, H., Wananabe, Y., Yamaguchi, K. E., Naraoka, A., Haruna, M., Kakegawa, R., Hayashi, K., and Kato, G., 2006, Chemical constraints and biological evolution of the early earth: constraints from banded iron formations, *in* Kesler, R., and Ohmoto, H. *eds.*, Evolution of the Early Earth's Atmosphere, Hydrosphere, and Biosphere – Constraints from Ore Deposits: Geological Society of America Memoir 198, p. 291-331.

Pomar, L., 2001, Types of carbonate platforms: a genetic approach: Basin Research, v. 3, p. 313-334.

Reading, H.G., *ed.*, 1986, Sedimentary Environments and Facies: Blackwell Scientific Publications, Oxford, 615 p.

Reijmer, J.J.G., Betzler, C, and Mutti, M., *eds.,* 2003, New perspectives in carbonate sedimentology, International Journal of Earth Sciences v. 92, p. 433-660.

Scholle, P.A., and Ulmer-Scholle, D.S., 2003, A Colour Guide to the Petrography of Carbonate Rocks: American Association of Petroleum Geologists Memoir 77, 474 p.

Sonnenfeld, P., 1984, Brines and Evaporites: Academic Press, Orlando, 613 p.

Stanley, S.M., and Hardie, L.A., 1998, Secular oscillations in the carbonate mineralogy of reef-building and sediment-producing organisms driven by tectonically forced shifts in seawater chemistry: Palaeogeography, Palaeoclimatology, Palaeoecology, v. 144, p. 3-19.

Trendall, A. F., and Blockley, J. G., 2004, Precambrian iron-formation, *in* Ericsson, P.G., Altermann, W., Nelson, D.R., Meuller, W.U., and Cantaneau, O., *eds.*, The Precambrian Earth: Tempos and Events: Elsevier, Amsterdam, p. 403-421.

Tucker, M.E., 1991, Sequence stratigraphy of carbonate-evaporite basins: models and application to the Upper Permian (Zechstein) of northeast England and adjoining North Sea: Journal Geological Society of London, v. 148, p. 1019-1036.

Wilson, J. L., 1975, Carbonate Facies in Geologic History: Springer-Verlag, Berlin, 471 p.

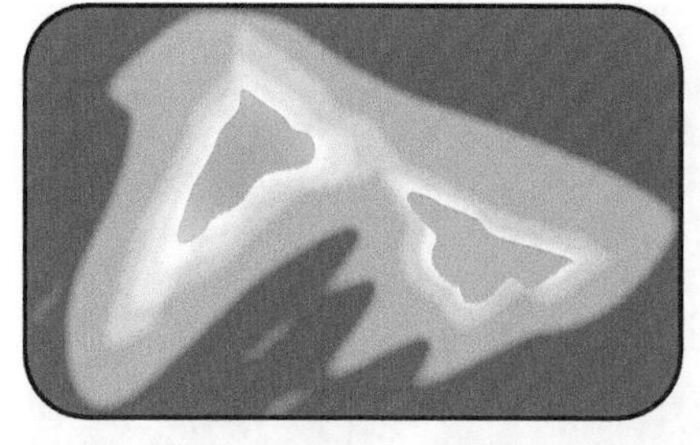

14. Warm-Water Neritic Carbonates

Brian Jones, Department of Earth and Atmospheric Sciences,
University of Alberta, Edmonton, AB, T6G 2E3, Canada

INTRODUCTION

Warm-water carbonate factories (Fig. 1) are vast, wondrous edifices where interactive processes produce a spectrum of diverse carbonate sediments. Sedimentation dynamics in these factories are controlled by many factors, including climate, water temperature and salinity, water depth, water turbidity, and resident biota. In the absence of siliciclastic sediments, carbonates form and accumulate on banks, rimmed and unrimmed shelves, ramps, atolls, and around isolated oceanic islands. Factory input is Ca and CO_3 from seawater. Output, mediated by numerous abiotic and biotic processes, is the carbonate sediment that varies spatially in accord with environmental parameters. The animals and plants that thrive in well-illuminated waters of the warm-water factory are multifaceted processors that variously contribute sediment and influence deposition by trapping and binding of sediment. Facies architecture is commonly intricate and controlled by subtle environmental factors that are difficult to monitor. Extrinsic changes (*e.g.,* tectonics, eustasy, climate) can affect any of the environmental parameters that control factory operations. Vertical accumulations of carbonate will reflect temporal changes to the depositional systems of the carbonate factory.

The first part of this chapter provides an overview of the carbonate factory and its operation. The second part presents examples of modern warm-water depositional systems with the view of constructing facies models that can be used as guides in the interpretation of ancient successions. The third part examines the factors that control the temporal development of carbonate successions in the face of eustatic, climatic, and tectonic changes.

THE CARBONATE FACTORY

Warm-water carbonate factories are today located between ~30°N and ~30°S latitude, where there is little or no influx of siliciclastic sediment that can:

1. Volumetrically mask carbonate production;
2. Increase water turbidity, thereby reducing light penetration which, in turn, reduces growth of photozoans; and
3. Curtail the growth of many animals, including corals, by impeding their feeding mechanisms.

Carbonate accumulation is therefore favoured where hinterlands have little fluvial input (*e.g.*, Florida Bay) or banks are surrounded by deep oceanic water (*e.g.*, Great Bahama Bank; Fig. 1).

The strong link between sediment production and photozoan growth means that the depth of the photic zone (80–100 m in clear water) generally controls the lower limit of the benthos and hence the basal depth of the carbonate factory. Accordingly, the aerial extent of the carbonate factory is the area of substrate that lies between sea level and the base of the photic zone. This is readily apparent from the satellite image of the Great Bahama Bank–Florida Shelf area, where the carbonate factories are sharply delimited by 'drop-offs' (Fig. 1). Carbonate factories can extend across large areas, *i.e.*, the Great Bahama Bank covers an area of ~ 96 000 km^2 and the Great Barrier Reef complex (Australia) is 2600 km long and covers an area of ~ 344 400 km^2.

Sediment Composition

Carbonate sediments are formed of aragonite, low-magnesium calcite (LMC: < 4 mol % $MgCO_3$), and high-magnesium calcite (HMC: > 4 mol % $MgCO_3$). Compositional heterogeneity is the norm because the skeletal components from different animals and plants (Fig. 2) are mixed. The different chemical characteristics of the aragonite, LMC, and HMC will influence the diagenetic evolution of the sediments.

Sediment Components

Sediment production includes *precipitates*, *skeletal grains*, *pellets* (and *peloids*), *compound grains, intraclasts*, and *coated grains*; all are superbly illustrated in Scholle and Ulmer-Scholle (2003). Combined, in accord with local conditions, these components form the plethora of facies found in modern and ancient carbonate successions.

Precipitates

Proving abiotic precipitation is difficult because macro- and micro-organisms are typically nearby and the spectre of their involvement is ever-present. Nevertheless, ooids (spherical to elliptical coated grains, < 2 mm in diameter) are commonly but not always attributed to abiogenic precipitation. Although typically formed in high-energy, shallow-marine environments like those around Joulters Cay, Bahamas, they can also form in low-energy embayments such as the Laguna Madre, Texas. Those ooids, which usually have a large nucleus (commonly pellets) encased by only one or two cortical lamina, have also been called superficial ooids or egg-shell ooids.

Calcareous mud (micrite) can be formed by abiotic precipitation, bio-mediated precipitation, or through the mechanical breakdown of larger grains. The mystery of 'whitings', which are clouds of calcareous mud suspended in the water column that appear almost instantaneously, remains open. Shinn *et al.* (1989) suggested that some whitings represent bottom sediment that has been temporarily re-suspended by physical phenomena, whereas other whitings

Figure 1. Satellite image of the Florida–Bahamas–Cuba area showing large areas of carbonate sedimentation (white) in Florida Bay, Florida Shelf, Little Bahama Bank, Great Bahama Bank, Cay Sal Bank, and along north and south coasts of Cuba, separated by deep water zones (dark blue). Image courtesy of the Image Science and Analysis Laboratory, NASA Johnson Space Center. [http://veimages.gsfc.nasa.gov/6417/Bahamas.A2004024.1600.1km.jpg].

originate from local *in situ* inorganic precipitation. In contrast, Robbins and Blackwelder (1992) proposed that whitings resulted from biological precipitation of $CaCO_3$ that is induced by picoplankton with the cells acting as nucleation sites for calcite precipitated from seawater that is supersaturated with respect to calcite. Since then, whitings have been variously attributed to rapid resuspension of micrite from the seafloor, possibly by fish, pseudo-homogeneous precipitation caused by microbial activity, and repetitive epitaxial growth of carbonate on re-suspended bottom material. The lack of whitings in any of the Indo-Pacific reef areas adds further to the 'whitings mystery'.

Skeletal Grains

Skeletal grains (bioclasts, biofragments) are produced by the disarticulation, biological destruction, and physical breakage of animal and plant skeletons. The size and morphology of such grains ultimately depends on the skeletal architecture and degree of breakdown. For example, with continued physical and biological breakdown, the -sized grains produced from the initial breakdown will ultimately be reduced to small, needle-shaped crystals.

Pellets and Peloids

A pellet (or faecal pellet) is the byproduct of invertebrate animals (*e.g.,* worms, shrimp, holothurians) that indiscriminately ingest sediment in their search for food. Pellet morphology is a function of the size of the animal's intestinal system. Although typically < 0.3 mm in diameter, large sea cucumbers (holothurians) can produce faecal pellets ~ 1 cm in diameter. Faecal pellets, bound solely by mucus will rapidly disintegrate to mud unless cemented quickly. A peloid, formed by micritization of other allochems such as ooids, intraclasts, and skeletal fragments, may appear identical to faecal pellets. Micritization is achieved as microbial borings are filled with micrite.

Compound Grains

Also known as 'grapestones', they are formed of clusters of carbonate sand grains (ooids, pellets, skeletal grains), commonly micritized, that are bound together by micrite cement or filamentous microbes.

Intraclasts

Intraclasts result from physical breakage of penecomptemporaneously lithified seafloor or early lithification of grain aggregates. They typically accumulate close to their place of origin.

Coated Grains

Coated grains include ooids (= ooliths, oolites), pisoliths (= pisoids, pisolites), oncoliths (= oncoid, oncolite), and rhodoliths (= rhodoids, rhodolites). Ooids and pisoids are commonly attributed to abiogenic precipitation whereas oncoliths and rhodoliths are formed biogenically. Oncoids commonly > 2 mm diameter, are usually found in shallow waters where phototrophic filamentous microbes trap and bind sediment particles to a substrate, and mediate calcite precipitation. Rhodoliths formed by coralline algae and associated encrusting foraminifera can grow to depths of ~ 90 m because the coralline algae utilize a different part of the light spectrum than green algae.

AUTOGENIC CONTROLS IN THE CARBONATE FACTORY

Most sediment in the warm-water neritic carbonate factory is produced in the photic zone with maximum productivity in water < 20 m deep (Fig. 3A): the depositional window. Given that much of this production is inexorably linked to photophilic organisms, Pomar (2001) defined the euphotic, oligophotic, and aphotic zones and labelled biotic elements according to the zone in which they lived (Fig. 3B–D). A mesophilic zone can also be defined, if desired. The depth of these zones depends on water clarity. In clear ocean waters the base of the photic zone is usually at ~ 100 m. It is also important to note that these zones are not linked solely to water depth, *e.g.,* aphotic or sciaphilic (shade-loving) organisms can thrive in shallow water providing they live in the shadows.

Light penetration is not the sole control on photosynthetic calcareous invertebrates. Other factors such as water temperature (T), salinity (Fig. 4), light conditions, energy levels, water clarity, substrate conditions, nutrient requirements, and safety are just as important. Each organism has its own environmental preferences. In addition, for each parameter there will be a tolerance range and an optimal range (Fig. 4); modern hermatypic corals, for example, can tolerate temperatures of 18–36°C and 22–40‰ salinity but prefer temperatures of 25–29°C and salinity between 25–35‰. Corals stressed by factors such as increased water temperature, increased sedimentation or nutrient overloading, commonly respond by expelling their symbiotic zooxanthellate. As a result, the corals turn white ('bleaching'), and may eventually die.

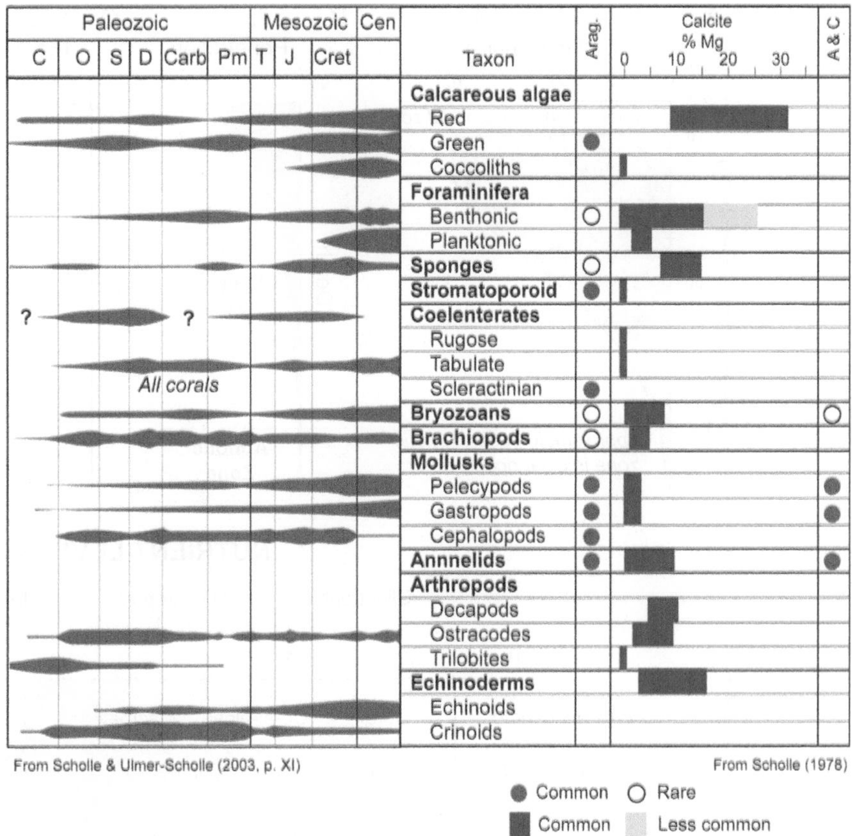

Figure 2. Geological ranges and composition of skeletal organisms that contribute to carbonate sediment formation.

Nutrient availability, often ignored in the context of carbonate systems, exerts a fundamental control over the biota distribution and hence, sediment production. Photozoans use nutrients (*e.g.*, N, P, Fe, Si) to synthesize proteins utilized in cell maintenance, growth, and reproduction (Mutti and Hallock, 2003). Such nutrients may be introduced into a carbonate factory by terrestrial runoff, leaching of coastal carbonates, storm disturbance of offshore oceanographic systems, volcanic eruptions, or upwelling from ocean depths. Establishment of nutrient gradients, ranging from oligotrophic to mesotrophic to eutrophic to hypertrophic, strongly influence the structure of the resident biota (Fig. 3E). The plankton biomass, which controls the amount of particulate organic matter in the water column, increases as the nutrient supply increases (Hallock and Schlager, 1986). Paradoxically, however, an increase in nutrients will encourage growth of organisms that feed on plankton during the larval and adult stages (*e.g.*, sponges), but curtail growth of photozoans by decreasing light penetrative levels (Fig. 3F), and in the case of zooxanthellate corals, cause feeding overstress (Hallock and Schlager, 1986).

In clear, oligotrophic tropical–subtropical ocean waters, zooaxanthellate corals thrive because they are superbly adapted to low-nutrient conditions. Under mesotrophic conditions, bioerosion increases and benthic macroalgae (including calcareous green algae) increase along with the echinoids and gastropods that graze on them. Most symbiotic organisms cannot compete with the faster growing macroalgae and sponges that thrive under mesotrophic conditions. Hypertrophic conditions, which leads to decreased light penetration because

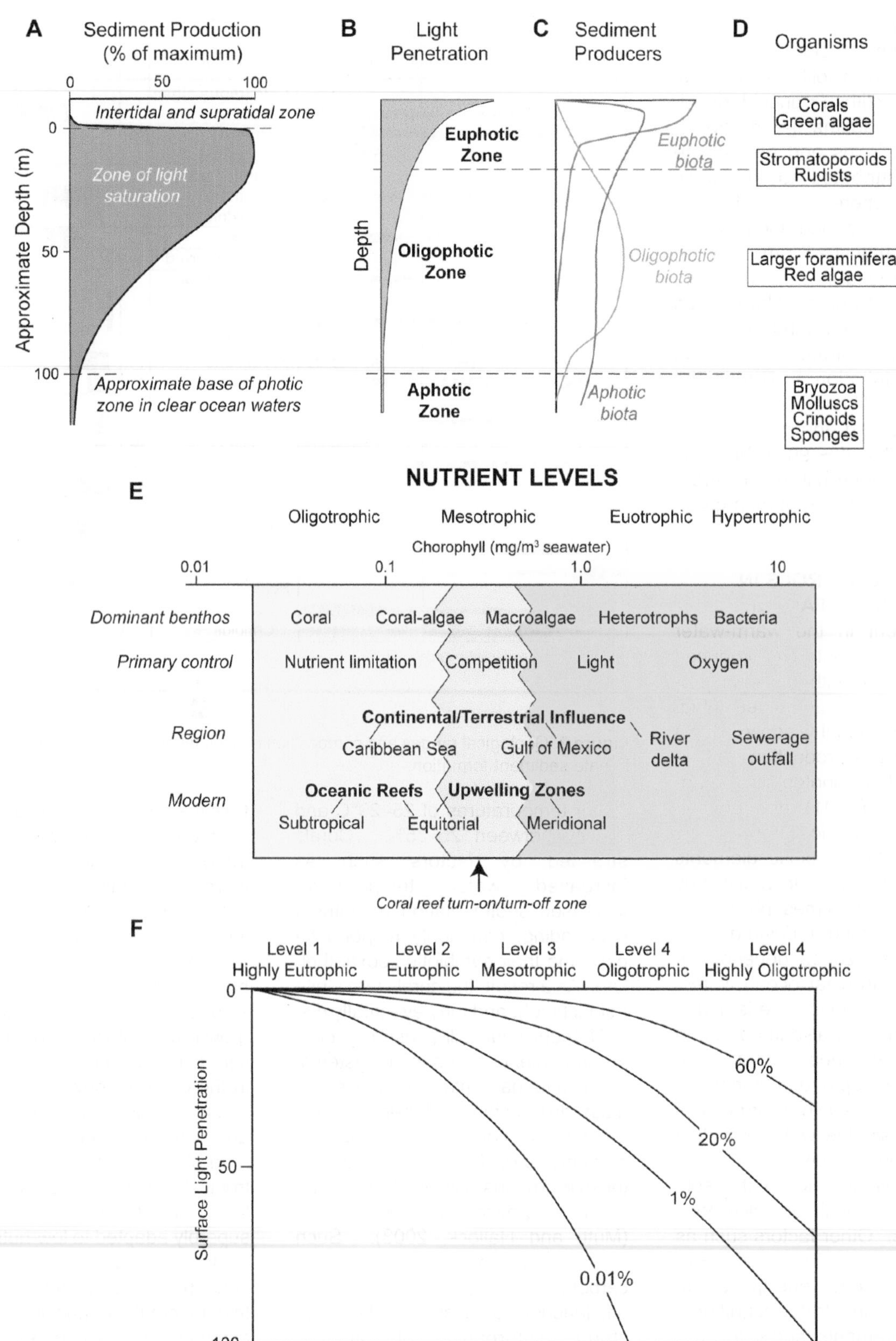

Figure 3. Sediment production in the carbonate factory. **A.** Schematic curve illustrating carbonate sediment production relative to depth based mainly on corals and green algae. After Schlager (1998, Fig. 1); **B.** Zonation of upper part of ocean column according to light penetration; **C.** Distribution of euphotic, oligophotic, and aphotic biotas; **D.** Distribution of biotic groups according to light penetration. Figures 3B, C, and D after Pomar (2001, Fig. 4); **E.** Nutrient gradients in low-latitude waters. After Mutti and Hallock (2003, Fig. 2); **F.** Estimated depths of photosynthetically active surface light penetration as a function of trophic resource decline. Isolines are for optimum growth of: 60% – branching corals; 20% – head corals; 1% – base of euphotic zone; 0.01% – maximum depth for zooxanthellate corals. After Hallock (1987, Fig. 2).

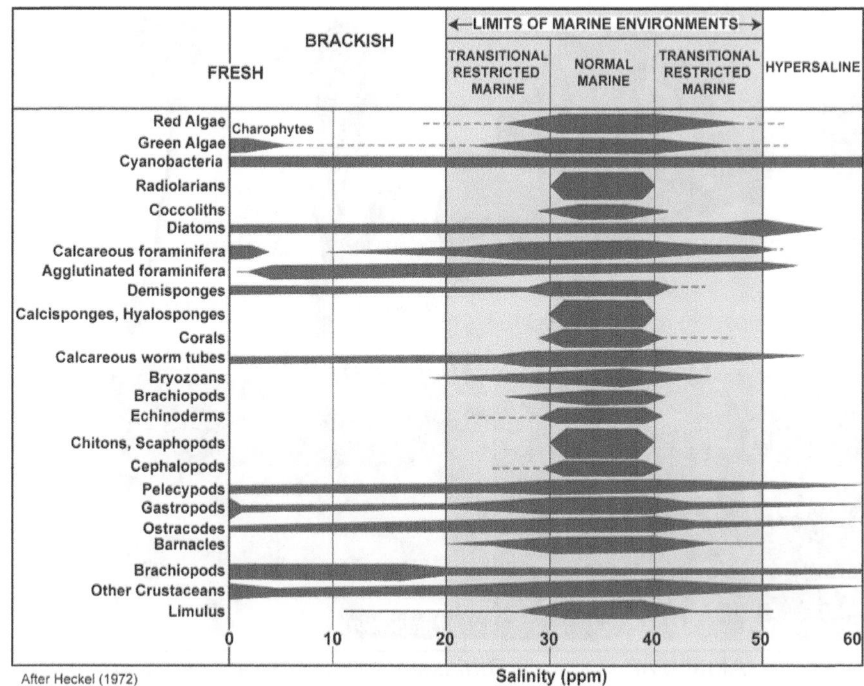

Figure 4. Salinity ranges for various animals and plants that contribute to carbonate sediment production. Thickness of line indicates relative abundance.

of the high chlorophyll levels in the water column (Fig. 3F), encourages colonization by filter-feeding benthos, including bivalves, sponges, and ascidians (Mutti and Hallock, 2003).

Little is known about the role that nutrient supply played in ancient oceans, largely because of the problem of assessing the nutrient levels (or a proxy) of ancient seas. However, Mutti and Hallock (2003) argued that various sedimentological, biogenic, and geochemical proxies can be used to establish nutrient fluxes in ancient systems. MacNeil and Jones (2008) suggested that the presence of stromatoporoid-microbe associations (mainly *Renalcis*), in parts of the Upper Devonian Alexandra Reef System in the southern part of the Northwest Territories, reflected higher nutrient gradients.

The Role of Algae and Plants in the Carbonate Factory

Algae and plants contribute to carbonate sedimentation by producing sediment, trapping sediment, binding sediment, and creating coated grains (Fig. 5). Calcareous green algae (mainly *Halimeda*, *Penicillus*, *Rhipocephalus*, and *Udotea*; Fig. 5D) can influence sedimentation in many ways. *Halimeda* mounds on the landward side of the Great Barrier Reef (Orme and Salama, 1988) and the Kalukalukang Bank in the eastern Java Sea (Roberts *et al.*, 1987) that occur in water 20–100 m deep, are up to 52 m thick with up to 12 m relief, cover hundreds of square metres, and accrete at rates of up to 5.9 m/1000^{-1} years. Hillis (1997) suggested that *Halimeda* species alone produce ~ 8% of the total worlds carbonate sediment. In the Bight of Abaco (Bahamas, Fig. 5G), the volume of sediment produced by calcareous algae, over the last 5500 years, has exceeded the volume of sediment found on the lagoon floor. The excess sediment was flushed out of the basin by storm activity. However, it is not known if analogous processes have been ongoing throughout the Phanerozoic or if they are only applicable to modern settings.

Sea-grass meadows commonly grow in shallow, near-shore waters (Fig. 5C and E). *Thalassia testudinum* (turtle grass), found throughout the Caribbean Sea, has leaves that baffle currents causing entrained sediment grains to be deposited around their root system, which also binds the sediment in place. The diverse epiphytic biota that thrives on the *Thalassia* leaves (*e.g.*, foraminifera, coralline algae, diatoms) locally contributes up to ~ 500 g of $CaCO_3$ m^{-2} yr^{-1} to the sediment budget of a lagoon. The combination of sediment production, trapping, and binding typically leads to the formation of aerially extensive sea grass banks that, in turn, lead to shallowing of the seafloor (Fig. 6). Sea grasses are, however, largely a post-Oligocene phenomenon.

Microbial (cyanobacterial) mats commonly trap and bind sediment on the seafloor and can develop into stromatolites, like those found in Shark Bay, Australia and Exuma Sound, Bahamas. By covering the sediment, microbial mats can also prevent sediment erosion and transportation in areas where currents are strong enough to move sediment grains.

Mangrove trees that fringe the landward margins of lagoons and shelves (Fig. 5) have dense entangled prop root systems that quickly baffle even the strongest currents and cause deposition of entrained sediment between their prop root systems that also bind the sediment in place. Although the trees themselves contain no $CaCO_3$, bivalves, calcareous worms, and barnacles commonly attach themselves to the prop roots and will, upon death, become part of the associated sediment.

Boring of Solid Substrates

Boring organisms, including animals (*e.g.*, bivalves, sponges) and micro-organisms (*e.g.*, fungi, bacteria) penetrate hard substrates (*e.g.*, coral skeletons, hardgrounds) in search of food or protection (Fig. 7A–C). Bivalves (Fig. 7B) and sponges commonly bore into coral skeletons. Boring is achieved by dissolution or mechanical excavation. Dissolution releases Ca and CO_3 back into the seawater, whereas excavation produces sediment grains that become part of the sediment budget – *e.g.*, the sponge *Cliona* produces up to 8 kgm^{-3}yr^{-1} of silt from the corals and hard substrates on Grand Cayman, and up to ⅓ of all sediment in some Pacific atoll lagoons. Boring also weakens lithified substrates thereby making them susceptible to physical destruction during cyclonic storms.

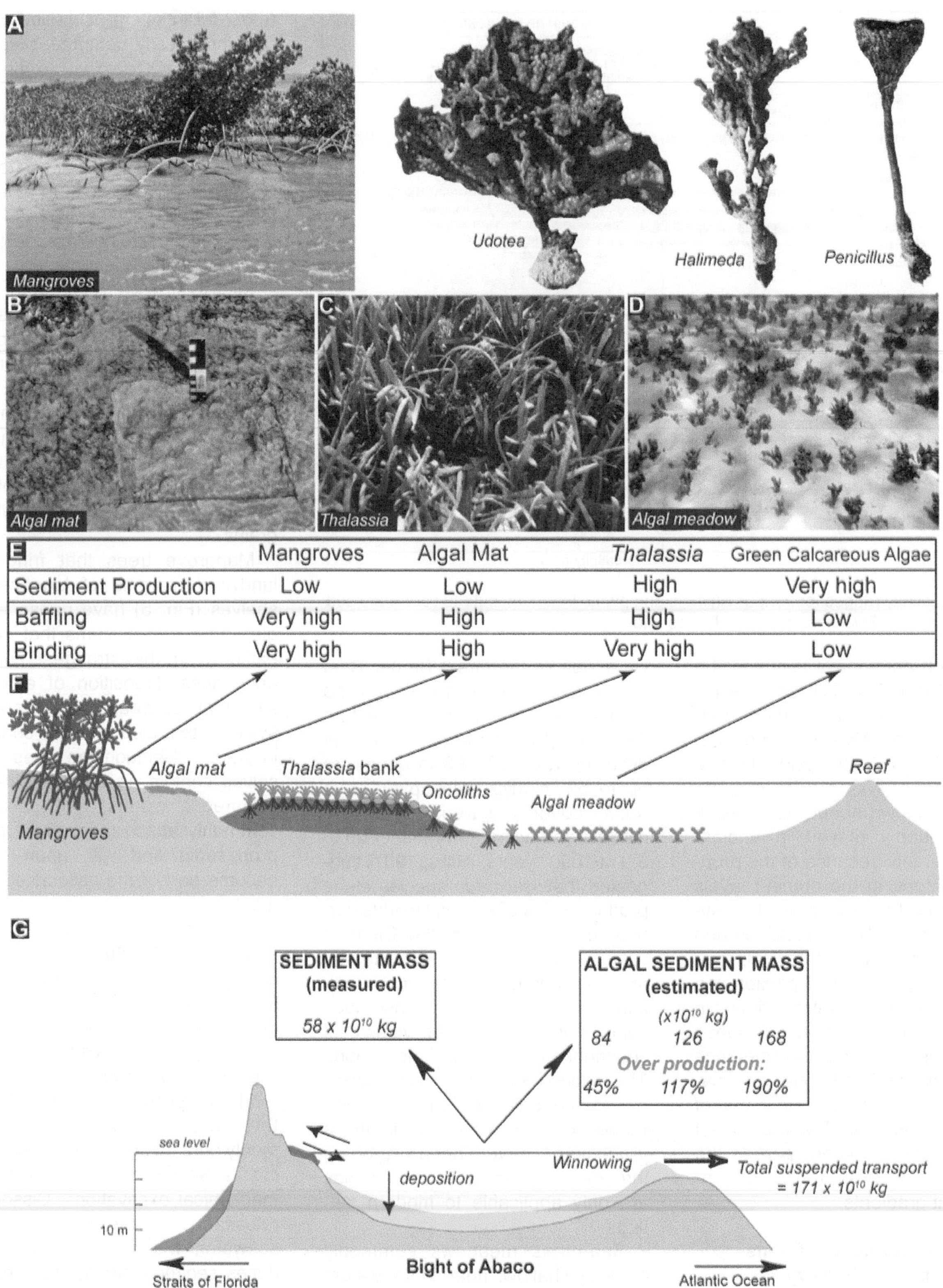

	Mangroves	Algal Mat	*Thalassia*	Green Calcareous Algae
Sediment Production	Low	Low	High	Very high
Baffling	Very high	High	High	Low
Binding	Very high	High	Very high	Low

Figure 5. Plants and algae in the carbonate factory – sediment producers, trappers, and binders. **A.** Sediment accumulating around prop root system of mangrove trees, Caicos Island; **B.** Algal mat covered calcareous mud, Frank Sound, Grand Cayman; **C.** *Thalassia* leaves covered with epibiota (white), Frank Sound, Grand Cayman; **D.** Calcareous green algae (mainly *Halimeda*) on floor of Frank Sound, Grand Cayman. Water ~ 3 m deep. Individual *Udotea*, *Halimeda*, and *Penicillus* plants shown above; **E.** Relative roles of mangroves, algal mats, *Thalassia*, and calcareous algae in the production, baffling, and binding of carbonate sediment; **F.** Hypothetical cross-section across a rimmed shelf or lagoon showing distribution of mangroves, *Thalassia*, algae mat, and calcareous green algae; **G.** Comparison of mass of mud-sized sediment (measured) and the estimated volume of mud-sized sediment produced from calcareous algae in the Bight of Abaco, Bahama Islands (after Neumann and Land, 1975, Fig. 5). The overproduction refers to the amount of mud-sized sediment produced by the algae relative to what is found in the bight.

Figure 6. Vertical section through a *Thalassia* bank in Prospect Point Lagoon, Grand Cayman showing height above floor of lagoon, extensive root system, and long leaves. Section created by storm waves, water ~ 0.4 m deep.

Burrowing Animals and Bioturbation

Carbonate sediments host a diverse infauna of worms, burrowing shrimps, and sea anemones (Fig. 7). Amply demonstrated by the ghost shrimp (*Callianassa*), these burrowers quickly overturn huge quantities of sediment as they seek food and protection from predators. *Callianassa*, for example, constructs large complex burrow systems (*Ophiomorpha*) to depths of 2 m, comprising shafts and chambers (Fig. 7F) that are lined with pellets produced by the shrimp. The entrance is rarely seen whereas the exit is obvious because of the mounds of sediment formed as the shrimp expels sediment from its galleries (Fig. 7E). Such burrowing homogenizes sediment by mixing different layers and destroying sedimentary structures. *Callianassa* ejects as much as 3.9 kg m^{-2} of sediment per day, some of which is carried away by local currents.

Lithification – Hardgrounds and Firmgrounds

Hardgrounds are hard, lithified marine substrates whereas *firmgrounds* have a consistency that is between that of soft sediment and hardground. Developed via submarine cementation, these layers are important because they foster paleoecological communities that differ from those associated with the precursor soft sediments. Burrowing animals (*e.g.*, shrimp) or plants roots (*e.g.*, *Thalassia*) are replaced by boring organisms (*e.g.*, worms, sponges), or invertebrate organisms with holdfasts requiring hard substrates (*e.g.*, corals, stromatoporoids), or encrusting organisms (*e.g.*, bryozoans, coralline algae). Such a succession can also herald a change in the type of biogenic sediment that is produced. Firmgrounds can be inhabited by burrowers and borers (Chapter 17).

Hardgrounds, important hiatal surfaces that are traceable over vast areas of the seafloor, develop where there is little or no sediment accumulation. As shown in Figure 8, they are characterized by various combinations of:

1. An irregular surface with overhangs and re-entrants that could not exist in soft sediment;
2. Encrusting organisms (*e.g.*, bryozoans);
3. Holdfasts;
4. Borings;
5. Reworked concretions, commonly bored and encrusted, formed of the same lithology as the hardground;
6. Pyrite, phosphate, and Mg oxide mineralization;
7. A contrast between the facies below and above the hard ground surface; and
8. Truncation of constituent grains and cement if erosive scouring took place.

Hardgrounds are known from successions of all ages (Taylor and Wilson, 2003) with those found in the Ordovician of North America and the Jurassic of Europe being particularly well-studied.

Storms and Hurricanes

Carbonate factories are usually tranquil with currents being generated solely by local winds or tides, and little water movement being the norm in protected lagoons. Tropical cyclonic storms, typically short-lived, profoundly affect sedimentation patterns because:

1. Associated horizontal gradients in atmospheric pressure cause water level to rise at the shoreline – coastal set-up;
2. Onshore winds accentuate coastal set-up and generate wind-drift currents;
3. Gradient currents compensate for the coastal set-up and flow offshore; and
4. Wave effects mobilize bottom sediments (Aigner, 1985).

Onshore-sediment transport dominates in shallow depositional systems whereas offshore transport dominates in deep-water systems. Onshore winds push water landward and large volumes of lagoonal or shelf sediments can be transported onto muddy tidal flats or beaches. As the storm wanes, the water piled onshore flows offshore with resultant strong currents carrying sediment to the fore reef and downslope.

Shelf deposits or lagoonal sediments can be affected by storm activity as:

1. Fine-grained sediments are preferentially winnowed;
2. Sediments are homogenized and sedimentary structures destroyed;
3. Coral reefs are severely damaged, especially if previously weakened by bioerosion;
4. Intraclasts are formed during hardground destruction; and
5. Sedimentary structures (*e.g.*, crossbedding) indicative of high-energy conditions are generated.

Storm deposits can be recognized by:

1. Geomorphic evidence such as spillover lobes and onbank sand lobes;
2. Stratigraphic evidence such as sharp-based beds overlying erosive surfaces or beds of high-energy sediments randomly distributed through a succession formed largely of quiet-water deposits; and
3. Biostrationomic evidence that includes coquina layers and condensed beds.

FACIES MODELS FOR WARM-WATER NERITIC CARBONATE DEPOSITION

A plethora of terms have been used to classify different warm-water depositional settings. Most terms, however, are descriptive and based on modern systems that commonly reflect the physiographic setting. Pomar (2001) argued that such a classification is fundamental to sequence stratigraphy, and suggested that a genetic classification of carbonate platforms was necessary with the principal classifiers being sediment type, locus of sediment production, and hydraulic energy. Invalidating the physiographic approach to depositional system classification is, however, unnecessary, as issues regarding sediment production are a perquisite to understanding how facies develop and respond to environmental change.

Facies models for warm-water carbonate factories are derived largely from modern settings where all aspects of the systems can, at least in theory, be determined. Maps of modern systems (*see* Figs. 10–18) clearly show that facies distribution is related to platform shape, size, geomorphology, location of islands relative to current directions, water depth, and bottom topography. Therefore, interpretation of ancient depositional systems becomes difficult in the face of limited outcrop, lack of seismic profiles, and biotas formed largely of extinct organ-

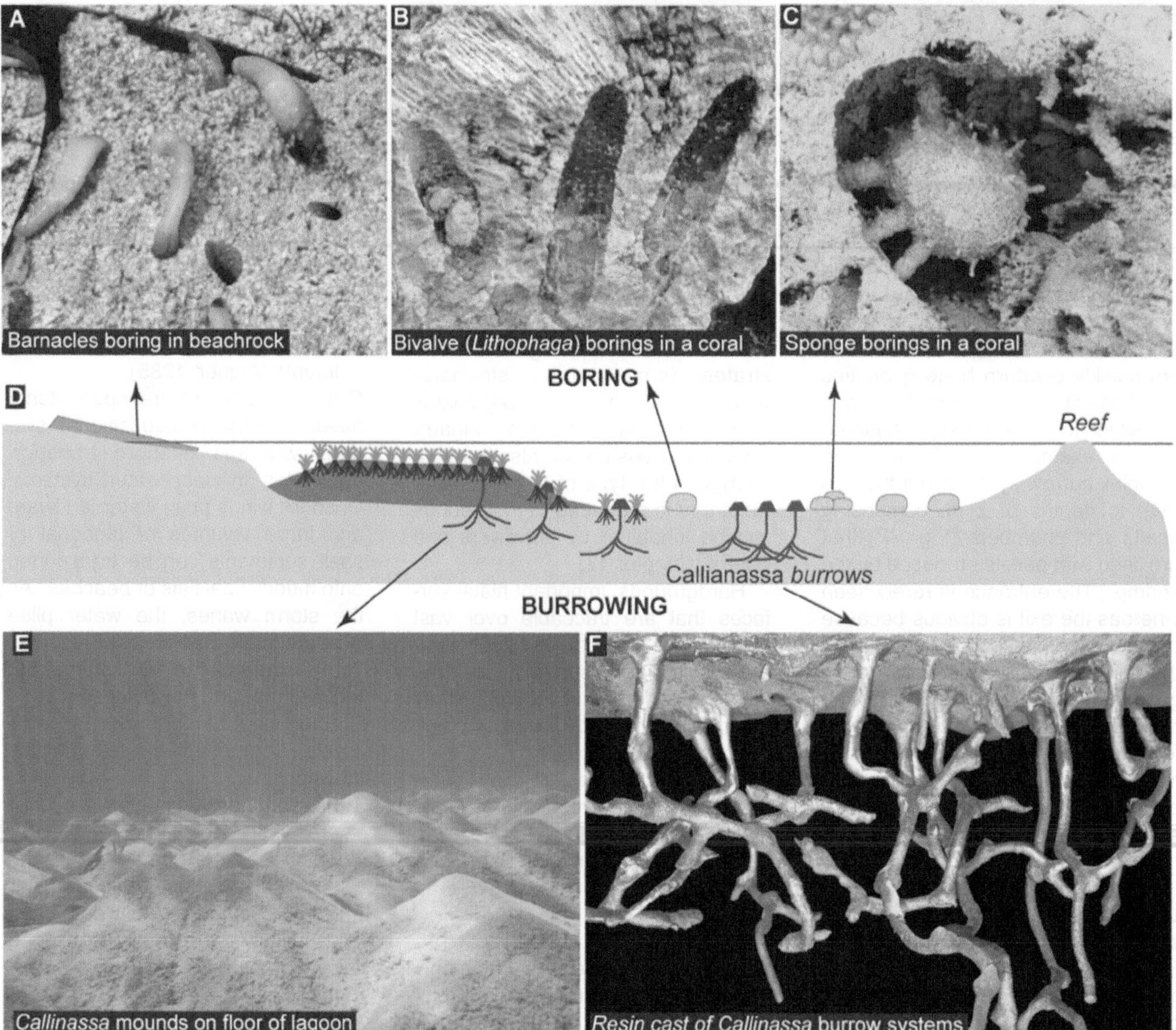

Figure 7. Animal modification of substrates in carbonate factory. (All images, except F, from the Grand Cayman, by author. Image F, courtesy Murray Gingras). **A.** Barnacles boring into beach rock, Bodden Town; **B.** *Lithophaga*-borings in large coral, Ironshore Formation; **C.** Cast of sponge boring in leached coral, Cayman Formation; **D.** Schematic cross-section of rimmed shelf/lagoon showing distribution of boring and burrowing organisms; **E.** Volcano-like mounds of sediment piled around exit holes of *Callianassa* burrows, East Sound, water ~ 6 m deep; **F.** Resin cast of modern *Callianassa* burrows.

isms that can be poorly understood in terms of their paleoecological preferences. A thorough understanding of modern carbonate sediments coupled with the perspective of their spatial arrays does, nevertheless, provide a solid foundation for the interpretation of ancient successions. Regardless, as stressed in Chapter 13, ancient systems may have behaved differently and this possibility must be considered before facies models based on modern systems can be uncritically applied to the older rock record.

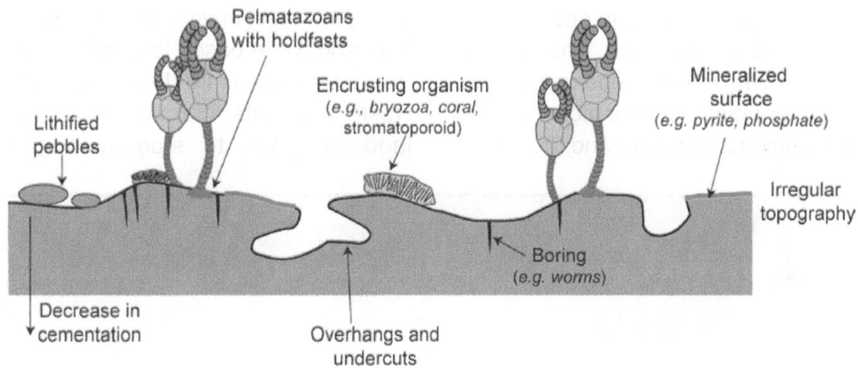

Figure 8. Schematic diagram showing main features of a hardground.

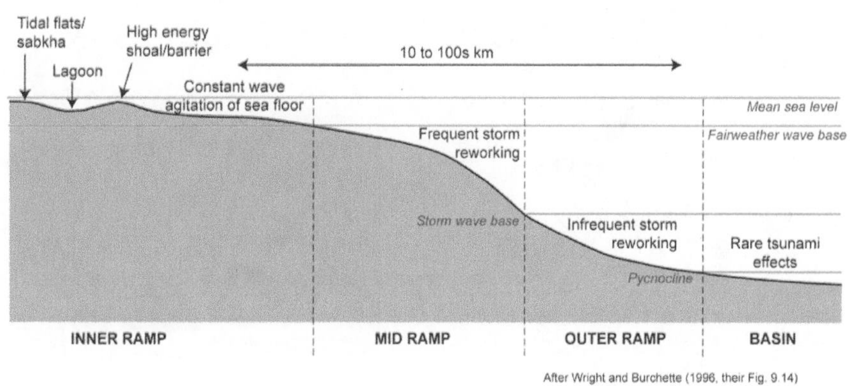

Figure 9. Main depositional regimes in a ramp system.

Open Shelves and Ramps (Figures 9, 10, and 11)

Open shelves, also known as open unrimmed shelves, range from ramps with uniform slopes (< 1°) that merge gradually with the open basin (Read, 1985) to open shelves (distally steepened ramp of Read, 1985) that have a higher gradient in its outer, deeper part. Ramps, which are common along passive margins and in foreland basins and possess a uniform slope from the shoreline to the open ocean, have been termed 'homoclinal ramps' whereas those with a slope break in deeper water have been called 'distally steepened ramps'. Features common to most ramps include:

1. An initial topography inherited from the underlying substrate;
2. The absence of reefs that can radically alter the profile of a shelf as they grow; and
3. The potential to morph into other types of platforms if sedimentation (*e.g.*, reefs, shoals) is strongly focused in one area.

By using the dominant sedimentary processes, Burchette and Wright (1992) divided this depositional regime into the inner, mid, and outer ramp (Fig. 9), as follows:

- The *inner ramp* for the zone above fair-weather wave base and therefore includes recognizable successions of sediments that accumulated in the shoreface, beach, lagoonal, and tidal flat settings.
- The *mid-ramp*, which lies between the fair-weather wave base and storm wave base, where storm deposits (tempestites) with graded bedding and hummocky cross-bedding are common.
- The *outer ramp* stretches from the base of the storm wave base to the open ocean basin; storm-generated currents can deposit graded beds and also rework some of the pre-existing sediments. In other settings, the outer ramp can experience restricted circulation such that the bottom waters may be suboxic or anoxic.

Ramps are generally devoid of large skeletal reefs. Nevertheless, small, isolated skeletal-microbial reefs and mud-dominated mounds can develop in the mid- and outer ramp (*see* Chapter 17). Such reefs commonly develop during sea-level rise with topographic highs being preferred sites of development.

Ramps and unrimmed shelves have:

1. A gently sloping seafloor, 10–300 km wide, adjacent to a continent;
2. Facies belts of variable width that parallel bathymetric contours;
3. Gradual transitions between neighbouring facies belts;
4. High-energy carbonate sand wave- and tide-agitated, shallow-water environments above fair-weather wave base;
5. Skeletal muddy sand to mud in quiet, deeper water areas below fair-weather wave base and only periodically disturbed by storms;
6. No continuous reef trends; and
7. Localized patch reefs and sand shoals.

Ramps

The Persian Gulf, a classic modern ramp, is an asymmetric foreland basin with its deep axis near the Iranian coast from which vast quantities of siliciclastic sediment are shed (Fig. 10). On the southern, cratonward side, carbonate sediments accumulate on a ramp with a seaward gradient of < 35 cm/km[1] because little siliciclastic sediment is derived from the hinterland. A transect across that ramp includes:

1. Microbial intertidal flats that pass landward into sabkhas and skeletal-pelletal lime muds in protected coastal lagoons;
2. High-energy skeletal–oolitic sand shoals, beach barrier systems, and scattered coral reefs;
3. A deep area, where molluscan and foraminiferal-dominated sands pass laterally into skeletal muddy

sands with molluscan debris; and
4. A gradual transition into bivalve-rich marls in the deeper axial zone.

Sediment distribution is controlled by the seasonal Shamal Wind (from NW) that generates waves several metres high and wave bases to depths of 30 m. Local tides are less important. Restricted water circulation in the lagoons leads to stagnation and evaporation that drives salinity as high as 70‰. Few animals and plants can tolerate such high salinities (Fig. 4).

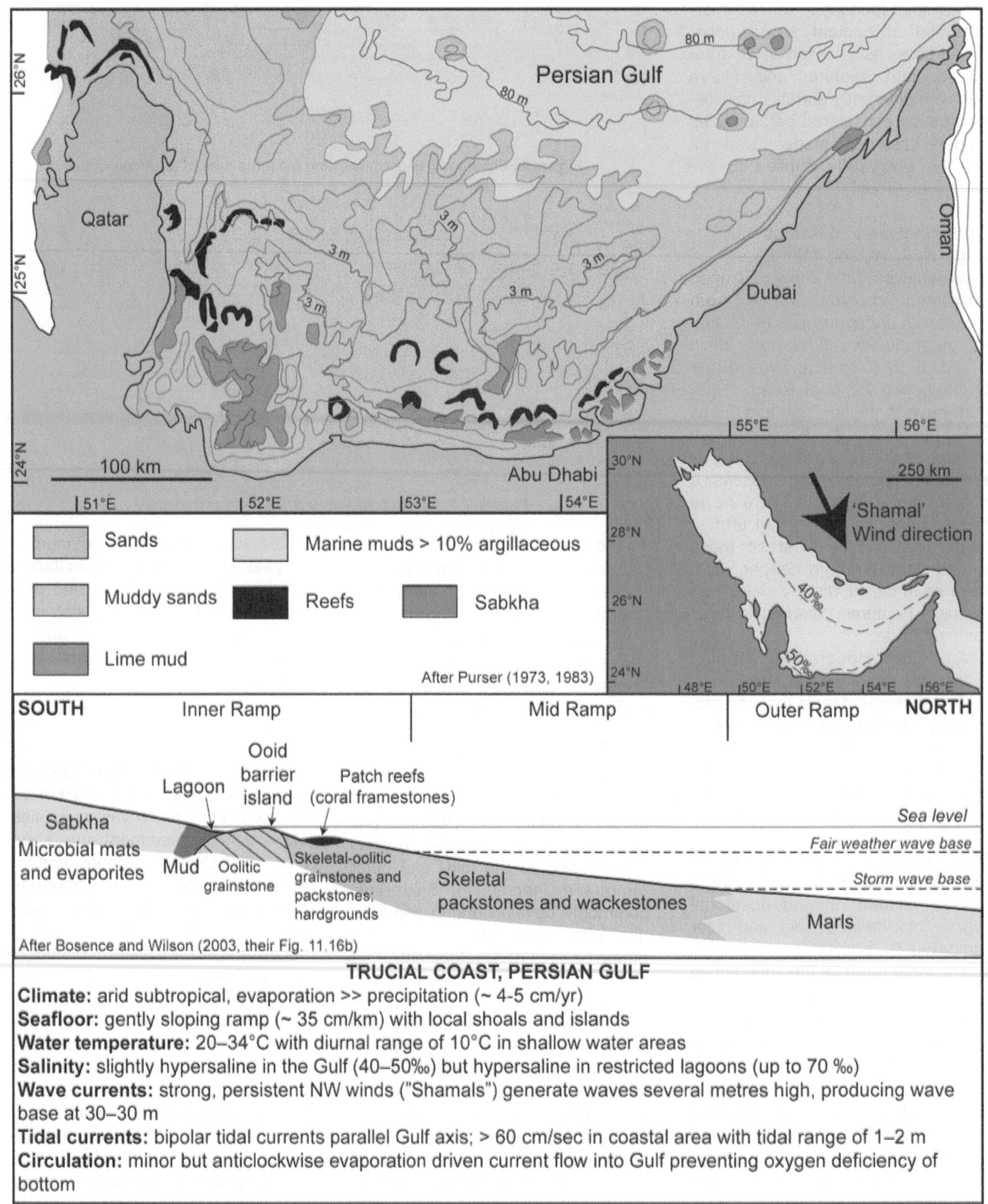

TRUCIAL COAST, PERSIAN GULF

Climate: arid subtropical, evaporation >> precipitation (~ 4-5 cm/yr)
Seafloor: gently sloping ramp (~ 35 cm/km) with local shoals and islands
Water temperature: 20–34°C with diurnal range of 10°C in shallow water areas
Salinity: slightly hypersaline in the Gulf (40–50‰) but hypersaline in restricted lagoons (up to 70 ‰)
Wave currents: strong, persistent NW winds ("Shamals") generate waves several metres high, producing wave base at 30–30 m
Tidal currents: bipolar tidal currents parallel Gulf axis; > 60 cm/sec in coastal area with tidal range of 1–2 m
Circulation: minor but anticlockwise evaporation driven current flow into Gulf preventing oxygen deficiency of bottom

Figure 10. Lithofacies map and cross-section for ramp off the Trucial Coast, Abu Dhabi, in the Persian Gulf.

Open Shelves

Open shelves characterize the west coast of Florida (Fig. 11A) and the northeast corner of the Yucatan Peninsula (Fig. 11B). They lie on opposite sides of the Gulf of Mexico, which is a partly enclosed, microtidal, low-energy sea that is largely wind dominated with most energy generated from waves that transverse the open gulf. Hurricanes, however, can quickly modify sedimentation patterns. Facies on the West Florida Shelf include:

1. Inner shelf quartz/molluscan sands;
2. Outer shelf coralline algal sands;
3. Shelf edge oolitic, peloidal, and

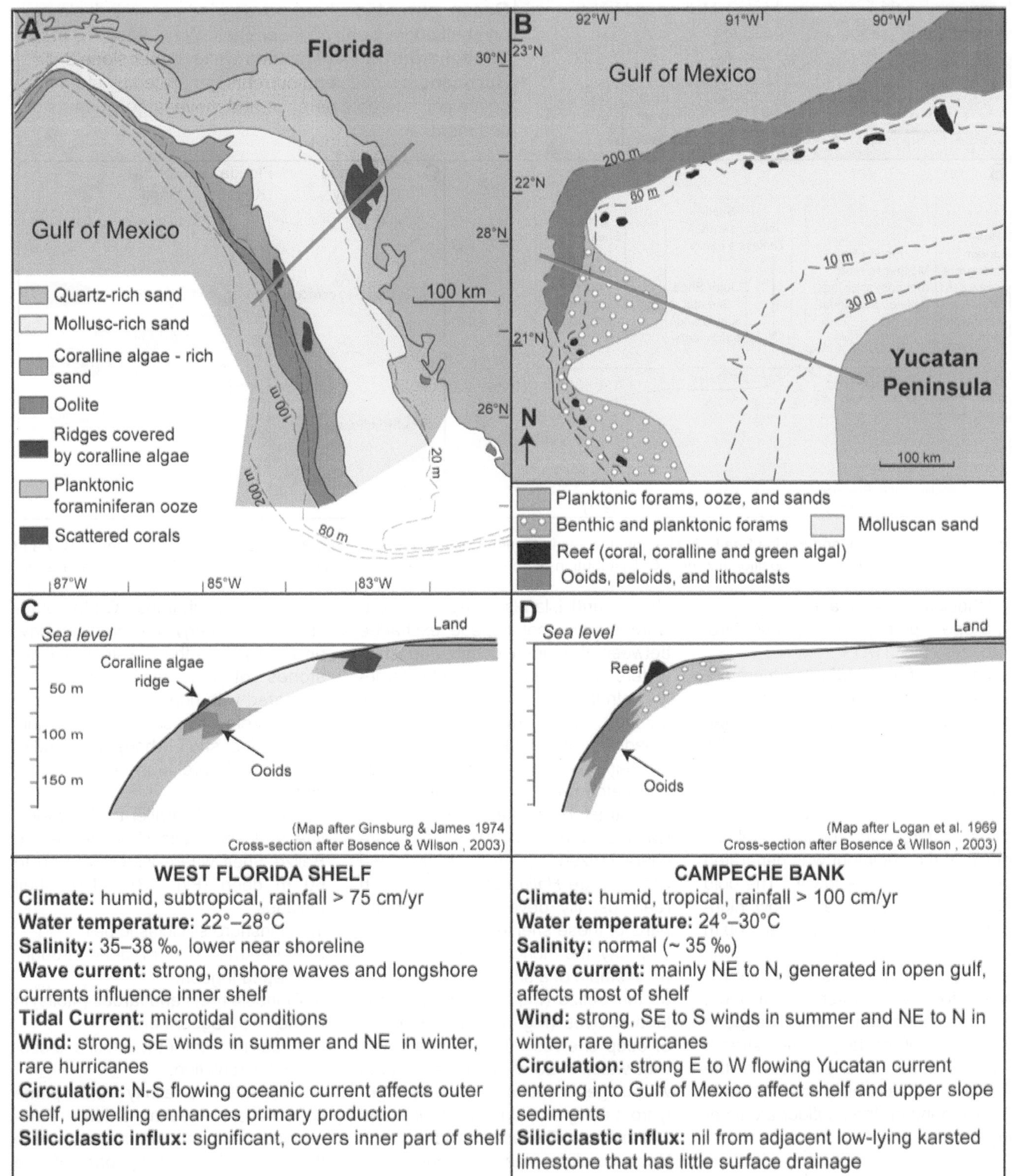

Figure 11. Lithofacies maps of unrimmed shelves. **A.** West Florida Shelf; **B.** Yucatan Shelf. The ooid facies are relict sediments that formed when sea level was lower than it is today.

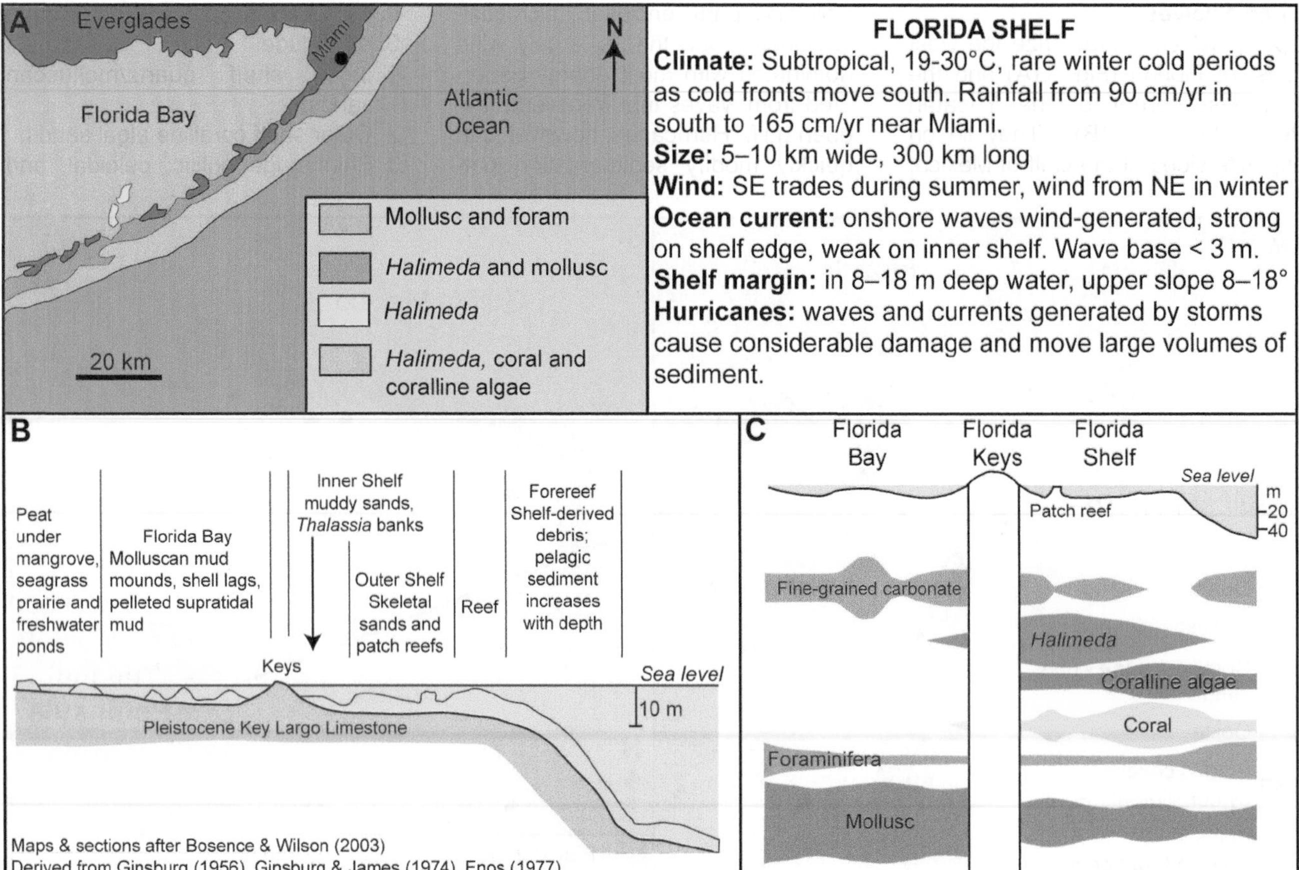

Figure 12. Florida Bay and Florida Shelf. **A.** Dominant facies in Florida Bay and Florida Shelf; **B.** Cross-section across Florida Bay and Florida Shelf showing main depositional zones; **C.** Distribution of dominant sediment contributors across Florida Bay and Florida Shelf.

lithoclastic sands; and

4. Slope planktonic foraminiferal oozes (Fig. 11A).

Landward of the shelf edge there are small reefs constructed of branching and massive corals, bryozoans, and *Halimeda*.

Facies belts on the Yucatan Shelf are similar but inshore quartz sands are absent (Fig. 11B). On both, the West Florida and Yucatan shelves, much of the sediment was generated when sea level was lower. The ooid facies that formed in a shallow, high-energy setting, occurs in water > 60 m deep on both shelves (Fig. 11). Similarly, sands on the West Florida Shelf, which formed in water ~ 5 m deep, are now found near the shelf edge in water 80–100 m deep. Such stranded sediments clearly illustrate the dangers of assuming that all sediments found on the seafloor are time-equivalent.

Rimmed Shelves (Figures 12, 13, and 14)

Rimmed shelves have reefs, sand shoals, and islands along their seaward margin that impede exchange between the open ocean and shelf waters. Facies development is largely controlled by water depth. Shallow shelves have muddy sea-grass meadows and biofragmental sands on their inner parts and skeletal sands and patch reefs on their outer parts. In contrast, deep shelves (covered by water as much as 30 m deep) are characterized by mud or, in some cases, large *Halimeda* banks. Patch reefs usually develop on the outer parts of these shelves.

Today, the Florida Shelf (Fig. 12) is an example of a shallow shelf, whereas the Belize (Fig. 13) and Queensland shelves (Fig. 14) are examples of deep shelves. The Florida Shelf, with water < 20 m deep, is covered with biogenic sediment derived largely from the green algae *Penicillus* and *Halimeda*. *Thalassia*, algae, molluscs, coralline algae, echinoids, and small corals grow in the fine-grained inner shelf sediments. Sediments are homogenized by burrowing worms, molluscs, and shrimps. On the outer shelf, high-energy onshore wind-driven waves keep the Pleistocene limestones of the seafloor bare or veneered with rippled skeletal sands. Patch reefs are scattered along this part of the shelf whereas low-relief reef mounds are found on the inner shelf.

Facies belts parallel to the shoreline and shelf axis characterize the Belize Shelf where waters are up to 60 m deep (Fig. 13). This shelf includes:

1. An inshore siliciclastic belt;
2. A central, 6-60 m deep, trough-shaped lagoon.
3. Pinnacle reefs and faroes (atolls) in the shelf lagoon; and
4. A barrier reef that inhibits cross-shelf circulation.

Water depth and salinity are the prime controls over sedimentation. Fine-grained sediments that accumulate in the deep, low-energy part of the lagoon are characterized by an increase from 30% carbonate content inshore (marl) to 90% carbonate con-

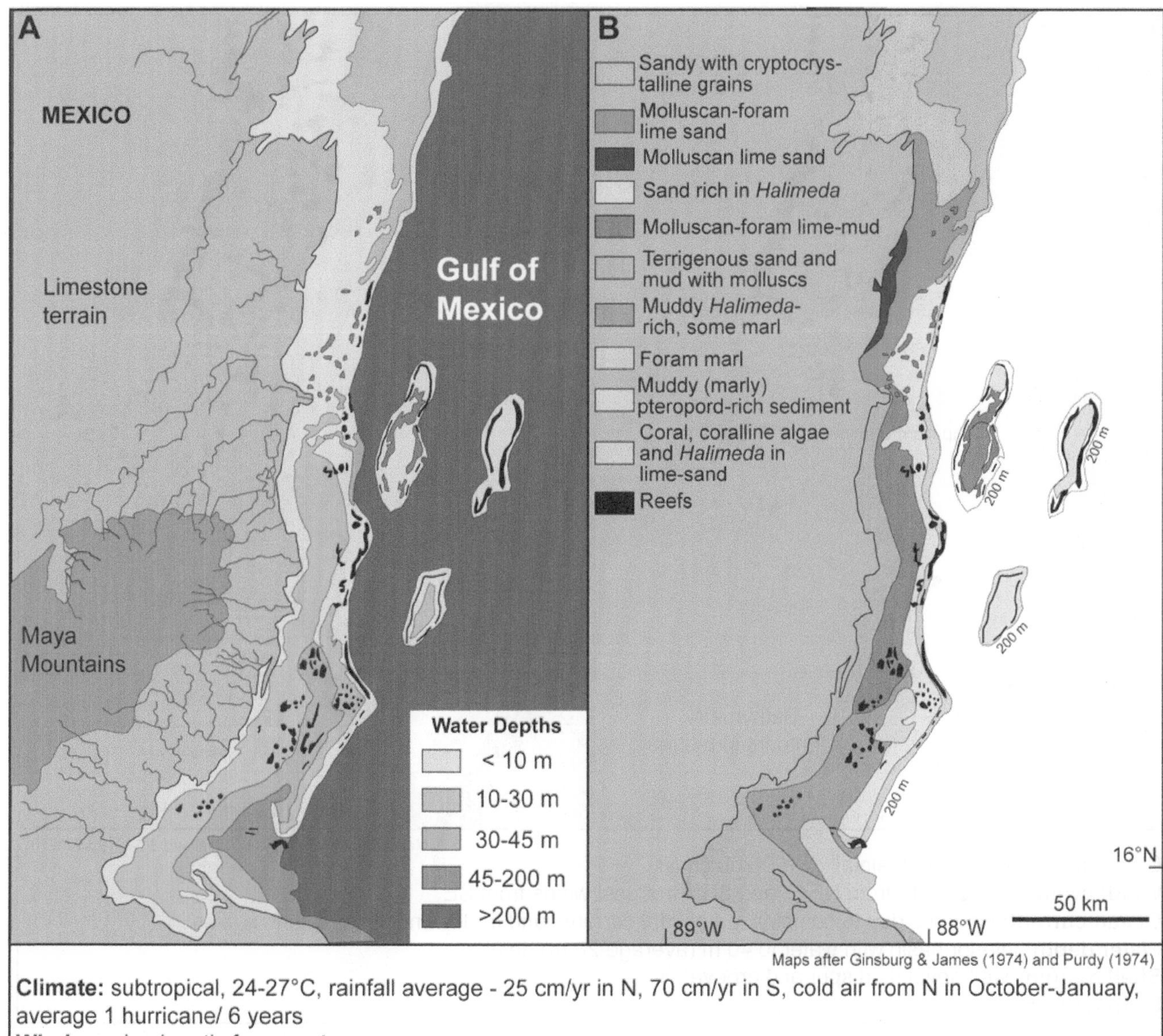

Climate: subtropical, 24-27°C, rainfall average - 25 cm/yr in N, 70 cm/yr in S, cold air from N in October-January, average 1 hurricane/ 6 years
Wind: predominantly from east
Ocean current: predominantly from E due to Caribbean current
Water depth: up to 50 m on shelf, < 10 m on N part of shelf
Shelf margin: rimmed by 3-10 km wide barrier (< 3 m water depth), coral reefs and islands, steep on ocean side
Salinity: ~ 36‰, well-mixed to 30-50 m, decreases to N and S due to freshwater influx, slightly higher in north lagoon due to lower rainfall

Figure 13. The Belize Shelf, Gulf of Mexico. **A.** Bathymetry; **B.** Lithofacies map.

tent near the barrier reef, and concomitant with a north to south increase in water depth; the dominant bioclasts change from molluscs and foraminifera to green algal (*Halimeda)* plates in the centre, to foraminifera in the deepest water. Blankets of coarse sediment (coral–green algae) accumulate on the leeward side of the barrier reef.

Lithofacies associated with the Great Barrier Reef (Fig. 14), controlled largely by water depth, include:

1. An inshore belt of terrigenous mud and sand;
2. A zone of calcareous muddy sand;
3. A zone of carbonate mud; and
4. An outer shelf edge zone formed of numerous reefs and inter-reef sands.

Halimeda banks develop in deeper waters. Shelf reefs, surrounded by biofragmental debris, are common throughout. Sands between the reefs are derived largely from *Halimeda*, molluscs, bryozoa, echinoids, corals, and foraminifera. Most of the calcareous mud results form the breakdown of skeletal material that originates around the reefs.

Banks
(Figure 15)

Bank sedimentation is controlled by water depth, temperature, and energy level. If present, islands can impact facies development by impeding wind-generated waves and cur-

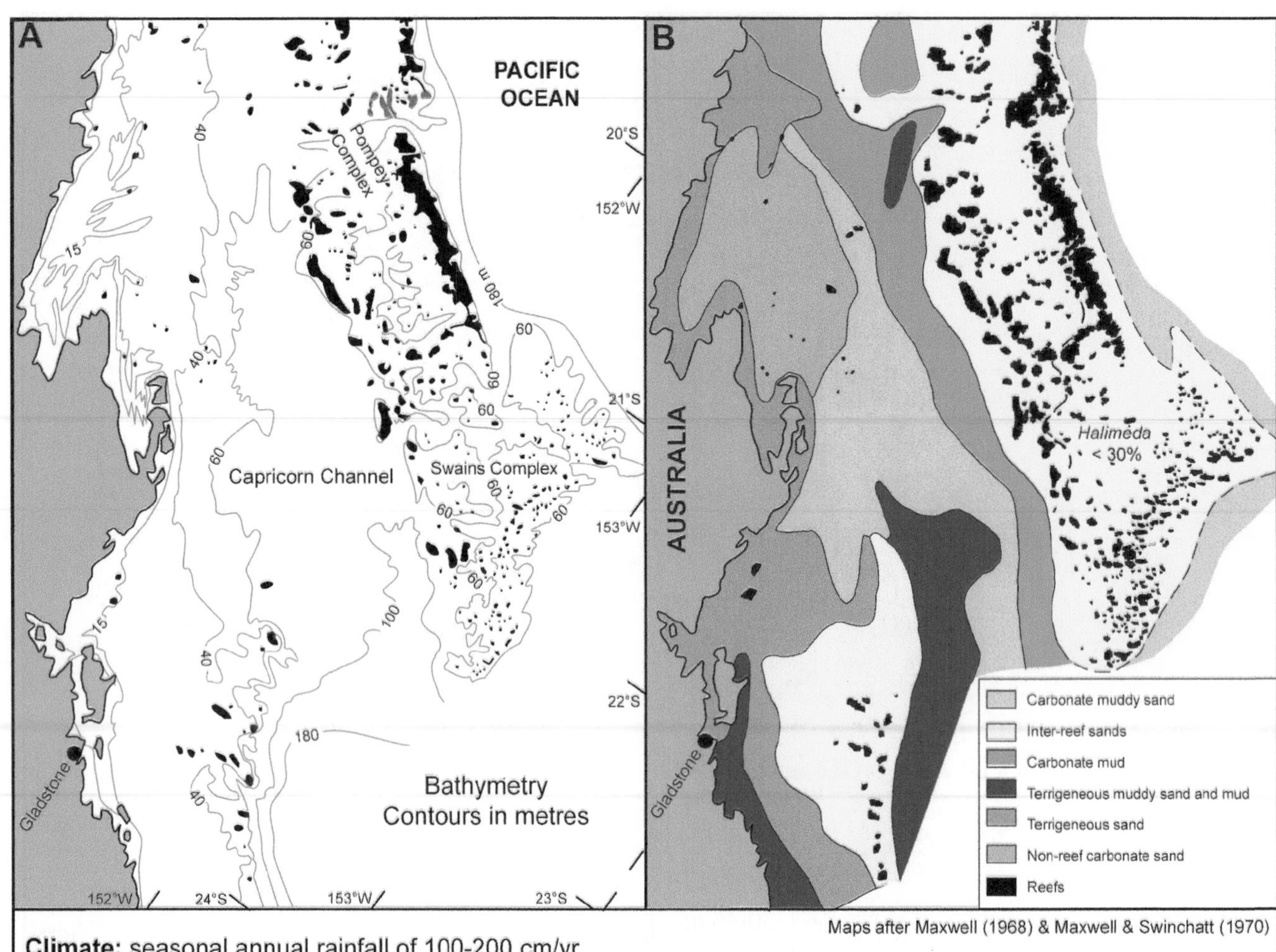

Climate: seasonal annual rainfall of 100-200 cm/yr
Wind: average 8-20 km/hr from NE,E, and SE: strongest winds from SE
Ocean current: vigorous due to combination of tidal currents and trade-wind currents and waves
Water depth: interior basin with water to 40 m (average 28 m) deep
Shelf margin: well-developed spur and groove
Salinity: ~ 35‰

Figure 14. The Northeast Australian shelf. **A.** Bathymetry; **B.** Lithofacies map.

rents. Storms and hurricanes have a significant impact. Based on banks found throughout the Caribbean Sea (*e.g.*, Pedro Bank), the Gulf of Mexico (*e.g.*, Florida Middle Ground), and the Bahamas (Cay Sal Bank – Fig. 1), the following attributes characterize most carbonate bank sedimentation:

- Banks < 10 m deep have moderate- to high-energy currents that sweep much of the sediment off the bank. Sea-grass communities can develop whereas corals are scattered. Coral reefs, if present, grow along the windward margin.
- Banks 10–45 m deep are characterized by a wide array of sediment types and a diverse, coral-dominated biota.
- Banks > 45 m deep are characterized by an azooxanthellate coral community.
- Some banks have depth-controlled biotic zones, whereas others do not; and
- High-energy conditions, created by geostrophic currents, wind-driven currents, or storms and hurricanes, have a profound impact on bank sedimentation. Currents up to 100 cm sec^{-1} have been recorded on Thunder Knoll in the southwest Caribbean Sea. As a result, fine-grained sediments are commonly washed off the bank, leaving gravels and sands as the dominant bank top sediments.

The Great Bahama Bank (Fig. 1), covering an area of 96 000 km^2, is bounded by steep (> 40°) faulted margins, surrounded by deep oceanic water, and has water circulation controlled by tides and winds (Fig. 1). Coarse-grained sediments occur around the bank edge (Fig. 15) where tides and wind-driven waves create high-energy conditions. Mud and pellet mud accumulate on the shallower bank interior, where energy levels are lower. Reefs grow along the eastern margin where energy levels are high. Andros Island blocks east–west water movement so that waters along the west coast of the island stagnate and become more saline during the summer months (up to 40 ‰). Sediments on the shallow bank interior are disturbed only during major storms.

Pedro Bank, located southwest of Jamaica, is 7900 km^2 in area and cov-

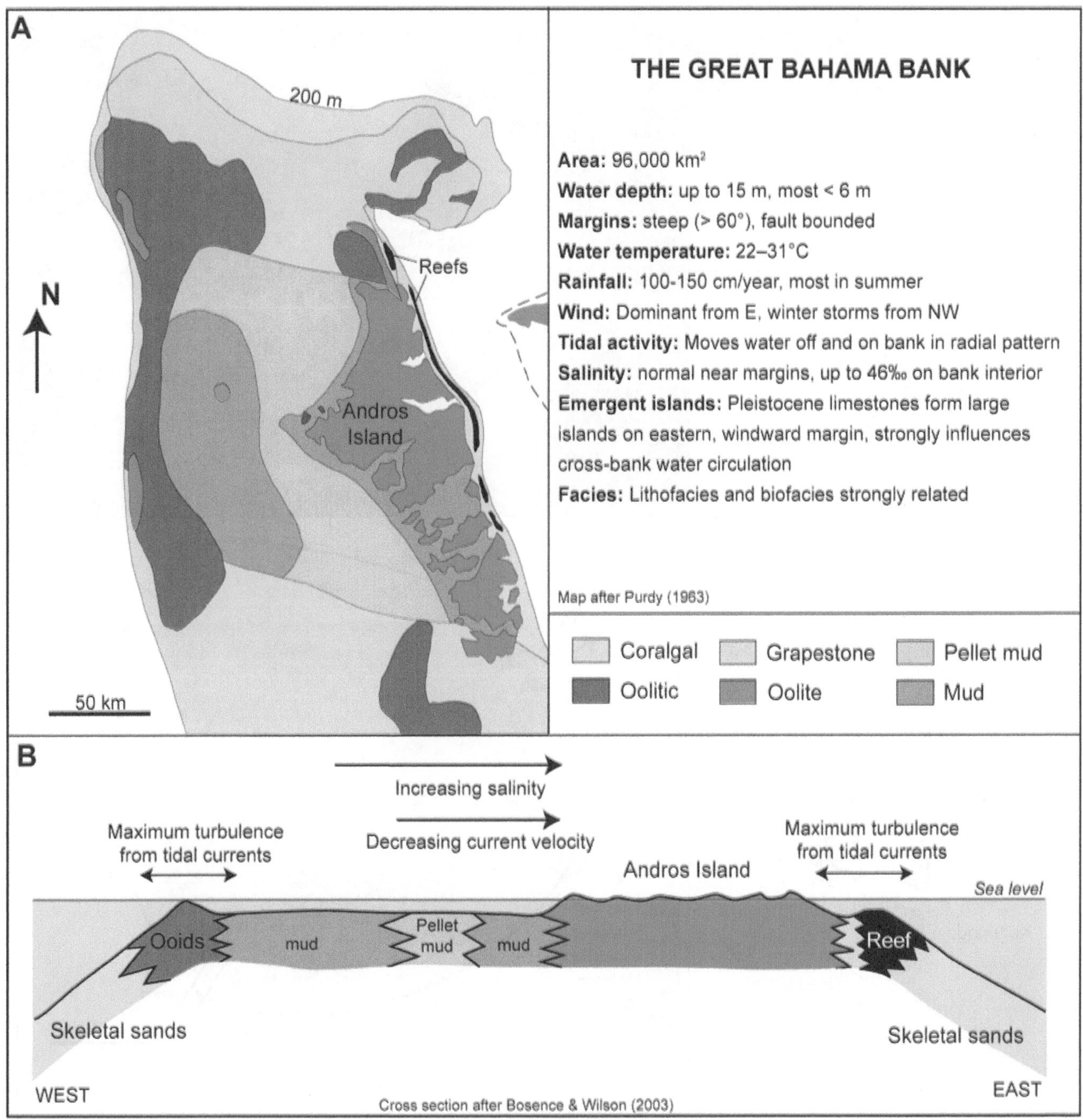

Figure 15. Great Bahama Bank. **A.** Lithofacies map (modified from Purdy, 1963a, b); **B.** Schematic cross-section (east–west) across bank showing distribution of facies (modified from Bosence and Wilson, 2003, their Fig. 11.14c).

ered by water that deepens to the southeast. Modern sediments are mainly poorly sorted, medium- to coarse-grained skeletal-pellet sands that contain numerous benthic and planktonic foraminifera and coccoliths. The sand, which forms in the shallower areas, is redistributed across the bank by strong currents.

Atolls
(Figure 16)

Atolls, found in the modern Atlantic, Pacific, and Indian oceans, typically have a circular to ovate reef enclosing a central lagoon and are surrounded by deep oceanic waters (Fig. 16). A volcanic island is located in the middle of most atolls. The Maldives, located on the Laccadive-Chagos submarine ridge in the Indian Ocean (Fig. 16A) include a double row of 22 atolls that are separated by the Inner Sea basin, which is up to 450 m deep.

Atolls, including Rasdhoo in the Maldives (Fig. 16C, D and E) and Kapingamarangi Atoll in the Pacific Ocean, have a concentric arrangement of facies in their deep lagoons. On the lagoon side of the reef, there is usually a blanket of reef-derived sands that are washed into the lagoon by onshore winds. In the central part of the lagoon, water depth exerts primary control over the sediments. In Kapingamarangi Atoll, *Halimeda* sands form below 30 m but at depths > 60 m, low light conditions

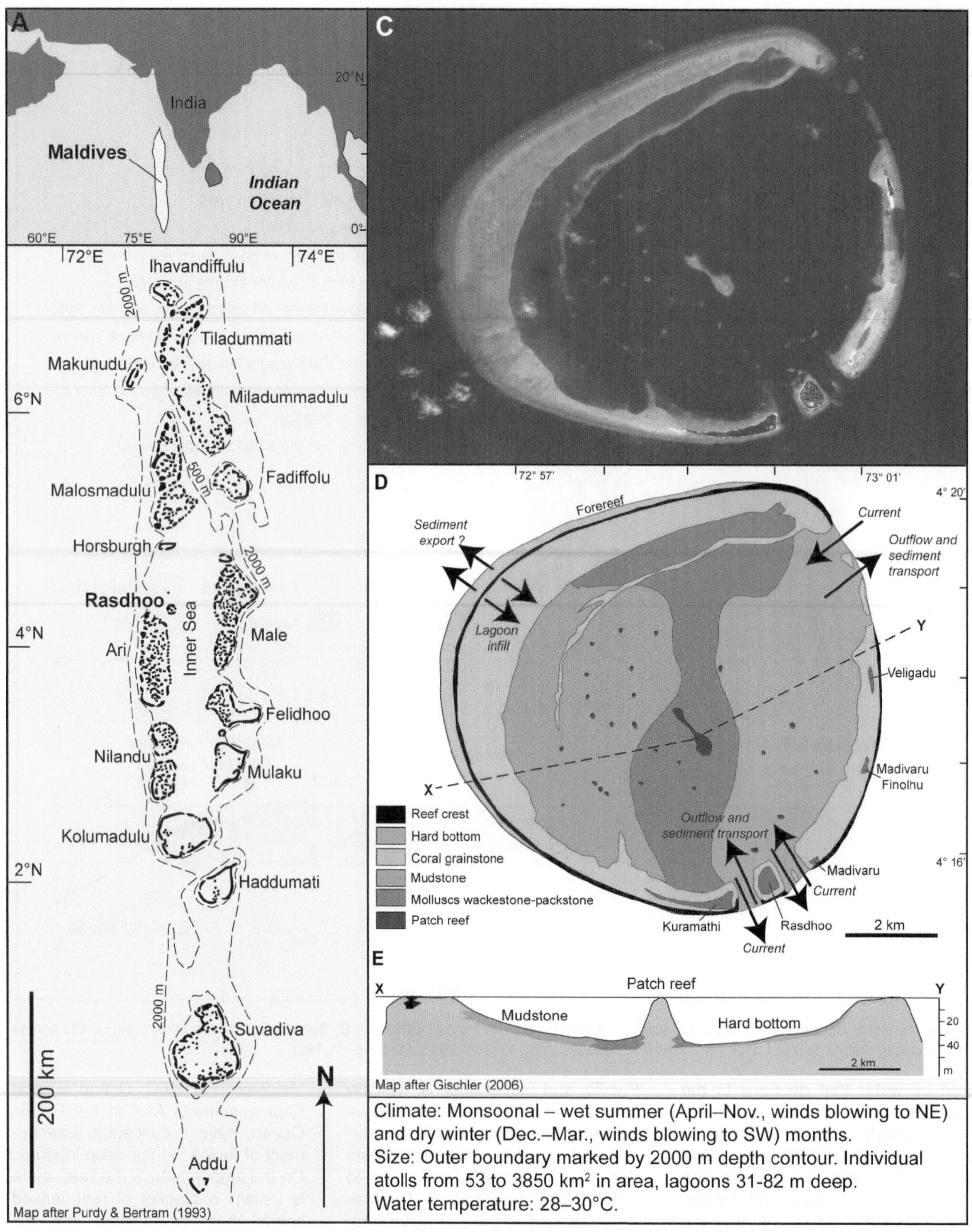

Figure 16. Maldives, Indian Ocean. **A.** Location of Maldives; **B.** Maldives atolls distributed along a north−south ridge; **C.** Satellite image of Rasdhoo Atoll – see Fig. 16B for location. Image courtesy of the Image Science and Analysis Laboratory, NASA Johnson Space Center [image ISS010-E-13155; http://eol.jsc.nasa.gov]; **D, E.** Lithofacies map and cross-section for Rasdhoo Atoll.

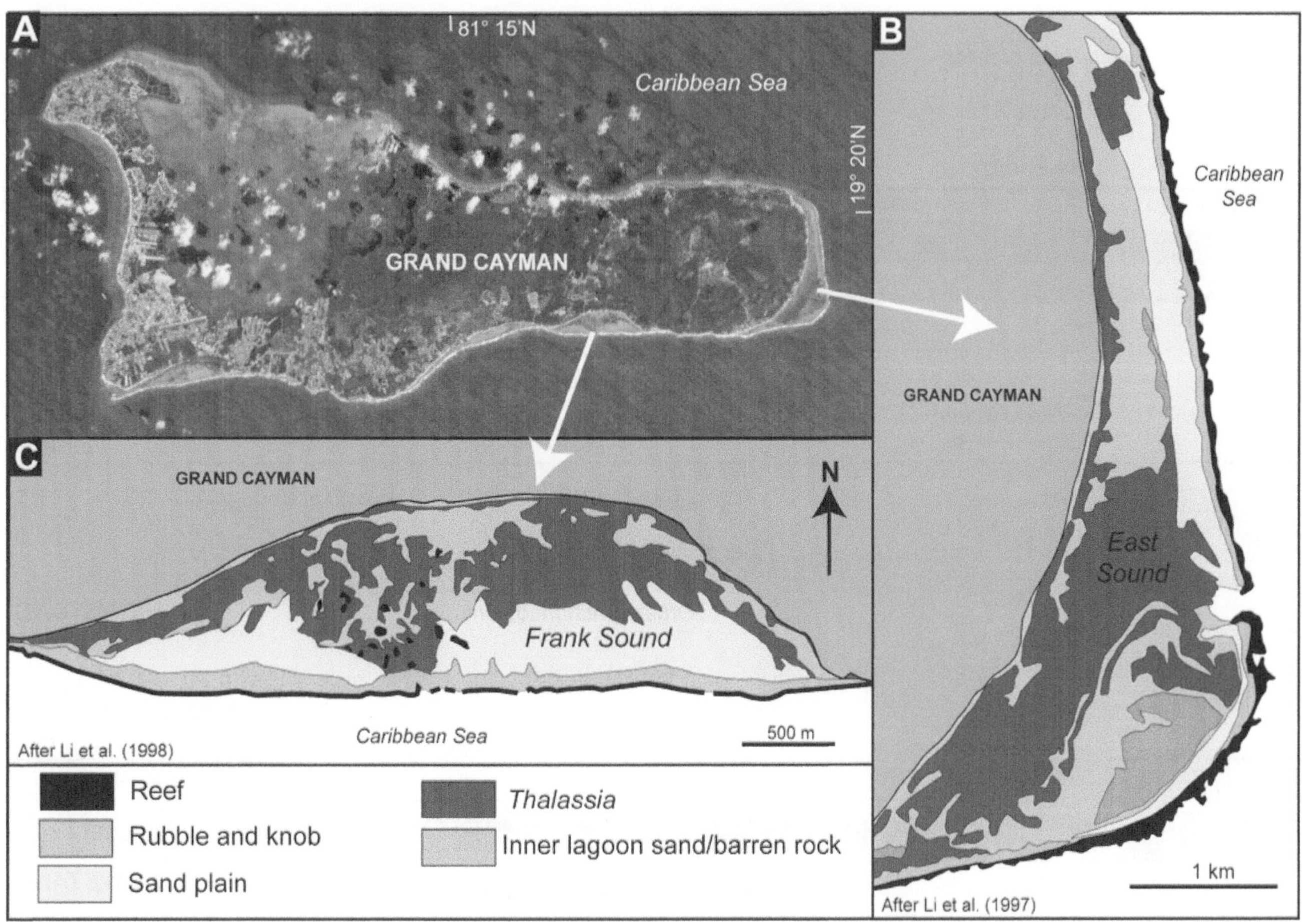

Figure 17. Lithofacies in small lagoons of an isolated oceanic island – Grand Cayman. **A.** Satellite image showing small lagoons around periphery of island. Image courtesy of the Image Science and Analysis Laboratory, NASA Johnson Space Center [image ISS014-E-20404; http://eol.jsc.nasa.gov]; **B.** Lithofacies map of East Sound; **C.** Lithofacies map of Frank Sound.

preclude *Halimeda* growth and benthic foraminifera become dominant. Channels through the outer reef have a significant impact on sedimentation. Rasdhoo Atoll (Fig. 16D) with channels in the NE and SE quadrants, allow ocean currents to enter the lagoon and sediment-laden currents to flow out. Thus, bare rock is found on the eastern part of the lagoon whereas mud accumulates in the low-energy regimes, in the western part of the lagoon.

Sediment in the lagoon of Rasdhoo Atoll is formed largely of skeletal grains derived from corals, *Halimeda*, molluscs, and foraminifera. Missing from the Maldives is *Penicillus* – a green calcareous algae from which so much fine-grained sediment is derived in Caribbean shallow-water environments. Faecal pellets are rare and ooids are absent, as they are in many other Indo-Pacific reefs and platform systems. Rankey and Reeder (2009) suggested that ooids will only form where there is a source of nuclei, agitation, and supersaturated ocean waters. By suggesting that the carbonate supersaturation state is the main reason for the paucity of ooids in Pacific depositional systems, Rankey and Reeder (2009) may have resolved the "oolite problem" as posed by Milliman (1974).

Small, Isolated Oceanic Islands (Figure 17)

Small islands, isolated by deep oceanic waters are common throughout the oceans of the world. In tropical realms, carbonate sediments commonly form on the narrow, shallow shelves and in the small lagoons that fringe these islands. Under everyday conditions, sedimentation operates as in other carbonate factories and facies patterns, akin to those found on their larger brethren, result. Grand Cayman (Fig. 17A), located in the north central part of the Caribbean Sea, is a good example of such an island system.

Storms and hurricanes radically impact sedimentation in the shallow depositional systems typically found around these islands. Storm generated waves and currents can entrain virtually all of the bottom sediments and transport them onshore or flush them out into the fore reef area. Mud- and silt-sized sediment is preferentially winnowed out, leaving behind homogenized sands and rubble. For example, sands and rubble dominate the lagoons around Grand Cayman and it is difficult to find any mud, although the green algae are present.

Epeiric Shelves and Seas (Figure 18)

Epeiric seas, also known as intracratonic and interior cratonic basins, are difficult to model because none exist today. Although the definition of an

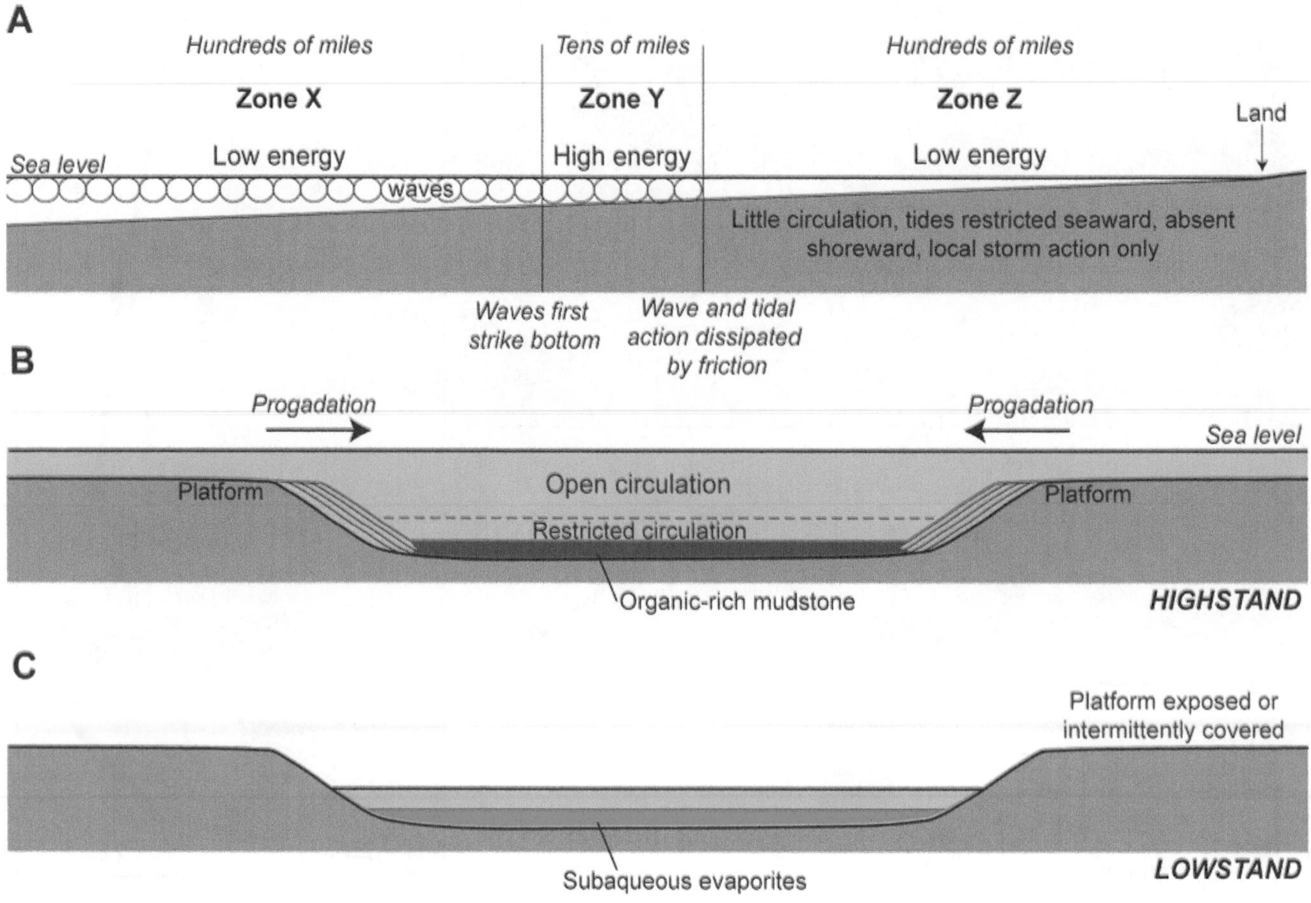

Figure 18. Depositional models for epeiric seas. **A.** Depositional scheme proposed by Irwin (1965, Fig. 3); **B, C.** Depositional schemes in epeiric seas under highstand and lowstand phases, from Wright and Burchette (1996, Fig. 9.8).

epeiric sea is open to some debate (Pratt and Holmden, 2008), it is generally considered to be a vast (100–1000 km wide), enclosed, shallow sea, commonly < 10 m deep. Some authors have suggested that Hudson Bay, Canada, the Baltic Sea, and the Gulf of Carpentaria, Australia, are modern examples of such basins – but even if so, they are not domains where carbonates are forming.

The interpretation of ancient carbonate successions that formed in these vast epeiric seas is largely based on ancient deposits, including those found in the Cambrian–Ordovician of North America and parts of the Permian and Tertiary succession of the Middle East. In general, epeiric seas were characterized by:

1. Vast, flat, shallow subtidal zones with the only relief being inherited from pre-existing topographic lows and scattered skeletal sand shoals;
2. Intertidal zones that may have been tens of kilometres wide; and
3. Wide supratidal zones.

It is doubtful that subsidence would have been uniform and local highs and lows probably developed in response to differential tectonic movement. Deep basins that may have existed in some areas were surrounded by rimmed shelves or ramps. Shaw (1964) and Irwin (1965) were the first to develop depositional schemes for these vast shallow seas, dividing them into:

1. Zone X for the open ocean below wave base;
2. Zone Y for the narrow zone that was under tidal influence; and
3. Zone Z for the platform beyond the effects of tidal exchange (Fig. 18A).

Storm-generated waves and currents would have played a major role in sedimentation on the epeiric platform. Water would have piled-up in a downwind direction, perhaps leaving upwind areas temporally barren of water. As storms abated, water would have sloshed back across the platform. Such high-energy periods would have led to homogenization of the sediment, in much the same way as hurricane-generated storms homogenize sediments on modern platforms. Other researchers have argued that deposition in these settings was controlled by storm and tidal activity with each island and bank giving rise to its own succession of prograding intertidal and supratidal deposits (Fig. 18B). Water depth is critical in this argument because it influenced water circulation and the type of biota that inhabited an area (Fig. 18B and C). During highstands, lateral progradation from coastal areas commonly overstepped the organic-rich mudstones that had accumulated in the central, deeper part of the sea where organic-rich mudstones accumulated (Fig. 18B).

Successions that formed in epeiric

seas commonly appear to have formed in 'normal' seawater. Evaporites may, or may not, be present in these successions. The Permian Khuff Formation, for example, located in the Arabian Gulf, is 1500 m thick. It is formed of stacked shallowing-upward cycles, each 2–10 m thick, that have oolitic limestone at their bases grading upward into laminite and sabkha deposits. This succession contains several evaporitic intervals, including some that probably formed under subaqueous conditions.

Much remains to be understood regarding the factors that controlled carbonate sedimentation across the vast expanses of epeiric seas. Although the lack of modern analogs is a hindrance, application of the principles derived from other modern depositional systems should provide insights into these complex depositional regimes.

CARBONATE SANDBODIES

Large sandbodies formed of bioclastic grains or ooids are common in many carbonate depositional systems. Despite their obvious importance and potential as hydrocarbon reservoirs in ancient successions, these sandbodies have received relatively little study. They are particularly common in high-energy environments, such as those found around platform margins and on the interior parts of ramps. Such high-energy conditions are generated by regular tidal oscillations, wind-driven waves, and aperiodic storms. Platform marginal sandbodies on the Great Bahama Bank are up to 2 m high, 75 km long, and 1–4 km wide. Ripples on the surfaces of the sandbodies are testimony to the constant movement of the sediment under high-energy conditions. Channels that cut through these banks are commonly floored by sea grass.

The northern margin of the Little Bahama Bank, characterized by marvellous examples of platform marginal sandbodies (Fig. 19A), has been the subject of numerous recent studies (Rankey *et al.*, 2006; Reeder and Rankey, 2008). The fascinating sedimentological quandaries that arise from these sandbodies are:

1. Their apparent stationary positions despite the fact that they are in high-energy settings;
2. The continual birth and growth of ooids; and
3. The fact that the ooids appear to be confined to the shoals and do not migrate into neighbouring facies.

These features have been explained by the 'spin cycle' model that relies on the fact, that in many tidal systems, tidal flows are mutually exclusive with the flood (ebb) tidal flows being focused along marginal channels, whereas the ebb (flood) flow is funnelled through the main channel (Fig. 20). These opposing flow directions set up a cyclical pattern of currents that allows ooid growth by keeping them constantly agitated while confining them to the shoal area (Fig. 20).

In some situations, ebb-tidal delta lobes will prograde across a lagoon, thereby overstepping the lagoonal facies and almost instantaneously causing a significant shallowing (Fig. 19E and F). Thus, the muds, seagrass banks, scattered corals, and patch reefs of the original lagoon are covered by cross-bedded coarse bioclastic/oolitic sands that have large subaqueous dunes with ripples superimposed on their surfaces (Fig. 19E and F).

FACIES MODELS AND ANCIENT SUCCESSIONS

Modern carbonate systems are the foundation for most facies models and underpin most principles used in the interpretation of ancient carbonate sequences (Figs. 9–18). Nevertheless, over reliance on uniformitarianism and the desire to link sedimentation to one parameter (*e.g.*, sea level) must be avoided because carbonate sediments are the product of many different factors that change with time. Application of facies models (Fig. 21) for the interpretation of ancient warm neritic carbonate deposits should be undertaken within the framework and limitations of the following considerations:

- Facies models based on modern tropical warm-water settings are, in essence, 'snapshots' that provide an overview of a depositional systems at the time of study. As such, they offer a static rather than dynamic view of a depositional system.
- Many aspects of carbonate sedimentation are poorly understood even in modern systems where they are actively forming today.
- The assumption that today's oceans include examples of all the depositional systems that have existed throughout geological history, despite the concerns of Wright and Burchette (1996) warning that some ancient carbonate depositional systems have no modern counterparts.
- The size of a depositional system can strongly influence sediment generation and accumulation; however, even the largest modern depositional systems (*e.g.*, Great Bahama Bank, northeastern Australian Shelf) are small when compared to some of the vast systems that existed in the past; and
- Sediment production is largely tied to the resident microbes, plants and animals. As they have evolved through geologic time, so the nature of sediment production has changed.

Providing these notes of caution are heeded, facies models like those in Figure 21 embody the processes that produce carbonate sediments in many different settings. As such, they provide a basis for the interpretation of ancient successions.

NERITIC WARM-WATER CARBONATE SYSTEMS

Throughout the world there are many examples of thick carbonate successions formed of sediments that originated in the shallow, warm-water environs of platforms, banks, ramps, atolls, or isolated oceanic islands. Selected examples include the 720-m thick Latemar Limestone (Middle Triassic) found in the Dolomites of northern Italy, 500 m (possible 1750 m) of shallow-water Tertiary carbonates found on Grand Cayman Island in the Caribbean, and 4500 m of Late Triassic to Paleogene shallow-water carbonates on the Abruzzi Platform of northern Italy. The Latemar Limestone comprises ~ 600 shallowing-upward cycles (average 0.65 m cycle^{-1}) that reflect the repeated creation of the same environment condi-

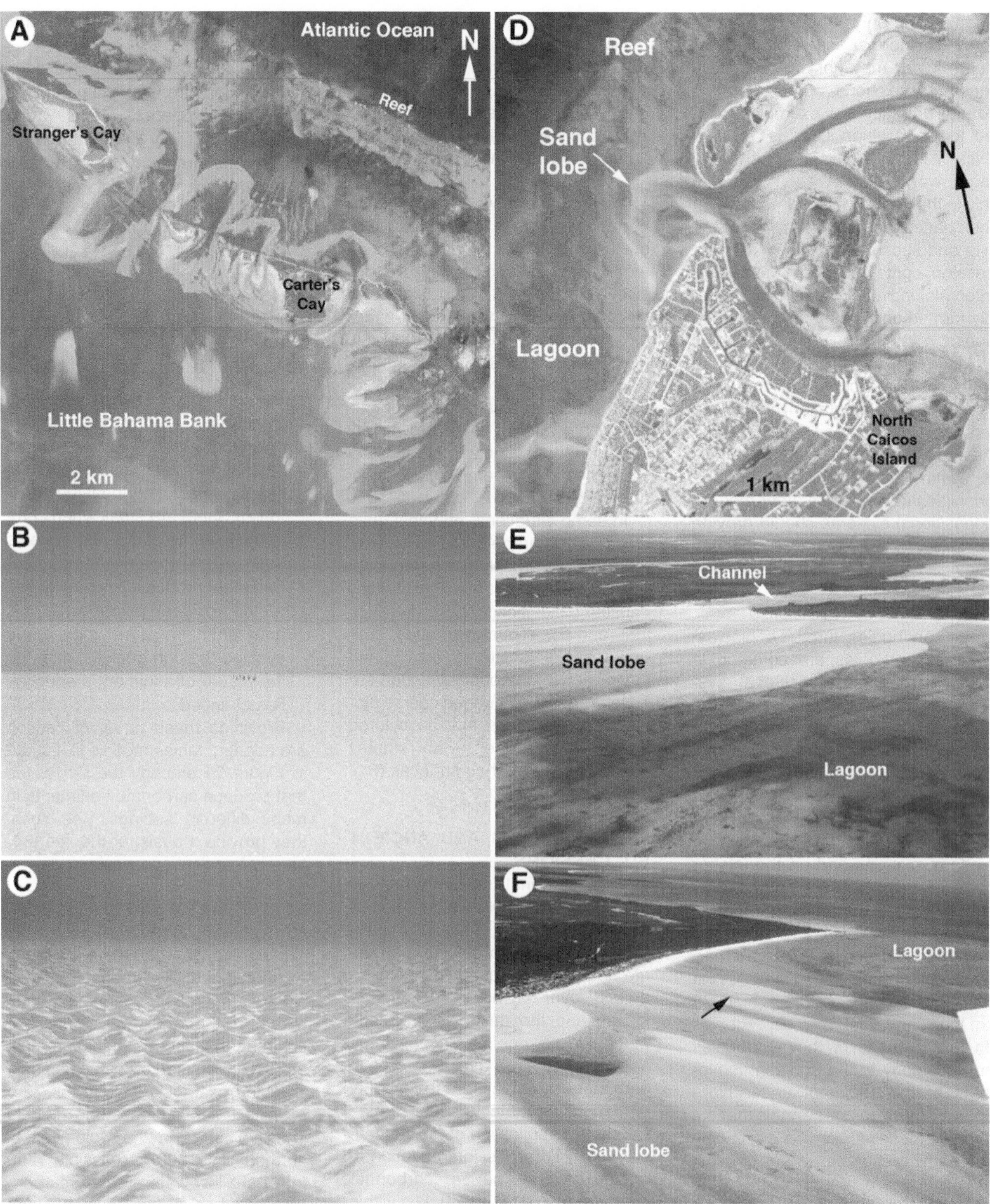

Figure 19. Sandbodies on shallow carbonate shelves. **A.** Ooid sand shoals along northern margin of Little Bahama Bank. Image from Google Earth; **B.** Extensive ooid shoal, Joulter's Cay, Bahamas; note people (centre) for scale; **C.** Rippled surface of ooid shoal indicative of sediment movement, Joulter's Cay, Bahamas; **D.** Ebb-tidal delta lobe prograding into lagoon, North Caicos Island. Image from Google Earth; **E.** Aerial view showing front edge of sand lobe shown in Fig. 19D as it progrades across the lagoonal sediments; **F.** Surface of sand lobe shown in Fig. 19D (arrow), lagoon in background. Note subaqueous dunes on upper surface of sand lobe.

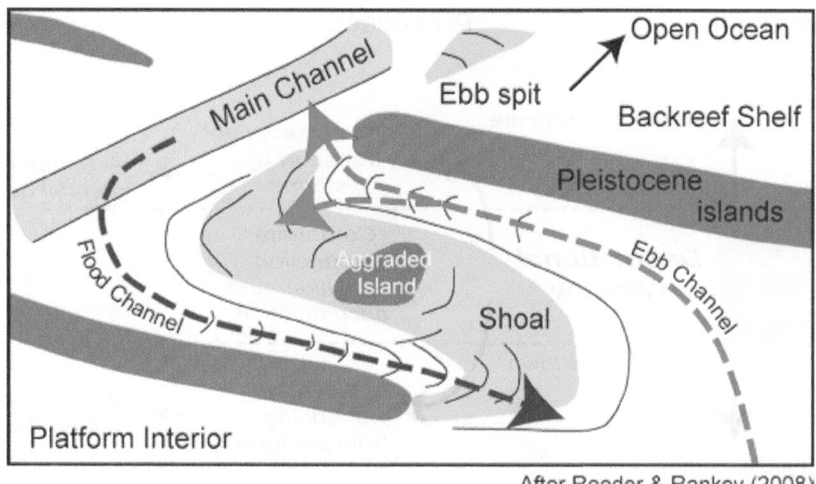

After Reeder & Rankey (2008)

Figure 20. The ‘spinning top’ model for development of ooid sand body near edge of Bahama Bank.

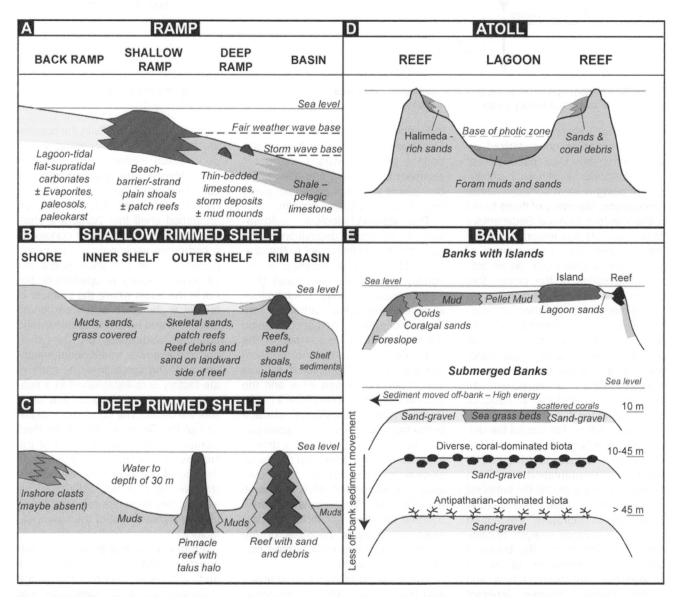

Figure 21. Idealized facies models for **A.** ramp; **B.** shallow rimmed shelf; **C.** deep rimmed shelf; **D.** atoll, and **E.** banks of various depths.

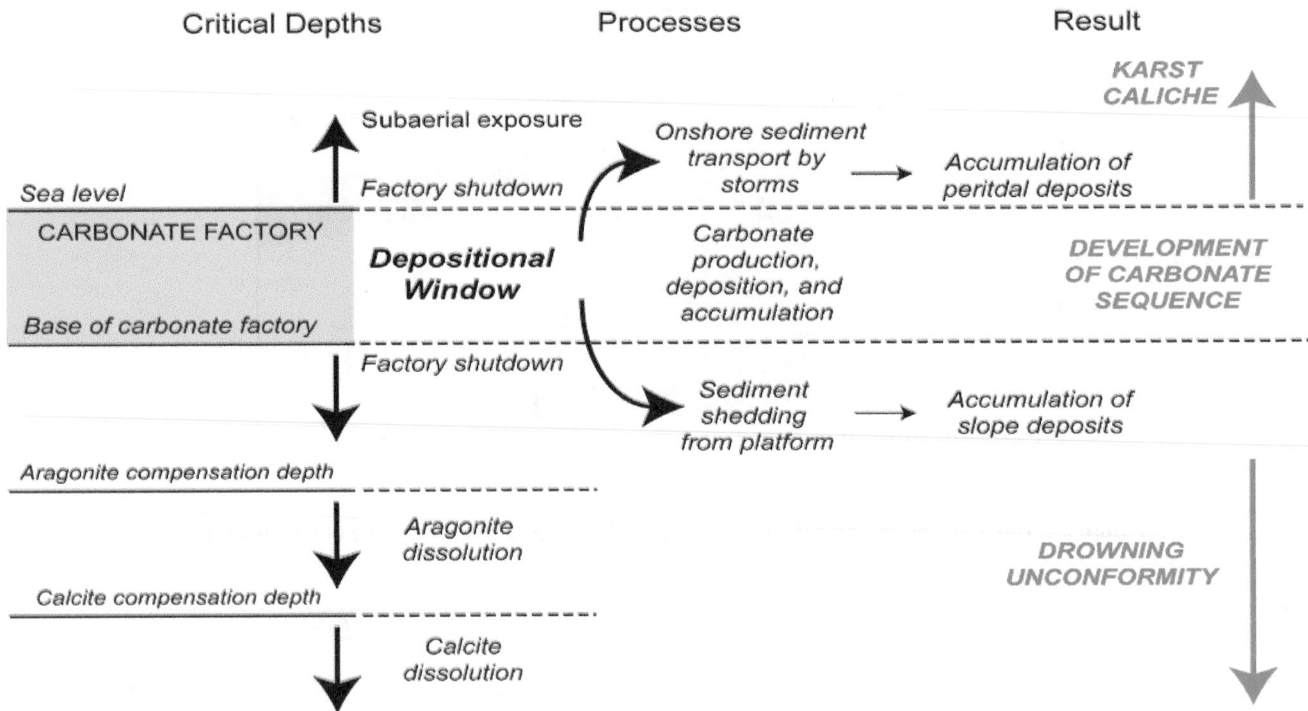

Figure 22. Relationships between the critical water depths, processes, and development of carbonate sequences and sequence boundaries. Base of carbonate factory = base of photic zone.

tions. The Tertiary succession on Grand Cayman is characterized by unconformity-bounded packages of strata (formations), with each unconformity being a karst surface. Other successions, like many of those found in the Western Canada Sedimentary Basin (WCSB) are important hydrocarbon reservoirs. Interpretation of these successions relies on an intimate knowledge of the factors that control their temporal development.

Sediment accumulation is controlled by the accommodation (= space between sea level and the depositional surface, or simply, water depth at any given time) and sediment production. Disequilibrium can be triggered by tectonic, eustatic, or climatic changes that disturb the autogenic parameters that control the delicate balance between sea level, the depositional surface, and the rate of sediment accumulation. Such dynamic changes dictate the temporal evolution of the facies architecture. Sequence stratigraphy, developed largely from seismic analysis of siliciclastic systems, traces the temporal and spatial migration of facies through the integration of time and relative sea-level changes. Inherent to the siliciclastic system is the notion that sediment supply, via rivers, is independent of sea-level change. This premise is far from true in carbonate systems where autogenic sediment is the norm and even slight changes in sea level can affect every part of the depositional system.

The geomorphological configuration of carbonate depositional systems is critically important to the spatial and temporal development of carbonate sedimentary sequences (Fig. 22). On isolated banks, which typically have steep sides, carbonate production, deposition, and accumulation will take place while the top of the bank (*i.e.*, the depositional surface) remains in the depositional window that is defined by sea level and the base of the photic zone (Figs. 21, 22). Dropping the depositional surface below the base of the photic zone can lead to the development of a drowning unconformity. Elevating the depositional surface above sea level allows development of a subaerial unconformity. Thus, isolated banks become excellent 'dipsticks' for past sea-level changes because of the unconformities that exist in their carbonate successions.

The situation of attached shelves, with land on one side and an ocean basin on the other, is far more complex because sea-level rise and fall can lead to lateral shifts in the carbonate factory and thus, changes in the depositional system (Figs. 21 and 22). Hence, the geomorphological configuration of the system becomes critical. Consider, for example, an unrimmed shelf that borders a shallow basin under highstand conditions – carbonate sediments will form and accumulate on the shelf where the carbonate factory is operational but not in the basin where the seafloor is far below the base of the carbonate factory. If sea level falls to a lowstand position, sediments on the shelf become exposed to subaerial weathering. At the same time, the carbonate factory is re-established in a seaward position in the area that once was the basin. The complex stratigraphy of the Devonian Alexandra Reef Complex in the southern part of the Northwest Territories of Canada developed through this mechanism (MacNeil and Jones, 2006). Conversely, sea-level rise can lead to the landward movement of the carbonate factory.

From a sequence stratigraphy perspective, drowning unconformities and subaerial unconformities are important 'sequence boundaries' that delineate the genetic packages within a succession. Such unconformities will, in the context of the ensuing highstand, be the 'antecedent topography' that will have a significant impact on the next phase of sedimentation. The 'facies architecture' within each sequence will reflect the temporal and spatial variations that took place in the carbonate factory because of changes in sea level, the depositional surface (*i.e.*, accommodation space), and sediment production.

Importance of Sea-level Change

Changes in sea level and tectonism are the main allocyclic controls that affect cycle development in shelf carbonate successions. Sea-level change, commonly the preferred explanation, is controlled by climate that dictates continental glaciation and hence the volume of water in the ocean. Tectonic adjustment of vertical and horizontal plate positions causes changes in the volume of the ocean basins. Sea-level changes caused by melting or freezing of the ice caps are generally more rapid than those caused by tectonism.

Milankovitch cyclicity that operates through the integrated effects of the astronomic precession of the equinoxes (23 000–19 000 years periodicity), obliquity of the elliptic (41 000 years), and eccentricity (100 000 years), cause periodic rapid rises in sea level that are followed by periods of stability. Milankovitch cycles have been detected in the oxygen isotope record of Pleistocene sediments from the deep sea and from Pleistocene coral reef terraces off Barbados and New Guinea. Although there is general agreement as to the timing of these cycles over the last 500 000 years or so, the positions of sea level during the various highstands and lowstands are open to debate. Allocyclic controls on sedimentation such as eustasy might be expected to produce similar cycles over widespread areas as each region should have been affected by the same rise in sea level. This is, however, not the case and it has been amply demonstrated that many sedimentary cycles are laterally variable. Such geographic differences may be due to the effects of local tectonism and differences in geodetic sea level (sea level that varies at any given time due to effects of gravity caused by mantle−core interactions).

Faulting can have a major impact on carbonate sedimentation because such movement can rapidly move a depositional surface into, or out of, the depositional window. The tectonic controls on carbonate cycles, however, are less well understood. This is because of the difficulty of invoking a recurring yet unpredictable, short-term mechanism that would account for the systematic repetition of depositional conditions needed to produce the similar small-scale sedimentary cycles that are common in shelf successions. This might result from 'in-plate stress', whereby the stress conditions common to all lithospheric plates is changed as a result of plate movement. An increase in pressure of ~ 1 kbar will lead to ~ 20 m of basin margin uplift and ~ 75 m subsidence in the basin centre. The reverse happens when stress is released. Such changes can potentially account for cyclic development of sedimentation on carbonate shelves.

Sequence Boundaries

Drowning unconformities, karst surfaces, and caliche horizons are sequence boundaries that delimit the bounding surfaces of genetically related sediment packages.

Drowning Unconformities

Drowning of the carbonate depositional surface causes the carbonate factory to shutdown because;

1. It deprives the photozoans of the light needed for photosynthesis; and
2. The demise of the benthos, especially if changes in upwelling and circulation patterns increase the amounts of nutrients and organic matter, or cause oxygen-deficiencies (Zempolich, 1993).

A drowning unconformity may be recognized by:

1. An abrupt change from a shallow- to a deep-water biota;
2. A change from shallow- to deep-water facies;
3. A hardground;
4. A mineralized surface;
5. An erosional surface, and
6. Dissolution of aragonite and concomitant calcite precipitation if water depths are between the aragonite and calcite compensation depths or dissolution of aragonite and calcite if depths are below the calcite compensation depth (Fig. 22).

During the Jurassic, isolated carbonate seamounts on the Italian continental crust underwent two stages of drowning: *incipient drowning* with water depths in the euphotic zone but with abnormally low sediment production and accumulation rates, and *complete (terminal) drowning* when the tops of the seamounts sank below the euphotic zone (Bice and Stewart, 1990). In this case, development of the drowning unconformity was controlled by subsidence and sea-level change, seamount configuration, lack of an elevated rim, absence of reef-building organisms, and the rapid formation of deep basins. Demise of the northern Tethyan Urgonian carbonate platform, in the Helvetic Alps of central Switzerland involved four stages of platform demise with each drowning phase being characterized by erosion and phosphogenesis (Föllmi and Gainon, 2008).

Subaerial Unconformities

During lowstands, aragonite-rich sediments exposed to the atmosphere and flushed by freshwater will undergo rapid lithification that may coincide with, or be followed by, weathering that, depending on climate, leads to caliche/calcrete crust (dry climates) or karst (wet climates) development. Lithification and weathering can be forestalled if the fall in sea level is accompanied by progradation of siliciclastic systems from the adjacent coast, as on the carbonate platforms of northeastern Australia.

Weathering typically produces a karst landscape characterized by rugged topography with significant relief. Today, karst towers, up to 35 m high, characterize the Stone Forest (China). On isolated islands, differential karst development between the rim and core can produce an 'atoll-shaped' topography with the coastal rim originally formed by erosion, not reef growth. In coastal situations, tower karst will develop in the marginal plain whereas conical karst will form in the more seaward areas. In the context of the next transgression, this topography is the antecedent surface that may significantly impact depositional patterns. Consider for example, a situation whereby a karst surface with towers up to 30 m high is rapidly inundated by a sea-level rise of 35 m. Tower tops would be covered by 5 m of water, whereas the inter-tower areas would be covered by 35 m of water. Photozoans would preferentially colonize the topographic highs whereas fine-grained sediment would accumulate in the low areas where fewer organisms lived.

The Tertiary succession on Grand Cayman Island is, for example, formed of a series of unconformity-bounded packages of carbonate rocks. The 'Cayman Unconformity' with a relief of ≥ 40 m (Fig. 23) separates Miocene from Pliocene carbonates and developed during the Messinian (terminal Miocene), when sea level was ≥ 40 m lower than today. Deposition began during the Early Pliocene when sea level rose to ~ 15 m above the present day sea level and sediment then started to fill in the topographic lows. That process continues today as modern sedimentation is still filling the basin on Grand Cayman.

In dry climates, calcrete (caliche) crusts (generally < 15 cm thick) can form on exposed surfaces and generally have subdued relief. The Pleistocene succession on Grand Cayman is formed of a series of depositional packages that are separated by unconformities marked by calcrete crusts.

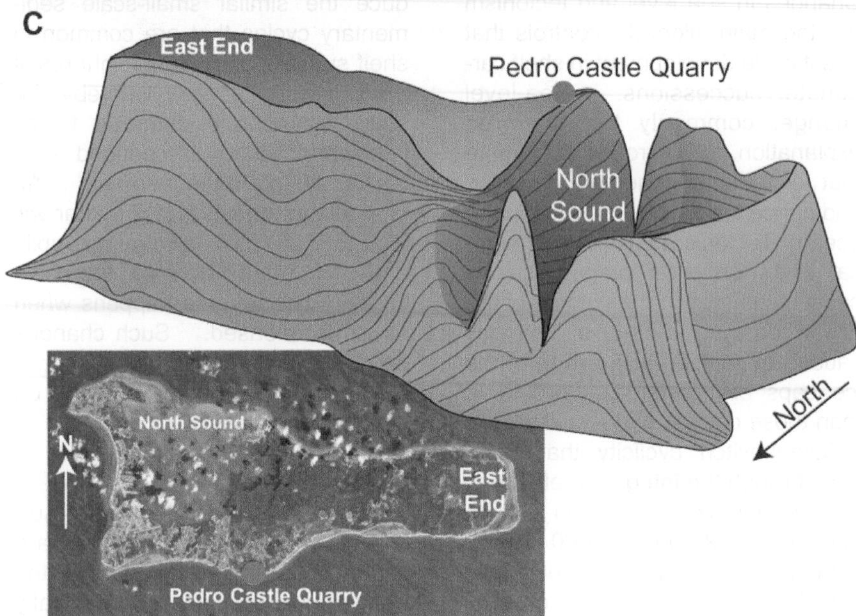

Figure 23. The Cayman Unconformity on Grand Cayman – an example of a paleokarst surface that forms a sequence boundary. **A.** Cayman Unconformity in Pedro Castle Quarry, separating Cayman Formation from Pedro Castle Formation; **B.** Vertical surface showing boring by lithophagid bivalves extending from unconformity surface; **C.** Schematic diagram showing topography on the Cayman Unconformity based on elevations derived from outcrops and wells (Jones and Hunter 1994, Fig. 12). Satellite image shows location of Pedro Castle Quarry and North Sound.

Facies Architecture of Unconformity-Bounded Sequences

Highstand conditions typically promote carbonate sediment production and accumulation because vast tracts of shallow-water environments exist on isolated platforms and attached shelves. A transgression, leading to the flooding of these areas, however, does not result in the immediate establishment of a carbonate factory because time (lag time) is required for the photozoans to colonize and become viable, long-standing sediment producers. A water depth (lag depth) of 1–2 m is probably needed for a fully functional carbonate factory to develop. The physical record of the lag time, which is generally poor, can be variable. Some transgressive Holocene successions have a basal mangrove peat overlying the unconformity. In other cases, flooding of lithified rock surfaces can promote colonization by boring organisms (*e.g.*, sponges and bivalves) that produce a bored surface similar to those associated with hardgrounds (Fig. 23B and C).

Once established, the carbonate factory will continue to produce sediment providing suitable environmental conditions are maintained. It must be remembered, however, that sediment production is not solely the function of sea level, as other factors such as salinity, water temperature, and turbidity also influence sediment production because of the control they exert on the resident photozoans. Such parameters operate over a preferred range, such that, 'too little-too low' is just as detrimental as, 'too much-too high'. Thus, for most parameters there are two 'off' switches – one at each end of the optimal range. For the animals and plants in the carbonate factory, high salinity is just as detrimental to their existence as low salinity (Fig. 4).

Vertical accretion of carbonate sediment depends largely on the balance between sediment production rate, sea-level position, subsidence rate, and lateral sediment transportation (Fig. 24A). This is immediately evident from considering the configuration of strata associated with each stage of a transgressive–regressive cycle. Under highstand (HST) conditions, progradation takes place due to highstand shedding of sediment (Fig. 24A). In contrast, during falling (FSST) and lowstand (LST) stages, previously deposited carbonate sediments will be exposed that may then, depending on climatic conditions, be modified by karst or caliche development. At the same time, the depositional systems will downstep with the configuration of accumulated sediments being a function of the amount of sediment being produced. During the transgressive (TST) stage, the balance between the rates of sediment accumulation (SA) and accommodation (AC) will control the geometry of the strata, as follows (Fig. 24A):

- If SA >> AC, a prograding TST will form.
- If SA = AC, an aggrading TST will form.
- If SA < AC, a series of backstepping carbonate accumulations may form; and
- If SA << AC, a drowning unconformity will form.

Peritidal and subtidal cycles are common features of many carbonate platform successions. The classic peritidal cycle (Fig. 24B), which evolves as a tidal flat progrades across a platform, is well known from many studies (*see* Chapter 16). The basal deep-marine subtidal deposits of these cycles grade upward into progressively shallower deposits before terminating in tidal-flat facies that display clear evidence of subaerial exposure (Fig. 24B). Osleger (1991), however, pointed out that many platform successions are composed of subtidal cycles, so named because they are formed entirely of subtidal lithofacies (Fig. 24C). An important aspect of subtidal cycles is that their constituent facies and the thicknesses of those facies change across the breadth of the shelf. In a Late Cambrian ramp system from Utah, shallow ramp cycles include oolitic grainstone and oncolitic–skeletal packstone–grainstone, whereas the mid- and deep-ramp segments are characterized by interbedded burrowed wackestone–packstone and nodular and argillaceous mudstone–wackestone (Fig. 24C). As subtidal cycles develop, it seems that the depositional system never becomes shallow enough for the depositional surface to be in the peritidal window of deposition. This may be due to reworking and redistribution of sediment once the depositional surface intersects an energy barrier, such as the normal fair-weather or storm-wave reworking. In the Persian Gulf, coral–algal sands of the Great Pearl Bank lie in water 8–20 m deep, suggesting that there is little aggradation above the depth of normal wave reworking. Similarly, the depositional surfaces of many banks in the Caribbean Sea never reach sea level because currents are constantly sweeping sediment off the bank top into the surrounding deep ocean waters. *Subtidal cycles appear to be common components of carbonate ramps whereas peritidal cycles seem to be a feature of flat-topped rimmed platforms.* This may be a function of the energy regimes that are associated with each of these depositional systems.

Cycles in platform carbonate successions, irrespective of their style, reflect the superimposition of high-frequency changes in accommodation space on a low-frequency change. The high-frequency changes are probably attributable to subsidence or eustatic changes in sea level. It should be noted, however, that cycle thickness and the proportions of the different types of facies will vary throughout a regressive–transgressive cycle as a result of changes in accommodation (Fig. 24D). During the transgressive and early part of the highstand stages, the increase in accommodation usually leads to the development of thick successions of subtidal deposits (Fig. 24D).

SUMMARY

Sediments in the warm, clear turquoise waters of tropical platforms and banks have been the focus of considerable research for many years because they are easy to access and the climate is generally enjoyable. Interpreters of ancient carbonate successions face the challenge of deciphering the factors that controlled sediment production, accumulation, and complex facies mosaics. This is no easy task because carbonate sediments form in response to a multitude of different but interrelated factors (Fig. 25). The biota is of fundamental importance as it contributes directly to the sediment and commonly mediates deposition. As the biota changes with evolution, climate, water temperature, salinity, nutrient level, light, and oxygenation so does the produced sediment change (Fig. 25). Extrinsic factors, such as sea-level change, platform geomorphology, subsidence, and uplift collectively control the manner in which carbonate successions develop through time. Although much is known about the factors that control carbonate sediment production in these settings, many issues remain unresolved and are in need of further research. The importance of nutrients, for example, has been largely ignored despite the fact that nutrients are fundamental to the health of the resident plants and animals that greatly impact sediment production and accumulation.

In looking at ancient carbonate successions, the challenge is to determine which parameters controlled formation of the different types of sediment and the overall architecture of the sequence. With our ever-increasing knowledge of modern carbonate sediments and sequence stratigraphy, such interpretations allow a better understanding of how carbonate sediments evolved through time. Such interpretations are playing an increasingly important role in the exploration of carbonate successions for the valuable hydrocarbon reserves and ore bodies that they commonly contain.

ACKNOWLEDGEMENTS

I am extremely grateful to the Natural Sciences and Engineering Research Council of Canada for funding that has permitted ongoing research on carbonate sediments and their environments and allowed the preparation of this manuscript. Brian Pratt kindly provided me with preprints of

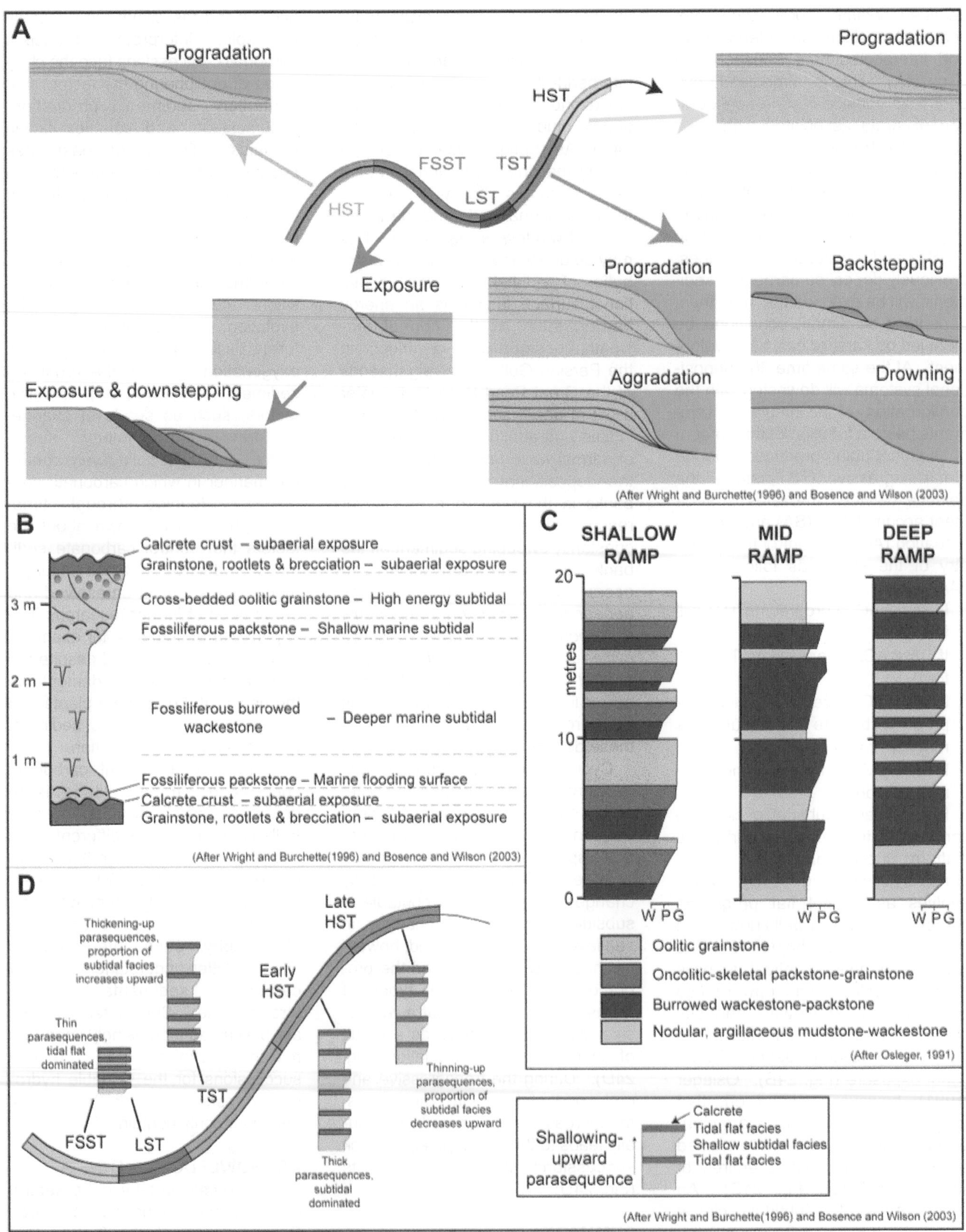

Figure 24. A. Development of carbonate depositional tracts during an idealized cycle of relative sea-level change. See text for factors that control processes during the TST; **B.** Idealized peritidal cycle; **C.** Subtidal cycles found on different parts of a ramp; **D.** Stacking patterns of peritidal cycles. Curve is long-term (3rd order) relative sea-level cycle; the sections show thickness and composition (subtidal vs. intertidal) of the 4th or 5th order cycles within each systems tract.

Biota	*In carbonate factory, composition (aragonite, HMC, LMC) and morphology of skeletal parts of calcareous microbes, plants, and calcifying plants exert primary control over sediment production and composition.*
Evolution	*Evolutionary changes meant that different plants and animals dominated sediment contributions at different times.*
Latitude	*Influences water temperature that controls plant and animal communities and hence, sediment production and accumulation. Basic division into warm water and cool warm realms.*
Eustatic sea level changes	*Controls sea level thereby exerting primary control over operations in the carbonate factory, accommodation space, development of cycles, and hence, overall configuration of carbonate succession.*
Temperature/salinity	*Control distribution of microbes, plants, and animals, hence controls production and distribution of sediment.*
Nutrient	*Nutrient levels control plant and animal communities and hence control production and distribution of sediment.*
Light penetration	*Light penetration controls depth to base of carbonate factory; influenced by nutrient levels and detritus in water column.*
Water circulation/ oxygenation	*Well-oxygenated waters essential for growth of all calcifying invertebrate animals and most plants. Circulation essential for benthos.*
Geomorphology	*Geomorphological configuration of depositional system exerts strong control over facies distribution (e.g., windward and leeward sides of islands).*
Siliciclastic sediment supply	*High influx of siliciclastic sediment reduces depth of light penetration, masks carbonate sediment production, and inhabits growth of most calcifying invertebrate animals.*
Basin margin subsidence	*Creates accommodation space needed for accumulation of thick carbonate successions. Too rapid subsidence may, however, produce drowning unconformity*
Fault-related uplift	*Uplift may halt sediment production; increase rate of siliciclastic sediment supply to shelf and/or lead to formation of calcrete crusts or karst terrains on earlier formed carbonates.*
Fault-related subsidence	*Rapid fault-related subsidence may drown carbonate platform, halting carbonate sediment production and leading to development of drowning unconformity.*

Figure 25. Summary of factors that control development and accumulation of carbonate sediments in warm-water neritic settings.

papers that are to appear in a book that he and Chris Holmden are assembling on the poorly understood epeiric seas of the past. Noel James encouraged me to write this chapter and then provided critical insights that greatly improved its final form and content.

REFERENCES

Basic sources of information

Aigner, T., 1985, Storm Depositional Systems. Dynamic Stratigraphy in Modern and Ancient Shallow-Marine Sequences: Lecture Notes in Earth Sciences, v. 3, Springer-Verlag, Berlin, Heidelberg, 174 p.

A readable process-oriented summary of modern and ancient storm deposits, concentrating on ramp systems.

Bosence, W.J., and Wilson, R.C.L., 2003, Carbonate depositional systems, *in* Coe, A.L., *ed.*, The Sedimentary Record of Sea-level Change: Cambridge University Press, Cambridge, p. 209-233.

This chapter and the companion chapter on sequence stratigraphy provide a concise, easily understandable summary of the main aspects of carbonate sedimentary and sequence development.

Handford, C.R., and Loucks, R.G., 1993, Carbonate depositional sequences and systems tracts – responses of carbonate platforms to relative sea-level changes, *in* Loucks, R.G., and Sarg, J.F., *eds.*, Carbonate Sequence Stratigraphy: Recent Developments and Applications. American Association of Petroleum Geologists Memoir 57, p. 3-41.

One of the first attempts to place carbonate depositional systems in the con-

text of sequence stratigraphy, it provides eloquent examination of the main principles involved.

Jones, B., and Desrochers, A., 1992, Shallow carbonate platforms, *in* Walker, R.G., and James, N.P., *eds.*, Facies Models. Response to Sea level Change: Geological Association of Canada, p. 277-301.

This chapter is the precursor to the current chapter.

Mutti, M., and Hallock, P., 2003, Carbonate systems along nutrient and temperature gradients: some sedimentological and geochemical constraints: International Journal of Earth Sciences, v. 92, p. 465-475.

A thought-provoking paper that examines the role that nutrients play in the development and growth of the sediment forming in carbonate systems.

Pomar, L., 2001, Types of carbonate platforms: a genetic approach: Basin Research, v. 13, p. 313-334.

An alternate approach to the classification of carbonate platforms is outlined in this very interesting paper.

Scholle, P.A., and Ulmer-Scholle, D.S., 2003, A color guide to the petrography of carbonate rocks: grains, textures, porosity, diagenesis: American Association of Petroleum Geologists Memoir 77, 474 p.

A magnificence reference book that superbly images the main components found in carbonate sediments and rocks.

Wright, V.P., and Burchette, T.P., 1996, Shallow-water carbonate environments, *in* Reading, H.G., *ed.*, Sedimentary Environments: Processes, Facies and Stratigraphy: Blackwell Science Ltd., Oxford, p. 325-394.

This excellent chapter provides a comprehensive overview of carbonate depositional systems and explains the development of carbonate successions in the context of sequence stratigraphy.

Wright, V.P., and Burchette, T. P., 1997, Carbonate Ramps: Geological Society of London, Special Publication 149, 328 p.

The chapters in this book present many case studies from many parts of the world of carbonate ramp systems of different geological ages.

Other references

Bice, D.M., and Stewart, K.G., 1990. The formation and drowning of isolated carbonate seamounts: tectonic and ecological controls in the northern Apennines, *in* Tucker, M.E., Wilson, J.L., Crevello, P.D., Sarg, J.R., and Read, J.F., *eds.*, Carbonate Platforms, Facies, Sequences and Evolution: International Association of Sedimentologists, Special Issue 9, p.145-168.

Burchette, T.P., and Wright, V.P., 1992, Carbonate ramp depositional systems: Sedimentary Geology, v. 79, p. 3-57.

Corlett, H., and Jones, B., 2007, Epiphyte communities on *Thalassia testudinum* from Grand Cayman, British West Indies: their composition, structure, and contribution to lagoonal sediments: Sedimentary Geology, v. 194, p. 245-262.

Enos, P., 1977, Holocene sediment accumulations of the South Florida Shelf Margin, *in* Enos, P., and Perkins, P.D., *eds.*, Quaternary Sedimentation in South Florida, Geological Society of America Memoir 147, p. 1-129.

Föllmi, K.B., and Gainon, F., 2008, Demise of the northern Tethyan Urgonian carbonate platform and subsequent transition towards pelagic conditions: The sedimentary record of the Col de la Plaine Morte area, central Switzerland: Sedimentary Geology, v. 205, p. 142-159.

Ginsburg, R.N., 1956, Environmental relationships of grain-size and constituent particles in some Florida carbonate sediments, American Association of Petroleum Geologists Bulletin, v. 40, p. 2384-2427.

Ginsburg, R.N., and James, N.P., 1974, Holocene carbonate sediments of continental shelves, *in* Burk, C.A., and Drake, C.L., *eds.*, The Geology of Continental Margins: Springer-Verlag, Berlin, p. 137-155.

Gischler, C., 2006, Sedimentation on Rasdhoo and Ari atools, Maldives, Indian Ocean: Facies, v. 52, p. 341-360.

Hallock, P., and Schlager, W., 1986, Nutrient excess and the demise of coral reefs and carbonate platforms: Palaios, v. 1, p. 389-398.

Hallock, P., 1987, Fluctuations in the trophic resource continuum: a factor in global diversity cycles?: Paleoceanography, v. 2, p. 457-471.

Heckel, P. H., 1972, Recognition of ancient shallow marine environments, *in* Rigby, J.K., and Hamblin, W.K., *eds.*, Recognition of Ancient Sedimentary Environments: SEPM Special Publication 16, p. 226-286.

Hillis, L., 1997, Corals reefs from a calcareous alga perspective and a first carbonate budget: Proceedings of the 8th International Coral Reef Symposium, p. 761-766.

Irwin, M.L., 1965, General theory of epeiric clear water sedimentation: American Association of Petroleum Geologists Bulletin, v. 49, p. 445-459.

Jones, B., and Hunter, I.G., 1994, Messinian (Late Miocene) karst on Grand Cayman, British West Indies: an example of an erosional sequence boundary: Journal of Sedimentary Research, v. 64, p. 531-541.

Li, C., Jones, B., and Blanchon, P., 1997, Lagoon-shelf sediment exchange by storms – evidence from foraminiferal assemblages, east coast of Grand Cayman, British West Indies: Journal of Sedimentary Research, v. 67, p.17-25.

Li, C., Jones, B., and Kalbfleisch, W.B.C., 1998, Carbonate sediment transport pathways based on foraminifera: case study from Frank Sound, Grand Cayman, British West Indies: Sedimentology, v. 45, p. 109-120.

Logan, B.W., Harding, J.L., Ahr, W.M., Williams, J.D., and Snead, R.G., 1969, Carbonate sediments and reefs, Yucatan Shelf, Mexico: Memoir American Association of Petroleum Geologists, v. 22, p. 1-198.

MacNeil, A.J., and Jones, B., 2006, Sequence stratigraphy of a Late Devonian ramp-situated reef system in the Western Canada Sedimentary Basin: dynamic responses to sea-level changes and regressive reef development: Sedimentology, v. 53, p. 321-359.

MacNeil, A. J., and Jones, B., 2008, Nutrient-gradient controls on Devonian reefs: insights from the ramp-situated Alexandra Reef system (Frasnian), Northwest Territories, Canada, *in* Lukasik, J., and Simo, J.A., *eds.*, Controls on Platform and Reef Development: SEPM Special Publication 89, p. 125-143.

Maxwell, W.G.H., 1968, Atlas of the Great Barrier Reef: Elsevier, Amsterdam, 258 p.

Maxwell, W.G.H., and Swinchatt, J.P., 1970, Great Barrier Reef: variation in a terrigenous carbonate province: Geological Society of America Bulletin, v. 81, p. 691-724.

Milliman, J.D., 1974, Marine Carbonates: Springer, New York.

Neumann, A.C., and Land, L.S., 1975, Lime mud deposition and calcareous algae in the Bight of Abaco, Bahamas: A Budget: Journal of Sedimentary Petrology, v. 45, p. 763-786.

Orme, G.R., and Salama, M.S., 1988, Form and seismic stratigraphy of *Halimeda* in part of the northern Great Barrier Reef Province: Coral Reefs, v. 6, p. 131-137.

Osleger, D., 1991, Subtidal carbonate cycles: implications for allocyclic vs. autocyclic controls: Geology, v. 19, p. 917-920.

Pratt, B.R., and James, N.P., 1986, The St. George Group (Lower Ordovician) of western Newfoundland: tidal flat island model for carbonate sedimentation in shallow epeiric seas: Sedimentology, v. 33, p. 313-343.

Pratt, B.R., and Holmden, C., 2008, Introduction, *in* Pratt, B.R., and Holmden, C., *eds.*, Dynamics of Epeiric Seas, Geological Association of Canada, Special Paper 48, p. 1-5.

Purdy, E.G., 1963a, Recent calcium carbonate facies of the Great Bahama Bank. I. Petrography and reaction groups: Journal of Geology, v. 71, p. 334-355.

Purdy, E.G., 1963b, Recent calcium carbonate facies of the Great Bahama Bank. II. Sedimentary facies: Journal of Geology, v. 71, p. 472-497.

Purdy, E.G., 1974a, Reef configurations:

cause and effect, *in* Laporte, L.F., *ed.*, Reefs in Time and Space, SEMP Special publication 18, p. 9-76.

Purdy, E.G., 1974b, Karst-determined facies patterns in British Honduras: Holocene carbonate sedimentation model: American Association of Petroleum Geologists Bulletin, v. 58, p. 825-855.

Purdy, E.G., and Bertram, G.T., 1993, Carbonate concepts from the Maldives, Indian Ocean: American Association of Petroleum Geologists, Studies in Geology # 34, p. 1-56.

Purser, B.H., 1973, The Persian Gulf: Springer-Verlag, Berlin, 471 p.

Purser, B.H., 1983, Sédimentation et diagenèse de carbonate néritiques récents, v. 1: Editions Technip, Paris, 389 p.

Rankey, E.C., 2006, Form, function and feedbacks in a tidally dominated ooid shoal, Bahamas: Sedimentology, v. 53, p. 1191-1210.

Rankey, E.C., and Reeder, S.L., 2009, Holocene ooids of Aitutaki Atoll, Cook Islands, South Pacific: Geology, v. 37, p. 971-974.

Rankey, E.C., Riegl, B., and Steffan, K., 2006, Form, function and feedbacks in a tidally dominated ooid shoal, Bahamas, Sedimentology, v. 53, p. 1191-1210.

Read, J.F., 1985, Carbonate platform facies models: American Association of Petroleum Geologists Bulletin, v. 69, p. 1-21.

Reeder, S.L., and Rankey, E.C., 2008, Interactions between tidal flows and ooid shoals, northern Bahamas: Journal of Sedimentary Research, v. 78, p. 175-186.

Robbins, L.L., and Blackwelder, P.L., 1992, Biochemical and ultrastructural evidence for the origin of whitings: A biologically induced calcium carbonate precipitation mechanism: Geology, v. 20, p. 464-468.

Roberts, H.H., Phipps, C.V., and Effendi, L., 1987, *Halimeda* bioherms of the eastern Java Sea, Indonesia: Geology, v. 15, p. 371-374.

Schlager, W., 1998, Exposure, drowning and sequence boundaries on carbonate platforms, *in* Camoin, G.F., and Davies, P.J., eds., Reefs and Carbonate Platforms in the Pacific and Indian Oceans: Blackwell Science, Oxford, England, p. 3-21.

Scholle, P., 1978, A color illustrated guide to carbonate rock constituents, textures, cements, and porosities: American Association of Petroleum Geologists, Memoir 27, 241 p.

Shaw, A.B., 1964, Time in Stratigraphy: McGraw-Hill, New York, 365 p.

Shinn, E.A., Steinen, R.P., Lidz, B.H., and Swart, P.K., 1989, Whitings, a sedimentologic dilemma: Journal of Sedimentary Petrology, v. 59, p. 147-161.

Taylor, P.D., and Wilson, M.A., 2003, Palaeocology and evolution of marine hard substrate communities: Earth-Science Reviews, v. 62, p. 1-103.

Tucker, M.E., and Wright, V.P., 1990, Carbonate Sedimentology: Blackwell Scientific Publications, Oxford, 282 p.

Zempolich, W.G., 1993, The drowning succession in Jurassic carbonates of the Venetian Alps, Italy: a record of supercontinent breakup, gradual eustatic rise, and eutrophication of shallow-water environments, *in* Loucks, R.G., and Sarg, J.F., *eds.*, Carbonate Sequence Stratigraphy. Recent Developments and Applications: American Association of Petroleum Geologists Memoir 57, p. 63-105.

15. Cool- and Cold-Water Neritic Carbonates

Noel P. James, Department of Geological Sciences, Queen's University, Kingston, ON, K7L 3N6, Canada

Jeff Lukasik, Global Exploration Technology, Statoil, Oslo, NO-0246, Norway

INTRODUCTION

Carbonate sediments are born in a spectrum of neritic environments from the tropics to the poles, and have been so throughout geologic time. Warm-water settings with their stunning, vibrant carbonate biota in turquoise, gin-clear waters was, for many years, the irresistible modern depositional analog for all ancient carbonate rocks. Doing comparative sedimentology in shallow-water areas that are pleasant to visit, easy to study, and require little equipment was rewarding and profoundly advanced our science (Chapter 14). As a result, for many years neritic carbonates were interpreted as deposits that formed only in tropical oceans.

But there were dissenting voices. Beginning with Chave (1967), Schlanger and Konishi (1975) and the seminal work of Lees and Buller (1972) and Lees (1975) all of these studies stressed that carbonate deposition was also taking place in mid- to high latitudes (Fig. 1) and was significantly different from that in tropical neritic environments. This was especially true in the southern hemisphere, adjacent to the Southern Ocean, as chronicled in studies of Cenozoic limestones (Nelson, 1978) and the compendia edited by Nelson (1988) and later by James and Clarke (1997). These volumes contained, for the first time, numerous studies of cool- and cold-water carbonates in the older Phanerozoic record. In the last decade it has become clear that the nature of these extra-tropical neritic carbonates is not simply a function of colder seawater temperatures, but is also intimately tied into the relative availability of nutrients and trophic resources. Thus, James (1997) divided the neritic carbonate system into the *Photozoan Association* and the *Heterozoan Association* (*see* Chapter 13 - Introduction). Recently, there has been a dramatic increase in the study of these modern and ancient deposits (*e.g.*, Pedley and Carannante, 2006) because the realization has grown that these sedimentary rocks contain essential information about ancient oceans outside the tropical realm.

Figure 1. Ice-rafted dropstone in bryozoan-rich rudstone, Lower Permian Darlington Limestone, Maria Island, Tasmania; scale intervals 10 cm.

DEFINITIONS

As outlined in Chapter 13, it seems that a minimum average surface seawater temperature of ~20°C provides a measurable boundary, albeit diffuse, between temperate and tropical carbonate depositional realms. Below this temperature, reef-building corals and precipitates such as ooids, lime mud, and prolific synsedimentary cements of the Photozoan Association become conspicuously absent and are replaced by filter-feeding biota of the Heterozoan Association. This temperature threshold as defined by organisms is, however, not always sharp due to its modification by nutrient and trophic resource availability. For example, some species of modern stony corals can thrive if winter temperatures do not fall below 18°C, and large benthic foraminifers with photosymbionts can exist in both temperate and tropical waters depending on the level of trophic resources.

The heterozoan biota comprise mostly benthic foraminifers, molluscs (bivalves and gastropods), echinoderms, barnacles, and bryozoans; the only phototrophs are coralline algae. These biota are not restricted to mid-latitudes but extend to the poles. As metabolic requirements of organisms decrease in progressively cooler waters, growth and calcification rates also slow down, such that these sediments form at much lower rates than do warm-water sediments. Such sediments are also called temperate carbonates, non-tropical carbonates, and extra-tropical carbonates.

The typical environments where

A

REALM	PROVINCE	ATTRIBUTES	TEMP
COOL WATER	WARM TEMPERATE	HETEROZOAN and TRANSITIONAL HETEROZOAN; open shelf & ramp minor photozoan elements (large benthic foraminifers, scattered zooxanthellate corals, poorly calcified green algae), minorcarbonate mud, local hardgrounds, abundant corallines, seagrasses or macrophytes inboard (post early Cretaceous), palimpsest sediment, extensive bioerosion and maceration, bryozoan, mollusc, foraminiferfactory.	15-20°C
	COOL TEMPERATE	HETEROZOAN: open shelf & ramp, interbedded glacigene deposits, minor carbonate mud, local hardgrounds, palimpsest sediment, extensive bioerosion & maceration, corallines and macrophytes inboard (post Cretaceous), conspicuous bivalves, & barnacles with bryozoans.	5-15°C
COLD WATER	POLAR	HETEROZOAN: open shelf or oceanic bank, seasonal ice cover, interbedded glacigene deposits, IRD, glendonites, both biogenic carbonate (bryozoans, molluscs, barnacles) & biosiliceous (diatoms & sponges) sediment.	<5°C

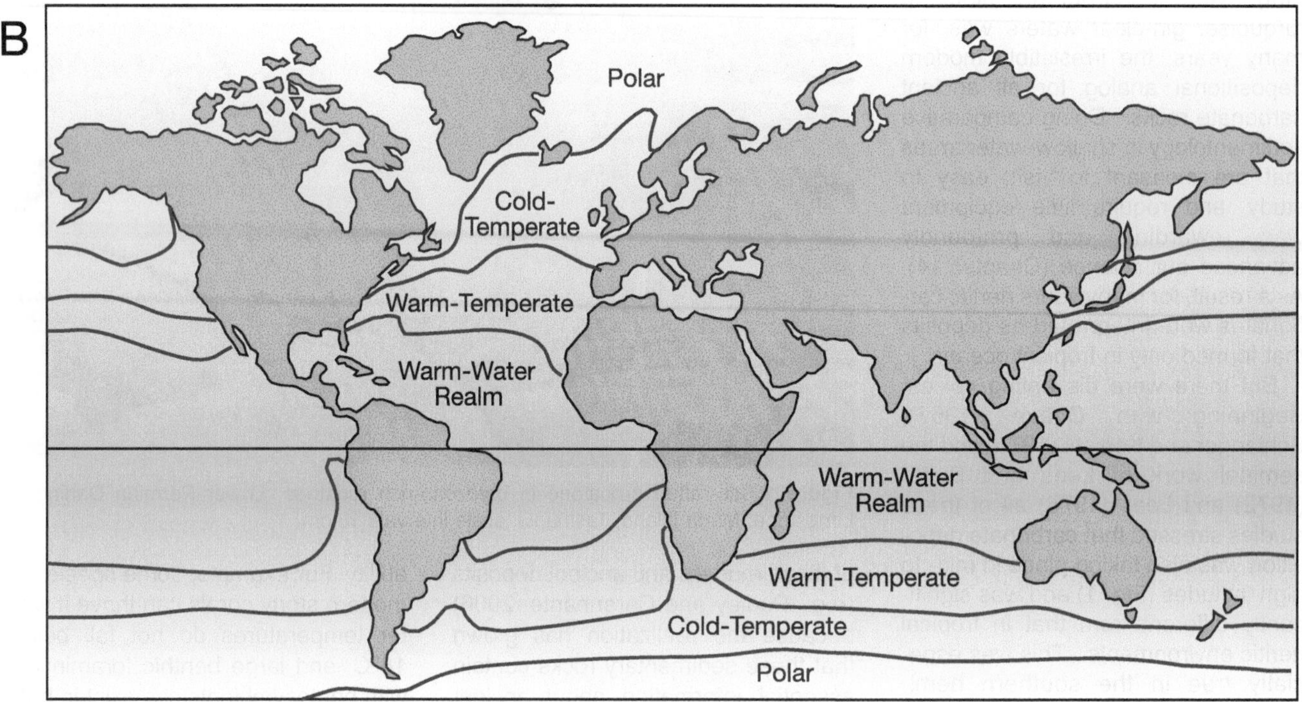

Figure 2. A. Terminology followed in this paper for subdivision of cool- and cold-water carbonate facies; **B.** Global map of average surface seawater isotherms delineating modern cool- and cold-water provinces.

such sediments accumulate are open shelves or ramps (Chapter 13) as there are no reefs or significant barriers present, except perhaps sea ice or an ice shelf, to keep open ocean waves and swells from sweeping across the entire neritic environment. Thus, sediments originate and accumulate across the depositional spectrum but are locally sorted, swept into sand shoals, deposited into topographic lows, and transported in the same way as neritic siliciclastic sediments. They may be born as carbonates but they accumulate as siliciclastics; they are clastic carbonates! The deposits commonly possess the best of both worlds. They have the compositional attributes of carbonates that permit environmental differentiation and the bedforms and physical sedimentary structures of terrigenous clastics that allow interpretation of paleohydrodynamics.

James (1997) defined different depositional provinces within the non-tropical depositional realm. In the last decade, research on modern and post-Paleozoic temperate carbonates has increased, such that this classification can now be refined. At the moment, the deposits would seem to be most usefully divided into cool-water and cold-water facies realms (Fig. 2A). Cool-water, temperate realm carbonates are largely restricted to the vast mid-latitudes while cold-water polar carbonates are clearly related to areas of glacigene sedimentation (Fig. 2B).

The cool-water realm is herein separated into two provinces; *warm temperate* and *cold temperate*. Although separated in the modern world by a minimum near-surface seawater temperature of 15°C, these different depositional provinces are defined most practically by sediment composition. Warm-temperate facies (*cf.* Betzler *et al.*, 1997) are those whose sediments are heterozoan in character but contain up to 20% hardy photozoan elements such as poorly calcified green algae, scattered zooxanthellate corals, and large benthic foraminifers with photosymbionts. As such, these facies are *transitional heterozoan* (Halfar *et al.*, 2006b). Cold-temperate facies are typified by wholly *heterozoan* sediments; photozoan calcareous components comprise <1%. In modern settings these facies can also accumulate on top of Pleistocene glacigene deposits. In the rock record, they can be interbedded with glacigene sedimentary rocks.

Polar carbonates of the cold-water realm accumulate in high-latitude environments with minimum surface water temperatures <5°C (Fig. 2B). As there is usually little carbonate sedimentation beneath permanent ice shelves or oceanic sea ice, these deposits are localized to ice-front environments or areas of winter-ice cover. Carbonates from these settings have a significant siliciclastic component, can contain conspicuous ice-rafted debris (IRD), and glendonites, and are interbedded with non-carbonate strata.

CONTROLS

Terrigenous Clastic Sedimentation

Carbonate sediments only accumulate in abundance where terrigenous clastic sedimentation is arrested and carbonate production can flourish. This balance is especially critical in cool- and cold-water settings where low seawater temperatures reduce the metabolism of ectothermic organisms slowing the rate of biogenic carbonate precipitation. Cool-water carbonates are thus restricted to areas of low siliciclastic input or, beyond the reach of terrigenous sediment transport, on the outer parts of platforms. In short – reduced dirt input!

Oceanography

Cool-water carbonates are generally located on the western sides of continents where arms of oceanic gyres advect cool waters equatorward. These are also areas of coastal upwelling where cool, nutrient-rich waters are brought up onto the shelf, promoting prolific benthic bioproductivity. Thus, heterozoan biogenic sediments can extend well into tropical latitudes. Carbonates from the cool-water realm can also form at unexpectedly high latitudes such as those generated under the influence of the Gulf Stream system that advects warm waters into the Norwegian polar realm.

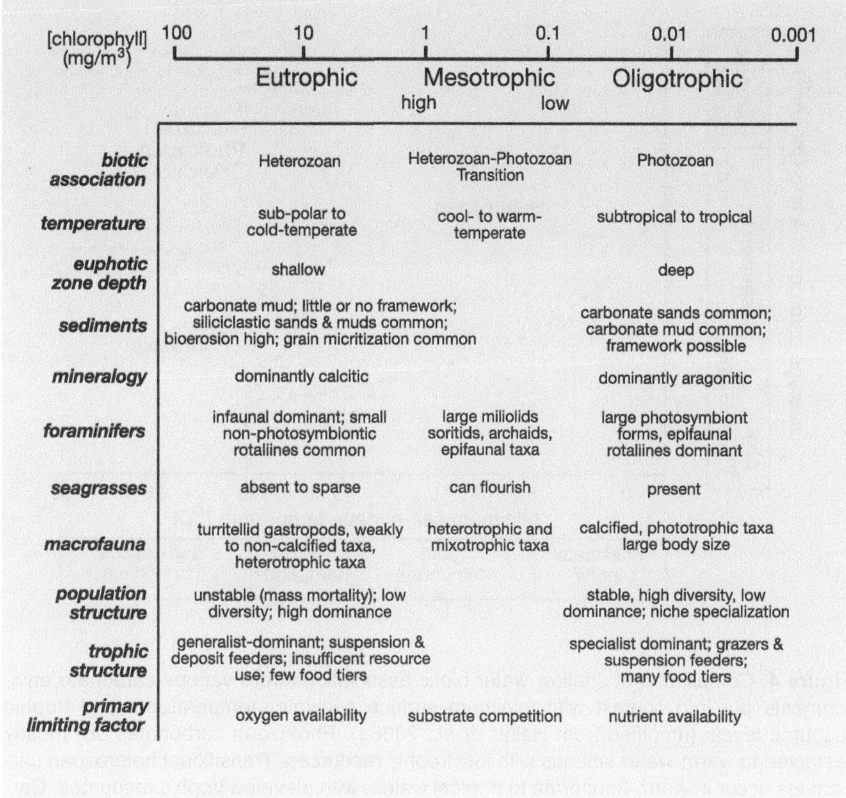

Figure 3. The Trophic Resource Continuum and correlative biotic, ecologic and sedimentologic information used to define trophic resource settings in the rock record.

Another major oceanographic control is the water column stratification across these platforms that make it possible to have shallow seafloors bathed in warm water with deeper parts of the same platform submersed in cool water, resulting in shallow-water photozoan sediments and deep-water heterozoan sediments. As cool-water platforms are typically deep, much of the seafloor will be within, or below, the seasonal thermocline leading to rapidly decreasing bottom-water temperatures with increasing water depth. The deepest, outermost regions of the platform can also come within the oxygen minimum zone. Seasonality, however, is important; summer waters are typically stratified but winter waters can be thoroughly mixed to depths of several hundred meters because of mid-latitude storms.

Nutrients and Trophic Resources

Elemental nutrients, especially nitrogen and phosphate, are critical to carbonate systems in that they stimulate growth of phytoplankton in the water column providing food for the benthos below. Areas of high nutrient flux have correspondingly high levels of sessile heterotrophic benthic productivity ultimately resulting in enhanced sediment production. Places of highest seasonal primary productivity are:

1. Zones of upwelling;
2. Oceanographic fronts and shear zones;
3. Areas of fluvial input; and
4. Ice margins.

Areas of downwelling and intervals of water column de-stratification are also important in the open ocean by transferring nutrients from surface plankton blooms to the seafloor.

Trophic resources, the sum of nutrients and organic matter, vary along a trophic resource continuum (Fig. 3). High trophic resource levels lead to eutrophy resulting in extensive phytoplankton bloom development, increased water column turbidity,

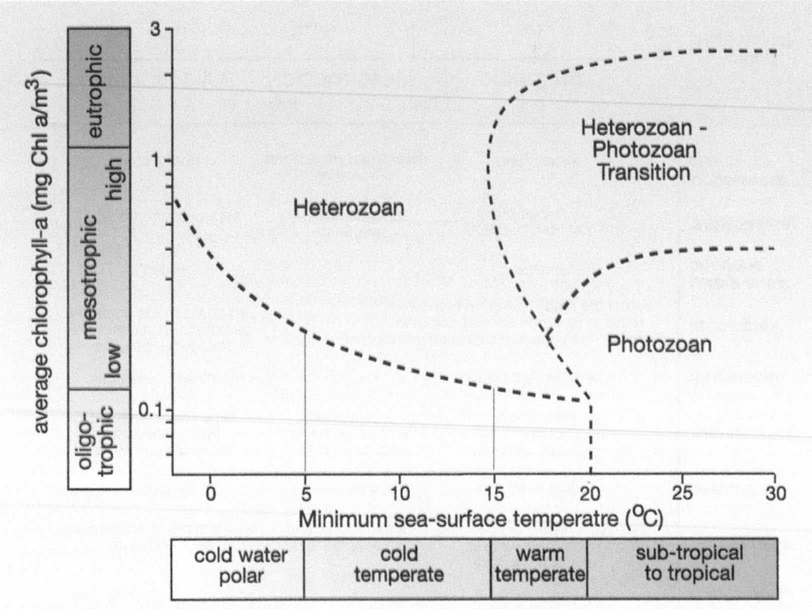

Figure 4. Compilation of shallow-water biotic associations from various carbonate environments put into context with minimum surface seawater temperatures and trophic resource levels (modified from Halfar *et al.*, 2006a). Photozoan carbonates are mostly restricted to warm-water settings with low trophic resources. Transitional heterozoan carbonates occur in warm-temperate to tropical waters with elevated trophic resources. Carbonates within the cold-temperate and cold-water provinces are heterozoan.

diminished seafloor illumination and a shallowing of the euphotic zone. Oligotrophic waters with low trophic resource levels are phytoplankton poor and thus clear with a deep euphotic zone. Changes in the trophic resource continuum from nutrient-rich, relatively eutrophic states to nutrient-poor, oligotrophic states affect the character of the substrate and turbidity of the water column to the point where they impact the nature and composition of the benthic assemblage. Biotic, sedimentologic, and ecologic information interpreted from the sediments can be used to identify the trophic resource levels present during their deposition (Fig. 3), especially when calibrated with nutrient and biotic information from modern cool-water environments (Lukasik *et al.*, 2000; Mutti and Hallock, 2003).

Temperature and Trophic Resources

Temperature and trophic resources are negatively correlated, meaning that warm waters with photozoan biota are mostly oligotrophic to low-mesotrophic (nutrient poor) whereas cool and cold waters yielding heterozoan biota are more mesotrophic to eutrophic (nutrient rich). Rising water temperatures stimulate the metabolism of organisms, forcing them to feed at higher rates thus cleaning the water column of suspended particulate matter. Clearer waters enable deeper sunlight penetration, in turn allowing more mixotrophic and phototrophic organisms to flourish at depth. Decreasing water temperature has the opposite effect, slowing metabolic activity and lowering the food intake of organisms, resulting in decreasing water clarity and a shallowing of the photic zone. Because of this interaction, most cool- and cold-water settings are almost never oligotrophic (Fig. 4).

There is, however, a caveat! Trophic resource levels can modify the influence water temperature has on the biota thus generating a gradient between Photozoan and Heterozoan associations within the warm- and cool-water realms (Fig. 4). Although biota from warm-water settings are overwhelmingly photozoan and those from cold-temperate to polar environments are heterozoan, it is mostly in the warm-temperate (15−20°C) and subtropical zones (>20°C) where trophic resources have the greatest impact in modifying the temperature-controlled biotic signature. It is within these zones that warm-water or warm, highly saline environments with high trophic resources can produce heterozoan carbonates similar to those generated in the colder waters of the cool-temperate realm and where relatively oligotrophic waters of the warm-temperate realm can produce photozoan accumulations. In short, whereas cool-water carbonates are almost always heterozoan in character, heterozoan carbonates are not always cool-water in origin (James, 1997).

Light

Water clarity and thus light penetration can be reduced by suspended mud, prolific plankton growth, and the angle of incident light, an especially important factor at high latitudes is where light vanishes for several months each year. Whereas heterozoan carbonates can accumulate in the absence of light and, thus, at all depths regardless of water clarity, biogenic components of the Photozoan association (including calcareous red algae) require light to grow. Seafloor illumination is important because it determines the depth to which sea grasses, non-calcareous algae, mixotrophs, and coralline algae can survive.

Hydrodynamics

Sedimentation depends upon the hydrodynamic regimen, frequency and intensity of storms as well as depth of the shelf or inner- to mid-ramp. Energy levels range from wave-swept, stormy platforms of the Southern Ocean to relatively quiet shelves of the Mediterranean Sea. Off the coast of southern Australia, sediment grains are moved constantly to depths of 60 m (zone of wave abrasion) and episodically to depths of 120 m, with large storms shifting sediments at depths of 250 m. The result is that all neritic facies are sandy, with mud and carbonate silt being exported off the platform to the adjacent slope or accumulating in intrashelf basins. As water energy lessens, the zones of disturbance shallow, abrasion decreases, and more mud remains in place. Because so many heterozoan cool-water carbonate shelves today are open-ocean and high energy, there is an impression that there is no mud accumulation, but that is not the case.

THE CARBONATE FACTORY

Introduction

These carbonates can be recognized in the rock record on the basis of their composition. Modern deposits allow us to relate workings of this cool-water carbonate factory to identifiable marine controls and oceanographic thresholds. Attributes of these deposits can then be translated into the older rock record (*see* section on Depositional Dynamics). Because cool- and cold-water carbonates accumulate slowly, the deposits are often palimpsest, a mixture of modern and older previously deposited sediment (*see* below).

At first glance, the sediments look like the leftovers from warm-water carbonates — those organisms that were otherwise overwhelmed by phototrophs and mixotrophs in the warm-water realm and so did not contribute much sediment — but they are not! Whereas warm-water environments are generally nutrient-depleted (oligotrophic), cool- and cold-water neritic environments are, in contrast, typically nutrient-rich (mesotrophic and eutrophic) such that the biomass is extraordinarily high and the benthic biota spectacular. Recall, for example, that these are the great fishing grounds of the world; the tropics are not!

Substrates

The sediment-producing organisms live on and within open rippled sands, on hard rocky substrates, and on ephemeral grasses and seaweeds. Numerous workers have commented that the diversity and abundance of calcareous epibenthos on open rippled sands is almost an order of magnitude less that that on hard rocky substrates such that there are really two sediment factories:

1. A low production sediment plain; and
2. A highly productive rocky seafloor (Fig. 5).

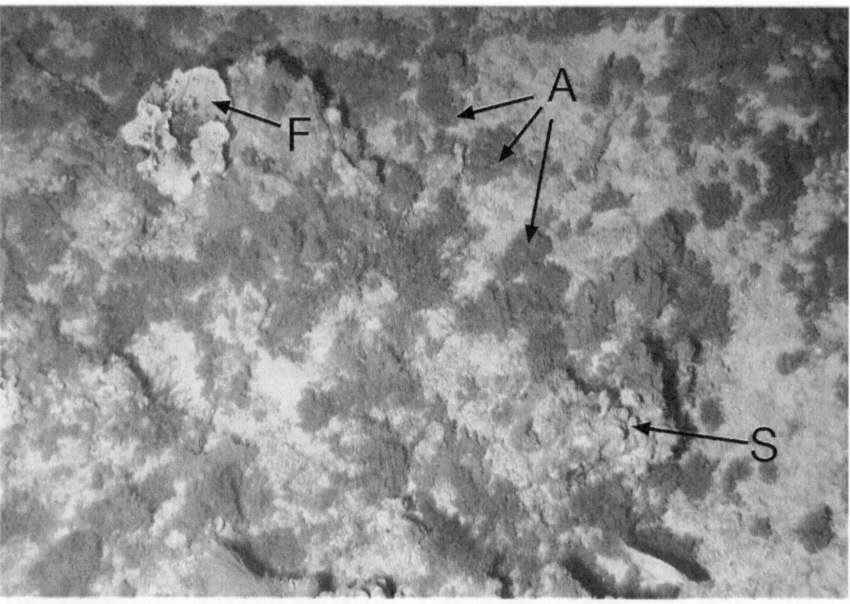

Figure 5. Seafloor off northwestern Tasmania (120 mwd) covered fenestrate (F) and articulated branching (A) bryozoans and sponges (S); field of view 1.0 m (image courtesy CSIRO, Australia).

Seaweeds such as marine grasses and macroalgae are not preserved in the rock record but are enormously important. Inner shelf (ramp) grasses can be prolific, outstripping their tropical cousins both in terms of diversity and abundance. Sea grasses are angiosperms and thus post Early Cretaceous. They are restricted to sediment substrates and warm-temperate environments. Their blades are substrates for growth of prolific calcareous epiphytes such as coralline algae (encrusting and articulated), small and large benthic foraminifers, serpulid worms, and bryozoans, and so are fundamental parts of the carbonate factory. At the same time their complex rhizome (root) systems bind the sediment and prevent marine erosion.

Macrophytes, the red, brown, and green soft non-calcareous macroalgae (Fig. 6) occur in both cool- and cold-water environments but are generally restricted to hard, rocky substrates and have fewer calcareous epiphytes. The most spectacular and arguably most important are kelp, large brown macroalgae (phaeophytes) that form marine forests in cool-temperate and cold, high-nutrient, high-energy, rocky seafloor environments (Freiwald, 1993; Henrich *et al.*, 1995). The dense, prolific growth of the kelp reduces water energy and the rigid but flexible thalli serve as substrates for some calcareous epiphytes. The rhizoid holdfast and proximal cauloid are encrusted by barnacles, erect-flexible bryozoans, serpulids, crustose corallines, sponges, ascidians, and hydrozoans. The phylloid or flat, broad strap algae, which in some species gets detached each winter, is encrusted by membraniporan bryozoans (Fig. 6 - inset), spirorbids, and epiphytic foraminifers. Although it is difficult to establish the age of the oldest significant kelp, it would seem to be mid-Cenozoic.

Coralline Algae

Coralline algae (Fig. 7) are the only benthic calcareous autotrophs. The oldest sedimentologically important coralline algae are Early Cretaceous but do not become prolific until the Paleogene. They are the most widespread of all calcareous algae, growing from tropical to polar latitudes, at depths from just over 250 meters water depth (mwd) to the surf zone. Whereas many genera inhabit shallow water, just as many are adapted to growing in dim light meaning they can thrive in nutrient-rich, turbid conditions. Their numbers are reduced under eutrophic conditions because of low seafloor illumination brought about by unrestrained phytoplankton growth. Thus, they are most abundant in the oligophotic zone under mesotrophic conditions.

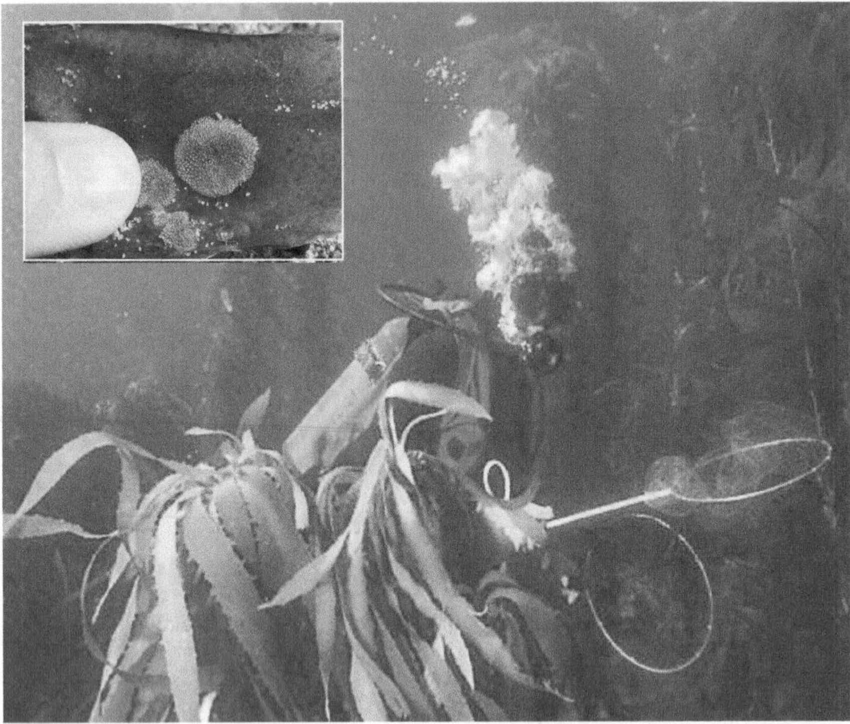

Figure 6. Diver amongst kelp off southern California (image courtesy D. Stokes); inset = encrusting bryozoan (*Membranipora* sp.) on stype (finger scale).

Corallines form granule- to cobble-size concentrically laminated nodules called rhodoliths (Fig. 7A). Such nodules can be smooth or branched wherein the branches break off during high-energy events to yield gravels of rod-shaped sticks (maerl). Geniculate or articulated branching corallines (Fig. 7B) are particularly numerous on broad-bladed sea grasses as well as on rocky substrates where they disintegrate to form innumerable rod-shaped, sand-size sediment grains. They can also encrust hard substrates and shells as well as living seaweeds. Intermixed encrusting and branching growths can also form banks up to several meters high.

Skeletal Invertebrates

The benthic calcareous invertebrates, which dominate neritic cool-water settings and produce sediment, are many (Fig. 8). The smallest are single-celled benthic foraminifers that, together with ostracode shells, generate much of the fine sand-size sediment fraction. Large benthic foraminifers are restricted to illuminated shallow seafloor environments. Foraminifers are augmented by numerous spicules from tunicates, especially ascidians. Since the Paleozoic, molluscs in the form of bivalves and gastropods have dominated this realm. Infaunal clams are generally most prolific inboard whereas epifaunal bivalves occur everywhere. Likewise, herbivorous gastropods are most abundant in shallow-water settings whereas predatory snails are present across the depositional spectrum. Although mollusc shells may be biofragmented on the seafloor (eaten by crabs and durophagus fishes), or broken and abraded in the surf, many remain as whole snails and clams. Brachiopods, although present, are not numerous except in some high-latitude settings. The most conspicuous and arguably most important components in this realm today, however, are cheilostome and cyclostome bryozoans (Fig. 9). These clonal organisms grow across the environmental spectrum (James and Bone, 2010) and contribute mud-, silt-, sand- and gravel-size particles to the sediment (Fig. 8). They, along with molluscs, form the core of the neritic temperate-water carbonate factory. Regular and irregular echinoids are also numerous both as infaunal ani-

Figure 7. Coralline algae. **A.** Buckets of rhodoliths 2 to 10 cm in diameter dredged from 51 mwd off southwestern Australia; **B.** delicate branching coralline algae from an offshore grass bed (scale 2 cm intervals).

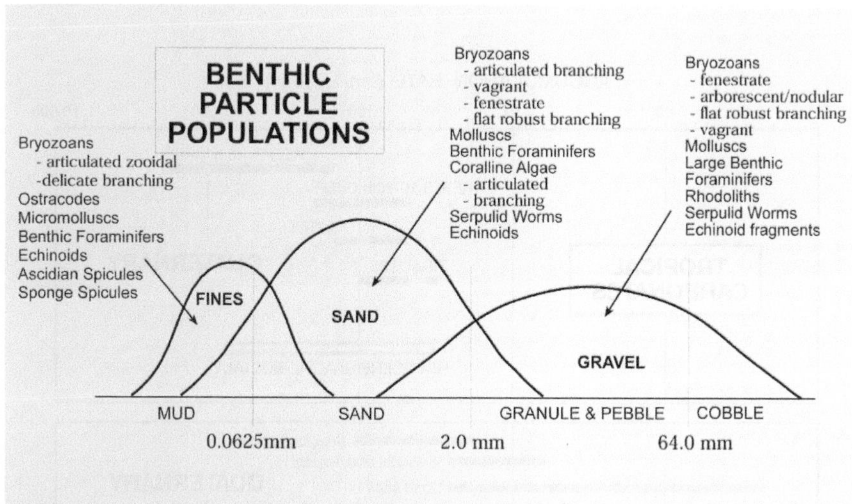

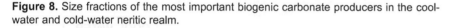

Figure 8. Size fractions of the most important biogenic carbonate producers in the cool-water and cold-water neritic realm.

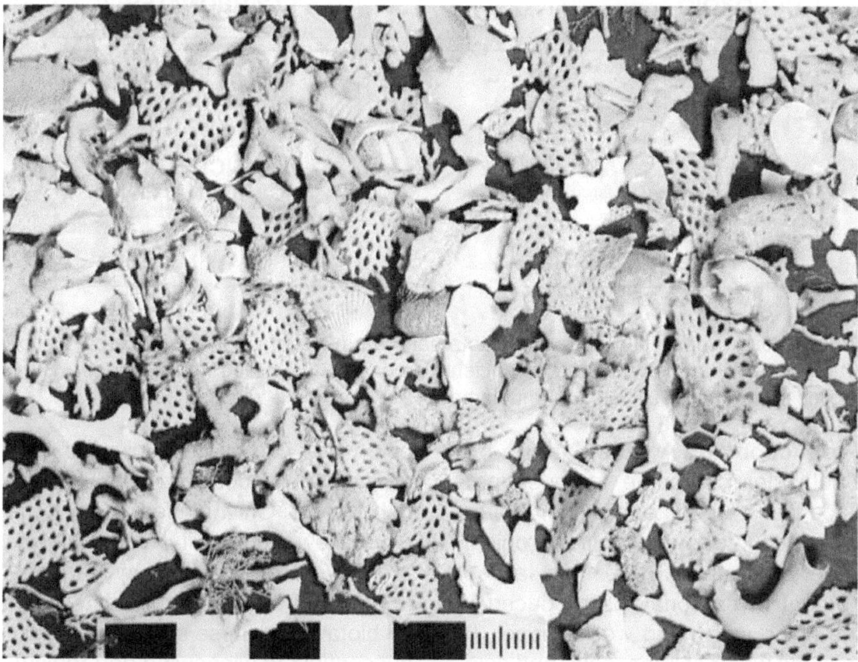

Figure 9. Sample of bryozoan-rich seafloor sand from 83 mwd on the Lacepede Shelf, southern Australia (cm scale).

mals that produce innumerable spines and plate fragments and epifaunal forms that generate more robust gravel-size particles (Fig. 8). Serpulid worms, both small isolated tubes and larger multi-tube clusters create a significant proportion of the sand and larger size fraction (Fig. 8). Barnacles generally disintegrate into many plates and are especially numerous in high-energy and high-latitude environments. Although carbonate mud is not generally a conspicuous part of these modern sediments, they are produced by pelagic calcareous plankton such as coccoliths and planktic foraminifers as well as through maceration. Finally, corals are present, but they are azooxanthellate, *i.e.* they do not have symbionts, and so are small skeletons that are not volumetrically important.

The relationship between nutrient availability, temperature, and light is complex but optimum growth conditions occur when a balance is met between plentiful suspended food supply and water clarity. Mixotrophs, such as large benthic foraminifers, are most numerous in warm-water euphotic, oligotrophic zones but become scarce in mesotrophic waters <15°C. However, molluscs and bryozoans are ubiquitous across the trophic resource spectrum but can be crowded out on hard substrates in the photic zone by numerous algae. Likewise, grasses grow across the photic zone but the broad-leaf-forms with numerous calcareous epiphytes are rare below ~15°C; in contrast macroalgae such as kelp seem to need higher nutrients (highly mesotrophic to eutrophic settings) to thrive and grow best where water temperatures are <15°C. Similarly, barnacles, serpulid worms, and echinoids are most prolific in mesotrophic and eutrophic environments.

Finally, although most of the sediments are calcareous, siliceous sponges are an integral part of the benthic biota. Their spicules are easily transported due to their hollow nature and low specific gravity and thus usually accumulate in the fine-grained fraction of the sediment mixture (Fig. 8). Such spicules become progressively more prolific in cooler, nutrient-rich waters and are especially numerous in deposits of the cold-water realm.

Bioerosion and Maceration

Early seafloor processes are especially important in cool-water environments (Smith and Nelson, 2003). With the high biomass, predation is rampant in cool-water environments; bryozoans and molluscs are eaten by crabs and durophagus fishes, bivalves are drilled and ingested by carnivorous gastropods, starfish prey upon epifaunal bivalves, phaeophytes are grazed by echinoids.

Agents of bioerosion (bivalves, gastropods, worms, sponges and microbes) are also profuse. Their actions lead to particle destruction through repeated macro- and micro-boring. This results in both production of mud by the boring process (especially by sponges) and grain disintegration into fine sand and mud fragments by repeated infestation. In warm-water environments endolith perforations are filled by sediment and/or marine cement, leading to the formation of micrite envelopes and hence particle preservation. In most cool-water environments, because cement is less abundant and fine-grained mud is scarce, the holes usually remain empty. This lack of micrite envelopes means that aragonitic grains are not preserved in the rock record, they vanish.

Particles also fall apart *in situ* after the disintegration of their organic matrix holding them together (maceration). This process is aided by the action of microbes that consume intercrystalline organics.

Chemical Diagenesis

Cementation

Absence of non-skeletal particles such as ooids, lack of obvious precipitated lime mud, and relative scarcity of marine cement leads to the notion that cool-water carbonate environments are not environments of early, abiotic Mg-calcite or aragonite precipitation. Whereas this seems to be largely true there is also contrary evidence for episodic hardground formation via intergranular precipitation of synsedimentary carbonate cement (*cf.* James and Bone, 1992; Nelson and James, 2000).

Dissolution

The role of carbonate dissolution on cool-water shelves is complex and poorly constrained. Aragonite skeletons seem to turn 'chalky' in many cool-water settings signaling partial dissolution, whereas Mg-calcite may 'leak' magnesium (Nelson, 1988). Recent work (James *et al.*, 2005; Wright and Cherns, 2008) has confirmed that even though overlying waters can be oversaturated with respect to aragonite there is profound loss of aragonite within sediment on the modern seafloor, likely via dissolution associated with oxidation of organic material. As a result, part of the biota may disappear soon after deposition leaving the sedimentary record strongly biased towards calcite components. It is as yet unclear how widespread this process is. Regardless, from a diagenetic standpoint, cool- and cold-water carbonates are calcite-dominated and thus well preserved in the rock record. In short, what you see on the seafloor is not what you get in the rock record; only part of the original sediment is preserved on the other side of the diagenetic curtain.

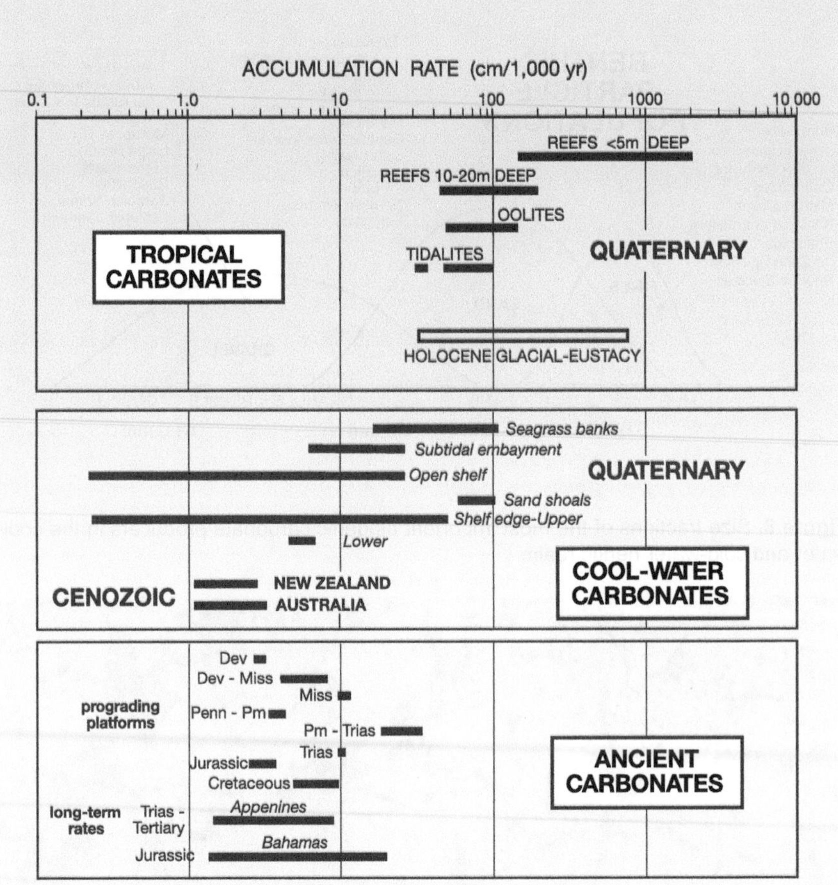

Figure 10. Accumulation rates for Quaternary warm-water, tropical, Quaternary cool-water, heterozoan, temperate, and ancient carbonates, (compiled from Schlager, 1981; James and Bone, 1991; Boreen and James, 1993 - modified from James, 1997).

Production and Accumulation

Cool-water carbonate factories produce less carbonate than their warm-water counterparts. Unlike tropical-water systems where the largest amount of carbonate is produced in the clear, warm and well-lit upper 15 m of the water column, cool- and cold-water systems are less dependent on depth for carbonate production given that most of the sediments are generated from heterozoan factories located across the depositional spectrum. In general, net rates of carbonate production in the temperate waters are roughly one order of magnitude less than those in warmer water settings, but there are exceptions (Schlager, 1981). The shallow sea grass, kelp, and coralline algal facies of the cool- and cold-water realm are, however, often prodigious sediment producers whose production rates rival those from the warm-water realm.

Low accumulation rates (Fig. 10) are a hallmark of neritic cool- and cold-water carbonates, linked to generally low production in cool, nutrient rich waters. On a platform scale, the overall low productivity, extensive bioerosion/maceration and low diagenetic potential of a predominantly skeletal biota make these sediments highly susceptible to winnowing and redistribution across the platform. As such, accumulation is highly variable, dependent mostly on the distribution of hydrodynamic energy. Thickest accumulations occur in areas of lowest hydrodynamic energy such as below wave base in outer-shelf settings or within shallow nearshore environments, or in shallow areas across the mid to inner shelf where epiphytic carbonate production is high and the substrates are protected by a thick cover of sea grass or kelp. Thinnest accumulations, with high sediment omission, occur in areas where waves and swells constantly sweep the seafloor pushing sediment elsewhere.

FACIES MODELS

Introduction

Facies models for heterozoan, neritic, cool-water carbonates are most usefully formulated on the basis of controls such as minimum surface seawater temperature, trophic resources, and geotectonic setting, as observed in the modern ocean. Water temperature modified by trophic resources largely determines the character of the sediment-producing biota and the rate at which carbonate sediment is produced. Geotectonic setting establishes the nature of the platform and intensity of the hydrodynamic regimen in which the sediments accumulate. Seawater temperature, as reflected by the biota, is difficult to quantify but on a broad scale (Fig. 2) the cool-water temperate realm falls into two natural facies, warm-temperate and cold-temperate. The most common geotectonic settings (Fig. 11) in which these sediments are found can be resolved into:

1. Open-ocean, as represented by continental margins or offshore banks;
2. Interior basins, as represented by seas like the Mediterranean, together with a spectrum of epicontinental, foreland, and intracratonic depocenters. The main difference between these two settings is that open-ocean neritic environments are generally wave- or tide-dominated open shelves with relatively deep hydrodynamic zones affecting large areas whereas interior basins generally form lower energy ramp or open-shelf systems with or without high-energy shorelines; and
3. The seaway setting is distinct from the others. Seaways are generally narrow marine passages that link larger bodies of water such as oceanic straits, elongate intra-platformal channels, or inter-island pathways.

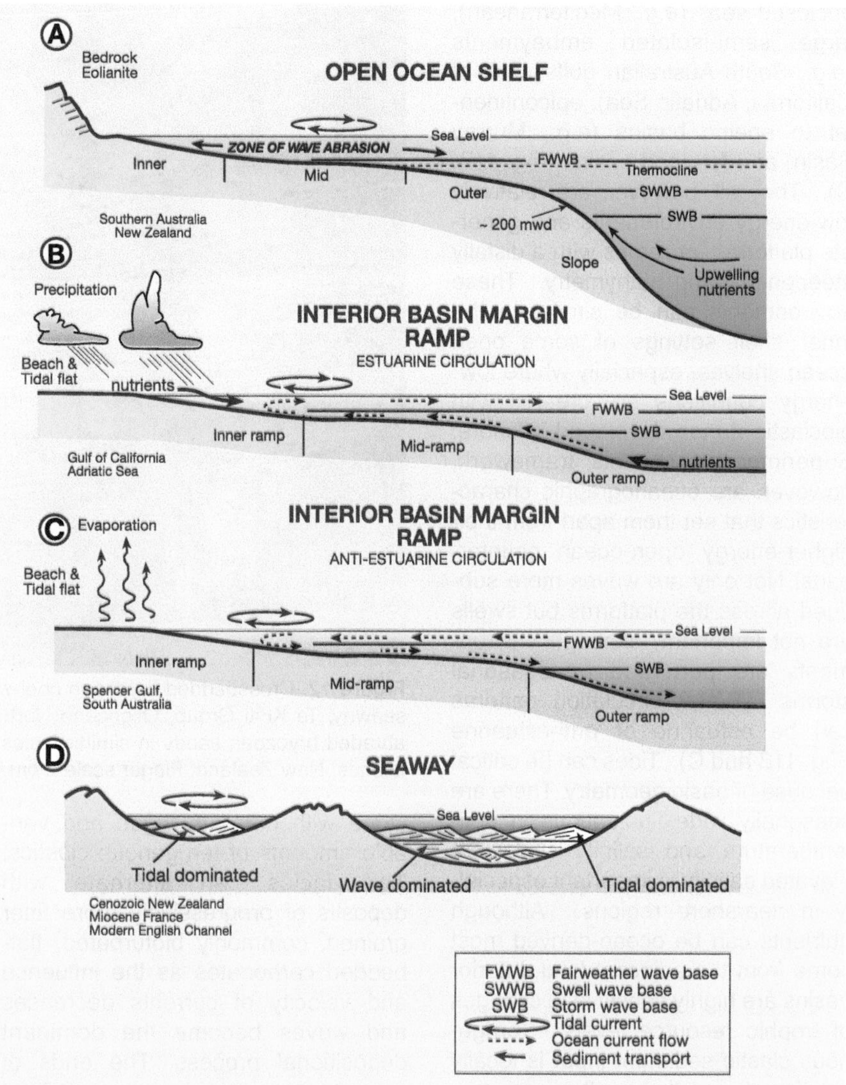

Figure 11. Geotectonic and oceanographic settings for cool-water carbonate facies; **A.** Open-ocean shelf; **B.** Interior basin ramp with a nutrient-rich, normal seawater inflow under a humid climate; **C.** Interior basin ramp with nutrient-poor normal seawater inflow and saline bottom- water outflow under a semi-arid climate; **D.** cross-section of the setting and general facies in a carbonate seaway.

Open Ocean

Open-ocean environments (Fig. 11A) are located along continental margins or on isolated banks and tend to be wave-dominated depositional systems due to their location in universally high-energy mid- to high-latitude settings. They are open shelves or distally steepened ramps with a steep shoreface, a broad, relatively flat or gently inclined ramp-like seafloor, 30–80 m deep with no outer-margin rim. Siliciclastic sands are common inboard but are kept there because of the high-energy system and long-shore drift processes. Fairweather wave base is deep, commonly 50–60 mwd such that sediment to these depths is moved almost constantly (sometimes called the zone of wave abrasion) and transported both offshore and onshore. The water column is well mixed in the winter but poorly stratified in summer with the shallow thermocline at 50–80 mwd. Water temperatures below the thermocline in warm-temperate neritic environments are usually between 10 and 15°C. Salinities are typically normal marine. Nutrients are mostly supplied via seasonal upwelling of mid-ocean waters onto the outer shelf or even to the shoreline. The mid- to outer-shelf depositional regions are largely similar because they all lie below the shallow thermocline and so in cool waters with roughly similar attributes, it is the shallow inboard facies that are different.

Interior Basin

Interior basin depocenters include a wide variety of settings such as enclosed seas (*e.g.*, Mediterranean), large semi-isolated embayments (*e.g.*, South Australian gulfs, Gulf of California, Adriatic Sea), epicontinental to epeiric basins (*e.g.*, Murray Basin) and foreland basins (Fig. 11B, C). They all, however, are relatively low-energy environments and generate platforms or ramps with a distally steepened ramp bathymetry. These environments can be similar to the inner shelf settings of some open ocean shelves, especially where low-energy conditions are created with bioclastic dunes developed offshore. Superimposed on this framework, however, are oceanographic characteristics that set them apart from their higher-energy open-ocean counterparts! Not only are waves more subdued across the platforms but swells are not important, although environments are perturbed by seasonal storms. Water circulation patterns can be estuarine or anti-estuarine (Fig. 11B and C). Tides can be critical because of basin geometry. There are seasonally wide fluctuations in both temperature and salinity such that elevated salinity is important especially in nearshore regions. Although nutrients can be ocean-derived most come from the adjacent land. Interior basins are highly sensitive to changes of trophic resource levels. Terrigenous clastic sediment input is locally significant and the seafloor environments are relatively shallow throughout (generally <100 mwd).

Figure 12. Crossbedded bryozoan cool-water limestones deposited in a tide-dominated seaway, Te Kuiti Group, Oligocene North Island, New Zealand; Inset = well-worn and abraded bryozoan sands in similar facies from Eocene Te One Limestone on Chatham Islands, New Zealand; Finger scale 2 cm wide.

Seaways

In some situations where conditions are optimum for seawater flow acceleration, facies are strongly current bedded (Figs. 11D and 12). Such currents are tidal or unidirectional ocean flow. These conditions can develop in oceanic straits, in passages between islands, or in elongate seaways that range from 10s to 100s of kilometers in scale. Cross-sectional geometry of seaways, together with the oceanic tide and wave climate, appears to determine whether the system will be wave- or current-dominated. When current-dominated, large-scale, simple and compound subaqueous crossbedded dunes will form in water depths of 40−60 m and be up to 9 m in height. The sediments are generally heterozoan grainstone and rudstone with coralline algae and variable amounts of terrigenous clastics. This facies can alternate with deposits of progressively more finer grained, commonly bioturbated, flat-bedded carbonates as the influence and velocity of currents decreases and waves become the dominant depositional process. The ends of seaways can also be sites of flood tidal delta deposition in the form of shallowing-upward cycles up to 50 m thick.

Facies are generally similar across depositional provinces and dominated by hydrodynamic processes, they are the most clastic of carbonates. Most documented examples to date are in open-ocean settings (*see* Anastas *et al.*, 2006) with the exception of flooded incised valleys (Reynaud *et al.*, 2006).

Approach

The modern world contains many environments in these depositional realms. The approach here is to formulate models based on the Cenozoic, combining observations from the modern seafloor and melding them with studies of the Pleistocene, older Neogene and Paleogene limestones. This approach is possible because the biogenic aspects of the deposits have changed relatively little over the last 90 m.y. Many of the models are also directly applicable to the Mesozoic but because this was largely a globally warm period, there cool-water carbonate deposits are relatively rare. The Paleozoic, divorced by time from the younger carbonates because of its biota, is treated separately (*see* below).

Although this approach has been successfully applied to tropical photozoan carbonates, it is slightly complicated in mid- and high-latitude carbonates. Modern marine sediments are today accumulating during an ice-house period in earth history that for the last 2.5 m.y. has been characterized by repeated continental glaciation (Fig 13). Such glaciation resulted in large fluctuations in sea level, profound climate change, cooling of seawater in temperate latitudes, and local glaciation of the shelves themselves. Against this background, the relatively slow rates of carbonate sedimentation in cool-water environments have led to several peculiar attributes of this system.

In warm-water tropical settings, when carbonate sediments are exposed to meteoric diagenesis they are typically cemented via meteoric diagenesis and so they are unavail-

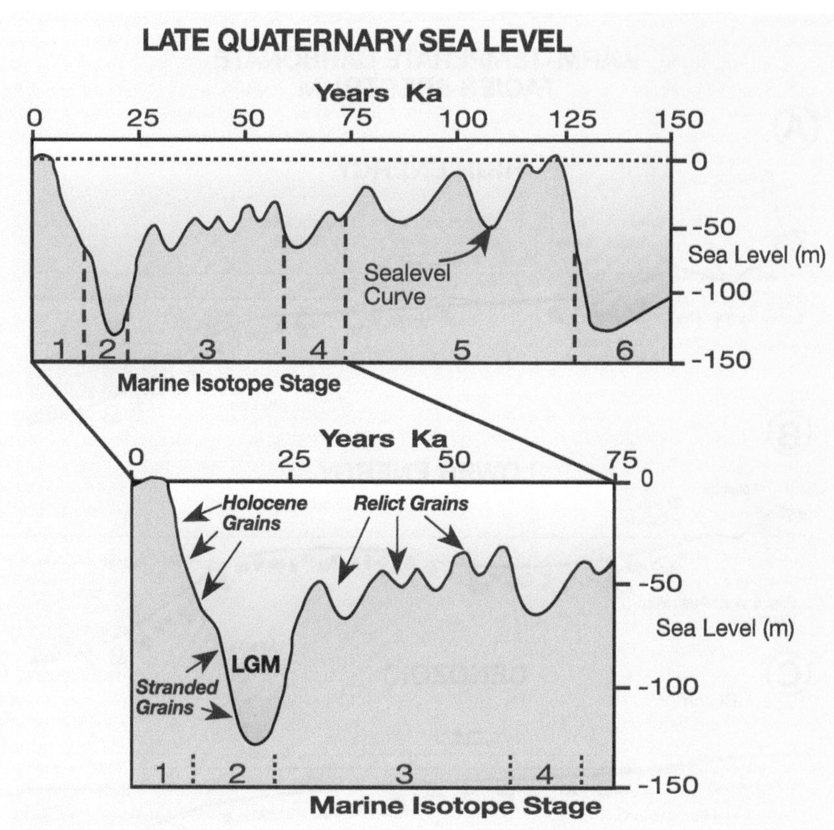

Figure 13. The Late Quaternary eustatic sea-level curve (Chapell and Shackleton, 1986) with the periods in which Relict, Stranded and Holocene particles would have formed in a cool-water carbonate system; LGM = Last Glacial Maximum.

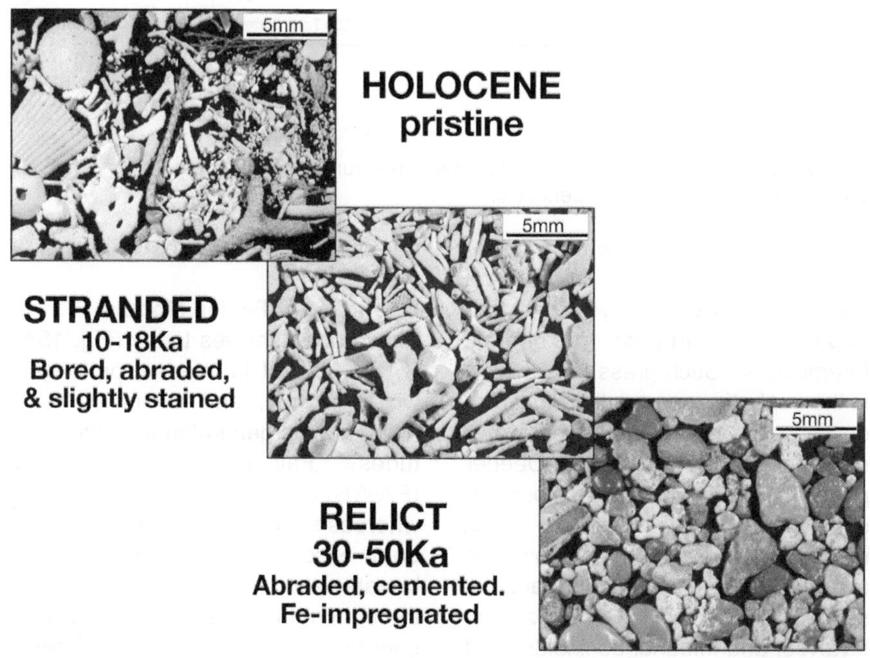

Figure 14. Images of Holocene, Stranded, and Relict particles from the south Australian continental margin.

able for re-sedimentation when sea level subsequently rises again. Cool-water carbonates are not so cemented (lack of aragonite) and thus older grains can be mixed in with new sediment to form a palimpsest accumulation (Fig. 14). The *relict carbonate grains* formed at lowstands, and interstadials whereas *stranded carbonate particles* formed during early stages of sea-level rise, and are now out of equilibrium with modern conditions (*e.g.,* coralline algae looking fresh but at depths well below the photic zone). Such sediment mixtures are *palimpsest.*

Deteriorating climate, especially increased rainfall, results in increased delivery of siliciclastic sediment to the shelf during glacial periods to the point where the entire shelf may be covered with terrigenous sands and gravels. Today such shelves are, because of relatively high sea levels, cut off from most terrigenous clastic sediment input. Modern carbonate is accumulating across the neritic environment, albeit slowly, and so the resultant deposit is a mixture of modern carbonate and older terrigenous clastic sediment.

Finally, in colder parts of the temperate province the shelves were actually glaciated during the last glacial maximum (LGM) and so the sediment is a variety of marine glacigene deposits on top of which cool-water carbonates are slowly accumulating or being admixed with glacigene sediments by modern burrowing and hydrodynamic processes.

In summary, not only can local depositional conditions alter the distribution of facies belts within similar settings, but many cool-water carbonates are mixtures of modern and older particles that are in some cases difficult to separate. This caveat should be remembered when generating facies models derived from the modern world that are then applied to the older rock record.

WARM-TEMPERATE CARBONATE FACIES MODELS

General Attributes

In the modern world, the warm-temperate carbonate facies lie mostly in the lower mid-latitude belts not directly affected by glaciation. They are the most well studied of any carbonates in the cool-water realm and include open ocean platforms and interior basins of southern Australia, western and northwestern Africa, the Gulf of California, southeast North America, southern Brazil and the Mediterranean Sea (Fig. 2B). The preservation potential of such carbonates is high and they form most of the temperate-water limestones found in the Cenozoic rock record. As such, warm temperate carbonates are what we now acknowledge them as the classic cool-water carbonates. This depositional province (Fig. 15) is dominated by heterozoan and transitional heterozoan facies defined by the presence and abundance of large benthic foraminifers (Fig. 16). Calcareous green algae are present but are usually poorly or non-calcified and thus sparsely represented in the sedimentary record, if at all. Likewise, large corals are typically present as isolated colonies confined to a few genera. They never form reefs and they are not significant in terms of overall sediment production.

The shallow inboard seafloor, especially in lower energy environments, is covered with broad-bladed seagrass meadows, generally more luxuriant and diverse than in similar tropical settings. Numerous seaweeds are intermixed with the sea grasses, especially on rocky and gravelly substrates. Attendant calcareous epiphytes together with prolific infaunal and epifaunal molluscs as well as bryozoans form an important carbonate factory. Regardless of setting there is a common inboard - outboard tripartite facies motif; shoreface sands (terrigenous or carbonate) quickly grade into luxuriant seagrass meadows that can in turn pass seaward into skeletal sands rich in bryozoans or coralline algae (commonly rhodolith) gravels. Bryozoan-rich, deeper water facies lie below the thermocline (generally <15°C) and so are similar to shallow-water cool-temperate facies (*see* below).

Inboard facies are also partitioned by trophic resources. Oligotrophic (rarely) to low mesotrophic settings provide the optimum balance between plentiful suspended food supply and water clarity (allowing sunlight penetration) and so are sites of prolific sea grass and attendant calcareous epiphyte growth together with a background of suspension feeders (bivalves and bryozoans) and large mixotrophic benthic foraminifers. Such grass banks within the euphotic zone are the most productive parts of the whole cool-water carbonate factory system. Deeper parts of the oligophotic seafloor are sites of rhodolite growth. Most mesotrophic environments, however, are characterized by an abundance of suspension-feeding taxa (bryozoans, bivalves, and infaunal echinoids) and an absence of large benthic foraminifers due to a reduced depth of the photic zone. Eutrophic settings are typified by a much reduced epibenthic suspension feeding population (bryozoans and bivalves) and an abundance of infaunal animals.

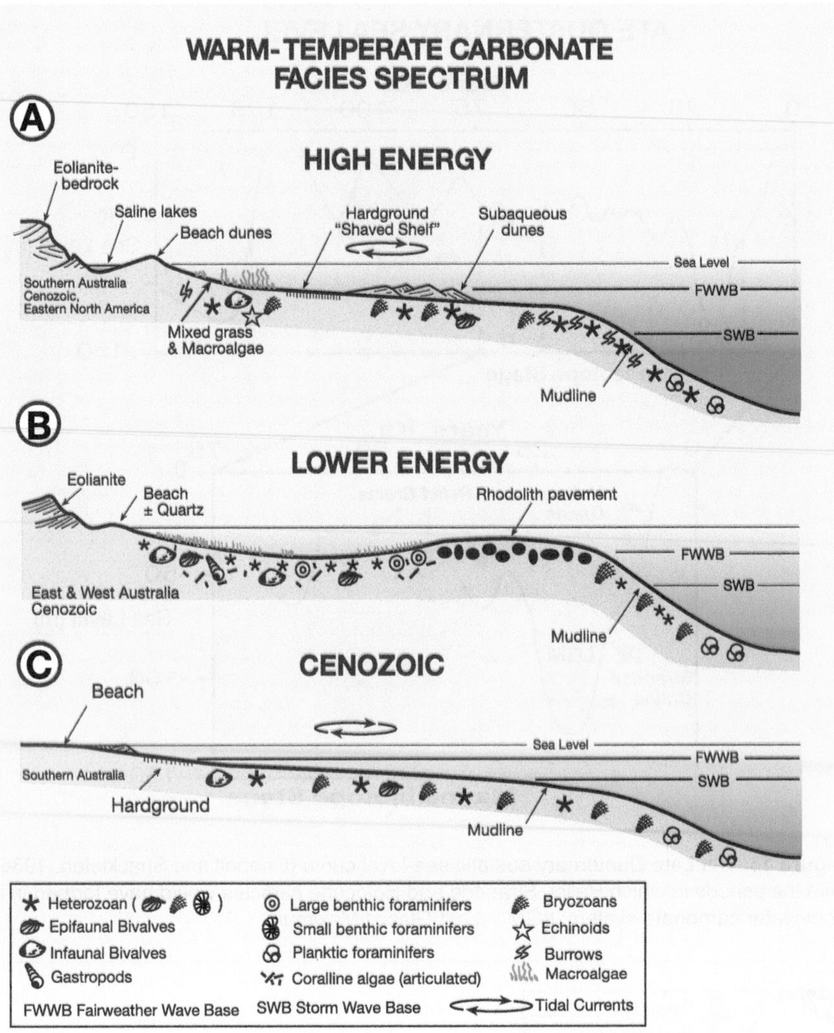

Figure 15. Warm-temperate carbonate facies spectrum. **A.** High-energy open shelf; **B.** Lower-energy open shelf; **C.** Cenozoic.

Open Ocean Facies Models

These open shelves today (Fig. 15A, B) are subject to intense, prolonged winter storms, and swept by long rolling swells generated in higher latitudes. Fair weather wave base (FWWB) can be as deep as 60 mwd whereas storm wave base (SWB) can be in excess of 150 mwd. The carbonate sediment factory operates throughout but it seems that the high energy, while resulting in shelf sediment accumulation, also transports prodigious amounts of carbonate onshore to form eolianites, and offshore to form prograding slope sediment wedges (Fig. 17). In many cases Cenozoic shelves and basins

Figure 16. Pliocene grassbed rudstone facies composed of bivalves, oysters, and large benthic foraminifers (*Marginopora vertebralis* - Roe Calcarenite, Western Australia).

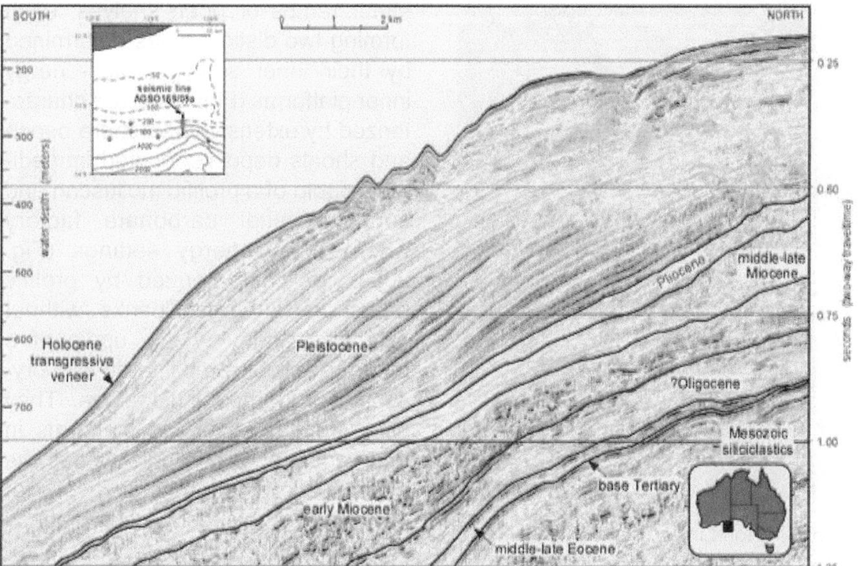

Figure 17. A north−south seismic image of the continental margin in the central part of the Great Australian Bight, Australia illustrating the thick Pleistocene prograding wedge of cool-water carbonates and embedded bryozoan mounds (green) (image courtesy Geoscience Australia).

are characterized by overall lower energy facies (Fig. 15C).

The *inner shelf* or shoreface lies within, or above, the thermocline, in low-mesotrophic to mesotrophic settings and entirely within the euphotic zone. It is high-energy throughout, except for protected small embayments. The strandline is in the form of rocky headlands, extensive eolianites cliffs, and prograding beach-dune complexes with hypersaline lagoon corridors between. There are *no* muddy peritidal complexes.

The steep shoreface descends to 10 mwd or more where the seafloor is open rippled sand, bedrock, or areas of thin-bladed to local broad-bladed sea grass. Bryozoans, molluscs (mostly infaunal bivalves), echinoids, and small benthic foraminifers produce the biofragmental sand and gravel (Fig. 19) whereas coralline red algae encrust most bedrock surfaces. Benthic growth is most abundant on hard substrates.

This zone has been called a *shaved shelf* (James *et al.*, 1994; Fig. 15A) because even though carbonate production is relatively high, most of the carbonate is razored off by the constant wave activity to accumulate as sediment on the beaches, only to be swept into dunes, or offshore in the mid-shelf or deeper. The geological record of this zone is one of hardgrounds that can be encrusted with corallines.

The wide *middle shelf* is everywhere characterized by wave-rippled sand (Fig. 18) that can be swept into large, crossbedded, subaqueous dunes. This is a zone of sediment production, but little *in situ* accumulation due to wave sweeping.

The *outer shelf* is universally a zone of carbonate production and accumulation. It is an environment little influenced by the day-to-day action of waves and swells but strongly affected by seasonal storms. Bryozoans, sponges and other heterozoan biota form a 'sticky zone' of prolific sediment production where the shelf rolls over onto the upper slope (Fig. 5). The 'mudline', or that portion of the seafloor below which mud accumulates in significant quantities, characterizes deeper parts of this environment. This bioturbated mud is a mixture of planktonic carbonate fallout and transported benthic skeletal fragments, clay (forming marl) and siliceous sponge spicules. Since it is an area distal from significant terrigenous sediment input, these are the most extensive temperate carbonate sediments in the modern ocean.

The *shelf edge*, because it is so deep (100−200 mwd), is poorly known. It is usually a gentle roll-over of the open shelf margin into progressively deeper and muddier sediments with a corresponding decrease in macrobiota (Fig. 15A). In most regions it is not an important sedimentological break as much as it is a gradual decrease of macrobiota

Figure 18. Seafloor off northwestern Tasmania (138 mwd) covered with wave-rippled, bryozoan-rich sand and gravel; field of view = 1.8 m (image courtesy CSIRO).

Figure 19. Bryozoan–pecten rudstone, Miocene Point Turton Limestone, Yorke Peninsula, Southern Australia; finger scale.

as the open shelf edge rolls-over into progressively deeper and muddier sediments. The *upper slope* is even more poorly documented. Much of the muddy sediment is silt-grade skeletal material, and includes carbonate skeletal fragments and siliceous spicules swept from the shelf and mixed with pelagic fallout, especially coccoliths. Overall, pelagic deposition during highstands of sea level is punctuated by redeposited shelf sediments during lowstands, when the shelves were narrower and the high-energy zones were located nearer to the shelf edge (*see* section on Depositional Dynamics). Regardless, sediment transport off shelf usually is so profound that spectacular seaward-prograding clinoforms have developed wherein up to 600 m of Quaternary slope sediment has accumulated in 2 m.y. (Fig. 17). This upper slope environment is also the location of bryozoan–sponge biogenic mounds that grew during periods of elevated trophic resources.

When the overall energy setting of these warm-temperate open-ocean platforms is relatively low, facies belts become blurred (Fig. 15B). This is especially common in the transition zone between tropical and temperate depositional systems. The outer shelf and shelf edge is typically covered with rhodolith gravel, local grasses and seaweeds together with their attendant epiphytes. These pass basinward into deeper water, muddy, upper slope carbonates.

Interior Basin Facies Models

These depositional environments are overall more tranquil than open-ocean settings wherein the depth of FWWB is negligible compared to the open ocean and winter storms, although violent, are generally local and of short duration. Such lower energy settings contain a similar facies spectrum to their higher-energy counterparts but are thus overall finer grained. Interior basin platforms are either ramps or open shelves, each forming two distinct types determined by their inner settings, high-energy inner platforms (Fig. 20A) are characterized by extensive nearshore beach and shoals deposits located immediately updip of a prolific molluscan and coralline algal carbonate factory whereas low-energy settings (Fig. 20B) are characterized by prolific nearshore seagrass banks without extensive grainy shoals updip of a muddy seafloor ramp littered with bryozoans and rhodolithic algae. They differ from open ocean platforms in that sedimentation occurs across the entire shallow platform, but are similar in that the thickest accumulations occur downdip of the carbonate factory.

Inner platform (ramp) settings bathed in low-mesotrophic waters such as the Mediterranean Sea are expectedly diverse; rocky shorelines, barrier islands with leeward muddy lagoons with numerous bivalves, ostracodes, and gastropods, or beaches; offshore crossbedded sand shoals (packstone and grainstone) composed of abraded biofragmental grains and extensively burrowed, muddy seagrass beds. In more protected settings like Spencer Gulf, South Australia (Gostin *et al.*, 1988)

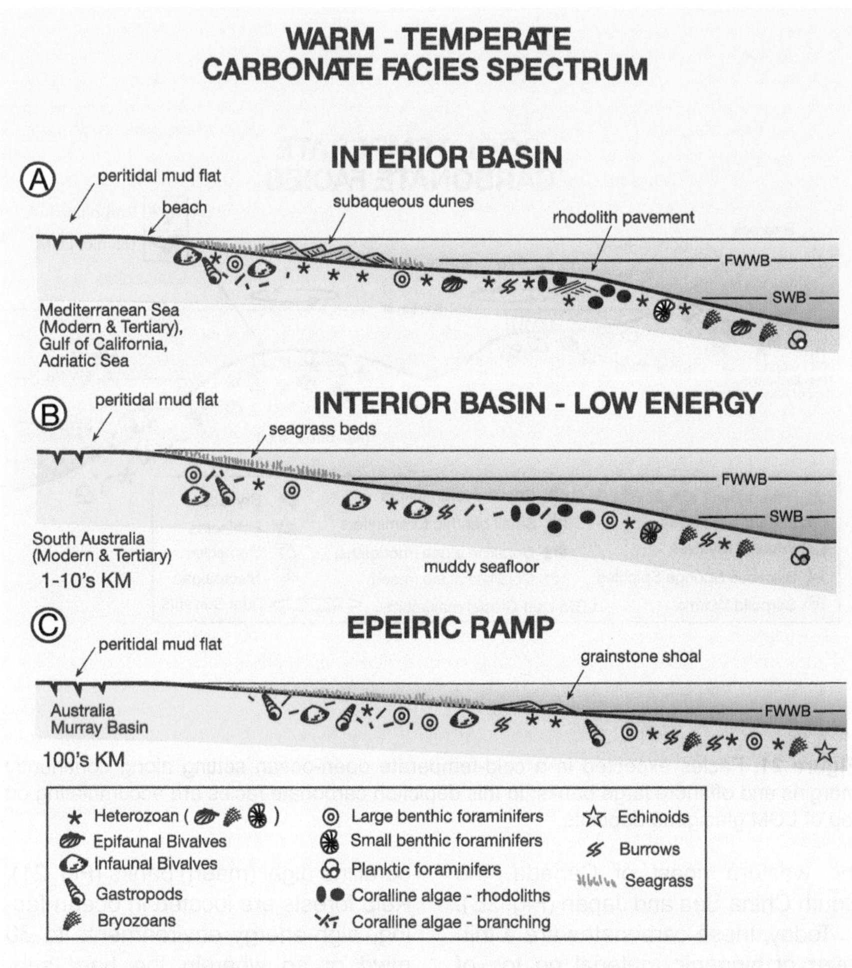

Figure 20. Warm-temperate carbonate facies spectrum. A. Interior Basin, high-energy inner ramp; **B.** Interior Basin, low-energy inner ramp; **C.** Epeiric Ramp; the core heterozoan assemblage is dominated by bryozoans, molluscs, and small benthic foraminifers.

there are also muddy tidal flats with all of the attributes of typical peritidal deposits (Chapter 16). Thus, the important concept that peritidal facies can be present in cool-water settings as long as there is mud produced in the offshore carbonate factory. Many Cenozoic ramps also have extensive regions of biofragmental subaqueous dunes that are laterally equivalent to the grass beds (Braga *et al.*, 2006). These large embayments are, however, variable. Spencer Gulf is very Mediterranean-like, with generally oligotrophic seawater and anti-estuarine circulation driven by intense summer evaporation and elevated salinity development at the head of the Gulf.

By contrast, in the Gulf of California (Halfar *et al.*, 2006a, b) year-round upwelling occurs, with extensive tidal mixing in the northern gulf. A strong nutrient gradient is present from oligotrophic at the gulf entrance in the south to eutrophic in north. Molluscs are ubiquitous and occur in all facies but zooxanthellate corals are only present in oligotrophic settings in the south; coralline algae dominate in the mesotrophic central part; bryozoans are most abundant in the more eutrophic head of the gulf. The Adriatic (McKinney, 2007) has an east-to-west trophic resources gradient with relatively warm oligotrophic− mesotrophic waters adjacent to the Croatian coast and eutrophic waters near the Italian coast due to high nutrients and counterclockwise circulation. The seafloor under oligotrophic conditions is dominated by an upright, filter-feeding, calcareous biota whereas the more terrigenous clastic muddy seafloor immersed in eutrophic waters is dominated by an infaunal biota. Bryozoans are particularly evident in oligotrophic waters where local meadows of upright articulated bryozoans (*Cellaria* sp.) form extensive muddy floatstones.

Mid-platform settings in mesotrophic waters contain sub-wave base facies typified by red algae in the form of encrusted pavements but more often as fields of rhodoliths and articulated coralline algae with associated bryozoans, molluscs and foraminifera. These calcareous algae thrive in the oligophotic zone, a window between FWWB and the limit of seafloor illumination, a dim but productive zone. Sediments accumulate within depressions on an uneven seafloor, *e.g.,* the détritique côtier sediments of the Mediterranean, within this carbonate factory area (Fig. 20A). Deposits here range from burrowed and thick-bedded to conspicuously cross-laminated grainstone and packstone beds. Rudstones can form buildups 4 m high and up to 50 m across but have low accretion rates of 0.1−0.4 m/ky. The seafloor is below significant sea grass growth but local seaweeds still persist.

Outer platform areas typically accumulate muddy facies with a typical cool-temperate biota of filter-feeding and infaunal calcareous organisms together with conspicuous planktonic foraminifers.

Broad, very shallow basins many hundreds of kilometers across with very shallow seafloor gradients and depths that are nowhere below storm wave base, as exemplified by the Cenozoic Murray Basin in Australia (Lukasik *et al.*, 2000), have been called *epeiric ramps* (Fig. 20C). Finely grained, muddy facies typical of the inner and outer ramp generate in overall low-energy settings separated by an offshore grainy skeletal facies belt developed within the zone of wave winnowing. All but the grainy offshore facies sediments possess a ubiquitous background assemblage of gastropods and bivalves, together with numerous bryozoans and echinoids, are pervasively bioturbated and are characterized by decimeter-scale firmground- and hardground-bounded sediment cycles. Deposits are storm modified throughout but textured by daily shallow-water processes. All facies are profoundly affected by changes in trophic resources as controlled by climate. They exhibit low-mesotrophic to mesotrophic attributes (abundant large benthic foraminifers) during arid periods and high-mesotrophic to eutrophic attributes (an infaunal heterotrophic biota) during humid periods when abundant nutrients and terrigenous clays are delivered by fluvial processes. These epeiric ramp platforms differ from the others in that they seem to record an extremely detailed record of climate and relative sea-level change.

COLD-TEMPERATE FACIES MODELS

General Attributes

Modern cold-temperate shelves have been episodically glaciated or influenced by glaciation throughout the Pleistocene. The shallow seafloor was exposed, veneered with siliciclastic sediments, or covered by glacigene deposits. Diamictites are most extensive in the northern hemisphere where continental glaciation was exceptionally widespread. This is not necessarily the case in the southern hemisphere because of fewer large continents at these attitudes and thus less glaciation. Following retreat of continental glaciers after the last glacial maximum many northern hemisphere shelves have become sites of heterozoan carbonate deposition, especially on their outboard extremities. By contrast, those in the southern hemisphere, particularly New Zealand, appear to have siliciclastic sediments covered by similar carbonates. The distribution of such sediments in the northern hemisphere is, however, complicated by oceanic current circulation patterns. This is especially evident along the extensive and well-studied continental shelves of NW Europe where relatively warm waters of the Gulf Stream system intrude well above the Arctic Circle creating a province of cold-temperate carbonates in ice-free waters at relatively high latitudes.

Minimum surface seawater temperatures on these cold-temperate, open shelves, regardless of location, ranges from ~5 to 15°C. Regions in this cold-temperate category are the platforms around New Zealand, United Kingdom, the northeast Atlantic, the western coast of Canada, the South China Sea and Japan (Fig. 2B).

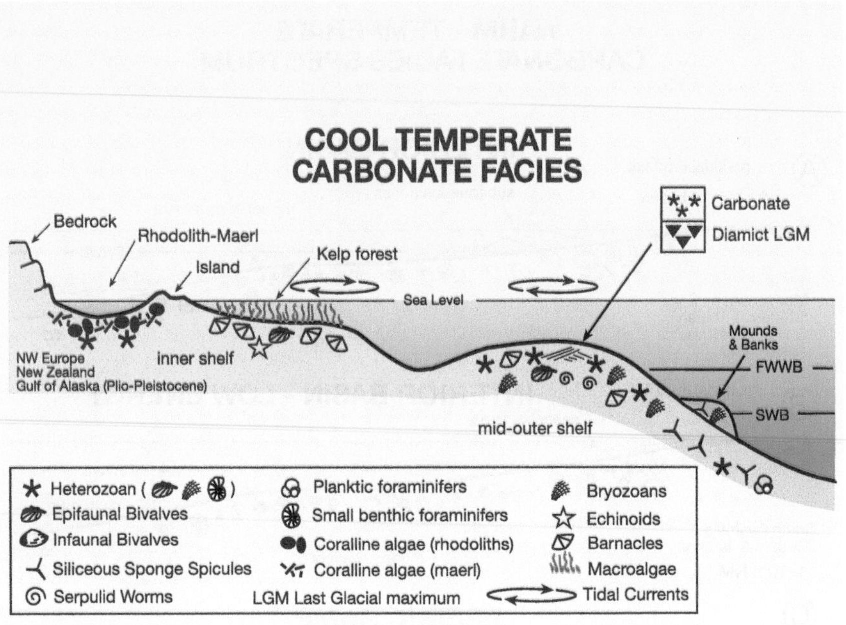

Figure 21. Facies expected in a cold-temperate open-ocean setting along continental margins and offshore large banks; in this depiction carbonate facies are accumulating on top of LGM glacigene deposits.

Today, these carbonates are a thin layer of biogenic material on top of LGM sediments. Carbonates are almost completely heterozoan, nearshore kelp forests are typical, and marine waters are overwhelmingly mesotrophic. While similar to those in non-glaciated regions, the facies motif in glaciated settings is somewhat different because of a dissimilar depositional history and higher trophic resource levels. Of particular importance is the presence of a hard substrate formed from boulder pavements resulting from winnowing of mud and sand during the Holocene transgression. Such successions in the rock record comprise interbedded glacigene or siliciclastic deposits and carbonates. The preservation potential of such carbonates is unpredictable because they can be removed during glacial re-advance or by terrestrial erosion during lowstands.

Open Ocean Facies Models

The *inner shelf* to ~50 mwd or less comprises two major facies that are dominated by algae, kelp forests and coralline algal (maerl) banks (Fig. 21). Kelp forests are located in ocean-facing, high-energy environments to 30 mwd or so wherein the hard substrates between phaeophyte holdfasts are colonized by innumerable barnacles that, together with epilithic bivalves and echinoderm debris, generate copious amounts of barnacle-rich sediment. This material covers the adjacent seafloor in the form of rippled sands that are generally barren but in less energetic settings are burrowed and populated by numerous epifaunal and innumerable infaunal bivalves.

The maerl facies in the NW Atlantic occupies somewhat protected environments behind islands. The depth limits of this red algal facies become shallower with increasing latitude, 20 mwd offshore Brittany (48°N) to 6–15 mwd offshore Norway (70°–80°N). Rigid branching corallines grow on hard surfaces or form banks (reefs) of interlocking branches and algal crusts. The branches are broken off during storms and together with less numerous barnacle plates form nuclei for the growth of rhodoliths that can in turn be fragmented to form gravels of coralline twigs. These particles in areas of active tidal flow are swept into banks and shoals of crossbedded maerl.

Sediments on the subphotic *mid- to outer shelf* are likewise mostly generated on hard substrates, either bedrock or on winnowed glacigene boulder lags. Bivalves, bryozoans, azooxanthellate corals, and serpulids, together with echinoids, and benthic foraminifers densely colonized these hard substrates. Their remains are again swept off to form rippled and burrowed sands that are habitats for infaunal and epifaunal bivalves; equivalent to the shaved shelf of the warm temperate realm. The seafloor in some areas is covered by thickets of articulated branching bryozoans (*Cellaria* sp.) that act as a source for subaqueous bryozoan sand dunes of up to 10 m relief above the seafloor. Gravels composed of serpulid worm tubes are widespread where currents are subdued along the shelf edge.

Sediments from two offshore banks in the NE Atlantic (the Porcupine and Rockall banks) illustrate a strong, depth-controlled facies pattern. The shallow, subphotic bank tops are variably outcrops and carbonate sand. In areas of upwelling (Rockall Bank), heterozoan carbonate sand is abundant, rippled, and dominated by bryozoans and serpulids. Sponges are prolific. Where there is no upwelling (Porcupine Bank), sediments are less abundant and composed mainly of bivalves with lesser bryozoans, gastropods, serpulids, echinoids and benthic foraminifers. Below *ca.* 200 m, all sediments are dominated by planktic foraminifers. Between 200 and 500 m there are numerous patches and mounds of the azooxanthellate coral *Lophelia*.

COLD-WATER, POLAR CARBONATE FACIES MODELS

Introduction

North of, and adjacent to, the polar front in the Barents Sea, eastward along the coasts of Greenland, and southward along the coastline of eastern Canada are areas that share similar environmental attributes; shelves are ice-covered during winter months, water temperatures rarely rise above 5°C, and icebergs

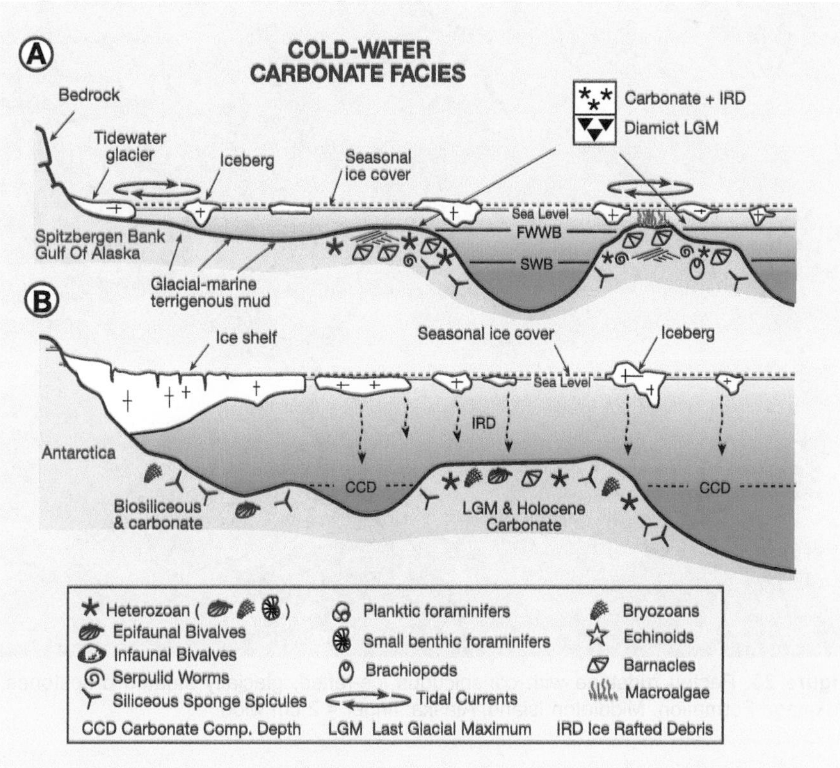

Figure 22. Cold-water carbonate facies in high-latitude environments with seasonal ice cover. **A.** Facies on continental shelves and offshore banks as represented in the northern hemisphere; **B.** Carbonate facies on the deep Antarctic continental shelf, much of which is relict and rich in biosiliceous material.

frequently scour their sediment-mantled shelves. This is the cold-water, polar realm that is also present around Antarctica (Fig. 2B). These heterozoan carbonates are differentiated by the presence of ice-rafted debris (IRD) and an abundant biosiliceous component composed of diatom frustules and sponge spicules.

Depositional conditions in these environments are extreme and characterized by strong seasonality. In many locations the ocean is ice-covered for many months and periods of winter darkness alternate with continuous summer light. The sea-ice front moves back and forth across the region during the year and, because of active water mixing, is a site of high trophic resources, especially during the spring melt. Periods of eutrophy alternate with periods of relative mesotrophy. When the sea is ice free it can be subjected to intense storm activity. Most carbonate deposition is localized to those areas that are seasonally ice-free; biogenic deposition below permanent ice shelves and sea ice is generally negligible and mostly biosiliceous. Recent work has, however, discovered that the bryozoan–sponge community can produce sediment beneath permanent ice cover, in perpetual darkness, as much as 100 km inboard from the ice front.

Open Ocean Facies

Carbonates accumulating on Pleistocene glacigene sediments but without any modern glacial sediment input are located in the open ocean away from continents on large banks in the Barents Sea and on relatively shallow banks at the outer edge of continental shelves where sea-ice contains little glacigene material (Gulf of Alaska, Antarctica). They are similar in style to cold-temperate facies but contain IRD (Figs. 1, 22, 23, 24, and 25).

The large Spitzbergen Bank and SW Svalbard Shelf are adjacent to the polar front. The post-glacial neritic environments were originally covered by muddy diamictite populated by an infaunal bivalve community. Subsequent increase in current strength winnowed the mud leaving

Figure 23. Pecten rudstone with conspicuous ice-rafted, glacially striated dropstones; Yakataga Formation, Middleton Island, Alaska; finger = 2 cm wide.

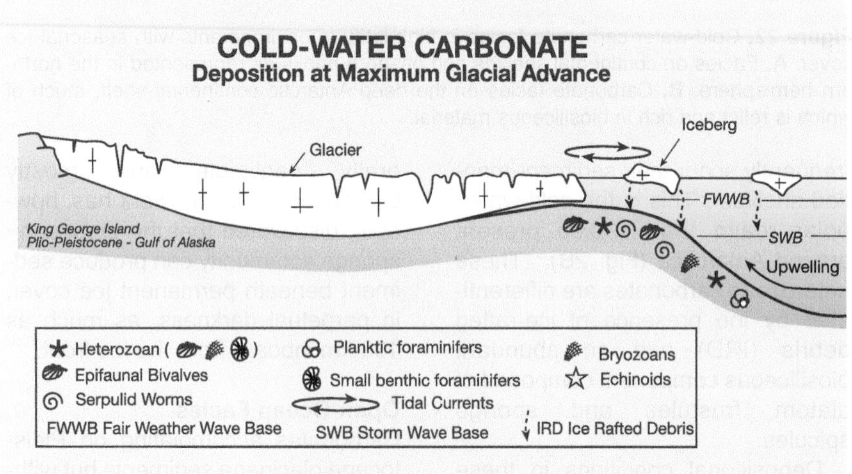

Figure 24. Cold-water, polar carbonate deposition near an ice front during maximum glacial advance; terrigenous clastic deposition is largely arrested by cold conditions.

behind a bivalve and boulder-cobble lag that was populated by a modern epibenthic community of barnacles, bivalves (especially pectens) bryozoans, brachiopods, and serpulids (Fig. 22A). Shallow bank tops from 20–60 mwd are mantled by extensive barnacle sands that are swept into subaqueous dunes up to 1 m high. This sediment is produced on hard substrates populated by kelp forests to depths of ~25 mwd. In deeper water, a boulder-pavement seafloor of winnowed Pleistocene diamict is colonized by patches of hydrozoans, bryozoans, soft corals, and large barnacles. Ice-rafted debris on the narrow SW Svalbard shelf comes from icebergs whereas on the more isolated Spitzbergen Bank it comes from passing ice floes.

Antarctica is, in contrast, quite a different setting (Fig. 22B). The perpetually ice covered continent has vast floating ice shelves that cover much of the continental shelf that is in turn much deeper (average 500 mwd) than any other continental margins. Icebergs are numerous and IRD plus windblown sediment form the major source of terrigenous sediment. The overall setting is one of extensive upwelling and high seasonal primary productivity. At and beyond the calving front, biogenic productivity is high and reflected by the formation of both calcareous and siliceous biogenic sediments (Fig. 26). Shallowest parts of the shelf at ~60 mwd are current-swept with the familiar epibenthic suspension feeders on hard substrates producing carbonates composed of bivalve, bryozoan, echinoid, gastropod, and benthic foraminifer remains.

There are however, two unique features. First, epibenthic suspensions feeders such as sponges, bryozoans, and ascidians together with brachiopods and alcyonarians across most of the deep shelf are growing on soft and not hard substrates. Many types of sponges can also cover more than 50% of the seafloor with their spicules forming dense mats 1–2 m thick. There is virtually no infauna of deposit-feeding or burrowing animals, and molluscs are conspicuously rare; it is a very Paleozoic-like biota. This situation is ascribed to very low sedimentation rates and lack of durophagus predators. Carbonate silt is not abundant.

Second, the relatively shallow outer parts of the continental shelf and shelf edge are covered in heterozoan carbonates resting on top of diamictite. Although familiar, the carbonates are predominantly relict, deposited during the last glacial maximum (LGM) when ice was at the shelf edge and the seafloor was ~165–200 m lower than today. These carbonates are not interglacial but were formed during the maximum glaciation – they are the coldest of all carbonates. Surprisingly, this does not appear to be unique. Plio-Pleistocene deposits of this facies (Eyles and Lagoe, 1989; Gazdzicki, 1984; James *et al.*, 2009) are astonishingly similar (Fig. 23). They are typically redeposited in the form of tempestites. Locally, accumulation rates can be high, on the order of 15 cm/1000 years, (Chapter 5).

Open-ocean seamounts in the same environments near the seasonal ice front have similar facies. The overall environment of one such bank in the Greenland Sea (Henrich *et al.*, 1992) is mesotrophic and ice-covered, with a two-month summer window of ice retreat, plankton bloom,

Figure 25. Late summer at Otto Fiord, Ellesmere Island, Arctic Canada with melting sediment-rich icebergs; image width in foreground = ~300 m.

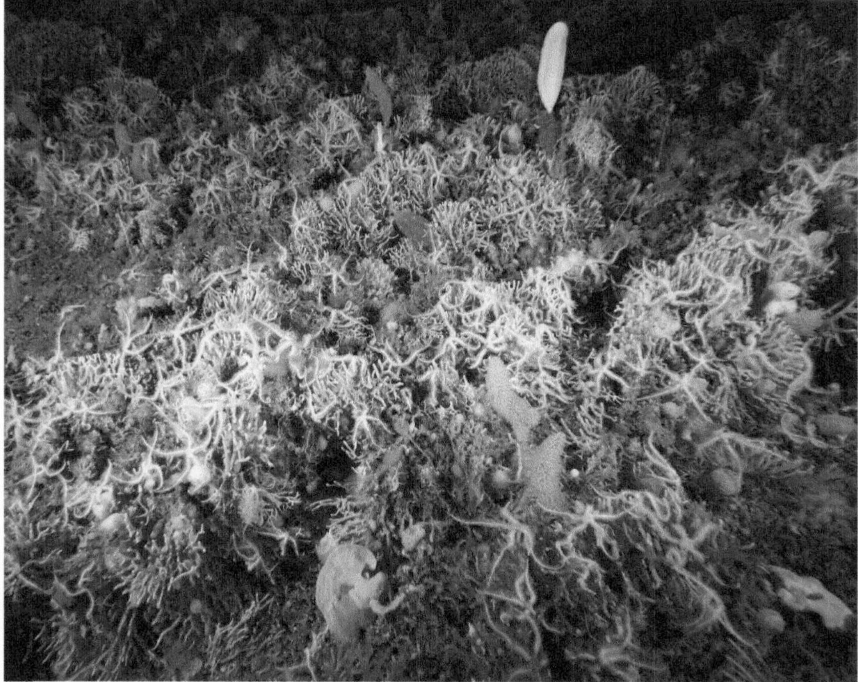

Figure 26. A rich calcareous invertebrate community dominated by hydrozoans (stylasterines–*Errina* sp.), soft corals, bryozoans, brittle stars, sponges, crinoids, alcyonarians, and gorgonians at a depth of ~800 m on the upper slope of George V Land, Antarctica. Sponges become more numerous with increasing depth. Copyright Australian Antarctic Division, Geoscience Australia.

and transfer of large quantities of food to the benthic community below. The shallowest seafloor (133–260 m) is mantled by a siliceous mat of microbes, sponge spicules, and bryozoans up to 10 cm thick, with meter-size bryozoan–hydrozoan–serpulid mounds and bryozoan–sponge thickets. The upper slopes (260–400 mwd) are veneered with bryozoan–bivalve–foraminifer and minor spiculite as well as numerous pectens, while small elevations are colonized by sponge–bryozoan buildups. Rocky slopes below 400 mwd have a variety of small buildups formed by bryozoans, sponges and crinoids, with crinoids being most abundant on soft bottoms.

PALEOZOIC COOL-WATER CARBONATES

Introduction

Recognition of cool-water carbonates and their facies is not straightforward in Paleozoic strata because the heterozoan benthic biota, upon which many facies are differentiated, is fundamentally different. Overall it is an epibenthic-dominated carbonate community with relatively little predation from which many post-Paleozoic cool-water heterozoan components are missing or not significant as sediment producers. Ephemeral components such as sea grass and large phaeophytes are absent, and the role of other macroalgae is unknown. Coralline algae, echinoids, bivalves, and serpulids although present were not significant carbonate sediment producers. On the other hand, pelmatozoans and brachiopods were fundamental elements of the depositional system. Bryozoans remained abundant although dominated by erect-rigid, non-articulated forms. Acorn barnacles had not evolved. The ecological attributes such as light, seawater temperature, and nutrient requirements of rugose corals, tabulate corals, and stromatoporoids are highly controversial (*see* Chapter 17). Whereas small benthic foraminifers are present throughout, large benthic photosymbiont-bearing foraminifers only appear in the late Paleozoic. Finally, ocean chemistry was different with higher silica contents because diatoms had not evolved, and higher carbonate saturation because calcareous plankton had not developed. On balance, the core of the heterozoan assemblage was brachiopods, bryozoans, pelmatozoans and small benthic foraminifers.

With the absence of so many critical components that can be related to modern environments, the character of cool-water carbonates relies heavily on stratal relationships, especially where carbonates are temporally and spatially related to confirmed cool-water strata.

Most of the Paleozoic cool-water carbonates identified are Mississippian to Permian, a global icehouse period related to global cooling and the late Paleozoic Ice Age (LPIA). The rarity of non-tropical carbonates in the early to mid-Paleozoic is likely because that period was dominated by greenhouse conditions, much like the Cretaceous. Those documented to date can be related to lower seawater temperature, stratification, or large-scale climatic change.

The Thermocline Facies Model

This model (Fig. 27) is most thoroughly studied in the Western Canada Sedimentary Basin (Martindale and Boreen, 1997). The setting is a westward-thickening and deepening Mississippian foreland basin succession wherein shallow-water sediments with ooids, calcareous green algae, and other photozoan components grade outboard into heterozoan facies. The succession is interpreted such that photozoan facies accumulated in warm waters above a thermocline with contemporaneous heterozoan facies below in waters so cold as to allow glendonite formation. Sub-thermocline crinoid shoals periodically transgress inboard signaling a breakdown of the thermocline and intrusion of cold waters into very shallow environments. Deep-water facies are bryozoan rich (Fig. 28), with bryozoan downslope mounds and deeper water spiculitic mudstones and shales. Similar facies disposition in the Middle Ordovician are likewise interpreted.

The Climate Change Facies Model

Climate change is recorded at the end of the LPIA by either regional cooling or regional warming (Fig. 29). The *climate change cooling model* is present in the Sverdrup Basin of the Canadian Arctic wherein photozoan carbonates and evaporites pass stratigraphically upward into cold-water deposits (Beauchamp and Desrochers, 1997). This transition appears to document changes in northern hemisphere global seawater circulation patterns that resulted in progressive refrigeration and attendant nutrient enrichment. Warm, shallow-water Pennsylvanian and earliest Permian photozoan limestones comprise the well-know phylloid algal, fusulinid, ooid facies that, like the Mississippian, grades basinward, perhaps across a thermocline, into heterozoan and deeper water biosiliceous chert facies. Seawater cooling and trophic resource increase during the Early Permian resulted in a progressively less diverse and more heterozoan facies succession in shallow water. By Late Permian, all facies across the storm-dominated ramps were biosiliceous chert. This situation is important because it reflects the higher silica content of Paleozoic seawater that allowed extensive sponge spicule sedimentation to occur (Fig.

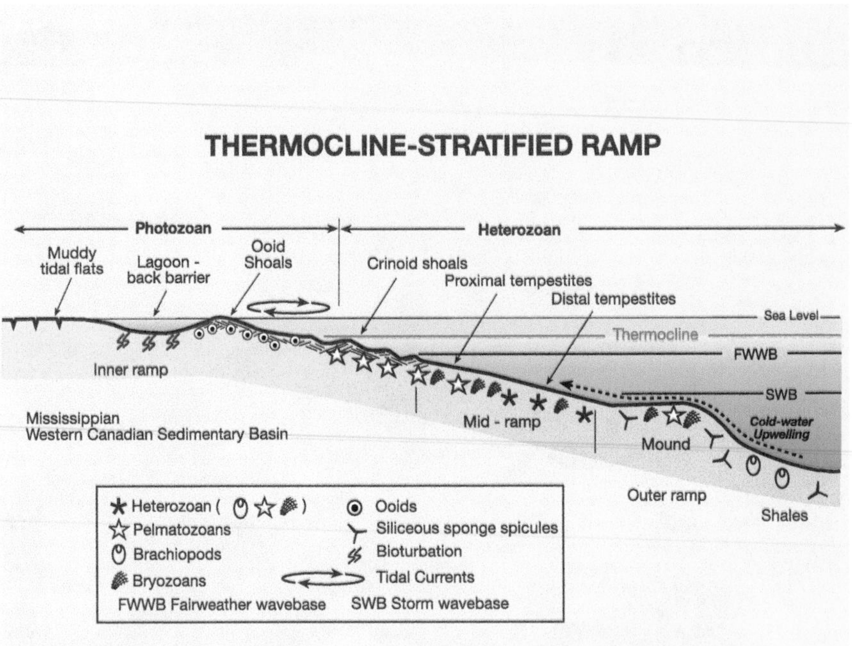

Figure 27. Carbonate facies on a Paleozoic temperature-stratified ramp as exemplified by the Mississippian in parts of the Western Canadian Sedimentary Basin (after Martindale and Boreen, 1997).

Figure 28. Bryozoan rudstone and floatstone; Early Permian Assistance Formation, Ellesmere Island, Arctic Canada; finger scale = 2 cm wide.

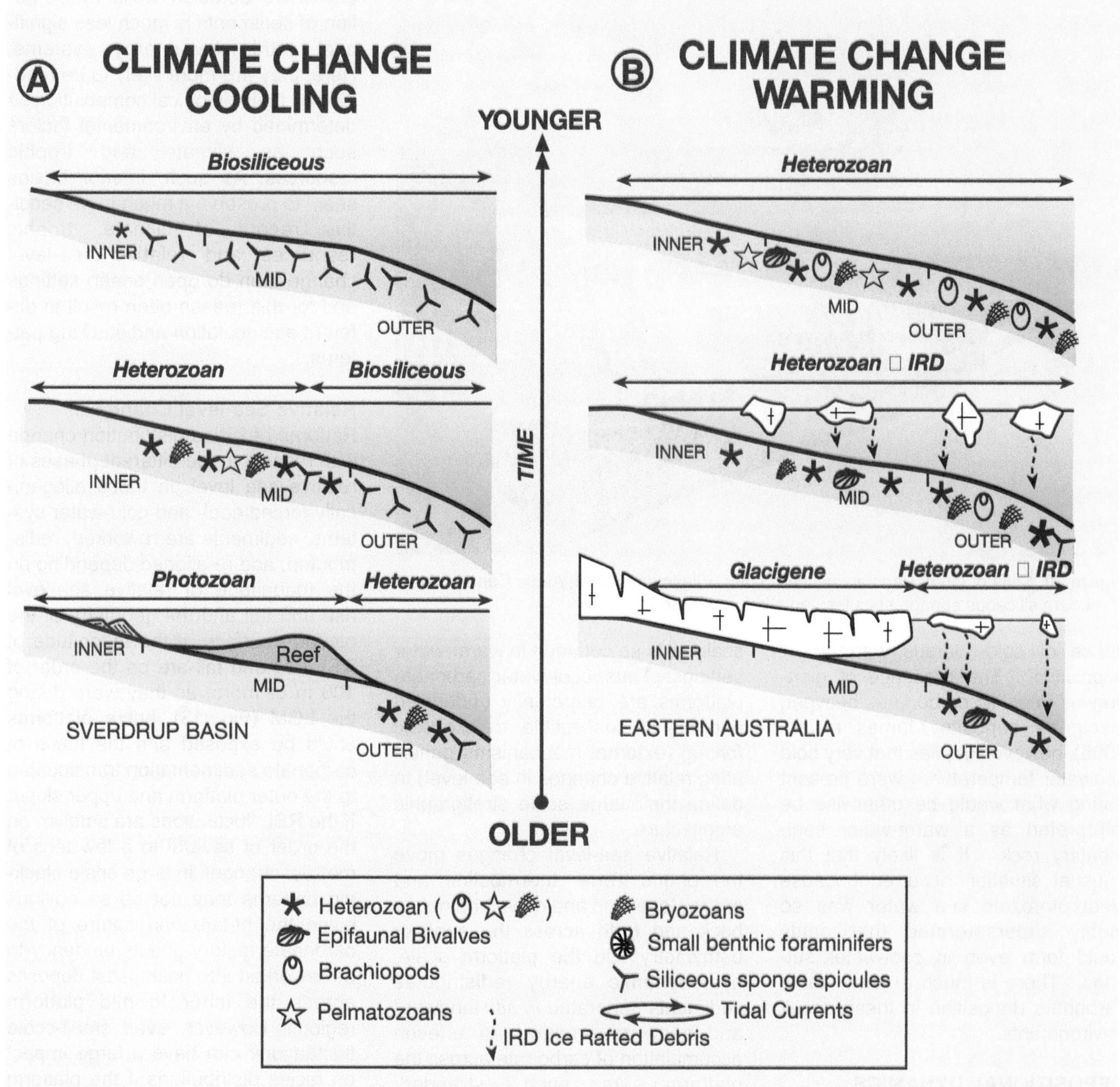

Figure 29. Climate change Permian carbonate facies as illustrated by cooling in what is now the Sverdrup Basin, Arctic Canada **(A)**, and warming in what is now eastern Australia, southern Gondwana **(B)**.

30), yet there is little evidence for the presence of freezing temperatures in the form of glendonites and significant IRD in the succession. This Paleozoic biosiliceous cooling motif is also present in the Late Ordovician just prior to the end-Ordovician Hirnantian glaciation.

The contrasting *climate change warming model* is present in the Lower Permian strata of the southern hemisphere, albeit at higher paleolatitudes. The succession, especially in eastern Australia and other parts of Gondwana (Rogala *et al.*, 2007, Fielding *et al.*, 2008), contains carbonates associated with continental deglaciation and warming and glacial deposits containing abundant IRD (Fig. 1) and glendonites. Carbonates occur in ramp-like settings away from areas of terrigenous clastic input, are heterozoan in composition and are characterized by low diversity, high-abundance heterozoan biotas. Coldest water carbonate facies contain IRD that become progressively less abundant up-section as seawater warmed and ice disappeared; there are no glendonites in the carbonate facies.

THE PROTEROZOIC

This is the most difficult period in earth history to assess carbonate facies in terms of seawater temperature because there are no calcareous invertebrates to act as proxies. Yet it is nevertheless clear, especially in the Neoproterozoic, that wide fluctuations in climate occurred, to the point that there may have been global glaciation. Glacigene sediments are typically overlain by cap carbonates that contain no clear evi-

Figure 30. Cliffs of latest Permian spiculitic chert, Ellesmere Island, Arctic Canada; inset of modern siliceous sponge spicules; tents in foreground for scale.

dence of cold- versus warm-water deposition. The presence of glendonites in oolitic carbonates between glacigene deposits (James *et al.*, 2005), however, implies that very cold seawater temperatures were present during what would be otherwise be interpreted as a warm-water sedimentary rock. It is likely that this unusual situation occurred because Neoproterozoic sea water was so highly supersaturated that ooids could form even in cool-water settings. There is much to learn about carbonate deposition in these occult environments.

DEPOSITIONAL DYNAMICS

General Attributes

Stratigraphic packaging is the stacking of facies through time in response to fluctuations in relative sea level (RSL). How these stratigraphic patterns develop is dependent on the creation and destruction of accommodation (the space in which sediments can accumulate) as defined by the interaction of tectonics, eustasy and biology (*see* Chapter 2). The generally low rates of carbonate production, lack of early marine cementation, and common hydrodynamic redistribution of sediments means that cool- and cold-water platforms are generally unable to aggrade to sea level or to form their own topography on a large scale as is so common in warm-water settings. Thus, cool-water carbonate platforms are perpetually underfilled and highly susceptible to *allogenic forcing* (external mechanisms generating relative changes in sea level) to define their large scale stratigraphic architecture.

Relative sea-level changes move the photic zone, thermocline and energy levels up and down and hence back and forth across the seafloor bathymetry. On the platform scale, hydrodynamic energy redistributes sediments generated *in situ* landward and seaward creating an uneven accumulation of carbonate across the platform. As such, stranded, palimpsest sediments and relatively thin stratigraphic successions on the upper platform are to be expected. Aggradation is limited and progradation is mostly confined to outer platform and, in low-energy settings, nearshore areas. Although production happens everywhere from shoreline to mudline, sediments accumulate most readily in low energy areas encompassed by the sheltered substrates of nearshore seagrass meadows and kelp forests, depressions on a wave swept seafloor and broad flat expanses below fair-weather wave base. Sedimentation patterns on high-energy platforms are heavily influenced by the depth of wave base and wave abrasion while redistribution of sediments is much less significant within low-energy systems. Here, they are more likely to be modified by their biological composition as determined by environmental factors such as climate and trophic resources. As such, interior basins seem to preserve a much more sensitive record of climate, trophic resources and relative sea-level change than do open ocean settings and for this reason often result in different accumulation and stacking patterns.

Relative Sea-level Change

Patterns of facies distribution change with respect to the different phases of relative sea level. In these allogenically-forced cool- and cold-water systems, sediments are reworked, redistributed, and re-aligned depending on the magnitude of relative sea-level rise and fall and the geometry of the platform surface. If the magnitude of RSL rise and fall are on the order of 100 m or more, as they were during the LGM (Fig. 13), entire platforms could be exposed and the focus of carbonate sedimentation translocated to the outer platform and upper slope. If the RSL fluctuations are smaller, on the order of several to a few tens of metres, changes in large scale stacking patterns may not be so obvious given the heterozoan nature of the carbonate factory and its tendency to be reworked into palimpsest deposits across the inner to mid platform regions. However, even small-scale fluctuations can have a large impact on facies distributions if the platform is flat (Fig. 31). This is especially true for epeiric platforms and broad open shelves where a 50 m drop of RSL can lead to 100 km or more of seafloor exposure!

Sequence Stratigraphy

Sequence stratigraphic principles applied to these platforms are best derived from the siliciclastic literature, but with a caveat. These are clastic carbonates (as stressed above), born *in situ*, but redistributed by hydrodynamics. Sedimentation below the zone of active wave base is slow, but nearly constant – an important difference to siliciclastics. Principles derived from the warm-water carbon-

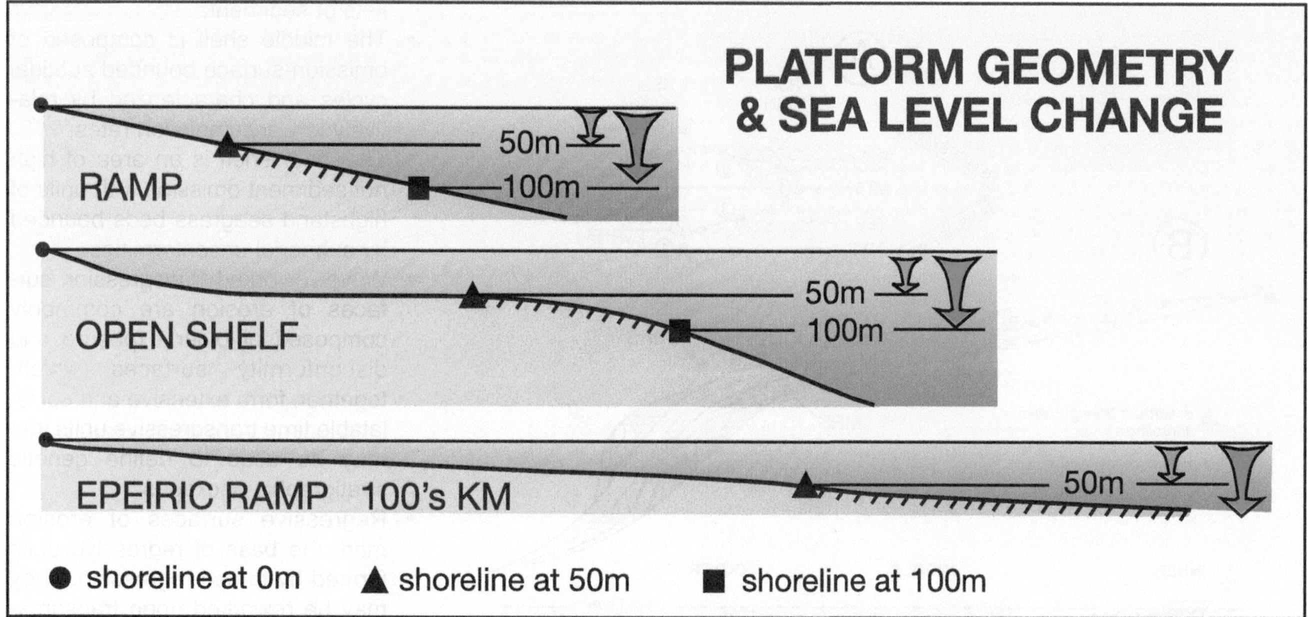

Figure 31. Effect of platform geometry and magnitude of relative sealevel fluctuation on the distance of basinward shifts of shorelines.

ate literature (*cf.* Sarg, 1988) are not so easily applied to these mostly drowned systems in the cool-water realm. Falls in RSL do not shut down the carbonate factory, only redistribute it. Updip, exposure surfaces can be cryptic due to their low diagenetic potential. From a terminology standpoint, the literature has a dizzying array of sequence stratigraphic terms for similar successions. They are simplified here, taking a common sense approach to separate stratigraphic packages as defined by changes in relative sea level from transgression (RSL rise) to regression (RSL fall) bounded by maximum transgressive and maximum regressive shorelines (Fig. 32A).

Although various platform types respond differently to changes in RSL, there are some particular attributes shared by most. The low overall productivity, nonphototrophic nature and easily reworked character of sediments generated on cool-water carbonate platforms means that stratigraphic packages composed of palimpsest sediment riddled with omission surfaces are to be expected in the inner to mid-platform regions affected by a fluctuating wave base. The packaging in these systems is forced by RSL change. As such, cool-water carbonates record periods of transgression and regression as reworked, disconformable surfaces created by a scouring of wave base (Fig. 33) that may be correlated over long distances (Fig. 32B). While the regressive surfaces of erosion (RSE) record critical downstepping of facies belts during a fall in relative sea level (Hunt and Tucker, 1992; Lukasik and James, 2006), the transgressive surfaces of erosion (TSE) tend to cut down and superimpose themselves onto previous RSE's often making them the most obvious and correlatable horizons on which to separate depositional sequences (Caron *et al.*, 2004). These inner and mid-platform regions are punctuated by a series of omission surfaces and development of subtidal cycles (James and Bone, 1994) while down dip, below the most basinward depth of wave base (time of maximum regressive shoreline) lies an intercalated package of time-continuous sediment (Fig. 32C).

How these packages and surfaces express themselves differs between open-ocean, interior basin and epeiric ramp successions. For simplicity, models of stratigraphic packaging in warm-temperate situations derived only through large changes in relative sea level are illustrated. There is not yet enough information to generate meaningful models for cold temperate and cold water successions. For detailed examples of smaller scale cyclicity in the cool-water realm, refer to the references at the end of this chapter.

Warm-temperate Open-ocean Shelf

During the period of *maximum regression* or most basinward movement of the shoreline, a large expanse of the broad, shallow inner and middle shelf is subaerially exposed, wave base affects a relatively narrow zone of seafloor on the outer shelf, and increased upwelling occurs offshore as a response to changing oceanography. As such, bryozoan mounds develop on the upper slope amidst a background of spicule-rich fine-grained carbonate sediment. Together they form a distinct *lowstand* wedge (Fig. 34A).

As RSL rises during *transgression*, slowly at first then rapidly, wave base scours and reworks the underlying subaerial exposure surface as a series of transgression surfaces of erosion as the carbonate factory steps landward. Wave sweeping starts to impinge upon progressively shallower regions of the newly flooded shelf moving sediment basinward where it accumulates as a blanket over the lowstand wedge. Transgressive erosion updip may also trigger downslope sediment and gravity flows of the weakly cemented sediment. Bryozoan mounds become smothered. In the middle to inner shelf, transgressive surfaces of erosion amalgamate forming a widespread, time transgressive interval that separates regressive units below

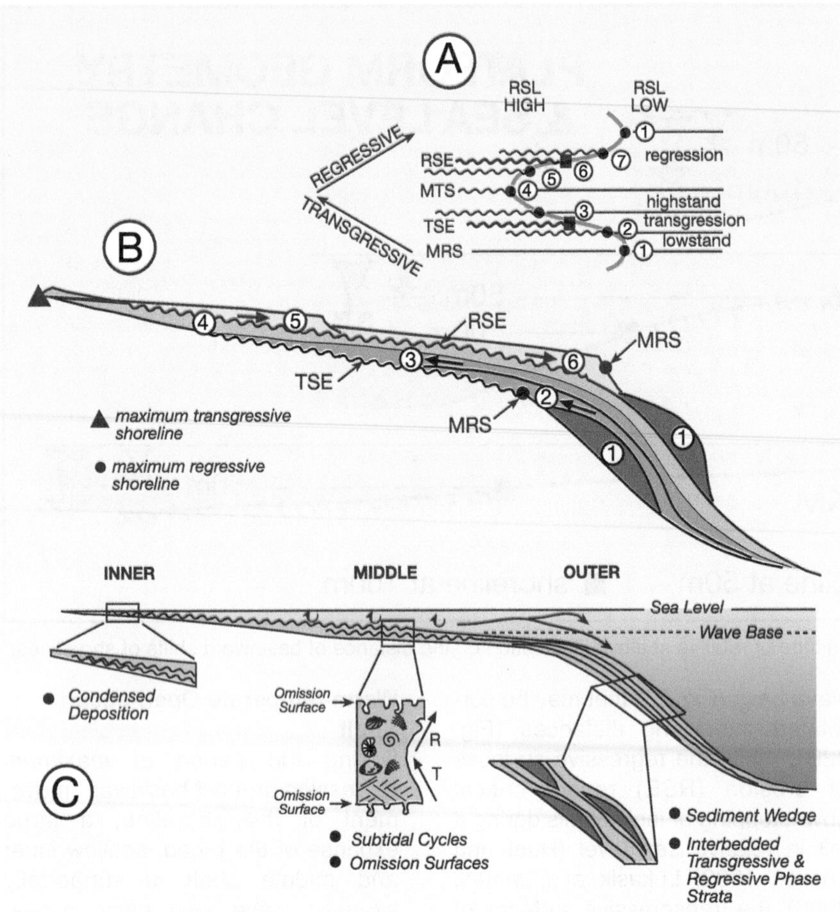

Figure 32. A. Common sequence stratigraphic terminology used in forced regressive, cool- and cold-water carbonate systems and their illustration in a general platform context; **B.** Periods of regression result in a shift of facies and wave base basinward towards a maximum regressive surface (MRS). Subsequent transgression forces facies and wave base landward, ultimately deepening to a maximum transgressive shoreline. Regressive surfaces of erosion (RSE) and transgressive surfaces of erosion (TSE) are characteristic surfaces often preserved in ancient successions; **C.** Inner platform areas are often characterized by low net accumulation; mid-platform regions by omission-bounded subtidal cycles generated under smaller scale fluctuations of RSL, and outer shelf to upper slope stratigraphic packages that are more conformable.

from transgressive above.

When the relative rise in sea level begins to slow during RSL *highstand,* the highly prolific nearshore seagrass beds, if present, have the opportunity to aggrade and even prograde. Across the inner to middle shelf, the zone of wave abrasion is firmly established sweeping sediment from the carbonate factory landwards into seagrass beds and basinward where it accumulates below wave base. The outer shelf is a zone of prolific bryozoan and sponge production where it rolls over onto the spiculitic muds of the upper slope.

As relative sea level starts to fall during *forced regression* (Fig. 34B), wave base steps basinward resulting in regressive surfaces of erosion carving into sediments deposited during the recent highstand. In steeper regions, downslope sediment movement in the form of bypass channels and mass sediment wasting may occur. Thin sediment wedges step progressively downwards until RSL stops falling. When it does, this depositional surface of maximum regression marks the boundary between RSL fall and subsequent RSL rise.

General Characteristics

- Outer shelf and upper slope are areas of constant aggradation and progradation with regressive and lowstand wedges intercalated with transgressive and highstand blankets of sediment.
- The middle shelf is composed of omission-surface bounded subtidal cycles and characterized by relatively low accumulation rates.
- The inner shelf is an area of high net sediment omission with units of highstand seagrass beds bounded by subaerial unconformities.
- Wave reworked transgression surfaces of erosion are commonly composed of older exposure and disconformity surfaces which, together, form extensive and correlatable time transgressive units that may be used to define genetic stratigraphic packages.
- Regressive surfaces of erosion mark the base of regressive units formed by forced regression. They may be reworked upon transgression in areas of thin sedimentary cover. Their maximum basinward extent marks the maximum regressive surface.

Warm Temperate Interior Basins

Interior basins with high-energy inner platforms respond to RSL change differently from their lower-energy ramp counterparts. High-energy systems develop grainy beaches and carbonate shoals inboard (Fig. 20A) that grade outboard into the highly productive carbonate factory from which their sediment is sourced. Upon *transgression*, these shoreface complexes climb updip as a series of wedges towards the ultimate maximum transgressive shoreline (Fig. 35A). Transgressive surfaces of erosion are present but may be hidden in later transgressive phases given the overall high-energy nature of the overlying *highstand* wedge that prograde out over the top. Progradation of these grainy systems occurs during highstands of sea level grading basinward into thinner and muddier carbonate sediments downdip. Grainy basin floor fans develop due to sediments bypassing the slope from shallower regions. Upon RSL fall, the forced *regressive* wedges of shoreface sands, downstep out over the highstand strata below. Regressive surfaces of erosion are common. Basin floor fans formed from the earlier highstand period may become buried by later regressive wedge deposition. The maximum regressive sur-

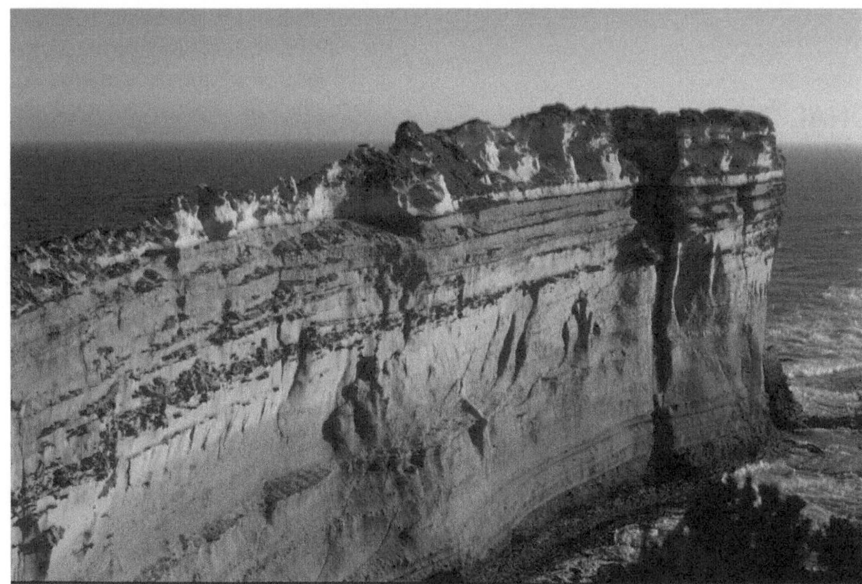

Figure 33. Oligocene–Miocene cool-water limestones, eastern Victoria, Australia (cliff ~40 m high). The shallowing-upward, outer- to mid-shelf (grey-brown) successions are sharply overlain by buff, shallower water, grainy limestones.

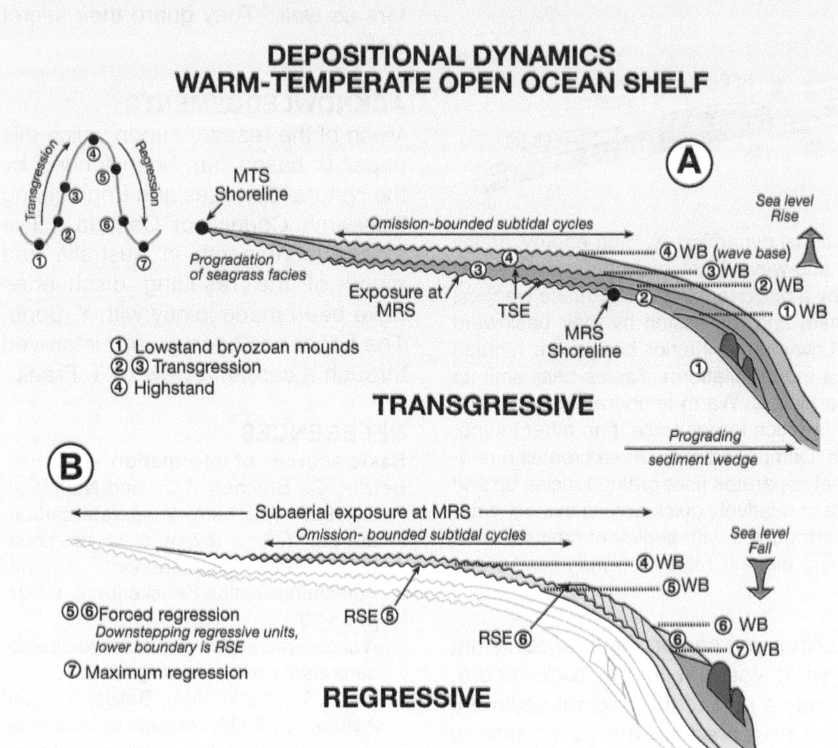

Figure 34. Warm-temperate open shelf depositional dynamics. **A.** Lowstand phases see widespread exposure of the platform updip and thick sediment wedge formation and bryozoan mound development downdip. Transgressive units are defined by TSE development and a backstepping of facies onto the platform. Highstands of relative sea level are periods of sediment generation and deposition in the outer shelf, wave sweeping and low net sediment accumulation in the mid-shelf, and aggradation to progradation of nearshore seagrass beds if present; **B.** Regression is marked by the development of RSE's and a fore-stepping of sedimentary facies belts basinward.

face develops at the lowest point of RSL fall resulting in extensive subaerial exposure updip.

Low-energy systems, however, tend to form ramps and as such have a much more subdued topography across their depositional profile (Fig. 20B). Changes in RSL cause the facies belts to shift up and down the ramp without the formation of distinctive, thick sediment wedges (Fig. 35B). During times of increased rates of RSL change, transgressive and regressive surfaces of erosion may form. Otherwise, facies shifts are most commonly recorded as condensed beds or expressed by firmgrounds indicating sediment omission. Prolific carbonate production and binding of sediment within the extensive seagrass deposits on the inner ramp allow for the progradation of seagrass facies during RSL highstands.

Warm Temperate Epeiric Ramps

Epeiric ramps respond to RSL changes in much the same way as low-energy systems interior basin systems but on a much broader and hence more subtle scale (Fig. 35C). Widespread sediment omission occurs in all phases of RSL change. Low carbonate production rates combined with widespread and rapid shifts of facies belts with small changes in RSL (Fig. 31) mean that sedimentation is highly condensed as reflected in the ubiquitous omission surfaces, thin beds, overprinted bioturbation fabrics and palimpsest sediment composition. *Transgressive* surfaces often juxtapose grainy, winnowed and otherwise open marine facies over muddy, nearshore sediments with a sharp solitary hardground or erosive surface. These surfaces are very sharp and form natural boundaries for separating genetic stratigraphic packages. Likewise, *regressive* surfaces separate open marine relatively grainy sediments below from nearshore facies above but they are often expressed slightly differently depending on the facies onto which they are forming or the rates of RSL fall.

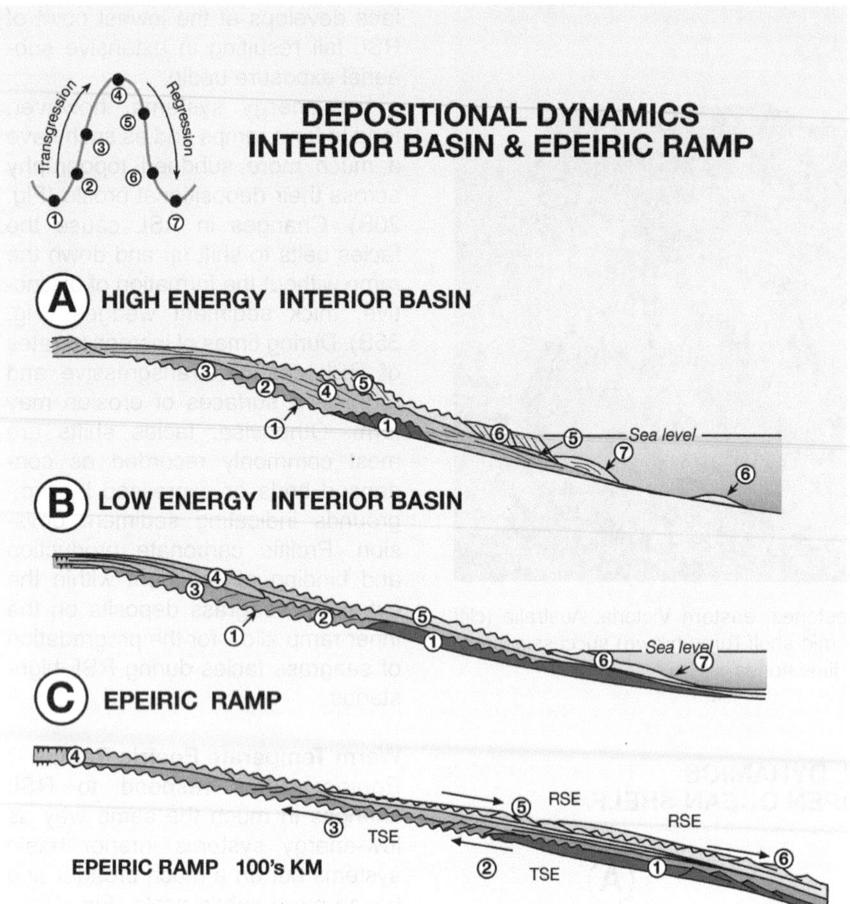

Figure 35. Warm-temperate interior basin depositional dynamics. **A.** High-energy, grainy shoreface association grades basinward into a sub-wave base carbonate factory and grainy basin floor fans. Transgression is marked by a backstepping of shoreface wedges, highstand by a progradation of beaches over them and regression by their basinward migration as discrete shoreface complexes; **B.** Low-energy interior basins are rimmed with shallow seagrass facies grading down into a muddy platform. Facies belts shift up and down across the ramp in response to RSL changes; **C.** Warm-temperate epeiric ramp depositional dynamics. Epeiric ramps operate on a much larger scale than other interior basins and are thus primarily low-energy systems. Dampened wave base creates a relatively narrow band of winnowed sands offshore that separates finer grained facies up and down dip from it. Facies belt migrations are large and relatively quick across this extremely shallow platform. As such, epeiric ramp strata are riddled with sediment omission surfaces, the most obvious of which are made during highest rates of transgression and regression.

SUMMARY

Some key points to remember about neritic carbonates of the cool- and cold-water realms are:

- They contain essential information about ancient oceans outside of the tropical realm.
- They may be born as carbonates but they accumulate as siliciclastics; they are clastic carbonates.
- The control of seawater temperature on the biotic character of the sediments may be strongly modified by the availability of sunlight and water clarity as determined by trophic resources.
- What you see on the seafloor is not what you get in the rock record; only a part of the original sediment is preserved on the other side of the diagenetic curtain.
- Peritidal carbonate facies are generally absent from these systems.
- Not only can local depositional conditions alter the distribution of facies belts within similar settings, but many of these carbonates are palimpsest, meaning that particles of different ages are mixed together.
- The stratigraphic packaging of these carbonates is overwhelmingly controlled by allogenic mechanisms due to their low productivity and inability of these systems to internally generate changes in accommodation.

Recognition of cool- and cold-water carbonates greatly increases our ability to understand the neritic realm of ancient oceans and interpret carbonate sedimentary rocks in a more imaginative fashion. The models presented here are the first organized attempt to bring together the overriding themes of sedimentation in this system. Modern environments are, however, difficult and expensive to study and so future refinements of these models will come slowly. Regardless, these carbonates will continue to be challenging to interpret because their compositions, upon which so many deductions are based, are the combined products of not only temperature, but many other parameters as well. They guard their secret well.

ACKNOWLEDGEMENTS

Much of the research upon which this paper is based has been funded by the Natural Sciences and Engineering Research Council of Canada. The extensive research in Australia and many of the resulting discoveries have been made jointly with Y. Bone. The paper has been greatly improved through a careful review by T. Frank.

REFERENCES

Basic sources of information

Betzler, C., Brachert, T.C., and Nebelsick, J., 1997, The warm temperate carbonate province: a review of facies, zonations and delineations: Courier Forschungsinstitut Senckenberg, v. 201, p. 83-99.

A critical paper defining warm-temperate carbonate depositional systems.

Braga, J.C., Martin, J.M., Betzler, C., and Aguirre, J., 2006, Models of temperate carbonate deposition in Neogene basins of SE Spain: a synthesis, *in* Pedley, H.M. and Carannante, G., *eds.*, Cool-water Carbonates; Depositional Systems and Palaeoenvironmental Controls: Geological Society of London, Special Publication 255, p. 121-136.

A summary of the numerous studies of the well exposed carbonates in Cenozoic basin in southern Spain.

James, N.P. and Clarke, A.D., 1997, Cool-water Carbonates: SEPM Special Publication 56, 440p.

An overview article and a series of

papers on modern and ancient cool-water carbonates - an excellent place to start; see particularly, *James, N.P. The Cool-water Carbonate Depositional Realm, p. 1-22.*

James, N.P. and Bone, Y., 2010, Neritic Carbonate Sediments in a Temperate Realm; Southern Australia; Springer, Berlin, 341p.
A synthesis of deposition on the largest cool-water carbonate shelf on the globe with a focus on the relationship between deposition and oceanography.

Lees, A. and Buller, A.T., 1972, Modern temperature-water and warm-water shelf carbonate sediments contrasted: Marine Geology, v. 13, p. M67-M73.
The seminal paper that defined carbonate grain associations on the modern seafloor and synthesized many cool-water carbonate occurrences.

Mutti, M. and Hallock, P., 2003, Carbonate systems along a nutrient and temperature gradients; some sedimentological and geochemical constraints: International Journal of Earth Sciences, v. 92, p. 465-475.
An overview of current thinking about the relationship between nutrients and styles of carbonate sedimentation.

Nelson, C.S., 1988, Non-tropical shelf carbonates - modern and ancient: Sedimentary Geology, v. 60, 367p.
The first real compilation of cool-water carbonates; see particularly *Nelson, C.S., An introductory perspective on non-tropical shelf carbonate, p. 3-12.*

Pedley, H.M. and Carannante, G., *eds.*, 2006, Cool-water Carbonates; Depositional Systems and Palaeoenvironmental Controls: Geological Society of London, Special Publication 255, 373 p.
Numerous articles on cool-water carbonates; the most current reference at the time of writing.

Other references - Modern, Cenozoic and Mesozoic

Anastas, A., Dalrymple, R.W., James, N.P. and Nelson C.S., 2006, Lithofacies and dynamics of a cool-water carbonate seaway: mid-Tertiary, Te Kuiti Group, New Zealand, *in* Pedley, H.M. and Carannante, G., *eds.*, Cool-water Carbonates: Depositional Systems and Palaeoenvironmental Controls: Geological Society of London, Special Publication 255, p. 245-268.

Andruleit, H., Freiwald, A. and Schafer, P., 1996, Bioclastic carbonate sediments on the southwestern Svalbard shelf: Marine Geology, v. 134, p. 163-182.

Bassi, J.C., Carannante, G., Murru, M., Simone, L. and Toscano, F., 2006, Rhodalgal/bryomol assemblages in temperate-type carbonate, channelized systems: the early Miocene of the Sarcidano area (Sardinia, Italy), *in* Pedley, H.M. and Carannante, G., *eds.*, Cool-water Carbonates; Depositional Systems and Paleoenvironmental Controls: Geological Society of London, Special Publication 255, p.35-52.

Boreen, T.D. and James, N.P., 1993, Holocene sediment dynamics on a cool-water carbonate shelf: Journal of Sedimentary Research, v. 63, p. 574-588.

Boreen, T.D., James, N.P., Wilson, C. and Heggie, D., 1993, Surficial cool-water carbonate sediments on the Otway continental margin, southeastern Australia: Marine Geology, v. 112, p. 35-56.

Bornhold, B.D. and Yorath, C.J., 1984, Surficial geology of the continental shelf, northwestern Vancouver Island: Marine Geology, v. 5, p. 89-112.

Bosence, D.W.J., 1985, The "coralligene" of the Mediterranean - a recent analog for Tertiary coralline algal limestones, *in* Toomey, D.F. and Nitecki, M.H., *eds.*, Paleontology, Contemporary Research and Applications: Springer-Verlag, Berlin, p. 216-225.

Burne, R.V. and Colwell, J.B., 1982, Temperate carbonate sediments of Northern Spencer Gulf, South Australia: a high salinity 'foramol' province: Sedimentology, v. 29, p. 223-238.

Carannante, G., Esteban, M., Milliman, J.D. and Simone, L., 1988, Carbonate lithofacies as paleolatitude indicators: problems and limitations: Sedimentary Geology, v. 60, p. 333-346.

Chave, K.E., 1967, Recent carbonate sediments - an unconventional view: Journal of Geological Education, v. 15, p. 200-204.

Coffey, B.P. and Read, J.F., 2004, Mixed carbonate–siliciclastic sequence stratigraphy of a Paleogene transition zone continental shelf, southeastern USA: Sedimentary Geology, v. 166, no.1-2, p. 21-57.

Collins, L.B., 1988, Sediments and history of the Rottnest Shelf, southwest Australia: a swell dominated, non-tropical carbonate margin: Sedimentary Geology, v. 60, p. 15-49.

Doyle, L.J. and Holmes, C.W., 1985, Shallow structure, stratigraphy and carbonate sedimentary processes of west Florida upper continental slope: American Association of Petroleum Geologists Bulletin, v. 69, p. 1133-1144.

Eyles, N. and Lagoe, M.B., 1989, Sedimentology of shell-rich deposits (coquinas) in the glaciomarine upper Cenozoic Yakataga Formation, Middleton Island, Alaska: Geological Society of America Bulletin, v. 101, p. 129-142.

Farrow, G.E. and Fyfe, J.A., 1988, Bioerosion and carbonate mud production on high-latitude shelves: Sedimentary Geology, v. 60, p. 281-297.

Farrow, G.E., Allen, N.H. and Akpan, E.B., 1984, Bioclastic carbonate sedimentation on a high-latitude tide-dominated shelf: northeast Orkney Islands, Scotland: Journal of Sedimentary Petrology, v. 54, no. 2, p. 373-393.

Fornos, J.J. and Ahr, W.M., 1997, Temperate carbonates on a modern, low-energy, isolated ramp: the Balearic platform, Spain: Journal of Sedimentary Research, v. 67, p. 364-373.

Freiwald, A., 1993, Coralline algal maerl frameworks - islands within the phaeophytic kelp belt: Facies, v. 29, p. 133-148.

Freiwald, A., 1998, Modern nearshore cold-temperature calcareous sediments in the Troms District, northern Norway: Journal of Sedimentary Research, v. 68, p. 763-776.

Gazdzicki, A., 1984, The *Chlamys* coquinas in glacio-marine sediments (Pliocene) of King George Island, West Antarctica: Facies, v. 10, p. 145-152.

Gillespie, J.L. and Nelson, C.S., 1996, Mixed siliciclastic carbonate facies on Wanganui Shelf, New Zealand: a contribution to the temperate carbonate model, *in* James, N.P. and Clarke, J.A.D., *eds.*, Cool-water Carbonates: SEPM Special Publication 56, p. 127-140.

Gostin, V.A., Belperio, A.P. and Cann, J.H., 1988, The Holocene non-tropical coastal and shelf carbonate province of southern Australia: Sedimentary Geology, v. 60, p. 51-70.

Halfar, J., Starasser, M., Riegel, B. and Godinez-Orta, L., 2006a, Oceanography, sedimentology, and acoustic mapping of a bryomol carbonate factory in the northern Gulf of California, Mexico, *in* Pedley, H.M., and Carannante, G., *eds.*, Cool-water Carbonates; Depositional Systems and Palaeoenvironmental Controls: Geological Society of London, Special Publication 255, p.197-216.

Halfar, J., Godinez-Orta, L., Mutti, M., Valdez-Holguin, J.E. and Borges, J.M., 2006b, Carbonates calibrated against oceanographic parameters along a latitudinal transect in the Gulf of California, Mexico: Sedimentology, v. 53, p. 297.

Hayton, S., Nelson, C.S. and Hood, S.D., 1995, A skeletal assemblage classification system for non-tropical deposits based on New Zealand Cenozoic limestones: Sedimentary Geology, v. 100, p. 123-141.

Henrich, R., Hartmann, M., Reitner, J., Shafer, P., Freiwald, A., Steinmetz, S., Dietrich, P. and Thiede, J., 1992, Facies belts and communities of the Arctic Vesterisbanken Seamount (Central Greenland Sea): Facies, v. 27, p. 71-104.

Henrich, R., Freiwald, A., Betzler, C., Bader, B., Schaefer, P., Samtleben, C., Brachert, T.C., Wehrmann, A., Zankl, H. and Kuehlmann, D.H.H., 1995, Controls on modern carbonate sedimentation on warm-temperate to arctic coasts, shelves and seamounts in the northern hemisphere: implications for fossil counterparts: Facies, v. 32, p. 71-108.

Henrich, R., Freiwald, A., Bickert, T. and Schäfer, P., 1997, Evolution of an Arctic open-shelf carbonate platform, Spitsbergen Bank (Barents Sea), *in* James, N.P. and Clarke, A.D., *eds*., Cool-water Carbonates: SEPM Special Publication 56, p. 163-181.

James, N.P. and Bone, Y., 1991, Sediment dynamics of an Oligo-Miocene cool water shelf limestone, Eucla Platform, southern Australia: Sedimentology, v. 38, p. 323-342.

James, N.P. and Bone, Y., 1992, Synsedimentary cemented calcarenite layers in Oligo-Miocene cool-water shelf limestones, Eucla Platform, southern Australia: Journal of Sedimentary Petrology, v. 62, p. 860-872.

James, N.P. and Bone, Y., 1994, Paleoecology of cool-water, subtidal cycles in Mid-Cenozoic limestones, Eucla Platform, Southern Australia: Palaios, v. 9, p. 457-477.

James, N.P., 1997, The cool-water carbonate depositional realm, *in* James, N.P. and Clarke, J.A.D., *eds*., Cool-water Carbonates: SEPM Special Publication 56, p. 1-22.

James, N.P. and Bone, Y., 2007, A Late Pliocene to early Pliocene, inner shelf, sub-tropical, seagrass-dominated carbonate: Roe Calcarenite, Great Australian Bight, Western Australia: Palaios, v. 22, p. 343-359.

James, N.P., Von der Borch, C.C., Gostin, V. and Bone, Y., 1991, Modern carbonate and terrigenous clastic sediments on a cool-water, high-energy, mid-latitude shelf: Lacepede Southern Australia: Sedimentology, v. 39, p. 877-904.

James, N.P., Boreen, T.D., Bone, Y. and Feary, D.A., 1994, Holocene carbonate sedimentation on the West Eucla Shelf, Great Australian Bight; a shaved shelf: Sedimentary Geology, v. 90, p. 161-177.

James, N.P., Collins, L.B., Bone, Y. and Hallock, P. 1999, Sub-tropical carbonates in a temperate realm: modern sediments on the southwest Australian Shelf: Journal of Sedimentary Research, v. 69, p. 1297-1321.

James, N.P., Bone, Y., Collins, L.B. and Kyser, T.K., 2001, Surficial sediments of the Great Australian Bight: facies dynamics and oceanography on a vast cool-water carbonate shelf: Journal of Sedimentary Research, v. 71, p. 549-568.

James, N.P., Feary, D.A.F., Betzler, C., Bone, Y., Holburn, A.E., Li, Q., Machiyama, H., Simo, J.A. and Surlyk, F., 2004, Origin of Late Pleistocene bryozoan reef-mounds: Great Australian Bight: Journal of Sedimentary Research, v. 74, p. 20-48.

James, N.P., Bone, Y. and Kyser, T.K., 2005, Where has all the aragonite gone? Mineralogy of neritic cool-water carbonates, southern Australia: Journal of Sedimentary Research, v. 75, p. 454-463.

James, N.P., Bone, Y., Brown, K.M. and Cheshire, N., 2007, Calcareous epiphyte production in cool-water carbonate depositional environments; Southern Australia, *in* Swart, P. and Eberli, G., *eds*., Advances in Carbonate Sedimentology: International Association of Sedimentologists Special Publication.

James, N.P., Eyles, C.H., Eyles, N., Hiatt, E.C., and Kyser, T.K., 2009 Oceanographic significance of an extreme cold-water carbonate environment: glaciomarine sediments of the Pleistocene Yakataga Formation, Middleton Island, Alaska: Sedimentology, v. 56, p.367-397.

Jones, H.A. and Davies, P.J., 1983, Superficial sediments of the Tasmanian continental shelf and part of Bass Strait: Bulletin of the Bureau of Mineral Resources, Geology and Geophysics, No. 218, 25 p.

Kamp, P.J., Harmsen, F.J., Nelson, C.S. and Boyle, S.F., 1988, Barnacle-dominated limestone with giant cross-beds in a non-tropical, tide-swept, Pliocene forearc seaway, Hawke's Bay, New Zealand: Sedimentary Geology, v. 60, p. 173-195.

Larsonneur, C., Bouysse, P. and Auffret, J.P., 1982, The superficial sediments of the English Channel and its Western Approaches: Sedimentology, v. 29, p. 851-864.

Lees, A., 1975, Possible influences of salinity and temperature on modern shelf carbonate sedimentation: Marine Geology, v. 19, p. 159-198.

Light, J.M. and Wilson, J.B., 1998, Cool-water carbonate deposition on the West Shetland Shelf: a modern distally steepened ramp, *in* Wright, V.P. and Burchette, T.P.; *eds*., Carbonate Ramps: Geological Society of London, Special Publication 149, p. 73-106.

Lukasik, J.J. and James, N.P., 2003, Deepening-upward carbonate subtidal cycles, Murray Basin, South Australia: Journal of Sedimentary Research, v. 73, p. 653-671.

Lukasik, J.J., James, N.P., McGowran, B. and Bone, Y., 2000, An epeiric ramp: Low-energy, cool-water carbonate facies in a Tertiary inland sea, Murray Basin, South Australia: Sedimentology, v. 47, p. 851-882.

Marshall, J. and Davies, P.J., 1978, Skeletal carbonate variation on the continental shelf of eastern Australia: Australian Bureau of Mineral Resources Journal of Geology and Geophysics, v. 3, p. 85-92.

Martindale, W., and Boreen, T.D., 1997, Temperature-stratified Mississippian carbonates as hydrocarbon reservoirs - examples from the Foothills of the Canadian Rockies, *in* James, N. P., and Clarke, M. J., *eds*., Cool-water Carbonates, SEPM, Special Publication 56, p. 391-409.

McKinney, F.K., 2007, The Northern Adriatic Ecosystem: Columbia University Press, New York, 299p.

Nelson, C.S., 1978, Temperate shelf carbonate sediments in the Cenozoic of New Zealand: Sedimentology, v. 25, p. 737-771.

Nelson, C.S., Keane, S.L. and Head, P.S., 1988a, Non-tropical carbonate deposits on the modern New Zealand shelf: Sedimentary Geology, v. 60, p. 71-90.

Nelson, C.S., Hyden, F.M., Keane, S.L., Leask, W.L. and Gordon, D.P., 1988b, Application of bryozoan zoarial growth-form studies in facies analysis of non-tropical carbonate deposits in New Zealand: Sedimentary Geology, v. 60, p. 301-322.

Nelson, C.S. and James, N.P. 2000 Marine cements in mid-Tertiary cool-water shelf limestones of New Zealand and southern Australia: Sedimentology v. 47, p. 609-630.

Pedley, M. and Grasso, M., 2002, Lithofacies modelling and sequence stratigraphy in microtidal cool-water carbonates: a case study from the Pleistocene of Sicily, Italy: Sedimentology, v. 49, p. 533.

Reynaud, J-Y., Dalrymple, R.W., Vennin, E., Parize, O., Besson, D. and Rubino, J.-L., 2006, Topographic controls on production and deposition of tidal cool-water carbonates, Uzes Basin, SE France: Journal of Sedimentary Research, v. 76, p. 117-130.

Rivers, J., James, N.P., Kyser, T.K. and Bone, Y., 2007, Genesis of palimpsest cool-water carbonate sediments on the southern Australian continental margin; Southern Australia: Journal of Sedimentary Research, v. 77, p. 480-494.

Schlanger, S.O. and Konishi, K., 1975, The geographic boundary between the coral-algal and the bryozoan-algal limestone facies: a paleolatitude indicator: IX International Congress of Sedimentology, Nice, p. 187-190.

Scoffin, T.P., 1988, The environments of production and deposition of calcareous sediments on the shelf west of Scotland: Sedimentary Geology, v. 60, p. 107-124.

Scoffin, T.P. and Bowes, G.E., 1988, The facies distribution of carbonate sediments on Porcupine Bank, northeast Atlantic: Sedimentary Geology, v. 60, p. 125-134.

Simone, L. and Carannante, G., 1988, The fate of foramol ('temperate-type') carbonate platforms: Sedimentary Geology, v. 60, p. 341-354.

Smith, A.M. and Nelson, C.S., 2003, Effects of early sea-floor processes on the taphonomy of temperate shelf skeletal carbonate deposits: Earth Science Reviews, v. 63, p. 1-31.

Wilson, J.B., 1988, A model for temporal changes in the faunal composition of shell gravels during a transgression on the continental shelf around the British

Isles: Sedimentary Geology, v. 60, p. 95-105.
Young, H.R. and Nelson, C.S., 1988, Endolithic biodegradation of cool-water skeletal carbonates on Scott shelf, northwestern Vancouver Island, Canada: Sedimentary Geology, v. 60, p. 251-267.

Other references - Paleozoic and Proterozoic

Beauchamp, B. and Desrochers, A., 1997, Permian warm- to very cold-water carbonates and cherts in northwest Pangea, *in* James, N.P. and Clarke, A.D., *eds*., Cool-water Carbonates: SEPM Special Publication 56, p. 327-347.
Brookfield, M.E., 1988, A mid-Ordovician temperate carbonate shelf - the Black River and Trenton Limestone Groups of southern Ontario, Canada: Sedimentary Geology, v. 60, p. 137-153.
Fairchild, I.J., 1993, Balmy shores and icy wastes: the paradox of carbonates associated with glacial deposits in Neoproterozoic times: *in* Wright, V.P. *ed*., Sedimentology Review 1, Blackwells, Oxford, U.K., p. 1-16.
Fielding, C. R., Frank, T. D., and Isbell, J., 2008, The late Paleozoic ice age - a review of current understanding and synthesis of global patterns, *in* Frank, T. D., and Isbell, J., *eds*., Resolving the Late Paleozoic Ice Age in Time and Space: Geological Society of America, Special Paper v. 441, p. 343-354.
Gates, L.M., James, N.P. and Beauchamp, B., 2004, A glass ramp: shallow-water Permian spiculite chert sedimentation, Sverdrup Basin, Arctic Canada: Sedimentary Geology, v. 168, p. 125-147.
Lasemi, Z., Norby, R.D. and Treworgy, J.D., 1998, Depositional facies and sequence stratigraphy of a lower Carboniferous bryozoan-crinoidal ramp in the Illinois Basin, amid-continent USA, *in* Wright, V.P. and Burchette, T.P., *eds*., Carbonate Ramps: Geological Society of London, Special Publication 149, p. 369-396.
Lavoie, D., 1995, A Late Ordovician high-energy, temperate-water carbonate ramp, southern Quebec, Canada: implications for Late Ordovician oceanography: Sedimentology, v. 42, p. 95-116.
Martindale, W. and Boreen, T.D., 1997, Temperature-stratified Mississippian carbonates as hydrocarbon reservoirs-examples from the foothills of the Canadian Rockies; *in* James, N.P. and Clarke, J.A.D., *eds*., Cool-water Carbonates: SEPM Special Publication, 56, p. 391-410.
Pope, M.C. and Read, J.F., 1997, High-resolution stratigraphy of the Lexington Limestone (Late Middle Ordovician) Kentucky, U.S.A.: a cool-water carbonate-clastic ramp in a tectonically active foreland basin, *in* James, N.P. and Clarke, J.A.D., *eds*., Cool-water Carbonates: SEPM Special Publication 56, p. 411-429.
Rogala, B., James, N.P. and Reid, C., 2007, Deposition of polar carbonates during interglacial highstands on an early Permian shelf, Tasmania: Journal of Sedimentary Research, v. 77, p. 587-606.

Sequence stratigraphy references

Caron, V., Nelson, C.S. and Kamp, P.J.J., 2004, Transgressive surfaces of erosion as sequence boundary markers in cool-water shelf carbonates: Sedimentary Geology, v. 164, p. 179-189.
Lukasik, J.J. and James, N.P., 2006, Carbonate sedimentation, climate change, and stratigraphic completeness on a Miocene cool-water epeiric ramp, Murray Basin, South Australia, *in* Pedley, H.M. and Carannante, G., *eds*., Cool-water Carbonates; Depositional Systems and Palaeoenvironmental Controls: Geological Society of London, Special Publication 255, p. 217-244.
Plint, A.G. and Nummendal, D., 2000, The falling stage systems tract: recognition and importance in sequence stratigraphic analysis, *in* Hunt, D. and Gawthorpe, R.L., *eds*., Sedimentary Responses to Forced Regression: Geological Society of London, Special Publications 172, p. 1-17.

REFERENCES CITED

Chappell, J. and Shackleton, N.J., 1986, Oxygen isotopes and sea level: Nature, v. 324, p. 137-140.
Fielding, C. R., Frank, T. D., and Isbell, J., 2008, The late Paleozoic ice age - a review of current understanding and synthesis of global patterns, *in* Frank, T. D., and Isbell, J., *eds*., Resolving the Late Paleozoic Ice Age in Time and Space: Geological Society of America, Special Paper v. 441, p. 343-354.
Hunt, D. and Tucker, M.E., 1992, Stranded parasequences and the forced regressive wedge systems tract: deposition during base-level fall: Sedimentary Geology, v. 81, p. 1-9.
James, N.P., Narbonne, G.M., Dalrymple, R.W. and Kyser, T.K., 2005, Glendonites in Neoproterozoic low-latitude, interglacial, sedimentary rocks, northwest Canada: insights into the Cryogenian ocean and Precambrian cool-water carbonates: Geology, v. 33, p. 9-12.
Sarg, J.F., 1988, Carbonate sequence stratigraphy, *in* Wilgus, C.K., Hastings, B.S., Kendall, C.G.S.C., Posamentier, H.W., Ross, C.A. and Van Wagoner, J.C., *eds*., Sea-level Changes: An Integrated Approach: SEPM Special Publication 42, p. 155-182.
Schlager, W., 1981, The paradox of drowned reefs and carbonate platforms: Geological Society of America Bulletin, v. 92, p. 197-211.
Wright, V. P. and Cherns, L., 2008, The subtle thief: selective dissolution of aragonite during shallow burial and the implications for carbonate sedimentology, *in* Lukasik, J., and Simo, J.A.T., *eds*., Controls on carbonate platform and reef development: SEPM Special Publication No 89, p. 47-54.

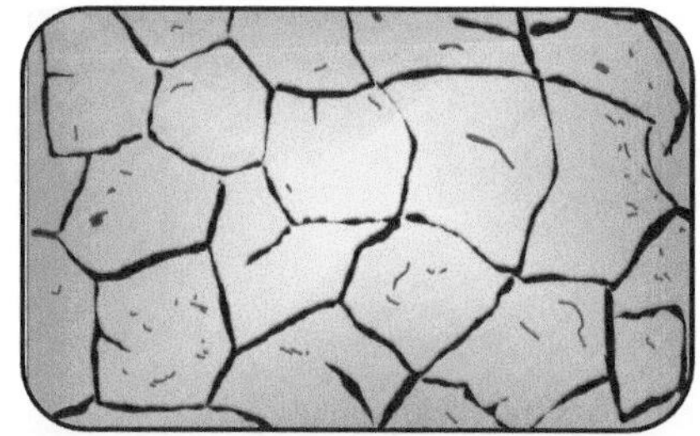

16. Peritidal Carbonates

Brian R. Pratt, Department of Geological Sciences, University of Saskatchewan, Saskatoon, Saskatchewan S7N 5E2, Canada

INTRODUCTION

Limestone and dolostone, deposited on muddy tidal flats and in adjacent shallow seawater, are one of the most conspicuous families of carbonate facies in the stratigraphic record (Fig. 1A and B). The term *peritidal* (from the Greek *peri*, meaning around or near, and tidal, relating to tides) was coined, in passing, by R. L. Folk in 1973, and has proven a useful adjective for the spectrum of these nearshore facies, even in regions where tides were probably so small as to have had only minor influence on deposition.

Tidal flats are low-energy, low-relief repositories of generally fine-grained sediment consisting mostly of allochthonous calcareous particles 'born' in the subtidal 'carbonate factory'. Beaches are higher energy and, therefore, are composed of well-sorted, coarser material. Because tidal flats comprise a geomorphologically variable setting, peritidal strata are typically heterogeneous, but one of the salient attributes that distinguish them is the many – but not all – features that can be compared directly with modern analogues (Fig. 2A and B); this makes these facies ostensibly easy to recognize in the field. They are critical paleobathymetric indicators, for they delineate the position of the sea level at certain points in the stratigraphy. They form around the interface between marine and subaerial processes. Hence, they are also indicative of early diagenetic phenomena to which they may have been subjected. If the subaerial setting is arid, evaporite minerals can precipitate; if the setting is humid, features can be preserved that indicate the presence of brackish or fresh water. They are often dolomitic because of the particular water chemistry and hydrology that may develop in the shallow subsurface.

The stratigraphy of peritidal car-

Figure 1. Peritidal carbonate rock exposures. **A.** Panorama of Middle Cambrian–Early Ordovician strata, east-central British Columbia, Canada. Distinctly bedded limestones and dolomites in the lower part (Snake Indian Formation) consist of shallow subtidal and peritidal facies. The massive unit (Eldon Formation) above the small icefield to the right of center consists of subtidal limestones. These are overlain, in turn, by shallow subtidal and peritidal limestones (Lynx Group) then by deeper water shales and limestones (Survey Peak Formation); **B.** Quarry wall ~7 m high showing successive shallowing-upward cycles of peritidal limestones. Supratidal tops of cycles exhibit dinosaur footprints. San Giovanni Rotondo limestone (Early Cretaceous), Gargano, Italy.

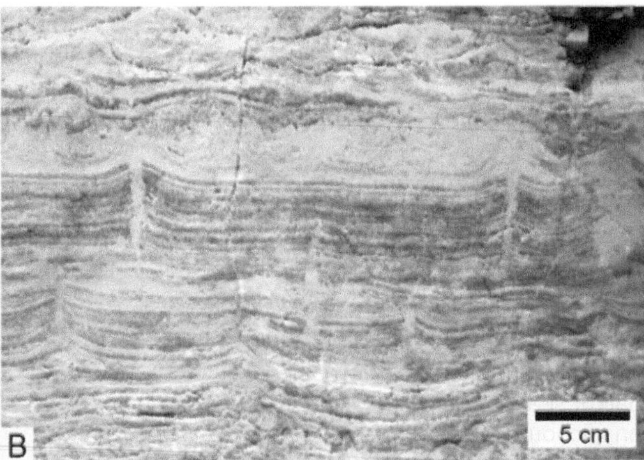

Figure 2. A. Shallow trench dug into the sabkha of Abu Dhabi, showing subtidal sediment at the base, overlain, in turn, by black intertidal microbially laminated muds having desiccation cracks, then light-colored supratidal muds, and finally by eolian quartz sand (Photograph courtesy of P.A. Scholle); **B.** Microbially laminated dolomite cut by desiccation cracks, overlain by intertidal or subtidal mudstone. Providence Island Formation (Early Ordovician), northern New York, USA (Photograph courtesy of N.P. James).

bonates is a combination of aggradation, progradation, erosion and retrogradation of bodies of sediment, and therefore is the register of the interaction of physical and biological processes, climate, sea-level change and so forth. This makes them seemingly simple on the one hand, but decidedly challenging on the other.

This chapter first summarizes tidal processes and then discusses how the setting has changed through geological time in response to organic evolution. The unique diagenetic regime of tidal flats is outlined and the sediments and their rock counterparts are described. Peritidal facies are treated in terms of how they can be recorded stratigraphically, thereby presenting current hypotheses to account for their distribution including perceived patterns of cyclicity. Also mentioned are a couple of examples that appear, at first glance, to be peritidal but prove not to be so – which signals the importance of careful documentation and evaluation of inferred processes, and consideration of alternative hypotheses.

THE PERITIDAL ENVIRONMENT

Some Historical Background

Geologists first took an interest in modern carbonate tidal flats with the publication of M. Black's 1933 study of supratidal 'algal marshes' of Andros Island, the largest island of the Bahamas archipelago. The understanding of peritidal carbonate facies then underwent a dramatic boost with intensive research efforts on Holocene tidal flats and associated shallow subtidal sediments during the 1960s and early 1970s. Seminal studies were carried out in southern Florida, Andros Island (Bahamas), Belize, the Arabian/Persian Gulf, and Shark Bay (Western Australia); other tropical and subtropical examples have since been described. These are fundamental reference points because they explain the processes operating in this environment and illustrate the variety of potentially preservable features, which is why some of these studies were sponsored by oil companies. Modern carbonate tidal flats are easily accessible and students are encouraged to explore one for themselves.

Ancient counterparts were quickly identified in a series of pioneering comparative studies. Since then, many examples of peritidal facies have been documented and the whole setting has been well reviewed in textbooks. Armed with confident facies interpretations, sedimentologists since the late 1980s have been focusing on what controls the stratigraphy of peritidal strata. Even so, it is important to remember that carbonate rocks can be capricious, not least because organisms are typically involved, and modern analogues are lacking for a number of ancient facies.

Tidal Processes

Oceanic tides result mostly from the gravitational attraction between the Earth and Moon. The tidal bulges are semi-diurnal but because of the slow orbit of the Moon around the Earth, their period is slightly longer than 12 hours. Tides are strongly influenced by the shape and bathymetry of the water bodies through which they pass, as well as bottom topography and coastal geomorphology. The advance of the tide over the shoreline is called the flood, and its retreat is called the ebb tide. Depending on tidal range and slope of the shore, tidal currents can be created and commonly the ebb tide drains via tidal channels or creeks cut into the tidal flat (*see* Chapter 9).

The difference in elevation between normal high-tide and normal low-tide is the tidal range or amplitude. A region is termed microtidal if the tidal range is <2 m, mesotidal if it is 2–4 m, and macrotidal if >4 m. The open ocean and unobstructed modern tropical coasts are microtidal; tidal height can increase across shallow shelves and in embayments. Enclosed seas and large lakes have minimal tides. Facies analysis and modeling suggest that ancient inland epeiric seas were mostly microtidal, with some being virtually non-tidal.

Three bathymetric zones are recognized in the nearshore setting: subtidal, intertidal and supratidal (Fig. 3A and B). The subtidal zone is permanently submerged and ranges from low-energy lagoonal environments to higher energy shoals and reefs. This zone is the carbonate factory where sediment is generated (*see* Chapters 14 and 15). Semi-monthly spring tides – when tidal range is at its maximum

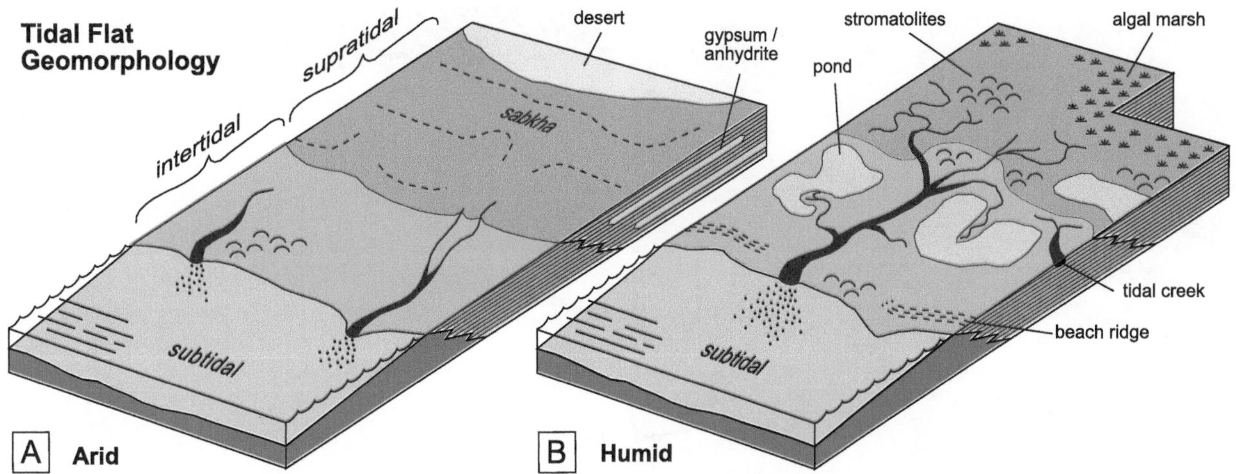

Figure 3. Block diagrams showing the main morphological elements of carbonate tidal flats. **A.** Arid setting, based on the sabkha of the southern shore of the Arabian/Persian Gulf; **B.** Humid setting, based on the western side of Andros Island, Bahamas.

– may briefly expose the shallowest subtidal areas. The intertidal zone (or littoral zone to marine biologists), lies between normal low-tide and high-tide levels and is therefore alternately exposed and submerged. Tidal creeks may be bordered by levees. The supratidal zone is above normal high-tide and is flooded only during spring tides, high winds and storms. The higher intertidal and supratidal zones may be dotted with brackish or saline ponds (Fig. 4A). Supratidal flats may become evaporitic in arid climates, and for these the Arabic word *sabkha* has been adopted by sedimentologists (*see* Chapter 20). Clearly, in tandem with the increasing degree of subaerial exposure, physical energy is also attenuated across the tidal flat.

The three-fold environmental subdivision is fairly well marked on modern tidal flats by a distinct lateral distribution of sedimentary features like desiccation cracks and the restricted benthic biota which is adapted to the degree of emergence (Fig. 4A–D). However, with regard to sediment deposition, the flood and ebb of the tide have little effect. Rather, most sediment is moved by high winds, storms and hurricanes which stir up the adjacent offshore areas and drive sediment-laden waters onto the tidal flat. Storms shape coarse particles into beach ridges on higher energy coasts. Tsunami impact is also possible, although as yet few examples have been recognized.

Where do Tidal Flats Form?

Modern examples have accreted as wedges along coasts and in the lee of rocky islands, spits, coral reefs and carbonate sand shoals (Fig. 5). These are low-energy areas that have collected sediment during the rapid Holocene sea-level rise. As well as comparable ancient coastal locales, tidal flats also formed as discrete islands and banks in carbonate platform interiors and in the broad epicontinental shelf seas and inland epeiric seas that existed frequently in the geological past (Fig. 6). These platforms and seas also commonly exhibit peritidal facies that appear to be laterally extensive on a regional scale.

Biological Evolution of Tidal Flats

Organic evolution has left its stamp on peritidal carbonate facies just as it has, under various guises, on almost all other sedimentary rock types. In the Precambrian, the subtidal carbonate factory generated abiotic and microbially influenced particles such as ooids, lime mud and intraclasts. Reefs were stromatolitic (*see* Chapter 17). The seabed was variably loosely covered by a flocculent microbial mat composed of cyanobacteria (photosynthetic) and bacteria (heterotrophic: degrading microbial organic matter). Intertidal and supratidal surfaces were covered with mats and biofilms undoubtedly consisting of a somewhat different microbial community, whose composition depended on decreasing moisture and increasing salinity. Laterally linked domical stromatolites were common.

Come the 'Cambrian explosion', the carbonate factory began to generate copious amounts of bioclasts derived from many different taxonomic groups (*see* Chapter 14). Ooids and intraclasts continued to form, but lime mud seems to have become more abundant, even though it probably was still mostly microbial in origin. Moreover, silt-sized micritic peloids also became increasingly common – some fecal, others probably microbial too. The shallow subtidal and perhaps the lowermost intertidal zone started to host bioturbating invertebrates like certain kinds of worms. These infaunal animals digested sediment as they exploited organic matter on, and in, the sediment, or lived in burrows and filter-fed from the water column (*see* Chapter 3).

By contrast, the microbial biota of tidal flats probably did not change appreciably, and stromatolite formation continued to take place. However, crusts composed of clotted and peloidal micrite with laminoid pores called 'fenestrae' appear to have become common. While these crusts are regarded as primarily microbial products, albeit diagenetically complex, they suggest that the beginning of the Phanerozoic did not just involve a simple displacement of microbial communities by multicellular organisms. That stromatolites became rarer after the Early Ordovician has often been ascribed to grazing pressure, especially by herbivo-

Figure 4. Geomorphology of modern tidal flats. **A.** Oblique, low-altitude aerial photograph of the northwestern coast of Andros Island, Bahamas. Offshore subtidal is to the left. The dark areas are microbial mat-covered intertidal areas; lighter colored areas are ponds. Tidal channels up to a few meters wide drain the tidal flat, and the white areas along the channel edges are supratidal levees (Photograph courtesy of N.P. James); **B.** View across intertidal muddy sands weakly bound by a light-colored microbial mat and littered with oncoids and intraclasts, beyond which is a protected lagoon with mangroves. Bonaire, Netherlands Antilles; **C.** View across dark-colored supratidal mat passing to light-colored, burrowed intertidal muddy carbonate sands and the protected lagoon. Bonaire; **D.** View across the supratidal microbial mat to a storm-generated carbonate sand beach, behind which are mangroves. Bonaire. Mangroves in **B–D** are 3–4 m high.

rous gastropods. This is too simplistic an explanation, however, because microbial mats continued to be ubiquitous, indeed to the present-day. Stromatolite formation typically requires concomitant permineralization of organic substances along with cementation, which are not directly related to faunal–microbial ecological interactions. The reasons are still unclear but differences in microbial communities and seawater chemistry seem implicated.

Later pulses of diversification and mass extinctions led to a changing cast of bioclast types that were swept onto tidal flats. Various microbial groups adapted to etching hard substrates began to produce microborings in bioclasts and cemented surfaces, even in the intertidal zone. Lime mud came from a variety of sources, especially via the breakdown of certain kinds of calcareous algae composed of easily disaggregated micritic crystals. Some ostracode, gastropod and foraminiferal taxa evolved to tolerate fluctuating salinity or hypersalinity, as well as elevated temperature, and were able to live on tidal flats.

With the diversification of crustaceans during the Mesozoic and Cenozoic, bioturbation not only became more intense and deeper but took place even in supratidal areas, because some species were now able to spend extended periods out of the water. In fact, these crab and shrimp taxa are so industrious that by homogenizing the sediment they effectively erase the kinds of features that act as key evidence of the peritidal environment. Interestingly, different kinds of burrowers operate in modern carbonate versus siliciclastic tidal flats—perhaps due to the contrasting sediment type.

Another important phenomenon is the diversification of plants. First, the rise of land plants in the middle Paleozoic led to changes in the nature of soil formation, including in low-lying areas adjacent to tidal flats. Later, the diversification of angiosperms, which began in the late Mesozoic, has had a profound effect: In modern settings seagrass meadows stabilize the shallow subtidal seabed, while mangroves colonize protected parts of tidal flats and lagoons. Both developments resulted, through the Cenozoic, in new ecological niches and substrates for mud-producing organisms.

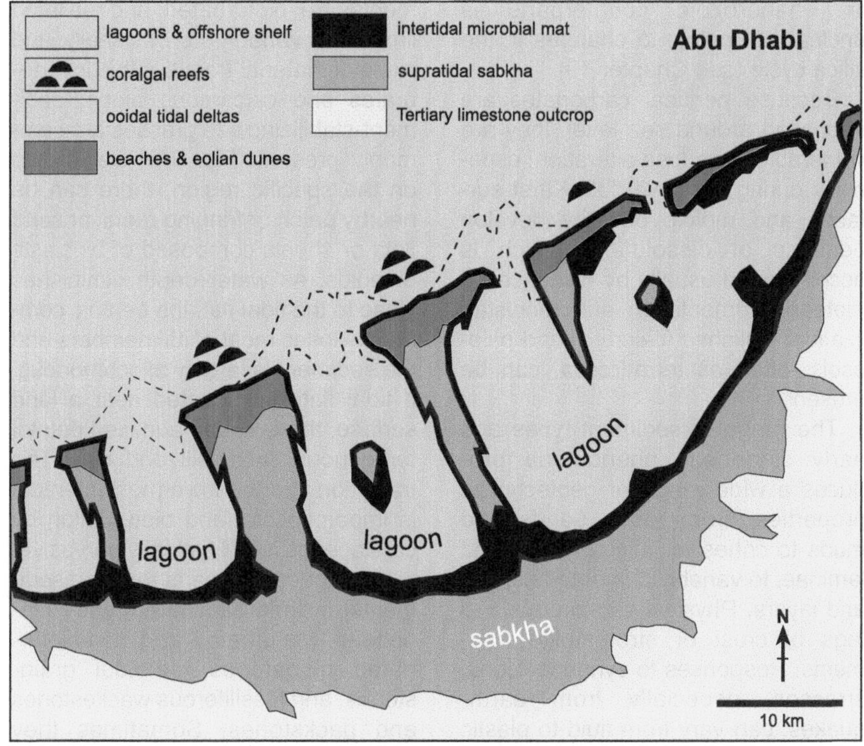

Figure 5. Simplified map of modern tidal flat near Abu Dhabi, Arabian/Persian Gulf. Coastal lagoons are floored with peloidal and bioclastic sands and muds, and are drained by subtidal channels.

Diagenetic Processes of Tidal Flats

Tidal-flat sediments accumulate at the interface between seawater and air which means that in arid areas, evaporation concentrates the pore-waters residing in the tidal flats, and if the appropriate level of hypersalinity is reached, gypsum and anhydrite precipitate as isolated crystals and layers (*see* Chapter 20). Halite can crystallize on the surface but this is typically ephemeral and no vestige is usually preserved. The sediments can also be affected by freshwater, as rainwater and groundwater.

In modern examples, the semi-diurnal flushing by seawater does not seem to cause carbonate cementation beneath the surface. However, this may not have been true in the geological past when indications are that seawater may have had a somewhat different composition. For example, flat-pebble intraclasts are common in Precambrian and early Paleozoic peritidal strata and indicate early lithification of shallow subtidal and tidal-flat sediment layers.

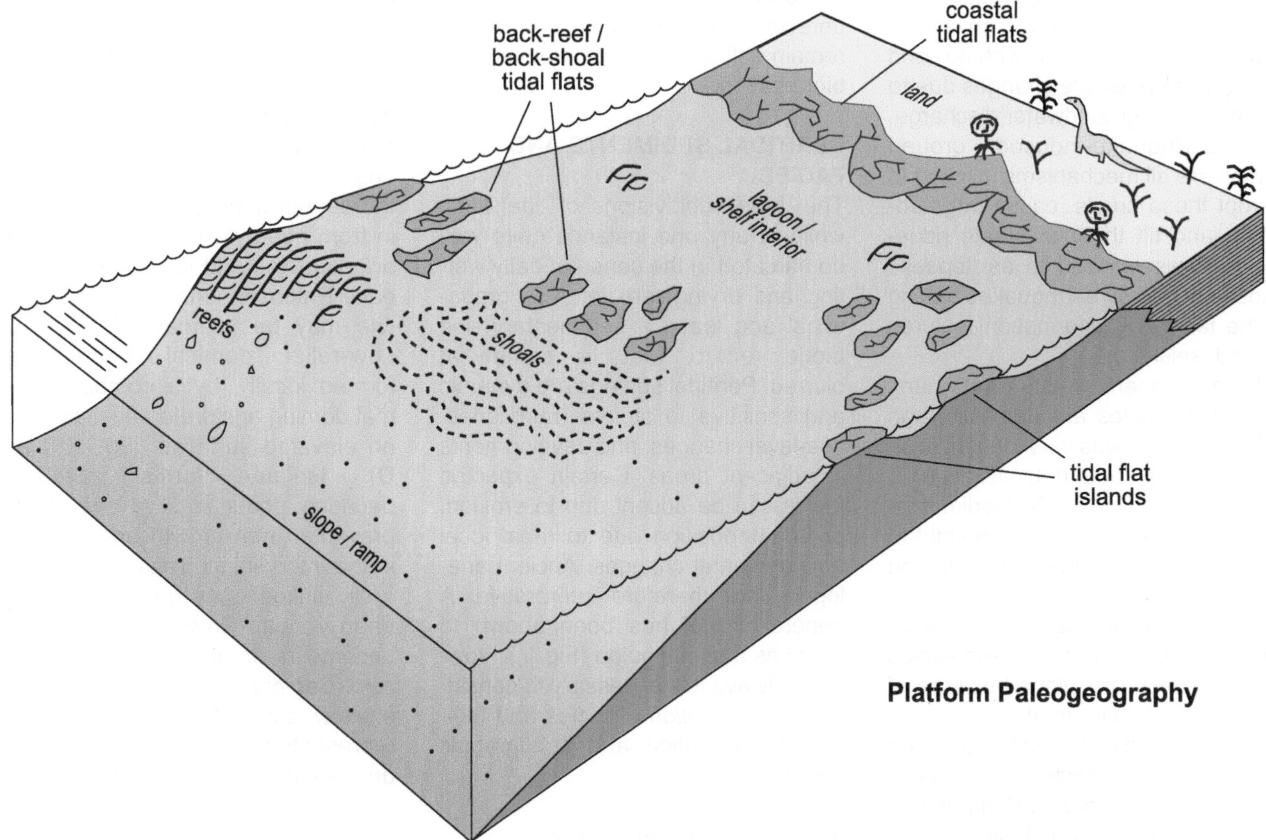

Figure 6. Block diagram of a hypothetical carbonate platform showing possible locations where tidal flats form: in the lee of reefs and carbonate sand shoals, as islands, and as coastal deposits.

Pendent fibrous cement can be present, a reliable indicator of percolation through air-filled pores like fenestrae. Sea spray under elevated air temperatures appears to cause precipitation of rinds or fringes of fibrous aragonite. Some ancient examples show the growth of pisoids, initially as mobile grains which eventually become stationary and continue to accrete.

Within microbial mats, especially those that form stromatolites, laminites and layers with fenestral porosity, photosynthesis can result in $CaCO_3$ precipitation because of the uptake of CO_2. Probably more important, however, is that biodegradation of cells and mucous secretions cause extra-cellular polymeric substances (exopolysaccharides) to become reactive due to CO_2 production and the release of Ca^{2+} that was bonded to ligands.

Modern tidal flats and beaches sometimes exhibit surface crusts and beachrock, respectively. Several factors appear to be involved in rapid, synsedimentary high-Mg calcite and aragonite precipitation, including elevated temperature, evaporation, fluctuating salinity, and microbial processes. Alternating wetting and drying, pore-pressure changes due to storm impact, groundwater discharge, and earthquake-induced ground motion are all mechanisms invoked to disrupt these crusts, cause fragmentation, and tilt them to create ridge-like features referred to as 'tepees'. Those caused by earthquakes belong to the family of deformation features termed 'seismites'.

Some Recent crusts and many ancient examples are dolomitic, and indeed this fact was very influential on early thinking about dolomitization, in the 1960s and 1970s. Synsedimentary dolomite replacement is thought to be due to evaporative concentration of Mg^{2+}.

Many Proterozoic tidal-flat dolomites are cherty, and sometimes these nodules are black because of contained organic matter. In some cases, they exhibit well-preserved remains of cyanobacterial cells silicified in various stages of degradation, and in fact, the paleontological record of Proterozoic microfossils is primarily from such peritidal facies. Why this phenomenon is essentially not seen in Phanerozoic counterparts is unclear but points to changes in the silica cycle (*see* Chapter 19).

Because peritidal carbonates are deposited around sea level, they are vulnerable to vadose alteration, especially during sea-level fall. Karst surfaces and moldic porosity develop because of dissolution, which is accompanied usually by evidence for meteoric cementation and recrystallization. Caliches (calcrete) and paleosols, often as terra rossa, can be present.

The range of sediment types and early diagenetic phenomena produces a wide range of geotechnical properties, from loose sands and muds to cohesive, microbially bound laminae, to variably cemented objects and layers. Physical erosion leads to lags of crust or stromatolite fragments. Responses to syndepositional stresses, especially from earthquakes, can vary from fluid to plastic to brittle, leading to a variety of deformation features such as shrinkage, injection, convolute bedding, folds, cracks, microfaults and wholesale brecciation. Such events typically go unrecorded in subtidal sediments that remain soft and prone to subsequent bioturbation.

PERITIDAL SEDIMENTS AND FACIES

The three subdivisions of tidal flats, while at any one instance quite well demarcated in the sense of daily wetting and drying, are laterally gradational and leave a sedimentary and biotic record that is somewhat blurred. Peritidal settings are dynamic and sensitive to storms, hurricanes, sea-level changes, and developments in adjacent areas. Certain expected facies can be absent due to erosion, or non-deposition due to rapid local environmental changes. Ancient analogues are therefore interpretive. A general model has been shown to serve as a useful guide (Fig. 7). However, always it is necessary to consider the *association* of facies and their lateral distribution and stratigraphic context.

Shallowest Subtidal Zone

Seaward of the tidal flat is the carbonate factory and because salinity is normal-marine (35‰), these sediments are bioturbated and pelleted lime mud variably rich in shelly and skeletal material from benthic invertebrates and calcareous algae; sediment-stabilizing sea grasses are commonly present (Fig. 8A). Depending on the specific region, there can be nearby patch or fringing reefs, or sand flats or shoals composed of bioclasts or ooids. As water depth diminishes close to the tidal flat, the benthic community loses most of its members and the sediment is largely allochthonous. If tidal flats are located near a land surface, there will be admixed detrital terrigenous sand, silt and clay. The transition into the lowermost intertidal is imperceptible, and bioturbation by crustaceans is still typically pervasive.

Ancient counterparts to these sediments include laminated and thin-bedded, bioturbated and non-bioturbated mudstones, peloidal grainstones, and fossiliferous wackestones and packstones. Sometimes they contain an autochthonous low-diversity fauna suggestive of ecological stress (Fig. 8B). Storm-generated interbeds of wave- and unidirectional current-ripple crosslaminated grainstones that have scoured bases, with flat-pebble intraclasts being common in Precambrian and early Paleozoic cases (Fig. 9A and B). Sand- and silt-sized particles consist of bioclasts and peloids mostly winnowed from the deeper subtidal, or ooids washed in from nearby shoals. In Precambrian examples and in increasingly few early Paleozoic ones, microbial laminite may be intercalated (Fig. 9C). Low-relief domical stromatolites formed locally, by a combination of mat doming and preferential accretion on elevated surfaces (Fig. 10C and D). Isolated forms developed because scour around the domes prevented mat growth and accretion in the intervening areas, whereas laterally linked ones developed where energy was lower. Stromatolites become rarer after the early Paleozoic. Beginning in the Mesozoic, the shallow subtidal zone is commonly represented by medium- to thick-bedded bioturbated packstones.

Intertidal Zone

Due to burrowing crustaceans, lower intertidal sediments and facies in non-arid settings, well back into the Meso-

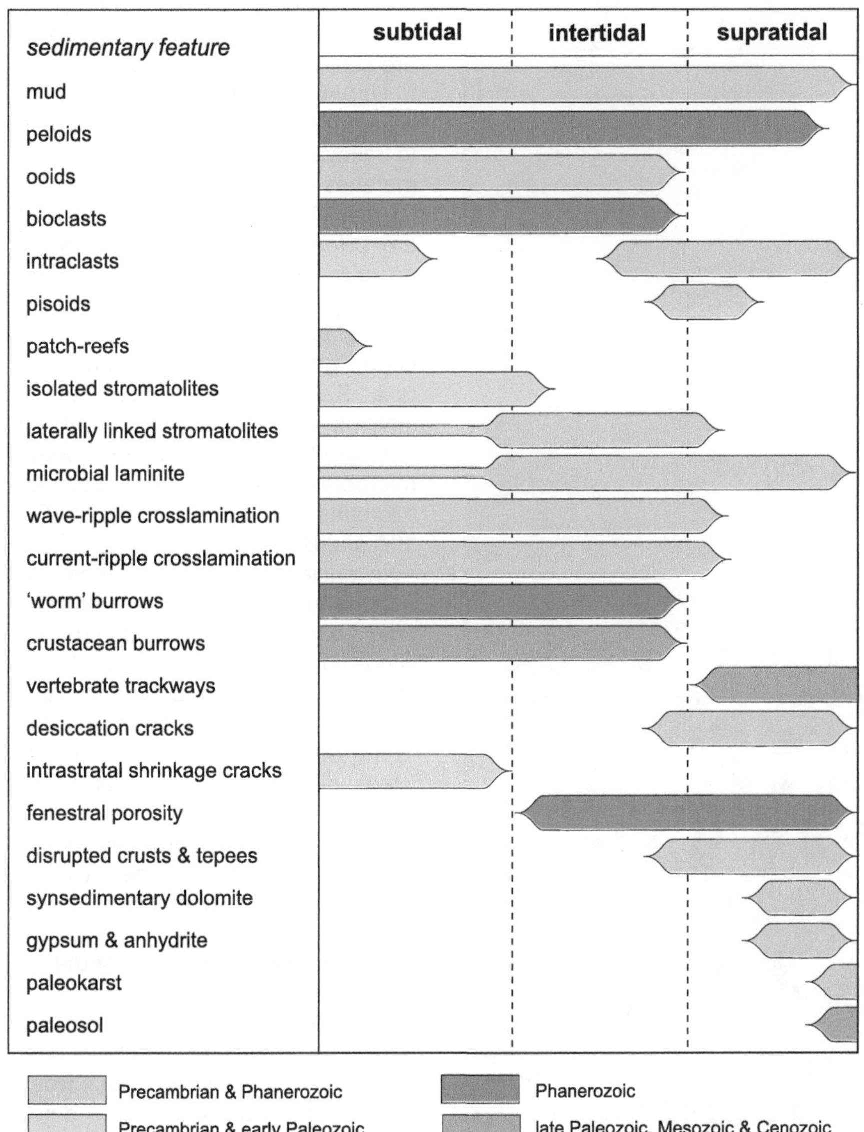

Figure 7. Generalized bathymetric distribution of key components and features. The width of individual spindles shows their approximate relative abundance across the three peritidal zones; spindles are, however, not scaled to the relative abundance of the different items. Spindles are color-coded with respect to broad intervals of geological time in which they are most commonly found. A typical supratidal distribution is depicted even though major storms, hurricanes and tsunamis can cast coarse-grained particles far landward. Not shown are bioclasts derived from invertebrates living on tidal flats, such as in ponds and marshes. Also not shown are features characteristic of beaches.

zoic, are typically extensively bioturbated, and thus difficult to distinguish from shallow subtidal deposits. Middle and late Paleozoic examples may contain burrows too. Key indicators are grain size – because the sediment is allochthonous – and identity of associated *in situ* invertebrate fauna – which should be suggestive of stressed conditions.

Mechanically deposited intertidal sediments are thin- and wavy-bedded mudstones and fine-grained grainstones, similar to shallow subtidal sediments (Fig. 9B). Unidirectional current- and wave-ripple crosslamination is common not just in the grainstones but also in the mudstones, because the latter were often composed of silt-sized peloids or tiny aggregates of micrite, coarse enough to move by traction. Plane laminae and drapes of mudstone (or calcareous shale) may be present, forming 'flaser' bedding and 'tidal' bedding, both of which are signature features of siliciclastic tidal flats (*see* Chapter 9). These couplets result from sediment movement by gentle tidal currents and waves alternating with slackwater, during which mud settles out of suspension.

In arid settings as well as those of early Paleozoic age and older, bioturbation is restricted and variably distinct wavy microbial laminite can be present, just as in the supratidal zone (*see* below). Stromatolites can also be present, just as they are in the shallowest subtidal environments (Fig. 10A–D). Beds composed of clotted and peloidal micrite and having fenestral porosity are a common limestone type in both the intertidal and supratidal zones (*see* Fig. 13A and B). There is some indication, however, that this fabric is more typical of humid regions.

Modern tidal creeks are floored with a lag of intraclasts and coarse-grained bioclasts overlain by sand-sized particles. Although their position tends to be more stable geographically than their siliciclastic counterparts because of localized synsedimentary cementation and the overall low-energy setting, they do migrate laterally during high-energy events, reworking tidal-flat sediment and leaving laterally extensive fining-upward beds. Ancient counterparts contain intraclastic rudstones and crosslaminated grainstones showing lateral-accretion bedding; tidal-bundled laminae indicative of neap-spring cyclicity has been reported. By contrast, beach deposits tend to consist of low-angle, seaward-dipping beds composed of well-rounded and well-sorted grainstones. Slabs of broken beachrock can be present.

Supratidal Zone

Supratidal areas receive fine-grained sandy and muddy sediment only during storms and hurricanes. Because they are so ecologically stressed, invertebrate animals are excluded and only microbial mats and biofilms are present (Fig. 11A–E), giving rise to microbially laminated mudstone. The various surface irregularities – domes, bumps and wrinkles – cause the laminae to be undulating or wiggly (Fig. 12A and B). Locally, there are storm-deposited grainstone and mudstone laminae (Fig. 12B). Some

A

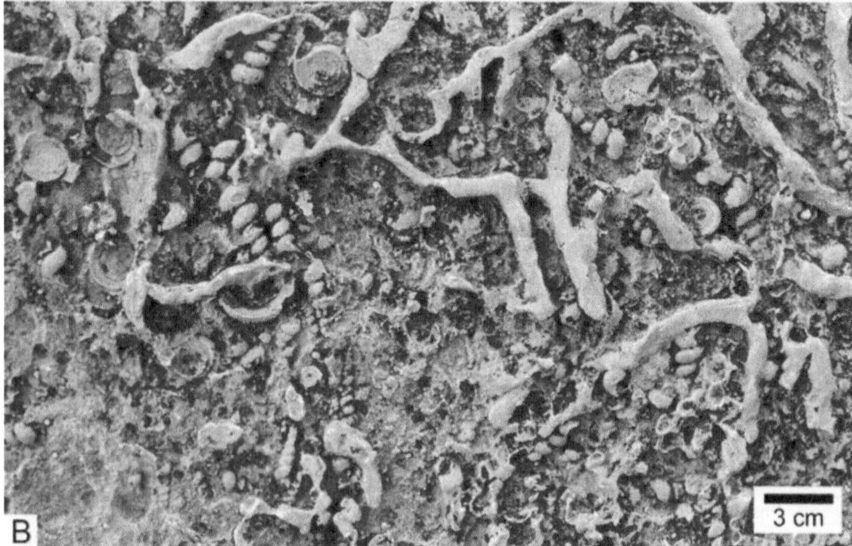

Figure 8. The shallow subtidal carbonate factory. **A.** Underwater photograph of a ~1 m deep lagoon showing sandy bottom supporting a benthic community that includes udoteacean calcareous green algae *Halimeda opuntia*, fleshy green algae and the sea grass *Thalassia testudinum*. The sediment also contains foraminifers and red algae as well as fragments of corals, gastropods, echinoids and other bioclasts washed in from the nearby reef to seaward. The mounds of sediment are produced by burrowing callianassid shrimp. Bonaire; **B.** Weathered bedding plane showing dolomitized gastropod steinkerns and horizontal burrows belonging to *Palaeophycus*. The distinctive fauna comprising large numbers of only two taxa of planispiral and high-spired gastropods is suggestive of a stressed setting perhaps close to low-tide level. Boat Harbour Formation (Early Ordovician), western Newfoundland, Canada.

microbial layers are composed of early-cemented clotted and peloidal micrite with fenestral pores, as in the intertidal zone (Figs. 13A, B, and 14). They can be fragmented into intraclasts or pushed upward into tepees (Fig. 15A–C). Cavities under these dislodged crusts typically contain geopetal sediment washed in or pisoids that accreted *in situ* (Fig. 15D).

Supratidal sediments are prone to desiccation which can produce polygonal mudcracks (Figs. 2A, B and 11E). Such cracks get filled with sediment that washes or blows in from above. Supratidal surfaces in Mesozoic and Cenozoic examples can preserve vertebrate trackways. Theoretically, they could preserve arthropod trackways and insect burrows like their siliciclastic counterparts, but none have been reported. Mangroves are common in modern settings.

Upper intertidal and supratidal areas in humid regions are exposed to rainwater, resulting in ponds and marshes exhibiting markedly fluctuating salinity. Marshes are muddy and are colonized by mangroves. Ponds are also muddy, and have flocculent microbial growth and bioclasts belonging to specially adapted ostracode, gastropod and foraminiferal taxa. Relatively few ancient equivalents have been specifically identified, but mudstones containing wispy microbial laminae or calcified microbes are likely candidates (Fig. 16A and B).

In ancient arid settings, there are occasionally crystals or beds of anhydrite (often replacing gypsum) and wind-blown terrigenous sand and clay, if adjacent to a land surface. Supratidal sediments are vulnerable to the effects of minor sea-level fall, which gives rise to subaerial alteration such as karst surfaces and moldic porosity. The diversification of land plants promoted the development of paleosols (Fig. 17).

Pitfalls in Recognizing Peritidal Facies

Because of the great advances in carbonate sedimentology that have been made over the past decades, it would seem quite straightforward to detect the precise record of the sediment–air interface in a stratigraphic section, and thus identify and evaluate a suite of peritidal facies; indeed it is often so. The sediment types, sedimentary structures and biotic characteristics, however, may be rather equivocal, because they are shared by other environments such as lakes (*see* Chapter 21).

Care must be taken in making paleoecological interpretations based on living descendents or presumed analogues to long-extinct organisms. Benthic fossils and bioturbation can be rare or absent in subtidal facies for various reasons. Even though this might give the impression of ecological stress, it may be unrelated to tidal exposure. Weak currents and gentle wave action at the limit of wave-base produce rippled beds that are similar

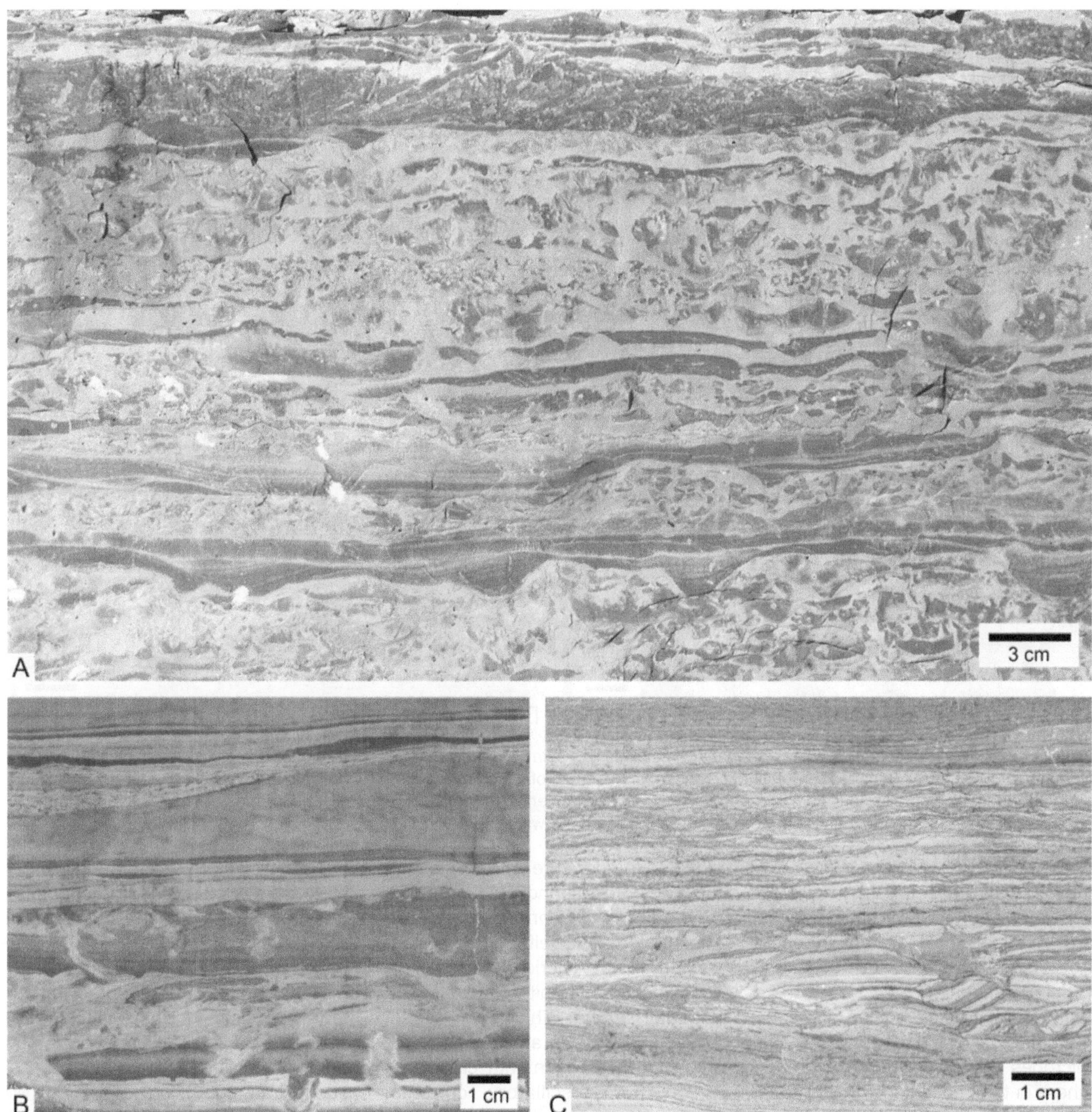

Figure 9. Shallowest subtidal and lower intertidal facies. **A.** Outcrop of dolomitic limestone consisting of interbedded burrow-mottled mudstone and crosslaminated peloidal grainstone overlying scoured surfaces. Lenticular thin bed near top is composed of flat-pebble intraclastic rudstone. Kindblade Formation (Late Cambrian), Oklahoma, USA; **B.** Polished surface of dolomite showing interbedded plane and lenticular laminae, scoured surfaces, and vertical U-shaped burrows belonging to *Diplocraterion*. Watts Bight Formation (Early Ordovician), western Newfoundland; **C.** Polished surface of dolomite showing plane and wavy laminae that are a combination of microbial binding and mechanical sedimentation. Pinching and swelling of laminae, low-angle microfaults and brecciation represent intrastratal deformation interpreted to have been caused by synsedimentary earthquake shaking. Altyn Formation (Mesoproterozoic), northwestern Montana, USA.

to tidal-flat counterparts, and intercalated mud laminae settle out during times of slackwater. Shallow-water sediments can be transported offshore far beyond their locus of formation, given the appropriate paleogeography and powerful agents such as off-surge from major storms and tsunamis.

Desiccation cracks need to be distinguished from intrastratal cracks and shrinkage features which go by names such as 'syneresis' and 'diastasis' cracks and, in Proterozoic limestones, 'molar tooth structure'. These are deformation features, some of which appear to be seismites, and while they often exhibit polygonal patterns on bedding planes, the polygons tend to be smaller and more angular in outline compared to true

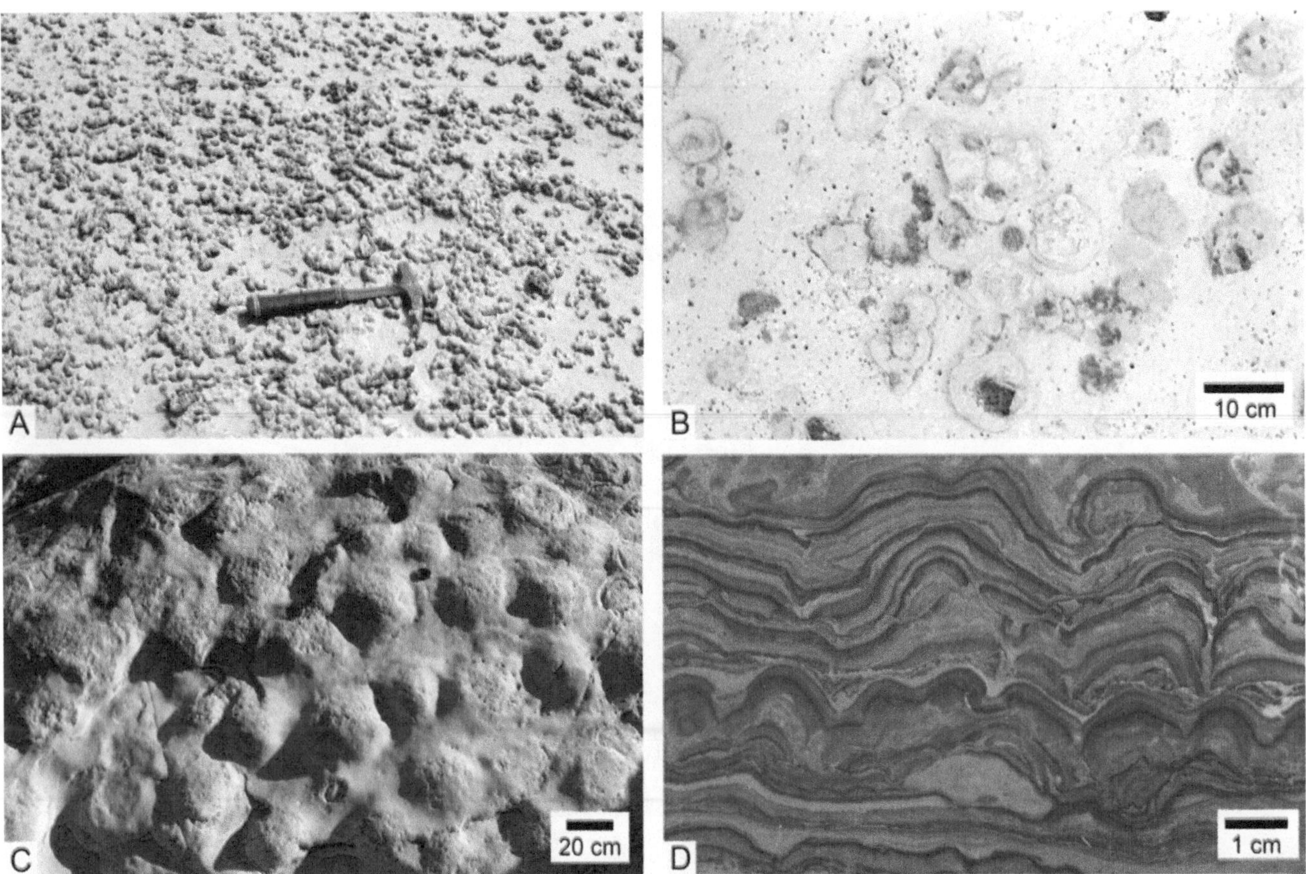

Figure 10. Shallow subtidal and lower intertidal stromatolites. **A.** Gelatinous microbial mat forming isolated and laterally linked domes. Hammer is 30 cm long. Bonaire; **B.** Gelatinous sandy domes, some of which are overturned. Coin is 2.5 cm wide. Bonaire; **C.** Bedding surface of domes. Boat Harbour Formation (Early Ordovician), western Newfoundland; **D.** Polished surface of wavy microbial laminite and laterally linked domes. March Point Formation (Late Cambrian), western Newfoundland.

desiccation cracks (Fig. 11F). Also, they are typically filled by injected sediment whereas desiccation cracks are V-shaped in cross-section, open to the surface, and contain sediment simply washed in.

Planar microbial laminae and stromatolites due to photosynthetic microbes can form, theoretically, anywhere in the photic zone, including in anomalously tranquil lagoonal settings. Examples include the famous Late Jurassic Solnhofen Limestone of southern Germany, as well as the Early Silurian of southern Ontario (Fig. 12C). That these are both fossil Lagerstätten – sites of extraordinary preservation of soft-tissues – points to the taphonomic implications of this phenomenon.

Some intervals of sparsely fossiliferous laminated mudstone in epeiric sea strata have been considered as recording intermittent subaerial exposure due to tidal effects, but some of these can be explained as due to either the onset of elevated salinity or elevated temperature while remaining submerged (Pratt and Haidl, 2008).

The Helena Formation of the Mesoproterozoic Belt Supergroup of western North America is an instructive example of a unit that mimics the peritidal setting. The 'background' facies is laminated, argillaceous, silty lime mudstone often with crosslamination from oscillatory currents. Superficially it does resemble lenticular and wavy tidal bedding. However, the laminae are pervasively graded and small-scale scoured surfaces are comparatively rare. These are both characteristics that are not typically the products from the hustle and bustle of tidal flats. Shrinkage cracks are common but have nothing to do with subaerial exposure. Domical stromatolites are sporadically present but are not intimately associated with the other facies as should be the case in a peritidal setting. Rather, they encrust surfaces scoured up to ~50 cm deep. These erosional features are not tidal creeks or sculpted on tidal flats by storms, but are interpreted to have been scoured by tsunamis. Interbedded, locally hummocky, crosslaminated grainstones are composed of ooids transported into deeper water by tsunami off-surge, rather than recording in situ shallow-water shoals. The Helena Formation is not peritidal at all, but masquerades as such (Pratt, 2001).

PERITIDAL STRATIGRAPHY

The overall stratigraphic record of peritidal facies is the spatio-temporal response to *autogenic* and *allogenic* forcing factors whose relative contribution needs to be distinguished. In essence, the former are intrinsic to the setting and specifically unpredictable in detail, whereas the latter are extrinsic and potentially predictable. Autogenic factors include the ecological characteristics that govern the carbonate factory, and hydrographic processes and paleoclimatic phenomena such as tides, winds and storms which govern the movement

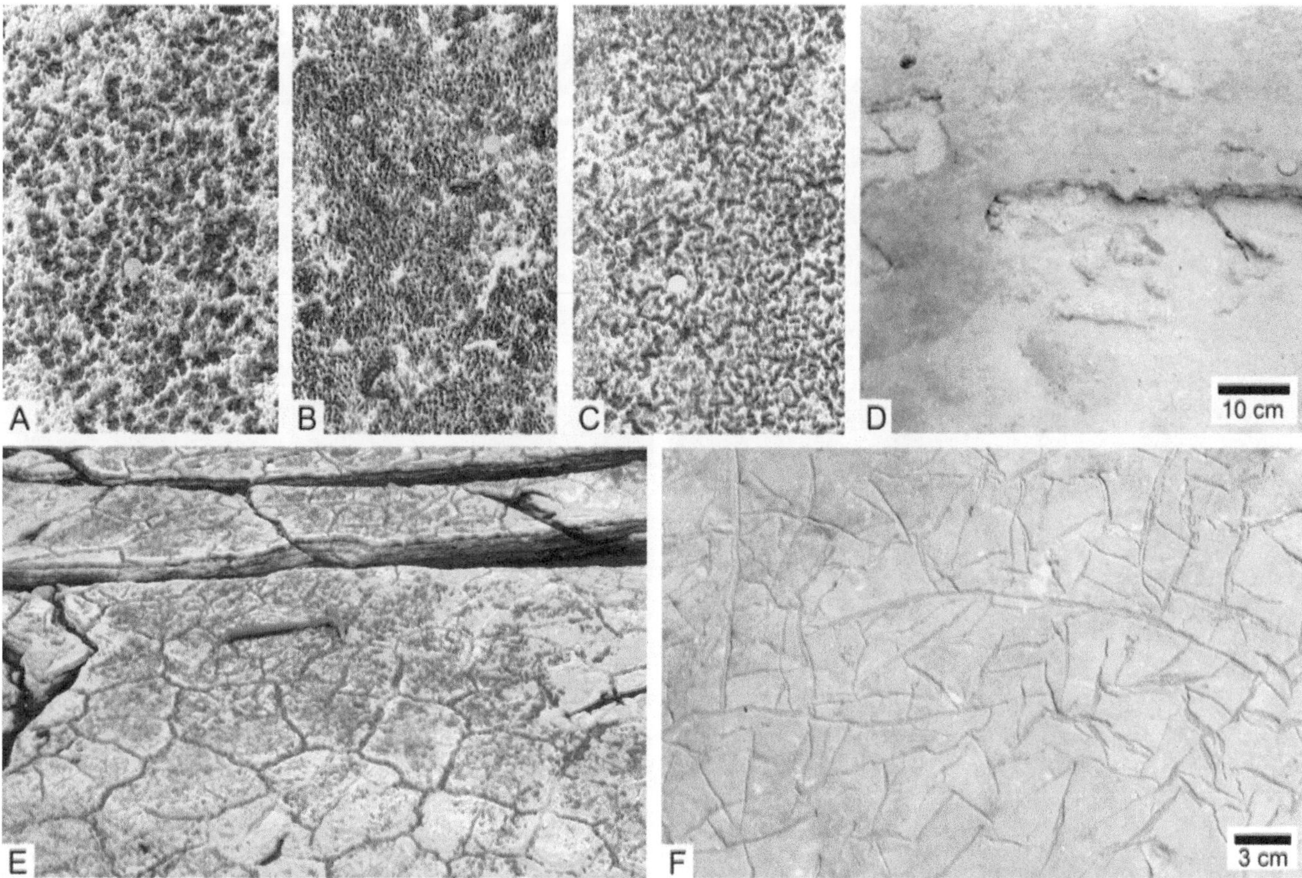

Figure 11. Supratidal surface details. **A–C.** Dark-colored microbial mat showing pustules and wrinkles. Bonaire; **D.** Partly eroded, light-colored smooth mat. Bonaire. Scale bar same for **A–D**; **E.** Bedding plane of desiccation-cracked microbial laminite. Polygons are irregular because of strong microbial binding. Hammer is 30 cm long. Watts Bight Formation (Early Ordovician), western Newfoundland; **F.** Bedding plane of shallow subtidal to intertidal dolomite laminae criss-crossed with intrastratal shrinkage cracks, a phenomenon more commonly seen in Proterozoic and early Paleozoic strata than in younger limestones. These are not due to desiccation from subaerial exposure. Boat Harbour Formation (Early Ordovician), western Newfoundland.

of sediment. Allogenic factors include regional tectonic mechanisms like the rate of subsidence and flexure and global sea-level change, known as eustasy.

One must keep in mind that, in general, the relative abruptness of the stratigraphic response to the various forcing factors does not necessarily equate to the rate at which they operated. Sedimentation rate varies and can cease altogether; there can be lag times before sedimentation resumes. Of course, many events escape being recorded at all.

Modern tidal flats reveal themselves to be remarkably dynamic and complex systems in that they can aggrade, prograde, retrograde and migrate laterally, through a combination of sediment addition and erosion. Geomorphological changes occur over short distances. For example, the western side of Andros Island shows that progradation and erosion can operate in the same general area during rising sea level. Changing hydrographic patterns seaward of the tidal flat caused, say, by the growth of a sand shoal or barrier reef, can have major local effects.

On the other hand, modern analogues are less insightful for peritidal stratigraphy because they are areally rather small and coastal, and they accreted during the latest phase of rapid sea-level rise, after wild eustatic swings through the Quaternary. Thus, it is the rock record alone that must be deciphered in order to learn how entire platforms and epeiric seas behave under less dramatic sea-level change.

Paleogeography plays an important role in where and how peritidal facies accumulate. The coastal areas of probably all carbonate platforms and epeiric seas were mantled by tidal flats even if tidal range was small, thanks to wind, storm and hurricane transport. These deposits are often not preserved because this region is vulnerable to later erosion as a consequence of sea-level fall and tectonic flexure. Peritidal facies are common in many platform interiors, indicating that a surplus of sediment was produced and stored in low-energy areas well seaward of the coast. However, in spite of the availability of accommodation, some platforms lack peritidal facies altogether. This shows how delicately tuned the carbonate factory is.

Cyclic versus Non-cyclic Successions

Portions of some carbonate platforms consist of stratigraphically and paleogeographically continuous, more or less uniform tidal-flat facies, implying that environmental conditions remained stable for extended periods of time. The effects of extrinsic factors like eustasy appear to be

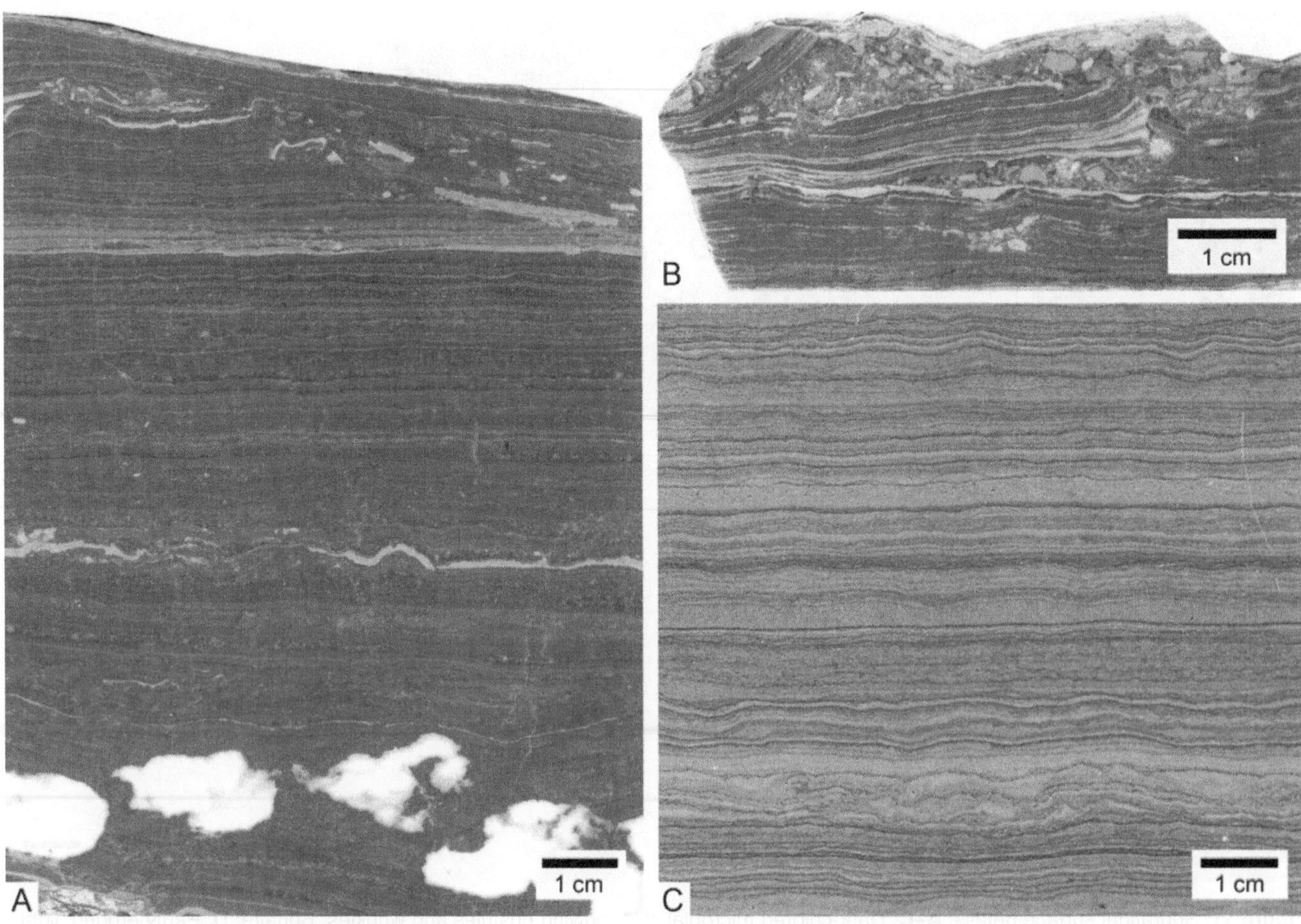

Figure 12. Intertidal and supratidal microbial laminites. Polished surfaces. **A.** Dolomite showing calcitized replacive anhydrite nodules at base and scoured surface at top overlain by intraclasts. Maywood Formation (Late Devonian), north-central Montana; **B.** Dolomitic limestone showing layers peeled back and overlain by coarse sand- and granule-sized grains. Boat Harbour Formation (Early Ordovician), western Newfoundland; **C.** Microbially laminated bituminous dolomite. This is not peritidal but a tranquil, restricted subtidal deposit. Eramosa Formation (Early Silurian), southern Ontario, Canada.

muted in such regions, possibly because they were small.

Most peritidal successions are, however, heterogeneous and individual facies, as beds typically 0.1–1 m thick, are repetitive. In many cases, this repetition is expressed as cycles, and the meter-scale, asymmetric shallowing-upward cycle comprises the most commonly observed motif (Fig. 18). On the other hand, in some other cases cycles are symmetric: each records a gradual shallowing to the supratidal zone followed by gradual deepening and a return to subtidal conditions. However, in still other cases the appearance of cyclicity is deceptive and, instead, stratigraphic patterns and thicknesses are irregular, that is to say, unordered. This suggests that accumulation involved an essentially random combination of accretion, interruption, reworking, migration and drowning. The record of transgressions and regressions indicates, then, just localized phenomena (*see Autocycles* below).

In the simplest case, a shallowing-upward cycle represents a sediment body that has passively filled its available accommodation, regardless of its lateral extent, and has aggraded from the highest subtidal level to just beyond high tide. Assuming sea level did not change during accumulation, the thickness of this stratigraphic package can theoretically yield an approximate measure of the tidal range. Given sufficient sediment supply, and once aggradation has reached the limit of depositional processes, the tidal flat may prograde or migrate laterally.

But, if subsidence, flexure and eustasy are superimposed on simple aggradation, gauging the tidal range may be thwarted. Nevertheless, meter-scale shallowing-upward cycles typically form anyway because the rate of tidal flat accretion normally outpaces externally driven increases in accommodation. Subaerial exposure horizons, manifested as low-relief karst surfaces or paleosols, indicate times of tectonic upwarping or episodes during which eustatic sea-level fall occurred faster than subsidence (Fig. 19).

If sedimentation ceases until, with continued subsidence, the tidal flat becomes subtidal again so that there is renewed accommodation, and this happens repeatedly, the whole succession is said to consist of shallowing-upward cycles, each one separated by a flooding surface. These flooding events – not the same as the diurnal advance of the tide – could well involve extended reworking and nondeposition before tidal flat accretion takes place again. However, if local sedimentation continues but slows

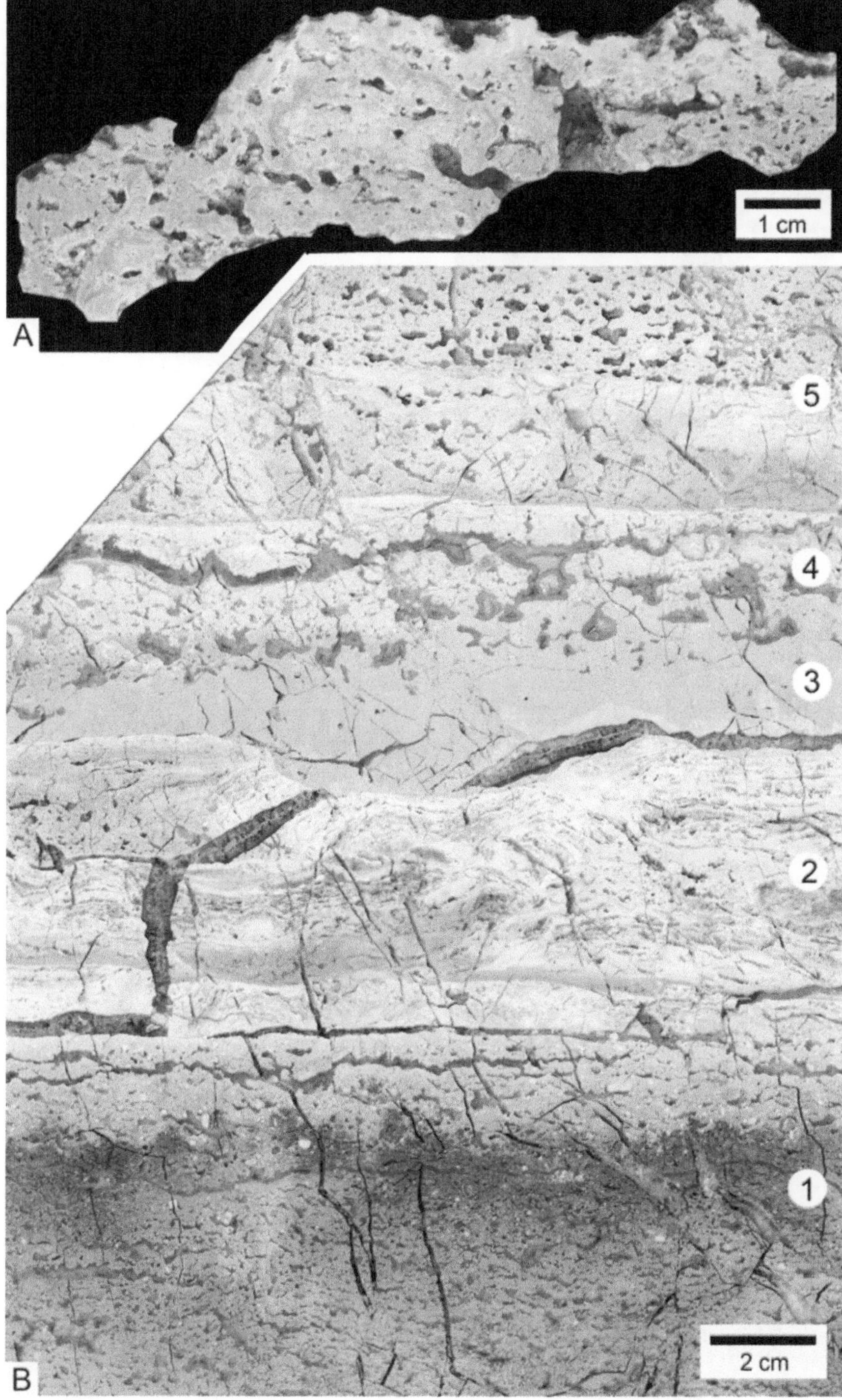

Figure 13. Intertidal and supratidal fenestral limestone. **A.** Polished surface through crust showing indistinct thrombolitic texture and fenestral pores. Bonaire; **B.** Polished surface through a succession of different facies with several small-scale hiatuses. 1-clotted mudstone with small fenestrae; 2-wavy microbial laminite–stromatolite; 3-mudstone; 4-clotted mudstone with large fenestrae; 5-two layers of clotted mudstone with fenestrae. Microbial laminite-stromatolite (2) is disrupted by a crack filled with fibrous calcite cement. Oblique cracks are post-depositional. Late Triassic–Early Jurassic, Peloponnesus, Greece.

down after the tidal flat has accreted to the supratidal zone, supratidal deposits are overlain, in turn, by intertidal then subtidal facies, and if this takes place repeatedly, the succession consists of symmetric cycles.

Causes of Shallowing-upward Cyclicity

Shallowing-upward peritidal cycles are regarded as 4th and 5th order phenomena in the standard hierarchy, and in sequence-stratigraphic parlance they are the parasequences that are bundled into larger scale sequences (Spence and Tucker, 2007). If a stratigraphy is truly cyclic it implies a pattern that is persistently organized in terms of controlling factors and the sedimentary response on a regional scale. Repetitive but non-cyclic patterns imply the dominance of localized phenomena. Thus, one of the problems that must be overcome for a confident interpretation of peritidal cyclicity in specific examples is the often limited information about lateral extent of cycles and their component facies.

Cycles that can be ascribed purely to intrinsic or autogenic controls are termed *autocycles*, whereas those ascribed to extrinsic or allogenic forces like eustasy and episodic subsidence are *allocycles*. Autogenic and allogenic mechanisms can operate concurrently because they are not mutually exclusive.

Autocycles

In many successions, peritidal facies are laterally discontinuous, vary in thickness, and lack a stratigraphic pattern. These are envisaged as recording tidal-flat islands that shift around by aggradation, erosion, progradation and retrogradation in a seemingly irregular fashion (Fig. 20A). This is ascribed to the vagaries of autogenic hydrographic factors that have geographically limited effects, leading to temporally and spatially heterogeneous sediment production and deposition. Lateral migration of tidal flats through time means that facies and contacts are typically diachronous. Nevertheless, shallowing-upward cycles are sometimes formed because after tidal flats accrete to the supratidal zone,

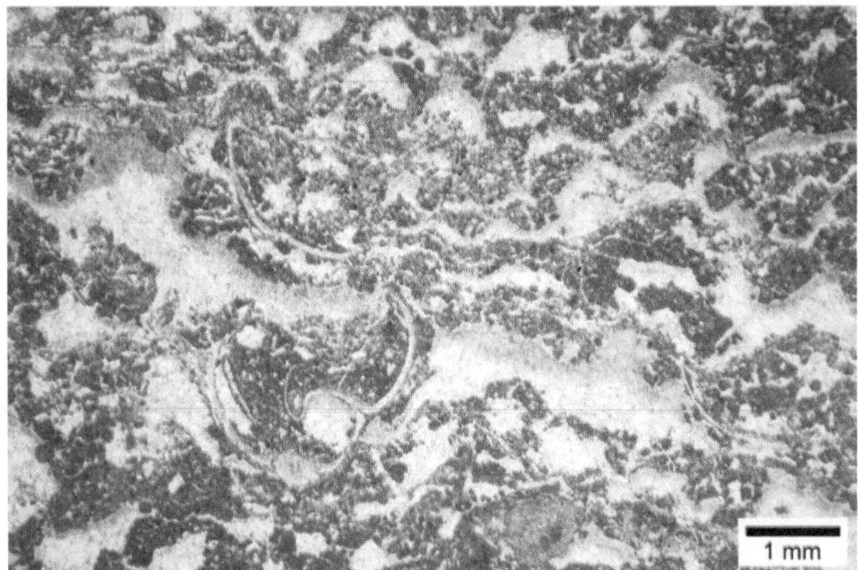

Figure 14. Intertidal and supratidal fenestral limestone. Thin section of peloidal and clotted micrite containing ostracode valves and corroded gastropod shells. Ceilings of fenestrae are coated by pendent fibrous calcite cement, indicating percolation of marine waters through air-filled pores. Table Point Formation (Middle Ordovician), western Newfoundland.

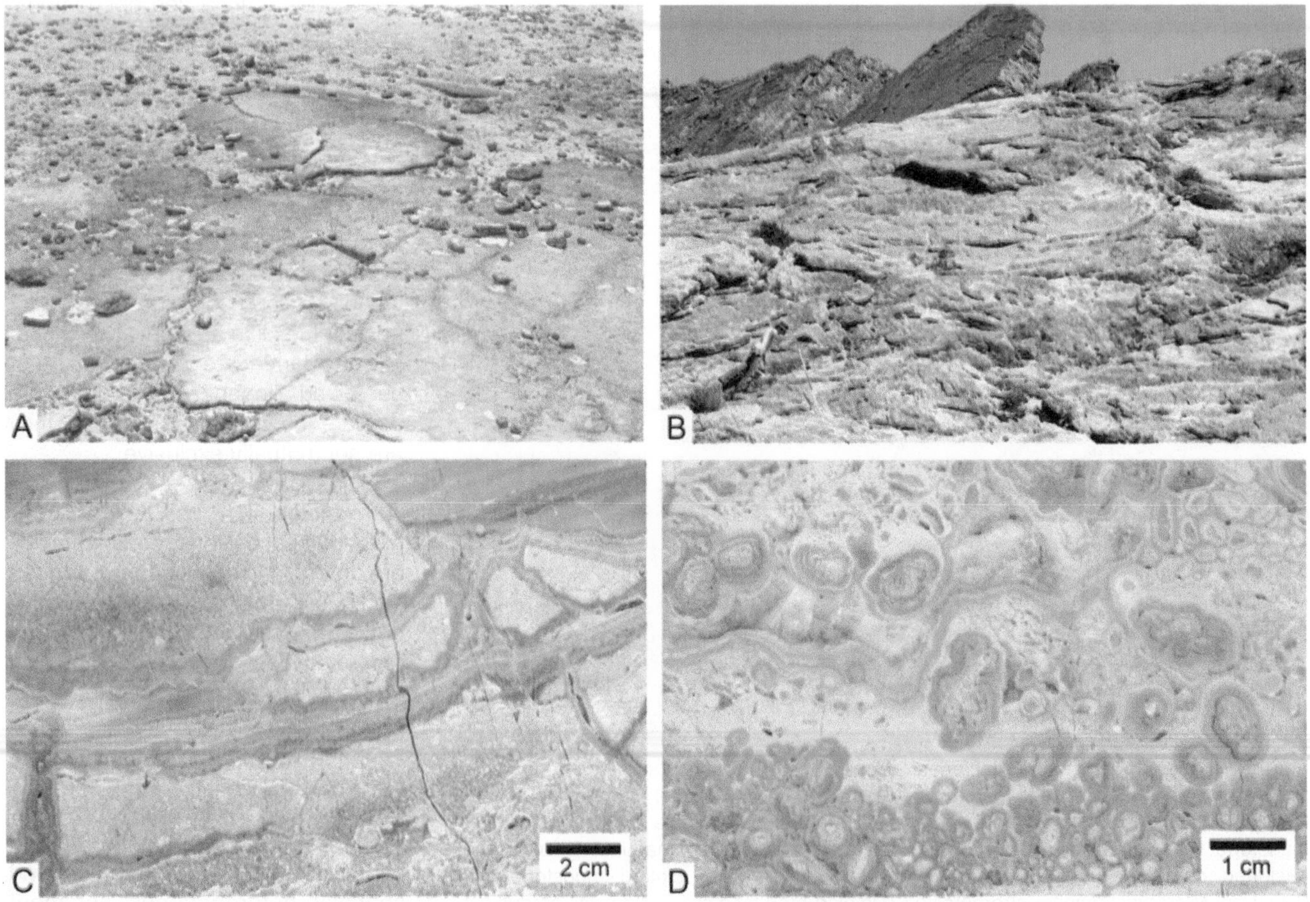

Figure 15. Supratidal crusts and tepees. **A.** Supratidal crusts disrupted, probably by storm-induced, elevated porewater pressure, and eroded into intraclasts. Hammer is 30 cm long. Bonaire; **B.** Bedding surface of tepees. Early Jurassic, High Atlas Mountains, Morocco. Person at lower left is ~175 m tall. Photograph courtesy of C.G.St.C. Kendall; **C.** Polished surface of disrupted crusts. Monte Bardia Limestone (Late Jurassic), Sardinia, Italy; **D.** Polished surface of pisoidal rudstone with fibrous high-Mg calcite cement rinds and geopetal mudstone in cavities. Capitan Formation (Late Permian), southeastern New Mexico, USA.

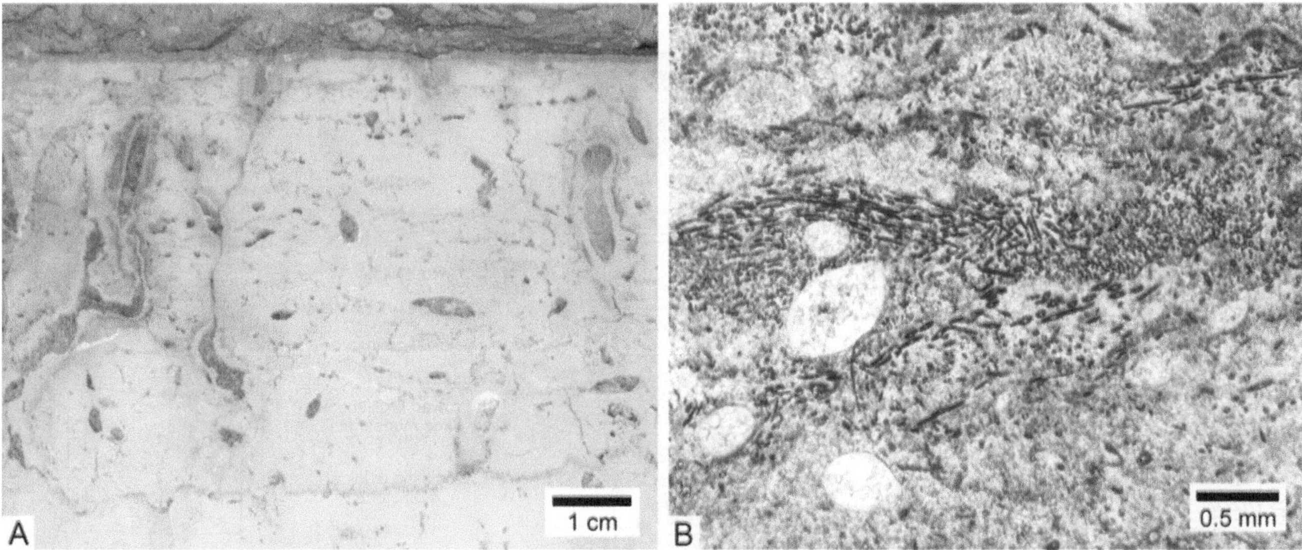

Figure 16. Tidal-flat pond (or restricted lagoon) facies. **A.** Polished surface of mudstone containing scattered ostracode valves, shrinkage cracks and burrow networks of small *Thalassinoides* type, overlain by storm-deposited grainstone which was pumped down into the burrows. Contact between Gull River and Bobcaygeon formations (Middle Ordovician), southern Ontario; **B.** Thin section of limestone composed of loosely packed micrite, peloids, articulated ostracodes and *Girvanella* (calcified filamentous cyanobacteria). This was deposited during the transgression over the unconformity on underlying Early Ordovician limestones. Table Point Formation (Middle Ordovician), western Newfoundland.

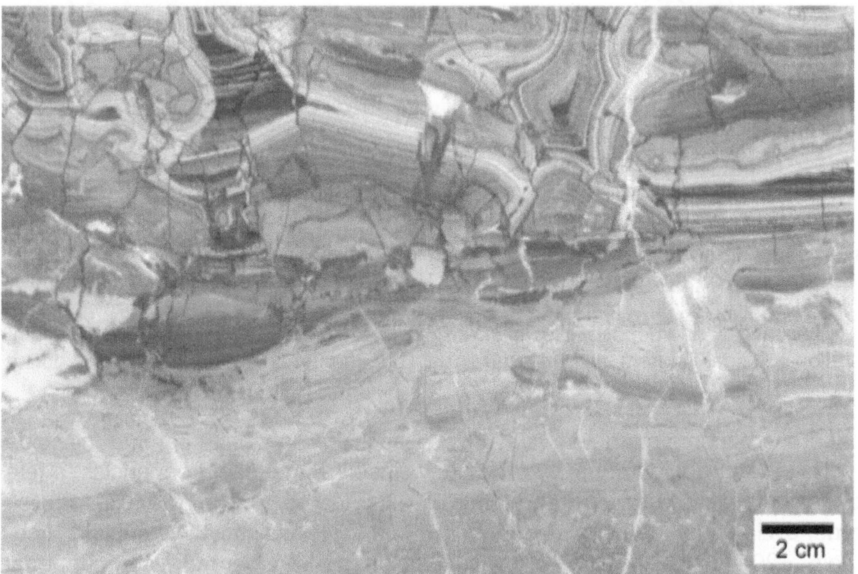

Figure 17. Polished surface of fenestral and microbially laminated mudstone whose upper layers are brecciated and red-stained from subaerial exposure above. The upper surfaces of fragments in the middle part are covered by microbially laminated mudstone. The fragments in the upper part are enveloped by multiple layers of fibrous calcite cement, and cavities are filled by Fe-rich geopetal laminated mudstone. The limestone is cut by several generations of fractures, beginning with synsedimentary vertical and horizontal dilatational cracks. Late Triassic–Early Jurassic, Peloponnesus, Greece.

aggradation can cease until they are flooded and sedimentation in that particular, now-subtidal, locality resumes. In the autocycle model, the peritidal parts of platforms are interpreted to be mosaics of tidal flats and shallow subtidal banks (Pratt and James, 1986). In successions where the lateral extent of peritidal facies cannot be determined but stratigraphic distribution appears to be statistically random, the autocycle explanation appears to be the most reasonable interpretation.

Small-amplitude sea-level oscillations may have occurred but are masked and indistinguishable from autogenic factors. However, more pronounced regional or global allogenic controls can still be superimposed. For example, autogenic processes can generate the unordered facies patterns but episodic subsidence events or eustatic sea-level rises can package them into laterally extensive allocycles.

Allocycles

The architecture of some platforms consists of shallowing-upward cycles that look laterally continuous on a regional scale because this can be observed on mountainsides or confidently determined by correlation between outcrops. Bounding surfaces are essentially time lines. This cyclicity seems to require extrinsic forcing, either by eustatic sea-level fluctuation or by subsidence, or both. These mechanisms lead to episodic creation of accommodation that hypothetically can be filled by regional progradation (Fig. 20B) or aggradation (Fig. 20C), as well as autocycles (*see* above). In the first two situations, it is predictable that as the carbonate factory becomes progressively reduced in size, its efficiency decreases. In purely aggradational scenarios, carbonate production also decreases or changes in character as circulation is hindered and subti-

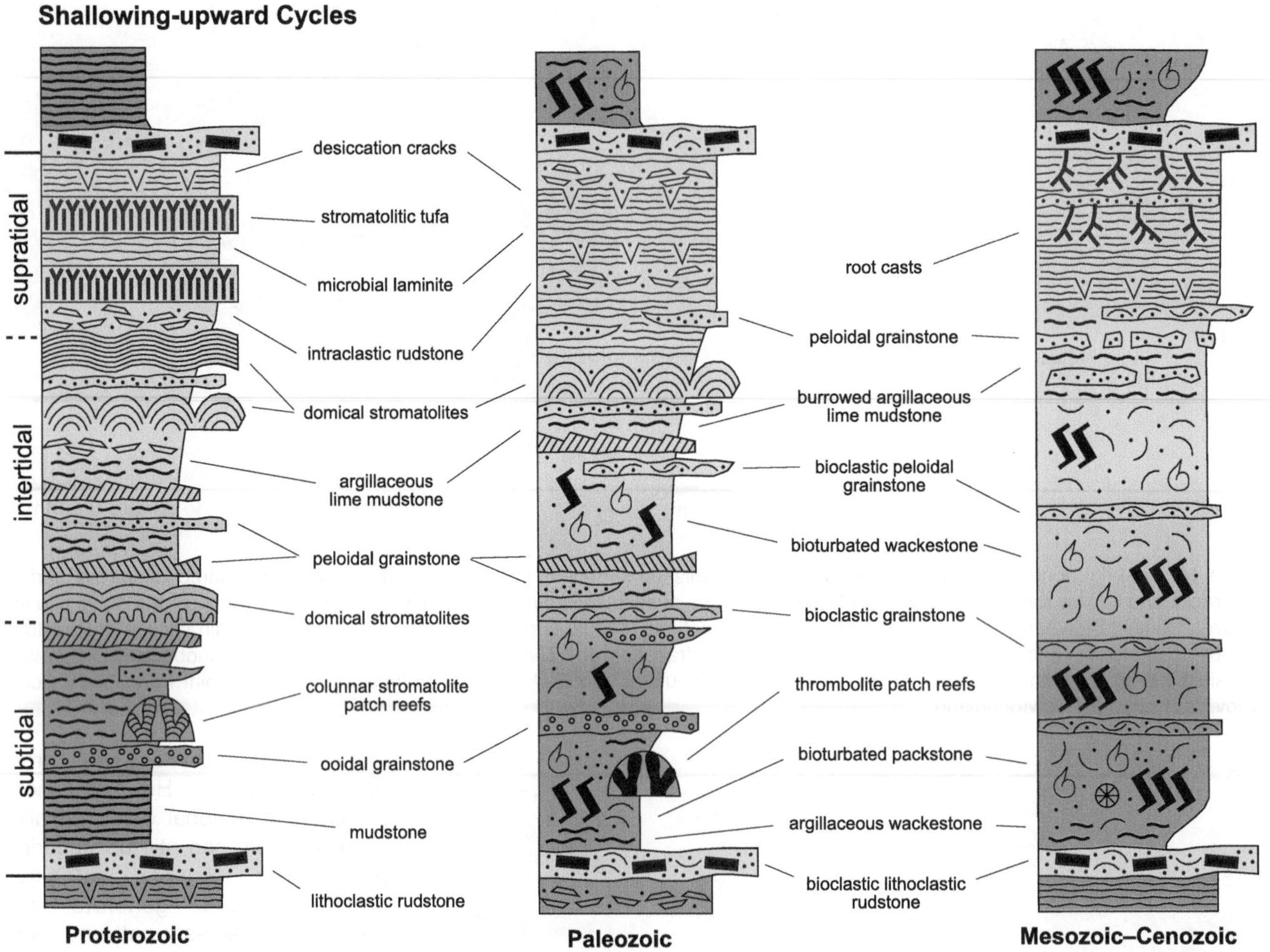

Figure 18. Hypothetical stratigraphic sections showing meter-scale, shallowing-upward peritidal cycles. The three sections illustrate the evolutionary differences in sediment types and facies between Proterozoic, Paleozoic and Mesozoic–Cenozoic times. Particularly noteworthy are the appearance of bioclasts in the early Paleozoic, the later decline of stromatolites, and the evolution of deep-burrowing crustaceans and angiosperms. Each cycle is shown to begin with a lag deposit reflecting transgression over the underlying cycle. The complete cycle records more or less continuous accretion of a tidal flat from the shallow subtidal zone through to the supratidal zone during uniform subsidence.

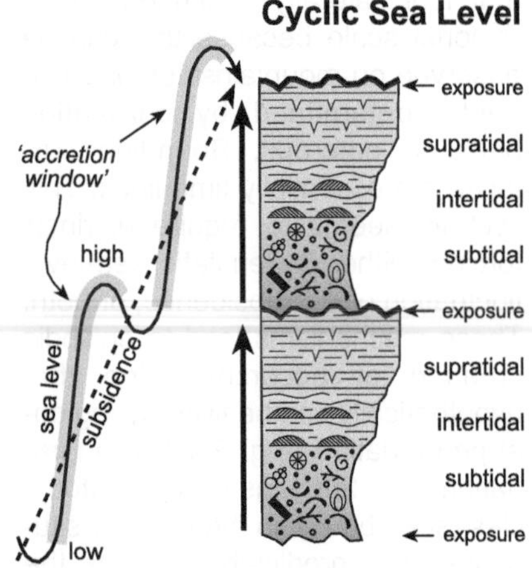

Figure 19. Simple case of cycle development during eustatic rises in sea level superimposed on uniform subsidence. Tidal flat aggradation, the 'accretion window', is followed by subaerial exposure and minor erosion during sea-level fall.

dal areas become restricted.

The stratigraphic frequency of allocycles and their bundling into higher order cycles has been ascribed to the main ~20–40–100 kyr Milankovitch rhythms which are orbital in origin (precessional, obliquity and eccentricity, respectively). While each of these frequencies has a different cause, in a sense they are hierarchical and, akin to interference of sound waves, the peaks and troughs can coincide, thus periodically enhancing their effect. Depending on background subsidence rate and magnitude of sea-level change, cycles of sea-level fall can be punctuated by subaerial exposure horizons (Fig. 19). Subsidence or tectonic flexure can also overprint the orbital controls.

Even if cycle periodicities and stacking patterns seem to coincide closely with orbital frequencies, a reliable temporal framework is crucial. Middle Triassic shallow subtidal limestones interspersed with subaerial exposure surfaces in the Latemàr platform of northern Italy were long considered a seminal example of Milankovitch cyclicity, but as biostratigraphic, geochronologic and magnetostratigraphic control has become more precise, a sub-Milankovitch frequency has been revealed.

In addition, Milankovitch rhythmicity likely did not operate very distinctively at all times in the geological past. During so-called 'greenhouse' intervals, orbital influences on ocean temperature and thermal expansion were relatively small compared to 'icehouse' times. The latter are char-

Origin of Peritidal Cycles

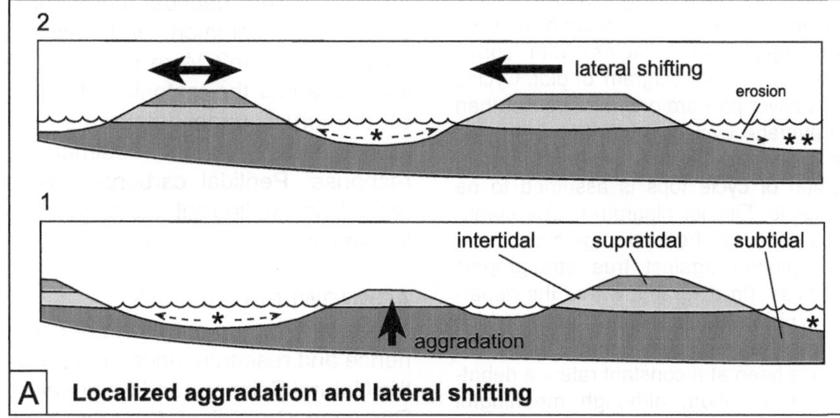

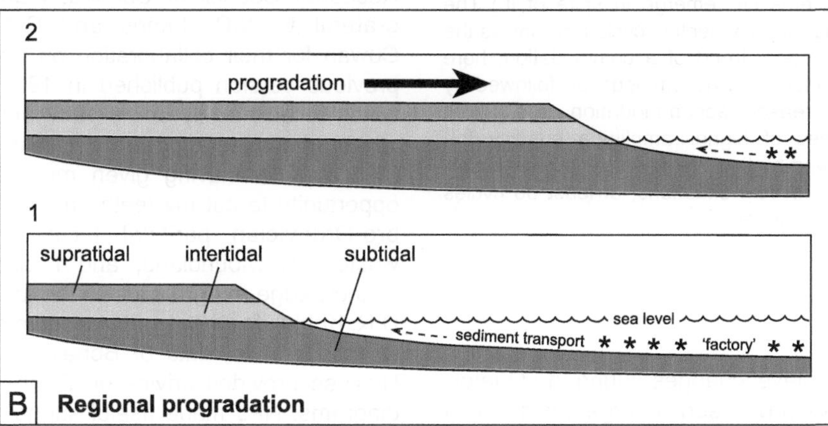

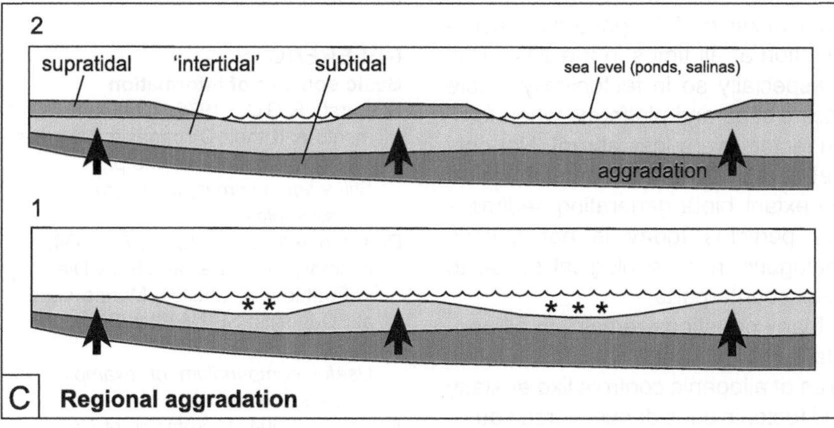

Figure 20. Models illustrating various ways in which meter-scale shallowing-upward peritidal cycles can theoretically form. Knowing the lateral extent of the cycles is key to determining which mechanism is the most likely. **A.** Laterally discontinuous cycles due to localized nucleation, aggradation and lateral shifting, creating islands; **B.** Laterally continuous cycle from regional progradation. Progradation eventually ceases as the subtidal carbonate factory reaches a point when it is too reduced to generate significant quantities of sediment; **C.** Laterally continuous cycle from regional aggradation. Intertidal-like facies form as shallow subtidal areas become restricted.

acterized by polar icecaps which lock up large amounts of the Earth's water, leading to sea-level fluctuations of much greater amplitude, as witnessed during the Quaternary. Thus, it has been suggested that these gave rise to laterally continuous allocycles whereas autocycles are typical of greenhouse times (Lehrmann and Goldhammer, 1999). But, if allocycle thickness is random and not bundled, then a tectonic control would appear to be dominant over eustasy (Bosence *et al.,* 2009).

One method to analyze peritidal successions is to graph the stratigraphic record of shallowing-upward cycles against cycle thickness as a 'Fischer diagram' or plot (Fig. 21). The cumulative departure from mean cycle thickness gives an idea of long-term changes in accommodation, higher order cycle stacking patterns caused by eustasy at lower frequencies, and the subsidence rate of the individual platform. Stratigraphic periodicities can also be compared mathematically with Milankovitch frequencies by time-series analysis in order to generate power spectra of cycle thicknesses. These various exercises seem to work well during 'greenhouse' times of low-amplitude eustasy, although they reveal that the stratigraphic rhythm can be complicated by non-deposition during 'missed beats', erosion and tectonic events affecting subsidence.

Pulsed and irregularly timed subsidence is probably to be expected in platforms in extensional tectonic settings. Nonetheless, one attempt to correlate cycles to seismites as a proxy for strong earthquakes, taken (hypothetically) as registering subsidence events, did not show a match (Pratt, 1994).

FINAL THOUGHTS AND FUTURE DIRECTIONS

Facies description and interpretation are still critically important geological tools, despite all the technical and conceptual advancements in sedimentology. A familiarity with the range of peritidal facies is vital in any petroleum exploration involving carbonate platforms. Armed with the criteria described here combined with disciplined and detailed observation, an open mind, a sound grasp of the literature and a store of field experience, the student is ready to tackle specific stratigraphic units and scenarios. Given the fact that carbonate facies analysis is only a few decades old, there is much scope to refine or correct prior interpretations. Owing to limitations of surface exposures and drilling density, the lateral distribution of peritidal facies has been confidently determined in relatively few cases.

Modern analogues are instructive, but it is important to remember that

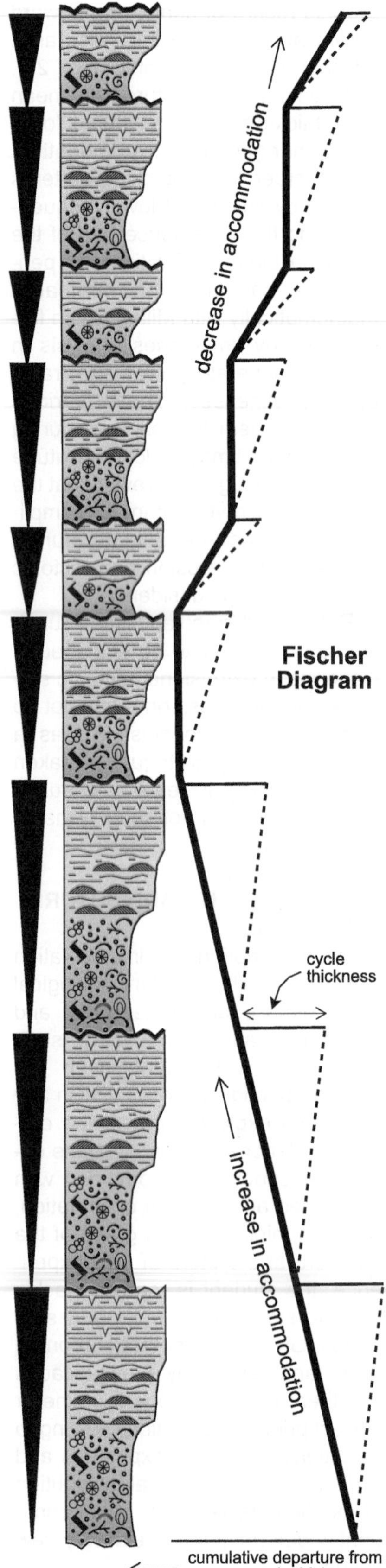

Figure 21. Hypothetical succession of meter-scale shallowing-upward cycles (shown by lithological column and by widening-upward triangles) with corresponding 'Fischer diagram' or plot. Cycles in the lower part are one-third thicker than the average cycle, and in the upper part the thinnest cycles are one-third thinner; erosion of cycle tops is assumed to be negligible. Fischer diagram shows cumulative departure from average cycle thickness plotted against true stratigraphic thickness. (In diagrams where the departure is plotted against cycle number, which is dimensionless, subsidence is assumed to have been at a constant rate – a debatable assumption, although meaningful patterns still emerge in spite of it.) The thick line connecting cycle tops shows the long-term trend of accommodation, here of increased accommodation followed by decreased accommodation (*e.g.,* from times of higher amplitude eustasy followed by times of lower amplitude eustasy). In real situations, at least 50 cycles are required.

sea-level changes during the Quaternary have been so large and so rapid that these sediments may not be all that indicative of the patterns of accumulation at all times in the past. This is especially so in tectonically stable epeiric seas and platforms when forcing factors were less abrupt. Not only that, it is important to remember that the extant biota generating sedimentary particles today is not directly analogous in an ecological sense to fossil counterparts.

Many questions remain. In all peritidal settings what were the relative roles of allogenic controls like eustasy and tectonic subsidence, versus autogenic factors like the vigor of the carbonate factory and the paleoclimatic context to the role of storms? How do changing rates of these processes affect the system? What do sharp boundaries, such as those between successive cycles, represent in terms of time and what may go unrecorded? What components of the observed stratigraphic patterns are predictable and what are likely to be random? Can Milankovitch rhythms be as widely recognized in cyclic successions as some suggest? How do patterns differ in detail between greenhouse versus icehouse times? If some epeiric seas of continental interiors had minimal tides, why do peritidal carbonates seem to be common, and how did they accumulate? More needs to be learned about the tectonic behaviour of the crust under platforms and epeiric seas and the sedimentary response. Peritidal carbonate rocks are still an exciting subject and a challenging storehouse of information!

ACKNOWLEDGEMENTS

This chapter is an outgrowth of experience and research funded mainly by the Natural Sciences and Engineering Research Council of Canada. I am grateful to N.P. James and C.A. Cowan for their collaboration on the previous version published in 1992, which relied heavily on its 1977 and 1983 antecedents by NPJ. I thank NPJ also for having given me the opportunity to cut my teeth on Cambro-Ordovician peritidal facies in western Newfoundland, and I also acknowledge my gratitude to the late D.R. Kobluk for having introduced me to modern tidal flats of Bonaire. A. Husinec provided advice on Fischer diagrams. NPJ made useful suggestions to an earlier draft.

REFERENCES

Basic sources of information

Bathurst, R.G.C., 1975, Carbonate Sediments and their Diagenesis, 2nd edition: Elsevier, Amsterdam, 658 p.
Still a fine summary of several key modern examples.

Demicco, R.V. and Hardie, L.A., 1994, Sedimentary Structures and Early Diagenetic Features of Shallow Marine Carbonate Deposits: SEPM Atlas Series No. 1, 265 p.
Useful compendium of examples with copious illustrations.

Flügel, E., 2004, Microfacies of Carbonate Rocks: Springer-Verlag, Berlin, 976 p.
Exceptionally detailed treatment of carbonate petrography, including that of peritidal facies.

Ginsburg, R.N., *ed.*, 1975, Tidal Deposits: A Casebook of Recent Examples and Fossil Counterparts: Springer, New York, 428 p.
Wide-ranging selection of peritidal sediments and facies, both carbonate and siliciclastic.

Hardie, L.A. and Shinn, E.A., 1986, Carbonate depositional environments, modern and ancient. Part 3: tidal flats: Colorado School of Mines Quarterly, v. 81, 74 p.
Authoritative descriptions of peritidal carbonate sediments and rocks.

Shinn, E.A., 1983a, Tidal flats, *in* Scholle, P.A., Bebout, D.G. and Moore, C.H., *eds.*, Carbonate Depositional Environments: American Association of Petroleum Geologists Memoir 33, p. 171-210.
Well illustrated and authoritative summary of both modern and ancient peritidal carbonates.

Tucker, M.E. and Wright, V.P., 1990, Carbonate Sedimentology: Blackwell Scientific Publications, Oxford, 428 p.
Advanced textbook that places peritidal carbonate facies in their overall context.

Wright, V.P. and Tucker, M.E., 1996, Shallow-water carbonate environments, *in* Reading, H.G., *ed.*, Sedimentary Environments: Processes, Facies and Stratigraphy, 3rd edition, Blackwell Science, Oxford, p. 325-394.
Basic textbook chapter on peritidal carbonates.

Modern examples

Alsharhan, A.S. and Kendall, C.G.St.C., 2003, Holocene coastal carbonates and evaporites of the southern Arabian Gulf and their ancient analogues: Earth-Science Reviews, v. 61, p. 191-243.

Enos, P. and Perkins, R.D., 1979, Evolution of Florida Bay from island stratigraphy: Geological Society of America Bulletin, v. 90, p. 59-83.

Gebelein, C.D., 1977, Dynamics of Recent Carbonate Sedimentation and Ecology, Cape Sable, Florida: E.J. Brill, Leiden, 120 p.

Hardie, L.A., *ed.*, 1977, Sedimentation on the Modern Carbonate Tidal Flats of Northwest Andros Island, Bahamas: Johns Hopkins University Studies in Geology 22, 202 p.

Logan, B.W., Davies, G.R., Read, J.F. and Cebulski, D.E., 1970, Carbonate Sedimentation and Environments, Shark Bay, Western Australia: American Association of Petroleum Geologists Memoir 13, 223 p.

Logan, B.W., Hoffman, P. and Gebelein, C.D., 1974, Evolution and Diagenesis of Quaternary Carbonate Sequences, Shark Bay, Western Australia: American Association of Petroleum Geologists Memoir 22, 358 p.

Lokier, S. and Steuber, T., 2008, Quantification of carbonate-ramp sedimentation and progradation rates for the late Holocene Abu Dhabi shoreline: Journal of Sedimentary Research, v. 78, p. 423-431.

Purser, B.H., *ed.*, 1973, The Persian Gulf: Holocene Carbonate Sedimentation and Diagenesis in a Shallow Epicontinental Sea: Springer, Heidelberg, 471 p.

Rankey, E.C. and Morgan, J., 2002, Quantified rates of geomorphic change on a modern carbonate tidal flat, Bahamas: Geology, v. 30, p. 583-586.

Shinn, E.A., Lloyd, R.M. and Ginsburg, R.N., 1969, Anatomy of a modern carbonate tidal flat, Andros Island, Bahamas: Journal of Sedimentary Petrology, v. 39, p. 1202-1228.

Wanless, H.R., Tyrrell, K.M., Tedesco, L.P. and Dravis, J., 1988, Tidal-flat sedimentation from Hurricane Kate, Caicos Platform, British West Indies: Journal of Sedimentary Petrology, v. 58, p. 724-738.

Wantland, K.F. and Pusey, W.C., *eds.*, 1975, Belize Shelf—Carbonate Sediments, Clastic Sediments, and Ecology: American Association of Petroleum Geologists, Studies in Geology, 2, 599 p.

Ancient examples

Adams, R.D. and Grotzinger, J.P., 1996, Lateral continuity of facies and parasequences in Middle Cambrian platform carbonates, Carrara Formation, southeastern California, U.S.A.: Journal of Sedimentary Research, v. 66, p. 1079-1090.

Bosence, D.W.J., Wood, J.L., Rose, E.P.F. and Qing, H., 2000, Low- and high-frequency sea-level changes control peritidal carbonate cycles, facies and dolomitization in the Rock of Gibraltar (Early Jurassic, Iberian Peninsula): Journal of the Geological Society, London, v. 157, p. 61-74.

Bosence, D., Procter, E., Aurell, M., Bel Kahla, A., Boudagher-Fadel, M., Casaglia, F., Cirilli, S., Mehdie, M., Nieto, L., Rey, J., Scherreiks, R., Soussi, M., Waltham, D., 2009, Dominant tectonic signal in high-frequency, peritidal carbonate cycles? A regional analysis of Liassic platforms from western Tethys: Journal of Sedimentary Research, v. 79, p. 389-415,

Brown, M.A., Archer, A.W. and Kvale, E.P., 1990, Neap-spring tidal cyclicity in laminated carbonate channel-fill deposits and its implications: Salem Limestone (Mississippian), south-central Indiana: Journal of Sedimentary Petrology, v. 60, p. 152-159.

Chen, D., Tucker, M.E., Jiang, M. and Zhu, J., 2001, Long-distance correlation between tectonic-controlled, isolated carbonate platforms by cyclostratigraphy and sequence stratigraphy in the Devonian of South China: Sedimentology, v. 48, p. 57-78.

Egenhoff, S.O., Peterhänsel, A., Bechstädt, T., Zühlke, R. and Grötsch, J., 1999, Facies architecture of an isolated carbonate platform: tracing the cycles of the Latemàr (Middle Triassic, northern Italy): Sedimentology, v. 46, p. 893-912.

Enos, P. and Samankassou, E., 1998, Lofer cyclothems revisited (Late Triassic, Northern Alps, Austria): Facies, v. 38, p. 207-228.

Grimwood, J.L., Coniglio, M. and Armstrong, D.K., 1999, Blackriveran carbonates from the subsurface of the Lake Simcoe area, southern Ontario; stratigraphy and sedimentology of a low-energy carbonate ramp: Canadian Journal of Earth Sciences, v. 36, p. 871-889.

Grotzinger, J.P., 1986, Cyclicity and paleoenvironmental dynamics, Rocknest platform, northwest Canada: Geological Society of America Bulletin, v. 97, p. 1208-1231.

Grötsch, J., 1996, Cycle stacking and long-term sea-level history in the Lower Cretaceous (Gavrovo Platform, NW Greece): Journal of Sedimentary Research, v. 66, p. 723-736.

Haas, J., 2004, Characteristics of peritidal facies and evidences for subaerial exposures in Dachstein-type cyclic platform carbonates in the Transdanubian Range, Hungary: Facies, v. 50, p. 263-286.

Haas, J., Lobitzer, H. and Monostori, M., 2007, Characteristics of the Lofer cyclicity in the type locality of the Dachstein Limestone (Dachstein Plateau, Austria): Facies, v. 53, p. 113-126.

Hamon, Y. and Merzeraud, G., 2008, Facies architecture and cyclicity in a mosaic carbonate platform: effects of fault-block tectonics (Lower Lias, Causses platform, south-east France): Sedimentology, v. 55, p. 155-178.

Hofmann, A., Dirks, P.H.G.M. and Jelsma, H.A., 2004, Shallowing-upward carbonate cycles in the Belingwe Greenstone Belt, Zimbabwe: a record of Archean sea-level oscillations: Journal of Sedimentary Research, v. 74, p. 64-81.

Husinec, A. and Read, J.F., 2007, The Late Jurassic Tithonian, a greenhouse phase in the Middle Jurassic–Early Cretaceous 'cool' mode: evidence from the cyclic Adriatic Platform, Croatia: Sedimentology, v. 54, p. 317-337.

Knoll, A.H., Swett, K. and Mark, J., 1991, Paleobiology of a Neoproterozoic tidal flat/lagoon complex: the Draken Conglomerate Formation, Spitsbergen: Journal of Paleontology, v. 65, p. 531-570.

Laporte, L.F., 1967, Carbonate deposition near mean sea-level and resultant facies mosaic: Manlius Formation of New York State: American Association of Petroleum Geologists Bulletin, v. 51, p. 73-101.

Lehmann, C., Osleger, D.A. and Montañez, I.P., 1998, Controls on cyclostratigraphy of Lower Cretaceous carbonates and evaporites, Cupido and Coahuila platforms, northeastern Mexico: Journal of Sedimentary Research, v. 68, p. 1109-1130.

Lehrmann, D.J. and Goldhammer, R.K., 1999, Secular variation in parasequence and facies stacking patterns of platform carbonates: a guide to application of stacking-pattern analysis in strata of diverse ages and settings, *in* Harris, P.M., Saller, A.H. and Simo, J.A., *eds.*, Advances in Carbonate Sequence Stratigraphy: Application to Reservoirs, Outcrops and Models: SEPM Special Publication 63, p. 187-225.

Masse, J.-P., Fenerci, M. and Pernarcic, E., 2003, Palaeobathymetric reconstruction of peritidal carbonates, Late Barremian, Urgonian, sequences of Provence (SE France): Palaeogeography, Palaeoclimatology, Palaeoecology, v. 200, p. 65-81.

Matter, A., 1967, Tidal flat deposits in the Ordovician of Maryland: Journal of Sedimentary Petrology, v. 37, p. 601-609.

Montañez, I.P. and Osleger, D.A., 1993, Parasequence stacking patterns, third-order accommodation events, and sequence stratigraphy of Middle to Upper Cambrian platform carbonates, Bonanza King Formation, southern Great Basin, *in* Loucks, R.G. and Sarg, J.F., *eds.*, Recent Advances and Applications of Carbonate Sequence Stratigraphy: American Association of Petroleum Geologists Memoir 57, p. 305-325.

Osleger, D.A. and Read, J.F., 1991, Relation of eustasy to stacking patterns of meter-scale carbonate cycles, Late Cambrian, USA: Journal of Sedimentary Petrology, v. 61, p. 1225-1252.

Peterhänsel, A. and Egenhoff, S.O., 2008, Lateral variation of cycle stacking patterns in the Latemàr, Triassic, Italian Dolomites, *in* Lukasik, J. and Simo, J.A., *eds.*, Controls on Carbonate Platform and Reef Development: SEPM Special Publication 89, p. 217-229.

Pratt, B.R., 1994, Seismites in the Mesoproterozoic Altyn Formation (Belt Supergroup), Montana: a test for tectonic control of peritidal carbonate cyclicity: Geology, v. 22, p. 1091-1094.

Pratt, B.R., 2001. Oceanography, bathymetry and syndepositional tectonics of a Precambrian intracratonic basin: integrating sediments, storms, earthquakes and tsunamis in the Belt Supergroup (Helena Formation, c. 1.45 Ga), western North America: Sedimentary Geology, v. 141/142, p. 371-394.

Pratt, B.R., 2002, Tepees in peritidal carbonates: origin via earthquake-induced deformation, with example from the Middle Cambrian of western Canada: Sedimentary Geology, v. 153, p. 57-64.

Pratt, B.R., and Haidl, F.M., 2008, Microbial patch reefs in Upper Ordovician Red River strata, Williston Basin, Saskatchewan: signal of heating in a deteriorating epeiric sea, *in* Pratt, B.R. and Holmden, C., *eds.*, Dynamics of Epeiric Seas: Geological Association of Canada, Special Paper 48, p. 315-352.

Pratt, B.R. and James, N.P., 1986, The St. George Group (Lower Ordovician) of western Newfoundland: tidal flat island model for carbonate sedimentation in epeiric seas: Sedimentology, v. 33, p. 313-343.

Preto, N., Hinnov, L.A., de Zanche, V., Mietto, P. and Hardie, L.A., 2004, The Milankovitch interpretation of the Latemar platform cycles (Dolomites, Italy): implications for geochronology, biostratigraphy, and Middle Triassic carbonate accumulation, *in* D'Argenio, B., Fischer, A.G., Premoli Silva, I., Weissert, H. and Ferreri, V., *eds.*, Cyclostratigraphy: Approaches and Case Histories: SEPM Special Publication 81, p. 167-182.

Riding, R., 2000, Microbial carbonates: the geological record of calcified bacterial–algal mats and biofilms: Sedimentology, v. 47 (Suppl. 1), p. 179-214.

Roehl, P.O., 1967, Stony Mountain (Ordovician) and Interlake (Silurian), facies analogs of Recent low-energy marine and subaerial carbonates, Bahamas: American Association of Petroleum Geologists Bulletin, v. 51, p. 1979-2032.

Salad Hersi, O. and Dix, G.R., 1999, Blackriveran (lower Mowhawkian, Upper Ordovician) lithostratigraphy, rhythmicity, and paleogeography: Ottawa Embayment, eastern Ontario, Canada: Canadian Journal of Earth Sciences, v. 36, p. 2033-2050.

Sami, T.T. and James, N.P., 1994, Peritidal carbonate platform growth and cyclicity in an early Proterozoic foreland basin, upper Pethei Group, northwest Canada: Journal of Sedimentary Research, v. 64, p. 111-131.

Satterley, A.K., 1996, Cyclic carbonate sedimentation in the Upper Triassic Dachstein Limestone, Austria: the role of patterns of sediment supply and tectonics in a platform–reef–basin system: Journal of Sedimentary Research, v. 66, p. 307-323.

Shinn, E.A., 1983b, Birdseyes, fenestrae, shrinkage pores and loferites: a reevaluation: Journal of Sedimentary Petrology, v. 53, p. 619-628.

Skompski, S., Luczyński. P., Drygant, D. and Kozłowski, W., 2008, High-energy sedimentary events in lagoonal successions of the Upper Silurian of Podolia, Ukraine: Facies, v. 54, p. 277-296.

Spence, G.H. and Tucker, M.E. 2007, A proposed integrated multi-signature model for peritidal cycles in carbonates. Journal of Sedimentary Research, v. 77, p. 797-808.

Strasser, A., Hillgärtner, H. and Pasquier, J.-B., 2004, Cyclostratigraphic timing of sedimentary processes: an example from the Berriasian of the Swiss and French Jura Mountains, *in* D'Argenio, B., Fischer, A.G., Premoli Silva, I., Weissert, H. and Ferreri, V., *eds.*, Cyclostratigraphy: Approaches and Case Histories: SEPM Special Publication 81, p. 135-151.

Warren, J., 2000, Dolomite: occurrence, evolution and economically important associations: Earth-Science Reviews, v. 52, p. 1-81.

Wells, M.R., Allison, P.A., Piggott, M.D., Pain, C.C., Hampson, G.J. and Dodman, A., 2008, Investigating tides in the Early Pennsylvanian seaway of northwest Europe using the Imperial College Ocean Model, *in* Pratt, B.R. and Holmden, C., *eds.*, Dynamics of Epeiric Seas: Geological Association of Canada, Special Paper 48, p. 375-400.

Wilkinson, B.H., Diedrich, N.W. and Drummond, C.N., 1996, Facies associations in peritidal carbonate sequences: Journal of Sedimentary Research, v. 66, p. 1065-1078.

Wilkinson, B.H., Drummond, C.N., Diedrich, N.W. and Rothman, E.D., 1997a, Biological mediation of stochastic peritidal carbonate accumulation: Geology, v. 25, p. 847-850.

Wilkinson, B.H., Drummond, C.N., Rothman, E.D. and Diedrich, N.W., 1997b, Stratal order in peritidal carbonate sequences: Journal of Sedimentary Research, v. 67, p. 1068-1082.

Yang, W., Harmsen, F. and Kominz, M.A., 1995, Quantitative analysis of a cyclic peritidal carbonate sequence, the Middle and Upper Devonian Lost Burro Formation, Death Valley, California—a possible record of Milankovitch climatic cycles: Journal of Sedimentary Research, v. B65, p. 306-322.

Yang, W. and Lehrmann, D.J., 2003, Milankovitch climatic signals in Lower Triassic (Olenekian) peritidal carbonate successions, Nanpanjiang Basin, South China: Palaeogeography, Palaeoclimatology, Palaeoecology, v. 201, p. 283-306.

Zühlke, R., 2004, Integrated cyclostratigraphy of a model Mesozoic carbonate platform—the Latemar (Middle Triassic, Italy), *in* D'Argenio, B., Fischer, A.G., Premoli Silva, I., Weissert, H. and Ferreri, V., *eds.*, Cyclostratigraphy: Approaches and Case Histories: SEPM Special Publication 81, p. 183-211.

17. Reefs

Noel P. James, Department of Geological Sciences and Geological Engineering, Queens University, Kingston, ON, K7L 3N6, Canada

Rachel Wood, School of Geosciences, University of Edinburgh, Edinburgh, EH9 3JW, UK; and Edinburgh Consortium for Subsurface Science and Engineering, UK

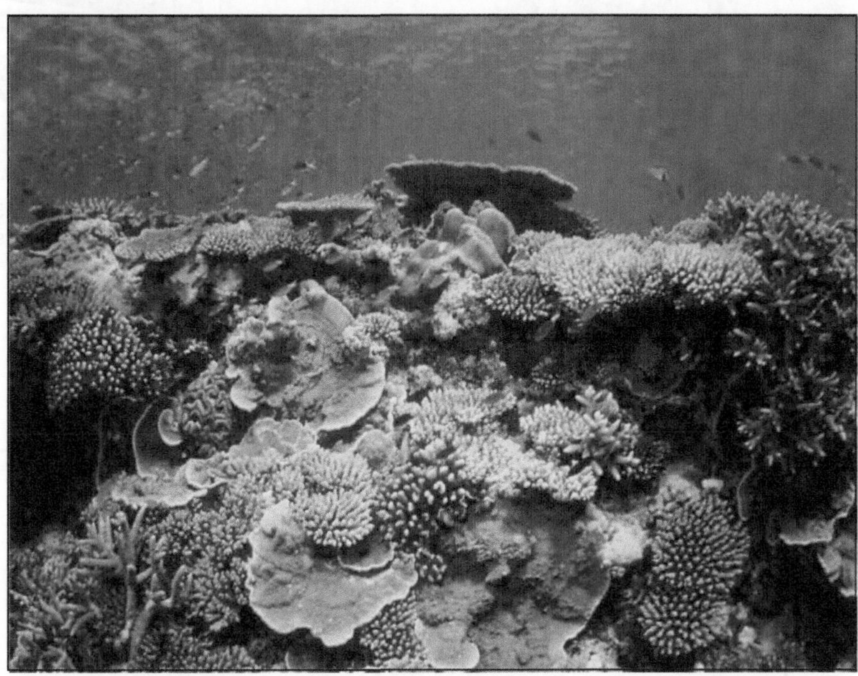

Figure 1. A living reef in ~2 m of water composed mainly of mixotroph scleractinian corals illustrating the wide variety of growth forms, Heron Island, Great Barrier Reef, Australia. Field of view ~ 3 meters. (image courtesy P. Pufahl).

INTRODUCTION

When we think of reefs, visions of clear tropical water, luxuriant coral growth and brightly colored fish spring to mind (Fig. 1). Today, such reefs are built by large corals and abundant algae, and stand as impressive wave-resistant structures above the surrounding seafloor. Our vision quickly becomes blurred, however, in quiet, murky, cold or deep water where these structures have varied relief, are constructed by different organisms, and are as much piles of sediment as hard rock.

Ancient reefs were *biologically constructed reliefs* that grew on the seafloor and are now rock bodies of massive carbonate, surrounded by bedded strata (Fig. 2). Not only are they our most important repositories of evolving marine biology and changing oceanography but they are of significant economic importance as hydrocarbon reservoirs. It is a challenge to formulate facies models for such features because their growth was governed as much by interactions within the evolving biosphere as by universal physical and chemical laws. This article is an integration of the themes that run through geological history and characterize reefs of all ages.

REEF STRUCTURE

Reefs are essentially biological constructions and their character, at any given time, is a product of the organisms alive at that moment. Within this phrase, however, lies a world of complexity that is difficult to synthesize mostly because of complex feedback loops between organisms, environments, and evolution (Fig. 3). A strict uniformitarian approach to understanding reefs has limited utility; we can gain insights into processes but predict the products with difficulty. Ancient reefs and their facies are not modern reefs just formed by different organisms: in many cases they had fundamentally different ecological demands and geological attributes that have no living counterparts!

Regardless, a reef is a complex, but integrated, community of calcareous and non-calcareous organisms, inorganic and organic carbonate precipitates, and sediment that rises above the seafloor. Some structures grow passively on broad open shelves whereas others become so large, as to change the environment itself (*e.g.,* the Great Barrier Reef). Principal reef building components range across the biological spectrum from invertebrate metazoans to microbes. This community produces a highly porous structure with growth cavities, and yet is commonly cemented early to form a hard limestone structure on the seafloor. The living surface is also continuously broken down by a variety of bioeroders and predators. The recurring processes that seem to be present and characterize these biological reliefs are (Fig. 3):

1. In-place fixation of carbonate by organic assemblages of microbes, algae, and metazoans;
2. Development of internal cavity systems during growth;
3. Synsedimentary lithification; and
4. Bioerosion.

These attributes have varied markedly through time (Fig. 4).

Biological Fixation of Carbonate

Calcareous Metazoans

Reef-building calcareous metazoans come from all invertebrate taxa but the most recurring and important groups are sponges, corals, bryozoans, and bivalves. Many reef metazoans are colonial or clonal, and are capable of surviving and re-growing after damage to their living tis-

Figure 2. A. Two Neoproterozoic reefs, the larger one of which is about 600 m wide and 300 m high, Mackenzie Mountains, NW Canada; **B.** Small tabulate coral reef, middle Ordovician, western Newfoundland, Canada. Person for scale.

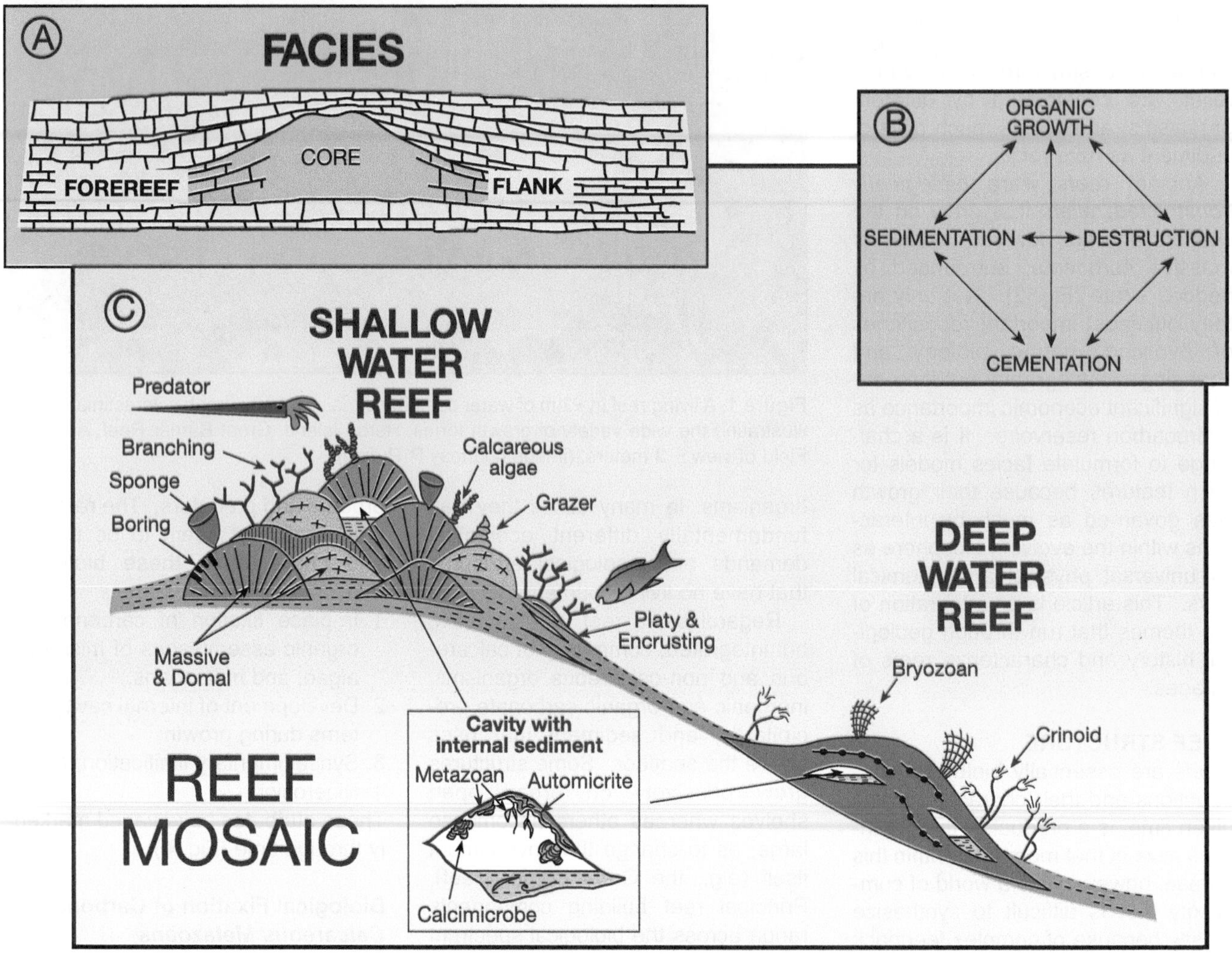

Figure 3. A series of sketches illustrating **A.** Cross-sectional geometry of a typical reef as exposed in outcrop; **B.** Complex interrelationship between processes that control reef composition; **C.** Main attributes of shallow-water and deep-water reefs.

sues. They are obligate calcifiers in the sense that the organism controls calcification (or biomineralization). Many of these robust invertebrates have been called 'hypercalcifiers' (Stanley and Hardie, 1998) because of their propensity to produce large skeletons. Such skeletons range in shape from domal and hemispherical (single coral colonies, and stromatoporoid sponges can be up to 3 m across), to laminar or encrusting, to delicate plate-like and branching. The more delicate growth types are readily fragmented to form sediment in shallow water.

Calcareous metazoans are either heterotrophs or mixotrophs. Heterotrophs consume other organisms whereas mixotrophs are heterotrophs that also contain symbiotic microorganisms, generally photosynthetic cyanobacteria or microalgae; they are a plant–animal mixture. The host organism ingests food whereas the symbionts produce carbohydrates and lipids for the host via photosynthesis. The most important mixotrophs in modern reefs are corals with symbionts called zooxanthellae. By providing almost limitless energy, zooxanthellae allow hermatypic (reef-building) corals to produce calcium carbonate several times faster than ahermatypic (non-reef building), azooxanthellate corals. But it is a tradeoff between symbiont and host; zooxanthellae are limited to the photic zone and do best in low-nutrient (oligotrophic) environments. Thus, luxuriant modern coral reefs are limited to shallow-water, nutrient-poor environments. Corals without symbionts form reefs in deep, dark environments and are very slow growing.

Sponges include hexactinellids (siliceous spicules), and those with calcareous skeletons - calcified sponges belonging to several different groups (stromatoporoids - Fig. 5A), demosponges (spiculate or non-spiculate, but with calcareous skeletons), archaeocyaths, and chaetetids. Corals comprise both tabulates (Fig. 5B) and rugosans in the Paleozoic, and scleractinians (Fig. 1) in the Mesozoic and Cenozoic. The other important organisms are bryozoans (Fig. 5C) that, even though relatively small and delicate compared to the foregoing, are critical elements of many reefs and act as scaffolds between which *in situ* microbially induced micrite (automicrite) can form and cements precipitate.

Although we cannot be certain about the physiological tolerances of all fossil reef-builders, some generalities are possible. Scleractinian corals, the modern reef-building corals, evolved in the Triassic and probably possessed zooxanthellae by the Late Triassic. Tabulate corals were a diverse group of organisms

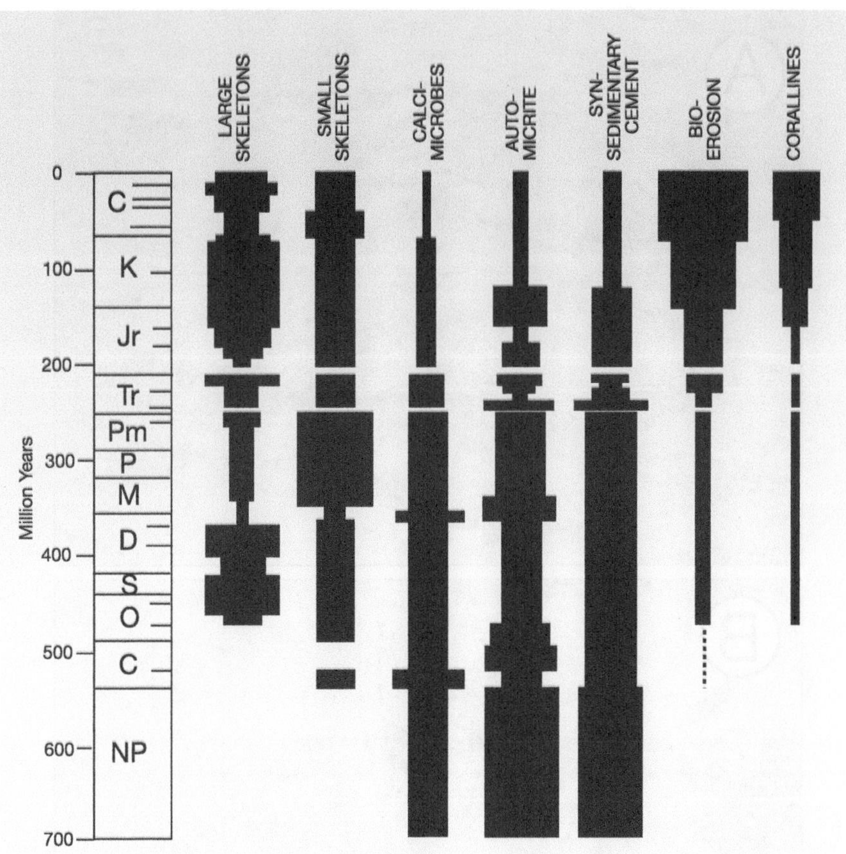

Figure 4. Generalized qualitative depictions of the relative importance of various components and processes of reef formation through the Neoproterozoic and Phanerozoic. Size of spindles is relative to those attributes of modern reefs, shallow and deep.

and it is not clear whether they contained similar symbionts. Rudist bivalves, which were mound builders and massive producers of sediment in Cretaceous seas, are similar but unrelated to modern, giant clams (*Tridacna*), although some of the most important groups did not appear to have symbionts. Fossil sponges, both spiculate and soft bodied, and calcified sponges (such as stromatoporoids and archaeocyaths) are poorly understood and their association with photosymbionts is uncertain (Wood, 1999) even though modern sponges on outer-shelf reefs commonly contain cyanobacterial symbionts. For those organisms that did contain phototrophic symbionts, light and nutrients would have been important limiting factors.

Calcareous Algae

Calcareous algae, both red (coralline and squamaracean) and green (codiacean and udoteacean), are prolific reef-associated sediment producers and can even form reefs on their own. They are also obligate calcifiers. Encrusting red algae are critically important accessory components in reefs dominated by calcareous metazoans because they bind and stabilize reef frameworks. Branching and segmented green algae are significant sediment producers on modern reefs, and red or green algae may have occupied a similar niche in the past. One group of leaf-like (phylloid) algae (Fig. 6A) are particularly conspicuous in the late Paleozoic. Calcareous algae have been part of the carbonate factory since the Ordovician but were not significant in reef facies until the Mississippian and have continued to be so ever since.

Microbes

Carbonate precipitation associated with microbes grades from peloidal, clotted or laminated micrite (automicrite) to calcified microbial sheaths (calcimicrobes) that have a spectrum of intermediate structures, whose attributes depend upon seawater

Figure 5. A. Stromatoporoid - Silurian, Anticosti Island, Canada: cm scale; **B.** Tabulate coral, Silurian, Anticosti Island, Canada: cm scale; **C.** Broken core illustrating fenestrate bryozoans and surrounding synsedimentary cement, Mississippian, Western Canadian Sedimentary Basin, cm scale (image courtesy W. Martindale).

carbonate saturation and post-mortem preservation.

Microbes are highly diverse single-celled prokaryotes and eukaryotes. These organisms are non-obligate calcifiers in that they induce carbonate precipitation by their metabolic processes or post-mortem chemistry; most such precipitates are microcrystalline carbonate (micrite). They are incapable of calcification under unfavorable conditions such as low carbonate saturation. Calcification takes place beneath or within a biofilm consisting of thin or thick layers of organic matter (EPS – extracellular polysaccharide) that can contain a variety of living and dead microbial communities within a matrix of degrading organic matter and mucilage. Such microcrystalline carbonate is called automicrite because it is produced in-place, generally on a substrate and it can be an important binding or cementing agent.

The most common calcimicrobes are tubular (*Girvanella*, *Rothpletzella*), subspherical (*Renalcis*; Fig. 6B), or branching (*Epiphyton*). Generally, they are interpreted as sheaths of bacteria that have undergone variable taphonomic alteration. Many other similar structures of complex affinity, such as *Tubiphytes* (a foraminifer–cyanobacteria association), have also been placed in this category.

Automicrite is also the main component of layered stromatolites and thrombolites (clotted rather than laminated structures). Stromatolites, composed of automicrite and synsedimentary cement, were the only reef-builders throughout the Precambrian and were augmented by thrombolites and calcimicrobes in the Neoproterozoic. They have continued to be a variable contributor throughout the Phanerozoic. Calcimicrobes, once established in the Neoproterozoic, have continued to be a significant and sometimes dominant element of reefs throughout the Phanerozoic. Whereas automicrite is essentially an encrusting or binding element in most reefs, often within cavities, it may nevertheless contribute substantially to the reef rock volume. Calcimicrobes, *e.g., Renalcis,* are also important sediment producers, similar to *Tubiphytes* in the Permian and the foraminifer *Homotrema* in the Quaternary.

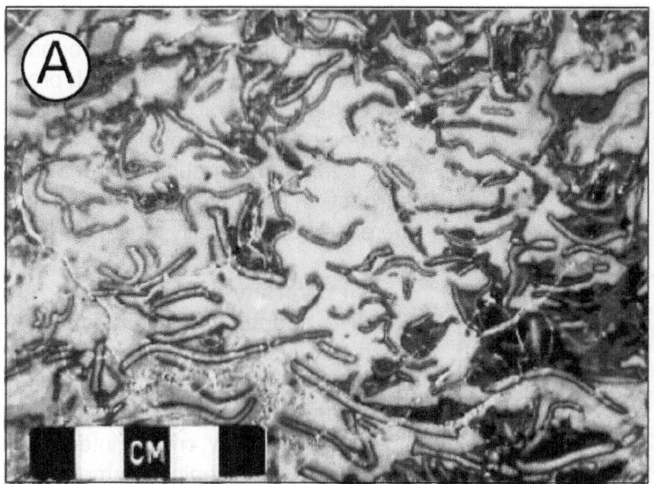

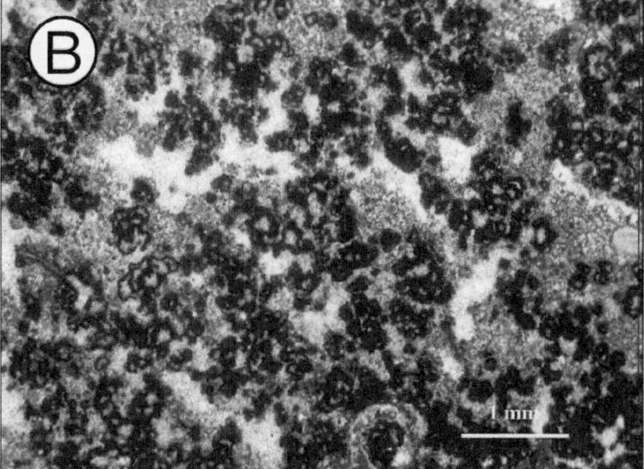

Figure 6. A. Polished core slab of phylloid algae (now preserved as leached molds) surrounded by synsedimentary marine cement; Pennsylvanian, Ellesmere Island, Arctic Canada (image courtesy G.R. Davies); **B.** Photomicrograph, plane polarized light of the calcimicrobe *Renalcis* (dark clustered lunate objects) surrounded by microclotted automicrite, Upper Devonian, Canning Basin, Australia (image courtesy J.L.Wray).

The critical point is that any one or any combination of these components can form a reef. The relative importance and abundance of each component has, however, changed through geologic time, leading to the bewildering array of ancient reef fabrics present in the stratigraphic record.

Synsedimentary Lithification

Most reefs are lithified immediately below the living surface by a variety of calcite or aragonite precipitates. Much of such cement is microcrystalline and so, not obvious, but other cements are spectacular, precipitating as crystal arrays from the walls and ceilings of cavities (Fig. 7) or between sediment grains. Remember, however, that much of the micrite cement may be automicrite. Such cements not only make reefs rigid objects on the seafloor but also occlude much original pore space.

Synsedimentary cementation is an attribute of reefs of all ages but is temporally and spatially controlled, in part, by changing seawater chemistry (*see,* Evolving Ocean Chemistry). The most intensive cementation is in Precambrian reefs, where some are dominated by cement. The amount of crystalline cement in reefs decreases somewhat with the appearance of calcareous organisms in the Cambrian and then again following the evolution of calcareous plankton in the Jurassic. Thus, many Paleozoic reefs contain conspicuously more synsedimentary cement than Jurassic and younger structures. Location of the reef also determines the amount of cementation. Platform-margin and down-slope reefs usually contain marine cements whereas lagoonal or inner-ramp reefs rarely do and so can be unlithified.

Internal Cavity Systems

A surprising amount of any reef is void space because the combination of metazoan–algal–microbe growth geometries produces cavities, even in those reefs almost wholly composed of mud. Voids (Figs. 3, and 8) can be populated by cryptic organisms that encrust walls and hang down from ceilings. They range from photic organisms near the openings to heterotrophs in the dark, lightless interiors. Automicrite often preferentially coats inner surfaces of cavities. Cavities are also sites of fine-grained sediment accumulation (internal sediment), material composed of tiny biofragments and mud that trickles into the holes from above or skeletons that fall from the walls and ceiling after death, to form geopetal sediment. Combinations of cavity dwellers, cement, and internal sediment can generate complex and puzzling textures (*e.g., Stromatactis,* Fig. 9).

Water flow is typically continuous through these interconnected reef voids allowing seawater and nutrients to circulate throughout the interior of the structure. These tranquil crypts can be quite different from the turbulent surface of the reef just above. Reef cavities first appeared in the Neoproterozoic coincident with calcimicrobes and thrombolites. Thereafter they were a significant element of most reefs.

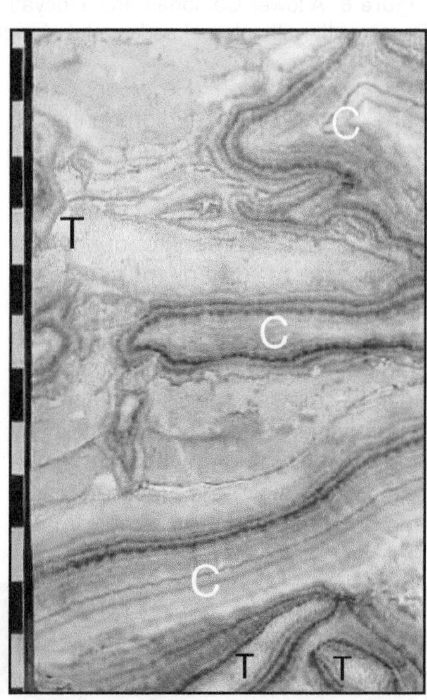

Figure 7. Layered fibrous calcite synsedimentary cements (C) and thamnoporid corals (T), Late Devonian, Western Canadian Sedimentary Basin; cm scale.

Bioerosion and Predation

Although a reef is a constructional edifice, a host of organisms erode the living structure via boring, graz-

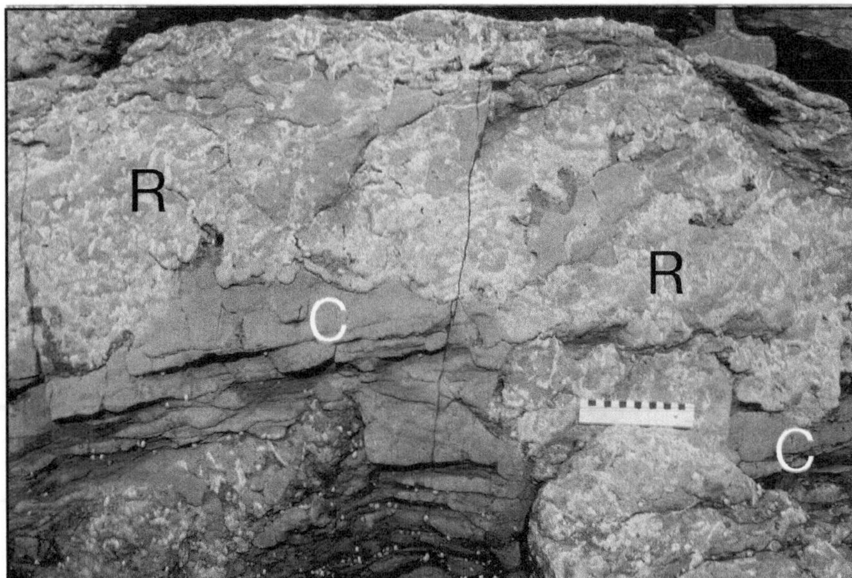

Figure 8. A lower Cambrian archaeocyath-calcimicrobial reef (R) and large cavities (C) filled mostly with internal sediment, Labrador, Canada; cm scale.

ing, and predation (Perry and Hepburn, 2008). These actions enhance organism diversity, weaken reef structure, and produce sediment.

Boring (Fig. 10) is defined as the excavation of carbonate during life processes of organisms such as sponges, bivalves, and worms. Boring has been taking place since the Cambrian but was not really important until the Jurassic, and since then has progressively increased in intensity (Fig. 4). Microbes have been significant borers since the Neoproterozoic and have become increasingly important ever since.

Grazing (scraping and rasping) by herbivores, particularly gastropods, fish, and echinoderms, removes seaweed, algal turf, and hard calcareous algae, and can significantly erode the actively growing reef. More importantly, herbivores reduce the proliferation of fast-growing soft algae and seaweed and thus create open space for the growth of competing reef-building calcareous metazoans.

Calcareous reef builders are preyed upon by fish that consume the fleshy part of the organism and excrete the ingested reef carbonate as sediment (Fig. 3C). Although present in the Paleozoic, this component of in-place reef removal, bioerosion and sediment production began in earnest in the Mesozoic and has increased dramatically since.

Fossil Diversity

Relative abundance of different fossils is potentially one of the most useful observations when interpreting reefs and mounds because such observations allow critical interpretation of oceanographic parameters at the time of growth. It must always be kept in mind, however, that only the calcareous part of the community, and moreover only part of that, is preserved. Organism diversity is, together with competitive interactions and predation, a product of all those factors that affect growth (*see*, The Reef Growth Window). Recent observations indicate that disturbance from hurricanes, outbreaks of predators (*e.g.*, crown-of-thorns starfish), die-off of important grazers (*e.g.*, echinoids), and major water temperature and salinity perturbations (*e.g.*, El Nino) all dramatically rearrange the modern reef community structure. Undisturbed buildups have relatively low diversity because competitive domi-

Figure 9. Irregular masses of calcite (*Stromatactis*) that were originally internal cavities partly filled with internal sediment and then occluded by synsedimentary cement in a Silurian mud mound, Gaspe, Québec, Canada; hammer scale upper right.

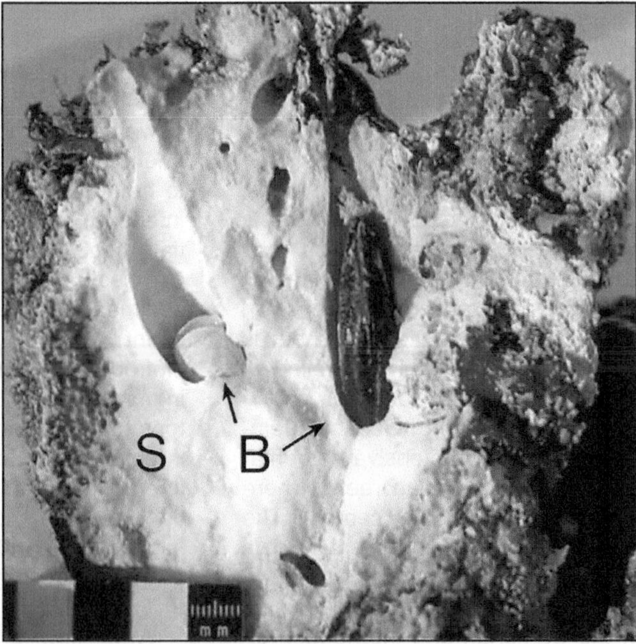

Figure 10. A broken piece of modern reef coral that was dead and overgrown by soft algae with extensive bioerosion by bivalves (B) and sponges (to the left of S), Bermuda; cm scale.

nants have eliminated inferior species. Structures subject to frequent or intense disturbance have higher species richness because competitive domination is repeatedly interrupted, giving inferior species a chance to persist. Reefs prone to the highest rates of disturbance are less diverse because only small numbers of resistant or rapidly colonizing species persist. Thus, patchiness and evidence of extensive fragmentation and debris formation should be expected in the most diverse fossil buildups. Low diversity is also a characteristic of newly established communities (those that have moved into a new environment) and those subject to severe and continuing physical/chemical stress.

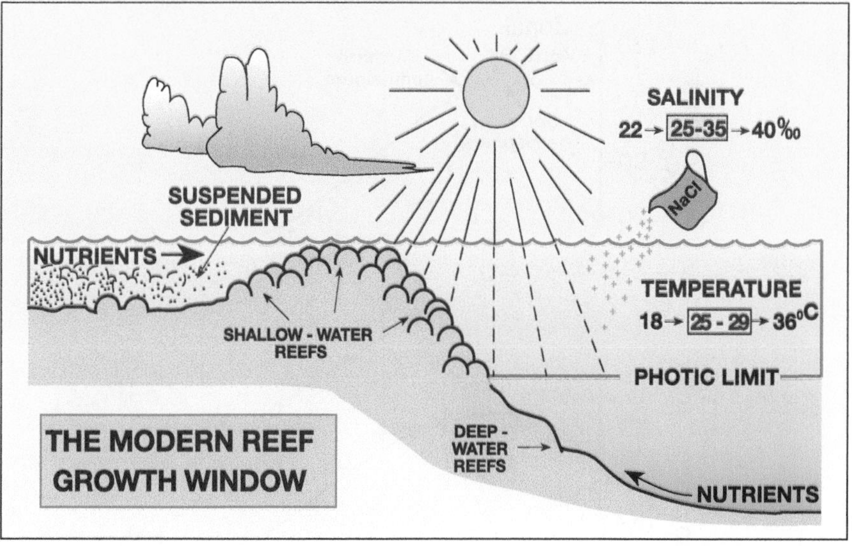

Figure 11. Sketch of the environmental parameters that define the growth window of modern mixotroph coral reefs.

Growth Form and Environment

The relationship between organism shape and environment is one of the oldest and most controversial topics in biology and paleontology. There are, however, limitations to the application of these concepts directly to fossil reefs and no general patterns are applicable to all reefs. For example, variations in shape are the result of interactions between environmental factors and the genetically dictated growth pattern of the organism. Nevertheless, the relationship between the organism and surrounding sediment in the rock record does allow us to make some useful generalizations.

Light intensity, rate of sedimentation, and water turbulence in particular exert an important influence on the growth pattern of an individual colony. The relationship between external shape and internal growth banding geometry of a fossil colony can be used to infer the relative rates of sedimentation and water roughness. Those skeletons whose base is cemented to a hard substrate (*e.g.*, scleractinian corals) are more stable in rough water than those that sit on or in soft sediment (*e.g.*, stromatoporoid sponges). Modern coral shape is also dependent on light. Domal and branching forms with their encompassing tissue are well adapted to relatively shallow water where light is refracted and comes from all directions. By contrast, deeper water, lamellar growth forms that have thin sub-horizontal plates maximize surface area relative to size and so are particularly adapted to lower light levels wherein all intercepted light is vertical. There is, however, a conundrum, as flat, encrusting growth forms are also adapted to high-energy conditions at the reef crest and so care must be taken when interpreting such relationships. Finally, those reefs that form in areas of strong wave and current activity are typically elongated parallel to the current flow.

AUTOGENIC CONTROLS–THE REEF GROWTH WINDOW

Individual reef communities grow best under a specific set of environmental conditions, hence each reef, both today and in the past, can be defined by a reef growth window (James and Bourque, 1992; Kiessling *et al.*, 2002) framed by a specific set of autogenic controls (Fig. 11). Modern coral reefs are only a small component of the Phanerozoic spectrum, yet because they are alive today, we can examine key factors that control their growth.

Modern Coral Reefs

The modern reef growth window (Fig. 11) is determined by a combination of factors that control growth of the dominant reef building organism, in this case, corals. These mixotrophs require specific conditions of temperature, nutrients and light as described below.

Light

A single coral colony can house several different types of zooxanthellae, each adapted to highly local light conditions. Light decreases exponentially with depth, and the lower limits of hermatypic coral and calcareous green algal growth today are 80–100 m. Vertical growth rates below 15 m decrease in a nonlinear, possibly exponential fashion.

Recent studies indicate that zooxanthellate corals, if stressed, either die or revert to being heterotrophs and have corresponding lower growth rates. Other important animal–plant mixotrophs are alcyonarians (soft corals), tridacnid bivalves (giant clams), cyanobacteria-associated sponges, ascidians, and large benthic foraminifers.

Carbonate Saturation

Carbonate saturation in modern seas is now known to be an important determinant not only of carbonate production rates, but also coral reef distribution, coral species diversity, and the rate of coral growth (Kleypas, 1997). For example, the latitudinal range of carbonate-producing species today is mostly governed by temperature and carbonate supersaturation (where concentrations of calcium and carbonate ions exceed the thermodynamic mineral solubility product). Coral reef growth and production rates increase three-fold for a four-fold increase in carbonate saturation. Reefs in well mixed, highly

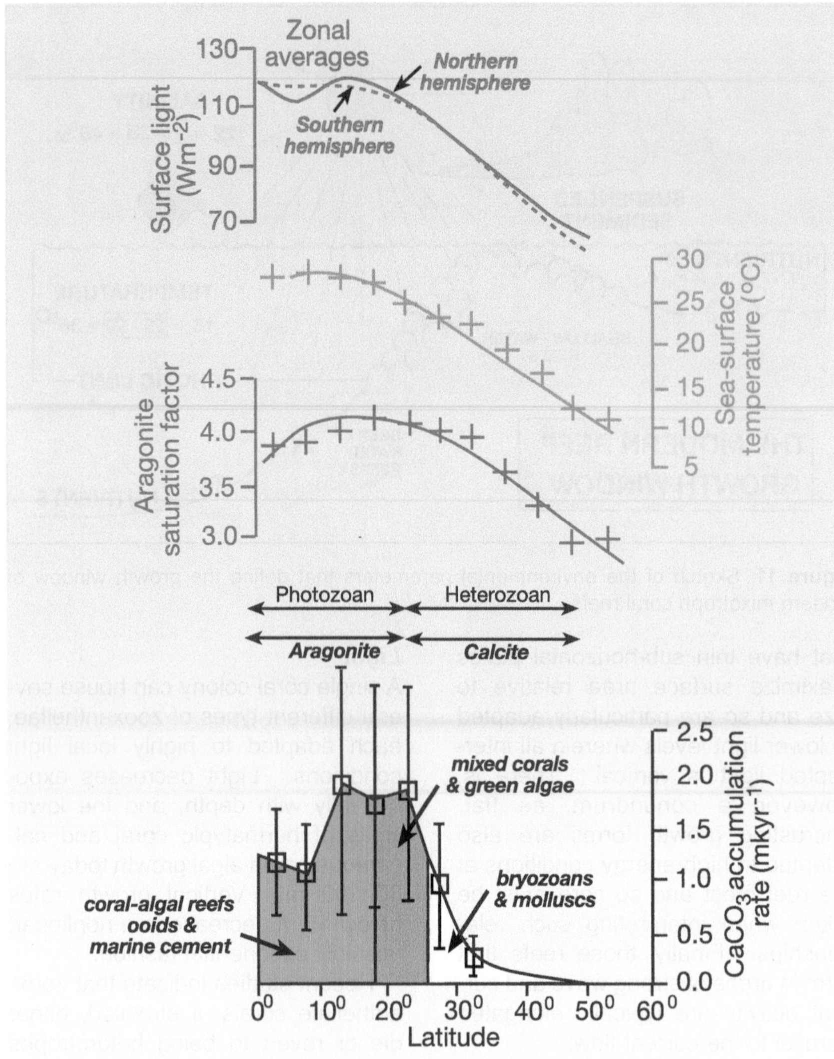

Figure 12. Latitudinal distributions of Holocene calcium carbonate accumulation rates and origin, compared with surface-water aragonite saturation ratio, temperature and light availability (modified from Buddemeier, 1997).

nutrient supply, by enabling other organisms to thrive, leads to modified or arrested reef growth. Importantly, however, if many of the hypercalcifiers in the past were not mixotrophs, then they may have grown in more nutrient-rich environments than they do today and thus were not light-limited as are modern shallow-water reefs.

supersaturated waters also tend to have abundant internal synsedimentary cements compared to those in low saturation settings.

Nutrients

Coral (mixotroph) reefs flourish in oligotrophic, nutrient-impoverished oceanic regions and seem to have done so since the Triassic largely because they retain and recycle nutrients very efficiently, using inorganic nutrients from the water and waste products (ammonia, organic phosphates) from the animal host (Pomar and Hallock, 2008). Even though the waters are low in nutrients, adequate amounts are supplied in energetic environments by high-water flux across the reef.

Increased nutrient levels, from upwelling on the outer platform and fluvial runoff on the inner platform, lead to dramatic changes in reef structure (Hallock and Schlager, 1986). In particular, active phytoplankton growth prevents light from reaching the seafloor, while at the same time energized benthic algae can overgrow and crowd out the calcareous reef-building benthos. At intermediate levels, the mixotrophs are replaced by more heterotrophic animals and 'fouling organisms' such as filamentous algae, fleshy algae, and small suspension-feeding animals (barnacles and bivalves). Reefs still grow in such regions where herbivores are present to graze back the algae. Thus, the paradox: increased

Temperature and Salinity

Zooxanthellate corals grow in waters between 18° and 36°C, but are best adapted to form reefs in waters between 25° and 29°C. Periodic exposure is not necessarily lethal and some intertidal corals are out of water for many hours each day. Modern reef organisms show a latitudinal zonation according to their ability to harvest calcium carbonate from seawater (Fig. 12). Aragonite sediments dominate in lower latitudes. Calcite sediment prevails in the cooler waters (*see* Chapter 15). Reefs formed by light-independent calcifiers (*e.g.*, bryozoans) are also more numerous in cooler, deeper waters (but can be important in waters just below coral reef growth). Most shallow reefs that are dominated by hypercalcifiers are today, and have been in the past, largely located in the tropical latitudes because calcification is metabolically easiest in warm tropical waters.

Most calcareous invertebrates today thrive and grow most rapidly in waters of normal salinity. The salinity window of modern reef-building corals ranges from 22 to 40‰, but most grow best in waters between 25 and 35‰. Elevated salinities progressively favor calcareous green algae with few organisms able to survive above 40‰.

Sedimentation

Whereas coarse sediment in high-energy settings may cause abrasion, it is the fine-grain suspended sediment that is most hostile to coral growth because it decreases light penetration and covers or clogs the polyps. It is difficult to decouple the effect of fine terrigenous sediment and nutrients on coral reef growth because they usually occur together. Fine sediment alone does not arrest coral reef growth but limits it to sediment-tolerant corals, which either grow rapidly to rise above the substrate or have the ability to remove sediment from their polyps. Modern reefs in such environments are distinguished by comparatively low diversity, exceptionally large corals, and patchy, coral covered substrates.

REEF FACIES

From the foregoing it is clear that reefs are infinitely variable and have formed an amazing suite of buildups throughout geologic history. Yet there are themes that run through this myriad of structures and we have arbitrarily divided them, for purposes of useful discussion, into different types (Fig. 13):

1. Skeletal reefs;
2. Skeletal-microbial reefs;
3. Microbial reefs; and
4. Mud mounds.

These are, however, merely end members in a continuous spectrum of biological reliefs that run the gamut from stacked large skeletons, to complex multi-organism constructions, to piles of mud. The current view is that all such structures are reefs! These different reefs, however, have varied remarkably through geologic time (*see* Fig. 33).

The reader should be aware, however, that there has been, for decades, an ongoing debate as to what constitutes an ancient reef (*see* James, 1983; Fagerstrom, 1987; Geldsetzer *et al.*, 1988; Wood, 1999; Stanley, 2001 for summaries). Nevertheless, two terms, bioherms and biostromes, are commonly used to define biogenically constructed geological structures. A *bioherm* is a lens-shaped reef; a *biostrome* is a tabular rock body, usually a single bed of reef-like composition. Another commonly used generic epithet with no compositional, size, or shape connotation is *carbonate buildup*. The preceding terms carry no implication of scale or composition (Fig. 2).

The shape of reefs in plan varies from linear fringing reefs or barrier reefs (Fig. 14A and B) to atoll reefs to subcircular cup-reefs (Fig. 14C) or patch reefs (Fig. 14D). During geological periods when large calcareous skeletal and microbial structures were common, reefs grew as fringing reefs inboard, as patch reefs across platforms, and as semi-continuous to continuous barrier reefs along platform margins. Coeval mounds developed in either calm-water settings across the platform, or on ramps, or on the slope. During geological periods when there were no large reef builders, reef mounds and mud mounds were the only buildups and either grew leeward of sand shoals but more commonly on deep-water ramps and slopes.

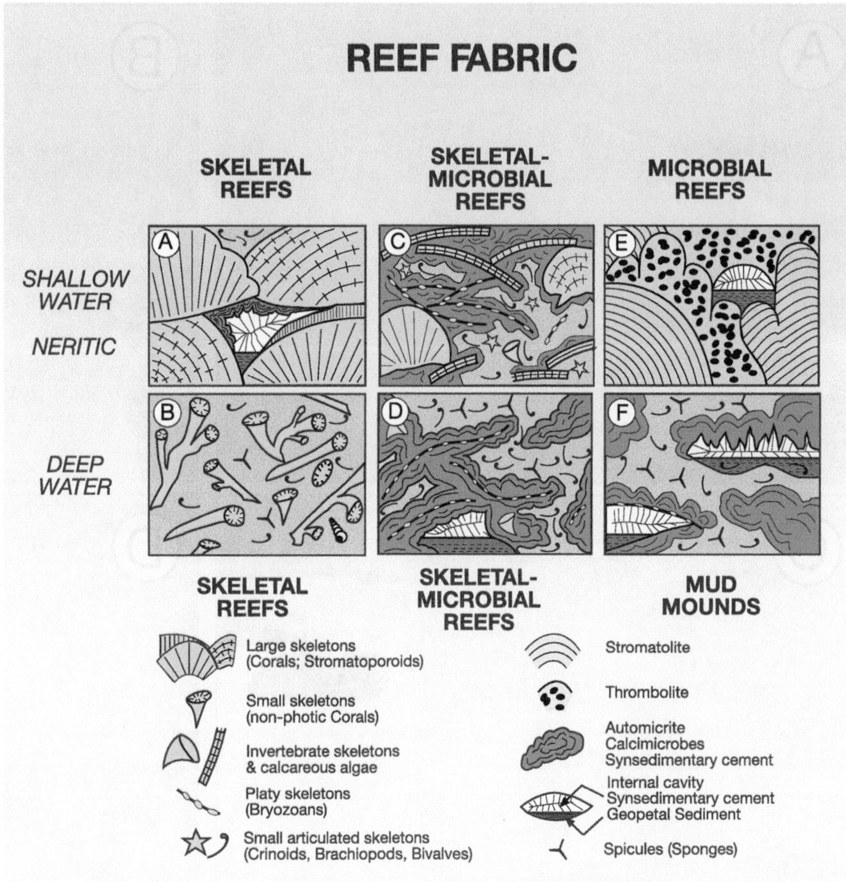

Figure 13. A sketch of the different rock fabrics encountered in shallow-water and deep-water reefs.

SKELETAL REEFS

Shallow Reefs

These are the beautiful shallow reefs of modern seas. They are restricted in geologic time to specific parts of the Phanerozoic, namely, the Middle Ordovician to Late Devonian (Fig. 15), the Late Jurassic, and the Cenozoic (*see* Fig. 33). Today, and in the rock record, they comprise a spectrum of skeletons from tiny to immense. Skeletons dominate the structures forming most of the rock volume (Fig. 16). Although synsedimentary cements, microbes, and automicrite are present they are usually of subsidiary importance, except in parts of the Paleozoic. Because of the wide variety of skeletal growth forms, the reef internal structure is complex, typically resulting in many open spaces that in some reefs comprise as much as 30% of the reef volume. Internal cavities are sites of a diverse cryptic biota, calcimicrobial growth, and internal sediment. Automicrite is present on, around, and between skeletons. Synsedimentary cement can be spectacular with the largest amounts found in reefs at the platform margin. Growth forms and skeletal diversity are strongly controlled by hydrodynamic energy and available space. Diversity is lowest at the reef crest and in deep-water, with both environments favoring sheet-like skeletal morphologies (Fig. 17A). The highest diversity of skeletons and shapes, and therefore rock types, occur at intermediate depths. These reefs, with their large skeletons and rela-

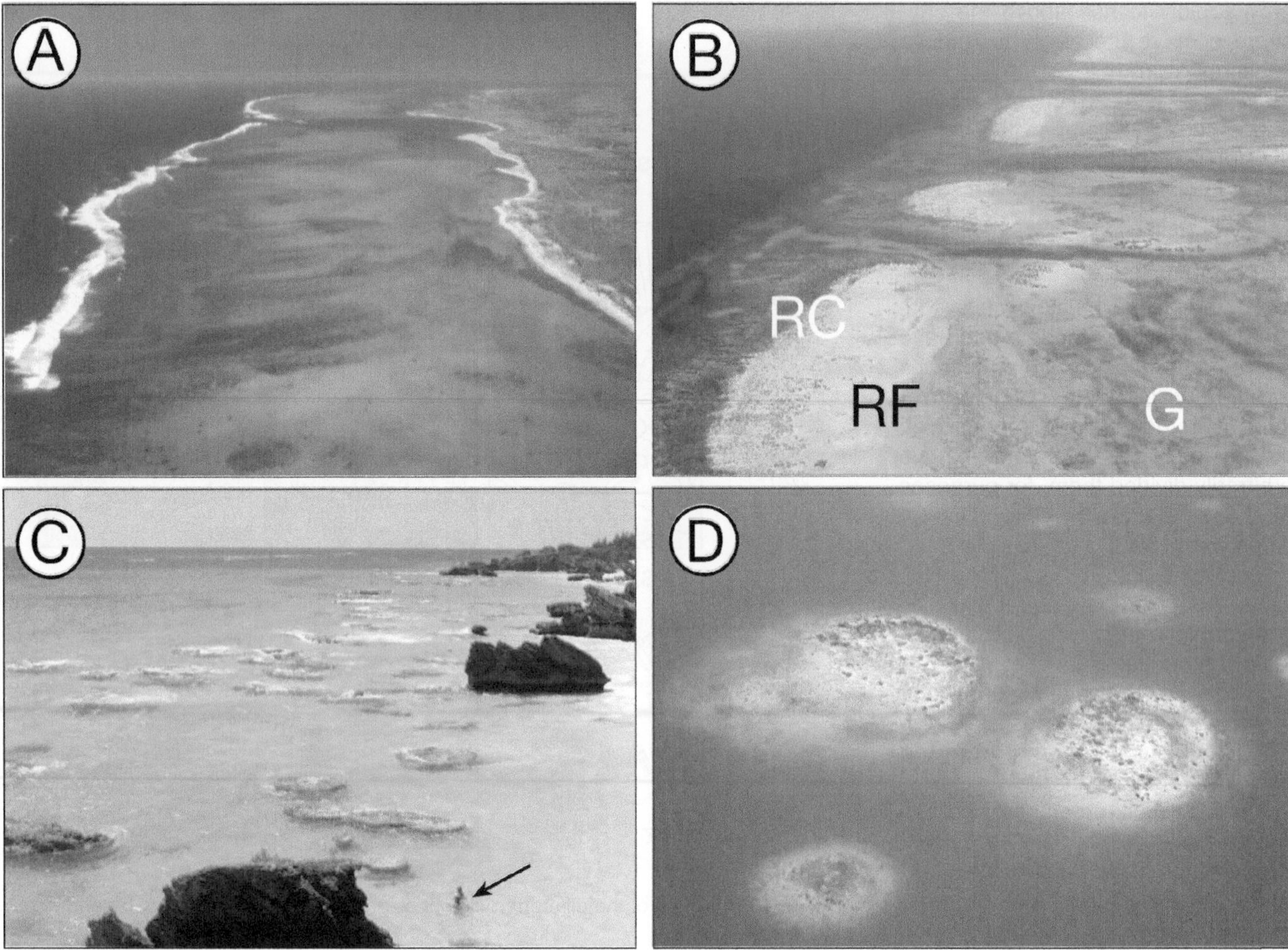

Figure 14. A. Aerial view of the Ningaloo fringing reef off northwestern Australia; the distance between the reef crest at the left and the shore at the right is ~4 kilometers; **B.** Aerial view, looking south, of the Belize Barrier Reef on a calm day with the Caribbean Sea to the left. The narrow dark linear band is the reef crest (RC), in the lee of which to the right is the shallow reef flat (RF), depth ~2 meters, width ~100 m, that grades in the far right to seagrass meadows (G). Deep channels cut the reef into discrete segments (image courtesy W. Martindale); **C.** Numerous cup reefs composed of encrusting coralline algae, sessile vermetid gastropods, a few corals, and extensive synsedimentary cement, south shore Bermuda, arrow - people at lower right for scale; **D.** Aerial view of modern coral patch reefs, each is ~100 m across in several meters of water in Glovers Atoll Reef lagoon (image courtesy W. Martindale).

tively rapid growth rates are amongst the most extensive in the rock record.

Such reefs were built by coralline sponges (stromatoporoids), tabulate corals (Figs. 15 and 16), or scleractinian corals. As scleractinian corals are tied to the photic zone, modern reefs only extend to a maximum depth of ~100 m. If Paleozoic corals and calcified sponges, such as stromatoporoids, were not mixotrophs or were less dependent on photosymbionts, then shallow-water skeletal reefs could have extended much deeper. Consequently, there are two potential facies profiles for skeletal reefs (Fig. 17). This concept has important facies analysis consequences, mainly that tabulates and stromatoporoids do not necessarily indicate shallow-water paleoenvironments.

Mixotroph Skeletal Reefs (Scleractinian Corals)

Depth Zonation

Modern coral reefs exhibit depth zonation (Fig. 17) because of decreasing wave energy, light intensity, and to a lesser extent temperature, and changing biological interactions (*e.g.*, Graus and Macintyre, 1989). The *Reef Front Facies* lies between about 10 m (the base of surface wave action) and 100 m. This is an environment of diverse reef-builders varying in shape from hemispherical to branching to columnar to dendroid to laminar. Accessory organisms and various niche dwellers such as bivalves, gastropods, coralline algae, segmented calcareous algae, and sponges are common. Below about 30 m, wave intensity is lower, light is attenuated and most reef corals are prone and plate-shaped (Fig. 17). Pockets, streams and chutes of skeletal sand, especially calcareous algae particles, accumulate seaward between ridges of dense coral growth. The *Fore Reef Facies* is generally gravel and sand comprising whole or fragmented skeletal debris, blocks of reef limestone, and skeletons of reef builders. Such deposits grade basinward into muds exhibiting many of the attributes described in the chapter on carbonate slopes (Chapter 18). This picture is particularly sensitive to perturbations of the growth window, especially a decrease in light, which leads to a shallowing of all zones.

Energy Zonation

Growing as they do in the trade-wind belts, the shallow parts of modern coral reefs have a strong windward–leeward zonation (Fig. 17). This is true whether they form the margins of large platforms or isolated structures inboard.

Zonation is best developed in windward locations (Fig. 14). The *Reef Crest Facies*, which extends to a depth of about 15 m, receives most of the wind and wave energy. Composition depends upon the degree of wind intensity; swell height, and periodicity of cyclonic storms. Only organisms that can encrust, which are generally sheet-like forms, are able to survive where wind and swell are constant and intense. If wave and swell intensity are more episodic or only moderate to strong, encrusting forms still dominate but can be bladed or have short, stubby branches. Where wave energy is moderate, robust branching corals proliferate. The large variety of branching corals on modern reefs is due to their rapid evolution in the late Cenozoic. Hemispherical to massive forms characterize this zone in older scleractinian reefs. The *Reef Flat Facies* varies form a pavement of cemented, large clasts, scattered rubble and coralline algal nodules in areas of intense waves and swell, to shoals of well-washed lime sand in areas of moderate wave energy. Most material comes from the reef crest and is swept onto the pavement during cyclonic storms. The vagaries of wave refraction may pile the sands into cays and islands that in turn protect small, calm-water environments adjacent to the reef crest. The shielded *Back Reef Facies* is where much of the mud formed on the reef comes out of suspension. This coupled with the prolific growth of sand- and mud-producing bottom fauna, such as calcareous green algae, results in mud-rich rocks. Corals here are stubby and dendroid, and/or large, globular forms that extend above the substrate to withstand episodic agitation and quiet muddy periods.

Figure 15. A Devonian skeletal reef composed mainly of stromatoporoids and tabulate corals, Ellesmere Island, Arctic Canada; pencil scale.

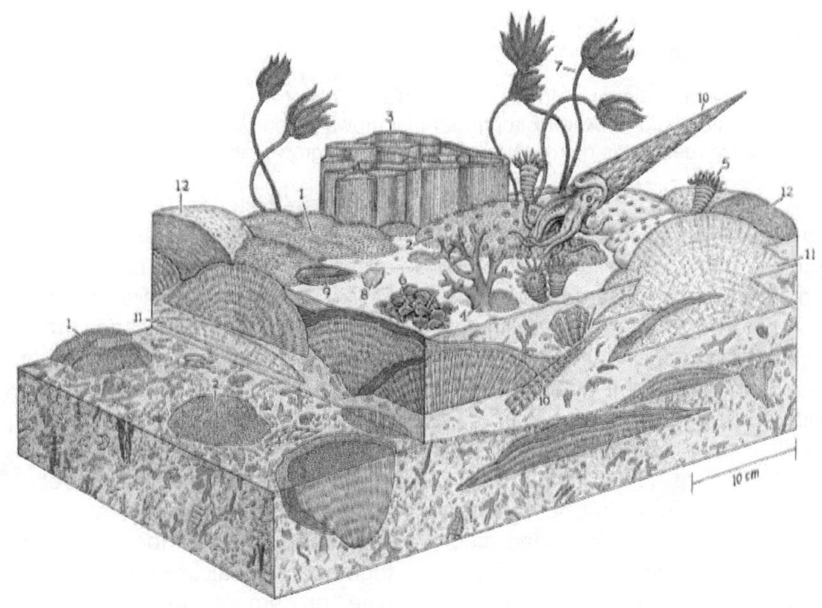

Figure 16. Reconstruction of a Silurian skeletal reef composed mostly of tabulate corals and stromatoporoids (courtesy J. Sibbick); 1. Tabulate coral (*Favosites*), 2. Tabulate coral (*Heliolites*), 3. Tabulate coral (*Halysites*), 4. Bryozoan, 5. Rugose coral, 6. Spiriferid brachiopod, 7. Crinoid, 8. Brachiopod, 9. Trilobite, 10. Orthocone nautiloid, 11. Stromatoporoid, 12. Thrombolite.

Reefs on the leeward side of banks and atolls can be strikingly different, either because of reduced wave activity or because of 'bad water'. Such water is generated on the platform by fine-grain sediment production, oxygen depletion, heating, or evaporation. It is usually driven off the bank downwind, across the leeward margin, inhibiting reef growth to a depth of several 10s of meters. Such shallow leeward margins can, therefore, be bare rock floors covered by soft fleshy algae or hard coralline algae with active reef growth taking place only in deeper water, below this interface.

Nutrient–Sediment Zonation

There is typically a cross-shelf zonation in reef composition that partly reflects the outboard decrease in fine sediment and nutrients.

Inner shelf reefs are characterized

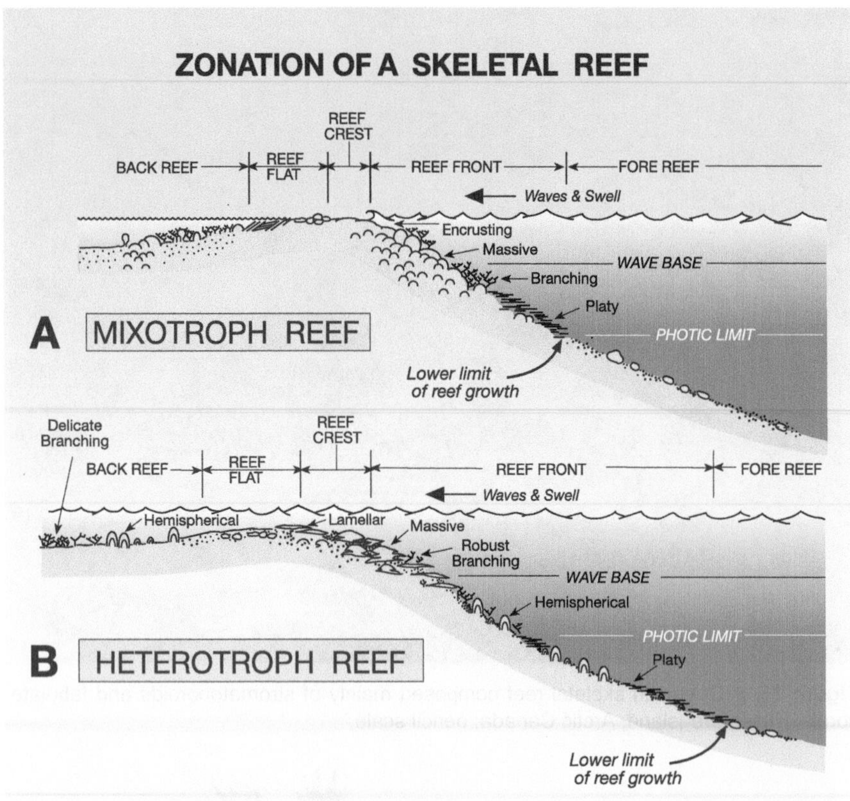

Figure 17. Sketch illustrating the facies of skeletal reefs growing on the windward margin of a carbonate platform. **A.** Mixotroph reef that grows to the base of the photic zone; **B.** Heterotrophic reef that is not constrained by light and grows into deeper water.

by i) quickly growing corals with a high tolerance for fine sediment, low salinities and an inability to withstand turbulent waters, ii) large and abundant heterotrophic sponges, iii) low epifaunal diversity, iv) few soft corals, and v) abundant soft algae and few calcareous algae.

Outer shelf reefs (in areas of little upwelling) are distinguished by i) slow-growing mixotroph corals that cannot withstand suspended sediment or low salinities but are adapted to high-energy seas, ii) reduced numbers of sponges, some of which contain photosynthetic symbionts, iii) common tridacnid bivalves in the Pacific, iv) high epifaunal diversity, and v) prolific calcareous algae.

Heterotroph Skeletal Reefs (Stromatoporoid Sponges and Tabulate Corals)

Stromatoporoids are a group of sponges of probably mixed affinity that are distantly related to various modern calcified sponges. As modern calcified sponges are not shallow-reef builders we must rely on ancient reef case histories (*e.g.*, Geldsetzer *et al.*, 1988; Wood, 1999) to formulate a general energy zonation (Fig. 17). Unlike modern corals, Paleozoic stromatoporoid sponges and tabulate corals, even though they often grew as laminar forms, sat on or were anchored within the sediment. Large, tabular forms that were solidly rooted in the sediment inhabited high-energy zones such as the reef crest. Tabular stromatoporoids, locally bound together by calcimicrobes, automicrite, or cement, are also known to have occurred in rough-water environments. Domal, bulbous, and dendroid forms occupied calm-water zones, either below wave base or in sheltered areas of the back reef and the lagoon (delicate stick-like amphiporoids). These skeletons were locally reworked during cyclonic storms and redeposited as rubble units associated with peritidal facies. As stromatoporoids did not typically have an encrusting habit, it is doubtful that they were successful builders in the surf zone. Sand- and gravel-rich facies containing reworked stromatoporoid skeletons, up to several 10s of centimeters in size are probably an expression of the inability of these animals to withstand surf conditions.

Deep-water Skeletal Reefs

Today such reefs grow worldwide, in perpetual darkness, well below the photic zone and below wave base (Freiwald and Roberts, 2005). They are constructed by a several calcareous heterotrophs, the most prominent of which are scleractinian corals that lack photosymbionts, especially the branching form *Lophelia pertusa* (Fig. 18). These reefs are most numerous between water depths of 250 and 1500 m where temperatures range from 4 to 12°C. Most reefs are up to 5 km long and 40 m high and have generally steep sides. Corals range from small cm-scale cups to m-scale dendroid bushes. Other important organisms are soft corals with spicules (alcyonarians), stylasterene hydrocorals, zoanthids (anemones), crinoids, spiculate and calcareous lithistid sponges. Most reefs are unlithified, with the exception of 'lithoherms' in the Straits of Florida. The reefs are somewhat self-organized; individual colonies coalesce to form thickets (an aggregate of closely spaced colonies); thickets merge to form coppices (piles of skeletons colonized by living corals); coppices join to form banks or reefs. The coppices are usually intensively bioeroded, are filled with muddy internal sediment, and act as a substrate for ancillary epibenthic growth leading to increased biotic diversity. A theme of all these reefs is one of numerous corals in a fine-grain matrix that can be either carbonate or siliciclastic mud, or both, and contains numerous planktic foraminifers and pteropods.

There is considerable debate as to the controls on the location of these reefs that are largely dependent upon plankton as a food source. Many, but not all, such reefs appear to thrive best on bathymetric highs such as seamounts, drowned glacial moraines, subaqueous dunes, and sediment drifts. They are also located in areas of elevated nutrients such as oceanic fronts or upwelling. Others are sited on top of cold hydrocarbon seeps. These cold-seep reefs are part of a lineage of similar structures that

Figure 18. Prolific growth of *Lophelia pertusa* on the up-current end of a lithoherm at ~600 meters water depth in the Straits of Florida (image courtesy A.C. Neumann).

extends back to the Silurian but there are very few corals from methane seep communities older than Late Eocene.

The rock record of such skeletal reefs is not extensive; the oldest seem to be Triassic. They are not numerous until the Cenozoic where an especially good record exists in NW Europe and the Mediterranean region. Pre-Cenozoic deep-water reefs seem to be either skeletal–microbial mounds or mud mounds.

MIXED SKELETAL–MICROBIAL REEFS

Attributes

Many ancient reefs bear two important constituents: small or delicate skeletal metazoans that are either heterotrophs (*e.g.,* sponges and bryozoans) or autotrophs (*e.g.*, phylloid algae), and a microbial component of calcimicrobes or inferred microbialite (Webb, 1996; Wood, 1999). These structures, which have a mixed consortium of organisms, are unlike most modern skeletal reefs and have been called reef mounds or biogenic mounds (James and Bourque, 1992). Such reefs often contain extensive framework cavities, with their own distinctive cryptic biotas and abundant synsedimentary cements (Fig. 19A).

Shallow-water Skeletal–Microbial Reefs

Either autotroph-microbial or heterotrophic-microbial communities dominated ancient mixed reefs in shallow waters of the photic zone. No direct modern analogues are known for either of these systems. Some modern coral reef cavities are reported to be encrusted by secondary microbialites crusts, but these do not seem to be widespread. Modern *Halimeda* algae form mounds in water depths below 50 m (Roberts and Macintyre, 1988), but not in known association with carbonate-producing microbial populations.

Autotroph-dominated

Ancient examples of mixed autotroph reefs include Pennsylvanian to Early Permian phylloid algal mounds (Fig. 6A). These algae are probably of mixed affinities and grew as laminar, cup-, bowl- or upright leaf-like forms that may have been encrusted by automicrite. Such reefs are usually low-relief isolated structures, appear to have grown in shallow waters, and are often associated with packstones and grainstones that indicate growth under fairly energetic conditions. These mounds were typically the source of significant skeletal debris that formed extensive flank beds. *Halimeda* mounds have been reported from the Late Miocene, and may have been stabilized by rapid lithification of micrite and microbial crusts (Martin *et al.*, 1997).

Heterotroph-dominated

Mixed heterotroph reefs are common in the ancient record. Lower Cambrian reefs occur as small isolated mounds, consisting of archaeocyath sponges and calcimicrobes, particularly *Renalcis* and *Epiphyton* (Fig. 8). Lower Ordovician reefs were likewise dominated by calcimicrobes, microbialite and, locally, corals.

Some back-reef shallow-water Frasnian (Late Devonian) reefs exhibit large skeletal metazoans (particularly stromatoporoid sponges, and subsidiary corals) as well as extensive fenestral microbialite and calcimicrobes both as free-standing mounds, and as secondary encrustations within cryptic habitats (Fig. 20). Succeeding Famennian reefs growing in the aftermath of the Late Devonian mass extinction were composed of framebuilding microbialite and calcimicrobes (*e.g., Renalcis* and *Rothplezella)* and a diverse community of spiculate sponges.

Shallow-water Mississippian reefs are formed by various endemic communities composed of laminated microbial mounds that have a rich encrusting open surface and cryptic fauna dominated by algae, bryozoans, corals, and sponges (Mundy, 1994; Webb, 1994; Ahr *et al.*, 2003).

The mid-to Late Permian is characterized by extensive fringing reefs that have a well-developed zonation. They consist, in part, of a primary framework of frondose bryozoans and calcified sponges (many of whom were cavity-dwellers), that were bound by extensive crusts of laminated automicrite (Fig. 19). *Tubiphytes* is common, together with various encrusting algae including *Archeolithoporella*. Volumetrically, many of these reefs were dominated by sediment and synsedimentary cement.

The first metazoan reefs to form

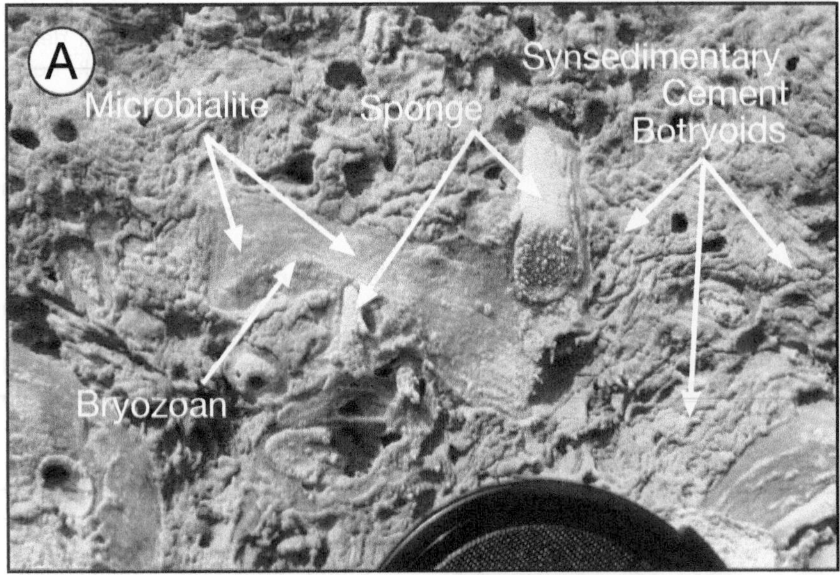

Figure 19. A. Pendant sponges attached to a bryozoan frond. The whole has been encrusted with microbialite. Remaining cavity space is filled with synsedimentary cement botryoids. Permian Capitan Reef, McKittrick Canyon, Guadalupe Mountains, Texas, USA, Lens cap = 52 mm; **B.** Reconstruction of bryozoan-sponge community, Permian Capitan Reef, USA (courtesy J. Sibbick); 1. Frondose bryozoans, 2. Solitary sphinctozoan sponges, 3. *Archaeolithoporella* (encrusting alga), 4. Automicrite, 5. Synsedimentary cement botryoids, 6. Sediment.

after the end-Permian extinction in the Anisian (Lower Triassic) were of a similar ecological cast and large framework cavities, with the addition of the calcified microbe *Cladogirvanella.* In these examples, automicrite is a secondary, often cryptic component, but may still contribute locally up to 70% of the reef rock volume. Such successions are inferred to have formed within open cavity systems with freely circulating seawater, in response to decreasing light and energy conditions as the reef was progressively buried. Upper Jurassic shallow-water reefs were dominated by hexactinellid and lithistid spiculate sponges, coated by automicrite and thrombolites, and encrusted by a variety of biota.

DEEP-WATER SKELETAL–MICROBIAL REEFS

Deep-water, heterotroph mixed skeletal–microbial reef communities are known from the Silurian to Devonian, the Jurassic, and the Cretaceous. They were very different from modern skeletal deep-water reefs. Muddy reefs with *Stromatactis* (Fig. 9), lithistid sponges, and calcimicrobes were widespread during the Silurian, and typically show a vertical zonation that culminates in shallow-water stromatoporoid sponge-rich reefs. Similar communities, with the addition of microbialites, stromatolites, and receptaculitids, formed deep-water reefs in the mid to Late Devonian.

Upper Jurassic deep-water mixed reefs were constructed by thrombolitic–stromatolitic columns or mounds associated with *Tubiphytes*, the worm *Terebella*, and hexactinellid and lithistid sponges. The platy scleractinian coral *Microsolena*, which is thought to have fed, in part, heterotrophically, is found in low light settings, such as shallow turbid or deep-water environments.

MICROBIAL REEFS

Archean and Proterozoic

Microbial reefs dominated much of Earth history, from the Archean until Middle Ordovician (*see* Fig. 33) but are only found in younger Phanerozoic facies where most invertebrates have been excluded by environmental stress. They are built primarily by stromatolites and thrombolites that have a wide variety of morphologies (*see* Fig. 23). Precambrian reefs throughout the Archean, early, and middle Proterozoic are essentially a series of stacked stromatolites and to a lesser extent, thrombolites (Grotzinger and James, 2000). They have neither cavities nor inter-microbial synsedimentary cements. Many reefs display a clear fractal organization. Nevertheless, the basic building blocks exhibit a wide variety of delicate branching, lamellar, hemispherical, and conical morphologies that are thought to be due to a combination of different microbial communities, hydrodynamics, and sedimentation rates. These microbialites range in size from cm to dm, with some large individual organo-sedimentary structures up to 8 m high by 30 m long.

Shallow-water, high-energy Proterozoic buildups were formed by isolated stromatolitic domes, linked domes, columnar stromatolites, and their elongate equivalents (Fig. 22). Stromatolite fragments that were ripped off during storms and swept into crossbedded gravels locally surrounded individual shallow reefs. Reefs at platform margins show a strong zonation of microbialite growth

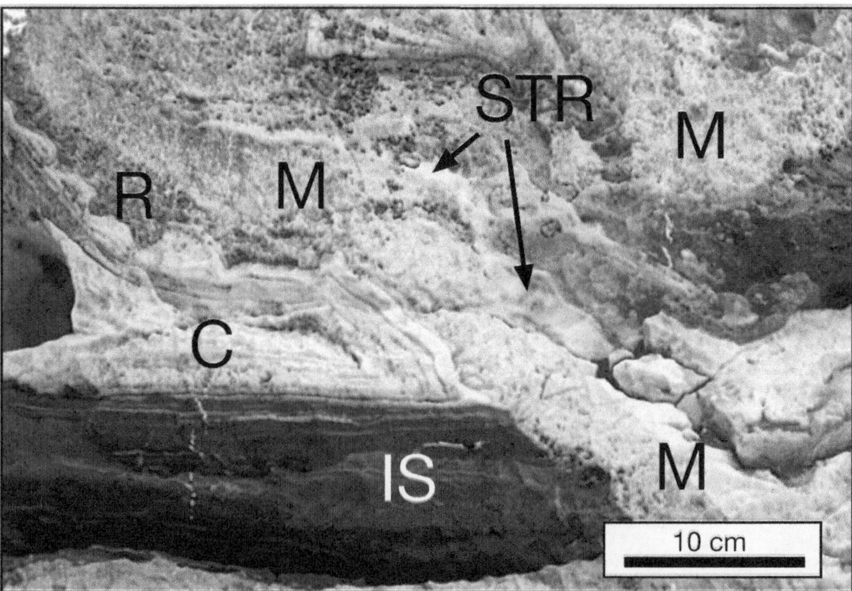

Figure 20. An Upper Devonian reef, Canning Basin, Western Australia. The robust platy stromatoporoid (STR) is encrusted first by microbialite (M). The so-formed internal reef cavity roof is encrusted by the calcimicrobe *Renalcis* (R), and the cavity is floored by internal geopetal sediment (IS). Remaining pore space is finally filled by isopachous layers of synsedimentary cement (C).

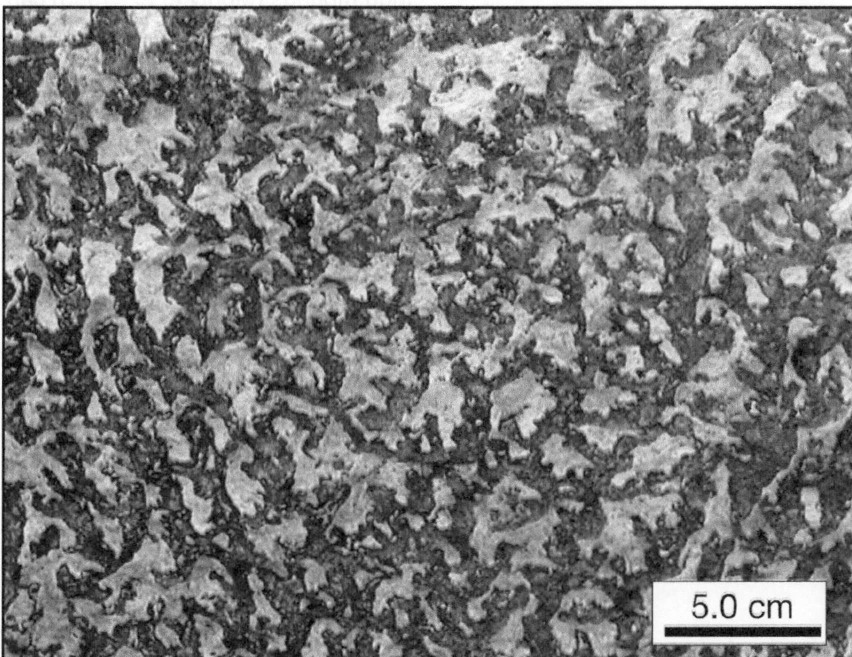

Figure 21. Neoproterozoic thrombolites, Nama Basin, Namibia.

forms that mimic facies zonations of metazoan reefs. Isolated deep subtidal and slope reefs (probably below fair-weather wave base) were built primarily by conical (conoform) stromatolites (Fig. 23).

There seems to be an evolution of microbial structures through the Precambrian (Fig. 23). Archean and Paleoproterozoic reefs were mainly synsedimentary precipitates (possibly microbially mediated) having microbial layers. Microbes appear to become more prominent in Paleoproterozoic and Mesoproterozoic structures. A profound change, however, took place in the Neoproterozoic with the increased importance of thrombolites and the appearance of calcimicrobes (Fig. 21) as reef builders. This resulted in the earliest formation of cavities, complete with internal sediment and synsedimentary cement, features that would become a prominent attribute of all subsequent Phanerozoic reefs. This community had the ability to construct reefs with up to 100 m of relief above the sea floor in deeper water environments (Turner *et al.*, 1997). Toward the end of the Neoproterozoic, the first skeletal invertebrates (*e.g., Cloudina*, *Namacalathus,* and *Namapoikia*) began to populate the surfaces of these reefs and were entombed by the microbial precipitates.

Modern stromatolites grow in areas of abundant calcium carbonate saturation, and where faster-growing algae and seaweeds are excluded due to environmental stresses, such as active tidal currents, low nutrients, or high salinity (Fig. 24).

Paleozoic

The late Paleozoic and early Mesozoic, a period when there were relatively few large skeletal reef-formers, is usually envisaged as a time without significant reefs, but a time of ramps (James and Bourque, 1992; Tucker and Wright, 1990). While this is generally true, this interval was also a period of significant carbonate platforms, many of which had steep flanks (Della-Porta *et al.,* 2004; Kenter *et al.*, 2005). The shallow platform itself was typically composed of ooid–pisoid–calcareous algal sand shoals and interbedded mudstones. The platform margin and upper slope, however, was a skeletal–microbialite facies with prolific synsedimentary cement that extended across a gentle roll-over (below wave base) and downslope to paleodepths of ~300 m or more. This reef-like facies passed in turn below 300 m into lower slope facies of debrites and sediment gravity-flow deposits of mixed shallow-water grains and microbial clasts eroded from the upper slope.

These skeletal–microbialite-cement rock types had no or minimal seafloor relief, and yet were critical and time-specific elements of these platform margins. They are composed of several rock types that are depth-related:

1. Shelf margin = structureless micrites with numerous synsedi-

Figure 22. Mesoproterozoic stromatolites, Pethei Formation (1.9 Ga) Great Slave Lake, Northwest Territories, Canada; cm scale.

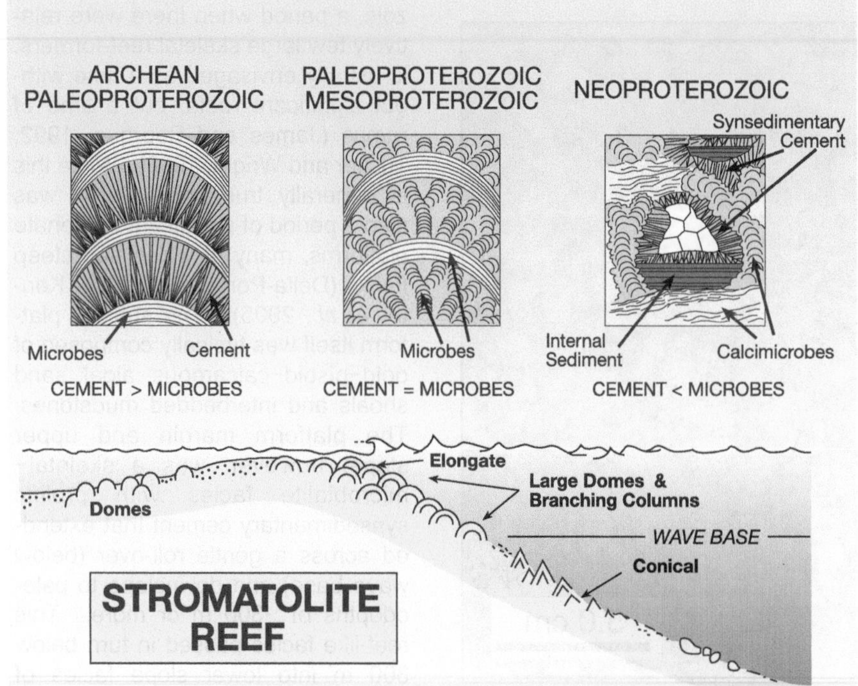

Figure 23. Sketch of a zoned Proterozoic marginal reef showing different stromatolite growth forms and their environment. Boxes are sketches of general microbial microstructure during different times during the Archean and Proterozoic. In the Archean-Paleoproterozoic the amount of synsedimentary cement was generally greater than the microbial content; in the latter part of the Paleoproterozoic and Mesoproterozoic they were about equal; in the Neoproterozoic microbes and calcimicrobes were volumetrically more abundant than synsedimentary cement.

mentary fibrous cement-filled cavities, occasional bryozoans and sponge spicules but plentiful calcimicrobes (*Donezella, Renalcis, Girvanella*, and *Ortonella*) as well as calcareous algae (phylloids and other types);

2. Upper slope (20–150 m depth) numerous microbial crusts, globose and laminar sheets, prolific synsedimentary cements but minor skeletal elements; and
3. Lower slope clotted micrites with rare voids and minor cements but common re-sedimented shallow-water grains.

These are reefs in composition but not in shape or relief; perhaps they are better described as slope biostromes. Regardless they were critical in generating steep-sided platforms.

MUD MOUNDS

Attributes

Mud mounds (Monty *et al.*, 1995) can form large (often over 100 m high and 400 m wide), isolated, steep-sided structures that consist of more than 80% fine-grained carbonate (micrite; Fig. 25A). This micrite has both *in situ* (autochthonous) and detrital (allochthonous) origin, but commonly shows accretionary structures constructed by successive phases, known as polygenetic muds ('polymuds') that form on open surfaces and within semi-enclosed cavities. Such polymud fabrics produce complex, three-dimensional accumulations, which, in turn, form open frameworks that can subsequently be filled with mud or synsedimentary cement. Many mounds also display a rich attached metazoan biota of crinoids, tabulate corals, brachiopods, trilobites, sponges, ostracodes, and bryozoans (Fig. 25B).

The clotted, peloidal or laminated textures, the encrusting or frame-forming habit, and the often-inferred high-Mg mineralogy argue for an *in situ* microbial or organomineralic origin augmented by rapid synsedimentary lithification. This origin is supported by moderate rates of carbonate accumulation (0.2–0.8 m/1000 yrs), and the steepness of (35–40°) of mound flanks. Abundant *Stromatactis* cavities that parallel the accretionary mound surfaces (Fig. 9)

Figure 24. Stromatolites ~0.5 m high growing today in Hamelin Pool, Shark Bay, Western Australia.

suggest a close relationship between mound formation, and internal sediment filled voids, cementation, and carbonate production (Monty *et al.*, 1995).

It is possible that mud-mound formation relies mainly on early diagenetic organomineralization processes to aid calcite nucleation on dead organic matter, particularly via molecules derived from decayed sponges. The presence of abundant, reactive Ca^{2+}-binding fulvic acids have been confirmed in well-preserved Cretaceous mud-mounds and modern coral reef caves.

Mud-mounds were often established in deep waters below the photic zone as indicated by the absence of algae and cyanobacterial activity (*e.g.*, micritic envelopes). Locally, however, they show shallowing-upward facies signatures characteristic of shallow depths and significant depositional energies. In the Frasnian 'récifs rouges' of the Belgian Ardennes (Boulvain, *in* Alvaro *et al.*, 2007), the basal facies is composed of *Stromatactis*-rich red lime mudstone to wackestone with abundant ubiquitous sponge spicules. This may have formed in hypoxic waters below the photic zone. Platy corals appear higher, but are overlain by a more diversified coral, stromatoporoid and skeletal algae wackestone to packstone with *Stromatactis*. Uppermost facies consist of a stromatoporoid, coral and bryozoan framework with various microbial and calcimicrobial encrusters, which formed in photic waters above fair-weather base, but sometimes in restricted or lagoonal environments (Fig. 26).

Early Mississippian reefs (Waulsortian facies) also show a similar progression of facies. The growth of the well-known Muleshoe Mound in New Mexico (Fig. 25A) records a shift from predominantly upward (aggradational) to lateral (progradational) growth. The largest volume of these buildups is composed of polymud fabrics and early cements, sponges, and large fenestrate bryozoans. The framework is composed of rigid micrite masses with rounded, bulbous shapes and thrombolitic fabrics that are lined by early marine cements. In the uppermost facies, both the microbialites and large bryozoans grew with a pronounced high-angle orientation into currents. Crinoid-rich flanking grainstone beds draped the reef slopes. In places, fenestrate bryozoans built a delicate framework that formed a limestone containing up to 90% early fibrous cement (Fig. 25B).

Initiation

Initiation of deep-water mud-mound growth remains a mystery, but they are commonly found in groups suggesting that their formation was environmentally mediated. Several environmental triggers have been proposed: the episodic formation of nutrified water masses, reduced sediment supply during platform drowning, or localized oxygen depletion of sea water. Some mud-mounds are aligned along faults or fissures which may have acted as conduits for hydrothermal fluids. For example, the microspars and brachiopods of Muleshoe Mound have been shown to have higher $\delta^{13}C$ values than non-mound samples, possibly related to methanogenic fermentation.

Distribution

Mud-mound formation began in the Paleoproterozoic (especially Neoproterozoic) and extended until the Miocene but they are mostly a Paleozoic phenomenon, with widespread mud-mound formation occurring during the Early Cambrian, Early Ordovician, Late Devonian and the Mississippian, which was dominated by Waulsortian mounds.

Summary

Mud mounds, mixed skeletal–microbial, and microbial reefs present a continuum of shared ecologies and sedimentary characteristics. However, the location and initiation of mud-mound formation does appear to be mediated by environmental factors that differ from those of shallow skeletal reefs, and there may be real differences in the style of primary production and organic matter recycling between these different reef systems.

ALLOGENIC CONTROLS

The allostratigraphy of reefs is a function of i) varying conditions in the growth window, ii) ecological succession, iii) antecedent topography, and iv) the rates of sea-level rise and organism growth.

Reef growth during the Holocene rise in sea level, as revealed by extensive scientific drilling (Montaggioni, 2005), illustrates several themes. Even though the rates may have been different in the ancient,

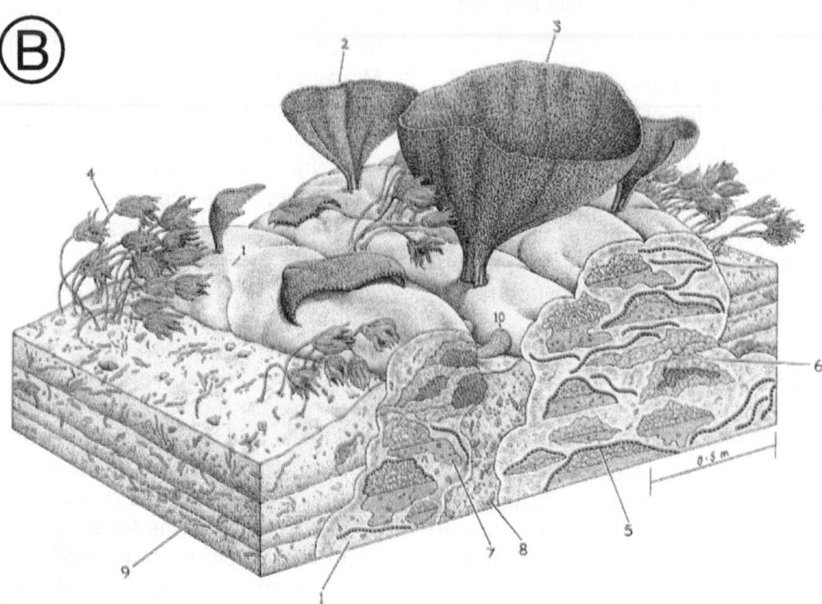

Figure 25. A. Muleshoe Mud Mound, Mississippian, Sacramento Mountains, New Mexico, USA; **B.** Reconstruction of Muleshoe Mud Mound community, New Mexico, USA (courtesy J. Sibbick); 1. Automicrite, 2.& 3. Fenestrate bryozoans, 4. Crinoid, 5. Fenestrate bryozoan roofing cavity, 6. Synsedimentary cement, 7. Geopetal internal sediment, 8. Neptunian dike filling, 9. Intermound sediment, 10. Sponge.

the stratigraphic record is still a measure of the interplay between them from which first-order interpretations can be made. Stratigraphy is critical in this analysis, it reveals the interplay between rates; rates of sea-level rise versus rates of reef growth.

Response to Relative Sea-level Change

Start-up

The initiation of reef growth can occur i) during a rise of sea level after subaerial exposure, ii) during sea-level fall when suitable seafloor falls within the reef growth window, iii) when reef growth creates an elevated seafloor for a new type of reef community, iv) when any local factor hostile to reef growth is eliminated, and v) a change in local tectonics. Where the initial reef grows will be determined by the ecological demands of the community.

Reef growth during the Holocene sea-level rise often began upon Pleistocene limestone (commonly karsted); some reefs started as shallow or intermediate depth communities; others experienced lag times of up to 2000 years and did not grow until the water depth was greater than 20 m. Mid- to inner-shelf reefs often show a delayed start-up whereas shelf margin reefs grew immediately. Accretion rates of inner shelf reefs, however, increased dramatically once the outer barrier reef reached sea level.

Reef Growth Strategies

Holocene sea level is characterized by an early rapid rise, later slowing and final stability. Although the response of reefs to this global trend was highly varied, some reactions were independent of reef community and setting (Fig. 27). *Keep-up reefs* are those that maintained their crests at or near sea level, so tracked sea-level rise. *Catch-up reefs* began in relatively deep water and grew upward faster than rising sea level. Once catch-up and keep-up reefs reached sea level, they maintained their near-surface crest facies or developed capping facies of subtidal sediment or storm-ridge deposits. Subsequent reef growth was limited to the seaward margin so that reefs prograded over reef front and fore-reef sediments. *Give-up reefs* initially grew as other reefs, but were subject to some change of conditions that caused the community to die and reef growth to cease. Examples included increased turbidity, nutrient excess or hot saline waters that swept off the shelves when sea level flooded over wide shallow platforms (the 'bad water' syndrome).

All responses can be found within one reef system. At Papeete, in French Polynesia, for example, keep-up reefs are found on the fast-growing windward margin, catch-up reefs grew on the leeward margin, and give-up reefs characterize the muddy patch reef margins.

Drowning

Reefs can turn off and die for a variety of reasons. A simple eustatic rise in sea level does not seem to be enough because the growth of most healthy reefs can keep pace. Instead reef

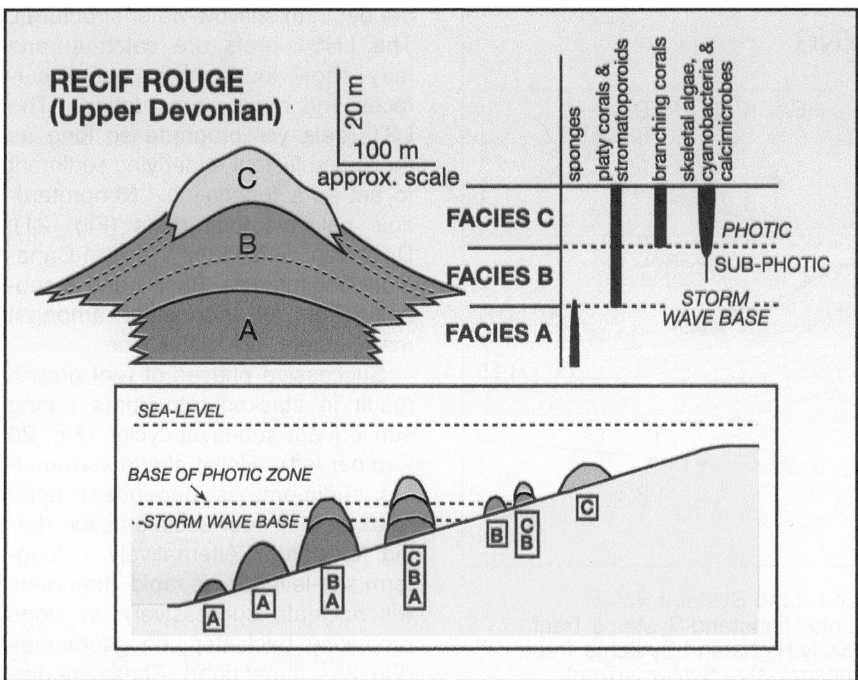

Figure 26. Facies mosaic and vertical succession of an Upper Devonian "récifs rouges", Belgium. (Top left) complete facies succession; dotted lines are growth surfaces. (Top right) ecological assemblages and depth zonation. (Inset) Conceptual model of distribution of various types of monogenetic and composite mounds on a slope with stacking order of facies for each mound. Facies A = red, stromatactis, sponge-rich mound; Facies B = pink coral/stromatoporoid-mud mound; Facies C = grey coral-thrombolite-calcimicrobe mound.

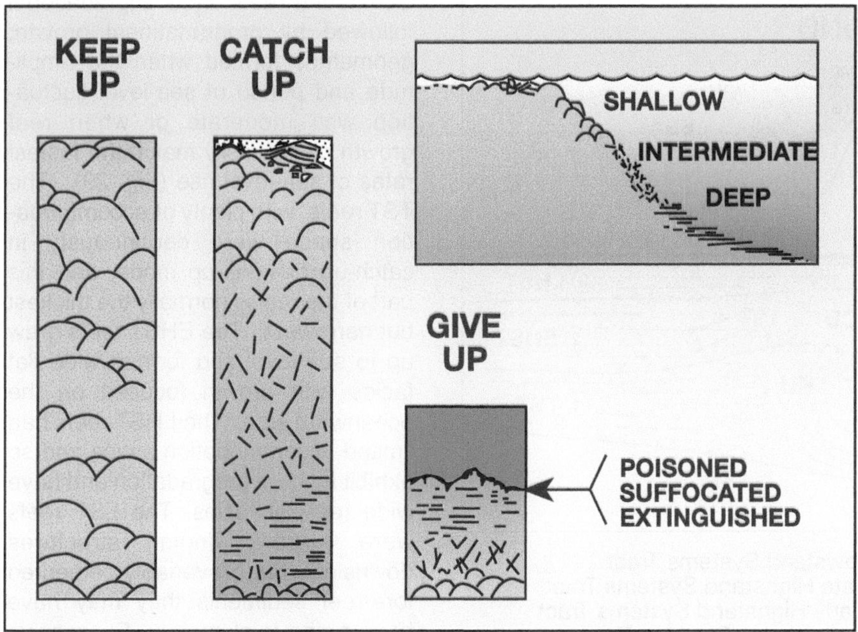

Figure 27. Growth strategies illustrated by reefs during a period of rising sea level; after Neumann and Macintyre (1985). (Inset) The shallow, intermediate and deep zones of reef growth; columns = the response of these zones to different rates of sea level change. Keep-up reefs track sea level and have uniform facies throughout (in this case, all shallow water facies). Catch-up reefs can have an initial shallow water facies, then lag behind sea level rise for a short period only to grow upwards through progressively shallower waters to reach sea level; such reefs display shallowing-upward facies trends. Give-up reefs start to grow but are killed off by excessive nutrients (poisoned), abundant fine sediment (suffocated) or rapid subsidence below the photic zone (extinguished). The top of a give-up reef can be a hardground or it can grade upward into a deep-water reef.

demise is usually the result of sea-level rise accompanied by shallowing of the photic zone, increased turbidity, nutrient excess, or hot saline waters. The submergence of a reef platform below the photic zone is often referred to as 'drowning', and is usually caused by a tectonically induced rapid rise in relative sea level.

Reef Body Geometry

The development of particular reef geometry is determined by changing accommodation space, and the ability of the community to respond to that change. Responses are hierarchical, and can be superimposed. Most reef structures are variations of either aggradation, aggradation then progradation, or progradation. The response of the reef is expressed as a transgressive systems tract (TST), an early highstand systems tract (EHST), a late highstand systems tract (LHST) and a lowstand systems tract (LST). An important caveat, however, is that the vertical exaggeration on seismic sections and geological cross-sections gives the impression of much more relief than originally present; many ancient reefs resemble pancakes more than they do cones. Some reef communities also have a naturally cavernous growth habit, such as scleractinian corals, which have secure and permanent attachment to a hard substrate and grow to substantial sizes. Other organisms with more delicate and small growth forms never formed substantial relief above the sea floor.

Aggrading Reefs

An aggrading reef shows continuous near-vertical growth as the shallow reef system constantly keeps up with sea-level rise and remains submerged (Fig. 28). This growth is dominant when the amplitude of sea-level fluctuations is large and the period short and when rates of production barely matched those of accommodation space increase. The TST reefs can be narrow but show thick accumulations, with near vertical margins and limited peri-reefal sediment. They aggraded in a keep-up mode, but continuously lagging sea-level rise and the reef community grew at an intermediate

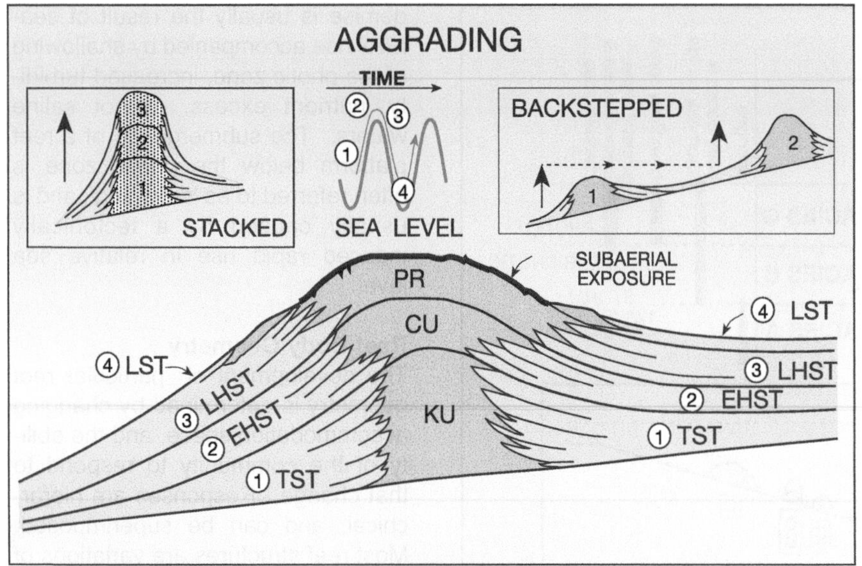

PR = Prograde
CU=Catch Up
KU=Keep Up
GU=Give Up

LST = Lowstand Systems Tract
LHST=Late Highstand Systems Tract
EHST=Early Highstand Systems Tract
TST=Transgressive Systems Tract

Figure 28. A sketch illustrating aggrading reef growth during a rise and fall in sea level when fluctuations were large and of short period (icehouse times) and/or the reef builders could barely match rates of sea level rise. In relatively shallow water aggradation is mainly during TST, EHST and LHST phases. Such reefs may be stacked to form large complexes or if sea level is very rapid may backstep upslope (insets).

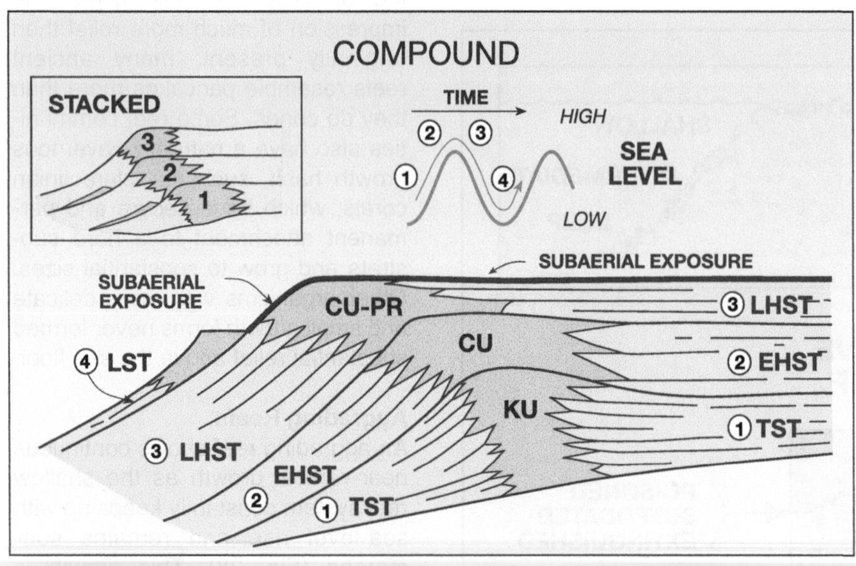

PR = Prograde
CU=Catch Up
KU=Keep Up
GU=Give Up

LST = Lowstand Systems Tract
LHST=Late Highstand Systems Tract
EHST=Early Highstand Systems Tract
TST=Transgressive Systems Tract

Figure 29. A sketch illustrating compound reef growth patterns during a rise and fall in sea level when the scale and period of sea level fluctuations were intermediate and/or reef builders could match the fastest rates of sea level rise. The motif is one of aggradation followed by progradation. TST reefs are mostly keep-up; EHST reefs are catch-up and LHST reefs are prograding. Inset shows stacking of reefs as the result of growth during several major fluctuations in sea level.

ate depth to shallow-water structures. The LHST reefs are catch-up and may show localized exposure surfaces and minor progradation. The LST reefs will prograde so long as there is sufficient underlying sediment to act as a foundation. Neoproterozoic calcimicrobial reefs (Fig. 21), Devonian reefs in the Western Canadian Sedimentary Basin, and Cenozoic reefs of Indonesia, amongst many others, are of this type.

Successive phases of reef growth result in stacked, structures during subsequent sea-level cycles (Fig. 28 – upper left). Relief above surrounding sediments is dependent upon rates of inter-reef sedimentation during lowstands. Alternatively, if long-term sea-level rise is rapid, then reefs will nucleate successively up slope producing back-stepping geometries (Fig. 28 – upper right). These are particularly important as hydrocarbon reservoirs because the reef reservoir facies can be buried by deep-water shales that are both source rocks and seals.

Compound Reefs

Compound reefs show aggradational followed by progradational growth; geometries formed when the amplitude and period of sea-level fluctuation was moderate or when reef growth could easily match the fastest rates of sea-level rise (Fig. 29). The TST reefs, with plenty of accommodation space were continuously in catch-up or keep-up mode, and this part of the reef is normally the thickest but narrowest. The EHST reefs grew up to sea level and formed wide flat facies with growth focused on the oceanward side. The LHST reefs had limited accommodation space and so exhibit marked progradation and have wide reef-flat facies. The LST reefs were narrow fringing structures downslope but on variably cemented fore-reef sediments they may have been prone to slumping. Successive reef phases of reef growth (Fig. 29 – inset) illustrate pronounced overall prograding geometry. Some Devonian reefs of the Canning Basin and Miocene reefs of Mallorca, for example, have this geometry.

Progradational Reefs

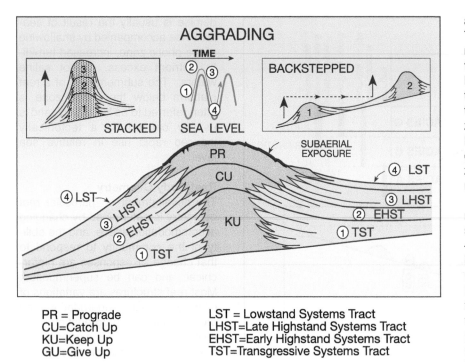

Progradation occurred when sea-level change was minimal and of long duration, and the reef community growth rate could exceed rates of sea-level change. These reefs grew into shallow water and were often exposed: facies expanded laterally due to restricted accommodation space (Fig. 30). Progradation was often established soon after reef initiation if growth was in shallow waters with minimal accommodation. The TST reefs were typically zoned structures but the style of EHST and LHST reefs depended on basin relief. If relief was large, then progradation could continue and wide reef flats would develop. If relief was minimal, then reef geometry would flatten and much fore-reef debris would be created. Successive phases of reef growth then resulted in relatively thin structures having pronounced progradational geometry. The LST development would be represented by a slight downshift in facies. Such reefs can be found around the shallow margins of the Michigan Basin, at the edge of the Permian Reef Complex and along the leeward margin of the Bahamas.

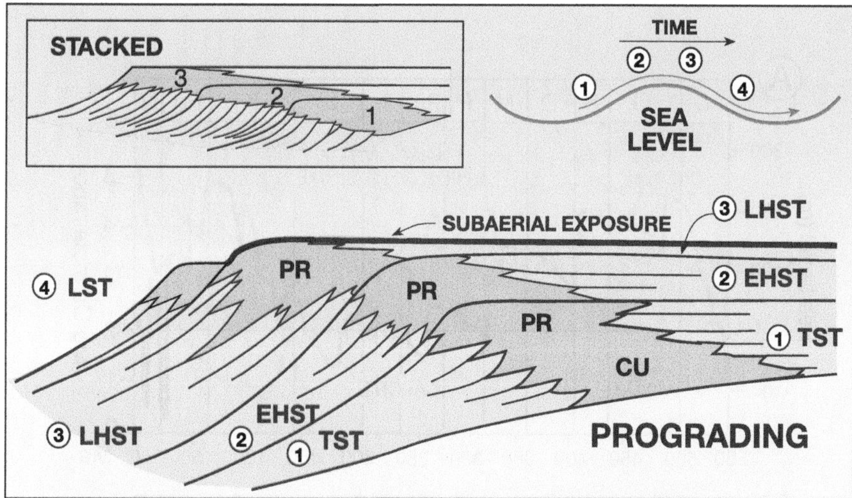

PR = Prograde
CU=Catch Up
KU=Keep Up
GU=Give Up

LST = Lowstand Systems Tract
LHST=Late Highstand Systems Tract
EHST=Early Highstand Systems Tract
TST=Transgressive Systems Tract

Figure 30. A sketch illustrating prograding reef growth during a rise and fall in sea level when changes in sea level were small and of long duration (greenhouse times) and/or the reef builders could easily exceed the rates of sea level change. If platform-basin relief was large progradation prevails during LHST and LST but if small then biostromes or sand shoals can develop. Inset shows the result of repeated progradation during several sea level cycles.

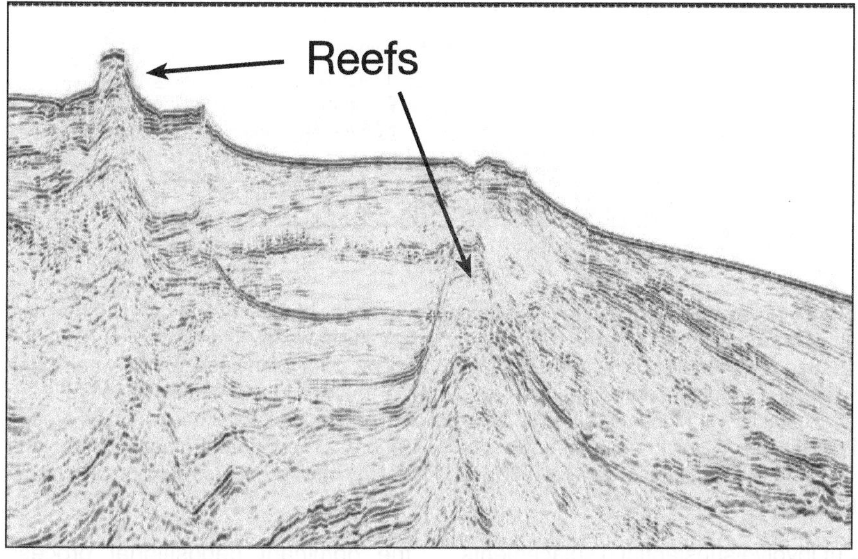

Figure 31. Seismic image of isolated Pleistocene aggrading carbonate reefs growing on prograding siliciclastics of the Mahakam Delta, Indonesia (image courtesy A. Saller, reproduced from AAPG 2004, reprinted by permission of the AAPG whose permission is required for further use).

Isolated Reefs

A common feature of isolated reefs is their well-defined vertical facies zonation and the near-absence of lateral zonation. Carbonate mounds are also able to develop (often on pre-existing topographic highs) in otherwise siliciclastic regimes during periods of minimal siliciclastic input, such as during transgressions when the locus of siliciclastic input shifted landward (Fig. 31). In contrast to shallow-water reefs, accommodation is not a problem and so they are generally equidimensional.

Response to Tectonic Change

Tectonics and time ultimately drive many environmental aspects, such as sea level, circulation patterns, climate, and evolving sea-water chemistry change (Fig. 32). Reefs and carbonate platforms commonly begin on pre-existing highs, such as horst blocks or salt domes during the early stages of rifting. Reefs are most numerous on passive continental margins at low latitudes where they typically form barrier or fringing reefs. Similar carbonate shelves with reefs can form around the margins of shallow intracratonic basins, where they can be interbedded with thin beds of evaporites. Such basins may be prone to isolation from the open ocean leading to infilling by extensive evaporites. Rates of thermal subsistence on passive margins are slow and predictable, and are similar to those found in intracratonic basins. Reef geometries are thus generally compound to progradational. By contrast, reef-formation toward the cratonward side of foreland basins is relatively rapid and reef geometries more aggradational. Finally, reefs forming in strike-slip basins and on thrust complexes are affected by the vagaries of local tectonic movements, which may be highly episodic

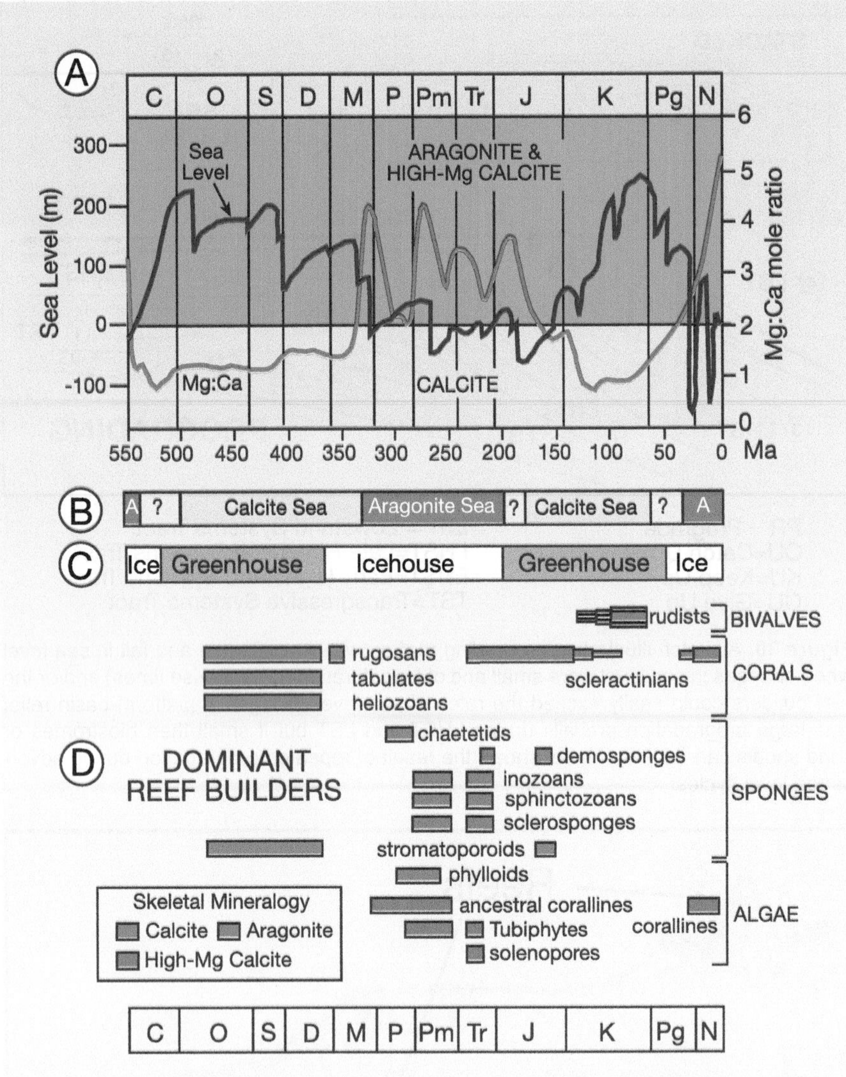

Figure 32. A. Correspondence between changing ocean chemistry and carbonate mineralogy through time as a function the of Mg:Ca ratio in seawater (after Stanley and Hardie, 1998) and general global sea level change (Vail *et al.*, 1977); the boundary between fields of calcite (<4 mole % $MgCO_3$), high-Mg Calcite (>4mole % $MgCO_3$), and aragonite is the horizontal line at Mg/Ca = 2; **B.** The different dominant non-skeletal mineralogies precipitated in sea water through time (Sandberg, 1983); **C.** The different global climatic and oceanographic periods (Fischer, 1982); **D.** Mineralogy of different reef-building organisms (Stanley and Hardie, 1998).

but involve substantial displacements, and are thus highly unpredictable.

Epeiric seas (Chapter 14), those that formed by the extensive flooding of continents during globally high sea levels, have negligible basin-floor topography. Water depths rarely exceeded 10 m, such that shallow subtidal and intertidal sediments dominate, with many episodes of exposure. Epeiric seas could support patch reefs, often elongated because of wind waves, storms, or tides. Due to the limited accommodation space on epeiric platforms, progradation is the dominant depositional process resulting in stacked shallowing-upward sequences or biostromes.

Carbonate Saturation

Living, shallow-water coral reefs are restricted to tropical and subtropical environments characterized by warm temperatures, high light intensities, and high aragonite supersaturation. The growth of coral skeletons and other calcifying organisms precipitates carbonate ions, forcing a re-equilibration of the bicarbonate-dominated marine inorganic carbon system, which also creates a source of carbon dioxide. A coral reef, therefore, represents the net accumulation of $CaCO_3$ and a source of CO_2, which, in turn, probably leads to feedback-controls on the global carbon cycle.

Coral reef calcification depends on the saturation state of aragonite in surface waters, and if calcification declines, then reef-building capacity also decreases. Predictions suggest that by the middle of the next century, carbon dioxide concentrations will rapidly rise to double pre-industrial levels. This will be catastrophic for reefs and the modern carbonate system because the time scales required for natural feedbacks to operate is far greater than the rate of greenhouse gas increase. Both experimental work and models suggest that an increased concentration of carbon dioxide will lower pH leading to ocean acidification, so decreasing the aragonite saturation state in the tropics by 30 percent and biogenic aragonite precipitation by 14 to 30% (Kleypas *et al.*, 2001). Indeed, biological calcification rates are already known to be 10 to 20% lower than under pre-industrial conditions. Coral reefs are particularly threatened, because most modern reef-building organisms secrete aragonite, a metastable form of $CaCO_3$.

Carbonate saturation has probably varied dramatically throughout geological time. It appears that Precambrian oceans were highly oversaturated, but this may have declined around the beginning of the Phanerozoic with the appearance of numerous calcareous invertebrates. Since then, Ca^{2+} in seawater has broadly followed sea-level change, such that Ca^{2+} levels were reasonably high during the Cambrian–Mississippian and the Jurassic to Cretaceous.

It has been suggested that the geological distribution of microbialites might be controlled by physicochemical factors, including the saturation state of sea water and/or global temperature distribution. This may also explain the decline in abundance of reef microbialite after the Jurassic due to the increase of pelagic carbonate that led to a reduced seawater saturation state. This would have lowered supersaturation levels below a threshold for abundant automicrite formation, thus restricting its formation to cryptic reef habitats where abnormal chemistries could have existed. Such a situation might also explain the absence of *Stromatactis* in the late Phanerozoic.

Evolving Ocean Chemistry

The dominant form of precipitated crystalline $CaCO_3$ has oscillated during the geological past, with both inorganic and organic production of aragonite and high-Mg calcite dominating carbonate formation during cool periods (icehouse), and low-Mg calcite predominating during warm periods (greenhouse; Fig. 32). Such mineralogical shifts are interpreted as markers for major changes in seawater chemistry.

Stanley and Hardie (1998) have proposed that shifts in the Mg:Ca ratio have controlled the predominance of calcite versus aragonite secretors, particularly reef builders, due to the inhibiting effect of high Mg^{2+} concentration on calcite secretion. They suggest that the changing Mg:Ca ratio of seawater has been controlled by variations in the rate of production of oceanic crust, because oceanic hydrothermal alteration is a major sink for Mg and an important source of Ca. Experimental work has subsequently confirmed the profound influence of Mg:Ca seawater ratios on modern reef builders, including scleractinian corals and the calcareous green alga *Halimeda*.

Scleractinian corals were, for example, important reef-builders in the Jurassic, but they did not build extensive reefs during the greenhouse period (calcite seas) of the Cretaceous. During this period, species diversity remained high but their abundance on carbonate platforms was low compared to the Jurassic. Distribution shifted to outer platform settings and higher latitudes (~35–45°N). Many hypotheses are offered to explain these observations, including the high temperatures, restricted circulation, and unstable sediment conditions of Cretaceous platforms, and, in particular, the favoring of calcite-producing rudist bivalves over aragonite corals (Wood, 1999). Rudist bivalves, with their outer calcitic skeletons, underwent a dramatic radiation in the Late Cretaceous when the Mg:Ca ratio of seawater reached exceptionally low values, so favoring calcite over aragonite secretors.

Major shifts in the dominant composition of carbonate skeletal particles through geological time also mirrors, in part, these proposed changes in seawater chemistry and climate (Fig. 32D), but mass extinctions also play a role by triggering changes in the predominant form of $CaCO_3$ produced by marine calcifiers (Kiessling *et al.,* 2008). Changes in the abundance of aragonitic organisms following mass extinction events appear to have been predominantly driven by selective recovery rather than selective extinction.

Evidence is also persuasive that changing seawater chemistry has influenced the style of early diagenesis in carbonate regimes, particularly in reefs. The mineralogy of early marine reef cements also seems to follow the same secular changes. For example, aragonitic botryoids are known exclusively from phases of aragonite seas (Early Cambrian, mid-Carboniferous to Early Jurassic, and mid-Late Cenozoic), whereas radiaxial fibrous calcite is common in reefs that grew in calcite seas (particularly the Ordovician to Devonian), it is virtually unknown from aragonite seas of the Cenozoic. Enhanced rates of calcite cementation during calcite seas, due to the elevated abundance of calcium ions, may have promoted rapid lithification of the reef framework and aided preservation of cryptic biota that were otherwise vulnerable to disturbance, but this has yet to be documented.

Climate

Climate is a major controlling force on the evolution of reefs on both short- and longtime scales, as the latitudinal range of carbonate-producing species is largely governed by temperature and carbonate supersaturation. Reef growth therefore shows cyclicity at all scales in response to short-term oscillations (*e.g.,* Milankovitch and glacial-interglacial cycles) as well as to longer term climatic intervals driven by slower, tectonically driven processes.

Global climate has oscillated through greenhouse and icehouse phases, in concert with aragonite and calcite seas, respectively (Fig. 32). During icehouse phases of continental glaciation (*e.g.,* Pennsylvanian–Early Permian, Miocene–Pleistocene), high-frequency sequences on carbonate platforms were generated by eustatic sea-level changes of 50–100 m. Subaerial exposure, unfilled accommodation space and conspicuous regional disconformities were common. Aggrading reef growth in a keep-up mode is dominant in icehouse times, as rates of production barely match those of accommodation space increase. These large changes in sea level caused ramps to have steep gradients, and platform tops to have significant depositional relief characterized by pinnacle reefs and erosional topography. Icehouse reefs are typically dominated by heterotrophs or autotrophs with aragonitic or High Mg Calcite mineralogies.

During greenhouse times with little global ice (*e.g.,* Late Cambrian–Early Ordovician, Devonian, Triassic, Cretaceous), reef sequences were generated by small (possibly <10 m or so) sea-level fluctuations. Reef cycles typically consist of very shallow-water facies and regional-scale tidal-flat caps and minor disconformities. Greenhouse reefs are often either compound or progradational as reef growth could easily match or outpace the fastest rates of sea-level rise. The TST reefs, with plenty of accommodation space were continuously in catch-up or keep-up mode, and this part of the reef is normally the thickest but narrowest. Ramps often have very low gradients and platforms tend to be progradational, with minor topography. Greenhouse reefs can also have a more extensive range as carbonate settings extend into higher latitudes due to elevated global temperatures. Composition may also be more uniform, often dominated by low-Mg calcite skeletal components.

GEOLOGIC HISTORY OF REEFS

The evolving history of reefs is one of the great topics in Earth science and has been tackled by numerous workers. Amongst the more recent summaries are those by Fagerstrom (1987), Wood, (1999), Stanley

(2001), and Kiessling *et al.* (2002). The record is one of recurring themes punctuated by episodes of mass extinction.

Themes

A simple plot of reef types and components against geologic time (Fig. 33) reveals several themes. The most obvious is the evolution from stromatolite-dominated reefs of the Archean and Proterozoic to more complex biotic structures of the Phanerozoic. Coincident with this change is the appearance of thrombolites and calcimicrobes in the Neoproterozoic that resulted in a fundamental change from dense microbial structures to more porous, open buildups that characterize much of the Phanerozoic.

The Phanerozoic has several superimposed themes. Skeletal-microbial reefs are generally universal, with the microbial component becoming less important since post-Jurassic time. Skeletal reefs are confined to specific periods, particularly when conditions were right for the growth of large clonal metazoans such as corals and stromatoporoid sponges. Likewise carbonate mud mounds are largely confined to the Late Cambrian to Early Cretaceous, with their acme in the middle and late Paleozoic.

The modern reef biota appeared in the Middle Jurassic but was largely suppressed by the prolific growth of calcitic rudist bivalves in the Cretaceous, only to rebound after the K–T extinction. There is considerable debate, however, as to whether this coral-dominated biota was able to construct spectacular large reefs like those of the modern day before the Cenozoic.

Climate and changing seawater chemistry are also important controls on the history of reef growth, both in terms of complex feedback mechanisms that govern skeletal mineralogy and hence promote some groups over others, as well as controlling the styles of early lithification.

A final theme concerns the relationship between carbonate platforms and skeletal reefs. Large, shallow-water, flat-topped platforms coincide with periods of skeletal reef development because skeletal reefs are char-

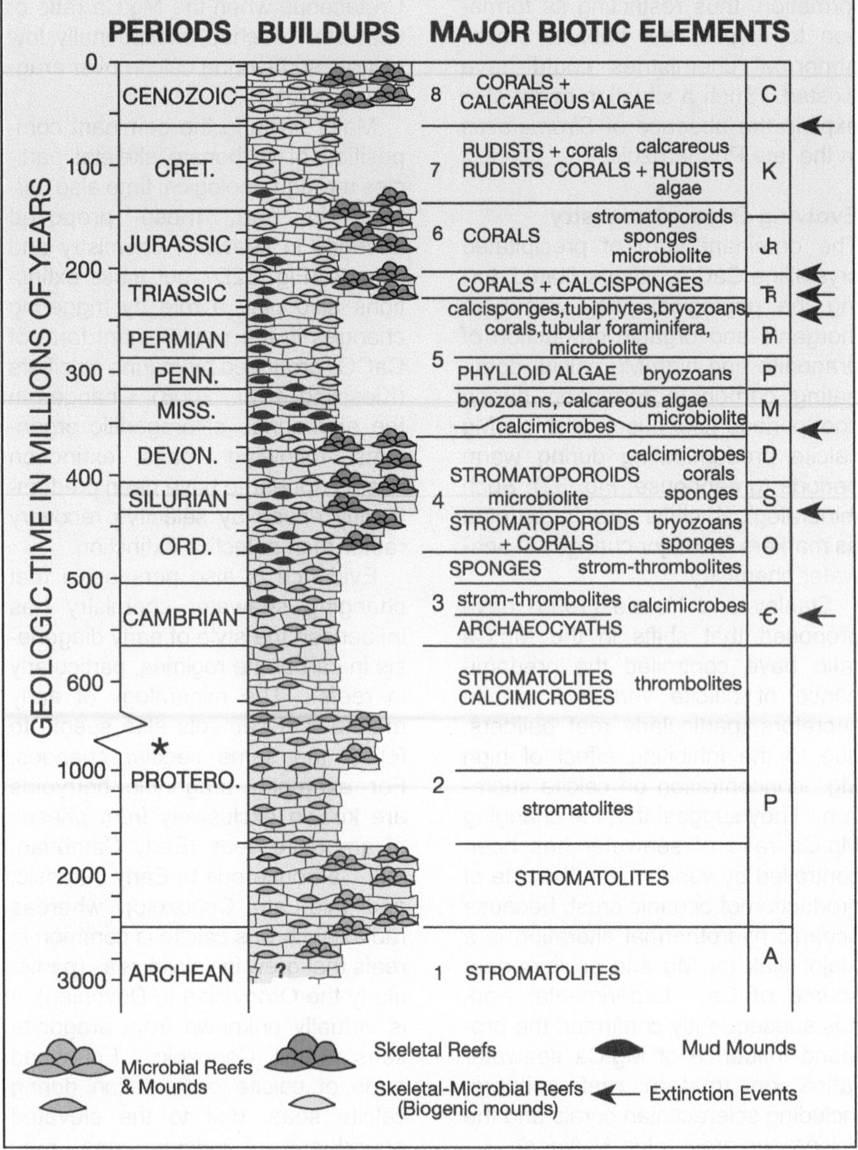

Figure 33. An idealized stratigraphic column representing geologic time and illustrating periods when there were only skeletal - microbial reefs (biogenic mounds) and those times when there were both skeletal-microbial reefs and skeletal reefs. The periods during which mud mounds were important are also highlighted. Numbers indicate different associations of reef- and mound-building biota. Arrows signal major extinction events; * = scale change.

acterized by well-cemented carbonate buttresses along their seaward margin. These ramparts absorbed waves and swells, thus allowing more tranquil, protected lagoonal facies to develop across the platform interiors. When skeletal reefs were absent carbonate ramps were the norm.

Mass Extinctions

Reefs have long been thought to be highly susceptible to mass extinction events, with major reef-building biota typically going largely or wholly extinct. Metazoan reefs may take some 2–8 Myr to recover after such events, longer than other communities, and post-extinction reefs are often composed of entirely new groups of skeletal metazoans. In some cases, either calcimicrobes (*e.g.*, the end Devonian) or microbialites (*e.g.*, the end Permian) can dominate post-extinction reefs, but it should be remembered that these elements were often already part of the pre-extinction reef community.

Tropical shallow-marine carbonate production has been noted to fall after some mass extinctions (*e.g.*, the Late

Devonian and end Permian), but not others (end-Cretaceous).

Changes in seawater saturation or chemistry may also explain the response of reef- and other carbonate-producing biota to mass extinctions. When skeletal groups first appear they tended to adopt the mineralogy favored by ambient seawater chemistry, but subsequent extinction events may have preferentially removed those groups and selected for new skeletal biota whose skeletal mineralogy reflected the new ocean chemistry.

SYNTHESIS

Reefs are organo-sedimentary seafloor reliefs that have persisted throughout Earth history, from the Archean to the present day. The basic constructional elements have been microbes, calcareous algae, metazoans, and synsedimentary carbonate cements. Alone or together these components have built a myriad of structures that range from dense carbonate mud mounds to skeletal aggregates of stunning diversity. They grew from the lightless depths to the sunlit shallows and across the environmental spectrum from tranquil lagoons to wave-swept platform margins. As biological structures subject to evolutionary change and innovation, reefs exhibit an endless complexity and their growth was particularly sensitive to variations in physical and chemical oceanography at all scales. The facies models synthesized herein should be viewed as a background against which to consider, and as a guide with which to interpret an almost infinite variety of reefs. Each geologic period and each environment had a specific style of reef, and each one is unique.

ACKNOWLEDGEMENTS

This paper is the result of research supported by the Natural Sciences and Engineering Research Council of Canada (NPJ) and the ECOSSE (RW). Bill Martindale and Brian Pratt critically assessed the manuscript and the final version reflects their questioning, careful analysis, and detailed editing. We dedicate this synthesis to Pierre-Andre Bourque who recently died in the field; he is sorely missed.

REFERENCES

Basic sources of information

Fagerstrom, J.A., 1987, The Evolution of Reef Communities: J. Wiley & Sons, New York, 600 p.
A sound introduction to the wide diversity of Phanerozoic reef communities emphasising the guild concept.

Flugel, E., 2004, Microfacies of Carbonate Rocks: Springer-Verlag, Berlin Heidelberg, 976 p.
An excellent source of information on reef components and reef fabrics.

Freiwald, A. and Roberts, J.M., *eds*, 2005, Cold-water Corals and Ecosystems: Springer-Verlag, Berlin Heidelberg, 1243 p.
A wide variety of important articles on Tertiary and Modern deep-water coral reefs.

Geldsetzer, H.H.J., James, N.P. and Tebbutt, G.E., 1988, Reefs, Canada and Adjacent Areas: Canadian Society of Petroleum Geologists, Memoir 13, 775 p.
A gold mine of information! A total of 115 short case studies of Precambrian, Paleozoic, and early Mesozoic reefs organized chronologically.

Grotzinger, J.P. and James, N.P., *eds.*, 2000, Carbonate Sedimentation and Diagenesis in the Evolving Precambrian World, SEPM Special Publication 65. (*see* introduction — Grotzinger, J.P. and James, N.P., Precambrian Carbonates: evolution of understanding, p.1-20.)
A compendium of studies on the Precambrian with several important studies of reefs.

James, N.P., 1983, Reef Environment, *in* Scholle, P.A., Bebout, D.G. and Moore, C.H., *eds.*, Carbonate Depositional Environments: American Association of Petroleum Geologists Memoir 33, p. 345-440.
An overview of modern and ancient reef sedimentology with numerous colour illustrations and photographs; read in conjunction with this paper.

James, N.P. and Bourque, P.-A., 1992, Reefs and Mounds, *in* Walker, R.G. and James, N.P., *eds.*, Facies Models, Response to sea level change; Geological Association of Canada, St. John's, Newfoundland, p. 323-347.
The precursor to this article summarizing understanding at that time.

Kiessling, W., Flugel, E. and Golonka, J., *eds.*, 2002, Phanerozoic Reef Patterns, SEPM Special Publication, 72, 775 p.
The most current summary of reefs through time with chapters devoted to each geologic period (including maps) as well as summary articles; the best detailed reference with much data on reefs of all ages; read in conjunction with Stanley (2001).

Monty, C.L.V., Bosence, D.W.J., Bridges, P.H. and Pratt, B.R., *eds.*,1995, Carbonate Mud Mounds: Their Origin and Evolution: International Association of Sedimentologists Special Publication 23: Blackwell Science Ltd., Oxford, p. 475-493.
A focused book with critical articles on the puzzling genesis of these largely enigmatic structures.

Stanley, G.D. Jr., *ed.*, 2001, The History and Sedimentology of Ancient Reef Systems: Kluwer Academic/Plenum Publishers, New York, 458 p. (*see* Chapter 1 - Stanley, G.D. Jr. Introduction to Reef Ecosystems and their Evolution).
Numerous articles on reefs through time; read in conjunction with Keissling et al., *2002).*

Stanley, G.D., Jr. and Fagerstrom, J.A., *eds.*, 1988, Ancient reef ecosystems: Palaios, v. 3, p. 110-250.
An issue with 12 papers devoted to the paleoecology of reefs and mounds; including several summary contributions.

Toomey, D.F., *ed.*, 1981, European Fossil Reef Models: SEPM Special Publication 30, 546 p.
A total of 17 papers with emphasis on latest Paleozoic, Mesozoic, and Cenozoic structures.

Wilson, J.L., 1975, Carbonate Facies in Geologic History: Springer-Verlag, Heidelberg, 471 p.
A classic text, an overall reference to fossil reefs with succinct descriptions and numerous references.

Wood, R., 1999, Reef Evolution, Oxford University Press, New York, 414 p.
A text devoted to the understanding of fossil reefs from a mostly geobiological viewpoint; a different and important perspective.

OTHER REFERENCES

The following are cited in the text or are important reading on specific aspects of reef attributes, modern and ancient.

Modern reefs

Blanchon, P., Jones, B. and Kalbfleisch, W., 1997, Anatomy of a fringing reef around Grand Cayman: storm rubble, not coral framework: Journal of Sedimentary Research, v. 67, p. 1-16.

Callender, W.R., Staff, G.M., Powell, E. and Macdonald, I.R., 1990, Gulf of Mexico hydrocarbon seep communities V. Biofacies and shell orientation of autochthonous shell beds below storm wave base: Palaios, v. 5, p. 2-14.

Camoin, G.F. and Montaggioni, L.F., 1994, High energy coralgal-stromatolite frameworks from Holocene reefs (Tahiti, French Polynesia): Sedimentology, v. 41, p. 655-676.

Cortes, J., Macintyre, I.G. and Glynn, P.W., 1994, Holocene growth history of an eastern Pacific fringing reef, Punta Islotes, Costa Rica: Coral Reefs, v. 13, p.

65-73.

Ginsburg, R.N. and Schroeder, J.H., 1973, Growth and submarine fossilization of algal cup reefs, Bermuda: Sedimentology, v. 20, p. 575-614.

Graus, R.R. and Macintyre, I.G., 1989, The zonation patterns of Caribbean coral reefs as controlled by wave and light energy input, bathymetric setting and reef morphology: computer simulation experiments: Coral Reefs, v. 8, p. 9-18.

Harris, P.M., 1996, Reef styles of modern carbonate platforms: Bulletin of Canadian Petroleum Geology, v. 44, p. 72-81.

Hovland, M., Mortensen, P.B., Brattegard, T., Strass, P. and Rokoengen, K., 1998, Ahermatypic coral banks off mid-Norway: evidence for a link with seepage of light hydrocarbons: Palaios, v. 13, p. 189-200.

James, N.P. and Ginsburg, R.N., 1979, The Seaward Margin of Belize Barrier and Atoll Reefs: International Association of Sedimentologists, Special Publication 3, 191 p.

Macintyre, I.G. and Glynn, P.W., 1976, Evolution of modern Caribbean fringing reef, Galeta Point, Panama: American Association of Petroleum Geologists, Bulletin, v. 60, p. 1054-1072.

Messing, C.G., Neumann, A.C. and Lang, J. C., 1990, Biozonation of deep-water lithoherms and associated hardgrounds in the Straits of Florida: Palaios, v. 5, p. 15-53.

Montaggioni, F.F., 2005, History of Indo-Pacific coral reef systems since the last glaciation: Development patterns and controlling factors: Earth Science Reviews, v. 71, p. 1-75.

Mullins, H.,T., Newton, C.R., Heath, K. and Vanburen, H.M., 1981, Modern deep-water coral mounds north of Little Bahama Bank: criteria for recognition of deep-water coral bioherms in the rock record: Journal of Sedimentary Petrology, v. 51, p. 999-1013.

Neumann, A.C. and Macintyre, I.G., 1985, Reef response to sea level rise: keep-up, catch-up or give-up: Proceedings of the Fifth International Coral Reef Congress, Tahiti, v. 3, p. 105-110.

Neumann, A.C., Kofoed, J.W. and Keller, G.H., 1977, Lithoherms in the Straits of Florida: Geology, v. 5, p. 4-10.

Perry, C.T. and Hepburn, L.J., 2008, Syndepositional alteration of coral reef framework through bioerosion, encrustation, and cementation: Taphonomic signatures of reef accretion and reef depositional events: Earth Science Reviews, v. 86, p. 106-144.

Reed, J.K., 2002, Deep-water *Oculina* coral reefs of Florida: biology, impacts and management: Hydrobiologica, v. 471, p. 43-55.

Reitner, J., 1993, Modern cryptic microbialite/metazoan facies from Lizard Island (Great Barrier Reef, Australia) - Formation and concepts: Facies, v. 29, p. 3-40.

Roberts, H.H. and Macintyre, I.G., *eds.*, 1988, Halimeda: Coral Reefs, v. 6, p. 121-271.

Taviani, M., Freiwald, A. and Zibrowius, H., 2005, Deep coral growth in the Mediterranean sea: an overview, *in* Cold water corals and ecosystems, part of Erlangen Earth conference series: Springer-Verlag, New York, p. 137-156.

Tudhope, A.W. and Scoffin, T.P., 1994, Growth and structure of fringing reefs in a muddy environment, south Thailand: Journal of Sedimentary Research, v. A64, p. 752-764.

Wanless, H.R. and Tagett, R.P., 1989, Origin and dynamic evolution of carbonate mudbanks in Florida Bay: Bulletin of Marine Science, v. 44, p. 454-489.

Ancient reefs

Ahr, W.M., Harris, P.M., Morgan, W.A. and Somerville, I.D., *eds.,* 2003, Permo-Carboniferous Carbonate Platforms and Reefs: SEPM Special Publication 78, 414 p.

Alvaro, J., J., Aretz, M., Boulvain, F., Munnecke, A., Vachard, D. and Vennin, E., *eds.*, 2007, Paleozoic Reefs and Bioaccumulations: Climatic and Evolutionary controls: Geological Society of London Special Publication 275, 291p.

Bernecker, M. and Weidlich, O., 1990, The Danian (Paleocene) coral limestone of Fakse, Denmark: A model for ancient aphotic, azooxanthallate coral mounds: Facies, v. 22, p. 103-138.

Brunton, F.R. and Dixon, O.A., 1994, Siliceous sponge-microbe biotic associations and their recurrence through the Phanerozoic as reef mound constructors: Palaios, v. 9, p. 370-387.

Crevello, P.D. and Harris, P.M., 1984, Depositional models for Jurassic reefal buildups, *in* Ventress, W.P.S., Bebout, D.G., Perkins, B.F. and Moore, C.H., *eds.*, Jurassic of the Gulf Rim: Gulf Coast Section, SEPM p. 57-102.

Fagerstrom, J.A. and Weidlich, O., 1999, Origin of the upper Capitan-massive limestone (Permian), Guadalupe Mountains, New Mexico-Texas; is it a reef?: Geological Society of America Bulletin, v. 111, p. 159-176.

James, N.P. and Gravestock, D., 1990, Lower Cambrian shelf and shelf margin buildups, Flinders Ranges, South Australia: Sedimentology, v. 37, p. 455-480.

James, N.P. and Kobluk, D.R., 1978. Lower Cambrian patch reefs and associated sediments, southern Labrador, Canada: Sedimentology, v. 25, p. 1-35.

James, N.P., Feary, D.A.F., Betzler, C., Bone, Y., Holburn, A.E., Li, Q., Machiyama, H., Simo, J.A. and Surlyk, F., 2004, Origin of Late Pleistocene bryozoan reef-mounds: Great Australian Bight: Journal of Sedimentary Research, v. 74, p. 20-48.

Kauffman, E.G. and Johnson, C.C., 1988, The morphological and ecological evolution of Middle and Upper Cretaceous reef-building rudistids: Palaios, v. 3, p. 194-216.

Kershaw, S. and Keeling, M., 1994, Factors controlling the growth of stromatoporoid biostromes in the Ludlow of Gotland, Sweden: Sedimentary Geology, v. 89, p. 325-335.

Kenter, J.A.M., Harris, P.M. and Della-Porta, G., 2005, Steep microbial boundstone-dominated platform margins-examples and implications: Sedimentary Geology, v. 178, p. 5-31.

Kirkby, K.C. and Hunt, D., 1996, Episodic growth of a Waulsortian buildup: the Lower Carboniferous Muleshoe Mound, Sacramento Mountains, New Mexico, USA, *in* Strogen, P., Somerville, I.D. and Jones, G.L., *eds.*, Recent Advances in Lower Carboniferous Geology: Geological Society of London, Special Publication 107, p. 97-110.

Kruse, P.D., Zhuravlev, A. Yu. and James, N.P., 1995, Primordial metazoan-calcimicrobial reefs: Tommotian (Early Cambiran) of the Siberian Platform: Palaios, v. 10, p. 291-321.

Leinfelder, R.R. and Keupp, H., 1995, Upper Jurassic mud mounds: Allochthonous sedimentation versus autochthonous carbonate production: Facies (Erlangen), v. 32, p. 17-26.

Martín, J.M., Braga, J.C. and Riding, R., 1997, Late Miocene *Halimeda* alga-microbial segment reefs in the marginal Mediterranian Sorbas Basin, Spain: Sedimentology, v. 44, p. 441-456.

Mundy, D.J.C., 1994, Microbiolite-sponge-bryozoan-coral framestones in Lower Carboniferous (Late Visean) buildups in northern England (UK); *in* Embry, A.F., Beauchamp, B. and Glass, D.J., *eds.*; Pangea: Global Environments and Resources: Canadian Society of Petroleum Geologists Memoir 17, p. 713-729.

Murillo-Muñetón, G. and Dorobek, S.L., 2003, Controls on the evolution of carbonate mud mounds in the Lower Cretaceous Cupido Formation, northeastern Mexico: Journal of Sedimentary Research, v. 73, p. 869-886.

Narbonne, G.M. and James, N.P.,1996, Mesoproterozoic deep-water reefs from Borden Peninsula, Arctic Canada: Sedimentology, v. 43, p. 827-848.

Palmer, T.J. and Fursich, F.T., 1981, Ecology of sponge reefs from the Upper Bathonian (Middle Jurassic) of Normandy: Palaeontology, v. 24, p. 1-25.

Playford, P.E., 1980, Devonian "Great Barrier Reef" of the Canning Basin, Western Australia: American Association of Petroleum Geologists, Bulletin, v. 64, p. 814-840.

Pomar, L., 1991, Reef geometries, erosion surfaces and high-frequency sea level changes, upper Miocene Reef complex, Mallorca, Spain: Sedimentology, v. 38, p. 243-270.

Pratt, B.R. and James, N.P., 1982, Cryptalgal-metazoan bioherms of Early Ordovician age in the St. George Group, western Newfoundland: Sedimentology, v. 29, p. 543-569.

Riding, R. and Zhuravlev, A.Y., 1995, Structure and diversity of oldest sponge-microbe reefs: Lower Cambrian, Aldan River, Siberia: Geology, v. 23, p. 649-652.

Riding, R., Martin, J.M. and Braga, J.C., 1991, Coral-stromatolite reef framework, Upper Miocene, Almeria, Spain: Sedimentology, v. 38, p. 799-819.

Saller, A.H., Harris, P.M., Kirkland, B.L. and Mazzullo, S.J., *eds.*, 1999, Geological Framework of the Capitan Reef: SEPM Special Publication 65, 224 p.

Scott, R.W., 1995, Global environmental controls on Cretaceous reef ecosystems: Palaeogeography, Palaeoclimatology, Palaeoecology, v. 119, p. 187-199.

Schmidt, D.-U. Leinfelder, R.R. and Nose, M., 2001, Growth dynamics and ecology of Upper Jurassic mounds, with comparisons to mid-Paleozoic mounds: Sedimentary Geology, v. 145, p. 343-376.

Stemmerik, L., Larson, P.A., Larssen, G.B., Mørk, A. and Simonsen, B.T., 1994, Depositional evolution of Lower Permian *Palaeoaplysina* build-ups, Kapp Duner Formation, Bjørnøya, Arctic Norway: Sedimentary Geology, v. 92, p. 161-174.

Stephens, N. P. and Sumner, D.Y., 2003, Famennian microbial reef facies, Napier and Oscar Ranges, Canning Basin, western Australia: Sedimentology, v. 50, p.1283-1302.

Turner, E.C., James, N.P. and Narbonne, G.M., 1997, Sea level dynamics and growth of Neoproterozoic deep-water reefs, Mackenzie Mountains, Canada: Journal of Sedimentary Research, v. 67, p. 437-450.

Watts, N. R. and Riding, R., 2000, Growth of rigid high-relief patch reefs, Mid-Silurian, Gotland, Sweden: Sedimentology, v. 47, p. 979-994.

Webb, G.E., 1994, Non-Waulsortian Missippian bioherms: a comparative analysis: *in* Embry, A.F., Beauchamp, B. and Glass, D.J., *eds.*, Pangea: Global Environments and Resources, Canadian Society of Petroleum Geologists Memoir 17, p. 701-712.

Webb, G.E., 2005, Quantitative analysis and paleoecology of earliest Mississippian microbial reefs, Gudman Formation, Queensland, Australia: not just post-disaster phenomena: Journal of Sedimentary Research, v. 75, p. 877-896.

Wendt, J., Belka, Z., Kaufmann, B., Kostrewa, R. and Hayer, J., 1997, The world's most spectacular carbonate mud mounds (Middle Devonian, Algerian Sahara): Journal of Sedimentary Research, v. 67, p. 424-436.

Wilson, M.E.J., 2005, Development of equatorial delta-front patch reefs during the Neogene, Borneo: Journal of Sedimentary Research, v. 75, p. 114-133.

Wood, R. A., Zhuravlev, Y. and Anaaz, C.T.,1993, The ecology of Lower Cambrian buildups from Zuune Arts, Mongolia: implications for early metazoan reef evolution: Sedimentology, v. 40, p. 829-858.

Wood, R., 1995, The changing biology of reef-building: Palaios, v. 10, p. 517-529.

Wood, R., Dickson, J.A.D. and Kirkland-George, B., 1994, Turning the Capitan Reef upside down: a new appraisal of the ecology of the Permian Capitan reef, Guadalupe Mountains, Texas and New Mexico: Palaios, v. 9, p. 422-427.

REFERENCES CITED

Buddemeier, R.W., 1997, Symbiosis; making light work of adaptation: Nature, v. 388, p. 229-230.

Della-Porta, G., Kenter, J.A.M., and Bahamonde, J.R., 2004, Depositional facies and stratal geometry of an Upper Carboniferous prograding and aggrading high-relief carbonate platform (Cantabrian Mountains, N Spain): Sedimentology, v. 51, p. 267-295.

Fischer, A.G., 1982, Long-term climatic scillations recorded in stratigraphy, *in* Berger, W., *ed.*, Climate in Earth History: National Research Council, Studies in Geophysics, National Academy Press, Washington D.C., p. 97-104.

Hallock, P. and Schlager, W., 1986, Nutrient excess and the demise of coral reefs and carbonate platforms: Palaios, v. 1, p. 389-398.

Kiessling, W., Aberhan, M. and Villier, L., 2008, Phanerozoic trends in skeletal mineralogy driven my mass extinctions: Nature Geoscience, v. 1, p. 527-530.

Kleypas, J.A., 1997, Modelled estimates of global reef habitat and carbonate production since the last glacial maximum: Paleoceanography, v. 12, p. 533-545.

Kleypas, J. A., Buddemeier, R. W. and Gattuso, J.-P., 2001, The future of coral reefs in an age of global change: Geologische Rundschau, International Journal of Earth Sciences (1999), v. 90, p. 426-437.

Lukasik, J. and Simo, J.A., 2008, Controls on Carbonate Platform and Reef Developmen: SEPM Special Publication 89, 364 p.

Pomar, L., and Hallock, P., 2008, Carbonate factories: a conundrum in sedimentary geology: Earth Science Reviews, v. 87, p.134-169.

Ries, J.B., Stanley, S.M. and Hardie, L.A., 2006, Scleractinian corals produce calcite, and grow more slowly, in artificial Cretaceous seawater: Geology, v. 34, p. 525-528.

Saller, A.H., Noah, J.T., Ruzuar, A.P. and Schneider, R., 2004, Linked lowstand delta to basin-floor fan deposition, offshore Indonesia: an analog for deepwater reservoir systems: American Association of Petroleum Geologists Bulletin, v. 88, p. 21-46.

Sandberg, P.A., 1983, An oscillating trend in Phanerozoic nonskeletal carbonate mineralogy: Nature, v. 305, p.19-22.

Stanley, S.M. and Hardie, L.A., 1998, Secular oscillations in the carbonate mineralogy of reef-building and sediment-producing organisms driven by tectonically forced shifts in seawater chemistry: Palaeogeography, Palaeoclimatology, Palaeoecology, v. 144, p. 3-19.

Tucker, M.E. and Wright, V.P., 1990, Carbonate Sedimentology: Blackwell Scientific Publications, Oxford, 482 p.

Vail, P.R., Mitchum, R.M., Jr. and Thompson, S., III, 1977, Seismic stratigraphy and global changes of sea level, Part four: global cycles of relative changes of sea level, *in* Payton, C.E., *ed.*, Seismic Stratigraphy-Applications to Hydrocarbon Exploration: American Association of Petroleum Geologists Memoir 26, p. 83-98.

Van de Poel, H.M. and Schlager, W., 1994, Variations in Mesozoic-Cenozoic skeletal carbonate minelogy: Geologie en Mijnbouw, v. 73, p. 31-51.

Webb, G.E., 1996, Was Phanerozoic reef history controlled by the distribution of non-enzymatically secreted reef carbonates (microbial carbonate and biologically induced cement)?: Sedimentology, v. 43, p. 947-972.

18. Carbonate Slopes

Ted E. Playton, Department of Geological Sciences, Jackson School of Geosciences, University of Texas at Austin, 1 University Station C1100, GEO 3.216, Austin, Texas, 78712 USA, and Chevron Energy Technology Company, 6001 Bollinger Canyon Road, D-1256; San Ramon, CA, 94583-2324, USA

Xavier Janson, Bureau of Economic Geology, Jackson School of Geosciences, University of Texas at Austin, University Station, Box X; Austin, Texas, 78713-8924, USA

Charles Kerans, Department of Geological Sciences, Jackson School of Geosciences, University of Texas at Austin, 1 University Station C1100, GEO 3.216; Austin, TX, 78712, USA

INTRODUCTION

Significance

Carbonate slopes are volumetrically significant parts of carbonate platforms and contain stratigraphic records which, although preserved differently than platform-top or platform-edge sediments, reflect the growth, evolution, and depositional conditions of the carbonate system. The broad grain size ranges, diversity of re-sedimentation processes initiated by different drivers, and complex stratal architecture of carbonate-slope systems result in high degrees of spatial and stratigraphic heterogeneity. This variability makes the development of depositional models and predictive relationships challenging. Improved understanding of carbonate-slope systems is of interest from both an academic and an applied standpoint because slope strata provide insights regarding carbonate platform accumulations, as well as predictive relationships for understanding coeval basinal and platform-top settings.

Table 1. Comparison of carbonate and siliciclastic slope characteristics

	Carbonate Slopes	Siliciclastic Slopes
Dominant Sediment Grain Size Range	mud to boulders (μms to 10s m)	mud to sand (μms to mms)
Sand Grain Characteristics	irregular to spherical shapes, primary intragranular and microporosity common	angular to spherical shapes, primary intragranular and microporosity uncommon
Mud Characteristics	aragonite needles, planktonic skeletal forms, less cohesive	platy micaceous forms, more cohesive
Dominant Sediment Sources	platform top, margin, slope, water column	hinterland, water column
Dominant Re-sedimentation/ Flow Processes	bedload and suspended load: rockfall, debris flow, (hyper) concentrated flow, turbidity flow, suspension	primarily suspended load: turbidity flow, suspension, lesser debris flow and concentrated flow
Early Lithification	common: submarine cementation and biological binding	uncommon
Potential for Brittle Failure and Gravitational Collapse	high: early lithification and high gradients, coarse debris common	low: lack of lithification and common regrading, coarse debris uncommon
Common Maximum Slope Declivities	re-sedimented: 35-40° autochthonous: 90°	all re-sedimented: 3-6°
Sediment Dispersal	inherently line-fed, requires sediment focusing mechanism for downslope point source	inherently point-sourced with modification from strike reworking
Depositional Patterns	strike-continuous fine- to coarse-grained aprons on slope or at toe-of-slope; fan-channel complexes less common	strike-discontinuous toe-of-slope or basinal fan-channel complexes and fine-grained bypass slopes common

Approach

When compared to deep-water siliciclastic or carbonate platform-top settings, carbonate slopes are poorly understood owing to lack of continuous, high-quality outcrops, seismic imaging limitations, and less research due, in part, to historically poor hydrocarbon production. Despite these limitations, past research on carbonate-slope and basin systems has provided fundamentals on:

1. The range of deposit types, margin styles, and depositional profiles;
2. An appreciation of the differences between carbonate and siliciclastic slopes (Table 1);
3. Relationships between sediment characteristics and the slope profile; and
4. Sequence stratigraphic concepts from the modern and their limitations.

The generalized variations, however, that exist in carbonate-slope-deposit types and bed- to platform-scale architecture remain unclear, as do the controlling drivers behind such variations. This chapter presents our approach to simplify the depositional and architectural characteristics of carbonate slopes using classification schemes for the deposit types, two-dimensional large-scale stratal patterns, internal stacking architecture, and three-dimensional sediment distribution patterns. Uniform description of these systems allows for meaningful comparison in terms of key depositional and architectural attributes, isolation of variables, and ultimately, recognition of the controlling intrinsic and extrinsic drivers.

Key Breakthroughs

Deposits and Processes

Building upon pioneering deep-water siliciclastic studies, early deep-water carbonate research documented the re-sedimentation of material derived from shallow-water environments to the slope and basin, and presented recognition criteria for autochthonous *versus* re-sedimented, and shallow-*versus* deep-depositional environment interpretations (*e.g.*, Cook *et al.*, 1972). With this understanding, subsequent case studies and syntheses catalogued the spectrum of deposit types, facies associations, and bedding architecture observed in carbonate slope and basin systems, and identified the affects of early lithification on slope development (*e.g.*, Davies, 1977; James, 1981). In particular, the recognition of very coarse, chaotic debris deposits [such as megabreccias] on the slope from gravitational collapse of early-lithified, commonly margin-derived material introduced a re-sedimentation process and resulting deposit type not present in deep-water siliciclastic systems (Table 1; Mountjoy *et al.*, 1972).

Margin and Slope Morphology

Following the advances in deposit description and recognition, some deep-water carbonate studies shifted in focus to the morphological aspects and large-scale stratal relationships of shelf-to-basin profiles. Two end-member margin transitions were first delineated by McIlreath and James (1978):

1. Clinoformal, accretionary (depositional) margins (Fig.1A), where re-sedimented slope and margin facies interfinger; and
2. Escarpment (bypass) margins, where coeval slope and margin facies are physically disconnected by a surface of non-deposition upon which slope strata onlap (Fig. 1B).

Schlager and Ginsburg (1981) reported on modern accretionary and escarpment profiles of the Bahamas Archipelago, Caribbean, and also identified 'erosional' profiles, interpreted here as escarpments that undergo prolonged net removal of margin and slope sediment from contour current modification and/or repeated collapse (Fig. 2). In addition to current modification, other parameters were identified that influence the depositional profile and margin morphology, such as prevailing wind direction in particular (*e.g.,* Mullins, 1983). Modern leeward margins of the Caribbean are mostly accretionary and net-progradational owing to offbank sediment transport, whereas modern windward margins are net-aggradational escarpments because of dominant onbank sediment transport.

Along strike, modern Bahamian slopes were shown to develop as laterally extensive sediment aprons composed of smaller scale sheets, lobes, and channelform sedimentary bodies, exemplifying line-fed systems and the *apron model* (Fig. 3; Mullins and Cook, 1986). This is due to the margin-parallel morphology of carbonate sediment factories that contribute material downslope, in contrast to inherently *point-sourced* siliciclastic systems that form deep-water channel–fan complexes rather than aprons.

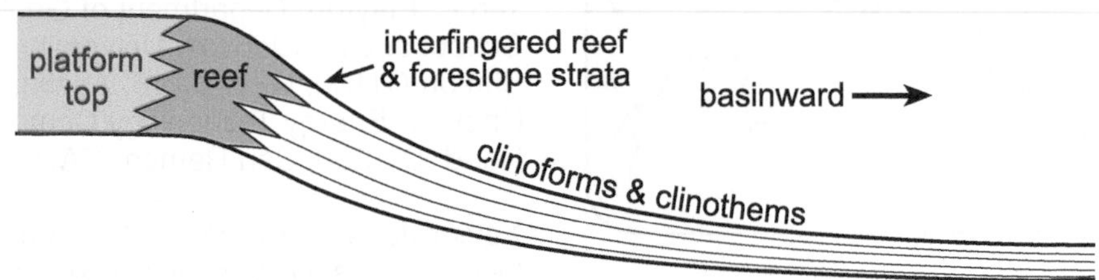

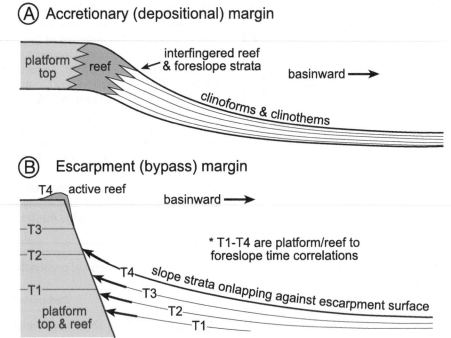

Figure 1. A. Accretionary (depositional); and **B.** escarpment (bypass) margins. **(A)** and **(B)** modified after McIlreath and James, 1978.

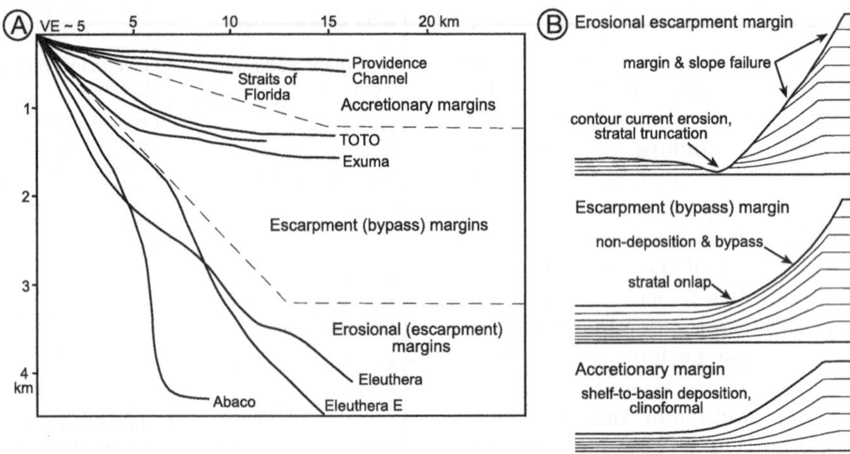

Figure 2. Slope profiles and stratal models of accretionary, escarpment, and erosional profiles of the Bahamas, Caribbean. **A.** Accretionary, escarpment, and erosional depositional profiles showing variations in slope angle and slope height; **B.** Models of accretionary, escarpment, and erosional profiles, highlighting variations in stratal architecture. **(A)** and **(B)** modified after Schlager and Ginsburg, 1981.

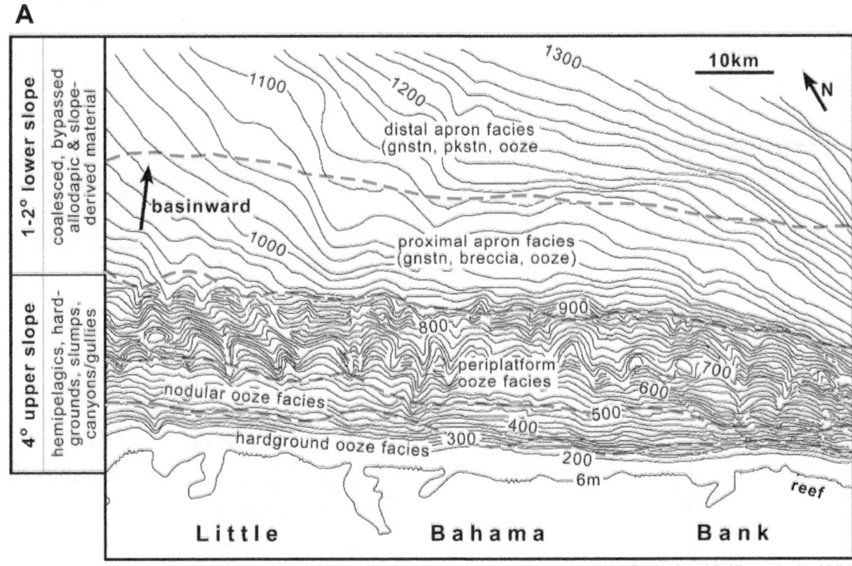

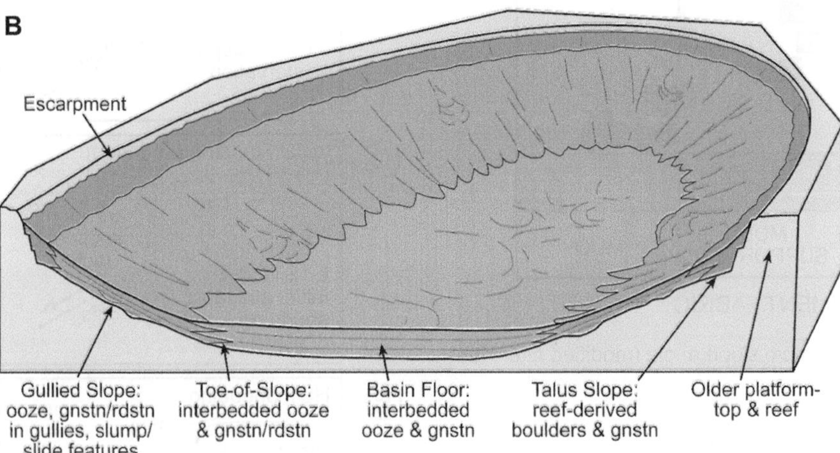

Figure 3. Modern strike-continuous aprons of the Bahamas, Caribbean. **A.** Plan view of gullied slope system with bathymetric contours and facies tract boundaries (dashed red lines), northern Little Bahama Bank, Bahamas (modified after Mullins *et al.,* 1984); **B.** Conceptual geologic block model of escarpment margins and strike-continuous foreslope aprons around the cul-de-sac of the Tongue of the Ocean, Bahamas (modified after Schlager and Chermak, 1979).

Among studies that quantified slope attributes (*e.g.*, Schlager and Camber, 1986), Kenter (1990) showed a correlation between slope-sediment fabric and slope angle where the angle-of-repose of deposits increases with decreasing mud content, except in cases of significant biological binding and slope channelization (Fig. 4). Later studies further related clinoform shape and curvature to sediment distribution along the slope profile (*e.g.,* Adams and Schlager, 2000).

Sequence Stratigraphic Concepts

Well-constrained Pleistocene–Holocene age dates and sea-level curves allowed for the incorporation of sequence stratigraphic principles into carbonate slope and basin models, and the goal of linking deposit types and stratal architecture to accommodation changes. Analyses from the Caribbean demonstrated the concept of *highstand shedding*, where platform-top sediment factories flourish and shed material downslope during platform submergence (highstands of eustatic sea level), and conversely are shut down during platform exposure (lowstands of eustatic sea level) leading to less sediment accumulation on the slope (Fig. 5; *e.g.*, Droxler and Schlager, 1985). Grammer and Ginsburg (1992) documented the high-frequency systems tract evolution over the last glacial–interglacial cycle of the Tongue of the Ocean proximal slope, Great Bahama Bank, and (in addition to further validating the highstand-shedding model) showed the effects of accommodation change on margin collapse and slope starvation (Fig. 6).

These examples from the modern, however, represent tropical carbonate platforms during peak icehouse conditions, and are not directly applicable to all examples, especially in the case of the timing and triggering mechanisms of gravitational collapse. Intrinsic triggers for collapse can include:

1. Basinward margin accretion beyond the angle of repose (*e.g.*, Ginsburg *et al.*, 1991);
2. Mechanical erosion and chemical weathering during exposure *(e.g.,* Grammer and Ginsburg, 1992);
3. Release of pore pressure related to exposure (Spence and Tucker, 1997);
4. Increased pore pressure during transgression (George *et al.*, 1995); and
5. Differential compaction of slope and basin sediments during progradation (Hunt and Fitchen, 1999).

Extrinsic collapse triggers, such as tectonic seismicity, tsunamis, and catastrophic storms, are independent of accommodation change (*e.g.*, Mutti *et al.*, 1984). Finally, oligophotic boundstone factories on the upper portions of the slope can extend well below the euphotic zone and influence of sea level (> 300 m water depth), operating dominantly through autogenic growth and collapse that is somewhat independent of high frequency accommodation changes (*e.g.,* Della Porta *et al.*, 2003).

Terminology

Nomenclature used in the description of carbonate slope and basin systems has been historically inconsistent due to the overlap of process-based and morphological attributes that characterize them. Thus, definition and standardization of terms are warranted and outlined below (Fig. 7):

- *Platform top, platform edge, slope*

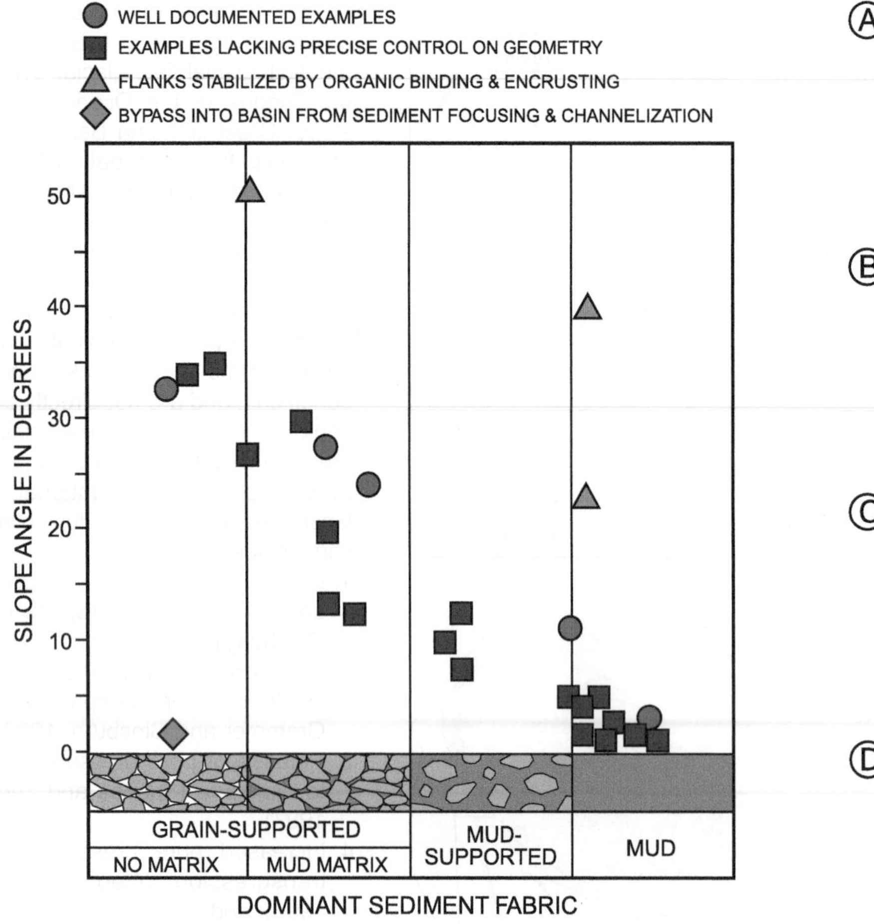

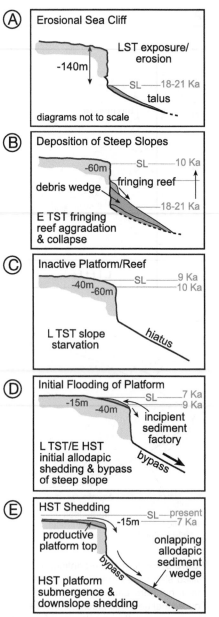

Figure 4. Plot of slope sediment fabric versus maximum slope angle (modified after Kenter, 1990).

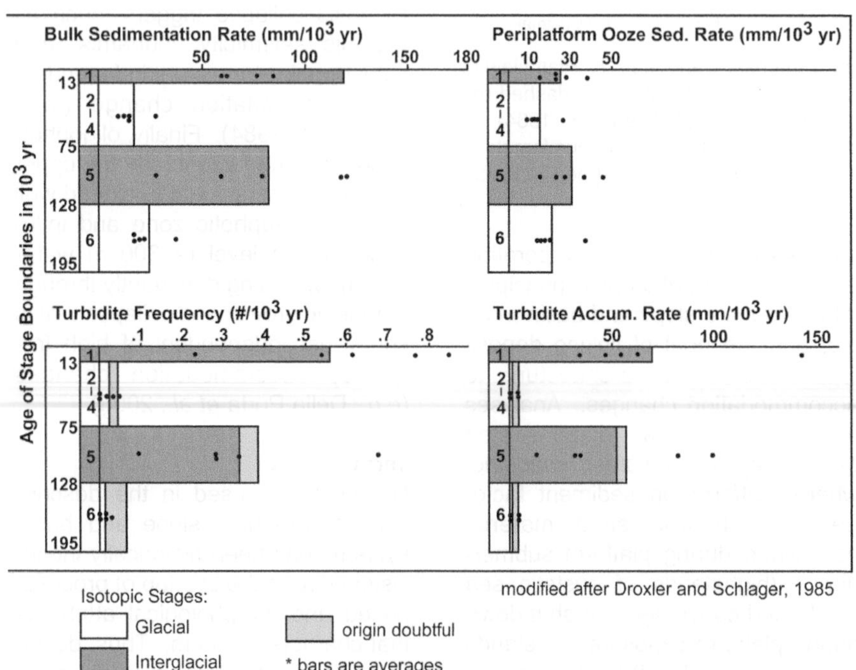

Figure 5. Graphs of sedimentation rates and bed frequency of resedimented platform-derived material during interglacial versus glacial periods. Age dates are constrained by stable isotopic analyses (modified after Droxler and Schlager, 1985).

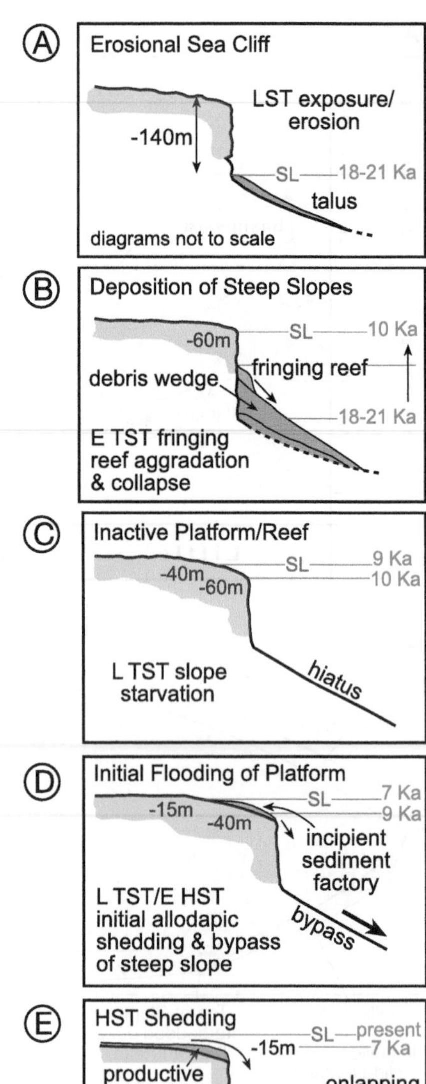

Figure 6. Schematic diagrams depicting the evolution of the platform margin and proximal foreslope during the last Holocene glacial-interglacial cycle, Tongue of the Ocean, Bahamas. Deposit types and their timing are linked to platform-top exposure/submergence and changing eustatic sea level (modified after Grammer and Ginsburg, 1992).

or slope profile, toe-of-slope, and basin are used here as spatial or geometrical divisions and positions along the depositional profile. In general terms, the *platform edge* (greatest seaward increase in gradient) and *toe-of-slope* (seaward transition from inclined to flat-lying strata) represent the major inflections of the profile. The *slope, or slope profile*, is the inclined part of

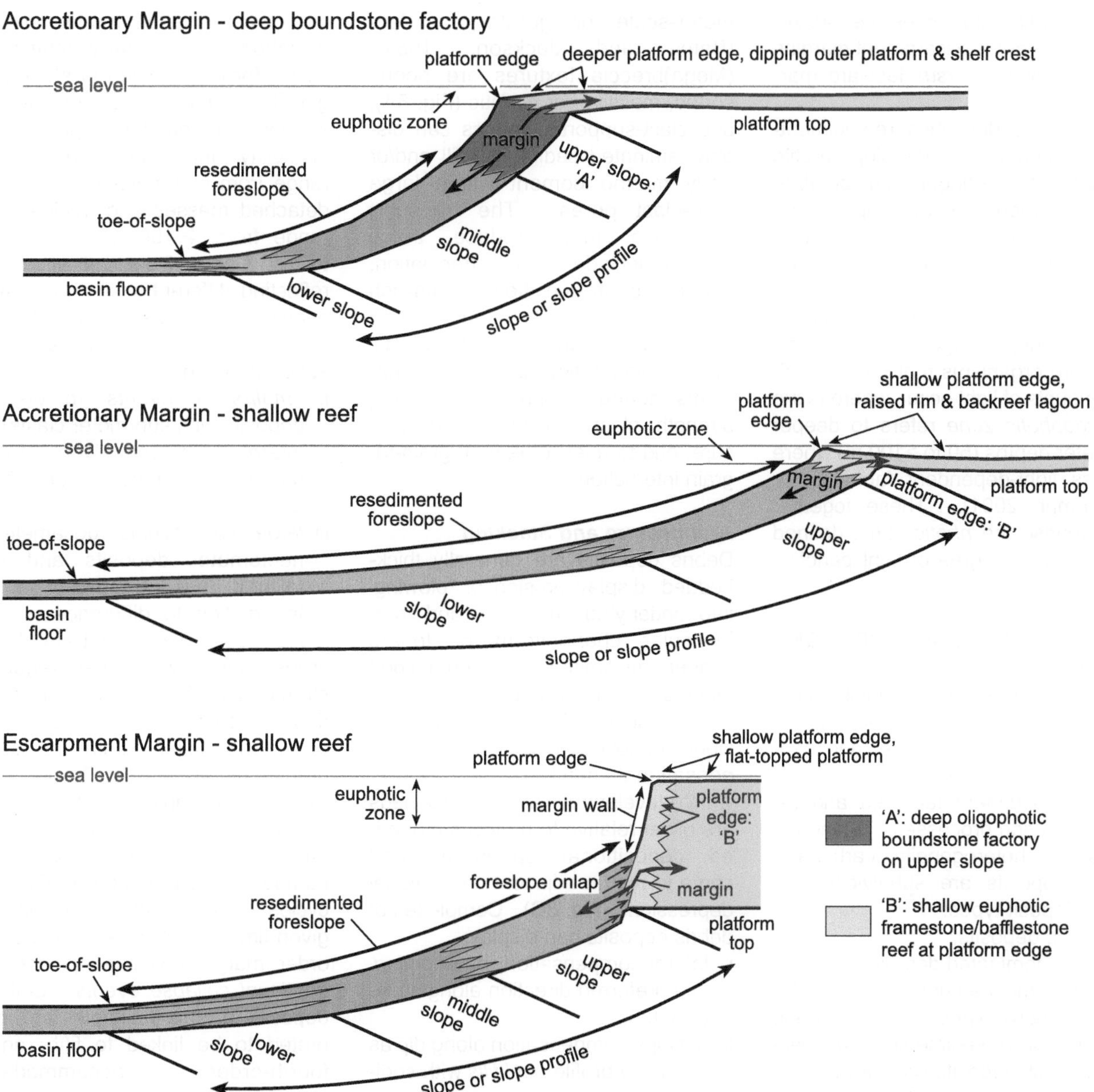

Figure 7. Terminological breakdown for some typical generalized carbonate shelf-to-basin depositional models. Foreslope facies colors (red, blue, and gray) are arbitrary.

the depositional profile (between the platform edge and toe-of-slope), and is best subdivided into facies tracts according to distinctive deposits, processes, and architecture (*i.e., upper-, middle-,* and/or *lower slope* facies tracts).

- *Reef* is used here as a relatively early-lithified, autochthonous biological accumulation that constructs relief above the seafloor and usually occurs around the upper slope, platform edge, or on the platform top (Chapter 17). *Reefs*, as used here, are dominantly restricted to shallow-water environments (<50 m water depth), and composed of skeletal frameworks producing framestone and bafflestone fabrics. Amalgamated biostromes of early-lithified oligophotic boundstone (typically bindstone and bound framestone fabrics) that occur in deeper upper slope settings (generally 30–300 m water depth) are not here considered '*true reefs*', and accordingly are referred to as *deep boundstone factories*.
- *Margin* encompasses the transition from platform-top to re-sedimented foreslope environments, often including the interfingering of outer-platform, reef, and upper-slope settings. Thus, the *platform edge* is considered to be the central feature of the *margin*. *Margins* can also be non-reefal and/or non-lithified (*i.e.*, shoal-dominated) and without seaward interfingering of environments (*i.e.*, erosional escarpments). *Margin wall* is used in cases where the uppermost portions of depositional profiles exhibit near-vertical declivities, which form through both depositional and erosional processes. Finally, types of *margins* describe configurations over time (*i.e.*, prograding *versus* retrograding margins) and/or pronounced geometrical

and stratal differences (*i.e.*, accretionary *versus* escarpment margins or windward *versus* leeward margins).
- *Foreslope* describes re-sedimented material along the slope profile that can interfinger with, or abut, environments on the upper slope and around the platform edge (boundstone factory, reefal, and non-reefal).
- The *euphotic zone* defines shallow-water depths (generally <50 m) where organisms that require high amounts of light occur, whereas the *oligophotic zone* refers to deeper water depths (50 to >100 m) where less light-dependent fauna occur (Pomar, 2001). These together comprise the *photic zone*, defined by 'some' degree of light penetration.

DEPOSIT TYPE CLASSIFICATION

Introduction

Carbonate foreslope deposits comprise a wide spectrum of sedimentological texture, grain size, and bedding geometry owing to a diverse suite of sediment factories and re-sedimentation processes. To encompass this heterogeneity, carbonate-slope deposits are subdivided into three deposit types:
1. Debris deposits;
2. Grain-dominated deposits; and
3. Mud-dominated deposits (Table 2).

Such deposit types include discrete families of re-sedimented sediment gravity-flow deposits with similar sedimentary characteristics, bedding styles, re-sedimentation processes, and sources.

Debris Deposits

Deposits

Debris deposits are generated from gravitational collapse and coherent mass wasting of early lithified material mostly derived from lithified margin environments (Table 2). Debris deposits (Fig. 8) range in dominant grain size from cobble to boulder (0.1 to > 50 m), and consist of clast- to matrix-supported (mega)breccia, isolated blocks, boulder complexes (blocks amassed together), and olistoliths (detached masses of strata generally decameters or greater across; Fig. 8). Megabreccias are deposits that have an abundance of meter-scale or greater boulders (Bates and Jackson, 1984). (Mega)breccia textures are poorly sorted, massive, or chaotic (Fig. 8A), and clast-supported fabrics can display laminated sediment infill and/or early marine cement within large inter-clast pores. The transport processes that produce these deposits include rockfall, translation, hyperconcentrated flow (grain-rich matrix), and debris flow (fine-grained matrix) (Mulder and Alexander, 2001). The dominant flow support mechanisms during transport are matrix strength, clast buoyancy, pore pressure, and to a lesser extent, grain-to-grain interaction.

Architecture and Stacking

Debris deposits are generally thick-bedded, display erosion or 'plowing' into underlying substrate, and form lenticular channelform or lobate shapes with limited strike and dip bed lengths (meters and decameters minimum, respectively; Table 2; Fig. 8C). Debris deposits typically display compensational stacking architecture, which develops when lenticular beds are offset relative to previous mounded, depositional topography, and come to rest in adjacent depositional depressions (Fig. 8C). Complexes of debris deposits can display:
i. lateral compensation, or *shingling*, in a preferred direction along strike (Fig. 8c);
ii. upslope compensation along dip as the slope profile *backfills* with sediment; and
iii. compensation with no particular organization.

Debris deposits can also occur as isolated, discontinuous tongues or wedges on the slope or basin floor, forming depositional topography that influences subsequent sedimentation. Debris deposits in debris-poor foreslopes typically consist of sub-meter lenses and boulders on the proximal foreslope that are sparsely intercalated within grain- or mud-rich strata.

Re-sedimentation - Collapse Scale and Frequency

Brittle collapse processes, and thus the source factories for debris deposits, are usually limited to lithified margin environments, where early lithification of sediment occurs from marine cementation and biologic stabilization. This early lithification allows for construction of steep margin declivities (up to vertical) that enhance the potential for gravitational instability. Individual debris deposits range from sub-meter scale beds to detached masses and boulders that range from 4–100 m across (olistoliths; Gomez-Perez *et al.*, 1999), reflecting different scales of collapse events. On the basis of field observation, collapse events can be usefully subdivided into:
i. *small-scale events* (meter-scale deposits with sub-meter clasts);
ii. *intermediate-scale events* (multi-meter-scale deposits and boulders); and
iii. *large-scale events* (decameter to hectometer deposits and olistoliths).

In addition to differences in collapse scale, well-constrained outcrops that have robust sequence stratigraphic frameworks show that debris deposits occur at different relative frequencies within the stratigraphic hierarchy (*e.g.*, Tinker, 1998; Kerans and Tinker, 1999). Some examples exhibit substantially greater numbers of collapse events than the number of high-frequency, fifth-order cycles on the platform-top within a given time period, reflecting sub-fifth-order margin collapse episodes, or *continual collapse events*. Collapse deposits in other examples are interpreted to be linked to fifth- and/or fourth-order accommodation changes, and represent *episodic collapse events*. Furthermore, several outcrop datasets record rare failure episodes that occur at third- or second-order-scale periodicities, and are referred to as *non-typical events*.

As observed from well-constrained outcrop data, a simple relationship exists between collapse event scale and relative frequency, reflecting different magnitudes of instability, temporal duration required to develop (and recover from) instability, and triggering mechanism. The general trend is as follows:
- *Continual collapse events* are generally *small-scale*, sub-fifth-order in periodicity, and tied to autogenic growth and failure processes (somewhat independent of relative sea-level changes).

Table 2. Characteristics of debris, grain-dominated, and mud-dominated foreslope deposits

	Debris Deposits	Grain-Dominated Deposits	Mud-Dominated Deposits
Fabrics/ Textures	**clast- & block-dominated**; v. coarse clast- and matrix-supported (mega) breccias, isolated boulders, boulder complexes, olistoltihs	**sand- and gravel-dominated;** fine to coarse gnstn and grainy pkstn; intraclastic and bioclastic gnstn and rdstn	**clay- and silt-dominated**; mudstone, siltstone, chalk, shale, marl; wackestones & muddy packstones (turbidites)
Dominant Grain Size Range	**cobble to boulder** (10s cm to 10s m; excluding incorporated matrix)	**very fine/fine sand to pebble** (100μm to cms)	**clay to silt** (< 100μm; excluding intermixed grains in turbidity flows)
Sedimentary Features	very poorly sorted; massive, chaotic or ungraded (except for floating clasts at bed cap); laminated sediment infill in clast-supported breccias	well- to poorly-sorted; normal- to inverse-grading or massive; low-angle to planar lamination; partial Bouma sequences (commonly Ta-Tb); imbricated clasts, aligned grains	burrowing/bioturbation; hardgrounds, firmgrounds, differential cementation, and encrustation; chert; partial to full Bouma sequences when sand grains present; parallel lamination
Bedding Style	thick-bedded; **lenticular, erosive**, channelform/lobate; lateral and upslope compensational stacking; strike and dip lengths <10m-100s m; aspect ratios can be 1:1	thin- to medium-bedded; slightly lenticular to **tabular sheets**; erosive to non-erosive; low-angle compensational stacking; strike and dip lengths 10s-100s m; aspect ratios 1:10s or 1:100s	laminated to thin-bedded; **drapes and blankets**; nodular/irregular; strike and dip lengths up to kms
Architecture and Facies Association 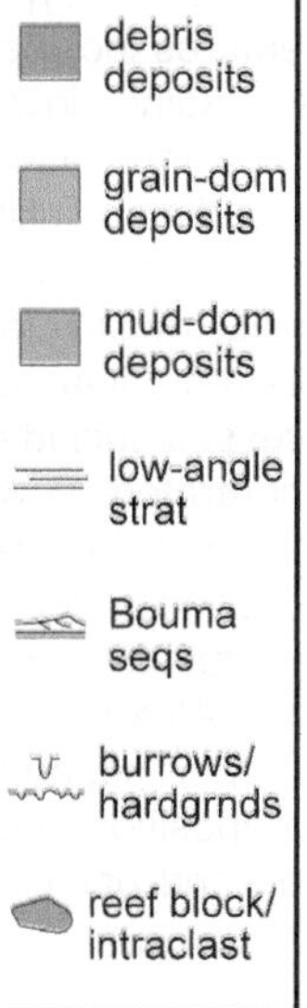 debris deposits grain-dom deposits mud-dom deposits low-angle strat Bouma seqs burrows/ hardgrnds reef block/ intraclast	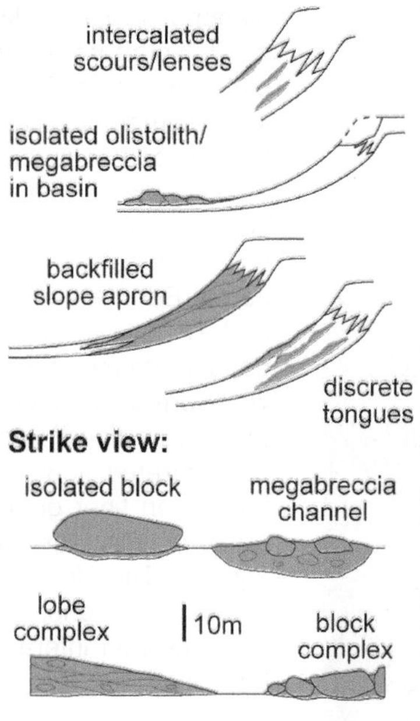 **Dip view:** intercalated scours/lenses isolated olistolith/ megabreccia in basin backfilled slope apron discrete tongues **Strike view:** isolated block megabreccia channel lobe complex 10m block complex	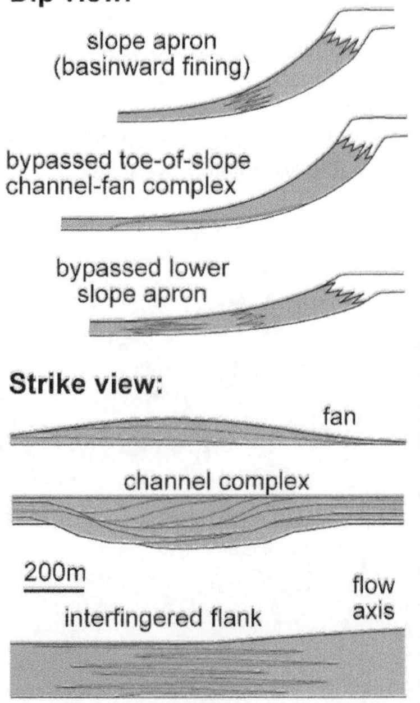 **Dip view:** slope apron (basinward fining) bypassed toe-of-slope channel-fan complex bypassed lower slope apron **Strike view:** fan channel complex 200m interfingered flank flow axis	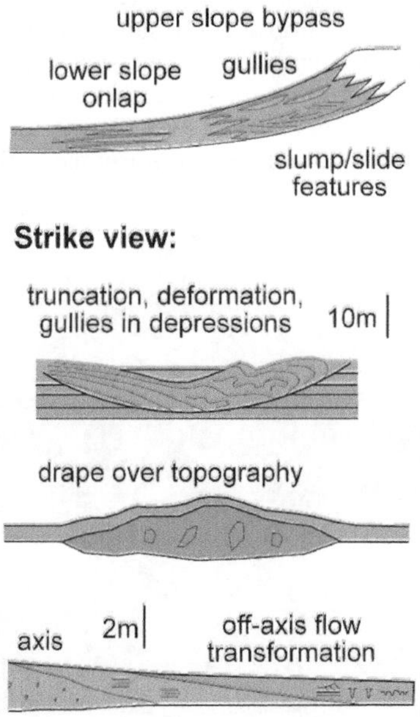 **Dip view:** upper slope bypass lower slope onlap gullies slump/slide features **Strike view:** truncation, deformation, gullies in depressions 10m drape over topography axis 2m off-axis flow transformation
Transport Processes	rockfall, hyperconcentrated flow, debris flow	hyperconcentrated (grain) and concentrated flow	suspension and turbidity flow
Flow Support	matrix strength, pore pressure, buoyancy	grain-to-grain interaction, buoyancy	fluid turbulence
Source Factory	early-lithified **platform edge and upper slope (margin) environments**, lesser outer platform material	non-skeletal & skeletal grains, bioclasts, and intraclasts from moderate to **high energy platform-top and reefal environments** (excluding peloids & pellets)	**low energy lagoonal or platform interior environments** and **water column**
Resedi-mentation Process	**brittle failure/gravitational collapse of early-lithified material**	**offbank sweeping from waves and tidal or storm currents**	**offbank sweeping from wave, tidal, and storm currents (periplatform); water column fallout (pelagic)**; flow transformation in off-axis positions

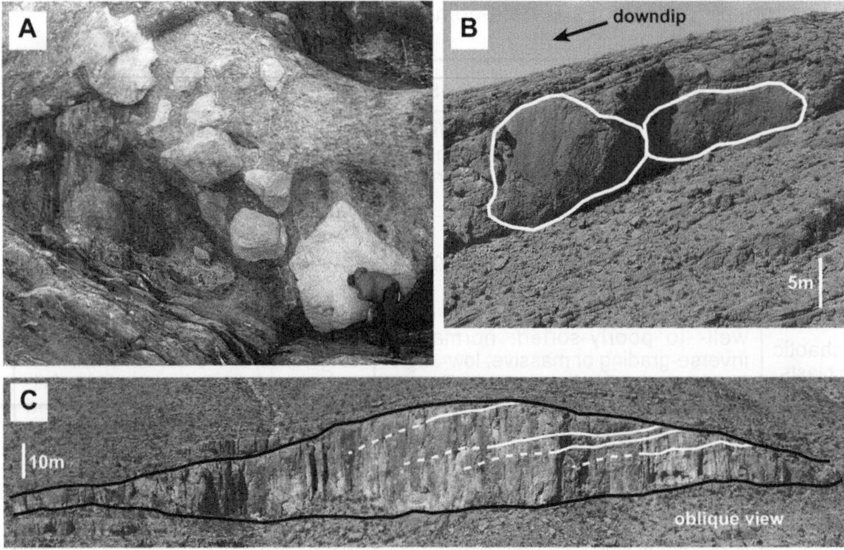

Figure 8. Debris deposit textures and morphology. **A.** Matrix-supported megabreccia with large margin-derived blocks overlying and eroding interbedded slope mudstones and rudstones, Upper Cambrian, Cow Head Group, western Newfoundland, Canada; **B.** Coral-sponge-microbial boundstone boulders (white outlines) encased in allodapic grainstone, Pliensbachian (Lower Jurassic), High Atlas, Morocco; **C.** Oblique strike view of lenticular margin-derived megabreccia complex with mounded cap, slightly erosive base, internal shingling geometries (white lines), and axial bedding amalgamation (white dashed lines), Lower Leonardian (Lower Permian), Sierra Diablo Mountains, west Texas.

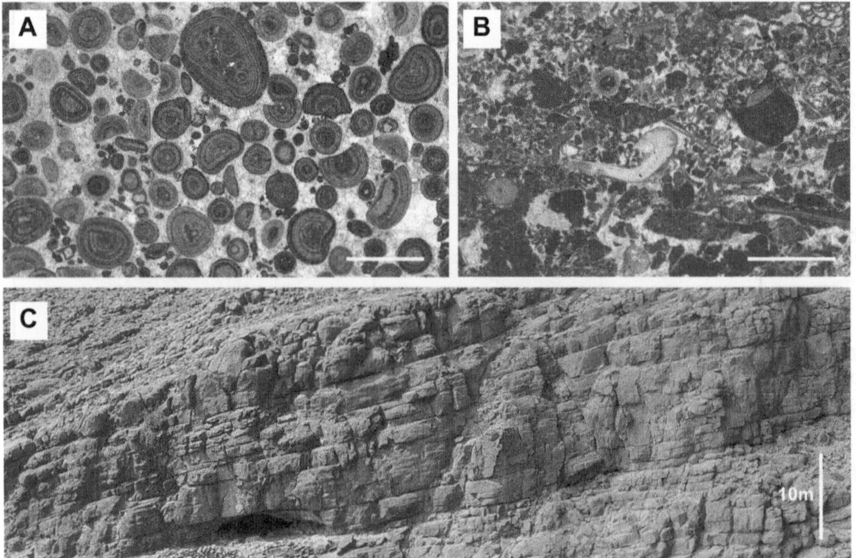

Figure 9. Grain-dominated deposit textures and bedding style. **A.** Moderately-sorted pisoid-peloid-ooid grainstone, Famennian (Upper Devonian), Canning Basin, Western Australia; **B.** Poorly-sorted peloid-intraclast-bioclast grainstone/ rudstone, Guadalupian (Upper Permian), Guadalupe Mountains, west Texas; **C.** Tabular bedding and sheet-stacked architecture within grain-dominated foreslope apron, Pliensbachian (Lower Jurassic), High Atlas, Morocco.

- *Episodic collapse events* are generally *small to intermediate scale*, occur at fifth- or fourth-order periodicities, and tied to relative sea-level changes.
- *Non-typical collapse events* are generally *intermediate to large scale*, occur at the greatest periodicities (third- to second-order), and are tied to either long-term changes in sequence architecture (*i.e.*, second-order maximum flooding surfaces) or extrinsic failure triggers.

Grain-Dominated Deposits

Deposits

Grain-dominated deposits comprise deposits generated by downslope transport of sand- to gravel-sized particles that are mostly derived from platform-top and platform-edge settings (Table 2). Grain-dominated deposits range in grain size from very fine sand to pebbles (100μm to cm), and consist of grain-dominated packstones, grainstones, and rudstones with variable degrees of sorting (Fig. 9A and B). Intraclasts and bioclasts up to decimeters across are common, and indicate reworking of lithified substrates and fragmentation of macrofaunal communities, respectively (Fig. 9B). Grain composition ranges from purely non-skeletal, derived from platform-top shoals or lagoons, to purely skeletal, often indicating platform-edge and/or outer platform origins (Fig. 9A and B). Typical sedimentological features include normal and/or inverse grading, low-angle to planar lamination, imbricated clasts, small-scale pore sediment infill [in rudstones], aligned grains, and partial Bouma sequences (mostly Ta and Tb divisions). Grain-dominated deposit transport processes include well-sorted hyperconcentrated flow (true grain flows), pebble-rich hyperconcentrated flow, and concentrated flow when finer grained fractions are present (Mulder and Alexander, 2001). The dominant flow support mechanisms prior to deposition are grain-to-grain interaction and particle buoyancy.

Architecture and Stacking

Grain-dominated deposits are thin- to medium-bedded, slightly- to non-erosive, tabular to slightly lenticular (Fig. 9C), and form sheets having relatively low aspect ratios of (1:10 to 1:200 thickness: width ratios) and greater dip and strike lengths (decameters to hectometers) than debris deposits (Table 2). Complexes of tabular grain-dominated deposits display broad, slightly compensational internal stacking (Fig. 9C), and occur as,

i. laterally unconfined aprons having gradual axis-to-flank compositional transitions; or
ii. laterally confined channel-fan complexes that have relatively abrupt axis-to-flank compositional transi-

tions.

Laterally unconfined aprons can span the entire slope profile and fine into the basin, or can occur only in lower slope to toe-of-slope settings indicating sediment bypass across the upper slope (slope apron *versus* base-of-slope apron; Mullins and Cook, 1986). By definition, grain-dominated aprons extend for great distances along strike (up to 10s of kilometers) reflecting the line-source nature of contributing sediment factories. However, subtle textural fining and interfingering relationships are observed internally, defining intra-apron axes and flanking positions. In contrast, channel–fan complexes occur in lower slope, toe-of-slope, or basinal settings (up to 15 km into the basin) indicating sediment bypass across the slope profile, and can exhibit sub-kilometer strike widths with sharp flanking contacts (*e.g.,* Savary and Ferry, 2004). Channel–fan complexes are also characterized by lower slope to toe-of-slope onlap, internal unidirectional shingling architecture, and erosive channelform morphologies on the slope that pass into mounded, non-erosive lobate bodies in distal settings (*e.g.*, Vigorito *et al.*, 2005). The strike-discontinuous nature of channel–fan complexes indicate updip sediment focal points, such as collapse-generated margin reentrants, that can funnel shallow-derived sediment downslope, promote channelization, and generate isolated bypassed accumulations in distal settings.

Re-sedimentation

Grain-dominated deposits are primarily derived from shallow platform-top and platform-edge settings that generate unconsolidated sand and gravel, such as high-energy shoals (Table 2). Shallow framestone and bafflestone reefs are also capable of generating considerable volumes of bioclastic to intraclastic rudstone and grainstone (in addition to debris deposits). Peloids (micritized grains and pellets) represent sand production in lower energy lagoon or platform interior environments that is mitigated by biogenic and chemical processes, rather than physical energy. Wave action, tidal currents, and storm currents, especially basinward tidal ebb flow and storm return flow, drive re-sedimentation of shallow-derived sand and gravel to the foreslope and basin. Such wave and tidal current processes are interlinked with the production and dispersal of sand and gravel through;

i. wave agitation required for coated non-skeletal grain factories;
ii. fragmentation of skeletal communities for bioclast production; and
iii. erosion of early-lithified substrates (*i.e.,* hardgrounds) for intraclast generation.

Morphological attributes of platforms that influence physical energy impingement, such as shelf width and platform-edge or shelf-crest barriers, are also interconnected with the production and downslope re-sedimentation of sand and gravel (discussed later).

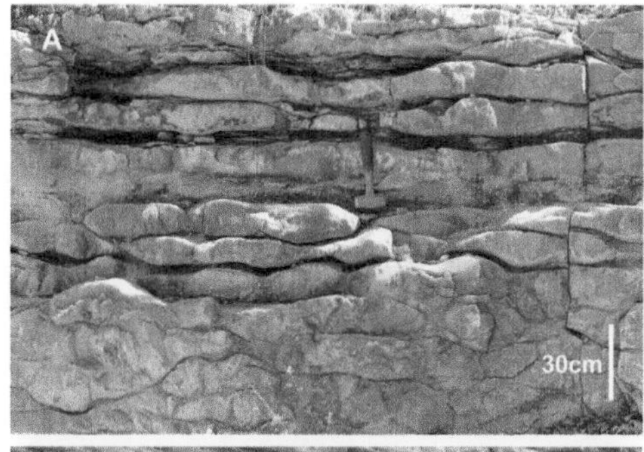

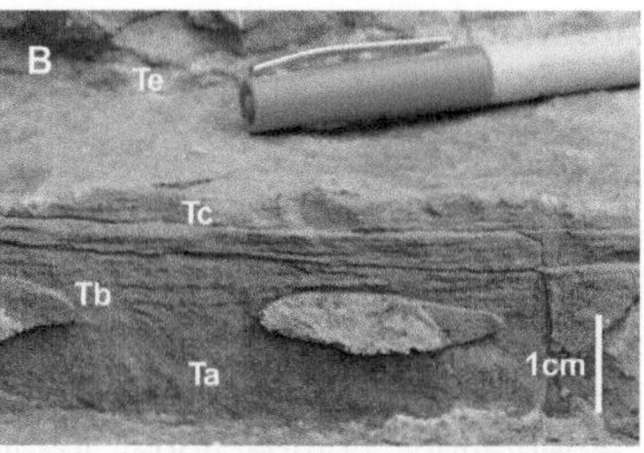

Figure 10. Mud-dominated deposit bedding styles and structures. **A.** Nodular bedding from bioturbation and differential cementation in skeletal-peloid wackestones, Guadalupian (Upper Permian), Guadalupe Mountains, west Texas; **B.** Peloidal packstone-to-wackestone turbidite with Bouma sequence divisions labeled, Lower Leonardian (Lower Permian), Sierra Diablo Mountains, west Texas. Resistant lenses are chert; **C.** Thin-bedded mudstone with recessive partings, Lower Ordovician, Cow Head Group, western Newfoundland, Canada; hammer is 30cm long.

Mud-Dominated Deposits

Deposits

Mud-dominated deposits consist of fine-grained deposits (dominated by clay- or silt-sized particles) that originate in protected platform settings (periplatform) and/or the water column (pelagic; Table 2). These deposits accumulate during,

i. periods of relative foreslope quiescence (background sedimentation);
ii. periods of fine-grained allodapic shedding (turbidity flows); and
iii. flow separation/transformation into off-axis, flanking positions.

Mud-dominated deposits range in dominant grain size from clay to silt (<60 μm), including mudstone, siltstone, chalk, and argillaceous (marly) textures (Fig. 10A). Wackestone and mud-dominated packstone textures are common as the gradational bases of calcareous turbidites (dominantly Ta divisions; Fig. 10B), and upright skeletal wackestone fabrics can develop from fine-grained blanketing and infiltration of autochthonous skeletal substrates. Mud-dominated deposits are generally;

i. parallel laminated or structureless;
ii. typically display signatures of omission (firmgrounds and hardgrounds), burrowing/ bioturbation, and/or chertification; and
iii. are associated with organic encrustation.

Turbidites can have partial to full Bouma sequences depending on the grain-size range (Fig. 10B). Mud-dominated deposit transport processes include pure suspension fallout, fluid turbulence, and combinations thereof, all of which entail suspension fallout as the dominant depositional mechanism (either from higher up in the water column or from turbulent plumes; Mulder and Alexander, 2001).

Architecture and Stacking

These deposits are thin-bedded or laminated (Fig.10C), often wavy- or nodular-bedded (Fig. 10A), can drape foreslope topography, and can blanket deep-water seascapes with extensive bed lengths along strike and dip (kilometers; Table 2). However, early cementation, compactional loading, and dewatering can result in foreslope readjustment and the formation of slump/slide complexes along distinct intraformational truncation or shear translation surfaces (*e.g.*, Davies, 1977; Coniglio, 1986). These failure features form topographic irregularities on the slope that can evolve into gullies or small canyons, and serve as conduits for subsequent bypassing material (*gullied slopes*; Fig. 3; Schlager and Chermak, 1979; Mullins *et al.*, 1984). These phenomena are well documented on modern slopes of the Bahamas where slump-prone mud-dominated facies and gullies dominate upper-slope environments, and bypassed aprons relatively enriched in grain-dominated deposits occur in lower- to toe-of-slope settings (Fig. 3).

Re-sedimentation

The clay- to silt-sized particles that dominate mud-dominated deposits are derived mostly from platform-top environments and the water column, forming periplatform and pelagic deposits, respectively (Table 2). Mud-producing platform-top factories are primarily biotic communities in low-energy, protected lagoon or platform-interior environments and, to a lesser extent, can include shallow-water chemical/microbial precipitates, submarine bioerosion, and mechanical reworking. Waves, tidal and storm currents transport allodapic material offbank either in suspension or as turbidity flows, depending on sediment concentration and the fractions of other incorporated grain sizes. Pelagic deposition indicates ambient suspension fallout from the water column of dominantly planktonic microfossil communities (carbonate and other), products from hinterland weathering, and fine volcaniclastics. There is a temporal constraint on carbonate pelagic communities because the dominant contributors, coccolithophorids and planktic foraminifera, evolved in the Mesozoic (Fischer and Arthur, 1977). Thus, the potential for mud-dominated deposits in the rock record is greater in Mesozoic to recent slopes than in those of the Paleozoic, when carbonate mud sources were primarily confined to protected platform environments, microbial settings, mechanical erosion, and chemical precipitation.

STRATAL PATTERNS AND SPATIAL ARCHITECTURE

Large-scale Stratal Patterns

Introduction

Differentiation of accretionary (depositional) and escarpment (bypass) margins (McIlreath and James, 1978) is an essential first-order observation in terms of large-scale geometries (Figs. 1 and 11). The two profiles display fundamentally different large-scale margin transitions and stratal patterns that indicate differing ratios of material exported downslope relative to the volume of material required to fill the slope profile. Despite these differences, accretionary and escarpment margins can pass along strike from one another (*e.g.*, windward and leeward margins of the Great Bahama Bank; Mullins, 1983), inferring significant lateral variation in the sediment factories and their degree of contribution to the foreslope. The terms 'accretionary' and 'escarpment' margins are preferred here, because both end-members involve some level of bypass and depositional processes. 'Erosional' margins are here considered as a specific type of escarpment with distinctive stratal truncation characteristics and implications regarding oceanographic effects (*e.g.,* Hooke and Schlager, 1980).

Escarpment Margins

Escarpment margins are decoupled systems where coeval margin and re-sedimented foreslope environments are separated by a surface of non-deposition, or a bypass surface (Fig. 11a and b). This configuration occurs when the volume of foreslope sediment required to fill the slope profile is not available and/or is not re-sedimented downslope, due to,

i. greater rates of vertical margin aggradation than foreslope accumulation; and/or
ii. large (hectometers to kilometers), inherited slope heights from platform nucleation on antecedent topography (Fig. 11a, b).

Escarpment margin foreslopes are characterized by stratal onlap against palimpsest, lithified surfaces (often former margins or foreslopes; Fig. 11a and b). Escarpment margin configurations are aggradational or retrogradational, and are 'unable to prograde' until sufficient foreslope substrate is present to accommodate an overlying, advancing load (Fig. 11a and b). Escarpments are subdivided into two classes that vary in terms of stratal evolution:

1. *Inherited escarpments*, defined by the progression from escarpment-to-accretionary configurations over time (Fig. 11a), and
2. *Growth escarpments*, defined by the progression from accretionary-to-escarpment configurations over time (Figs. 11b and 12).

Inherited escarpments are defined by continuous onlap of foreslope strata against a single, long-lived escarpment surface, indicating a sustained inability for the contributing sediment factories to completely fill the slope profile with re-sedimented material (Fig. 11a). These geometries result from platform nucleation on high-relief antecedent topography (slope heights up to a kilometer or more), such as fault blocks or relict carbonate platforms, whereby the system originates and operates as an escarpment until the basin fills. Prior to basin infill, the margin is unable to prograde (even under low accommodation conditions) due to the absence of foreslope substrate, and will consequently collapse producing aggradational or retrogradational margin configurations (some erosional) regardless of accommodation setting (Fig. 11a). Upon complete basin infill and a substantial decrease in original slope height, accretionary margin configurations and progradation are possible (Fig. 11a; Eberli *et al.*, 1993).

By contrast, *growth escarpments* display progression from accretionary to escarpment stratal patterns, and nucleate on gently dipping surfaces that commonly postdate substantial platform backstepping (Fig. 11b). This evolution requires sustained periods of upward-directed margin growth, such that

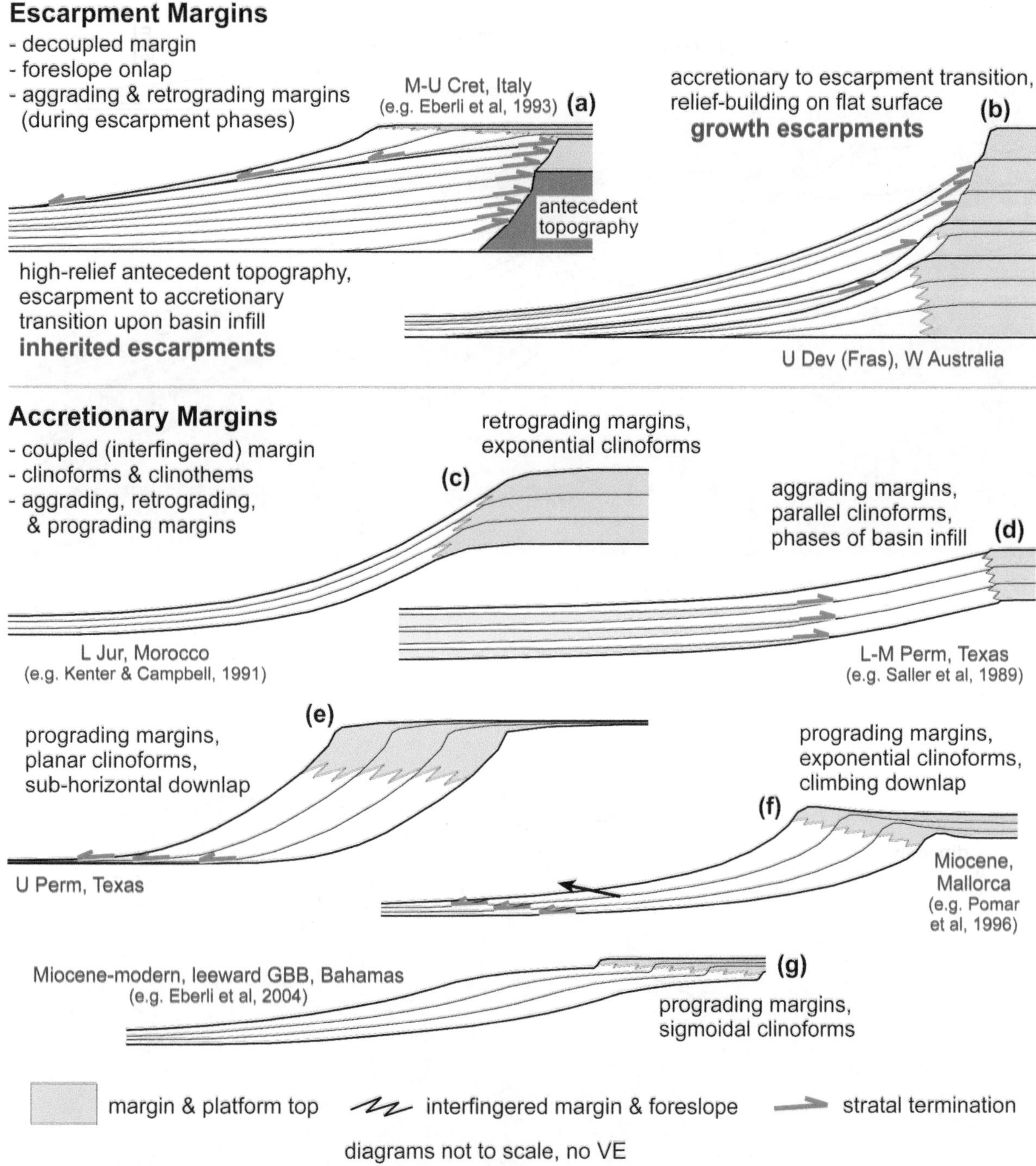

Figure 11. Large-scale stratal architecture and lap terminations of different escarpment and accretionary margins based on actual outcrop and subsurface examples. Escarpments (**a-b**) show variations in stratal evolution, and accretionary margins (**c-g**) show variations in margin configuration, clinoform shape/curvature, and downlap pattern.

i. initial synoptic relief is constructed; and
ii. continued upbuilding of the margin eventually outpaces coeval foreslope accumulation by increasing shelf-to-basin relief (Schlager, 1981).

Consequently, growth escarpments characteristically exhibit aggradational to retrogradational configurations, increasing degrees of foreslope onlap against older surfaces, and decreasing foreslope thickness over time (Figs. 11b and 12). The overall change from accretionary to escarpment configurations is gradational, resulting in smaller scale alternations and internal stratal complexity (Figs. 11b and 12). Margin and/or foreslope declivities generally increase throughout growth escarpment evolution, and margin walls with sub-vertical dips can develop during peak or terminal escarpment phases (Figs. 11b and 12).

Accretionary Margins

Accretionary margins are characterized by interfingering of relatively coeval margin and foreslope environ-

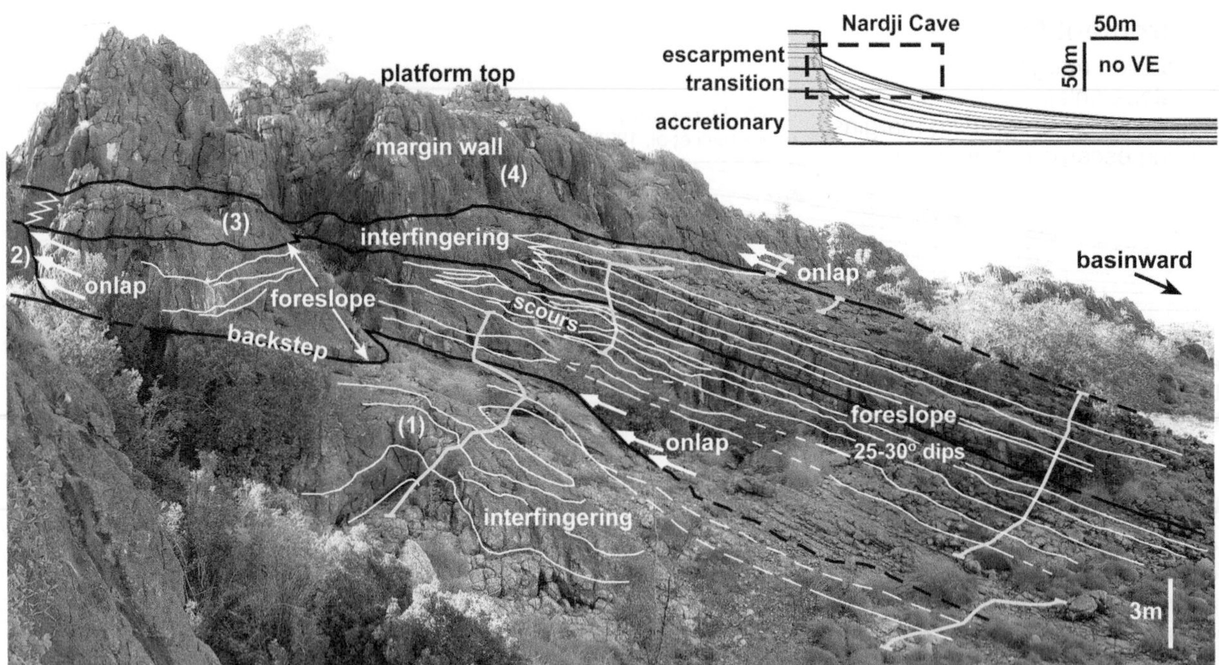

Figure 12. Stratal architecture at the accretionary-escarpment margin transition of the Nardji Cave growth escarpment, Lower Frasnian (Upper Devonian), Canning Basin, Western Australia. Discrete phases of margin-foreslope development and configuration are outlined in black and numbered in white. Phase 1 displays accretionary configurations with interfingering reefal boundstone and foreslope deposits. The margin in Phase 2 backstepped, aggraded, and thereby outpaced the coeval foreslope forming an onlapping, escarpment configuration. Phase 3 is characterized by slight margin progradation and accretionary margin-foreslope interfingering. Phase 4 is strongly aggradational and forms a sub-vertical escarpment margin wall, marking the end of the transition from accretionary geometries. Yellow lines are measured sections.

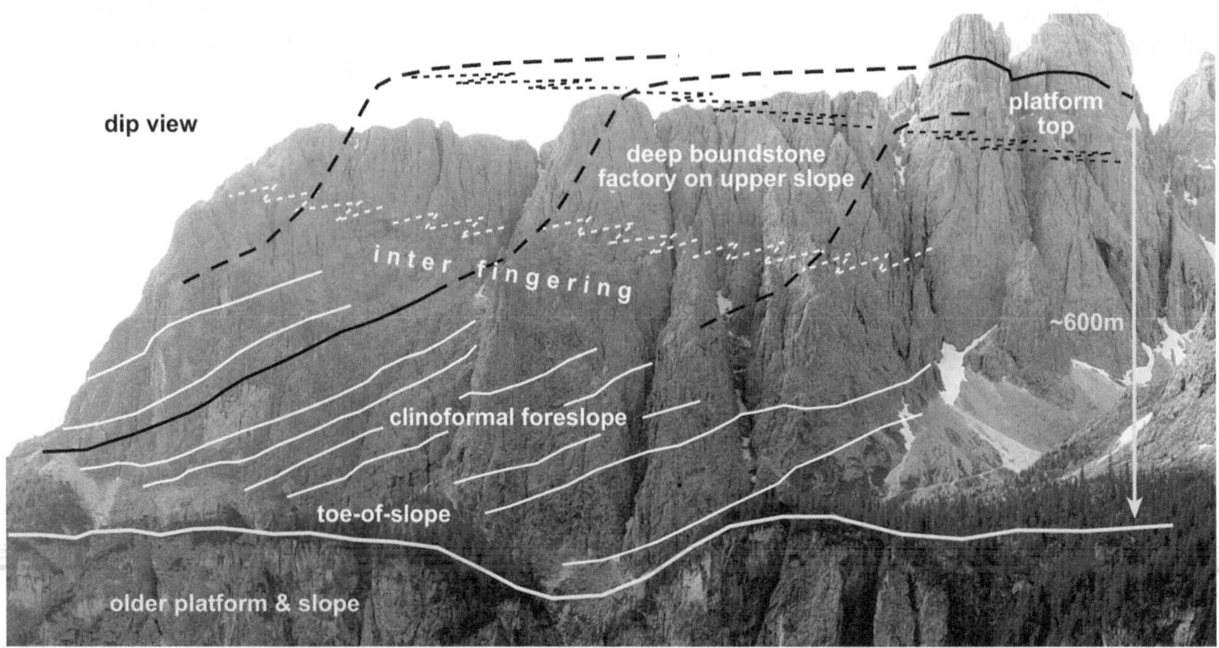

Figure 13. Stratal architecture of the Carnian (Upper Triassic) accretionary prograding Sella Platform, Dolomites, northern Italy, showing planar clinoforms, interfingered deep boundstone factory and foreslope strata, and sub-horizontal downlap at the toe-of-slope.

ments (Figs. 11c–g and 13). Thus, by definition, accretionary margins develop clinoforms, or depositional surfaces along contemporaneous platform-edge to toe-of-slope environments, and clinothems, the packages of sediment or rock that are separated by clinoforms (after Rich, 1951; Figs. 11c-g and 13). Unlike escarpments, clinoformal accretionary systems display progradational as well as retrogradational and aggradational margin configurations (Figs. 11c–g and 13). Clinoforms of prograding margins indicate the volume of re-sedimented material was sufficient to fill the slope profile at the angle of repose, thereby providing a foreslope

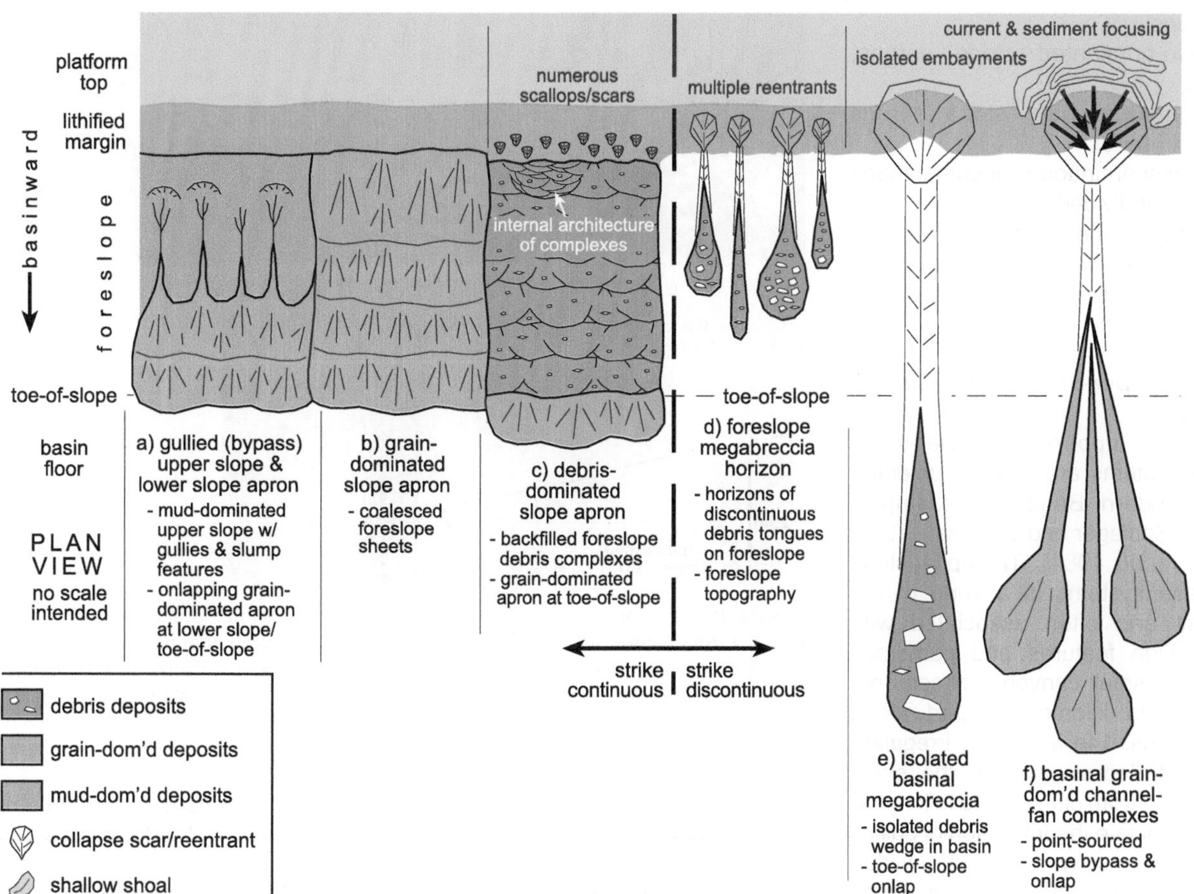

Figure 14. Plan view schematic diagrams of carbonate slope and basin spatial architecture showing range of strike variability from strike-continuous aprons (**a-c**) to discontinuous tongues and channel-fan complexes (**d-f**). Strike-discontinuous architectures often superimpose onto strike-continuous aprons during times of margin collapse and slope bypass.

substrate for the margin to advance basinward over. Prograding clinothems typically taper in thickness toward the toe-of-slope, and can downlap onto the basin floor abruptly, or gradationally, depending on the degree of basinal input (Bosellini, 1984; Figs. 11e–g and 13). Retrograding and aggrading accretionary margins require,

i. increasingly larger volumes of foreslope sediment over time, to keep the slope profile filled in pace with a continually upbuilding margin (Schlager, 1981), and/or
ii. margin sediment factories that are less responsive to sea-level rise and not as likely to outpace the coeval foreslope (*e.g.,* oligophotic boundstone on the upper slope).

Such systems typically display clinothem thinning in younger strata, owing to increased shelf-to-basin relief and volume to fill over time (Fig. 11c). Phases of basin fill, such as siliciclastic bypass episodes, decrease or sustain shelf-to-basin relief and increase sediment thickness in distal settings (Fig. 11d). Consequently, *parallel clinoforms* develop that bound isopachous clinothems with internal onlap wedges (Fig. 11d).

Individual clinoforms vary in terms of shape and curvature, reflecting the distribution of deposit types along the slope profile (*e.g.,* Adams and Schlager, 2000; Fig. 11c–g). Planar clinoforms (minimal changes in declivity along most of the slope profile) indicate relatively homogeneous foreslopes, and often exhibit abrupt or sub-horizontal downlap patterns (Bosellini, 1984; Figs. 11e and 13). In contrast, exponential clinoforms (progressive basinward decrease in declivity along the slope profile) display curvature due to gradational facies changes and basinward fining along dip (Fig. 11c and f). Sigmoidal clinoforms develop when sediment accumulation on the middle to upper slope relatively outweighs that of more distal settings (Fig. 11g), often as a result of sediment removal and strike reworking from deep (>100 m) contour currents (*e.g.,* Eberli and Ginsburg, 1989; Betzler *et al.*, 1999).

Spatial Architecture

Introduction

Spatial architecture describes strike variability and categorizes the plan-view distribution and types of carbonate slope-to-basin deposits and features (Fig. 14). Carbonate slope and basin configurations can be subdivided into

i. *strike-continuous* aprons; and
ii. *strike-discontinuous* accumulations, including channel–fan- and debris complexes (Fig. 14).

Strike-discontinuous deposits can superimpose onto aprons at frequent or infrequent periodicities during foreslope evolution, reflecting cyclic re-sedimentation to the slope or non-typical events, respectively. Strike-continuous aprons can be subdivided into:

1. Gullied upper slope/lower slope

aprons;
2. Grain-dominated slope aprons; and
3. Debris-dominated slope aprons (Fig. 14a–c).

Strike-discontinuous accumulations can be subdivided into:
1. Foreslope megabreccia horizons;
2. Isolated basinal megabreccia; and
3. Basinal channel-fan complexes (Fig. 14d–f).

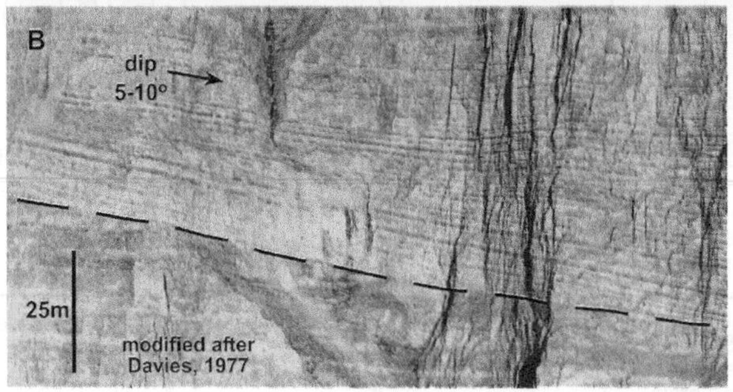

Figure 15. Outcrop photographs of mud-dominated gullied slopes and slump/slide features. **A.** Strike view of numerous closely spaced gullies and channels filled with allodapic grainstone/rudstone and encased in silty skeletal wackestone, Famennian (Upper Devonian), Canning Basin, Western Australia; **B.** Dip view of large-scale stratal truncation feature from slumping/sliding within bedded mudstones to packstones, Serpukhovian (Lower Carboniferous), Sverdrup Basin, Canadian Arctic.

Gullied Upper-Slope/Lower-Slope Aprons

Gullied upper-slope/lower-slope apron systems are well documented from modern Bahamian examples (Fig. 3; Schlager and Chermak, 1979; Mullins *et al.*, 1984). The upper slope of these systems is a zone of net bypass and often associated with slump/slide features and numerous gullies or small canyons spaced decameters to kilometers apart along strike (Figs. 14a and 15). Irregularities and depressions on the slope generated from slump/slide failure are likely to initiate the formation of gullies that subsequently deepen and widen with continued bypass of material. The upper-slope bypass zone is commonly mud-dominated (excluding grain-dominated gully fills), and prone to early cementation, hardground development, and readjustment through translational or rotational (slump/slide) shear failure processes (Fig. 15B) Due to early lithification, mud-dominated upper slopes can develop 10−15° declivities, which are greater than those of the lower slope, and thus represent an exception to Kenter's (1990) relationship. Sediment derived from the platform top, margin, and upper foreslope itself bypasses via multiple point source gullies along strike, thereby depositing in lower-slope to toe-of-slope settings and onlapping the middle slope (Figs. 4 and 14a). The bypassed lobes and sheets coalesce along strike forming a strike-continuous lower-slope apron that is typically enriched in grain-dominated- and debris deposits when compared to the upper-slope bypass zone (Fig. 14a).

Grain-dominated Slope Aprons

Strike-continuous grain-dominated slope aprons are composed of coalesced complexes of sheet-stacked grain-dominated deposits that extend from margin to toe-of-slope environments (Fig. 14b). Internally, individual tabular beds typically stack into slightly mounded lobe complexes that are decameters to hectometers across. These complexes coalesce laterally and along dip to form a larger scale apron that can extend for 10s to 100s of kilometers along strike. These aprons display basinward fining with the transition from grain-dominated deposits on the slope to mud-dominated deposits in the basin, and thus produce exponential clinoform curvature (*e.g.*, Adams and Schlager, 2000). Maximum declivities in proximal portions of the apron approach 30° without significant biological stabilization. In lower-slope to toe-of-slope environments, where grain- and mud-dominated deposits begin to interfinger, individual lobe complexes are better expressed as axial *versus* flanking positions are marked by gradational interfingering transitions.

Debris-dominated Slope Aprons

Laterally extensive debris-dominated slope aprons consist of amalgamated complexes of compensationally stacked debris deposits that extend from toe-of-slope to margin settings (Figs. 13, 14c and 16A). The lenticular shapes of debris deposits result in a characteristic compensational stacking style that forms accretionary geometries in an upslope direction, or backfilling architecture (Figs. 14c and 16A). Backfilling indicates stacking of deposits behind (immediately updip of) pre-existing depositional topography on the slope, such that the slope profile fills from the toe-of-slope to the margin over time forming a clinothem (Fig. 16A). Such clinothems are internally composed of multiple, nested scales of backfilled beds and bedsets, ranging from a meter to decameters, respectively (Figs. 14c and 16A). Due to the high angle of repose for coarse debris deposits and depositional homogeneity along the slope profile, these aprons produce 30−40° dipping, planar clinoforms that sharply downlap the toe-of-slope and interfinger with bypassed grain- or mud-dominated deposits (Figs. 14c and 16A). The development of a strike-continuous debris apron requires numerous phases of repeated small-scale gravitational collapse at multiple points along strike, thus the lithified margin source can display abundant meter-scale scars, scallops, and healed failure features (Fig. 14c).

Foreslope Megabreccia Horizons (Debris Deposits)

Foreslope megabreccia deposits

Figure 16. Outcrop photographs of debris deposit architecture. **A.** Dip view of prograding foreslope debris apron from the Guadalupian (Upper Permian) of the Guadalupe Mountains, west Texas, showing internal debris backfilling architecture (white lines) within bounding 20-30° dipping clinoforms (black lines); **B.** Strike view looking downdip of debris horizons within 20-30° dipping foreslope strata, Famennian (Upper Devonian), Canning Basin, Western Australia. Mud- and grain-dominated deposits (bedded) infill strike topography generated from discontinuous debris deposits (blue). Yellow lines are measured sections.

defined here represent phases of intermediate- to large-scale collapse that generate horizons of discontinuous debris deposits on the slope, such as megabreccia channels/lobes, isolated blocks, and boulder complexes (Figs. 14d and 16B). In this case, debris deposits do not dominate slope deposition (as with debris-dominated slope aprons), but represent periods of instability interspersed within otherwise 'typical' apron development (Fig. 16B). Megabreccias are lenticular and discontinuous along debris horizons forming lateral depositional topography (meters to decameters of synoptic relief spaced decameters to hectometers apart) that influences and redirects subsequent grain- or mud-dominated sediment gravity flows (Figs. 14d and 16B). Thus, deposits that immediately postdate collapse events occur adjacent to megabreccia accumulations, sidelap megabreccia margins, and may be erosional from redirection and sediment focusing on the slope (Fig. 16B). Despite the great densities and volumes that megabreccias entail, they are capable of coming to rest on the slope at high angles (up to 35°), often related to friction from plowing or erosion into the underlying semi-lithified, compactable substrate (Fig. 16B). The discontinuous nature of debris deposits indicates strike-limited points of instability and failure of the margin, implying the presence of multiple considerable reentrants (decameters across) along strike that can potentially serve as downslope sediment focal points (Fig. 14d).

Isolated Basinal Megabreccia (Debris Deposits)

Isolated basinal megabreccias represent phases of extreme instability and large-scale collapse of the outer platform, margin, and portions of the slope (Fig. 14e). Such events produce voluminous debris deposits (*i.e.*, olistoliths decameters or more across that accumulate in distal toe-of-slope and basinal settings (up to 10–15 km from the margin; Fig. 14e). In general, these deposits are isolated spatially (spaced 1 to 10s of kilometers apart) and occur infrequently within the overall foreslope stratigraphy, reflecting 'non-typical collapse events' that punctuate ambient apron development. Strike and dip lengths and thickness of megabreccias vary depending on the type of deposit (*i.e.*, coherent blocks *versus* debris-flow tongues). They are, however, overall laterally discontinuous and form marked irregularities at the toe-of-slope or on the basin floor that require time to bury and anneal. Perhaps the most significant products of large-scale collapse events are the pronounced, isolated embayments generated at the margin and the potential for associated slope canyon development (Fig. 14e). Although difficult to observe in outcrop, these embayments can be reconstructed to span hundreds of meters across, and thus can serve as strike-limited, long-lived downslope sediment feeders for subsequently re-sedimented material (Fig. 14e).

Basinal Grain-dominated Channel–Fan Complexes

Grain-dominated channel–fan complexes develop in distal settings when sediment generated from line source carbonate factories is focused downslope through an irregularity in the margin, such as an embayment (Fig. 14f). This sediment focusing promotes channelization, bypass of the slope profile, and accumulation of strike-discontinuous, shallow-derived sediment at the toe-of-slope or on the basin floor (Fig. 14f). Large (hundreds of meters across) margin embayments serve as long-lived downslope sediment focal points, and can be inherited/fault-controlled, or form due to large-scale gravitational failure resulting from intrinsic (*i.e.*, margin accretion) or extrinsic (*i.e.*, seismicity) triggers. Thus, basinal megabreccia deposition (Fig. 14e) may precede the emplacement of bypassed channel–fan complexes. In addition to sediment focusing, large embayments concentrate current energy that in turn promotes the development of high-energy shoal rims around the collapsed perimeter, providing a local sediment source for downslope re-sedimentation (Ball, 1967; Fig. 14f). For this reason, bypassed channel–fan complexes are commonly grain-dominated (*e.g.*, Savary and Ferry, 2004), and mark an exception to Kenter's (1990) relationship (Figs. 4 and 14f).

These grain-dominated systems are fed through channel networks that are confined to lower-slope and toe-of-slope settings or extend onto the basin floor for kilometers (*e.g.*, Braga *et al.*, 2001). Channelization indicates confined flow and the potential for sediment to transport over great distances. In the most distal environments where flow becomes unconfined, fans accumulate and channel throats backfill to the toe-of-slope or lower slope to produce onlap stratal terminations (Fig. 14f). The internal architecture of carbonate channel complexes is similar to that of their siliciclastic counterparts, with scour-filled thalwegs, hierarchical erosional surfaces, bed shingling, and complex compensational or crosscutting relationships (*e.g.*, Savary and Ferry, 2004). Channels are commonly decameters to hectometers wide, decameters deep, and linear in plan view (Fig. 14f), although documentation of levees and meander bars indicate local sinuosity.

Fans comprise the distal components of channel systems and reflect the change from confined to unconfined flow processes at the channel head (Fig. 14f). Fans can develop at the toe-of-slope or deep into the basin (up to 35 km), can be up to hectometers or kilometers across, and are typically decameters thick (*e.g.*, Payros *et al.*, 2007). Grain-dominated fans, however, generally fall within the lower ranges in terms of size and distance traveled (hectometers and 1–10 kilometers, respectively). They are characterized by broad, lobate morphologies with gradational textural fining at the flanks, and exhibit well-developed shingling geometries internally.

CARBONATE SLOPE AND BASIN MODELS

Introduction

Carbonate-slope and basin-depositional systems can be broken down by deposit type, large-scale stratal pattern, and spatial architecture. Deposit-type proportions (pie charts; Table 3) provide a quantitative estimate of the gross deposit type ratios within the foreslope, and have been calculated by taking transects within a genetic package along the slope profile, converting facies descriptions into the deposit-type scheme, and weighting the calculations on vertical thickness. Groups of examples with similar deposit types and architectural attributes, integrated with morphological, faunal, oceanographic, and extrinsic characteristics of the entire carbonate accumulation, define categories and subcategories of carbonate slopes. As a first order grouping, all carbonate slopes can be subdivided by dominant deposit type (debris-, grain-, and mud-dominated), each of which has distinctive overarching characteristics and are linked to a particular morphological or biotic setting (Table 3). In terms of strike variability, significant differences in the spatial distribution and proportions of debris deposits are connected to collapse scale, frequency, and triggering mechanism. Growth and inherited escarpments represent architecturally based subdivisions of carbonate slopes that have distinctly different deposit-type proportions, stratal characteristics, scales, and controls on stratal development.

Debris-Dominated Foreslopes

Debris-dominated foreslopes consist of greater than 50% debris deposits and record collapse of early-lithified margin environments as the dominant process during foreslope development (Table 3; Fig. 17A-1; *e.g.*, Della Porta *et al.*, 2003). These systems form strike-continuous, debris-dominated aprons that have internal complexes of compensational, lenticular-debris deposits that backfill the slope profile and interfinger at the toe-of-slope with grain- or mud-dominated deposits (Figs. 14c, 16A, 17A-1). Debris-dominated foreslopes are amongst the largest accretionary margin systems that have slope heights approaching a kilometer (Fig. 18), and are observed in aggradational and progradational configurations. Debris-dominated escarpments do not occur. In general, debris-dominated foreslopes exhibit planar clinoforms and are associated with deeper platform edges (20–80 m water depth), seaward-dipping outer platform strata, and shelf crest development (*e.g.,* Tinker, 1998).

The overriding control on debris-dominated foreslope development is the presence of deep oligophotic boundstone factories on the upper slope that extend down the slope profile up to 300 m (water depth) below the influence of sea-level fluctuations and wave-base (Table 3; Fig. 17A-1; *e.g.*, Weber *et al.,* 2003; Della Porta *et al.*, 2003). Such boundstone factories are capable of continual growth, collapse, and production of debris deposits on the foreslope without interruption from high-frequency accommodation changes. As deep boundstones are seldom exposed, gravitational collapse is driven by autogenic basinward accretion on a steeply inclined surface. Thus, these systems represent autonomous slope-sediment factories that are disconnected from the platform top and largely driven by nutrient levels and circulation patterns, and less so by relative sea-level changes (Della Porta *et al.*, 2003; 2004). Long-term accommodation, however, does affect the development of interfingering toe-of-slope accumulations by controlling platform sediment production and re-

Table 3. Chart of carbonate foreslope examples from outcrop work and literature arranged by deposit types (debris-, grain-, and mud-dominated; horizontal axis) and large-scale stratal architecture (accretionary and escarpment margins; vertical axis). General characteristics and controls are outlined

		Debris-Dominated	Grain-Dominated	Mud-Dominated
Textures & Style		cobbles & boulders; lenticular megabreccia & blocks; backfilled aprons	sand, pebbles, & boulders; gnstn & rdstn; tabular beds; coalesced aprons & lobes	clay & silt; mdstn & wkstn; nodular beds; gullied upper slopes/bypassed aprons
Key Controls		extensive deep boundstone factories on upper slope	waveswept, narrow shelves and/or shallow euphotic framestone/bafflestone reefs	protected, broad shelves and/or external fine-grained sediment
Accretionary Margins - interfingered margin & foreslope	Prograding (strong → weak)	10%, 20%, 70% - L Carb (Serp), Tengiz Field, KZ (e.g. Weber et al, 2003); - U Carb (Bash), Asturias, Spain (e.g. Della Porta et al, 2003); - U Perm (Guad), Yates Fm, W TX; - M-U Tri (Carn), Dolomites, Italy (e.g. Kenter, 1990) 10%, 30%, 60% U Perm (Guad), 7-Rivers Fm, W TX	30%, 10%, 60% U Miocene, Mallorca, Spain (shallow euphotic reef system) (e.g. Pomar et al, 1996) 15%, 25%, 60% - U Dev (Fam), Canning Basin, W Australia; - L Perm (Leon), W TX	5%, 30%, 65% U Miocene to present, western-leeward Great Bahama Bank (shallow euphotic reef system) (e.g. Eberli et al, 2004) 55%, 20%, 25% M Cret (Alb), Gorbea Platform, Basque, Spain (e.g. Gomez et al, 1999)
	Aggrading/ Retrograding	20%, 15%, 65% - U Carb (Mosc), Asturias, Spain (e.g. Della Porta et al, 2003); - M Tri (Lad), Dolomites, Italy (e.g. Harris, 1994)	15%, 5%, 80% L Jur (Pliens), High Atlas, Morocco (e.g. Kenter & Campbell, 1991)	5%, 80%, 15% L Perm (U Leon), W TX (e.g. Saller et al, 1989) 5%, 75%, 20% - U Dev (Fras), Canning Basin, W Australia; - mud mounds
Escarpment Margins	Growth (foreslope onlap & aggrading/retrograding margins; accretionary-escarpment transition)	debris-dominated escarpments not observed	Stratal Controls: increasing accommodation & shallow euphotic reefs 20%, 5%, 75% U Dev (Fras), Canning Basin, W Australia	Stratal Controls: increasing accommodation & shallow euphotic reefs 5%, 25%, 70% U Dev (Fras), Alberta Basin, Canada (e.g. Whalen et al, 2000)
	Inherited (escarpment-accretionary transition)	Pie Chart Legend: mud-dom'd; debris; grain-dom'd	Stratal Control: high-relief antecedent topography/large slope height 20%, 20%, 60% M-U Cret (Cen-Camp), Maiella Platform, Italy (e.g. Eberli et al, 1993)	Stratal Control: high-relief antecedent topography/large slope height 15%, 25%, 60% Tertiary to modern windward margins, Caribbean (e.g. Grammer & Ginsburg, 1992)

sedimentation. Another important consideration is that deep oligophotic boundstone factories are mostly restricted to middle Paleozoic through middle Mesozoic time (see Chapter 17), thus debris-dominated foreslopes are more likely to occur during some geologic ages than others.

Grain-Dominated Foreslopes

Grain-dominated foreslopes consist of greater than 50% grain-dominated deposits and record offbank transport of sand- to gravel-sized material, from platform-top and platform-edge environments, as the dominant process during foreslope development (Table 3; Fig. 17A-2). These systems develop strike-continuous, grain-dominated aprons composed of tabular-bedded, coalesced lobes and sheets that fine downslope into mud-dominated deposits, producing exponential clinoform curvature (Figs. 14b and 17A-2). Grain-dominated foreslopes are observed in prograding, aggrading, and retrograding settings, and as the flanks for both accretionary and escarpment systems. As a result, depositional profile scales are highly variable, ranging from <100 to >1000 m in slope height, and <3 km in slope width (with exception to inherited escarpments; Fig. 18). Grain-domi-

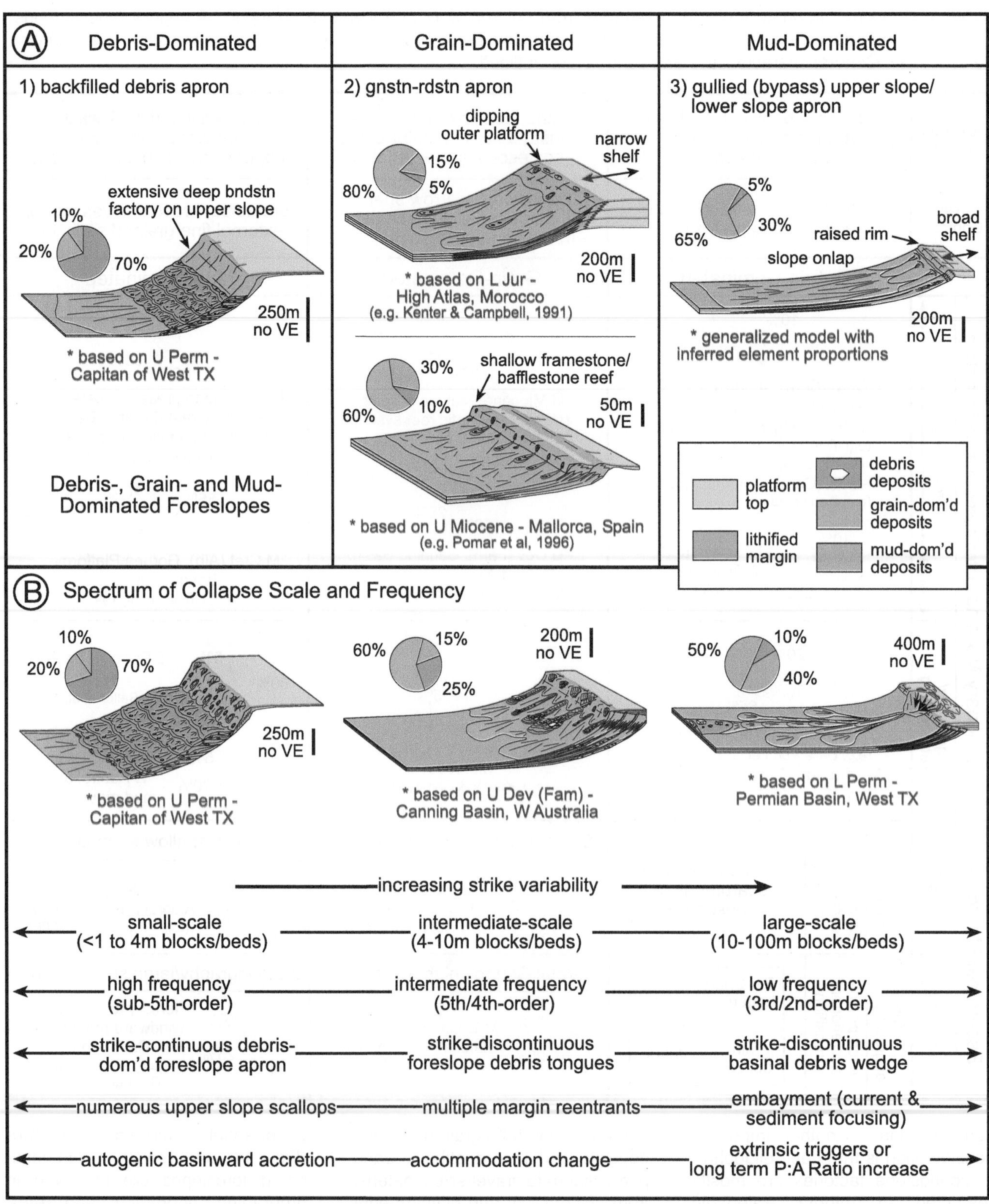

Figure 17. Conceptual carbonate slope and basin block models based on outcrop and subsurface examples. **A.** Deposit type proportions and strike variability of debris-, grain-, and mud-dominated foreslopes (excluding episodic and non-typical collapse events). 1) Debris-dominated foreslopes are associated with deep oligophotic boundstone factories. 2) Grain-dominated foreslopes are associated with wave-swept platform tops and/or shallow skeletal reefs. 3) Mud-dominated (gullied) foreslopes are associated with broad and/or protected platform interiors; **B.** Models showing spectrum of collapse scale and frequency, and associated characteristics. In general, the degree of strike variability of the margin, foreslope, and basin increases with increasing collapse scale, decreasing collapse frequency, and extrinsic triggering mechanisms.

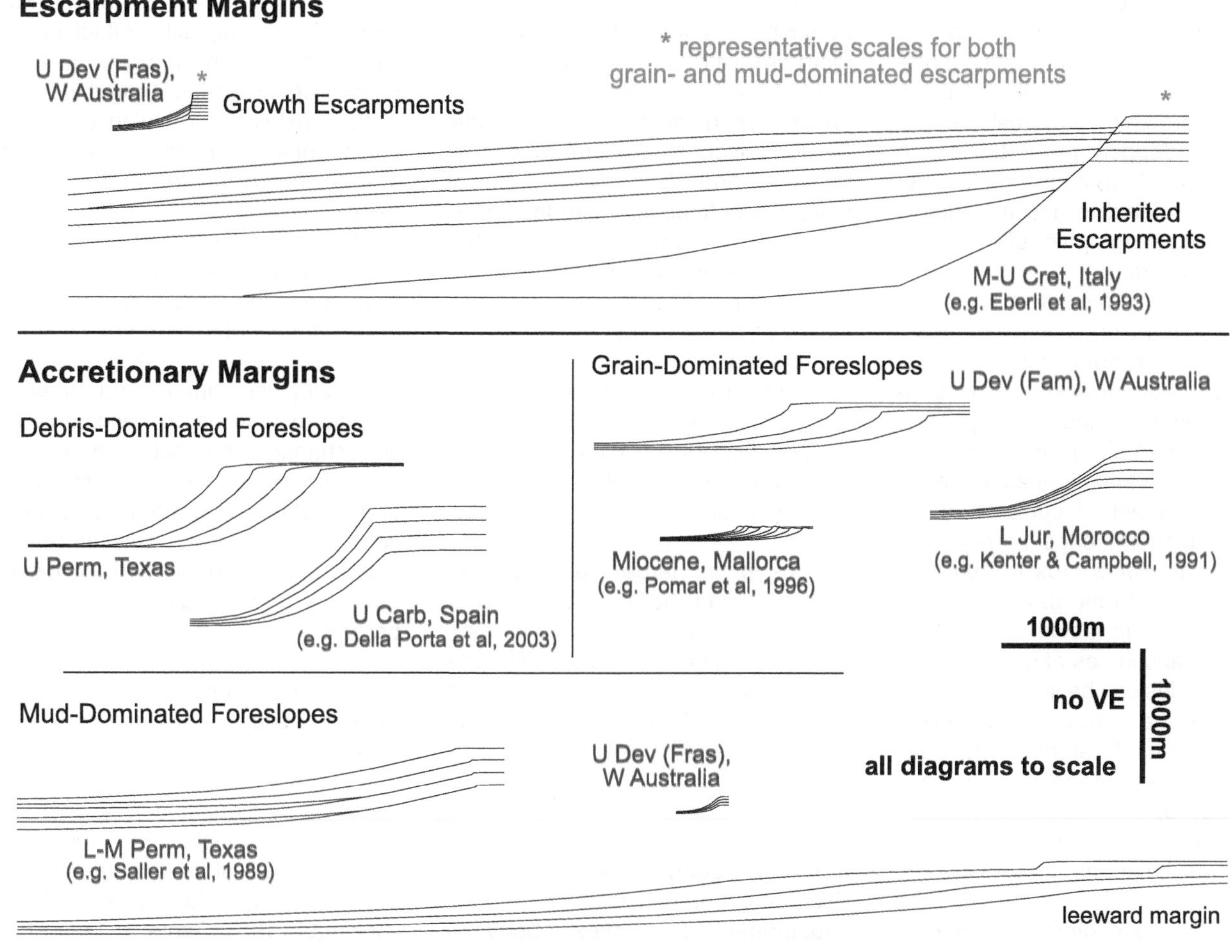

Figure 18. Simplified stratal architectures at same scale and with no vertical exaggeration for debris-, grain-, and/or mud-dominated foreslopes of escarpment and accretionary margin systems, based on outcrop and subsurface examples.

nated foreslopes are most variable in terms of debris deposit proportions and superimposed spatial architecture, reflecting differences in accommodation setting, margin sediment type, collapse frequency, and extrinsic factors.

The overriding controls on the development of grain-dominated foreslopes are platform morphological features and margin faunal assemblage or fabric (Table 3; Fig. 17A-2). Morphological features that are amenable to prolific current sweeping of the platform-top, such as open rims (absence of barriers), dipping- or flat-topped outer platform strata, and narrow shelf widths (<10 km), promote the production and downslope re-sedimentation of sand- and gravel-sized particles (Fig. 17A-2). Other grain-dominated foreslopes are associated with shallow, euphotic framestone to bafflestone reefs that are capable of producing considerable volumes of bioclastic sand and gravel, in addition to debris deposits from gravitational collapse. This is dominantly due to

i. the likelihood of fragmentation from direct interaction with wave-base and shallow currents; and
ii. the abundance of associated 'accessory' skeletal organisms that are not bound into the reef framework and available for downslope re-sedimentation (Fig. 17A-2).

Mud-dominated Foreslopes

Mud-dominated foreslopes consist of greater than 50% mud-dominated deposits, and dominantly record off-bank transport and/or water column fallout of clay- and silt-sized particles, from protected platform-top environments and pelagic sources, respectively (Table 3, Fig. 17A-3). These systems develop early-cemented, mud-dominated upper-slope bypass zones that have numerous gullies along strike and slump/slide features (Figs. 14a, 15 and 17A-3). Re-sedimented material bypasses the upper slope and accumulates as an onlapping, strike-continuous lower slope to toe-of-slope apron composed of coalesced lobes and sheets that are rich in grain-dominated or debris deposits (Figs. 14a and 17A-3). As in grain-dominated foreslopes, mud-dominated foreslopes exhibit prograding, aggrading, and retrograding morphologies, and flank both accretionary and escarpment margin systems. Consequently, there is a wide range of depositional profile scales encompassing slope heights from <100 to >1000 m, and slope widths from <1 to >10 km (Fig. 18). The smallest mud-dominated foreslopes (which include 'mud mounds') are those with largely non-reefal autochthonous skeletal communities that occupy upper- to middle-slope settings (instead of gully systems), and produce *in situ* floatstone fabrics upon blanketing by fine-grained material. The most recognizable variations within mud-dominated foreslopes are in debris deposit proportions and superimposed spatial architecture, which are related to accommodation setting, oceanographic effects, and extrinsic factors.

The overriding controls on mud-dominated foreslope development are platform morphological features, external sediment input, oceanographic effects, and geologic age (Table 3, Fig. 17A-3). Platforms with characteristics that dissipate or hinder the impingement of current energy, such as platform edge barriers and broad shelf widths (10s to 100 km), are conducive to the development of platform-top mud factories (Fig. 17A-3). External sources of clay- and silt-sized particles can be substantial contributors to the foreslope when mixed with fine-grained carbonate material. In particular, prevailing wind direction is a factor, as

i. onbank sediment transport along windward margins reduces the volumes of sand- and gravel-sized particles re-sedimented downslope; and
ii. offbank-sediment transport, inherent to leeward margins, enhances the downslope re-sedimentation of periplatform mud.

Lastly, other important oceanographic parameters, such as circulation and nutrient levels, together with geologic age can partially control the relative proportions of mud-dominated deposits on the foreslope by dictating the productivity and/or presence of carbonate planktonic communities and resulting pelagic sedimentation (Fischer and Arthur, 1977).

Strike Variability and Collapse

Strike variability in carbonate slope and basin systems is closely linked to different styles of spatial architecture generated from margin collapse (Figs. 14d–f and 17B). Foreslope spatial architecture becomes more laterally heterogeneous with increasing scale and decreasing frequency of collapse, as debris accumulations transition from aprons to isolated wedges, depositional topography is more pronounced, and failure features at the margin are larger and require longer durations to anneal (Fig. 17b). Furthermore, with larger collapse-generated margin irregularities (*i.e.*, embayments) comes greater potential for long-lived downslope sediment focusing and the development of strike-discontinuous, grain-dominated channel–fan complexes (Figs. 14f, and 17b). Dominant triggering mechanisms for failure change with collapse scale and frequency, where

i. continual, small-scale collapse events are induced from autogenic basinward accretion;
ii. episodic, intermediate-scale collapse events are linked to high-frequency accommodation changes; and
iii. non-typical, large-scale collapse events can be extrinsically triggered (*i.e.*, seismicity) or coincident with increases in long-term Progradation: Aggradation Ratio of the margin (P: A Ratio; Tinker, 1998; Kerans and Tinker, 1999; Fig. 17B).

Escarpment Margins

Inherited Escarpments

Escarpment margin systems, and the substantial differences between inherited and growth escarpments, warrant a separate discussion because they have distinctly different deposit-type proportions, stratal configurations, depositional profile scales, and controls on stratal evolution (Table 3; Figs. 11a, 18, and 19;). Inherited escarpments are characterized by escarpment to accretionary stratal evolution and aggrading or retrograding margin configurations, because margins are unable to prograde prior to basin infill due to large initial slope heights (*e.g.*, Eberli *et al.*, 1993; Figs. 11a, and 19a). As a result, inherited escarpments are among the largest carbonate slope and basin systems with slope heights >1 km and slope widths >10 km (Figs. 18, and 19a). Grain- and mud-dominated versions of these systems exist due to morphological, faunal, oceanographic, and extrinsic controls. The most distinctive deposit types of inherited escarpments, however, are extensive debris-deposit tongues interspersed throughout foreslope strata that comprise ~1/4 of the slope sediment volume and extend for many kilometers into the basin (Fig. 19a). These deposits reflect unstable, collapse-prone margins inherent to the highly disequilibrated depositional profiles that develop when the volume of foreslope substrate is inadequate for a given shelf-to-basin relief.

The overriding control on inherited escarpment development and stratal evolution is the slope height inherited from antecedent topography (Table 3; Figs. 18, and 19a). Regardless of morphological, faunal, and oceanographic conditions, or accommodation changes that occur during the escarpment phase, the resulting stratal patterns and margin configurations are predictable, reflecting large slope heights and the inability of platform-top and margin sediments to prograde (Figs. 11a, 19a).

Growth Escarpments

Growth escarpments develop from building of synoptic relief over time due to preferential growth of the shallow-water carbonate factory, and sustained margin upbuilding such that the coeval foreslope accumulation is 'outpaced' and onlaps older surfaces (Schlager, 1981; Table 3; Figs. 11b, and 19b). Thus, stratal patterns evolve from accretionary to escarpment configurations throughout development [with an intervening transitional phase], and show increasing shelf-to-basin relief and degrees of foreslope onlap over time (Figs. 11b, 12, and 19b). Margins of growth escarpments are inherently aggradational or retrogradational due to upward-directed growth (Figs. 11b, and 19b). As these systems build relief from initially flat-lying surfaces, depositional profile scales are small, especially when compared to inherited escarpments, thus slope heights are generally <300 m and slope widths <5 km (Figs. 18, and 19b). Foreslope beds decrease in thickness and steepen over time as shelf-to-basin relief and foreslope aerial extent increase (Figs. 11b, and 19b). Similar to inherited escarpments, growth escarpment foreslopes can be grain- or mud-dominated depending on morphological, faunal, oceanographic, and extrinsic variables. Debris deposits within growth escarpment foreslopes are volumetrically

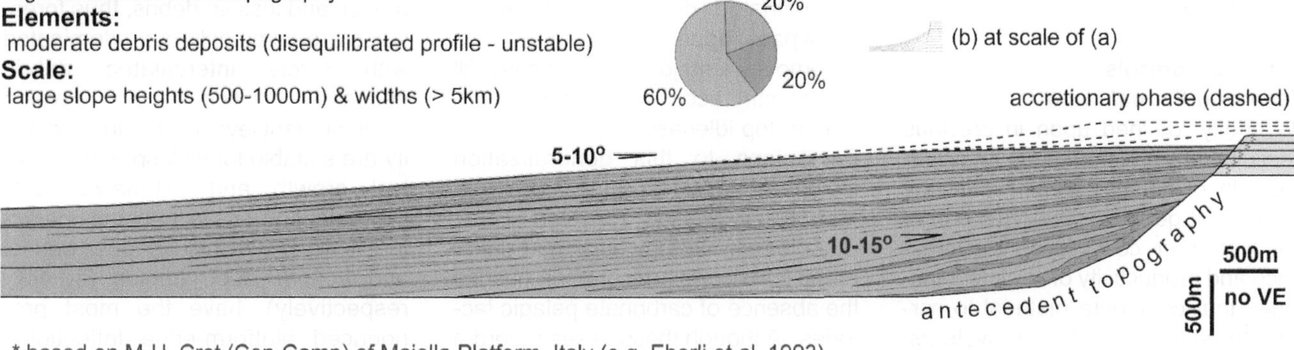

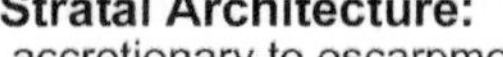

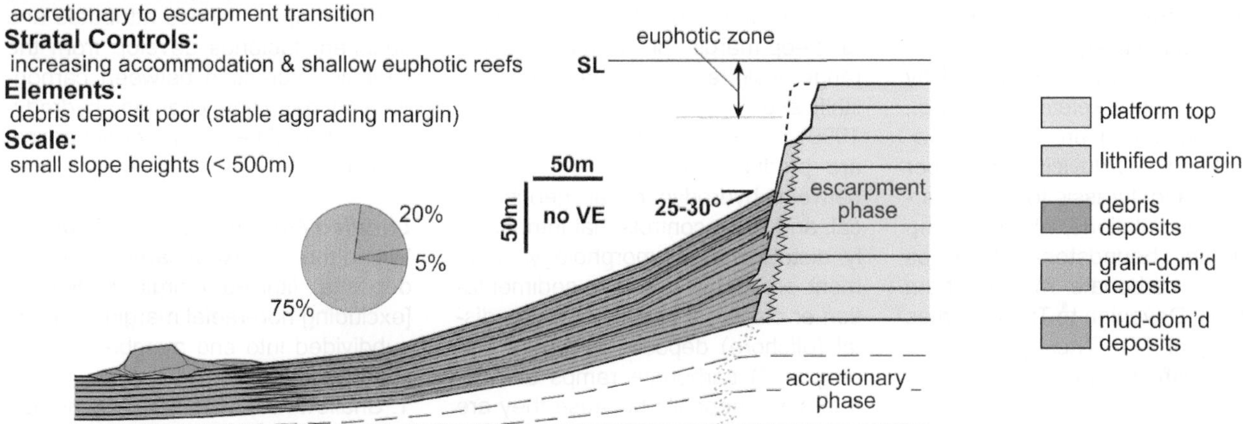

Figure 19. Inherited **(a)** and growth **(b)** escarpment margins based on outcrops, showing generalized deposit type proportions, deposit type distribution, stratal evolution, and difference in scale.

minor (<10%), especially when compared to those of inherited escarpments (up to 20%), and most commonly occur as small-scale breccia lenses intercalated within proximal foreslope strata (Fig. 19b). This relative paucity of collapse and subsequent debris deposit deposition reflects the stable configuration of an aggrading margin, where upward-directed margin growth occurs over a solid underlying foundation usually consisting of lithified, older margins, and therefore is not prone to collapse (Figs. 11b, and 19b). However, large-scale (non-typical) collapse events occasionally occur toward the end of escarpment growth, when the depositional profile is disequilibrated and most sensitive to collapse triggers (Fig. 19b).

The overriding controls on growth escarpment development are twofold:

1. Increasing accommodation conditions; and
2. Euphotic reef assemblages that build to sea level and are capable of tracking rising sea level (Table 3; Fig. 19b).

The coeval effects of these dual parameters allow for the building of relief and sustained upward growth of the margin required to evolve from accretionary to escarpment configurations (Schlager, 1981). Deeper boundstone factories that are less sensitive to changes in relative sea level are unlikely to outpace the contemporaneous foreslope and favor accretionary margin development (Fig. 11c).

CONTROLS ON CARBONATE SLOPE TO BASIN SYSTEMS

Introduction

Carbonate-slope and basin systems are highly variable with respect to deposit-type proportions and distribution, large-scale stratal patterns, and spatial architecture. These variations are driven by several controls that affect the production/availability and re-sedimentation of contributing sediment factories and sources. No single control exists that is expressed consistently or dominantly across the spectrum of carbonate slopes because these systems are driven by the interplay of multiple intrinsic (linked to carbonate system) and extrinsic (external to carbonate system) controls.

Intrinsic Controls

Temporal Faunal Variations

As briefly touched upon in previous sections, it is important to consider the role of secular faunal variations throughout geologic history as one of the first-order controls on the presence and productivity of sediment factories that contribute material to carbonate slope and basin systems. Among the expansive list of carbonate faunal changes throughout the Phanerozoic, the temporally constrained occurrences of skeletal reefs and carbonate plankic communities have significant impacts on the generation of debris and mud-dominated deposits, respectively.

In general, margins dominated by shallow euphotic skeletal reefs (*e.g.*, coral framestones) are most prevalent in Middle Ordovician to Upper Devonian, and Jurassic to recent carbonate accumulations, whereas deep oligophotic boundstone factories dominated carbonate margins from uppermost Devonian to Triassic times (*see* Chapter 17). These contrasting styles of lithified margin sediment type are associated with different collapse characteristics (*i.e.*, potential, frequency, trigger), and thus have substantial implications on resulting debris deposit proportions and architecture within foreslope strata (discussed later). Transitional periods, such as the Late Devonian and points within the middle to late Mesozoic (*e.g.*, Gomez-Perez *et al.*, 1999), exhibit coexisting shallow reefs and deep boundstones along a single margin, and have collapse characteristics common to both styles of lithified sediment types.

Carbonate (nanno)planktic communities are not documented in the rock record prior to the Middle Jurassic (Fischer and Arthur, 1977), when platform-top- to margin-derived carbonate mud and external mud input (*e.g.*, hinterland siliciclastics, volcaniclastics) accounted for the clay- to silt-sized sediment fraction of slope and basin environments. As a result, a potentially significant source of pelagic carbonate mud was not present before the middle Mesozoic, which, for those geologic periods,

i. reduces the likelihood for mud-dominated deposits and gullied (bypass) upper slopes; and
ii. removes a major supply of basin fill that can accumulate during platform-top idleness.

Exceptions to this generalization include the Middle–Late Devonian and Early–Middle Permian (*e.g.*, Saller *et al.*, 1989), where mud-dominated foreslopes are common despite the absence of carbonate pelagic factories. Although these cases record a considerable contribution from non-carbonate mud sources, the origin of the remainder carbonate mud fraction is debatable and/or unclear.

Ramp to Rim Evolution

The evolution from low-angle ramps to steep-margin platforms covers a range of slope angles and degrees of lithification (*e.g.,* Kerans and Tinker, 1999). Slope angles and lithification are products of the feedback loop between hydrodynamical, geochemical, and biotic controls that intrinsically dictate margin morphology, sediment productivity, and re-sedimentation or dispersal processes. The distal (offshore) deposits of gently dipping (<2°) carbonate ramps are not true slope deposits because they are generated by traction and suspension processes from within storm wave base, as opposed to sediment gravity flow processes. Sediment gravity flow deposition begins when carbonate accumulations build shelf-to-basin relief and slope angles approach 2-5°. In many such cases, mud-dominated turbidite deposition results because of prolific production of clay- and silt-sized particles and downslope re-sedimentation, often forming mud-dominated gullied slope systems (Schlager and Chermak, 1979; Mullins *et al.*, 1984).

When platforms generate enough relief to form a distinguishable platform-edge inflection and slope angles from 5–15°, an energy gradient threshold is reached where erosional scouring, offbank transport, and current activity are favored. These conditions develop current-sculpted hydrodynamic accumulations (banks) and *in situ* organic biohermal buildups at the platform edge and upper slope, forming an incipient lithified margin. In such cases, the margin banks and buildup flanks are sources for sand, gravel, and less so debris, thus foreslopes are commonly grain-dominated with rarely intercalated debris deposits. Lithified margins develop when nutrient levels and current activity are suitable for widespread biological growth and extensive early cementation. These systems characteristically build the steepest detrital and *in situ* slope angles (40–90°, respectively), have the most pronounced platform-edge inflections, and display the highest proportions of debris deposits due to elevated propensity for collapse. In summary, morphological and faunal variations ultimately govern the re-sedimentation process and deposits on carbonate foreslopes. These attributes and sediment factories develop from the intrinsic interaction between carbonate sedimentation, changing hydrodynamical regimes, and geochemical conditions.

Lithified Margin Sediment Type

Within the context of carbonate-slope deposits, lithified margin sediments [excluding non-reefal margins] can be subdivided into end members as follows:

1. Shallow euphotic framestone/bafflestone reefs (*shallow reefs*); and
2. *Deep oligophotic boundstone factories* (Table 4).

Shallow reefs consist primarily of light-dependent macro-skeletal communities, such as coralgal, rudistid, and stromatoporoid assemblages, and are centered at the platform edge generally between 0 and 50 m of water depth (Pomar, 2001). Deep boundstone factories, however, are dominated by less-light-dependent microbial or ahermatypic organisms, and are centered on the upper slope typically between 30 and 300 m of water depth (Pomar, 2001).

Generally, euphotic framestone/bafflestone reefs are prone to the production of sand and gravel for grain-dominated deposits, as well as collapsed material for debris deposits (Table 4). Conversely, oligophotic boundstone factories are dominantly prone to cobble and boulder production for debris deposits. Excluding inherited escarpments, foreslopes associated with shallow reefs consist of 10% or less debris deposits, whereas those associated with deep boundstone factories contain up to 70% debris deposits (Table 3). Additionally, if oceanographic effects and external sediment input are not significant factors, foreslopes associated with shallow reefs contain 60–75% grain-dominated deposits. This is dominantly due to differences in the degree of interaction with waves and currents, and the abundance of associated accessory organisms.

In addition to debris deposit production from early lithification and collapse, shallow reefs are also prone to mechanical erosion and fragmentation into sand- and gravel-sized particles due to continual interaction with wave-base and shallow currents. Shallow-reef factories also tend to share environments with diverse populations of skeletal organisms that are not necessarily bound within the reefal framework, and thus available for downslope transport as skeletal sand and bioclasts. In contrast, deep boundstone factories,

i. dominantly accumulate below the effects of agitation and significant mechanical erosion; and
ii. prefer depositional environments where unbound accessory skeletal organisms are not as plentiful.

For the abovementioned reasons,

i. deep boundstone factories principally donate cobbles and boulders to the foreslope (not sand and gravel) from their own autogenic basinward accretion and collapse; whereas
ii. shallow reefs are likely to contribute a range of sediment (from bioclastic sand to framestone boulders) to the foreslope due to a variety of processes.

As a result of considerable differences in depth range, the surface area of oligophotic boundstone factories (for a given width along strike) is an order of magnitude larger than that of euphotic framestone/bafflestone reefs (Table 4). The extensive sediment production areas of deep boundstone factories, coupled with the dissociation from high-frequency accommodation changes, allow for uninterrupted (continual) growth, collapse, and generation of large volumes of debris deposits, processes that can substantially outweigh others contributing to foreslope development (as for debris-dominated foreslopes) (Table 3; Fig. 17A-1). Conversely, shallow reefs have relatively smaller production areas and are sensitive to high-frequency relative sea level changes, thus the generation of debris deposits is partitioned within accommodation cycles, potentially less in volume, and less likely to be a dominating foreslope process.

Table 4. Characteristics of end-member lithified margin sediment types and implications on foreslope deposit type production and resedimentation

Lithified Margin Sediment Type	Shallow Euphotic Framestn/Bafflestn Reef	Deep Oligophotic Boundstone Factory
Dominant Faunal Assemblage	**skeletal** (i.e. coralgal, rudistid, stromatoporoid)	**microbial** & ahermatypic fauna
Depth Range	centered at platform edge; generally **0-50 m (euphotic)**; within influence of wave base and relative sea level changes	centered on upper slope; generally **30-300 m (oligophotic)**; large portions of factory are often below the influence of wave base and relative sea level changes
Surface Area of Sediment Factory	**10,000s m^2** per km along strike; comprises < 20% of slope surface area	**100,000s m^2** per km along strike; comprises up to 30-40% of slope surface area
Dominant Sediment Production	**grain-dominated and debris deposit production**; skeletal sand, pebbly bioclasts, & cobbles/boulders	**debris deposit production**; minimal sand production; pebbles, cobbles, and boulders from cms to 10s m
Collapse Controls	**interlinked with relative sea level**; sensitive to exposure/ submergence & degree of current energy; episodic collapse	**autogenic basinward accretion**, local instability & collapse; tied to nutrient levels, much less so relative sea level changes; +/- continual collapse
Morphological Features	associated with **raised rims and flat-topped platforms** built to sea level; backreef aprons and lagoons	associated with **dipping outer platforms**, deeper platform edges (20-50 m), and shelf crests
Accessory Skeletal Organisms	**abundant**; significant fractions both bound within reef framework & not bound	**moderate to rare**; most skeletal material bound within factory and/or volumetrically minor

In summary, lithified margin sediment type has significant implications regarding sediment production potential and the propensity for resedimentation of particular deposit types onto the foreslope. Deep boundstone factories are associated with debris-dominated foreslopes due to uninterrupted, continual donation of debris to the foreslope and unlikelihood for production of sand- and gravel-sized particles (Table 4). Shallow-reef factories, however, have not been documented with associated debris-dominated foreslopes, and conversely tend to generate both grain-dominated and debris deposits owing to the shallow, agitated, skeletal-rich environments they inhabit. Furthermore, it is important to recognize that these are generalized statements and characteristics for end-member styles of reef systems that are partitioned temporally throughout the geologic record (*see* Chapter 17). In particular, and as mentioned previously, deep boundstone factories are not widespread throughout the Phanerozoic, but mostly confined to the late Paleozoic through early Mesozoic (Late Devonian through Triassic), and seemingly straddle global mass extinction events (*i.e.,* Frasnian–Famennian extinction and Permian–Triassic extinction).

Platform Morphology

Morphological characteristics of the platform and margin, such as the presence of platform-edge barriers, shelf width, and lagoon depth, affect the interaction and impingement of waves and tidal or storm currents.

Wave action and currents [or lack thereof] are especially important factors on the production and re-sedimentation of material that comprise grain- and mud-dominated deposits. Mud-dominated foreslopes are associated with protected platforms, platform-edge energy barriers (raised rims), backreef lagoons, and/or broad (10s to 100 km) platform interiors (Table 3; Fig. 17A-3). These characteristics provide favorable environments for clay and silt production, and greater likelihood of sequestering grain-rich sediment on the platform-top instead of downslope re-sedimentation. In contrast, grain-dominated foreslopes are associated with open platforms, including flat-topped- and seaward-dipping outer-platform configurations, and/or narrow (<5 km) shelf widths (Fig. 17A-2). These features are conducive to high-energy, current-swept platform tops, where wave agitation and mechanical erosion assist in grain-rich sediment production and offbank currents drive downslope re-sedimentation. Exceptions to this include grain-dominated foreslopes associated with euphotic framestone/bafflestone reefs (*e.g.*, Pomar *et al.*, 1996), which can produce larger volumes of sand and gravel than platform-derived clay and silt, despite the presence of a raised rim.

Extrinsic Controls

Accommodation Change

Interestingly, various scales of relative sea-level change do not significantly affect the development of debris-, grain-, or mud-dominated foreslopes, as prograding and aggrading versions of each exist (Table 3). However, changes in accommodation have impacts on margin stability, timing of large-scale collapse events, development of growth escarpments, and bypassed, toe-of-slope accumulations.

Margin Stability Prograding margins in low-accommodation settings are prone to collapse from basinward accretion beyond the angle of repose without sufficient foreslope substrate to advance over. Conversely, aggrading or retrograding margins in high-accommodation settings are stable and not as prone to collapse, owing to solid underlying foundations (often former lithified margins) for upward-directed growth. Thus, all else being equal, foreslopes of prograding margins are expected to contain higher proportions of debris deposits than foreslopes of retrograding or aggrading margins (Table 3; Fig. 17A-2 and B).

Timing of Large-scale Collapse Events Aside from extrinsic triggers like seismicity and hurricanes, large-scale collapse events [and the potential for sediment focal point generation] also occur across long-term (*i.e.*, second- or third-order) changes in accommodation from relatively more aggradational to progradational sequence architecture. This is due to increased shelf-to-basin relief and potential profile disequilibration during aggradation, followed by severe instability of the margin upon attempted progradation over uncemented or absent substrate.

Development of Growth Escarpments As discussed previously, high and increasing accommodation conditions, coupled with euphotic reefs that are sensitive to sea-level change, are required for the development of growth escarpments and the evolution from accretionary to escarpment stratal configurations (Figs. 11b, 12, and 19b; Table 3). These represent the combined effects of intrinsic (lithified margin sediment type) and extrinsic (accommodation) controls to produce distinctive stratal patterns.

Bypassed Toe-of-Slope Accumulations Variable downlap patterns can reflect the size, productivity, and re-sedimentation of platform-top sand factories. Slightly higher accommodation conditions can result in broad, prolific sand-producing environments, favoring the re-sedimentation and accumulation of grain-dominated deposits at the toe-of-slope and development of climbing downlap patterns (Bosellini, 1984). In contrast, very low-accommodation conditions can constrict sand-producing environments spatially and temporally, resulting in relatively less accumulation of grain-dominated deposits at the toe-of-slope and the formation of sub-horizontal downlap patterns.

Oceanographic Effects

Oceanographic characteristics, such as prevailing wind direction (windward–leeward effects), deep contour currents, circulation patterns, and nutrient flux, are important factors that affect lithified margin sediment factories, sediment distribution patterns, and stratal architecture.

- Foreslopes of leeward margins are typically poor in debris deposits due to continual burying or poisoning of the margin from offbank sediment transport, and meager or interrupted development of lithified, collapsible material (Table 3; *e.g.*, Mullins, 1983).
- The foreslopes of windward margins are likely to be mud-dominated, as margin-derived sand and gravel are swept to backreef environments despite the presence of shallow euphotic framestone reefs (Table 3; *e.g.*, Mullins, 1983).
- Deep contour currents can significantly affect
  i. sediment distribution patterns from strike reworking,
  ii. primary depositional texture through winnowing, and
  iii. stratal architecture from erosion or sculpting, such as for sigmoidal clinoforms and erosional escarpments (Figs. 2, and 11g; *e.g.*, Hooke and Schlager, 1980; Eberli and Ginsburg, 1989).
- Nutrient levels and flux, interlinked with large-scale circulation patterns, are key factors for planktic and oligophotic boundstone productivity (or demise), and thus impact the generation of mud-dominated and debris deposits, respectively (Fischer and Arthur, 1977; Della Porta *et al.*, 2003).
- The large-scale type of margin, such as intracontinental *versus* continental (passive) margins, is linked with circulation patterns and the resulting productivity of contributing sediment factories. Intracontinental basins are more likely to be restricted, poorly circulated, and evaporitic, whereas open-ocean, passive margins are prone to the contrary.

Slope Height

As previously discussed for inherited escarpments, large slope heights (kilometer or more) are inherited when platforms nucleate on high-relief antecedent topography, resulting in continuous foreslope onlap against the escarpment surface and aggradational or retrogradational margin configurations (Figs. 11a, and 19a: Table 3). Slope height acts as an independent control in these cases, where the above-mentioned geometries develop regardless of accommodation setting, lithified margin sediment type, and platform morphology due to the imbalance of sediment available for re-sedimentation and the sediment volume required to fill the entire slope profile. Additionally, depositional profiles that are out of equilibrium (due to large slope heights) have an increased sensitivity to collapse triggers, thus the associated foreslopes contain relatively large proportions of debris deposits (Fig. 19a).

Tectonic Activity and Setting

The regional tectonic setting of a basin is closely linked to subsidence patterns and long-term accommodation changes that can determine margin configuration and stratal patterns. The style and stage of tectonic activity also can dictate the presence and size of antecedent topography that is suitable for carbonate-platform nucleation. For example, the inherited escarpments observed are similar in age and rooted on antecedent topography that was either directly formed by, or indirectly resulted from, Mesozoic rifting and fault-block movement (*e.g.*, Eberli *et al.*, 2004). Active tectonic seismicity (*i.e.*, fault movement and earthquakes) is especially critical as a large-scale collapse trigger, and can produce sizeable margin embayments and basinal megabreccias (Fig. 14e), with little or no linkage to other controlling variables (Borgomano, 2000). Thus, tectonically active periods are likely to develop long-lived sediment focusing mechanisms at the margin that facilitate preferential deposition of bypassed, channel–fan complexes in toe-of-slope or basin-floor settings (Fig. 14f).

External Sediment Input

The input of external sediment sources can also affect foreslope deposit-type proportions and stratal patterns. Mud derived from hinterland weathering, volcanic activity, and non-carbonate pelagic communities increase the likelihood for development of mud-dominated foreslopes (Table 3). The rates and thickness of accumulation of fine-grained material in toe-of-slope or proximal basin-floor settings also influences downlap patterns in progradational systems (climbing *versus* sub-horizontal; Bosellini, 1984; Fig. 11e-f).

Terrigenous siliciclastic input affects carbonate-slope and basin systems by

i. impacting margin growth; and
ii. adding to sediment volume on the foreslope and basin floor.

Clay- and silt-sized siliciclastics and turbid-water conditions can disrupt filter feeding and decrease light penetration ('poisoning'), dramatically hindering margin growth and the potential for generation of debris deposits. Siliciclastic intermixing with carbonate-foreslope deposits and basin infill from bypass contribute to the volume of downslope substrate, which is linked to margin stability, slope height and progradation rates [or P: A Ratios], and accretionary *versus* escarpment margin configurations. For example, thick, onlapping basinal wedges of siliciclastics are observed in accretionary, aggrading Leonardian (Lower Permian) systems of west Texas (Saller *et al.*, 1989; Table 3; Fig. 11d), reflecting substantial phases of siliciclastic bypass. These wedges most likely played a role in upholding accretionary margin relationships by infilling the basin, sustaining slope height during margin aggradation, and maintaining the balance between sediment availability and sediment volume required to fill the slope profile.

SUMMARY OF CARBONATE SLOPE SYSTEMS

Past chapters of *Carbonate Slopes* in *Facies Models* have underscored the facies types and margin styles observed (McIlreath and James, 1978), and the importance of modern examples in our understanding of carbonate slope and basin systems (Coniglio and Dix, 1992). This edition reflects the foundation of knowledge built by the above-mentioned, and numerous other integral studies over the past four decades, as well as concepts developed from recent and ongoing field or subsurface research. As repeatedly concluded from many diverse studies, carbonate slopes are highly variable compositionally, architecturally, and spatially due to a spectrum of sediment sources, re-sedimentation processes, and controlling factors. The vast amount of sediment stored in slope systems offers an insightful and voluminous stratigraphic record that reflects the depositional conditions and characteristics of the platform-top, margin, and basin. Provided herein is a characterization approach that uniformly subdivides carbonate slopes in terms of their deposit types, large-scale stratal patterns, and spatial architecture. Consistent description of these systems allows for meaningful comparison, identification of existing variations, and interpretation of the dominant drivers behind the observed variations, as well as establishes groundwork for predictive capability.

Carbonate-slope deposits can be categorized into debris, grain-dominated, and mud-dominated deposits, which simplify complex heterogeneity by linking depositional characteristics to their sources and re-sedimentation processes. Escarpment *versus* accretionary margin configurations are characterized by fundamentally different stratal patterns that serve as geometrical end members and embody a host of implications regarding evolution and controls. Spatial architecture accounts for the distribution, internal geometries, and temporal superimposition of strike-continuous *versus* strike-discontinuous carbonate-slope accumulations, providing an overall measure for strike variability and the types of seascapes partitioned over time. Combinations of these characteristics reflect the interplay between mulitple intrinsic and extrinsic controls, all of which can be minor *or* overarching influences on carbonate slope and basin development.

Further work is necessary to characterize carbonate-slope systems not contained herein (*e.g.*, empty bucket margins, cool-water settings, *etc.*), identify effects of different diagenetic overprints or processes, and test hypotheses with forward stratigraphic models that incorporate slope processes. As carbonate-slope and basin accumulations continue to show value as economic resources and/or applied tools, available concepts and models need tailoring to clearly outline predictive relationships and rule sets for implementation into hydrocarbon exploration and production strategies.

ACKNOWLEDGEMENTS

The preparation of this manuscript was made possible by the Reservoir Characterization Research Laboratory, the Bureau of Economic Geology, the Department of Geological Sciences at the Jackson School of Geosciences, University of Texas at Austin, and Chevron Energy Technology Company in San Ramon, California. Special thanks to Noel James, Scott Tinker, Ron Steel, Mitch Harris, Jeroen Kenter, and Phillip Playford for research contributions and opportunities. Funding, fieldwork logistics, and collaborations were made possible by Chevron Australia, RCRL, BEG, JSG Geology Foundation, Shell International E&P, Chevron ETC, ConocoPhillips SPIRIT Program, AAPG-GIA, SWAAPG, Geological Survey of Western Australia, Vrije Universiteit, Windjana Gorge and Guadalupe Mountains National Parks, Mount Pierre and Leopold Downs Stations, and the Sierra Diablo and Corn Ranches. Thanks to Meghan Playton, Jerome Bellian, and Beatriz Garcia-Fresca for field support.

REFERENCES

Basic sources of information

Coniglio, M. and Dix, G.R., 1992, Carbonate slopes, *in*, Walker, R.G. and James, N.P., *eds.*, Facies Models: Response to Sea Level Change: Geological Association of Canada, Geoscience Canada Series, p. 349-374.
An excellent synthesis of carbonate slope deposit types, processes, vertical successions, and end-member models, with particular focus on the recent systems of Florida and the Bahamas.

Cook, H.E., and Enos, P., *eds.*, 1977, Deep-water Carbonate Environments: SEPM Special Publication 25, 336 p.
A collection of early, yet classic, carbonate slope and basin outcrop case studies, reservoir studies, and syntheses. These works represent many of the first detailed descriptions and interpretations of carbonate slope deposits, processes, margin styles, and secular changes.

Cook, H.E. and Mullins, H.T., 1983, Basin margin environment, *in,* Scholle, P.A., Bebout, D.G., and Moore, C.H., *eds.*, Carbonate Depositional Environments: American Association of Petroleum Geologists Memoir 33, p. 539-618.
A comprehensive, visual account of carbonate slope and toe-of-slope deposit textures, fabrics, and features, with numerous color outcrop photographs and photomicrographs.

Enos, P. and Moore, Jr., C.H., 1983, Forereef-slope environment, *in*, Scholle, P.A., Bebout, D.G., and Moore, C.H., *eds.*, Carbonate Depositional Environments: American Association of Petroleum Geologists Memoir 33, p. 507-538.
A comprehensive, visual account of carbonate proximal slope deposit textures, fabrics, and features, with numerous color outcrop photographs, photomicrographs, and special emphasis on debris aprons immediately seaward of reefal margins.

McIlreath, I.A. and James, N.P., 1978, Carbonate slopes, *in*, Walker, R.G., *ed.*, Facies Models: Geological Association of Canada, Geoscience Canada Series, p. 245-257.
The first comprehensive synthesis of carbonate slope and basin systems, including deposit type identification, carbonate sediment gravity flow processes, examples from ancient and modern systems, and delineation of depositional (accretionary) versus bypass (escarpment) margins.

Mulder, T. and Alexander, J., 2001, The physical character of subaqueous sedimentary density flows and their deposits: Sedimentology, v. 48, p. 269-299.
An excellent review, compilation, condensation, and quantitatively-based reclassification of sediment gravity flow types and processes, encompassing more than forty years of cumulative research and applicable for both siliciclastic and carbonate deep-water systems.

Tucker, M.E., and Wright, V.P., 1990, Carbonate Sedimentology: Blackwell Science, 421 p.
A comprehensive, yet detailed, overview of most aspects of carbonates along the depositional profile, including facies, depositional environments, sedimentation processes, intrinsic and extrinsic controls, and examples from ancient and modern systems.

Other references

Adams, E.W., and Schlager, W., 2000, Basic types of submarine slope curvature: Journal of Sedimentary Research, v. 70, p. 814-828.

Ball, M.M., 1967, Carbonate sand bodies of Florida and the Bahamas: Journal of Sedimentary Petrology, v. 37, (2), p. 556-591.

Betzler, C., Reijmer, J.J. G., Barnet, K., Eberli, G.P., and Anselmetti, F.S., 1999, Sedimentary patterns and geometries of the Bahamian outer carbonate ramp (Miocene-lower Pliocene, Great Bahama Bank): Sedimentology, v. 46, (6), p. 1127-1143.

Blomeier, D.P.G. and Reijmer, J.J.G., 2002, Facies architecture of an early Jurassic carbonate platform slope (Jbel Bou Dahar, High Atlas, Morocco): Journal of Sedimentary Research, v. 72, (4), p. 462-475.

Borgomano, J.R.F., 2000, The Upper Cretaceous carbonates of the Gargano-Murge region, southern Italy: A model of platform-to-basin transition: American Association of Petroleum Geologists Bulletin, v. 84, p. 1561-1588.

Bosellini, A., 1984, Progradation geometries of carbonate platforms: Examples from the Triassic of the Dolomites, northern Italy: Sedimentology, v. 31, p. 1-24.

Braga, J., Martin, J.M., and Wood, J.L., 2001, Submarine lobes and feeder channels of redeposited, temperate carbonate and mixed siliciclastic-carbonate platform deposits (Vera Basin, Almeria, southern Spain): Sedimentology, v. 48, no. 1, p. 99-116.

Brown, A.A., and Loucks, R.G., 1993, Influence of sediment type and depositional processes on stratal patterns in the Permian Basin-margin Lamar Limestone, McKittrick Canyon, Texas, *in,* Loucks, R.G. and Sarg, J.F., *eds.*, Carbonate Sequence Stratigraphy: American Association of Petroleum Geologists Memoir 57, p. 133-156.

Coniglio, M., 1986, Synsedimentary submarine slope failure and tectonic deformation in deep-water carbonates, Cow Head Group, western Newfoundland: Canadian Journal of Earth Science, v. 23, p. 476-490.

Cook, H.E., McDaniel, P.N., Mountjoy, E.W., and Pray, L.C., 1972, Allochthonous carbonate debris flows at Devonian bank ('reef') margins, Alberta, Canada: Bulletin of Canadian Petroleum Geology, v. 20, (3), p. 439-497.

Crevello, P.D. and Schlager, W., 1980, Carbonate debris sheets and turbidites, Exuma Sound, Bahamas: Journal of Sedimentary Petrology, v. 50, no. 4, p. 1121-1148.

Davies, G.R., 1977, Turbidites, debris sheets, and truncation structures in Upper Paleozoic deep-water carbonates of the Sverdrup Basin, Arctic Archipelago, *in*, Cook, H.E. and Enos, P., *eds.*, Deep-water carbonate environments: SEPM Special Publication 25, p. 221-247.

Della Porta, G., Kenter, J.A.M., and Bahamonde, J.R., 2004, Depositional facies and stratal geometry of an Upper Carboniferous prograding and aggrading high-relief carbonate platform (Cantabrian Mountains, N. Spain): Sedimentology, v. 51, (2), p. 267-295.

Della Porta, G., Kenter, J.A.M., Bahamonde, J.R., Immenhauser, A., and Villa, E., 2003, Microbial boundstone dominated carbonate slope (Upper Carboniferous, N Spain): Microfacies, lithofacies distribution and stratal geometry: Facies, v. 49, p. 175-208.

Droxler, A.W. and Schlager, W., 1985, Glacial versus interglacial sedimentation rates and turbidite frequency in the Bahamas: Geology, v. 13, p. 799-802.

Eberli, G.P., and Ginsburg, R.N., 1989, Cenozoic progradation of northwestern Great Bahama Bank, a record of lateral platform growth and sea level fluctuations, *in*, Crevello, P., Wilson, J.L., Sarg, J.F., and Read, J.F., *eds.*, Controls on Carbonate Platform to Basin Development: SEPM Special Publication 44, p. 339-352.

Eberli, G.P., Anselmetti, F.S., Betzler, C., Van Konijnenburg, J.H., and Bernoulli, D., 2004, Carbonate platform to basin transitions on seismic data and in outcrops: Great Bahama Bank and the Maiella platform margin, Italy, *in,* Eberli, G.P., Masaferro, J.L., and Sarg, J.F., *eds.,* Seismic Imaging of Carbonate Reservoirs and Systems: American Association of Petroleum Geologists Memoir 81, p. 207-250.

Eberli, G.P., Bernoulli, D., Sanders, D., and Vecsei, A., 1993, From aggradation to progradation; the Maiella Platform, Abruzzi, Italy, *in*, Simo, J.A.T., Scott, R.W., and Masse, J.P., *eds.,* Cretaceous Carbonate Platforms: American Association of Petroleum Geologists Memoir 56, p. 213-232.

Enos, P., 1977, Tamabra limestone of the Poza Rica Trend, Cretaceous, Mexico, *in*, Cook, H.E. and Enos, P., *eds.*, Deep-water Carbonate Environments: SEPM Special Publication 25, p. 273-314.

Fitchen, W.M., Starcher, M.A., Buffler, R.T., and Wilde, G.L., 1995, Sequence stratigraphic framework and facies models of early Permian carbonate platform margins, Sierra Diablo, West Texas, *in*, Garber, R.A., and Lindsay, R.F., *eds.*, Wolfcampian-Leonardian shelf margin facies of the Sierra Diablo – Seismic scale models for subsurface exploration: West Texas Geological Society Annual Symposium, no. 95-97, p. 23-66.

Garber, R.A., Grover, G.A., and Harris, P.M., 1989, Geology of the Capitan shelf margin – subsurface data from the northern Delaware Basin, *in,* Harris, P.M. and Grover, G.A., *eds.*, Subsurface and outcrop examination of the Capitan shelf margin, northern Delaware Basin: SEPM Core Workshop No. 13, p. 3-269.

George, A.D., Playford, P.E., and Powell, C.McA., 1995, Platform margin collapse during Famennian reef evolution, Canning Basin, Western Australia: Geology, v. 23, (8), p. 691-694.

Ginsburg, R.N., Harris, P.M., Eberli, G.P., and Swart, P.K., 1991, The growth potential of a bypass margin, Great Bahama Bank: Journal of Sedimentary Petrology, v. 61, (6), p. 976-987.

Glaser, K.S., and Droxler, A.W., 1991, High production and highstand shedding from deeply submerged carbonate banks, northern Nicaragua Rise: Journal of Sedimentary Petrology, v. 61, p. 128-142.

Gomez-Perez, I., Fernandez-Mendiola, P.A., and Garcia-Mondejar, J., 1999, Depositional architecture of a rimmed carbonate platform (Albian, Gorbea, western Pyrenees): Sedimentology, v. 46, p. 337-356.

Grammer, G.M., and Ginsburg, R.N., 1992, Highstand versus lowstand deposition on carbonate platform margins: Insight from Quaternary foreslope in the Bahamas: Marine Geology, v. 103, p. 125-136.

Grammer, G.M., Ginsburg, R.N., and Harris, P.M., 1993, Timing of deposition, diagenesis, and failure of steep carbonate slopes in response to a high-amplitude/high frequency fluctuation in sea level, Tongue of the Ocean, Bahamas, *in,* Loucks, R.G. and Sarg, J.F., *eds.*, Carbonate Sequence Stratigraphy: American Association of Petroleum Geologists Memoir 57, p. 107-131.

Harris, M.T., 1994, The foreslope and toe-of-slope facies of the Middle Triassic Latemar Buildup (Dolomites, northern Italy): Journal of Sedimentary Research, v. B64,(2), p. 132-145.

Hooke, R.LeB. and Schlager, W., 1980, Geomorphic evolution of the Tongue of Ocean and the Providence Channels, Bahamas: Marine Geology, v. 35, p. 343-366.

Hunt, D. and Fitchen, W.M., 1999, Compaction and the dynamics of carbonate platform development: Insights from the Permian Delaware and Midland Basins; southeastern New Mexico and west Texas, USA, *in*, Harris, P.M., Saller, A.H., and Simo, J.A., *eds.*, Advances in Carbonate Sequence Stratigraphy: Application to Reservoirs, Outcrops, and Models: Society for Sedimentary Geology Special Publication No. 63, p. 75-106.

James, N.P., 1981, Megablocks of calcified algae in the Cow Head breccia, western Newfoundland: Vestiges of a Cambro-Ordovician platform margin: Geological Society of America Bulletin, v. 92, (11), p. 1799-1811.

James, N.P. and Ginsburg, R.N., 1979, The seaward margin of Belize barrier and atoll reefs: Special Publication of the International Association of Sedimentologists, 3, 191 p.

James, N.P. and Mountjoy, E.W., 1983, Shelf-slope break in fossil carbonate platforms: An overview, *in*, Stanley, D.J. and Moore, G.T., *eds.*, The Shelf Break; Critical Interface on Continental margins: SEPM Special Publication 33, p. 189-206.

Kenter, J.A.M., 1990, Carbonate platform flanks: Slope angle and sediment fabric: Sedimentology, v. 37, p. 777-794.

Kenter, J.A.M. and Campbell, A.E., 1991, Sedimentation on a Lower Jurassic carbonate platform flank: Geometry, sediment fabric and related depositional structures (Djebel Bou Dahar, High Atlas, Morocco): Sedimentary Geology, v. 72, p. 1-34.

Kenter, J.A.M., Harris, P.M., and Della Porta, G., 2005, Steep microbial boundstone-dominated platform margins; Examples and implications: Sedimentary Geology, v. 178, nos. 1-2, p. 5-30.

Kenter, J.A.M., Van Hoeflaken, F., Bahamonde, J.R., Bracco Gartner, G.L., Keim, L., and Besems, R.E., 2002, Anatomy and lithofacies of an intact and seismic-scale Carboniferous carbonate platform (Asturias, NW Spain): Analogues of hydrocarbon reservoirs in the Pricaspian Basin (Kazakhstan), *in*, Zempolich, W.G. and Cook, H.E., *eds.*, Paleozoic Carbonates of the Commonwealth of Independent States (CIS); Subsurface Reservoirs and Outcrop Analogs: Society for Sedimentary Geology Special Publication 74, p. 183-205.

Kerans, C., and Tinker, S.W., 1999, Extrinsic stratigraphic controls on development of the Capitan reef complex, *in*, Saller, A.H., Harris, P.M., Kirkland, B.L., and Mazzullo, S.J., *eds.*, Geologic Framework of the Capitan Reef: Society for Sedimentary Geology Special Publication 65, p. 15-36.

McDaniel, P.N. and Pray, L.C., 1969, Bank-to-basin transition in Permian (Leonardian) carbonates, Guadalupe Mountains, Texas, *in*, Depositional Environments in Carbonate Rocks; A symposium: SEPM Special Publication 14, p. 79-104.

McIlreath, I.A., 1977, Accumulation of a Middle Cambrian, deep-water limestone debris apron adjacent to a vertical, submarine canyon escarpment, southern Rocky Mountains, Canada, *in*, Cook, H.E. and Enos, P., *eds.*, Deep-water Carbonate Environments: SEPM Special Publication 25, p. 113-124.

Moore, Jr., C.H., Graham, E.A., and Land, L.S., 1976, Sediment transport and dispersal across the deep fore-reef and island slope (-55m to -305m), Discovery Bay, Jamaica: Journal of Sedimentary Petrology, v. 46, no.1, p. 174-187.

Mountjoy, E.W., Cook, H.E., Pray, L.C., and McDaniel, P.N., 1972, Allochthonous carbonate debris flows: Worldwide indicators of reef complexes, banks or shelf margins: Report of the Session - International Geological Congress, Stratigraphy and Sedimentology, Section 6, v. 24, Issue 6, p. 172-189.

Mullins, H.T., 1983, Modern carbonate slopes and basins of the Bahamas, *in,* Cook, H.E., Hine, A.C., and Mullins, H.T., *eds*., Platform Margin and Deep Water Carbonates: SEPM Short Course Notes 12, p. 4.1-4.138.

Mullins, H.T. and Cook, H.E., 1986, Carbonate apron models: Alternatives to the submarine fan model for paleoenvironmental analysis and hydrocarbon exploration: Sedimentary Geology, v. 48, p. 37-79.

Mullins, H.T., Heath, K.C., Van Buren, H.M., and Newton, C.R., 1984, Anatomy of an open-ocean carbonate slope: northern Little Bahama Bank: Sedimentology, v. 31, p. 141-168.

Payros, A., Pujalte, V, and Orue-Etxebarria, X., 2007, A point-sourced calciclastic submarine fan complex (Eocene Anotz Formation, western Pyrenees); Facies architecture, evolution and controlling factors: Sedimentology, v. 54, (1), p. 137-168.

Pomar, L., Ward, W.C., and Green, D.G., 1996, Upper Miocene reef complex of the Llucmajor area, Mallorca, Spain, *in*, Franseen, E.K., Esteban, M., Ward, W.C., and Rouchy, J.M., *eds*., Models for Carbonate Stratigraphy from Miocene Reef Complexes of Mediterranean Regions: Society for Sedimentary Geology, Concepts in Sedimentology and Paleontology, v. 5, p. 191-225.

Saller, A.H., Barton, J.W., and Barton, R.E., 1989, Slope sedimentation associated with a vertically building shelf, Bone Spring Formation, Mescalero Escarpe Field, southeastern New Mexico, *in,* Crevello, P.D., Wilson, J.L., Sarg, J.F., and Read, J.F., *eds*., Controls on Carbonate Platform and Basin Development: SEPM Special Publication 44, p. 275-288.

Savary, B., and Ferry, S., 2004, Geometry and petrophysical parameters of a calcarenitic turbidite lobe (Barremian-Aptian,Pas-de-la-Cluse, France): Sedimentary Geology, v. 168, p. 281-304.

Schlager, W., 1981, The paradox of drowned reefs and carbonate platforms: Geological Society of America Bulletin, v. 92, (4), p. 1197-1211.

Schlager, W. and Camber, O., 1986, Submarine slope angles, drowning unconformities, and self-erosion of limestone escarpments: Geology, v. 14, p. 762-765.

Schlager, W. and Chermak, A., 1979, Sediment facies of platform-basin transition, Tongue of the Ocean, Bahamas, *in*, Doyle, L.J. and Pilkey, O.H., *eds*., Geology of continental slopes: SEPM Special Publication 27, p. 193-208.

Schlager, W. and Ginsburg, R.N., 1981, Bahama carbonate platforms – the deep and the past: Marine Geology, v. 44, p. 1-24.

Spence, G.H. and Tucker, M.E., 1997, Genesis of limestone megabreccias and their significance in carbonate sequence stratigraphic models: A review: Sedimentary Geology, v.112, p. 163-193.

Tinker, S.W., 1998, Shelf-to-basin facies distribution and sequence stratigraphy of a steep-rimmed carbonate margin: Capitan depositional system, McKittrick Canyon, New Mexico and Texas: Journal of Sedimentary Research, v. 68, p. 1146-1174.

Van Konijnenburg, J.H., Bernoulli, D., and Mutti, M., 1999, Stratigraphic architecture of a lower Cretaceous-Lower Tertiary carbonate base-of-slope succession: Gran Sasso d'Italia (Central Appenines, Italy), *in,* Harris, P.M., Saller, A.H., and Simo, T., *eds*., Advances in Carbonate Sequence Stratigraphy-Application to Reservoirs, Outcrops, and Models: Society for Sedimentary Geology Special Publication 63, p. 291-315.

Vigorito, M., Murru, M., and Simone, L., 2005, Anatomy of a submarine channel system and related fan in a foramol/rhodalgal carbonate sedimentary setting; A case history from the Miocene syn-rift Sardinia Basin, Italy: Sedimentary Geology, v. 174, (1-2), p. 1-30.

Weber, L.J., Francis, B.P., Harris, P.M., and Clark, M., 2003, Stratigraphy, lithofacies, and reservoir distribution, Tengiz Field, Kazakhstan, *in*, Ahr, W.M., Harris, P.M., Morgan, W.A., and Somerville, I.D., *eds*., Permo-Carboniferous Carbonate Platforms and Reefs: Society for Sedimentary Geology Special Publication 78, p. 351-394.

Wilson, J.L., 1974, Characteristics of carbonate-platform margins: American Association of Petroleum Geologists Bulletin, v. 58, (5), p. 810-824.

Whalen, M.T., Eberli, G.P., van Buchem, F.S.P., Mountjoy, E.W., and Homewood, P.W., 2000, Bypass margins, basin-restricted wedges, and platform-to-basin correlation, Upper Devonian, Canadian Rocky Mountains: Implications for sequence stratigraphy of carbonate platform systems: Journal of Sedimentary Research, v. 70, (4), p. 913-936.

References cited

Bates, R.L. and Jackson, J.A., *eds*., 1984, Dictionary of Geological Terms, 3rd Edition: American Geological Institute, Anchor Books, New York, 571 p.

Fischer, A.G. and Arthur, M.A., 1977, Secular variations in the pelagic realm, *in*, Cook, H.E. and Enos, P., *eds*., Deepwater Carbonate Environments: SEPM Special Publication 25, p. 19-50.

Mutti, E., Lucchi, F.R., Seguret, M., and Zanzucchi, G., 1984, Seismoturbidites: A new group of resedimented deposits: Marine Geology, v. 55, p. 103-116.

Pomar, L., 2001, Types of carbonate platforms: A genetic approach: Basin Research, v. 13, p. 313-334.

Rich, J.L., 1951, Three critical environments of deposition, and criteria for recognition of rocks deposited in each of them: Geological Society of America Bulletin, v. 62, (1), p. 1-19.

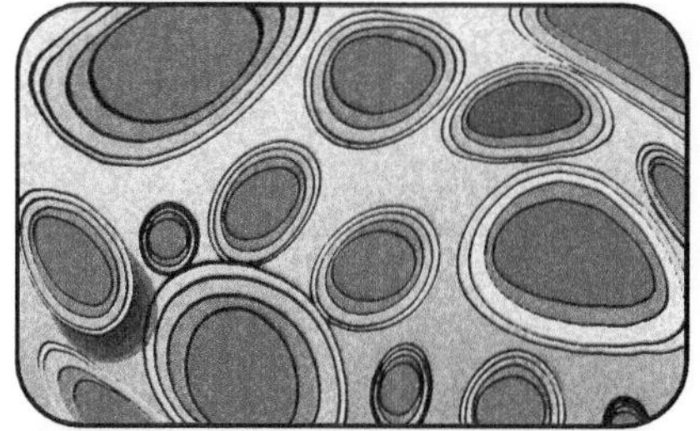

19. Bioelemental Sediments

Peir K. Pufahl, Department of Earth and Environmental Science,
Acadia University, Wolfville, Nova Scotia, B4P 2R6, Canada

INTRODUCTION

Iron formation, chert, and phosphorite are bioelemental sediments because they are composed of the nutrient elements Fe, Si and P. These bioessential elements are required for numerous life processes that ultimately lead to their precipitation as stable minerals. Iron is an important micronutrient and is necessary for photosynthesis. Silicon is used by organisms to construct siliceous shells and skeletons. Phosphorus is a component of many skeletal systems, is a building block of DNA, is essential for cellular energy transfer, and limits biological productivity over geologic timescales. As the precipitation of these elements is so closely linked to biology, bioelemental sediments are simply not recorders of geological processes, but are intimately involved in the evolution of life. It is, therefore, a challenge to develop general facies models because, although bioelemental sediments are primarily biologic, their accumulation has been governed, as much by the evolving biosphere, as by seawater chemistry and physical sedimentologic processes.

Nearly all iron formation is Precambrian. This is the result of direct or indirect biologic formation that operated in conjunction with a host of abiotic precipitation processes (Fig. 1). Chert occurs in both the Precambrian and Phanerozoic (Fig. 1). Precambrian chert is purely abiotic, having formed in a silica-saturated ocean, whereas Phanerozoic chert is the consequence of biotic precipitation in silica-undersaturated seawater. Phosphorite is almost entirely a Phanerozoic phenomenon of which precipitation is dependent on microbial regulation of P in porewater (Fig. 1). This shift from predominantly abiotic precipitation in the Precambrian to biotic accumulation in the Phanerozoic reflects biologic evolution and the ever increasing role life played in the biogeochemical cycling of Fe, Si, and P.

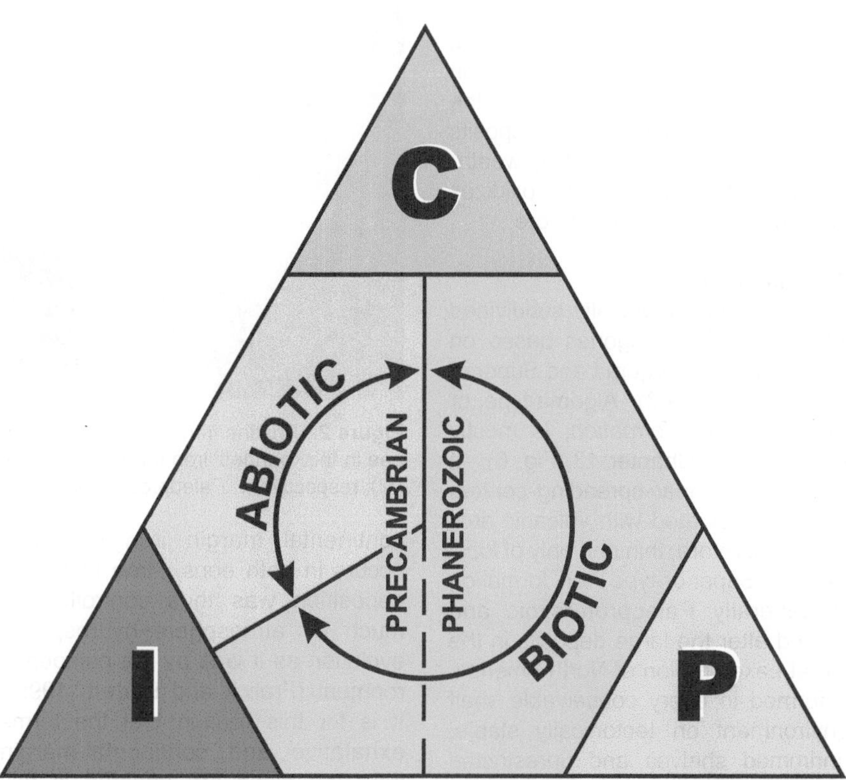

Figure 1. Bioelemental sediment types. Most iron formation (I) is restricted to the Precambrian; its precipitation was controlled by biotic and abiotic precipitation processes in the water column and sediment. Chert (C) occurs in both the Precambrian and Phanerozoic. Precambrian chert is abiotic; Phanerozoic chert is the result of biologic precipitation. Almost all phosphorite (P) is Phanerozoic in age; precipitation is governed by microbial mediation of the concentration of phosphorus in porewater.

This chapter provides an integrated view of bioelemental sediments through time with particular focus on the interplay between biologic, chemical, and sedimentologic processes. The emphasis is on facies that result from direct chemical and biochemical precipitation, as well as those formed through physical and biologic reworking. Finally, discussion focuses on the autogenic and allogenic controls governing the stratigraphic stacking of lithofacies.

IRON FORMATION

Iron formation is a Precambrian, Fe-rich, marine chemical sedimentary rock. The original definition also stipulates a minimum Fe concentration of at least 15 wt. % (James, 1954), but this arbitrary lower limit is often too restrictive when describing the full range of iron formation facies (*e.g.*, Klein, 2005). Iron formation is not the same as *ironstone*, which is generally granular, less siliceous, more aluminous, and Phanerozoic in age. Distinguishing Fe-rich sedimentary rocks in this way highlights important compositional and genetic differences. Most large iron formations resulted from the initial oxygenation of the ocean (Cloud, 1973) and the evolution of Fe-precipitating bacteria (*e.g.*, Konhauser *et al.*, 2002). Thus, it is a distinct class of sedimentary rock restricted primarily to the Precambrian. Ironstone is exclusively authigenic and limited to certain oxygenated Phanerozoic environments.

Aside from the importance of iron formation as a recorder of early Earth history, it is also economically significant because it contains most of the world's Fe ore. It is mined on every continent except Antarctica and the largest deposits occur in Australia, Brazil, and Russia. Other large iron formations occur in Canada and the United States. The richest deposits are naturally concentrated by weathering, which leaches Si and oxidizes Fe to produce a high-grade ore.

Figure 2. Pristine iron formation. Suboxic shelf environment. Red and dark gray laminae in this 'banded' iron formation are red hematite/chert (jasper; H) and gray magnetite (M), respectively. Paleoproterozoic Sokoman Formation, Labrador, Canada.

Classification

Iron formation is typically subdivided into two broad categories based on tectonic regime: Algoma and Superior types (Gross, 1983). Algoma-type, or *exhalative iron formation*, is mostly Archean (*see* Chapter 13, Fig. 6). It accumulated near spreading centers or was associated with volcanic arcs and is, therefore, thin and only of local extent. Superior-type iron formation is generally Paleoproterozoic and named after the large deposits in the Great Lakes region of North America. It formed in every conceivable shelf environment on tectonically stable, unrimmed shelves and unrestricted epeiric platforms (*see* Chapter 13, Fig 6). The appearance of *continental margin iron formation* coincides approximately with the development of extensive shelves and a rapid rise in oxygen known as the Great Oxidation Event (Holland, 2002). These events, together with the upwelling of anoxic, deep-ocean water rich in hydrothermal derived Fe and Si, led to the widespread precipitation of iron formation (Cloud, 1973). The use of non-geographic terms to classify iron formation, such as 'exhalative' and 'continental margin', is preferable because such a classification is based on sedimentologic setting, rather than comparing one deposit type to another.

The change in depositional style of iron formation through the Precambrian conveys the idea that processes causing precipitation differed substantially from the Archean to the Proterozoic. This perception, however, is not accurate because exhalative and continental margin iron formation occurs in both eons. Iron formation deposition was thus controlled as much by atmosphere–hydrosphere evolution as it was by the paleoenvironment (Fralick and Barrett, 1995). It is for this reason that the terms exhalative and continental-margin iron formations are adopted in this chapter instead of using the traditional terminology of Algoma- and Superior-type. Emphasizing depositional context rather than geographic locality is much less restrictive and certainly more useful to the sedimentologist.

As in terrigenous clastic and carbonate depositional systems, autogenic processes produced a diverse array of iron formation lithofacies that can be subdivided into laminated and granular varieties. Laminated, *pristine iron formation* was originally a chemical mud that precipitated in calm-water settings when ferrous Fe transported in anoxic bottom waters was either mixed with oxygenated seawater or oxidized during anoxygenic, bacterial photosynthesis (Konhauser *et al.*, 2002). It was 'pristine' in the sense that it was autochthonous and the site of chemical mud accumulation with little evidence of reworking (Fig. 2). The 'banding' that often characterizes pristine iron formation reflects compositional differences in laminae produced by discrete precipitation events within the water column. This texture is where the commonly used term 'banded iron formation' or BIF originated.

Granular, *reworked iron formation* is a high-energy deposit composed of intraclasts and coated grains (Fig. 3). Waves and currents ripped up fresh precipitates or pristine iron formation in shallow energetic environments to produce cross-stratified and graded beds. Such sediment is sometimes referred to as a granular iron formation (GIF; Simonson, 2003) or simply a grainstone. The latter is borrowed from the modified Dunham Classification for carbonate rocks. As for limestones, a grain type term can be used to further differentiate iron formation grainstones. For example, a grainstone composed primarily of chert grains would be a chert grainstone, whereas one formed mainly of ankerite would be an ankerite grainstone.

Pristine Iron Formation

Understanding how pristine iron formation was generated has been an ongoing challenge. Although modern hot-springs and hydrothermal vents have provided clues to understanding

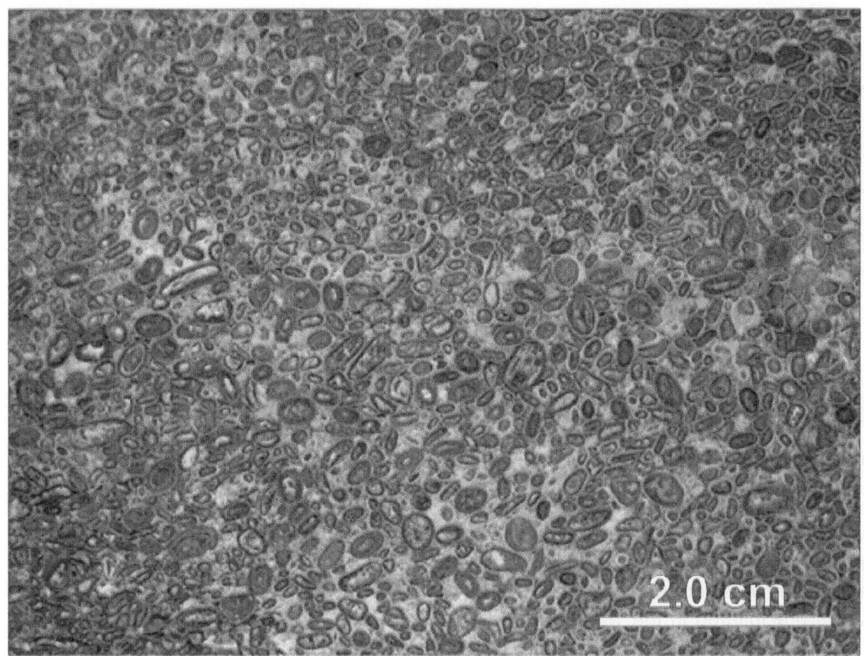

Figure 3. Reworked iron formation. Shallow-shelf environment. Rounded, sand and granule-sized hematite and chert intraclasts in a self-supporting framework. Paleoproterozoic Gunflint Formation, Ontario, Canada. Photo courtesy of Bruce M. Simonson.

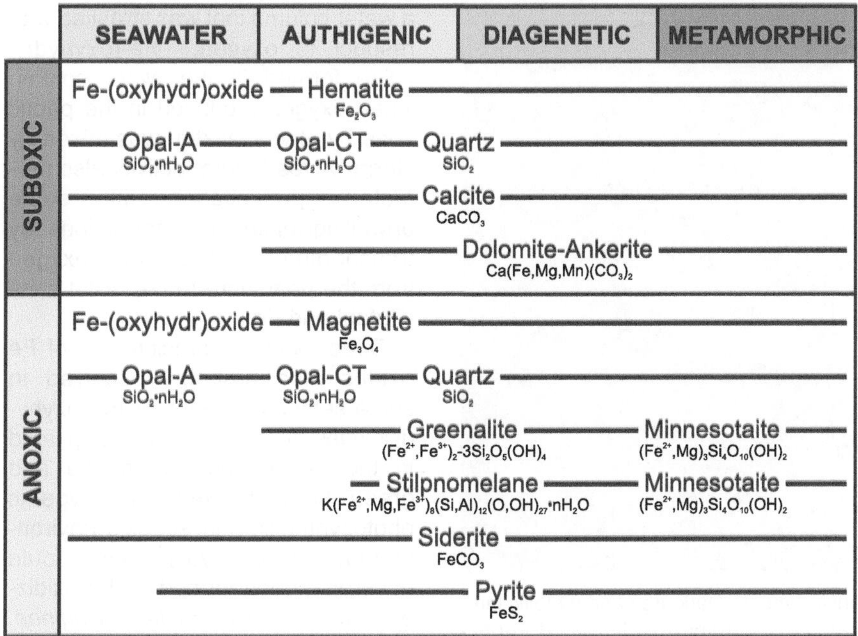

Figure 4. Paragenesis typical of pristine iron formation in suboxic and anoxic paleoenvironments.

the accumulation of laminated, Fe-rich sediment, their mineralogy and areal extent are much different than those of many iron formations. Metamorphism has also altered primary precipitates to a complex assortment of Fe-rich minerals, making it difficult to interpret depositional processes. What has become clear in recent years, however, is that pristine iron formation is the product of both physiochemical and bacterial precipitation in quiet water settings. The wide range of laminated pristine iron formation facies reflects the unique combination of precipitation pathways within such environments.

Suboxic Environments

Low-energy suboxic environments are characterized by hematite (Fe_2O_3) and chert composed primarily of quartz (SiO_2; Fig. 4). Hematite formed during burial via the transformation of a Fe-(oxyhydr)oxide precursor that precipitated in the water column or at the seafloor (Fig. 5A). This precursor was likely ferrihydrite ($Fe_2O_3 \cdot 0.5H_2O$) produced by mixing dissolved ferrous Fe and oxygenated seawater. Such precipitation is kinetically favored in solutions with high Fe concentrations and elevated temperatures, conditions typical of the Precambrian ocean.

Chert likely precipitated as a gel from silica-saturated seawater and authigenically in the sediment (Fig. 5A). Such chert is strictly abiogenic and reflects the high dissolved silica concentrations of the Precambrian ocean prior to the evolution of silica-secreting organisms. Calcite ($CaCO_3$) and members of the dolomite–ankerite ($CaMg(CO_3)_2$–$Ca(Fe, Mg, Mn)(CO_3)_2$) series were common co-precipitates in suboxic environments that were also supersaturated with respect to carbonate.

Anoxic Environments

Anoxic quiet-water environments are distinguished by the presence of magnetite (Fe_3O_4), greenalite ($(Fe^{2+}, Fe^{3+})_2\text{-}3Si_2O_5(OH)_4$), and stilpnomelane ($K(Fe^{2+},Mg,Fe^{3+})_8(Si,Al)_{12}(O, OH)_{27} \cdot nH_2O$; Figs. 4, 5B and 6). All of these minerals contain some ferrous Fe, and reflect precipitation under extremely low oxygen concentrations. Experimental work shows that dissolved oxygen levels were probably no higher than ~10^{-20} $pO_{2\text{-water}}$ and were likely as low as 10^{-70} $pO_{2\text{-water}}$.

Magnetite formed authigenically and directly in seawater (Fig. 5B). Iron isotope data suggest that Fe-(oxyhydr)oxide precipitated in suboxic surface water was altered to magnetite when accumulating on an anoxic seafloor (Johnson *et al.*, 2008). The transformation of Fe-(oxyhydr)oxide to magnetite likely occurred through the addition of ferrous Fe either by diffusion from overlying seawater or via bacterial dissimilatory iron reduction. Dissimilatory iron reduction involves the oxidation of sedimentary organic matter by heterotrophic bacteria that use

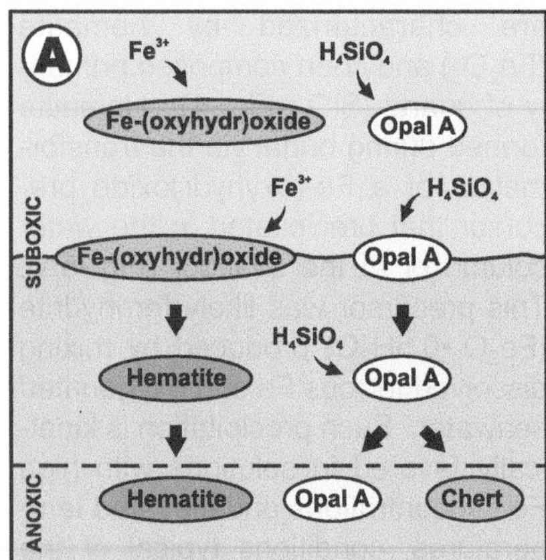

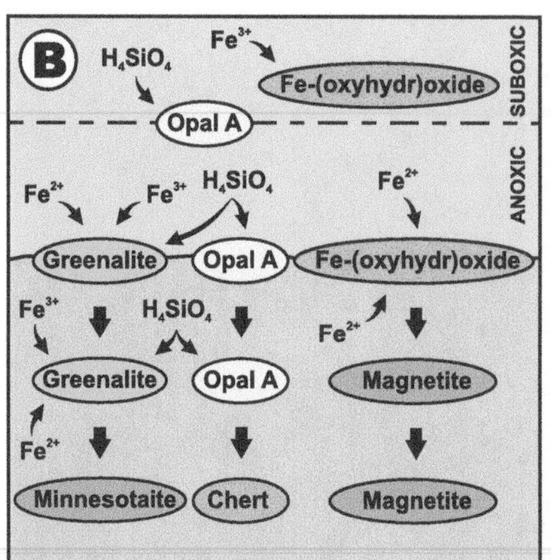

Figure 5. Precipitation of pristine iron formation. **A.** Suboxic environment. Fe-(oxyhydr)oxide precipitated in the water column or at the seafloor was authigenically converted to hematite during burial. Chert initially precipitated as opal-A from silica-saturated seawater or porewater; **B.** Anoxic environment. Magnetite formed by the addition of Fe^{2+} to Fe-(oxyhydr)oxide. Fe-(oxyhydr)oxide precipitated in a suboxic surface ocean and was converted to magnetite as it fell through anoxic seawater or after deposition on an anoxic seafloor. As in suboxic settings opal-A was precipitated in the water column or formed authigenically.

Figure 6. Pristine iron formation. Anoxic shelf environment. Dark gray and metallic laminae are chert (C) and magnetite (M), respectively. Paleoproterozoic Ironwood Formation, Wisconisn, USA. Lens cap is 6 cm in diameter.

Fe-(oxyhydr)oxides as an electron acceptor. Sedimentologic evidence shows that the conversion of Fe-(oxyhydr)oxide to magnetite and the direct precipitation of magnetite also occurred in the water column. Magnetite laminae forming sand shadows behind pebbles and draping reactivation surfaces of ripples and dunes indicate that magnetite formed in seawater accumulated rapidly through suspension rain (Fralick and Pufahl, 2006).

Greenalite and stilpnomelane were created authigenically from original Fe-rich silica gels (Fig. 5B). During metamorphism, greenalite and stilpnomelane change to minnesotaite ($(Fe^{2+},Mg)_3Si_4O_{10}(OH)_2$; Klein, 2005) a common alteration mineral in many iron formations (Fig. 4). With increasing metamorphic grade amphiboles, pyroxenes, and fayalite are high-temperature reaction products. Chert formed where seawater was silica-saturated. Siderite ($FeCO_3$) also precipitated in areas saturated with respect to carbonate.

Pyrite commonly occurs with magnetite and chert in anoxic exhalative paleoenvironments. In such settings hydrogen sulfide and Fe combined near the hydrothermal vent to produce pyrite (FeS_2; Fig. 7).

Bacterial Involvement

There is no doubt that bacteria were essential in precipitating pristine iron formation, however, the nature and extent of their role is not well understood. Indirect physiochemical precipitation of Fe oxides using photosynthetic oxygen has long been thought to be an important factor (Cloud, 1973). Such precipitation occurred in shallow-water environments where photosynthesizing stromatolites generated suboxic conditions. In deeper water settings, abiogenic precipitation likely took place in a water column that was stratified with respect to oxygen. Fe-(oxyhydr)oxide formed in the water column when oxygen produced in the photic zone by phytoplankton was mixed by storm waves. Storm activity also promoted precipitation in some upwelling-related iron formations by transporting photosynthetic oxygen from the nearshore to the distal shelf (Pufahl and Fralick, 2004).

Direct bacterial precipitation of Fe oxides also probably occurred in some settings. Precursor Fe-(oxyhydr)oxides may have been precipitated in the photic zone by purple and green bacteria during anoxygenic photosynthesis. In suboxic environments, Fe-(oxyhydr)oxides could have also been formed by Fe-oxidizers such as *Gallianella ferruginea*. Other bio-oxidizing bacteria presumably precipitated hematite and magnetite directly. In some anoxic paleoenvironments, heterotrophic, Fe-reducing bacteria probably assisted with the conversion of Fe-(oxyhydr)oxides to magnetite.

The conspicuous absence of fossil bacteria and organic-matter in many iron formation lithofacies suggests that direct bacterial precipitation was environment specific. This observation further suggests that the array of pristine iron formation facies formed primarily by indirect precipitation.

Reworked Iron Formation

The accumulation of granular, reworked iron formation is generally well understood because familiar sedimentary structures pinpoint physical processes of deposition. Grainstone production was greatest in shallow-water regions where the seafloor was continually swept by waves and coastal currents.

Shallow-marine Environments

The shallow, high-energy seafloor was the *grainstone factory*. Coarse sand-sized intraclasts (Figs. 3 and 8) were born here by reworking of fresh precipitates, especially in the zone of wave abrasion (*see* Chapter 8). Fine-grained sediments were winnowed and coarser grains were concentrated, producing widespread, shallow-water grainstone. Along wave dominated shorelines, strong currents and prolific grain production produced offshore bar complexes.

Shallow-water grainstone consists of a well-cemented framework of rounded, coarse-grained intraclasts (Fig. 9). The relative proportion of grain types reflects the redox conditions and degree of reworking of the source area. For example, grainstone derived from a suboxic environment is composed primarily of chert and hematite, whereas grainstone from anoxic settings contains Fe-silicates and magnetite. Unlike terrigenous clastic sediments, the rounded nature of individual grains does not reflect transport distance, but rather the reworking of precipitates often with a gel-like consistency.

Depending on paleoenvironment, a variety of coated grain types can also occur. Ooids accumulated in energetic environments where seawater was carbonate saturated (*see* Chapter 16). Oncoids formed in high-energy peritidal settings having plentiful cyanobacteria (Fig. 10; *see* Chapter 16). Hematite-coated chert grains were produced when freshly precipitated silica gel was rolled through ferrihydrite mud (Fig. 9).

Microcrystalline and mesocrystalline chert are the most common authigenic cements in shallow-water grainstone. Greenalite and stilpnomelane are important pore fillers in anoxic environments. Common meteoric and burial diagenetic cements include blocky calcite and sucrosic dolomite or ankerite.

Figure 7. Photomicrograph of pristine iron formation. Anoxic exhalative environment. Laminated pyrite (P; gold) and chert (C). Reflected light. Neoarchean Morley iron formation, Ontario, Canada. Photo courtesy of Philip W. Fralick.

Figure 8. Reworked iron formation. Shallow-water shelf environment. Cross-stratified, jasperlitic grainstone. Paleoproterozoic Sokoman Formation, Labrador, Canada.

Deep-marine Environments

In deep, generally anoxic environments, primary grain production occurred only on the middle shelf in areas above storm wave base (*see* Chapter 8). Intermittent storm reworking of laminated pristine iron formation produced thin lags, graded tempestites, and cross-stratified grainstones (Fig. 11) composed of chert, magnetite and Fe-silicates. Grainstone beds are exceedingly rare in distal shelf settings, having been deposited during the height of

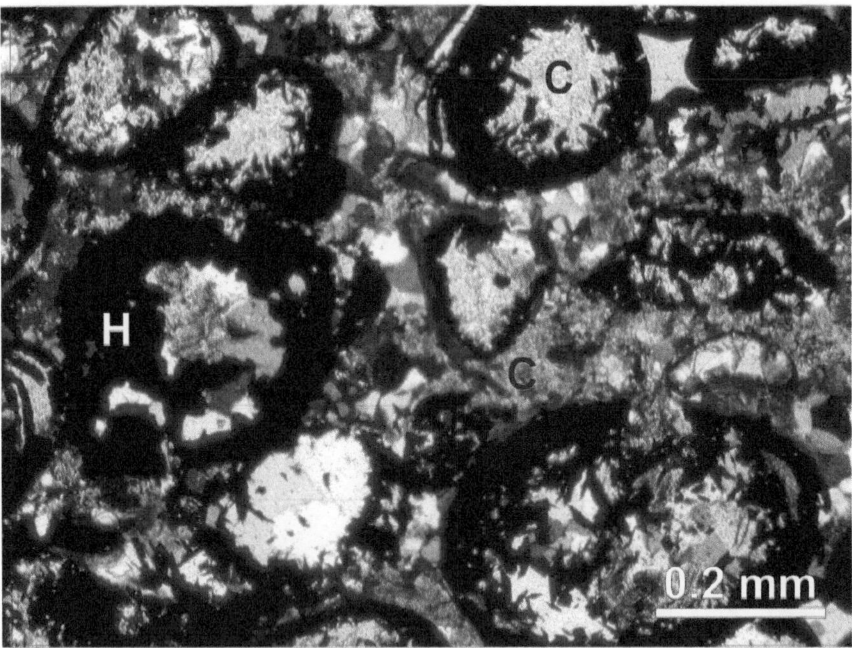

Figure 9. Photomicrograph of reworked iron formation. Shallow-water shelf environment. Grainstone composed of hematite coated (H) chert grains cemented by mesocrystalline chert. Iron carbonate replacement (C) of some grains and cement. Cross-polarized light. Paleoproterozoic Sokoman Formation, Québec, Canada.

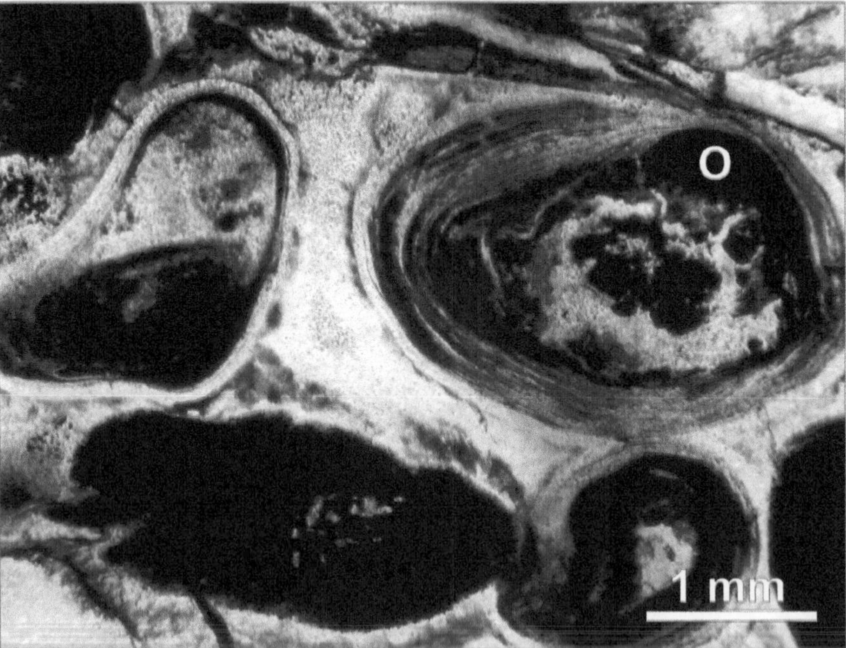

Figure 10. Photomicrograph of reworked iron formation. Shallow-water shelf environment. Chert grainstone with granule-sized oncolites (O). Plane-polarized light. Paleoproterozoic Gunflint Formation, Ontario, Canada.

the largest and most severe storms. Cements in deep-water grainstones are typical of anoxic environments and include greenalite, stilpnomelane, chert, and local siderite. Granular, reworked iron formation does not occur in exhalative settings because accumulation was well below storm wave base.

CHERT

Chert occurs in both Precambrian and Phanerozoic sedimentary rocks and is composed almost entirely of silica (~90–99 wt. % SiO_2). Chert is important because it often encapsulates fossils and sediment, forming a geochemically resistant time capsule (Behl and Garrison, 1994). In addition, its stratigraphic distribution provides a record of elevated surface productivity and, in some cases, deposition below the carbonate compensation depth. Fractured chert is also a significant petroleum reservoir in many Phanerozoic upwelling successions.

Nearly all Precambrian chert is formed by abiogenic precipitation, wherever seawater was saturated with respect to silica. Phanerozoic chert is primarily biogenic and forms almost exclusively in middle- and distal-shelf environments. The shift from abiotic to biotic precipitation is linked to the evolutionary radiation of siliceous microorganisms that occurred in the late Neoproterozoic. This biologic event marks the transition from a silica-saturated to undersaturated ocean and forever changed the global silica cycle. The evolution of sponges in the Cambrian produced widespread subtidal chert; however, they were not abundant enough to prevent supersaturation and continued precipitating chert in shallow-marine environments. The rise of radiolarians in the Ordovician resulted in the near cessation of abiogenic silica deposition in shallow-water settings. A major radiation of diatoms in the Late Cretaceous caused a precipitous decline in the production of sponge spiculites that permanently shifted the locus of chert formation to deeper water environments.

Classification

The term 'chert' has been applied to siliceous rocks spanning a wide range of compositions and textures reflecting a variety of authigenic and diagenetic processes. Chert in its purest form is a fine-grained siliceous sedimentary rock that is hard, dense, vitreous, and breaks with a concoidal fracture (Isaacs, 1982). It is composed of either opal-CT or quartz. The former contains disordered cristobalite and tridymite, which are quartz polymorphs (Calvert, 1974). Less pure siliceous rocks such as *porcelanite* have different textural characteristics than pure chert. Porcelanite contains more clay, is less hard and dense, has a blocky to splintery fracture, and a matte appearance like unglazed porcelain. Chert and porcelanite can appear similar in the field, hence siliceous rocks should always be examined petrographically or by X-ray diffraction (Behl and Garrison, 1994).

Chert can be further subdivided based on its sedimentologic attributes. *Pristine chert* is generally well laminated and characteristic of calm-water settings (Fig. 12). *Reworked chert* is formed of grains and pebble-sized clasts developing in areas of constant wave agitation. Chert conglomerate is produced in environments where intense silicification is interrupted by frequent storms (Fig. 13).

Pristine Chert

Pristine chert generally forms through the authigenic and diagenetic alteration of laminated siliceous sediment. In low-energy environments where bottom and/or porewater become silica saturated *replacive chert* is also produced. Such overlapping processes create a complex array of primary and secondary textures.

Shallow-marine Environments

Most shallow-water pristine chert is a co-precipitate with iron formation or a replacement in Precambrian peritidal carbonates. This is because orthosilicic acid (H_4SiO_4) was easily concentrated by evaporation of silica-saturated seawater along shorelines that were free of diluting siliciclastics. Phanerozoic, shallow-water pristine chert is only common in the Cambrian when supersaturation of orthosilicic acid in neritic settings was still possible.

Pristine chert is generally well laminated having precipitated directly from seawater in low-energy environments such as lagoons and sheltered bays (Fig. 12). The initial precipitate was likely a metastable opal-CT ($SiO_2 \cdot nH_2O$) gel before recrystallizing to quartz (SiO_2) during diagenesis (Fig. 4; Maliva *et al.*, 2005). In seawater containing very high concentrations of orthosilicic acid the precursor was presumably hydrous, amorphous opal-A ($SiO_2 \cdot nH_2O$). The presence of organic matter probably enhanced precipitation by providing abundant suitable nucleation sites.

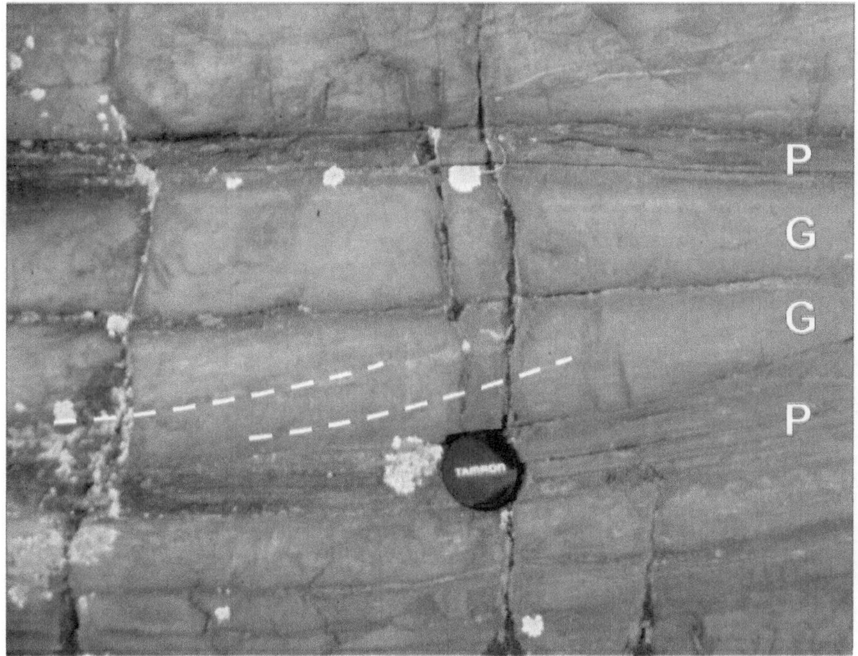

Figure 11. Reworked iron formation. Deep shelf environment. Trough cross-stratified chert grainstone (G) interbedded with laminated magnetite-rich pristine iron formation (P). Paleoproterozoic Ironwood Formation, Wisconsin, USA. Lens cap is 6 cm in diameter.

Figure 12. Pristine chert. Lagoon environment. Well-laminated Precambrian chert (C) and iron carbonate (I). Paleoproterozoic Gunflint Formation, Ontario, Canada.

Many pristine cherts possess a splintered, 'jig-saw puzzle' fabric (Fig. 14). A process called *autobrecciation* produces this characteristic texture that reflects the diagenetic volume loss when opal-CT and opal-A dehydrate and recrystallize to quartz. Intense post-brecciation silicification preserves this texture by cementing *in situ* angular clasts before significant seafloor reworking can occur. Petrographically, pristine chert is composed of micro- and mesocrystalline quartz, except in intensely evaporitic environments where length-slow chalcedony is a co-precipitate (Fig. 15).

Replacive chert is also common in many shallow-water Archean and

Figure 13. Reworked chert. Upwelling environment. Storm reworked chert conglomerate. Campanian Amman Silicified Limestone Formation, Jordan. Hammer is 35 cm in length.

Figure 14. Pristine chert. Lagoon environment. Autobrecciated chert with 'jig-saw' puzzle fit of fragments. Paleoproterozoic Gunflint Formation, Ontario, Canada.

Proterozoic carbonates. Such *peritidal chert* was formed at least, in part, by the replacement of carbonate minerals. It too is composed primarily of micro- and mesocrystalline quartz. Although the replacement processes are not especially well understood, it is apparent that silicification happened very soon after deposition. Textural evidence suggests that replacement of carbonate occurred at the seafloor as silica-saturated seawater percolated down through the sediment (Fig. 16). Other known mechanisms of replacive chert formation include permineralization of bacterial mats and replacement of evaporite minerals.

Deep-marine Environments

Deep-water pristine chert was produced below wave base in middle and distal shelf environments. Direct precipitation from seawater, authigenesis, and replacement were also important chert-forming processes in Archean and Proterozoic subtidal settings. Hydrothermal chert precipitated in Precambrian exhalative systems well below storm wave base, when silica-saturated vent fluids cooled upon mixing with seawater. As in shallow environments, siliceous layers first precipitated as opal-A and converted to quartz during diagenesis.

Most Phanerozoic, deep-water pristine chert is part of the *upwelling triad* of sediments that includes chert, phosphorite, and black shale. Chert also forms in open-ocean and spreading ridge environments, well below the carbonate compensation depth. Such chert is either diagenetic or replacive, forming from orthosilicic acid derived from the dissolution of siliceous microfossils. Because Phanerozoic seawater is so undersaturated with respect to silica, it is extremely corrosive to biosiliceous detritus. Opalline silica is also easily remobilized in sediment making it available for secondary precipitation as discrete laminae or nodules (Calvert 1974; Murray *et al.*, 1992). Abiogenic Phanerozoic chert is far less abundant and is strictly hydrothermal in origin.

In upwelling environments, chert primarily forms from the remobilization of silica derived from diatoms or siliceous sponge spicules. Diatoms thrive in high-nutrient settings and are common in late Mesozoic and Cenozoic upwelling regimes (Fig. 17). Sponges characterize Paleozoic upwelling environments where they filter fed on copious phytoplankton (*see* Chapter 15, Fig. 30). Radiolarians are the dominant source of silica in deeper ocean settings.

Diatoms, sponge spicules, and radiolarians consist of opal-A, which with time and burial will transform to metastable opal-CT and eventually to diagenetic quartz (Fig. 18A). This recrystallization process occurs via two discrete dissolution/precipitation steps that expel water to produce the well-ordered crystal structure of

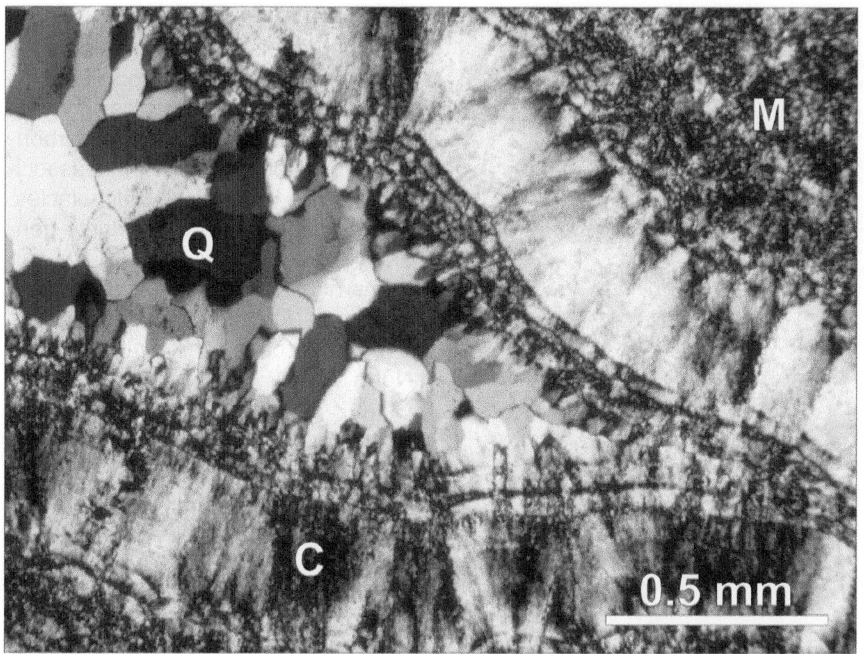

Figure 15. Chert cements. Peritidal environment. Common cement types include microcrystalline chert (M), mesocrystalline or blocky quartz (Q), and fibrous length slow chalcendony (C). Cross-polarized light. Paleoproterozoic Fleming Formation, Labrador, Canada.

Figure 16. Replacive chert. Shallow platform environment. Chert layer (C) replaces carbonate (P). Silicification occurred as silica-saturated seawater percolated from the seafloor down through the sediment. The layer's sharp upper contact is the paleoseafloor. The undulatory lower contact records the maximum depth to which the silicification front penetrated. Paleoproterozoic Denault Formation, Labrador, Canada.

quartz (Isaacs, 1982). The kinetics of these phase changes is controlled by increased temperature and sediment composition (Fig. 19). Clay and organic matter inhibit the rate of opal-A to opal-CT transformation. The presence of calcium carbonate seems to increase the rate of opal-CT nucleation and may also accelerate quartz formation. Opal-CT generally precipitates as blades and needles, whereas diagenetic quartz is either microcrystalline, mesocrystalline, or chalcedonic. Porcelanite forms instead of chert in organic-rich sediments with appreciable amounts of clay. The main difference being that porcelanite consists of opal-CT whereas chert is microcrystalline quartz. As in the Precambrian, these phase transitions result in diagenetic volume loss and in some cases, autobrecciation (Fig. 20).

Reworked chert

Reworked chert only forms in coastal areas where constant wave agitation creates abundant grains from freshly precipitated opal-CT and/or opal-A. Thus, the widespread accumulation of granular chert is restricted to the Precambrian when shallow environments were bathed in silica-saturated seawater. Chert grainstone is most common in iron formation where cross-stratified beds were produced in the grainstone factory. In tide-dominated environments bidirectional paleocurrents are common, as are flat-pebble conglomerates (*see* Chapters 9 and 16) shaped by storm reworking of peritidal chert. Silicified hardgrounds also developed in some wave-swept peritidal settings, providing a firm and stable substrate for stromatolitic growth. Low sedimentation rates in these areas led to rapid and intense silicification of the seafloor producing hardgrounds.

The precipitation of ubiquitous authigenic chert cement in shallow-water Precambrian marine environments was probably enhanced within suboxic segments of the coastline where Fe-(oxyhydr)oxides precipitated. Here, silica was likely enriched in porewater through Fe-redox pumping (Fig. 18B). Fe-redox pumping is a cyclic mechanism that concentrated dissolved silica in porewater by releasing silica adsorbed onto freshly precipitated Fe-(oxyhydr)oxides. During burial dissolution of Fe-(oxyhydr)oxides below the suboxic–anoxic redox boundary liberated sorbed silica to porewater. The escape of silica out of the sediment was prevented by re-adsorption of silica onto Fe-(oxyhydr)oxides just above this redox interface. Fe-redox cycling would have been very efficient at transferring silica from the

Figure 17. Diatomite. Upwelling environment. Diatomaceous laminae are opal-A that are offset slightly by microfaulting. Miocene Monterey Formation, California, USA. Hammer is 35 cm in length.

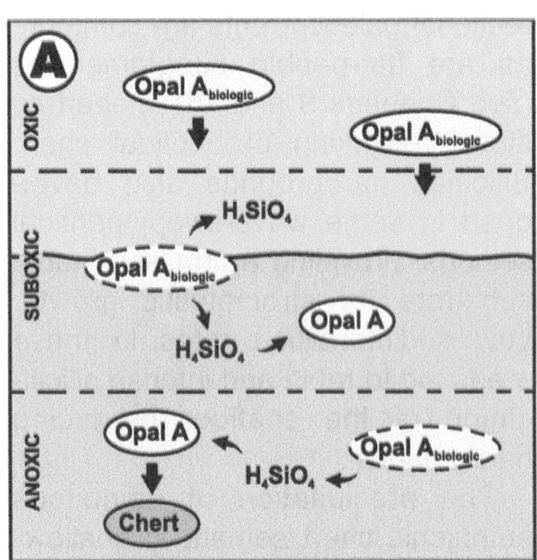

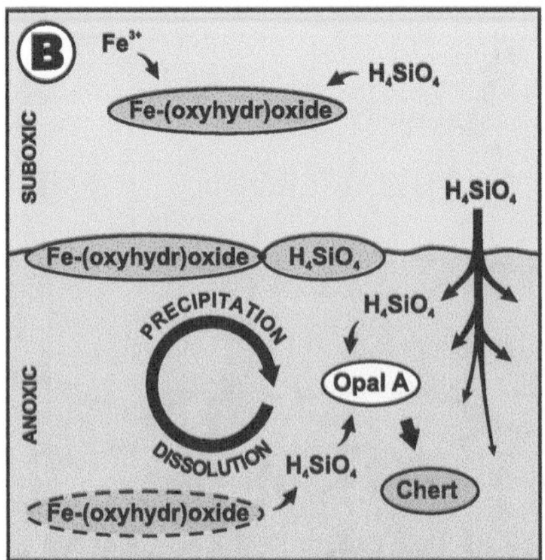

Figure 18. Precipitation of chert. **A.** Upwelling environment. Chert precipitates first as opal-A from remobilized orthosilicic acid (H_4SiO_4) derived from dissolution of siliceous tests such as diatom frustules (opal A-biogenic); **B.** Iron redox pumping of orthosilicic acid in porewater. In suboxic settings the cyclic precipitation and dissolution of Fe-(oxyhydr)oxide within sediment likely maintained the high orthosilicic acid concentrations necessary for authigenic chert precipitation.

water column to the sediment because of the high affinity silica has for Fe-(oxyhydr)oxides (Konhauser *et al.*, 2007). This process was likely important in producing the abundant chert cements that characterize granular iron formation.

Reworked chert is rare in the Phanerozoic because most chert-forming environments are below wave-base. It occurs only in some upwelling environments where autobrecciation of biosiliceous sediment preconditioned the seafloor with abundant, easily reworked intraclasts. The combination of autobrecciation and storm reworking of pristine chert produces spectacular chert conglomerates (Fig. 13). Chert grainstone is uncommon because the seafloor is not continually agitated by fair weather waves, but is reworked only during the largest storms. Such intermittent reworking allows the seafloor to become well silicified between storms, producing much larger clasts.

PHOSPHORITE

Phosphorite is P-rich, bioelemental sediment commonly associated with coastal upwelling. As it can form in both terrigenous clastic and carbonate depositional systems, it is associated with a wide array of lithofacies. The amount of P in phosphorite generally exceeds 18 wt. % P_2O_5, but can be as high as 40 wt. %, making it an important fertilizer ore. Economic deposits occur in the United States, North Africa, the Middle East, and China. The North African and Middle Eastern phosphorites are part of the Late Cretaceous South Tethyan Phosphogenic Province (STPP), the single largest accumulation of phosphorite on Earth.

Apart from the economic value of phosphorite, it is also the most important long-term sink for the global P cycle. In coastal upwelling environments, P is extracted from the surface ocean by phytoplankton and authigenically converted to carbonate fluorapatite (francolite) in the accumulating organic-rich mud below. This process governs primary productivity on geologic time scales, and ultimately controls the rate at which carbon dioxide is removed from the atmosphere and deposited as sedimentary organic matter. Large-scale deposition of organic-rich, phosphatic strata is thus an important feedback mechanism that regulates climate (Föllmi *et al.*, 1993). Because such feedback did not operate in the Precambrian, phosphorite is primarily a Phanerozoic phenomenon.

Classification

Phosphorite is classified based on tectonic and oceanographic setting. *Insular phosphorite* forms on carbonate atolls or islands (*see* Chapter 13, Fig. 6) where lagoonal organic matter and/or guano from nesting seabirds are altered to francolite. Through leaching by meteoric water, these deposits can form thick crusts and in some cases even phosphatize underlying limestone (Fig 21). The geothermal circulation of nutrient-rich seawater upward through the atoll may also help drive phosphatization in these settings. *Seamount phosphorite* is commonly associated with Fe–Mn crusts and may simply be submerged insular phosphorite or the result of

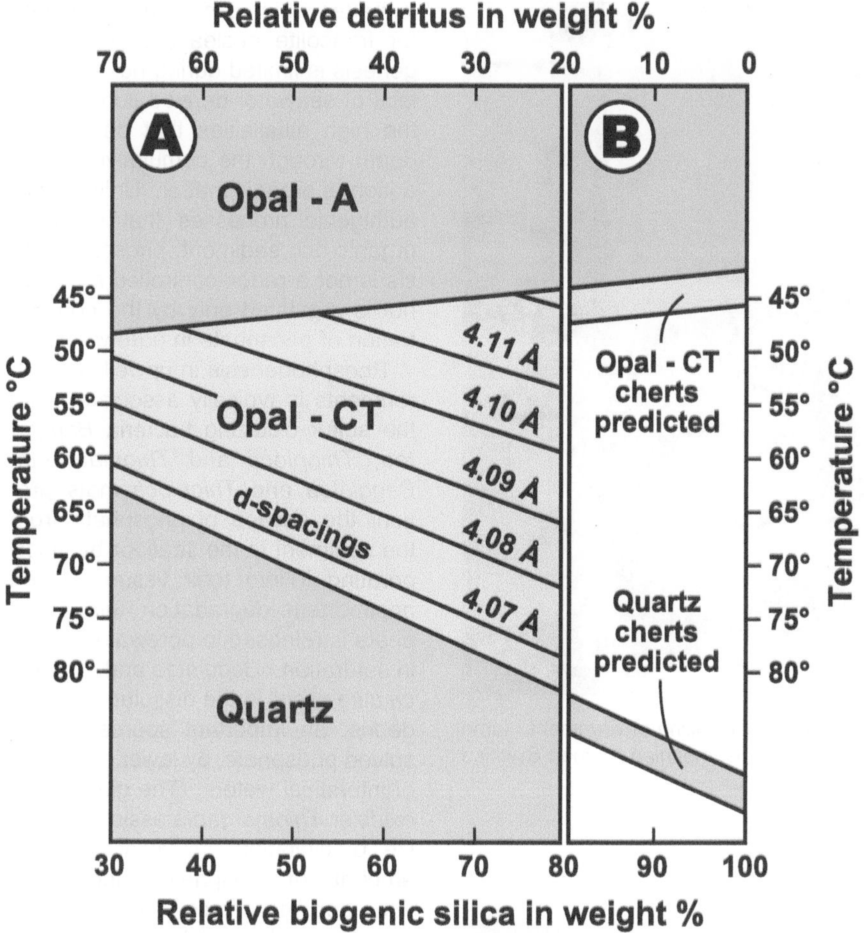

Figure 19. A. Silica phase changes with respect to composition of detritus-rich, diatomaceous sediment from the Monterey Formation, California; **B.** Linear prediction of phase changes in more pure biosiliceous sediment. After Isaacs (1982) and Behl and Garrison (1994). The mobilization of silica and shallow formation of chert potentially occurs over a wider temperature range than illustrated.

francolite precipitation under low nutrient conditions. *Continental margin phosphorite* forms on the slope and shelf within organic-rich sediment beneath sites of active coastal upwelling. Modern upwelling-related phosphorite occurs along western South America, Baja California, southwestern Africa, and western India. In these settings, postmortem accumulation of blooming phytoplankton creates the organic-rich mud necessary for francolite precipitation. Aerially extensive *epeiric sea phosphorite* was deposited in ancient marginal seas and epeiric platforms during sea-level highstands. Much of what is known about shallow phosphatic strata comes from these 'phosphorite giants'. No such deposits are forming today.

Authigenic, biological and hydrodynamic processes work together in all of these settings to form either pristine or reworked phosphorite. *Pristine phosphorite* lacks any reworking or winnowing and consists of phosphatic laminae, nodules, or *in situ* peloids within organic-rich mud (Figs. 22 and 23). *Reworked phosphorite* is a granular deposit created by hydraulically reworking pristine facies (Fig. 24).

Like iron formation and chert, pristine and reworked phosphatic facies can further be subdivided using a naming system adapted from the modified Dunham Classification for carbonate rocks. Phosphatic sediments have also been categorized based on their degree of induration (Garrison and Kastner, 1990). This system breaks phosphorite into three categories: F-, D-, and P-phosphates. F-phosphate is essentially pristine phosphorite and consists of light colored phosphatic laminae and peloids. The D and P-phosphates are reworked facies composed of phosphatic gravels and sands, respectively.

Pristine Phosphorite

Pristine phosphorite forms through phosphogenesis, which is the authigenic precipitation of francolite within sediment or at the seafloor. Francolite is a highly substituted carbonate fluorapatite with the general formula $(Ca_{10-a-b}Na_aMg_b(PO_4)_{6-x}(CO_3)_{x-y-z}(CO_3{\cdot}F)_{x-y-z}(SO_4)_zF_2)$. Its precipitation is a microbially mediated process influenced by bottom- and porewater-redox potential and acidity, chemical gradients, and sedimentation rate (Glenn *et al.*, 1994). Phosphogenesis is distinct from the hydraulic and biological reworking processes that concentrate phosphatic sediment into granular deposits.

Phosphogenesis usually produces pristine facies containing discrete phosphatic layers and local *in situ* phosphatic nodules and peloids (Fig. 25). Peloids are subspherical, silt- to granule-sized francolite grains, which, unlike carbonate peloids, are authigenic and not fecal in origin. Nodules are pebble-sized. Phosphatized particles within pristine phosphorite can include francolite-replaced foraminifera, bivalves, brachiopods, fish debris, and vertebrae bone fragments. Under the right conditions, phosphogenesis can even preserve soft tissue. Diatoms and sponge spicules are the most common siliceous microfossils associated with phosphorites.

Upwelling Environments

The organic-rich seafloor beneath coastal upwelling cells is the *phosphorite factory*. Pristine phosphorite produced in this environment is usually unbioturbated because of the lack of oxygen at the seafloor. Oxygen is rapidly depleted in bottom water by microbes that degrade accumulating organic matter, resulting in an oxygen minimum zone.

Phosphogenesis occurs when porewater becomes supersaturated with phosphate ($H_2PO_4^-$, HPO_4^{2-}, PO_4^{3-}; Fig. 26A). Phosphate is produced when organic matter is degraded through a sequence of microbially mediated redox reac-

Figure 20. Laminated (L) and autobrecciated (A) chert. Upwelling environment. Lime mudstone was silicified with remobilized Si from diatoms. Campanian Amman Silicified Limestone Formation, Jordan. Scale is 3 cm in length.

Figure 21. Insular phosphorite. Phosphatized limestone is dark gray (P). Note karst surface (dashed line) above. Christmas Island, Australia. Photo courtesy of Yvonne Bone.

tions. In order of decreasing energy yield, these reactions include oxic respiration, denitrification, metal oxide reduction, sulfate reduction, and methanogenesis. They define distinct authigenic zones in the sediment that correspond to the profiles of observed porewater concentrations of O_2, NO_3^-, Mn^{2+}, Fe^{2+}, and SO_4^{2-}. Francolite precipitation occurs just beneath the seafloor, within 5–20 cm of the sediment–water interface, in association with the microbial reduction of nitrate, Mn-oxides, Fe-oxides, and sulfate. Formation is aided by the release of polyphosphate from degrading organic matter, which can act as a template for francolite nucleation. Phosphogenesis is limited during burial by the lack of seawater-derived fluorine and the high alkalinities that develop at depth through the cumulative degradation of organic matter. Unlike other authigenic processes that occur in organic-rich sediment, phosphogenesis is not a redox-controlled reaction, but is regulated only by the concentration of phosphate in porewater.

Phosphogenesis in upwelling environments is typically associated with the sulfur oxidizing bacteria *Beggiatoa*, *Thioploca* and *Thiomargarita*. *Beggiatoa* and *Thioploca* mats prevent the escape of phosphate from the sediment to the seafloor by incorporating it into their tissue. Upon postmortem degradation this phosphate is released to porewater, aiding in saturation. *Beggiatoa* and *Thioploca* also assist in the dissolution of fish debris, an important source of dissolved phosphate, by lowering the pH of interstitial waters. The giant sulfur oxidizer *Thiomargarita* assists francolite precipitation by releasing large amounts of phosphate when conditions become anoxic. The microbes can themselves act as nucleation sites and be coated with francolite.

Non-upwelling environments

In areas not associated with prominent upwelling, such as seamounts, the eastern margin of Australia and the continental slope off southern Baja California, the concentration of phosphate is regulated by Fe-redox pumping across a suboxic seafloor (Fig. 26B). Here, phosphate readily adsorbs onto Fe-(oxyhydr)oxide in much the same way orthosilicic acid was scavenged from the Precambrian ocean. Such preferential adsorption of phosphate is kinetically favored because Phanerozoic seawater is severely silica-undersaturated. Upon deposition, the phosphate is released to porewater when Fe-(oxyhydr)oxide dissolves below the Fe-redox boundary (Heggie *et al.*, 1990). The authigenic transformation of primary ferrihydrite (which adsorbs phosphate) to goethite (FeO(OH)) also introduces dissolved phosphate to porewater (Poulton and Canfield, 2006). Although the conversion of ferrihydrite to hematite is thermodynamical-

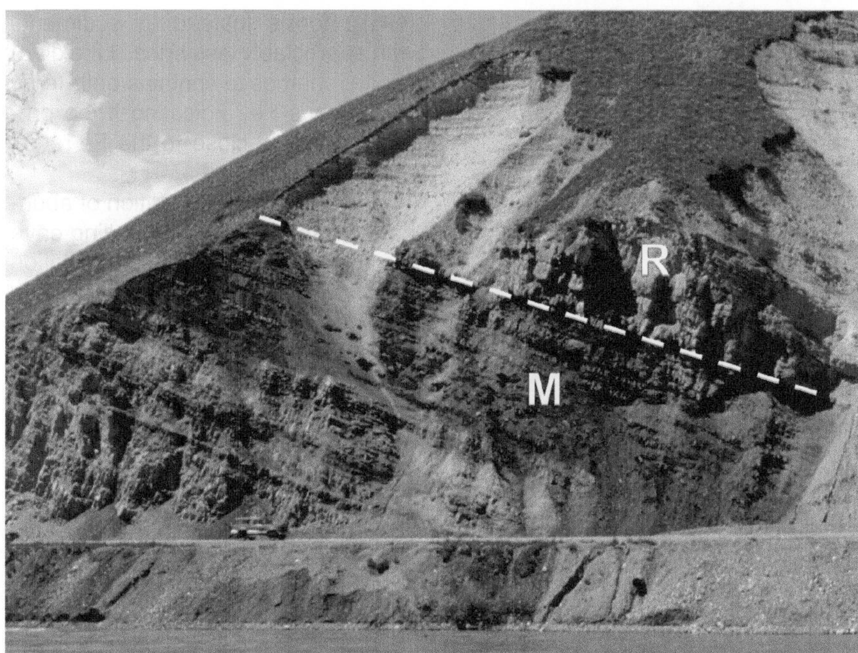

Figure 22. Pristine phosphorite. Middle platform environment. Phosphorite-rich, laminated black shale interbedded with phosphorite grainstone of the Meade Peak Member (M), which is overlain by the Rex Chert Member (R). Organic-rich intervals in the Meade Peak contain up to 15 wt. % total organic carbon. Permian Phosphoria Formation, Wyoming, USA. Photo courtesy of Eric E. Hiatt. Truck in foreground is 4.5 m long.

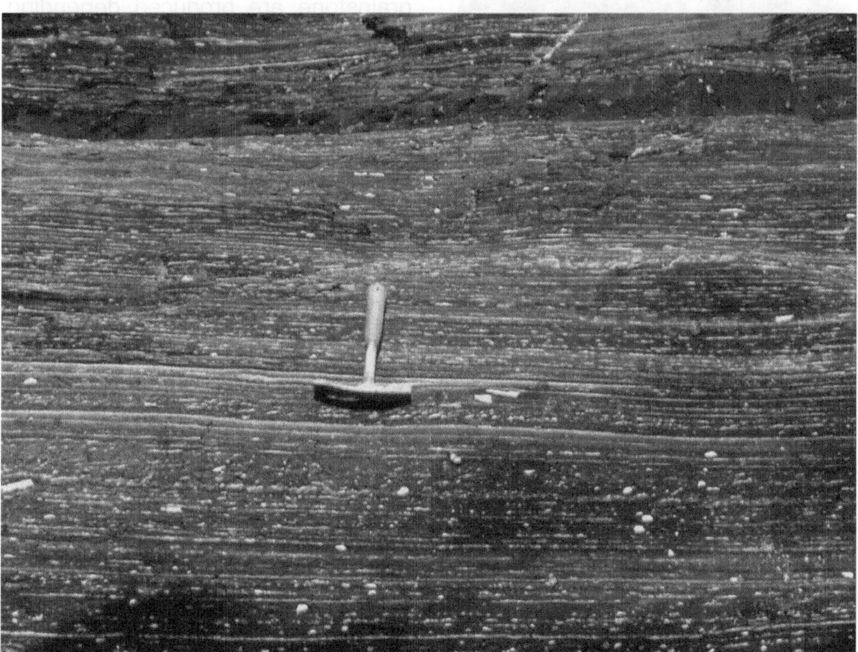

Figure 23. Pristine phosphorite. Distal shelf environment. *In situ* francolite nodules (white specs) within laminated black shale. Miocene Monterey Formation, California, USA. Photo courtesy of Karl B. Föllmi. Hammer is 35 cm in length.

ly favored, the low dissolved Fe content of Phanerozoic porewater and the presence of certain trace elements are thought to promote the precipitation of goethite instead.

Fe-redox pumping to operate efficiently as a P pump requires either repeated mixing of Fe-(oxyhydr)-oxide below the Fe-redox boundary through bioturbation or else a Fe-redox boundary that oscillates vertically with time. Such oscillations in porewater Eh are attributed to seasonal fluctuations in the deposition of sedimentary organic matter. Increased delivery of organic matter produces a rise in the biological oxygen demand at the seafloor, which causes the Fe-redox boundary to move upward through the sediment dissolving Fe-(oxyhydr)oxide. Once the bulk of the organic carbon is microbially respired, oxygen diffuses deeper into the sediment causing this redox boundary to move downward. This allows Fe-(oxyhydr)-oxides to re-precipitate and re-adsorb phosphate before it diffuses out of the sediment (Fig. 26B).

Stratigraphic Condensation

Regardless of the mechanism releasing phosphate to porewater, supersaturation is only achieved during stratigraphic condensation. Periods of low or net negative deposition stabilize the zone of phosphogenesis within the sediment, allowing francolite to precipitate. The thickness and type of pristine phosphorite that develops depends on the stability of this zone. For example, intense phosphogenesis with diminished sedimentation produces continuous phosphatic laminae, whereas francolite precipitation associated with slightly higher sedimentation rates produces discontinuous layers.

Pristine phosphorite is a component of condensed facies, hence it typically contains other features indicative of arrested sedimentation. Such features include carbonate concretion horizons, firmgrounds, and hardgrounds (Fig. 24). When oxygen is plentiful, substrate-controlled trace fossils of the *Glossifungites* and *Trypanites* ichnofacies can penetrate these surfaces (*see* Chapter 3). Multi-generational phosphatic hardgrounds develop in areas that are severely sediment starved by repeated episodes of intense phosphogenesis and submarine erosion (Fig. 27).

Glauconite, Pyrite and Dolomite

There is a depth-stratified sequence of authigenic minerals that co-precipitate with francolite (Glenn and Arthur, 1988). Glauconite ($(K,Na,Ca)_{1.2\text{-}2.0}(Fe^{3+},Al,Fe^{2+},Mg)_{4.0}[Si_{7\text{-}7.6}Al_{1.0\text{-}0.4}O_{20}](OH)_4 \bullet n(H_2O)$) precipitates first followed by pyrite and then dolomite at progressively

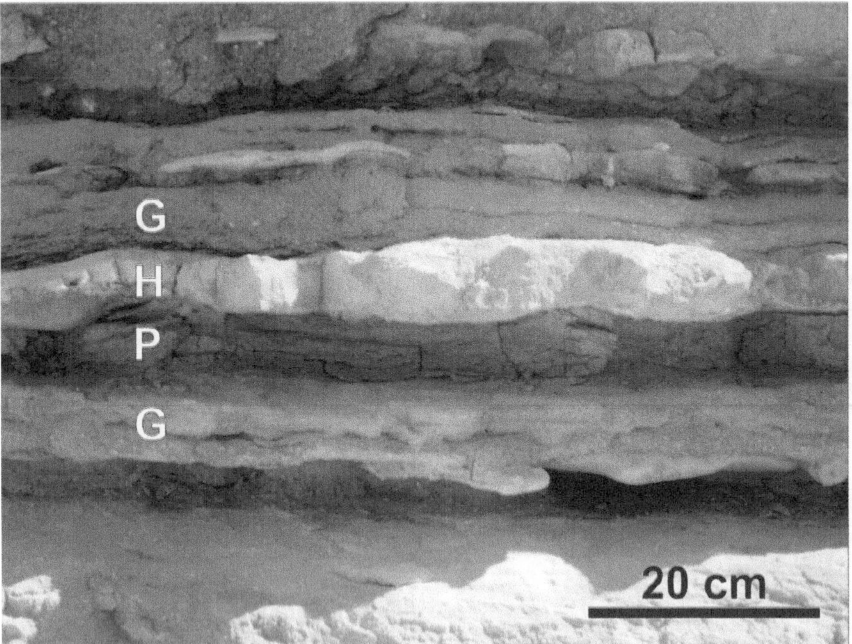

Figure 24. Reworked phosphorite. Middle platform environment. Granular phosphorite beds (G) formed by winnowing pristine phosphorite (P). Winnowing is generally associated with stratigraphic condensation, which also produces hardgrounds (H). Campanian Alhisa Phosphorite Formation, Jordan.

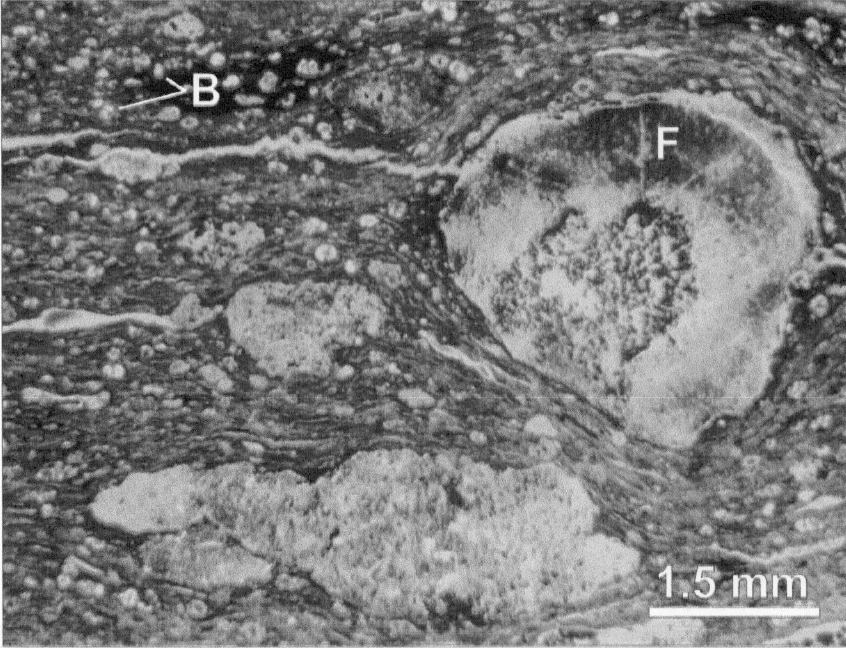

Figure 25. Photomicrograph of pristine phosphorite. Middle platform environment. *In situ* francolite peloids (F) and abundant benthic foraminifers (B). Black areas are organic matter. Plane-polarized light. Campanian Alhisa Phosphorite Formation, Jordan.

deeper levels in the sediment. Glauconite occurs as authigenic peloids, coatings, or cement. Pyrite precipitates as disseminated framboids or as cortical layers on coated phosphate grains. Dolomite is generally a microcrystalline cement that binds detrital grains and earlier authigenic minerals together.

Glauconite forms in upwelling and non-upwelling phosphogenic environments. It precipitates within suboxic sediments following the partial microbial reduction of Fe-(oxyhydr)oxide (Glenn and Arthur, 1988). Authigenic chamosite ($Fe^{2+}_3Mg_{1.5}AlFe^{3+}_{0.5}Si_3AlO_{12}(OH)_6$) forms instead in sediments with appreciably less silica. Like glauconite, chamosite contains both ferric and ferrous Fe, indicating it too precipitates early, close to the Fe-redox boundary.

The microbial degradation of abundant organic matter in upwelling environments also fuels the precipitation of pyrite and dolomite. Pyrite precipitates in the zone of sulfate reduction when ferrous Fe from Fe-(oxyhydr)-oxide dissolution combines with bacterially produced sulfide. Dolomite forms when sulfate, a kinetic inhibitor to dolomite precipitation, is exhausted. The increase in alkalinity at depth created by the cumulative degradation of organic matter and the release of Mg from clay minerals also favor dolomite precipitation.

Reworked Phosphorite

Reworked, granular phosphorite consists of beds or lags that are syndepositionally winnowed from pristine facies. Phosphatic wackestone to grainstone are produced depending on the degree of hydraulic reworking and the intensity of bioturbation. Such deposits consist of phosphatic intraclasts, coated grains, vertebrate bone fragments, and peloids (Fig. 28). Carbonate allochems can include mollusc and brachiopod shell fragments as well as foraminifera. Matrix material can be silt and clay in terrigenous clastic systems or lime mudstone when interbedded with carbonates. Cements are typically calcareous, siliceous, or phosphatic.

Shallow Environments

Shallow-water, granular phosphorite is rare in modern marine environments, but dominates ancient, wave-swept epeiric platforms with prominent coastal upwelling. For reasons that are not well understood, the phosphorite factory in these epeiric systems operated across a wide depositional spectrum to produce phosphorite giants. Storms were apparently the most important agent in reworking pristine phosphorite into granular deposits.

In the nearshore sandy, cross-stratified phosphatic grainstone records the combination of storm waves and coastal currents (*see* Chapter 8). Here, constant syndepo-

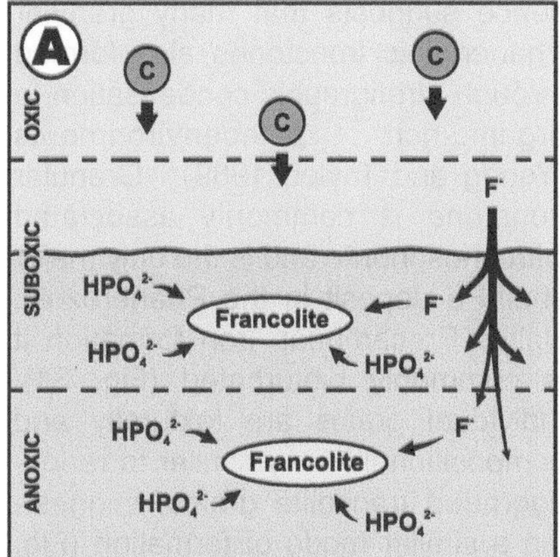

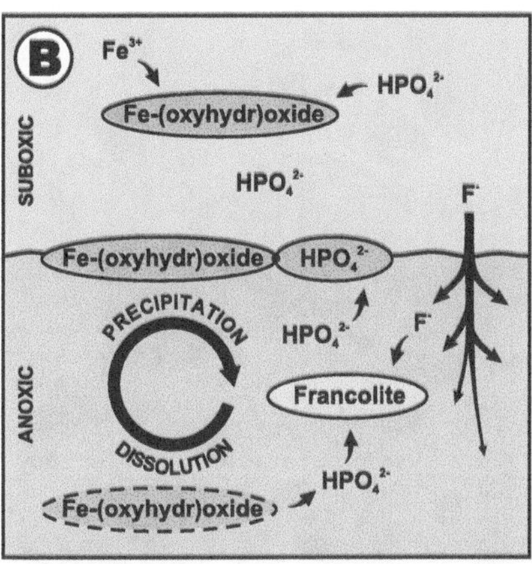

Figure 26. A. Upwelling environment. Microbial degradation of sedimentary organic matter (C) increases the dissolved phosphate concentration of porewater causing francolite to precipitate. Phosphogenesis is aided by the release of polyphosphate, a template for nucleation. Diffusion from overlying seawater provides the necessary F. Phosphate is shown as hydrogen phosphate (HPO_4^{2-}), which is most abundant at the pH of normal seawater (8.0-8.3). Porewater will also contain dihydrogen phosphate ($H_2PO_4^-$) and, at higher pH's, a very small amount of orthophosphate (PO_4^{3-}); **B.** Non-upwelling environment. Iron redox cycling of phosphate in porewater. Cyclic precipitation of Fe-(oxyhydr)oxide and its dissolution below the Fe-redox boundary during burial maintains the high phosphate concentrations necessary for phosphogenesis. Phosphate is also released to porewater when primary ferrihydrite authigenically transforms to goethite.

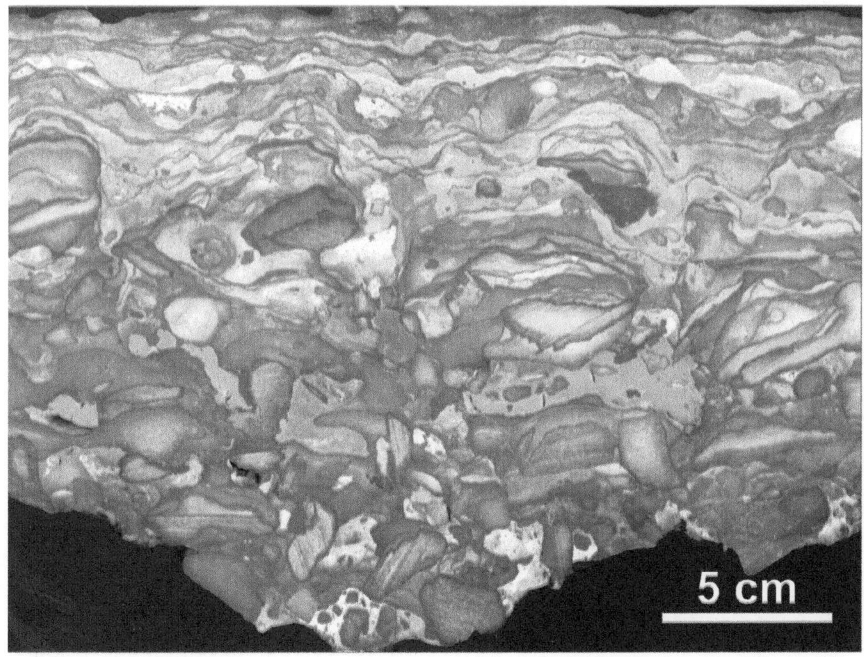

Figure 27. Stratigraphic condensation. Phosphatic hardground. Discontinuous layers record multiple episodes of phosphogenesis and hydraulic reworking. Miocene Monterey Formation, California, USA. Photo courtesy of Karl B. Föllmi.

sitional reworking of the seafloor did not allow pristine facies to accumulate. The shallow middle platform is closer to the upwelling center and characterized by winnowed lags and phosphatic tempestites interbedded with pristine facies. Entrapment of terrigenous clastics in nearshore settings led to stratigraphic condensation and widespread phosphogenesis in this organic-rich environment. Phosphatic lags were generated through storm-wave winnowing of pristine phosphorite. Winnowing was most effective on topographic highs where phosphatic grainstone was shed to accumulate in surrounding lows. Storm-generated currents also transported and redeposited winnowed phosphatic clasts to form massive, coarse-tail graded, and hummocky cross-stratified tempestites. Amalgamation of tempestites produced granular phosphorite beds several meters thick (Fig. 29). The relative thickness of amalgamated beds reflects variations in storm frequency.

Deep Environments

Beneath modern and ancient upwelling centers, the only granular deposits are thin phosphatic lags and distal tempestites (Garrison and Kastner, 1990). Storm-generated phosphatic turbidites dominate farther outboard, where there is a prominent increase in shelf inclination. In some phosphatic depositional systems, turbulent suspension entrained live infaunal crustacea from neritic settings. Upon deposition in deep anoxic environments, these allochthonous trace makers reworked large volumes of sediment, but quickly perished in the absence of oxygen.

Coated Francolite Grains and Phanerozoic Ironstones

Coated francolite grains are ubiquitous in many reworked phosphorites. Debate about their origins has, at times, led to misinterpretations about how some granular deposits formed. Unlike ooids, which precipitate while in turbulent suspension, coated francolite grains are authigenic, precipitating just beneath the seafloor within the zone of phosphogenesis (Pufahl and Grimm, 2003), analogous to soil glaebules and calcrete pisolites. They are considered the granular equivalent of condensed beds, and, as such, preserve a record of physical and chemical change in bottom and porewater.

There are two types of coated francolite grains: redox aggraded and unconformity-bounded. Redox-aggraded grains consist of concordant, concentric francolite laminae interlayered with redox sensitive minerals such as chamosite and pyrite (Fig. 30A). Cortical laminae record changes in porewater Eh during authigenesis that result from variations in the sedimentation rate of organic matter. Redox-aggraded

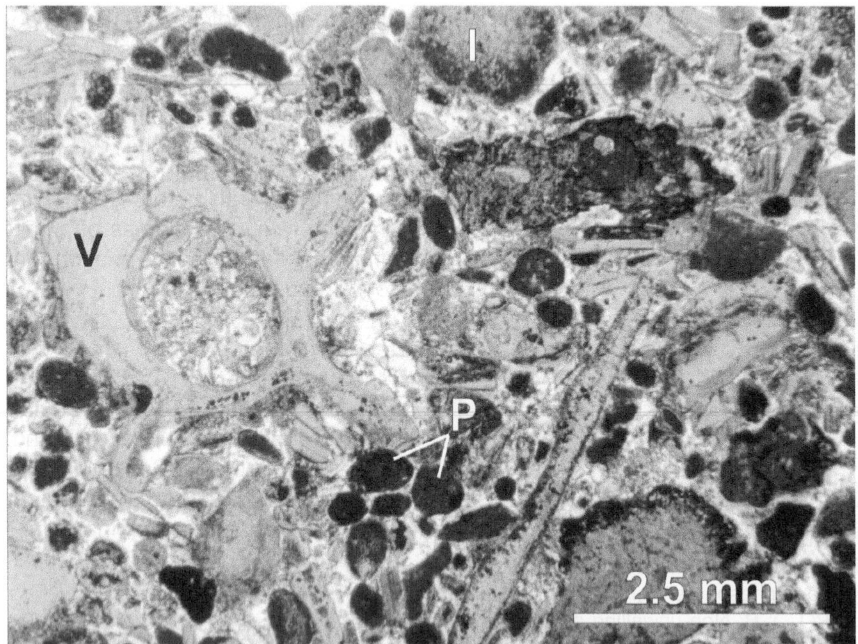

Figure 28. Photomicrograph of reworked phosphorite. Middle platform environment. Phosphatic peloids (P), phosphatic intraclasts (I), vertebrate bone fragments (V). Plane-polarized light. Campanian Alhisa Phosphorite Formation, Jordan.

Figure 29. Amalgamated granular phosphorite. Middle platform environment. Recessive layers are pristine phosphorite and carbonate concretionary horizons. Individual amalgamated phosphorite beds are 0.5 to 2 m thick. One of the thickest economic phosphorite deposits in the world. Campanian Alhisa Phosphorite Formation, Jordan.

grains are thus sensitive indicators of surface ocean productivity and biological oxygen demand at the seafloor.

Unconformity-bounded grains exhibit internal discordances and erosional surfaces, attributable to multiple episodes of phosphogenesis, exhumation, and erosion, followed by reburial into the zone of phosphogenesis (Fig. 30B). Therefore, they contain a record of substrate reworking/winnowing and indicate breaks in calm-water deposition caused by storms and episodic undercurrents.

Textural and mineralogical evidence suggests that many granular Phanerozoic ironstones also formed through stratigraphic condensation in organic-rich paleoenvironments (Young and Taylor, 1989). Granular ironstone is commonly associated with phosphorite and is the only major shelf-Fe deposit in the Phanerozoic. Unlike Precambrian iron formation it is commonly bioturbated (Fig. 31). Individual grains are texturally and compositionally very similar to redox-aggraded francolite grains, suggesting a similar mode of formation (Fig. 32). Finely laminated cortical layers are generally composed of berthierine ($Fe^{2+}_{1.5}AlFe^{3+}_{0.2}\ Mg_{0.2}Si_{1.1}Al_{0.9}O_5(OH)_2$) and/or chamosite. Both contain Fe^{2+} and Fe^{3+} indicating formation close to the Fe-redox boundary during suboxic authigenesis. Siderite, pyrite and, in some cases, francolite are important cements and replacement products.

BIOELEMENTAL DEPOSITIONAL SYSTEMS

Like carbonate depositional systems, the production and accumulation of bioelemental sediment is controlled by subsidence rate, terrigenous clastic input, and position of sea level. Most important, however, are the effects of seawater chemistry, oceanography, and microbial biology. As bioelemental sediments are precipitates, seawater chemistry is a fundamental control on deposit type. Ocean circulation patterns such as upwelling are also critical because they deliver Fe, Si, and P. Bacterial processes promote deposition by increasing precipitation rate.

Sediment production is at a maximum when these parameters are finely tuned, and this situation generally occurs during sea-level rise. Increased shelf accommodation traps diluting terrigenous clastics in nearshore environments. It also shifts the locus of upwelling from the shelf–slope break to more in-board positions, where the reduction in slope allows bioelemental sediments to accumulate over a much greater area. In addition, wave-induced or other cross-shelf currents can also form along flooded margins and contribute to winnowing and reworking of pristine facies into granular deposits.

Bioelemental facies that develop

on unrimmed platforms with prominent upwelling are texturally similar to those that accumulate on high-energy terrigenous clastic shelves and carbonate ramps. This is because bioelemental facies are hybrid sediments that consist of chemical precipitates with a strong hydrodynamic overprint. Deposits such as exhalative iron formation and the more exotic phosphatic accumulations, such as insular or seamount phosphorite, have no terrigenous clastic or carbonate counterparts. As chert occurs in such close affinity with both iron formation and phosphorite it is not treated separately, but is discussed in joint context.

Exhalative Iron Formation

Exhalative iron formation formed around Archean and Proterozoic hydrothermal vents along spreading centers or within back-arc basins and extensional forearcs. Metal sulfides and Fe-(oxyhydr)oxides precipitated when acidic vent fluids containing dissolved Fe, Si and S were mixed with cooler, more oxygenated and alkaline bottom waters. Such precipitation produced a distinct set of laterally gradational facies that reflect changing plume composition during dispersal (Fig. 33). Walther's Law is generally not applicable because accumulation occurred primarily through vertical accretion, well below storm wave base.

Metal sulfides accumulated rapidly close to the vent during initial mixing of the hydrothermal plume, in many cases forming a sulfide lens (Fig. 33). With enough sulfide, a volcanogenic massive sulfide deposit was produced, which can contain significant quantities of Cu, Zn, Pb, Au, and Ag. Precipitation at the discharge point of the vent created a poorly sorted, sulfide-cemented breccia composed of toppled hydrothermal chimneys. In contrast, Fe-(oxyhydr)oxides that precipitated in the hydrothermal plume, dispersed particles for tens to hundreds of kilometers away from the vent (Fig. 33). Discrete pulses of hydrothermal activity and gravity settling produced pristine iron formation with well-formed laminae. Laminae consist of either hematite or magnetite,

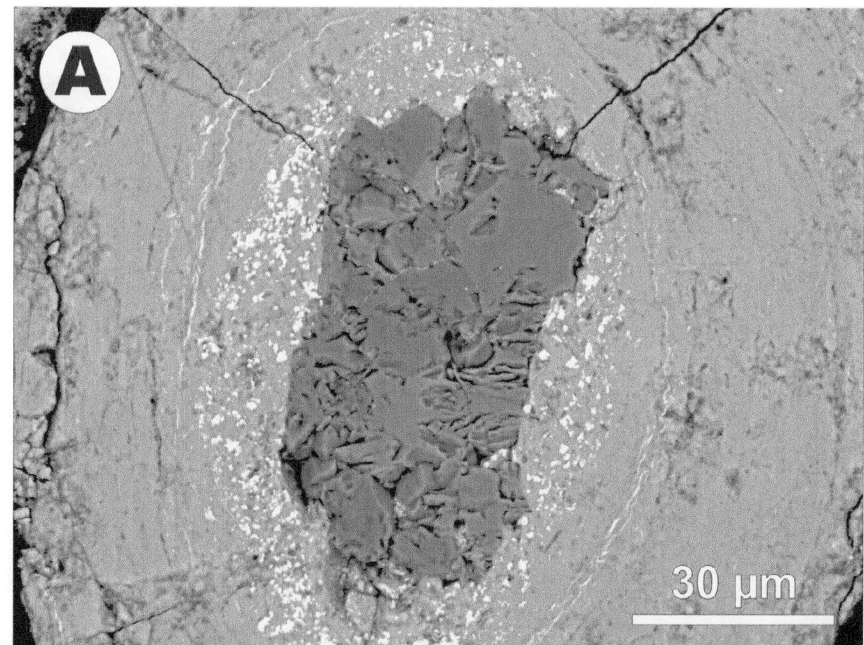

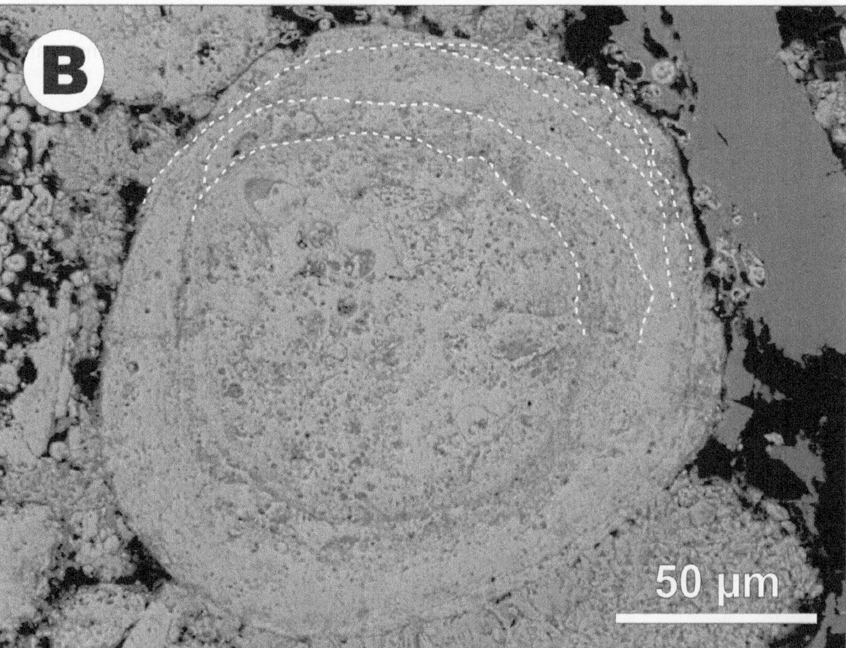

Figure 30. Coated francolite grains. **A.** Redox-aggraded grain with silt-sized feldspar nucleus. Compositionally distinct innermost cortex laminae contain abundant framboidal pyrite (white specs). Contacts marked by thin, discontinuous pyrite laminae. Backscattered electron image. Quaternary, Peru; **B.** Unconformity-bounded grain. Nucleus is a phosphatic intraclast. Dashed lines highlight irregular, unconformable contacts between laminae. Back-scattered electron image. Campanian Alhisa Phosphorite Formation, Jordan.

depending on the redox conditions at the seafloor. Chert accumulated relatively close to the vent in association with Fe-(oxyhydr)oxides (Fig. 33). In suboxic settings the formation of chert was likely enhanced through the deposition of Fe-(oxyhydr)oxides that scavenged orthosilicic acid from the hydrothermal plume (Fischer and Knoll, 2009).

Excellent examples of Archean exhalative iron formations occur throughout the Canadian Shield, many of which are economically important. All have magnetite and are interbedded to varying degrees with submarine volcanic rocks and black shale. Some Proterozoic

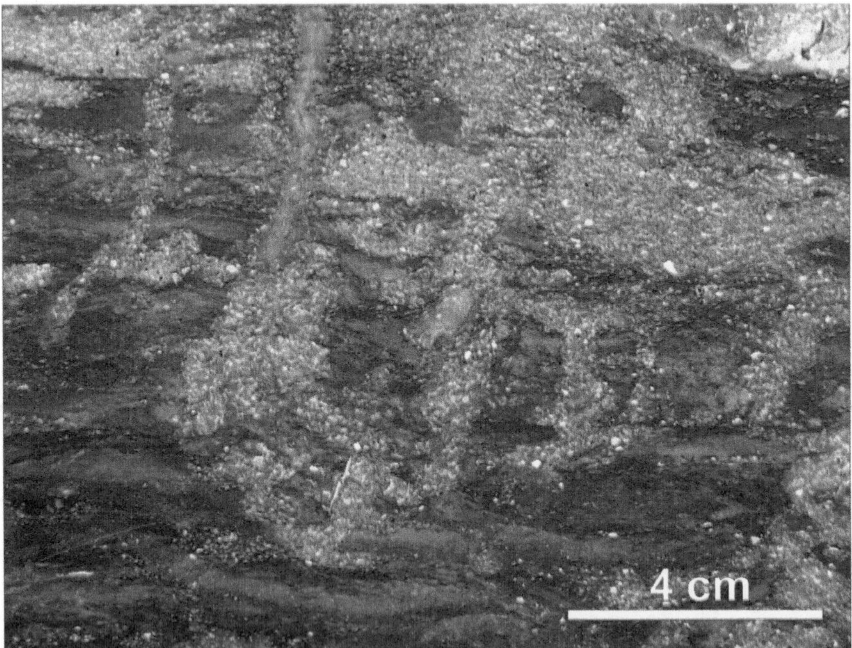

Figure 31. Granular ironstone. Shallow shelf environment. Ironstone grains passively infill burrows in bioturbated mudstone. Devonian Mahantango Formation, Pennsylvania, USA. Photo courtesy of Bruce M. Simonson.

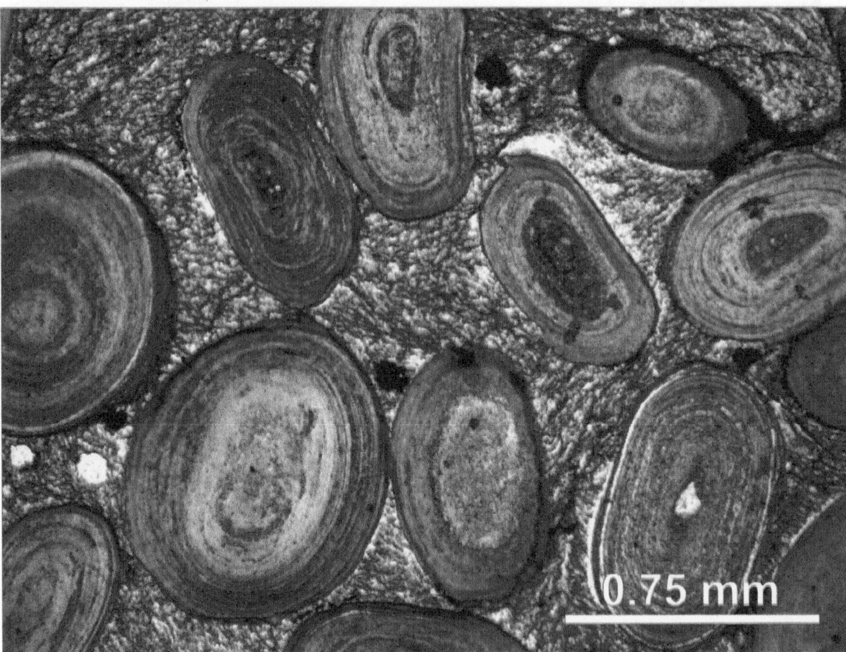

Figure 32. Coated ironstone grains. Finely laminated cortices are similar to redox-aggraded francolite grains suggesting a similar mode of formation. Plane-polarized light. Late Ordovician to Silurian Don Braulio Formation, Sierra de Villicúm, Argentina. Photo courtesy of Bruce M. Simonson.

exhalative iron formations, like those in the Jerome mining district of Arizona contain hematite, suggesting precipitation from a deep ocean that was, at times, suboxic.

Continental Margin Iron Formation

Continental margin iron formation accumulated on the extensive unrimmed platforms that developed at the end of the Archean. Excellent examples include the Paleoproterozoic iron formations of the Animikie Basin in the Lake Superior region of North America, the Labrador Trough of eastern Canada, the Hamersley Basin of Western Australia, and the Transvaal Basin of South Africa. The oldest true continental margin iron formations are those from the Hamersley and Transvaal basins (~2.5 Ga). The abundance of pristine iron formation implies deposition in deep shelf or possibly slope environments. Younger iron formations (~1.9 Ga) from North America accumulated in the full spectrum of shelf environments.

Upwelling is interpreted to have provided a sustained supply of anoxic bottom water rich in dissolved Fe and Si (Fig. 34) in such settings. Rare-earth-element concentration data and Nd isotopic data indicate a hydrothermal origin for the Fe and Si before they were mixed with seawater. Germanium and Si ratios imply that at least some Si originated from continental weathering. High Fe concentrations were probably maintained in the water column through diagenetic recycling of Fe at the seafloor. Modeling suggests that within anoxic paleoenvironments microbial reduction of Fe-(oxyhydr)oxides in the sediment returned a significant portion of Fe back to the overlying water column, keeping ambient Fe concentrations high.

Like all unrimmed shelf systems, lithofacies stacking patterns within continental margin iron formation comprise a sedimentary wedge that fines and thickens basinward. Nearshore facies typically consist of cross-stratified grainstones produced by a variety of shelf currents (Fig. 34). Pristine iron formation accumulated along low-energy shorelines and in intertidal areas. Lithofacies generally possess the same textural attributes as carbonate sediments deposited in similar environments. Pristine iron formation is interbedded to varying degrees with granular tempestites in offshore areas (Fig. 34). These distal facies, because of significant hydrodynamic overprinting, are texturally more akin to sediments on high-energy terrigenous clastic shelves.

Although there is a consensus that iron formation precipitation occurred in a stratified ocean, a great deal of uncertainty exists regarding the character and causes of stratification. The lateral distribution of iron formation lithofacies does, however, suggest

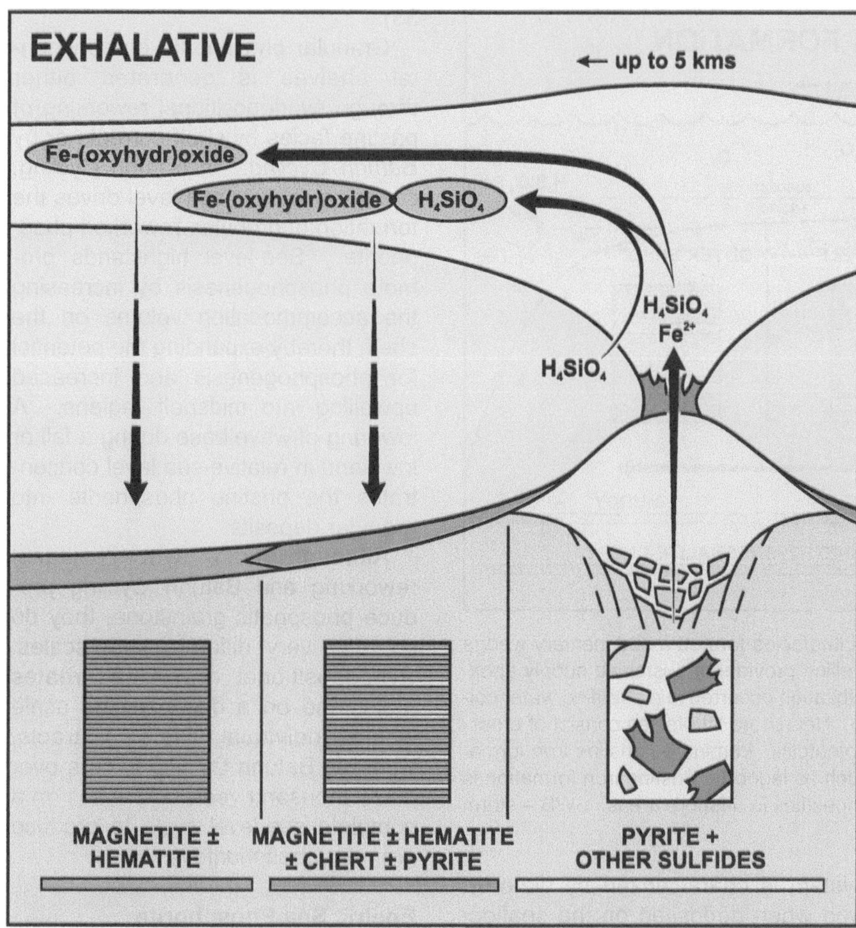

Figure 33. Exhalative iron formation. Mixing of the hydrothermal plume with seawater produces a distinct set of laterally gradational lithofacies. Metal sulfides precipitated rapidly close to the vent forming a sulfide lens and breccia composed of toppled hydrothermal chimneys. Iron-(oxyhydr)oxides precipitated as the hydrothermal plume dispersed. Magnetite or hematite laminae formed depending on seafloor redox conditions. Chert also accumulated relatively close to the vent in association with Fe-(oxyhydr)oxides. The formation of chert was enhanced in suboxic bottom water via the precipitation of Fe-(oxyhydr)oxides, which scavenged orthosilicic acid from the hydrothermal plume.

that oxygen stratification was the primary control on facies disposition (Fig. 34). The sharp transition from hematitic, pristine and granular iron formation in nearshore environments to laminated, magnetite-rich deposits in middle-shelf settings implies a surface ocean that was suboxic to at least fair-weather wave base. The presence of stromatolites within some shallow-water facies further suggests that precipitation of Fe-(oxyhydr)oxides could have been induced by mixing of photosynthetic oxygen with Fe-rich seawater. Oxygen production by cyanobacterial phytoplankton in offshore areas probably produced a suboxic surface ocean having a prominent chemocline.

Destruction of this layer during storm mixing, and the offshore transport of photosynthetic oxygen during storms is interpreted to have caused precipitation of Fe-(oxyhydr)oxides in middle- and distal-shelf environments. Such a process likely produced the intimate interlayering of magnetite and chert that characterizes pristine iron formation in these settings (Fig. 6; Pufahl and Fralick, 2004). Storm mixing and transport of oxygen caused rapid precipitation of Fe-(oxyhydr)oxides, stripping seawater of dissolved ferrous Fe. The resultant decrease in dissolved Fe, together with the background precipitation of abiogenic silica, formed magnetite–chert couplets as well as chemically graded laminae with Fe-rich bases and siliceous tops. Changes in the rate of photosynthetic oxygen production may have also produced magnetite–chert couplets. As in modern environments, variations in upwelling and thus, nutrient delivery to middle- and distal-shelf environments probably caused phytoplankton blooms that increased surface ocean oxygen levels. This rapid increase in oxygen also would have stripped seawater of dissolved Fe and led to the formation of interlaminated magnetite and chert. In addition, changes in surface ocean temperature brought about by either storm mixing or variations in upwelling intensity of cool intermediate waters could have also formed some magnetite–chert couplets. Recent laboratory experiments demonstrate that the rate of Fe precipitation by photosynthesizing, anoxygenic bacteria, reaches a maximum between 20 and 25°C (Posth *et al.*, 2008). Decreasing or increasing water temperatures apparently slows biogenic Fe precipitation and may promote the precipitation of abiogenic, opalline silica by temperature related-changes in silica solubility. Proposed photochemical precipitation processes, utilizing sunlight to directly oxidize ferrous to ferric Fe, were probably unimportant in producing Fe-rich laminae because they have been shown to be much too slow to produce significant amounts of sediment.

As on modern shelves, chemoclines, other than oxygen stratification, likely existed in specific environments. The presence of such steep chemical gradients may have contributed to sediment production because many biomineralizing bacteria make a living bridging these redox interfaces.

In a sequence stratigraphic context, the same surfaces used for correlation in Phanerozoic siliciclastic successions are also present in continental margin iron formation (Fig. 35). There are, however, two main differences; i) the lack of bioturbation, macrofossils, and terrestrial vegetation can make recognition of key surfaces difficult; and ii) the maximum flooding surface is not generally marked by a prominent depositional hiatus because of the accumulation of copious precipitates in middle- and distal-shelf environments. Con-

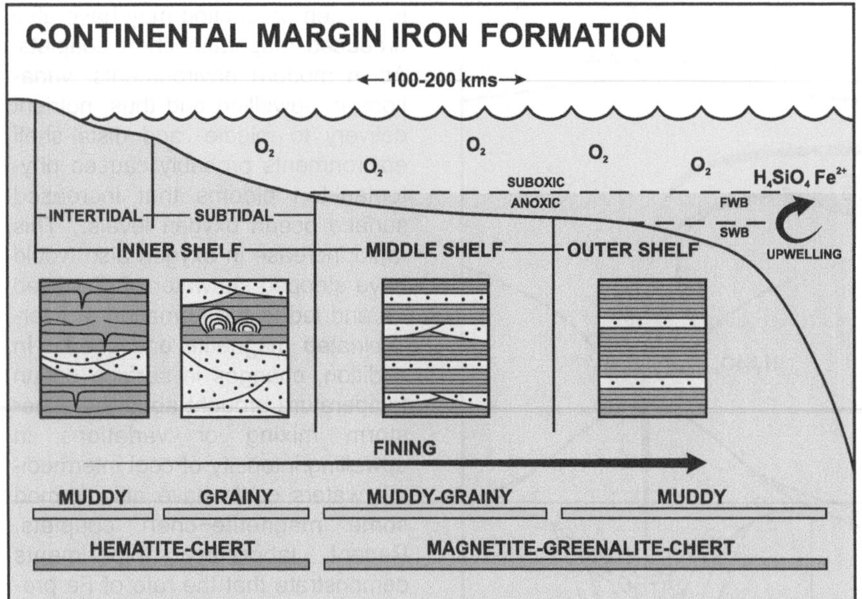

Figure 34. Continental margin iron formation. Lithofacies formed a sedimentary wedge that fines and thickens basinward. Coastal upwelling provided a sustained supply anoxic bottom water rich in dissolved Fe and Si. Precipitation occurred in a stratified water column that was suboxic to fair-weather wave base. Nearshore lithofacies consist of cross-stratified grainstones that are in some cases stromatolitic. Laminated pristine iron formation accumulated in low energy environments such as lagoons. Pristine iron formation is interbedded to varying degrees with granular tempestites in offshore areas. SWB = storm wave base; FWB = fair-weather wave base.

densed intervals instead formed in shallow, wave-swept environments where low sedimentation rates allowed hardgrounds to develop (Fig. 35).

Continental Margin Phosphorite

Continental margin phosphorite forms on the distal shelf beneath coastal upwelling cells (Fig. 36). On modern shelves it accumulates along western continental margins because favorable trade winds induce upwelling. Continental margin phosphorite is part of the upwelling triad of sediments that characterize all Phanerozoic upwelling regimes.

Associated lithofacies can be quite variable because continental margin phosphorite forms in both terrigenous clastic and carbonate depositional systems. However, it is more common in clastic successions because periodic eutrophic nutrient levels prevent the development of even the most robust heterozoan carbonate producers (*see* Chapter 15). Nevertheless, a distinct set of diachronous facies belts develop that reflect the lateral distribution of upwelling-related, planktic ecosystems. The central and most active part of the upwelling system is characterized by diatoms, who when deposited on the seafloor eventually produce biogenic chert. Less nutrient-rich margins of the upwelling front are dominated by nanoplankton and autotrophic dinoflagellates. Their deposition creates mirrored facies belts of organic-rich, hemipelagic ooze.

Phosphogenesis occurs within any of these facies belts in association with the microbial respiration of accumulating organic matter. Such degradation produces a prominent oxygen minimum zone that impinges on the shelf where pristine phosphorite is actively forming (Fig. 36). Phosphogenesis is greatest during marine transgression because sediment starvation in offshore areas creates widespread stratigraphic condensation. Such conditions are also a prerequisite for the formation of granular ironstone, providing there is sufficient Fe supplied by continental weathering. Continental margin phosphorite, and, in some cases granular ironstone, are important sequence stratigraphic markers because they form best along maximum flooding surfaces where terrigenous clastics were trapped in shallow environments (Fig. 35).

Granular phosphorite on continental shelves is generated either through syndepositional reworking of pristine facies by shelf currents or by *Baturin Cycling*. In Baturin Cycling, changing relative sea level drives the formation of granular, reworked phosphorite. Sea-level highstands promote phosphogenesis by increasing the accommodation volume on the shelf, thereby expanding the potential for phosphogenesis and increased upwelling into midshelf regions. A lowering of wave base during a fall or lowstand in relative sea level concentrates the pristine phosphorite into granular deposits.

Although both syndepositional reworking and Baturin Cycling produce phosphatic grainstone, they do so over very different time scales. Syndepositional reworking creates grainstone on a decadal time scale within individual systems tracts, whereas Baturin Cycling occurs over a few thousand years and relies on a complete sea-level cycle to produce granular phosphorite.

Epeiric Sea Phosphorite

Epeiric sea phosphorites can form 'phosphorite giants' that accumulated in ancient marginal seas and epeiric platforms. Outstanding examples include the Late Cretaceous STPP in North Africa and the Middle East as well as the Permian Phosphoria Formation in the western United States. Because they lack modern analogues, there is uncertainty as to whether these phosphorite giants resulted from global variations in P cycling or from local sedimentologic and tectonic controls on P burial. Lithofacies are coarse grained because most sediment accumulated in shallow water above storm weather wave-base (Figs. 35 and 37). Phosphorite is generally associated with carbonates because terrigenous clastics were trapped in nearshore environments. As is typical in epeiric seas, frictional dampening of tide-generated currents prevented the widespread formation of tidal deposits.

As in continental margin phosphorite, phosphogenesis was stimulated by the accumulation of upwelling-related organic matter during sea-

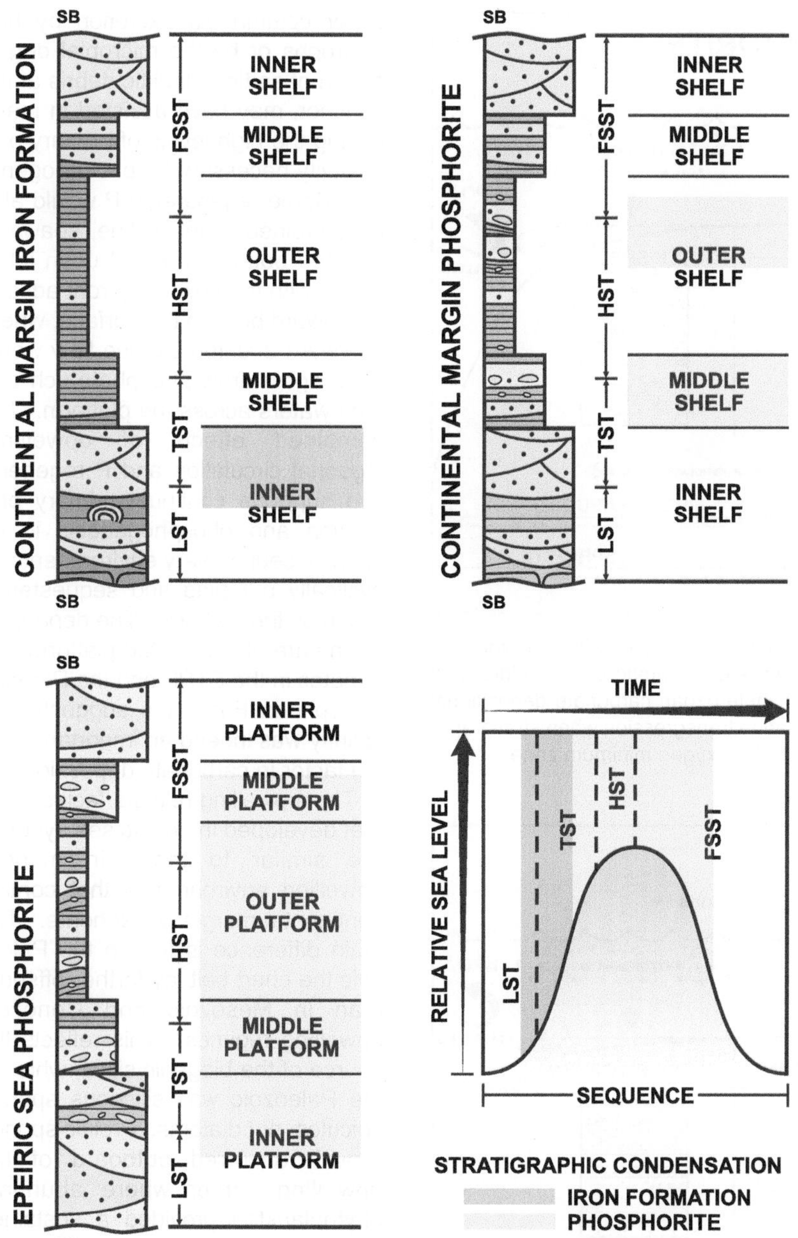

Figure 35. Idealized stacking of iron formation and phosphorite lithofacies through complete sea-level cycles. In continental margin iron formation stratigraphic condensation is restricted to the LST and TST in the shallow grainstone factory because of high distal accumulation rates. Stratigraphic condensation in phosphatic successions accompanies phosphogenesis and occurs preferentially at inflection points on the relative sea-level curve. Phosphogenesis is most pronounced in continental margin environments along the maximum flooding surface at the inflection between the TST and HST, when terrigenous clastics are trapped in nearshore environments. Phosphogenesis can also occur during the change in accommodation between the HST and FSST. The condensed interval in epeiric sea phosphorite is much thicker because the shallow seafloor is continually reworked by wave activity. The thickness of each sequence can range between one hundred and several hundred meters depending on the magnitude of relative sea level rise. Lithofacies are keyed to those depicted in Figures 34, 36, and 37. Exhalative iron formation is not shown because its stratigraphy is not governed by sea-level induced changes in accommodation. SB = sequence boundary; LST = lowstand systems tract; TST = transgressive systems tract; HST = highstand systems tract; FSST = falling stage systems tract.

level rise. The lack of bioturbation in many pristine facies implies the development of an oxygen minimum zone during francolite precipitation. Most phosphatic grainstones formed through the successive winnowing, transport and redeposition of phosphatic grains and intraclasts derived from pristine phosphorite facies via storm-generated currents. The thickest granular deposits are interpreted to have accumulated through event-driven amalgamation during episodes of heightened storm activity.

Unlike modern environments, where upwelling-related phosphogenesis is restricted to the outer shelf, francolite precipitation in epeiric seas occurred across the entire platform, wherever suitable conditions prevailed (Fig. 37). This *phosphorite nursery* may reflect the combined effects of nutrient transport away from the locus of upwelling, and the cyclic regeneration of P across the platform. Surface waters in modern upwelling systems are rapidly depleted in nutrients and productivity is greatly diminished within a relatively short distance away from the upwelling center. In epeiric seas that had highly seasonal or low net precipitation rates, such as the STPP and the Permian Phosphoria Formation, dissolved phosphate may have been transported away from the upwelling front to nearshore environments by lagoonal circulation. Lagoonal (anti-estuarine) circulation is characterized by the shoreward inflow of surface water and the outflow of saline water at depth, and is common within shallow basins with low or seasonal precipitation, high evaporation rates, and low riverine input. High evaporation rates cause an increase in the density of surface water in nearshore environments by increasing its salinity. This water sinks and flows below the surface in a seaward direction, resulting in an accompanying onshore-directed surface current.

Such circulation may have led to the shoreward flow of P-rich surface water from the upwelling center, thus stimulating primary production and phosphogenesis over the entire platform. The reintroduction of P back into solution at depth, either in the

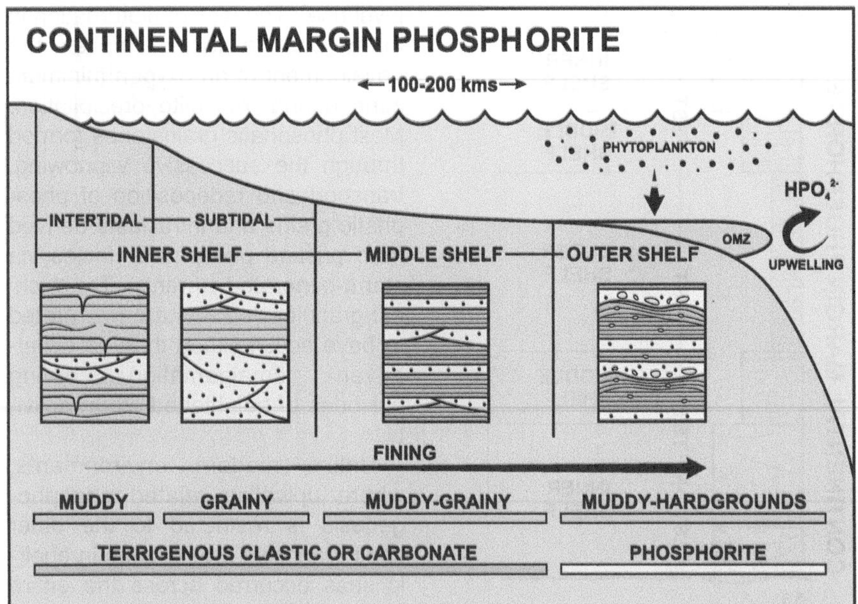

Figure 36. Continental margin phosphorite. Phosphorite accumulates within organic-rich sediment on the distal shelf beneath coastal upwelling cells. Nearshore and middle shelf lithofacies are those typical of terrigenous clastic or heterozoan carbonate depositional systems. Phosphogenesis is most intense during marine transgression when stratigraphic condensation in offshore areas is widespread. OMZ = oxygen minimum zone.

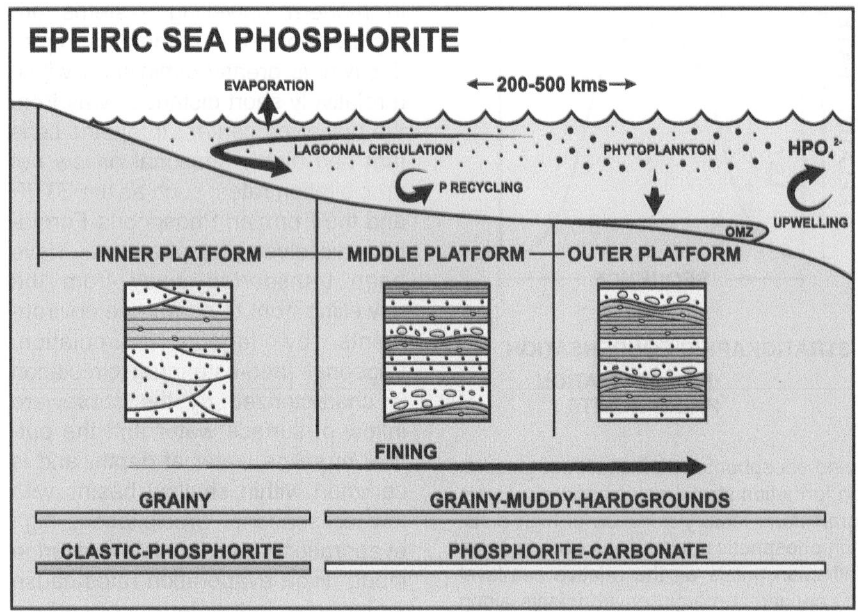

Figure 37. Epeiric sea phosphorite. As with continental margin phosphorite, phosphogenesis was stimulated by the accumulation of upwelling-related organic matter during relative sea level rise. High surface ocean productivity was maintained across the platform by lagoonal circulation, which drew dissolved phosphate from the upwelling centre to shallow water settings. This, in conjunction with the cyclic regeneration of P back to the water column, sustained the high nutrient levels necessary for primary production and thus, phosphogenesis. Lithofacies are grainy because they accumulated in relatively shallow-water environments above storm wave base. Nearshore deposits consist of cross-stratified grainstones. Amalgamated beds are interbedded with pristine phosphorite in middle platform environments. Laminated hemipelagic sediments are interbedded thin, granular phosphatic tempestites on the distal platform.

water column via excretion by heterotrophs or by the microbial degradation of settled planktic debris at the seafloor, may have assisted in maintaining the high levels of primary productivity necessary for phosphogenesis. Some regenerated P would also be entrained in the saline, seaward-directed bottom flow and upon interaction with the upwelling front advected upward back to the surface, where landward directed surface flow would once again draw phosphate-rich surface waters across the platform. The combined effects of upwelling, lagoonal circulation and P regeneration supports continual primary production and phosphogenesis in an array of sedimentary environments by cyclically pumping and sequestering P across the platform. The depauperate nature of associated platform carbonates in the STPP and the Permian Phosphoria Formation suggests that salinity was indeed an important limiting factor in carbonate deposition.

The upwelling-related facies belts that developed in epeiric sea systems are similar to those in modern upwelling environments that contain continental margin phosphorite. The main difference is that in the Paleozoic the chert belt lay farther offshore than in Mesozoic and Cenozoic upwelling regimes. This reflects the source of the biogenic silica, which in the Paleozoic was siliceous sponge spicules, not diatoms. Prolific sponge growth occurred outboard of the upwelling center where abundant phytoplankton provided a sustained food supply.

TEMPORAL DISTRIBUTION

The temporal occurrence of bioelemental sediment reflects variations in ocean chemistry linked to tectonic processes, biologic evolution, and climate change (Fig. 38). Each of these factors has had a profound influence on the biogeochemical cycling of Fe, P and Si and the types of bioelemental sediments produced.

Iron Formation

The Archean is characterized by deep-water exhalative iron formation deposited in tectonically active areas around spreading centers associated with volcanic arcs. Such deposition reflects the lack of shallow shelves

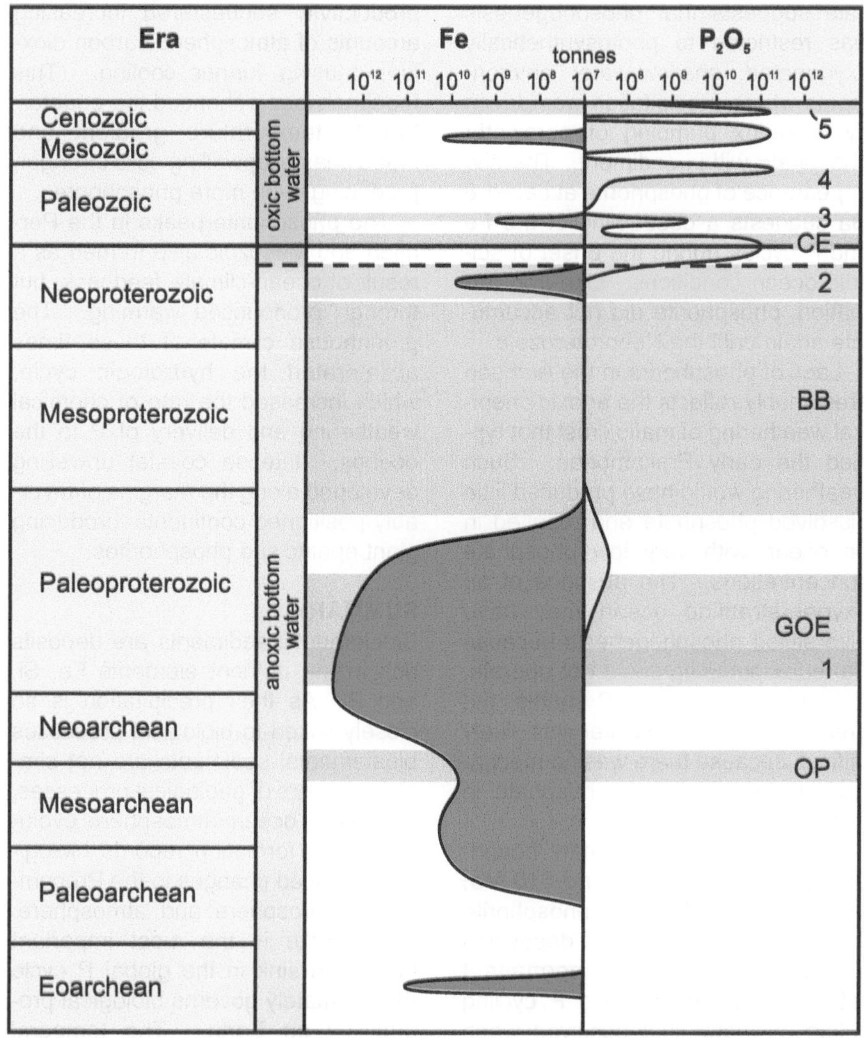

Figure 38. Temporal distribution of iron formation (red), ironstone (purple), and phosphorite (yellow). Curves are based on deposit age and resource estimates in Glenn *et al.* (1994), Kholodov and Butuzova (2001), and Klein (2005). The plateau in iron formation abundance in the Paleoproterozoic between ~2.5 and 2.2 Ga may consist of two peaks (Isley and Abbott, 1999). Events: OP = oxygenic photosynthesis; GOE = Great Oxidation Event; BB = Boring Billion; CE = Cambrian Explosion. Glaciations: 1 = Huronian; 2 = Snow Ball; 3 = Ordovician; 4 = Permian; 5 = Neogene.

and the influence of long-lasting mantle plumes that dominated Archean tectonics. Although shallow-water, non-exhalative varieties exist, they appear to be far less numerous and restricted to the Neoarchean, when narrow cratonic margins formed during the onset of Phanerozoic-style plate tectonics.

The dramatic rise in iron formation at 2.9 Ga probably corresponds to the evolution of oxygenic photosynthesis and resulting precipitation of ferric Fe from the Archean ocean (Fig. 38). Although some evidence suggests that iron formation prior to this time was also linked to photosynthetic oxygen, most data indicate deposition through a combination of anoxygenic photosynthesis, dissimilatory iron reduction, oxygen produced via non-phototrophic sources, and episodic increases in the input of hydrothermal Fe and Si during mantle plume events.

The broad peak between ~2.5 and 2.2 Ga signals a shift from deep-water iron formation deposition to upwelling-driven neritic accumulation on the expansive, unrimmed platforms that developed at the end of the Archean. Such areal extensive Paleoproterozoic iron formation formed in the full spectrum of shelf environments from an oxygen-stratified ocean born during the Great Oxidation Event (Fig. 38). This rapid rise in oxygen has been attributed to a major, tectonically induced change in the oxidation state of volcanic gases. The switch to expelling oxidizing rather than reducing gases is thought to have permitted, for the first time, the accumulation of abundant photosynthetic oxygen in the surface ocean.

Recent geochemical studies indicate that the decline and disappearance of iron formation at *ca.* ~1.8 Ga was probably also a consequence of the Great Oxidation Event. Oxic chemical weathering of the continents delivered ever-increasing amounts of sulfate to the ocean, which, in turn, promoted increased rates of bacterial sulfate reduction. The sulfide that was generated combined with ferrous Fe to form pyrite, ostensibly causing the cessation of iron formation deposition. Such a process apparently titrated the ocean of available Fe that would have otherwise precipitated as the Fe-(oxyhydr)oxides and Fe-silicates in iron formation.

A major shift in the Precambrian Si cycle also occurred at *ca.* ~1.8 Ga when subtidal cherts disappeared. A decrease in the input of hydrothermal Si, and Si derived from chemical weathering are thought to have caused this change. With the onset of sulfidic ocean conditions it is also possible that the Fe-redox pump that helped to concentrate Si at the seafloor stopped operating.

The period that follows is called the 'Boring Billion' because apparently little happened during this protracted interval of Earth history. Evidence suggests that the sulfidic ocean that prevailed for nearly a billion years perturbed the cycling of biologically important elements, causing a long lull in eukaryote evolution. It is not until the Neoproterozic that iron formation reappears (Fig. 38), when the Earth was shocked out of this state by the extensive 'snowball' glaciations.

Neoproterozoic iron formation is, by far, the least understood. Ice cover that shrouded much of the Earth between 740 and 630 Ma may have slowed chemical weathering and delivery of sulfate to the oceans. Such conditions would produce less sulfide in deep, oxygen-free portions of the oceans, allowing ferrous Fe concentrations in seawater to rise once more. Iron formation is thought to have precipitated in coastal basins when the ice melted during interglacials and oxygen re-entered the marine system. Neoproterozoic iron formation differs from its earlier Precambrian counterparts in that it is associated with glacial deposits and has a much simpler Fe mineralogy that is dominated by hematite.

Iron formation nearly disappeared from the stratigraphic record just after the last snowball glaciation at *ca.* 580 Ma. At this time, the advent of an oxygenated deep ocean, changes in ocean productivity, and possible differences in the cycling of S had a profound effect on seawater chemistry. This, in combination with an oceanic reservoir that became depleted in Fe sometime in the Early Cambrian, restricted iron formation to hydrothermal systems. These Phanerozoic exhalative iron formations generally contain more hematite than those in the Precambrian because most precipitated from oxygenated bottom waters.

All other large marine iron deposits in the Phanerozoic are granular ironstones (Fig. 38). Major peaks in the Ordovician and Jurassic have been attributed to transgressive events that produced widespread stratigraphic condensation in a variety of basin types. These occurrences also correspond to major phosphogenic episodes.

Phosphorite

Phosphorite is mostly a Phanerozoic phenomenon and only a single depositional episode occurring entirely within the Precambrian (Fig. 38). Little is known about this Precambrian interval, but it appears to span between 2.2 and 1.8 Ga, beginning in the middle of the Great Oxidation Event just after the Huronian Glaciation. The occurrence of phosphorite at this time likely records an abrupt increase in the delivery of dissolved phosphate to the oceans. This pulse may reflect the switch from mechanical weathering during the Huronian Glaciation to post-glacial chemical weathering of continental crust under an oxygenated atmosphere. Limited data suggests that phosphogenesis was restricted to photosynthetically oxygenated, shallow-water environments where precipitation was driven by Fe-redox pumping of porewater phosphate within sediment. The disappearance of phosphorite at *ca.* ~1.8 Ga suggests a decoupling of the Fe and P cycles during the onset of sulfidic ocean conditions. Like iron formation, phosphorite did not accumulate again until the Neoproterozoic.

Lack of phosphorite in the Archean presumably reflects the anoxic chemical weathering of mafic crust that typified the early Precambrian. Such weathering would have produced little dissolved phosphate and resulted in an ocean with very low phosphate concentrations. The absence of an oxygen-stratified ocean may have also stifled phosphogenesis because Fe-redox pumping could not operate. Without this important P shuttle, the precipitation of francolite was likely difficult because there was no mechanism to supersaturate phosphate in porewater.

Neoproterozoic–Cambrian boundary phosphorites (~700 and 510 Ma) are the Earth's first true phosphorite giants (Fig. 38). Their deposition probably signals the permanent return of coupled Fe and P cycling within seawater that occurred when widespread sulfidic ocean conditions finally ceased. Once the ocean became fully oxygenated at ~580 Ma, Fe-redox pumping could produce phosphorite in all shelf environments. Biologic evolution in the Ediacaran and Cambrian also played an important role promoting phosphogenesis. Concentration of P in the biological cycle by way of microbes, fecal pellets, filter feeding, bioturbation, and development of phosphatic shells during the Cambrian Explosion all led to increased P levels within sediment.

Phosphorite giants in the Ordovician and Miocene were related to the invigorated coastal upwelling during glaciations (Fig. 38). The pronounced equator-to-pole temperature gradient that develops during the buildup of high-latitude glaciers leads to more vigorous atmospheric circulation and thus, coastal upwelling, resulting in the widespread accumulation of organic-rich sediment and phosphorite. This increase in surface ocean productivity sequestered increasing amounts of atmospheric carbon dioxide, causing further cooling. This feedback loop enhanced the equator-to-pole temperature gradient and made coastal upwelling cells stronger, producing even more phosphorite.

The phosphorite peaks in the Permian and Mesozoic also formed as a result of ocean-climate feedback, but through pronounced warming. The greenhouse climate of these times accelerated the hydrologic cycle, which increased the rate of chemical weathering and delivery of P to the oceans. Intense coastal upwelling developed along the margins of favorably positioned continents, producing giant epeiric sea phosphorites.

SUMMARY

Bioelemental sediments are deposits rich in the nutrient elements Fe, Si, and P. As their precipitation is so closely linked to biological processes bioelemental sediments are not simply recorders of geological processes, but reflect ocean–atmosphere evolution. Iron formation records biologically induced changes in the Precambrian hydrosphere and atmosphere. Phosphorite is the most important long term sink in the global P cycle that ultimately governs biological productivity on Earth. The temporal abundance of chert reflects differences in the cycling of Si in the Precambrian and Phanerozoic oceans.

Two broad lithofacies categories exist in iron formation, chert and phosphorite: *pristine* and *reworked*. Pristine bioelemental facies are generally well laminated and form through the accumulation of primary precipitates in calm-water environments. Granular reworked deposits are created through hydraulic reworking and winnowing of primary precipitates in energetic settings. The observed array of pristine and reworked lithofacies reflects the unique combination of precipitation pathways and hydrodynamics that occur in different depositional environments.

The bioelemental sediment factory can produce these facies in the full spectrum of ocean environments, wherever there is a sustained supply of Fe, Si, and P. The most production occurs on unrimmed platforms with prominent upwelling during relative sea-level rise. Elevated sea-level increases the accommodation volume on shelves and traps diluting terrigenous clastics in nearshore environments. The locus of upwelling and delivery of nutrient elements is also shifted from the shelf–slope break to more proximal shelf positions where the reduction in slope allows bioelemental sediments to accumulate over a much greater area. Currents that rework pristine facies into granular deposits also develop during platform flooding.

The bioelemental facies that form are similar to those that accumulate on high-energy siliciclastic shelves and carbonate ramps. This is because bioelemental sediments are hybrid deposits that consist of chemical precipitates and have a strong hydrodynamic overprint. Like carbonate depositional systems, lithofacies stacking patterns are controlled as much by subsidence rate, terrigenous clastic input, and position of sea level as by oceanography and microbial biology.

Sedimentologists, microbiologists and geochemists have their work cut out for them as they strive to grasp the exact role microbes play in forming bioelemental deposits. Such a collaborative approach is not only necessary for refining current facies models, but is also essential to understanding the evolution of life.

ACKOWLEDGEMENTS

This chapter draws substantially on my research supported by the Natural Sciences and Engineering Research Council of Canada. Thoughtful discussions with E.E. Hiatt greatly improved the manuscript. E.E. Hiatt and P.W. Fralick reviewed an early version of the chapter. C.F. Koebernick assisted with drafting.

REFERENCES

Basic sources of information

Behl, R.J. and Garrison, R.E., 1994, The origin of chert in the Monterey Formation of California (USA), *in* Iijima, A., Abed, A. and Garrison, R.E., *eds*., Siliceous, Phosphatic and Glauconitic Sediments of the Tertiary and Mesozoic: Proceedings of the 29th International Geological Congress, Part C, p. 101-132.
Although concentrating on the Monterey Formation, this paper provides an excellent overview of chert-forming processes.

Bentor, Y.K., *ed*., 1980, Marine Phosphorite, Geochemistry, Occurrence, Genesis: SEPM Special Publication 29, 249 p.
An excellent source of general information on a variety of phosphorite types that summarized the state of knowledge at the time.

Calvert, S.E., 1974, Deposition and diagenesis of silica in marine sediments: International Association of Sedimentologists Special Publication, v. 1, p. 273-299.
Still one of the best overall references on biosiliceous sediments.

Canfield, D.E., 2005, The early history of atmospheric oxygen: Homage to Robert M. Garrels: Annual Review of Earth and Planetary Science, v. 33, p.1-36.
A detailed account of the evolution of atmospheric oxygen with reference to the deposition of iron formation.

Chauvel, J.J., Yuki, C., El Shazly, E.M., Gross, G.A., Laajoki, K., Markov, M.S., Rai, K.L., Stulchikov, V.A., Augustithis, S.S., 1990, Ancient Banded Iron Formations (Regional Presentations): Theophrastus Publications, S.A., Greece, 463 p.
Articles on a wide variety of iron formations from across the globe.

Clout, J.M. and Simonson, B.M., 2005, Precambrian iron formations and iron formation-hosted iron ore deposits: Economic Geology, 100th Anniversary Volume, p. 643-679.
A sound synopsis of iron formation and their economic importance.

Cook, P.J. and Shergold, J.H., *eds*., 1986, Phosphate Deposits of the World, Volume 1, Proterozoic and Cambrian Phosphorites: Cambridge University Press, New York, 386 p.
A true gold mine of information! Contains an exhaustive survey of Precambrian and Cambrian phosphorite deposits.

Föllmi, K.B., *ed*., 1994, Concepts and controversies in phosphogenesis: Eclogae Geologicae Helvetiae, v. 87, p. 639-788.
A collection of three papers focusing on phosphorite geochemistry, phosphogenesis and the deposition of phosphatic deposits.

Föllmi, K.B., Garrison, R.E. and Grimm, K.A., 1991, Stratification in phosphatic sediments: illustrations from the Neogene of California, *in* Einsele, G., Ricken, W. and Seilacher, A., *eds*., Cycles and Events in Stratigraphy, p. 492-507.
An excellent overview of phosphorite depositional processes.

Fralick, P.W. and Barrett, T.J., 1995, Depositional controls on iron formation associations in Canada: International Association of Sedimentologists – Special Publication, v. 22, p. 137-156.
Although focusing on Canadian iron formation, this paper summarizes important attributes of exhalative and continental margin iron formation.

Glenn, C.R. and Arthur, M.A., 1988, Petrology and major element geochemistry of Peru Margin phosphorites and associated diagenetic minerals: authigenesis in modern organic-rich sediments: Marine Geology, v. 80, p. 231-267.
Summarizes the paragenesis of francolite and associated authigenic minerals in phosphatic depositional systems.

Glenn, C.R., Prévot-Lucas, L. and Lucas, J., *eds*., 2000, Marine Authigenesis: from Global to Microbial: SEPM Special Publication 66, 536 p.
An outstanding collection of articles on a myriad of authigenic sediment types with emphasis on phosphorus cycling and phosphorite.

Gross, G.A., 1983, Tectonic systems and the deposition of iron-formation: Precambrian Research, v. 20, p. 171-187.
An historic review paper on the classification and deposition of iron formation.

Heggie, D.T., Skyring, G.W., O'Brien, G.W., Reimers, C., Herczeg, A., Moriarty, J.W., Burnett, W.C. and Milnes, A.R., 1990, Organic carbon cycling and modern phosphorite formation on the East Australian continental margin: an overview *in* Notholt, A.J.G. and Jarvis, I., *eds*., Phosphorite Research and Development: Geological Society of London, Special Publication 52, p. 87-117.
An important contribution to understanding phosphogenesis in non-upwelling environments.

Holland, H.D., 2004, The Geologic History of Seawater, *in* Elderfield, H., *ed*., Treatise on Geochemistry: The Oceans and Marine Geochemistry: Elsevier, p. 583-625.
A detailed account of changing seawater chemistry through time with reference to iron formation and phosphorus cycling.

Isaacs, C.M., 1982, Influence of rock composition on kinetics of silica phase changes in the Monterey Formation, Santa Barbara area, California: Geology, v. 10, p. 304-308.
A succinct explanation of the diagenesis of biosiliceous sediment.

Klein, C., 2005, Some Precambrian banded iron-formations (BIFs) from around the world: their age, geologic setting, mineralogy, metamorphism, geochemistry, and origin: American Mineralogist, v. 90, p. 1473-1499.
A detailed account of iron formation paragenesis and metamorphism.

Konhauser, K., 2007, Introduction to Geomicrobiology: Blackwell, Oxford, 425 p.
A treasure trove of information on

microbial precipitation processes with emphasis on iron and silica mineralizing bacteria.

Maliva, R.G., Knoll, A.H. and Siever, R., 1989, Secular change in chert distribution: a reflection of evolving biological participation in the silica cycle: Palaios, v. 4, p. 519-532.

Links the evolution of silica secreting organisms to the temporal distribution of Phanerozoic chert.

Maliva, R.G., Knoll, A.H. and Simonson, B.M., 2005, Secular change in the Precambrian silica cycle: insights from chert petrology: Geological Society of America Bulletin, v. 117, p. 835-845.

An important contribution to understanding the accumulation of Precambrian chert.

Mel'nik, Y.P., 1982, Precambrian Banded Iron-formations, Physiochemical Conditions of Formation: Elsevier, New York, 310 p.

The most comprehensive treatment of abiotic iron formation precipitation.

Murray, R.W., Jones, D.L., Buchholtz ten Brink, M.R., 1992, Diagenetic formation of bedded chert: evidence from chemistry of the chert-shale couplet: Geology, v. 20, p. 271-274.

Discusses the constraints on the mobilization and precipitation of silica in fine-grained sediment.

Notholt, A.J.G., Sheldon, R.P. and Davidson, D.F., *eds.*, 1989, Phosphate Deposits of the World, Volume 2, Phosphate Rock Resources; Cambridge University Press, New York, 563 p.

A comprehensive synthesis of most Phanerozoic phosphorite deposits. A true gem!

Ohmoto, H., Watanabe, Y., Yamaguchi, K.E., Naraoka, Haruna, M., Kakegawa, Hayashi, K. and Kato, 2006, Chemical constraints and biological evolution of early Earth: constraints from banded iron formations, *in* Kesler, E. and Ohmoto, H., *eds.*, Evolution of Early Earth's Atmosphere, Hydrosphere, and Biosphere – Constraints from Ore Deposits: Geological Society of America Memoir 198, p. 291-331.

A sound introduction to iron formation and its place in Earth history.

Peter, J.M., 2003, Ancient iron formations: their genesis and use in the exploration for stratiform base metal sulphide deposits, with examples from the Bathurst Mining Camp, *in* Lentz, D.R., Geochemistry of Sediments and Sedimentary Rocks: Evolutionary Considerations to Mineral Deposit-Forming Environments: Geological Association of Canada, GeoText 4, p. 145-176.

One of the best overviews on hydrothermal processes and exhalative iron formation.

Pufahl, P.K. and Grimm, K.A., 2003, Coated phosphate grains: proxy for physical, chemical, and ecological changes in seawater: Geology, v. 31, p. 801-804.

Highlights the numerous phosphogenic processes in upwelling environments.

Simonson, B.M., 2003, Origin and evolution of large Precambrian iron formations, *in* Chan, M.A. and Archer, A.W., *eds.*, Extreme Depositional Environments: Mega End Members in Geologic Time: Geological Society of America Special Paper 370, p. 231 -244.

An excellent overview of giant continental margin iron formation with a historical account of various depositional models.

Trendall, A.F. and Blockley, J.G., 2004, Precambrian iron-formation, *in* Erisson, P.G., Altermann,W., Nelson, D.R., Mueller, W.U. and Catuneau, O., *eds.*, The Precambrian Earth, Tempos and Events: Elsevier, Amsterdam, p. 403-421.

A solid synthesis of iron formation with particular emphasis on those from the Hamersley Basin.

Uitterdijk Appel, P.W. and Laberge, G.L., 1987, Precambrian Iron-formations: Theophrastus Publications, S.A., Greece, 674 p.

Another good general reference on iron formations from across the globe.

Other references

Anbar, A.D. and Knoll, A.H., 2002, Proterozoic ocean chemistry and evolution: a bioinorganic bridge?: Science, v. 297, p. 1137-1142.

Barber, R.T., and Smith, R.L., 1981, Coastal Upwelling Ecosystems, *in* Longhurst, A.R., *ed.*, Analysis of Marine Ecosystems: Academic Press, New York, p. 31-68.

Barrett, T.J., Fralick, P.W. and Jarvis, I., 1988, Rare-earth-element geochemistry of some Archean iron formations north of Lake Superior, Ontario: Canadian Journal of Earth Sciences, v. 25, p. 570-580.

Baturin, G.N., 1971, Stages of phosphorite formation on the ocean floor: Nature, v. 232, p. 61-62.

Bekker, A., Holland, H.D., Wang, P.L., Rumble III, D., Stein, H.J., Hannah, J.L., Coetzee, L.L. and Beukes, N.J., 2004, Dating the rise in atmospheric oxygen: Science, v. 427, p. 117-120.

Benninger, L.M. and Hein, J.R., 2000, Diagenetic evolution of seamount phosphorite, *in* Glenn, C.R., Prévot-Lucas, L. and Lucas, J., *eds.*, Marine Authigenesis: from Global to Microbial: SEPM Special Publication 66, p. 245-256.

Canfield, 1998, A new model for Proterozoic ocean chemistry: Nature, v. 396, p. 450-453.

Canfield, D.E., Poulton, S.W., Knoll, A.H., Narbonne, G.M., Ross, G., Goldberg, T, and Strauss, H., 2008, Ferruginous conditions dominated later Neoproterozoic deep-water chemistry: Science, v. 321, p. 949-952.

Cloud, P., 1973, Paleoecological significance of the banded iron-formation: Economic Geology, v. 68, p. 1135-1143.

Derry, L.A. and Jacobsen, S.B., 1990, The chemical evolution of Precambrian seawater: evidence from REEs in banded iron formation: Geochemica et Cosmochemica Acta, v. 54, p. 2965-2977.

Diaz, J., Ingall, E., Benitez-Nelson, C., Paterson, D., de Jonge, M.D., McNulty, I. and Branders, J.A., 2008, Marine polyphosphate: a key player in geologic phosphorus sequestration: Science, v. 320, p. 652-655.

Filippelli, G.M., 1997, Controls on phosphorus concentration and accumulation in oceanic sediments: Marine Geology, v. 139, p. 231-240.

Fischer, W.W. and Knoll, A.H., 2009, An iron shuttle for deepwater silica in Late Archean and early Paleoproterozoic iron formation: Geological Society of America Bulletin, v. 121, p. 222-235.

Föllmi, K.B., Weissert, H. and Lini, A., 1993, Nonlinearities in phosphogenesis and phosphorus-carbon coupling and their implications for global change, *in* Wollast, R., Mackenzie, F.T. and Chou, L., *eds.*, Interactions of C, N, P and S Biogeochemical Cycles and Global Change, Springer, p. 447-474.

Föllmi, K.B., Badertscher, C., de Kaenel, E., Stille, P., John, C.M., Adatte, T. and Steinmann, P., 2005, Phosphogenesis and organic-carbon preservation in the Miocene Monterey Formation at Naples Beach, California – the Monterey hypothesis revisited: Geological Society of America Bulletin, v. 117, p. 589-619.

Fralick, P.W. and Pufahl, P.K., 2006, Iron formation in Neoarchean deltaic successions and the microbially mediated deposition of transgressive systems tracts: Journal of Sedimentary Research, v. 76, p. 1057-1066.

Garrison, R.E. and Kastner, M., 1990. Phosphatic sediments and rocks recovered from the Peru margin during ODP Leg 112, *in* Suess, E., von Heune R., *et al.*, *eds.*, Proceedings of the Ocean Drilling Program, Scientific Results. Ocean Drilling Program, College Station, p. 111-134.

Glenn, C.R., Föllmi, K.B., Riggs, S.R., Baturin, G.N., Grimm, K.A., Trappe, J., Abed, A.M., Galli-Oliver, C., Garrison, R.E., Ilyin, A.V., Jehl, C., Rohrlich, V., Sadaqah, R.M.Y., Schidlowski, M., Sheldon, R.E., and Siegmund, H., 1994, Phosphorus and phosphorites: Sedimentology and environments of formation. Eclogae Geologicae Helveticae. 87, p. 747-788.

Grimm, K.A. and Föllmi, K.B., 1994, Doomed pioneers: allochthonous crustacean tracemakers in anaerobic basinal strata, Oligo-Miocene San Gregorio Formation, Baja California Sur, Mexico: Palaios, v. 9, p. 313-334.

Hamide, T., Konhauser, K.O., Raiswell, R., Goldsmith, S. and Morris, R.C., 2003,

Using Ge/Si ratios to decouple iron and silica fluxes in Precambrian banded iron formations: Geology, v. 31, p. 35-38.

Hansel, C. M., Brenner, S.G., Nico, P., Fendorf, S., 2004, Structural constraints of ferric (hydr)oxides on dissimilatory iron reduction and the fate of Fe(II): Geochimica et Cosmochimica Acta, v. 68, p. 3217-3229.

Hiatt, E.E. and Budd, D.A., 2001, Sedimentary phosphate formation in warm shallow waters: new insights into the palaeoceanography of the Permian Phosphoria Sea from analysis of phosphate oxygen isotopes: Sedimentary Geology, v. 145, p. 119-133.

Hiatt, E.E. and Budd, D.A., 2003, Extreme paleoceanographic conditions in a Paleozoic oceanic upwelling system: Organic productivity and widespread phoshogenesis in the Permian Phosphoria sea, *in* Chan, M.A. and Archer, A.W., *eds*., Extreme Depositional Environments: Mega End Members in Geologic Time: Geological Society of America Special Paper 370, p. 245-264.

Hoashi, M., Bevacqua, D.C., Otake, T., Watanabe, Y., Hickman, A.H., Utsunomiya, S., and Ohmoto, H., 2009, Primary haematite, formation in an oxygenated sea 3.46 billion years ago: Nature Geoscience, v. 2, p. 301-306.

Holland, H.D., 2002, Volcanic gases, black smokers, and the Great Oxidation Event: Geochemica et Cosmochemica Acta, v. 66, p. 3811-3826.

Huston, D.L., and Logan, G.A., 2004, Barite BIFS and bugs: evidence for the evolution of the Earth's early hydrosphere: Earth and Planetary Science Letters, v. 220, p. 41-55.

Isley, A.E. and Abbott, D.H., 1999, Plume-related mafic volcanism and the deposition of banded iron formation: Journal of Geophysical Research, v. 104, p. 15,461-15,477.

Jahnke, R.A., Emerson, S.R., Roe, K.K. and Burnett, W.C., 1983, The present day formation of apatite in Mexican continental margin sediments. Geochimica et Cosmochimica Acta, v. 47, 259-266.

James, H.L., 1954, Sedimentary facies of iron formation: Economic Geology, v. 49, p. 235-293.

Johnson, C.M., Beard, B.L., Klein, C., Beukes, N.J. and Roden, E.E., 2008, Iron isotopes constrain biologic and abiologic processes in banded iron formation genesis: Geochimica et Cosmochimica Acta, v. 72, p. 151-169.

Kastner, M., Keene, J.B. and Gieskes, J.M., 1977, Diagenesis of siliceous oozes: I. Chemical controls on the rate of opal-A diagenesis - an experimental study: Geochimica et Cosmochimica Acta, v. 40, p. 1041-1059.

Kholodov, V.N. and Butuzova, G.Y., 2001, Problems of iron and phosphorus geochemistry in the Precambrian: Lithology and Mineral Resources, v. 36, p. 339-352.

Kimberley, M.M., 1978, Paleoenvironmental classification of iron formations: Economic Geology, v. 73, p. 215-229.

Konhauser, K.O., Hamade, T., Raiswell, R., Morris, R.C., Ferris, F.G., Southam, G. and Canfield, D.E., 2002, Could bacteria have formed the Precambrian banded iron formations?: Geology, v. 30, p. 1079-1082.

Konhauser, K.O., Lalonde, S.V., Amskold, L. and Holland, H.D., 2007, Was there really an Archean phosphate crisis?: Science, v. 315, p. 1234.

Krapez, B., Barley, M.E. and Pickard, A.L., 2003, Hydrothermal and resedimented origins of the precursor sediments to banded iron formation: sedimentological evidence from the Early Palaeoproterozoic Brockman Supersequence of Western Australia: Sedimentology, v. 50, p. 979-1011.

Morris, R.C., 1993, Genetic modeling for banded iron-formation of the Hamersley Group, Pilbara Craton, Western Australia: Precambrian Research, v. 60, p. 243-286.

Murray, J.W., 1979, Iron oxides, *in* Burns, R.G., *ed*., Marine Minerals: Reviews in Mineralogy, v. 6, p. 47-91.

Nelson, G.J., Pufahl, P.K. and Hiatt, E.E., 2010, Paleoceanographic constraints on Precambrian phosphorite accumulation, Baraga Group, Michigan, USA: Sedimentary Geology, v. 226, p. 9-21.

Nisbet, E.G., Grassineau, N.V., Howe, C.J., Abell, P.I., Regelous, M. and Nisbet, R.E.R., 2007, The age of Rubisco: the evolution of oxygenic photosynthesis: Geobiology, v. 5, p. 311-335.

Posth, N.R., Hegler F., Konhauser, K.O., and Kappler, A., 2008, Alternating Si and Fe deposition caused by temperature fluctuations in Precambrian oceans: Nature Geoscience, v. 1, p. 703-708.

Poulton, S.W. and Canfield, D.E., 2006, Co-diagenesis of iron and phosphorus in hydrothermal sediments from the southern East Pacific Rise: implications for the evaluation of paleoseawater phosphate concentrations: Geochimica et Cosmochimica Acta, v. 70, p. 5883-5898.

Prévot, L., El Faleh, E.M. and Lucas, J., 1989, Details on synthetic apatites formed through bacterial mediation mineralogy and chemistry of the products: Sciences Geologiques (Bulletin), v. 42, p. 237-254.

Pufahl, P.K., Grimm, K.A., Abdulkader, A.M. and Sadaqah, R.M.Y., 2003, Upper Cretaceous (Campanian) phosphorites in Jordan: implications for the formation of a South Tethyan phosphorite giant: Sedimentary Geology, v. 161, p. 175-205.

Pufahl, P.K. and Fralick, P.W., 2004, Depositional controls on Paleoproterozoic iron formation accumulation, Gogebic Range, Lake Superior Region, USA: Sedimentology, v. 51, p. 791-808.

Raiswell, R., 2006, An evaluation of diagenetic recycling as a source of iron for banded iron formations *in* Kesler, S.E. and Ohmoto, H., *eds*., Evolution of Early Earth's Atmosphere, Hydrosphere, and Biosphere – Constraints from Ore Deposits: Geological Society of America Memoir 198, p. 223-238.

Reimers, C.E., Kastner, M. and Garrison, R.E., 1990, The role of bacterial mats in phosphate mineralization with particular reference to the Monterey Formation, *in* Burnett, W.C., and Riggs, S.R., *eds*., Phosphate Deposits of the World, Neogene to Modern Phosphorites: Cambridge University Press, New York, p. 300-311.

Schen, Y., Schidlowski, M. and Chu, X., 2000, Biochemical approach to understanding phosphogenic events of the terminal Proterozoic to Cambrian: Palaeogeography, Palaeoclimatology, Palaeocology, v. 158, p. 99-108.

Schulz, H.N. and Schulz, H.D., 2005, Large sulfur bacteria and the formation of phosphorite: Science, v. 307, p. 416-418.

Simonson, B.M. and Hassler, S.W., 1996, Was the deposition of large Precambrian iron formations linked to major marine transgressions?: Journal of Geology, v. 104, p. 665-676.

Slack, J.F., Grenne, T., Bekker, A., Pouxel, O.J. and Lindberg, P.A., 2007, Suboxic deep seawater in the late Paleoproterozoic: evidence from hematitic chert and iron formation related to seafloor-hydrothermal sulfide deposits, central Arizona, USA: Earth and Planetary Science Letters, v. 255, p. 243-256.

Soudry, D., 2000, Microbial phosphate sediment, *in* Riding, R.E. and Awramik, S.M., *eds*., Microbial Sediments, p. 12-136.

Trappe, J., 2001, A nomenclature system for granular phosphate rocks according to depositional texture: Sedimentary Geology, v. 145, p. 135-150.

Young, T.P. and Taylor, W.E.G., 1989, Phanerozoic ironstones: Geological Society of London Special Publication 46, 251 p.

20. Marine Evaporites

Alan C. Kendall, School of Environmental Sciences, University of East Anglia, Norwich, NR4 7TJ, England, UK

INTRODUCTION

Evaporites are chemical sediments that precipitate from brines where there is a water budget deficit (potential evaporation losses exceed atmospheric precipitation). Concentration occurs by solar evaporation at an air-water interface, even if this occurs *within* sediments. The feedstock for most large evaporite deposits was seawater, but this may be augmented by continental and/or hydrothermal waters. Ancient marine evaporites are those precipitated from marine waters or those diagenetically emplaced within marine sediments. Some, however, may actually have been non-marine or precipitated from mixed water sources but may not be identified as such.

Evaporites are minor components in some sediment sequences and then are parts of other depositional models. Elsewhere evaporites are the major, or only, component of sedimentary bodies which, in extreme cases, are vast (covering millions of square kilometers), thick (several kilometers), and basin-central in location. These require separate models.

Because evaporites display both detrital and crystalline textures, and can also be substantially diagenetically modified, they constitute one of the most variable of sedimentary rock groups. A diverse group of minerals occurs (Table 1), but most are rare. Carbonates are commonly closely associated with more soluble minerals, and then should be considered evaporites that form in the initial stages of brine concentration. They commonly contain large amounts of preserved organic matter and are important hydrocarbon source rocks.

Evaporite minerals crystallize at brine/air interfaces (less commonly within the brine column), on the floor of brine pools, within brine-soaked sediments as displacive or cement crystals, or as efflorescent crusts (Fig. 1a). Efflorescent crusts have low preservation potential (although evidence for their former existence remains); consequently, most evaporite deposits can be described as;

1. Subaqueous accumulations of surface-nucleated crystals (*crystal cumulates* Fig. 1b);
2. Subaqueous bottom precipitates (*crystal crusts* Fig. 1c);
3. *Intra-sediment precipitates* that replace, displace or incorporate host sediments, (Fig. 1d); or
4. *Clastic accumulations* of transported evaporite particles (Fig. 1e).

After deposition, evaporites are diagenetically changed, altering the original mineralogy and sediment textures (often obliterating all original characters), or they may completely dissolve away. Many diagenetic changes are early, therefore depositional evaporite models must consider and incorporate these changes.

Evaporites only accumulate where the depositional and the burial environment are settings where brine dilution or dissolution of salts by excessive influx of less saline waters fails to happen. This requirement confines marine evaporites to locations isolated from, or having greatly restricted access to, the ocean. Today, marine evaporites are confined to coastal supratidal settings and locations where marine waters usually seep (rather than flow) into low-lying pools and small basins. Some ancient evaporites formed in similar environments, but these are minor. Thick and extensive evaporites ('saline giants') formed throughout the Phanerozoic, and have no modern equivalents. Some contain so much salt that their formation must have caused significant reductions of oceanic salinity and thereby affected oceanic circulation (Hay *et al.*, 2006).

Table 1. Marine evaporite minerals (the more common in bold font)

Anhydrite	**$CaSO_4$**	**Halite**	**NaCl**
Aragonite	$CaCO_3$	Kainite	$MgSO_4.KCl.1\frac{1}{4}H_2O$
Bassanite	$CaSO_4.\frac{1}{2}H_2O$	**Kieserite**	**$MgSO_4.H_2O$**
Bischofite	$MgCl_2.6H_2O$	Langbeinite	$2MgSO_4.K_2SO_4$
Calcite	**$CaCO_3$**	Leonite	$MgSO_4.K_2SO_4.4H_2O$
Carnallite	**$MgCl_2.KCl.6H_2O$**	Loewite	$2MgSO_4.2Na_2SO_4.5H_2O$
Dolomite	**$CaCO_3.MgCO_3$**	**Polyhalite**	**$2CaSO_4.MgSO_4.K_2SO_4.2H_2O$**
Epsomite	$MgSO_4.7H_2O$	**Sylvite**	**KCl**
Glauberite	$CaSO_4.Na_2SO_4$	Tachyhydrite	$CaCl_2.2MgCl_2.12H_2O$
Gypsum	**$CaSO_4.2H_2O$**		

CONTROLS ON EVAPORITE PRECIPITATION

Brine Concentration

The order in which evaporite minerals precipitate from modern seawater is determined by their relative solubilities: the least soluble precipitating first (McCaffrey *et al.*, 1987). Calcium carbonate begins to precipitate from seawater concentrated 1.8 times, usually as aragonite. Gypsum begins precipitating at 3.8 times seawater concentration, and halite only when the brine reaches a concentration 10.6 times that of seawater. Magnesium sulfates should first appear at brine concentrations about 70 times seawater, whereas a 90 times concentration is needed to precipitate potassium-bearing phases. By this concentration stage, the brine volume is so small that it can lodge within the pore-space of earlier-precipitated evaporites rather than being a surface brine. Evaporite sequences exhibit major compositional differences from the theoretical seawater evaporation sequence (more soluble materials being depleted or absent)

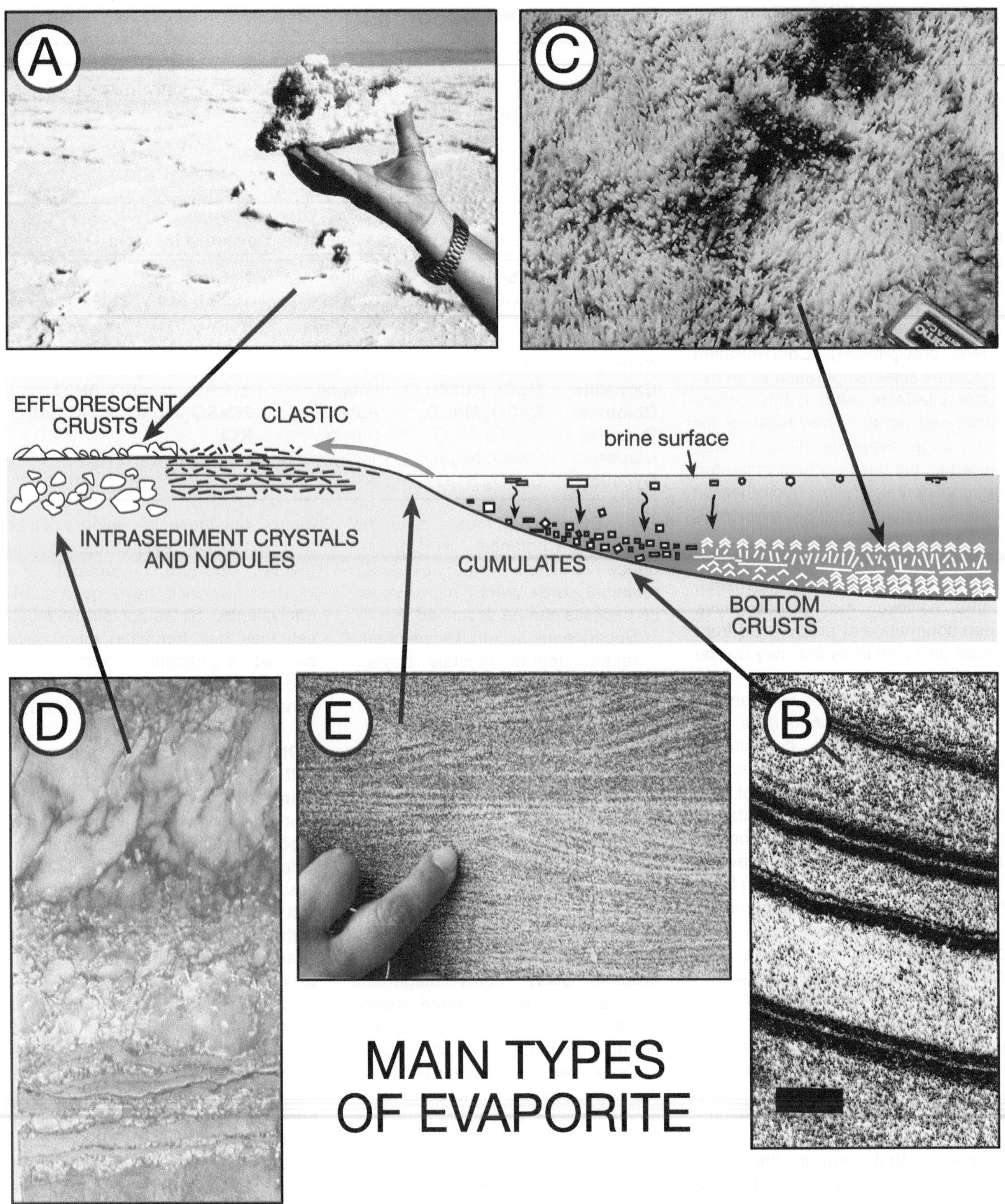

Figure 1. Main types of evaporite. **A.** Efflorescent crusts (usually ephemeral): modern saltpan, west Texas, **B.** Accumulations of cumulate (halite) crystals: Messinian, Mediterranean (photo courtesy T. Lowenstein), **C.** Bottom-grown crystal crusts, Howz-e-Soltan Lake, Iran (photo courtesy F. Fayazi), **D.** Intra-sediment precipitates, displacive anhydrite nodules – some enterolithic, Upper Jurassic, southern England (photo courtesy D. Shearman), and **E.** Clastic evaporites – cross-bedded gypsum: Messinian, Sicily, (photo courtesy B.C. Schreiber). This diagram does not imply that cumulate evaporites accumulate in shallower water than bottom crusts.

and thus are not products of simple seawater evaporation models.

If brine and air are at the same temperature, water only evaporates when its vapor pressure exceeds the partial pressure of water in the atmosphere (the relative humidity, Kinsman, 1976). The vapor pressure of a brine varies with the chemical activity of water in that brine, itself a function of its ionic strength (or salinity). Low humidities are needed to evaporate brines to the concentrations required to precipitate high-solubility minerals. Mean relative humidities are high over oceans, whereas along low-latitude coasts, they only fall to 70–80%, which should be insufficient to allow halite to be precipitated. Humidities that are sufficiently low to allow primary precipitation of potassium and magnesium salts are confined to regions far distant from oceans. These salts should not occur in marginal-marine evaporitic environments.

However, absolute humidity and temperature contrasts can be more important than relative humidity (Walton, 1978). When temperatures of brine and the overlying air are unequal, air will absorb or release water, and temperatures of brine and water approach each other. If brine temperatures are higher and they warm the overlying air, the air is able to absorb water, even if it originally had a high relative humidity. Brine evaporation will then occur with concurrent cooling of the brine. This cooling, however, is more than offset by solar heating. Offshore winds have low relative humidities and high temperatures, favoring evaporation. Onshore winds have high relative humidities, but the air temperatures, controlled by the ocean, will be lower than that of the brine, causing the air to become warmed and thus promoting evaporation. Halite precipitation and accumulation is therefore not confined to continental settings.

Brine Composition

As each mineral phase is precipitated, specific chemical components are removed from solution, so modifying the remaining brine composition. Evaporation pathways of brines are determined by the original Ca^{2+} to SO_4^{2-} ratio. This ratio, important at the stage of gypsum saturation, determines whether more evolved brines will become depleted in calcium or sulfate. Evaporation of modern seawater produces alkaline carbonates, gypsum and brines depleted in calcium but enriched in sulfate. This was not always true of ancient seawater.

The most common bittern salt deposits are composed of halite+sylvite+carnallite±bischoffite (*i.e.,* magnesium sulfate-poor evaporites), but cannot be derived from simple evaporation of modern seawater. They may have formed by evaporation of hydrothermal or basin groundwaters (Hardie, 1990), but might also have formed from ancient seawaters with compositions different to modern seawater (Hardie, 1996). Seawater compositions are determined by the mixing of waters from continental runoff and waters (commonly brines) that have been substantially modified at mid-oceanic ridges by geothermal circulation. The geothermal waters are calcium enriched, but impoverished in sulfate. During geological periods, when mid-oceanic circulation was greater than today, seawaters had lower relative sulfate contents and, upon evaporation, would precipitate $MgSO_4$-impoverished bitterns.

Basin Hydrology

Evaporite depositional systems are dynamic and open, and must be considered in terms of inter-related water and salt budgets. The hydrologic framework during evaporite deposition consists of i) gravity-driven groundwaters (marine or continental) in aquifers that discharge at the edges of the depositional basin, ii) surface flows (marine or continental) that episodically transport water and salts toward any depositional depressions, and iii) dense brines beneath lower parts of the depositional basin that accumulate as a result of density-driven brine reflux.

Inflow of Water and Salts

Large volumes of seawater are needed to generate evaporites. A single batch of seawater, even in an initially deep basin, cannot produce a significant deposit (complete evaporation of a kilometer of seawater will only precipitate ~14 m of evaporites, mostly halite). Some salt sequences are several kilometers thick.

Basin Restriction

Although large water volumes are needed, the depositional site cannot be too open or else brines will constantly be diluted by inflows. Water losses by evaporation are unable to offset dilution by an unrestricted surface-water inflow (Maiklem, 1971). This means that evaporite basins must be physically restricted, or even isolated, from the sea. The ratio between the surface area of a basin and the cross-sectional area of any inlet must be at least $>10^6$:1 for gypsum to form, and for halite the inlet must be at least eight orders of magnitude smaller than the basin area (Lucia, 1972).

Disconnected, but marine-fed, basins must gain their water by seepage through permeable barriers, and might even receive a significant proportion of their water from continental sources. Seepage through barriers, even if supplemented by short-lived flooding events, is usually unable to offset evaporative losses in the basin. When evaporation losses exceed inflow, basins desiccate. The loss of brine volume, and a consequent lowering of the basin brine level, is termed *evaporative drawdown* (Maiklem, 1973).

Preservation of perennial brines in basins (required for 'deep-water' marine evaporites) requires;

i. Episodic seawater input into the basin over the barrier; or
ii. A reduction in the evaporation rate; or
iii. Additional seepage inflow from the basin periphery or through the basin floor.

All marine-derived evaporite sequences have lower amounts of the more soluble phases than simple seawater evaporation would produce. Thus brine components of the missing more-soluble phases must have been exported from the depositional basin. The size of any inlet into a basin is, however, so small during halite precipitation that there must be complete surface disconnection of the basin from the open ocean. In isolated basins there is no opportunity for any *surface* reflux of the remaining brines out of the basin.

Brine Reflux in Isolated Basins

Loss of more soluble brine components must be by reflux through the basin floor or sides. However, it is unreasonable to expect that brine reflux can easily occur over the main part of an evaporite basin floor. If the floor were permeable enough to allow wholesale brine export, then this same permeability would also allow upward seepage of more dilute formation waters. Waters beneath basins affected by evaporative drawdown will be affected by a vertical hydraulic gradient that should drive groundwaters upward (Kendall, 1989). Brines within the basin would then be constantly diluted by this seepage, and concentrated brines could never form. As fast as they formed, they would also sink away through the permeable basin floor, or be diluted by upward inflow. Thus, a major prerequisite for evaporite deposition in basins is the presence of a basal aquitard (one capable of confining brines within basins) or a plume of subsurface waters that have already been concentrated – so preventing density-driven brine convection (the 'brine curtain' of Warren 1999). Basal seals in large basins are unlikely to be everywhere complete or effective. Permeable parts are potential sites for brine reflux or for upward seepage of more dilute groundwaters.

Large evaporite basins are capable of generating and exporting (by reflux) large brine volumes – volumes capable of filling rock porosity to depths of many kilometers beneath the evaporite basin (Brantley and Donovan, 1990). The geochemistry of many subsurface brines suggests that they were generated and exported from evaporite basins.

Evaporite deposition in apparently hydrologically-closed basins is usually considered to be simply a product of evaporation without any involvement of the subsurface. However, if pore-waters in underlying sediments are less dense, the denser surface brines (generated by surface evaporation) sink into the underlying sediments long before the surface brines can reach sufficiently high salinities to precipitate salts. The water body disappears without precipitating any salt.

Transfer of brine into the shallow subsurface (by reflux), however, gradually increases the salinity of the shallow groundwaters (Fig. 2). Eventually surface brines are unable to sink and their evaporation causes evaporite precipitation at the surface. Reflux only occurs when the density contrast between surface brines and underlying groundwaters exceeds a certain threshold (perhaps with differences of 50-100g.l^{-1}, Bowler, 1986). This implies that salt crusts and subaqueous evaporites commonly lie above sediments that contain groundwaters that are less saline and possibly even undersaturated with respect to the more soluble surface salts. Thus, even if salts precipitate at the surface, they are not necessarily preserved when buried. Preservation will only occur when pore-waters in underlying sediments have been concentrated to near-saturation levels for the salts in question.

Creation of a plume of dense brines beneath evaporitic mudflats or salt pans also generates a subsurface hydrological structure that refracts flow lines of up-flowing (and potentially unsaturated) groundwaters (Fig. 2). This focuses groundwater discharge away from central salt-accumulating areas toward the margins of the basin, promoting preservation of basin-central salts.

The presence of subsurface brine plumes, known from present-day continental evaporite basins, must also have occurred in association with ancient marine evaporitic shelves and basins. These brine plumes can be inferred by the very fact that evaporites have been preserved, and by the occurrence of cement overgrowths on primary evaporite crystals (and other diagenetic changes) that require the presence of supersaturated pore-filling brines.

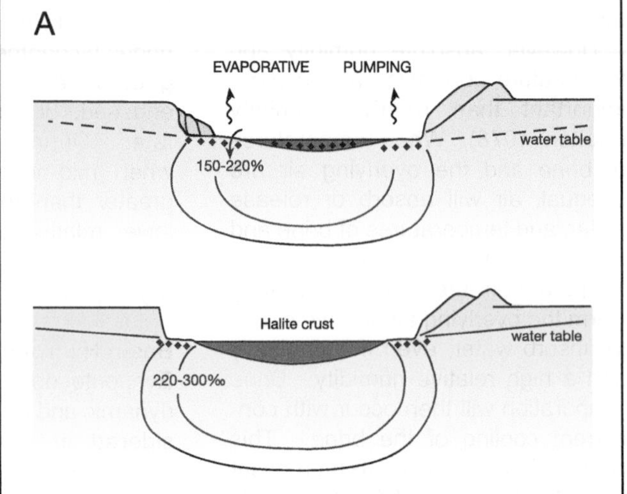

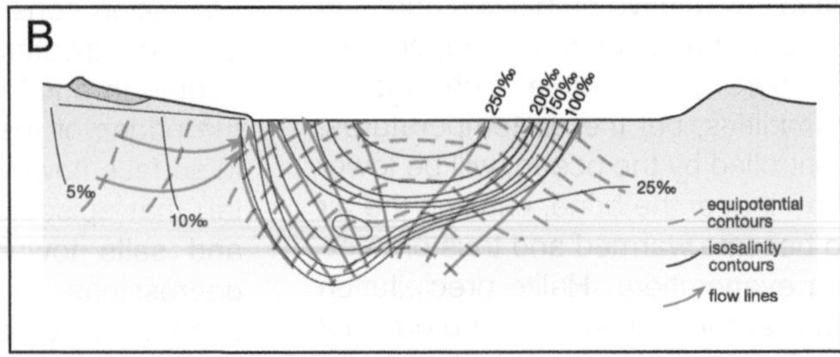

Figure 2. A. Stages in the development of brine-plume beneath a saline lake; increasing density of brine-plume allows more saline and denser surface brines to accumulate at the surface. Sequence may reflect changes over time or a change towards increasing aridity. **B.** Cross-section of brine-plume beneath the salt crust of Lake Frome (Australia) showing salinity variations, equipotential isolines and inferred groundwater flowlines that define the refracting character of the saline lens. The lens focuses groundwater discharge towards playa margins. Both diagrams modified from Bowler (1986).

Inflow-outflow Ratios in Evaporite Basins

Surface brines can only remain within a basin where water losses (evaporation + brine reflux) are more or less balanced by inflow through the barrier (+ basin-floor seepage inflow, rainfall, and additions of continental surface waters). If inflows exceed water losses, the basin fills, becoming more dilute. If losses exceed inflow, the basin desiccates (Fig. 3).

The balance between losses and inputs also determines the salinity of basin brines (Fig. 3). Brine salinity is partially a function of the residence

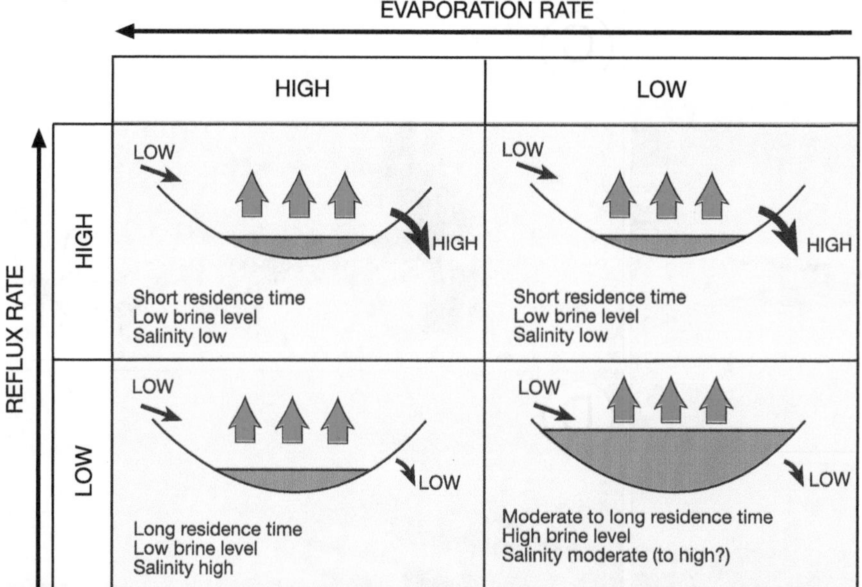

Figure 3. Models of isolated evaporite basins. Evaporative drawdown occurs in response to low rates of influx (relative to evaporative and seepage losses). Deep-water evaporites only form in basins with low rates of reflux and evaporation losses; all others experience significant desiccation. However, with low relative rates of evaporation, long time periods are required to concentrate the entire brine volume in the basin to the point where evaporites accumulate.

time of water in the basin (Logan, 1987). With low reflux rates, residence times are long, and evaporation has time (with sufficient aridity) to concentrate brines to high salinities. When residence times are short (high reflux rates), the brines spend insufficient time in the basin to be concentrated to high salinities before being exported from the basin. With these short residence times, even when the climate is hyper-arid, only evaporite minerals of relatively low solubility (carbonates, gypsum) precipitate.

Neither influx nor reflux rates are stable during a basin's history. Sea-level changes increase or decrease the hydraulic head between ocean and basin-brine level, so increasing or decreasing the amount of seepage inflow and residence times. Sediment aggradation causes concurrent rises in brine levels, reducing hydraulic head and the rate of seepage influx. Sediment aggradation is also associated with lateral onlap of basin flanks, expanding the area of possible brine reflux. Sites of influx can also become choked by precipitates, progressively reducing or stopping inflow. Variations in the inflow-reflux ratio cause changes in the volume of surface brine retained in the basin and thus changes in the type of evaporite mineral precipitated. Such changes can occur without any climate change. Halite deposits thus need not imply more aridity than gypsum accumulations.

EVAPORITE FACIES

Evaporites form in a variety of modern geographic settings (Fig. 4). Settings inter-grade, are affected by similar depositional processes, and so can produce similar facies. A coastal brine-pan deposit is almost identical to a brine pan facies in a continental setting. Identification of the geographic setting of ancient evaporites commonly depends more upon geochemical information (identifying the origin of the precipitating brines) and upon associated facies, than on the evaporite facies themselves. An absence of Recent deep-basinal and widespread shallow-marine evaporitic settings (Fig. 5) means that ancient evaporites of these types must be identified by comparing their facies with those of continental salt lakes and with relatively small marine-fed salinas. Models explaining the distribution and extent of these ancient facies must, however, be determined by deductive reasoning, coupled with information about the physical and chemical behavior of brine systems.

One of the most important environmental parameters to be established for ancient evaporites is the depth of brine during their deposition.

Brine Depth

Three environmental settings for subaqueous evaporites (Fig. 6), based upon sedimentary structures, have been recognized (Schreiber *et al.*, 1976). Criteria used include:

1. Structures indicating wave and current activity, identifying high-energy intertidal and shallow-subtidal environments;
2. Algal structures (in the absence of wave- and current-induced structures), indicating a deeper environment, but one still within the photic zone; and
3. Widespread, evenly laminated sediments, lacking evidence of current and algal activity (but associated with gravity-displaced sediments) characterizing the deepest, subphotic environment.

There are problems with these criteria. Some stromatolites grow in protected shallow, quiet-water settings; thus an absence of current structures from microbial-laminated sediments is not necessarily a criterion of greater depth. Brines can be turbid with suspended crystallites, flocculating clays, organic matter, and colored micro-organisms. The photic limit in such brines is considerably less than a meter, rendering the criterion of little use. Finally, discrimination between planar stromatolites and films formed from detrital organic detritus (with no depth significance) may not be possible.

Widely correlated lamination is not unique to deep-water environments. It occurs within evaporites deposited in episodically flooded environments (*see* Fig. 20).

Dissolution surfaces and bottom-grown crusts are, however, useful in assessing depositional depth in evaporites (Hovorka *et al.*, 2007)(Fig. 7). Frequent dissolution surfaces imply that frequent water influxes into the depositional site dissolved uppermost parts of previously precipitated evaporites. Each flooding event diluted the entire brine column, and this can only happen when

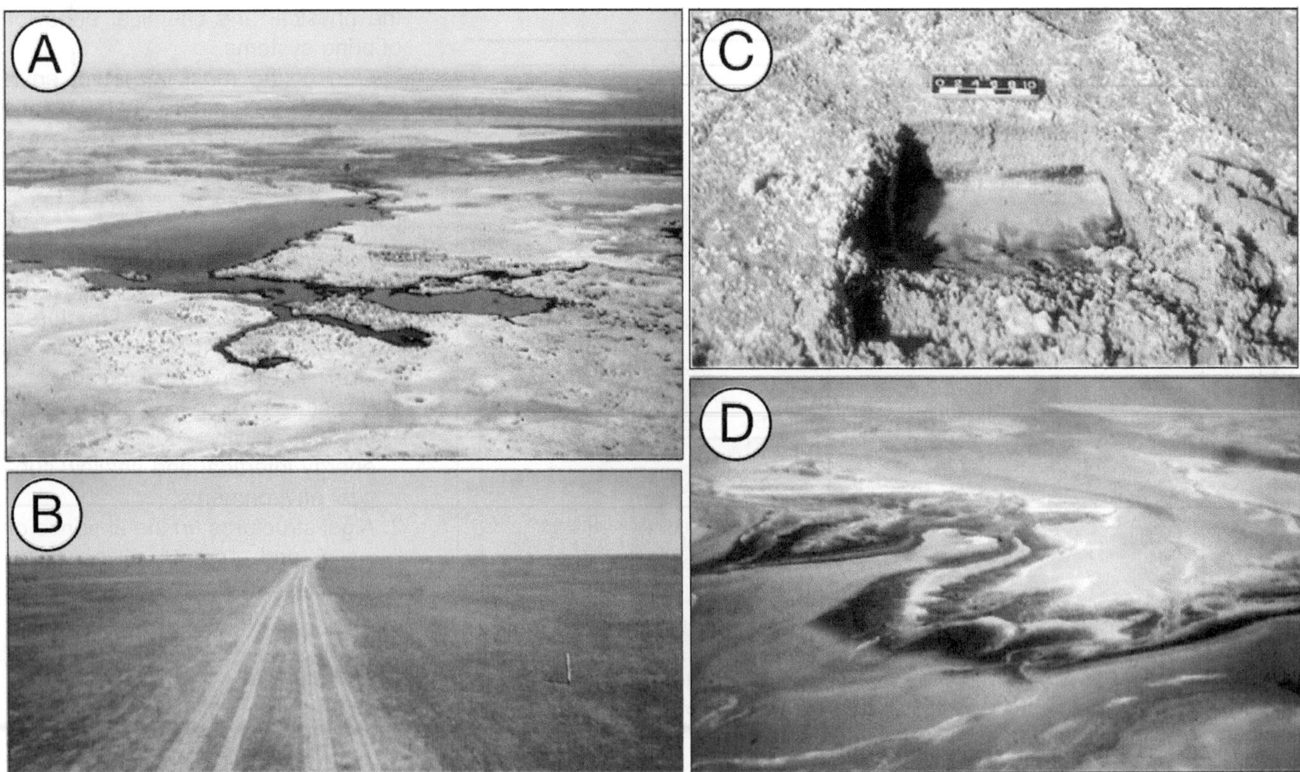

Figure 4. A. Aerial view of a salina: Lake MacLeod (Western Australia) with a marginal perennial marine lake (~2 km across) fed by seepage through a barrier (part of seepage face in foreground) that passes into evaporite flats in far distance, **B.** Ground-level view of Lake Macleod evaporite flat, **C.** Detail of shallow pit (Lake MacLeod evaporite flat), with efflorescent salt crust and shallow brine table, **D.** Oblique view of Abu Dhabi coastline with subtidal marine lagoon (20 km across), dark intertidal microbial mat (1-2 km wide) and sabkha surface in foreground; barrier islands and open Persian Gulf in far distance. (Photo courtesy R.K. Park.)

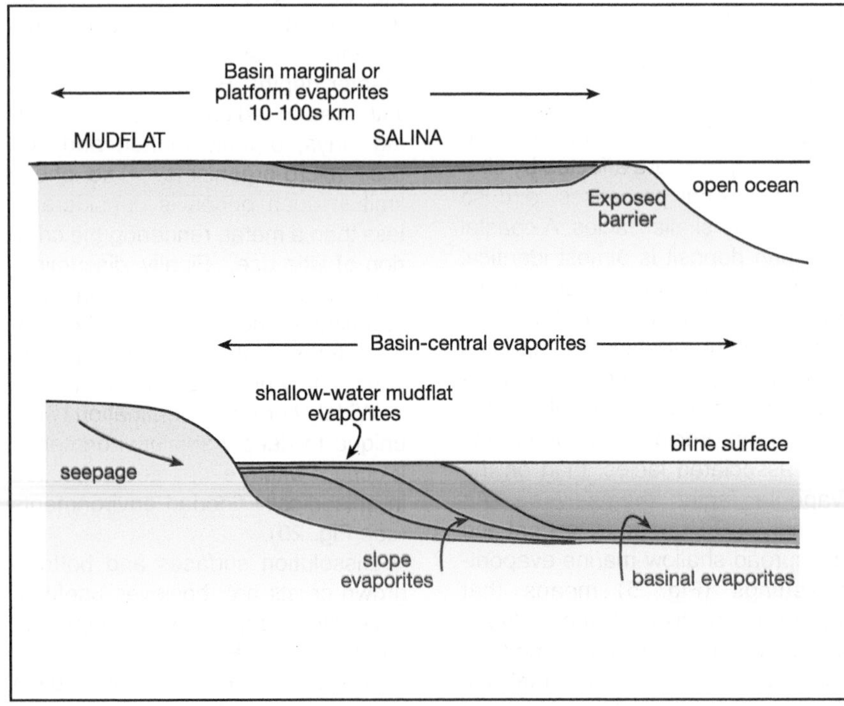

Figure 5. Diagrams of ancient evaporitic settings.

brine volumes are small, implying shallow brine depths. Other surfaces have karstic and related features, indicating depositional surfaces became subaerially exposed.

Growth of subaqueously-precipitated evaporite crystal crusts requires that they be in contact with supersaturated brines. Evaporative concentration leading to supersaturation, however, only occurs at the air-water interface. In deep bodies of stratified brine, descent of supersaturated brines is usually prevented by brine stratification, and bottom-grown crusts should not form. Advection of supersaturated brines occurs down to a salinocline so that only where basin floors lay at shallower depths can bottom crusts potentially grow. Even when brine stratification is absent, as it is for brief periods in the present-day Dead Sea, a drive to transport supersaturated brines down to a deep-basin floor may be absent (but see later).

Basin Hydrodynamics

Different evaporite minerals precipi-

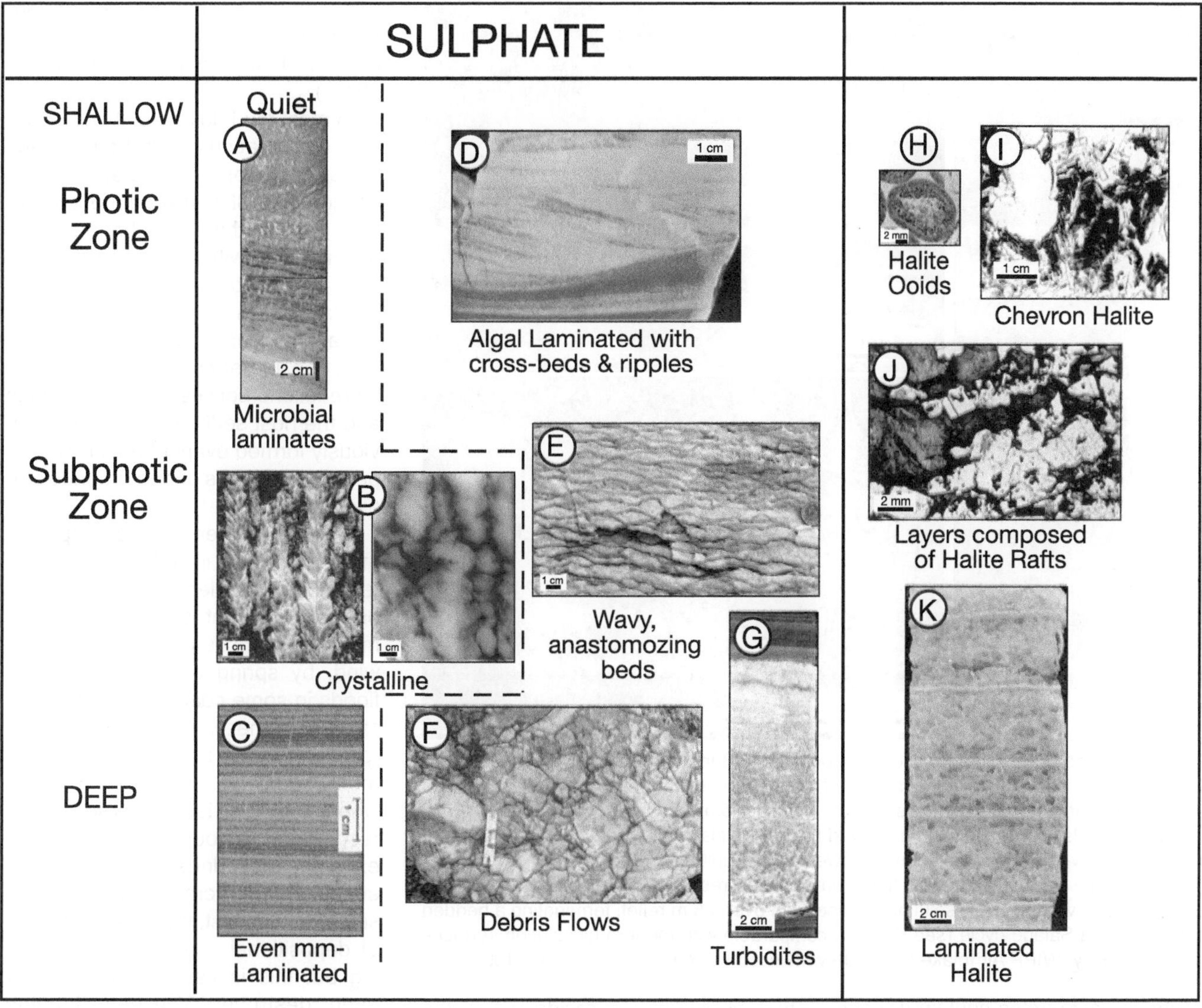

Figure 6. Main depth environments of subaqueous evaporite deposition and the chief sulfate and halite facies present in each (adapted from Schrieber *et al.*, 1973). **A.** Microbial laminated sulfates, anhydrite after gypsum crystals that grew within microbial mat (Souris River Formation, Saskatchewan), **B.** Gypsum crusts, Miocene, near Paphos, Cyprus (photo courtesy B.C. Schreiber) and anhydrite replacements of similar crystals, Buckner Formation, Florida, **C.** Even-laminated sulfates, Castile Formation, New Mexico, **D.** Crossbedded sulfates, Miocene, Sicily (photo courtesy B.C. Schreiber), **E.** Wavy beds (flaser-bedding), Zechstein, Harz Mountains, Germany, **F.** Debris flows, Zechstein, Harz Mountains, Germany (scale bar 15 cm long), **G.** Turbidites, Zechstein, north Germany, **H.** Halite oolites, Tuz Gölü Lake, Turkey (photo courtesy E. Tekin), **I.** chevron halite, modern salt crust, Death Valley (photo courtesy T. Lowenstein), **J.** Thin-section of halite rafts (with sylvite cement) Salado Formation, New Mexico (photo courtesy T. Lowenstein) and **K.** Laminated halite, Permian Castile Formation, New Mexico.

tate from brines of very different densities. Some evaporite depositional models i) suggest lateral variation in evaporite mineralogy occurs, and ii) that this reflects a lateral variability of brine salinity across the basin. These models imply the brines in the basin exhibited a marked lateral change in density. But, where brine depths exceed a few meters, any substantial horizontal variation in brine density is hydrodynamically unstable. The brine body should rapidly reach equilibrium by becoming vertically stratified and horizontally more uniform. In this situation, the same evaporite mineral precipitates over the entire depositional area and changes in mineralogy will also be simultaneous across the basin. Vertical changes in evaporite mineralogy should reflect basin-wide changes in brine density and concentration. Apparent lateral changes in evaporite facies either indicate, i) deposition occurred in shallow environments where bottom friction prevent hydrodynamic equilibrium conditions from being reached, or ii) the different facies are not, in fact, time equivalents of each other.

Main Evaporite Facies

Three evaporite facies groups are identified:

1. *Mud-flat ('sabkha') facies,* in which evaporites are ephemeral or accumulate within older sediments in episodically flooded, essentially subaerial environments - less commonly efflorescent surface crusts are preserved;
2. *Shallow-water facies*, characterized by evaporites precipitated from thin, desiccating brine

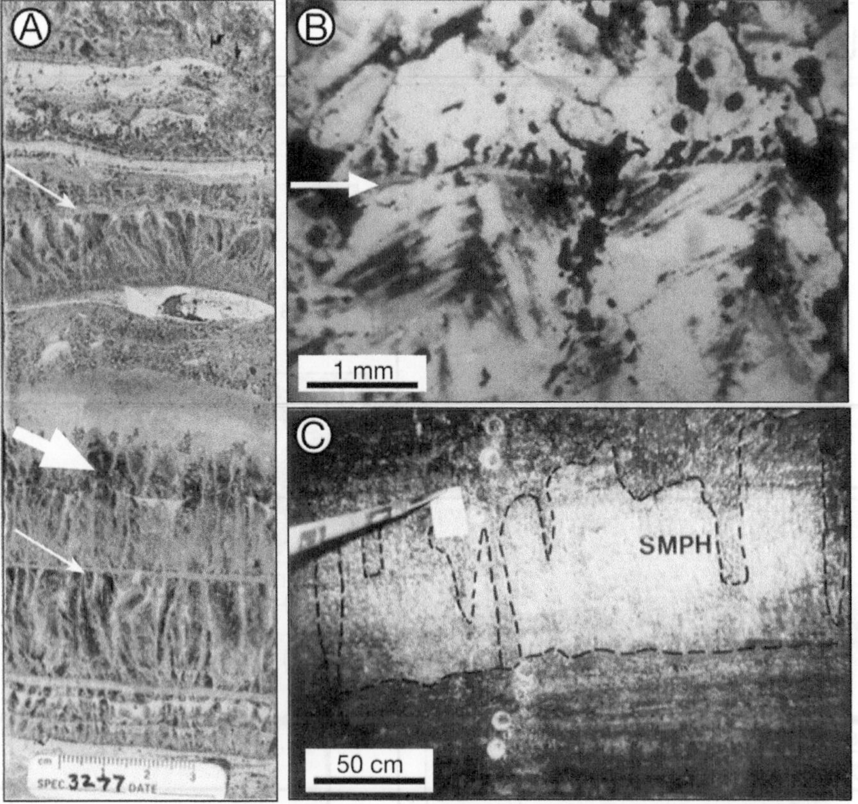

Figure 7. Types of bedding surface in evaporites. **A.** Buried, unmodified crystal terminations (thick arrow), indicating absence of dissolution events, and truncated crystals (inclined thin arrows) indicating dissolution affecting former gypsum crusts, Salado Formation, New Mexico, **B.** Solution-truncated halite crystals (white arrow), indicating dissolution by a low-salinity water influx; followed by precipitation of small gypsum crystals (now anhydrite and halite) before halite precipitation resumed. Palo Duro Basin (photo courtesy S. Hovorka), **C.** Dissolution surface, with over 0.5 m relief, terminating a bedded halite unit in a Salado cycle (*see* Fig. 36), comparable with modern dissolution surfaces in Death Valley; WIPP Air Intake Shaft, New Mexico. From Holt and Powers (1990).

sheets; evaporites contain dissolution surfaces and/or abundant bottom-grown crusts (ephemeral lakes and salt pans); and

3. *'Deep'-water facies* characterized by laminated and gravity-displaced sediments, but lacking evidence for shallower-water environments. 'Deep-water' evaporites can, however, accumulate at depths of only a few tens of meters: the essential features of laminated evaporites only requiring the presence of brine stratification. Stratification occurs in brine bodies only ten meters (or less) deep.

MUD-FLAT EVAPORITES

These grow in subaerially exposed, but occasionally flooded, saline mud (or sand-) flats around playa lakes and salina ponds, and in supratidal flats of coastal sabkhas. Brine surfaces lie within the sediment, commonly <1 m beneath the surface, so that all sediments reside within the phreatic or capillary zones and remain moist.

Sedimentary Processes

Saline mud flats are equilibrium deflation-sedimentation surfaces (Stokes surfaces) that have a topography controlled by the brine table and its gradients. Wind removes particulate material that dries out on the flats, but this process can be inhibited by hard efflorescent crusts or by covering coarse lag deposits. Beach ridges, for example, rise above the general level of a coastal sabkha because they are covered by shell lags. On the other hand, disruption of uppermost sediments by salts can create materials very susceptible to removal by wind action.

Evaporation concentrates pore fluids, and saline mud flats are sites of brine formation. Evaporation of subsurface brines only occurs in the uppermost parts of the capillary zone where at least some pore-throats are not brine filled (so allowing upward escape of water vapor to the atmosphere).

In most mud-flats little, if any, evaporite precipitation occurs from surface brines (except ephemeral efflorescences, Fig. 8B); instead the evaporites grow within soft, brine-soaked sediments or within the overlying capillary zone to form intra-sediment crystals and nodules (Fig. 8A). Precipitation occurs by brine evaporation from the uppermost capillary fringe; by reactions between refluxing brines and the host sediment (including previously formed evaporites); or by temperature changes during groundwater ascent.

Host sediments are typically muddy, but also include organic matter, sands or earlier-precipitated evaporites. Sediment is transported onto mud-flat surfaces episodically by the wind, by spring tide or storm-sheet floods in some coastal sabkhas or by storm-fed sheets moving downslope (or down-wind) from salinas. Sediment additions are only preserved where the brine table is raised a corresponding amount: otherwise the extra sediment dries out and is deflated away. Where flood-introduced sediments persist, the episodic nature of depositional events imparts layering but this is partially disrupted or even destroyed between flooding events by the growth and dissolution of ephemeral evaporite crystals, and by drying and/or thermal volume changes causing extensive, multiple mud-cracking.

Efflorescent crusts of saline minerals temporarily accumulate on mud flats i) during groundwater discharge and evaporation (Fig. 8B), or ii) by the evaporation to dryness of ponded floodwaters. Because evaporation is both rapid and complete, crusts contain metastable and highly soluble salts. Rain and floodwaters dissolve the more soluble constituents to form concentrated but chemically 'simple' brines which, ultimately, reach basin-center ponds or salt pans.

Thicker saline crusts are similar to those of salt pans. Growth of evaporite crystals causes volume increases and formation of salt blisters, salt polygonal crack systems, commonly with over-thrust rims (Chapter 21) or,

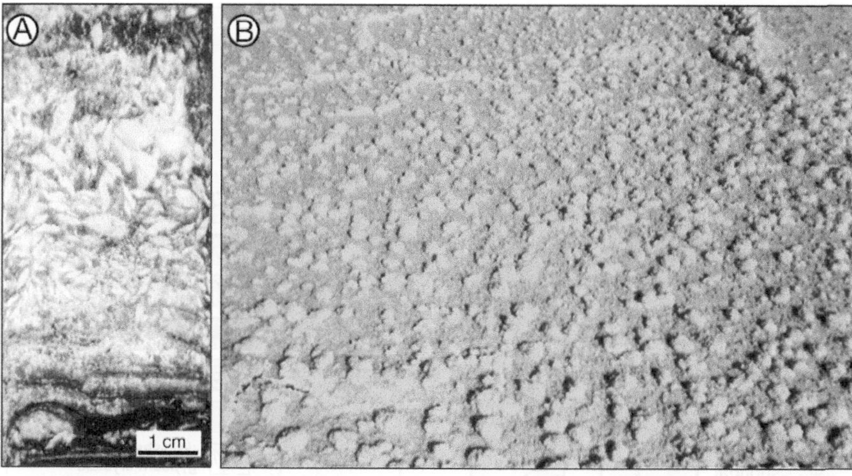

Figure 8. Recent mud-flat evaporites. **A.** Displacive lenticular gypsum crystals forming 'gypsum mush' layer, Abu Dhabi sabkha. (Photo courtesy B.C. Schreiber), **B.** Ephemeral halite crust with 'pop-corn' like growths (10 cm across), Howz-e-Soltan salt Lake, Iran. Although halite is the most conspicuous surface evaporite, gypsum is the only phase accumulating in the sediments. (Photo courtesy F. Fayazi.)

when affected by rain, highly irregular dissolution surfaces, sometimes with considerable relief (Fig. 9).

Even salt crusts that survive dissolution by floodwaters and become buried are not necessarily preserved. Underlying groundwaters are commonly unsaturated. When groundwaters move upward, drawn by evaporative losses near the surface, they dissolve buried remnants of the salt crust and re-precipitate them at the sediment surface. Farther away from the water input sites, however, groundwaters can become increasingly more saline and calcium sulfates, or exceptionally even halite, become stable within the sediment because the underlying groundwaters are already saturated with respect to those minerals. Minerals in surface crusts thus do not usually reflect the character of any evaporite minerals accumulating within the sediment. Many dolomitic, redbed sequences reflect deposits of these evaporitic, but essentially non-evaporite-preserving, environments (Fig. 10). Thick, well-preserved intervals of efflorescent halite, however, have been preserved in the rock record (Fig. 11) in exceptional circumstances (Lowenstein *et al.* 2003)

Previously, coastal sabkha brines were considered marine and supplied either by episodic, wind-driven seawater flooding, or by *lateral* flow of either seawater or continental-shallow groundwaters; movement being caused by evaporative losses on the sabkha surface. Wood *et al.* (2002), however, demonstrate that 95% of the solutes precipitated in the capillary zone of Trucial Coast sabkhas come from *ascending* deep, continental brines. Precipitation of gypsum and anhydrite within sabkha sediments does not occur by brine evaporation (Wood *et al.*, 2005). As brines are transported upward, drawn by water evaporative losses at the surface, they warm (especially during summer months) and minerals with retrograde solubility (such as gypsum and anhydrite) precipitate.

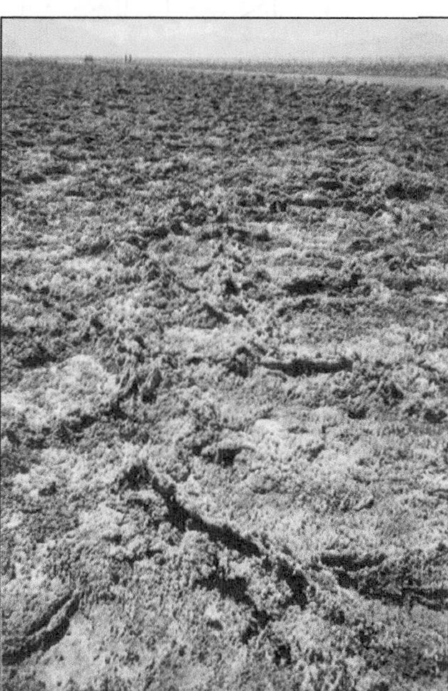

Figure 9. Markedly upthrust polygons, partially dissolved by rain to form irregular, 0.5 m relief. Devil's Golf Course, Death Valley, California.

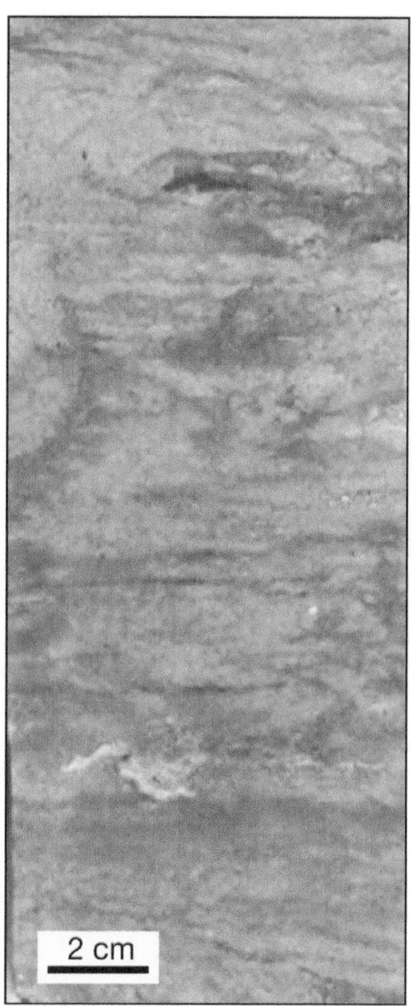

Figure 10. Silty and sandy mudstone with discontinuous and disturbed silt/sand layers: Lower Watrous redbeds, Saskatchewan.

The reader should be aware that the ascending groundwater re-interpretation of the Trucial Coast sabkhas has yet to be integrated with other, and perhaps inconsistent, data. These include the presence of two generations of calcium sulfate (one transgressive, the other regressive, Fig. 12) and interpretations of much of the regressive anhydrite as a replacement of earlier salina gypsum (Hardie, 1986; Kirkham, 1997).

Sedimentary Facies

Two types of mud flat are distinguished in continental playas (Hardie *et al.*, 1978), and similar concepts can be adopted for marine-marginal sabkhas and salinas. *Saline mud flats* occur closer to saline lakes, salt pans or the sea, whereas *dry mud*

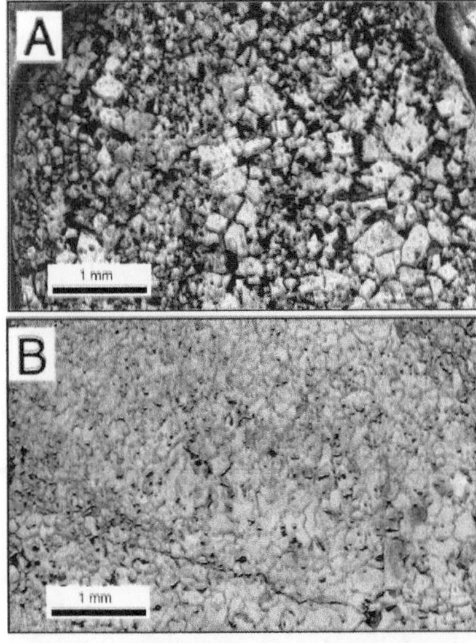

Figure 11. Preserved efflorescent halite in subaerial halite crusts: thin sections A. above 50 m with halite cubes and abundant porosity (dark grey areas), B. below 50 m, now a tight interlocking mosaic caused by cement overgrowth on original crystals; Salar de Atacama, Chile; (photos courtesy T. Lowenstein.)

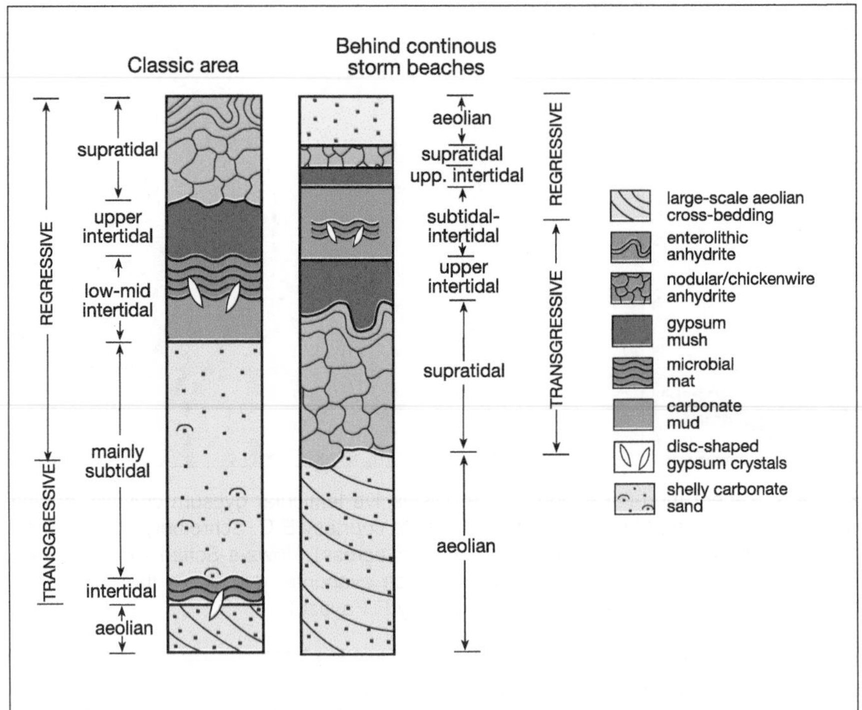

Figure 12. Diagrams showing anhydrite occurrence in Abu Dhabi sabkha (after Kirkham (1997). In the shoreward 'classic' area subtidal sediments are overlain by a 'regressive sequence' of intertidal and supratidal sediments, the last mentioned containing anhydrite. Farther inland and behind beach ridges, there is a basal 'transgressive' sequence (with thick anhydrite) beneath the transgressive deposit (with its own, lesser developed, anhydrite).

flats appear in more distal locations and pass laterally into coarser sand flats, or interfinger with aeolian, fluvial braidplain or sheetwash sediments.

Dry Mud Flats

Depositional and desiccation structures (Fig. 13) are usually preserved. Mud and sandy laminae form during and after flooding, but are disrupted by mud-cracks, sheet-cracks (Fig. 13B) and by growth and subsequent dissolution of saline minerals. Dry mud flats commonly are covered by efflorescent saline crusts between flooding events. When saline crusts begin to dissolve in floodwaters, introduced coarser sediment is trapped in depressions on the dissolving crust. Upon burial and complete dissolution of any remaining saline crusts, these coarser sediments are preserved as irregular coarser grained patches and layers (Smoot and Castens-Seidell, 1994) (Fig. 13C).

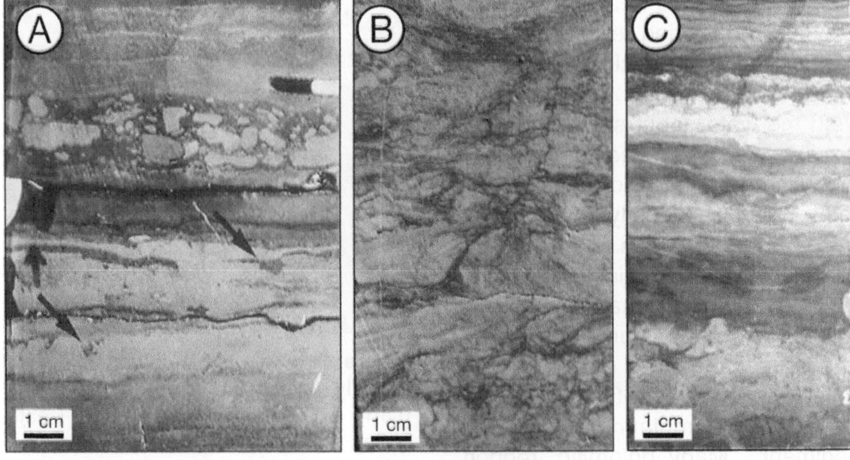

Figure 13. Ancient dry mud-flat dolomites A. Centimeter-bedded silty dolomites with erosive bases and sediment-filled moulds after displacive halite hopper crystals (arrowed), B. Silty dolomite, originally laminated but now disrupted by superimposed generations of desiccation cracks and probable repeated growth and dissolution of evaporites, C. laminated dolomites with discontinuous silt layers mimicking efflorescent evaporite crusts. A and B from Paradox Formation (Pennsylvanian) Utah, C from Lake Alma anhydrite (Ordovician), Saskatchewan.

Saline Mud Flats

Primary sedimentary structures in this facies are destroyed by growth of intra-sediment evaporite crystals. A chaotic mix of evaporite crystals within an unlaminated matrix forms as enormous numbers and volumes of displacive evaporite crystals slice into the original sediment, pushing it around as they grow (Figs. 14 and 15).

Evaporite Facies in Mud Flats

Intra-sediment evaporite crystals grow by i) incorporating sediment as inclusions within the growing crystals, often along crystal growth layers (Fig. 15A) or ii) displacively, by pushing aside the sediment, or by both processes.

Intra-sediment gypsum crystals typically are discoidal (flattened nor-

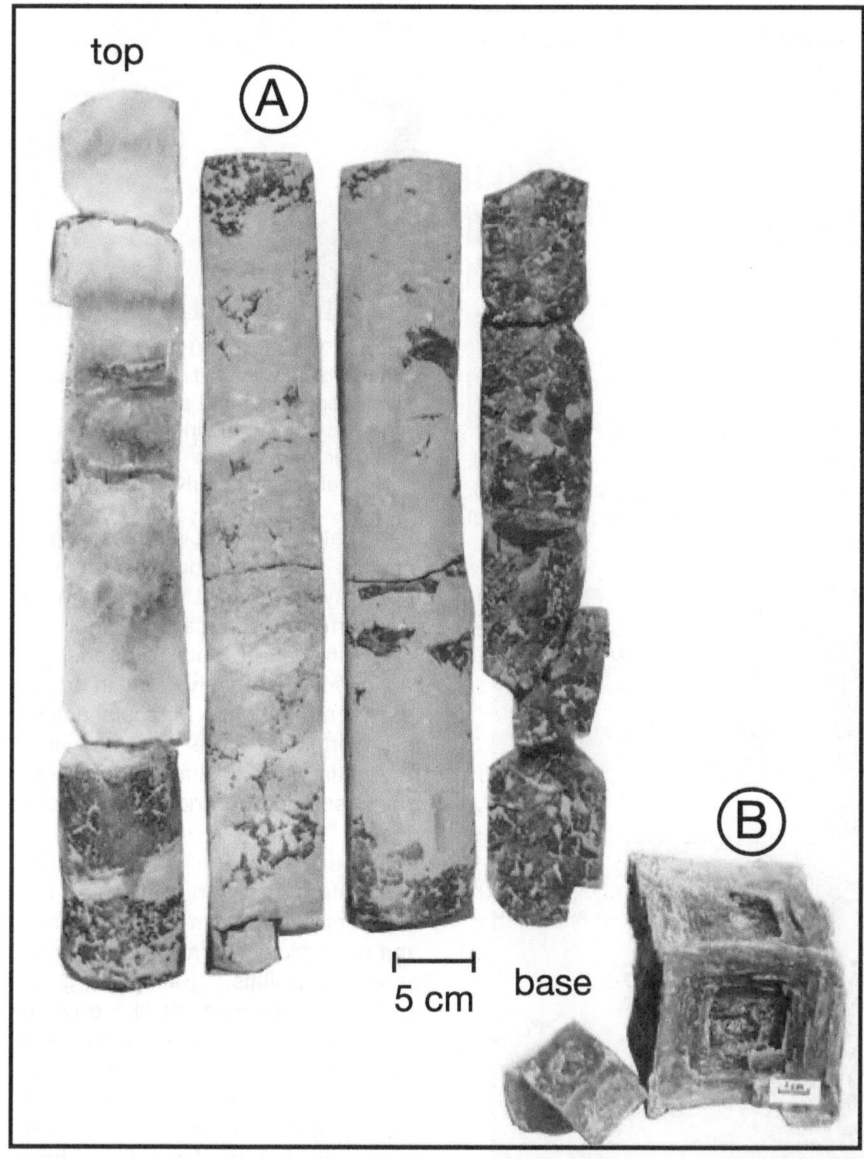

Figure 14. A. Upward passage from red silty and anhydritic mudstone with abundant, displacive halite (haselgebirge facies) via similar sediments but with lesser amounts of halite (crystals with hopper morphology) into bedded halite with thin anhydrite and clay laminae. Sequence marks change from saline mud flat into a salt-pan deposit. Lotzberg Salt (M. Devonian), N. Alberta, **B.** Single displacive hopper crystals of halite, Dead Sea (photo courtesy of B.C. Schreiber).

mal to c-axis, Fig. 8A) and may be complexly twinned, forming rosettes (desert roses). Gypsum grows displacively within microbial mats or fine-grained sediments (sometimes forming a crystal mush), but grows as a cement in sandy sediments. Gypsum precipitation causes calcium depletion in groundwaters, and the increased Mg/Ca ratio in brines promotes dolomitization of pre-existing aragonite and/or magnesite precipitation.

Anhydrite has a more restricted distribution and is confined to the capillary zone. It replaces earlier gypsum or is newly formed. Nucleation of anhydrite requires a seasonal high temperature above 35°C and the mineral is only preserved where mean annual temperatures exceed 20°C. It forms discrete nodules and bands of coalesced nodules (Figs. 16 and 17). Displacement, replacement and dilution of the host sediment can occur to such an extent that it becomes relocated to internodule areas or is reduced to mere partings between the anhydrite nodules. Much of this *mosaic anhydrite*, however, has replaced pre-existing gypsum, rather than being displacive. Pseudomorphs of gypsum crystals are commonly absent because of adjustment during compaction, and by continued growth of primary anhydrite. Composite anhydrite nodules are remains of gypsum crystal clusters, and massive anhydrite forms from gypsum mush layers.

Gypsum nodules also occur, and elsewhere gypsum and anhydrite regularly alter backward and forward, depending on temperature changes and flooding or rainfall events.

Displacive halite in modern playa sediments has been described, but it is uncertain whether precipitation occurred beneath a shallow brine or within mud-flat capillary zones. Crystals commonly grow preferentially along cube corners and edges (especially when precipitated from highly supersaturated brines) forming cubes with hopper-like pyramidal hollows on each crystal face (Fig. 14B). In extreme cases, skeletal/dendritic halite crystals form (Fig. 15B). Halite beneath and on coastal sabkha surfaces is not usually an accumulative phase, but dissolves diurnally in morning dew; is blown away, or dissolves in floodwaters or upon burial.

Preserved efflorescent crusts (also termed subaerial salts) are uncommon and are composed of an interlocking aggregate of mm-sized (and smaller) cubic halite crystals (Lowenstein *et al.*, 2003)(Fig. 11). Vugs, created by partial dissolution during rainfall events, are partially filled by wind-blown dust. The same detrital material is also trapped on the surface by films of hydroscopic brine creating silty lenses or discontinuous laminae. Where subaerial halite is preserved in the subsurface it is massive. Porosity occurs down to depths of 50 m (Fig. 11A), but is eliminated at deeper levels by halite cement overgrowths that create a tight, interlocking, polygonal crystal aggregate (Fig. 11B). Many ancient massive halite deposits might represent diagenetic replacements of similar buried efflorescent crusts.

SHALLOW-WATER MARINE EVAPORITES

Deposition of most shallow-water evaporites occurs in brines at, or

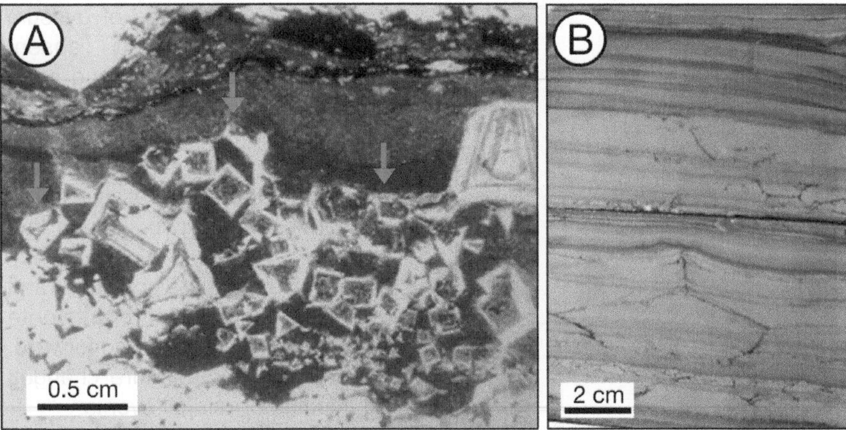

Figure 15. Displacive halite. **A.** With zonally arranged sediment inclusions. Arrows mark erosion surface truncating halite crystals, Permian of West Texas (photo courtesy of S. Horvoka), **B.** Skeletal halite (now replaced by dark pyritic films) displacing storm-sheet dolomite layers, Lake Alma Anhydrite (Ordovician) of Saskatchewan.

Figure 16. Sabkha cycles in Frobisher Evaporite (Mississippian of Saskatchewan). **A.** Laminated dolomite (probably hypersaline lagoon) overlain by microbial mat containing anhydrite pseudomorphs of vertically oriented lenticular gypsum crystals, and nodular mosaic anhydrite. **B.** Upper part of sabkha cycle with inclined rows of nodular mosaic anhydrite (after subaqueous gypsum), truncated by erosion surface and overlain by laminated micrites of succeeding lagoonal member. Nature of anhydrite suggests it has overprinted subaqueous (salina) gypsum deposits.

Figure 17. Trench ~50 cm deep in Abu Dhabi sabkha with diapiric layers of anhydrite (after gypsum). Uppermost layer that truncates the anhydrite is a storm-washover and aeolian carbonate-clastic unit. Photo courtesy R.K. Park.

near, saturation with respect to gypsum or halite, and in environments that may be subject to strong wave and current action, causing sediment scour, transport and re-deposition. Most facies probably form in brines less than 5 m deep. Microbial activity is important in more protected environments, and sediments are commonly subject to periodic or episodic drying.

Modern shallow-water marine evaporites occur in perennial and ephemeral (salt pan) salina ponds, and are typically surrounded by mud- or evaporite-flats. Salt pans grade insensibly into mud-flat like environments where evaporites accumulate as detrital sediments or by surface precipitation. These evaporite-flats are occupied for only brief periods (days, weeks) by unconfined brine sheets, usually only a few centimeters deep.

Evaporites having shallow-water features occur in ancient deposits where they form laterally extensive deposits unlike those of modern examples. Some formed by lateral accretion in depositional environments similar to those of the present, but most require explanations that are drastic departures from present-day settings.

Shallow-water evaporites are commonly cyclic, composed of shoaling (or brining-upward) sequences and capped by mud-flat deposits. Evaporite beds within cycles often can sometimes be correlated for distances of tens to hundreds of kilometers.

Sedimentary Processes

Perennial Marine Lakes

Perennial marine lakes persist for tens to thousands of years without desiccating. They require substantial perennial water inflow – a restricted inlet to the sea or a highly permeable barrier. During wet periods they can expand greatly, flooding surrounding flats.

For soluble minerals to be precipitated, bottom brines must be sufficiently dense to support the surface brine for the time necessary to allow it to become supersaturated with respect to the saline mineral. When bottom brines are insufficiently dense, surface brines sink before supersaturation is reached.

Evaporation at the lake surface generates the concentrated brines in which saline minerals nucleate and grow. Brines and crystals sink but, unless the entire brine column is saturated, the crystals dissolve before reaching the lake bottom. Their dissolution progressively increases brine concentrations until crystals no longer dissolve and they begin to accumulate on the lake floor (Hardie *et al.*, 1978).

Lakes experiencing progressive reduction in their inflow exhibit a progressive change from precipitating low- to high-solubility salts. However, lakes affected by short-term decreases in inflow might display no change in the minerals precipitated because the change in inflow/evaporation ratio is not immediately matched by a sufficient change in the density of the bottom brine.

Dissolution of the uppermost evaporite deposits in shallow perennial lakes occurs during freshening events (storms, wet seasons or periods) when mixing of the brine column is complete. Wind turbulence causes mixing. During strong persistent winds there can be wind set-up whereby brines are driven and maintained up-slope. Migration of lake margins for tens of kilometers is controlled by changes in wind conditions, and brine sheets can become disconnected from the main lake to move over neighboring mud- or evaporite-flats.

Wave action is limited by the restricted depth and/or fetch. Maximum expenditure of wave energy occurs at lake margins where evaporite crystals can be continuously moved and abraded *as they grow*. This forms rippled sands and gravels composed of rounded crystal clasts, and sometimes formation of halite or gypsum ooids.

Figure 18. Rubbery, wrinkled microbial mat, deformed into polygons and encrusted with gypsum crystals: Hyeres salt lagoons, S. France. Lens cap is 30 cm in diameter. Photos courtesy G.M. Harwood.

Crystal growth at brine surfaces produces i) small, thin and platy or needle-like crystals or ii) brine-displacing, hollow hopper-shaped crystals. Crystals also coalesce into floating rafts. Crystals and rafts are held up by surface tension until they grow sufficiently large to sink, or they are spilled by waves. They are sometimes transported downwind to accumulate at shorelines, but most accumulate on lake floors forming loosely packed layers of cumulate crystals.

Crystal crusts precipitate on lake bottoms from well-mixed supersaturated brines. They overgrow cumulate crystals or older crusts, or nucleate on stable substrates (*e.g.*, microbial mats, Fig. 18). Crusts deposited at greater depths are protected from dissolution during flood events by the density stratification that develops between the more dilute inflow and the denser bottom brine, and the crusts become draped by detrital or chemical laminae. Crusts in shallow areas, however, are partially dissolved during freshening events and can be difficult to distinguish from salt-pan deposits.

Shallow lakes with bottom brines at, or below, halite super-saturation commonly support luxuriant microbial mats. Gypsum and carbonates precipitate in, and upon, the mats.

Marine Salt Pans

Marine salt pans are occupied by ephemeral bodies of brine that form when inflow temporarily exceeds outflow. A return to normal evaporative conditions causes these ephemeral lakes to shrink and disappear. Repeated floods generate sequences composed of thin salt crusts of halite, or gypsum, separated by mud partings and dissolution surfaces (Fig. 19) (Chapter 21).

During floods, shallow brackish lakes form, and fine-grained sediment (introduced by floodwaters) is deposited as thin layers. Floodwaters also dissolve i) efflorescent saline crusts on surrounding flats, and ii) previously precipitated salt crusts of the salt pan. After floods, lake waters concentrate by evaporation. At first, they precipitate cumulate crystals, but eventually crystals grow on the lake floor as bottom crusts. Newly exposed salt crusts have high porosities with pores filled with dense brine. Exposed salt surfaces are kept moist by evaporative

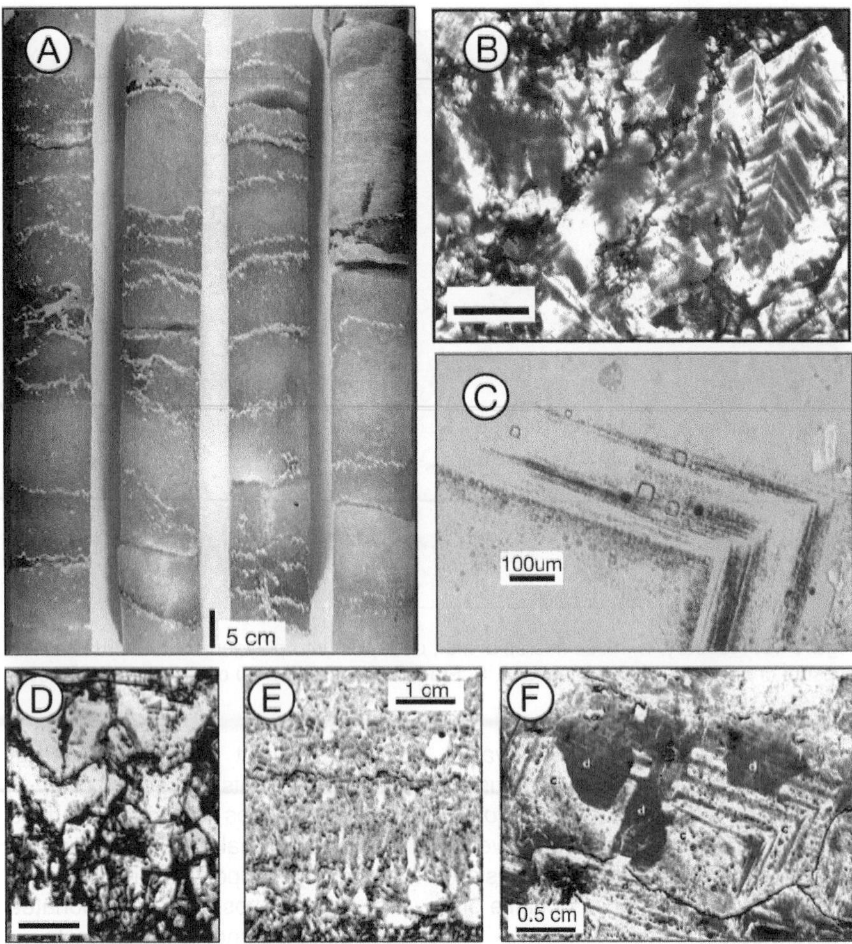

Figure 19. Brine-pan halite. **A.** Crusts of chevron halite separated by anhydrite laminae, Lower Prairie Evaporite (M. Devonian) Saskatchewan. **B.** Thin-section of modern chevron halite, salt crust, Death Valley (scale bar = 1mm). **C.** Fluid inclusions within chevron halite, Miocene, Spain. **D.** Cornet halite, modern salt crust, Death Valley, (scale bar = 1mm) **E.** Halite crust with numerous vertical dissolution tubes. **F.** Irradiated thin-section (creates blue color within halite) showing primary crystals with growth layers (c) cut by dissolution cavities, later filled by inclusion-free halite (d). Rot Salt, The Netherlands, photo courtesy J.R. Urai. Photos B-E courtesy T. Lowenstein

draw and by precipitation of dew on hydrophilic salt surfaces during cold nights. The evaporation rate falls to values as low as 1/170th of the rate from standing bodies of the same brine, so that the brine level rarely drops more than a few meters beneath the surface. Nevertheless brines continue to concentrate by evaporation, with development of overgrowth cements on salt crystals (Fig. 19F), precipitation of more soluble minerals within porosity, and growth of displacive salt crystals in clastic mud layers.

Continued crystal growth causes crust expansion and formation of polygonal fractures with overthrust edges (*see* Chapter 21). Preferred brine evaporation from fractures allows precipitation of efflorescent salts that widen cracks, and continues the upturn of polygon margins to form relief up to 50 cm (Fig. 9).

Each flood introduces new salts (from surrounding flats and the hinterland), so that less salt in the pan is usually dissolved than was precipitated during preceding lake concentration phases, and net accumulation is possible. Each salt layer is composed of material that has been repeatedly dissolved and re-precipitated, but salts can also be lost by wind deflation or during brine reflux.

Gypsum pans differ from more saline salt pans. During floods, clastic and displacive gypsum crystals are more commonly reworked than completely dissolved, and accumulate as crystal lags and intraclasts at the base of mud layers. Also, after mud layers are deposited, they are colonized by microbial mats. Their contorted surfaces are templates for later gypsum precipitation as surface crystal crusts (Fig. 18) or as cements that grow beneath the blistered surfaces. Surface topography is exaggerated by continued gypsum crystal growth that buckles, domes and fractures the layers even more.

Marine Evaporite Flats

Marine evaporite flats are only occupied ephemerally by brine sheets. These are only a few centimeters deep but can cover areas of several thousand square kilometers. They form as outflows from adjacent ephemeral rivers or from saline lakes, either seasonally (when lowered evaporation rates allow lakes to expand), or episodically, when wind stress detaches brine sheets. Floods dissolve earlier-formed evaporites. This material then re-precipitates as floodwaters evaporate, and these materials are reworked by the flood sheets to form laterally persistent detrital layers of subaqueous gypsum or halite (Fig. 20), some traceable over the entire extent of the flooding, perhaps hundreds or thousands of square kilometers. Brine sheets dry up or move elsewhere, leading to exposure of the evaporite flat surface. Exposure may cause i) growth of additional intra-sediment evaporites (which, if extensive, converts evaporite flats into saline mud flats), ii) alteration of gypsum to anhydrite, or iii) partial to complete dissolution of the just-precipitated evaporite crusts. Surfaces are also affected by influxes of wind-blown dust. Microbial mats are likely to be ephemeral features, especially where exposure allows formation of salt crusts. Most microbial-laminated evaporites probably form beneath near-permanent shallow brine bodies rather than upon evaporitic flats where conditions are too harsh.

Evaporite flats occur essentially in the same geographic settings as mud flats. If displacive evaporite growth in mud flats (sabkhas) is generally a product of upward groundwater flux, then evaporite flats possibly form where such groundwater fluxes are low.

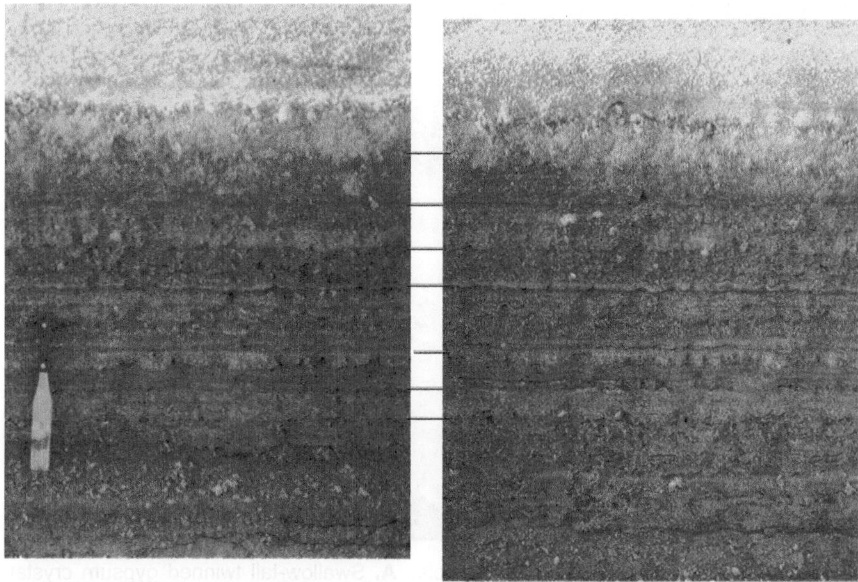

Figure 20. Thin-bedded to laminated clastic gypsum from evaporite flat of Lake MacLeod (Western Australia). Beds are laterally persistent (photos were taken ~100 m apart) and record flooding and precipitation events. Spatula is 30 cm long.

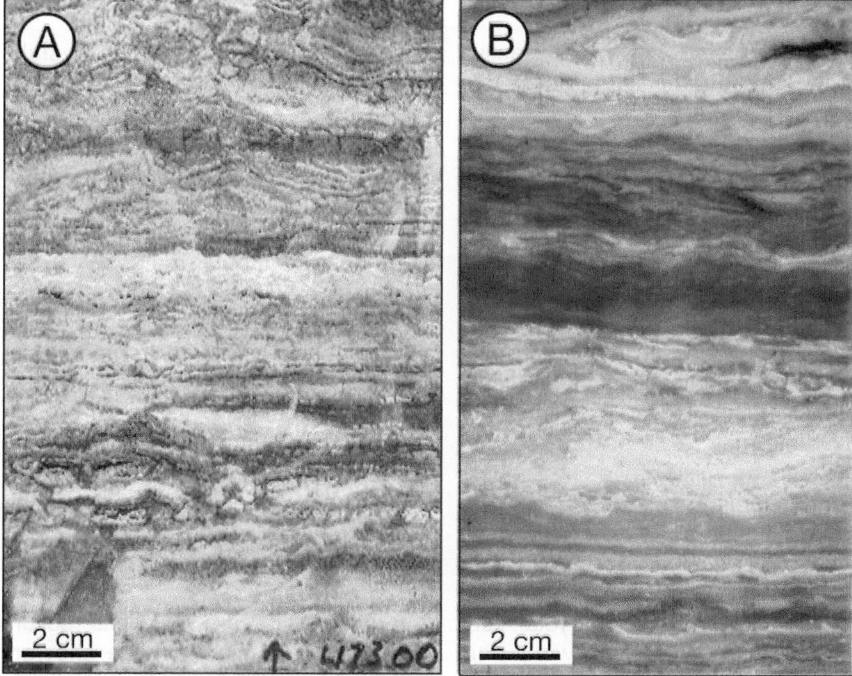

Figure 21. Comparison between **A)** modern laminated gypsum deposits from Lake MacLeod (*see* fig. 20) and **B)** laminated anhydrites from Dawson Bay Formation, Saskatchewan.

Evaporite Facies in Shallow-water Marine Settings

An abundance of shallow-water clastic textures and structures, and the presence of desiccation cracks, crystal crusts and microbial mats make identification of shallow-water evaporites relatively straightforward. Difficulties arise when depositional features are lost or obscured by later diagenesis, or when the sediments are modified by early diagenetic over-printing, as when desiccation subjects subaqueous evaporites to mud- flat processes.

Laminated Sulfates

Shallow-water laminites consist of current-deposited carbonate micrite and clastic silt- and sand-sized gypsum crystals/ cleavage fragments in reverse- and normally-graded laminae (Hardie and Eugster, 1971)(Figs. 20 and 21). Crystals originally grow i) as bottom crusts that are then broken and reworked or ii) were precipitated brine-surface crystals that sank. Yet other crystals grow displacively within bottom sediments and may then be reworked. All particles eventually become overgrown by cement; laminae converting to interlocking gypsum-crystal mosaics. Crossbedding, ripple-drift bedding, basal scoured surfaces and rip-up clasts testify to environments with periodic high-energy events such as floods and storms. Each layer forms during times when evaporitic flats and bodies of shallow brine are temporarily flooded by sediment-charged water. Microbial mats collect and bind evaporite sediment and, as the flood subsides, the coarser load is deposited as a traction layer or as a settle-out producing a normally-graded lamina. Cyanobacteria grow through the new lamina, re-establish themselves on the surface and protect the underlying sediment from erosion. Reverse-graded beds possibly form when an upward segregation of coarser particles occurs in highly concentrated, flowing sand sheets during storm surges, a feature characteristic of evaporite flats. Alternatively, episodes of brine freshening induce selective recrystallization of the upper parts of gypsum laminae. Some asymmetric ripple-marks with over-steepened sides represent adhesion ripples, indicating deposition of wind-blown gypsum onto moist surfaces. Shallow-water deposition is also demonstrated by the presence of stromatolites, by bird or dinosaur footprints, or by fossil brine shrimp or their fecal pellets.

Other laminated gypsums are composed of crystals that suffered little transport. Crystals displace/ enclose microbial mat carbonates and organic material (Fig. 6A; Fig. 23C). Lamination here reflects original microbial mat lamination.

Conversion of gypsum to anhydrite obscures differences between the different lamina types, so that ancient anhydrite laminites rarely provide sufficient evidence for precise environmental reconstruction. Some laminites even become replaced by nodular anhydrite, pro-

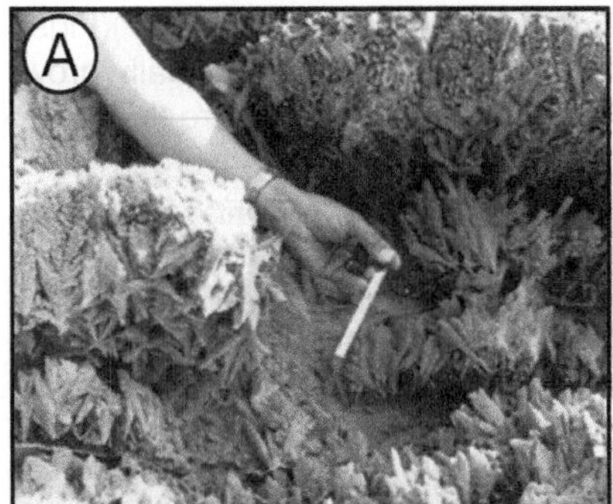

Figure 22. Coarsely crystalline gypsum facies. A. Palisades of gypsum crystals (scale bar 6 cm long), Miocene of Sicily. (Photo courtesy B.C. Schrieber). B. Layered anhydrite with pseudomorphs after gypsum crystals (scale bar divisions in cm), Otto Fiord Formation (Pennsylvanian), Ellesmere Island. (Photo courtesy N.C. Wardlaw).

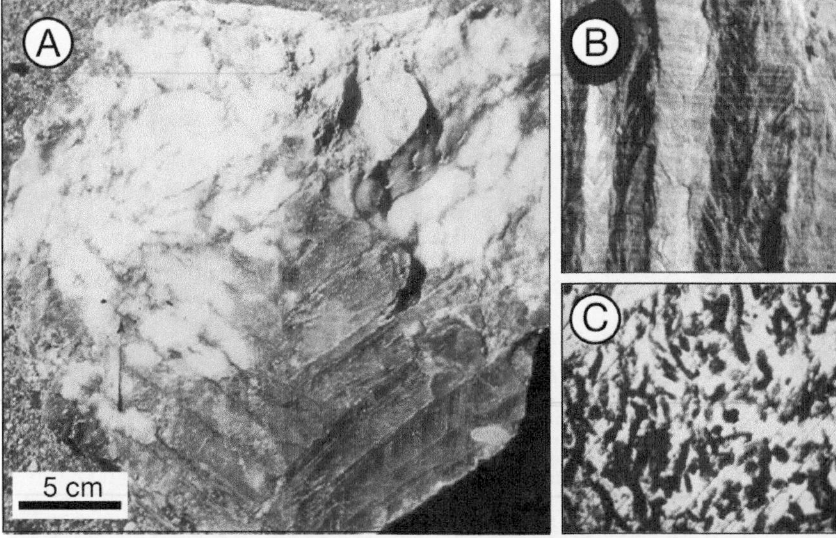

Figure 23. Coarsely crystalline gypsum facies A. Swallow-tail twinned gypsum crystal (partially replaced by nodular anhydrite) (photo supplied J. Melvin). B. Inclusion-defined dissolution surfaces cross-cutting laterally-coalescent gypsum crystals, Marian Lake, Australia, lens cap is 30 cm in diameter; (photo courtesy B.C., Schreiber). C. Microbial filaments within gypsum crystal, Messinian, Sicily; field of view is 3.2 mm wide; (photo courtesy B.C. Schreiber).

ducing rocks that resemble sabkha anhydrite.

Gypsum Crusts

This facies occurs as beds of vertically standing, elongate and commonly swallow-tail twinned crystals (Fig. 22), less than a centimeter to over five meters high. Crystals define a vertical palisade fabric or radiating upward-conical clusters. Other crystals exhibit bizarre growth and twinning patterns and suffer crystal-splitting to generate palmate clusters of sub-parallel sub-crystals (Fig. 23A). Twinning occurs at a variety of angles and sometimes only one twin grows, producing curved sabre-shaped crystals.

Crystals contain faint to clearly-defined lamination defined by carbonate and anhydrite inclusions (Warren, 1982). Inclusions may parallel crystal faces (marking growth increments) or they define horizontal dissolution surfaces (Fig. 23B). Crystals also invade microbial mats, incorporating microbial filaments (Fig. 23C).

Larger gypsum crystals form in environments of constant and active brine flow – artificial brine ponds drained during wet seasons only develop clastic gypsum layers or crusts of cm-sized crystals.

Beds composed of gypsum crystals can be difficult to identify when converted to anhydrite. During burial they are replaced by nodular mosaic anhydrite (Fig. 24A). Where original gypsum crystals enclosed carbonate laminae associated with planar dissolution surfaces (Fig. 23B), the nodular anhydrite may retain such lamination and can be mistaken as a replacement of laminated gypsum. Boundaries between anhydrite nodules in replacements of large palmate gypsum crystals are commonly controlled by the original gypsum crystal sub-crystal boundaries, allowing this facies to be identified (Shearman, 1983)(Figs. 23A and 24A). However, when gypsum dehydration occurs in low-permeability sediments, the expelled structural water cannot easily escape: the sulfate becomes thixotropic, flows, and original gypsum crystal forms are lost (Fig. 16).

Gypsum crystals can also be replaced by halite, sylvite or polyhalite pseudomorphs (Fig. 24B) outlined by anhydrite. Such replacements apparently only occur in the shallowest of brine ponds where there is a large temperature contrast between the brine and the gypsum crystals beneath on the pond floor.

Coarse Clastic Gypsum

Gypsum sands and pebbly sands are composed of worn gypsum cleavage fragments with variable amounts of carbonate and other materials. They indicate that gypsum can be transported and deposited by brines in the same manner and environments as other clastic sediments. Clastic gypsum commonly exhibits current or wave-activity structures and occurs as shoestring sands or as sand sheets and represent channel, beach, offshore shoal or spit deposits.

Halite Crusts

Various halite growth-habits occur, but the most important are chevron halite (Fig. 19B) and crusts composed of cornet-shaped crystals (Fig. 19D). Fluid inclusions are concentrated into growth layers parallel to cube faces (Fig. 19C). In crystals with corners uppermost, the zoning appears in vertical sections as chevrons with upwardly pointing apices (Wardlaw and Schwerdtner, 1966; Lowenstein and Hardie, 1985; Brodylo and Spencer, 1987).

Upper surfaces of halite layers are either crystal growth faces (characteristic of deeper salt lakes) or truncation surfaces associated with cavities

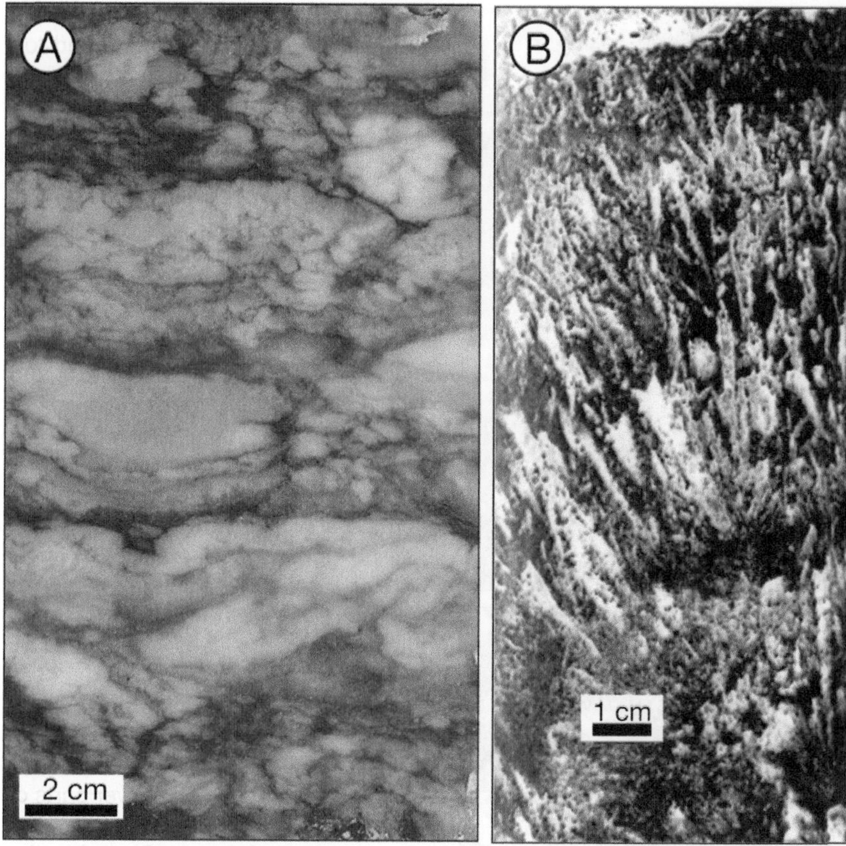

Figure 24. A. Nodular mosaic anhydrite pseudomorphic after swallow-tailed twinned gypsum, Lake Alma Anhydrite (Ordovician), Saskatchewan. **B.** Gypsum crystals replaced by anhydrite and halite, Permian, Yorkshire (photo supplied by D. Smith).

(Figs. 7B, and 19E) recording episodes of brine undersaturation and halite dissolution (Arthurton, 1973), and are characteristic of salt pans. Each layer is composed of primary inclusion-zoned halite and clear halite which fills the inter-crystalline porosity between zoned crystals or dissolution cavities (Fig. 19F).

Detrital Halite

Detrital halite is probably more important than published studies suggest: it is very susceptible to recrystallization. Where well-preserved, it is composed of fragmentary surface-grown hopper crystals and small cubes that represent overgrown hoppers, foundered crystal rafts, crystals that can have precipitated during brine mixing, or reworked material from bottom-grown crusts. Commonly it is ripple-marked, exhibits cross-bedding and can include other detrital material. The hydraulic behavior of halite in brine is similar to that of the quartz/water system. Halite crystal growth commonly continues after deposition as overgrowths upon original crystals and the detrital origin is easily obscured.

Detrital halite dominates higher energy environments – situations where crystals are subject to constant bottom motion and crust development is curtailed. Constant motion also promotes halite precipitation onto grain surfaces forming halite ooids (Handford, 1990).

Potash Salts

Less is known about origins of potash-magnesia salts. The primary nature of sylvites in varved sylvinites (Fig. 25A) is indicated by their intimate association with halite that preserves subaqueous textures (Lowenstein and Spencer, 1990). Sylvite layers were crystal cumulates that crystallized as a result of surface brine cooling. Crusts of bottom-grown carnallite are also known, interbedded with layers of detrital halite (Fig. 25B). The presence of tachyhydrite in some evaporites – a mineral that cannot survive exposure to the atmosphere – indicates evaporite deposition was entirely subaqueous.

Most potash-magnesia salts, however, are early diagenetic replacements of, or are additions to, earlier halite or sulfate deposits. All textures and structures can be diagenetic, but less altered beds can still exhibit surprisingly well-preserved depositional features. Some sylvite and carnallite are early diagenetic cements that fill intercrystalline porosity (Fig. 25C) and dissolution cavities created in brine-pan halite crusts. Carnallite precipitates when warm, potassium-rich surface brines descend below the pan surface during desiccation phases. Cooling of the brines causes them to become supersaturated with respect to carnallite.

Other diagenetic changes may remove all traces of original depositional features, converting many evaporites into essentially metamorphic rocks. Facies modeling of such rocks is especially difficult, if not impossible.

DEEP-WATER EVAPORITES

Deep-water evaporites are controversial. Almost every deposit identified as deep water has also been, at some point, interpreted as having formed in shallow-water or supratidal settings. For some people, evaporite basins were deep enough to sink battleships in, but for others the floors of the same basins would have been barely moist.

Modern analogues for deep-water evaporites include laminated carbonates and bottom-grown halite of the Dead Sea (~400 m deep) and small, ~30 m deep, sodium sulfate lakes of western Canada. They are not particularly representative of the inferred environments where ancient deep-water marine evaporites formed. Because there are no fully suitable modern analogues, identification of deep-water evaporite facies is instead based entirely upon ancient examples. These facies occur only in deeper (and commonly central) parts of basin-central evaporites, and can include slope and basin-floor deposits. Nucleation and crystal growth occurs at the brine surface and crystals settle through the brine column as a pelagic rain to form cumulate deposits.

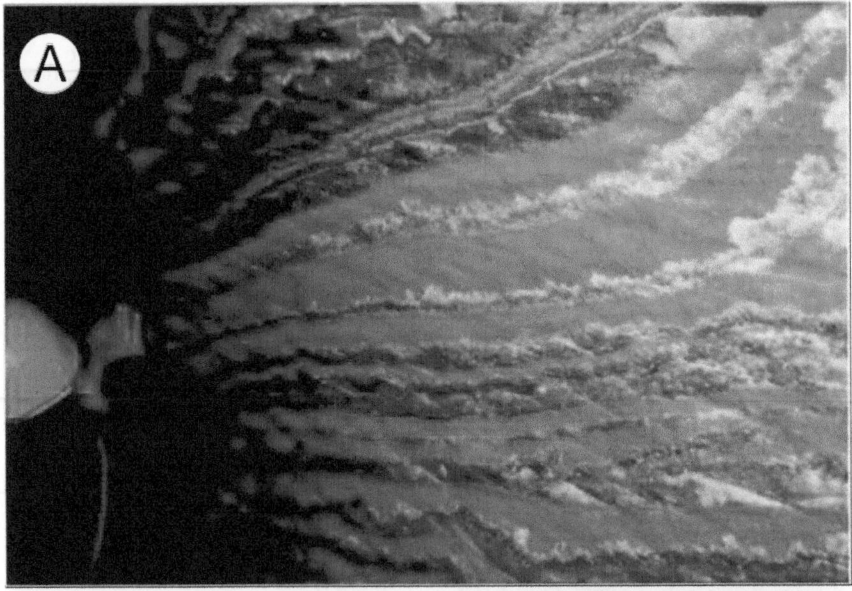

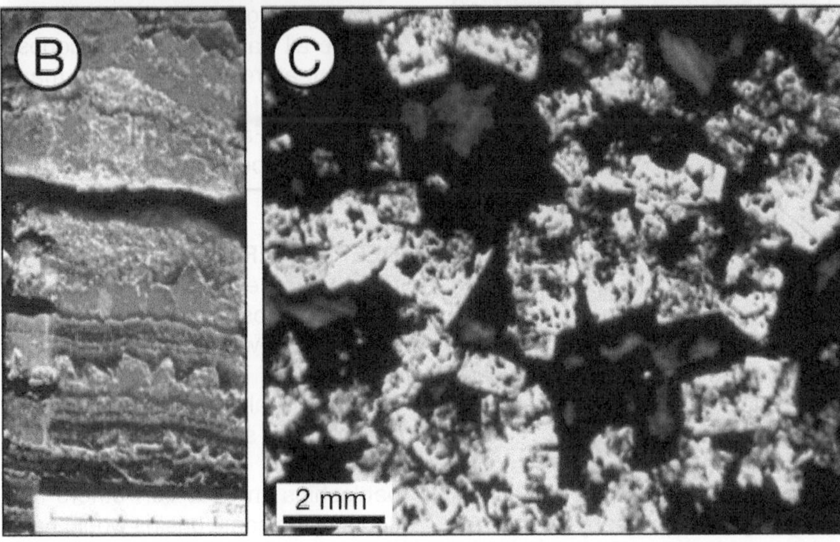

Figure 25. Potash facies. **A.** Layered sylvinites, successive crusts of chevron halite, each overlain by red layers of cumulate sylvite crystals, Permian Evaporites, Solikamsk (Russia). **B.** Bottom-grown carnallite crystals, Aptian evaporites, Brazil. **C.** Thin-section photograph of halite cubes surrounded by void-filling carnallite cement and mudstone; McNutt Potash zone, Salado Formation (Permian), New Mexico (Photo courtesy T.K. Lowenstein).

The distinctive features of deep-water evaporites reflect the relatively large size and volume of the brine body that generates vertical and lateral facies continuity. Large bodies of brine do not, as a whole, fluctuate in composition in response to short-term changes – the large brine body acts as a buffer. However, regular interlaminations of minerals with different solubilities commonly do occur, but indicate fluctuations of the **surface** brine salinity – not the entire brine body. Most of the inferred deep-water evaporites are laminar to thin-bedded, and individual laminae are traceable for long distances and can be basin-wide (Fig. 26; Anderson *et al.*, 1972). Stratigraphic units composed of laminar evaporitic carbonates, sulfates and halite can extend for thickness of tens to hundreds of meters, and correlate over entire basins (tens to hundreds of kilometers). This vertical and lateral constancy is the main argument for the presence of a substantial body of brine.

Laminated evaporites form as pelagic rain, with crystals precipitating at the brine surface and settling down to the sediment surface, but could also represent distal puffs of evaporite turbidites, the more proximal parts of which occur on basin-slopes and toe of slopes. Laminae reflect changes (perhaps seasonal) in the saturation states of surface brines or, if they are distal turbidites, variations in the rate of sediment supply from basin-marginal slopes.

Comparable laminated calcium sulfates or halite have not been identified from modern deep-water evaporitic settings, although similar laminated evaporitic carbonates occur in some deep-water saline lakes (Garber *et al.*, 1987). When the bottom brines in modern saline lakes do precipitate basin-floor salts, the products are crusts and masses of coarsely crystalline and variably oriented crystals of gypsum and halite (in the Dead Sea) or halite and mirabilite (Canadian Prairies brine lakes; *see* Fig. 31A).

Turbidites and mass-flow deposits composed entirely or, in part, of resedimented evaporites can also accumulate in deep-water environments. Originally formed in shallower parts of the basin, their lateral transport implies the existence of relief in the basin and deposition within the deep-water setting.

Sedimentary Processes

Depositional processes in deep-water evaporitic environments are believed to have been similar to those in the larger and deeper, continental perennial salt lakes. Perennial saline lakes more than a few meters deep are density or temperature stratified. Ancient deep-water evaporites also precipitated from stratified brine systems. This has important implications:

1. The persistent, denser, bottom brine protects previously precipitated evaporites from dissolution during flooding/dilution events that only modify surface brines;
2. Persistent stratification prevents surface brines from reaching the basin floor, so growth of bottom crusts may be prevented; and
3. Bottom brines become anoxic, allowing organic matter to be preserved and bacterial reduction of sulfate.

In deep evaporite basins a yearly influx of fresher water, especially if it is seasonal, generates a buoyant layer floating on the denser bottom

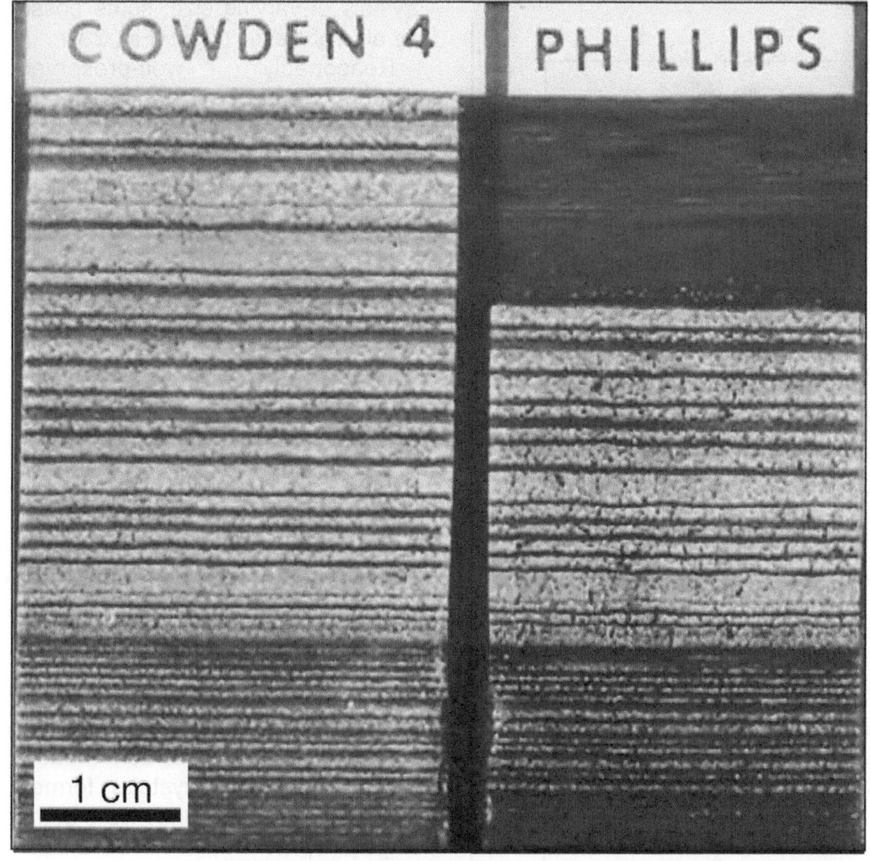

Figure 26. Correlation of organic-rich calcite and anhydrite laminae; Permian Castile Formation of west Texas and New Mexico; wells are 30 km apart. (Photo courtesy W.E. Dean).

brine. When the influx exceeds (or more or less equals) yearly evaporative losses, a stable salinocline forms at a depth between 5 and 15 m, and this prevents downward penetration of turbulence and downward transport of supersaturated brines. Solar heating of deep brine bodies (especially if the brines are colored) reinforces this water stratification stability because the heat is trapped in the upper water layer, increasing its temperature and making it even less dense. In shallower settings, however, heat can become trapped in the lower brine, raising its temperature and perhaps imparting instability to the brine column.

If the seasonal(?) water influx exceeds the amount that will be evaporated during the succeeding year, the brine body remains strongly stratified (meromictic condition with a sharp boundary (pycnocline) between brine layers) with both temperature and salinity stabilizing the density stratification (Anati *et al.*, 1987). In those years bottom brine salinity does not increase.

When influxes are less than evaporative losses, the buoyant surface layer evaporates away to create a near-uniform brine column. This holomictic condition lasts for several months until a new seasonal pycnocline begins to form. However, during many essentially holomictic years the stratification does not entirely disappear and slightly more saline surface brines come to lie above less saline (Fig. 27). This unexpected situation occurs because the density increase caused by a higher salinity is offset by a density increase produced by the higher temperature of the surface-heated brines (Anati *et al.*, 1987). The upper warmer brine, however, becomes denser as it loses heat by diffusion more rapidly than diffusive salt transfer downward into the deeper brine. When the density contrast between the two brine layers becomes zero, the upper brine (which retains its higher salinity) becomes unstable, and further cooling causes it to descend. This double diffusive mechanism and associated instability may allow transport of supersaturated brines down to deep evaporite basin floors, there to precipitate crystal crusts.

Changes between holomictic and meromictic conditions may explain bottom-grown crystal crusts (Fig. 28) in ancient deep-water evaporites (Kendall and Harwood, 1996). Laminated intervals record meromictic years (all evaporites formed in upper brine layers). Bottom-grown crystal crusts, however, form under holomictic conditions, either during short periods of overturn, or from supersaturated fingers of descending brine generated during diffusive mixing in hot, dry seasons. Alternations between bottom crusts and intervening laminated intervals would reflect variations in the amount of fresher water influx between different years: larger influxes inducing meromictic conditions and deposition of laminae; smaller influxes causing holomictic conditions and formation of bottom crusts.

Deep-water Evaporite Facies

Laminar Sulfates and Carbonates

Laminar sulfate (originally gypsum), either alone or in couplets or triplets with carbonates and/or organic matter, is the most common deep-water evaporite facies (Fig. 26). It passes vertically up or down into laminar carbonates that are probably also of evaporitic origin, or upward into laminar halite. It can also include re-sedimented layers. Sulfate laminae are thin, usually only 1–2 mm thick, and are typically bounded by smooth, flat surfaces that contrast with the more irregular lamination of shallow-water laminites. Within short vertical sequences, laminae tend to be of near uniform thickness, and individual laminae are traceable over long distances (commonly tens to hundreds of km).

Laminar sulfates accumulate as pelagic rain, with crystals forming at the surface of the brine column and sinking down to the deep brine sediment surface (cumulate deposits). Some, however, might represent distal evaporite turbidites (or storm deposits) that characterize basin-slopes.

Laminar cumulate deposits com-

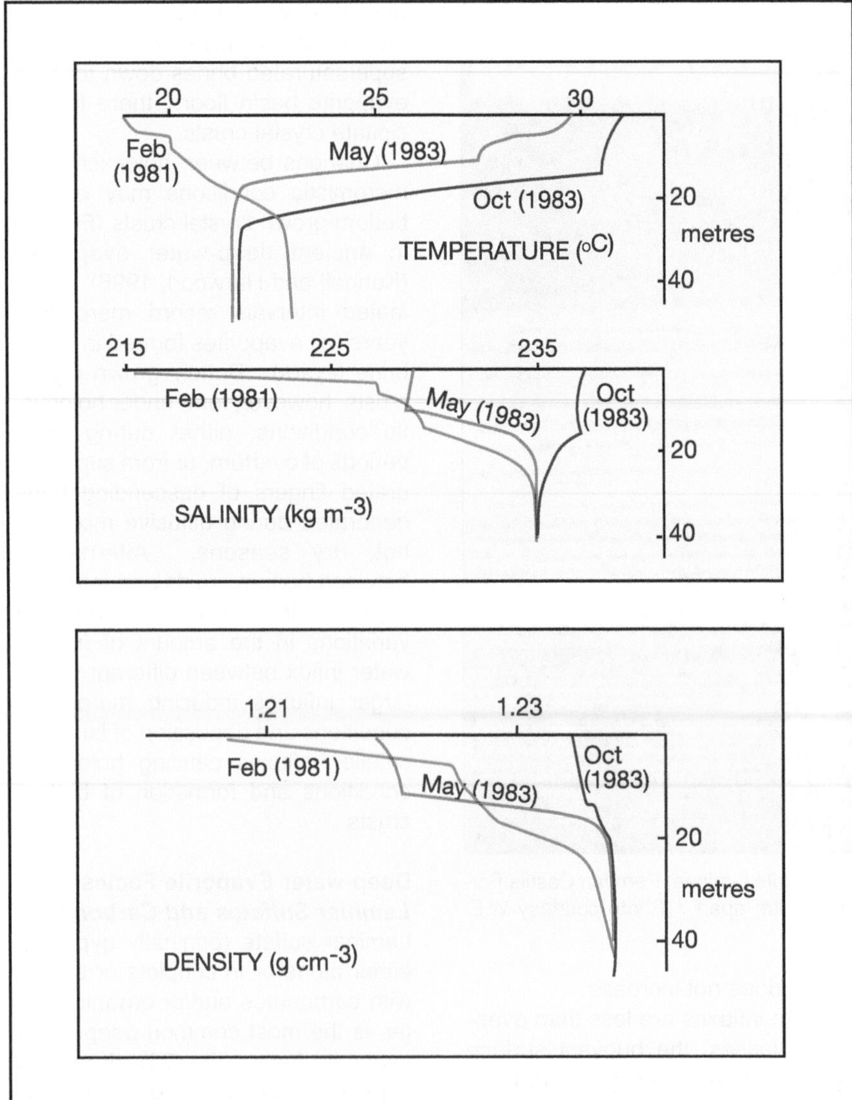

Figure 27. Density stratification in Dead Sea. Spring fresh-water influx creates marked density stratification with less saline brine floating upon dense bottom brine; summer heating and evaporation eliminates upper brine to create a surface brine that is more saline but warmer than the bottom brine; late autumn situation where the summer surface brine has cooled, become denser and has either mixed with or descended through the lower brine to create an essentially holomictic brine body (after Anati *et al.*, 1987).

monly are interpreted as seasonal or annual increments – varves. Usually it cannot be conclusively demonstrated that evaporite laminae are truly annual, but it is commonly stated that no other hypothesis explains all the features of these evaporites. However, evaporitic carbonate laminae in Searles Lake and the modern Dead Sea, are seasonal, but are not necessarily annual – only one laminae being deposited every three or four years.

Laminar carbonates and sulfates record deposition in a brine body with a bottom largely unaffected by wave or current activity. Such stagnant, permanently stratified bodies of brine need not, however, be particularly deep. Well-developed brine stratification dampens wave motion at shallow depths, leading to a false impression of greater depth.

Bottom-growth Gypsum (and Anhydrite Replacements)

Basin-floor laminated sulfates are commonly interbedded with various nodular, micro-nodular, sluggy, or flaser-like anhydrites, some of which can be identified as poorly-preserved replacements of either i) bottom-grown gypsum crystal crusts (Fig. 28) or ii) sediment displacing(?) gypsum crystals of varying size, some possibly also of clastic origin (Fig. 29).

Reasonably well-preserved pseudomorphs of gypsum crusts are identified in the Castile Evaporite of the Delaware Basin. Well-preserved pseudomorphs pass, by a complete transition, into sluggy to flaser-like anhydrite, much of which is inferred to have originally been bottom grown gypsum crusts (Fig. 28). Other flaser-anhydrite (Fig. 29B) appears to pass, also by a series of transitional stages, into thicker than normal laminated anhydrite containing micro-nodules, some of which are pseudomorphs of gypsum crystals, interpreted to be sediment-displacive (although some may be clastic; Fig. 29A).

The various 'nodular' anhydrites constitute part of basin-wide, millennial cycles, either forming i) tops of incomplete cycles, or ii) transitions between laminated anhydrite below and halite above. Their position indicates the gypsum precursors (bottom crusts, displacive crystals) formed when bottom basin brines were at the stage just prior to halite net accumulation. Seemingly identical flaser- and related anhydrite occurs in the Zechstein of northern Europe although well-preserved pseudomorphs after gypsum crusts have not been identified. Intervals of nodular anhydrite are traceable laterally across the Zechstein basin, and can be traced upslope, passing laterally into thicker platform anhydrites pseudomorphing shallow-water gypsum.

The dramatic thickening of nodular sulfates from basins onto platforms might reflect a preferred precipitation of gypsum within shallow, warm waters. The various non-laminar anhydrite facies in basinal sequences then might represent times when bottom brines were warmer. Identification of them as bottom-grown gypsum deposits, however, suggests they represent episodes when some form of brine overturn of more saline-than-normal brines affected basin floors. This conclusion is important because much nodular anhydrite in basinal evaporite sequences has been misidentified as sabkha deposits. This misidentification requires unnecessary and complex sequences of brine-level changes (or rapid differential subsidence) to have taken place.

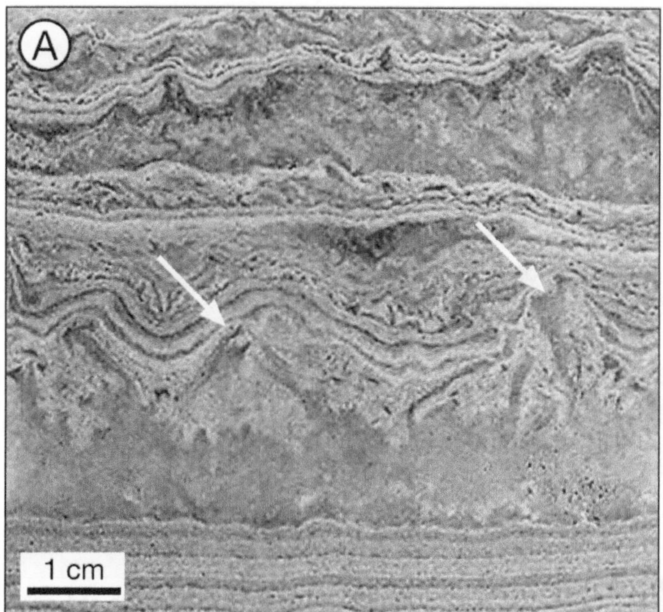

Figure 28. Anhydrite pseudomorphs of bottom-grown gypsum-crystal crusts. **A.** With draping anhydrite-calcite laminae and swallow-tail twinning evident in some former crystals at arrows, gypsum now replaced by nodular anhydrite, **B.** Superimposed former gypsum crusts. Both A and B from the lowest anhydrite member of Castile Formation, West Texas.

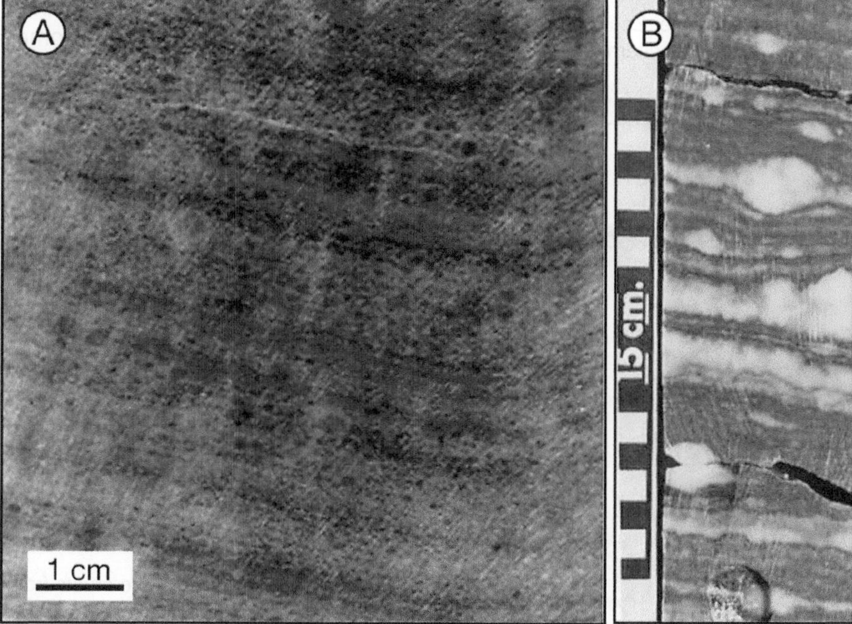

Figure 29. Anhydrite pseudomorphs after displacive(?) gypsum crystals. **A.** Anhydrite blebs in thick anhydrite laminae associated with replaced gypsum crusts, Castile Formation, West Texas. **B.** Flaser and nodular anhydrite within laminated anhydrite, much of which appears to have replaced gypsum crystals (some retaining angular shapes), Z1 Anhydrite (Permian Zechstein), Germany.

Laminated and Bedded Halites

Deep-water halite is difficult to recognize because most examples have recrystallized. Even so, inferred deep-water halite is invariably finely laminated (Fig. 30A) containing anhydrite–carbonate laminae similar to those of deep-water laminated sulfates. Even finer lamination within salt layers occurs, defined by variations in inclusion content. The fineness and perfection of this lamination (Fig. 1B) indicates the original halite were accumulations of very small (cumulate) crystals (Fig. 30B).

Some presumed deeper-water halites are composed of large bottom grown crystals (Fig. 31B and C). A deeper-water setting is suggested by i) clear, inclusion-free halite crystals, and ii) an absence of dissolution surfaces associated with anhydrite laminae. Instead, anhydrite buries and defines the crystal terminations of bottom-grown crystals (Fig. 31C). Bottom growth of halite in deeper water settings has yet to be satisfactorily explained. Possibly much of the facies is deposited in only a few tens of meters of brine, but this cannot be assumed, nor does it explain away deep-water halite of the modern Dead Sea. In the Silurian Salina A salts of the Michigan Basin, this facies only occurs in the basin center, suggesting deeper brines there (Nurmi and Friedman, 1977).

Polyhalite and Bittern Salts

Almost all high-solubility salts were deposited within previously deposited mud-flat or shallow-water deposits or in shallow brines. However, some polyhalite and kieserite from the Upper Permian Zechstein (Fig. 32) occurs within deeper-water slope deposits (Colter and Reed, 1980). Polyhalite is an early replacement of anhydrite (that itself replaced gypsum syndepositionally) by halite-precipitating brines rich in potassium and magnesium ions. The brines formed in platform salt-depositing areas and flowed down over the platform edge and across depositional slope deposits (Peryt *et al.*, 1998).

Gravity-displaced Evaporites

Clastic evaporite intervals within deep-water laminated evaporites are slump, collapse, mass-flow, density

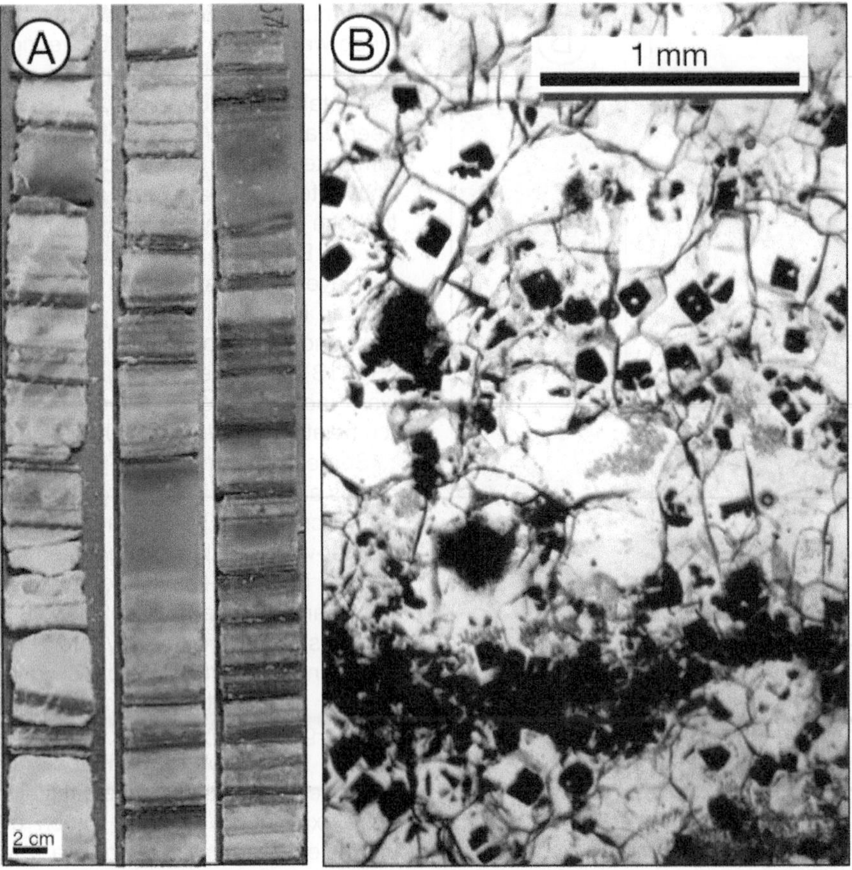

Figure 30. Laminated, deep-water halite. A. With anhydrite laminae; Zechstein Z2 evaporite, Yorkshire, England. B. Cumulate halite beds, Miocene, Mediterranean. (Photo courtesy T. Lowenstein.)

Figure 32. Bedded halite with red polyhalite (and minor kieserite) slope deposits; Forden Evaporites (Permian), Yorkshire.

current or turbidity current deposits. Their occurrence is probably the best indication of a large body of brine during deposition. They also imply the existence of unstable slope conditions. Gypsum and anhydrite turbidites (Fig. 33) are similar to their non-evaporite equivalents. Sometimes entire Bouma sequences occur, but most beds are composed only of graded units, or possess poorly-developed parallel laminae in their uppermost parts. Mass-flow deposits are represented by breccias composed of blocks of reworked sulfate or crystal fragments (Fig. 34A), either alone or with carbonate clasts. They are intimately associated with units affected by slumping (Fig. 34B).

Three main facies associations of re-sedimented evaporites are recognized (Manzi *et al.*, 2005). **Chaotic**

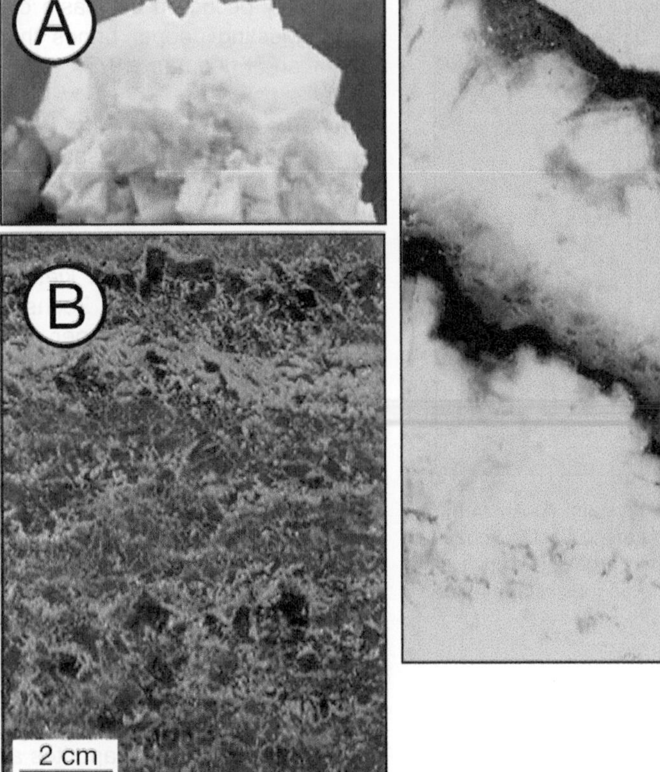

Figure 31. 'Deep-water' halite crusts from A. Modern, Saskatchewan lake (photo courtesy W. Last). B. Silurian A1 salt, Michigan (photo courtesy R. Nurmi), vertical scale = 1 cm. C. Paradox (Pennsylvanian) Formation, Utah. Cubic terminations of clear halite crystals are outlined by overlying anhydrite laminae. The absence of dissolution at the top of crystal crusts indicates deposition occurred in brines deep enough to allow crusts to escape the effects of brine-dilution events (also responsible for the anhydrite precipitation).

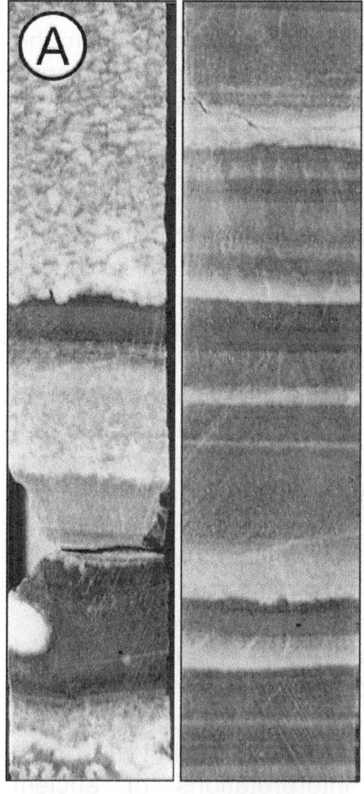

Figure 33. Sulfate turbidites. **A.** Graded clastic anhydrite beds (with white and dark bituminous clasts) overlying dark laminated anhydrite (with other layers with a finely clastic texture). Zechstein-2 anhydrite; Germany. **B.** Amalgamated gypsum turbidites (now anhydrite), some of which display pronounced grading (scale in cm); Z-1 Anhydrite, S. Harz (Germany).

deposits are proximal complexes composed of evaporite slabs, boulders and blocks, debris flows and hyper-concentrated flow deposits composed of coarse, and sometimes pebbly, gypsum sand deposits. **Lobe deposits** are products of high- to low-density gravity flows composed of medium- to fine-grained gypsum sands, silts and shales. They are interbedded with thin-bedded, fine-grained gypsum sands and silts. **Drape deposits** consist of gypsum laminae interbedded with bituminous shales. Originally considered primary (cumulate) deep-water deposits; their close association with clastic deposits suggests they are also re-sedimented, forming from more dilute flows.

The Messinian salinity crisis occurred during an intense phase of geodynamic re-organization of Mediterranean area. Shallow-water evaporites formed in semi-closed thrust-top basins and their subsequent uplift and subaerial exposure led to widespread collapse and re-sedimentation of gypsum into deeper basins. It is debatable whether these deep-water deposits are strictly evaporites – they are re-sedimented deposits that just happen to be composed of gypsum. The basin into which they were deposited need not even have been particularly saline, although this would have aided gypsum preservation.

Re-sedimented halite deposits are only rarely identified but do form spectacular slump deposits with clasts exceeding 10 000 m^3. Their nature is only clearly established when they contain entrained non-halite clasts.

EVAPORITE FACIES MODELS

Facies models have been devised for modern evaporitic settings – continental playa basins (*see* Chapter 21) and marginal marine sabkhas and salinas. They are deduced for ancient deep-water and widespread shelf evaporites. The occurrence, however, of similar depositional facies within different geographic settings, and the vast size of many ancient evaporites (with no modern parallel), makes it difficult to use geography-based and/or modern depositional models directly for ancient evaporites. The distribution of (presumed) marine evaporite facies within depositional basins is judged to be more important.

The need to establish a non-marine origin before continental evaporites are positively identified is obvious, but this is not easy (Taberner *et al.*, 2000). Models of continental evaporites deposited in continental saline lakes are considered in Chapter 21.

Basin-central and basin-marginal marine evaporite distributions are distinguished. All modern marine-fed evaporite environments (sabkhas and salinas) are basin-marginal; there being no modern examples of marine basin-central evaporites (although similar depositional patterns can occur in larger salinas). Basin-central evaporites occupy the entire depositional basin, although this can be considerably shrunken compared with the depositional area when underlying non-evaporite sediments accumulated. Marginal shelf evaporites pass laterally into marine strata elsewhere in the basin with an intervening barrier between evaporites and open marine environments. However, the barrier might be difficult to identify if it was narrow or ephemeral. For basin-central evaporites, it is the entire exposed rim of the basin that forms the necessary restricting barrier. Thus basin-central evaporites should be represented on basin rims by non-sequences.

Finally, almost all basin-central evaporites are lowstand system tract deposits, filling the entire central basin floor and forming at times when sea levels drop below basin rims (Tucker, 1991; Kendall and Harwood, 1996). In contrast, basin-marginal evaporites are commonly components of transgressive and highstand system tracts. If the barrier occurs at the shelf edge, almost the entire shelf might become evaporitic and evaporite-accumulating. There are no modern wide, evaporitic shelves. If pre-evaporite deposits formed on a ramp, then the required barrier for evaporite deposition forms in a more shoreward position than on shelves, restricting the lateral extent of the back-barrier evaporites.

There is criticism of the uncritical use of stratigraphic concepts, conceived for siliciclastics, to evaporite

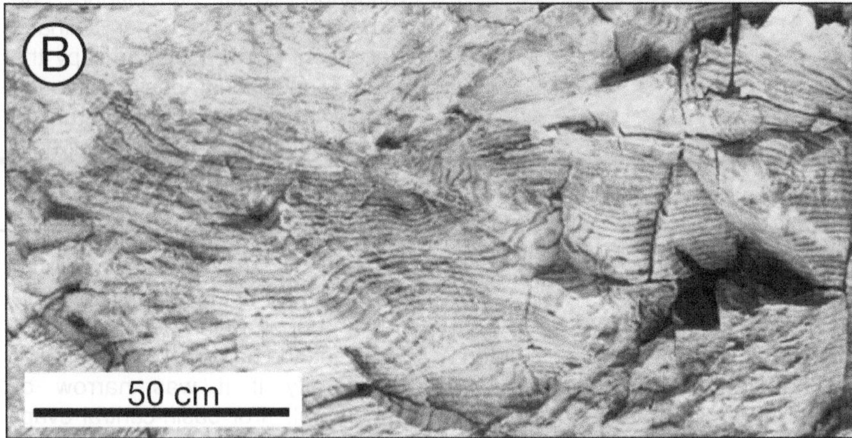

Figure 34. Sulfate slope deposits. **A.** Mass-flow deposit composed of broken and transported gypsum crystals (now anhydrite). **B.** Slumped and contorted, laminated anhydrite, both from Z-1 Anhydrite of S. Harz.

successions. In particular, i) the necessity for severe basin restriction ii) rapid water-level shifts caused by desiccation not necessarily related to sea-level changes and iii) the rapidity with which sediment aggradation occurs mean that, for basin-central evaporites, 'conventional' sequence-stratigraphy is impossible, or at best can be extraordinarily difficult.

BASIN-MARGINAL (SHELF) EVAPORITE MODELS

Coastal Sabkha Models

Carbonate examples are described in Chapter 16. Coastal sabkhas can merge insensibly landwards into continental sabkhas and all parts of coastal sabkhas may be affected by continental groundwaters. Typical non-marine evaporites (*e.g.*, trona), as well as calcium sulfates and halite, can thus precipitate in the marine-derived sediments.

Coastal sabkhas are products of depositional and diagenetic processes, the most important being emplacement of early diagenetic calcium sulfate (less commonly halite). Host sediments reflect the more mobile fraction of the offshore sediment mosaic as well as detrital sediment from the hinterland. Offshore sediment is washed over the sabkha surface during storms.

Groundwaters beneath sabkhas become concentrated toward sabkha interiors, and all but the very seaward and landward margins can be halite saturated. Concentration occurs by evaporation from the top of the capillary fringe and by dissolution of earlier-formed evaporites (particularly halite). In the conventional view, evaporative losses are replenished by downward seepage of storm-driven floodwaters and/or by gradual intra-sediment flow, fluxing either from the seaward margin or from a continental groundwater source affecting landward parts of the sabkha. The water table inclines seawards, but lateral brine migration is very slow.

Sabkhas formed by coastal progradation have a characteristic vertical sequence (Figs. 12 and 16): subtidal sediments (commonly restricted lagoonal) at the base; intertidal sediments including microbial mats above; and a capping supratidal deposit with abundant displacive gypsum ± anhydrite (Warren and Kendall, 1985). Displacive gypsum grows within intertidal and subtidal sediments once these become located beneath the sabkha environment.

Sabkhas are deflation surfaces. Identification of an ancient sabkha sequence requires recognition of this disconformity surface, but most published interpretations of ancient sabkha sequences fail to identify this essential element.

Variations in Sabkha Sequences

1. *Diastrophic control* Shallowing-upward sequences terminated by marine sabkhas might form in three different ways: by i) sediment progradation, ii) eustatic falls in sea level, or iii) brine-level drops caused by evaporative drawdown. The first two also affect non-evaporite sequences and Chapter 16 discusses different hypotheses for shallowing-upward carbonate successions, of which evaporite-capped successions form an arid-zone variant.

Problems arise when successions are interpreted only as a result of sediment progradation (Hardie, 1986). Modern sabkhas are protected from the full impact of onshore storms by beach barriers or offshore shoals/islands. This protection allows the sabkhas to prograde, but once the lagoon behind the barrier becomes sediment-filled, further progradation requires formation of a new barrier, behind which progradation can be renewed. Widespread cycles that are composed of fine-grained sediments capped by sabkhas thus should not accrete laterally in a uniform manner; instead they should migrate laterally in a series of jumps, as offshore barriers develop into barriers. The entire sediment body should be an arid-zone equivalent of a chenier plain – an environment that occurs in Kuwait, the Mediterranean coast of Egypt, and from Abu Dhabi, but which appears not to have been documented in ancient examples.

One problem affecting many ancient evaporites that are interpreted to have been generated by prograding sabkhas is that they overlie open marine carbonates rather than restricted lagoonal deposits. Here we must assume that either i) barriers were absent and mud flats were protected from wave attack by the extensiveness and shallowness of the offshore environment, or ii) the prograding sabkha interpretation is incorrect.

A second problem concerns the preservation potential of wide prograding sabkhas. These require low gradient, stable shelves – locations where active lateral flow of groundwater is unlikely. Sabkha surfaces are subject to continual evaporative losses and, unless this loss is replenished, the supratidal environment dries out and is subject to wind erosion. Wide sabkhas therefore require a regional groundwater flow system that allows the water table to rise as the sabkhas prograde. Without such a system, the uppermost evaporite-bearing sediments beyond seawater flood-recharge would be removed by wind deflation. Dolomitized subtidal carbonates with erosive upper surfaces and scattered nodular anhydrite after displacive gypsum (but lacking overlying supratidal units), could have formed this way.

Any seaward-dipping groundwater table would also cause additional sediment (mainly eolian) to be deposited as the brine-surface rises, but requires a large neighboring landmass with sufficient relief to act as a recharge area for the groundwater system. For many widespread evaporites, interpreted as prograding sabkha deposits, this requirement is not substantiated. Neither have the thicker eolian supratidal sequences of more landward parts of the inferred widespread sabkhas complexes been identified. For these reasons, interpretations of many (all?) widespread sabkha evaporites as simple progradational sequences are suspect.

Evaporites precipitated within marine-marginal sabkhas have a very low preservation potential after they are buried. They form in settings with ascending groundwaters. Burial transfers sabkha evaporites into a regime where the porewaters are unsaturated, so they begin to dissolve. Complete dissolution would leave, at most, a thin horizon of dissolution collapse. On the other hand, burial of sabkhas that form on the margins of a hypersaline body of brine, with its underlying plume of saline groundwaters (necessary to prevent massive reflux loss of the surface brines), should allow preservation of the less soluble sabkha salts.

The uppermost parts of vertically repeated sediment cycles, each of which is capped by evaporites, have commonly been interpreted as the products of prograding marine-marginal sabkha deposits, the arid-zone equivalents of coal measures. But, if sabkha brines are being fed by ascending unsaturated groundwaters that pass through older sabkha evaporites, the underlying evaporites will dissolve. If it were to be argued that the older evaporites were protected by their low permeability, then the groundwaters would not been able to ascend to the surface to generate the younger sabkha evaporites. Gypsum at West Caicos Island accumulates in a sabkha by evaporation (and possibly by brine warming) of ascending marine-derived groundwaters (Perkins *et al.*, 1994). These groundwaters have dissolved gypsum of an earlier sabkha deposit and are currently dissolving the lower parts of the presently accumulating surface gypsum deposit.

If the hydrologic model of Wood *et al.* (2002) is widely applicable, then the concept of vertically stacked sabkha measures appears fundamentally flawed. Only the uppermost evaporite deposit should survive. More likely, vertically stacked deposits are subaqueous evaporites that became exposed and converted into mud-flat (sabkha) deposits (see below).

Sediment emergence as a result of forced regression, with formation of widespread exposure surfaces and the overprinting of formerly subaqueous evaporites by mud-flat processes (generating displacive/replacive evaporite nodules and crystals similar to those of prograding sabkhas; Fig. 35), occurs independently of any sediment progradation. Exposure occurs by global external events (glacial sea-level drops) or by increased evaporative drawdown.

Most (and possibly) all sabkha evaporites result from a combination of causes - even in the classic Abu Dhabi sabkha, anhydrite occurs as an overprint of earlier subaqueous gypsum in a prograding sequence, some of which has been forced by a slight sea-level drop. Widespread sabkha evaporites are probably always overprints of shallow-water facies caused by externally-forced regressions or evaporative drawdowns. If this is correct, then modern prograding coastal sabkhas are not good analogues of ancient deposits.

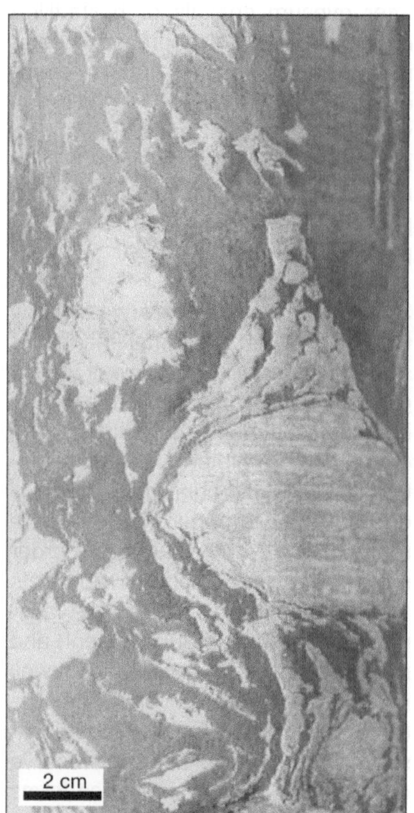

Figure 35. Nodular anhydrite overprinting and distorting subaqueous gypsum crystals (now also anhydrite) within a dolomite matrix, Frobisher Evaporite (Mississippian) of Saskatchewan.

2. *Nature of the host sediment* This determines permeability and thus the amount of drainage in the sabkha sediments and the thickness of the capillary fringe. In turn, permeability controls further evolution of sabkha brines and subsequent compactional history. Impermeable sediments inhibit brine reflux.

Carbonates, particularly aragonite, in host sediments are important. Their dolomitization releases calcium that reacts with sulfate to form additional gypsum or anhydrite. Reflux of sabkha-generated brines that dolomitize deeper carbonates causes additional gypsum precipitation in those sediments. In non-carbonate sediments, dolomitization does not occur and sabkhas brines retain 60–70% of their sulfate; much less gypsum is precipitated and brines remain magnesium-rich. Sulfate-rich brines formed in non-carbonate sabkha sediments can also react with earlier formed gypsum to form polyhalite.

Lithified or cohesive sediments preserve anhydrite pseudomorphs after gypsum crystals or halite-filled crystal moulds. Compressible sediments (including finer-grained evaporites, but particularly organic-rich varieties such as microbial mats), instead, allow gypsum and anhydrite nodules to grow, coalesce and compact, perhaps even to form sluggy or pseudo-layered anhydrites.

3. *Nature of the offshore water body* Commonly this is normal marine to slightly hypersaline and well below gypsum saturation. Thus subtidal and intertidal sediments beneath sabkha evaporites are bioturbated and skeletal-rich, with microbial mats (if present) confined to upper intertidal environments. Where sabkhas border more saline water bodies, sediments beneath sabkha evaporites are laminated (burrowing infauna absent) and microbial mats can extend down well into deeper water environments (browsing biota absent). Where offshore waters precipitate and preserve gypsum, the sabkha sequence forms the uppermost unit of a largely subaqueous evaporite. Fragments of the subaqueous evaporite are transported onto the sabkha surface by storms, there to form clastic beds of gypsum, anhydrite or even halite debris. Such sabkha sequences are largely composed of these clastic beds, with nodular or displacive additions, and are difficult to distinguish from shallow subaqueous evaporites.

4. *Transgressive evaporites* Nodular earlier Holocene anhydrite in the Abu Dhabi occurs in a zone parallel to and landward of, continuous storm beaches (Kirkham, 1997). It occurs within Pleistocene eolian sands and can occur up to 30 cm below the sabkha surface where it is overlain by layers of gypsum mush and subtidal lagoonal microbial laminated muds (Fig. 12). This vertical transgressive sequence is the reverse of that of the idealized regressive sabkha succession. A gypsum mush is commonly also present above the subtidal muds, indicating subsequent regression (progradation). The importance of the two calcium sulfate deposits is that i) not all nodular anhydrite caps regressive sequences, ii) transgressive evaporites can be more widespread and abundant than regressive evaporites, and iii) preservation of a transgressive evaporite beneath subtidal sediments (overlain by a regressive anhydrite/gypsum sequence) suggests that the uppermost (regressive) evaporite has not formed from ascending, undersaturated groundwaters (*cf.* Wood *et al.,* 2002) – because the lower (transgressive) evaporite should then have dissolved.

Marginal Marine Salina Models

Relatively small marginal marine salinas occur in depressions on sabkhas, between coastal dunes or in ephemeral stream deltas or fan deltas. Those lying below sea level are of greater size, are more persistent, and have thicker evaporite sequences. They occur in tectonic depressions and behind coastal sedimentary barriers. Marine inflow can be via minor inlets through the barrier (*e.g.,* inlet between evaporitic Kara-Bogaz Gol and the Caspian Sea), surface flow across a supratidal flat (ephemeral brine pans on supratidal mudflats, Colorado River delta, Baja California), or by passage through the barrier, discharging into the salina as springs, brine sheets or as seepages (Lake MacLeod, Western Australia). Salinas are covered by water for brief or long periods, but in all cases evaporite deposition is primarily by precipitation from a surface brine, rather than by precipitation within the sediment. Salinas associated with wide, high-relief barriers differ markedly from sabkhas in that they are unaffected by tide/storm surges. They therefore lack transported marine sediment (unless this is wind transported). Salinas can be large and comparable with small ancient evaporite basins, *e.g.*, the Rann of Kutch (India) 30 000 km^2 and Lake MacLeod (Western Australia) 2000 km^2.

Salinas and continental playa lake complexes are similar. Both contain subaqueous and subaerial sub-environments in which evaporites accumulate. These can be arranged in similar patterns, and depositional processes and products are comparable. They commonly consist of an ephemeral or perennial saline lake located in the lowest part of a depression, surrounded by saline mud flats or evaporite flats that pass outward into peripheral dry mud flats.

Variations in Salina Sequences

The variation in mineralogy and facies of salina evaporites is determined by factors similar to those controlling continental playas. Climate controls the maximum salinity to which brines can be concentrated, but more commonly inflow-reflux ratios actually determine the concentrations reached. Ratios are determined by the presence/absence of surface inflow, the nature (width, permeability) of the barrier, the hydraulic head between sea level and the brine level in the salina (these all control inflow rates), size of the salina (determines amounts of water loss by evaporation) and the effectiveness of any basal aquitard (controls brine seepage losses). These aspects are further discussed in the section *Basin-central evaporites.*

Ancient Widespread Shelf Evaporite Models

Many evaporites are composed of shallow-water and/or mud-flat facies, but must have accumulated on a scale different from those occurring today. The main problem in interpreting these evaporites is to determine if they were i) generated during progradation of marginal marine evaporitic environments comparable with those of the present day (in which case they are diachronous), or ii) they formed simultaneously in evaporitic settings significantly larger than any today. This is a major unresolved problem for many evaporite deposits and commonly the question simply has not been considered. Detailed stratigraphic studies indicate that the second explanation is the rule. Individual beds of gypsum crusts are traceable for tens or hundreds of kilometers or can be traced into adjacent units of microbially laminated evaporite-flat deposits. They must have formed within vast expanses of evaporitic lagoons reaching tens of thousands of square kilometers, over which brine depths were only a few meters deep (or even shallower) and flanked by equally impressive evaporitic flats. Tidal influences must have been minimal and deposition was affected more by storms that created laminated gypsum sands and silts. Storms reworked subaqueous sediments from shallow evaporitic lagoons and transferred the resultant carbonate and clastic evaporite sands onto neighboring flooded evaporitic flats. Microbial mats were ubiquitous in shallow subaqueous environments where brines were below halite saturation.

Stratiform evaporite units, 5 to 50 m thick, of mixed subaerial and shallow-water evaporites were generated in these settings. The lateral persistence of thin beds over large areas with only minor changes in thickness, mineralogy or facies indicates they formed on broad, flat shelf areas – areas which could be affected by rapid transgressions or evaporitic drawdowns and by seasonal changes in brine stratification (Babel, 2007).

The evaporites commonly form the upper parts of shallowing-upward cycles, with underlying open-marine or restricted-marine carbonates. These, in areas of tectonic subsidence, can be vertically stacked to generate thick sedimentary packages. Landward they interfinger with continental siliciclastics and seawards they apparently pass into marine deposits. The lateral equivalency of marine sediments and platform evaporites can be more apparent than real, however, in that evaporites accumulate so rapidly that they can form during short-lived regressions that cannot be easily identified in the marine strata.

Internal characters of shelf mudflat evaporites suggest they formed in mud- and evaporite-flats. Mud- and evaporite-flat sediments form shallowing upward cycles but correlation of individual facies across a platform may be difficult because of their limited continuity. They formed as a facies mosaic of saline and mud-flats that separated local brine pans. Here the different sub-environments can be comparable in size and facies to modern coastal marine equivalents. However, the lateral extent of the evaporite body has no modern counterpart.

Widespread areas of shallow subaqueous evaporite precipitation also have no modern counterparts, although the facies developed can be matched in modern perennial coastal salina and continental playa lakes. Commonly, these evaporites are cyclic and composed of shallowing-upward units 2–50 m thick, underlain by transgressive open-marine sediments and perhaps terminated by thin sabkha sequences (overprinting the uppermost subaqueous evaporites). Units are traceable over distances of tens to hundreds of kilometers, *e.g.*, San Andres Formation of west Texas, Yates Formation anhydrites of west Texas and the Ferry Lake Anhydrite of east Texas. Much of this evaporite, previously interpreted as widespread sabkha deposits, was originally deposited subaqueously in wide evaporitic shelves. Relicts of subaqueously formed gypsum are abundant in many 'classic' ancient sabkha evaporites; *e.g.*, almost all nodular anhydrites from the Paleozoic of Saskatchewan (*e.g.*, Figs. 16 and 24A), most of which, at one time or another, have been interpreted as sabkha deposits, are now identified as replacements of subaqueous gypsum (Kent, 2003). Attempts to discriminate between nodular anhydrites formed in sabkhas (nodular-mosaic with nodules flattened parallel to bedding) and replacements of salina gypsum crusts (vertically elongate nodules) are probably incorrect. The shape and orientations of gypsum-replacing anhydrite nodules commonly are controlled by impurity distributions (Shearman, 1983) and, if the original gypsum crystals were cut by carbonate/organic laminae (marking crystal-growth interruptions, Fig. 23B), the replacement anhydrite will consist of nodules elongated parallel to this lamination.

In the absence of a modern analogue, it is difficult to account for evaporites deposited in continuous shallow brine bodies. Comparison with the large salina of Lake MacLeod (Western Australia) suggests that evaporites composed of laminated clastic gypsum (Figs. 20 and 21) are deposited on evaporite flats rather than from perennial surface brines. Other evaporites, however, such as bottom-grown gypsum or cumulate halite, were deposited in more continuously subaqueous settings. Those composed of large gypsum crystals indicate deposition in relatively stable bodies of brine; this facies being confined to deeper parts of modern salinas or to salt ponds that were not drained nor subject to brine mixing. Brine-flow directions have been determined where oriented sabre-shaped gypsum crystals grew toward the brine flow (Babel and Becker, 2006).

Bodies of relatively low-salinity brine might precipitate gypsum in dry holomictic seasons, only to have this gypsum dissolve away during wetter meromictic seasons; this generates laminated carbonates. It is also possible that records of apparent unchanging gypsum deposition are misleading. Halite layers could have been precipitated above gypsum crusts, but dissolved during succeeding brine dilution events before immediately succeeding gypsum layers formed.

Widespread layers of shallow-water halite are usually subdivided by dissolution surfaces and layers or laminae of anhydrite, carbonate or detrital material – all indicating episodes of brine freshening. Cycles (Fig. 36) within shelf salts exhibit an upward change in fabrics and sedimentary structures reflecting variations in the paleo-water-table position (Holt and Powers, 1990). Com-

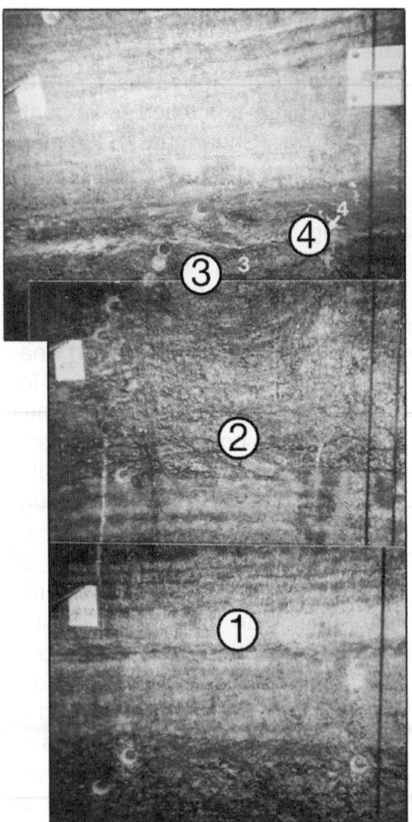

Figure 36. Photomosaic of a complete halite cycle composed of 1) bedded mud-poor halite at the base, terminated by a major dissolution surface, 2) 'podular' muddy halite, 3) 'dilated' mud-rich halite and 4) halitic mudstone. The cycle is underlain by 'dilated' halite of the underlying cycle and is overlain by bedded halite of the overlying cycle. Salado Formation, WIPP Air Intake Shaft, New Mexico. From Holt and Powers (1990).

plete cycles contain four facies. At the base are stratified, mud-poor halites with bottom-growth fabrics containing upward-increasing amounts of passive pore-filling and displacive halite cements. An irregular, hummocky surface occurs at the top of this facies (Fig. 7C). Overlying 'podular' muddy halites contain lenses of halite with anastomosing layers of clay, and pass up into the third facies – 'dilated' mud-rich facies containing abundant displacive halite cements. The uppermost facies is a halite-containing mudstone having clastic depositional structures markedly affected by haloturbation and displacive halite.

The basal stratified facies precipitated beneath shallow brines or on salt pans. Vadose and phreatic alterations increased upward indicating increasing and more pronounced water-table drops. Pits, pipes and large pores developed during dissolution events and were subsequently filled passively by halite cements in the phreatic zone. 'Podular' muddy halite formed in salt pans subject to intense vadose alteration. This created irregular dissolution surfaces (hummocky salt pans similar to the Devil's Golf Course at Death Valley, California (Figs. 9). The increased mud content occurs because the mud becomes concentrated as significantly more halite dissolved. 'Dilated' mud-rich halites record extensive alteration within phreatic zones. The mechanical integrity of salt following extensive vadose dissolution was low and subsequent growth of displacive halite during phreatic conditions dilated the sediments. The relief on the depositional surface was subdued and flooding events introduced large volumes of mud. Mud layers were then broken and disrupted by haloturbation. The uppermost halitic mudstones record deposition on saline mud flats. The entire cyclic sequence records increasing desiccation of an intermittently flooded salt pan.

BASIN-CENTRAL EVAPORITE MODELS

There are no modern equivalents of *marine* basin-central evaporites. Basin-wide evaporites can be thin (<5 m) or thick (hundreds to thousands of meters) and consist of shallow-water/subaerial evaporites, with or without evaporites of deep-water aspect. Basins that were filled by deep brines commonly contain marginal slope and deeper basin-central facies.

Evaporites deposited in low-relief basins are similar to shelf evaporites, except for their basin-central distributions. They are exclusively composed of shallow-water or mud-flat facies. In basins that had high relief, shallow-water/subaerial facies formed on basin-marginal shelves, are the terminal phases of basin filling (after depositional relief was eliminated), or they accumulated on the floors of largely desiccated basins during periods of evaporative drawdown. Evaporites of deep-water aspect are confined to basin centers. Near basin edges they commonly pass laterally into shallow-water facies, perhaps with an intermediate facies belt of slope deposits characterized by reworked evaporites – mass-flow deposits, slumps and turbidites.

Thick basin-center evaporites are composite stratigraphic units, intercalated with marine or restricted marine carbonates, and some evaporites may be non-marine. Carbonates represent less-saline evaporitic intervals or episodes when the entire evaporite basin was flooded by normal-marine waters (transgressive or highstand system-tract deposits). Non-marine evaporites (and associated continental deposits) record episodes when the formerly marine-fed basin became completely disconnected from any seawater supply.

Three main types of evaporite depositional basin are distinguished (Fig. 37). In most basins, however, these basin types represent developmental stages in the basin's history. This is not surprising because brine levels in a basin are most unlikely to have been constant. Evaporite deposition implies surface disconnection from the ocean, thus brine levels rapidly change or fluctuate in response to even slight changes in the rates of inflow, outflow and evaporation. These rates are controlled by climate and the degree of basin restriction. Basin restriction, in turn, is controlled, in part, by changes in tectonism or sea level.

Isostatic compensation should be important in the formation of evaporite depositional basins because of the relatively high density of anhydrite (especially if this is formed by early diagenetic alteration of gypsum). Many large evaporite deposits form within extensional basins that are sited above thin, fractured crust that rapidly subsides. Also important is that partial or complete desiccation of evaporite basins should lead to isostatic uplift, such that evaporites might well form in basinal settings that were shallower than expected from the stratigraphic relationships of underlying strata. Rapid accumulation of thick and dense evaporite deposits drives subsidence. Thick evaporite deposits thus require an initial basin depth that can be substantially less than their final thickness.

Van den Belt and de Boer (2007)

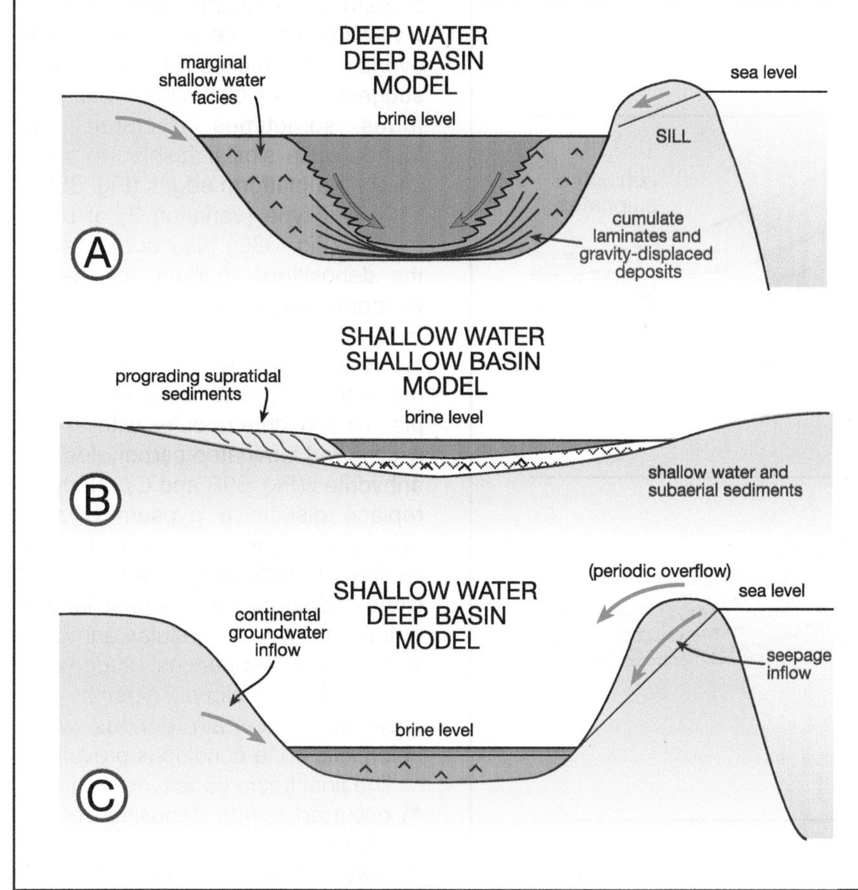

Figure 37. Depositional models for basin-central evaporites.

argue that rapid basin subsidence caused by isostatic adjustment (itself caused by rapid evaporite precipitation) eliminates the need for deep-basin desiccation as an explanation for thick evaporite sequences composed of shallow-water facies. This is undoubtedly correct for some sequences, but formation of thick sequences still requires the existence of an initial depression, which by Van den Belt and de Boer's calculations can be several hundred meters deep. This, for evaporite deposition, is 'deep'.

Shallow-water, Shallow-basin Model

This model accounts for relatively thin (<10 m) evaporites with basin-central distributions. This last feature alone should be sufficient to prevent these evaporites being compared with modern prograding coastal sabkhas, but a similarity in evaporite facies has commonly resulted in this misinterpretation. The evaporites develop within cratonic sag basins that did not subside rapidly and were never deep.

Evaporites of this type are composed of shallow-water and mud-flat facies and commonly form in the upper parts of carbonate cycles exhibiting signs of upward-increasing restriction. They have usually been interpreted as 'shallowing-upward' cycles but are better identified as 'brining-upward' cycles particularly when the evidence is equivocal. In evaporite-terminated cycles of the Ordovician Herald Formation (Upper Red River) of the Williston Basin (Kendall, 1976) and the Middle Devonian Lucas Formation of the Michigan Basin, thin-bedded laminated evaporitic dolomites containing desiccation cracks, microbial mats and evidence of storm deposition pass up into nodular anhydrite, the majority of which replaces vertically-aligned complex-twinned gypsum crystals (Fig. 23A). The more saline parts of these cycles were apparently deposited in more stable, subaqueous environments than the less saline carbonates beneath, some of which resemble mud flat deposits (or shallow saline lake deposits). Thus some brining-upward cycles can also be deepening-upward cycles, or cycles with little evidence for water-level change.

Deep-water, Deep-basin Model

Evaporite basins that were deep depressions and at least partly filled with deep brines are dominated by evaporites of deep-water aspect (at least in their lowermost parts). On basin floors these evaporites are laminated/thin- bedded sulfates or halite, whereas on marginal slopes, reworked platform evaporites occur (Peryt *et al.*, 1993). The presence of a high-relief basin is usually demonstrated by i) deep-water siliciclastics or carbonates beneath the evaporites – possibly with starved basin features, ii) depositional slope facies in both older sediments and the evaporites, or iii) high, pre-evaporite, carbonate buildups. When initial basin-marginal slopes are gentle, the basin periphery becomes the depositional site of thick, shallow-water gypsum deposits (Fig. 38A) that build upward and outward into the basin, and a platform is constructed. In large part, the difference in sedimentation rate between basin and flanks is a product of the greater degree to which brines can be evaporated in shallower waters. Because deposition is slower in basin centers, steep depositional slopes develop at the platform edge. Upper parts of slopes are sites of slumping and mass flow, whereas lower parts contain graded beds deposited by turbidity currents. Basins are largely filled by laminated deposits which are (annual?) precipitates or distal evaporite fines (or both).

Four main variations upon this platform-to-basin theme occur.

The first (variation 1) is the basic model with a shallow-water platform dominated by shallow-water facies that passes insensibly into slope deposits (commonly with turbidites and less common mass-flow deposits) that, in turn, pass into basal laminated facies (Fig. 38A) (Schlager and Bolz, 1977; Peryt *et al.*, 1993).

A variant of this basic model (variation 2) is seen in Zechstein of the Harz area (Germany) where quarries reveal huge blocks (10s of meters

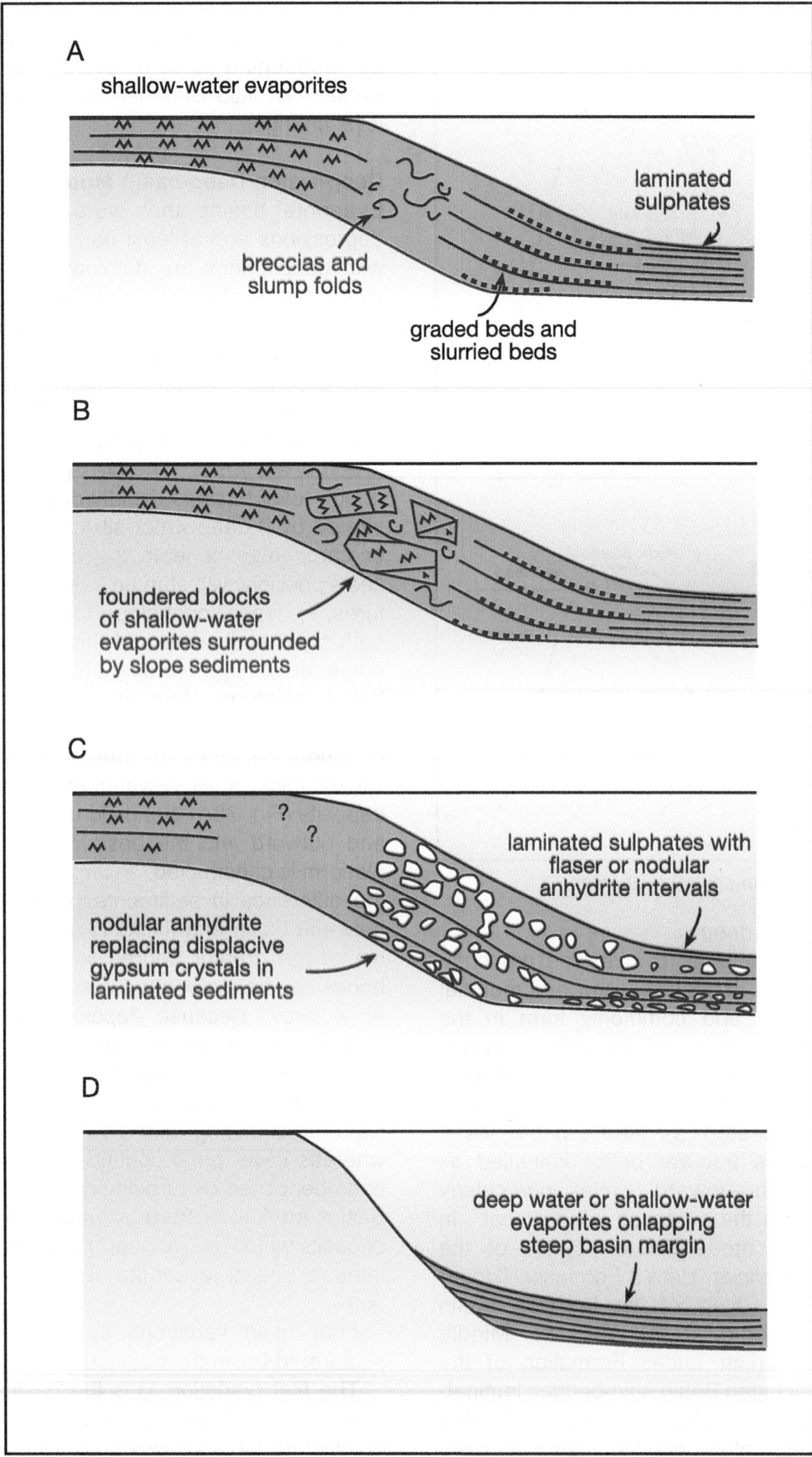

Figure 38. Schematic diagrams of deep-water and slope evaporitic environments. **A.** Basic model (after Schlager and Bolz 1977). **B.** Margin affected by block failure and collapse. **C.** Slope facies dominated by displacive gypsum crystal growth that greatly thickens the toe of slope 'basinal' laminite sequence, details of passage into shallow-water platform facies not known. **D.** Basin margin lacking marginal facies.

across) of anhydrite with shallow-water features that are flanked and overlain by turbidites (Fig. 39A). This suggests edges of shallow-water platforms sometimes fractured and foundered in similar fashion to some carbonate platform edges (Fig. 38B).

A third type (variation 3) of basin margin (Fig. 38C) also occurs along the depositional margins of the Z2 evaporite (especially on the English side of the North Sea). Here, the increasing thickness of the flanking depositional slopes consists of displacive anhydrite nodules within highly distorted laminated carbonates and anhydrites (Fig. 39B and C). Nodules replace displacive gypsum crystals that grew within essentially basinal laminated sediments. Intervals of slope nodular anhydrites pass laterally into intervals of nodular anhydrite within basin sequences, suggesting the displacive slope gypsum only grew during certain periods when holomictic brine conditions prevailed.

The final basin edge type (variation 4) occurred during deposition of the Permian Castile Formation in the Delaware Basin and perhaps during some of the time that deeper water evaporites formed in the Otto Fiord Formation of Arctic Canada. There is no marginal thickening of the evaporites, and passage into peripheral shallow-water platform sequences does not occur (Fig. 38D). In the Otto Fiord Formation, evaporites form upward thinning wedges as they are traced up steep carbonate slopes of the Nansen Formation, whereas in the Castile, laminated sulfates of deep-water aspect occur within a few hundred meters of the steep carbonate margin.

The same basic basin-filling pattern during gypsum deposition also occurs when some deep-water halites (and even some potash salts) are precipitated. Stratigraphic units in Permian Zechstein 2 evaporites (Figs. 30A and 32) of the North Sea thicken from the basin toward a marginal depositional shelf, where they thin again onto a shallow-water platform (Colter and Reed, 1980). This indicates the lower parts of thick marginal and the basin-floor salts were deep-water deposits. Deep-water potash salts constitute a major problem. In order that a basin fill with brine capable of

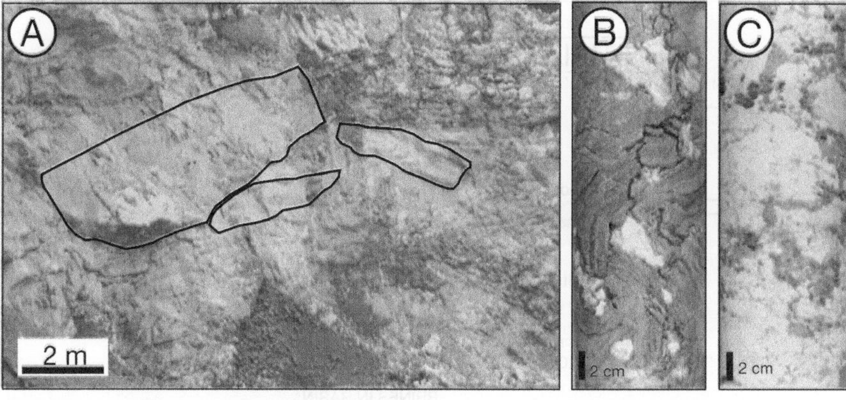

Figure 39. Marginal facies. **A.** Foundered blocks of shallow-water sulfates (some outlined), flanked and overlain by contorted laminated sulfate mass-flow deposits and turbidites, S. Harz, Germany (*see* Fig. 38B). **B.** Laminated slope dolomites and anhydrites distorted by addition of nodular anhydrite (here interpreted as replaced displacive gypsum), which in **C)** has pushed host sediment into thin selvages, producing a sediment that could easily be mistaken for a sabkha evaporite. (**B** and **C** from Z1 Anhydrite: Bates 8 well; UK North Sea).

precipitating potash salts, each batch of seawater first has to precipitate the greater part of its sodium chloride as halite. The amount of halite that needs to be precipitated should equal in volume the entire volume of the depositional basin and thus no deep basin would be left in which to precipitate the potash salts. Possibly the deep-water potash salts initially formed on the marginal platforms and the marginal slope deposits are either reworked salts or precipitated when supersaturated brines, created on high-salinity marginal shelves, flowed down depositional slopes. Alteration of pre-existing calcium sulfates into polyhalite by such descending brine flows also occurs. The only modern analogue might be salts in the Dead Sea that form when end-brines (generated in artificial potash ponds) are pumped out and descend down the slopes of the northern, deep part of the Dead Sea.

Depositional models for 'deep-water' evaporite basins were reviewed by Logan (1987). *Stratopycnal basins* have sub-horizontal density layering and, when they have surface inflow, are the classic evaporite basins of early workers. In shallow basins this model is unsound hydrodynamically. Strong winds (needed to promote evaporation) mix waters potentially down to depths of several tens of meters, so that upper parts of the brine develop a vertical density structure (*i.e.*, an *isopycnal basin*) similar to that of the modern Hamelin Pool (Shark Bay, Western Australia). Here salinities reach 72‰. For gypsum to be precipitated in Hamelin Pool, inflow must be reduced to 10% of its already restricted rate. This would confine influx to channels through a permanently emergent sill, and brine levels in the basin would fall below sea level. Seepage inflow through the barrier would occur, but no surface reflux. Thus the basin would be a variant of a *seepage inflow basin*: one modified by some surface inflow. Halite precipitation in Hamelin Pool would require complete surface disconnection.

Isolated or highly restricted basins experience drawdown and thus differ only in degree from *shallow-water, deep basins*. The main problem (discussed earlier) is to explain how any substantial brine column can be retained in desiccating basins. One mechanism is that, during formation of deep-water laminated evaporites, part of each year was more humid (Kirkland, 2003). During such seasons, the brine did not evaporate and inflowing seawater replenishes the brine column that had been depleted (and lowered) by the previous dry season.

Hydrodynamic considerations suggest deep-water evaporites were precipitated from brines that were laterally rather uniform, although they were probably density stratified for much of the time. This means that at any given time, evaporites of the same mineral facies are deposited throughout the basin.

Shallow-water, Deep-basin Model

Many thick basin-central evaporites seem to have been deposited in deep basins subject to significant evaporative drawdown. They were therefore isolated or extremely restricted basins and obtained most of their inflow by seepage. Like those of the previous model, these evaporites are underlain by deep-water marine sediments. Unlike them, they are composed of shallow-water and/or mud-flat facies. The facies developed depends upon the depth of the brine retained in the basin, and the inflow/reflux ratio. A shallow-water, deep-basin model was originally proposed for the Devonian Elk Point Basin, for the Miocene of the Mediterranean, and the Silurian of the Michigan Basin. More recent work has thrown considerable doubt upon the desiccated Mediterranean hypothesis. This is because deep depressions (hundreds of meters) are likely to be the focus toward which groundwaters flow. If underlying sediments and rocks become brine-saturated, there will then be no sink for surface-generated brines. After a period of time, such basins should stop desiccating and substantial depths of brine persist. The brine level of the Dead Sea, for example, will never completely desiccate even if inflow was reduced to zero.

Evaporite accumulation in evacuated deep basins is only possible where there is a basin aquitard, either a pre-existing impermeable layer on the basin floor, or one created by evaporite cementation during initial (and leaky) stages of brine concentration. Without the aquitard, the evacuated depression should act as a discharge area for underlying groundwaters. An aquitard lines at least part of the basin floor enclosing a topographically closed volume. Within this *hydro-sealed* part of the basin, long-term storage of brine occurs.

Basin margins, as well as localized regions within basins (such as earlier carbonate buildups) act as locations of basin brine reflux or as sites of groundwater discharge.

This probably occurred in the Devonian Elk Point Basin and the Silurian Michigan Basin and is shown where anhydrite occurs preferentially around former reefs and passes laterally into basin-floor brine-pan halites (Kendall, 1989). The anomalous anhydrite precipitated when calcium-rich groundwaters issued from the exposed reefs and mixed with sulfate-rich (but calcium-depleted) marine-derived brines in the basin.

When considering the effect of basal aquitards, two different situations can be contrasted (Fig. 40). When brine tables lie within the hydro-sealed part of the basin and seepage discharge is less than evaporation losses, the basin desiccates and water flows into the basin. Evaporites and brines are confined to areas underlain by the hydroseal so that reflux only occurs i) at slow rates through the aquitard or ii) during short episodes when floods raise brine-levels above the hydroseal. Consequently, brine residence times in the basin are long, and evaporation is able, with appropriate arid climatic conditions, to concentrate brines. Highly saline evaporites accumulate when climates are sufficiently arid.

In contrast, when brine levels lie above the hydroseal the basin is more open. Brines are lost by seepage outflow in marginal areas where the basal aquitard is absent. This loss reduces the residence times of brines in the basin and brines are exported before they can be fully concentrated to levels the arid climate would otherwise permit. Only salts with lower solubility are precipitated, regardless of how arid the climate was. If groundwaters become progressively more saline, reflux of surface brines is diminished and more saline evaporites then can accumulate.

In isolated basins there is no simple relationship between the type of evaporite deposited and sea-level change. Brine salinities (and the minerals they precipitate) are controlled by brine residence times in the basin. These are determined primarily by the absolute and the relative rates of water flow into, and brine flux out of, the basin. One of the more direct methods of changing the inflow, however, is to change sea level. When this rises (but not high enough to drown the basin rim which would terminate evaporite deposition), the hydraulic head between the sea and the depressed brine level in the basin is increased. This increases water seepage through the permeable barrier. If reflux rates are unchanged, the enhanced influx increases brine residence times and promotes higher brine salinities. In disconnected basins therefore, sea-level rises are marked by deposition of more saline evaporites. Lowered sea levels cause reduction in influx rates, decreasing brine residence times and deposition of evaporites that have lower solubilities. These effects are opposite to those expected in basins having free connection with the ocean. The upward sequence from anhydrites to carbonates in the Middle Devonian of northern Alberta (Muskeg and Sulfur Point Formations), usually interpreted as reflecting a sea-level rise, can instead be seen as being due to a sea-level fall that terminated with deposition of an overlying non-marine Watt Mountain unit (Kendall, 1988).

Inflow/reflux is more likely to change because of sediment aggradation in the basin (Fig. 40), especially because evaporites can accrete so rapidly. Initially, evaporite deposition is confined to bottom and central parts of a basin floor where reflux is severely limited by an underlying aquitard (or by a plume of subsurface brine). Residence times of surface brines are long; they become concentrated to high salinities and saline salts precipitate. With continued

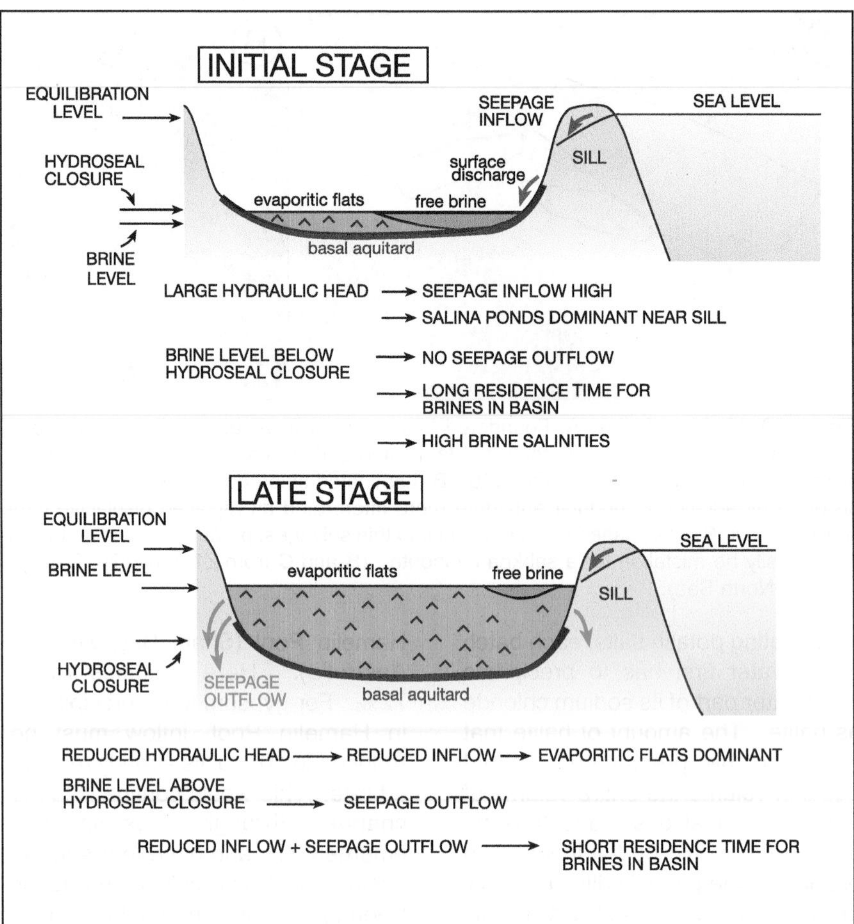

Figure 40. Diagrams illustrating two stages in the development of seepage basins. Initial stage has a large hydraulic head (between equilibration- and basin-brine-levels) and brine levels located below hydroseal closure (low reflux). Late stage has a reduced hydraulic head and brine levels above the hydroseal closure that allows seepage outflow of the basin brines. The consequences of these variations are shown with respect to brine residence times and brine salinities. Diagrams based on Logan (1987). In the late stage it is assumed that no diagenetic seal is formed that curtails brine reflux (*see* Fig. 41).

evaporite accretion, however, the depositional area expands onto basin flanks where the aquitard (or the brine plume) is absent (or has yet to develop). Increasing amounts of brine reflux occur, progressively reducing residence times of brines. As surface brine salinities drop, the more soluble phases cease precipitating (some, already precipitated, can even begin to dissolve) and an upward sequence of decreasing mineral solubility is generated (Fig. 40). This model can explain symmetrical cycles in the Pennsylvanian Paradox evaporites of Utah and Colorado.

Most evaporite sequences, however, even those that onlap basin flanks, have continued upward increases in mineral solubility. Here, lateral expansion of depositional areas was not coupled with decreases in brine residence times. This implies the presence of efficient seals or saline groundwaters on basin flanks. Commonly, however, flanks are composed of only slightly older carbonates that undoubtedly were still permeable at the beginning of the evaporite depositional episode. This suggests that evaporite basins can be self-sealing – porosity in basin flanking rocks, upon contact with widening brine plumes, becomes occluded by evaporite cements (Fig. 41). In particular, dolomitization of flanking carbonates (almost universally associated with evaporite basins) is accompanied by precipitation of calcium sulfates as replacive nodules or cements that eliminate porosity.

Evaporite sequences exhibiting upward decreases in mineral solubility are possibly confined to basins with basin flanks composed of siliciclastics (less susceptible to evaporite cementation related to dolomitization), or in which conduits for refluxing brines were repeatedly opened by syndepositional tectonism. Both explanations apply, for instance, to Paradox Basin evaporite-carbonate cycles.

Finally, changes in aridity cause variations in evaporite mineralogy and facies in isolated and desiccated basins. These changes can be seasonal or long term. The presence of free brine in a basin indicates inflow exceeds or equals losses from reflux and evaporation. When evaporative losses exceed inflow, surface brines disappear. Increases in aridity thus promote development of evaporite-flat environments where, if brine residence times are long (low-reflux rates), evaporites of high solubility accumulate in the sediment. Decreases in aridity allow perennial or seasonal surface brines to occur, but can prevent accumulation of the more saline mineral phases.

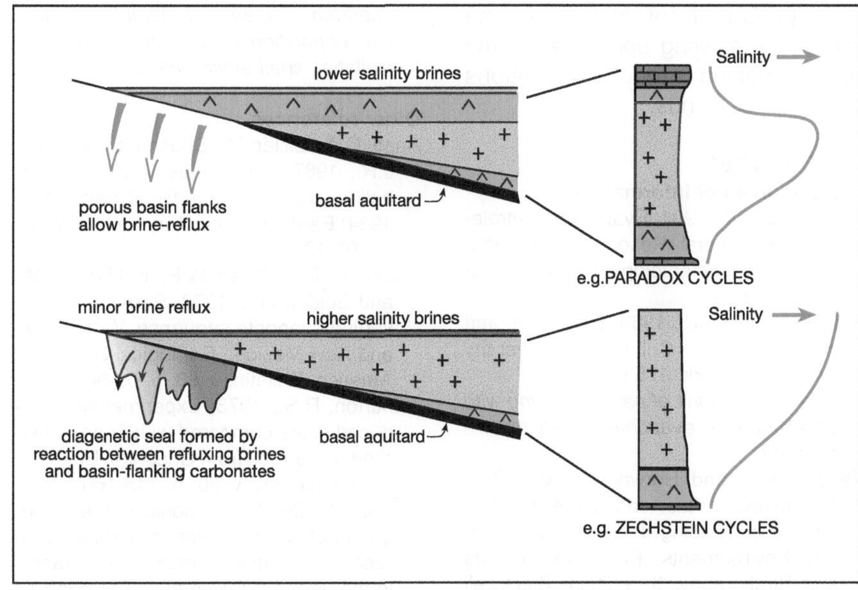

Figure 41. Diagrams showing inferred depositional sequences in basins without a hydroseal at bank-flanking locations (top) where brine reflux reduces brine residence times and evaporite sequence becomes first more saline then less saline upwards, and (below) basins that develop a diagenetic hydroseal as a result of a reaction between refluxing brines and flanking carbonates. Impeded brine reflux here causes brine residence times to remain long, allowing continued brine concentration and deposition of evaporite sequences with high-solubility salts in their uppermost parts.

SUMMARY

Evaporite facies are best defined by internal sedimentary and early diagenetic characters rather than by inferred geographic environments. Three main facies occur: *mud-flat*, *shallow-water* and *deep-water* evaporites, and two assemblages of marine evaporites: *basin-marginal* – largely marine and deposited as parts of transgressive and highstand system tracts, and *basin-central* – believed to be largely marine, but which i) can be influenced by continental groundwaters and thus ii) grade insensibly into entirely continental deposits. Basin-central evaporites only form when the entire depositional basin becomes restricted or isolated and are thus lowstand system tract deposits. Evaporative drawdown in these basins, however, greatly amplifies the effects of sea-level falls.

Since the 3rd edition of Facies Models was published, much has been learned. The sabkha paradigm continues to diminish. Sequence-stratigraphy has probably proven less successful than might otherwise have been expected, but ideas about hydrogeological settings of evaporites and the consequences for their preservation have proven to be increasingly important. Too often, however, depositional interpretations and models are based on inadequate criteria (*e.g.*, nodular anhydrites equating to sabkha deposits regardless of other considerations) and models are based upon inadequate numbers of case histories (sometimes only one). Some recent re-interpretations have yet to be integrated with previous models.

DEDICATION and ACKNOWLEDGEMENTS

This paper is dedicated to Doug Shearman and Gill Harwood who were always willing to discuss all matters evaporitic with me and whom I sorely miss. I am grateful to all who provided unaccredited material and the illustrations I have used (too many to list). Noel James kept me honest by insisting upon brevity and clarity, making useful suggestions towards those ends.

REFERENCES

Basic sources of information

Melvin, J.L., *ed.*, 1991, Evaporites, Petroleum and Mineral Resources: Developments in Sedimentology: Elsevier, Amsterdam, 556 p.

Schreiber, B.C., *ed.*, 1988, Evaporites and Hydrocarbons: Columbia University Press, New York, 475 p.

Two compilations of papers dealing with all aspects of evaporites; both recommended.

Kendall, A.C. and Harwood, G.M., 1996, Marine evaporites: arid shorelines and basins, *in* Reading, H.G. *ed.*, Sedimentary Environments: Processes, Facies and Stratigraphy, 3rd edition: Blackwell Science, Oxford, p. 281-324.

Contains similar material to this Facies Models paper but with greater emphasis upon case histories and sequence stratigraphic aspects.

Hardie, L.A., 1984, Evaporites: marine or non-marine?: American Journal of Science, v. 284, p. 193-240.

Criteria employed in distinguishing marine and non-marine evaporites; dispels previous views that non-marine evaporites are insignificant and that they always differ mineralogically and geochemically from marine evaporites.

Schreiber, B.C. and El Tabakh, M., 2000, Deposition and early diagenetic of evaporites: *Sedimentology,* v. 47 (Suppl. 1), p. 215-238.

Short review of evaporite depositional environments and diagenetic changes, giving more emphasis to geochemical evidence than this paper.

Warren, J.K., 1989, Evaporite Sedimentology: Importance in Hydrocarbon Accumulation: Prentice Hall, Englewood Cliffs, 285 p.

Warren, J.K., 1999, Evaporites: their Evolution and Economics: Blackwell Sciences, Oxford, 438 p.

Warren, J. K., 2006, Evaporites: Sediments, Resources, and Hydrocarbons: Springer, Berlin, 1035 p.

Successive reviews of evaporite science: the 1989 book discusses evaporite sedimentology and its importance to hydrocarbons; the 1999 version examines sequence-stratigraphic aspects and links to mineral deposits, and the 2006 magnum opus *updates and better illustrates almost all matters evaporitic. A good book to dip into; has a very extensive bibliography.*

Schreiber, B.C., Lugli, S. and Babel, M., 2007, Evaporites through space and time: The Geological Society of London, Special Publication 285, 373 p.

The most recent compilation, with important papers on depth indicators and deposition in ancient gypsum-precipitating shelf environments.

Other references

Anati, D.A., Stiller, M., Shasha, S. and Gat, J.R., 1987, Changes in the thermohaline structure of the Dead Sea: 1979-1984: Earth and Planetary Letters, v. 84, p. 109-121.

Anderson, R.Y., Dean, W.E., Kirkland, D.W. and Snider, H.I., 1972, Permian Castile varved evaporite sequence, West Texas and New Mexico: Geological Society of America, Bulletin, v. 83, p. 59-86.

Arthurton, R.S., 1973, Experimentally produced halite compared with Triassic layered halite-rock from Cheshire, England: Sedimentology, v. 20, p. 145-160.

Babel, M., 2007, Depositional environments of a salina-type evaporite basin recorded in the Badenian gypsum facies in the northern Carpathian Foredeep, *in* Schreiber, B.C., Lugli, S., and Babel, M. *eds.*, Evaporites through Space and Time: Geological Society of London Special Publication 285, p. 107-142.

Babel, M. and Becker, A., 2006, Cyclonic brine-flow pattern recorded by oriented gypsum crystals in the Badenian Evaporite Basin of the northern Carpathian foredeep: Journal of Sedimentary Petrology, v. 76, p. 996-1011.

Bebout, D.G. and Maiklem, W.R., 1973, Ancient anhydrite facies and environments, Middle Devonian Elk Point Basin, Alberta: Bulletin Canadian Petroleum Geologists, v. 21, p. 287-343.

Brantley, S.L. and Donovan, B., 1990, Marine evaporites, bittern seepage, and the genesis of subsurface brines: Chemical Geology, v. 84, p. 187-189.

Bowler, J.M., 1986, Spatial variability and hydrologic evolution of Australian lake basins: analogue for Pleistocene hydrologic change and evaporite formation: Palaeogeography. Palaeoclimatology. Palaeoecology, v. 54, p. 21-41.

Brodylo, L.A. and Spencer, R.J., 1987, Depositional environment of the Middle Devonian Telegraph salts, Alberta, Canada: Bulletin Canadian Petroleum Geologists, v. 35, p. 186-196.

Colter, V.S. and Reed, G.E., 1980, Zechstein 2 Fordon Evaporites of the Atwick No 1 borehole, surrounding areas of N.E. England and the adjacent southern North Sea, *in* Füchtbauer, H. and Peryt, T., *eds.*, The Zechstein Basin with Emphasis on Carbonate Sequences: Contributions to Sedimentology v. 9, p. 115-129.

Garber, R.A., Levy, Y. and Friedman, G.M., 1987, The Sedimentology of the Dead Sea: Carbonates and Evaporites, v. 2, p. 43-57.

Handford, C.R., 1990, Halite depositional facies in a solar salt pond: a key to interpreting physical energy and water depth in ancient deposits?: Geology, v. 18, p. 691-694.

Hardie, L.A., 1986, Ancient carbonate tidal-flat deposits: Colorado School Mines Quarterly, v. 81, p. 37-57.

Hardie, L.A., 1990, The roles of rifting and hydrothermal $CaCl_2$ brines in the origin of potash evaporites: an hypothesis: American Journal of Science, v. 290, p. 43-106.

Hardie, L.A., 1996, Secular variation in seawater chemistry; an explanation for the coupled secular variation in the mineralogies of marine limestones and potash evaporites over the past 600 m.y.: Geology, v. 24, p. 279-283.

Hardie, L.A. and Eugster, H.P., 1971, The depositional environment of marine evaporites: a case for shallow, clastic accumulation: Sedimentology, v. 16, p. 187-220.

Hardie, L.A., Smoot, J.P. and Eugster, H.P., 1978, Saline lakes and their deposits: a sedimentological approach, *in* Matter, A. and Tucker, M.E., *eds.*, Modern and Ancient Lake Sediments: International Association of Sedimentologists, Special Publication 2, p. 7-41.

Hardie, L.A. and Lowenstein, T.K., 2004, Did the Mediterranean Sea dry out during the Miocene? A reassessment of the evaporite evidence from DSDP legs 13 and 42A cores: Journal of Sedimentary Research, v. 74, p. 453-461.

Hardie, L.A., Lowenstein, T.K. and Spencer, R.J., 1985, The problem of distinguishing between primary and secondary features in evaporites, *in* Schreiber, B.C. and Harner, H.L., *eds.*, Sixth Symposium on Salt, Alexandria, VA: Salt Institute, v. 1, p. 11-40.

Hay, W.W., Migdisov, A., Balukhovsky, A.N., Wold, C.N., Flögel, S. and Söding, E., 2006, Evaporites and the salinity of the ocean during the Phanerozoic: Implications for climate, ocean circulation and life: Palaeogeography, Palaeoclimatology, Palaeoecology, v. 240, p. 3-46.

Holt, R.M and Powers, D.W., 1990, Geological mapping of the air intake shaft at the Waste Isolation Pilot Plant: DOE-WIPP 90-051.

Hovorka, S.D., Holt, R.M. and Powers, D. W., 2007, Depth indicators in Permian Basin evaporites, *in* Schreiber, B.C., Lugli, S., and Babel, M., *eds.*, Evaporites Through Space and Time: Geological Society of London, Special Publication 285, p. 335-364

Kendall, A.C., 1976, The Ordovician carbonate succession (Bighorn Group) of southeastern Saskatchewan: Saskatchewan Dept. Mineral Resources, Report 180, 185 p.

Kendall, A. C., 1988, Aspects of evaporite basin stratigraphy, *in* Schreiber, B. C., *ed.*. Evaporites and Hydrocarbons: Columbia University Press, New York, p. 11-65.

Kendall, A.C., 1989, Brine mixing in the Middle Devonian of western Canada and its possible significance to regional dolomitization: Sedimentary Geology, v. 64, p. 271-285.

Kent, D.M., 2003, Some thoughts on the formation of calcium sulphate deposits in Paleozoic rocks of southern Saskatchewan, *in* Summary of Investigations, 2003,1: Saskatchewan Geological Survey, Sask. Industry and Resources, Misc. Rep. 2003-$.1, CD-ROM, Paper A-8.

Kinsman, D.J.J, 1976, Evaporites: relative humidity control on primary mineral facies: Journal of Sedimentary Petrology, v. 45, p. 273-279.

Kirkham, A., 1997, Shoreline evolution, aeolian deflation and anhydrite distribution of the Holocene, Abu Dhabi: GeoArabia, v. 2, p. 403-416.

Kirkland, D.W., 2003, An explanation for the varves of the Castile Evaporite (Upper Permian) Texas and New Mexico, USA: Sedimentology, v. 50, p. 899-920.

Logan, B.W., 1987, The Lake MacLeod Evaporite Basin, Western Australia: American Association of Petroleum Geologists, Memoir 44, 140 p.

Lowenstein, T.K. and Hardie, L.A., 1985, Criteria for the recognition of salt pan evaporites: Sedimentology, v. 32, p. 627-644.

Lowenstein, T.K., Hein, M.C., Bobst, A.L., Jordan, T.E., Ku, T-L. and Luo, S., 2003, An assessment of stratigraphic completeness in climate-sensitive, closed-basin lake sediments: Salar de Atacama, Chile: Journal of Sedimentary Research, v. 73, p. 91-104.

Lowenstein, T.K. and Spencer, R.J., 1990, Syndepositional origin of potash evaporites: petrographic and fluid inclusion evidence: American Journal of Science, v. 290, p. 1-42.

Lucia, F.J., 1972, Recognition of evaporite-carbonate shoreline sedimentation, *in* Rigby, J.D. and Hamblin, W.K. *eds*., Recognition of Ancient Sedimentary Environments, SEPM Special Publication 16, p. 160-191.

McCaffrey, M.A., Lazar, B. and Holland, H.D., 1987, The evaporation path of seawater and the coprecipitation of Br^- and K^+ with halite: Journal of Sedimentary Petrology, v. 57, p. 928-937.

Maiklem, W.R., 1971, Evaporative drawdown – a mechanism for water level lowering and diagenesis in the Elk Point Basin: Bulletin Canadian Petroleum Geologists, v. 19, p. 487-503.

Manzi, V., Anzi, V., Lugli, S., Ricci Lucchi, F. and Roveri, M., 2005, Deep-water clastic evaporites deposition in the Messinian Adriatic foredeep (northern Apennines, Italy): did the Mediterranean ever dry out?: Sedimentology, v. 52, p. 875–902.

Nurmi, R.D. and Friedman, G.M., 1977, Sedimentology and depositional environments of basin center evaporites, Lower Salina Group (Upper Silurian), Michigan Basin, *in* Fisher, J.H., *ed*., Reefs and Evaporites – Concepts and Depositional Models, American Association of Petroleum Geologists, Studies in Geology, v. 5, p. 23-52.

Orti-Cabo, F. and Busson, G., eds., 1984, Introduction to the Sedimentology of the coastal salinas of Santa Pola (Alicante, Spain): Revista d'Investgacions Geologiques, v. 38/39, p. 1-235.

Perkins, R.D., Dwyer, G.S., Rosoff, D.B., Fuller, J., Baker, P.A. and Lloyd, R.M., 1994, Salina sedimentation and diagenesis: West Caicos Island, British West Indies, *in* Purser, B., Tucker, M., and Zenger, D., *eds*. Dolomites, a volume in honour of Dolomieu, International Association of Sedimentologists, Special Publication 21, p. 37-54.

Peryt, T.M., Orti, F. and Rosell, L., 1993, Sulfate platform-basin transition of the Lower Werra Anhydrite (Zechstein, Upper Permian), western Poland: facies and petrography: Journal of Sedimentary Petrology, v. 63, p. 646-658.

Peryt, T.M., Pierre, C., and Gryniv, S.P., 1998, Origin of polyhalite deposits in the Zechstein (Upper Permian) Zdrada platform (northern Poland): Sedimentology, v. 45, p. 565-578

Schlager, W. and Bolz, H., 1977, Clastic accumulation of sulfate evaporites in deep water: Journal of Sedimentary Petrology, v. 47, p. 600-609.

Schreiber, B.C., Friedman, G.M., Decima, A., and Schreiber, E., 1976, Depositional environments of Upper Miocene (Messinian) evaporite deposits of the Sicilian Basin: Sedimentology, v. 23, p. 729-760.

Shearman, D.J., 1983, Syndepositional and late diagenetic alteration of primary gypsum to anhydrite, Sixth International Symposium on Salt: Salt Institute, 1, p. 41-50.

Smoot, J.P. and Castens-Seidell, B., 1994, Sedimentary features produced by efflorescent salt crusts, Saline Valley and Death Valley, California, *in* Sedimentology and Geochemistry of Modern and Ancient Saline Lakes: SEPM Special Publication 50, p. 73-90.

Taberner, C. Cendón, D.I., Pueyo, J.J. and Ayora, C., 2000, The use of environmental markers to distinguish marine from continental deposition and to quantify the significance of recycling in evaporite basins: Sedimentary Geology, v. 137, p. 213-240.

Tucker, M.E., 1991, Sequence stratigraphy of carbonate-evaporite basins: models and application to the Upper Permian (Zechstein) of northeast England and adjoining North Sea: Journal Geololgical Society of London, v. 148, p. 1019-1036.

Van den Belt, F.J.G. and de Boer, P.L., 2007, A shallow basin model for 'saline giants' based on isostacy-driven subsidence, *in* Nicholls, G., Williams, E. and Paola, C., *eds*., Sedimentary Processes, Environments and Basins: A tribute to Peter Friend: Blackwell, p. 241-252.

Walton, A.W., 1978, Evaporites: relative humidity control of primary mineral facies: a discussion: Journal of Sedimentary Petrology, v. 48, p. 1357-1378.

Wardlaw, N.C. and Schwerdtner, W.M., 1966, Halite-anhydrite seasonal layers in Middle Devonian Prairie Evaporite, Saskatchewan, Canada: Geological Society of America, Bulletin, v. 77, p. 331-342.

Warren, J.K., 1982, The hydrological setting, occurrence and significance of gypsum in late Quaternary salt lakes in South Australia: Sedimentology, v. 29, p. 609-637.

Warren, J.K. and Kendall, C.G.S.,1985, Comparison of sequences formed in marine sabkha (subaerial) and salina (subaqueous) settings – modern and ancient: American Association of Petroleum Geologists, v, 89, p. 1013-1023.

Wood, W.W., Sanford, W.E. and Al Habshi, A.R.S., 2002, Sources of solutes to the coastal sabkhas of Abu Dhabi: Geological Society America, Bulletin, v. 114, p. 259-268.

Wood, W.W., Sanford, W.E. and Frape, S.K., 2005, Chemical openness and potential for misinterpretation of the solute environment of coastal sabkhat: Chemical Geology, v. 215, p. 361-372.

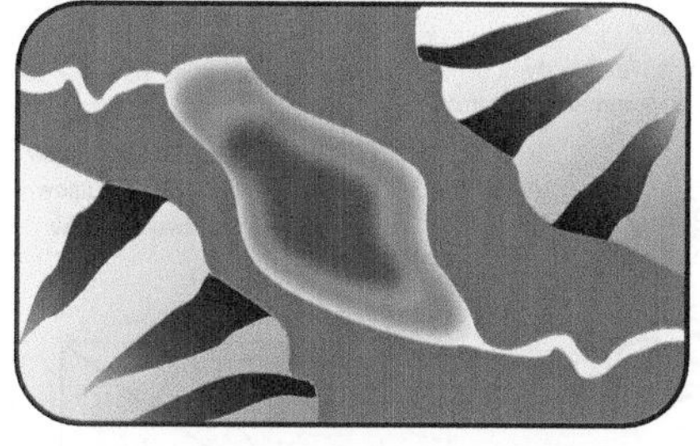

21. Lakes

Robin W. Renaut, Department of Geological Sciences, University of Saskatchewan, Saskatoon, SA, S7N 5E2, Canada

Elizabeth H. Gierlowski-Kordesch, Department of Geological Sciences, Ohio University, Athens, Ohio 45701-2979, USA

INTRODUCTION

The diversity of lakes on Earth is almost endless. From the long, narrow, deep (>1640 m) Lake Baikal in Siberia to the hundreds of thousands of small, shallow lakes scattered across the North American prairies, to the broad, shallow Lake Victoria in East Africa, to the ice-covered hypersaline lakes of the Dry Valleys region in Antarctica, lakes show extraordinary variety (Fig. 1).

A lake is "an inland body of standing water occupying a depression in the Earth's crust" (Kelts, 1988). The depressions occupied by lakes have many origins, but about 90% of all modern lake basins originated by tectonic and glacial processes (Cohen, 2003). Tectonic processes produce rift (*e.g.*, Lake Tanganyika, East Africa), strike-slip to pull-apart (*e.g.*, Dead Sea, Israel), foreland basin (*e.g.*, Eocene Green River Formation, USA), and intra-arc lakes (*e.g.*, Lake Titicaca, Bolivia). Glacial erosion and damming by glacial deposits give rise to large proglacial (*e.g.*, Lake Agassiz, central North America), glacial valley (*e.g.*, Wastwater, Cumbria, UK), and 'kettle' lakes (*e.g.*, the shallow fresh and saline lakes of Saskatchewan, Canada). The remaining 10% of lakes include those from dissolution in karstic settings (*e.g.*, Florida lakes, USA), those produced by volcanic and hydrothermal activity (*e.g.*, maar and crater lakes such as Lake Taupo, New Zealand), and those from many other origins (deflation, coastal lakes, meteorite impact, *etc.*). Each lake is a unique environment with sediments that reflect the local hydrology, bedrock, and biology of the drainage basin, and the regional climate. This makes their sediment records valuable archives of environmental change (Cohen, 2003; Smol, 2008).

The diversity of lacustrine depositional systems necessitates a different approach to the development of facies models from the others presented in this book. Lake sediments

Figure 1. The diversity of lakes. **A.** Coniston Water, a freshwater open lake in a glaciated valley, Cumbria, England. **B.** Lake Bogoria, a saline, alkaline (pH: 10.3) meromictic lake in the Kenya Rift Valley. **C.** Salar de Atacama, a saline pan (halite) in Chile. **D.** Frying Pan Lake, an acidic (pH: 2.4) lake, fed by sublacustrine hot springs, in a hydrothermal eruption crater, Waimangu, North Island, New Zealand.

can be siliciclastic, volcaniclastic, carbonate, evaporitic, or organic in almost any combination, and lake systems incorporate elements of many other depositional systems, such as deltas, coastal siliciclastic systems, and turbidites. This chapter is a short overview of lacustrine environments and facies models, emphasizing natural facies relationships that are based mainly upon the available accommodation space and the amount of annual recharge. More detailed general discussion of lacustrine facies is given in Anadón *et al.* (1991a), Talbot and Allen (1996), Cohen (2003), Verrecchia (2007), and Gierlowski-Kordesch (2010).

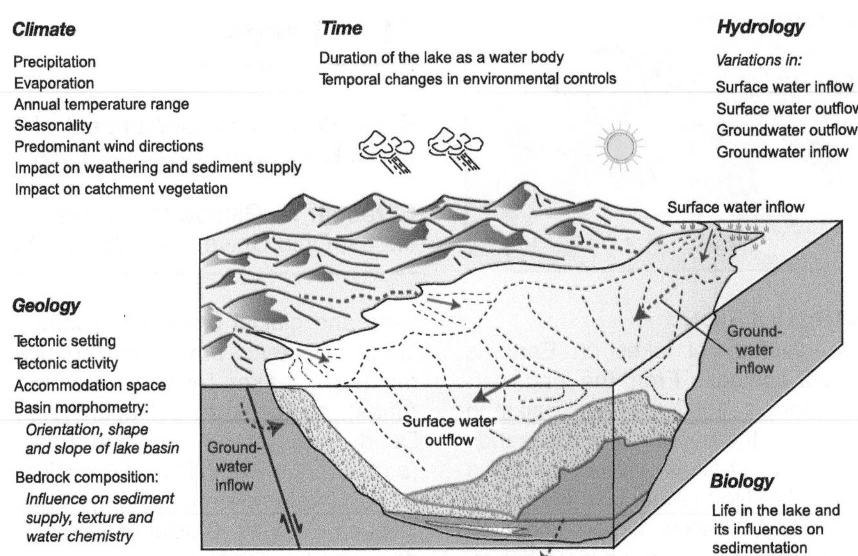

Figure 2. The main controls of lacustrine sedimentation.

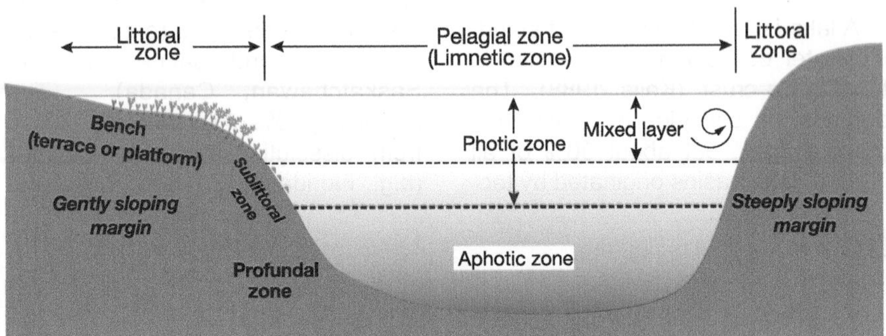

Figure 3. The zonation of lakes.

CONTROLS OF LACUSTRINE SEDIMENTATION

Lake sedimentation (Fig. 2) is controlled mainly by tectonics and climate, which together regulate the types of water input (surface runoff, groundwater, direct precipitation) and their sediment load (bedload, suspended clays and silts, and the dissolved solute load). The tectonic setting and hydrology of a lake basin, as well as climate and biologic processes (*e.g.*, plankton productivity), affect the mosaic of possible subenvironments within any lacustrine system (*see* Bohacs *et al.*, 2000, 2003; Cohen, 2003).

Basin morphometry and the internal dynamics of lake waters, including wave activity and current flow, stratification of the water column, and the hydrodynamics of the inflow (*e.g.*, sheetfloods, density currents) influence the sediment influx and lead to reworking of sediments within the lake. Lake watershed drainage and hydrology, whether open and through-flowing or hydrologically closed and terminal, control both clastic and chemical sedimentation, and influence organic deposition.

Climate determines the evaporation/precipitation ratio, the volume of surface water and its sediment load, the delivery of sedimentary particles by wind and rain, and the influx of atmospheric solutes (*e.g.*, chloride). Climate also influences lake-water circulation patterns. In contrast, tectonics can control accommodation and the general drainage within a lake basin, including hydrologic inputs and outputs from groundwater, and determines the bedrock composition and its spatial relationships. These in turn affect the composition and texture of sediment input and the lake hydrochemistry. The effects of tectonics and climate on lake sedimentation are intimately interwoven. These primary controls also impact the organic productivity and life in a lake, and can influence the organic components of lake sediments, their distribution and type (detrital versus autochthonous), and the facies. Compared to many other depositional systems, lakes are both diverse and highly complex.

LACUSTRINE ENVIRONMENTS AND PROCESSES

Lacustrine environments are defined by energy levels, as in marine settings, and extend from the shoreline to deeper water offshore regions (Fig. 3). The shape of the lake basin and the gradient of the basin slopes control the distribution of facies across a lake. The shallow-water zone around a lake margin is called the *littoral zone*; the *sublittoral zone* is the transition to deeper water. Light can still penetrate down to the sediments in the sublittoral zone. Gently sloping platforms called *benches* or *terraces* can extend the littoral zone, resulting in steep transitions into deeper water. The offshore area of deepest water with minimal to no sunlight penetration is termed the *profundal zone*. The upper photic zone of the water column above the profundal zone is the *pelagic or pelagial zone*. Shallow lakes have no profundal zone.

Physical Processes in Lakes

Physical processes in lakes are dominated by waves and currents (Imboden, 2004). Tidal effects are negligible even in large lakes. Waves move coarse sediment along the shoreline and transport fine sediment offshore by processes analogous to those along marine shorelines. Their effectiveness varies according to lake orientation and fetch with respect to dominant winds. Waves also mix the upper layers of lake water, which promotes circulation of nutrients and helps to prevent thermal stratification.

Currents and water circulation in lakes are mainly wind driven or induced by the inertial flow of rivers, density currents, and atmospheric pressure inequalities. Lake currents are rarely strong enough to move coarse bedload, but are important in dispersing fine sediment within the lake. Wind shear and low atmospheric pressure can lead to a pile up of water at one end of the lake, generating a large-scale wave-like motion of the lake body termed a *seiche*, analogous to the sloshing of water in a bathtub. With amplitudes sometimes exceeding a meter, large waves (*internal seiches*) can also form at the contact between the upper and lower water masses during density stratification. In large lakes, geostrophic effects deflect water currents to the right (N hemisphere) or left (S hemisphere), similar to their effects in the oceans. These can induce circulation, upwellings, and cause underflows capable of redistributing sediment. *Sediment focusing*, a process where sedimentation rates are higher in different areas of a lake, attests to the effects of sediment re-suspension and movement by surface waves, currents, and internal waves (Lamoureux, 1999; Horppila and Niemisto, 2008).

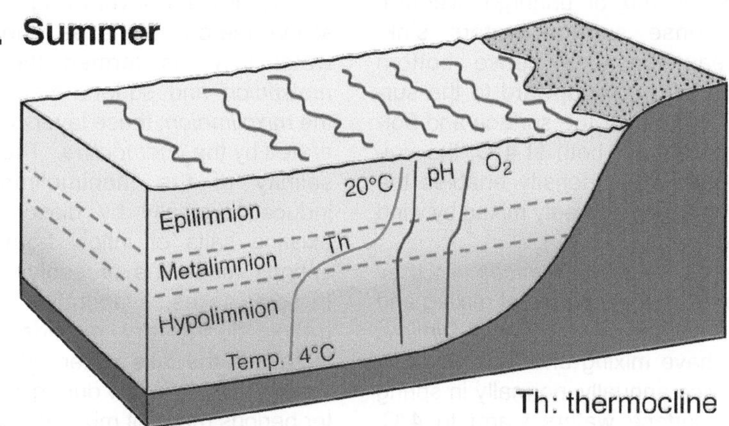

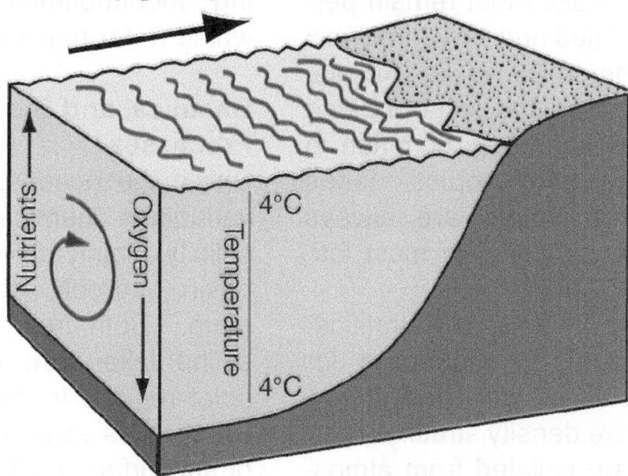

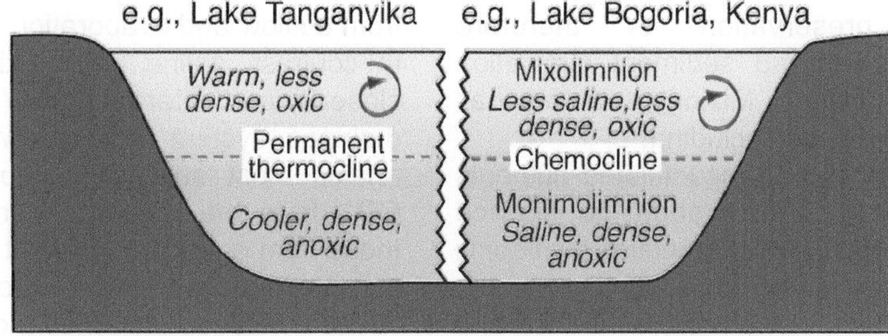

Figure 4. Stratification in lakes. **A.** Vertical profile of temperature, pH and oxygen in a typical perennial lake during warm summer months. Warm, less dense surface waters lie upon denser cooler waters, separated by the thermocline. **B.** Breakdown of stratification during winter months when surface waters cool to 4°C and attain their maximum density. Surface and bottom waters can be mixed by wind induced currents, waves and gravity. **C.** Meromixis (stable stratification) can develop when surface waters are consistently warmer than the cooler, dense bottom waters or when relatively less saline surface waters lie upon denser more saline brines.

Lake water Stratification and Mixing

Lake sedimentation and facies are strongly influenced by the lacustrine thermal regime and development of lake water stratification (Fig. 4). Two different water layers develop in many lakes due to differential radiative heating. The upper warmer, surface water mass, termed the *epilimnion*, is the least dense. The colder denser bottom water is termed the *hypolimnion*. These water masses are separated by the *metalimnion* or *thermocline*, the water layer or boundary where the temperature transition occurs, leading to a stratified water body (Fig. 4A). Limnologists classify the thermal regime of a lake by how often these water masses mix during the year (Wetzel, 2001). Wind and currents induced by inflow cause mixing of water masses, but temperature-controlled variations in water density account for most mixing. The maximum density of water occurs at 4°C. Whenever epilimnion waters cool to 4°C (or warm up if < 4°C) in a stratified lake, they can become as dense or denser than the underlying water mass, creating instability that allows the water column to mix or undergo overturn. The dense surface waters sink, whereas the less dense bottom waters can move upward to the surface. Whenever the surface and bottom waters are both at 4°C, the uniform lake-water density enables the water mass to be freely mixed by wind and currents (Fig. 4B).

Lakes are *monomictic* if they undergo a single phase of mixing and water circulation annually. *Dimictic* lakes have mixing and water circulation twice annually, normally in spring when surface waters warm to 4°C, and in autumn when they cool to 4°C. *Polymictic* lakes are shallow lakes that temporarily stratify for short periods, but otherwise mix freely throughout the year. Lakes that remain permanently stratified and very rarely mix are termed *meromictic* lakes (Hakala, 2004). *Meromixis*, the state of being meromictic, is especially common in deep, steep-sided tropical lakes where surface waters are always warm, but can develop in most latitudes (Fig. 4C).

The thermal regime and stratification have major consequences for lacustrine sedimentation and facies. When lakes are density stratified, bottom waters are isolated from atmospheric oxygenation and become dysoxic or anoxic when bacteria have depleted their oxygen. Organic matter preservation is therefore enhanced and sediment lamination can survive. Mixing of waters has many effects, including:

i. recycling of accumulated nutrients (N, P) from bottom to surface waters, which stimulates planktonic productivity;
ii. cooling or warming of surface and bottom waters, which can impact carbonate equilibria and biologic activity; and
iii. oxygenation of bottom waters, which can promote mineral precipitation (*e.g.*, ferric iron and Mn minerals) and allow bioturbation by benthic organisms.

Records of such seasonal and aperiodic mixing events and long-term meromixis are mostly preserved in the deeper water, offshore lacustrine facies.

Temperature is not the only control of meromixis. Density stratification can also develop due to differences in salinity of water layers (Fig. 4C). In saline meromictic lakes, the basal water layer is termed the *monimolimnion* and surface waters form the *mixolimnion*; these layers are separated by the *chemocline*. The higher salinity of the monimolimnion is induced typically by dissolution of existing salts or inflow from saline springs (shoreline or sublacustrine). In some lakes, evaporative concentration during dry periods initially increases the lake water salinity and density. Dilute inflow during later wetter periods may not mix with the existing saline water mass. This leads to a stable stratification unless, with evaporation, the mixolimnion density increases to equal that of the underlying monimolimnion, allowing the whole water mass to mix.

Chemical and Biochemical Processes

Unlike the oceans, lakes are highly variable in chemical composition and salinity, mainly reflecting the drainage basin bedrock, hydrology, and climate. Both freshwater (dilute) and saline lakes are present from the polar regions to the equator. Freshwater lakes exist wherever a surface or subsurface outflow maintains open drainage and the annual recharge from precipitation, runoff, and groundwater is balanced by the water loss from outflow and evaporation (Fig. 5). In contrast, saline lakes exist in a closed or intermittently closed drainage system that receives and retains most solutes (except some CO_2) from the watershed, including those from groundwater input. When the rate of evaporation or groundwater loss exceeds the amount of annual recharge, the lake can shrink in volume, fall below the height of its surface outlet, become progressively more saline, and eventually precipitate evaporites (Fig. 5).

The dominant cations in most dilute lakes are typically Na^+ and Ca^{2+}, with more variable K^+ and Mg^{2+} concentrations, depending mainly on bedrock composition in the drainage basin. Anions show greater variability. Carbonate species (HCO_3^-, CO_3^{2-}) dominate in limestone regions. Igneous and metamorphic terranes can also produce some bicarbonate ions through silicate mineral weathering by hydrolysis. Sulfate ions are mainly products of sulfide mineral oxidation (*e.g.*, pyrite), but are also released by weathering of sulfate evaporites (gypsum, anhydrite), and are present in rainfall (from marine aerosols or pollution). Most chloride originates from marine aerosols and connate groundwaters, but can also derive from halite dissolution and fluid inclusions within evaporites. Other common solutes include silica, fluoride, and nutrients (N, P species).

Chemical precipitates within dilute lakes include carbonates, minerals linked to changing redox conditions such as Fe- and Mn-oxyhydroxides, and some clay minerals and phosphates. In hydrologically closed basins, evaporite minerals can precipitate from lake brines. Saline lake brines and sediment pore waters also interact with detrital grains and earlier authigenic minerals to produce a wide range of early diagenetic minerals including clays, zeolites, and sodium silicates. Mixing of inflow waters with brines also can induce mineral precipitation in the water column by the common ion effect (*e.g.*, gaylussite: $Na_2Ca(CO_3)_2 \cdot 5H_2O$, which forms when dilute Ca-bearing river water mixes with sodium carbonate brines). Solutes acquired by weathering reactions and from rainfall are transported to the sites of mineral precipitation at various locations in the basin by runoff and groundwater, undergoing many fractionation processes along their flow paths that selectively remove some ions and dissolved species. The remaining solutes progressively increase in concentration, both relatively and in total abundance, until some in turn are also removed from the fluid, which ultimately simplifies the final lake-brine composition (Fig. 6). The principal mechanisms of solute removal and fractionation include:

1. Mineral precipitation when the waters become sufficiently supersaturated, mainly by evaporation or cooling, in both the drainage basin and the lake itself;
2. Degassing of CO_2 from surface water and shallow groundwater;
3. Selective dissolution of previously precipitated efflorescent salts by dilute surface runoff; and
4. Sulfate reduction together with

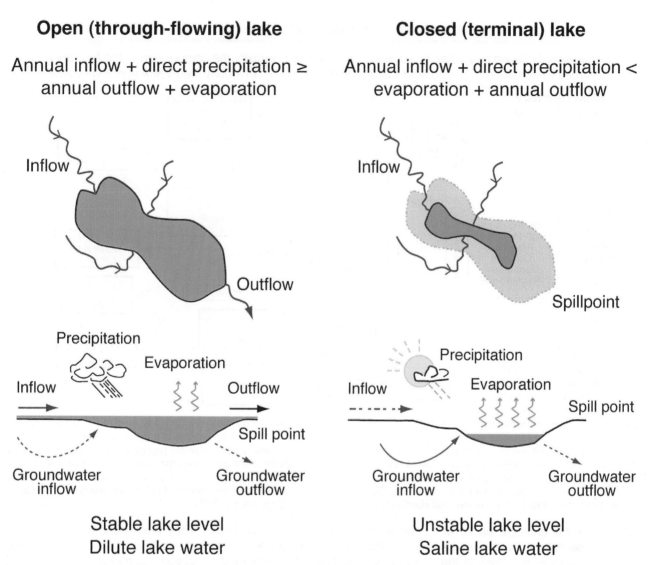

Open (through-flowing) lake

Annual inflow + direct precipitation ≥ annual outflow + evaporation

Inflow

Outflow

Precipitation

Evaporation

Inflow

Outflow

Spill point

Groundwater inflow

Groundwater outflow

Stable lake level

Dilute lake water

Closed (terminal) lake

Annual inflow + direct precipitation < evaporation + annual outflow

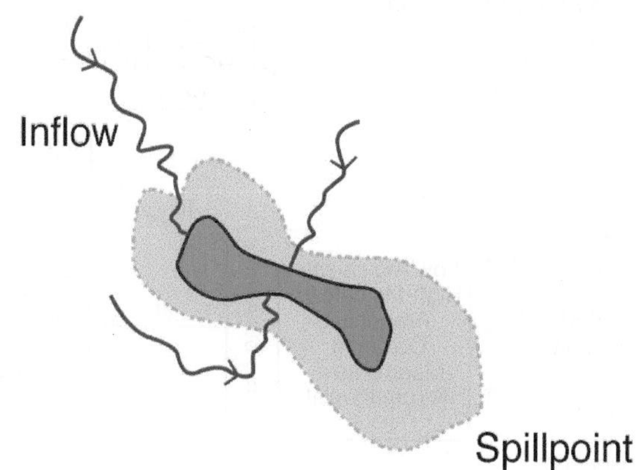

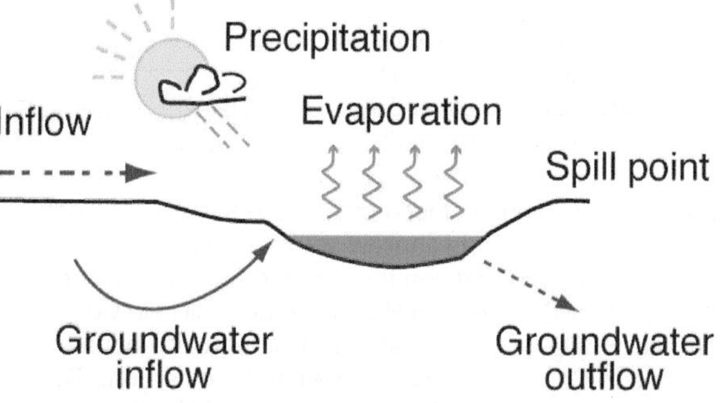

Unstable lake level

Saline lake water

Figure 5. Hydrology of open (through-flowing) and closed (terminal) lakes.

other chemical and biochemical processes.

By these mechanisms, the inflow waters and lake water progressively change in chemical composition and increase in salinity, allowing many different evaporite mineral suites to form (Fig. 6). This pattern contrasts with the relatively simple and more predictable sequence of evaporitic mineral precipitation and evaporite facies in marine waters (*see* Chapter 20). Evaporite mineralogy and facies are controlled mainly by the original ratios of the major ions in the dilute inflow waters before evaporation, the subsequent fractionation processes (Fig. 6), and the amount of leakage that occurs in a lake basin (Eugster and Hardie, 1978; Rosen, 1994; Jones and Deocampo, 2003).

Inflow from deep hydrothermal fluids can sometimes modify brine compositions causing deviations from the predictive models, and can promote precipitation of exotic minerals such as borates and lithium salts. Highly acidic lakes (pH < 3) occur in volcanic regions where H_2S and other gases condense in lake water and are oxidized to produce sulfate ions and protons (Fig. 1D). Many small, acidic saline lakes are present in topographically closed, but groundwater-connected, basins in arid Western Australia. Several minerals, including halite, gypsum, iron oxides, alunite, and jarosite, precipitate in these basins where the brines undergo strong evapoconcentration, but few minerals are present in the bedrock to buffer the solutions and neutralize the acidity (Benison *et al.*, 2007).

Some sedimentary processes in lakes are predominantly biochemical and are mediated by microbes, macroflora and fauna, particularly the precipitation of carbonate minerals, which occurs in both shallow marginal water and offshore in open water (Gierlowski-Kordesch, 2010). Lakes in almost any climate or tectonic regime can precipitate or deposit carbonates. Accumulation of carbonate sediment depends mainly on the presence and abundance of carbonate rocks or Ca-bearing volcanic rocks in the drainage area, or in the bedrock through which groundwater migrates. Even relatively dilute lake waters attain supersaturation with respect to $CaCO_3$, but without the influence of biologic processes, precipitation of carbonate minerals may not occur. The shells of bivalves, gastropods, and ostracodes, together with carbonates precipitated by charophytes, and detrital carbonate

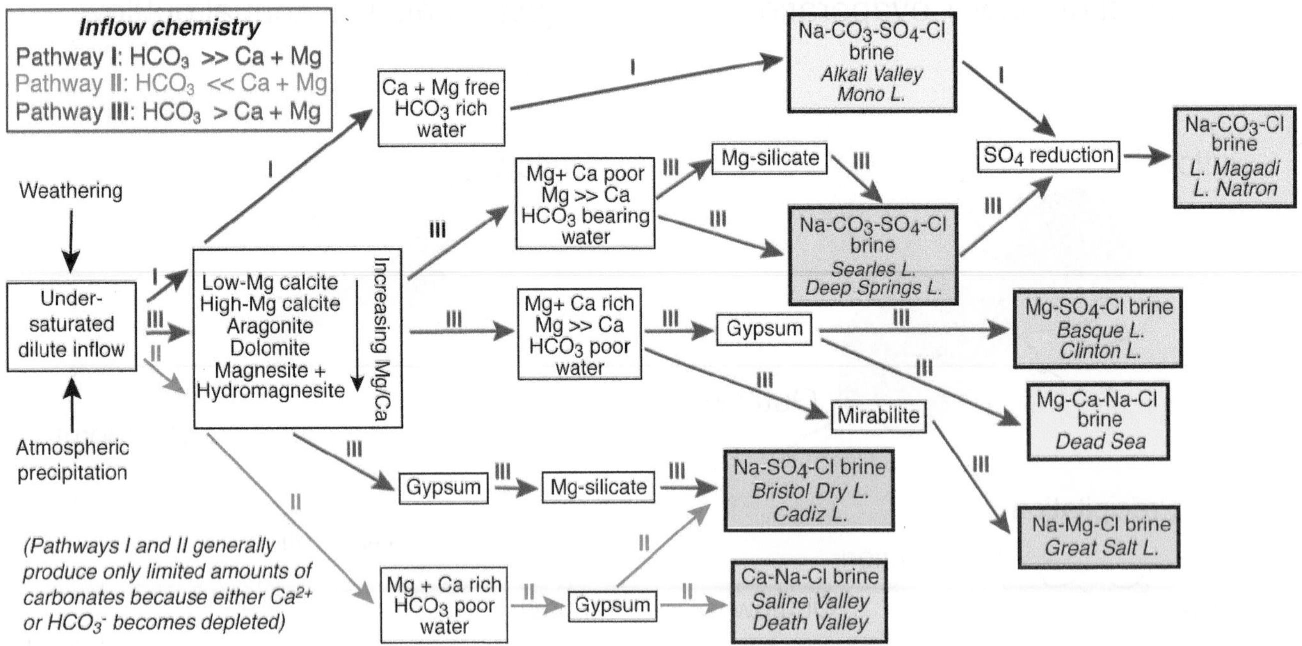

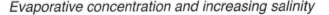

Figure 6. Evolution of closed basin waters to produce brines of different chemical composition. Ca is lost from the dilute fluid by $CaCO_3$ precipitation in soils and shallow sediments (cements, calcrete), by formation of tufa and travertine at springs, and in lakes. This increases the Mg/Ca ratio and can then lead to Mg-carbonate precipitation. After alkaline earth carbonate precipitation, the fluid is Ca-rich and HCO_3-poor or Ca-poor and HCO_3-rich and will follow one of several evolutionary paths with progressive evaporation. For example, if Ca-rich and HCO_3- poor after any alkaline-earth carbonate precipitation (paths II or III), gypsum will precipitate, leaving a fluid relatively enriched in Ca or SO_4 and depleted in the other ion. By these fractionation mechanisms, brines of different compositions can be generated. For further details, see Eugster and Hardie (1978) and Jones and Deocampo (2003).

grains from the drainage basin, all contribute detrital calcite and aragonite to lake sediments. Much of the carbonate deposited in lakes, however, consists of minute (< 5 µm) $CaCO_3$ crystals that nucleate in the upper water column and settle by gravity. In many lakes, pelagic calcite or aragonite *whitings* form periodically by processes analogous to those on shallow-marine carbonate platforms. Crystal nucleation in the water column can be induced by evaporative concentration, seasonal warming, or degassing of CO_2, all of which decrease $CaCO_3$ solubility. Most carbonate precipitation in lakes, however, is probably bio-induced by photosynthetic algae and bacteria that remove CO_2 (Thompson *et al.*, 1997).

LACUSTRINE FACIES MODELS

Shorelines

Lake shorelines are highly variable depending mainly on their slope, energy levels in the littoral zone, and the type, supply, and texture of the sediment. Some lakes, including many playa lakes, are bordered by low-energy mudflats or sandflats with little vegetation. In contrast, others are rimmed by littoral wetlands or bedrock that provides steep lake margins. Where the littoral gradient and energy levels are low, there may be no net accumulation of coarse sediments to record the location of the shoreline during any time interval. In some lakes, however, shoreline deposits, including beaches, are well developed. The character of the shoreline environments and facies differs in lakes that are actively precipitating carbonates from those with siliciclastic sedimentation.

Siliciclastic-dominated Shorelines

Siliciclastic shoreline deposits accumulate when sand and coarser sediment is available in the littoral zone, and energy levels are high enough to sort it (Fig. 7). Shoreline sediments in lakes are typically of small scale compared to their marine equivalents. They are similar to microtidal siliciclastic-shoreline deposits along ocean margins, but the onshore–offshore subenvironments are generally poorly defined except in large lakes. Sand or gravel beaches, spits, and small offshore barriers are present in many large lakes, and in smaller elongate lakes where a predominant wind direction sets up effective longshore currents. Characteristic facies include swash crossbedded sandstone, washover sandstone and siltstone, and back-barrier, low-energy lagoonal mudstone (Fig. 7; Table 1). Gravel beaches form imbricated orthoconglomerate, but most examples are thin and poorly sorted except in large lakes. Regressive lacustrine beach sandstone and conglomerate are important indicators of temporary paleoshorelines, especially in late Quaternary successions where many are preserved in their original context.

The shallow offshore zones in large lakes with shallow bench slopes have similar facies to those of shallow

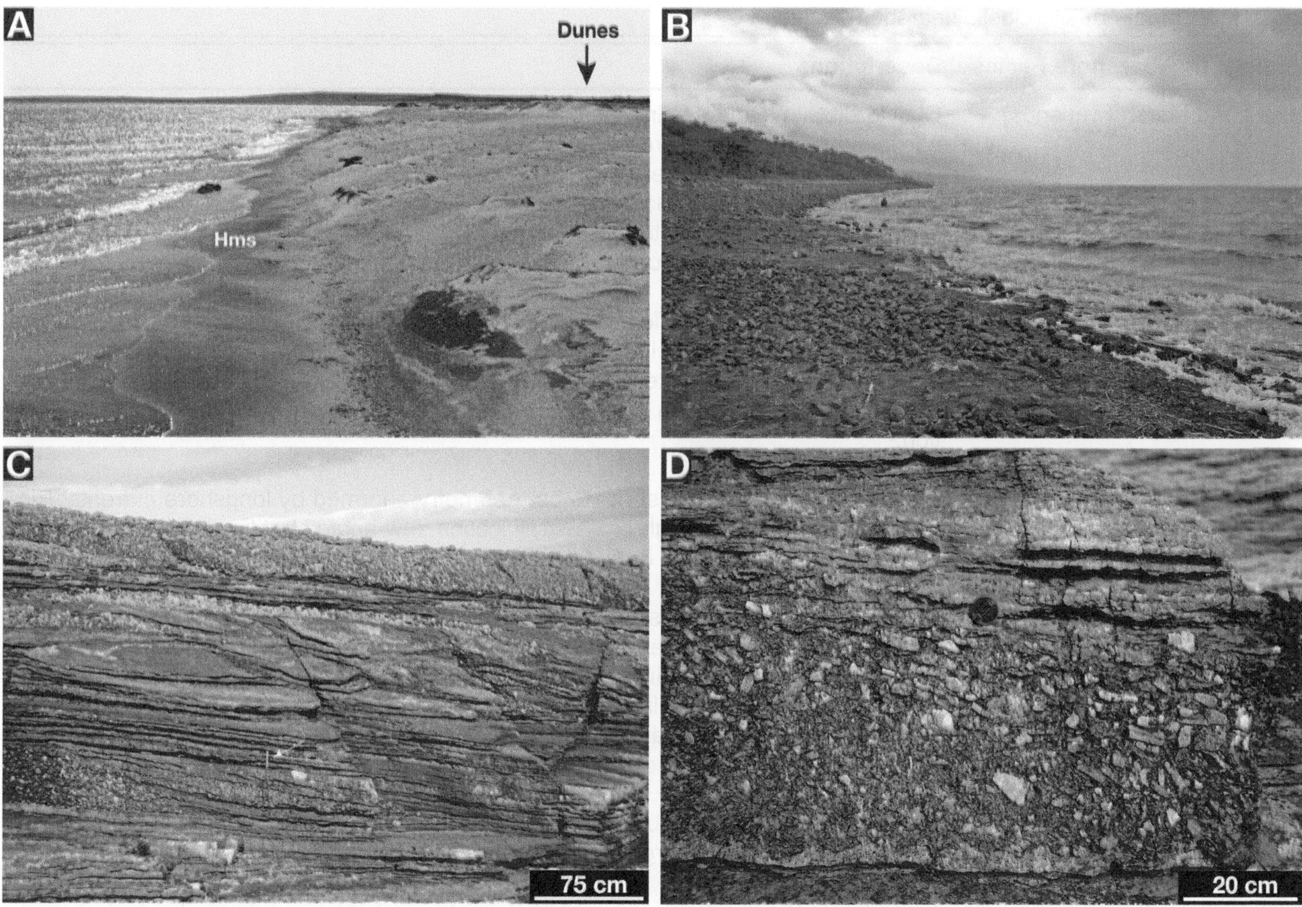

Figure 7. Siliciclastic shoreline sediments. **A.** Modern beach, berm, and eolian dunes at Koobi Fora, Lake Turkana, Kenya. Hms: heavy minerals that produce well defined laminae in the littoral sands. **B.** Modern, poorly sorted, angular beach gravels at Lake Bogoria, Kenya. Although a small lake, strong winds blowing axially along the lake generate waves and longshore currents strong enough to sort coarse gravels. **C, D.** Sandstone **(C)** showing low-angle parallel laminae and beds (lacustrine shoreface deposits) and steeply dipping, imbricated angular conglomerate **(D),** representing a former gravelly lacustrine shoreline. Middle Devonian Orcadian basin, Orkney, Scotland.

marine shelves. Interbedded mudstone and wave-rippled sandstone, some with swaley and hummocky cross-stratification, have been reported from several large paleolake successions.

Carbonate Shorelines

Carbonate shorelines develop in lakes where the inflow waters have been in contact with a significant proportion of carbonate rocks in the drainage basin or in the subsurface. Environments and facies of lacustrine carbonate shorelines vary mainly with slope angle and littoral energy level (Table 2). Lakes with moderately steep margins along their shorelines can develop carbonate bench profiles, whereas those with gentler slopes develop ramp profiles. The sub-environments and facies of benches and ramps, in turn, differ according to littoral energy levels (Figs. 8, and 9A), which vary to a great extent with lake size.

Low-energy carbonate benches are composed mainly of marl (> 50% carbonate mixed with siliciclastic mud) and carbonate sand that prograde lakewards. Much of the carbonate originates from mollusc and ostracode debris, together with carbonate sand and mud precipitated by charophytes, and bio-induced carbonate mud that forms in open water. Charophytes, which are rooted green algae that calcify externally by encrustation and internally as reproductive oogonia and gyrogonites, form dense 'forests' in the photic zone of the lower bench and slope (Figs. 8, and 9B). The shallow proximal bench can be covered by oncoids and calcified microbial mats (stromatolites: Fig. 9C, D, and E) that pass landwards into calcareous peats with rooted macrophytes.

In contrast, high-energy carbonate benches, which are most common in large lakes, are typically sites of ooid formation. Ooid sand (grainstone) forms large benches of crossbedded shoals, in which the prograding ooids periodically slump or form sediment gravity flows into deeper water, or are reworked along shorelines to form beaches and, in arid climates, subaerial eolian dunes. In some freshwater lakes, molluscs are sorted by wave action to form beaches composed of shell debris that can be variably mixed with ooid sand or fine gravel. Reworking of the shells during transgression or regression produces thin beds (cm to dm thick) and lenticular lags of shelly grainstone (coquina).

Gently sloping ramps and low wave energy typify shallow lakes, littoral wetlands, and the margins of some playas. These are sites of carbonate mud deposition (micrite, wackestone) with evidence of recurrent subaerial exposure, such as

Table 1. Summary of siliciclastic nearshore facies

Siliciclastic Nearshore Facies	Descriptions
Ponds and marsh (wetlands)	Clay to silt with planar lamination or completely structureless. Pebbles to sand with planar lamination to ripple cross-lamination sourced from landward areas or washover fans. Massive mud texture from bioturbation to pedoturbation. Invertebrate and plant fossils dispersed or along laminae. Peat, lignite, or coal layers or carbonaceous shales and siltstones. Gley soils to more dry soils. Vertebrate, invertebrate, plant and planktonic fossils. Rhizoliths, root marks commonly abundant.
Backshore	Mudflats of exposed lake floor; sandflats of exposed beaches and shoreface/foreshore sands. Sand dunes from wind reworking of sandy lakeshore sediments. Strandplains of beach ridges from high lake levels composed of mostly sand to silt sediments with low-angle stratification, trough crossbedding, planar lamination, and ripple cross-lamination.
Beaches, spits, bars	Silt to sand to gravel with sheet to bar geometry – formed by longshore currents. Ripple cross-lamination, trough cross-stratification, low-angle stratification, planar lamination or crude stratification. Best developed in large lakes.
Upper shoreface	Many sediment types possible. High-energy deposits of sand and gravel. Sands include ripple cross-lamination, planar lamination, graded bedding, lenticular to flaser bedding, trough crossbedding, and hummocky cross-stratification. Low energy deposits of silt and clay include planar lamination to massive (structureless) textures. Shell coquinas and sandy to gravelly shoals can accumulate. Wider zones develop in large lakes.
Lower shoreface	Same as upper shoreface but more finer grained deposits possible, depending on slope gradient, high vs. low energy lake margin, and size of lake.
Offshore transition	*Large lakes*: Contourites composed of planar laminated to ripple cross-laminated sands from wind-induced bottom currents. Also storm-induced hummocky cross-stratified sands. These sand deposits are interbedded with structureless and laminated clays and muds. *Small lakes*: Transition to offshore sediments is gradual or abrupt depending on slope gradient and types of shoreline sediments.

Key references: Johnson *et al.*, 1980, Johnston *et al.*, 2007; Renaut and Owen, 1991; Deocampo, 2002.

paleosols, calcite or silica rhizoliths from rooted plants, calcrete, and a wide range of early diagenetic features, shallow erosional channels, and microkarstification. Such carbonates are locally stained red and mottled due to iron oxidation. Bioturbation and ichnofossils are common, as are calcified microbial mats (thin stromatolites), charophytes, ostracodes, gastropods, and bivalves. Such low-energy, marshy carbonate environments prone to repeated exposure and early diagenesis are termed *palustrine* environments (Platt and Wright, 1992; Alonso-Zarza and Wright, 2010).

High-energy ramps are localities with oolite formation and stromatolites in nearshore areas, and carbonate muds in deeper water. Ooid shoals are typically planar or lenticular, with ooids reworked into beaches along the shoreline. Bioturbation is common in low energy parts of the ramp and can produce massive mudstone.

Carbonate bioherms, found in many lakes, range from encrusted stable substrates (stromatolites) to large buildups (stromatolitic and thrombolitic; Fig. 9F). Molluscan biostromes form in some high-energy settings (Fig. 9G). Beachrock, with early cementation by calcite or aragonite, is formed along some carbonate and siliciclastic shorelines (Fig. 9H). Many carbonate build-ups and beachrock are associated with sites of groundwater discharge and springs (*see* below).

Deltas, Fan Deltas, Sublacustrine Fans, and Splays

Lacustrine deltas are diverse although most are small compared to their marine counterparts. In some lakes (*e.g.*, groundwater-fed lakes and ponds) deltas are non-existent; in others, deltaic deposits are major contributors to the lacustrine sedimentary record and provide valuable indicators of changing shoreline locations and lake levels. Perennial lakes, both fresh and saline, are most likely to have well-defined deltaic environments and associated facies. Deltas in large lakes (*e.g.*, Lake Tanganyika, Lake Baikal) are as large as many marine deltas.

River currents flowing into lakes deliver siliciclastic and sometimes carbonate sediment as bedload (mainly sand and gravel) and suspended load (mainly silt and clay). Unlike in marine environments, where dilute river water normally flows into

Table 2. Summary of carbonate nearshore facies

Carbonate Nearshore Facies	Descriptions
Beachrock	Both siliciclastic and carbonate layers from earlier sedimentation cemented by calcite and aragonite and exposed with lowering of lake levels. Often related to groundwater inputs at lake margins.
Palustrine marsh	Marl/micrite exhibiting pedoturbation and other early diagenetic features from episodic subaerial exposure, including brecciation, karstification, marmorization, gleying, sparmicritization, rhizoliths, nodularization, and clay illuviation around roots. Peat and coal or evaporites possible, depending on hydrology and climate. Channels filled with rudstone, grainstone, and oncoids.
Shoals/beaches	Ripple cross-lamination to trough and low-angle cross-stratification. Grains can be composed of biogenic clasts of shell material, oncoids, as well as ooids. Also, coquinas of shell material, including gastropods, bivalves, and ostracodes.
Marl/micrite benches	Both low- and high-energy, steep lake margins. Marl and carbonate silt containing charophytes, molluscs, ostracodes, and plant material. Carbonate crusts common on all types of clasts. Also oncoids, ooids, and microbialites. Also grainstone and rudstone. Structureless (massive) to horizontal lamination as well as large-scale trough and low-angle cross-stratification. Bioturbation. Slumping, grain-flow processes, and avalanching produce turbidites offshore from bench.
Marl/micrite ramps	Both low- and high-energy gentle lake margins. Lake level changes can produce palustrine marshes. Shoals/beaches with wave-influenced shorelines. Structureless (massive) fine-grained carbonates to carbonate silts and sands with shell material debris and coquinas (molluscs, ostracodes), ooids, oncoids, charophytes, intraclasts, and plant material. Macrophytes can be coated with carbonates. Microbialites. Intraclastic rudstone (gravels) to grainstone from palustrine marshes. Carbonate crusts on many grain sizes.
Microbialites/bioherms	Microbial buildups from algae and bacteria, often related to spring and seep inputs, up to decameters in scale. Diverse morphologies include hemispherical to sub-hemispherical mounds, domes, bushes, columns, flat-laminated structures, and crusts.

Key references: Platt and Wright, 1991, 1992; Gierlowski-Kordesch *et al.*, 1991; Bertrand-Sarfati *et al.*, 1994; Casanova, 1994.

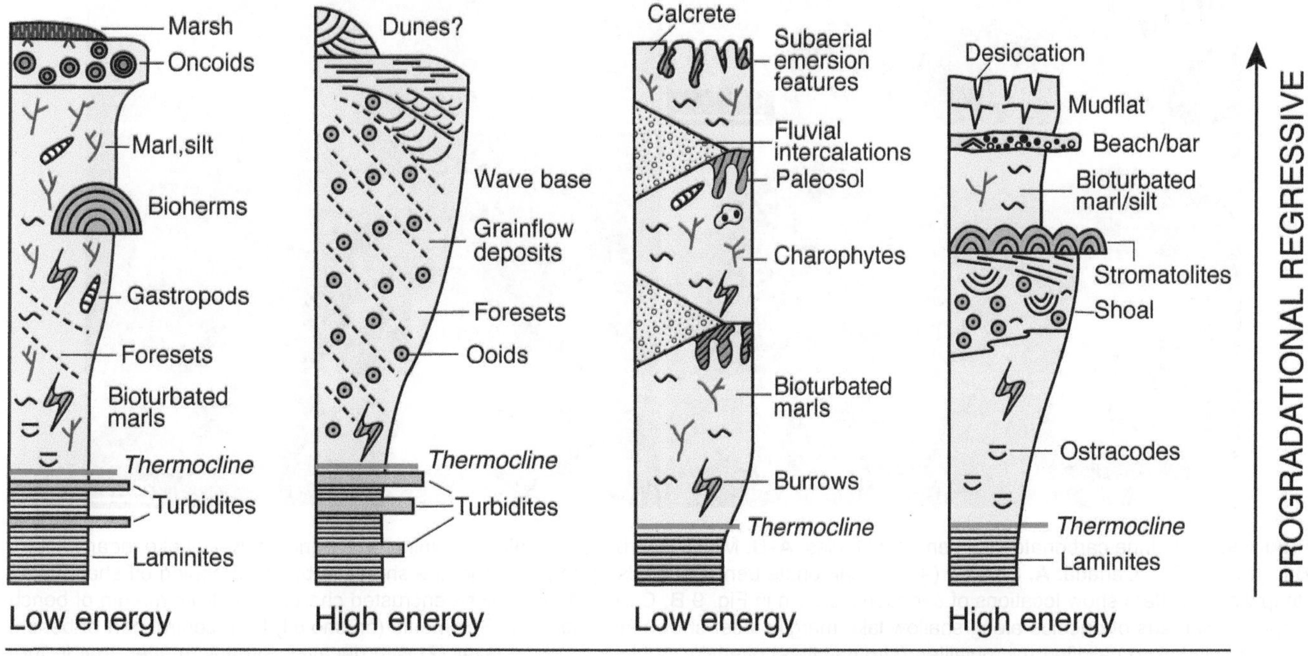

Figure 8. General vertical sequence of carbonates and associated facies that accumulate in marginal lacustrine ramp and bench settings (after Platt and Wright, 1991). (Reproduced with permission by Wiley-Blackwell.)

Figure 9. Lacustrine carbonate sediments and facies. **A–D.** Modern carbonates (calcite) forming on a bench at Kelly Lake, near Clinton, British Columbia, Canada. **A.** Shallow (< 1 m) carbonate bench extends up to 25 m from the shoreline before dropping off sharply into deep water. Letters show locations of sediments shown in Fig. 9 **B**, **C**, and **D**. **B.** Calcite encrusted charophytes from margin of bench slope. **C.** Spreads of oncoids along shallow lake margin; inset shows lamination. **D.** Sediments (on shovel) from central part of bench. (1) Laterally linked calcified microbialites (stromatolites) overlying a black reduced layer (2) with decaying organic matter, that in turn overlies charophyte sand and mud (3). **E.** Platform of very large (10–50 cm diameter) oncoids, early Holocene Galana Boi Formation, Lake Turkana, Kenya. **F.** Large Pleistocene stromatolite mound, Olorgesailie, Kenya. **G.** Biostrome of cemented oysters (*Etheria elliptica*), early Holocene Galana Boi Formation, Lake Turkana, Kenya. Hammer for scale. **H.** Recent lacustrine beachrock (calcite-cemented beach sands), Ileret, northeastern Lake Turkana, Kenya.

saline water, lake waters range from fresh to hypersaline. The style of sedimentation and resultant facies depend not only on the density of inflow (stream water plus its sediment load) but also on the density of the lake water (salinity plus suspended load; Fig. 10). The latter can vary through time, so the style of deltaic sedimentation and the associated facies relationships can vary accordingly (Renaut and Tiercelin, 1994; Table 3).

Where river and lake waters have similar densities, deceleration of inflow and fluid mixing near the river mouth results in deposition of the coarsest sediment near the shoreline (homopycnal flow; Fig. 10A). In shallow lakes, sand and gravel bars form that radiate from the river mouth. If the water at the proximal delta front is moderately deep, the bars form lakeward-dipping foresets. In *Gilbert-type deltas*, which are common in glacial lakes (Fig. 11A) and lakes in extensional basins, steep foreset beds prograde lakeward in sets up to tens of meters thick (Fig. 11B, C). Topset and bottomset beds sometimes are preserved. Steep foresets can become unstable and avalanche downslope as sediment gravity flows. Foresets in deltas with sandy bedload have gentler slopes and are more prone to reworking by waves and currents.

When the density of the inflow is less than that of the lake water, suspended silt- and clay-sized particles are transported as buoyant plumes far into the lake (hypopycnal flow; Fig. 10B). Some clay particles flocculate and accumulate near the delta front, whereas silt- to clay-sized particles can also settle in deeper water. Clay mineralogy, salinity, and organic matter influence the amount of flocculation. In lakes experiencing hypopycnal flow, any coarse bedload generally is deposited near the river mouth forming river-mouth bars and linear sand bodies. Progradation is enhanced under these conditions.

When rivers that have a high sediment load enter freshwater lakes, or cold streams (*e.g.*, from montane glaciers) flow into warm water lakes, underflows can occur that set up sediment-laden density currents (hyperpycnal flow; Fig. 10C). In

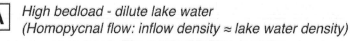

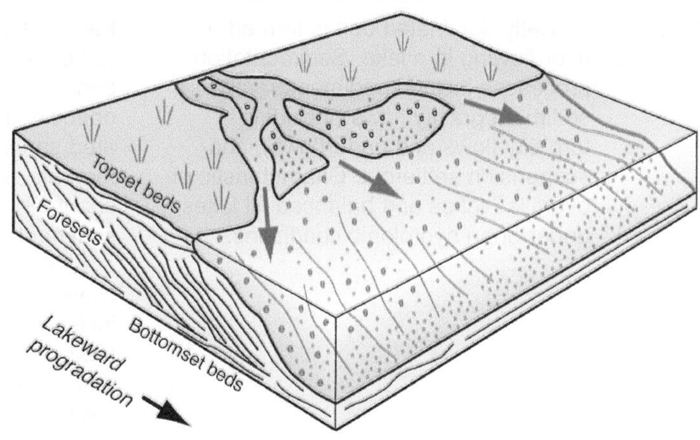

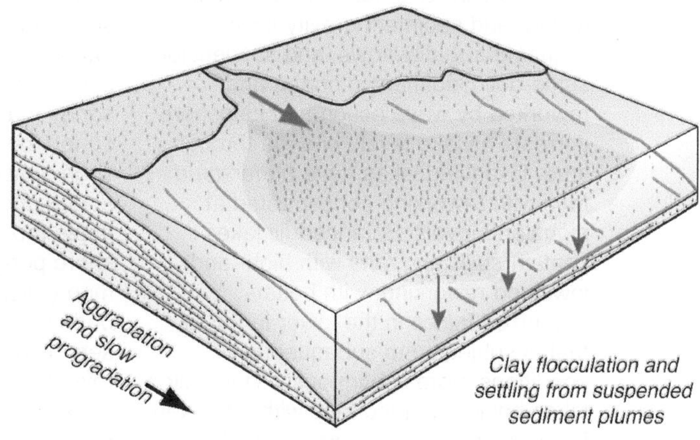

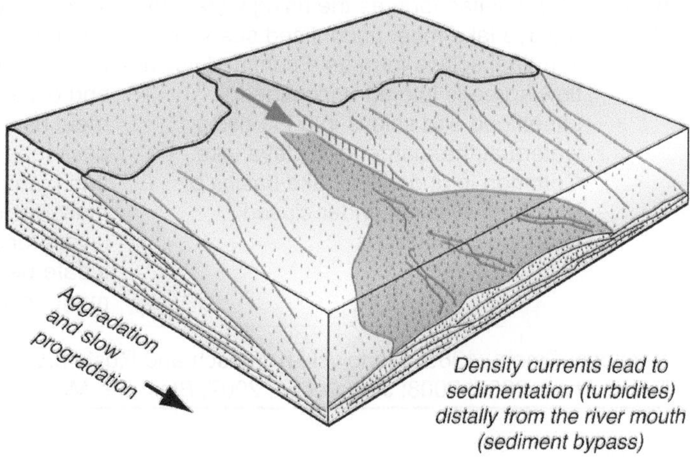

Figure 10. Styles of deltaic sedimentation in lakes that depend on the relative densities of the inflow and lake water. For explanation, see text.

Table 3. Summary of deltaic facies

Lake Deltas	Form and Interpretation	Facies
River delta	Fluvially dominated deltas formed by a river flowing in a lake. Sedimentation processes include sediment traction, sediment gravity flows (including turbidity flows and underflows), and suspension settle-out. Characteristic of overfilled lakes and balanced-fill lakes under open hydrologic conditions.	Facies resemble Mississippi delta facies. Fine-grained prodelta or bottomset deposits composed of suspension settle-out and beds of turbidites or underflows. Grading upward, delta front deposits of graded, parallel laminated, to ripple cross-laminated sand/silt/clay sediments or large-scale steeply dipping foresets. Very deep lakes can develop complex delta-front channels and levees with overbank settle-out areas with high sediment input from underflows or turbidites. Proximally, gravels are mixed with sands while distally sands, silts, and clays are present with typical turbiditic features. Nearshore incised canyons in rift lakes accumulate sands and gravels to feed delta channels. Topset beds comprise mouth bar deposits that comprise flaser to wavy bedding, scour-and-fill structures, mud lenses, and sand exhibiting trough cross-stratification and ripple cross-lamination. Other fluvial environments and associated sediments include floodplain lakes, marshes, overbank areas, levees, splays, and interdistributary troughs.
Fan delta	Subaqueous portions of alluvial fans extending into lakes. Sedimentation processes include debris flows, grain rolling, and sediment gravity flows, including turbidity flows, and suspension settle-out. Sediment is sorted and reworked by wave action and longshore currents as lake levels change through time. Overfilled lakes have *bufferstand fans* with reworking concentrated at highstand levels. Balanced-fill lakes contain *spillstand fans* with high- and lowstand reworking, with better-organized facies in the lower fan area that is subjected to subaqueous conditions more continuously. Underfilled lakes have *freestand* fans with reworked levels limited to lower areas as lake levels fluctuate under closed hydrologic conditions.	Matrix- and clast-supported conglomerate and rudstone (boulder to pebble grain sizes) that are coarsely to horizontally stratified or inversely graded are modified into (1) poorly sorted lags in high energy beaches, (2) moderately sorted, low-angle beds dipping lakeward of a beachface, (3) poorly sorted, low-angle to irregular bedding dipping lakeward with sandy matrix of a gravelly upper shoreface, and (4) poorly sorted low-angle beds dipping landward of a proximal backshore washover zone. Finer sediment is re-transported and sorted into (1) sandy lower shoreface with massive texture, ripple cross-lamination, or irregularly bedding, (2) distal backshore pond washover zone with poorly sorted pebbly sand and mud (siliciclastic and carbonate) in horizontal or low-angle beds dipping landward, (3) sandy to cobbly, large-scale foresets dipping lakeward, related to spit development on reworked terraces, and (4) poorly sorted pebbly sand in topset beds, identified as a spit-front apron. Bottomset beds are generally horizontally laminated clays, silts, and muds (carbonate and siliciclastic) with underflow/turbidite sands, containing slump structures and dewatering features. Topset beds are the subaerial alluvial fan facies.
Sheet delta/ terminal splay	Sheet deltas form as the rising waters of a playa lake meet unconfined sheetfloods. Terminal splays form as a confined fluvial flow spreads into an unconfined sheetflood and splays into the rising waters at the margin of a playa lake. Common in underfilled lake basins or balanced-fill lakes under closed hydrologic conditions.	When sand distributary channels with lag deposits of gravels and mud clasts reach base level flow becomes an unconfined sheetflood with deposition on playa lake floor as proximal and distal splay deposits. When large unconfined sheetfloods meet rising lake waters, sheet deltas develop. Decimeter-scale packages of sandstone/siltstone/mudrock exhibiting trough and planar cross-stratification, ripple cross-lamination, and massive (structureless) textures as well as normal graded layers form proximally from decelerating flows forming sheet to lobate bed geometry. Distally, structureless fine-grained sediment accumulates from suspension-settle out.

Key references: Smoot, 1985; Gierlowski-Kordesch and Rust, 1994; Back *et al.*, 1998; Soreghan *et al.*, 1999; Wells *et al.*, 1999; Fisher *et al.*, 2007; 2008; Jiang *et al.*, 2007; Blair and McPherson, 2008.

Figure 11. Deltaic sedimentation in lakes. **A.** Large coarse-grained Gilbert-type delta at Peyto Lake, Alberta, Canada fed by glacial outwash. Underflow occurs offshore in deeper water. **B.** Small Gilbert-type delta sandstone lobes (D) interbedded with lacustrine shales (S). Towards the top of the photograph the interbedding is most closely spaced (I), reflecting shallowing of the water. Miocene Tambach Formation, Kerio Valley, Kenya. Hammer for scale. **C.** Pleistocene lake marginal deposits, Searles Lake, California, USA. Gilbert-type foreset beds (gf) overlain by flaser-bedded sandstone and siltstone (fl), wave-rippled sandstone (wr), and transgressive laminated siltstone (ls). **D.** Low-density debris-flow breccia (df) showing imbricated clasts (ic) in laminated diatomaceous silts (d). Miocene, Mytilini basin, Samos, Greece. **E.** Thin upward-fining turbidite sandstone (t) in lacustrine siltstone, Miocene, Mytilini basin, Samos, Greece. **F.** Small (110 m radius) lacustrine fan-delta, Nasikie Engida, near Lake Magadi, Kenya. Efflorescent salts (trona) form a white rim along the margin of the fan delta.

stratified lakes, interflow also occurs at density boundaries in the water column (Sturm and Matter, 1978). If sediment load is high, channeling can take place on the delta front, and much of the sediment is then swept downslope into deeper water, sometimes almost continuously, blanketing the prodelta with clay and silt-sized particles. By this mechanism, lamination can form that reflects regular (*e.g.*, seasonal) or irregular fluvial floods. High-density turbidity flows, grain flows, and debris flows (Fig. 11D) occur from rapid increases in sediment delivery and oversteepening on the delta front. Small, graded, cm-scale turbidite sequences of predominantly upward-fining sand and mud can potentially accumulate in deeper water (Fig. 11E); most have partial Bouma sequences, including rippled cross-laminae and planar laminae. Rates of delta progradation can be slow in lakes dominated by underflows.

Fan deltas form at sites where alluvial fans discharge water and sediment directly into lakes (Fig. 11F). In shallow lakes, gradual progradation occurs as fan-derived sediments are deposited and spread laterally across the lake floor. In deep lakes, however, subaerial fan channels mostly serve as feeder conduits for subaqueous sediment transport by underflow into deeper water, thereby limiting fan delta progradation. Similar processes occur offshore from prograding deltas where there is an abrupt steepening of the lake-floor topography (*e.g.*, a fault at the margin of a submerged delta platform).

Sublacustrine fans, best known from deep rift lakes, are analogous to their marine counterparts. Incised channel networks up to tens of meters deep occur in their proximal parts and are usually filled by coarse siliciclastic or carbonate sands and gravels. Channel fills in deeper water include normally graded turbidites. Marginal levees defined by laminated and massive mud can form. Distal interchannel areas contain overbank deposits including finer grained turbidites and hemipelagic deposits, exhibiting progressive fining away from the fan.

Terminal splays and sheet deltas form where streams become unconfined or where unconfined flows reach lake base level at the shore of shallow playa lakes, spreading out lobes and sheets of sheetflood deposits. The sediments forming these splays and sheets, and their geometry, depend on the catchment area, vegetation patterns, discharge, grain size of transported sediment, and lake-level changes during flooding events. Proximal facies include planar and trough crossbedded, ripple cross-laminated, and massive to graded sandstone and muddy siltstone, while distal facies comprise thin beds of structureless claystone, and siliciclastic or carbonate siltstone and sandstone, interpreted as deposits settling from suspension in a decelerating flow.

Lake Center and Pelagic Environments

The 'offshore' zone in perennial lakes lies beyond the direct influence of shoreline processes, deltas and sublacustrine fans, and is characterized by predominantly fine-grained, relatively slow, sedimentation mainly from suspension (pelagic and hemipelagic deposits) or thin rippled silts and muds from bedload transport. The offshore sediments come from several sources (Figs. 12, and 13), some allochthonous (detrital) and others essentially autochthonous. Detrital, siliciclastic and carbonate silt particles and clay derived from fluvial inflow and littoral erosion are transported offshore by currents. These sediments are supplemented by eolian grains and pollen, and in some regions, ice-rafted particles or tephra (airfall tuffs and pumice), which act as stratigraphic marker horizons in some lacustrine successions (Fig. 13A). Episodic sediment-gravity flows can also deliver coarse sediment into deep water and the resultant deposits will become intercalated with the predominant basinal muds.

Much of the open-water sediment in the central parts of perennial lakes is biologic in origin. These deposits include both mineralized skeletal remains (microfossils) and organic (carbonaceous and fecal) debris from the plankton, which can be supplemented by organic matter from macrophytes (*e.g.*, reworked plant debris) and the mineralized and organic debris of nektonic (*e.g.*, fish) and benthonic organisms. In shallow lakes, the entire lake floor lies within the photic zone. Rooted plants can then cover much of the lake floor, inhibiting laminae formation. Pelagic oozes in lakes can be mainly silica, mainly carbonate, or a mixture of both in composition, and range in purity according to their degree of mixing with detrital lithic and mineral grains. Planktonic oozes are common in central parts of some large deep lakes, but also accumulate in shallow lakes and in protected areas that receive little detrital sediment influx.

Diatoms, which are algae that secrete skeletal frustules composed of opaline silica (opal-A: $SiO_2{\cdot}nH_2O$), can be planktonic, benthic, or epiphytic, and are found in Neogene to modern lake sediments from the littoral zone into the deepest water, and in waters ranging from acidic (pH < 3.5) to alkaline (up to ~ pH 9.5). Where diatoms form > 90% of the sediment mass, the rock is termed diatomite (Owen, 2002). Diatom frustules can be mixed with silt, clay, tephra, and other detrital grains in variable proportions, forming diatomaceous silts and clays. After burial and diagenesis, the diatom-rich sediments alter to opal-CT and quartzose chert through dissolution−reprecipitation reactions that eventually destroy most of the diatoms (Fig. 13B). Lacustrine diatomite deposition has been significant only since the Miocene when grasslands began to contribute silica into lake systems (Kidder and Gierlowski-Kordesch, 2005).

The fine-grained carbonates that accumulate in offshore sediments have several origins. Some fine-grained carbonates are skeletal (*e.g.*, from ostracodes and reworked mollusc and charophyte debris), some are detrital grains from the catchment (up to ~50%), and the rest of the pelagic carbonate comes from bio-induced calcite and aragonite precipitation. Carbonates may or may not reach or be preserved in deep offshore areas, depending on the level of anoxia determined by stratification properties of the lake (permanent *vs.* episodic), the temperature of bottom waters, and the amount of organic matter available for degradation and consequent CO_2 production. The primary carbonate mineralogy is determined by many factors including water temperature, prevailing Mg/Ca ratio, dissolved organic-matter content, level of carbonate supersaturation, and other controls. Although calcite mud predominates, aragonite mud can form in waters with high (> 1) Mg/Ca ratios or high temperatures. Hydromagnesite ($Mg_5(CO_3)_4(OH)_2{\cdot}4H_2O$), magnesite ($MgCO_3$) and dolomite ($CaMg(CO_3)_2$) mud forms in Mg-enriched lake waters, including examples in British Columbia, Canada. Metastable ikaite ($CaCO_3{\cdot}6H_2O$) mud sometimes forms in cold (< 10°C) waters. Early diage-

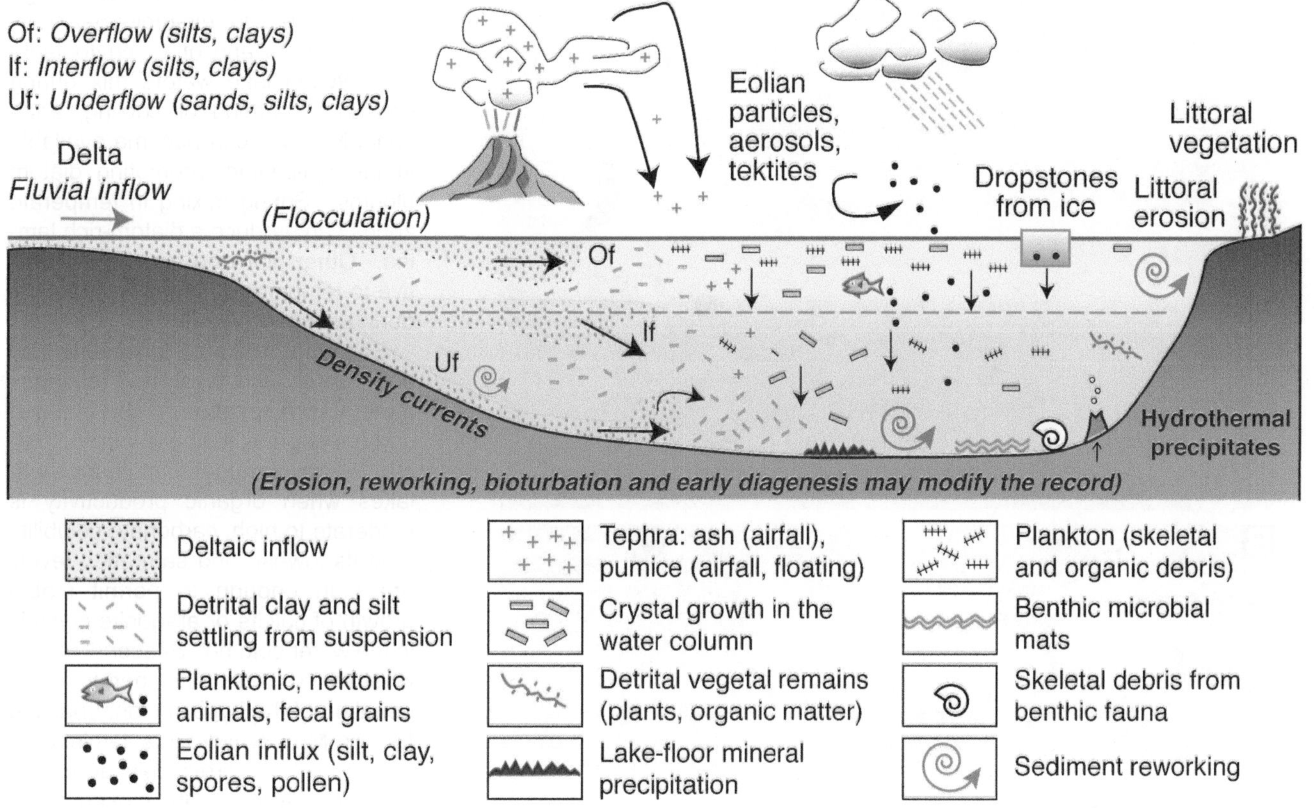

Figure 12. Potential sources of sediment in the offshore (open water) zone of lakes. This is generalized and is not intended to represent any particular type of lake.

nesis affects the metastable carbonate phases, so that calcite and dolomite are the most commonly preserved lacustrine carbonates in the sedimentary record. Some dolomite precipitation in lake sediments has been linked to bacterial mediation (*e.g.*, Bréhéret et al, 2008; Sanz-Montero *et al.*, 2008).

Chemical precipitates dominate the offshore zone in some perennial lakes. These include evaporites in saline lakes (*discussed below*), Fe- and Mn-oxyhydroxides and phosphates in lakes subject to changes in Eh, and several authigenic minerals that precipitate on the lake floor or form during early diagenesis (Fig. 13C).

Open-water Sedimentary Facies

Open-water lacustrine sediments from the central parts of lakes have long been a target for paleolimnologic studies because their slow sedimentation rates offer good potential for preserving long, high-resolution stratigraphic and paleoclimatic records. Many factors, however, control the offshore facies (Fig. 12). Some open-water sediments are laminated or varved, whereas others appear structureless in both core and outcrop.

Fine-grained, open-water lake sediments are commonly laminated (Figs. 14, and 15; Table 4). In many lacustrine successions, millimeter to sub-millimeter scale laminae occur in repeating cycles of two (couplets), three (triplets) or more laminae. Stacks of these repeating units, termed rhythmites, can be tens of meters thick. Some couplets and triplets are true ‘varves’ when each repeating unit represents an annual cycle of sedimentation, but caution is needed because field experiments with sediment traps have shown that this is not always the case.

Individual laminae are highly variable in composition. They can be detrital fine sand, silt, and clay, precipitated or detrital carbonates, other minerals, organic matter, microfossils, and in saline perennial lakes, fine cumulate evaporite minerals. Many variations in the stacking patterns of the laminae that form one rhythmic unit are found. For example, detrital silt can alternate with detrital clay (Fig. 15A); carbonate with clay or organic matter (Fig. 15B, C); diatom frustules with clay; evaporites with carbonates, and many other possible combinations (Fig. 14). The patterns record both seasonal (in true annual varves) and irregular changes in lake sedimentation that occurred repeatedly during that period of lake history.

Many laminae couplets and triplets record seasonal sedimentation. Seasonal variations in runoff and floods can produce alternations of relatively coarser and finer silt and clay. In cool climates, increased discharge can reflect snow and ice melt during spring: the coarser lamina forms during the spring runoff, whereas finer organic mud settles slowly during winter when the lake and drainage basin are variably ice covered. High-magnitude floods lead to rare beds or intercalations of turbidites and other coarser sediments in finely laminated sequences (Fig. 15B).

Some laminae are controlled primarily by seasonal changes in water

Figure 13. Offshore sediment facies. **A.** Bedded silt, sand and clay. Late glacial outwash deposits of Kamloops Lake, British Columbia, Canada. **B.** Laminated diatomaceous siltstones and diatomites of Miocene age, Mytilini basin, Samos, Greece. The original diatom opal-A has partly altered to opal-CT. **C.** Organic-rich lacustrine shales (> 5% TOC: total organic matter) deposited in a meromictic alkaline lake that have been altered to zeolites by saline, alkaline pore fluids. Miocene Ngorora Formation, Tugen Hills, Kenya Rift Valley.

temperature and stratification. High nutrient levels, for example, generally favor diatoms and other phytoplankton. When lake waters mix during seasonal warming or cooling, additional N and P can become available in the epilimnion, promoting diatom blooms. Spring mixing in temperate zones can produce a diatom-rich lamina. Other plankton are more sensitive to temperature and blooms occur during summer warming.

Seasonal changes in stratification patterns or lake level also influence mineral precipitation. Bio-induced carbonate precipitation is favored during warm summers in temperate lakes when organic productivity is moderate to high, carbonate solubility is at its lowest, and saturation levels are high enough to permit stable growth of calcite or aragonite crystals in the water column. Seasonal mixing in stratified lakes also permits oxygenation of the hypolimnion, enabling Fe- or Mn-oxyhydroxide laminae to form rapidly in previously anoxic or dysoxic waters. In closed-basin lakes, carbonate and evaporite laminae can reflect seasonal changes in salinity with the more soluble minerals forming during periods of increased evaporation and reduced water volume.

Tropical and subtropical lakes can develop annual laminae even though air and water temperatures are fairly uniform throughout the year. Monsoons, for example, can affect runoff and the wind direction and strength. The latter can affect the plankton distribution and lower the depth of mixing in the epilimnion. This, in turn, can modify the patterns of sedimentation.

Not all open-water sediments are laminated (Table 4). In some large lakes, uniform texture and composition can make laminae difficult to see (although some are revealed by X-ray imaging). However, non-laminated sediments in lake centers are also attributed to current-induced erosion, re-suspension and mixing, and to bioturbation at times when the bottom waters are oxygenated (Glenn and Kelts, 1991; Lamoureux, 1999; Gruszka, 2007).

Lacustrine Organic Sediments and Oil Shales

Sediments rich in organic matter form

EXAMPLES OF GLACIAL AND NON-GLACIAL ANNUAL VARVES

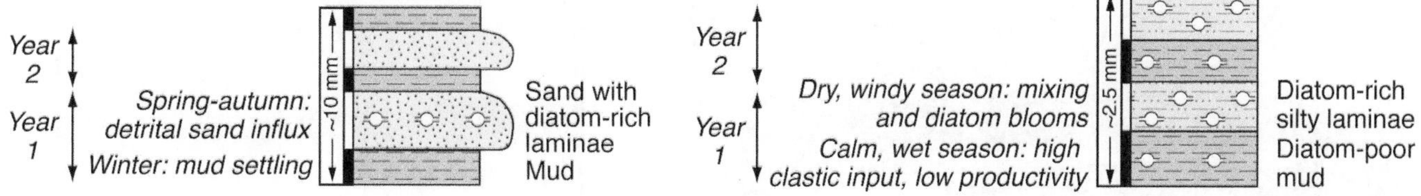

EXAMPLES OF LAMINAE LINKED TO BREAKDOWN OF STRATIFICATION AND OVERTURN

Temperate hardwater lakes: Non-glacial varves (Lake Zürich)

Fe-rich northern temperate lakes (Saskatchewan, Minnesota)

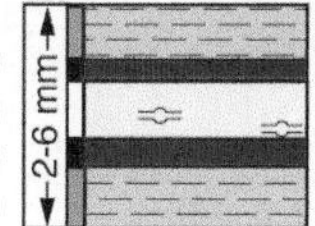

EXAMPLES OF LAMINAE LINKED TO CHEMICAL-BIOCHEMICAL PROCESSES

Deep stratified saline lake (e.g., Dead Sea in 1970s)

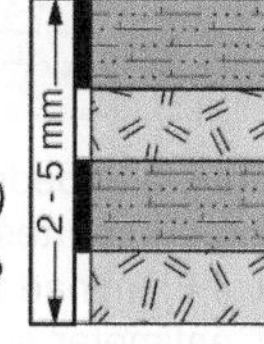

Organic-rich laminites (e.g., Orcadian Basin, Devonian)

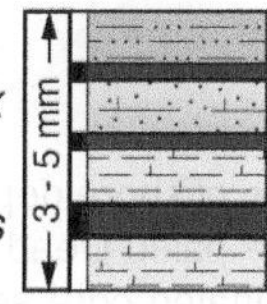

EXAMPLES OF IRREGULAR-EVENT LAMINAE (NON-RHYTHMITES)

Dust storm (waterlain loess) — Silt
Toxic event that kills biota — Phosphatic (fish-bone) lamina
Diatom bloom stimulated by Si-rich ash fall — Diatomite lamina
Airfall volcanic ash into lake — Tuff
Distal turbidite — Graded mud lamina
Scale variable

Light laminae
Intermediate laminae
Dark laminae

Figure 14. Examples of different types of lamination in lake sediments and their diverse origins. Expanded and adapted from Talbot and Allen (1996).

in both shallow lake-marginal and offshore environments. In very shallow (< ~2 m deep) lakes and along the swampy margins of some large lakes, organic-rich lacustrine peat forms mainly from *in situ* littoral macrophytes where the waters and sediments become oxygen-poor during and after burial (Figs. 16A, and 17A). Thin coal seams and lignite deposits, some with rooted plant remains, are then preserved in lacustrine successions. Some lignite originates from detrital vegetation washed into the lake (Fig. 17B).

In open-water settings, organic-rich mud (sapropel) that accumulates on the lake floor is the precursor to carbonaceous shale and oil shale (Fig. 16B). Organic-rich shale and kerogenite, recognized by their biomarkers, preserve a record of productivity. Favorable conditions for organic matter accumulation in lake sediments are where i) there are moderate to high rates of organic (plankton) productivity, ii) the organic

Figure 15. Laminae in lacustrine offshore deposits. **A.** Lacustrine sequence of alternating laminae and beds of silty sandstone and detrital claystone from the Holocene paleo-Dead Sea deposits (Ze'elim site), Dead Sea basin, Israel (courtesy of Yuval Bartov and Enzel Yehouda). **B.** Rhythmically laminated core section from Lake Zug, Switzerland, comprising light laminae rich in carbonate, alternating with dark clay laminae punctuated by thin turbiditic layers of fine sand and organic debris. Centimeter scale at left (courtesy of Kerry Kelts). **C.** Lamination in the third black shale of the Jurassic East Berlin Formation, Hartford basin, Connecticut, USA. The white laminae are composed of carbonates and the dark laminae are clays. Disrupted carbonate lamina indicates early diagenetic cementation and subsequent brecciation and transport of carbonate clasts.

matter content is undiluted by detrital siliciclastic and carbonate sediments, and iii) the organic matter is not rapidly destroyed by microbial decay processes after deposition. Both allochthonous (allogenic) and autochthonous (autogenic) organic matter contribute to the sediments, but the latter usually generates more oil after burial and maturation (Figs. 16C and 17C). The relative proportions of autogenic, microbial organic matter (mainly phytoplankton) and allogenic organic matter from the drainage basin (leaves, stems, pollen, spores) determine the potential quality of any future hydrocarbon deposits (Smith and Gibling, 1987; Talbot, 1988). Lacustrine organic-rich mud can form in lakes that are well oxygenated provided that the accumulation rates of organic matter exceed the rates of its dilution by clastic and carbonate sediments and decay by microbes. Stratified lakes, especially alkaline meromictic lakes, are considered particularly favorable types (Figs. 13C, 16B, and 17C). The lack of oxygen in the monimolimnion or hypolimnion eliminates benthic organisms, reduces bacterial decay, and enables laminated organic-rich mud to form that after burial and maturation becomes oil shale. Some organic-rich laminites, however, have been attributed to periodic growth of benthic microbial mats in shallow saline water and did not require deep anoxic water for their preservation.

Saline Lakes

Setting and Hydrology

For a lake to become saline, annual water loss by evaporation should exceed the water received annually from runoff, groundwater, and atmospheric precipitation under mostly closed basin conditions. Many saline lakes are present in dry semi-arid to arid environments, where annual evaporation is high, but as emphasized by Bohacs *et al.* (2000), the relationships among lake water salinity, annual precipitation, and annual evaporation are complex. Lake morphometry (*i.e.,* depth, surface area), the volume of annual recharge from runoff and groundwater, the presence or absence of water loss by lake-floor seepage, and the salinity of the inflow waters, also control or influence lake water salinity. Saline lakes and dilute lakes can co-exist in the same region. For example, Lake Baringo (dilute with < 2 g/l total dissolved salts) and Lake Bogoria (saline with > 60 g/l total

Table 4. Summary of open water - offshore facies

Open Water/Offshore Facies		**Descriptions**
Shallow	Massive (structureless)	Siliciclastic, diatomaceous or carbonate muds with or without dispersed fossils (*e.g.,* gastropods, bivalves, ostracodes, diatoms, pollen, *etc.*); bioturbation features common.
	Laminated	Fine to thicker lamination of siliciclastic, diatomaceous or carbonate muds and silts with sand layers and laminae to pebble- to gravel-sized grains. Saline lakes and playas contain laminae of evaporites, including gypsum, halite, ulexite, trona, gaylussite, *etc.* Preserved lamination of fine-grained sediments can occur when lake is meromictic, salinity is high, or microbial mats stabilize the substrate; invertebrate fossils are rare.
	Transported	Coquinas (grainstone, packstone) of invertebrate shells. Gravelly to sandy to oolitic shoals, with or without dispersed invertebrate shells, containing trough cross-stratification to ripple cross-lamination and structureless bedding and flat lamination. Traction-load sediments are associated with longshore currents or river inflow.
Deep	Massive (structureless)	Siliciclastic, diatomaceous or carbonate muds (thick laminae to thick bedding) with or without dispersed fossils (*e.g.,* gastropods, bivalves, ostracodes, diatoms, pollen, *etc.*); bioturbation features common.
	Laminated	Fine to thicker lamination of siliciclastic, diatomaceous or carbonate muds and silts with sand to grainstone laminae to layers. Laminae can include ostracodes, diatoms, and other small invertebrates, as well as plant remains, carbonate and chert clasts, rare pebble- to gravel-sized grains, and organic detritus. 'Varves' refers to regular lamination that is deposited on an annual basis within a glacial lake. 'Non-glacial varves' refers to similar lamination in a non-glacial lake. 'Rhythmites' have a regular lamination but an unknown timeframe for deposition. 'Laminites' simply describes regular to non-regular lamination. Laminae and high organic-matter preservation often favored by meromixis.
	Transported	Underflows and turbidites composed of siliciclastic, carbonate, or both kinds of grains as well as shell debris. Sedimentary structures include grading, ripple cross-lamination, flat lamination, massive textures, and slumping structures.

Key references: Anadón *et al.*, 1998; 2000; Anderson and Dean, 1988; Glenn and Kelts, 1991; Owen, 2002.

dissolved salts), which lie only 22 km apart in the Kenya Rift Valley, share almost identical annual precipitation and potential evaporation, but differ in hydrology, morphometry, and sedimentation. Saline lakes usually occupy topographically closed basins of interior drainage with no surface water outflow. Many are also hydrologically closed or 'endoreic' (Fig. 5), but some saline lakes have partially open drainage through lake-floor seepage, which controls both the salinity and facies distribution (Table 5).

Saline lakes can be perennial, maintaining surface water throughout the year, or ephemeral, drying up annually or every few years. Ephemeral saline lakes are termed playas, playa lakes, saline pans, salt pans, or one of many regional geographic names (salar, chott, *etc.*). Use of the terms 'sabkha' and 'salina', which typically imply a coastal marine influence or marine recharge, is not recommended for continental saline lakes.

Saline lakes are present in several physical and tectonic settings. The classic models, developed in the 1970s, are based on the block-faulted interior basins of the western United States (Fig. 18A) and the rift basins of East Africa (*e.g.*, Eugster and Hardie, 1978). There are many other types including, for example, the coastal and inland lakes of Australia in regions of low topography,

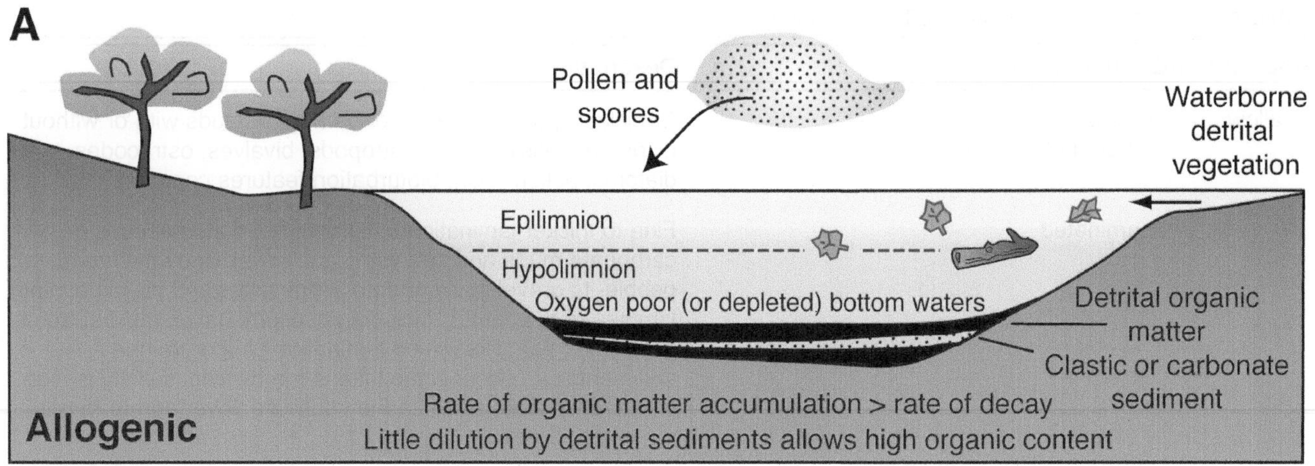

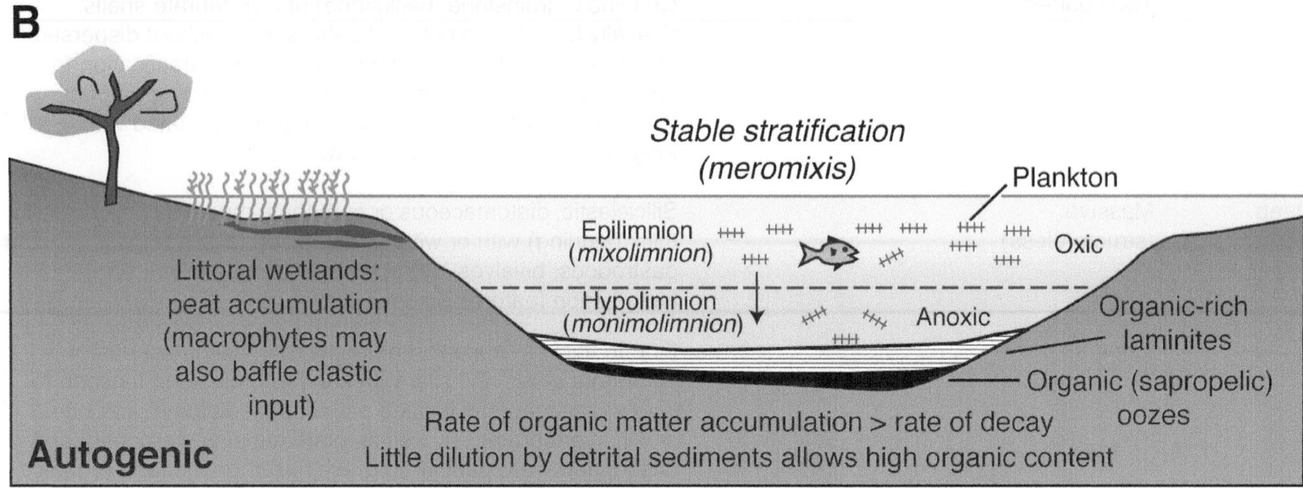

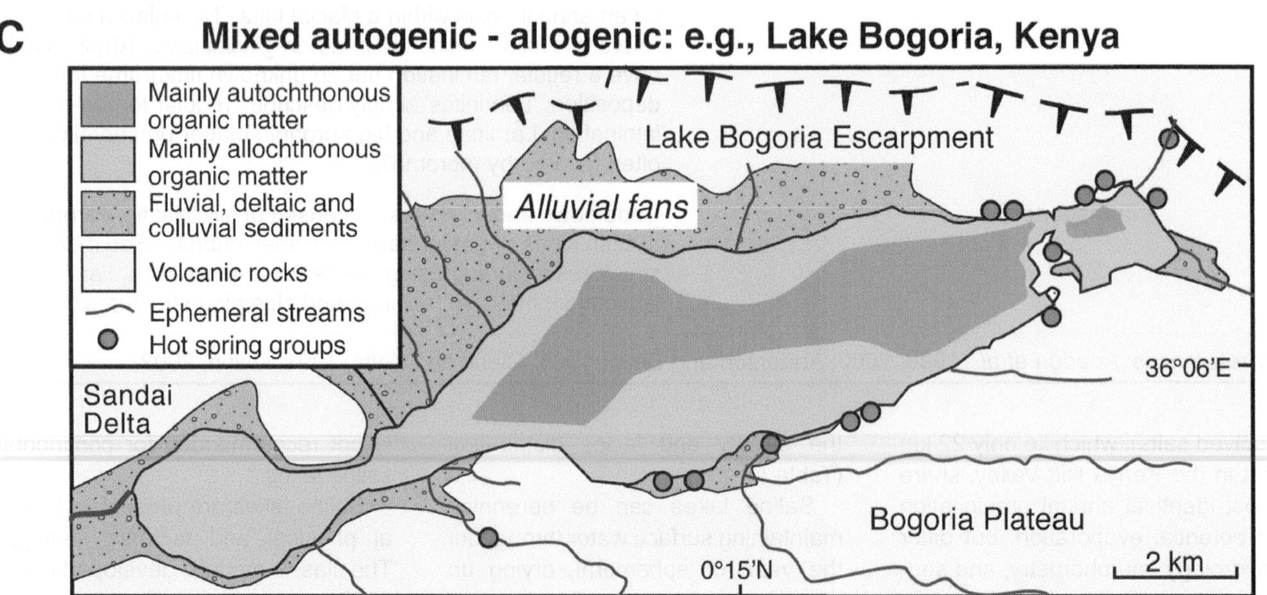

Figure 16. Organic sedimentation in lakes. **A.** Controls of allogenic sedimentation in lakes. **B.** Controls of autogenic sedimentation in lakes. **C.** Distribution of organic matter types in Lake Bogoria, Kenya. Allogenic organic matter (plant and soil detritus) from catchment vegetation dominates the lake margin; autogenic microbial organic matter from plankton accumulates in the central part of the lake (*see* Fig. 17C).

and the several hundred thousand, small saline lakes in glaciated terrain of the Canadian Prairies and in Interior British Columbia, many of which occupy glacial paleomeltwater channels (Last and Ginn, 2005).

The hydrology of saline lake basins is quite variable and is a major control on the mineralogy and facies development in continental closed basins (Fig. 18B). Some ephemeral lakes are fed primarily by surface runoff (streams, sheetfloods, atmospheric precipitation), whereas others rely mainly on groundwater discharge (including springs) or a combination of runoff and groundwater. To precipitate thick lacustrine evaporites, however, some groundwater discharge is normally required and is typically the dominant form of lake water recharge in regions of high aridity and low topography.

Rosen (1994) classified ephemeral lakes into three main groups based on their hydrology (Fig. 19). *Recharge playas* have a relatively deep, water table. The temporary lake receives fine-grained clastic sediments during ephemeral flooding and can precipitate thin salt crusts when runoff evaporates. However, some of those waters sink into the substrate and recharge a deeper aquifer. The playa surface normally dries to a hard mud-cracked surface that can undergo deflation. *Throughflow playas* leak, but their net outflow is less than their net inflow, so they sometimes accumulate evaporites. *Discharge playas* have almost no subsurface leakage, are the terminus of both local and regional groundwater systems, and are the closest to true hydrologically closed drainage basins. Discharge playas have an ephemeral lake or saline pan in their center that can accumulate thick evaporites.

Changing climate and changing levels of the water table can lead to the development of different playa types in a basin through time that will be recorded in the basin stratigraphy. In terrain with low topography, where the water table fluctuates in level near the land surface, playas and shallow lakes can form in shallow depressions. Termed *boinkas,* after Australian examples, these shallow saline lakes effectively become out-

Figure 17. Lacustrine organic matter. **A.** Rooted macrophytes in Lake Baringo, Kenya. Preservation potential is low in this shallow, polymictic freshwater lake because the water and sediments are freely oxidizing. **B.** Lignites (L) alternating with lacustrine diatomite (D), Neogene Kozani basin, northern Greece. The absence of rooted horizons shows that the lignites are allogenic and are composed of wood and leaf debris washed into the paleolake. **C.** Modern organic scum (planktonic cyanobacteria: *Arthrospira fusiformis*) up to 5 cm thick along the shoreline of saline, alkaline Lake Bogoria, Kenya. The high productivity in this tropical lake produces organic-rich muds and oozes that are a precursor of lacustrine oil shales.

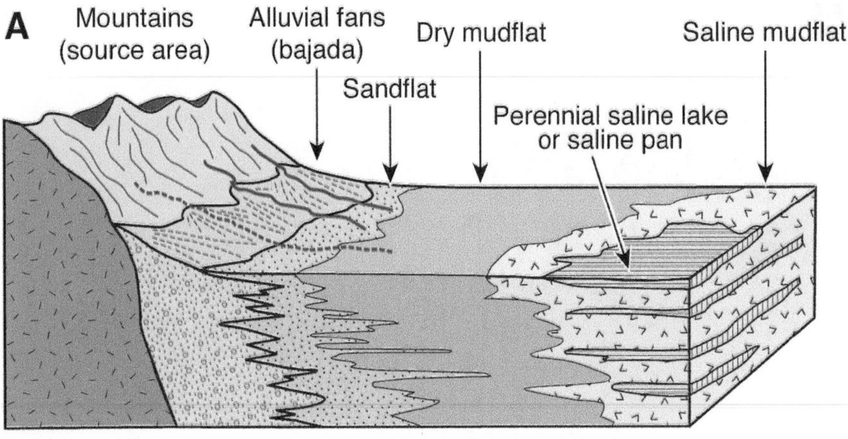

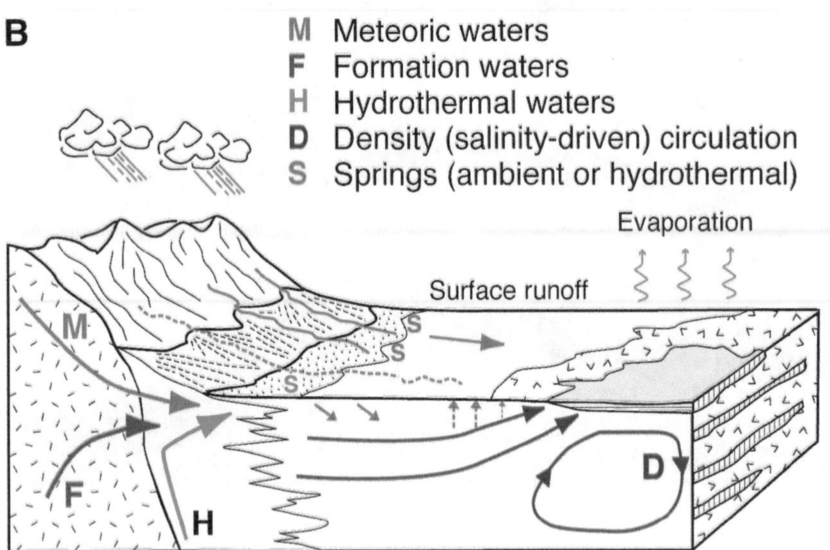

Figure 18. The playa lake model. **A.** General model for closed-basin sedimentation, based on the intermontane basins of the western United States (after Eugster and Hardie, 1978). **B.** Hydrology of intermontane closed basins (modified from Jones and Deocampo, 2003).

crops of the local water table. They are not fully hydrologically closed but some precipitate evaporites.

Peripheral Clastic Aprons and Dry Mudflats

Saline lakes in steep-sided, block-faulted tectonic basins are mostly bordered by alluvial fans that pass distally into sandflats and then into marginal lacustrine mudflats (Fig. 18). Similar fans form along the hangingwall of rift basin half-grabens. Clastic sediments are transported towards the lake during sporadic floods mainly as unchanneled sheetflow; most of this sediment is fine sand, silt, and clay deposited as sheets or lobes. In steep-sided tectonic basins, fan deltas can transport coarse sediments directly into perennial saline lakes (Fig. 11F). Peripheral mudflats are sometimes submerged when the lake is relatively high and receiving fine clastic input.

In proximal areas, mudflats dry out to form hard flat surfaces during the dry seasons and the water table is relatively deep. Dry mudflat sediments are typically graded silt and clay laminae, a few mm thick, produced by decelerating unchanneled flow (sheetwash or sheetflood; Figs. 20, and 21A). Rippled to cross-stratified coarser sands fill ephemeral shallow (< 1 m) washes incised into the proximal areas of mudflats. In some playas, carbonate mud (micrite, dolomicrite) or mixed carbonate–siliciclastic mud is deposited on the mudflats. Desiccation produces extensive crack networks and numerous mud chips that become reworked as intraclasts and are deposited to become intraformational breccia and conglomerate.

Saline Mudflats and Saline Pans

The dry mudflats pass lakewards into saline mudflats (Fig. 18) where saline groundwater lies very close to the surface and saturates the sediments in brine (Fig. 11F). The saline mudflat is often a site of evaporite mineral precipitation. Salts crystallize at the surface as efflorescent crusts of many types (Fig. 21B) and, depending on groundwater salinity, as intrasedimentary (typically displacive) salts both above and below the shallow water table forming individual crystals, crystal aggregates and nodules. These salts originate from capillary evaporation of near-surface brines, evapotranspiration by salt-tolerant plants if present, and evaporative pumping. Salts also precipitate directly from evaporating surface water when the playa is flooded.

Saline mudflat sediments are typically mixtures of mud (including carbonate mud in some basins) and salts. Bedding is less evident than in dry mudflats due to frequent disruption by salt growth and dissolution, wetting-drying cycles, and development of desiccation cracks at many scales (Fig. 20). Saline mudflat sediment is generally dark gray when wet and, upon drying, can be grayish green to reddish depending on the oxidation status of iron (i.e. Fe^{2+} or Fe^{3+}).

Compositional zoning of precipitated minerals is common across mudflats and reflects the progressive evaporative concentration of surface fluids and groundwater from basin margin to basin center. The more soluble salts tend to be dominant in the central lowest parts of the mudflats producing a 'bull's eye' pattern of precipitated minerals in some basins. Local variations in the sites of groundwater discharge and topography can, however, cause anomalous salt distributions.

In some basins the lowest, central part of the basin is occupied by a saline pan, composed of thick dry salt that is broken by large meter-scale polygons (Figs. 20, and 21C, D). Such saline pan salts vary in compo-

Table 5. Summary of saline lake and playa facies

Saline Lake/Playa Environments	Descriptions
Dry mudflats	Three types: (1) subaerially exposed perennial lake deposits, (2) slowly aggrading mudflats, and (3) rapidly aggrading mudflats; disruption by burrows, roots, and polygonal mudcracks of exposed perennial lake sediments; slowly aggrading mudflats consist of mm to sub-mm silt and mud layers reworked by deflation and desiccation; rapidly aggrading mudflats consist of sheetflood deposits reworked by deflation, desiccation, pedoturbation, and bioturbation; some surface efflorescent crusts; lenses and distorted layers of sand and silt floating in mud (sand patch fabric) from eolian and fluvial input; size of mudcrack polygons depends on thickness of cracked sediment.
Saline mudflats	Clay to sandy mud with intrasediment evaporite growth (many crystal habits as isolated crystals or crystal aggregates) or evaporite layers; salts can be void-filling cement as well; evaporite growth dependent on position of groundwater table; diagenetically-altered evaporite layers and crystals from groundwater; phreatic vs. vadose textures in evaporites; efflorescent crusts on surface from groundwater evaporation (thickness and texture dependent on groundwater depth); lenses and distorted layers of sand and silt floating in mud (sand patch fabric) from eolian and fluvial input; graded mud and silt laminae from sheetfloods; if developed on perennial to ephemeral lake sediments, disruption by mudcracks and roots can be preserved; sheetflood deposits of sand, silt, and mud become distorted with evaporite growth or dissolution of underlying crystal layer.
Ephemeral saline lake or saline pan	Couplets of thin mud layer(s) and thick crystal salt layer(s); polygonal cracks of varying sizes and depths; local disruption by roots and burrows; intrasediment evaporite growth from evaporative pumping, depending on depth to water table; diagenetically-altered evaporite layers from groundwater; evaporitic crusts.
Perennial saline lake	Offshore laminites and littoral lenticular to irregular layers composed of evaporites and mud/silt (evaporite geochemistry variable); turbidites with marginal evaporite intraclasts; shallow shoals of reworked evaporite intraclasts; sublittoral laminae and beds of silt and sand grains (siliciclastic, carbonate, evaporite) with continuous flat lamination to ripple cross-lamination or flaser to lenticular bedding; muds containing intrasediment growths and cements or simply structureless; evaporitic crusts on shallow and deep substrates; crystal cumulus layers (forming at air-brine interface, then settling); microbial structures and organic laminae; zeolites and cherts in alkaline lakes.

Key references: Eugster and Hardie, 1978; Hardie *et al.*, 1978; Lowenstein and Hardie, 1985; Smoot and Lowenstein, 1991; Rosen, 1994; Schubel and Lowenstein, 1997; Jones and Deocampo, 2003.

sition with basinal brine evolution (Fig. 6), but are commonly halite (NaCl), trona ($NaHCO_3 \cdot Na_2CO_3 \cdot 2H_2O$), gypsum ($CaSO_4 \cdot 2H_2O$), or sulfates such as epsomite ($MgSO_4 \cdot 7H_2O$), mirabilite ($Na_2SO_4 \cdot 10H_2O$), and bloedite ($Na_2SO_4 \cdot MgSO_4 \cdot 4H_2O$).

Salts in the saline pan can be bedded and show cycles (*e.g.*, annual) that reflect flooding and subsequent evaporation (Fig. 22). Flooding events introduce fine-grained clastic sediments and dissolve the uppermost salts of the saline pan surface. With evaporation, the brine reaches saturation with respect to a salt. Individual crystals and crystal rafts form at the air-water interface and subsequently sink (Fig. 21E). Some dissolve in the brine, increasing its salinity. Eventually the brine body becomes saturated with respect to one or more salts and bottom-nucleated crystals grow upward from the substrate upon the fine clastic layer, if present, incorporating sunken rafts. Once the brine is fully saturated, rapid crystal growth produces crystal beds several cm thick (Fig. 21F). With complete evaporation, the water table withdraws below the surface leaving a dry pan. Salt crystallization occurs interstitially from more concentrated subsurface pore fluids, sometimes producing more soluble salts than those of the main pan (*e.g.*, bittern salts such Mg- and K-salts) or diagenetically modifying previously precipitated salts. Polygonal crack networks and pressure ridges typically develop at the pan surface (Figs. 21D, and 22).

Perennial saline lakes, such as the large, deep Dead Sea and smaller shallow Lake Bogoria in Kenya are sites of periodic evaporite formation. Surface-nucleated cumulate crystals sometimes dissolve while sinking

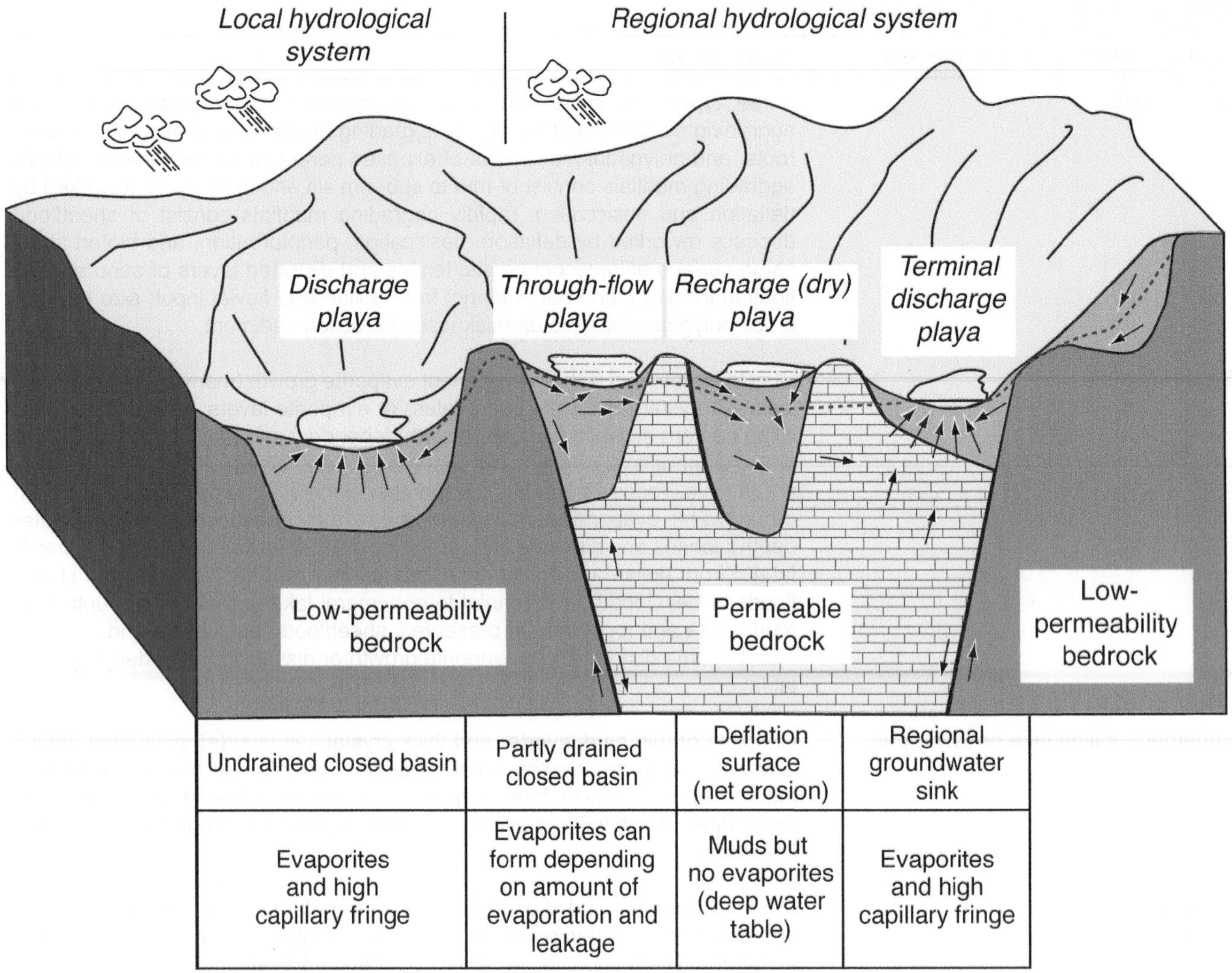

Undrained closed basin	Partly drained closed basin	Deflation surface (net erosion)	Regional groundwater sink
Evaporites and high capillary fringe	Evaporites can form depending on amount of evaporation and leakage	Muds but no evaporites (deep water table)	Evaporites and high capillary fringe

Figure 19. Hydrogeology and classification of playas and closed lacustrine basins (modified from Rosen, 1994).

because bottom waters initially can be undersaturated. Physical reworking of cumulate crystals by currents also produces detrital evaporites. In some shallow perennial saline lakes, bottom nucleated crystals grow vertically to form crystal beds. Perennial lakes that undergo periodic freshening from dilute inflow can have periods of high organic productivity. The sediments of such lakes are typically organic-rich laminated muds or carbonates that alternate with the bedded evaporites (Smoot and Lowenstein, 1991).

Springs

Springs and groundwater seepage zones provide a significant portion of the inflow in some lakes, especially those located in volcanic and karstic regions. Discharge can occur subaerially near the lake margins or subaqueously through the lake floor. The fluids discharged can be dilute or saline, and have temperatures ranging from just above the local freezing point to boiling.

Lakes fed mainly by springs and groundwater often have little siliciclastic sediment input except for eolian dust. Their sediments are mainly autochthonous, consisting of water-column and lake-floor precipitates (*e.g.*, carbonates, evaporites, clay minerals, sulfides), planktonic remains (skeletal and organic), and any mineral and organic deposits produced by benthic microbial mats or rooted plants if the lake is shallow. In such spring-fed lakes, lacustrine sediments pass laterally into subaerial deposits (*e.g.*, alluvial plain, sand dunes) or soils with little or no siliciclastic shoreline or deltaic facies. Rhizoliths and other wetland indicators are associated with the margins of lakes that were fed by groundwater seepage (Fig. 23A, B). Some springs leave a sedimentary record, whereas many do not and their former presence can only be inferred from indirect evidence such as facies relationships or trace fossils.

Most spring deposits in lake basins are carbonates, which can be subaerial deposits around the lake margins or subaqueous precipitates. Subaerial travertine and tufa deposits are found in the discharge zone near the base of alluvial fans in closed basins, and occur along fault lines in many locations (Fig. 23C). They are important sites of early Ca removal from closed-basin waters. Other travertine and tufa deposits form around the vents of sublacustrine springs and seepages, which can produce mounds or microbial bioherms. Hot springs can produce chimney-like structures, such as those in Mono Lake (Fig. 23D) and Searles Lake in

California, and Lake Abbe in Djibouti. Precipitation is abiotic through CO_2 degassing or is mediated by bacteria. In some examples, mixing of fluids of different composition promotes carbonate precipitation at the vent (*e.g.*, Ca-bearing springs entering a saline $Na\text{-}CO_3$ lake).

Very hot and boiling springs can precipitate sublacustrine sinter deposits composed of opaline silica (opal-A) that later alters diagenetically to quartz. Examples in the form of chimneys are known from crater lakes and as local areas of lake-floor cementation in several Kenya Rift lakes. Some sublacustrine sinters contain planktonic diatoms. Sulfide-rich mineral chimneys occur in Lake Tanganyika and Lake Kivu in East Africa, and native sulfur is precipitated around the vents in some acidic hot spring-fed lakes in New Zealand.

LAKE CLASSIFICATION AND LACUSTRINE SEQUENCE STRATIGRAPHY

Modern lakes are classified mainly by the seasonal stratification properties of their water columns. This approach is impractical for lake deposits in the geologic record because the sediments, not the lake waters, are preserved. Lake-level fluctuations and basin character influence the deposits of lacustrine basins and their stratal stacking patterns. Sedimentation patterns are controlled by allogenic factors such as tectonics and climate through interactions of four main variables: sediment supply, water supply, basin-sill height (spillpoint), and basin-floor depth (Bohacs *et al.*, 2000; 2003). Tectonic movements, which occur mainly over long time scales, affect basin hydrology by controlling basin shape and depth (*i.e.*, accommodation), watershed geology (sediment input), and drainage characteristics by defining the sill height or spillpoint of the basin and the presence or absence of groundwater inflow and outflow. Climate can govern surface water and precipitation influx, evaporation rate, and sediment supply over both short- and long-term time scales. Groundwater input is influenced more by geologic controls on hydrology (*i.e.*, structure and tectonics) than climate.

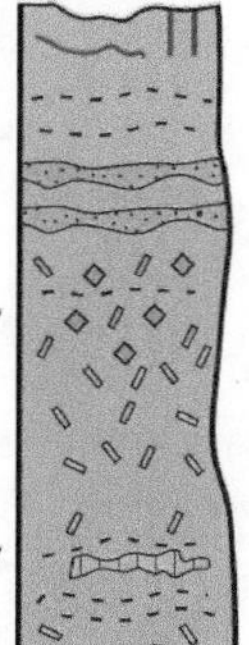

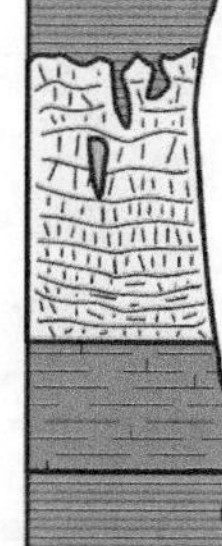

Figure 20. Typical facies of dry mudflats, saline mudflats, and saline pans (modified from Kendall, 1984).

Thus, tectonics and climate exert their influence together, not independently, to produce sequence stratigraphic patterns that define three main, lake-type end members: overfilled, balanced-fill, and underfilled (Figs. 24–26; Bohacs *et al.*, 2000). This tripartite lacustrine classification, previously recognized by other lake researchers (*e.g.*, Bradley, 1931; Schlische and Olsen, 1990; Glenn and Kelts, 1991) but named differently, is based on specific types of lithologic successions and their stacking patterns, and geometry, facies associations, fossils, and geochemical indicators.

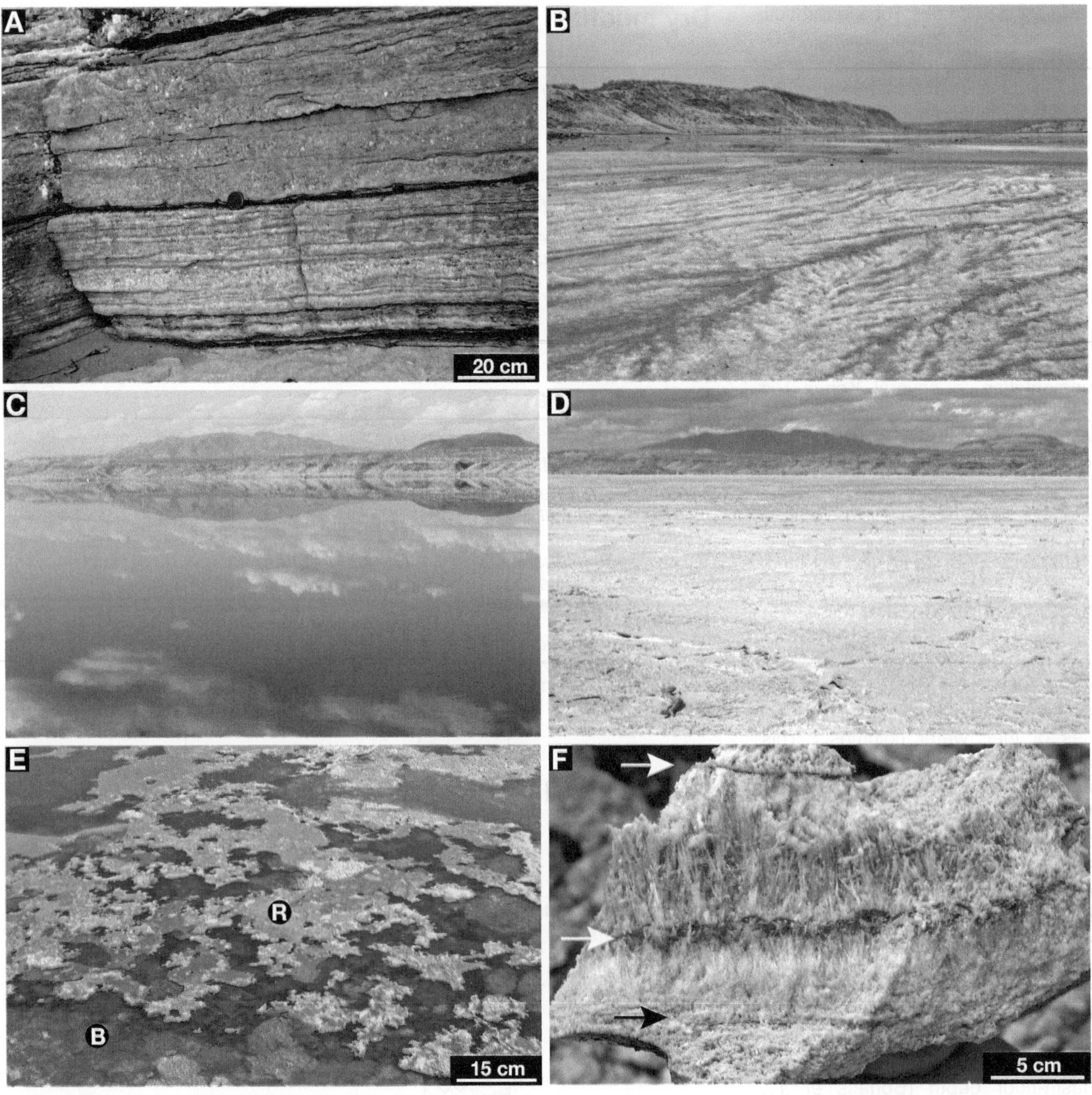

Figure 21. Saline lake environments and facies. **A.** Thin-bedded sandstones and siltstones deposited on sandflats-mudflats adjacent to a saline lake. Mid-Devonian Orcadian basin, Orkney, Scotland. **B.** Modern saline mudflats, Lake Magadi, Kenya. **C, D.** Lake Magadi saline pan during rainy season **(C)** with shallow (30 cm brine) and dry season **(D)** with polygonally-cracked trona flats. **E.** Crystal rafts (R) and bottom-nucleated crystal bushes (B) of nahcolite ($NaHCO_3$), Nasikie Engida, Lake Magadi, Kenya. **F.** Bedded trona cycles from the center of Lake Magadi saline pan, Kenya. Arrows indicate thin detrital layers overlying corroded salts that are linked to dilution of the brine during the rainy season(s).

Lacustrine Facies Associations

Overfilled Lakes

Overfilled lake basins occur where the water and sediment supply exceed the low rate of accommodation and where the watershed drainage remains open (Fig. 24). Such open lakes can fill a basin to its sill height to form a relatively deep lake or the lake surface can fall temporarily below the sill level to produce a shallower lake. Shoreline progradational sequences are common but lie within indistinctly defined stacks of parasequences. System tracts and sequence boundaries of deep overfilled lakes resemble classic Type 2 marine siliciclastic sequences that are composed of shelf-margin, transgressive, and high-stand system tracts. Those in shallow overfilled lakes resemble classic Type 1 marine siliciclastic sequences that comprise low-stand, transgressive, and high-stand systems tracts. Erosion and incision features can develop with a significant drop in lake level.

A mixed fluvial–lacustrine facies association exemplifies overfilled lake deposits, which are dominated by deep-water shale or micrite with turbiditic layers near basin centers, well-defined shoreline deposits, including

FLOODING STAGE

Dissolution of older saline crust by floodwaters (unchannelled wash and ephemeral streams)

Shallow lake (cm - m deep) becomes saline from dissolved salts. Springs (seasonal or perennial) supplement runoff in some lake basins

Suspended sediment (clastic and/or carbonate) is deposited on the corroded salt surface and may contain detrital and autochthonous (planktonic) organic matter

Benthic microbial mats (cyanobacteria and archaea) may form in alkaline lakes

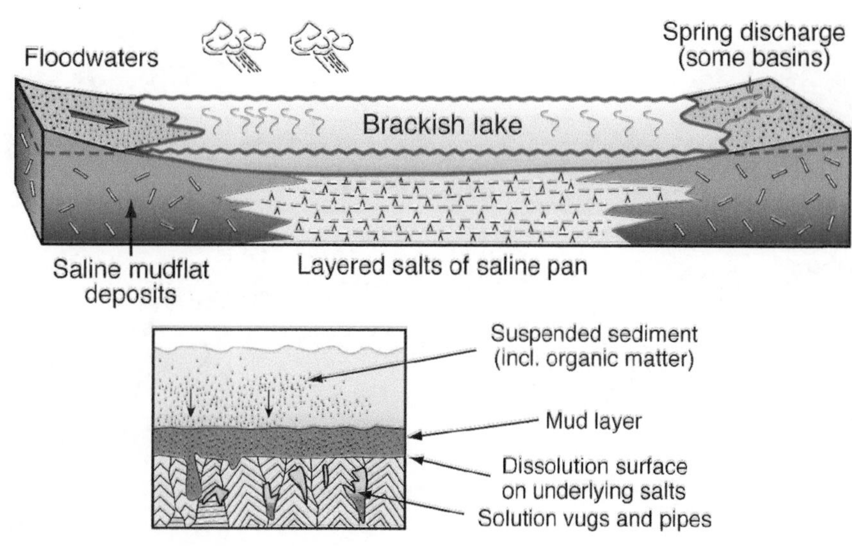

EVAPORATIVE CONCENTRATION

Evaporation leads to shrinking lake volume and increasing salinity

Saline minerals nucleate and grow at the air/water interface

Surface nucleated salts sink; initially many dissolve, increasing salinity and density of lower brine

When bottom waters are saturated, crystal overgrowths occur from foundered rafts, halite hoppers and other particles

Bottom-nucleated salts grow competitively, forming crystal beds

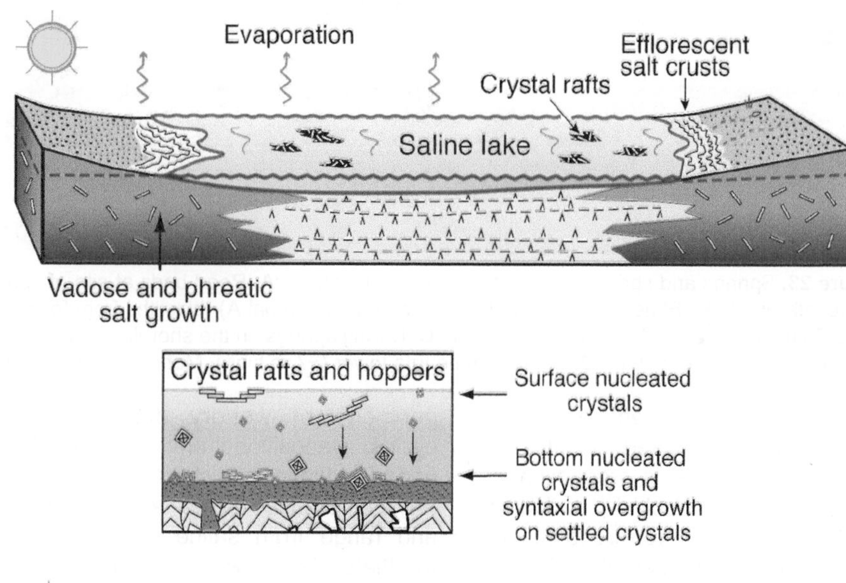

DESICCATION

Surface dries out and hypersaline brine withdraws below the exposed salt crust

Surface crust is disrupted by formation of polygons

Early diagenetic growth of salts occurs within the salt crust and mud layers

Periodic rain showers can dissolve surface crusts followed by formation of new thin crusts

Late interstitial salts can include more soluble species than the main salt crust (e.g. K, Mg-salts)

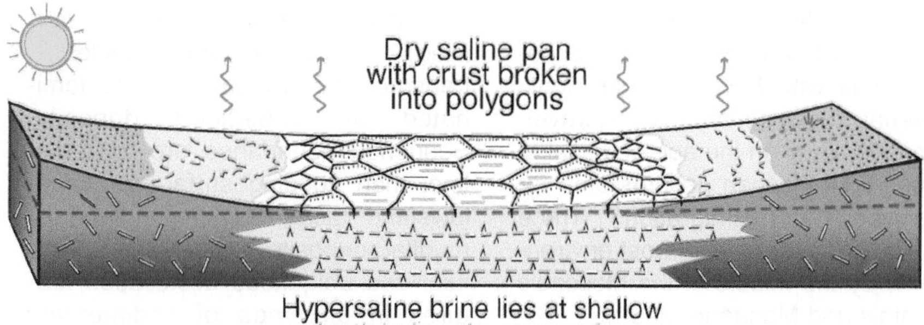

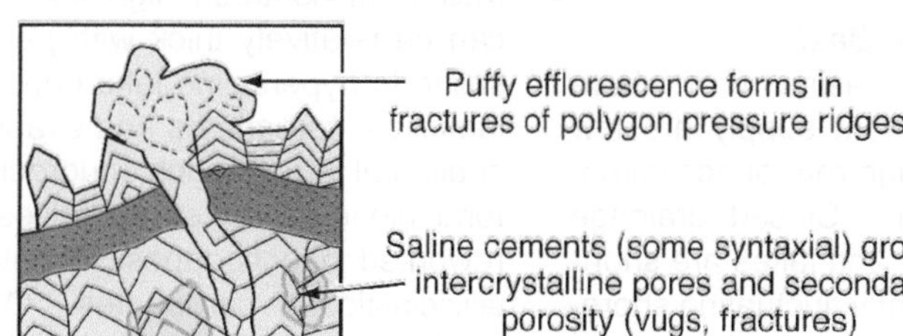

Figure 22. The saline pan cycle (modified from Lowenstein and Hardie, 1985).

Figure 23. Springs and spring deposits associated with lakes. **A.** Reed beds at site of spring discharge on the southeastern margin of saline, alkaline Lake Elmenteita, Kenya. **B.** Calcite and silica (opal-A) rhizoliths from former lake-marginal wetland, Pleistocene Olorgesailie Formation, Kenya. Marker is 13 cm long. **C.** Boiling springs on the shoreline of Lake Bogoria, Kenya. Travertine records the spring activity. Pool diameter is 5 m. **D.** Sublacustrine tufa tower (~7 m high), Mono Lake, California, USA.

proximal deltaic sediments and coals, and channel-dominated fluvial sediments. Lake water is predominantly fresh or slightly brackish, so indicators of high salinity are rare to absent. Organic matter preservation is normally low unless anoxic environments are present, and terrestrial components mix with lacustrine planktonic material. Species diversity, however, depends partly on the age of the lake: a long lake lifespan assures time for speciation and immigration. Trace fossils are preserved in most environments by the middle of the Mesozoic (Buatois and Mangáno, 2004).

Underfilled Lake Basins

Underfilled lake basins form where water and sediment supply are low compared to a high rate of accommodation (Fig. 25). Closed drainage usually develops and lakes are short-lived with frequently fluctuating shorelines. The stacking of thin parasequences will record an aggradational geometry of wet/dry depositional cycles. Depositional environments of underfilled lakes are characterized by an evaporative facies association, and range from saline playas and mudflats (shallow underfilled) to perennial saline lakes (deep underfilled). Associated alluvial facies in transgressive system tracts are dominated by sheetflood deposits. Sequence boundaries are less distinct than flooding surfaces. Lake-lowstand facies include evaporites, mudflat deposits, paleosols, eolianites, and ubiquitous desiccation features in all types of sedimentary rocks. In contrast, highstand facies can be relatively thick with perennial saline to hypersaline lake deposits of laminites, carbonates and evaporites, sublittoral organic-rich mudrocks, littoral bioherms or stromatolites, and reworked beach deposits containing siliciclastics or carbonates. Organic productivity can be high in this lake type, but desiccation, oxidation, and erosion can reduce the preservation potential for organic matter. High salinities limit the diversity of organisms and trace fossils in the underfilled ecosystem. Fauna and flora intolerant of high salinity are typically absent except near sites of dilute inflow such as springs and groundwater seepage zones.

Balanced-fill Basins

Balanced-fill basins (Fig. 26) are intermediate in water and sediment supply and accommodation rates. Both progradational and aggradational stacking patterns develop as lake levels and general lake characteristics oscillate with periodic changes between open and closed drainage. The 'fluctuating profundal' facies association of balanced-fill lakes differs between shallow and deep examples. For shallow lakes or lowstands of deeper lakes, both carbonate and siliciclastic rocks accumulate in a variable range of depositional environ-

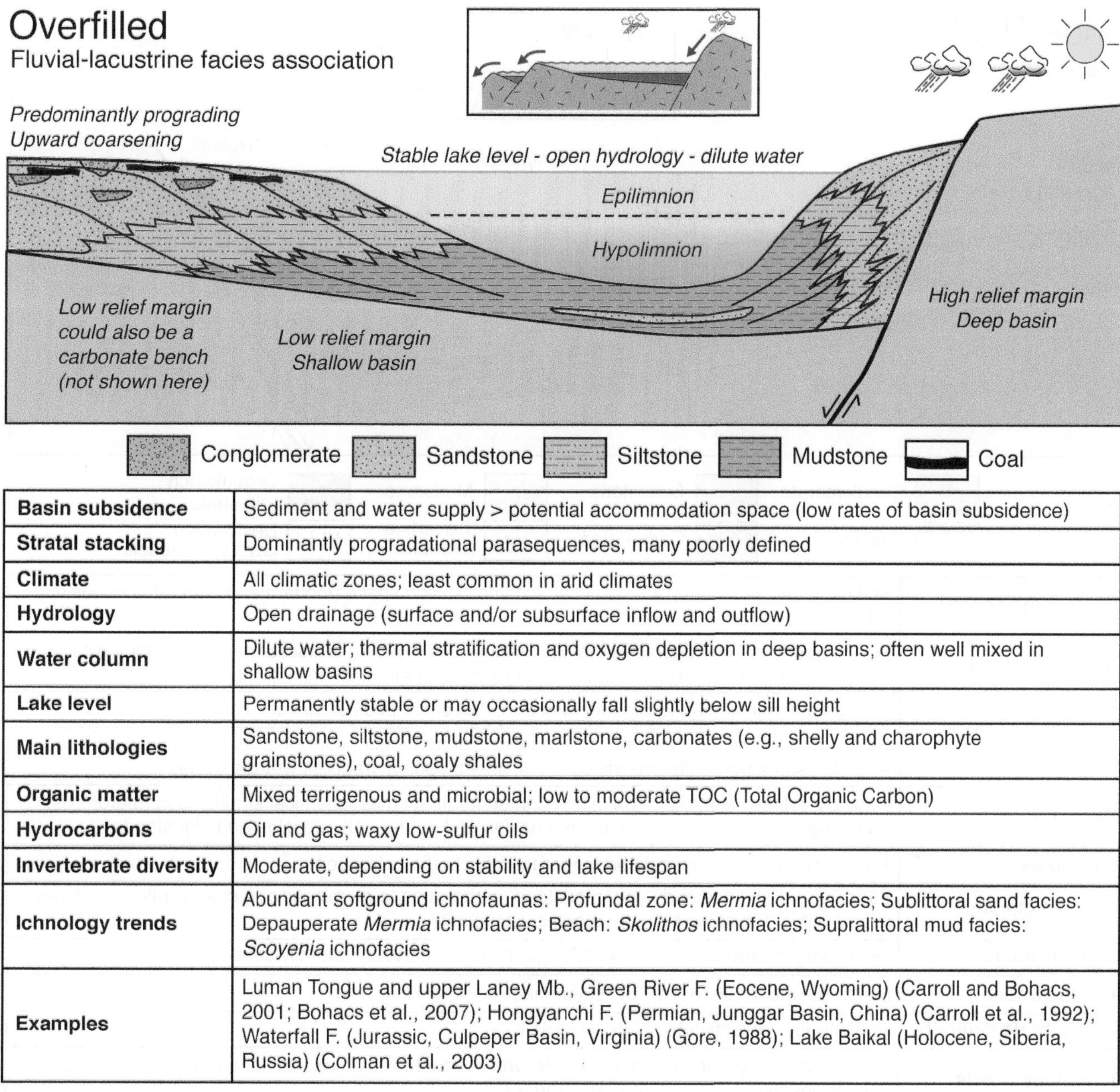

Basin subsidence	Sediment and water supply > potential accommodation space (low rates of basin subsidence)
Stratal stacking	Dominantly progradational parasequences, many poorly defined
Climate	All climatic zones; least common in arid climates
Hydrology	Open drainage (surface and/or subsurface inflow and outflow)
Water column	Dilute water; thermal stratification and oxygen depletion in deep basins; often well mixed in shallow basins
Lake level	Permanently stable or may occasionally fall slightly below sill height
Main lithologies	Sandstone, siltstone, mudstone, marlstone, carbonates (e.g., shelly and charophyte grainstones), coal, coaly shales
Organic matter	Mixed terrigenous and microbial; low to moderate TOC (Total Organic Carbon)
Hydrocarbons	Oil and gas; waxy low-sulfur oils
Invertebrate diversity	Moderate, depending on stability and lake lifespan
Ichnology trends	Abundant softground ichnofaunas: Profundal zone: *Mermia* ichnofacies; Sublittoral sand facies: Depauperate *Mermia* ichnofacies; Beach: *Skolithos* ichnofacies; Supralittoral mud facies: *Scoyenia* ichnofacies
Examples	Luman Tongue and upper Laney Mb., Green River F. (Eocene, Wyoming) (Carroll and Bohacs, 2001; Bohacs et al., 2007); Hongyanchi F. (Permian, Junggar Basin, China) (Carroll et al., 1992); Waterfall F. (Jurassic, Culpeper Basin, Virginia) (Gore, 1988); Lake Baikal (Holocene, Siberia, Russia) (Colman et al., 2003)

Figure 24. Schematic cross section and summary of the overfilled lake type and fluvial–lacustrine facies association (modified from Bohacs *et al.*, 2000).

ments within relatively thin aggradational parasequence sets that contain wet/dry cycles. Deep lakes or highstands can also accumulate thicker aggradational parasequence sets associated with turbidite deposition in profundal areas and progradational stacking associated with littoral zones. Limited incision and erosion events occur at sequence boundaries that are delineated by large basinward shifts of environments. Retrogradational stacking thus can also occur in transgressive system tracts. Lake levels in this basin type are typically related to sediment supply. Deposits in balanced-fill lakes exhibit facies and trace fossil distributions similar to both underfilled and overfilled lakes. Faunal diversity, based on lake age, can be the highest in this lake type because of the changing stresses in physical conditions through time. Organic matter is dominated by microbial production with relatively high preservation potential because of the meromixis that commonly is induced by dilute runoff during the highstands of relatively saline lakes.

CONCLUDING COMMENTS

When studying a lake or lacustrine sediment sequence (Fig. 27), it is important to recognize where it exists within its evolutionary life span because lake types evolve throughout the existence of a lake basin. In tectonic lake basins, sedimentation typically begins with fluvial deposition, reflecting the through-flowing drainage as tectonic movements open a depression in the crust. As a tectonic depression develops and deepens with time, the type of lake residing in the depression evolves as

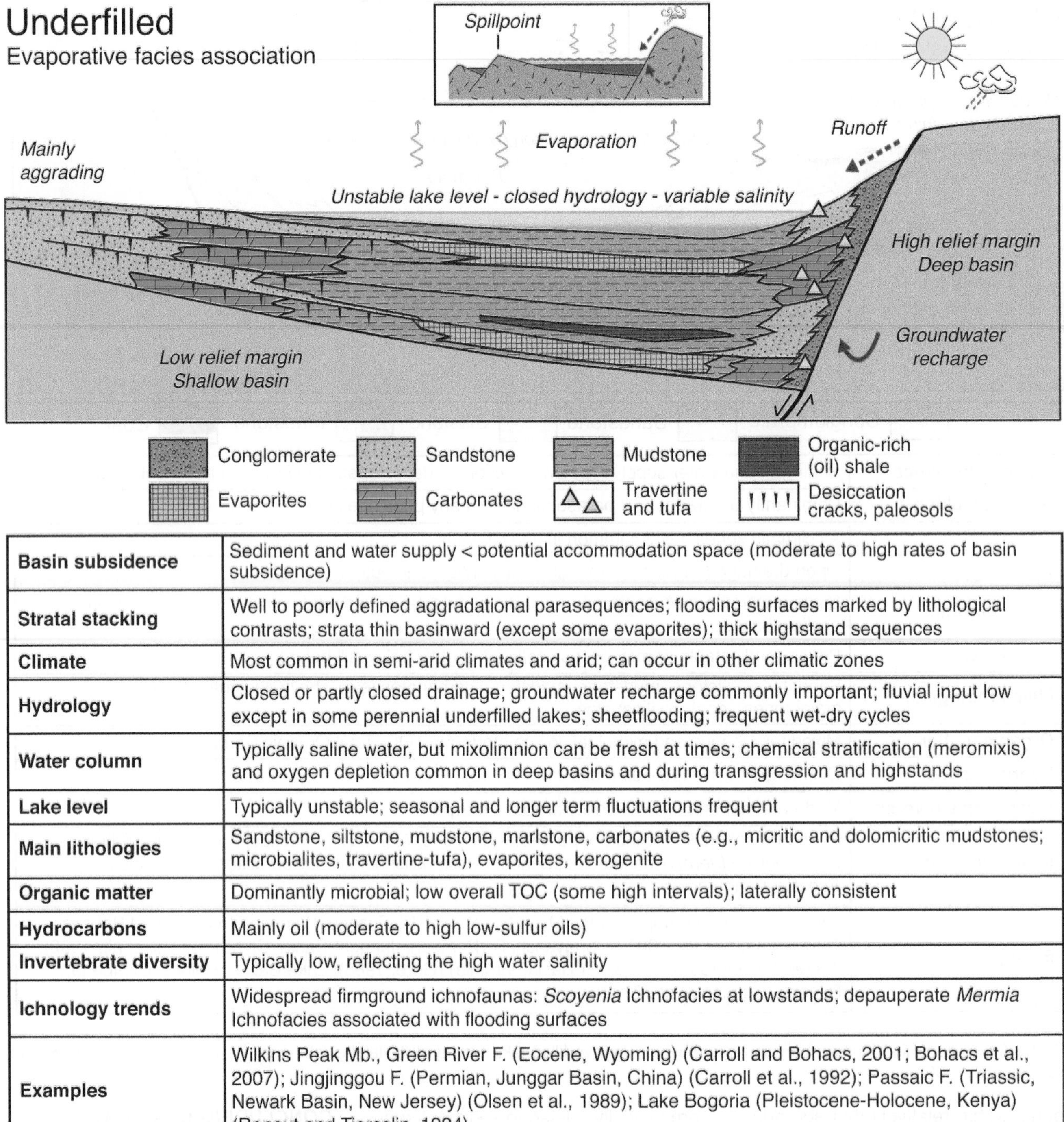

Basin subsidence	Sediment and water supply < potential accommodation space (moderate to high rates of basin subsidence)
Stratal stacking	Well to poorly defined aggradational parasequences; flooding surfaces marked by lithological contrasts; strata thin basinward (except some evaporites); thick highstand sequences
Climate	Most common in semi-arid climates and arid; can occur in other climatic zones
Hydrology	Closed or partly closed drainage; groundwater recharge commonly important; fluvial input low except in some perennial underfilled lakes; sheetflooding; frequent wet-dry cycles
Water column	Typically saline water, but mixolimnion can be fresh at times; chemical stratification (meromixis) and oxygen depletion common in deep basins and during transgression and highstands
Lake level	Typically unstable; seasonal and longer term fluctuations frequent
Main lithologies	Sandstone, siltstone, mudstone, marlstone, carbonates (e.g., micritic and dolomicritic mudstones; microbialites, travertine-tufa), evaporites, kerogenite
Organic matter	Dominantly microbial; low overall TOC (some high intervals); laterally consistent
Hydrocarbons	Mainly oil (moderate to high low-sulfur oils)
Invertebrate diversity	Typically low, reflecting the high water salinity
Ichnology trends	Widespread firmground ichnofaunas: *Scoyenia* Ichnofacies at lowstands; depauperate *Mermia* Ichnofacies associated with flooding surfaces
Examples	Wilkins Peak Mb., Green River F. (Eocene, Wyoming) (Carroll and Bohacs, 2001; Bohacs et al., 2007); Jingjinggou F. (Permian, Junggar Basin, China) (Carroll et al., 1992); Passaic F. (Triassic, Newark Basin, New Jersey) (Olsen et al., 1989); Lake Bogoria (Pleistocene-Holocene, Kenya) (Renaut and Tiercelin, 1994)

Figure 25. Schematic cross section and summary of the underfilled lake type and evaporative facies association (modified from Bohacs *et al.*, 2000).

accommodation space and climate also change with time. Then, as development of accommodation slows and decreases, the basin is eventually infilled. The lake types and their respective facies associations reflect this evolution. For example the lake types of the Green River Formation in Wyoming, USA, evolved from overfilled to balanced-filled to underfilled, back to balanced-filled, and finally to overfilled (Carroll and Bohacs, 2001). This pattern is also recognized in many other tectonic lake basins worldwide. In contrast, volcanic crater lakes can begin with deep water soon after an eruption and gradually fill with sediment eroded from the crater rim until they eventually overflow and become part of an open drainage system. The lake type and facies associations need to be considered when comparing lake successions in the geologic record (Gierlowski-Kordesch and Park, 2004; Bohacs *et al.*, 2007).

Much more work is needed to test the patterns of lake type changes throughout the life of a lake basin and their impact on life, sedimentation and

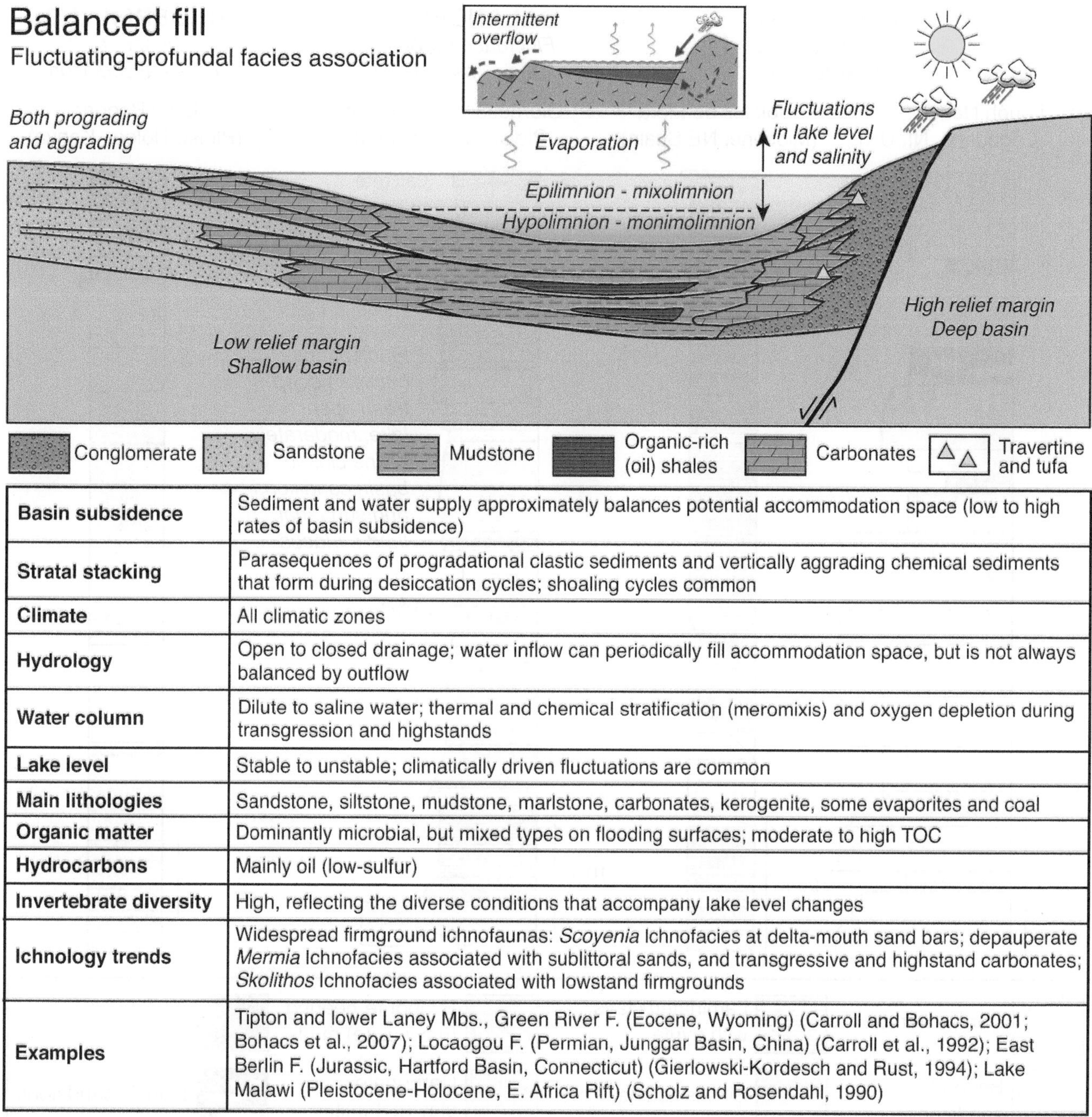

Basin subsidence	Sediment and water supply approximately balances potential accommodation space (low to high rates of basin subsidence)
Stratal stacking	Parasequences of progradational clastic sediments and vertically aggrading chemical sediments that form during desiccation cycles; shoaling cycles common
Climate	All climatic zones
Hydrology	Open to closed drainage; water inflow can periodically fill accommodation space, but is not always balanced by outflow
Water column	Dilute to saline water; thermal and chemical stratification (meromixis) and oxygen depletion during transgression and highstands
Lake level	Stable to unstable; climatically driven fluctuations are common
Main lithologies	Sandstone, siltstone, mudstone, marlstone, carbonates, kerogenite, some evaporites and coal
Organic matter	Dominantly microbial, but mixed types on flooding surfaces; moderate to high TOC
Hydrocarbons	Mainly oil (low-sulfur)
Invertebrate diversity	High, reflecting the diverse conditions that accompany lake level changes
Ichnology trends	Widespread firmground ichnofaunas: *Scoyenia* Ichnofacies at delta-mouth sand bars; depauperate *Mermia* Ichnofacies associated with sublittoral sands, and transgressive and highstand carbonates; *Skolithos* Ichnofacies associated with lowstand firmgrounds
Examples	Tipton and lower Laney Mbs., Green River F. (Eocene, Wyoming) (Carroll and Bohacs, 2001; Bohacs et al., 2007); Locaogou F. (Permian, Junggar Basin, China) (Carroll et al., 1992); East Berlin F. (Jurassic, Hartford Basin, Connecticut) (Gierlowski-Kordesch and Rust, 1994); Lake Malawi (Pleistocene-Holocene, E. Africa Rift) (Scholz and Rosendahl, 1990)

Figure 26. Schematic cross section and summary of the balanced-fill lake type and fluctuating-profundal facies association (modified from Bohacs *et al.*, 2000).

facies patterns, and resources. There is no universal facies model for lake sedimentation. Each lake is unique, composed of a mosaic of possible subenvironments that evolve through time within a specific tectonic setting and under the influence of variable climate. We now recognize the tripartite pattern of lake development, showing a spectrum of possible facies models. Limnogeology, the study of lake sedimentation in the geologic record, is a discipline still in its infancy – more data are needed to understand the vast complex array of lakes and lacustrine environments, and to develop and refine lacustrine facies models.

ACKNOWLEDGMENTS

Our research on lakes and lake sedimentation has been supported by grants from the Natural Sciences and Engineering Research Council (Canada), the International Development Research Centre (Canada), the European Community, the Natural Environment Research Council (UK), the Royal Society (UK), the National Science Foundation, the Petroleum Research Fund of the American Chemical Society, and the U.S. Geological Survey, which we gratefully acknowledge. We thank Kevin Bohacs, Bernie Owen, and Jenni Scott for their helpful comments on the original manuscript.

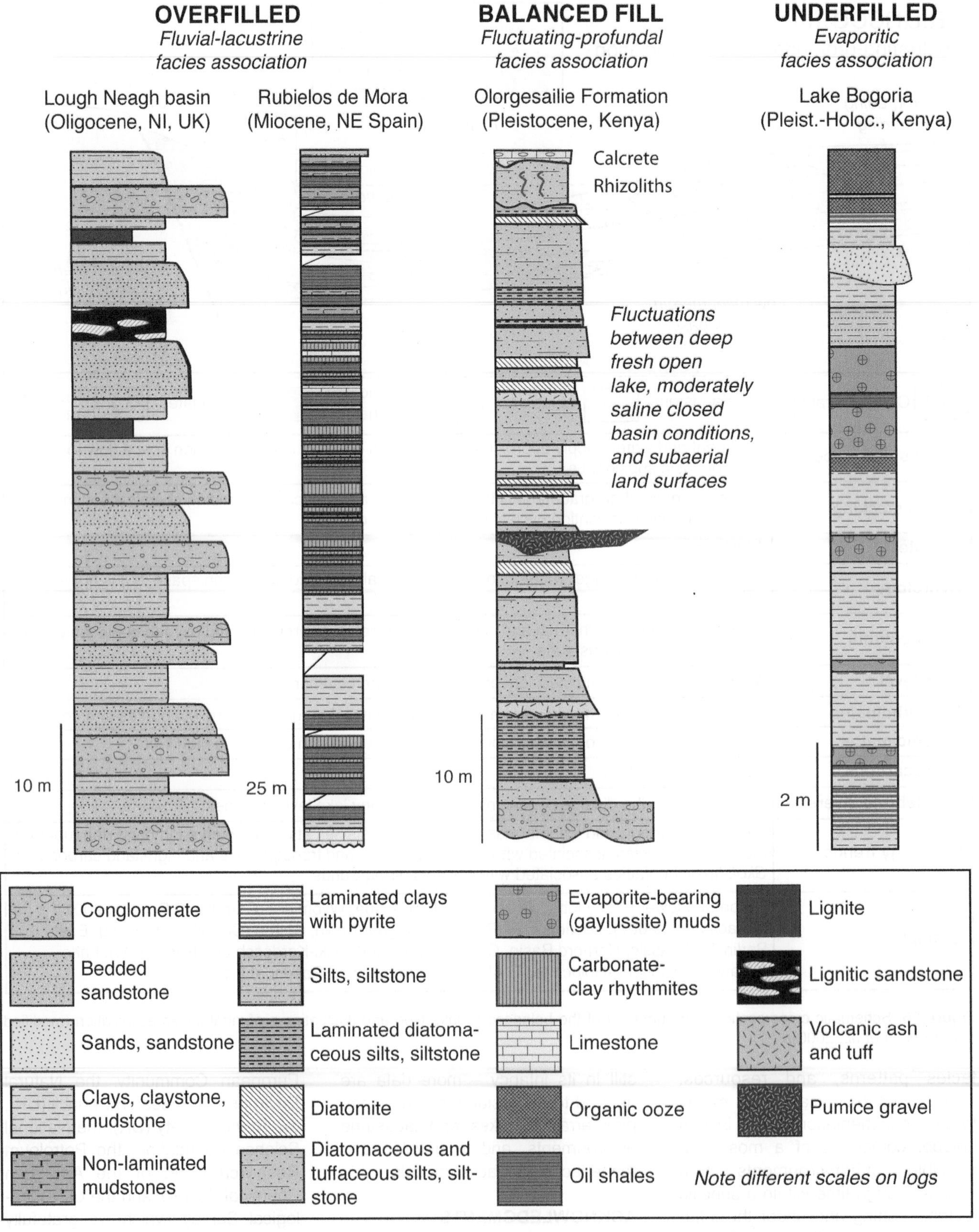

Figure 27. Examples of overfilled, balanced-fill and underfilled lake types (sources: Parnell and Shukla, 1994; Anadón *et al.*, 1991b; Owen and Renaut, 2000; Renaut and Tiercelin, 1994). For other examples, see Bohacs *et al.* (2000, 2003).

REFERENCES

Lakes – General accounts and overview papers

Anadón, P., Cabrera, Ll. and Kelts, K., *eds.*, 1991a, Lacustrine Facies Analysis: International Association of Sedimentologists Special Publication 13, 318 p.
A collection of papers documenting facies and depositional environments in lakes.

Cohen, A.S., 2003, Paleolimnology: Oxford University Press, Oxford, 500 p.
The most detailed scholarly account of lake sedimentology, lake paleocology, and techniques for paleoenvironmental analysis of lacustrine sediments.

Gierlowski-Kordesch, E. and Kelts, K., *eds.*, 1994, Global Geological Record of Lake Basins, v. 1: Cambridge University Press, Cambridge, 427 p.
Collection of short papers summarizing lacustrine basins throughout the geologic record.

Gierlowski-Kordesch, E.H. and Kelts, K.R., *eds.*, 2000, Lake Basins through Space and Time: American Association of Petroleum Geologists Studies in Geology 46, 648 p.
A second collection of short papers summarizing lacustrine basins throughout the geologic record.

Håkanson, L. and Jansson, M., 1983, Principles of Lake Sedimentology: Springer, Berlin, 316 p.
Detailed account emphasizing the physical processes of lake sedimentation, now re-issued.

Katz, B.J., *ed.*, 1991, Lacustrine Basin Exploration: Case Studies and Modern Analogs: American Association of Petroleum Geologists Memoir 50, 340 p.
Papers that document depositional environments, facies, and organic matter accumulation in several types of lacustrine basin (mainly rifts).

Kelts, K., 1988, Environments of deposition of lacustrine petroleum source rocks: an introduction, *in* Fleet, A.J., Kelts, K. and Talbot, M.R., *eds.*, Lacustrine Petroleum Source Rocks: Geological Society of London Special Publication 40, p. 3-26.
Overview of the depositional setting and facies of lacustrine source rocks.

Lerman, A., *ed.*, 1978, Lakes: Geology, Chemistry, Physics (1st Edition): Springer, Berlin, 363 p.
Although superseded by a later edition, this book contains several "classic" papers on lake sedimentology.

Matter, A. and Tucker, M.E., *eds.*, 1978, Modern and Ancient Lake Sediments: International Association of Sedimentologists Special Publication 2, 290 p.
A collection of papers documenting different facies and depositional environments in lakes.

O'Sullivan, P.E. and Reynolds, C.S., *eds.*, 2004, The Lakes Handbook, Volume 1 – Limnology and Limnetic Ecology: Blackwell, Malden, MA, 699 p.
Account of physical and biologic aspects of modern lakes.

Renaut, R.W. and Last, W.M., *eds.*, 1994, Sedimentology and Geochemistry of Modern and Ancient Saline Lakes: SEPM Special Publication 50, 334 p.
A collection of papers dealing specifically with the sedimentology of saline lakes and nonmarine evaporites.

Smol, J.P., 2008, Pollution of Lakes and Rivers: A Paleoenvironmental Perspective (2nd Edition): Blackwell, Oxford, 383 p.
An overview of how paleolimnologic approaches can be used to interpret the physical, chemical, and biologic information stored in lake and river sediments.

Smol, J.P. and Last, W.M., *eds.*, 2001-2008, Developments in Paleoenvironmental Research Book Series: Now published by Springer, Dordrecht, The Netherlands.
Now up to volume 12, this book series began as a five book series entitled "Tracking Environmental Change Using Lake Sediments". The rest of the series focuses on lake research in addition to other paleoclimatic topics.

Talbot, M.R. and Allen, P.A., 1996, Lakes, *in* Reading, H.G., *ed.*, Sedimentary Environments: Processes, Facies and Stratigraphy, 3rd Edition: Blackwell, Oxford, p. 83-124.
A useful summary of the main depositional environments in lakes and their characteristic facies.

Wetzel, R.G., 2001, Limnology: Lake and River Ecosystems, 3rd Edition: Academic Press, San Diego, 1006 p.
Overview of limnology, including the physical and chemical properties of lakes and the ecology of modern lake systems.

White, J.D.L. and Riggs, N.R., *eds.*, 2001, Volcaniclastic Sedimentation in Lacustrine Settings: International Association of Sedimentologists Special Publication 30, 309 p.
Collection of papers that consider the impact of volcanism on lacustrine processes and sediments.

Other references (including those cited in the text) by category

Carbonates

Alonso-Zarza, A.M and Wright, V.P., 2010, Palustrine carbonates, *in* Alonso-Zarza, A.M. and Tanner, L., *eds.*, Carbonates in Continental Settings: Facies, Environments and Processes: Developments in Sedimentology Vol. 61: Elsevier, Amsterdam, p. 103-131.

Anadón, P, Robles, F., Roca, E., Utrilla, R. and Vázquez, A., 1998, Lacustrine sedimentation in the diapir-controlled Miocene Bicorb Basin, eastern Spain: Palaeogeography, Palaeoclimatology, Palaeoecology, v. 140, p. 217-243.

Anadón, P., Utrilla, R. and Vázquez, A., 2000, Use of charophyte carbonates as proxy indicators of subtle hydrological and chemical changes in marl lakes: example from the Miocene Bicorb Basin, eastern Spain: Sedimentary Geology, v. 133, p. 325-347.

Bertrand-Sarfati, J., Freytet, P. and Plaziat, J.C., 1994, Microstructures in the Tertiary nonmarine stromatolites (France) – comparison with Proterozoic, *in* Bertrand-Sarfati, J. and Monty, C., *eds.*, Phanerozoic Stromatolites II: Kluwer, Dordrecht, p. 155-191.

Bréhéret, J.-G., Fourmont, A., Macaire, J.-J. and Négrel, P., 2008, Microbially mediated carbonates in the Holocene deposits from Sarliève, a small ancient lake of the French Massif Central, testify to the evolution of a restricted environment: Sedimentology, v. 55, p. 557-578.

Casanova, J., 1994, Stromatolites from the East African rift: a synopsis, *in* Bertrand –Sarfati, J. and Monty, C., *eds.*, Phanerozoic Stromatolites II: Kluwer, Dordrecht, p. 193-226.

Gierlowski-Kordesch, E.H., 1998, Carbonate deposition in an ephemeral siliciclastic alluvial system: Jurassic Shuttle Meadow Formation, Newark Supergroup, Hartford Basin, USA: Palaeogeography, Palaeoclimatology, Palaeoecology, v. 140, p. 161-184.

Gierlowski-Kordesch, E.H., 2010, Lacustrine carbonates, *in* Alonso-Zarza, A.M. and Tanner, L., *eds.*, Carbonates in Continental Settings: Facies, Environments and Processes: Developments in Sedimentology Vol. 61: Elsevier, Amsterdam, p. 1-101.

Platt, N.H. and Wright, V.P., 1991, Lacustrine carbonates: facies models, facies distributions and hydrocarbon aspects, *in* Anadón, P., Cabrera, Ll. and Kelts, K., *eds.*, Lacustrine Facies Analysis: International Association of Sedimentologists Special Publication 13, p. 57-74.

Platt, N.H. and Wright, V.P., 1992, Palustrine carbonates and the Florida Everglades: towards an exposure index for the fresh-water environment: Journal of Sedimentary Petrology, v. 62, p. 1058-1071.

Sanz-Montero, M.E., Rodríguez-Aranda, J.P. and García del Cura, M.A., 2008, Dolomite-silica stromatolites in Miocene lacustrine deposits from the Duero Basin, Spain: the role of organotemplates in the precipitation of dolomite: Sedimentology, v. 55, p. 729-750.

Thompson, J.B., Schultze-Lam, S., Beveridge, T.J. and Des Marais, D.J., 1997, Whiting events: biogenic origin due to the photosynthetic activity of cyanobacterial picoplankton: Limnology and Oceanography, v. 42, p. 133-141.

Verrecchia, E.P., 2007, Lacustrine and palustrine geochemical sediments, *in* Nash, D.J. and McLaren, S.J., *eds.*, Geochemical Sediments and Landscapes: Blackwell, Oxford, p. 298-330.

Coal and Oil

Anadón, P., Cabrera, Ll., Julía, R., Marzo, M., 1991b, Sequential arrangement and asymmetrical fill in the Miocene Rubielos de Mora Basin (northeast Spain), *in* Anadón, P., Cabrera, Ll., Kelts, K., *eds.*, Lacustrine Facies Analysis: International Association of Sedimentologists Special Publication 13, p. 257-275.

Bradley, W.H., 1931, Origin and microfossils of the oil shale of the Green River Formation of Colorado and Utah: U.S. Geological Survey Professional Paper 168, 58 p.

Carroll, A.R. and Bohacs, K.M., 2001, Lake-type controls on petroleum source rock potential in nonmarine basins: American Association of Petroleum Geologists Bulletin, v. 85, p. 1033-1053.

Carroll, A.R., Brassell, S.C., and Graham, S.A., 1992, Upper Permian lacustrine oil shales, southern Junggar basin, northwest China: American Association of Petroleum Geologists Bulletin, v. 76, p. 1874-1902.

Gierlowski-Kordesch, E.H., Gómez-Fernández, J.C. and Meléndez, N., 1991, Carbonate and coal deposition in an alluvial lacustrine setting: Lower Cretaceous (Weald) in the Iberian Range (east-central Spain), *in* Anadón, P., Cabrera, Ll. and Kelts, K., *eds.*, Lacustrine Facies Analysis: International Association of Sedimentologists Special Publication 13, p. 109-125.

Parnell, J. and Shukla, B, 1994. The Lough Neagh Basin, Northern Ireland, U.K., *in* Gierlowski-Kordesch, E. and Kelts, K., *eds.*, Global Geological Record of Lake Basins: Cambridge University Press, Cambridge, p. 277-279.

Smith, W.D. and Gibling, M.R., 1987, Oil shale composition related to depositional setting: a case study from the Albert Formation, New Brunswick, Canada: Bulletin of Canadian Petroleum Geology, v. 35, p. 469-487.

Talbot, M.R., 1988, The origins of lacustrine oil source rocks: evidence from the lakes of tropical Africa, *in* Fleet, A.J., Kelts, K. and Talbot, M.R., *eds.*, Lacustrine Petroleum Source Rocks. Geological Society of London, Special Publication 40, p. 29-43.

Deltas

Back, S., De Batist, M., Kirillov, P., Strecker, M.R. and Vanhauwaert, P., 1998, The Frolikha fan: a large Pleistocene glaciolacustrine outwash fan in northern Lake Baikal, Siberia: Journal of Sedimentary Research, v. 68, p. 841-849.

Blair, T.C. and McPherson, J.G., 2008, Quaternary sedimentology of the Rose Creek fan delta, Walker Lake, Nevada, USA, and implications to fan-delta facies models: Sedimentology, v. 55, p. 579-615.

Fisher, J.A., Waltham, D., Nichols, G.J., Krapf, C.B.E. and Lang, S.C., 2007, A quantitative model for deposition of thin fluvial sand sheets: Journal of the Geological Society, London, v. 164, p. 67-71.

Fisher, J.A., Krapf, C.B.E., Lang, S.C., Nichols, G.J., and Payenberg, T.H.D., 2008, Sedimentology and architecture of the Douglas Creek terminal splay, Lake Eyre, central Australia: Sedimentology, v. 55, p. 1915-1930.

Jiang, Z., Chen, D., Qiu, L., Liang, H. and Ma, J., 2007, Source-controlled carbonates in a small Eocene half-graben basin (Shulu Sag) in central Hebei Province, North China: Sedimentology, v. 54, p. 265-292.

Smoot, J.S., 1985, The closed-basin hypothesis and its use in facies analysis of the Newark Supergroup: U.S. Geological Survey Bulletin 1176, p. 4-10.

Evaporites, Saline Lakes, and Playas

Benison, K.C., Bowen, B.B., Oboh-Ikuenobe, F.E., Jagniecki, E.A., LaClair, D.A., Story, S.L., Mormile, M.R. and Hong, B.Y., 2007, Sedimentology of acid saline lakes in southern Western Australia: newly described processes and products of an extreme environment: Journal of Sedimentary Research, v. 77, p. 366-388.

Eugster, H.P. and Hardie, L.A., 1978, Saline lakes, *in* Lerman, A., *ed.*, Lakes: Chemistry, Geology, Physics: Springer, Berlin, p. 238-293.

Gierlowski-Kordesch, E. and Rust, B.R., 1994, The Jurassic East Berlin Formation, Hartford Basin, Newark Supergroup (Connecticut and Massachusetts): a saline lake-playa-alluvial plain system, *in* Renaut, R.W. and Last, W.M., *eds.*, Sedimentology and Geochemistry of Modern and Ancient Saline Lakes: SEPM Special Publication 50, p. 249-265.

Hardie, L.A., Smoot, J.P. and Eugster, H.P., 1978, Saline lakes and their deposits: a sedimentological approach, *in* Matter, A. and Tucker, M.E., *eds.*, Modern and Ancient Lake Sediments: International Association of Sedimentologists Special Publication 2, p. 7-41.

Jones, B.F. and Deocampo, D.M., 2003, Geochemistry of saline lakes, in Drever, J.I., *ed.*, Surface and Ground Water, Weathering, and Soils: Treatise on Geochemistry, Volume 5, Elsevier, Amsterdam, p. 393-424.

Kendall, A.C., 1984, Evaporites, *in* Walker, R.G., *ed.*, Facies Models, 2nd Edition: Geoscience Canada Reprint Series, v. 1, p. 259-296.

Last, W.M. and Ginn, F.M., 2005, Saline systems of the Great Plains of western Canada: an overview of the limnogeology and paleolimnology: Saline Systems, 1:10 doi:10.1186/1746-1448-1-10.

Lowenstein, T.K. and Hardie, L.A., 1985, Criteria for the recognition of salt-pan evaporites: Sedimentology, v. 32, p. 627-644.

Rosen, M.R., 1994, The importance of groundwater in playas: a review of playa classifications and the sedimentology and hydrology of playas, *in* Rosen, M.R., *ed.*, Paleoclimate and Basin Evolution of Playa Systems: Geological Society of America Special Paper 289, p. 1-18.

Schubel, K.A. and Lowenstein, T.K., 1997, Criteria for the recognition of shallow-perennial-saline-lake halites based on Recent sediments from the Qaidam Basin, western China: Journal of Sedimentary Research, v. 67, p. 74-87.

Smoot, J.P. and Lowenstein, T.K., 1991, Depositional environments of non-marine evaporites, *in* Melvin, J.L., *ed.*, Evaporites, Petroleum and Mineral Resources. Elsevier, Amsterdam, p. 189-347.

Facies Distribution and Sequence Stratigraphy

Bohacs, K.M., Carroll, A.R., Neal, J.E. and Mankiewicz, P.J., 2000, Lake-basin type, source potential, and hydrocarbon character: an integrated sequence-stratigraphic-geochemical framework, *in* Gierlowski-Kordesch, E.H. and Kelts, K.R., *eds.*, Lake Basins Through Space and Time: American Association of Petroleum Geologists, Studies in Geology 46, p. 3-33.

Bohacs, K.M., Carroll, A.R. and Neal, J.E., 2003, Lessons from large lake systems-thresholds, nonlinearity, and strange attractors, *in* Chan, M.A. and Archer, A.W., *eds.*, Extreme Depositional Environments: Mega End Members in Geologic Time: Geological Society of America Special Paper 370, p. 75-90.

Bohacs, K.M., Grabowski, Jr., G. and Carroll, A.R., 2007, Lithofacies architecture and variation in expression of sequence stratigraphy within representative intervals of the Green River Formation, Greater Green River Basin, Wyoming: The Mountain Geologist, v. 44, p. 39-58.

Buatois, L.A. and Mangáno, M.G., 2004, Animal-substrate interactions in freshwater environments; applications of ichnology in facies and sequence stratigraphic analysis of fluvio-lacustrine successions, *in* McIlroy, D., *ed.*, The Application of Ichnology to Palaeoenvironmental and Stratigraphic Analysis: Geological Society of London Special Publication 228, p. 311-333.

Gierlowski-Kordesch, E.H. and Park, L.E., 2004, Comparing species diversity in the modern and fossil records of lakes: Journal of Geology, v. 112, p. 703-717.

Smith M.E., Carroll, A.R., and Singer, B.S.,

2008, Synoptic reconstruction of a major lake system, Eocene Green River Formation, western United States: Geological Society of America Bulletin, v. 120, p. 54-84.

Lake Processes

Anderson, R.Y. and Dean, W.E., 1988, Lacustrine varve formation through time: Palaeogeography, Palaeoclimatology, Palaeoecology, v. 62, p. 215-235.

Glenn, C.R. and Kelts, K., 1991, Sedimentary rhythms in lake deposits, *in* Einsele, G., Ricken, W. and Seilacher, A., *eds.*, Cycles and Events in Stratigraphy: Springer, New York, p. 188-221.

Gruszka, B., 2007, The Pleistocene glaciolacustrine sediments in the Belchatów mine (central Poland): endogenic and exogenic controls: Sedimentary Geology, v. 193, p. 149-166.

Hakala, A., 2004, Meromixis as part of lake evolution - observations and a revised classification of true meromictic lakes in Finland. Boreal Environment Research, v. 9, p. 37-53.

Horppila, J. and Niemisto, J., 2008, Horizontal and vertical variations in sedimentation and resuspension rates in a stratifying lake – effects of internal seiches: Sedimentology, v. 55, p. 1135-1144.

Imboden, D.M., 2004, The motion of lake waters, *in* O'Sullivan, P.E., Reynolds, C.S., *eds.*, The Lakes Handbook, Volume 1 – Limnology and Limnetic Ecology: Blackwell Publishing, Malden, MA, p. 115-152.

Johnson, T.C., Carlson, T.W. and Evans, J.E., 1980, Contourites in Lake Superior: Geology, v. 8, p. 437-441.

Johnston, J.W., Thompson, T.A., Wilcox, D.A. and Baedke, S.J., 2007, Geomorphic and sedimentologic evidence for the separation of Lake Superior from Lakes Michigan and Huron: Journal of Paleolimnology, v. 37, p. 349-364.

Kidder, D.L. and Gierlowski-Kordesch, E.H., 2005, Impact of grassland radiation on the nonmarine silica cycle and Miocene diatomite: Palaios, v. 20, p. 198-206.

Lamoureux, S., 1999, Spatial and interannual variations in the sedimentation patterns, recorded in nonglacial varved sediments from the Canadian High Arctic: Journal of Paleolimnology, v. 21, 73-84.

Sturm, M. and Matter, A., 1978, Turbidites and varves in Lake Brienz (Switzerland): deposition of clastic detritus by density currents, *in* Matter, A. and Tucker, M.E., *eds.*, Modern and Ancient Lake Sediments: International Association of Sedimentologists Special Publication 2, p. 147-168.

Rift Lakes

Colman, S.M., Karabanov, E.B. and Nelson III, C.H., 2003, Quaternary sedimentation and subsidence history of Lake Baikal, Siberia, based on seismic stratigraphy and coring: Journal of Sedimentary Research, v. 73, p. 941-956.

Deocampo, D.M., 2002, Sedimentary processes and lithofacies in lake-margin groundwater-fed wetlands in East Africa, *in* Renaut, R.W. and Ashley, G.M., *eds.*, Sedimentation in Continental Rifts: SEPM Special Publication 73, p. 295-308.

Gore, P.J.W., 1988, Lacustrine sequences in an early Mesozoic rift basin: Culpeper Basin, Virginia, USA, *in* Fleet, A.J., Kelts, K., and Talbot, M.R., *eds.*, Lacustrine Petroleum Source Rocks: Geological Society of London Special Publication 40, p. 247-278.

Olsen, P.E., Schlische, R.W., and Gore, P.J.W., 1989, Tectonic, depositional, and paleoecological history of Early Mesozoic rift basins, eastern North America: 28th International Geological Congress Field Trip Guidebook T351, Washington, D.C., American Geophysical Union, 158 p.

Owen, R.B., 2002, Sedimentological characteristics and origins of diatomaceous deposits in the East African Rift System, *in* Renaut, R.W. and Ashley, G.M., *eds.*, Sedimentation in Continental Rifts: SEPM Special Publication 73, p. 233-246.

Owen, R.B. and Renaut, R.W., 2000, Spatial and temporal facies variations in the Pleistocene Olorgesailie Formation, southern Kenya rift valley, *in* Gierlowski-Kordesch, E.H. and Kelts, K.R., *eds.*, Lake Basins Through Space and Time: American Association of Petroleum Geologists Studies in Geology 46, p. 553-560.

Renaut, R.W. and Owen, R.B., 1991, Shore-zone sedimentation and facies in a closed rift lake: the Holocene beach deposits of Lake Bogoria, Kenya, *in* Anadón, P., Cabrera, Ll. and Kelts, K., *eds.*, Lacustrine Facies Analysis: International Association of Sedimentologists Special Publication 13, p. 175-198.

Renaut, R.W. and Tiercelin, J.-J., 1994, Lake Bogoria, Kenya Rift Valley – a sedimentological overview, *in* Renaut, R.W. and Last, W.M., *eds.*, Sedimentology and Geochemistry of Modern and Ancient Saline Lakes: SEPM Special Publication 50, p. 101-123.

Schlische, R.W. and Olsen, P.E., 1990, Quantitative filling model for continental extensional basins with applications to Early Mesozoic rifts of eastern North America: Journal of Geology, v. 98, p. 135-155.

Scholz, C.A. and Rosendahl, B.R., 1990, Coarse-clastic facies and stratigraphic sequence models from Lakes Malawi and Tanganyika, east Africa, *in* Katz, B.J., *ed.*, Lacustrine Basin Exploration: American Association of Petroleum Geologists Memoir 50, p. 151-168.

Soreghan, M.J., Scholz, C.A. and Wells, J.T., 1999, Coarse-grained, deep-water sedimentation along a border fault margin of Lake Malawi: seismic stratigraphic analysis: Journal of Sedimentary Research, v. 69, p. 832-846.

Wells, J.T., Scholz, C.A., and Soreghan, M.J., 1999, Processes of sedimentation on a lacustrine border-fault margin: interpretation of cores from Lake Malawi, East Africa: Journal of Sedimentary Research, v. 69, p. 816-831.

INDEX

Page numbers in *italics* indicate figures; in **bold** indicate tables

A

B

C

N

O

S

T